STANDARD NORMAL DISTRIBUTION

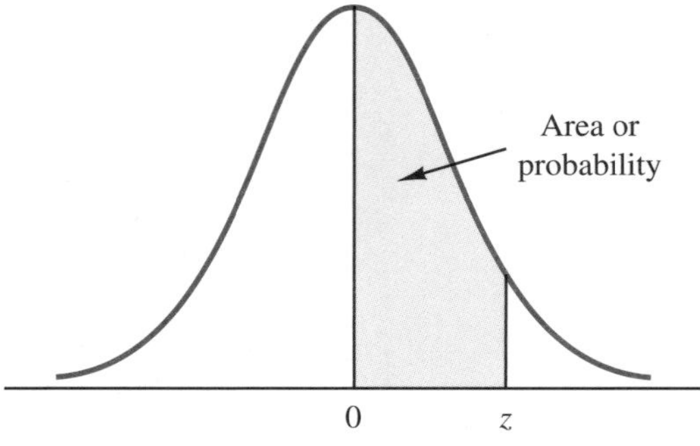

Entries in the table give the area under the curve between the mean and z standard deviations above the mean. For example, for $z = 1.25$ the area under the curve between the mean and z is .3944.

z	.00	.01	.02	.03	.04	.05	.06	.07	.08	.09
.0	.0000	.0040	.0080	.0120	.0160	.0199	.0239	.0279	.0319	.0359
.1	.0398	.0438	.0478	.0517	.0557	.0596	.0636	.0675	.0714	.0753
.2	.0793	.0832	.0871	.0910	.0948	.0987	.1026	.1064	.1103	.1141
.3	.1179	.1217	.1255	.1293	.1331	.1368	.1406	.1443	.1480	.1517
.4	.1554	.1591	.1628	.1664	.1700	.1736	.1772	.1808	.1844	.1879
.5	.1915	.1950	.1985	.2019	.2054	.2088	.2123	.2157	.2190	.2224
.6	.2257	.2291	.2324	.2357	.2389	.2422	.2454	.2486	.2518	.2549
.7	.2580	.2612	.2642	.2673	.2704	.2734	.2764	.2794	.2823	.2852
.8	.2881	.2910	.2939	.2967	.2995	.3023	.3051	.3078	.3106	.3133
.9	.3159	.3186	.3212	.3238	.3264	.3289	.3315	.3340	.3365	.3389
1.0	.3413	.3438	.3461	.3485	.3508	.3531	.3554	.3577	.3599	.3621
1.1	.3643	.3665	.3686	.3708	.3729	.3749	.3770	.3790	.3810	.3830
1.2	.3849	.3869	.3888	.3907	.3925	.3944	.3962	.3980	.3997	.4015
1.3	.4032	.4049	.4066	.4082	.4099	.4115	.4131	.4147	.4162	.4177
1.4	.4192	.4207	.4222	.4236	.4251	.4265	.4279	.4292	.4306	.4319
1.5	.4332	.4345	.4357	.4370	.4382	.4394	.4406	.4418	.4429	.4441
1.6	.4452	.4463	.4474	.4484	.4495	.4505	.4515	.4525	.4535	.4545
1.7	.4554	.4564	.4573	.4582	.4591	.4599	.4608	.4616	.4625	.4633
1.8	.4641	.4649	.4656	.4664	.4671	.4678	.4686	.4693	.4699	.4706
1.9	.4713	.4719	.4726	.4732	.4738	.4744	.4750	.4756	.4761	.4767
2.0	.4772	.4778	.4783	.4788	.4793	.4798	.4803	.4808	.4812	.4817
2.1	.4821	.4826	.4830	.4834	.4838	.4842	.4846	.4850	.4854	.4857
2.2	.4861	.4864	.4868	.4871	.4875	.4878	.4881	.4884	.4887	.4890
2.3	.4893	.4896	.4898	.4901	.4904	.4906	.4909	.4911	.4913	.4916
2.4	.4918	.4920	.4922	.4925	.4927	.4929	.4931	.4932	.4934	.4936
2.5	.4938	.4940	.4941	.4943	.4945	.4946	.4948	.4949	.4951	.4952
2.6	.4953	.4955	.4956	.4957	.4959	.4960	.4961	.4962	.4963	.4964
2.7	.4965	.4966	.4967	.4968	.4969	.4970	.4971	.4972	.4973	.4974
2.8	.4974	.4975	.4976	.4977	.4977	.4978	.4979	.4979	.4980	.4981
2.9	.4981	.4982	.4982	.4983	.4984	.4984	.4985	.4985	.4986	.4986
3.0	.4986	.4987	.4987	.4988	.4988	.4989	.4989	.4989	.4990	.4990

Preface

The purpose of *STATISTICS FOR BUSINESS AND ECONOMICS* is to give students, primarily those in the fields of business administration and economics, a conceptual introduction to the field of statistics and its many applications. The text is applications oriented and written with the needs of the non-mathematician in mind; the mathematical prerequisite is knowledge of algebra.

Applications of data analysis and statistical methodology are an integral part of the organization and presentation of the text material. The discussion and development of each technique is presented in an application setting, with the statistical results providing insights to decisions and solution to problems.

Although the book is applications oriented, we have taken care to provide sound methodological development and to use notation that is generally accepted for the topic being covered. Hence, students will find that this text provides good preparation for the study of more advanced statistical material. A revised and updated bibliography to guide further study is included as an appendix.

CHANGES IN THE EIGHTH EDITION

We appreciate the acceptance and positive response to the previous editions of *STATISTICS FOR BUSINESS AND ECONOMICS.* Accordingly, in making modifications for this new edition, we have maintained the presentation style and readability of those editions. The significant changes in the new edition are summarized here.

Content Revisions

The following list summarizes selected content revisions for this edition.

- A discussion of levels of measurement for data (Chapter 1)
- More material on graphical and tabular descriptive statistics (Chapter 2)
- The use of Microsoft® Excel for random sampling (Chapter 7)
- An earlier discussion of the term *margin of error* and additional information on when to use z and when to use t for interval estimation
- More emphasis on computing and interpreting p-values (Chapters 9 through 12)
- Formulas for regression computations have been modified to emphasize intuition and deemphasize computational issues (Chapter 14)
- Updated index numbers including the CPI and DJIA (Chapter 17)

New Examples and Exercises Based on Real Data

We have added approximately 200 new examples and exercises based on real data and recent reference sources of statistical information. Using *The Wall Street Journal, USA Today, Fortune, Barron's,* various websites, and a variety of other sources, we have drawn upon actual studies to develop explanations and to create exercises that demonstrate many uses of statistics in business and economics. We believe that the use of real data helps generate more student interest in the material and enables the student to learn about both the statistical methodology and its application. The eighth edition of the text contains approximately 350 examples and exercises based on real data.

New Case Problems

We have added four new case problems to this edition, bringing the total number of case problems in the text to twenty-six. The new case problems, which are based on referenced data sets, appear in the chapters on interval estimation and regression. These case problems provide students with the opportunity to analyze somewhat larger data sets and prepare managerial reports based on the results of the analysis.

New Statistics in Practice

Each chapter begins with a statistics in practice article that describes an application of the statistical methodology to be covered in the chapter. Statistics in practice have been provided by practitioners at companies such as Procter & Gamble, Mead, Dollar General, Colgate-Palmolive, Polaroid, Monsanto, and others. This edition includes two new statistics in practice: Small Fry Design (Chapter 3) and Citigroup (Chapter 5).

MINITAB and Excel Appendixes

MINITAB and Excel Spreadsheet appendices appear at the end of most chapters. These appendixes provide step-by-step procedures that make it easy for the student to use MINITAB or Excel to conduct the statistical analysis presented in the chapter. All appendixes have been updated for the latest versions of MINITAB and Excel. The Excel appendix for descriptive statistics (Chapter 2) now includes the use of the function wizard, the chart wizard, and the pivot table report. These procedures expose the student to the extensive capabilities of Excel for creating graphical presentations and for developing crosstabulations. A total of ten new and/or revised appendices appear at the end of the chapters on descriptive statistics, sampling, interval estimation, hypothesis testing, and tests of independence.

FEATURES AND PEDAGOGY

We have continued many of the features that appeared in previous editions. Some of the important ones are noted here.

Methods Exercises and Applications Exercises

The end-of-section exercises are split into two parts, Methods and Applications. The Methods exercises require students to use the formulas and make the necessary computations. The Applications exercises require students to use the chapter material in real-world situations. Thus, students first focus on the computational "nuts and bolts," then move on to the subtleties of statistical application and interpretation.

Self-Test Exercises

Certain exercises are identified as self-test exercises. Completely worked-out solutions for those exercises are provided in Appendix E at the back of the book. Students can attempt the self-test exercises and immediately check the solution to evaluate their understanding of the concepts presented in the chapter.

Notes and Comments

At the end of many sections, we provide Notes and Comments designed to give the student additional insights about the statistical methodology and its application. Notes and Comments include warnings about or limitations of the methodology, recommendations for application, brief descriptions of additional technical considerations, and other matters.

Data Sets Accompany the Text

Over 100 data sets are now available on the CD-rom that is packaged with the text. The data sets are available in both MINITAB and Excel formats. Data set logos are used in the text to identify the data sets that are available on the CD. Data sets for all case problems as well as data sets for larger exercises are included on the CD.

ANCILLARY TEACHING AND LEARNING MATERIALS

WebTutor™, a brand new electronic ancillary, is available with the new edition of *STATISTICS FOR BUSINESS AND ECONOMICS.* There are two main formats for this product:

- WebTutor is used by an entire class under the directions of the instructor. It provides Web-based learning resources to students as well as powerful communication and other course management tools including course calendar, chat, and e-mail for instructors. WebTutor is available on WebCT and Blackboard. See *http://webtutor.thomsonlearning.com* for more information.
- Personal WebTutor provides the learning resources for individual students to purchase and use for study and review. See *http://pwt.swcollege.com* for more information about this product.

Three print ancillaries are available to students either through their bookstore or for direct purchase through the on-line catalog at *www.swcollege.com.*

- Prepared by Mohammad Ahmadi of the University of Tennessee-Chattanooga, the ***Work Book*** (ISBN: 0-324-06676-7) will provide the student with significant supplementary study materials. It contains an outline, review, and list of formulas each text chapter, sample exercises with step-by-step solutions, exercises with answers, and a series of self-testing questions with answers.
- Another student ancillary is the ***Microsoft® Excel Companion for Business Statistics*** (ISBN: 0-324-06898-0) by David Eldredge of Murray State College. This manual provides step-by-step instructions for using Excel to solve many of the problems included in introductory business statistics. Directions for Excel 2000 and Excel 97 are included.
- **Solutions Manual** At the request of the instructor, a print version of the Solutions Manual can be packaged with the text for student purchase.

The following instructor support materials are available to adopters from the Thomson Learning™ Academic Resource Center at 800-423-0563 or through **www.swcollege.com:**

- **Instructor's Resource CD** (ISBN: 0-324-06674-0)—All instructor ancillaries are provided on a single CD-rom. Included in this convenient format are:
  - **Solutions Manual**—The Solutions Manual, prepared by the authors, includes solutions for all problems in the text.
  - **Instructor's Manual**—The Instructor's Manual, also prepared by the authors, contains solutions to all case problems presented in the text.
  - **PowerPoint™ Presentation Slides**—Prepared by John Loucks of St. Edward's University, the presentation slides contain a teaching outline that incorporates graphics to help instructors create even more stimulating lectures. The PowerPoint 97 slides may be adapted using PowerPoint software to facilitate classroom use.
  - **Test Bank** and **ExamView™**—Prepared also by Mohammad Ahmadi, the Test Bank includes true/false, multiple choice, short answer questions, and problems for each chapter. ExamView computerized testing software allows instructors to create, edit, store, and print exams.

ACKNOWLEDGMENTS

We would like to acknowledge the work of our reviewers who provided comments and suggestions of ways to continue to improve our text. Thanks to:

Lari Arjomand, Clayton College and State University
James Brannon, University of Wisconsin–Oshkosh
Donald Gren, Salt Lake Community College
Clifford Hawley, West Virginia University
Ronald Klimberg, St. Joseph's University
Bala Maniam, Sam Houston State University
Ceyhun Ozgur, Valparaiso University
H. V. Ramakrishna, Penn State University at Great Valley

We continue to owe a debt to our many colleagues and friends for their helpful comments and suggestions in the development of this and earlier editions of our text. Among them are:

Mohammad Ahmadi, University of Tennessee at Chattanooga
Robert Balough, Clarion University
Mike Bourke, Houston Baptist University
John Bryant, University of Pittsburgh
Peter Bryant, University of Colorado
Terri L. Byczkowski, University of Cincinnati
Robert Carver, Stonehill College
Robert Cochran, University of Wyoming
David W. Cravens, Texas Christian University
Robert Carver, Stonehill College
Robert Collins, Marquette University
Tom Dahlstrom, Eastern College
Gopal Dorai, William Patterson University
Nicholas Farnum, California State University–Fullerton
Paul Guy, California State University–Chico
Alan Humphrey, University of Rhode Island
Ann Hussein, Philadelphia College of Textiles and Science
C. Thomas Innis, University of Cincinnati
Ben Isselhardt, Rochester Institute of Technology
Jeffery Jarrett, University of Rhode Island
David Krueger, St. Cloud State University
Martin S. Levy, University of Cincinnati
David Lucking-Reiley, Vanderbilt University
Don Marx, University of Alaska, Anchorage
Tom McCullough, University of California–Berkeley
Glenn Nilligan, Ohio State University
Mitchell Muesham, Sam Houston State University
Roger Myerson, Northwestern University

Richard O'Connell, Miami University of Ohio
Alan Olinsky, Bryant College
Tom Pray, Rochester Institute of Technology
Harold Rahmlow, St. Joseph's University
Tom Ryan, Case Western Reserve University
Alan Smith, Robert Morris College
Bill Seaver, University of Tennessee
Willbann Terpening, Gonzaga University
David Tufte, University of New Orleans
Ted Tsukahara, St. Mary's College of California
Hroki Tsurumi, Rutgers University
Victor Ukpolo, Austin Peay State University
Ebenge Usip, Youngstown State University
Cindy VanEs, Cornell University
Jack Vaughn, University of Texas–El Paso
Andrew Welki, John Carroll University
Ari Wijetunga, Morehead State University
J. E. Willis, Louisiana State University
Mustafa Yilmaz, Northeastern University
Gary Yoshimoto, St. Cloud State University
Charles Zimmerman, Robert Morris College

A special thanks is owed to our associates from business and industry that supplied the Statistics in Practice features. We recognize them individually by a credit line in each of the articles. Finally, we are also indebted to our senior acquisitions editor Charles McCormick, Jr., our senior developmental editor Alice Denny, our senior production editor Deanna Quinn, our senior marketing manager Joe Sabatino, and others at South-Western/Thomson Learning for their editorial counsel and support during the preparation of this text.

David R. Anderson
Dennis J. Sweeney
Thomas A. Williams

About the Authors

David R. Anderson. David R. Anderson is Professor of Quantitative Analysis in the College of Business Administration at the University of Cincinnati. Born in Grand Forks, North Dakota, he earned his B.S., M.S., and Ph.D. degrees from Purdue University. Professor Anderson has served as Head of the Department of Quantitative Analysis and Operations Management and as Associate Dean of the College of Business Administration. In addition, he was the coordinator of the College's first Executive Program.

At the University of Cincinnati, Professor Anderson has taught introductory statistics for business students as well as graduate-level courses in regression analysis, multivariate analysis, and management science. He has also taught statistical courses at the Department of Labor in Washington, D.C. He has been honored with nominations and awards for excellence in teaching and excellence in service to student organizations.

Professor Anderson has coauthored nine textbooks in the areas of statistics, management science, linear programming, and production and operations management. He is an active consultant in the field of sampling and statistical methods.

Dennis J. Sweeney. Dennis J. Sweeney is Professor of Quantitative Analysis and Director of the Center for Productivity Improvement at the University of Cincinnati. Born in Des Moines, Iowa, he earned a B.S.B.A. degree from Drake University and his M.B.A. and D.B.A. degrees from Indiana University where he was an NDEA Fellow. During 1978–79, Professor Sweeney worked in the management science group at Procter & Gamble; during 1981–82, he was a visiting professor at Duke University. Professor Sweeney served as Head of the Department of Quantitative Analysis and as Associate Dean of the College of Business Administration at the University of Cincinnati.

Professor Sweeney has published more than 30 articles and monographs in the area of management science and statistics. The National Science Foundation, IBM, Procter & Gamble, Federated Department Stores, Kroger, and Cincinnati Gas & Electric have funded his research, which has been published in *Management Science*, *Operations Research*, *Mathematical Programming*, *Decision Sciences*, and other journals.

Professor Sweeney has coauthored nine textbooks in the areas of statistics, management science, linear programming, and production and operations management.

Thomas A. Williams. Thomas A. Williams is Professor of Management Science in the College of Business at Rochester Institute of Technology. Born in Elmira, New York, he earned his B.S. degree at Clarkson University. He did his graduate work at Rensselaer Polytechnic Institute, where he received his M.S. and Ph.D. degrees.

Before joining the College of Business at RIT, Professor Williams served for seven years as a faculty member in the College of Business Administration at the University of Cincinnati, where he developed the undergraduate program in Information Systems and then served as its coordinator. At RIT he was the first chairman of the Decision Sciences Department. He teaches courses in management science and statistics, as well as graduate courses in regression and decision analysis.

Professor Williams is the coauthor of ten textbooks in the areas of management science, statistics, production and operations management, and mathematics. He has been a consultant for numerous *Fortune* 500 companies and has worked on projects ranging from the use of data analysis to the development of large-scale regression models.

DATA AND STATISTICS

Chapter 1

CONTENTS

STATISTICS IN PRACTICE

BUSINESS WEEK*
New York, New York

With a global circulation of more than 1 million, *Business Week* is the most widely read business magazine in the world. More than 200 dedicated reporters and editors in 26 bureaus worldwide deliver a variety of articles of interest to the business and economic community. Along with feature articles on current topics, the magazine contains regular sections on International Business, Economic Analysis, Information Processing, and Science & Technology. Information in the feature articles and the regular sections help readers stay abreast of current developments and assess the impact of those developments on business and economic conditions.

Most issues of *Business Week* provide an in-depth report on a topic of current interest. Often, the in-depth reports contain statistical facts and summaries that help the reader understand the business and/or economic information. For instance, the January 17, 2000, issue contained a report on who gets hurt most by higher interest rates; the January 24, 2000, issue contained a report on what is wrong with mutual funds; and the February 28, 2000, issue contained a report on stock options. In addition, the weekly *Business Week Investor* contains statistics about the state of the economy including production indexes, stock prices, mutual funds, and interest rates.

Business Week also uses statistics and statistical information to help manage its own business. For example, an annual survey of subscribers helps the company learn about subscriber demographics, reading habits, likely purchases, lifestyles, and so on. *Business Week* managers use the statistical summaries from the survey to provide better services to subscribers and advertisers. For instance, a recent North American subscriber survey indicated that 90% of *Business Week* subscribers have a personal computer at home and that 64% of *Business Week* subscribers are involved with computer purchases at work. Such statistics alert *Business Week* managers to subscriber interest in articles about new developments in computers. The results of the survey are also made available to potential advertisers. The high percentage of subscribers using personal computers at home and the high percentage of subscribers involved with computer purchases at work would be an incentive for a computer manufacturer to consider advertising in *Business Week.*

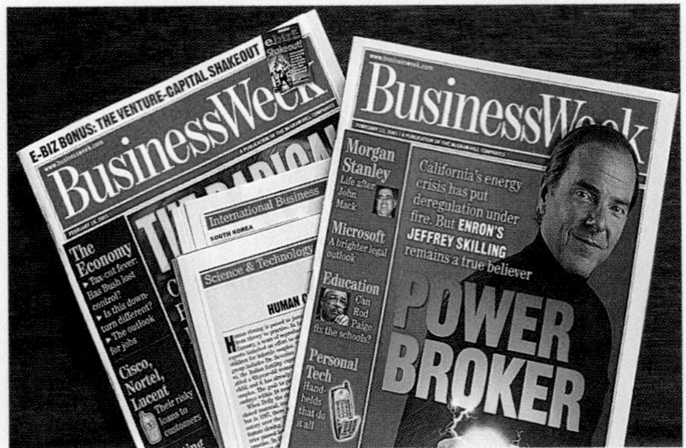

Business Week uses statistical facts and summaries in many of its articles. © Joe Higgins/South-Western.

In this chapter, we discuss the types of data that are available for statistical analysis and describe how the data are obtained. We introduce descriptive statistics and statistical inference as ways of converting data into meaningful and easily interpreted statistical information.

*The authors are indebted to Charlene Trentham, Research Manager at *Business Week,* for providing this Statistics in Practice.

Frequently, we see the following kinds of statements in newspaper and magazine articles:

- Cisco, which makes equipment that routes traffic across the Internet, was valued at $555 billion at Monday's Nasdaq close. (*USA Today,* March 28, 2000)
- The average executive compensation package for a chief executive at 50 of the largest corporations is $9.3 million. (*Forbes,* April 3, 2000)
- E-commerce sites spend an average of $108 to acquire each customer. (*Business 2.0,* March 2000)
- The *Washington Post* reaches 46% of the region's households on weekdays and 61% on Sundays, tops among big-city newspapers. (*Fortune,* January 10, 2000)

- Stocks account for 75% of the average investor's portfolio. (*The Wall Street Journal,* March 27, 2000)
- The average price of a ticket to a major league baseball game during the 1999 season was $14.91. (*USA Today,* April 15, 1999)

The numerical facts in the preceding statements ($555 billion, $9.3 million, $108, 46%, 61%, 75%, and $14.91) are called **statistics.** Thus, in everyday usage, the term *statistics* refers to numerical facts. However, the field, or subject, of statistics involves much more than numerical facts. In a broad sense, statistics is the art and science of collecting, analyzing, presenting, and interpreting data. Particularly in business and economics, a major reason for collecting, analyzing, presenting, and interpreting data is to give managers and decision makers a better understanding of the business and economic environment and thus enable them to make more informed and better decisions. In this text, we emphasize the use of statistics for business and economic decision making.

Chapter 1 begins with some illustrations of the applications of statistics in business and economics. In Section 1.2 we define the term **data** and introduce the concept of a data set. This section also introduces key terms such as **variables** and **observations,** discusses the difference between quantitative and qualitative data, and illustrates the difference between cross-sectional and time series data. Section 1.3 discusses how data can be obtained from existing sources or through survey and experimental studies designed to obtain new data. The important role that the Internet now plays in obtaining data is also highlighted. The use of data in developing descriptive statistics and in making statistical inferences is described in Sections 1.4 and 1.5.

1.1 APPLICATIONS IN BUSINESS AND ECONOMICS

In today's global business and economic environment, vast amounts of statistical information are available. The most successful managers and decision makers are the ones who can understand the information and use it effectively. In this section, we provide examples that illustrate some of the uses of statistics in business and economics.

Accounting

Public accounting firms use statistical sampling procedures when conducting audits for their clients. For instance, suppose an accounting firm wants to determine whether the amount of accounts receivable shown on a client's balance sheet fairly represents the actual amount of accounts receivable. Usually the number of individual accounts receivable is so large that reviewing and validating every account would be too time-consuming and expensive. The common practice in such situations is for the audit staff to select a subset of the accounts called a sample. After reviewing the accuracy of the sampled accounts, the auditors draw a conclusion as to whether the accounts receivable amount shown on the client's balance sheet is acceptable.

Finance

Financial advisors use a variety of statistical information to guide their investment recommendations. In the case of stocks, the advisors review a variety of financial data including price/earnings ratios and dividend yields. By comparing the information for an individual stock with information about the stock market averages, a financial advisor can begin to draw a conclusion as to whether an individual stock is over- or undervalued. For example, *Barron's* (January 10, 2000) reported that the average price/earnings ratio for the 30 stocks in the Dow Jones Industrial Average was 24.7. Philip Morris had a price/earnings ratio of

9. In this case, the statistical information on price/earnings ratios showed that Philip Morris had a lower price in comparison to its earnings than the average for the Dow Jones stocks. Therefore, a financial advisor might have concluded that Philip Morris was currently underpriced. This and other information about Philip Morris would help the advisor make buy, sell, or hold recommendations for the stock.

Marketing

Electronic scanners at retail checkout counters are being used to collect data for a variety of marketing research applications. For example, data suppliers such as ACNielsen and Information Resources, Inc., purchase point-of-sale scanner data from grocery stores, process the data, and then sell statistical summaries of the data to manufacturers. Manufacturers spent an average of $387,325 per product category to obtain this type of scanner data (Scanner Data User Survey, Mercer Management Consulting, Inc., April 1997). Manufacturers also purchase data and statistical summaries on promotional activities such as special pricing and the use of in-store displays. Brand managers can review the scanner statistics and the promotional activity statistics to better understand the relationship between promotional activities and sales. Such analyses are helpful in establishing future marketing strategies for the various products.

Production

With today's emphasis on quality, quality control is an important application of statistics in production. A variety of statistical quality control charts are used to monitor the output of a production process. In particular, an x-bar chart is used to monitor the average output. Suppose, for example, that a machine is being used to fill containers with 12 ounces of a soft drink. Periodically, a sample of containers is selected and the average number of ounces in the sample containers is computed. This average, or x-bar value, is plotted on an x-bar chart. A plotted value above the chart's upper control limit indicates overfilling, and a plotted value below the chart's lower control limit indicates underfilling. The process is "in control" and allowed to continue as long as the plotted x-bar values are between the chart's upper and lower control limits. Properly interpreted, an x-bar chart can help determine when adjustments are necessary to correct a production process.

Economics

Economists are frequently asked to provide forecasts about the future of the economy or some aspect of it. They use a variety of statistical information in making such forecasts. For instance, in forecasting inflation rates, economists use statistical information on such indicators as the Producer Price Index, the unemployment rate, and manufacturing capacity utilization. Often these statistical indicators are entered into computerized forecasting models that predict inflation rates.

Applications of statistics such as those described in this section are an integral part of this text. Such examples provide an overview of the breadth of statistical applications. To supplement these examples, we have asked practitioners in the fields of business and economics to provide chapter-opening Statistics in Practice articles that introduce the material covered in each chapter. The Statistics in Practice applications show the importance of statistics in a wide variety of decision-making situations.

1.2 DATA

Data are the facts and figures that are collected, analyzed, and summarized for presentation and interpretation. Together, the data collected in a particular study are referred to as the **data set** for the study. Table 1.1 shows a data set containing information for 25 of the shadow stocks tracked by the American Association of Individual Investors. Shadow stocks are common stocks of smaller companies that are not closely followed by Wall Street analysts.

Elements, Variables, and Observations

Elements are the entities on which data are collected. For the data set in Table 1.1, each individual company's stock is an element. With 25 stocks, the data set contains 25 elements. A **variable** is a characteristic of interest for the elements. The data set in Table 1.1 has the following five variables:

- *Exchange:* Where the stock is traded—NYSE (New York Stock Exchange), AMEX (American Stock Exchange), and OTC (Over-the-Counter).
- *Ticker Symbol:* The abbreviation used to identify the stock on the exchange listing.
- *Annual Sales:* Total sales for the company for the most recent 12 months in millions of dollars.
- *Earnings per Share:* Total earnings per share for the latest 12 months in dollars.
- *Price/Earnings Ratio:* Market price per share divided by the most recent 12 months' earnings per share.

TABLE 1.1 A DATA SET FOR 25 SHADOW STOCKS

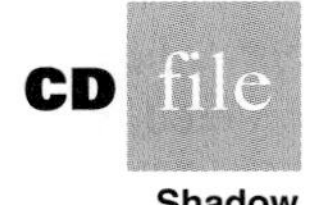
CD file
Shadow

Company	Exchange	Ticker Symbol	Annual Sales ($ millions)	Earnings per Share ($)	Price/ Earnings Ratio
Advanced Comm. Systems	OTC	ACSC	75.10	0.32	39.10
Ag-Chem Equipment Co.	OTC	AGCH	321.10	0.48	23.40
Aztec Manufacturing Co.	NYSE	AZZ	79.70	1.18	7.80
Cal-Maine Foods, Inc.	OTC	CALM	314.10	0.38	11.70
Chesapeake Utilities	NYSE	CPK	174.50	1.13	16.20
Dataram Corporation	AMEX	DTM	73.10	0.86	11.00
EnergySouth, Inc.	OTC	ENSI	74.00	1.67	13.20
Gencor Industries, Inc.	AMEX	GX	263.30	1.96	4.70
Industrial Scientific	OTC	ISCX	43.50	2.03	11.50
Keystone Consolidated	NYSE	KES	365.70	0.86	9.40
LandCare USA, Inc.	NYSE	GRW	111.40	0.33	29.40
Market Facts, Inc.	OTC	MFAC	126.70	0.98	26.50
Meridian Diagnostics, Inc.	OTC	KITS	36.30	0.46	14.70
Merit Medical Systems	OTC	MMSI	67.20	0.27	24.50
Met-Pro Corporation	NYSE	MPR	61.90	1.01	12.40
Nobility Homes, Inc.	OTC	NOBH	45.80	0.87	14.70
Omega Research, Inc.	OTC	OMGA	27.60	0.11	27.30
Point of Sale Limited	OTC	POSIF	12.30	0.28	25.40
Psychemedics Corp.	AMEX	PMD	17.60	0.13	39.40
Roadhouse Grill, Inc.	OTC	GRLL	118.40	0.26	20.80
Selas Corp. of America	AMEX	SLS	97.10	0.77	10.70
Toymax International, Inc.	OTC	TMAX	104.50	1.08	4.70
VSI Holdings, Inc.	AMEX	VIS	166.8	0.25	21
Warrantech Corporation	OTC	WTEC	207.30	0.13	29.80
Webco Industries, Inc.	AMEX	WEB	153.50	0.88	7.50

Source: American Association of Individual Investors web site, March 1999.

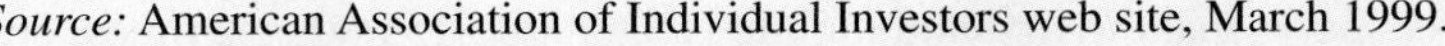

Data are obtained by collecting measurements on each variable for every element in a study. The set of measurements obtained for a particular element is called an **observation.** Referring to Table 1.1, we see that the set of measurements for the first observation (Advanced Comm. Systems) is OTC, ACSC, 75.10, 0.32, and 39.10. The set of measurements for the second observation (Ag-Chem Equipment Co.) is OTC, AGCH, 321.10, 0.48, and 23.40; and so on. Because the data set contains 25 elements, it has 25 observations.

Scales of Measurement

Data are collected using one of the following scales of measurement: nominal, ordinal, interval, and ratio. The scale of measurement determines the amount of information contained in the data and indicates the data summarization and statistical analyses that are most appropriate.

The scale of measurement for a variable is a **nominal scale** when the data are labels or names used to identify an attribute of the element. For example, referring to the data in Table 1.1, we see that the scale of measurement for the exchange variable is nominal because NYSE, AMEX, and OTC are labels used to identify where the company's stock is traded. In cases where the scale of measurement is nominal, a numeric code as well as nonnumeric labels may be used. For example, to facilitate data collection and to prepare the data for entry into a computer database, we might use a numeric code by letting 1 denote the New York Stock Exchange, 2 denote the American Stock Exchange, and 3 denote over-the-counter. In this case the numeric values, 1, 2, and 3 are the labels used to identify where the stock is traded. The scale of measurement is nominal even though the data are shown as numeric values.

The scale of measurement for a variable is an **ordinal scale** if the data have the properties of nominal data and the order or rank of the data is meaningful. For example, Eastside Automotive sends customers a questionnaire designed to obtain data on the quality of its automotive repair service. Each customer provides a repair service rating of excellent, good, or poor. Because the data obtained are the labels—excellent, good or poor—the data have the properties of nominal data. In addition, the data can be ranked, or ordered, with respect to the service quality. Data recorded as excellent indicates the best service, followed by good and then poor. Thus, the scale of measurement is ordinal. Note that the ordinal data can also be recorded using a numeric code. For example, we could use 1 for excellent, 2 for good, and 3 for poor to maintain the properties of ordinal data. Thus, data for an ordinal scale may be either nonnumeric or numeric.

The scale of measurement for a variable is an **interval scale** if the data have the properties of ordinal data and the interval between observations is expressed in terms of a fixed unit of measure. Interval data are always numeric. Scholastic Aptitude Test (SAT) scores are an example of interval-scaled data. For example, three students with SAT scores of 1120, 1050, and 970 can be ranked or ordered in terms of best performance to poorest performance. In addition, the differences between the scores are meaningful. For instance, student 1 scored $1120 - 1050 = 70$ points more than student 2, while student 2 scored $1050 - 970 = 80$ points more than student 3.

The scale of measurement for a variable is a **ratio scale** if the data have all the properties of interval data and the ratio of two values is meaningful. Variables such as distance, height, weight, and time use the ratio scale of measurement. A requirement of this scale is that it must contain a zero value that indicates that nothing exists for the variable at the zero point. For example, consider the cost of an automobile. A zero value for the cost would indicate that the automobile has no cost and is free. In addition, if we compare the cost of \$30,000 for one automobile to the cost of \$15,000 for a second automobile, the ratio property shows that the first automobile is \$30,000/\$15,000 = 2 times, or twice, the cost of the second automobile.

Qualitative and Quantitative Data

Data can be further classified as being either qualitative or quantitative. **Qualitative data** are labels or names used to identify an attribute of each element. Qualitative data use either the nominal or ordinal scale of measurement and may be nonnumeric or numeric. **Quantitative data** are numeric values that indicate how much or how many. Quantitative data use either the interval or ratio scale of measurement.

The statistical method used to summarize data depends upon whether the data are qualitative or quantitative.

A **qualitative variable** is a variable with qualitative data, and a **quantitative variable** is a variable with quantitative data. The statistical analysis that is appropriate for a particular variable depends upon whether the variable is qualitative or quantitative. If the variable is qualitative, the statistical analysis is rather limited. We can summarize qualitative data by counting the number of observations in each qualitative category or by computing the proportion of the observations in each qualitative category. However, even when the qualitative data have a numeric code, arithmetic operations such as addition, subtraction, multiplication, and division do not provide meaningful results. Section 2.1 provides ways for summarizing qualitative data.

On the other hand, arithmetic operations often provide meaningful results for a quantitative variable. For example, for a quantitative variable, the data may be added and then divided by the number of observations to compute the average value. This average is usually meaningful and easily interpreted. In general, more alternatives for statistical analysis are possible when the data are quantitative. Section 2.2 and Chapter 3 provide ways of summarizing quantitative data.

Cross-Sectional and Time Series Data

For purposes of statistical analysis, distinguishing between cross-sectional data and time series data is important. **Cross-sectional data** are data collected at the same or approximately the same point in time. The data in Table 1.1 are cross-sectional because they describe the five variables for the 25 shadow stocks at the same point in time. **Time series data** are data collected over several time periods. For example, Figure 1.1 is a graph of the

FIGURE 1.1 U.S. CITY AVERAGE PRICE PER GALLON FOR UNLEADED REGULAR GASOLINE

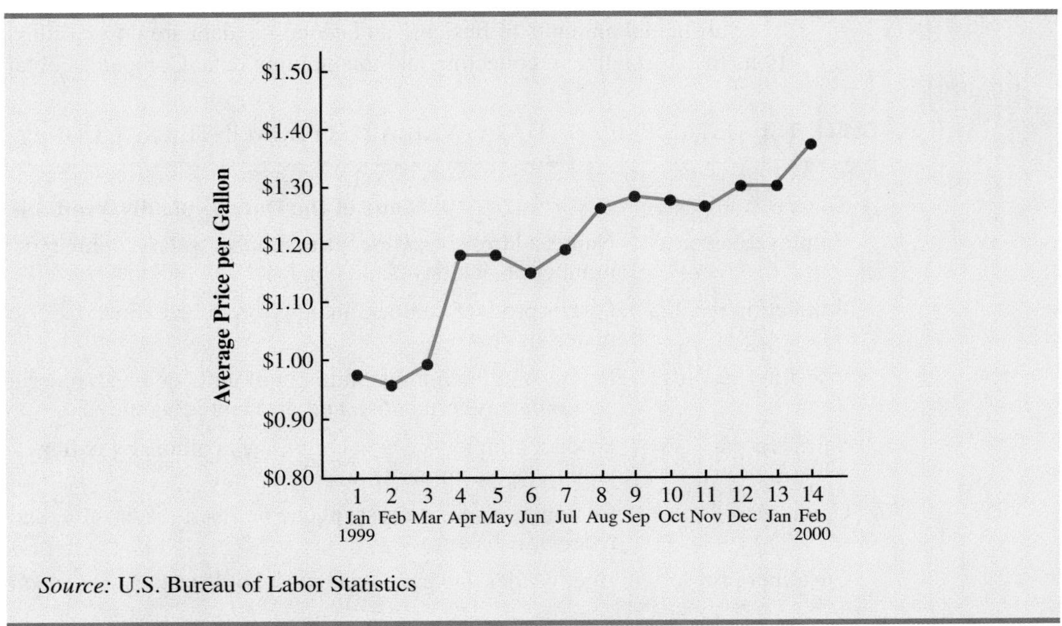

Source: U.S. Bureau of Labor Statistics

U.S. city average price per gallon for unleaded regular gasoline. It shows that the average price per gallon has grown from approximately $0.97 per gallon in January 1999 to $1.37 per gallon in February 2000. Most of the statistical methods presented in this text apply to cross-sectional data.

NOTES AND COMMENTS

1. An observation is the set of measurements obtained for each element in a data set. Hence, the number of observations is always the same as the number of elements. The number of measurements obtained on each element is equal to the number of variables. Hence, the total number of data items is the number of observations multiplied by the number of variables.
2. Quantitative data may be discrete or continuous. Quantitative data that measure how many are discrete. Quantitative data that measure how much are continuous because there is no separation between the possible values for the data.

1.3 DATA SOURCES

Data can be collected from existing sources or from surveys and experimental studies designed to obtain new data.

Existing Sources

In some cases, data needed for a particular application may already exist within a firm or organization. All companies maintain a variety of databases about their employees, customers, and business operations. Data on employee salaries, ages, and years of experience can usually be obtained from internal personnel records. Data on sales, advertising expenditures, distribution costs, inventory levels, and production quantities are generally available from other internal records. Most companies also maintain detailed data about their customers. Table 1.2 shows some of the data commonly available from the internal information sources of most companies.

Substantial amounts of business and economic data are now available from organizations that specialize in collecting and maintaining data. Companies obtain access to these

TABLE 1.2 EXAMPLES OF DATA AVAILABLE FROM INTERNAL COMPANY RECORDS

Source	Some of the Data Typically Available
Employee records	Name, address, social security number, salary, number of vacation days, number of sick days, and bonus
Production records	Part or product number, quantity produced, direct labor cost, and materials cost
Inventory records	Part or product number, number of units on hand, reorder level, economic order quantity, and discount schedule
Sales records	Product number, sales volume, sales volume by region, and sales volume by customer type
Credit records	Customer name, address, phone number, credit limit, and accounts receivable balance
Customer profile	Age, gender, income level, household size, address, and preferences

external data sources through leasing arrangements or by purchase. Dun & Bradstreet, Bloomberg, and Dow Jones & Company are three firms that provide extensive business database services to clients. ACNielsen and Information Resources, Inc., have built successful businesses collecting and processing data that they sell to advertisers and product manufacturers.

Data are also available from a variety of industry associations and special interest organizations. The Travel Industry Association of America maintains travel-related information such as the number of tourists and travel expenditures by states. Such data would be of interest to firms and individuals in the travel industry. The Graduate Management Admission Council maintains data on test scores, student characteristics, and graduate management education programs. Most of the data from these types of sources are available to qualified users at a modest cost.

In recent years, the Internet has become an important source of data. Almost all companies have Internet web sites and provide public access to them. Figure 1.2 shows the Internet addresses for a variety of companies, including the companies that contributed the Statistics in Practice applications that appear at the beginning of each chapter. By accessing

FIGURE 1.2 INTERNET ADDRESSES FOR SELECTED COMPANIES

ACNielsen Corporation
http://www.acnielsen.com

Alcoa Inc.
http://www.alcoa.com

Amazon.com, Inc.
http://www.amazon.com

American Airlines Inc.
http://www.aa.com

American Association of Individual Investors
http://www.aaii.com

American Express Company
http://www.americanexpress.com

Andersen Consulting
http://www.ac.com

America Online, Inc.
http://www.aol.com

AT&T
http://www.att.com

BMW of North America
http://www.bmw.com

The Boeing Company
http://www.boeing.com

Business Week
http://www.businessweek.com

BUY.COM, Inc.
http://www.buy.com

Caterpillar, Inc.
http://www.caterpillar.com

Charles Schwab & Co.
http://www.charlesschwab.com

Chemdex Corporation
http://www.chemdex.com

Chevron Corporation
http://www.chevron.com

Cisco Systems, Inc.
http://www.cisco.com

Citibank
http://www.citibank.com

The CocaCola Company
http://www.cocacola.com

Colgate-Palmolive Company
http://www.colgate.com

Dell Computer Corporation
http://www.dell.com

Delta Air Lines
http://www.delta.com

Disney
http://www.disney.com

Dollar General Corporation
http://www.dollargeneral.com

Dow Chemical Company
http://www.dowchemical.com

E.I. DuPont de Nemours and Company
http://www.dupont.com

E*TRADE Securities, Inc.
http://www.etrade.com

Ford Motor Company
http://www.ford.com

General Electric Company
http://www.ge.com

General Motors Corporation
http://www.gm.com

The Goodyear Tire & Rubber Company
http://www.goodyear.com

Harris Corporation
http://www.harris.com

Hewlett-Packard Company
http://www.hp.com

The Home Depot, Inc.
http://www.homedepot.com

Honeywell
http://www.honeywell.com

IBM Corporation
http://www.ibm.com

International Paper
http://www.ipaper.com

Johnson & Johnson
http://www.johnsonjohnson.com

L.L. Bean, Inc.
http://www.llbean.com

Lucent Technologies
http://www.lucent.com

McDonald's Corporation
http://www.mcdonalds.com

Mead Corporation
http://www.mead.com

Microsoft Corporation
http://www.microsoft.com

Minnesota Mining and Manufacturing (3M)
http://www.mmm.com

Morton International, Inc.
http://www.morton.com

Motorola, Inc.
http://www.motorola.com

Novell, Inc.
http://www.novell.com

Oracle Corporation
http://www.oracle.com

Polaroid Corporation
http://www.polaroid.com

Procter & Gamble Co.
http://www.pg.com

The Prudential Insurance Company of America
http://www.prudential.com

Quicken.com
http://www.quicken.com

SAP America, Inc.
http://www.sap.com

Sears Roebuck and Company
http://www.sears.com

Siebel Systems, Inc.
http://www.siebel.com

Sprint Corporation
http://www.sprint.com

Toyota Motor Corporation
http://www.toyota.com

The Travelers Insurance Company
http://www.travelers.com

Union Carbide
http://www.unioncarbide.com

United Way of America
http://www.unitedway.org

US Airways, Inc.
http://www.usair.com

Visa International
http://www.visa.com

Wal-Mart Stores, Inc.
http://www.walmart.com

Xerox Corporation
http://www.xerox.com

The Yankee Group
http://www.yankeegroup.com

these web sites, one can easily obtain a variety of product and other types of information concerning these companies. In addition, a number of companies now specialize in making information available over the Internet. As a result, one can obtain access to stock quotes, meal prices at restaurants, salary data, and an almost infinite variety of information.

Government agencies are another important source of existing data. For instance, the U.S. Department of Labor maintains considerable data on employment rates, wage rates, size of the labor force, and union membership. Table 1.3 lists selected governmental agencies and some of the data they provide. Most government agencies that collect and process data also make the results available through a web site. For instance, the Bureau of the Census has a wealth of data at its web site, *http://www.census.gov*. Figure 1.3 shows the home page for the Bureau of the Census.

Statistical Studies

Sometimes the data needed for a particular application are not available through existing sources. In such cases, the data can often be obtained by conducting a statistical study. Statistical studies can be classified as either *experimental* or *observational*.

The largest experimental statistical study ever conducted is believed to be the 1954 Public Health Service experiment for the Salk polio vaccine. Nearly 2 million children in grades 1, 2, and 3 were selected from throughout the United States.

In an experimental study, a variable of interest is first identified. Then one or more other variables are identified and controlled so that data can be obtained about how they influence the variable of interest. For example, a pharmaceutical firm might be interested in conducting an experiment to learn about how a new drug affects blood pressure. Blood pressure is the variable of interest in the study. The dosage level of the new drug is another variable that is hoped to have a causal effect on blood pressure. To obtain data about the effect of the new drug, a sample of individuals is selected. The dosage level of the new drug is controlled, with different groups of individuals being given different dosage levels. Data on blood pressure are collected for each group. Statistical analysis of the experimental data can help determine how the new drug affects blood pressure.

Studies of smokers and nonsmokers are observational studies because researchers do not determine or control who will smoke and who will not smoke.

In nonexperimental, or observational, statistical studies, no attempt is made to control the variables of interest. A survey is perhaps the most common type of observational study. For instance, in a personal interview survey, research questions are first identified. Then a questionnaire is designed and administered to a sample of individuals. Some restaurants use observational studies to obtain data about their customers' opinions of the quality of food, service, atmosphere, and so on. A questionnaire used by the Lobster Pot Restaurant in Redington Shores, Florida, is shown in Figure 1.4 (page 12). Note that the customers completing the questionnaire are asked to provide ratings for five variables: food quality, friendliness

TABLE 1.3 EXAMPLES OF DATA AVAILABLE FROM SELECTED GOVERNMENT AGENCIES

Government Agency	Some of the Data Available
Bureau of the Census *http://www.census.gov*	Population data and their distribution, data on number of households and their distribution, data on household income and their distribution
Federal Reserve Board *http://www.bog.frb.fed.us*	Data on the money supply, installment credit, exchange rates, and discount rates
Office of Management and Budget *http://www.whitehouse.gov/omb*	Data on revenue, expenditures, and debt of the federal government
Department of Commerce *http://www.doc.gov*	Data on business activity, value of shipments by industry, level of profits by industry, and growing and declining industries

FIGURE 1.3 U.S. BUREAU OF THE CENSUS HOMEPAGE

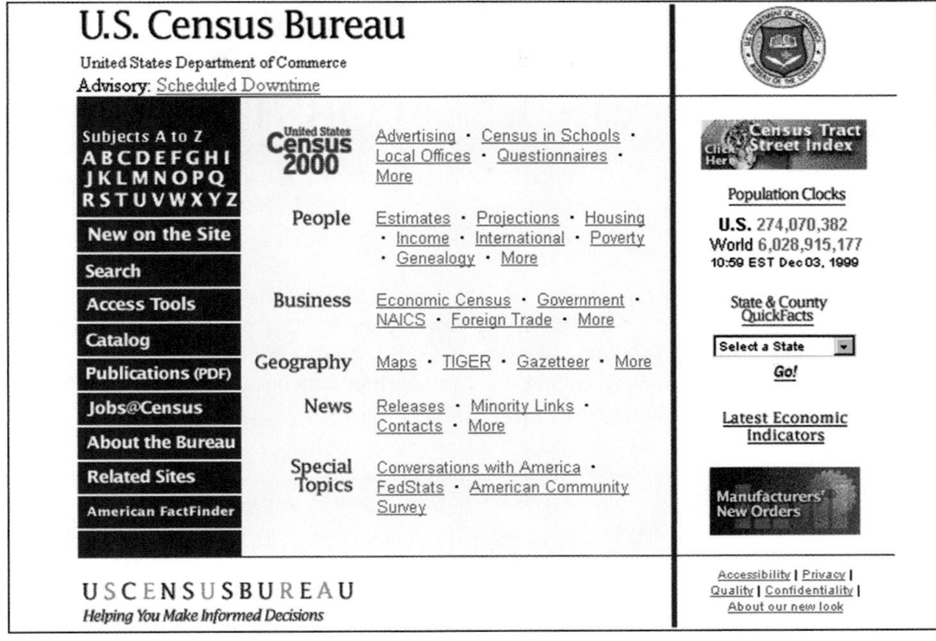

of service, promptness of service, cleanliness, and management. The response categories of excellent, good, satisfactory, and unsatisfactory provide data that enable Lobster Pot's managers to assess the quality of the restaurant's operation.

Managers wanting to use data and statistical analyses as an aid to decision making must be aware of the time and cost required to obtain the data. The use of existing data sources is desirable when data must be obtained in a relatively short period of time. If important data are not readily available from an existing source, the additional time and cost involved in obtaining the data must be taken into account. In all cases, the decision maker should consider the contribution of the statistical analysis to the decision-making process. The cost of data acquisition and the subsequent statistical analysis should not exceed the savings generated by using the information to make a better decision.

Data Acquisition Errors

Managers should always be aware of the possibility of data errors in statistical studies. Using erroneous data can be worse than not using any data at all. An error in data acquisition occurs whenever the data value obtained is not equal to the true or actual value that would have been obtained with a correct procedure. Such errors can occur in a number of ways. For example, an interviewer might make a recording error, such as a transposition in writing the age of a 24-year-old person as 42, or the person answering an interview question might misinterpret the question and provide an incorrect response.

Experienced data analysts take great care in collecting and recording data to ensure that errors are not made. Special procedures can be used to check for internal consistency of the data. For instance, such procedures would indicate that the analyst should review the accuracy of data for a respondent who is shown to be 22 years of age but who reports 20 years of work experience. Data analysts also review data with unusually large and small values,

FIGURE 1.4 CUSTOMER OPINION QUESTIONNAIRE USED BY THE LOBSTER POT RESTAURANT, REDINGTON SHORES, FLORIDA (*Used with permission*)

The LOBSTER Pot
RESTAURANT

We are happy you stopped by the Lobster Pot Restaurant and want to make sure you will come back. So, if you have a little time, we will really appreciate it if you will fill out this card. Your comments and suggestions are extremely important to us. Thank you!

Server's Name ____________________

	Excellent	Good	Satisfactory	Unsatisfactory
Food Quality	❑	❑	❑	❑
Friendly Service	❑	❑	❑	❑
Prompt Service	❑	❑	❑	❑
Cleanliness	❑	❑	❑	❑
Management	❑	❑	❑	❑

Comments ____

What prompted your visit to us? ____________________________

Please drop in suggestion box at entrance. Thank you.

called outliers, which are candidates for possible data errors. In Chapter 3 we present some of the methods statisticians use to identify outliers.

Errors often occur during data acquisition. Blindly using any data that happen to be available or using data that were acquired with little care can lead to misleading information and bad decisions. Thus, taking steps to acquire accurate data can help ensure reliable and valuable decision-making information.

1.4 DESCRIPTIVE STATISTICS

Most of the statistical information in newspapers, magazines, company reports, and other publications consists of data that are summarized and presented in a form that is easy for the reader to understand. Such summaries of data, which may be tabular, graphical, or numerical, are referred to as **descriptive statistics.**

Refer again to the data set in Table 1.1 where data on 25 shadow stocks are presented. Methods of descriptive statistics can be used to provide summaries of the information in this data set. For example, a tabular summary of the data for the qualitative variable Exchange is shown in Table 1.4. A graphical summary of the same data, called a bar graph, is shown in Figure 1.5. The purpose of these types of tabular and graphical summaries is to

TABLE 1.4 FREQUENCIES AND PERCENT FREQUENCIES FOR THE EXCHANGE VARIABLE

Exchange	Frequency	Percent Frequency
New York Stock Exchange (NYSE)	5	20
American Stock Exchange (AMEX)	6	24
Over-the-counter (OTC)	14	56
Totals	25	100

make the data easier to interpret. Referring to Table 1.4 and Figure 1.5, we can see easily that the majority of the stocks in the data set are traded over-the-counter. On a percentage basis, 56% are traded over-the-counter, 24% are traded on the American Stock Exchange, and 20% are traded on the New York Stock Exchange.

A graphical summary of the data for the quantitative variable price/earnings ratio for the shadow stocks, called a histogram, is provided in Figure 1.6. From the histogram, it is easy to see that the price/earnings ratios range from 0.0 to 39.9, with the highest concentrations between 8.0 and 15.9.

In addition to tabular and graphical displays, numerical descriptive statistics are used to summarize data. The most common numerical descriptive statistic is the average, or mean. Using the data on annual sales for the shadow stocks in Table 1.1, we can compute the average annual sales by adding the annual sales for all 25 stocks and dividing the sum by 25. Doing so provides average annual sales of $125.54 million. This average is taken as a measure of the central tendency, or central location, of the data.

In recent years interest has grown in statistical methods that can be used for developing and presenting descriptive statistics. Chapters 2 and 3 are devoted to the tabular, graphical, and numerical methods of descriptive statistics.

FIGURE 1.5 BAR GRAPH OF THE EXCHANGE VARIABLE

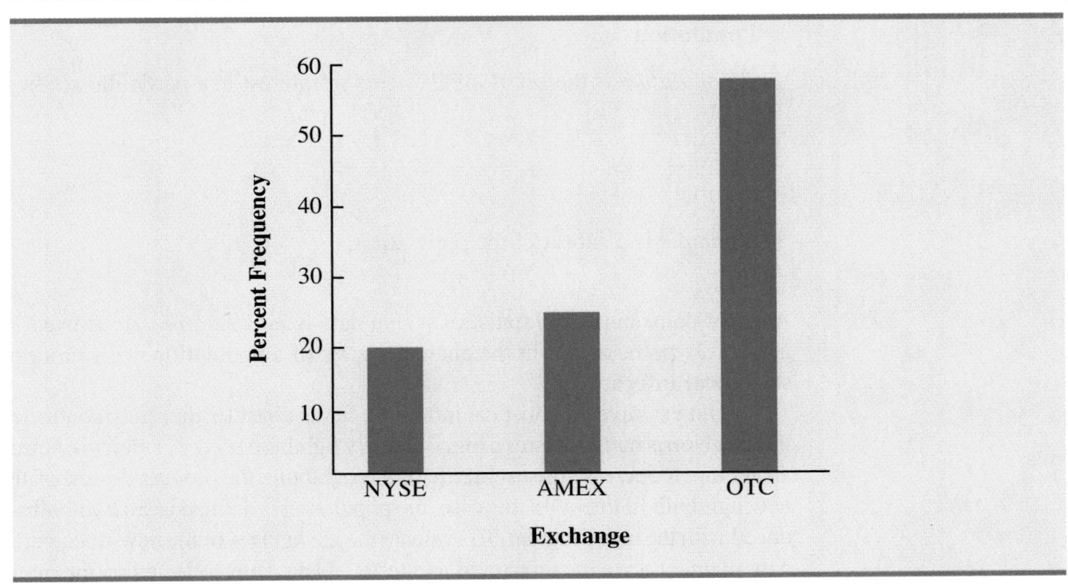

FIGURE 1.6 HISTOGRAM FOR 25 SHADOW STOCKS

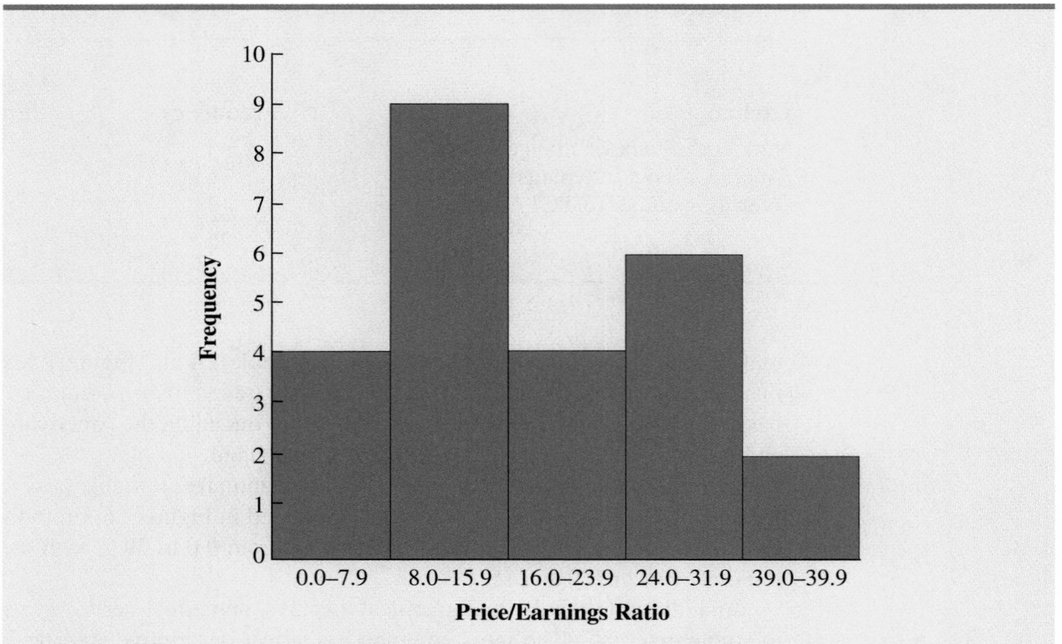

1.5 STATISTICAL INFERENCE

In many situations, data are sought for a large group of elements (individuals, companies, voters, households, products, customers, and so on). Because of time, cost, and other considerations, data are collected from only a small portion of the group. The larger group of elements in a particular study is called the **population,** and the smaller group is called the **sample.** Formally, we use the following definitions.

Population

A *population* is the set of all elements of interest in a particular study.

Sample

A *sample* is a subset of the population.

A major contribution of statistics is that data from a sample can be used to make estimates and test hypotheses about the characteristics of a population. This process is referred to as **statistical inference.**

As an example of statistical inference, let us consider the study conducted by Norris Electronics. Norris manufactures a high-intensity lightbulb used in a variety of electrical products. In an attempt to increase the useful life of the lightbulb, the product design group has developed a new lightbulb filament. In this case, the population is defined as all lightbulbs that could be produced with the new filament. To evaluate the advantages of the new filament, 200 bulbs with the new filament were manufactured and tested. Data were collected on the number of hours each lightbulb operated before filament burnout. The data from this sample are reported in Table 1.5.

TABLE 1.5 HOURS UNTIL BURNOUT FOR A SAMPLE OF 200 LIGHTBULBS FOR NORRIS ELECTRONICS

CD file
Norris

107	73	68	97	76	79	94	59	98	57
54	65	71	70	84	88	62	61	79	98
66	62	79	86	68	74	61	82	65	98
62	116	65	88	64	79	78	79	77	86
74	85	73	80	68	78	89	72	58	69
92	78	88	77	103	88	63	68	88	81
75	90	62	89	71	71	74	70	74	70
65	81	75	62	94	71	85	84	83	63
81	62	79	83	93	61	65	62	92	65
83	70	70	81	77	72	84	67	59	58
78	66	66	94	77	63	66	75	68	76
90	78	71	101	78	43	59	67	61	71
96	75	64	76	72	77	74	65	82	86
66	86	96	89	81	71	85	99	59	92
68	72	77	60	87	84	75	77	51	45
85	67	87	80	84	93	69	76	89	75
83	68	72	67	92	89	82	96	77	102
74	91	76	83	66	68	61	73	72	76
73	77	79	94	63	59	62	71	81	65
73	63	63	89	82	64	85	92	64	73

Suppose Norris is interested in using the sample data to make an inference about the average hours of useful life for the population of all lightbulbs that could be produced with the new filament. Adding the 200 values in Table 1.5 and dividing the total by 200 provides the sample average lifetime for the lightbulbs: 76 hours. We can use this sample result to estimate that the average lifetime for the lightbulbs in the population is 76 hours. Figure 1.7 is a graphical summary of the statistical inference process for Norris Electronics.

FIGURE 1.7 THE PROCESS OF STATISTICAL INFERENCE FOR THE NORRIS ELECTRONICS EXAMPLE

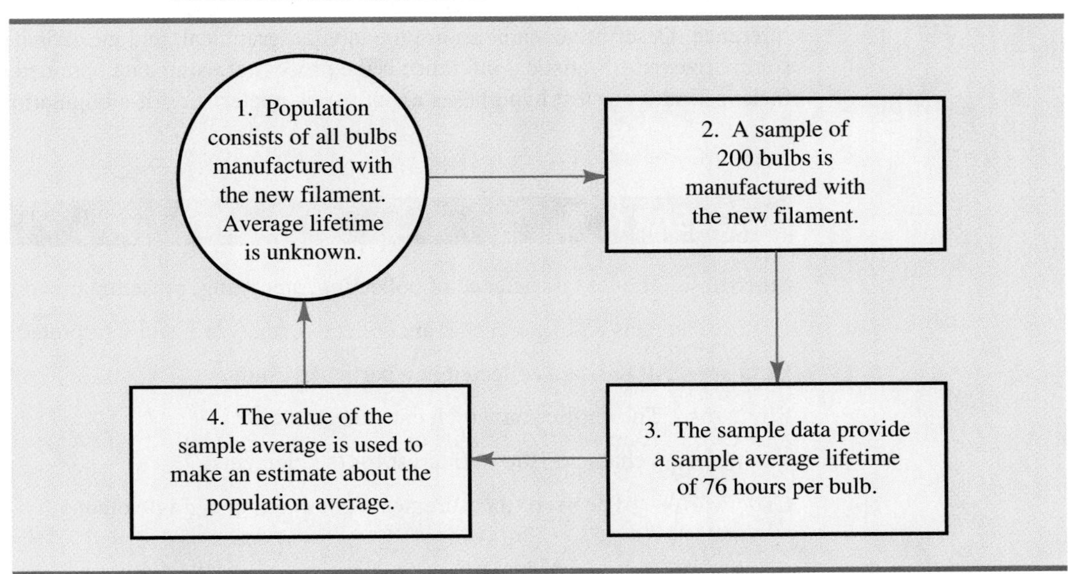

Whenever statisticians use a sample to estimate population characteristics of interest, they usually provide a statement of the quality, or precision, associated with the estimate. For the Norris example, the statistician might state that the estimate of the average lifetime for the population of new lightbulbs is 76 hours with a margin of error of ±4 hours. Thus, an interval estimate of the average lifetime for all lightbulbs produced with the new filament is 72 hours to 80 hours. The statistician can also state how confident he or she is that the interval from 72 hours to 80 hours contains the population average.

SUMMARY

Statistics is the art and science of collecting, analyzing, presenting, and interpreting data. Nearly every college student majoring in business or economics is required to take a course in statistics. We began the chapter by describing typical statistical applications for business and economics.

Data are the facts and figures that are collected, analyzed, presented, and interpreted. Four scales of measurement are available for obtaining data on a particular variable: nominal, ordinal, interval, and ratio. The scale of measurement for a variable is nominal when the data are labels or names used to identify an attribute of an element. The scale is ordinal if the data have the properties of nominal data and the order or rank of the data is meaningful. The scale is interval if the data have the properties of ordinal data and the interval between observations is expressed in terms of a fixed unit of measure. Finally, the scale of measurement is ratio if the data have all the properties of interval data and the ratio of two values is meaningful.

For purposes of statistical analysis, data can be classified as qualitative or quantitative. Qualitative data are labels or names used to identify an attribute of each element. Qualitative data use either the nominal or ordinal scale of measurement and may be nonnumeric or numeric. Quantitative data are numeric values that indicate how much or how many. Quantitative data use either the interval or ratio scale of measurement. Ordinary arithmetic operations are meaningful only if the data are quantitative. Therefore, statistical computations used for quantitative data are not always appropriate for qualitative data.

In Sections 1.4 and 1.5 we introduced the topics of descriptive statistics and statistical inference. Descriptive statistics are the tabular, graphical, and numerical methods used to summarize data. Statistical inference is the process of using data obtained from a sample to make estimates or test hypotheses about the characteristics of a population.

GLOSSARY

Statistics The art and science of collecting, analyzing, presenting, and interpreting data.

Data The facts and figures that are collected, analyzed, and interpreted.

Data set All the data collected in a particular study.

Elements The entities on which data are collected.

Variable A characteristic of interest for the elements.

Observation The set of measurements obtained for a single element.

Nominal scale A scale of measurement for a variable that uses a label or name to identify an attribute of an element. Nominal data may be nonnumeric or numeric.

Ordinal scale A scale of measurement for a variable that has the properties of nominal data and can be used to rank or order the data. Ordinal data may be nonnumeric or numeric.

Interval scale A scale of measurement for a variable that has the properties of ordinal data and the interval between observations is expressed in terms of a fixed unit of measure. Interval data are always numeric.

Ratio scale A scale of measurement for a variable that has all the properties of interval data and the ratio of two values is meaningful. Ratio data are always numeric.

Qualitative data Data that are labels or names used to identify an attribute of each element. Qualitative data use the nominal or ordinal scale of measurement and may be nonnumeric or numeric.

Quantitative data Data that indicate how much or how many of something. Quantitative data use the interval or ratio scale of measurement and are always numeric.

Qualitative variable A variable with qualitative data.

Quantitative variable A variable with quantitative data.

Cross-sectional data Data collected at the same or approximately the same point in time.

Time series data Data collected at several successive periods of time.

Descriptive statistics Tabular, graphical, and numerical methods used to summarize data.

Population The set of all elements of interest in a particular study.

Sample A subset of the population.

Statistical inference The process of using data obtained from a sample to make estimates or test hypotheses about the characteristics of a population.

EXERCISES

1. Discuss the differences between statistics as numerical facts and statistics as a discipline or field of study.

SELF test

2. *Condé Nast Traveler* conducts an annual poll of subscribers in order to determine the best places to stay throughout the world. Table 1.6 is a sample of nine European hotels from their most recent poll. The price of a standard double room during the hotel's high season ranges from $ (lowest price) to $$$$ (highest price). The overall score includes subscribers' evaluations of each hotel's rooms, service, restaurants, location/atmosphere, and public areas; higher overall scores correspond to a higher level of satisfaction.

a. How many elements are in this data set?

b. How many variables are in this data set?

c. Which variables are qualitative and which variables are quantitative?

d. What type of measurement scale is being used for each of the variables?

3. Refer to Table 1.6.

a. What is the average number of rooms for the nine hotels?

b. Compute the average overall score.

c. What is the percentage of hotels located in England?

d. What is the percentage of hotels with a room rate of $$?

TABLE 1.6 RATINGS FOR PLACES TO STAY IN EUROPE

Name of Property	Country	Room Rate	Number of Rooms	Overall Score
Graveteye Manor	England	$$	18	83.6
Villa d'Este	Italy	$$$$	166	86.3
Hotel Prem	Germany	$	54	77.8
Hotel d'Europe	France	$$	47	76.8
Palace Luzern	Switzerland	$$	326	80.9
Royal Crescent Hotel	England	$$$	45	73.7
Hotel Sacher	Austria	$$$	120	85.5
Duc de Bourgogne	Belgium	$	10	76.9
Villa Gallici	France	$$	22	90.6

Source: Condé Nast Traveler, January 2000.

SELF test

4. *Fortune* magazine provides data on how the 500 largest U.S. industrial corporations rank in terms of revenues and profits. Data for a sample of *Fortune* 500 companies are given in Table 1.7.
   a. How many elements are in this data set?
   b. What is the population?
   c. Compute the average revenue for the sample.
   d. Using the results in part (c), what is the estimate of the average revenues for the population?
5. Consider the data set for the sample of *Fortune* 500 companies in Table 1.7.
   a. How many variables are in the data set?
   b. Which of the variables are qualitative and which are quantitative?
   c. Compute the average profit for the sample.
   d. What percentage of the companies earned a profit over $100 million?
   e. What percentage of the companies have an industry code of 3?
6. Columbia House provides CDs and tapes to its mail-order club members. A Columbia House Music Survey asked new club members to complete an 11-question survey. Some of the questions asked were:
   a. How many CDs and tapes have you bought in the last 12 months?
   b. Are you currently a member of a national mail-order book club? (Yes or No)

TABLE 1.7 A SAMPLE OF 10 *FORTUNE* 500 COMPANIES

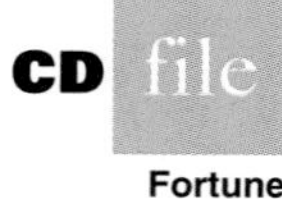

Company	Revenue ($ millions)	Profit ($ millions)	Industry Code
US Airways Group	8688.0	538.0	3
International Paper	19500.0	213.0	23
Tyson Foods	7414.1	25.1	20
Hewlett-Packard	47061.0	2945.0	13
Intel	26273.0	6068.0	49
Northrup Grumman	8902.0	214.0	2
Seagate Technology	6819.0	−530.0	11
Unisys	7208.4	387.0	10
Westvaco	2904.7	132.0	23
Campbell Soup	7505.0	660.0	20

Source: Fortune, April 26, 1999.

c. What is your age?
d. Including yourself, how many people (adults and children) are in your household?
e. What kind of music are you interested in buying? (15 categories were listed, including hard rock, soft rock, adult contemporary, heavy metal, rap, and country)

Comment on whether each question provides qualitative or quantitative data.

7. A California state agency classifies worker occupations as professional, white collar, or blue collar. The data are recorded with 1 denoting professional, 2 denoting white collar, and 3 denoting blue collar.
  a. The variable is worker occupation. Is it a qualitative or quantitative variable?
  b. What type of measurement scale is being used for this variable?

8. A *Wall Street Journal*/NBC News poll asked 2013 adults, "How satisfied are you with the state of the U.S. economy today?" (*The Wall Street Journal,* December 12, 1997). Response categories were Dissatisfied, Satisfied, and Not Sure.
  a. What was the sample size for this survey?
  b. Are the data qualitative or quantitative?
  c. Would it make more sense to use averages or percentages as a summary of the data for this question?
  d. Of the respondents, 28% said that they were dissatisfied with the state of the U.S. economy. How many individuals provided this response?

9. The Commerce Department reported receiving the following applications for the Malcolm Baldrige National Quality Award: 23 from large manufacturing firms, 18 from large service firms, and 30 from small businesses.
  a. Is type of business a qualitative or quantitative variable?
  b. What percentage of the applications came from small businesses?

10. State whether each of the following variables is qualitative or quantitative and indicate the measurement scale that is appropriate for each.
  a. Age
  b. Gender
  c. Class rank
  d. Make of automobile
  e. Number of people favoring the death penalty

11. State whether each of the following variables is qualitative or quantitative and indicate the measurement scale being used.
  a. Annual sales
  b. Soft-drink size (small, medium, large)
  c. Employee classification (GS1 through GS18)
  d. Earnings per share
  e. Method of payment (cash, check, credit card)

12. The Hawaii Visitors Bureau collects data on visitors to Hawaii. The following questions were among 16 asked in a questionnaire handed out to passengers during incoming airline flights in June.
  - This trip to Hawaii is my: 1st, 2nd, 3rd, 4th, etc.
  - The primary reason for this trip is: (10 categories including vacation, convention, honeymoon)
  - Where I plan to stay: (11 categories including hotel, apartment, relatives, camping)
  - Total days in Hawaii

  a. What is the population being studied?
  b. Is the use of a questionnaire a good way to reach the population of passengers on incoming airline flights?
  c. Comment on each of the four questions in terms of whether it will provide qualitative or quantitative data.

13. Figure 1.8 is a bar graph providing data on the number of riverboat casinos over the years 1991–1997.
    a. Are the data qualitative or quantitative?
    b. Are the data time series or cross-sectional?
    c. What is the variable of interest?
    d. Comment on the trend over time. Would you expect to see an increase or a decrease in 1998?

14. The following data set provides a snapshot of the financial performance of Ameritech Corporation (*Barrons,* December 29, 1997).

	1993	1994	1995	1996
Earnings per Share	\$2.78	\$2.13	\$3.41	\$3.83
Revenues (billions)	\$11.87	\$12.57	\$13.43	\$14.92
Net Income (billions)	\$1.51	\$1.17	\$1.89	\$2.12
Book Value per Share	\$14.35	\$10.98	\$12.67	\$13.98

    a. How many variables are there?
    b. Are the data qualitative or quantitative?
    c. Are they cross-sectional or time series data? Why?

15. Refer again to the data in Table 1.7 for *Fortune* 500 companies. Are they cross-sectional or time series data? Why?

16. The marketing group at your company has come up with a new diet soft drink that it claims will capture a large share of the young adult market.
    a. What data would you want to see before deciding to invest substantial funds in introducing the new product into the marketplace?
    b. How would you expect the data mentioned in part (a) to be obtained?

FIGURE 1.8 RIVERBOAT CASINOS IN THE UNITED STATES

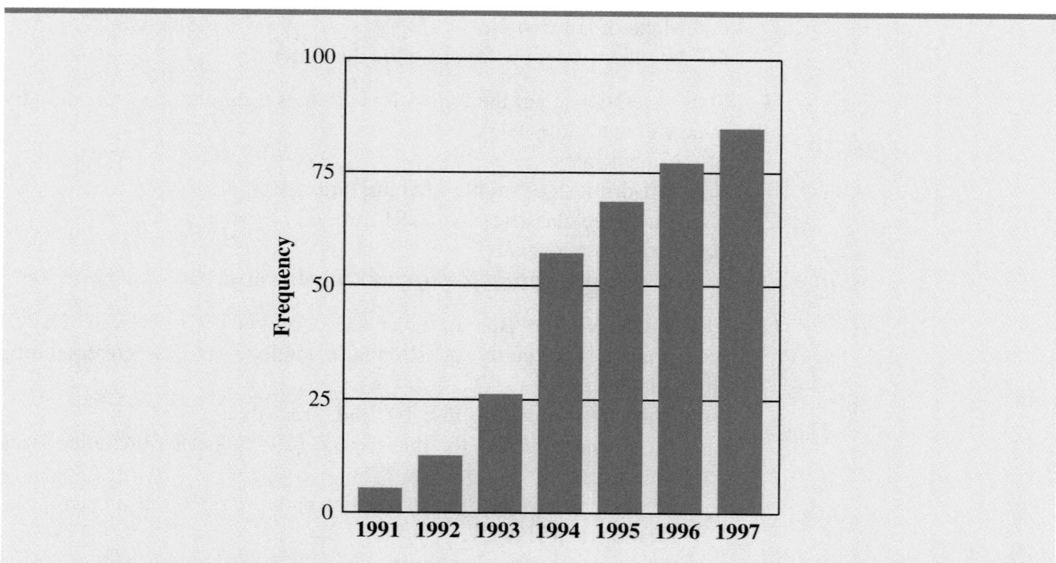

Source: Reprinted by permission of The *Wall Street Journal,* ©1997 Dow Jones & Company, Inc. All rights reserved worldwide.

17. A manager of a large corporation has recommended that a $10,000 raise be given to keep a valued subordinate from moving to another company. What internal and external sources of data might be used to decide whether such a salary increase is appropriate?

18. In a recent study of causes of death in men 60 years of age and older, a sample of 120 men indicated that 48 died as a result of some form of heart disease.

a. Develop a descriptive statistic that can be used as an estimate of the percentage of men 60 years of age or older who die from some form of heart disease.

b. Are the data on cause of death qualitative or quantitative?

c. Discuss the role of statistical inference in this type of medical research.

19. The *Business Week* 1996 North American subscriber study collected data from a sample of 2861 subscribers. Fifty-nine percent of the respondents indicated that their annual income was $75,000 or more and 50% reported having an American Express credit card.

a. What is the population of interest in this study?

b. Is annual income a qualitative or quantitative variable?

c. Is ownership of an American Express card a qualitative or quantitative variable?

d. Does this study involve cross-sectional or time series data?

e. Describe any statistical inferences *Business Week* might make on the basis of the survey.

20. A Scanner Data User Survey of 50 companies provided the following findings (Mercer Management Consulting, Inc., April 24, 1997):

- ACNielson earned 56% of the dollar-share of the market.
- The average amount spent on scanner data per category of consumer goods was $387,325.
- On a scale of 1 (very dissatisfied) to 5 (very satisfied), the average level of overall satisfaction with scanner data was 3.73.

a. Cite two descriptive statistics.

b. Make an inference of the overall satisfaction in the population of all users of scanner data.

c. Make an inference about the average amount spent per category for scanner data for consumer goods.

21. A 7-year medical research study reported that women whose mothers took the drug DES during pregnancy were *twice* as likely to develop tissue abnormalities that might lead to cancer as were women whose mothers did not take the drug.

a. This study involved the comparison of two populations. What were the populations?

b. Do you suppose the data were obtained in a survey or an experiment?

c. For the population of women whose mothers took the drug DES during pregnancy, a sample of 3980 women showed 63 developed tissue abnormalities that might lead to cancer. Provide a descriptive statistic that could be used to estimate the number of women out of 1000 in this population who have tissue abnormalities.

d. For the population of women whose mothers did not take the drug DES during pregnancy, what is the estimate of the number of women out of 1000 who would be expected to have tissue abnormalities?

e. Medical studies often use a relatively large sample (in this case, 3980). Why?

22. A firm is interested in testing the advertising effectiveness of a new television commercial. As part of the test, the commercial is shown on a 6:30 P.M. local news program in Denver, Colorado. Two days later, a market research firm conducts a telephone survey to obtain information on recall rates (percentage of viewers who recall seeing the commercial) and impressions of the commercial.

a. What is the population for this study?

b. What is the sample for this study?

c. Why would a sample be used in this situation? Explain.

23. The ACNielsen organization conducts weekly surveys of television viewing throughout the United States. The ACNielsen statistical ratings indicate the size of the viewing audience for each major network television program. Rankings of the television programs and of the viewing audience market shares for each network are published each week.

a. What is the ACNielsen organization attempting to measure?
b. What is the population?
c. Why would a sample be used for this situation?
d. What kinds of decisions or actions are based on the ACNielsen studies?

24. A sample of midterm grades for five students showed the following results: 72, 65, 82, 90, 76. Which of the following statements are correct, and which should be challenged as being too generalized?

a. The average midterm grade for the sample of five students is 77.
b. The average midterm grade for all students who took the exam is 77.
c. An estimate of the average midterm grade for all students who took the exam is 77.
d. More than half of the students who take this exam will score between 70 and 85.
e. If five other students are included in the sample, their grades will be between 65 and 90.

DESCRIPTIVE STATISTICS: TABULAR AND GRAPHICAL METHODS

CONTENTS

STATISTICS IN PRACTICE

COLGATE-PALMOLIVE COMPANY*
New York, New York

The Colgate-Palmolive Company started as a small soap and candle shop in New York City in 1806. Today, Colgate-Palmolive is a $9 billion company whose products can be found in more than 200 countries and territories around the world. While best known for its brand names of Colgate, Palmolive, Ajax, and Fab, the company also markets Mennen, Hill's Science Diet, and Hill's Prescription Diet products.

The Colgate-Palmolive Company uses statistics in its quality assurance program for home laundry detergent products. One concern is customer satisfaction with the quantity of detergent in a carton. Every carton in each size category is filled with the same amount of detergent by weight, but the volume of detergent is affected by the density of the detergent powder. For instance, if the powder density is on the heavy side, a smaller volume of detergent is needed to reach the carton's specified weight. As a result, the carton may appear to be underfilled when opened by the consumer.

To control the problem of heavy detergent powder, limits are placed on the acceptable range of powder density. Statistical samples are taken periodically, and the density of each powder sample is measured. Data summaries are then provided for operating personnel so that corrective action can be taken if necessary to keep the density within the desired quality specifications.

A frequency distribution for the densities of 150 samples taken over a one-week period and a histogram are shown in the accompanying table and figure. Density levels above .40 are unacceptably high. The frequency distribution and histogram show that the operation is meeting its quality guidelines with all of the densities less than or equal to .40. Managers viewing these statistical summaries would be pleased with the quality of the detergent production process.

In this chapter, you will learn about tabular and graphical methods of descriptive statistics such as frequency distributions, bar graphs, histograms, stem-and-leaf displays, crosstabulations, and others. The goal of these methods is to summarize data so that they can be easily understood and interpreted.

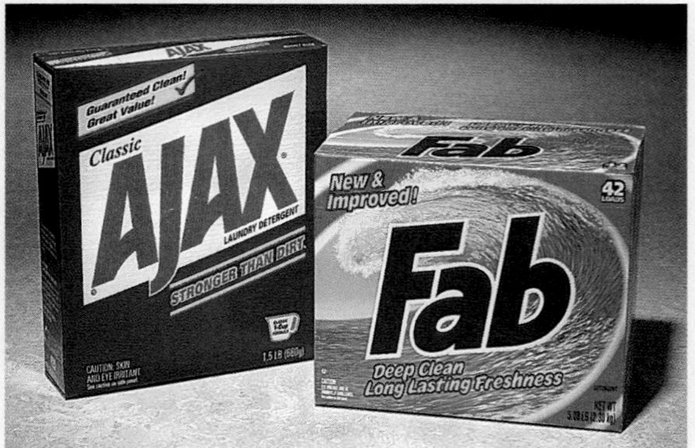

Statistical summaries help maintain the quality of these Colgate-Palmolive products. © Joe Higgins/South-Western.

Frequency Distribution of Density Data

Density	Frequency
.29–.30	30
.31–.32	75
.33–.34	32
.35–.36	9
.37–.38	3
.39–.40	1
Total	150

HISTOGRAM OF DENSITY DATA

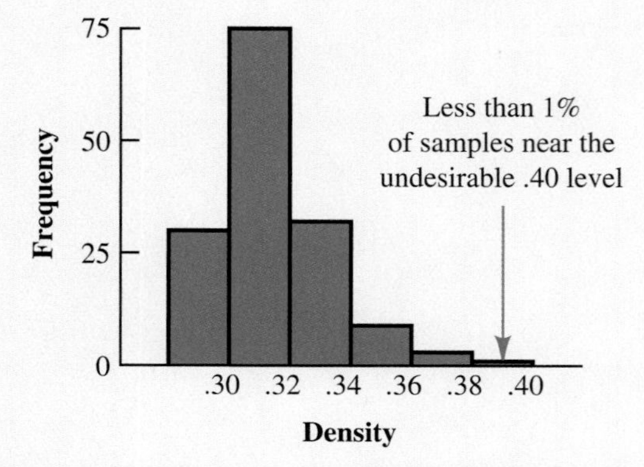

*The authors are indebted to William R. Fowle, Manager of Quality Assurance, Colgate-Palmolive Company, for providing this Statistics in Practice.

As indicated in Chapter 1, data can be classified as either qualitative or quantitative. **Qualitative data** are labels or names used to identify categories of like items. **Quantitative data** are numerical values that indicate how much or how many.

The purpose of this chapter is to introduce tabular and graphical methods commonly used to summarize both qualitative and quantitative data. Tabular and graphical summaries of data can be found in annual reports, newspaper articles, and research studies. Everyone is exposed to these types of presentations. Hence, it is important to understand how they are prepared and how they should be interpreted. We begin with tabular and graphical methods for summarizing data concerning a single variable. The last section introduces methods for summarizing data when the relationship between two variables is of interest.

Modern statistical software packages provide extensive capabilities for summarizing data and preparing graphical presentations. Minitab and Microsoft Excel are two packages that are widely available. In the chapter appendixes, we show some of their capabilities.

2.1 SUMMARIZING QUALITATIVE DATA

Frequency Distribution

We begin the discussion of how tabular and graphical methods can be used to summarize qualitative data with the definition of a **frequency distribution.**

Frequency Distribution

A frequency distribution is a tabular summary of data showing the number (frequency) of items in each of several nonoverlapping classes.

Let us use the following example to demonstrate the construction and interpretation of a frequency distribution for qualitative data. According to *Beverage Digest,* Coke Classic, Diet Coke, Dr. Pepper, Pepsi-Cola, and Sprite are the five top-selling soft drinks (*The Wall Street Journal Almanac,* 1998). Assume that the data in Table 2.1 show the soft drink selected for a sample of 50 soft drink purchases.

TABLE 2.1 DATA FROM A SAMPLE OF 50 SOFT DRINK PURCHASES

Coke Classic	Sprite	Pepsi-Cola
Diet Coke	Coke Classic	Coke Classic
Pepsi-Cola	Diet Coke	Coke Classic
Diet Coke	Coke Classic	Coke Classic
Coke Classic	Diet Coke	Pepsi-Cola
Coke Classic	Coke Classic	Dr. Pepper
Dr. Pepper	Sprite	Coke Classic
Diet Coke	Pepsi-Cola	Diet Coke
Pepsi-Cola	Coke Classic	Pepsi-Cola
Pepsi-Cola	Coke Classic	Pepsi-Cola
Coke Classic	Coke Classic	Pepsi-Cola
Dr. Pepper	Pepsi-Cola	Pepsi-Cola
Sprite	Coke Classic	Coke Classic
Coke Classic	Sprite	Dr. Pepper
Diet Coke	Dr. Pepper	Pepsi-Cola
Coke Classic	Pepsi-Cola	Sprite
Coke Classic	Diet Coke	

TABLE 2.2
FREQUENCY DISTRIBUTION OF SOFT DRINK PURCHASES

Soft Drink	Frequency
Coke Classic	19
Diet Coke	8
Dr. Pepper	5
Pepsi-Cola	13
Sprite	5
Total	50

To develop a frequency distribution for these data, we count the number of times each soft drink appears in Table 2.1. Coke Classic appears 19 times, Diet Coke appears 8 times, Dr. Pepper appears 5 times, Pepsi-Cola appears 13 times, and Sprite appears 5 times. These counts are summarized in the frequency distribution in Table 2.2.

This frequency distribution provides a summary of how the 50 soft drink purchases are distributed across the five soft drinks. This summary provides more insight than the original data shown in Table 2.1. Viewing the frequency distribution, we see that Coke Classic is the leader, Pepsi-Cola is second, Diet Coke is third, and Sprite and Dr. Pepper are tied for fourth. The frequency distribution has provided information about the relative popularity of the five best-selling soft drinks.

Relative Frequency and Percent Frequency Distributions

A frequency distribution shows the number (frequency) of items in each of several nonoverlapping classes. However, we are often interested in the proportion, or percentage, of items in each class. The relative frequency of a class is the fraction or proportion of items belonging to a class. For a data set with n observations, the relative frequency of each class is as follows:

Relative Frequency

$$\text{Relative Frequency of a Class} = \frac{\text{Frequency of the Class}}{n} \tag{2.1}$$

The *percent frequency* of a class is the relative frequency multiplied by 100.

A **relative frequency distribution** is a tabular summary of data showing the relative frequency for each class. A **percent frequency distribution** is a tabular summary of data showing the percent frequency for each class. Table 2.3 shows a relative frequency distribution and a percent frequency distribution for the soft drink data. In Table 2.3 we see that the relative frequency for Coke Classic is 19/50 = .38, the relative frequency for Diet Coke is 8/50 = .16, and so on. From the percent frequency distribution, we see that 38% of the purchases were Coke Classic, 16% of the purchases were Diet Coke, and so on. We can also note that 38% + 26% + 16% = 80% of the purchases were of the top three soft drinks.

Bar Graphs and Pie Charts

A **bar graph** is a graphical device for depicting data that have been summarized in a frequency, relative frequency, or percent frequency distribution. On one axis of the graph (usually the horizontal axis), we specify the labels that are used for the classes (categories) of data. A frequency, relative frequency, or percent frequency scale can be used for the other axis of the graph (usually the vertical axis). Then, using a bar of fixed width drawn above

TABLE 2.3 RELATIVE AND PERCENT FREQUENCY DISTRIBUTIONS OF SOFT DRINK PURCHASES

Soft Drink	Relative Frequency	Percent Frequency
Coke Classic	.38	38
Diet Coke	.16	16
Dr. Pepper	.10	10
Pepsi-Cola	.26	26
Sprite	.10	10
Total	1.00	100

FIGURE 2.1 BAR GRAPH OF SOFT DRINK PURCHASES

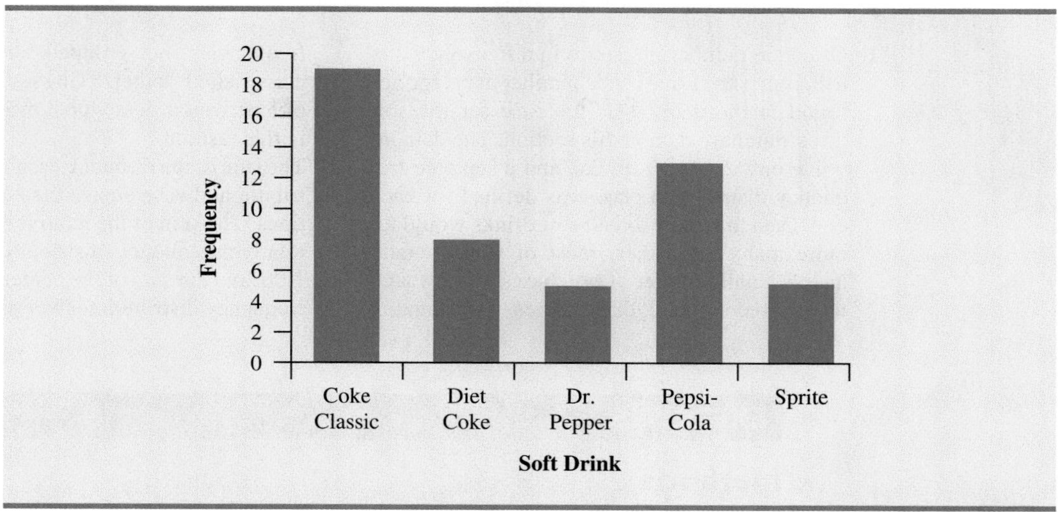

In quality control applications, bar graphs are used to identify the most important causes of problems. When the bars are arranged in descending order of height from left to right with the most frequently occurring cause appearing first, the bar graph is called a pareto diagram. *This diagram is named for its founder, Vilfredo Pareto, an Italian economist.*

each class label, we extend the length of the bar until we reach the frequency, relative frequency, or percent frequency of the class. For qualitative data, the bars should be separated to emphasize the fact that each class (category) is separate. Figure 2.1 is a bar graph of the frequency distribution for the 50 soft drink purchases. Note how the graphical presentation shows Coke Classic, Pepsi-Cola, and Diet Coke to be the most preferred brands.

The **pie chart** is another graphical device for presenting relative frequency and percent frequency distributions. To construct a pie chart, we first draw a circle to represent all of the data. Then we use the relative frequencies to subdivide the circle into sectors, or parts, that correspond to the relative frequency for each class. For example, because a circle has 360 degrees and Coke Classic has a relative frequency of .38, the sector of the pie chart labeled Coke Classic consists of .38(360) = 136.8 degrees. The sector of the pie chart labeled Diet Coke consists of .16(360) = 57.6 degrees. Similar calculations for the other classes yield the pie chart in Figure 2.2. The numerical values shown for each sector can be frequencies, relative frequencies, or percent frequencies.

FIGURE 2.2 PIE CHART OF SOFT DRINK PURCHASES

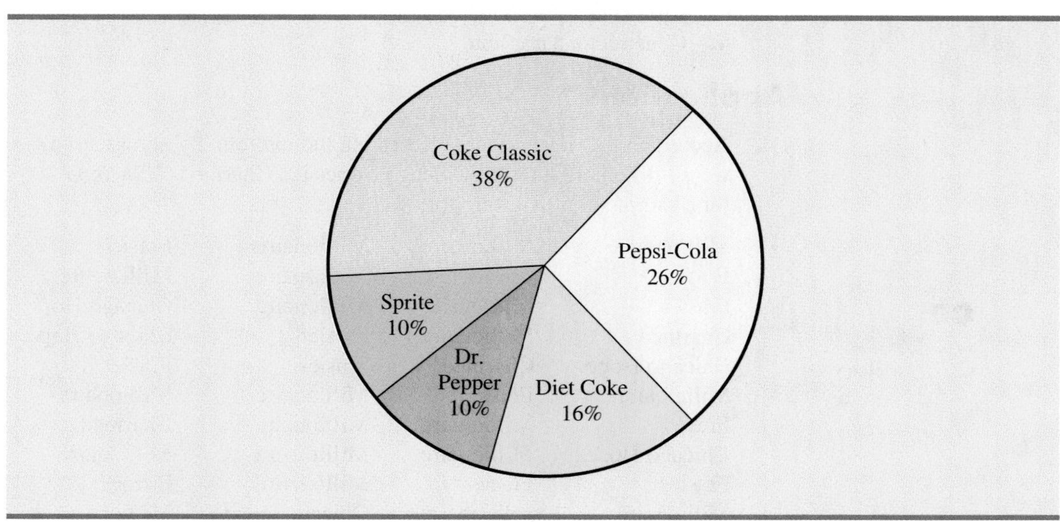

NOTES AND COMMENTS

1. Often the number of classes in a frequency distribution is the same as the number of categories found in the data, as is the case for the soft drink purchase data in this section. The data involve only five soft drinks, and a separate frequency distribution class was defined for each one. Data that included all soft drinks would require many categories, most of which would have a small number of purchases. Most statisticians recommend that classes with smaller frequencies be grouped into an aggregate class called "other." Classes with frequencies of 5 percent or less would most often be treated in this fashion.
2. The sum of the frequencies in any frequency distribution always equals the number of observations. The sum of the relative frequencies in any relative frequency distribution always equals 1.00, and the sum of the percentages in a percent frequency distribution always equals 100.

EXERCISES

Methods

1. The response to a question has three alternatives: A, B, and C. A sample of 120 responses provides 60 A, 24 B, and 36 C. Show the frequency and relative frequency distributions.
2. A partial relative frequency distribution is given.

Class	Relative Frequency
A	.22
B	.18
C	.40
D	

   a. What is the relative frequency of class D?
   b. The total sample size is 200. What is the frequency of class D?
   c. Show the frequency distribution.
   d. Show the percent frequency distribution.

3. A questionnaire provides 58 yes, 42 no, and 20 no-opinion answers.
   a. In the construction of a pie chart, how many degrees would be in the section of the pie showing the yes answers?
   b. How many degrees would be in the section of the pie showing the no answers?
   c. Construct a pie chart.
   d. Construct a bar graph.

Applications

4. According to Nielsen Media Research, the top four TV shows at 8:00 P.M., April 6, 2000, were Millionaire, Frasier, Chicago Hope, and Charmed (*USA Today,* April 13, 2000). Data for a sample of 50 viewers follow.

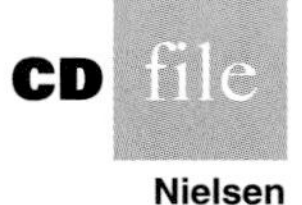

Nielsen

Millionaire	Millionaire	Millionaire	Frasier	Charmed
Frasier	Frasier	Millionaire	Millionaire	Frasier
Frasier	Millionaire	Millionaire	Chicago Hope	Millionaire
Charmed	Millionaire	Frasier	Chicago Hope	Millionaire
Chicago Hope	Charmed	Frasier	Frasier	Millionaire
Millionaire	Frasier	Millionaire	Millionaire	Chicago Hope
Frasier	Millionaire	Millionaire	Charmed	Chicago Hope
Chicago Hope	Millionaire	Millionaire	Millionaire	Millionaire
Frasier	Frasier	Millionaire	Frasier	Frasier
Millionaire	Millionaire	Chicago Hope	Millionaire	Frasier

a. Are these qualitative or quantitative data?
b. Provide frequency and percent frequency distributions for the data.
c. Construct a bar graph and a pie chart for the data.
d. On the basis of the sample, which show has the largest market share? Which one is second?

5. Freshmen entering the College of Business at Eastern University were asked to indicate their preferred major. The following data were obtained.

Major	Number
Management	55
Accounting	51
Finance	28
Marketing	82

Summarize the data by constructing:
a. Relative and percent frequency distributions
b. A bar graph
c. A pie chart

TABLE 2.4
THE EIGHT BEST-SELLING PAPERBACK BUSINESS BOOKS

- *The 7 Habits of Highly Effective People*
- *Investing for Dummies*
- *The Ernst & Young Tax Guide 2000*
- *The Millionaire Next Door*
- *The Motley Fool Investment Guide*
- *Rich Dad, Poor Dad*
- *The Wall Street Journal Guide to Understanding Money and Investing*
- *What Color is Your Parachute? 2000*

6. The eight best-selling paperback business books in February 2000 are listed in Table 2.4 (*Business Week,* April 3, 2000). Suppose a sample of book purchases in the Denver, Colorado, area provided the following data for these eight books.

7 Habits	Dad	7 Habits	Millionaire	Millionaire	WSJ Guide
Motley	Millionaire	Tax Guide	7 Habits	Dad	Dummies
Millionaire	Motley	Dad	Dad	Parachute	Dad
Dad	7 Habits	WSJ Guide	WSJ Guide	WSJ Guide	7 Habits
Motley	WSJ Guide	Millionaire	7 Habits	Millionaire	Millionaire
Millionaire	7 Habits	Millionaire	7 Habits	Motley	Motley
Motley	7 Habits	Dad	Dad	Dad	Dad
7 Habits	WSJ Guide	Tax Guide	Millionaire	Motley	Tax Guide
Motley	Motley	Millionaire	Millionaire	Dad	Dummies
Millionaire	Millionaire	Millionaire	Dad	Millionaire	Dad

a. Construct frequency and percent frequency distributions for the data. Group any books with a frequency of 5% or less in an "other" category.
b. Rank the best-selling books.
c. What percentage of the sales are represented by *The Millionaire Next Door* and *Rich Dad, Poor Dad?*

7. Leverock's Waterfront Steakhouse in Maderia Beach, Florida, uses a questionnaire to ask customers how they rate the server, food quality, cocktails, prices, and atmosphere at the restaurant. Each characteristic is rated on a scale of outstanding (O), very good (V), good (G), average (A), and poor (P). Use descriptive statistics to summarize the following data collected on food quality. What is your feeling about the food quality ratings at the restaurant?

G	O	V	G	A	O	V	O	V	G	O	V	A
V	O	P	V	O	G	A	O	O	O	G	O	V
V	A	G	O	V	P	V	O	O	G	O	O	V
O	G	A	O	V	O	O	G	V	A	G		

8. Position-by-position data for a sample of 55 members of the Baseball Hall of Fame in Cooperstown, New York, are shown here. Each observation indicates the primary position played by the Hall of Famers: pitcher (P), catcher (H), 1st base (1), 2nd base (2), 3rd base (3), shortstop (S), left field (L), center field (C), and right field (R).

L	P	C	H	2	P	R	1	S	S	1	L	P	R	P
P	P	P	R	C	S	L	R	P	C	C	P	P	R	P
2	3	P	H	L	P	1	C	P	P	P	S	1	L	R
R	1	2	H	S	3	H	2	L	P					

a. Use frequency and relative frequency distributions to summarize the data.
b. What position provides the most Hall of Famers?
c. What position provides the fewest Hall of Famers?
d. What outfield position (L, C, or R) provides the most Hall of Famers?
e. Compare infielders (1, 2, 3, and S) to outfielders (L, C, and R).

9. Employees at Electronics Associates are on a flextime system; they can begin their working day at 7:00, 7:30, 8:00, 8:30, or 9:00 A.M. The following data represent a sample of the starting times selected by the employees.

7:00	8:30	9:00	8:00	7:30	7:30	8:30	8:30	7:30	7:00
8:30	8:30	8:00	8:00	7:30	8:30	7:00	9:00	8:30	8:00

Summarize the data by constructing the following:
a. A frequency distribution
b. A percent frequency distribution
c. A bar graph
d. A pie chart
e. What do the summaries tell you about employee preferences in the flextime system?

10. Students in the College of Business Administration at the University of Cincinnati are asked to fill out a course evaluation questionnaire upon completion of their courses. It consists of a variety of questions that have a five-category response scale. One of the questions follows.

Compared to other courses you have taken, what is the overall quality of the course you are now completing?

____	____	____	____	____
Poor	Fair	Good	Very Good	Excellent

A sample of 60 students completing a course in business statistics during the spring quarter of 2000 provided the following responses. To aid in computer processing of the questionnaire results, a numeric scale was used with 1 = poor, 2 = fair, 3 = good, 4 = very good, and 5 = excellent.

3	4	4	5	1	5	3	4	5	2	4	5	3	4	4
4	5	5	4	1	4	5	4	2	5	4	2	4	4	4
5	5	3	4	5	5	2	4	3	4	5	4	3	5	4
4	3	5	4	5	4	3	5	3	4	4	3	5	3	3

a. Comment on why these are qualitative data.
b. Provide a frequency distribution and a relative frequency distribution summary of the data.
c. Provide a bar graph and a pie chart summary of the data.
d. On the basis of your summaries, comment on the students' overall evaluation of the course.

2.2 SUMMARIZING QUANTITATIVE DATA

Frequency Distribution

As defined in Section 2.1, a frequency distribution is a tabular summary showing the number (frequency) of items in each of several nonoverlapping classes. This definition holds for quantitative as well as qualitative data. However, with quantitative data we have

TABLE 2.5

YEAR-END AUDIT TIMES (IN DAYS)

12	14	19	18
15	15	18	17
20	27	22	23
22	21	33	28
14	18	16	13

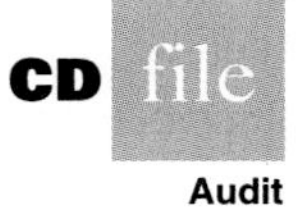

Audit

to be more careful in defining the nonoverlapping classes to be used in the frequency distribution.

For example, consider the quantitative data in Table 2.5. These data provide the time in days required to complete year-end audits for a sample of 20 clients of Sanderson and Clifford, a small public accounting firm. The three steps necessary to define the classes for a frequency distribution with quantitative data are as follow:

1. Determine the number of nonoverlapping classes.
2. Determine the width of each class.
3. Determine the class limits.

Let us demonstrate these steps by developing a frequency distribution for the audit-time data in Table 2.5.

Number of Classes. Classes are formed by specifying ranges that will be used to group the data. As a general guideline, we recommend using between 5 and 20 classes. For a small number of data items, as few as five or six classes may be used to summarize the data. For a larger number of data items, a larger number of classes is usually required. The goal is to use enough classes to show the variation in the data, but not so many classes that several contain only a few data items. Because the number of data items in Table 2.5 is relatively small ($n = 20$), we chose to develop a frequency distribution with five classes.

Width of the Classes. The second step in constructing a frequency distribution for quantitative data is to choose a width for the classes. As a general guideline, we recommend that the width be the same for each class. Thus the choices of the number of classes and the width of classes are not independent decisions. A larger number of classes means a smaller class width, and vice versa. To determine an approximate class width, we begin by identifying the largest and smallest data values. Then, once the desired number of classes has been specified, we can use the following expression to determine the approximate class width.

Making the classes the same width reduces the chance of inappropriate interpretations by the user.

$$\text{Approximate Class Width} = \frac{\text{Largest Data Value} - \text{Smallest Data Value}}{\text{Number of Classes}} \quad \textbf{(2.2)}$$

The approximate class width given by equation (2.2) can be rounded to a more convenient value based on the preference of the person developing the frequency distribution. For example, an approximate class width of 9.28 might be rounded to 10 simply because 10 is a more convenient class width to use in constructing and presenting a frequency distribution.

For the data involving the year-end audit times, the largest value is 33 and the smallest value is 12. Because we decided to summarize the data with five classes, using equation (2.2) provides an approximate class width of $(33 - 12)/5 = 4.2$. We therefore decided to use a class width of 5 days in the frequency distribution.

There is no one best frequency distribution for a data set. Different people may construct different, but equally acceptable, frequency distributions. The goal is to reveal the natural grouping and variation in the data.

In practice, the number of classes and the appropriate class width are determined by trial and error. Once a possible number of classes is chosen, equation (2.2) is used to find the approximate class width. The process can be repeated for a different number of classes. Ultimately, the analyst uses judgment to determine the combination of the number of classes and class width that provides the best frequency distribution for summarizing the data.

For the audit-time data in Table 2.5, after deciding to use five classes, each with a width of 5 days, the next task is to specify the class limits for each of the classes.

Class Limits. Class limits must be chosen so that each data item belongs to one and only one class. The *lower class limit* identifies the smallest possible data value assigned to the class. The *upper class limit* identifies the largest possible data value assigned to the class. In developing frequency distributions for qualitative data, we did not need to specify class limits because each data item naturally fell into a separate class (category). But with quantitative data, such as the audit times in Table 2.5, class limits are necessary to determine where each data value belongs.

For the audit-time data in Table 2.5 we defined the class limits as 10–14, 15–19, 20–24, 25–29, and 30–34. The smallest data value, 12, is included in the 10–14 class. The largest data value, 33, is included in the 30–34 class. For the 10–14 class, 10 is the lower class limit and 14 is the upper class limit. The difference between the lower class limits of adjacent classes is the class width. Using the first two lower class limits of 10 and 15, we see that the class width is $15 - 10 = 5$.

Once the number of classes, class width, and class limits have been determined, a frequency distribution can be obtained by *counting* the number of data values belonging to each class. For example, the data in Table 2.5 show that four values—12, 14, 14, and 13—belong to the 10–14 class. Thus, the frequency for the 10–14 class is 4. Continuing this counting process for the 15–19, 20–24, 25–29, and 30–34 classes provides the frequency distribution in Table 2.6. Using this frequency distribution, we can observe that:

TABLE 2.6

FREQUENCY DISTRIBUTION FOR THE AUDIT-TIME DATA

Audit Time (days)	Frequency
10–14	4
15–19	8
20–24	5
25–29	2
30–34	1
Total	20

1. The most frequently occurring audit times are in the class of 15–19 days. Eight of the 20 audit times belong to this class.
2. Only one audit required 30 or more days.

Other conclusions are possible, depending on the interests of the person viewing the frequency distribution. The value of a frequency distribution is that it provides insights about the data that are not easily obtained by viewing the data in their original unorganized form.

Class Midpoint. In some applications, we want to know the midpoints of the classes in a frequency distribution for quantitative data. The **class midpoint** is the value halfway between the lower and upper class limits. For the audit-time data, the five class midpoints are 12, 17, 22, 27, and 32.

Relative Frequency and Percent Frequency Distributions

We define the relative frequency and percent frequency distributions for quantitative data in the same manner as for qualitative data. First, recall that the relative frequency is simply the proportion of the observations belonging to a class. With n observations,

$$\text{Relative Frequency of Class} = \frac{\text{Frequency of the Class}}{n}$$

The percent frequency of a class is the relative frequency multiplied by 100.

Based on the class frequencies in Table 2.6 and with $n = 20$, Table 2.7 shows the relative frequency distribution and percent frequency distribution for the audit-time data. Note that .40 of the audits, or 40%, required from 15 to 19 days. Only .05 of the audits, or 5%, required 30 or more days. Again, additional interpretations and insights can be obtained by using Table 2.7.

Dot Plot

One of the simplest graphical summaries of data is a **dot plot.** A horizontal axis shows the range of values for the observations. Each observation is represented by a dot placed above the axis. Figure 2.3 is the dot plot for the audit-time data in Table 2.5. The three dots located above 18 on the horizontal axis indicate that there are three observations with a value

TABLE 2.7 RELATIVE AND PERCENT FREQUENCY DISTRIBUTIONS FOR THE AUDIT-TIME DATA

Audit Time (days)		Relative Frequency	Percent Frequency
10–14		.20	20
15–19		.40	40
20–24		.25	25
25–29		.10	10
30–34		.05	5
	Total	1.00	100

of 18. Dot plots show the details of the data and are useful for comparing the distribution of the data for two or more variables.

Histogram

A common graphical presentation of quantitative data is a **histogram.** This graphical summary can be prepared for data that have been previously summarized in either a frequency, relative frequency, or percent frequency distribution. A histogram is constructed by placing the variable of interest on the horizontal axis and the frequency, relative frequency, or percent frequency on the vertical axis. The frequency, relative frequency, or percent frequency of each class is shown by drawing a rectangle whose base is the class interval on the horizontal axis and whose height is the corresponding frequency, relative frequency, or percent frequency.

Figure 2.4 is a histogram for the audit-time data. Note that the class with the greatest frequency is shown by the rectangle appearing above the class of 15–19 days. The height of the rectangle shows that the frequency of this class is 8. A histogram for the relative or percent frequency distribution of this data would look the same as the histogram in Figure 2.4 with the exception that the vertical axis would be labeled with relative or percent frequency values.

As Figure 2.4 shows, the adjacent rectangles of a histogram touch one another. Unlike a bar graph, a histogram has no natural separation between the rectangles of adjacent classes. This format is the usual convention for histograms. Because the class limits for the audit-time data are stated as 10–14, 15–19, 20–24, 25–29, and 30–34, there appear to be one-unit intervals of 14 to 15, 19 to 20, 24 to 25, and 29 to 30 between the classes. These spaces are eliminated by drawing the vertical lines of the histogram halfway between the class limits. The vertical lines separating the classes for the histogram in Figure 2.4 are at 9.5, 14.5, 19.5, 24.5, 29.5, and 34.5. This minor adjustment to eliminate the spaces between

FIGURE 2.3 DOT PLOT FOR THE AUDIT-TIME DATA

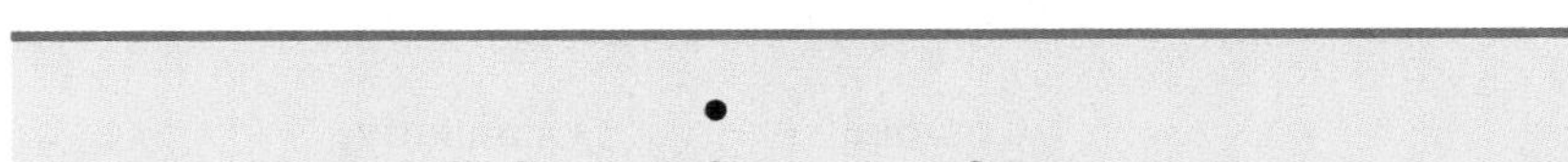
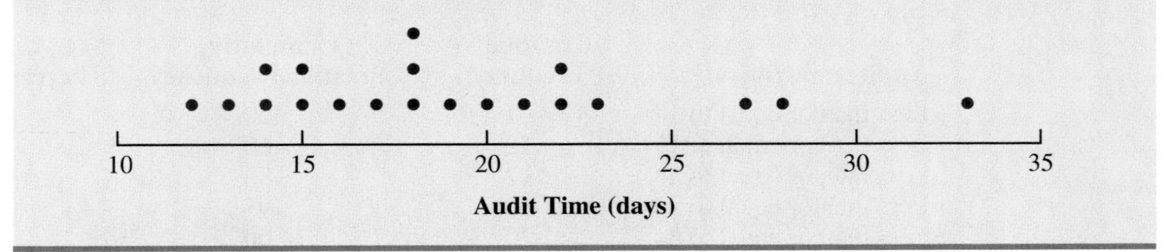

FIGURE 2.4 HISTOGRAM FOR THE AUDIT-TIME DATA

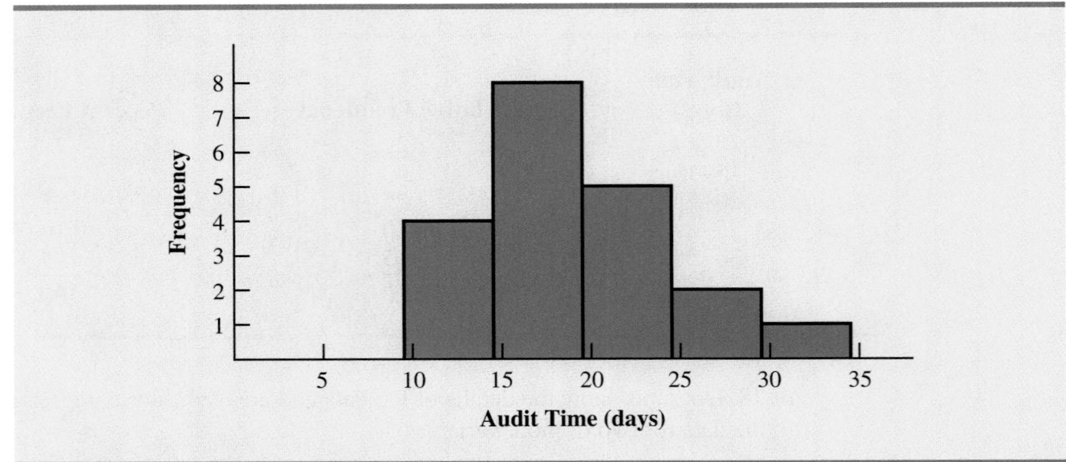

classes in a histogram helps show that, even though the data are rounded, all values between the lower limit of the first class and the upper limit of the last class are possible.

Cumulative Distributions

A variation of the frequency distribution that provides another tabular summary of quantitative data is the **cumulative frequency distribution.** The cumulative frequency distribution uses the number of classes, class widths, and class limits that were developed for the frequency distribution. However, rather than showing the frequency of each class, the cumulative frequency distribution shows the number of data items with values *less than or equal to the upper class limit* of each class. The first two columns of Table 2.8 provide the cumulative frequency distribution for the audit-time data.

To understand how the cumulative frequencies are determined, consider the class with the description "less than or equal to 24." The cumulative frequency for this class is simply the sum of the frequencies for all classes with data values less than or equal to 24. For the frequency distribution in Table 2.6, the sum of the frequencies for classes 10–14, 15–19, and 20–24 indicates that there are $4 + 8 + 5 = 17$ data values less than or equal to 24. Hence, the cumulative frequency for this class is 17. In addition, the cumulative frequency distribution in Table 2.8 shows that 4 audits were completed in 14 days or less and 19 audits were completed in 29 days or less.

TABLE 2.8 CUMULATIVE FREQUENCY, CUMULATIVE RELATIVE FREQUENCY, AND CUMULATIVE PERCENT FREQUENCY DISTRIBUTIONS FOR THE AUDIT-TIME DATA

Audit Time (days)	Cumulative Frequency	Cumulative Relative Frequency	Cumulative Percent Frequency
Less than or equal to 14	4	.20	20
Less than or equal to 19	12	.60	60
Less than or equal to 24	17	.85	85
Less than or equal to 29	19	.95	95
Less than or equal to 34	20	1.00	100

As a final point, we note that a **cumulative relative frequency distribution** shows the proportion of data items, and a **cumulative percent frequency distribution** shows the percentage of data items with values less than or equal to the upper limit of each class. The cumulative relative frequency distribution can be computed either by summing the relative frequencies in the relative frequency distribution or by dividing the cumulative frequencies by the total number of items. Using the latter approach, we found the cumulative relative frequencies in column 3 of Table 2.8 by dividing the cumulative frequencies in column 2 by the total number of items ($n = 20$). The cumulative percent frequencies were again computed by multiplying the relative frequencies by 100. The cumulative relative and percent frequency distributions show that .85 of the audits, or 85%, were completed in 24 days or less, .95 of the audits, or 95%, were completed in 29 days or less, and so on.

Ogive

A graph of a cumulative distribution is called an **ogive.** The data values are shown on the horizontal axis and either the cumulative frequencies, the cumulative relative frequencies, or the cumulative percent frequencies are shown on the vertical axis. Figure 2.5 is the ogive for the cumulative frequencies of the audit-time data in Table 2.8.

The ogive is constructed by plotting a point corresponding to the cumulative frequency of each class. Because the class limits for the audit-time data are 10–14, 15–19, 20–24, and so on, there appear to be one-unit gaps from 14 to 15, 19 to 20, and so on. As with the histogram, these gaps are eliminated by plotting points halfway between the class limits. Thus, 14.5 is used for the 10–14 class, 19.5 is used for the 15–19 class, and so on. The "less than or equal to 14" class with a cumulative frequency of 4 is shown on the ogive in Figure 2.5 by the point located at 14.5 on the horizontal axis and 4 on the vertical axis. The "less than or equal to 19" class with a cumulative frequency of 12 is shown by the point located at 19.5 on the horizontal axis and 12 on the vertical axis. Note that one additional point is plotted at the left end of the ogive. This point starts the ogive by showing that no data values fall below the 10–14 class. It is plotted at 9.5 on the horizontal axis and 0 on the vertical axis. The plotted points are connected by straight lines to complete the ogive.

FIGURE 2.5 OGIVE FOR THE AUDIT-TIME DATA

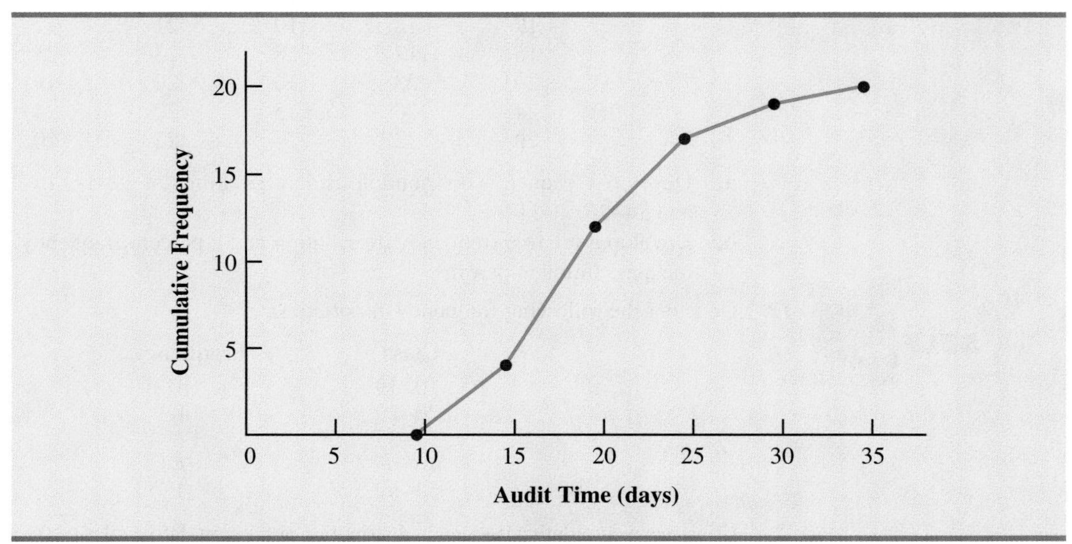

NOTES AND COMMENTS

1. The appropriate values for the class limits with quantitative data depend on the level of accuracy of the data. For instance, with the audit-time data of Table 2.5 the limits used were integer values because the data had been rounded to the nearest day. If the data were rounded to the nearest tenth of a day (e.g., 12.3, 14.4, and so on), then the limits would have been stated in tenths of days. For instance, the first class limits would have been 10.0–14.9. If the data were rounded to the nearest hundredth of a day (e.g., 12.34, 14.45, and so on), the limits would have been stated in hundredths of days. For instance, the first class limits would have been 10.00–14.99.
2. An *open-end* class is one that has only a lower class limit or an upper class limit. For example, in the audit-time data of Table 2.5, suppose two of the audits had taken 58 and 65 days. Rather than continue with the classes of width 5 with class intervals 35–39, 40–44, 45–49, and so on, we could simplify the frequency distribution to show an open-end class of "35 or more." This class would have a frequency of 2. Most often the open-end class appears at the upper end of the distribution. Sometimes an open-end class appears at the lower end of the distribution, and occasionally such classes appear at both ends.
3. The last entry in a cumulative frequency distribution is always the total number of observations. The last entry in a cumulative relative frequency distribution is always 1.00 and the last entry in a cumulative percent frequency distribution is always 100.
4. A bar graph and a histogram are essentially the same thing; both are graphical presentations of the data in a frequency distribution. A histogram is just a bar graph with no separation between bars. The separation between bars is appropriate for qualitative data because the data are discrete; no intermediate values are possible. For some quantitative data, a separation between bars is appropriate. Consider, for example, the number of classes in which a college student is enrolled. The data may only assume integer values. Intermediate values such as 1.5, 2.73, and so on are not possible. With continuous quantitative data, such as the audit times in Table 2.5, a separation between bars is not appropriate.

EXERCISES

Methods

11. Consider the following data.

14	21	23	21	16
19	22	25	16	16
24	24	25	19	16
19	18	19	21	12
16	17	18	23	25
20	23	16	20	19
24	26	15	22	24
20	22	24	22	20

a. Develop a frequency distribution using class limits of 12–14, 15–17, 18–20, 21–23, and 24–26.

b. Develop a relative frequency distribution and a percent frequency distribution using the class limits in part (a).

12. Consider the following frequency distribution.

Class	Frequency
10–19	10
20–29	14
30–39	17
40–49	7
50–59	2

Construct a cumulative frequency distribution and a cumulative relative frequency distribution.

13. Construct a histogram and an ogive for the data in Exercise 12.

14. Consider the following data.

8.9	10.2	11.5	7.8	10.0	12.2	13.5	14.1	10.0	12.2
6.8	9.5	11.5	11.2	14.9	7.5	10.0	6.0	15.8	11.5

a. Construct a dot plot.
b. Construct a frequency distribution.
c. Construct a percent frequency distribution.

Applications

15. A doctor's office staff has studied the waiting times for patients who arrive at the office with a request for emergency service. The following data were collected over a 1-month period (the waiting times are in minutes).

2 5 10 12 4 4 5 17 11 8 9 8 12 21 6 8 7 13 18 3

Use classes of 0–4, 5–9, and so on.
a. Show the frequency distribution.
b. Show the relative frequency distribution.
c. Show the cumulative frequency distribution.
d. Show the cumulative relative frequency distribution.
e. What proportion of patients needing emergency service have a waiting time of 9 minutes or less?

16. A sample of 25 computer hardware companies taken from the Stock Invester Pro database is shown in Table 2.9.
a. Develop tabular summaries and a histogram for the stock price data. Comment on typical stock prices and the distribution of stock prices.
b. Develop tabular summaries and a histogram for the earnings per share data. Comment on your results.

17. National Airlines accepts flight reservations by telephone. The following data show the call durations (in minutes) for a sample of 20 telephone reservations. Construct frequency and relative frequency distributions for the data. Also provide a histogram.

2.1	4.8	5.5	10.4
3.3	3.5	4.8	5.8
5.3	5.5	2.8	3.6
5.9	6.6	7.8	10.5
7.5	6.0	4.5	4.8

18. Wageweb conducts surveys of salary data and presents summaries on its web site. Using salary data as of January 1, 2000, Wageweb reported that salaries of marketing vice presidents ranged from \$85,090 to \$190,054 (*Wageweb.com,* April 12, 2000). Assume the following data are a sample of the annual salaries for 50 marketing vice presidents. Data are in thousands of dollars.

Wageweb

145	95	148	112	132
140	162	118	170	144
145	127	148	165	138
173	113	104	141	142
116	178	123	141	138
127	143	134	136	137
155	93	102	154	142
134	165	123	124	124
138	160	157	138	131
114	135	151	138	157

TABLE 2.9 DATA SET FOR 25 COMPUTER HARDWARE COMPANIES

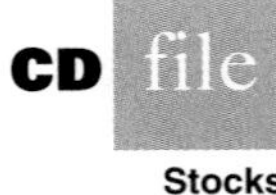

Stocks

Company	Stock Price	Institutional Ownership (%)	Price/Book Value	Earnings per Share (Annual $)
Amdahl	12.31	45.4	2.49	−2.49
Auspex Systems	11.00	66.1	2.22	0.85
Compaq Computer	65.50	83.0	6.84	2.01
Data General	35.94	91.5	4.25	1.15
Digi International	15.00	33.4	2.04	−0.89
Digital Equipment Corp.	43.00	58.8	1.92	−2.93
En Pointe Technologies	14.25	11.8	3.47	0.80
Equitrac	16.25	20.9	2.38	0.76
Franklin Electronic Pbls.	12.88	30.8	1.41	0.82
Gateway 2000	39.13	36.0	6.45	1.74
Hewlett-Packard	61.50	50.2	4.35	2.64
Ingram Micro	28.75	14.4	4.53	1.01
Maxwell Technologies	30.50	26.5	8.07	0.46
MicroAge	27.19	76.6	2.16	1.25
Micron Electronics	16.31	18.8	4.48	1.06
Network Computing Devices	11.88	39.8	3.34	0.15
Pomeroy Computer Resource	33.00	56.9	3.29	1.81
Sequent Computer Systems	28.19	57.0	2.65	0.36
Silicon Graphics	27.44	63.0	3.01	0.44
Southern Electronics	15.13	41.9	2.46	0.99
Stratus Computer	55.50	77.2	2.48	2.52
Sun Microsystems	48.00	59.3	7.50	1.67
Tandem Computers	34.25	61.3	3.61	1.02
Tech Data	38.94	82.3	3.80	1.50
Unisys	11.31	34.8	16.64	0.08

Source: Stock Investor Pro, American Association of Individual Investors, August 31, 1997.

a. What are the lowest and highest salaries?
b. Use a class width of $15,000 and prepare tabular summaries of the annual salary data.
c. What proportion of the annual salaries are $135,000 or less?
d. What percentage of the annual salaries are more than $150,000?
e. Prepare a histogram of the data.

19. The data for the numbers of units produced by a production employee during the most recent 20 days are shown here.

160	170	181	156	176
148	198	179	162	150
162	156	179	178	151
157	154	179	148	156

Summarize the data by constructing the following:
a. A frequency distribution
b. A relative frequency distribution
c. A cumulative frequency distribution
d. A cumulative relative frequency distribution
e. An ogive

20. The U.S. Bureau of the Census publishes a variety of information on the U.S. population. Following is the percent frequency distribution of the U.S. population by age as of July 1, 2000 (*The World Almanac and Book of Facts 2000*).

Age	Percent Frequency
0–13	20.0
14–17	5.7
18–24	9.6
25–34	13.6
35–44	16.3
45–54	13.5
55–64	8.7
65 or over	12.6
	100.0

a. What percentage of the population is 34 years old or less?
b. What percentage of the population is between 25 and 54 years old inclusively?
c. What percentage of the population is over 34 years old?
d. The total population is 275 million. How many people are less than 25 years old?
e. Suppose you believe that half the people in the 55–64 class are retired and that approximately all of the people 65 or older are retired. Estimate the number of retired people in the population.

Computer

21. The *Nielsen Home Technology Report* (February 20, 1996) reported on home technology and its usage by persons aged 12 and older. The following data are the hours of personal computer usage during one week for a sample of 50 persons.

4.1	1.5	10.4	5.9	3.4	5.7	1.6	6.1	3.0	3.7
3.1	4.8	2.0	14.8	5.4	4.2	3.9	4.1	11.1	3.5
4.1	4.1	8.8	5.6	4.3	3.3	7.1	10.3	6.2	7.6
10.8	2.8	9.5	12.9	12.1	0.7	4.0	9.2	4.4	5.7
7.2	6.1	5.7	5.9	4.7	3.9	3.7	3.1	6.1	3.1

Summarize the data by constructing the following:
a. A frequency distribution (with class width of 3 hours)
b. A relative frequency distribution
c. A histogram
d. An ogive
e. Comment on what the data indicate about personal computer usage at home.

2.3 EXPLORATORY DATA ANALYSIS: THE STEM-AND-LEAF DISPLAY

The techniques of **exploratory data analysis** consist of simple arithmetic and easy-to-draw graphs that can be used to summarize data quickly. One technique—referred to as a **stem-and-leaf display**— can be used to show both the rank order and shape of a data set simultaneously.

To illustrate the use of a stem-and-leaf display, consider the data in Table 2.10. These data are the result of a 150-question aptitude test given to 50 individuals who were recently interviewed for a position at Haskens Manufacturing. The data values indicate the number of questions answered correctly.

To develop a stem-and-leaf display, we first arrange the leading digits of each data value to the left of a vertical line. To the right of the vertical line, we record the last digit for each data value as we pass through the observations in the order they were recorded. The last digit for each data value is placed on the line corresponding to its first digit as follows:

TABLE 2.10 NUMBER OF QUESTIONS ANSWERED CORRECTLY ON AN APTITUDE TEST

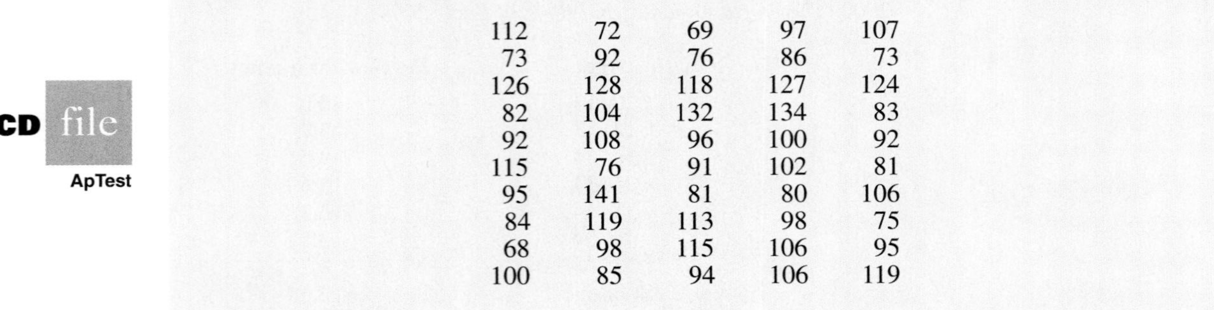

112	72	69	97	107
73	92	76	86	73
126	128	118	127	124
82	104	132	134	83
92	108	96	100	92
115	76	91	102	81
95	141	81	80	106
84	119	113	98	75
68	98	115	106	95
100	85	94	106	119

```
 6 | 9 8
 7 | 2 3 6 3 6 5
 8 | 6 2 3 1 1 0 4 5
 9 | 7 2 2 6 2 1 5 8 8 5 4
10 | 7 4 8 0 2 6 6 0 6
11 | 2 8 5 9 3 5 9
12 | 6 8 7 4
13 | 2 4
14 | 1
```

With this organization of the data, sorting the digits on each line into rank order is simple. Doing so leads to the stem-and-leaf display shown here.

```
 6 | 8 9
 7 | 2 3 3 5 6 6
 8 | 0 1 1 2 3 4 5 6
 9 | 1 2 2 2 4 5 5 6 7 8 8
10 | 0 0 2 4 6 6 6 7 8
11 | 2 3 5 5 8 9 9
12 | 4 6 7 8
13 | 2 4
14 | 1
```

The numbers to the left of the line (6, 7, 8, 9, 10, 11, 12, 13, and 14) form the *stem,* and each digit to the right of the line is a *leaf.* For example, consider the first line with a stem value of 6 and leaves of 8 and 9.

```
6 | 8 9
```

This indicates that there are two data values that have a first digit of six. The leaves show that the data values are 68 and 69. Similarly, the second line

```
7 | 2 3 3 5 6 6
```

indicates that six data values have a first digit of seven. The leaves show that the data values are 72, 73, 73, 75, 76, and 76.

To focus on the shape indicated by the stem-and-leaf display, let us use a rectangle to contain the leaves of each stem. Doing so, we obtain the following.

```
 6 | 8 9
 7 | 2 3 3 5 6 6
 8 | 0 1 1 2 3 4 5 6
 9 | 1 2 2 2 4 5 5 6 7 8 8
10 | 0 0 2 4 6 6 6 7 8
11 | 2 3 5 5 8 9 9
12 | 4 6 7 8
13 | 2 4
14 | 1
```

Rotating this page counterclockwise onto its side provides a picture of the data that is similar to a histogram with classes of 60–69, 70–79, 80–89, and so on.

Although the stem-and-leaf display may appear to offer the same information as a histogram, it has two primary advantages.

1. The stem-and-leaf display is easier to construct by hand.
2. Within a class interval, the stem-and-leaf display provides more information than the histogram because the stem-and-leaf shows the actual data values.

In a stretched stem-and-leaf display, whenever a stem value is stated twice, the first value corresponds to leaf values of 0–4, and the second value corresponds to leaf values of 5–9.

Just as a frequency distribution or histogram has no absolute number of classes, neither does a stem-and-leaf display have an absolute number of rows or stems. If we believe that our original stem-and-leaf display has condensed the data too much, we can easily stretch the display by using two or more stems for each leading digit(s). For example, to use two stems for each leading digit(s), we would place all data values ending in 0, 1, 2, 3, and 4 on one line and all values ending in 5, 6, 7, 8, and 9 on a second line. The following stretched stem-and-leaf display illustrates this approach.

```
 6 | 8 9
 7 | 2 3 3
 7 | 5 6 6
 8 | 0 1 1 2 3 4
 8 | 5 6
 9 | 1 2 2 2 4
 9 | 5 5 6 7 8 8
10 | 0 0 2 4
10 | 6 6 6 7 8
11 | 2 3
11 | 5 5 8 9 9
12 | 4
12 | 6 7 8
13 | 2 4
13 |
14 | 1
```

Note that values 72, 73, and 73 have leaves in the 0–4 range and are shown with the first stem value of 7. The values 75, 76, and 76 have leaves in the 5–9 range and are shown with the second stem value of 7. This stretched stem-and-leaf display is similar to a frequency distribution with intervals of 65–69, 70–74, 75–79, and so on.

The preceding example showed a stem-and-leaf display for data having up to three digits. Stem-and-leaf displays for data with more than three digits are possible. For example, consider the following data on the number of hamburgers sold by a fast-food restaurant for each of 15 weeks.

1565	1852	1644	1766	1888	1912	2044	1812
1790	1679	2008	1852	1967	1954	1733	

A stem-and-leaf display of these data follows.

Leaf Unit = 10

```
15 | 6
16 | 4  7
17 | 3  6  9
18 | 1  5  5  8
19 | 1  5  6
20 | 0  4
```

A single digit is used to define each leaf in a stem-and-leaf display. The leaf unit indicates how to multiply the stem-and-leaf numbers in order to approximate the original data. Leaf units may be 100, 10, 1, 0.1, and so on.

Note that a single digit is used to define each leaf and that only the first three digits of each observation have been used to construct the display. At the top of the display we have specified Leaf Unit = 10. To illustrate how to interpret the values in the display, consider the first stem, 15, and its associated leaf, 6. Combining these, we obtain the number 156. To reconstruct an approximation of the original observation, we must multiply this number by 10, the value of the *leaf unit.* Thus, $156 \times 10 = 1560$ is an approximation of the original observation used to construct the stem-and-leaf display. Although it is not possible to reconstruct the exact data from this stem-and-leaf display, the convention of using a single digit for each leaf enables stem-and-leaf displays to be constructed for data having a large number of digits. For stem-and-leaf displays where the leaf unit is not shown, it is assumed to equal 1.

EXERCISES

Methods

22. Construct a stem-and-leaf display for the following data.

70	72	75	64	58	83	80	82
76	75	68	65	57	78	85	72

SELF test

23. Construct a stem-and-leaf display for the following data.

11.3	9.6	10.4	7.5	8.3	10.5	10.0
9.3	8.1	7.7	7.5	8.4	6.3	8.8

24. Construct a stem-and-leaf display for the following data. Use a leaf unit of 10.

1161	1206	1478	1300	1604	1725	1361	1422
1221	1378	1623	1426	1557	1730	1706	1689

Applications

25. A psychologist developed a new test of adult intelligence. The test was administered to 20 individuals, and the following data were obtained.

114	99	131	124	117	102	106	127	119	115
98	104	144	151	132	106	125	122	118	118

Construct a stem-and-leaf display for the data.

26. The earnings per share data for a sample of 20 companies from the *Business Week* Corporate Scoreboard follow (*Business Week,* November 17, 1997):

Company	Earnings per Share ($)	Company	Earnings per Share ($)
Barnes & Noble	0.78	Hershey Foods	1.97
Citicorp	7.10	Hewlett-Packard	2.82
Compaq Computer	2.16	Humana	0.89
Dana	3.42	Microsoft	2.66
Dell Computer	2.03	Procter & Gamble	2.53
Digital Equipment	1.28	Quaker State	0.41
General Dynamics	4.82	Sara Lee	2.08
Goodyear	0.94	Snap-On Tools	2.38
Harley-Davidson	1.11	Sunstrand	2.53
Heinz	0.98	Xerox	3.95

Develop a stem-and-leaf display for the data. Use a leaf unit of 0.1. Comment on what you learned about the earnings per share for these companies.

27. In a study of job satisfaction, a series of tests was administered to 50 subjects. The following data were obtained; higher scores represent greater dissatisfaction.

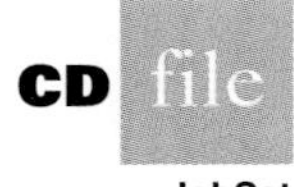

JobSat

87	76	67	58	92	59	41	50	90	75	80	81	70
73	69	61	88	46	85	97	50	47	81	87	75	60
65	92	77	71	70	74	53	43	61	89	84	83	70
46	84	76	78	64	69	76	78	67	74	64		

Construct a stem-and-leaf display for the data.

28. Periodically *Barron's* publishes earnings forecasts for the companies listed in the Dow Jones Industrial Average. The following are the 2000 forecasts of price/earnings (P/E) ratios for these companies implied by *Barron's* earnings forecasts (*Barron's,* February 14, 2000).

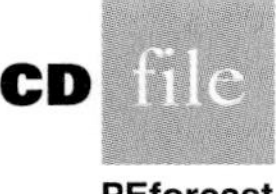

PEforcast

Company	2000 P/E Forecast	Company	2000 P/E Forecast
AT&T	23	Honeywell	13
Alcoa	15	IBM	28
American Express	25	Intel	37
Boeing	16	International Paper	14
Caterpillar	13	Johnson & Johnson	23
Citigroup	17	McDonald's	23
Coca-Cola	39	Merck	25
Disney	47	Microsoft	60
Dupont	18	Minnesota Mining	19
Eastman Kodak	11	J. P. Morgan	11
Exxon/Mobil	22	Philip Morris	5
General Electric	37	Procter & Gamble	26
General Motors	8	SBC Comm.	19
Hewlett-Packard	36	United Technologies	14
Home Depot	48	Wal-Mart	40

a. Develop a stem-and-leaf display for the data.

b. Use the results of the stem-and-leaf display to develop a frequency distribution and percent frequency distribution for the data.

2.4 CROSSTABULATIONS AND SCATTER DIAGRAMS

Crosstabulations and scatter diagrams are used to summarize data in a way that reveals the relationship between variables.

Thus far in this chapter, we have focused on tabular and graphical methods that are used to summarize the data for *one variable at a time.* Often a manager or decision maker is interested in tabular and graphical methods that will assist in the understanding of the *relationship between two variables.* Crosstabulation and scatter diagrams are two such methods.

Crosstabulation

Crosstabulation is a method that can be used to summarize the data for two variables simultaneously. Let us illustrate the use of a crosstabulation by considering the following application. Zagat's Restaurant Review is a service that provides data on restaurants located throughout the world. Data on a variety of variables such as the restaurant's quality rating and typical meal price are reported. Quality rating is a qualitative variable with rating categories of good, very good, and excellent. Meal price is a quantitative variable that generally ranges from \$10 to \$49. The quality rating and the meal price data were collected for a sample of 300 restaurants located in the Los Angeles area. Table 2.11 shows the data for the first 10 restaurants.

A crosstabulation of the data for this application is shown in Table 2.12. The left and top margin labels define the classes for the two variables. In the left margin, the row labels (good, very good, and excellent) correspond to the three classes of the quality rating variable. In the top margin, the column labels (\$10–19, \$20–29, \$30–39, and \$40–49) correspond to the four classes of the meal price variable. Each restaurant in the sample provides a quality rating and a meal price. Thus, each restaurant in the sample is associated with a cell appearing in one of the rows and one of the columns of the crosstabulation. For example, restaurant 5 is identified as having a very good quality rating and a meal price of \$33. This restaurant belongs to the cell in row 2 and column 3 of Table 2.12. In constructing a crosstabulation, we simply count the number of restaurants that belong to each of the cells in the crosstabulation table.

In reviewing Table 2.12, we see that the greatest number of restaurants in the sample (64) have a very good rating and a meal price in the \$20–29 range. Only two restaurants have an excellent rating and a meal price in the \$10–19 range. Similar interpretations of the other frequencies can be made. In addition, note that the right and bottom margins of the crosstabulation provide the frequency distributions for quality rating and meal price separately. From the frequency distribution in the right margin, we see that data on quality rat-

TABLE 2.11 QUALITY RATING AND MEAL PRICE FOR LOS ANGELES RESTAURANTS

Restaurant

Restaurant	Quality Rating	Meal Price (\$)
1	Good	18
2	Very Good	22
3	Good	28
4	Excellent	38
5	Very Good	33
6	Good	28
7	Very Good	19
8	Very Good	11
9	Very Good	23
10	Good	13
.	.	.
.	.	.
.	.	.

TABLE 2.12 CROSSTABULATION OF QUALITY RATING AND MEAL PRICE FOR 300 LOS ANGELES RESTAURANTS

	Meal Price				
Quality Rating	$10–19	$20–29	$30–39	$40–49	Total
Good	42	40	2	0	84
Very Good	34	64	46	6	150
Excellent	2	14	28	22	66
Total	78	118	76	28	300

ings show 84 good restaurants, 150 very good restaurants, and 66 excellent restaurants. Similarly, the bottom margin shows the frequency distribution for the meal price variable.

The value of a crosstabulation is that it provides insight about the relationship between the variables. From the results in Table 2.12, higher meal prices appear to be associated with the higher quality restaurants and the lower meal prices appear to be associated with the lower quality restaurants.

Converting the entries in the table into row percentages or column percentages can afford additional insight about the relationship between the variables. For row percentages, the results of dividing each frequency in Table 2.12 by its corresponding row total are shown in Table 2.13. For example, the percentage in the first row and first column, 50.0, is calculated by dividing 42 by 84 and multiplying by 100 ($42/84 \times 100 = 50.0\%$). Of the restaurants with the lowest (good) quality rating, we see that the greatest percentages are for the less expensive restaurants (50.0% have $10–19 meal prices and 47.6% have $20–29 meal prices). Of the restaurants with the highest (excellent) quality rating, we see that the greatest percentages are for the more expensive restaurants (42.4% have $30–39 meal prices and 33.4% have $40–49 meal prices). Thus, we continue to see that the more expensive meals are associated with the higher quality restaurants.

Crosstabulation is widely used for examining the relationship between two variables. In practice, final reports for many statistical surveys include a large number of crosstabulation tables. In the Los Angeles restaurant sample, the crosstabulation is based on one qualitative variable (quality rating) and one quantitative variable (meal price). Crosstabulations can also be developed when both variables are qualitative and when both variables are quantitative.

Scatter Diagram

A **scatter diagram** is a graphical presentation of the relationship between two quantitative variables. As an illustration of a scatter diagram, consider the situation of a stereo and sound equipment store in San Francisco. On 10 occasions during the past three months, the store has used weekend television commercials to promote sales at its stores. The managers want to investigate whether a relationship can be demonstrated between the number

TABLE 2.13 ROW PERCENTAGES FOR EACH QUALITY RATING CATEGORY

	Meal Price				
Quality Rating	$10–19	$20–29	$30–39	$40–49	Total
Good	50.0	47.6	2.4	0.0	100
Very Good	22.7	42.7	30.6	4.0	100
Excellent	3.0	21.2	42.4	33.4	100

TABLE 2.14 SAMPLE DATA FOR THE STEREO AND SOUND EQUIPMENT STORE

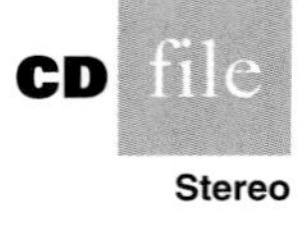

Week	Number of Commercials x	Sales ($100s) y
1	2	50
2	5	57
3	1	41
4	3	54
5	4	54
6	1	38
7	5	63
8	3	48
9	4	59
10	2	46

of commercials shown and the sales at the store during the following week. Sample data for the 10 weeks with sales in hundreds of dollars are shown in Table 2.14.

Figure 2.6 is the scatter diagram for the data in Table 2.14. The number of commercials (x) is shown on the horizontal axis and the sales (y) are shown on the vertical axis. For week 1, $x = 2$ and $y = 50$. A point with those coordinates is plotted on the scatter diagram. Similar points are plotted for the other 9 weeks. Note that on two of the weeks, one commercial was shown, on two of the weeks two commercials were shown, and so on.

The completed scatter diagram in Figure 2.6 indicates a positive relationship between the number of commercials and sales. Higher sales are associated with a higher number of commercials. The relationship is not perfect in that all points are not on a straight line. However, the general pattern of the points suggests that the overall relationship is positive.

Some general scatter diagram patterns and the types of relationships they suggest are shown in Figure 2.7. The top left panel depicts a positive relationship similar to the one we saw for the number of commercials and sales example. In the top right panel, the scatter diagram shows no apparent relationship between the variables. The bottom panel depicts a negative relationship where y tends to decrease as x increases.

FIGURE 2.6 SCATTER DIAGRAM FOR THE STEREO AND SOUND EQUIPMENT STORE

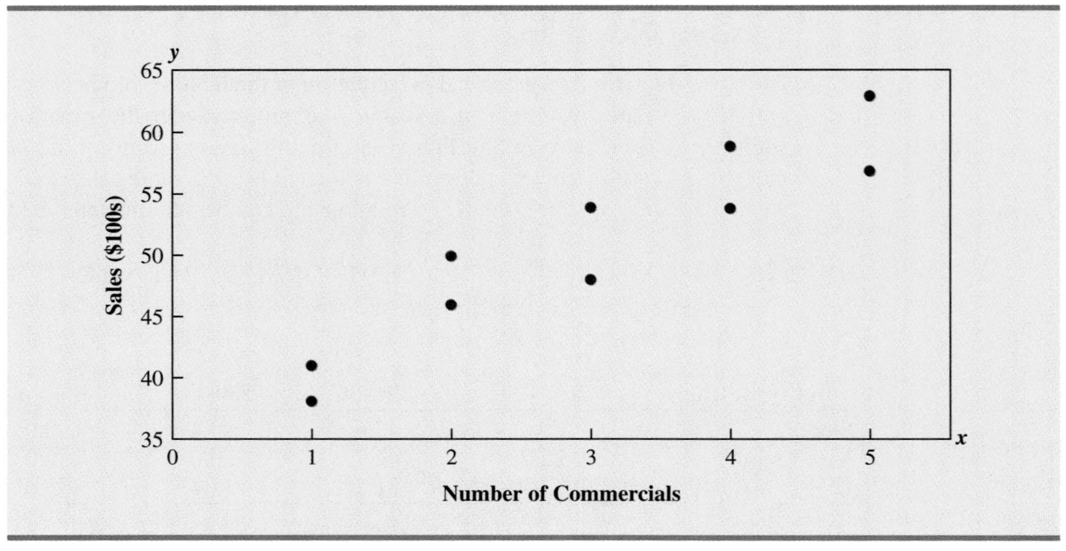

FIGURE 2.7 TYPES OF RELATIONSHIPS DEPICTED BY SCATTER DIAGRAMS

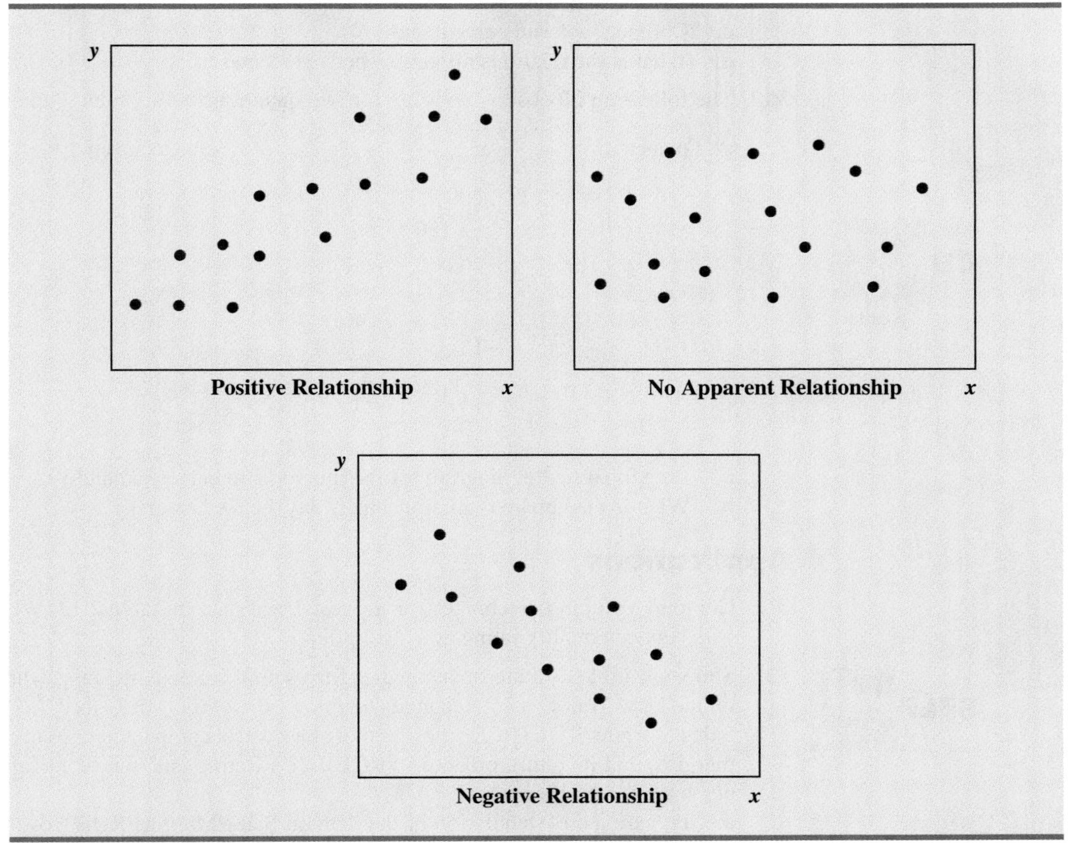

EXERCISES

Methods

29. The following data are for 30 observations on two qualitative variables, x and y. The categories for x are A, B, and C; the categories for y are 1 and 2.

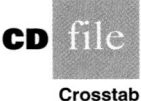

CD file Crosstab

Observation	x	y	Observation	x	y
1	A	1	16	B	2
2	B	1	17	C	1
3	B	1	18	B	1
4	C	2	19	C	1
5	B	1	20	B	1
6	C	2	21	C	2
7	B	1	22	B	1
8	C	2	23	C	2
9	A	1	24	A	1
10	B	1	25	B	1
11	A	1	26	C	2
12	B	1	27	C	2
13	C	2	28	A	1
14	C	2	29	B	1
15	C	2	30	B	2

a. Develop a crosstabulation for the data, with x in the rows and y in the columns.
b. Compute the row percentages.
c. Compute column percentages.
d. What is the relationship, if any, between x and y?

30. The following 20 observations are for two quantitative variables, x and y.

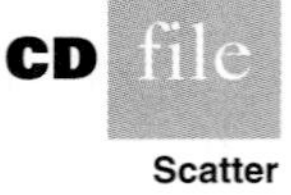
Scatter

Observation	x	y	Observation	x	y
1	−22	22	11	−37	48
2	−33	49	12	34	−29
3	2	8	13	9	−18
4	29	−16	14	−33	31
5	−13	10	15	20	−16
6	21	−28	16	−3	14
7	−13	27	17	−15	18
8	−23	35	18	12	17
9	14	−5	19	−20	−11
10	3	−3	20	−7	−22

a. Develop a scatter diagram for the relationship between x and y.
b. What is the apparent relationship, if any, between x and y?

Applications

31. Compute column percentages for the restaurant data in Table 2.12. What is the relationship between quality rating and meal price?

32. Shown in Table 2.15 are financial data for a sample of 36 companies whose stock is traded on the New York Stock Exchange (*Investor's Business Daily,* April 7, 2000). The data on Sales/Margins/ROE is a composite rating based on a company's sales growth rate, its profit margins, and its return on equity (ROE). EPS Rating is a measure of growth in earnings per share for the company.

a. Prepare a crosstabulation of the data on Sales/Margins/ROE (rows) and EPS Rating (columns). Use classes of 0–19, 20–39, 40–59, 60–79, and 80–99 for EPS Rating.
b. Compute row percentages and comment on any relationship that you see between the variables.

33. Refer to the data in Table 2.15.

a. Prepare a crosstabulation of the data on Sales/Margins/ROE and Industry Group Relative Strength.
b. Prepare a frequency distribution for the data on Sales/Margins/ROE.
c. Prepare a frequency distribution for the data on Industry Group Relative Strength.
d. How has the crosstabulation helped in preparing the frequency distributions in parts (c) and (d)?

34. Refer to the data in Table 2.15.

a. Prepare a scatter diagram of the data on EPS Rating and Relative Price Strength.
b. Comment on the relationship, if any, between the variables. The meaning of the EPS Rating is described in Exercise 32. Relative Price Strength is a measure on the change in the stock's price over the past 12 months. Higher values indicate greater strength.

35. The following data show the hotel revenue and the gaming revenue, in millions of dollars, for 10 Las Vegas casino hotels (*Cornell Hotel And Restaurant Administration Quarterly,* October 1997).

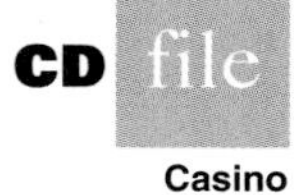
Casino

Company	Hotel Revenue ($millions)	Gaming Revenue ($millions)
Boyd Gaming	303.5	548.2
Circus Circus Enterprises	664.8	664.8
Grand Casinos	121.0	270.7

Company	Hotel Revenue ($millions)	Gaming Revenue ($millions)
Hilton Corp. Gaming Div.	429.6	511.0
MGM Grand, Inc.	373.1	404.7
Mirage Resorts	670.9	782.8
Primadonna Resorts	66.4	130.7
Rio Hotel & Casino	105.8	105.5
Sahara Gaming	102.4	148.7
Station Casinos	135.8	358.5

a. Develop a scatter diagram of the data on Hotel Revenue and Gaming Revenue.
b. Comment on the relationship, if any, between the variables.

TABLE 2.15 FINANCIAL DATA FOR A SAMPLE OF 36 COMPANIES

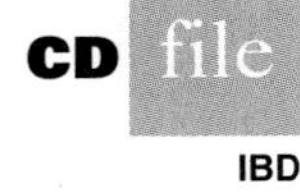

IBD

Company	EPS Rating	Relative Price Strength	Industry Group Relative Strength	Sales/Margins/ ROE
Advo	81	74	B	A
Alaska AirGp	58	17	C	B
Alliant Tech	84	22	B	B
Atmos Engy	21	9	C	E
Bank of Am.	87	38	C	A
Bowater PLC	14	46	C	D
Callaway Golf	46	62	B	E
Central Parking	76	18	B	C
Dean Foods	84	7	B	C
Dole Food	70	54	E	C
Elec. Data Sys	72	69	A	B
Fed. Dept. Stor.	79	21	D	B
Gateway	82	68	A	A
Goodyear	21	9	E	D
Hanson PLC	57	32	B	B
ICN Pharm.	76	56	A	D
Jefferson plt	80	38	D	C
Kroger	84	24	D	A
Mattel	18	20	E	D
McDermott	6	6	A	C
Monaco	97	21	D	A
Murphy Oil	80	62	B	B
Nordstrom	58	57	B	C
NYMAGIC	17	45	D	D
Office Depot	58	40	B	B
Payless Shoes	76	59	B	B
Praxair	62	32	C	B
Reebok	31	72	C	E
Safeway	91	61	D	A
Teco Energy	49	48	D	B
Texaco	80	31	D	C
US West	60	65	B	A
United Rental	98	12	C	A
Wachovia	69	36	E	B
Winnebago	83	49	D	A
York Intl.	28	14	D	B

Source: Investor's Business Daily, April 7, 2000.

SUMMARY

A set of data, even if modest in size, is often difficult to interpret directly in the form in which it is gathered. Tabular and graphical methods provide procedures for organizing and summarizing data so that patterns are revealed and the data are more easily interpreted. Frequency distributions, relative frequency distributions, percent frequency distributions, bar graphs, and pie charts were presented as tabular and graphical procedures for summarizing qualitative data. Frequency distributions, relative frequency distributions, percent frequency distributions, dot plots, histograms, cumulative frequency distributions, cumulative relative frequency distributions, cumulative percent frequency distributions, and ogives were presented as ways of summarizing quantitative data. A stem-and-leaf display was presented as an exploratory data analysis technique that can be used to summarize quantitative data. Crosstabulation was presented as a tabular method for summarizing data for two variables. The scatter diagram was introduced as a graphical method for showing the relationship between two quantitative variables. Figure 2.8 is a summary of the tabular and graphical methods presented in this chapter.

With large data sets, computer software packages are essential in constructing tabular and graphical summaries of data. In the two chapter appendixes, we show how Minitab and Microsoft Excel can be used for this purpose.

FIGURE 2.8 TABULAR AND GRAPHICAL METHODS FOR SUMMARIZING DATA

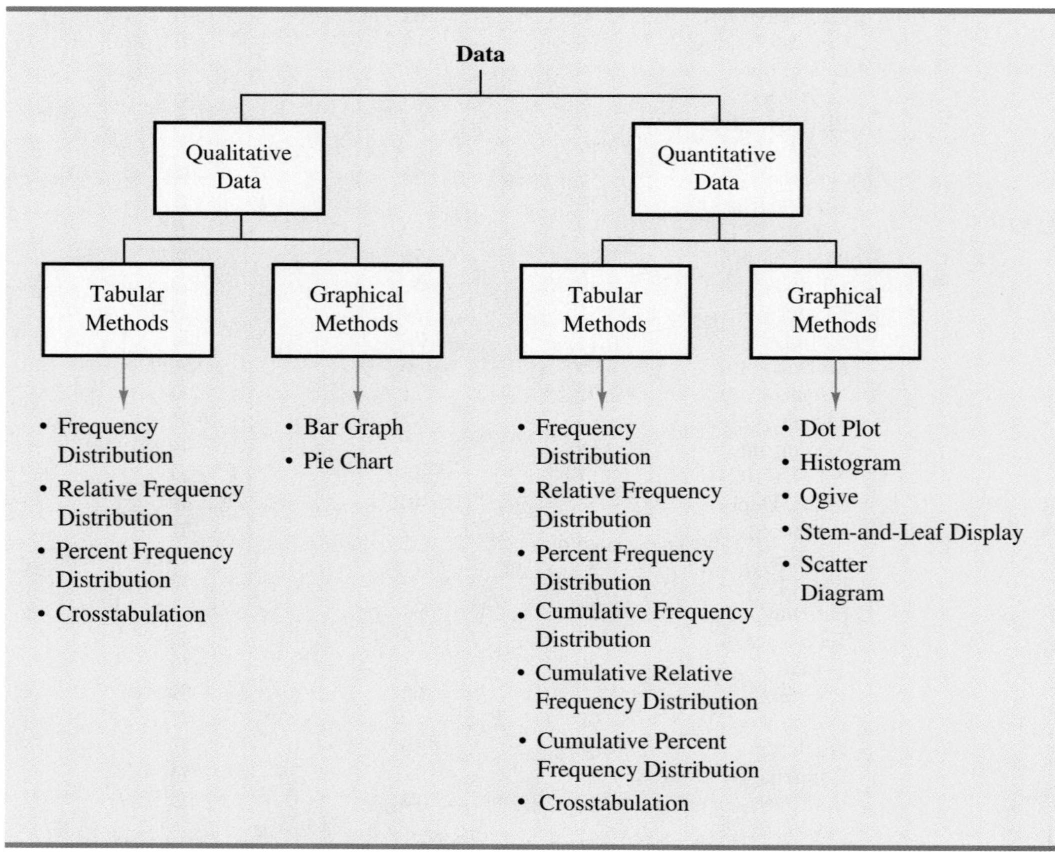

GLOSSARY

Qualitative data Data that are labels or names used to identify categories of like items.

Quantitative data Data that indicate how much or how many.

Frequency distribution A tabular summary of data showing the number (frequency) of items in each of several nonoverlapping classes.

Relative frequency distribution A tabular summary of data showing the fraction or proportion (relative frequency)—of data items in each of several nonoverlapping classes.

Percent frequency distribution A tabular summary of data showing the percentage of items in each of several nonoverlapping classes.

Bar graph A graphical device for depicting data that have been summarized in a frequency distribution, relative frequency distribution, or percent frequency distribution.

Pie chart A graphical device for presenting data summaries based on subdivision of a circle into sectors that correspond to the relative frequency for each class.

Class midpoint The point in each class that is halfway between the lower and upper class limits.

Dot plot A simple graphical summary of data with each observation represented by a dot placed above a horizontal axis that shows the distribution of the data values.

Histogram A graphical presentation of a frequency distribution, relative frequency distribution, or percent frequency distribution of quantitative data constructed by placing the class intervals on the horizontal axis and the frequencies on the vertical axis.

Cumulative frequency distribution A tabular summary of quantitative data showing the number of items with values less than or equal to the upper class limit of each class.

Cumulative relative frequency distribution A tabular summary of quantitative data showing the fraction or proportion of items with values less than or equal to the upper class limit of each class.

Cumulative percent frequency distribution A tabular summary of quantitative data showing the percentage of items with values less than or equal to the upper class limit of each class.

Ogive A graph of a cumulative distribution.

Exploratory data analysis Methods that use simple arithmetic and easy-to-draw graphs to summarize data quickly.

Stem-and-leaf display An exploratory data analysis technique that simultaneously rank orders quantitative data and provides insight about the shape of the distribution.

Crosstabulation A tabular summary of data for two variables. The classes for one variable are represented by the rows; the classes for the other variable are represented by the columns.

Scatter diagram A graphical presentation of the relationship between two quantitative variables. One variable is shown on the horizontal axis and the other variable is shown on the vertical axis.

KEY FORMULAS

Relative Frequency

$$\frac{\text{Frequency of the Class}}{n} \tag{2.1}$$

Approximate Class Width

$$\frac{\text{Largest Data Value} - \text{Smallest Data Value}}{\text{Number of Classes}} \tag{2.2}$$

SUPPLEMENTARY EXERCISES

36. *Autodata* and *USA Today* research provided data on the best-selling vehicle models for March 2000 (*USA Today,* April 4, 2000). Among those listed were Chevrolet Silverado/C/K pickup, Ford F-Series pickup, Ford Taurus, Honda Accord, and Toyota Camry. Data from a sample of 50 vehicle purchases are presented in Table 2.16.

TABLE 2.16 DATA FOR 50 VEHICLE PURCHASES

Silverado	Taurus	Accord	F-Series	Silverado
F-Series	Accord	Silverado	Camry	Taurus
Silverado	F-Series	Camry	Silverado	F-Series
Taurus	F-Series	Taurus	F-Series	Camry
Taurus	F-Series	F-Series	Accord	Camry
Silverado	Silverado	Silverado	Silverado	Taurus
Camry	Silverado	Accord	F-Series	F-Series
Taurus	F-Series	Accord	Camry	Accord
F-Series	Silverado	F-Series	F-Series	Taurus
Camry	Silverado	F-Series	F-Series	F-Series

a. Develop a frequency and percent frequency distribution.
b. What are the two top-selling vehicles?
c. Show a pie chart.

37. Each of the *Fortune* 1000 companies is classified as belonging to one of several industries (*Fortune,* April 17, 2000). A sample of 20 companies with their corresponding industry classification follows.

Company	Industry Classification	Company	Industry Classification
IBP	Food	Borden	Food
Intel	Electronics	McDonnell Douglas	Aerospace
Coca-Cola	Beverage	Morton International	Chemicals
Union Carbide	Chemicals	Quaker Oats	Food
General Electric	Electronics	Pepsico	Beverage

Company	Industry Classification	Company	Industry Classification
Motorola	Electronics	Maytag	Electronics
Kellogg	Food	Textron	Aerospace
Dow Chemical	Chemicals	Sara Lee	Food
Campbell Soup	Food	Harris	Electronics
Ralston Purina	Food	Eaton	Electronics

a. Provide a frequency distribution showing the number of companies in each industry.
b. Provide a percent frequency distribution.
c. Provide a bar graph for the data.

38. During 1998 and 1999, *Time* magazine carried cover stories on the following movies: *Blair Witch Project, Phantom Menace, Beloved, Primary Colors,* and *Truman Show. Time* keeps track of mail received on its cover stories; the number of pieces of mail received on these movies was 159, 89, 85, 57, and 51 respectively (*Time* September 27, 1999).
a. Develop a percent frequency distribution for the mail data.
b. Develop a pie chart for the data.
c. The cover stories on the *Blair Witch Project* and *Phantom Menace* were run in 1999; the others were run in 1998. What percent of the mail pertained to 1999 cover stories?

39. The data in Table 2.17 represent sales in millions of dollars for 20 companies in the health care services industry during third quarter 1997 (*Business Week,* November 17, 1997).
a. Construct a frequency distribution to summarize the data. Use a class width of 500.
b. Develop a relative frequency distribution for the data.
c. Construct a cumulative frequency distribution for the data.
d. Construct a cumulative relative frequency distribution for the data.
e. Construct a histogram as a graphical representation of the data.

40. The closing prices of 40 common stocks follow (*The Wall Street Journal,* March 17, 2000).

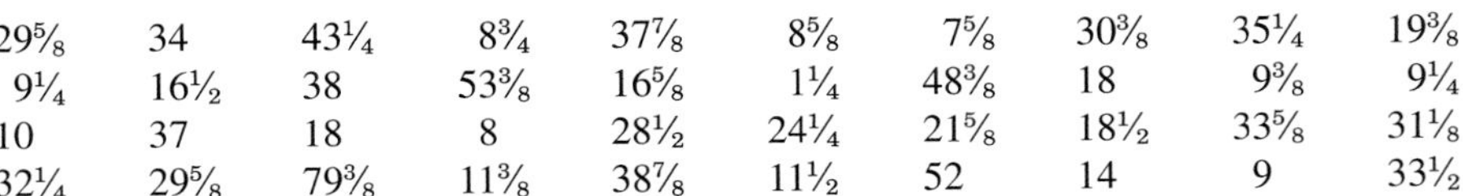

$29\frac{5}{8}$	34	$43\frac{1}{4}$	$8\frac{3}{4}$	$37\frac{7}{8}$	$8\frac{5}{8}$	$7\frac{5}{8}$	$30\frac{3}{8}$	$35\frac{1}{4}$	$19\frac{3}{8}$
$9\frac{1}{4}$	$16\frac{1}{2}$	38	$53\frac{3}{8}$	$16\frac{5}{8}$	$1\frac{1}{4}$	$48\frac{3}{8}$	18	$9\frac{3}{8}$	$9\frac{1}{4}$
10	37	18	8	$28\frac{1}{2}$	$24\frac{1}{4}$	$21\frac{5}{8}$	$18\frac{1}{2}$	$33\frac{5}{8}$	$31\frac{1}{8}$
$32\frac{1}{4}$	$29\frac{5}{8}$	$79\frac{3}{8}$	$11\frac{3}{8}$	$38\frac{7}{8}$	$11\frac{1}{2}$	52	14	9	$33\frac{1}{2}$

a. Construct frequency and relative frequency distributions for the data.
b. Construct cumulative frequency and cumulative relative frequency distributions for the data.

TABLE 2.17 SALES IN HEALTH CARE SERVICES FOR THIRD QUARTER 1997

Company	Sales	Company	Sales
Beverly Ent.	805	Novacare	357
Coventry	307	Phycor	284
Express Scripts	320	Quest Diag.	374
Healthsouth	748	Quorum Health	393
Horizon	445	Sun Healthcare	486
Humana	1968	Tenet Healthcare	2331
Int. Health	472	U. Wisconsin	389
Lab. Corp. Am.	377	Univ. Health	362
Manor Care	274	Vencor	845
Medpartners	1614	Wellpoint	1512

Source: Business Week, November 17, 1997.

c. Construct a histogram for the data.

d. Using your summaries, make comments and observations about the price of common stock.

41. Ninety-four new shadow stocks were reported by the American Association of Individual Investors (*AAII Journal,* February 1997). The term *shadow* indicates stocks for small to medium-size firms not followed closely by the major brokerage houses. Information on where the stock was traded—New York Stock Exchange (NYSE), American Stock Exchange (AMEX), and over-the-counter (OTC)—the earnings per share, and the price/earnings ratio was provided for the following sample of 20 shadow stocks.

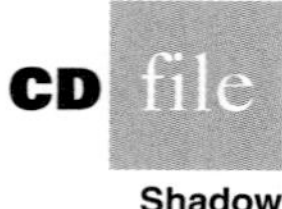

Stock	Exchange	Earnings per Share ($)	Price/ Earnings Ratio
Chemi-Trol	OTC	.39	27.30
Candie's	OTC	.07	36.20
TST/Impreso	OTC	.65	12.70
Unimed Pharm.	OTC	.12	59.30
Skyline Chili	AMEX	.34	19.30
Cyanotech	OTC	.22	29.30
Catalina Light.	NYSE	.15	33.20
DDL Elect.	NYSE	.10	10.20
Euphonix	OTC	.09	49.70
Mesa Labs	OTC	.37	14.40
RCM Tech.	OTC	.47	18.60
Anuhco	AMEX	.70	11.40
Hello Direct	OTC	.23	21.10
Hilite Industries	OTC	.61	7.80
Alpha Tech.	OTC	.11	34.60
Wegener Group	OTC	.16	24.50
U.S. Home & Garden	OTC	.24	8.70
Chalone Wine	OTC	.27	44.40
Eng. Support Sys.	OTC	.89	16.70
Int. Remote Imaging	AMEX	.86	4.70

a. Provide frequency and relative frequency distributions for the exchange data. Where are most shadow stocks listed?

b. Provide frequency and relative frequency distributions for the earnings per share and price/earnings ratio data. Use class limits of .00–.19, .20–.39, and so on, for the earnings per share data and class limits of 0.0–9.9, 10.0–19.9, and so on for the price/earnings ratio data. What observations and comments can you make about the shadow stocks?

42. A state-by-state listing of per capita personal income for 1998 follows.

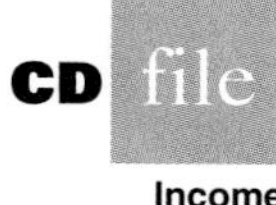

Ala.	21,500	Ga.	25,106	Md.	30,023
Alaska	25,771	Hawaii	26,210	Mass.	32,902
Ariz.	23,152	Idaho	21,080	Mich.	25,979
Ark.	20,393	Ill.	28,976	Minn.	27,667
Calif.	27,579	Ind.	24,302	Miss.	18,998
Colo.	28,821	Iowa	24,007	Mo.	24,447
Conn.	37,700	Kan.	25,049	Mont.	20,247
Del.	29,932	Ky.	21,551	Neb.	24,786
D.C.	37,325	La.	21,385	Nev.	27,360
Fla.	25,922	Maine	23,002	N.H.	29,219

N.J.	33,953	Ore.	24,775	Utah	21,096
N.M.	20,008	Penn.	26,889	Vt.	24,217
N.Y.	31,679	R.I.	26,924	Va.	27,489
N.C.	24,122	S.C.	21,387	Wash.	28,066
N.D.	21,708	S.D.	22,201	W. Va.	19,373
Ohio	25,239	Tenn.	23,615	Wis.	25,184
Okla.	21,056	Texas	25,028	Wyo.	23,225

Source: Bureau of Economic Analysis, *Current Population Survey,* March 2000.

Develop a frequency distribution, a relative frequency distribution, and a histogram for the data.

43. The conclusion from a 40-state poll conducted by the Joint Council on Economic Education is that students do not learn enough economics. The findings were based on test results from 11th- and 12th-grade students who took a 46-question, multiple-choice test on basic economic concepts such as profit and the law of supply and demand. The following table gives sample data on the number of questions answered correctly.

12	10	16	24	12	14	18	23
31	14	15	19	17	9	19	28
24	16	21	13	20	12	22	18
22	18	30	16	26	18	16	14
8	25	22	15	33	24	17	19

Summarize these data using the following:

a. A stem-and-leaf display
b. A frequency distribution
c. A relative frequency distribution
d. A cumulative frequency distribution
e. On the basis of these data, do you agree with the claim that students are not learning enough economics? Explain.

44. The daily high and low temperatures for 20 cities follow.

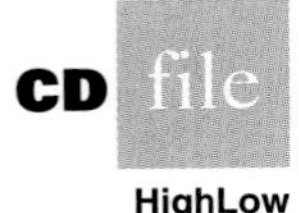

HighLow

City	High	Low	City	High	Low
Athens	75	54	Melbourne	66	50
Bangkok	92	74	Montreal	64	52
Cairo	84	57	Paris	77	55
Copenhagen	64	39	Rio de Janeiro	80	61
Dublin	64	46	Rome	81	54
Havana	86	68	Seoul	64	50
Hong Kong	81	72	Singapore	90	75
Johannesburg	61	50	Sydney	68	55
London	73	48	Tokyo	79	59
Manila	93	75	Vancouver	57	43

Source: USA Today, May 9, 2000.

a. Prepare a stem-and-leaf display for the high temperatures.
b. Prepare a stem-and-leaf display for the low temperatures.
c. Compare the stem-and-leaf displays from parts (a) and (b) and make some comments about the differences between daily high and low temperatures.
d. Use the stem-and-leaf display from part (a) to determine the number of cities having a high temperature of 80 degrees or above.
e. Provide frequency distributions for both high and low temperature data.

45. Refer to the data set for high and low temperatures for 20 cities in Exercise 44.

a. Develop a scatter diagram to show the relationship between the two variables, high temperature and low temperature.

b. Comment on the relationship between high and low temperature.

46. A study of job satisfaction was conducted for four occupations. Job satisfaction was measured using an 18-item questionnaire with each question receiving a response score of 1 to 5. The sum of the 18 scores provides the job satisfaction score for each individual in the sample. The data obtained is given. Higher scores indicate greater satisfaction.

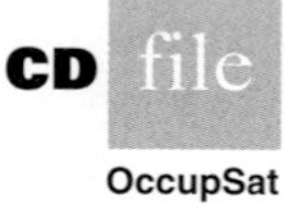

OccupSat

Occupation	Satisfaction Score	Occupation	Satisfaction Score	Occupation	Satisfaction Score
Lawyer	42	Physical Therapist	78	Systems Analyst	60
Physical Therapist	86	Systems Analyst	44	Physical Therapist	59
Lawyer	42	Systems Analyst	71	Cabinetmaker	78
Systems Analyst	55	Lawyer	50	Physical Therapist	60
Lawyer	38	Lawyer	48	Physical Therapist	50
Cabinetmaker	79	Cabinetmaker	69	Cabinetmaker	79
Lawyer	44	Physical Therapist	80	Systems Analyst	62
Systems Analyst	41	Systems Analyst	64	Lawyer	45
Physical Therapist	55	Physical Therapist	55	Cabinetmaker	84
Systems Analyst	66	Cabinetmaker	64	Physical Therapist	62
Lawyer	53	Cabinetmaker	59	Systems Analyst	73
Cabinetmaker	65	Cabinetmaker	54	Cabinetmaker	60
Lawyer	74	Systems Analyst	76	Lawyer	64
Physical Therapist	52				

a. Provide a crosstabulation of occupation and satisfaction score.

b. Compute the row percentages for your crosstabulation in part (a).

c. What observations can you make concerning the level of job satisfaction for these occupations?

47. Do larger companies generate more revenue? The following data show the number of employees and annual revenue for a sample of 20 *Fortune* 1000 companies (*Fortune,* April 17, 2000).

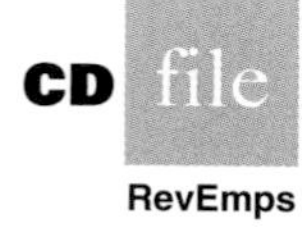

RevEmps

Company	Employees	Revenue $mil	Company	Employees	Revenue $mil
Sprint	77,600	19,930	American Financial	9,400	3,334
Chase Manhattan	74,801	33,710	Fluor	53,561	12,417
Computer Sciences	50,000	7,660	Phillips Petroleum	15,900	13,852
Wells Fargo	89,355	21,795	Cardinal Health	36,000	25,034
Sunbeam	12,200	2,398	Borders Group	23,500	2,999
CBS	29,000	7,510	MCI Worldcom	77,000	37,120
Time Warner	69,722	27,333	Consolidated Edison	14,269	7,491
Steelcase	16,200	2,743	IBP	45,000	14,075
Georgia-Pacific	57,000	17,796	Super Value	50,000	17,421
Toro	1,275	4,673	H&R Block	4,200	1,669

a. Prepare a scatter diagram to show the relationship between the variables Revenue and Employees.

b. Comment on any relationship that is apparent between the variables.

48. A survey of commercial buildings served by the Cincinnati Gas & Electric Company was concluded in 1992 (CG&E Commercial Building Characteristics Survey, November 25, 1992). One question asked what main heating fuel was used and another asked the year the commercial building was constructed. A partial crosstabulation of the findings follows.

Year Constructed	Fuel Type				
	Electricity	Natural Gas	Oil	Propane	Other
1973 or before	40	183	12	5	7
1974–1979	24	26	2	2	0
1980–1986	37	38	1	0	6
1987–1991	48	70	2	0	1

a. Complete the crosstabulation by showing the row totals and column totals.
b. Show the frequency distributions for year constructed and for fuel type.
c. Prepare a crosstabulation showing column percentages.

49. Table 2.18 contains a portion of the data on the file named Fortune on the CD ROM at the back of the book. It provides data on stockholders' equity, market value, and profits for a sample of 50 *Fortune* 500 companies (*Fortune,* April 26, 1999).
a. Prepare a crosstabulation for the variables stockholders' equity and profit. Use classes of 0–200, 200–400, . . . , 1000–1200 for profit, and classes of 0–1200, 1200–2400, . . . , 4800–6000 for stockholders' equity.
b. Compute the row percentages for your crosstabulation in part (a).
c. What relationship, if any, do you notice between profit and stockholders' equity?

50. Refer to the data set in Table 2.18.
a. Prepare a crosstabulation for the variables market value and profit.
b. Compute the row percentages for your crosstabulation in part (a).
c. Comment on any relationship that is apparent between the variables.

51. Refer to the data set in Table 2.18.
a. Prepare a scatter diagram to show the relationship between the variables profit and stockholders' equity.
b. Comment on any relationship that is apparent between the variables.

TABLE 2.18 DATA FOR A SAMPLE OF 50 *FORTUNE* 500 COMPANIES

Fortune

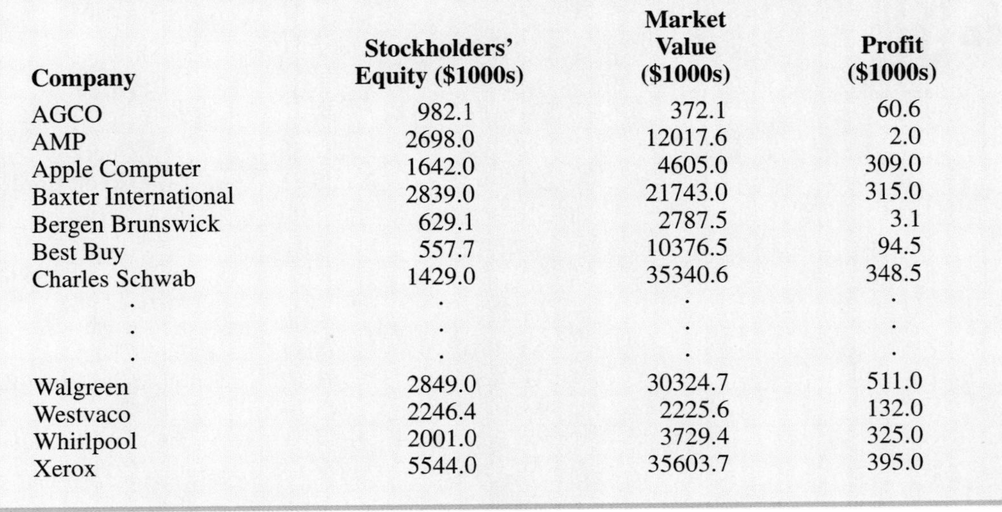

Company	Stockholders' Equity ($1000s)	Market Value ($1000s)	Profit ($1000s)
AGCO	982.1	372.1	60.6
AMP	2698.0	12017.6	2.0
Apple Computer	1642.0	4605.0	309.0
Baxter International	2839.0	21743.0	315.0
Bergen Brunswick	629.1	2787.5	3.1
Best Buy	557.7	10376.5	94.5
Charles Schwab	1429.0	35340.6	348.5
.	.	.	.
.	.	.	.
.	.	.	.
Walgreen	2849.0	30324.7	511.0
Westvaco	2246.4	2225.6	132.0
Whirlpool	2001.0	3729.4	325.0
Xerox	5544.0	35603.7	395.0

52. Refer to the data set in Table 2.18.
    a. Prepare a scatter diagram to show the relationship between the variables market value and stockholders' equity.
    b. Comment on any relationship that is apparent between the variables.

Case Problem CONSOLIDATED FOODS

Consolidated Foods operates a chain of supermarkets in New Mexico, Arizona, and California. A promotional campaign has advertised the chain's offering of a credit card policy whereby Consolidated Foods' customers have the option of paying for their purchases with credit cards, such as Visa and MasterCard, in addition to the usual options of cash or personal check. The policy is being implemented on a trial basis with the hope that the credit card option will encourage customers to make larger purchases.

After the first month of operation, a random sample of 100 customers was selected over a one-week period. Data were collected on the method of payment and how much was spent by each of the 100 customers. A portion of the sample data are shown in Table 2.19. Prior to the new credit card policy, approximately 50% of Consolidated Foods' customers paid in cash and approximately 50% paid by personal check.

Managerial Report

Use the tabular and graphical methods of descriptive statistics to summarize the sample data in Table 2.19. Your report should contain summaries such as the following.

1. A frequency and relative frequency distribution for the method of payment.
2. A bar graph or pie chart for the method of payment.

TABLE 2.19 PURCHASE AMOUNT AND METHOD OF PAYMENT FOR A SAMPLE OF 100 CONSOLIDATED FOODS CUSTOMERS

Consolid

Customer	Amount ($)	Method of Payment
1	28.58	Check
2	52.04	Check
3	7.41	Cash
4	11.17	Cash
5	43.79	Credit Card
6	48.95	Check
7	57.59	Check
8	27.60	Check
9	26.91	Credit Card
10	9.00	Cash
.	.	.
.	.	.
.	.	.
95	18.09	Cash
96	54.84	Check
97	41.10	Check
98	43.14	Check
99	3.31	Cash
100	69.77	Credit Card

3. Frequency and relative frequency distributions for the amount spent in each method of payment.
4. Histograms and/or stem-and-leaf plots for the amount spent in each method of payment.
5. A crosstabulation for the variables method of payment and amount spent.

What preliminary insights do you have about the amounts spent and method of payment at Consolidated Foods?

Appendix 2.1 USING MINITAB FOR TABULAR AND GRAPHICAL METHODS

Minitab has extensive capabilities for constructing tabular and graphical summaries of data. In this appendix we show how Minitab can be used to construct several graphical summaries and the tabular summary of a crosstabulation. The graphical methods presented are the dot plot, the histogram, the stem-and-leaf display, and the scatter diagram.

Dot Plot

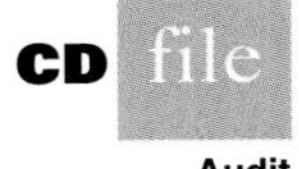

Audit

We use the audit-time data in Table 2.5 to demonstrate. The data have been entered into column C1 of a Minitab worksheet. The following steps will generate the dot plot shown in Figure 2.3.

Step 1. Select the **Graph** pull-down menu
Step 2. Choose **Dotplot**
Step 3. When the Dotplot dialog box appears:
Enter C1 in the **Variables** box
Click **OK**

Histogram

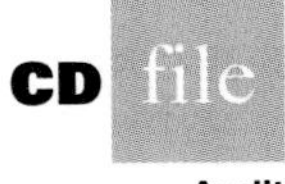

Audit

We show how to construct a histogram with frequencies on the vertical axis using the audit-time data in Table 2.5. The data have been entered into column C1 of a Minitab worksheet. The following steps will generate the histogram shown in Figure 2.4.

Step 1. Select the **Graph** pull-down menu
Step 2. Choose **Histogram**
Step 3. When the Histogram dialog box appears:
Enter C1 in row 1 of the **Graph variables** box
Select **Bar** under **Display** and **Graph** under **For each** in the **Data display** box
Select **Options**
Step 4. When the Histogram Options dialog box appears:
Select **Frequency** under **Type of Histogram**
Select **CutPoint** under **Type of Intervals**
Select **Midpoint/cutpoint positions** under **Definition of Intervals** and enter 10:35/5 in the box[1]
Click **OK**
Step 5. When the Histogram dialog box appears:
Click **OK**

1. The entry 10:35/5 indicates that 10 is the lowest value for the histogram, 35 is the highest value for the histogram, and 5 is the class width.

Stem-and-Leaf Display

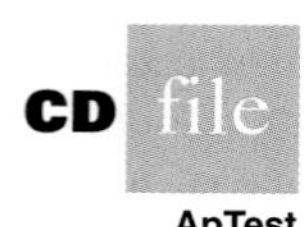
ApTest

We use the aptitude test data in Table 2.10 to demonstrate the construction of a stem-and-leaf display. The data have been entered into column C1 of a Minitab worksheet. The following steps will generate the stretched stem-and-leaf display shown in Section 2.3.

Step 1. Select the **Graph** pull-down menu
Step 2. Choose **Stem-and-Leaf**
Step 3. When the Stem-and-Leaf dialog box appears:
Enter C1 in the **Variables** box
Click **OK**

Scatter Diagram

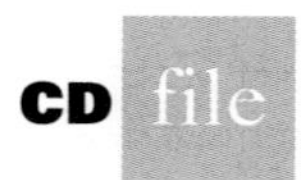
Stereo

We use the stereo and sound equipment store data in Table 2.14 to demonstrate the construction of a scatter diagram. The weeks are numbered from 1 to 10 in column C1, the data for number of commercials have been entered into column C2, and the data for sales have been entered into column C3 of a Minitab worksheet. The following steps will generate the scatter diagram shown in Figure 2.6.

Step 1. Select the **Graph** pull-down menu
Step 2. Choose **Plot**
Step 3. When the Plot dialog box appears:
Enter C3 under **Y** and C2 under **X** in the **Graph variables** section
Select **Symbol** under **Display** and **Point** under **For each** in the **Data display** section
Click **OK**

Crosstabulation

Restaurant

We use the data from Zagat's restaurant review, part of which is shown in Table 2.11, to demonstrate. The restaurants are numbered from 1 to 300 in column C1 of the Minitab worksheet. The quality ratings are in column C2, and the meal prices are in column C3.

Minitab can only create a crosstabulation for qualitative variables and meal price is a quantitative variable. So we have to first code the meal price data by specifying the category (class) to which each meal price belongs. The following steps will code the meal price data to create four categories of meal price in column C4: \$10–19, \$20–29, \$30–39, and \$40–49.

Step 1. Select the **Manip** pull-down menu
Step 2. Choose **Code**
Step 3. Choose **Numeric to Text**
Step 4. When the Code-Numeric to Text dialog box appears:
Enter C3 in the **Code data from columns** box
Enter C4 in the **Into columns** box
Enter 10:19 in the first **Original values** box and \$10–19 in the adjacent **New** box
Enter 20:29 in the second **Original values** box and \$20–29 in the adjacent **New** box
Enter 30:39 in the third **Original values** box and \$30–39 in the adjacent **New** box
Enter 40:49 in the fourth **Original values** box and \$40–49 in the adjacent **New** box
Click **OK**

For each meal price in column C3 the associated meal price category will now appear in column 4. We can now develop a crosstabulation for quality rating and the meal price categories by using the data in columns 2 and 4. The following steps will create a crosstabulation containing the same information as shown in Table 2.12.

Step 1. Select the **Stat** pull-down menu
Step 2. Choose **Tables**
Step 3. Choose **Cross Tabulation**
Step 4. When the Cross Tabulation dialog box appears:
Enter C2 C4 in the **Classification variables** box
Select **Counts**
Click **OK**

Appendix 2.2 USING EXCEL FOR TABULAR AND GRAPHICAL METHODS

Excel has extensive capabilities for constructing tabular and graphical summaries of data. Three of the most powerful tools available are the Function Wizard, The Chart Wizard, and the PivotTable Report.

Functions and the Function Wizard

Excel provides a variety of functions that are useful for statistical analysis. If you know what function you want and how to use it, you can simply enter the function directly into a cell of an Excel worksheet. If not, Excel provides a Function Wizard to help in identifying the functions available and in using them.

Function Wizard. To access the Function Wizard click f_x on the standard toolbar or select the **Insert** pull-down menu and choose f_x Function. The **Paste Function** dialog box will then appear (See Figure 2.9). The **Function category** box shows a list of the categories of Excel functions; we have selected **Statistical** in Figure 2.9. With statistical highlighted, a list of all the statistical functions available is displayed in the **Function name** box. Here, we have highlighted the COUNTIF function. Once a function has been highlighted, the proper form for the function along with a brief description appears below the Function category and Function name boxes. To obtain assistance in properly using the function, click **OK.**

SoftDrink

Frequency Distributions. We show how the COUNTIF function can be used to construct a frequency distribution for the data on soft drink purchases in Table 2.1. Refer to Figure 2.10 as we describe the tasks involved. The formula worksheet (shows the functions and formulas used) is in the background and the value worksheet (shows the results obtained using the functions and formulas) is in the foreground.

The label "Brand Purchased" and the data for the 50 soft drink purchases have been entered into cells A1:A51. We have also entered a label and the soft drink names into cells C1:C6. Excel's COUNTIF function can be used to count the number of times each soft drink appears in cells A2:A51. The following steps utilize the Function Wizard to produce the frequency distribution in the foreground of Figure 2.10.

Step 1. Select cell **D2,** access the Function Wizard, and choose **COUNTIF** from the list of statistical functions
Step 2. Click **OK**

FIGURE 2.9 EXCEL'S PASTE FUNCTION DIALOG BOX

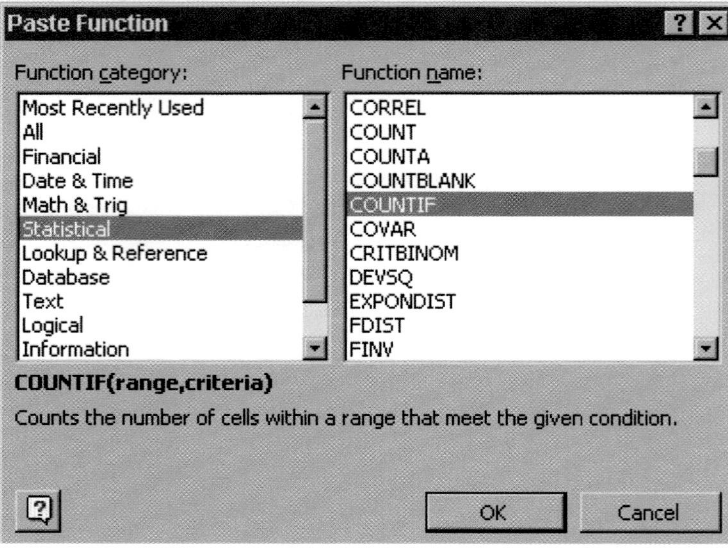

FIGURE 2.10 FREQUENCY DISTRIBUTION FOR SOFT DRINK PURCHASES CONSTRUCTED USING EXCEL'S COUNTIF FUNCTION

Note: Rows 11–44 are hidden.

	A	B	C	D	E
1	**Brand Purchased**		**Soft Drink**	**Frequency**	
2	Coke Classic		Coke Classic	=COUNTIF($A$2:$A$51,C2)	
3	Diet Coke		Diet Coke	=COUNTIF($A$2:$A$51,C3)	
4	Pepsi-Cola		Dr. Pepper	=COUNTIF($A$2:$A$51,C4)	
5	Diet Coke		Pepsi-Cola	=COUNTIF($A$2:$A$51,C5)	
6	Coke Classic		Sprite	=COUNTIF($A$2:$A$51,C6)	
7	Coke Classic				
8	Dr. Pepper				
9	Diet Coke				
10	Pepsi-Cola				
45	Pepsi-Cola				
46	Pepsi-Cola				
47	Pepsi-Cola				
48	Coke Classic				
49	Dr. Pepper				
50	Pepsi-Cola				
51	Sprite				
52					

	A	B	C	D	E
1	**Brand Purchased**		**Soft Drink**	**Frequency**	
2	Coke Classic		Coke Classic	19	
3	Diet Coke		Diet Coke	8	
4	Pepsi-Cola		Dr. Pepper	5	
5	Diet Coke		Pepsi-Cola	13	
6	Coke Classic		Sprite	5	
7	Coke Classic				
8	Dr. Pepper				
9	Diet Coke				
10	Pepsi-Cola				
45	Pepsi-Cola				
46	Pepsi-Cola				
47	Pepsi-Cola				
48	Coke Classic				
49	Dr. Pepper				
50	Pepsi-Cola				
51	Sprite				
52					

Step 3. When the COUNTIF dialog box appears:
Enter $A$2:$A$51 in the **Range** box
Enter C2 in the **Criteria** box
Click **OK**
Step 4. Copy cell D2 to cells D3:D6

The formula worksheet in the background of Figure 2.10 shows the cell formulas inserted by applying these steps. The value worksheet, in the foreground of Figure 2.10, shows the values computed using these cell formulas; we see that the Excel worksheet shows the same frequency distribution that we developed in Table 2.2.

If you are familiar with the COUNTIF function, and do not need the assistance of the Function Wizard, then you could simply enter the formulas directly into cells D2:D6. For instance, to count the number of times that Coke Classic appears, we could enter the following formula into cell D2:

=COUNTIF($A$2:$A$51,C2)

To count the number of times the other soft drinks appear, copy the formula into cells D3:D6.

Many more of Excel's functions will be demonstrated in future chapter appendixes. Depending on the complexity of the function, we will either enter it directly into the appropriate cell or utilize the Function Wizard.

Chart Wizard

Excel's Chart Wizard has extensive capabilities for developing graphical presentations. It is a tool that allows us to go beyond what can be done with functions and formulas alone. We show how it can be used to construct bar graphs, histograms, and scatter diagrams.

SoftDrink

Bar Graphs and Histograms. Here we show how the Chart Wizard can be used to construct bar graphs and histograms. Let us start by developing a bar graph for the soft drink data; we constructed a frequency distribution in Figure 2.10. The chart we are going to develop is an extension of that worksheet. Refer to Figure 2.11 as we describe the tasks involved. The value worksheet from Figure 2.10 is in the background; the chart developed using the Chart Wizard is in the foreground.

The following steps describe how to use Excel's Chart Wizard to construct a bar graph for the soft drink data using the frequency distribution appearing in cells C1:D6.

Step 1. Select cells C1:D6
Step 2. Select the **Chart Wizard** button on the Standard toolbar (or select the **Insert** pull-down menu and choose the **Chart** option)
Step 3. When the Chart Wizard–Step 1 of 4–Chart Type dialog box appears:
Choose **Column** in the **Chart type** list
Choose **Clustered Column** from the **Chart sub-type** display
Select **Next**
Step 4. When the Chart Wizard–Step 2 of 4–Chart Source Data dialog box appears:
Select **Next>**
Step 5. When the Chart Wizard–Step 3 of 4–Chart Options dialog box appears:
Select the **Titles** tab
Enter **Bar Graph of Soft Drink Purchases** in the **Chart title** box
Enter **Soft Drink** in the **Category (X) axis** box

FIGURE 2.11 BAR GRAPH OF SOFT DRINK PURCHASES CONSTRUCTED USING EXCEL'S CHART WIZARD

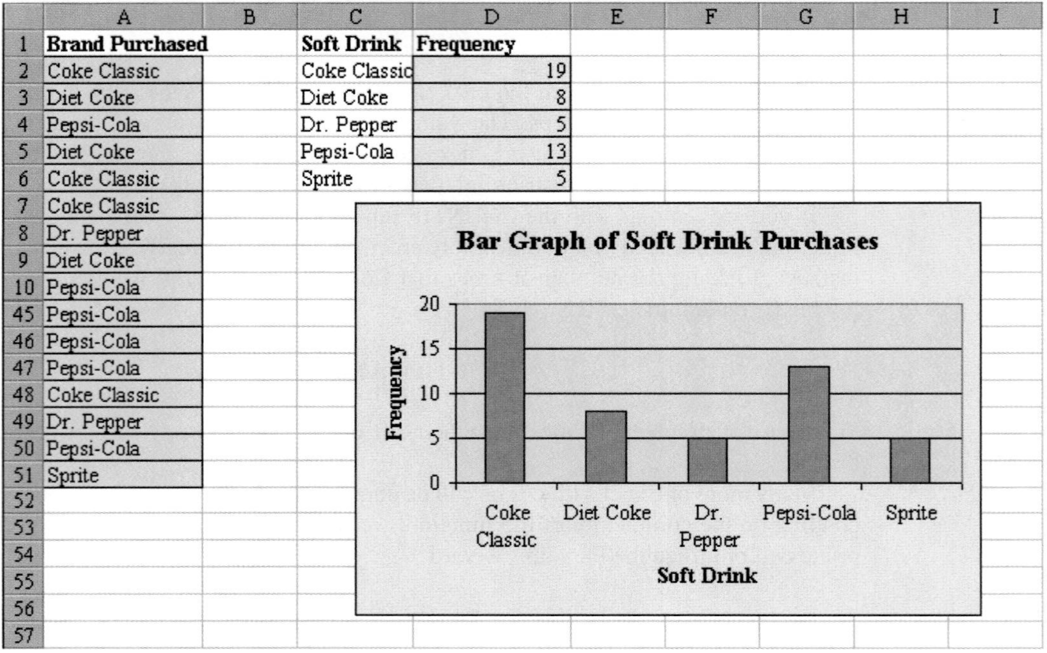

Enter **Frequency** in the **Values (Y) axis** box
Select the **Legend** tab and then
Remove the check in the **Show legend** Box
Select **Next>**

Step 6. When the Chart Wizard–Step 4 of 4–Chart Location dialog box appears:
Specify a location for the new chart (We used the current worksheet by selecting **As object in**)
Select **Finish**

The resulting bar graph (chart) is shown in Figure 2.11.[2]

Excel's Chart Wizard can produce a pie chart for the soft drink data in a similar fashion. The major difference is that in Step 3 we would choose Pie in the Chart type list.

As we stated in a note and comment at the end of Section 2.2, a histogram is essentially the same as a bar graph with no separation between the bars. Figure 2.12 shows the audit-time data with a frequency distribution in the background and a bar graph developed using the Chart Wizard (using the same steps just described) in the foreground. Because the adjacent bars in a histogram must touch, we need to edit the column chart (the bar graph) in order to eliminate the gap between each of the bars. The following steps accomplish this process.

Step 1. Right click on any bar in the column chart to produce a list of options
Step 2. Choose **Format Data Series**

2. The bar graph in Figure 2.11 is slightly larger than what was provided by Excel after selecting **Finish.** Resizing an Excel chart is not difficult. First, select the chart. Small black squares, called sizing handles, will appear on the chart border. Click on the sizing handles and drag them to resize the figure to your preference.

FIGURE 2.12 HISTOGRAM CONSTRUCTED USING EXCEL FOR THE AUDIT-TIME DATA

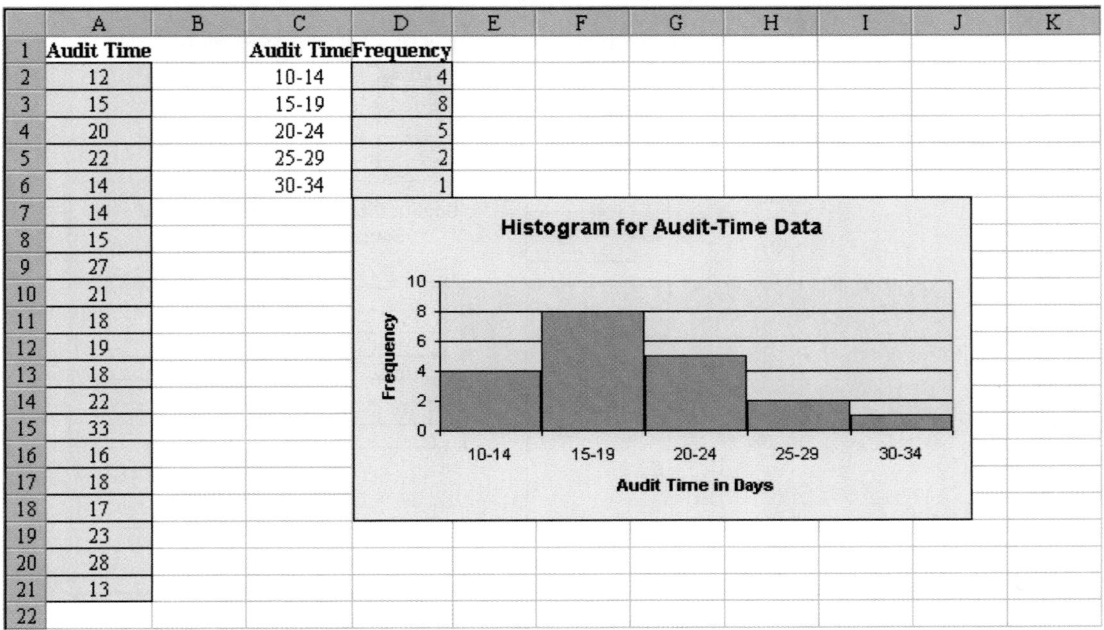

	A	B	C	D	E	F	G	H	I	J	K
1	Audit Time		Audit Time	Frequency							
2	12		10-14	4							
3	15		15-19	8							
4	20		20-24	5							
5	22		25-29	2							
6	14		30-34	1							
7	14										
8	15										
9	27										
10	21										
11	18										
12	19										
13	18										
14	22										
15	33										
16	16										
17	18										
18	17										
19	23										
20	28										
21	13										
22											

Step 3. When the Format Data Series dialog box appears:
Select the **Options** tab
Enter **0** in the **Gap width** box
Click **OK**

Stereo

Scatter Diagram. We use the stereo and sound equipment store data in Table 2.14 to demonstrate the use of Excel's Chart Wizard to construct a scatter diagram. Refer to Figure 2.13 as we describe the tasks involved. The data are in the background and the scatter diagram produced by the Chart Wizard is in the foreground. The following steps will produce the scatter diagram.

Step 1. Select cells B1:C11
Step 2. Select the **Chart Wizard** button on the standard toolbar (or select the **Insert** pull-down menu and choose the **Chart** option)
Step 3. When the Chart Wizard–Step 1 of 4–Chart Type dialog box appears:
Choose **XY (Scatter)** in the **Chart type:** display
Select **Next>**
Step 4. When the Chart Wizard–Step 2 of 4–Chart Source Data dialog box appears:
Select **Next>**
Step 5. When the Chart Wizard–Step 3 of 4–Chart Options dialog box appears:
Select the **Titles** tab
Enter **Scatter Diagram for the Stereo and Sound Equipment Store** in the **Chart title** box
Enter **No. of Commercials** in the **Value (X) axis** box
Enter **Sales Volume** in the **Value (Y) axis** box

FIGURE 2.13 SCATTER DIAGRAM FOR STEREO AND SOUND EQUIPMENT STORE USING EXCEL'S CHART WIZARD

	A	B	C	D	E	F	G
1	Week	No. of Commercials	Sales Volume				
2	1	2	50				
3	2	5	57				
4	3	1	41				
5	4	3	54				
6	5	4					
7	6	1					
8	7	5					
9	8	3					
10	9	4					
11	10	2					
12							
13							
14							
15							
16							
17							
18							
19							
20							

Scatter Diagram for the Stereo and Sound Equipment Store

Sales Volume (0–70); No. of Commercials (0–6)

Select the **Legend** tab
Remove the check in the **Show legend** box
Select **Next>**

Step 6. When the Chart Wizard–Step 4 of 4–Chart Location dialog box appears:
Specify a location for the new chart (We used the current worksheet by selecting **As object in**)
Select **Finish**

PivotTable Report

Excel's PivotTable Report provides a valuable tool for managing data sets involving more than one variable. We will illustrate its use by showing how to develop a crosstabulation.

Restaurant

Crosstabulation. We will illustrate the construction of a crosstabulation using the restaurant data in Figure 2.14. Labels have been entered in row 1 and the data for each of the 300 restaurants have been entered into cells A2:C301.

The crosstabulation shown in Table 2.12 has three rows under the heading Quality Rating, corresponding to the three quality categories, Good, Very Good, and Excellent. Unless we specify differently, the PivotTable Report will order the labels alphabetically causing the quality ratings to be listed in the order Excellent, Good, and Very Good. Because we want the quality ratings in the order Good, Very Good, Excellent, we must change the default order for the PivotTable Report. The following steps will do so:

Step 1. Select the **Tools** pull-down menu
Step 2. Choose **Options**

FIGURE 2.14 EXCEL WORKSHEET CONTAINING RESTAURANT DATA

Restaurant

Note: Rows 12–291 are hidden.

	A	B	C	D
1	Restaurant	Quality Rating	Meal Price ($)	
2	1	Good	18	
3	2	Very Good	22	
4	3	Good	28	
5	4	Excellent	38	
6	5	Very Good	33	
7	6	Good	28	
8	7	Very Good	19	
9	8	Very Good	11	
10	9	Very Good	23	
11	10	Good	13	
292	291	Very Good	23	
293	292	Very Good	24	
294	293	Excellent	45	
295	294	Good	14	
296	295	Good	18	
297	296	Good	17	
298	297	Good	16	
299	298	Good	15	
300	299	Very Good	38	
301	300	Very Good	31	
302				

Step 3. When the Options dialog box appears:
Select the **Custom lists** tab (see Figure 2.15)
In the **List entries** box, type **Good** and press Enter; type **Very Good** and press Enter; and type **Excellent**
Select **Add**
Click **OK**

We are now ready to use the PivotTable Report to construct a crosstabulation of the data for quality rating and meal price. Starting with the worksheet in Figure 2.14, the following steps are necessary:

Step 1. Select the **Data** pull down menu
Step 2. Choose the **PivotTable and PivotChart** Report
Step 3. When the PivotTable and PivotChart Wizard–Step 1 of 3 dialog box appears:
Choose **Microsoft Excel list or database**
Choose **PivotTable**
Select **Next>**
Step 4. When the PivotTable and PivotChart Wizard–Step 2 of 3 dialog box appears:
Enter A1:C301 in the **Range** box
Select **Next>**
Step 5. When the PivotTable and PivotChart Wizard–Step 3 of 3 dialog box appears:
Select **New Worksheet**
Click the **Layout** button
When the PivotTable and PivotChart Wizard–Layout diagram appears (See Figure 2.16):
Drag the **QualityRating** field button to the **ROW** section of the diagram

FIGURE 2.15 DIALOG BOX FOR CHANGING SORT ORDER IN EXCEL PIVOTTABLE

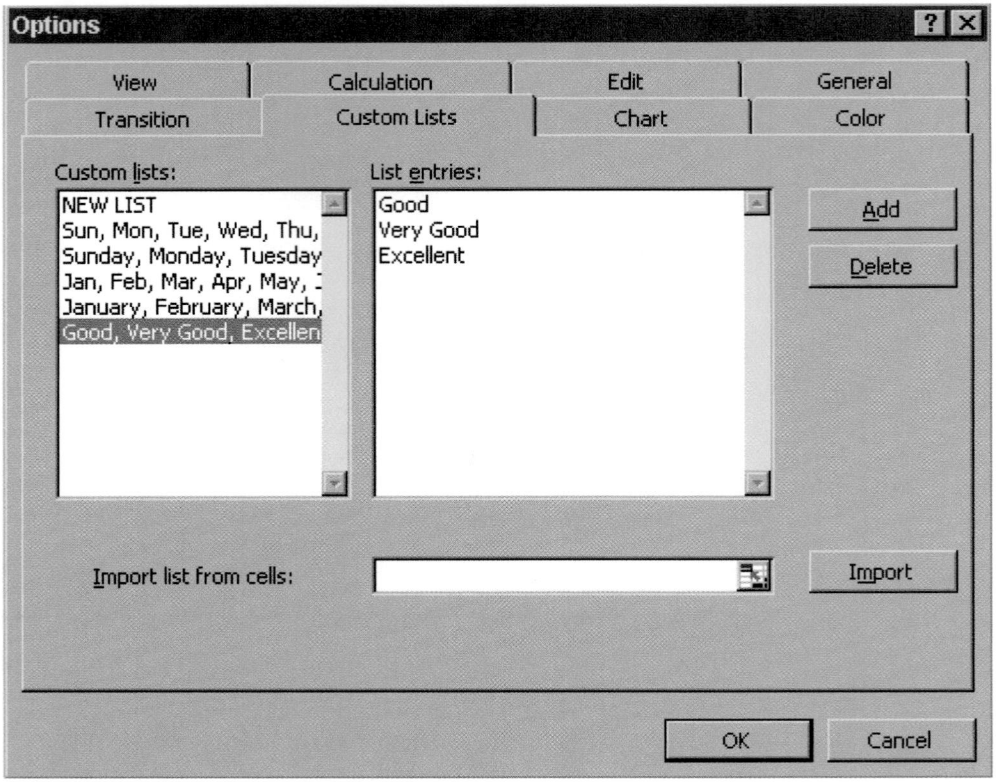

Drag the **Meal Price ($)** field button to the **COLUMN** section of the diagram
Drag the **Restaurant** field button to the **DATA** section of the diagram
Double click the **Sum of Restaurant** field button in the DATA section
When the PivotTable Field dialog box appears:
Choose **Count** under **Summarize by**
Click **OK** (Figure 2.17 shows the completed layout diagram)
Click **OK**
When the PivotTable and PivotChart Wizard–Step 3 of 3 dialog box reappears:
Select **Finish>**

A portion of the output generated by Excel is shown in Figure 2.18. Note that the output that appears in columns D through AK has been hidden so the results can be shown in a reasonably sized figure. The row labels (Good, Very Good, and Excellent) and row totals (84, 150, 66, and 300) that appear in Figure 2.18 are the same as the row labels and row totals shown in Table 2.12. But, in Figure 2.18, one column is designated for each possible value of meal price. For example, column B contains a count of restaurants with a $10 meal price, column C contains a count of restaurants with an $11 meal price, and so on. To view the PivotTable Report in a form similar to that shown in Table 2.12, we must group the

FIGURE 2.16 PIVOTTABLE AND PIVOTCHART WIZARD-LAYOUT DIAGRAM

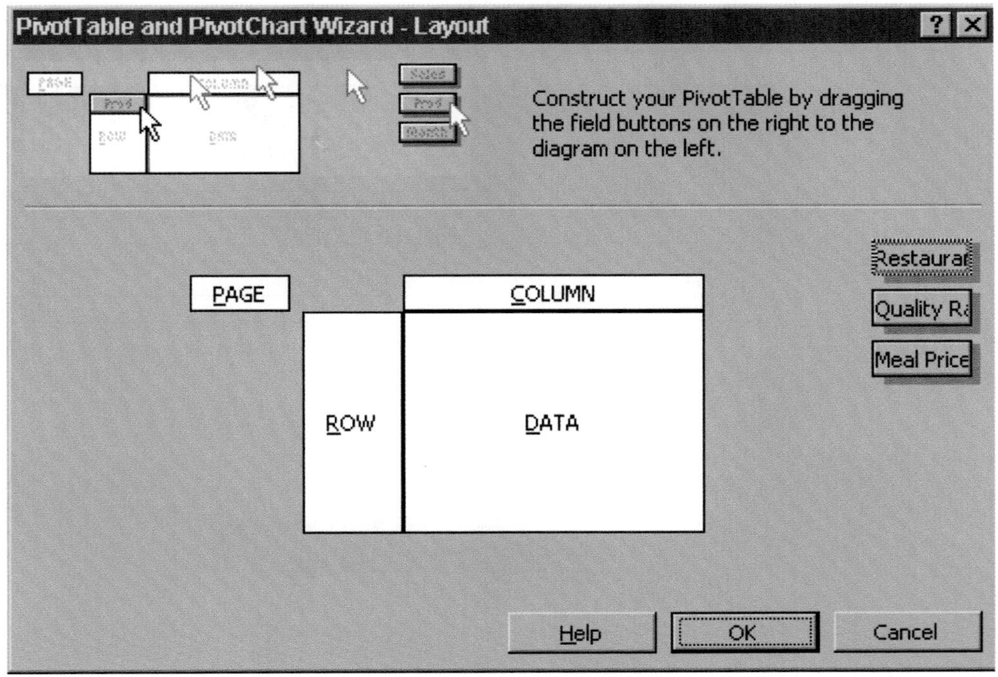

FIGURE 2.17 COMPLETED LAYOUT DIAGRAM

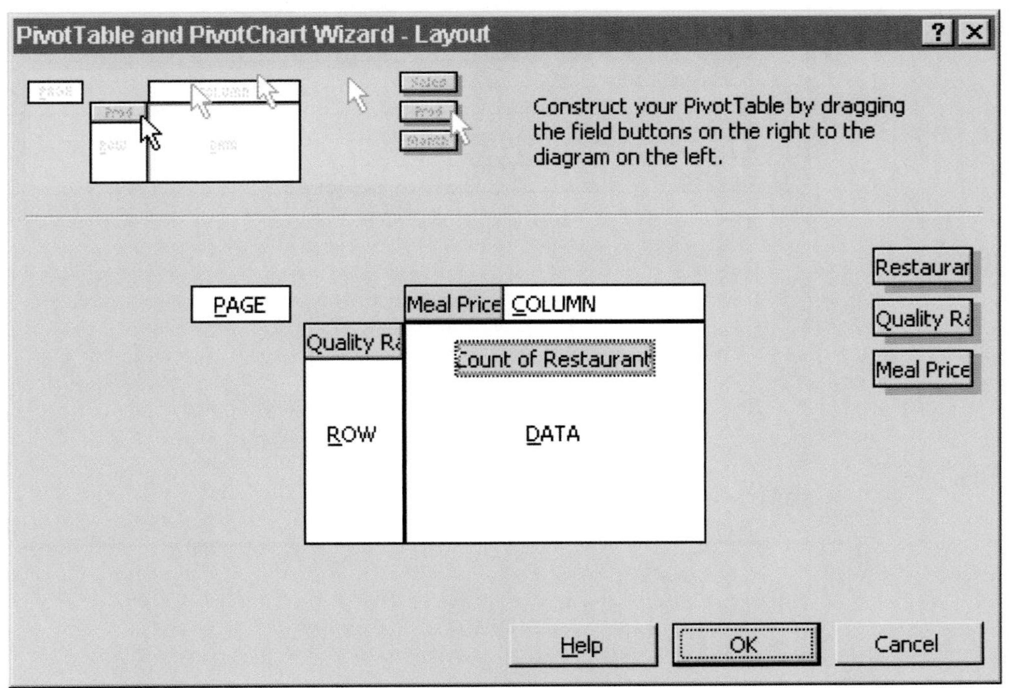

FIGURE 2.18 INITIAL PIVOTTABLE REPORT OUTPUT (COLUMNS D:AK ARE HIDDEN.)

	A	B	C	AL	AM	AN	AO
1							
2							
3	Count of Restaurant	Meal Price ($)					
4	Quality Rating	10	11	47	48	Grand Total	
5	Good	6	4			84	
6	Very Good	1	4		1	150	
7	Excellent			2	2	66	
8	Grand Total	7	8	2	3	300	
9							
10							
11							
12							
13							
14							
15							
16		PivotTable					
17		PivotTable ▾					
18							
19							
20							

FIGURE 2.19 FINAL PIVOTTABLE REPORT FOR RESTAURANT DATA

	A	B	C	D	E	F	G
1							
2							
3	Count of Restaurant	Meal Price ($)					
4	Quality Rating	10-19	20-29	30-39	40-49	Grand Total	
5	Good	42	40	2		84	
6	Very Good	34	64	46	6	150	
7	Excellent	2	14	28	22	66	
8	Grand Total	78	118	76	28	300	
9							
10							
11							
12							
13							
14							
15							
16		PivotTable					
17		PivotTable ▾					
18							
19							
20							

columns into four price categories: \$10–19, \$20–29, \$30–39, and \$40–49. The steps necessary to group the columns for the worksheet shown in Figure 2.18 follow.

Step 1. Right click on Meal Price (\$) in cell B3 of the PivotTable
Step 2. Select **Group and Outline**
Choose **Group**
Step 3. When the **Grouping** dialog box appears
Enter 10 in the **Starting at** box
Enter 49 in the **Ending at** box
Enter 10 in the **By** box
Click **OK**

The revised PivotTable output is shown in Figure 2.19. It is the final PivotTable. Note that it provides the same information as the crosstabulation shown in Table 2.12.

Chapter 3

DESCRIPTIVE STATISTICS: NUMERICAL METHODS

CONTENTS

STATISTICS IN PRACTICE

SMALL FRY DESIGN*
Santa Ana, California

Founded in 1997, Small Fry Design is a toy and accessory company that designs and imports products for infants. The company's product line includes teddy bears, mobiles, musical toys, rattles, and security blankets, and features high-quality soft toy designs with an emphasis on color, texture, and sound. The products are designed in the United States and manufactured in China.

Small Fry Design uses independent representatives to sell the products to infant furnishing retailers, children's accessory and apparel stores, gift shops, upscale department stores, and major catalog companies. Currently, Small Fry Design products are distributed in more than 1000 retail outlets throughout the United States.

Cash flow management is one of the most critical activities in the day-to-day operation of this young company. Ensuring sufficient incoming cash to meet both current and ongoing debt obligations can mean the difference between business success and failure. A critical factor in cash flow management is the analysis and control of accounts receivable. By measuring the average age and dollar value of outstanding invoices, management can predict cash availability and monitor changes in the status of accounts receivable. The company has set the following goals: the average age for outstanding invoices should not exceed 45 days and the dollar value of invoices more than 60 days old should not exceed 5% of the dollar value of all accounts receivable.

In a recent summary of accounts receivable status, the following descriptive statistics were provided for the age of outstanding invoices:

*The authors are indebted to John A. McCarthy, president of Small Fry Design, for providing this Statistics in Practice.

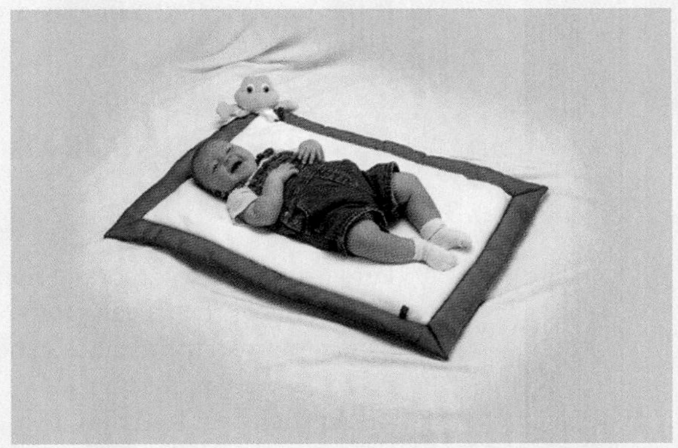

Some of the Small Fry Design products. © Photo courtesy of Small Fry Design.

Mean	40 days
Median	35 days
Mode	31 days

Interpretation of these statistics shows that the mean or average age of an invoice is 40 days. The median shows that half of the invoices have been outstanding 35 days or more. The mode of 31 days is the most frequent invoice age indicating that the most common length of time an invoice has been outstanding is 31 days. The statistical summary also showed that only 3% of the dollar value of all accounts receivable was over 60 days old. Based on the statistical information, management was satisfied that accounts receivable and incoming cash flow were under control.

In this chapter, you will learn how to compute and interpret some of the statistical measures used by Small Fry Design. In addition to the mean, median, and mode, you will learn about other descriptive statistics such as the range, variance, standard deviation, percentiles, and correlation. These numerical measures will assist in the understanding and interpretation of data.

In Chapter 2 we discussed tabular and graphical methods used to summarize data. These procedures are effective in written reports and as visual aids for presentations to individuals or groups. In this chapter, we present several numerical methods of descriptive statistics that provide additional alternatives for summarizing data.

We start by considering data sets consisting of a single variable. The numerical measures of location and dispersion are computed by using the n data values. If the data set contains more than one variable, each single variable numerical measure can be computed separately. In the two-variable case, we will also develop measures of the relationship between the variables.

Several numerical measures of location, dispersion, and association are introduced. If the measures are computed using data from a sample, they are called **sample statistics.** If the measures are computed using data from a population, they are called **population parameters.**

3.1 MEASURES OF LOCATION

Mean

Perhaps the most important numerical measure of location is the **mean,** or average value, for a variable. The mean provides a measure of central location for a data set. If the data are from a sample, the mean is denoted by $\bar{x}$; if the data are from a population, the mean is denoted by the Greek letter μ.

In statistical formulas, it is customary to denote the value of variable x for the first observation by x_1, the value of x for the second observation by x_2, and so on. In general, the value of x for the ith observation is denoted by x_i. For a sample with n observations, the formula for the sample mean is as follows.

Sample Mean

$$\bar{x} = \frac{\Sigma x_i}{n} \tag{3.1}$$

In the preceding formula, the numerator is the sum of the values of the n observations. That is,

$$\Sigma x_i = x_1 + x_2 + \cdots + x_n$$

The Greek letter Σ is the summation sign.

To illustrate the computation of a sample mean, let us consider the following class-size data for a sample of five college classes.

46 54 42 46 32

We use the notation x_1, x_2, x_3, x_4, x_5 to represent the number of students in each of the five classes.

$$x_1 = 46 \qquad x_2 = 54 \qquad x_3 = 42 \qquad x_4 = 46 \qquad x_5 = 32$$

Hence, to compute the sample mean, we can write

$$\bar{x} = \frac{\Sigma x_i}{n} = \frac{x_1 + x_2 + x_3 + x_4 + x_5}{5} = \frac{46 + 54 + 42 + 46 + 32}{5} = 44$$

The sample mean class size is 44 students.

Another illustration of the computation of a sample mean is given in the following situation. Suppose that a college placement office sent a questionnaire to a sample of business school graduates requesting information on starting salaries. Table 3.1 shows the data that

TABLE 3.1 MONTHLY STARTING SALARIES FOR A SAMPLE OF 12 BUSINESS SCHOOL GRADUATES

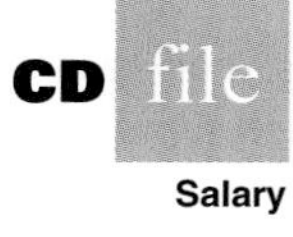

Salary

Graduate	Monthly Starting Salary ($)	Graduate	Monthly Starting Salary ($)
1	2850	7	2890
2	2950	8	3130
3	3050	9	2940
4	2880	10	3325
5	2755	11	2920
6	2710	12	2880

have been collected. The mean monthly starting salary for the sample of 12 business college graduates is computed as

$$\begin{aligned}\bar{x} &= \frac{\Sigma x_i}{n} = \frac{x_1 + x_2 + \cdots + x_{12}}{12}\\ &= \frac{2850 + 2950 + \cdots + 2880}{12}\\ &= \frac{35{,}280}{12} = 2940\end{aligned}$$

Equation (3.1) shows how the mean is computed for a sample with n observations. The formula for computing the mean of a population is the same, but we use different notation to indicate that we are working with the entire population. The number of observations in the population is denoted by N and the symbol for the population mean is μ.

Population Mean

$$\mu = \frac{\Sigma x_i}{N} \tag{3.2}$$

Median

The **median** is another measure of central location for data. The median is the value in the middle when the data are arranged in ascending order. With an odd number of observations, the median is the middle value. An even number of observations has no middle value. In this case, we follow the convention of defining the median to be the average of the values for the middle two observations. For convenience the definition of the median is restated as follows.

Median

Arrange the data in ascending order (smallest value to largest value).

(a) For an odd number of observations, the median is the middle value.
(b) For an even number of observations, the median is the average of the two middle values.

Let us apply this definition to compute the median class size for the sample of five college classes. Arranging the data in ascending order provides the following list.

32 42 46 46 54

Because $n = 5$ is odd, the median is the middle value. Thus the median class size is 46 students. Even though this data set has two values of 46, each observation is treated separately when we arrange the data in ascending order.

Suppose we also compute the median starting salary for the business college graduates. We first arrange the data in Table 3.1 in ascending order.

$$2710 \quad 2755 \quad 2850 \quad 2880 \quad 2880 \quad \underbrace{2890 \quad 2920}_{\text{Middle Two Values}} \quad 2940 \quad 2950 \quad 3050 \quad 3130 \quad 3325$$

Because $n = 12$ is even, we identify the middle two values: 2890 and 2920. The median is the average of these values.

$$\text{Median} = \frac{2890 + 2920}{2} = 2905$$

The median is the measure of location most often reported for annual income and property value data because a few extremely large incomes or property values can inflate the mean. In such cases, the median is a better measure of central location.

Although the mean is the more commonly used measure of central location, in some situations the median is preferred. The mean is influenced by extremely small and large values. For instance, suppose that one of the graduates (see Table 3.1) had a starting salary of \$10,000 per month (maybe the individual's family owns the company). If we change the highest monthly starting salary in Table 3.1 from \$3325 to \$10,000 and recompute the mean, the sample mean changes from \$2940 to \$3496. The median of \$2905, however, is unchanged, since \$2890 and \$2920 are still the middle two values. With the extremely high starting salary included, the median provides a better measure of central location than the mean. We can generalize to say that whenever a data set has extreme values, the median is often the preferred measure of central location.

Mode

A third measure of location is the **mode**. The mode is defined as follows.

Mode

The mode is the value that occurs with greatest frequency.

To illustrate the identification of the mode, consider the sample of five class sizes. The only value that occurs more than once is 46. Because this value, occurring with a frequency of 2, has the greatest frequency, it is the mode. As another illustration, consider the sample of starting salaries for the business school graduates. The only monthly starting salary that occurs more than once is \$2880. Because this value has the greatest frequency, it is the mode.

Situations can arise for which the greatest frequency occurs at two or more different values. In these instances more than one mode exists. If the data have exactly two modes, we say that the data are *bimodal.* If data have more than two modes, we say that the data are *multimodal.* In multimodal cases the mode is almost never reported; listing three or more modes would not be particularly helpful in describing a location for the data.

Percentiles

A **percentile** provides information about how the data are spread over the interval from the smallest value to the largest value. For data that do not have numerous repeated values, the pth percentile divides the data into two parts. Approximately p percent of the observations have values less than the pth percentile; approximately $(100 - p)$ percent of the observations have values greater than the pth percentile. The pth percentile is formally defined as follows.

Percentile

The pth percentile is a value such that *at least p* percent of the observations are less than or equal to this value and *at least* $(100 - p)$ percent of the observations are greater than or equal to this value.

Admission test scores for colleges and universities are frequently reported in terms of percentiles. For instance, suppose an applicant obtains a raw score of 54 on the verbal portion of an admission test. How this student performed in relation to other students taking the same test may not be readily apparent. However, if the raw score of 54 corresponds to the 70th percentile, we know that approximately 70% of the students had scores lower than this individual's and approximately 30% of the students had scores higher than this individual's.

The following procedure can be used to compute the pth percentile.

Following these steps makes it easy to calculate percentiles.

Calculating the pth Percentile

Step 1. Arrange the data in ascending order (smallest value to largest value).
Step 2. Compute an index i

$$i = \left(\frac{p}{100}\right)n$$

where p is the percentile of interest and n is the number of observations.
Step 3. (a) If *i is not an integer, round up.* The next integer *greater* than i denotes the position of the pth percentile.
(b) If *i is an integer,* the pth percentile is the average of the values in positions i and $i + 1$.

As an illustration of this procedure, let us determine the 85th percentile for the starting salary data in Table 3.1.

Step 1. Arrange the data in ascending order.

2710 2755 2850 2880 2880 2890 2920 2940 2950 3050 3130 3325

Step 2.

$$i = \left(\frac{p}{100}\right)n = \left(\frac{85}{100}\right)12 = 10.2$$

Step 3. Because i is not an integer, *round up.* The position of the 85th percentile is the next integer greater than 10.2, the 11th position.

Returning to the data, we see that the 85th percentile is the value in the 11th position, or 3130.

As another illustration of this procedure, let us consider the calculation of the 50th percentile. Applying step 2, we obtain

$$i = \left(\frac{50}{100}\right)12 = 6$$

Because i is an integer, step 3(b) states that the 50th percentile is the average of the sixth and seventh values; thus the 50th percentile is (2890 + 2920)/2 = 2905. Note that the *50th percentile is also the median.*

Quartiles

Quartiles are just specific percentiles; thus, the steps for computing percentiles can be applied directly in the computation of quartiles.

It is often desirable to divide data into four parts, with each part containing approximately one-fourth, or 25%, of the observations. Figure 3.1 shows a data set divided into four parts. The division points are referred to as the **quartiles** and are defined as

$$Q_1 = \text{first quartile, or 25th percentile}$$
$$Q_2 = \text{second quartile, or 50th percentile (also the median)}$$
$$Q_3 = \text{third quartile, or 75th percentile.}$$

The monthly starting salary data are again arranged in ascending order. Q_2, the second quartile (median), has already been identified as 2905.

2710 2755 2850 2880 2880 2890 2920 2940 2950 3050 3130 3325

The computations of Q_1 and Q_3 require the use of the rule for finding the 25th and 75th percentiles. Those calculations follow.

For Q_1,

$$i = \left(\frac{p}{100}\right)n = \left(\frac{25}{100}\right)12 = 3$$

Because i is an integer, step 3(b) indicates that the first quartile, or 25th percentile, is the average of the third and fourth values; thus, Q_1 = (2850 + 2880)/2 = 2865.

For Q_3,

$$i = \left(\frac{p}{100}\right)n = \left(\frac{75}{100}\right)12 = 9$$

FIGURE 3.1 LOCATION OF THE QUARTILES

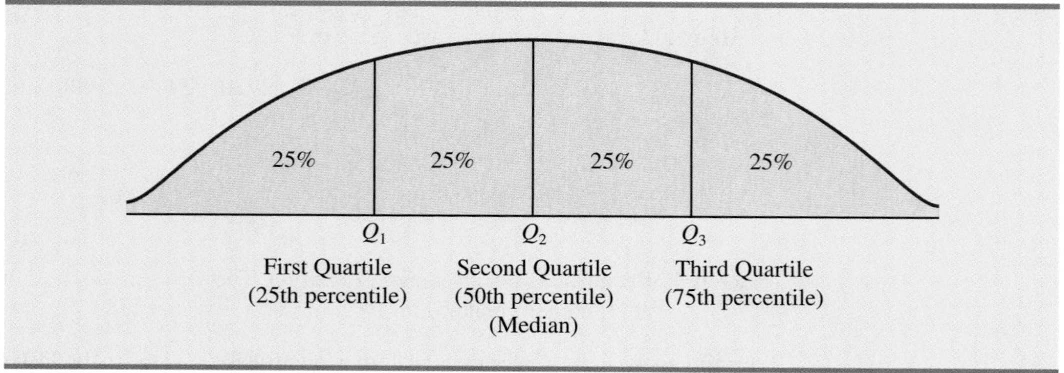

Again, because i is an integer, step 3(b) indicates that the third quartile, or 75th percentile, is the average of the ninth and tenth values; thus, $Q_3 = (\$2950 + \$3050)/2 = \$3000$.

The quartiles have divided the values into four parts, with each part consisting of 25% of the observations.

2710 2755 2850 | 2880 2880 2890 | 2920 2940 2950 | 3050 3130 3325

$Q_1 = 2865$ $Q_2 = 2905$ (Median) $Q_3 = 3000$

We have defined the quartiles as the 25th, 50th, and 75th percentiles. Thus, we have computed the quartiles in the same way as the other percentiles. However, other conventions are sometimes used to compute quartiles and the actual values reported may vary slightly depending on the convention used. Nevertheless, the objective of all procedures for computing quartiles is to divide data into roughly four equal parts.

NOTES AND COMMENTS

1. It is better to use the median than the mean as a measure of central location when a data set contains extreme values. Another measure, sometimes used when extreme values are present, is the *trimmed mean.* It is obtained by deleting the smallest and largest values from a data set and then computing the mean of the remaining values. For example, the 5% trimmed mean would be obtained by removing the smallest 5% and the largest 5% of the data values from a data set and then computing the mean of the remaining values. The 5% trimmed mean for the starting salaries in Table 3.1 is 2924.50.
2. An alternative to the quartile for dividing a data set into four equal parts has been developed by proponents of exploratory data analysis. The lower **hinge** corresponds to the first quartile, and the upper **hinge** corresponds to the third quartile. Because of different computational procedures, the values of the hinges and the quartiles may differ slightly. But, they can both be correctly interpreted as dividing a data set into approximately four equal parts. For the starting salary data in Table 3.1, the hinges and quartiles provide the same values.

EXERCISES

Methods

1. Consider the sample of size 5 with data values of 10, 20, 12, 17, and 16. Compute the mean and median.
2. Consider the sample of size 6 with data values of 10, 20, 21, 17, 16, and 12. Compute the mean and median.

3. Consider the sample of size 8 with data values of 27, 25, 20, 15, 30, 34, 28, and 25. Compute the 20th, 25th, 65th, and 75th percentiles.
4. Consider a sample with the data values of 53, 55, 70, 58, 64, 57, 53, 69, 57, 68, and 53. Compute the mean, median, and mode.

Applications

5. According to a salary survey conducted by the National Association of Colleges and Employers, bachelor's degree candidates in accounting received starting offers averaging $34,500 per year in 1999 (Bureau of Labor Statistics, *Occupational Outlook Handbook,*

2000–01 Edition). A sample of 30 students who graduated in 2000 with a bachelor's degree in accounting resulted in the following starting salaries. Data are in thousands of dollars.

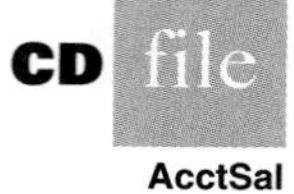

AcctSal

36.8	34.9	35.2	37.2	36.2
35.8	36.8	36.1	36.7	36.6
37.3	38.2	36.3	36.4	39.0
38.3	36.0	35.0	36.7	37.9
38.3	36.4	36.5	38.4	39.4
38.8	35.4	36.4	37.0	36.4

a. What is the mean starting salary?
b. What is the median starting salary?
c. What is the mode?
d. What is the first quartile?
e. What is the third quartile?

6. More and more investors are turning to discount brokers to save money when buying and selling shares of stock. The American Association of Individual Investors conducts an annual survey of discount brokers. Shown in Table 3.2 are the commissions charged by a sample of 20 discount brokers for two types of trades: 500 shares at $50 per share and 1000 shares at $5 per share.
a. Compute the mean, median, and mode for the commission charged on a trade of 500 shares at $50 per share.
b. Compute the mean, median, and mode for the commission charged on a trade of 1000 shares at $5 per share.
c. Which costs the most: trading 500 shares at $50 per share or trading 1000 shares at $5 per share?
d. Does the cost of a transaction seem to be related to the amount of the transaction? For example, the amount of the transaction when trading 500 shares at $50 per share is $25,000.

TABLE 3.2 COMMISSIONS CHARGED BY DISCOUNT BROKERS

Discount

	Commission ($)	
Broker	**500@$50**	**1000@$5**
AcuTrade	38.00	48.00
Bank of San Francisco	140.00	79.50
Burke Christensen & Lewis	34.00	34.00
Bush Burns Securities	35.00	35.00
Charles Schwab	155.00	90.00
Downstate Discount	55.00	60.00
Dreyfus Lion Account	154.50	88.50
First Union Brokerage	140.00	90.00
Levitt & Levitt	35.00	70.00
Max Ule	195.00	70.00
Mongerson & Co	95.00	66.00
Quick & Reilly	119.50	60.50
Scottsdale Securities, Inc.	50.00	63.00
Seaport Securities Corp.	50.00	70.00
St. Louis Discount	66.00	64.00
Summit Financial Services	95.00	60.50
T. Rowe Price Brokerage	134.00	80.00
Unified Financial Services	154.00	90.00
Wall Street Access	45.00	45.00
Your Discount Broker	55.00	70.00

Source: AAII Journal, January 2000.

7. The average person spends 45 minutes a day listening to recorded music (*The Des Moines Register,* December 5, 1997). The following data were obtained for the number of minutes spent listening to recorded music for a sample of 30 individuals.

Music

88.3	4.3	4.6	7.0	9.2
0.0	99.2	34.9	81.7	0.0
85.4	0.0	17.5	45.0	53.3
29.1	28.8	0.0	98.9	64.5
4.4	67.9	94.2	7.6	56.6
52.9	145.6	70.4	65.1	63.6

a. Compute the mean and mode.
b. Do these data appear to be consistent with the average reported by the newspaper?
c. Compute the median.
d. Compute the first and third quartiles.
e. Compute and interpret the 40th percentile.

8. Millions of Americans get up each morning and go to work in their offices at home. The growing use of personal computers is suggested to be one of the reasons more people can operate at-home businesses. Following is a sample of age data for individuals working at home.

22	58	24	50	29	52	57	31	30	41
44	40	46	29	31	37	32	44	49	29

a. Compute the mean and mode.
b. The median age of the population of all adults is 35.1 years (U.S. Census Bureau, November 1, 1997). Use the median age of the preceding data to comment on whether the at-home workers tend to be younger or older than the population of all adults.
c. Compute the first and third quartiles.
d. Compute and interpret the 32nd percentile.

9. Media Matrix collected data showing the most popular web sites when browsing at home and at work (*Business 2.0,* January 2000). The following data show the number of unique visitors (in 1000s) for the top 25 web sites when browsing at home.

CD file
Websites

Website	**Unique Visitors (in 1000s)**
about.com	5538
altavista.com	7391
amazon.com	7986
angelfire.com	8917
aol.com	23863
bluemountainarts.com	6786
ebay.com	8296
excite.com	10479
geocities.com	15321
go.com	14330
hotbot.com	5760
hotmail.com	11791
icq.com	5052
looksmart.com	5984
lycos.com	9950
microsoft.com	15593
msn.com	23505
netscape.com	14470
passport.com	11299
real.com	6785
snap.com	5730

Website	Unique Visitors (in 1000s)
tripod.com	7970
xoom.com	5652
yahoo.com	26796
zdnet.com	5133

a. Compute the mean and median.
b. Do you think it would be better to use the mean or the median as the measure of central tendency for these data? Explain.
c. Compute the first and third quartiles.
d. Compute and interpret the 85th percentile.

10. The *Los Angeles Times* regularly reports the air quality index for various areas of Southern California. Index ratings of 0–50 are considered good, 51–100 moderate, 101–200 unhealthy, 201–275 very unhealthy, and over 275 hazardous. Recent air quality indexes for Pomona were 28, 42, 58, 48, 45, 55, 60, 49, and 50.

a. Compute the mean, median, and mode for the data. Should the Pomona air quality index be considered good?
b. Compute the 25th percentile and 75th percentile for the Pomona air quality data.

11. The following data represent the number of automobiles arriving at a toll booth during 20 intervals, each of 10-minute duration. Compute the mean, median, mode, first quartile, and third quartile for the data.

26	26	58	24	22	22	15	33	19	27
21	18	16	20	34	24	27	30	31	33

12. In automobile mileage and gasoline-consumption testing, 13 automobiles were road tested for 300 miles in both city and country driving conditions. The following data were recorded for miles-per-gallon performance.

City:	16.2	16.7	15.9	14.4	13.2	15.3	16.8	16.0	16.1	15.3	15.2	15.3	16.2
Country:	19.4	20.6	18.3	18.6	19.2	17.4	17.2	18.6	19.0	21.1	19.4	18.5	18.7

Use the mean, median, and mode to make a statement about the difference in performance for city and country driving.

13. A sample of 15 college seniors showed the following credit hours taken during the final term of the senior year:

15 21 18 16 18 21 19 15 14 18 17 20 18 15 16

a. What are the mean, median, and mode for credit hours taken? Compute and interpret.
b. Compute the first and third quartiles.
c. Compute and interpret the 70th percentile.

14. Because of recent technological advances, today's digital cameras produce better-looking pictures than did their predecessors a year ago. The following data show the street price, maximum picture capacity, and battery life (minutes) for 20 of the latest models (*PC World,* January 2000).

Cameras

Camera	Price ($)	Maximum Picture Capacity	Battery Life (minutes)
Agfa EPhoto CL30	349	36	25
Canon PowerShot A50	499	106	75
Canon PowerShot Pro70	999	96	118
Epson PhotoPC 800	699	120	99
Fujifilm DX-10	299	30	229
Fujifilm MX-2700	699	141	124

Camera	Price ($)	Maximum Picture Capacity	Battery Life (minutes)
Fujifilm MX-2900 Zoom	899	141	88
HP PhotoSmart C200	299	80	68
Kodak DC215 Zoom	399	54	159
Kodak DC265 Zoom	899	180	186
Kodak DC280 Zoom	799	245	143
Minolta Dimage EX Zoom 1500	549	105	38
Nikon Coolpix 950	999	32	88
Olympus D-340R	299	122	161
Olympus D-450 Zoom	499	122	62
Richo RDC-500	699	99	56
Sony Cybershot DSC-F55	699	63	69
Sony Mavica MVC-FD73	599	40	186
Sony Mavica MVC-FD88	999	40	88
Toshiba PDR-M4	599	124	142

a. Compute the mean price.
b. Compute the mean maximum picture capacity.
c. Compute the mean battery life.
d. If you had to select one camera from this list, what camera would you choose? Explain.

3.2 MEASURES OF VARIABILITY

In addition to measures of location, it is often desirable to consider measures of variability, or dispersion. For example, suppose that you are a purchasing agent for a large manufacturing firm and that you regularly place orders with two different suppliers. After several months of operation, you find that the mean number of days required to fill orders is indeed about 10 days for both of the suppliers. The histograms summarizing the number of working days required to fill orders from the suppliers are shown in Figure 3.2. Although the mean number of days is roughly 10 for both suppliers, do the two suppliers have the same degree of reliability in terms of making deliveries on schedule? Note the dispersion, or variability, in the histograms. Which supplier would you prefer?

FIGURE 3.2 HISTORICAL DATA SHOWING THE NUMBER OF DAYS REQUIRED TO FILL ORDERS

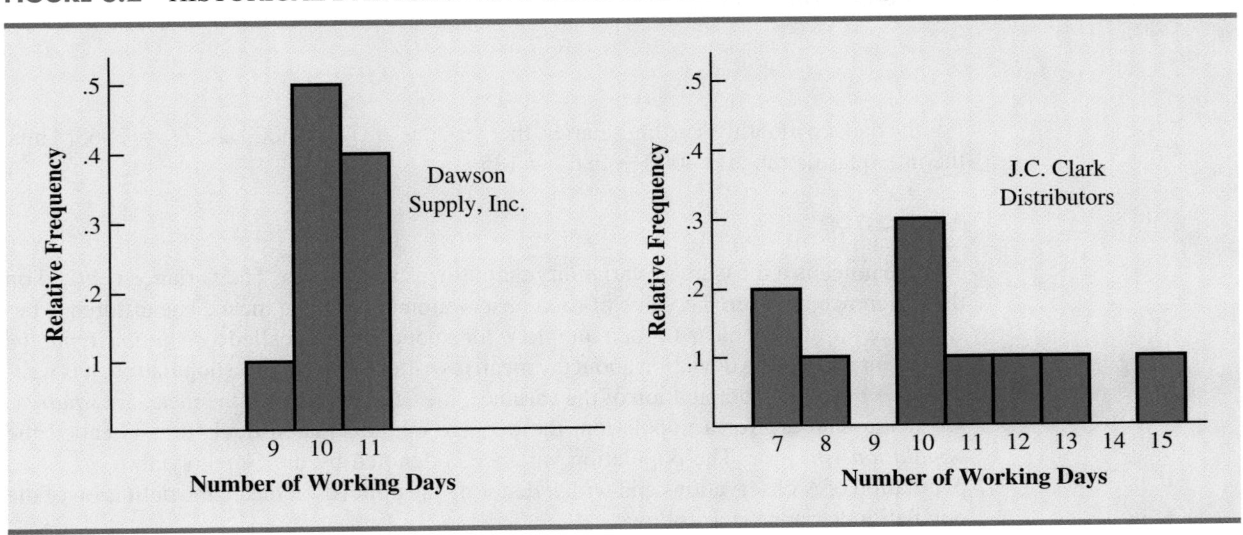

For most firms, receiving materials and supplies on schedule is important. The 7- or 8-day deliveries shown for J. C. Clark Distributors might be viewed favorably; however, a few of the slow 13- to 15-day deliveries could be disastrous in terms of keeping a workforce busy and production on schedule. This example illustrates a situation in which the variability in the delivery times may be an overriding consideration in selecting a supplier. For most purchasing agents, the lower variability shown for Dawson Supply, Inc., would make Dawson the preferred supplier.

We turn now to a discussion of some commonly used measures of variability.

Range

Perhaps the simplest measure of variability is the **range.**

Range

$$\text{Range} = \text{Largest Value} - \text{Smallest Value}$$

Let us refer to the data on monthly starting salaries for business school graduates in Table 3.1. The largest starting salary is 3325 and the smallest is 2710. The range is 3325 − 2710 = 615.

The range is easy to compute, but it is sensitive to just two data values: the largest and smallest.

Although the range is the easiest of the measures of variability to compute, it is seldom used as the only measure. The reason is that the range is based on only two of the observations and thus is highly influenced by extreme values. Suppose one of the graduates had a starting salary of $10,000. In this case, the range would be 10,000 − 2710 = 7290 rather than 615. This large value for the range would not be particularly descriptive of the variability in the data, because 11 of the 12 starting salaries are closely grouped between 2710 and 3130.

Interquartile Range

A measure of variability that overcomes the dependency on extreme values is the **interquartile range (IQR).** This measure of variability is simply the difference between the third quartile, Q_3, and the first quartile, Q_1. In other words, the interquartile range is the range for the middle 50% of the data.

Interquartile Range

$$IQR = Q_3 - Q_1 \qquad \textbf{(3.3)}$$

For the data on monthly starting salaries, the quartiles are $Q_3 = 3000$ and $Q_1 = 2865$. Thus, the interquartile range is 3000 − 2865 = 135.

Variance

The **variance** is a measure of variability that utilizes all the data. The variance is based on the difference between the value of each observation (x_i) and the mean. The difference between each x_i and the mean ($\bar{x}$ for a sample, μ for a population) is called a *deviation about the mean.* For a sample, a deviation about the mean is written $(x_i - \bar{x})$; for a population, it is written $(x_i - \mu)$. In the computation of the variance, the deviations about the mean are *squared.*

If the data are for a population, the average of the squared deviations is called the *population variance.* The population variance is denoted by the Greek symbol σ^2. For a population of N observations and with μ denoting the population mean, the definition of the population variance is as follows.

Population Variance

$$\sigma^2 = \frac{\Sigma(x_i - \mu)^2}{N} \tag{3.4}$$

In most statistical applications, the data being analyzed is a sample. When we compute a sample variance, we are often interested in using it to estimate the population variance σ^2. Although a detailed explanation is beyond the scope of this text, it can be shown that if the sum of the squared deviations about the sample mean is divided by $n - 1$, and not n, the resulting sample variance provides an unbiased estimate of the population variance. For this reason, the *sample variance,* denoted by s^2, is defined as follows.

Sample Variance

$$s^2 = \frac{\Sigma(x_i - \bar{x})^2}{n - 1} \tag{3.5}$$

To illustrate the computation of the variance for a sample, we use the data on class size for the sample of five college classes as presented in Section 3.1. A summary of the data, including the computation of the deviations about the mean and the squared deviations about the mean, is shown in Table 3.3. The sum of squared deviations about the mean is $\Sigma(x_i - \bar{x})^2 = 256$. Hence, with $n - 1 = 4$, the sample variance is

$$s^2 = \frac{\Sigma(x_i - \bar{x})^2}{n - 1} = \frac{256}{4} = 64$$

The variance is useful in comparing the variability of two variables.

Before moving on, let us note that the units associated with the sample variance often cause confusion. Because the values being summed in the variance calculation, $(x_i - \bar{x})^2$, are squared, the units associated with the sample variance are also squared. For instance, the sample variance for the class-size data is $s^2 = 64$ (students)^2. The squared units associated with variance make it difficult to obtain an intuitive understanding and interpretation of the numerical value of the variance. We recommend that you think of the variance as a

TABLE 3.3 COMPUTATION OF DEVIATIONS AND SQUARED DEVIATIONS ABOUT THE MEAN FOR THE CLASS-SIZE DATA

Number of Students in Class (x_i)	Mean Class Size ($\bar{x}$)	Deviation About the Mean ($x_i - \bar{x}$)	Squared Deviation About the Mean $(x_i - \bar{x})^2$
46	44	2	4
54	44	10	100
42	44	−2	4
46	44	2	4
32	44	−12	144
		0	256
		$\Sigma(x_i - \bar{x})$	$\Sigma(x_i - \bar{x})^2$

measure useful in comparing the amount of variability for two or more variables. In a comparison of the variables, the one with the larger variance has the most variability. Further interpretation of the value of the variance may not be necessary.

As another illustration of computing a sample variance, consider the starting salaries listed in Table 3.1 for the 12 business school graduates. In Section 3.1, we showed that the sample mean starting salary was 2940. The computation of the sample variance ($s^2 =$ 27,440.91) is shown in Table 3.4.

Note that in Tables 3.3 and 3.4 we show both the sum of the deviations about the mean and the sum of the squared deviations about the mean. For any data set, the sum of the deviations about the mean will *always equal zero.* Note that in Tables 3.3 and 3.4, $\Sigma(x_i - \bar{x}) = 0$. The positive deviations and negative deviations always cancel each other, causing the sum of the deviations about the mean to equal zero.

Standard Deviation

The **standard deviation** is defined as the positive square root of the variance. Following the notation we adopted for a sample variance and a population variance, we use s to denote the sample standard deviation and σ to denote the population standard deviation. The standard deviation is derived from the variance in the following way.

Standard Deviation

$$\text{Sample Standard Deviation} = s = \sqrt{s^2} \tag{3.6}$$

$$\text{Population Standard Deviation} = \sigma = \sqrt{\sigma^2} \tag{3.7}$$

TABLE 3.4 COMPUTATION OF THE SAMPLE VARIANCE FOR THE STARTING SALARY DATA

Monthly Salary (x_i)	Sample Mean ($\bar{x}$)	Deviation About the Mean ($x_i - \bar{x}$)	Squared Deviation About the Mean ($(x_i - \bar{x})^2$)
2850	2940	−90	8,100
2950	2940	10	100
3050	2940	110	12,100
2880	2940	−60	3,600
2755	2940	−185	34,225
2710	2940	−230	52,900
2890	2940	−50	2,500
3130	2940	190	36,100
2940	2940	0	0
3325	2940	385	148,225
2920	2940	−20	400
2880	2940	−60	3,600
		0	301,850
		$\Sigma(x_i - \bar{x})$	$\Sigma(x_i - \bar{x})^2$

Using equation (3.5),

$$s^2 = \frac{\Sigma(x_i - \bar{x})^2}{n-1} = \frac{301{,}850}{11} = 27{,}440.91$$

Recall that the sample variance for the sample of class sizes in five college classes is $s^2 = 64$. Thus the sample standard deviation is $s = \sqrt{64} = 8$. For the data set on starting salaries, the sample standard deviation is $s = \sqrt{27{,}440.91} = 165.65$.

The standard deviation is easier to interpret than the variance because standard deviation is measured in the same units as the data.

What is gained by converting the variance to its corresponding standard deviation? Recall that the units associated with the variance are squared. For example, the sample variance for the starting salary data of business school graduates is $s^2 = 27{,}440.91$ (dollars)^2. Because the standard deviation is simply the square root of the variance, the units of the variance, dollars squared, are converted to dollars in the standard deviation. Thus, the standard deviation of the starting salary data is \$165.65. In other words, the standard deviation is measured in the same units as the original data. For this reason the standard deviation is more easily compared to the mean and other statistics that are measured in the same units as the original data.

Coefficient of Variation

The coefficient of variation is a relative measure of variability; it measures the standard deviation relative to the mean.

In some situations we may be interested in a descriptive statistic that indicates how large the standard deviation is in relation to the mean. This measure is called the **coefficient of variation** and is computed as follows.

Coefficient of Variation

$$\frac{\text{Standard Deviation}}{\text{Mean}} \times 100 \tag{3.8}$$

For the class-size data, we found a sample mean of 44 and a sample standard deviation of 8. The coefficient of variation is $(8/44) \times 100 = 18.2$. In words, the coefficient of variation tells us that the sample standard deviation is 18.2% of the value of the sample mean. For the starting-salary data with a sample mean of 2940 and a sample standard deviation of 165.65, the coefficient of variation, $(165.65/2940) \times 100 = 5.6$, tells us the sample standard deviation is only 5.6% of the value of the sample mean. In general, the coefficient of variation is a useful statistic for comparing the variability of variables that have different standard deviations and different means.

NOTES AND COMMENTS

1. Statistical software packages and spreadsheets can be used to develop the descriptive statistics presented in this chapter. After the data have been entered into a worksheet, a few simple commands can be used to generate the desired output. In Appendixes 3.1 and 3.2, we show how Minitab and Excel can be used to develop descriptive statistics.
2. The standard deviation is a commonly used measure of the risk associated with investing in stock and stock funds (*Business Week*, January 17, 2000). It provides a measure of how monthly returns fluctuate around the long-run average return.
3. Rounding the value of the sample mean $\bar{x}$ and the values of the squared deviations $(x_i - \bar{x})^2$ may introduce errors when a calculator is used in the computation of the variance and standard deviation. To reduce rounding errors, we recommend carrying at least six significant digits during intermediate calculations. The resulting variance or standard deviation can then be rounded to fewer digits.
4. An alternative formula for the computation of the sample variance is

$$s^2 = \frac{\Sigma x_i^2 - n\bar{x}^2}{n - 1}$$

where $\Sigma x_i^2 = x_1^2 + x_2^2 + \cdots + x_n^2$.

EXERCISES

Methods

15. Consider the sample of size 5 with data values of 10, 20, 12, 17, and 16. Compute the range and interquartile range.

16. Consider the sample of size 5 with data values of 10, 20, 12, 17, and 16. Compute the variance and standard deviation.

SELF test

17. Consider the sample of size 8 with data values of 27, 25, 20, 15, 30, 34, 28, and 25. Compute the range, interquartile range, variance, and standard deviation.

Applications

18. A bowler's scores for six games were 182, 168, 184, 190, 170, and 174. Using these data as a sample, compute the following descriptive statistics.
    a. Range
    b. Variance
    c. Standard deviation
    d. Coefficient of variation

19. *PC World* provided ratings for the top 15 notebook PCs (*PC World,* February 2000). A 100-point scale was used to provide an overall rating for each notebook tested in the study. A score in the 90s is exceptional, while one in the 70s is above average. The overall ratings for the 15 notebooks that were tested are shown here.

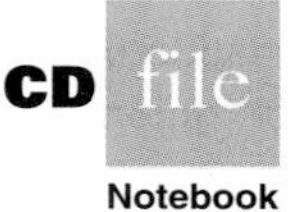

Notebook

Notebook	Overall Rating
AMS Tech Roadster 15CTA380	67
Compaq Armada M700	78
Compaq Prosignia Notebook 150	79
Dell Inspiron 3700 C466GT	80
Dell Inspiron 7500 R500VT	84
Dell Latitude Cpi A366XT	76
Enpower ENP-313 Pro	77
Gateway Solo 9300LS	92
HP Pavillion Notebook PC	83
IBM ThinkPad I Series 1480	78
Micro Express NP7400	77
Micron TransPort NX PII-400	78
NEC Versa SX	78
Sceptre Soundx 5200	73
Sony VAIO PCG-F340	77

Compute the range, interquartile range, variance, and standard deviation.

20. The *Los Angeles Times* regularly reports the air quality index for various areas of Southern California. A sample of air quality index values for Pomona provided the following data: 28, 42, 58, 48, 45, 55, 60, 49, and 50.
    a. Compute the range and interquartile range.
    b. Compute the sample variance and sample standard deviation.
    c. A sample of air quality index readings for Anaheim provided a sample mean of 48.5, a sample variance of 136, and a sample standard deviation of 11.66. What comparisons can you make between the air quality in Pomona and that in Anaheim on the basis of these descriptive statistics?

21. The Davis Manufacturing Company has just completed 5 weeks of operation using a new process that is supposed to increase productivity. The numbers of parts produced each week are 410, 420, 390, 400, and 380. Compute the sample variance and sample standard deviation.

22. Assume that the following data were used to construct the histograms of the number of days required to fill orders for Dawson Supply, Inc., and J. C. Clark Distributors (see Figure 3.2).

Dawson Supply Days for Delivery:	11	10	9	10	11	11	10	11	10	10
Clark Distributors Days for Delivery:	8	10	13	7	10	11	10	7	15	12

Use the range and standard deviation to support the previous observation that Dawson Supply provides the more consistent and reliable delivery times.

23. Police records show the following numbers of daily crime reports for a sample of days during the winter months and a sample of days during the summer months.

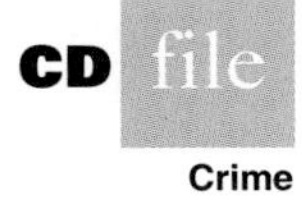
Crime

Winter	Summer
18	28
20	18
15	24
16	32
21	18
20	29
12	23
16	38
19	28
20	18

a. Compute the range and interquartile range for each period.
b. Compute the variance and standard deviation for each period.
c. Compute the coefficient of variation for each period.
d. Compare the variability of the two periods.

Discount

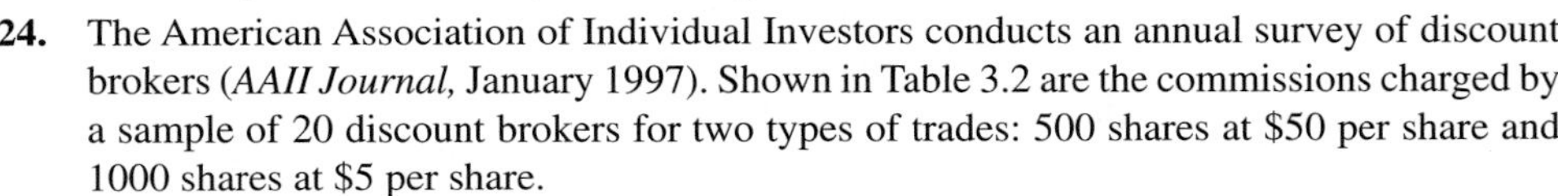

24. The American Association of Individual Investors conducts an annual survey of discount brokers (*AAII Journal,* January 1997). Shown in Table 3.2 are the commissions charged by a sample of 20 discount brokers for two types of trades: 500 shares at \$50 per share and 1000 shares at \$5 per share.
a. Compute the range and interquartile range for each type of trade.
b. Compute the variance and standard deviation for each type of trade.
c. Compute the coefficient of variation for each type of trade.
d. Compare the variability of cost for the two types of trades.

25. A production department uses a sampling procedure to test the quality of newly produced items. The department employs the following decision rule at an inspection station: If a sample of 14 items has a variance of more than .005, the production line must be shut down for repairs. Suppose the following data have just been collected:

3.43	3.45	3.43	3.48	3.52	3.50	3.39
3.48	3.41	3.38	3.49	3.45	3.51	3.50

Should the production line be shut down? Why or why not?

26. The following times were recorded by the quarter-mile and mile runners of a university track team (times are in minutes).

Quarter-mile Times:	.92	.98	1.04	.90	.99
Mile Times:	4.52	4.35	4.60	4.70	4.50

After viewing this sample of running times, one of the coaches commented that the quarter-milers turned in the more consistent times. Use the standard deviation and the coefficient of variation to summarize the variability in the data. Does the use of the coefficient of variation indicate that the coach's statement should be qualified?

3.3 MEASURES OF RELATIVE LOCATION AND DETECTING OUTLIERS

We have described several measures of location and variability for data. The mean is the most widely used measure of location, whereas the standard deviation and variance are the most widely used measures of variability. Using only the mean and the standard deviation, we also can learn much about the relative location of items in a data set.

z-Scores

By using the mean and standard deviation, we can determine the relative location of any observation. Suppose we have a sample of n observations, with the values denoted by x_1, $x_2, \ldots, x_n$. In addition, assume that the sample mean, $\bar{x}$, and the sample standard deviation, s, have been computed. Associated with each value, x_i, is another value called its *z*-score. Equation (3.9) shows how the *z*-score is computed for each x_i.

z-Score

$$z_i = \frac{x_i - \bar{x}}{s} \tag{3.9}$$

where

z_i = the *z*-score for x_i
$\bar{x}$ = the sample mean
s = the sample standard deviation

The *z*-score is often called the *standardized value*. The standardized value or *z*-score, z_i, can be interpreted as the *number of standard deviations x_i is from the mean $\bar{x}$*. For example, $z_1 = 1.2$ would indicate that x_1 is 1.2 standard deviations greater than the sample mean. Similarly, $z_2 = -.5$ would indicate that x_2 is .5, or 1/2, standard deviation less than the sample mean. Note also that *z*-scores greater than zero occur for observations with values greater than the mean, and *z*-scores less than zero occur for observations with values less than the mean. A *z*-score of zero indicates that the value of the observation is equal to the mean.

The *z*-score for any observation can be interpreted as a measure of the relative location of the observation in a data set. Thus, observations in two different data sets with the same *z*-score can be said to have the same relative location in terms of being the same number of standard deviations from the mean.

The *z*-scores for the class-size data are listed in Table 3.5. Recall that the sample mean, $\bar{x} = 44$, and sample standard deviation, $s = 8$, have been computed previously. The *z*-score of -1.50 for the fifth observation shows it is farthest from the mean; it is 1.50 standard deviations below the mean.

TABLE 3.5 *z*-SCORES FOR THE CLASS-SIZE DATA

Number of Students in Class (x_i)	Deviation About the Mean ($x_i - \bar{x}$)	*z*-score $\left(\frac{x_i - \bar{x}}{s}\right)$
46	2	2/8 = .25
54	10	10/8 = 1.25
42	−2	−2/8 = −.25
46	2	2/8 = .25
32	−12	−12/8 = −1.50

Chebyshev's Theorem

Chebyshev's theorem enables us to make statements about the proportion of data values that must be within a specified number of standard deviations from the mean.

Chebyshev's Theorem

At least $(1 - 1/z^2)$ of the data values must be within z standard deviations of the mean, where z is any value greater than 1.

Some of the implications of this theorem, with $z = 2$, 3, and 4 standard deviations, follow.

- At least .75, or 75%, of the data values must be within $z = 2$ standard deviations of the mean.
- At least .89, or 89%, of the data values must be within $z = 3$ standard deviations of the mean.
- At least .94, or 94%, of the data values must be within $z = 4$ standard deviations of the mean.

For an example using Chebyshev's theorem, assume that the midterm test scores for 100 students in a college business statistics course had a mean of 70 and a standard deviation of 5. How many students had test scores between 60 and 80? How many students had test scores between 58 and 82?

For the test scores between 60 and 80, we note that 60 is two standard deviations below the mean and 80 is two standard deviations above the mean. Using Chebyshev's theorem, we see that at least .75, or at least 75%, of the observations must have values within two standard deviations of the mean. Thus, at least 75 of the 100 students must have scored between 60 and 80.

Chebyshev's theorem requires $z > 1$, but z need not be an integer. Exercise 27 involves noninteger values of z greater than 1.

For the test scores between 58 and 82, we see that $(58 - 70)/5 = -2.4$ indicates 58 is 2.4 standard deviations below the mean and that $(82 - 70)/5 = +2.4$ indicates 82 is 2.4 standard deviations above the mean. Applying Chebyshev's theorem with $z = 2.4$, we have

$$\left(1 - \frac{1}{z^2}\right) = \left[1 - \frac{1}{(2.4)^2}\right] = .826$$

At least 82.6% of the students must have test scores between 58 and 82.

Empirical Rule

The empirical rule is based on the normal probability distribution, which will be introduced in Chapter 6 and used extensively throughout the text.

One of the advantages of Chebyshev's theorem is that it applies to any data set regardless of the shape of the distribution of the data. In practical applications, however, it has been found that many data sets have a mound-shaped or bell-shaped distribution like the one shown in Figure 3.3. When the data are believed to approximate this distribution, the **empirical rule** can be used to determine the percentage of data values that must be within a specified number of standard deviations of the mean.

Empirical Rule

For data having a bell-shaped distribution:

- Approximately 68% of the data values will be within one standard deviation of the mean.
- Approximately 95% of the data values will be within two standard deviations of the mean.
- Almost all of the data values will be within three standard deviations of the mean.

FIGURE 3.3 A MOUND-SHAPED OR BELL-SHAPED DISTRIBUTION

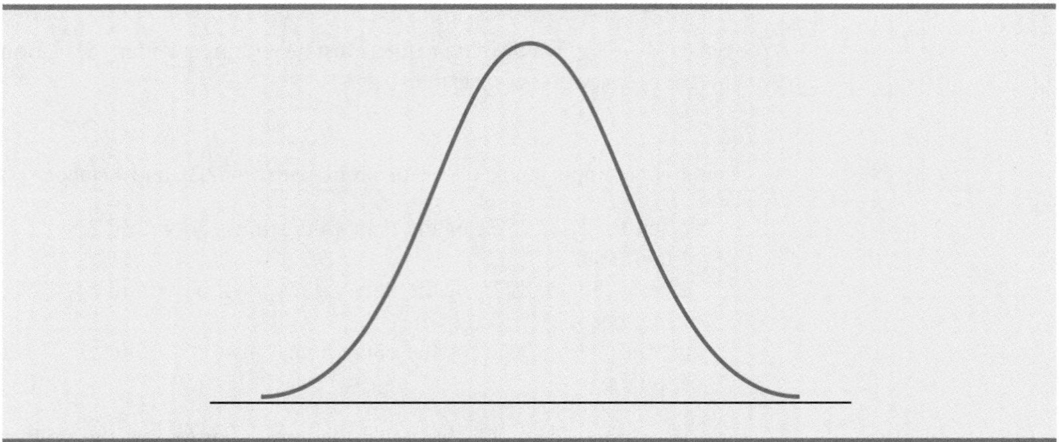

For example, liquid detergent cartons are filled automatically on a production line. Filling weights frequently have a bell-shaped distribution. If the mean filling weight is 16 ounces and the standard deviation is .25 ounces, we can use the empirical rule to draw the following conclusions.

- Approximately 68% of the filled cartons will have weights between 15.75 and 16.25 ounces (that is, within one standard deviation of the mean).
- Approximately 95% of the filled cartons will have weights between 15.50 and 16.50 ounces (that is, within two standard deviations of the mean).
- Almost all filled cartons will have weights between 15.25 and 16.75 ounces (that is, within three standard deviations of the mean).

Detecting Outliers

Sometimes a data set will have one or more observations with unusually large or unusually small values. Extreme values such as these are called **outliers.** Experienced statisticians take steps to identify outliers and then review each one carefully. An outlier may be a data value for which the data have been incorrectly recorded. If so, it can be corrected before further analysis. An outlier may also be from an observation that was incorrectly included in the data set; if so, it can be removed. Finally, an outlier may just be an unusual data value that has been recorded correctly and does belong in the data set. In such cases the item should remain.

It is a good idea to check for outliers before making decisions based on data analysis. Errors are often made in recording data and entering it into the computer. Outliers should not necessarily be deleted, but their accuracy and appropriateness should be verified.

Standardized values (z-scores) can be used to help identify outliers. Recall that the empirical rule allows us to conclude that for data with a bell-shaped distribution, almost all the data values will be within three standard deviations of the mean. Hence, in using z-scores to identify outliers, we recommend treating any data value with a z-score less than -3 or greater than $+3$ as an outlier. Such items can then be reviewed for accuracy and to determine whether they belong in the data set.

Refer to the z-scores for the class-size data in Table 3.5. The z-score of -1.50 shows the fifth item is farthest from the mean. However, this standardized value is well within the -3 to $+3$ guideline for outliers. Thus, the z-scores show that outliers are not present in the class-size data.

NOTES AND COMMENTS

1. Chebyshev's theorem is applicable for any data set and can be used to state the minimum number of data values that will be within a certain number of standard deviations of the mean. If the data set is known to be approximately bell-shaped, more can be said. For instance, the empirical rule allows us to say that *approximately* 95% of the data values will be within two standard deviations of the mean; Chebyshev's theorem allows us to conclude only that at least 75% of the data values will be in that interval.
2. Before analyzing a data set, statisticians usually make a variety of checks to ensure the validity of data. In a large study it is not uncommon for errors to be made in recording data values or in entering the values at a computer. Identifying outliers is one tool used to check the validity of data.

EXERCISES

Methods

27. Consider a sample with a mean of 30 and a standard deviation of 5. Use Chebyshev's theorem to determine the proportion, or percentage, of the data within each of the following ranges.

a. 20 to 40. **b.** 15 to 45. **c.** 22 to 38. **d.** 18 to 42. **e.** 12 to 48.

28. Data that have a bell-shaped distribution have a mean of 30 and a standard deviation of 5. Use the empirical rule to determine the proportion, or percentage, of data within each of the following ranges.

a. 20 to 40. **b.** 15 to 45. **c.** 25 to 35.

29. Consider the sample of size 5 with data values of 10, 20, 12, 17, and 16. Compute the z-score for each of the five data values.

30. Consider a sample with a mean of 500 and a standard deviation of 100. What is the z-score for each of the data values 520, 650, 500, 450, and 280?

Applications

31. The results of a national survey of 1154 adults showed that on average, adults sleep 6.9 hours per day during the workweek (2000 Omnibus Sleep in America Poll). Suppose that the standard deviation is 1.2 hours.

a. Use Chebyshev's theorem to calculate the percentage of individuals that sleeps between 4.5 and 9.3 hours per day.

b. Use Chebyshev's theorem to calculate the percentage of individuals that sleeps between 3.9 and 9.9 hours per day.

c. Assume that the number of hours of sleep is bell-shaped. Use the empirical rule to calculate the percentage of individuals that sleeps between 4.5 and 9.3 hours per day. How does this result compare to the value that you obtained using Chebyshev's theorem in part (a)?

32. According to ACNielsen, kids aged 12 to 17 watched an average of 3 hours of television per day for the broadcast year that ended in August (*Barron's,* November 8, 1999). Suppose that the standard deviation is 1 hour and that the distribution of the time spent watching television has a bell-shaped distribution.

a. What percentage of kids aged 12 to 17 watches television between 2 and 3 hours per day?

b. What percentage of kids aged 12 to 17 watches television between 1 and 4 hours per day?

c. What percentage of kids aged 12 to 17 watches television more than 4 hours per day?

33. Suppose that IQ scores have a bell-shaped distribution with a mean of 100 and a standard deviation of 15.

a. What percentage of people should have an IQ score between 85 and 115?
b. What percentage of people should have an IQ score between 70 and 130?
c. What percentage of people should have an IQ score of more than 130?
d. A person with an IQ score of more than 145 is considered a genius. Does the empirical rule support this statement? Explain.

34. The average labor cost for color TV repair in Chicago is $90.06 (*The Wall Street Journal,* January 2, 1998). Suppose the standard deviation is $20.

a. What is the z-score for a repair job with a labor cost of $71?
b. What is the z-score for a repair job with a labor cost of $168?
c. Interpret the z-scores in parts (a) and (b). Comment on whether either should be considered an outlier.

35. Wageweb conducts surveys of salary data and presents summaries on its web site. Using salary data as of January 1, 2000, Wageweb reported that salaries of benefits managers ranged from $50,935 to $79,577 (Wageweb.com, April 12, 2000). Assume the following data are a sample of the annual salaries for 30 benefits managers (data are in thousands of dollars).

CD file
WageWeb

57.7	64.4	62.1	59.1	71.1
63.0	64.7	61.2	66.8	61.8
64.2	63.3	62.2	61.2	59.4
63.0	66.7	60.3	74.0	62.8
68.7	63.8	59.2	60.3	56.6
59.3	69.5	61.7	58.9	63.1

a. Compute the mean and standard deviation for the sample data.
b. Using the mean and standard deviation computed in part (a) as estimates of the mean and standard deviation of salary for the population of benefits managers, use Chebyshev's theorem to determine the percentage of benefit managers with an annual salary between $55,000 and $71,000.
c. Develop a histogram for the sample data. Does it appear reasonable to assume that the distribution of annual salary can be approximated by a bell-shaped distribution?
d. Assume that the distribution of annual salary is bell-shaped. Using the mean and standard deviation computed in part (a) as estimates of the mean and standard deviation of salary for the population of benefits managers, use the empirical rule to determine the percentage of benefits managers with an annual salary between $55,000 and $71,000. Compare your answer with the value computed in part (b).
e. Do the sample data contain any outliers?

36. A sample of 10 National Basketball Association (NBA) scores provided the following data (*USA Today,* April 14, 2000).

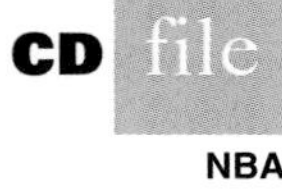

NBA

Winning Team	Points Scored	Losing Team	Points Scored	Winning Margin
Philadelphia	93	Washington	84	9
Charlotte	119	Atlanta	87	32
Milwaukee	101	Cleveland	100	1
Indiana	77	Toronto	73	4
Seattle	110	Minnesota	83	27
Boston	95	Orlando	91	4
Detroit	90	Miami	73	17
New York	91	New Jersey	89	2
Utah	102	L.A. Clippers	93	9
Phoenix	122	Vancouver	116	6

a. Compute the mean and standard deviation for the number of points scored by the winning team.

b. Assume that the number of points scored by the winning team for all NBA games is bell-shaped. Using the mean and standard deviation computed in part (a) as estimates of the mean and standard deviation of the points scored for the population of all NBA games, estimate the percentage of all NBA games in which the winning team will score 100 or more points. Estimate the percentage of games in which the winning team will score more than 114 points.

c. Compute the mean and standard deviation for the winning margin. Do the winning margin data contain any outliers? Explain.

37. *Consumer Review* posts reviews and ratings of a variety of products on the Internet. The following is a sample of 20 speaker systems and the ratings posted on January 2, 1998 (see *http://www.audioreview.com*). The ratings are on a scale of 1 to 5, with 5 being best.

Speakers

Speaker	Rating	Speaker	Rating
Infinity Kappa 6.1	4.00	ACI Sapphire III	4.67
Allison One	4.12	Bose 501 Series	2.14
Cambridge Ensemble II	3.82	DCM KX-212	4.09
Dynaudio Contour 1.3	4.00	Eosone RSF1000	4.17
Hsu Rsch. HRSW12V	4.56	Joseph Audio RM7si	4.88
Legacy Audio Focus	4.32	Martin Logan Aerius	4.26
Mission 73li	4.33	Omni Audio SA 12.3	2.32
PSB 400i	4.50	Polk Audio RT12	4.50
Snell Acoustics D IV	4.64	Sunfire True Subwoofer	4.17
Thiel CS1.5	4.20	Yamaha NS-A636	2.17

a. Compute the mean and the median.

b. Compute the first and third quartiles.

c. Compute the standard deviation.

d. What are the z-scores associated with the Allison One and the Omni Audio SA 12.3?

e. Do the data contain any outliers? Explain.

3.4 EXPLORATORY DATA ANALYSIS

In Chapter 2 we introduced the stem-and-leaf display as a technique of exploratory data analysis. Recall that exploratory data analysis enables us to use simple arithmetic and easy-to-draw pictures to summarize data. In this section we continue exploratory data analysis by considering five-number summaries and box plots.

Five-Number Summary

In a five-number summary, the following five numbers are used to summarize the data.

1. Smallest value
2. First quartile (Q_1)
3. Median (Q_2)
4. Third quartile (Q_3)
5. Largest value

The easiest way to develop a five-number summary is to first place the data in ascending order. Then it is easy to identify the smallest value, the three quartiles, and the largest value. The monthly starting salaries shown in Table 3.1 for a sample of 12 business school graduates are repeated here in ascending order.

2710 2755 2850 | 2880 2880 2890 | 2920 2940 2950 | 3050 3130 3325

$Q_1 = 2865$ $Q_2 = 2905$ (Median) $Q_3 = 3000$

The median of 2905 and the quartiles $Q_1 = 2865$ and $Q_3 = 3000$ were computed in Section 3.1. A review of the preceding data shows a smallest value of 2710 and a largest value of 3325. Thus the five-number summary for the salary data is 2710, 2865, 2905, 3000, 3325. Approximately one-fourth, or 25%, of the observations are between adjacent numbers in a five-number summary.

Box Plot

A **box plot** is a graphical summary of data based on a five-number summary. A key to the development of a box plot is the computation of the median and the quartiles, Q_1 and Q_3. The interquartile range, IQR $= Q_3 - Q_1$, is also used. Figure 3.4 is the box plot for the monthly starting salary data. The steps used to construct the box plot follow.

Box plots provide another way to identify outliers. But they do not necessarily identify the same values as those with a z-score less than −3 or greater than +3. Either, or both, procedures may be used. All you are trying to do is identify values that may not belong in the data set.

1. A box is drawn with the ends of the box located at the first and third quartiles. For the salary data, $Q_1 = 2865$ and $Q_3 = 3000$. This box contains the middle 50% of the data.
2. A vertical line is drawn in the box at the location of the median (2905 for the salary data). Thus the median line divides the data into two equal parts.
3. By using the interquartile range, IQR $= Q_3 - Q_1$, limits are located. The limits for the box plot are located 1.5(IQR) below Q_1 and 1.5(IQR) above Q_3. For the salary data, IQR $= Q_3 - Q_1 = 3000 - 2865 = 135$. Thus, the limits are $2865 - 1.5(135) = 2662.5$ and $3000 + 1.5(135) = 3202.5$. Data outside these limits are considered *outliers.*
4. The dashed lines in Figure 3.4 are called *whiskers.* The whiskers are drawn from the ends of the box to the smallest and largest data values *inside the limits* computed in step 3. Thus the whiskers end at salary values of 2710 and 3130.
5. Finally, the location of each outlier is shown with the symbol *. In Figure 3.4 we see one outlier, 3325.

FIGURE 3.4 BOX PLOT OF THE STARTING SALARY DATA WITH LINES SHOWING THE LOWER AND UPPER LIMITS

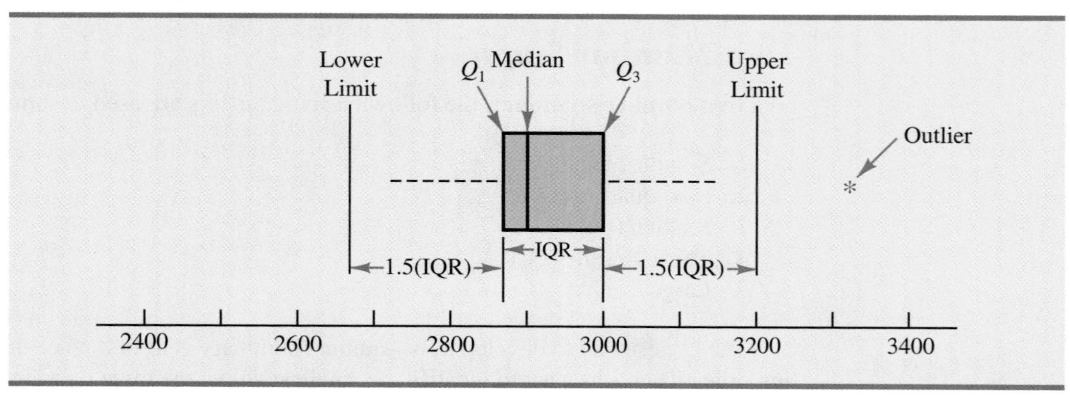

In Figure 3.4 we have included lines showing the location of the limits. These lines were drawn to show how the limits are computed and where they are located for the salary data. Although the limits are always computed, generally they are not drawn on the box plots. Figure 3.5 shows the usual appearance of a box plot for the salary data.

FIGURE 3.5 BOX PLOT OF THE STARTING SALARY DATA

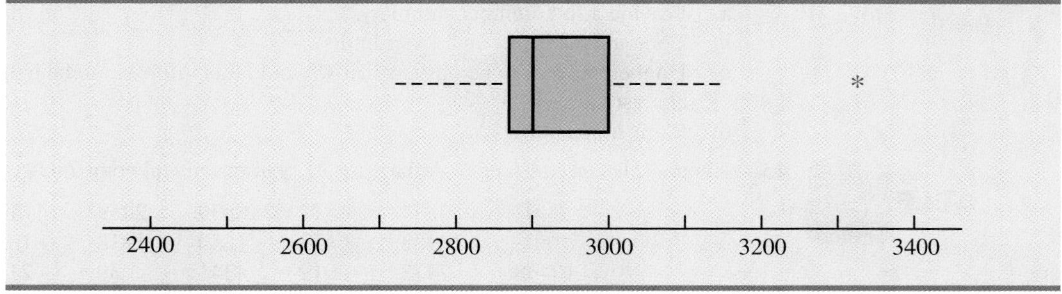

NOTES AND COMMENTS

1. When using a box plot, we may or may not identify the same outliers as the ones we select when using *z*-scores less than −3 and greater than +3. However, the objective of both approaches is simply to identify items that should be reviewed to ensure the validity of the data. Outliers identified by either procedure should be reviewed.
2. An advantage of the exploratory data analysis procedures is that they are easy to use; few numerical calculations are necessary. We simply sort the items into ascending order and identify the median and quartiles Q_1 and Q_3 to obtain the five-number summary. The limits and the box plot can then easily be determined. It is not necessary to compute the mean and the standard deviation for the data.
3. In Appendix 3.1, we show how to construct a box plot for the starting salary data using Minitab. The box plot obtained looks just like the one in Figure 3.5, but turned on its side.

EXERCISES

Methods

38. Consider the sample of size 8 with data values of 27, 25, 20, 15, 30, 34, 28, and 25. Provide the five-number summary for the data.

39. Show the box plot for the data in Exercise 38.

40. Show the five-number summary and the box plot for the following data: 5, 15, 18, 10, 8, 12, 16, 10, and 6.

41. A data set has a first quartile of 42 and a third quartile of 50. Compute the lower and upper limits. Should a data value of 65 be considered an outlier?

Applications

42. A goal of management is to earn as much as possible relative to the capital invested in their company. One measure of success in this effort is return on equity—the ratio of net income to stockholders' equity. Shown here are the return on equity percentages for 25 companies (*Standard & Poor's Stock Reports,* November 1997).

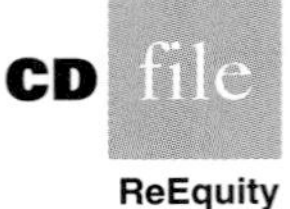
ReEquity

9.0	19.6	22.9	41.6	11.4
15.8	52.7	17.3	12.3	5.1
17.3	31.1	9.6	8.6	11.2
12.8	12.2	14.5	9.2	16.6
5.0	30.3	14.7	19.2	6.2

a. Provide a five-number summary.
b. Compute the lower and upper limits.
c. Do there appear to be outliers? How would this information be helpful to a financial analyst?
d. Show a box plot.

43. Annual sales, in millions of dollars, for 21 pharmaceutical companies follow.

8408	1374	1872	8879	2459	11413
608	14138	6452	1850	2818	1356
10498	7478	4019	4341	739	2127
3653	5794	8305			

a. Provide a five-number summary.
b. Compute the lower and upper limits.
c. Do there appear to be outliers?
d. Johnson & Johnson's sales are the largest in the list at $14,138 million. Suppose a data entry error (a transposition) had been made and the sales had been entered as $41,138 million. Would the method of detecting outliers in part (c) have identified the problem and allowed correction of the data entry error?
e. Show a box plot.

44. Corporate share repurchase programs are often touted as a benefit for shareholders. But, Robert Gabele, director of insider research for First Call/Thomson Financial, has noted that many of these have been undertaken solely to acquire stock for the companies' incentive options for top managers. Across all companies, existing stock options in 1998 represented 6.2% of all common shares outstanding. The following data show the number of shares covered by option grants and the number of shares outstanding for 15 companies. *Bloomberg* identified these companies as the ones that would need to repurchase the highest percentage of outstanding shares to cover their option grants (*Bloomberg Personal Finance,* January/February 2000).

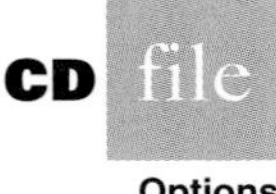
Options

Company	Shares of Option Grants Outstanding (millions)	Common Shares Outstanding (millions)
Adobe Systems	20.3	61.8
Apple Computer	52.7	160.9
Applied Materials	109.1	375.4
Autodesk	15.7	58.9
Best Buy	44.2	203.8
Cendant	183.3	718.1
Dell Computer	720.8	2540.9
Fruit of the Loom	14.2	66.9
ITT Industries	18.0	87.9
Merrill Lynch	89.9	365.5
Novell	120.2	335.0
Parametric Technology	78.3	269.3
Reebok International	12.8	56.1
Silicon Graphics	52.6	188.8
Toys R Us	54.8	247.6

a. What are the mean and median number of shares of option grants outstanding?
b. What are the first and third quartiles for the number of shares of option grants outstanding?
c. Are there any outliers for the number of shares of option grants outstanding? Show a box plot.
d. Compute the mean percentage of the ratio of the number of shares of option grants outstanding to the number of common shares outstanding. How does this percentage compare to the 1998 percentage of 6.2% reported for all companies?

45. The Highway Loss Data Institute's Injury and Collision Loss Experience report rates car models on the basis of the number of insurance claims filed after accidents. Index ratings near 100 are considered average. Lower ratings are better, indicating a safer car model. Shown are ratings for 20 midsize cars and 20 small cars.

CD file
Injury

Midsize cars:	81	91	93	127	68	81	60	51	58	75
	100	103	119	82	128	76	68	81	91	82
Small cars:	73	100	127	100	124	103	119	108	109	113
	108	118	103	120	102	122	96	133	80	140

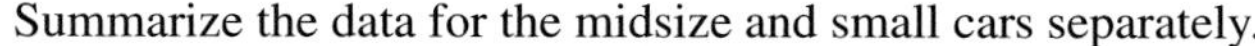

Summarize the data for the midsize and small cars separately.
a. Provide a five-number summary for midsize cars and for small cars.
b. Show the box plots.
c. Make a statement about what your summaries indicate about the safety of midsize cars in comparison to small cars.

46. Birinyi Associates, Inc., conducted a survey of stock markets around the world to assess their performance during 1997. Table 3.6 summarizes the findings for a sample of 30 countries.
a. What are the mean and median percentage changes for these countries?
b. What are the first and third quartiles?
c. Are there any outliers? Show a box plot.
d. What percentile would you report for the United States?

TABLE 3.6 1997 PERCENT CHANGE IN VALUE FOR WORLD STOCK MARKETS

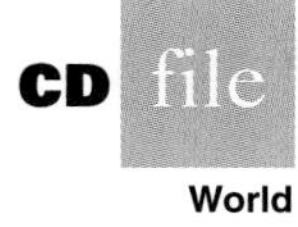

World

Country	Percent Change	Country	Percent Change
Argentina	24.70	Australia	7.91
Bahrain	49.67	Barbados	48.29
Bermuda	51.92	Brazil	44.84
Chile	12.80	Colombia	69.60
Croatia	−1.07	Czech Republic	−3.25
Ecuador	5.37	Estonia	62.34
Finland	32.31	Germany	47.11
Greece	59.19	India	18.60
Israel	27.91	Japan	−21.19
Lithuania	16.82	Mexico	54.92
Namibia	6.52	Nigeria	−7.97
Panama	59.40	Poland	2.27
Russia	125.89	Slovenia	18.71
Sri Lanka	15.49	Taiwan	18.08
Turkey	254.45	United States	22.64

Source: The Wall Street Journal, January 2, 1998.

3.5 MEASURES OF ASSOCIATION BETWEEN TWO VARIABLES

Thus far we have examined numerical methods used to summarize the data for *one variable at a time.* Often a manager or decision maker is interested in the *relationship between two variables.* In this section we present covariance and correlation as descriptive measures of the relationship between two variables.

We begin by reconsidering the application concerning a stereo and sound equipment store in San Francisco as presented in Section 2.4. The store's manager is interested in investigating the relationship between the number of weekend television commercials shown and the sales at the store during the following week. Sample data with sales expressed in hundreds of dollars are provided in Table 3.7 with one observation for each week ($n = 10$). The scatter diagram in Figure 3.6 shows a positive relationship, with higher sales (y) associated with a greater number of commercials (x). In fact, the scatter diagram suggests that a straight line could be used as a linear approximation of the relationship. In the following discussion, we introduce **covariance** as a descriptive measure of the linear association between two variables.

TABLE 3.7 SAMPLE DATA FOR THE STEREO AND SOUND EQUIPMENT STORE

CD file
Stereo

Week	Number of Commercials x	Sales Volume ($100s) y
1	2	50
2	5	57
3	1	41
4	3	54
5	4	54
6	1	38
7	5	63
8	3	48
9	4	59
10	2	46

FIGURE 3.6 SCATTER DIAGRAM FOR THE STEREO AND SOUND EQUIPMENT STORE

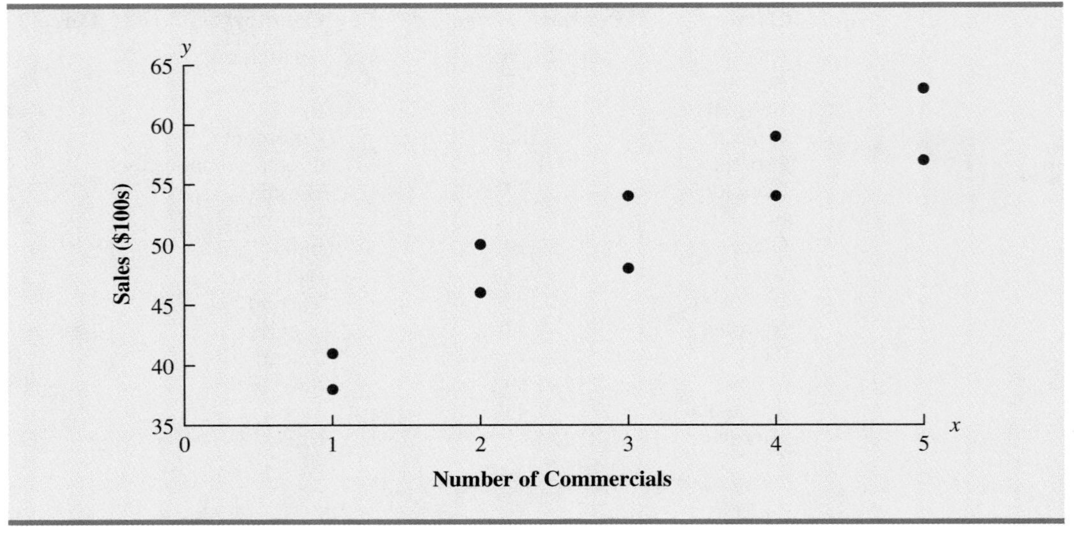

Covariance

For a sample of size n with the observations (x_1, y_1), (x_2, y_2) and so on, the *sample covariance* is defined as follows:

Sample Covariance

$$s_{xy} = \frac{\Sigma(x_i - \bar{x})(y_i - \bar{y})}{n - 1} \tag{3.10}$$

In this formula each x_i is paired with a y_i. We then sum the products obtained by multiplying the deviation of each x_i from its sample mean $\bar{x}$ by the deviation of the corresponding y_i from its sample mean $\bar{y}$; this sum is then divided by $n - 1$.

To measure the strength of the linear relationship between the number of commercials x and the sales volume y in the stereo and sound equipment store problem, we use equation (3.10) to compute the sample covariance. The calculations in Table 3.8 show the computation of $\Sigma(x_i - \bar{x})(y_i - \bar{y})$. Note that $\bar{x} = 30/10 = 3$ and $\bar{y} = 510/10 = 51$. Using equation (3.10), we obtain a sample covariance of

$$s_{xy} = \frac{\Sigma(x_i - \bar{x})(y_i - \bar{y})}{n - 1} = \frac{99}{9} = 11$$

The formula for computing the covariance of a population of size N is similar to equation (3.10), but we use different notation to indicate that we are working with the entire population.

Population Covariance

$$\sigma_{xy} = \frac{\Sigma(x_i - \mu_x)(y_i - \mu_y)}{N} \tag{3.11}$$

TABLE 3.8 CALCULATIONS FOR THE SAMPLE COVARIANCE

	x_i	y_i	$x_i - \bar{x}$	$y_i - \bar{y}$	$(x_i - \bar{x})(y_i - \bar{y})$
	2	50	−1	−1	1
	5	57	2	6	12
	1	41	−2	−10	20
	3	54	0	3	0
	4	54	1	3	3
	1	38	−2	−13	26
	5	63	2	12	24
	3	48	0	−3	0
	4	59	1	8	8
	2	46	−1	−5	5
Totals	30	510	0	0	99

$$s_{xy} = \frac{\Sigma(x_i - \bar{x})(y_i - \bar{y})}{n - 1} = \frac{99}{10 - 1} = 11$$

In equation (3.11) we use the notation μ_x for the population mean of the variable x and μ_y for the population mean of the variable y. The population covariance σ_{xy} is defined for a population of size N.

Interpretation of the Covariance

To aid in the interpretation of the sample covariance, consider Figure 3.7. It is the same as the scatter diagram of Figure 3.6 with a vertical dashed line at $\bar{x} = 3$ and a horizontal dashed line at $\bar{y} = 51$. Four quadrants have been identified on the graph. Points in quadrant I correspond to x_i greater than $\bar{x}$ and y_i greater than $\bar{y}$, points in quadrant II correspond to x_i less than $\bar{x}$ and y_i greater than $\bar{y}$, and so on. Thus, the value of $(x_i - \bar{x})(y_i - \bar{y})$ must be positive for points in quadrant I, negative for points in quadrant II, positive for points in quadrant III, and negative for points in quadrant IV.

The covariance is a measure of the linear association between two variables.

If the value of s_{xy} is positive, the points that have had the greatest influence on s_{xy} must be in quadrants I and III. Hence, a positive value for s_{xy} is indicative of a positive linear association between x and y; that is, as the value of x increases, the value of y increases. If the value of s_{xy} is negative, however, the points that have had the greatest influence on s_{xy} are in quadrants II and IV. Hence, a negative value for s_{xy} is indicative of a negative linear association between x and y; that is, as the value of x increases, the value of y decreases. Finally, if the points are evenly distributed across all four quadrants, the value of s_{xy} will be close to zero, indicating no linear association between x and y. Figure 3.8 shows the values of s_{xy} that can be expected with three different types of scatter diagrams.

Referring again to Figure 3.7, we see that the scatter diagram for the stereo and sound equipment store follows the pattern in the top panel of Figure 3.8. As we should expect, the value of the sample covariance is positive with $s_{xy} = 11$.

From the preceding discussion, it might appear that a large positive value for the covariance is indicative of a strong positive linear relationship and that a large negative value is indicative of a strong negative linear relationship. However, one problem with

FIGURE 3.7 PARTITIONED SCATTER DIAGRAM FOR THE STEREO AND SOUND EQUIPMENT STORE

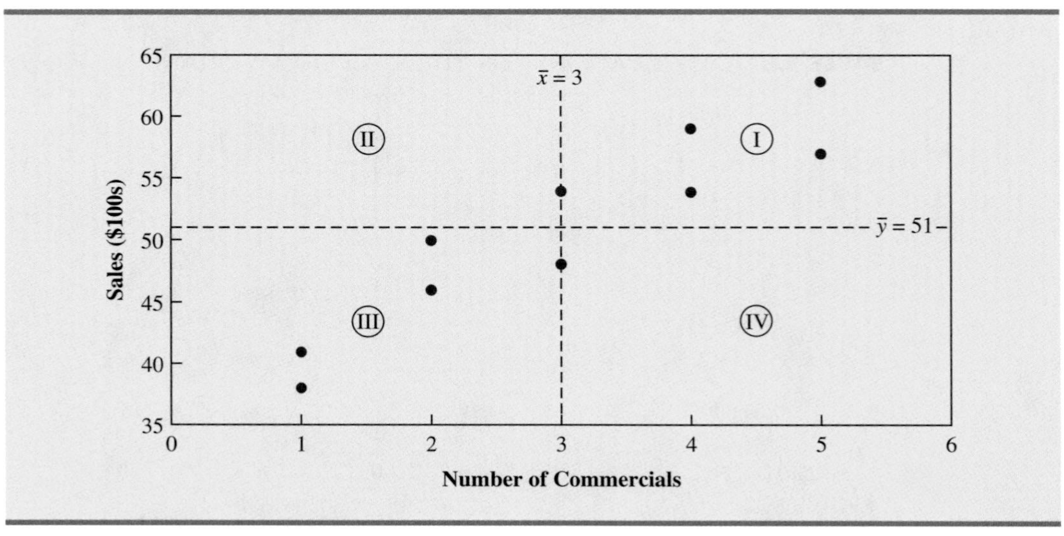

FIGURE 3.8 INTERPRETATION OF SAMPLE COVARIANCE

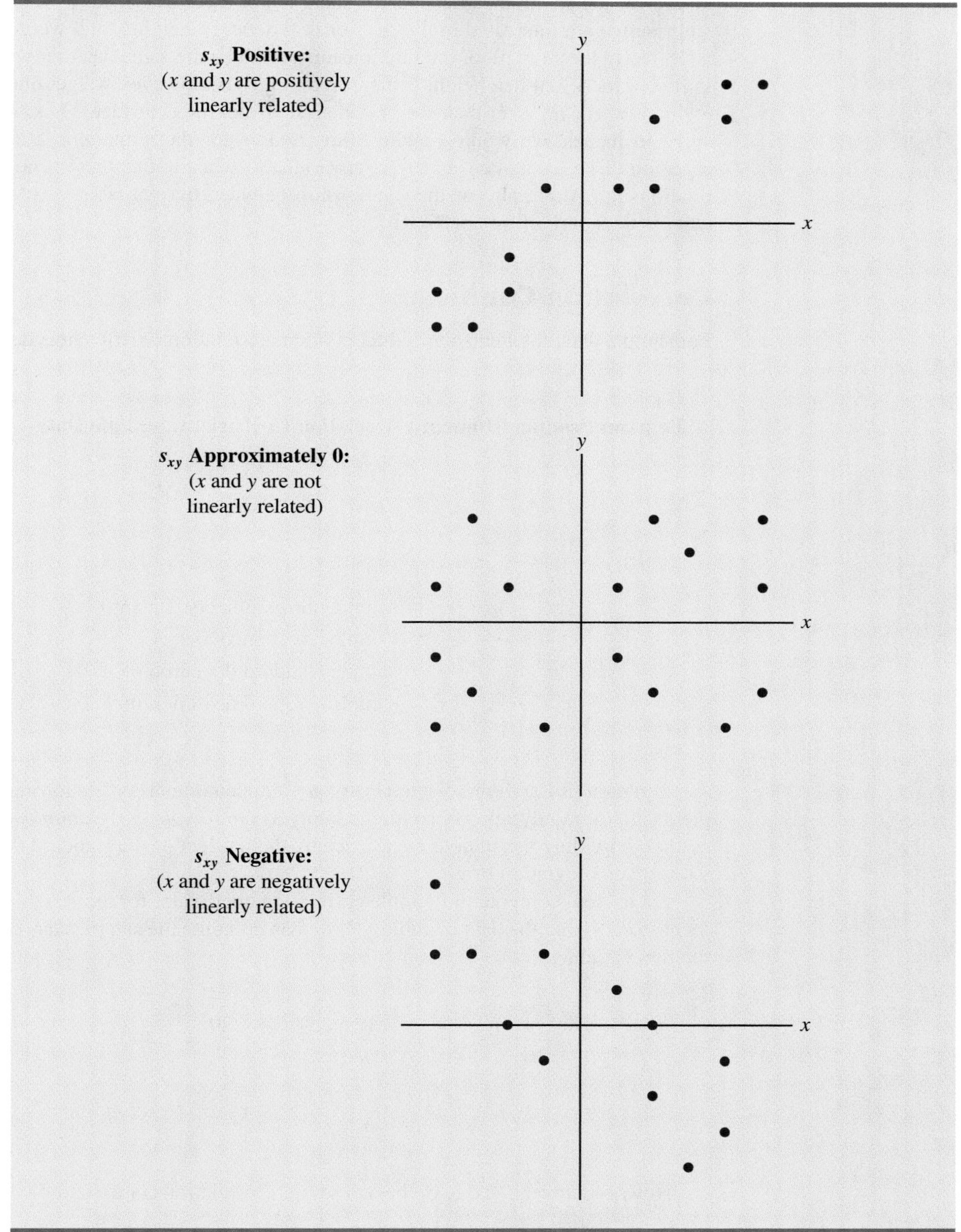

using covariance as a measure of the strength of the linear relationship is that the value we obtain for the covariance depends on the units of measurement for x and y. For example, suppose we are interested in the relationship between height x and weight y for individuals. Clearly the strength of the relationship should be the same whether we measure height in feet or inches. When height is measured in inches, however, we get much larger numerical values for $(x_i - \bar{x})$ than we get when it is measured in feet. Thus, with height measured in inches, we would obtain a larger value for the numerator $\Sigma(x_i - \bar{x})(y_i - \bar{y})$ in equation (3.10)—and hence a larger covariance—when in fact there is no difference in the relationship. A measure of the relationship between two variables that avoids this difficulty is the **correlation coefficient.**

Correlation Coefficient

For sample data, the Pearson product moment correlation coefficient is defined as follows.

Pearson Product Moment Correlation Coefficient: Sample Data

$$r_{xy} = \frac{s_{xy}}{s_x s_y} \tag{3.12}$$

where

r_{xy} = sample correlation coefficient
s_{xy} = sample covariance
s_x = sample standard deviation of x
s_y = sample standard deviation of y

Equation (3.12) shows that the Pearson product moment correlation coefficient for sample data (commonly referred to more simply as the *sample correlation coefficient*) is computed by dividing the sample covariance by the product of the standard deviation of x and the standard deviation of y.

Let us now compute the sample correlation coefficient for the stereo and sound equipment store. Using the data in Table 3.7, we can compute the sample standard deviations for the two variables.

$$s_x = \sqrt{\frac{\Sigma(x_i - \bar{x})^2}{n - 1}} = \sqrt{\frac{20}{9}} = 1.4907$$

$$s_y = \sqrt{\frac{\Sigma(y_i - \bar{y})^2}{n - 1}} = \sqrt{\frac{566}{9}} = 7.9303$$

Now, because $s_{xy} = 11$, we have a sample correlation coefficient of

$$r_{xy} = \frac{s_{xy}}{s_x s_y} = \frac{11}{(1.4907)(7.9303)} = +.93$$

The formula for computing the correlation coefficient for a population, denoted by the Greek letter ρ_{xy} (rho, pronounced "row"), follows.

Pearson Product Moment Correlation Coefficient: Population Data

$$\rho_{xy} = \frac{\sigma_{xy}}{\sigma_x \sigma_y} \tag{3.13}$$

where

ρ_{xy} = population correlation coefficient
σ_{xy} = population covariance
σ_x = population standard deviation for x
σ_y = population standard deviation for y

The sample correlation coefficient r_{xy} is an estimate of the population correlation coefficient ρ_{xy}.

Interpretation of the Correlation Coefficient

First let us consider a simple example that illustrates the concept of a perfect positive linear relationship. The scatter diagram in Figure 3.9 depicts the relationship between x and y based on the following sample data.

x_i	y_i
5	10
10	30
15	50

The straight line drawn through each of the three points shows a perfect linear relationship between x and y. In order to apply equation (3.12) to compute the sample correlation

FIGURE 3.9 SCATTER DIAGRAM DEPICTING A PERFECT POSITIVE LINEAR RELATIONSHIP

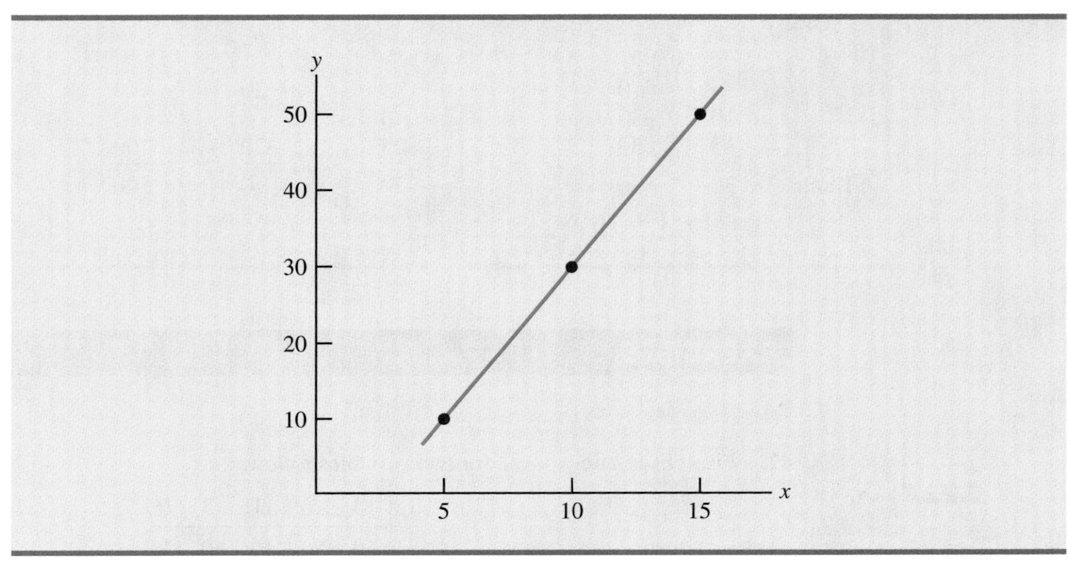

we must first compute s_{xy}, s_x, and s_y. Some of the necessary computations are contained in Table 3.9. Using the results in Table 3.9, we find

$$s_{xy} = \frac{\Sigma(x_i - \bar{x})(y_i - \bar{y})}{n - 1} = \frac{200}{2} = 100$$

$$s_x = \sqrt{\frac{\Sigma(x_i - \bar{x})^2}{n - 1}} = \sqrt{\frac{50}{2}} = 5$$

$$s_y = \sqrt{\frac{\Sigma(y_i - \bar{y})^2}{n - 1}} = \sqrt{\frac{800}{2}} = 20$$

$$r_{xy} = \frac{s_{xy}}{s_x s_y} = \frac{100}{5(20)} = 1$$

Thus, we see that the value of the sample correlation coefficient is 1.

In general, it can be shown that if all the points in a data set are on a straight line having positive slope, the value of the sample correlation coefficient is $+1$; that is, a sample correlation coefficient of $+1$ corresponds to a perfect positive linear relationship between x and y. Moreover, if the points in the data set are on a straight line having negative slope, the value of the sample correlation coefficient is -1; that is, a sample correlation coefficient of -1 corresponds to a perfect negative linear relationship between x and y.

The correlation coefficient ranges from -1 to $+1$. Values close to -1 or $+1$ indicate a strong linear relationship. The closer the correlation is to zero the weaker the relationship.

Let us now suppose that a certain data set shows a positive linear relationship between x and y, but the relationship is not perfect. The value of r_{xy} will be less than 1, indicating that the points in the scatter diagram are not all on a straight line. As the points in a data set deviate more and more from a perfect positive linear relationship, the value of r_{xy} becomes smaller and smaller. A value of r_{xy} equal to zero indicates no linear relationship between x and y, and values of r_{xy} near zero indicate a weak linear relationship.

For the data set involving the stereo and sound equipment store, recall that $r_{xy} = +.93$. Therefore, we conclude that there is a strong positive linear relationship between the number of commercials and sales. More specifically, an increase in the number of commercials is associated with an increase in sales.

TABLE 3.9 COMPUTATIONS USED IN CALCULATING THE SAMPLE CORRELATION COEFFICIENT

	x_i	y_i	$x_i - \bar{x}$	$(x_i - \bar{x})^2$	$y_i - \bar{y}$	$(y_i - \bar{y})^2$	$(x_i - \bar{x})(y_i - \bar{y})$
	5	10	−5	25	−20	400	100
	10	30	0	0	0	0	0
	15	50	5	25	20	400	100
Totals	30	90	0	50	0	800	200
	$\bar{x} = 10$	$\bar{y} = 30$					

EXERCISES

Methods

SELF test

47. Five observations taken for two variables follow.

x_i	4	6	11	3	16
y_i	50	50	40	60	30

a. Develop a scatter diagram with x on the horizontal axis.
b. What does the scatter diagram developed in part (a) indicate about the relationship between the two variables?
c. Compute and interpret the sample covariance for the data.
d. Compute and interpret the sample correlation coefficient for the data.

48. Five observations taken for two variables follow.

x_i	6	11	15	21	27
y_i	6	9	6	17	12

a. Develop a scatter diagram for these data.
b. What does the scatter diagram indicate about a possible relationship between x and y?
c. Compute and interpret the sample covariance for the data.
d. Compute and interpret the sample correlation coefficient for the data.

Applications

49. A high school guidance counselor collected the following data about the grade point averages (GPA) and the SAT mathematics test scores for six seniors.

GPA	2.7	3.5	3.7	3.3	3.6	3.0
SAT	450	560	700	620	640	570

a. Develop a scatter diagram for the data with GPA on the horizontal axis.
b. Does there appear to be any relationship between the GPA and the SAT mathematics test score? Explain.
c. Compute and interpret the sample covariance for the data.
d. Compute the sample correlation coefficient for the data. What does this value tell us about the relationship between the two variables?

50. A department of transportation's study on driving speed and mileage for midsize automobiles resulted in the following data.

Driving Speed	30	50	40	55	30	25	60	25	50	55
Mileage	28	25	25	23	30	32	21	35	26	25

Compute and interpret the sample correlation coefficient for these data.

51. *PC World* provided ratings for the top 15 notebook PCs (*PC World,* February 2000). The performance score is a measure of how fast a PC can run a mix of common business applications in comparison to their baseline machine. For example, a PC with a performance score of 200 is twice as fast as the baseline machine. A 100-point scale was used to provide an overall rating for each notebook tested in the study. A score in the 90s is exceptional, while one in the 70s is above average. The performance scores and the overall ratings for the 15 notebooks are shown.

PCs

Notebook	**Performance Score**	**Overall Rating**
AMS Tech Roadster 15CTA380	115	67
Compaq Armada M700	191	78
Compaq Prosignia Notebook 150	153	79
Dell Inspiron 3700 C466GT	194	80
Dell Inspiron 7500 R500VT	236	84
Dell Latitude Cpi A366XT	184	76
Enpower ENP-313 Pro	184	77
Gateway Solo 9300LS	216	92
HP Pavillion Notebook PC	185	83
IBM ThinkPad I Series 1480	183	78
Micro Express NP7400	189	77

Notebook	Performance Score	Overall Rating
Micron TransPort NX PII-400	202	78
NEC Versa SX	192	78
Sceptre Soundx 5200	141	73
Sony VAIO PCG-F340	187	77

a. Compute the sample correlation coefficient.
b. What does the sample correlation coefficient tell about the relationship between the performance score and the overall rating?

52. The Dow Jones Industrial Average (DJIA) and the Standard & Poor's 500 (S&P) Index are both used as measures of overall movement in the stock market. The DJIA is based on the price movements of 30 large companies; the S&P 500 is an index composed of 500 stocks. Some say the S&P 500 is a better measure of stock market performance because it is broader based. The closing price for the DJIA and the S&P 500 for 10 weeks, beginning with February 11, 2000, are shown (*Barron's,* April 17, 2000).

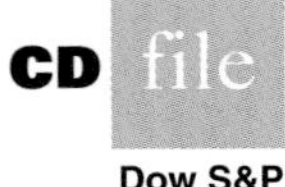

Dow S&P

Date	Dow Jones	S&P 500
February 11	10425	1387
February 18	10220	1346
February 25	9862	1333
March 3	10367	1409
March 10	9929	1395
March 17	10595	1464
March 24	11113	1527
March 31	10922	1499
April 7	11111	1516
April 14	10306	1357

a. Compute the sample correlation coefficient for these data.
b. Are they poorly correlated, or do they have a close association?

53. The daily high and low temperatures for 20 cities follow (*USA Today,* May 9, 2000).

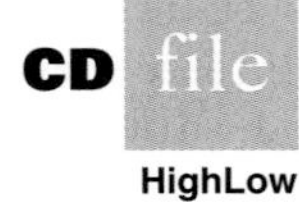

HighLow

City	High	Low
Athens	75	54
Bangkok	92	74
Cairo	84	57
Copenhagen	64	39
Dublin	64	46
Havana	86	68
Hong Kong	81	72
Johannesburg	61	50
London	73	48
Manila	93	75
Melbourne	66	50
Montreal	64	52
Paris	77	55
Rio de Janeiro	80	61
Rome	81	54
Seoul	64	50
Singapore	90	75
Sydney	68	55
Tokyo	79	59
Vancouver	57	43

What is the correlation between the high and low temperatures?

3.6 THE WEIGHTED MEAN AND WORKING WITH GROUPED DATA

In Section 3.1, we presented the mean as one of the most important measures of descriptive statistics. The formula for the mean of a sample with n observations is restated as follows.

$$\bar{x} = \frac{\Sigma x_i}{n} = \frac{x_1 + x_2 + \cdots + x_n}{n} \tag{3.14}$$

In this formula, each x_i is given equal importance or weight. Although this practice is most common, in some instances, the mean is computed by giving each observation a weight that reflects its importance. A mean computed in this manner is referred to as a **weighted mean.**

Weighted Mean

The weighted mean is computed as follows:

Weighted Mean

$$\bar{x} = \frac{\Sigma w_i x_i}{\Sigma w_i} \tag{3.15}$$

where

x_i = value of observation i
w_i = weight for observation i

When the data are from a sample, (3.15) provides the weighted sample mean. When the data are from a population, μ replaces $\bar{x}$ and (3.15) provides the weighted population mean.

As an example of the need for a weighted mean, consider the following sample of five purchases of a raw material over the past three months.

Purchase	Cost per Pound ($)	Number of Pounds
1	3.00	1200
2	3.40	500
3	2.80	2500
4	2.90	1000
5	3.25	800

Note that the cost per pound has varied from $2.80 to $3.40 and the quantity purchased has varied from 500 to 2500 pounds. Suppose that a manager has asked for information about the mean cost per pound of the raw material. Because the quantities ordered vary, we must use the formula for a weighted mean. The five cost-per-pound data values are $x_1 = 3.00$, $x_2 = 3.40$, $x_3 = 2.80$, $x_4 = 2.90$, and $x_5 = 3.25$. The mean cost per pound is found by weighting each cost by its corresponding quantity. For this example, the weights are $w_1 = 1200$, $w_2 = 500$, $w_3 = 2500$, $w_4 = 1000$, and $w_5 = 800$. Using (3.15), the weighted mean is calculated as follows:

$$\bar{x} = \frac{1200(3.00) + 500(3.40) + 2500(2.80) + 1000(2.90) + 800(3.25)}{1200 + 500 + 2500 + 1000 + 800}$$

$$= \frac{17{,}800}{6000} = 2.967$$

Thus, the weighted mean computation shows that the mean cost per pound for the raw material is \$2.967. Note that using (3.14) rather than the weighted mean formula would have provided misleading results. In this case, the mean of the five cost-per-pound values is (3.00 + 3.40 + 2.80 + 2.90 + 3.25)/5 = 15.35/5 = \$3.07, which overstates the actual mean cost per pound.

Computing a grade point average is a good example of the use of a weighted mean.

The choice of weights for a particular weighted mean computation depends upon the application. An example that is well known to college students is the computation of a grade point average (GPA). In this computation, the data values generally used are 4 for an A grade, 3 for a B grade, 2 for a C grade, 1 for a D grade, and 0 for an F grade. The weights are the number of credits hours earned for each grade. Exercise 56 at the end of this section provides an example of this weighted mean computation. In other weighted mean computations, quantities such as pounds, dollars, and/or volume are frequently used as weights. In any case, when data values vary in importance, the analyst must choose the weight that best reflects the importance of each data value in the determination of the mean.

Grouped Data

In most cases, measures of location and variability are computed by using the individual data values. Sometimes, however, we have data in only a grouped or frequency distribution form. In the following discussion, we show how the weighted mean formula can be used to obtain approximations of the mean, variance, and standard deviation for **grouped data.**

In Section 2.2 we provided a frequency distribution of the time in days required to complete year-end audits for the public accounting firm of Sanderson and Clifford. The frequency distribution of audit times based on a sample of 20 clients is shown again in Table 3.10. Based on this frequency distribution, what is the sample mean audit time?

To compute the mean using only the grouped data, we treat the midpoint of each class as being representative of the items in the class. Let M_i denote the midpoint for class i and let f_i denote the frequency of class i. The weighted mean formula (3.15) is then used with the data values denoted as M_i and the weights given by the frequencies f_i. In this case, the denominator of (3.15) is the sum of the frequencies, which is the sample size n. That is, $\Sigma f_i = n$. Thus, the equation for the sample mean for grouped data is as follows.

Sample Mean for Grouped Data

$$\bar{x} = \frac{\Sigma f_i M_i}{n} \tag{3.16}$$

where

$$\begin{aligned} M_i &= \text{the midpoint for class } i \\ f_i &= \text{the frequency for class } i \\ n = \Sigma f_i &= \text{the sample size} \end{aligned}$$

With the class midpoints, M_i, halfway between the class limits, the first class of 10–14 in Table 3.10 has a midpoint at (10 + 14)/2 = 12. The five class midpoints and the weighted mean computation for the audit time data are summarized in Table 3.11. As can be seen, the sample mean audit time is 19 days.

To compute the variance for grouped data, we use a slightly altered version of the formula for the variance provided in equation (3.5). In equation (3.5), the squared deviations of the data about the sample mean $\bar{x}$ were written $(x_i - \bar{x})^2$. However, with grouped data, the values are not known. In this case, we treat the class midpoint, M_i, as being representative of the x_i values in the corresponding class. Thus the squared deviations about the sample mean, $(x_i - \bar{x})^2$, are replaced by $(M_i - \bar{x})^2$. Then, just as we did with the sample mean calculations for grouped data,

TABLE 3.10 FREQUENCY DISTRIBUTION OF AUDIT TIMES

Audit Time (days)	Frequency
10–14	4
15–19	8
20–24	5
25–29	2
30–34	1
Total	20

TABLE 3.11 COMPUTATION OF THE SAMPLE MEAN AUDIT TIME FOR GROUPED DATA

Audit Time (days)	Class Midpoint (M_i)	Frequency (f_i)	f_iM_i
10–14	12	4	48
15–19	17	8	136
20–24	22	5	110
25–29	27	2	54
30–34	32	1	32
		20	380

$$\text{Sample mean } \bar{x} = \frac{\Sigma f_iM_i}{n} = \frac{380}{20} = 19 \text{ days}$$

we weight each value by the frequency of the class, f_i. The sum of the squared deviations about the mean for all the data is approximated by $\Sigma f_i(M_i - \bar{x})^2$. The term $n - 1$ rather than n appears in the denominator in order to make the sample variance the estimate of the population variance σ^2. Thus, the following formula is used to obtain the sample variance for grouped data.

Sample Variance for Grouped Data

$$s^2 = \frac{\Sigma f_i(M_i - \bar{x})^2}{n - 1} \tag{3.17}$$

The calculation of the sample variance for audit times based on the grouped data from Table 3.10 is shown in Table 3.12.

The standard deviation for grouped data is simply the square root of the variance for grouped data. For the audit time data, the sample standard deviation is $s = \sqrt{30} = 5.48$.

Before closing this section on computing measures of location and dispersion for grouped data, we note that the formulas (3.16) and (3.17) are for a sample. Population summary measures are computed similarly. The grouped data formulas for a population mean and variance follow.

Population Mean for Grouped Data

$$\mu = \frac{\Sigma f_iM_i}{N} \tag{3.18}$$

Population Variance for Grouped Data

$$\sigma^2 = \frac{\Sigma f_i(M_i - \mu)^2}{N} \tag{3.19}$$

TABLE 3.12 COMPUTATION OF THE SAMPLE VARIANCE OF AUDIT TIMES FOR GROUPED DATA (SAMPLE MEAN $\bar{x} = 19$)

Audit Time (days)	Class Midpoint (M_i)	Frequency (f_i)	Deviation ($M_i - \bar{x}$)	Squared Deviation $(M_i - \bar{x})^2$	$f_i(M_i - \bar{x})^2$
10–14	12	4	−7	49	196
15–19	17	8	−2	4	32
20–24	22	5	3	9	45
25–29	27	2	8	64	128
30–34	32	1	13	169	169
		20			570
					$\Sigma f_i(M_i - \bar{x})^2$

$$\text{Sample variance } s^2 = \frac{\Sigma f_i(M_i - \bar{x})^2}{n - 1} = \frac{570}{19} = 30$$

NOTES AND COMMENTS

1. An alternative formula for the computation of the sample variance for grouped data is

$$s^2 = \frac{\Sigma f_i M_i^2 - n\bar{x}^2}{n - 1}$$

where $\Sigma f_i M_i^2 = f_1 M_1^2 + f_2 M_2^2 + \cdots + f_k M_k^2$ and k is the number of classes used to group the data. Using this formula may ease the computations slightly.

2. In computing descriptive statistics for grouped data, the class midpoints are used to approximate the data values in each class. As a result, the descriptive statistics for grouped data are approximations of the descriptive statistics that would result from using the original data directly. We therefore recommend computing descriptive statistics from the original data rather than from grouped data whenever possible.

EXERCISES

Methods

54. Consider the following data and corresponding weights.

x_i	Weight (w_i)
3.2	6
2.0	3
2.5	2
5.0	8

a. Compute the weighted mean for the data.

b. Compute the sample mean of the four data values without weighting. Note the difference in the results provided by the two computations.

55. Consider the sample data in the following frequency distribution.

Class	Midpoint	Frequency
3–7	5	4
8–12	10	7
13–17	15	9
18–22	20	5

a. Compute the sample mean.
b. Compute the sample variance and sample standard deviation.

Applications

56. The grade point average for college students is based on a weighted mean computation. For most colleges, the grades are given the following data values: A (4), B (3), C (2), D (1), and F (0). After 60 credit hours of course work, a student at State University has earned 9 credit hours of A, 15 credit hours of B, 33 credit hours of C, and 3 credit hours of D.
   a. Compute the student's grade point average.
   b. Students at State University must have a 2.5 grade point average for their first 60 credit hours of course work in order to be admitted to the business college. Will this student be admitted?

57. The dividend yield is the percentage of the value of a share of stock that will be paid as an annual dividend to the stockholder. A sample of eight stocks held by Innis Investments had the following dividend yields (*Barron's,* January 5, 1998). The amount Innis has invested in each stock is also shown. What is the mean dividend yield for the portfolio?

Company	Dividend Yield	Amount Invested ($)
Apple Computer	0.00	37,830
Chevron Corp.	2.98	27,667
Eastman Kodak	2.77	31,037
ExxonMobil	2.65	27,336
Merck & Co.	1.58	37,553
Franklin Resources	0.57	17,812
Sears	2.00	32,660
Woolworth	0.00	17,775

58. A service station has recorded the following frequency distribution for the number of gallons of gasoline sold per car in a sample of 680 cars.

Gasoline (gallons)	Frequency
0–4	74
5–9	192
10–14	280
15–19	105
20–24	23
25–29	6
Total	680

Compute the mean, variance, and standard deviation for these grouped data. If the service station expects to service about 120 cars on a given day, what is an estimate of the total number of gallons of gasoline that will be sold?

59. In a survey of subscribers to *Fortune* magazine, the following question was asked: "How many of the last four issues have you read or looked through?" Suppose that the following frequency distribution summarizes 500 responses.

Number Read		Frequency
0		15
1		10
2		40
3		85
4		350
	Total	500

a. What is the mean number of issues read by a *Fortune* subscriber?
b. What is the standard deviation of the number of issues read?

SUMMARY

In this chapter we introduced several descriptive statistics that can be used to summarize the location and variability of data. Unlike the tabular and graphical procedures, the measures introduced in this chapter summarize the data in terms of numerical values. When the numerical values obtained are for a sample, they are called sample statistics. When the numerical values obtained are for a population, they are called population parameters. Some of the notation used for sample statistics and population parameters follow.

	Sample Statistic	Population Parameter
Mean	$\bar{x}$	μ
Variance	s^2	σ^2
Standard deviation	s	σ
Covariance	s_{xy}	σ_{xy}
Correlation	r_{xy}	ρ_{xy}

As measures of central location, we defined the mean, median, and mode. Then the concept of percentiles was used to describe other locations in the data set. Next, we presented the range, interquartile range, variance, standard deviation, and coefficient of variation as measures of variability or dispersion. We then described how the mean and standard deviation could be used together, applying the empirical rule and Chebyshev's theorem, to provide more information about the distribution of data and to identify outliers.

In Section 3.4 we showed how to develop a five-number summary and a box plot to provide simultaneous information about the location, variability, and shape of the distribution. In Section 3.5 we introduced covariance and the correlation coefficient as measures of association between two variables. In the final section, we showed how to compute a weighted mean and how to calculate a mean, variance, and standard deviation for grouped data.

The descriptive statistics we have discussed can be developed using statistical software packages and spreadsheets. In Appendix 3.1 we show how to develop most of the descriptive statistics introduced in the chapter using Minitab. In Appendix 3.2, we demonstrate the use of Excel for the same purpose.

GLOSSARY

Sample statistic A numerical value used as a summary measure for a sample (e.g., [the] sample mean, $\bar{x}$, the sample variance, s^2, and the sample standard deviation, s).

Population parameter A numerical value used as a summary measure for a population of data (e.g., the population mean, μ, the population variance, σ^2, and the population standard deviation, σ).

Mean A measure of central location for a data set. It is computed by summing all the data values and dividing by the number of observations.

Median A measure of central location. It is the value in the middle when the data are arranged in ascending order.

Mode A measure of location, defined as the value that occurs with greatest frequency.

Percentile A value such that at least p percent of the observations are less than or equal to this value and at least $(100 - p)$ percent of the observations are greater than or equal to this value. The 50th percentile is the median.

Quartiles The 25th, 50th, and 75th percentiles are the first quartile, the second quartile (median), and third quartile, respectively. The quartiles can be used to divide the data set into four parts, with each part containing approximately 25% of the data.

Hinges The value of the lower hinge is approximately the first quartile, or 25th percentile. The value of the upper hinge is approximately the third quartile, or 75th percentile. The values of the hinges and quartiles may differ slightly because of differing computational conventions, but their objective is to divide the data into four equal parts.

Range A measure of variability, defined to be the largest value minus the smallest value.

Interquartile range (IQR) A measure of variability, defined to be the difference between the third and first quartiles.

Variance A measure of variability based on the squared deviations of the data values about the mean.

Standard deviation A measure of variability computed by taking the positive square root of the variance.

Coefficient of variation A measure of relative variability computed by dividing the standard deviation by the mean and multiplying by 100.

z-score A value computed by dividing the deviation about the mean $(x_i - \bar{x})$ by the standard deviation s. A z-score is referred to as a standardized value and denotes the number of standard deviations x_i is from the mean.

Chebyshev's theorem A theorem applying to any data set that can be used to make statements about the proportion of observations that must be within a specified number of standard deviations of the mean.

Empirical rule A rule that can be used to compute the percentage of data values that must be within one, two, and three standard deviations of the mean for data having a bell-shaped distribution.

Outlier An unusually small or unusually large data value.

Five-number summary An exploratory data analysis technique that uses five numbers to summarize the data: smallest value, first quartile, median, third quartile, and largest value.

Box plot A graphical summary of data. A box, drawn from the first to the third quartiles, shows the location of the middle 50% of the data. Dashed lines, called whiskers, extending from the ends of the box show the location of data values greater than the third quartile and data values less than the first quartile. The locations of any outliers are also noted.

Covariance A numerical measure of linear association between two variables. Positive values indicate a positive relationship; negative values indicate a negative relationship.

Correlation coefficient A numerical measure of linear association between two variables that takes values between -1 and $+1$. Values near $+1$ indicate a strong positive linear relationship, values near -1 indicate a strong negative linear relationship, and values near zero indicate lack of a linear relationship.

Weighted mean The mean for a data set obtained by assigning each observation a weight that reflects its importance within the data set.

Grouped data Data available in class intervals as summarized by a frequency distribution. Individual values of the original data are not available.

KEY FORMULAS

Sample Mean

$$\bar{x} = \frac{\Sigma x_i}{n} \tag{3.1}$$

Population Mean

$$\mu = \frac{\Sigma x_i}{N} \tag{3.2}$$

Interquartile Range

$$\text{IQR} = Q_3 - Q_1 \tag{3.3}$$

Population Variance

$$\sigma^2 = \frac{\Sigma(x_i - \mu)^2}{N} \tag{3.4}$$

Sample Variance

$$s^2 = \frac{\Sigma(x_i - \bar{x})^2}{n - 1} \tag{3.5}$$

Standard Deviation

$$\text{Sample Standard Deviation} = s = \sqrt{s^2} \tag{3.6}$$

$$\text{Population Standard Deviation} = \sigma = \sqrt{\sigma^2} \tag{3.7}$$

Coefficient of Variation

$$\left(\frac{\text{Standard Deviation}}{\text{Mean}}\right) \times 100 \tag{3.8}$$

z-score

$$z_i = \frac{x_i - \bar{x}}{s} \tag{3.9}$$

Sample Covariance

$$s_{xy} = \frac{\Sigma(x_i - \bar{x})(y_i - \bar{y})}{n - 1} \tag{3.10}$$

Pearson Product Moment Correlation Coefficient: Sample Data

$$r_{xy} = \frac{s_{xy}}{s_x s_y} \tag{3.12}$$

Weighted Mean

$$\bar{x} = \frac{\Sigma w_i x_i}{\Sigma w_i} \tag{3.15}$$

where w_i = weight for observation i

Sample Mean for Grouped Data

$$\bar{x} = \frac{\Sigma f_i M_i}{n} \tag{3.16}$$

Sample Variance for Grouped Data

$$s^2 = \frac{\Sigma f_i (M_i - \bar{x})^2}{n - 1} \tag{3.17}$$

Population Mean for Grouped Data

$$\mu = \frac{\Sigma f_i M_i}{N} \tag{3.18}$$

Population Variance for Grouped Data

$$\sigma^2 = \frac{\Sigma f_i (M_i - \mu)^2}{N} \tag{3.19}$$

SUPPLEMENTARY EXERCISES

60. The average American spends \$65.88 per month dining out (*The Des Moines Register,* December 5, 1997). A sample of young adults provided the following dining out expenditures (in dollars) over the past month.

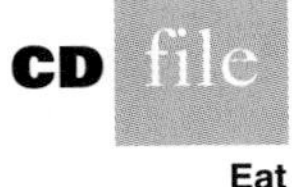

Eat

253	101	245	467	131	0	225
80	113	69	198	95	129	124
11	178	104	161	0	118	151
55	152	134	169			

a. Compute the mean, median, and mode.
b. Considering your results in part (a), do these young adults seem to spend about the same as an average American eating out?
c. Compute the first and third quartiles.
d. Compute the range and interquartile range.
e. Compute the variance and standard deviation.
f. Are there any outliers?

61. The total annual compensation for a board member at one of the nation's 100 biggest public companies is based in part on the cash retainer, an annual payment for serving on the board. In addition to the cash retainer, a board member may receive a stock retainer, a stock grant, a stock option, and a fee for attending board meetings. The total compensation can easily exceed $100,000 even with an annual retainer as low as $15,000. The following data show the cash retainer for a sample of 20 of the nation's biggest public companies (*USA Today,* April 17, 2000).

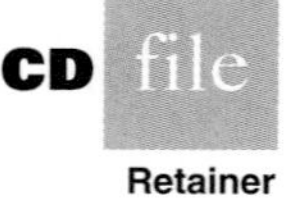

Retainer

Company	Cash Retainer
American Express	64
Bank of America	36
Boeing	26
Chevron	35
Dell Computer	40
DuPont	35
ExxonMobil	40
Ford Motor	30
General Motors	60
International Paper	36
Kroger	28
Lucent Technologies	50
Motorola	20
Procter & Gamble	55
Raytheon	40
Sears Roebuck	30
Texaco	15
United Parcel Service	55
Wal-Mart Stores	25
Xerox	40

Compute the following descriptive statistics.

a. Mean, median, and mode
b. The first and third quartiles
c. The range and interquartile range
d. The variance and the standard deviation
e. Coefficient of variation

62. A survey was conducted concerning the ability of computer manufacturers to handle problems quickly (*PC Computing,* November 1997). The following results were obtained.

Company	Days to Resolve Problems	Company	Days to Resolve Problems
Compaq	13	Gateway	21
Packard Bell	27	Digital	27
Quantex	11	IBM	12
Dell	14	Hewlett-Packard	14
NEC	14	AT&T	20
AST	17	Toshiba	37
Acer	16	Micron	17

a. What are the mean and median number of days needed to resolve problems?
b. What is the variance and standard deviation?
c. Which manufacturer has the best record?
d. What is the z-score for Packard Bell?
e. What is the z-score for IBM?
f. Are there any outliers?

63. The following data show home mortgage loan amounts handled by one loan officer at the Westwood Savings and Loan Association. Data are in thousands of dollars.

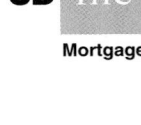

CD file
Mortgage

52.0	68.5	63.0	57.5	64.0	42.5	55.9	73.2	67.5	66.2
55.2	60.9	53.8	58.4	43.0	61.0	63.5	55.4	63.5	50.2
69.0	68.1	60.5	75.5	60.5	82.0	70.5	81.6	72.5	74.8

a. Find the mean, median, and mode.
b. Find the first and third quartiles.

64. According to Forrester Research, Inc., approximately 19% of Internet users play games online. The following data show the number of unique users (in thousands) for the month of March for 10 game sites (*The Wall Street Journal*, April 17, 2000).

Site	Unique Users
AOLGames.aol	9,416
extremelotto.com	3,955
freelotto.com	12,901
gamesville.com	4,844
iwin.com	7,410
prizecentral.com	4,899
shockwave.com	5,582
speedyclick.com	6,628
uproar.com	8,821
webstakes.com	7,499

Using these data, compute the mean, median, variance, and standard deviation.

65. The typical household income for a sample of 20 cities follow (*Places Rated Almanac*, 2000). Data are in thousands of dollars.

City	Income
Akron, OH	74.1
Atlanta, GA	82.4
Birmingham, AL	71.2
Bismark, ND	62.8
Cleveland, OH	79.2
Columbia, SC	66.8
Danbury, CT	132.3
Denver, CO	82.6
Detroit, MI	85.3
Fort Lauderdale, FL	75.8
Hartford, CT	89.1
Lancaster, PA	75.2
Madison, WI	78.8
Naples, FL	100.0
Nashville, TN	77.3
Philadelphia, PA	87.0
Savannah, GA	67.8
Toledo, OH	71.2
Trenton, NJ	106.4
Washington, DC	97.4

a. Compute the mean and standard deviation for the sample data.
b. Using the mean and standard deviation computed in part (a) as estimates of the mean and standard deviation of household income for the population of all cites, use Chebyshev's theorem to determine the range within which 75% of the household incomes for the population of all cities must fall.

c. Assume that the distribution of household income is bell-shaped. Using the mean and standard deviation computed in part (a) as estimates of the mean and standard deviation of household income for the population of all cites, use the empirical rule to determine the range within which 95% of the household incomes for the population of all cities must fall. Compare your answer with the value in part (b).

d. Does the sample data contain any outliers?

66. Public transportation and the automobile are two methods an employee can use to get to work each day. Samples of times recorded for each method are shown. Times are in minutes.

Public Transportation:	28	29	32	37	33	25	29	32	41	34
Automobile:	29	31	33	32	34	30	31	32	35	33

a. Compute the sample mean time to get to work for each method.

b. Compute the sample standard deviation for each method.

c. On the basis of your results from parts (a) and (b), which method of transportation should be preferred? Explain.

d. Develop a box plot for each method. Does a comparison of the box plots support your conclusion in part (c)?

67. Final examination scores for 25 statistics students follow.

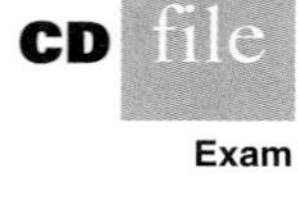

Exam

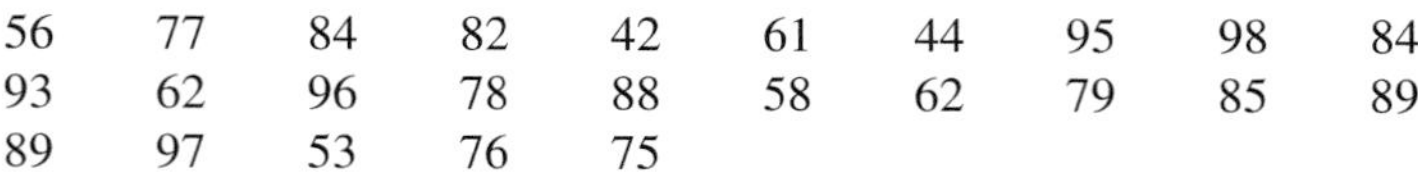

56	77	84	82	42	61	44	95	98	84
93	62	96	78	88	58	62	79	85	89
89	97	53	76	75					

a. Provide a five-number summary.

b. Provide a box plot.

68. The following data show the total yardage accumulated during the NCAA college football season for a sample of 20 receivers.

744	652	576	1112	971	451	1023	852	809	596
941	975	400	711	1174	1278	820	511	907	1251

a. Provide a five-number summary.

b. Provide a box plot.

c. Identify any outliers.

69. The typical household income and typical home price for a sample of 20 cities follow (*Places Rated Almanac,* 2000). Data are in thousands of dollars.

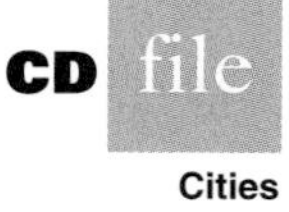

Cities

City	Income	Home Price
Bismark, ND	62.8	92.8
Columbia, SC	66.8	116.7
Savannah, GA	67.8	108.1
Birmingham, AL	71.2	130.9
Toledo, OH	71.2	101.1
Akron, OH	74.1	114.9
Lancaster, PA	75.2	125.9
Fort Lauderdale, FL	75.8	145.3
Nashville, TN	77.3	125.9
Madison, WI	78.8	145.2
Cleveland, OH	79.2	135.8
Atlanta, GA	82.4	126.9
Denver, CO	82.6	161.9
Detroit, MI	85.3	145.0
Philadelphia, PA	87.0	151.5
Hartford, CT	89.1	162.1
Washington, DC	97.4	191.9
Naples, FL	100.0	173.6
Trenton, NJ	106.4	168.1
Danbury, CT	132.3	234.1

a. What is the value of the sample covariance? Does it indicate a positive or a negative linear relationship?

b. What is the sample correlation coefficient?

70. *Road & Track* provided the following sample of the tire ratings and load-carrying capacity of automobiles tires.

Tire Rating	Load-Carrying Capacity
75	853
82	1047
85	1135
87	1201
88	1235
91	1356
92	1389
93	1433
105	2039

a. Develop a scatter diagram for the data with tire rating on the x-axis.

b. What is the sample correlation coefficient and what does it tell you about the relationship between tire rating and load-carrying capacity?

71. In Exercise 6, we computed a variety of descriptive statistics for two types of trades made by discount brokers: 500 shares at $50 per share, and 1000 shares at $5 per share. Table 3.2 shows the commissions charged on each of these trades by a sample of 20 discount brokers (*AAII Journal,* January 1997). Compute the covariance and the correlation coefficient for the two types of trades. What did you learn about the relationship?

72. The following data show the Dow stocks' trailing 52-weeks primary share earnings and book values as reported by 10 companies (*The Wall Street Journal,* March 13, 2000).

Company	Book Value	Earnings
Am Elec	25.21	2.69
Columbia En	23.2	3.01
Con Ed	25.19	3.13
Duke Energy	20.17	2.25
Edison Int'l	13.55	1.79
Enron Cp.	7.44	1.27
Peco	13.61	3.15
Pub Sv Ent	21.86	3.29
Southn Co.	8.77	1.86
Unicom	23.22	2.74

a. Develop a scatter diagram for the data with book value on the x-axis.

b. What is the sample correlation coefficient and what does it tell you about the relationship between the earnings per share and the book value?

73. The days to maturity for a sample of five money market funds are shown here. The dollar amounts invested in the funds are provided. Use the weighted mean to determine the mean number of days to maturity for dollars invested in these five money market funds.

Days to Maturity	Dollar Value ($million)
20	20
12	30
7	10
5	15
6	10

74. A forecasting technique referred to as moving averages uses the average or mean of the most recent n periods to forecast the next value for time series data. With a three-period moving average, the most recent three periods of data are used in the forecast computation. Consider a product with the following demand for the first three months of the current year: January (800 units), February (750 units), and March (900 units).

a. What is the three-month moving average forecast for April?

b. A variation of this forecasting technique is called weighted moving averages. The weighting allows the more recent time series data to receive more weight or more importance in the computation of the forecast. For example, a weighted three-month moving average might give a weight of 3 to data one month old, a weight of 2 to data two months old, and a weight of 1 to data three months old. Use the data above to provide a three-month weighted moving average forecast for April.

75. A frequency distribution for the duration of 20 long-distance telephone calls in minutes follows. Compute the mean, variance, and standard deviation for the data.

Call Duration (minutes)	Frequency
4–7	4
8–11	5
12–15	7
16–19	2
20–23	1
24–27	1
Total	20

76. Dinner check amounts at La Maison French Restaurant have the frequency distribution shown as follows. Compute the mean, variance, and standard deviation for the data.

Dinner Check ($)	Frequency
25–34	2
35–44	6
45–54	4
55–64	4
65–74	2
75–84	2
Total	20

77. Automobiles traveling on the New York State Thruway are checked for speed by a state police radar system. Following is a frequency distribution of speeds.

Speed (miles per hour)	Frequency
45–49	10
50–54	40
55–59	150
60–64	175
65–69	75
70–74	15
75–79	10
Total	475

a. What is the mean speed of the automobiles traveling on the New York State Thruway?

b. Compute the variance and the standard deviation.

Case Problem 1 CONSOLIDATED FOODS, INC.

Consolidated Foods, Inc., operates a chain of supermarkets in New Mexico, Arizona, and California. (See Case Problem, Chapter 2). Table 3.13 shows a portion of the data on dollar amounts and method of payment for a sample of 100 customers. Consolidated's managers requested the sample be taken to learn about payment practices of the stores' customers. In particular, managers were interested in learning about how a new credit card payment option was related to the customers' purchase amounts.

Managerial Report

Use the methods of descriptive statistics presented in Chapter 3 to summarize the sample data. Provide summaries of the dollar purchase amounts for cash customers, personal check customers, and credit card customers separately. Your report should contain the following summaries and discussions.

1. A comparison and interpretation of means and medians.
2. A comparison and interpretation of measures of variability such as the range and standard deviation.
3. The identification and interpretation of the five-number summaries for each method of payment.
4. Box plots for each method of payment.

Use the summary section of your report to provide a discussion of what you have learned about the method of payment and the amounts of payments for Consolidated Foods customers.

TABLE 3.13 PURCHASE AMOUNT AND METHOD OF PAYMENT FOR A SAMPLE OF 100 CONSOLIDATED FOODS CUSTOMERS

Consolid

Customer	Amount ($)	Method of Payment
1	28.58	Check
2	52.04	Check
3	7.41	Cash
4	11.17	Cash
5	43.79	Credit Card
6	48.95	Check
7	57.59	Check
8	27.60	Check
9	26.91	Credit Card
10	9.00	Cash
.	.	.
.	.	.
.	.	.
95	18.09	Cash
96	54.84	Check
97	41.10	Check
98	43.14	Check
99	3.31	Cash
100	69.77	Credit Card

Case Problem 2 NATIONAL HEALTH CARE ASSOCIATION

The National Health Care Association is concerned about the shortage of nurses the health care profession is projecting for the future. To learn the current degree of job satisfaction among nurses, the association has sponsored a study of hospital nurses throughout the country. As part of this study, a sample of 50 nurses were asked to indicate their degree of satisfaction in their work, their pay, and their opportunities for promotion. Each of the three aspects of satisfaction was measured on a scale from 0 to 100, with larger values indicating higher degrees of satisfaction. The data collected also showed the type of hospital employing the nurses. The types of hospitals were private (P), Veterans Administration (VA), and University (U). A portion of the data is shown in Table 3.14. The complete data set is on the data disk in the file named Health.

Managerial Report

Use methods of descriptive statistics to summarize the data. Present the summaries that will be beneficial in communicating the results to others. Discuss your findings. Specifically, comment on the following questions.

1. On the basis of the entire data set and the three job satisfaction variables, what aspect of the job is most satisfying for the nurses? What appears to be the least satisfying? In what area(s), if any, do you feel improvements should be made? Discuss.
2. On the basis of descriptive measures of variability, what measure of job satisfaction appears to generate the greatest difference of opinion among the nurses? Explain.
3. What can be learned about the types of hospitals? Does any particular type of hospital seem to have better levels of job satisfaction than the other types? Do your results suggest any recommendations for learning about and/or improving job satisfaction? Discuss.
4. What additional descriptive statistics and insights can you use to learn about and possibly improve job satisfaction?

TABLE 3.14 SATISFACTION SCORE DATA FOR A SAMPLE OF 50 NURSES

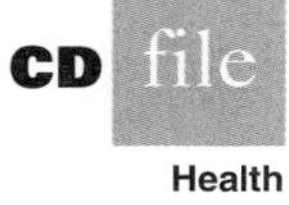

Health

Nurse	Hospital	Work	Pay	Promotion
1	Private	74	47	63
2	VA	72	76	37
3	University	75	53	92
4	Private	89	66	62
5	University	69	47	16
6	Private	85	56	64
7	University	89	80	64
8	Private	88	36	47
9	University	88	55	52
10	Private	84	42	66
.	.	.	.	.
.	.	.	.	.
.	.	.	.	.
45	University	79	59	41
46	University	84	53	63
47	University	87	66	49
48	VA	84	74	37
49	VA	95	66	52
50	Private	72	57	40

Case Problem 3 BUSINESS SCHOOLS OF ASIA-PACIFIC

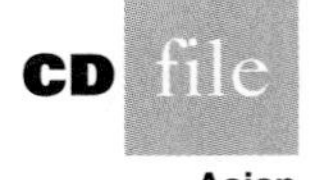

Asian

The pursuit of a higher education degree in business is now international. A survey shows that more and more Asians are choosing the Master of Business Administration degree route to corporate success (*Asia, Inc.,* September 1997). The number of applicants for MBA courses at Asia-Pacific schools is rising by about 30% a year. In 1997, the 74 business schools in the Asia-Pacific region reported a record 170,000 applications for the 11,000 full-time MBA degrees to be awarded in 1999. A main reason for the surge in demand is that an MBA can greatly enhance earning power.

Across the region, thousands of Asians are showing an increasing willingness to temporarily shelve their careers and spend two years in pursuit of a theoretical business qualification. Courses in these schools are notoriously tough and include economics, banking, marketing, behavioral sciences, labor relations, decision making, strategic thinking, business law, and more. *Asia, Inc.* provided the data set in Table 3.15, which shows some of the characteristics of the leading Asia-Pacific business schools.

Managerial Report

Use the methods of descriptive statistics to summarize the data in Table 3.15. Discuss your findings.

1. Include a summary for each variable in the data set. Make comments and interpretations based on maximums and minimums, as well as the appropriate means and proportions. What new insights do these descriptive statistics provide concerning Asia-Pacific business schools?
2. Summarize the data to compare the following:
   a. Any difference between local and foreign tuition costs.
   b. Any difference between mean starting salaries for schools requiring and not requiring work experience.
   c. Any difference between starting salaries for schools requiring and not requiring English tests.
3. Present any additional graphical and numerical summaries that will be beneficial in communicating the data in Table 3.15 to others.

Appendix 3.1 DESCRIPTIVE STATISTICS WITH MINITAB

In this appendix, we describe how to use Minitab to develop descriptive statistics. Table 3.1 listed the starting salaries for 12 business school graduates. Panel A of Figure 3.10 shows the descriptive statistics obtained by using Minitab to summarize these data. Definitions of the headings in Panel A follow.

N	number of data values
Mean	mean
Median	median
StDev	standard deviation
Min	minimum data value
Max	maximum data value
Q1	first quartile
Q3	third quartile

TABLE 3.15 DATA FOR 25 ASIA-PACIFIC BUSINESS SCHOOLS

Business School	Full-Time Enrollment	Students per Faculty	Local Tuition ($)	Foreign Tuition ($)	Age	%Foreign	GMAT	English Test	Work Experience	Starting Salary ($)
Melbourne Business School	200	5	24,420	29,600	28	47	Yes	No	Yes	71,400
University of New South Wales (Sydney)	228	4	19,993	32,582	29	28	Yes	No	Yes	65,200
Indian Institute of Management (Ahmedabad)	392	5	4,300	4,300	22	0	No	No	No	7,100
Chinese University of Hong Kong	90	5	11,140	11,140	29	10	Yes	No	No	31,000
International University of Japan (Niigata)	126	4	33,060	33,060	28	60	Yes	Yes	No	87,000
Asian Institute of Management (Manila)	389	5	7,562	9,000	25	50	Yes	No	Yes	22,800
Indian Institute of Management (Bangalore)	380	5	3,935	16,000	23	1	Yes	No	No	7,500
National University of Singapore	147	6	6,146	7,170	29	51	Yes	Yes	Yes	43,300
Indian Institute of Management (Calcutta)	463	8	2,880	16,000	23	0	No	No	No	7,400
Australian National University (Canberra)	42	2	20,300	20,300	30	80	Yes	Yes	Yes	46,600
Nanyang Technological University (Singapore)	50	5	8,500	8,500	32	20	Yes	No	Yes	49,300
University of Queensland (Brisbane)	138	17	16,000	22,800	32	26	No	No	Yes	49,600
Hong Kong University of Science and Technology	60	2	11,513	11,513	26	37	Yes	No	Yes	34,000
Macquarie Graduate School of Management (Sydney)	12	8	17,172	19,778	34	27	No	No	Yes	60,100
Chulalongkorn University (Bangkok)	200	7	17,355	17,355	25	6	Yes	No	Yes	17,600
Monash Mt. Eliza Business School (Melbourne)	350	13	16,200	22,500	30	30	Yes	Yes	Yes	52,500
Asian Institute of Management (Bangkok)	300	10	18,200	18,200	29	90	No	Yes	Yes	25,000
University of Adelaide	20	19	16,426	23,100	30	10	No	No	Yes	66,000
Massey University (Palmerston North, New Zealand)	30	15	13,106	21,625	37	35	No	Yes	Yes	41,400
Royal Melbourne Institute of Technology Business Graduate School	30	7	13,880	17,765	32	30	No	Yes	Yes	48,900
Jamnalal Bajaj Institute of Management Studies (Bombay)	240	9	1,000	1,000	24	0	No	No	Yes	7,000
Curtin Institute of Technology (Perth)	98	15	9,475	19,097	29	43	Yes	No	Yes	55,000
Lahore University of Management Sciences	70	14	11,250	26,300	23	2.5	No	No	No	7,500
Universiti Sains Malaysia (Penang)	30	5	2,260	2,260	32	15	No	Yes	Yes	16,000
De La Salle University (Manila)	44	17	3,300	3,600	28	3.5	Yes	No	Yes	13,100

FIGURE 3.10 DESCRIPTIVE STATISTICS AND BOX PLOT PROVIDED BY MINITAB

Panel A: Descriptive Statistics

N	Mean	Median	TrMean	StDev	SEMean
12	2940.0	2905.0	2924.5	165.7	47.8

Min	Max	Q1	Q3
2710.0	3325.0	2857.5	3025.0

Panel B: Box Plot

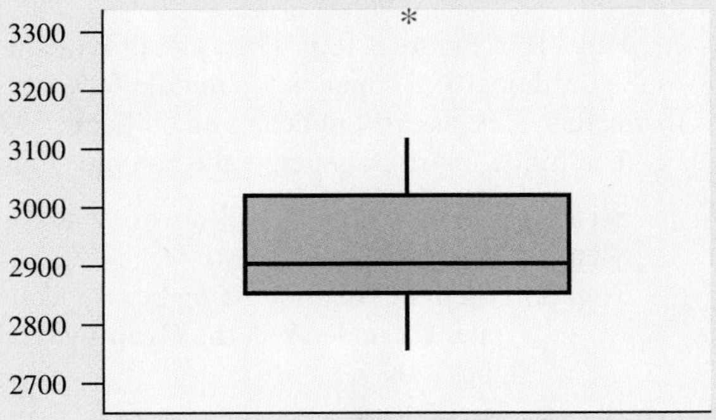

We have discussed, in a note, the numerical measure labeled TrMean and not discussed SEMean. TrMean refers to the *trimmed mean.* The trimmed mean indicates the central location of the data after removing the effect of the smallest and largest data values in the data set. Minitab provides the 5% trimmed mean; the smallest 5% of the data values and the largest 5% of the data values are removed. The 5% trimmed mean is found by computing the mean for the middle 90% of the data. SEMean, which is the *standard error of the mean,* is computed by dividing the standard deviation by the square root of N. The interpretation and use of this measure are discussed in Chapter 7 when we introduce the topics of sampling and sampling distributions.

Although the numerical measures of range, interquartile range, variance, and coefficient of variation do not appear on the Minitab output, these values can be easily computed if desired from the results in Figure 3.10 by the following formulas.

$$\text{Range} = \text{Max} - \text{Min}$$
$$\text{IQR} = \text{Q3} - \text{Q1}$$
$$\text{Variance} = (\text{StDev})^2$$
$$\text{Coefficient of Variation} = (\text{StDev}/\text{Mean}) \times 100$$

Finally, note that Minitab's quartiles $Q_1 = 2857.5$ and $Q_3 = 3025$ are slightly different than the quartiles $Q_1 = 2865$ and $Q_3 = 3000$ that we computed in Section 3.1. The reason is that different conventions* can be used to identify the quartiles. Hence, the values of Q_1 and Q_3 provided by one convention may not be identical to the values of Q_1 and Q_3 provided by another convention. Any differences tend to be negligible, however, and the results provided should not mislead the user in making the usual interpretations associated with quartiles.

*With the n data values arranged in ascending order (smallest value to largest value), Minitab uses the positions given by $(n + 1)/4$ and $3(n + 1)/4$ to locate Q_1 and Q_3, respectively. When a position is fractional, Minitab interpolates between the two adjacent ordered data values to determine the corresponding quartile.

Salary

Let us now see how the statistics in Figure 3.10 are generated. The starting salary data have been entered into column C2 (the second column) of a Minitab worksheet. The following steps will then generate the descriptive statistics.

Step 1. Select the **Stat** pull-down menu
Step 2. Choose **Basic Statistics**
Step 3. Choose **Display Descriptive Statistics**
Step 4. When the Display Descriptive Statistics dialog box appears:
Enter C2 in the **Variables** box
Click **OK**

Panel B of Figure 3.10 is a box plot provided by Minitab. The box drawn from the first to the third quartiles contains the middle 50% of the data. The line within the box locates the median. The asterisk indicates an outlier at 3325.

The following steps generate the box plot shown in panel B of Figure 3.10.

Step 1. Select the **Graph** pull-down menu
Step 2. Choose **Boxplot**
Step 3. When the Boxplot dialog box appears:
Enter C2 under **Y** in the **Graph variables** box
Click **OK**

Figure 3.11 shows the covariance and correlation output that Minitab provided for the stereo and sound equipment store data in Table 3.7. In the covariance portion of the figure, Commerci denotes the number of weekend television commercials and Sales denotes the sales during the following week. The value in column Commerci and row Sales, 11, is the sample covariance as computed in Section 3.5. The value in column Commerci and row Commerci, 2.22222, is the sample variance for the number of commercials, and the value in column Sales and row Sales, 62.88889, is the sample variance for sales. The sample correlation coefficient, 0.930, is shown in the correlation portion of the output. Note: The interpretation and use of the *p*-value provided in the correlation output are discussed in Chapter 14.

Stereo

Let us now describe how to obtain the information in Figure 3.11. We entered the data for the number of commercials into column C2 and the data for sales into column C3 of a Minitab worksheet. The steps necessary to generate the covariance output in the first three rows of Figure 3.11 follow.

Step 1. Select the **Stat** pull-down menu
Step 2. Choose **Basic Statistics**
Step 3. Choose **Covariance**

FIGURE 3.11 COVARIANCE AND CORRELATION PROVIDED BY MINITAB FOR THE NUMBER OF COMMERCIALS AND SALES DATA

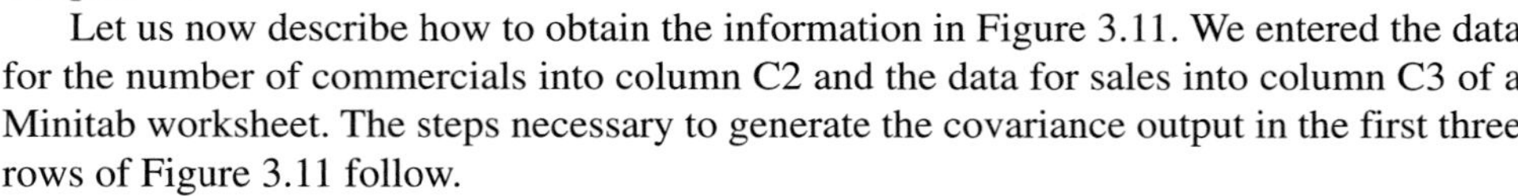

```
Covariances: Commercials, Sales

            Commerci     Sales
Commerci     2.22222
Sales       11.00000  62.88889

Correlations: Commercials, Sales

Pearson correlation of Commercials and Sales = 0.930
P-Value = 0.000
```

Step 4. When the Covariance dialog box appears:
Enter C2 C3 in the **Variables** box
Click **OK**

To obtain the correlation output in Figure 3.11, only one change is necessary in the steps for obtaining the covariance. In step 3, the **Correlation** option is selected.

Appendix 3.2 DESCRIPTIVE STATISTICS WITH EXCEL

Excel can be used to generate the descriptive statistics discussed in this chapter. We show how Excel can be used to generate several measures of location and variability for a single variable and to generate the covariance and correlation coefficient as measures of association between two variables.

Using Excel Functions

Excel provides functions for computing the mean, median, mode, sample variance, and sample standard deviation. We illustrate the use of these Excel functions by computing the mean, median, mode, sample variance, and sample standard deviation for the starting salary data in Table 3.1. Refer to Figure 3.12 as we describe the steps involved. The data have been entered in column B.

Salary

Excel's AVERAGE function can be used to compute the mean by entering the following formula into cell E1:

=AVERAGE(B2:B13)

FIGURE 3.12 USING EXCEL FUNCTIONS FOR COMPUTING THE MEAN, MEDIAN, MODE, VARIANCE, AND STANDARD DEVIATION

	A	B	C	D	E	F
1	Graduate	Starting Salary		Mean	=AVERAGE(B2:B13)	
2	1	2850		Median	=MEDIAN(B2:B13)	
3	2	2950		Mode	=MODE(B2:B13)	
4	3	3050		Variance	=VAR(B2:B13)	
5	4	2880		Standard Deviation	=STDEV(B2:B13)	
6	5	2755				
7	6	2710				
8	7	2890				
9	8	3130				
10	9	2940				
11	10	3325				
12	11	2920				
13	12	2880				
14						

	A	B	C	D	E	F
1	Graduate	Starting Salary		Mean	2940	
2	1	2850		Median	2905	
3	2	2950		Mode	2880	
4	3	3050		Variance	27440.91	
5	4	2880		Standard Deviation	165.65	
6	5	2755				
7	6	2710				
8	7	2890				
9	8	3130				
10	9	2940				
11	10	3325				
12	11	2920				
13	12	2880				
14						

Similarly, the formulas =MEDIAN(B2:B13), =MODE(B2:B13), =VAR(B2:B13), and =STDEV(B2:B13) are entered into cells E2:E5, respectively, to compute the median, mode, variance, and standard deviation. The worksheet in the foreground shows the values computed using the Excel functions are the same as we computed earlier in the chapter.

Excel also provides functions that can be used to compute the covariance and correlation coefficient. But, you must be careful when using these functions because the covariance function treats the data as if it were a population and the correlation function treats the data as if it were a sample. Thus, the result obtained using Excel's covariance function must be adjusted to provide the sample covariance. We show here how these functions can be used to compute the sample covariance and the sample correlation coefficient for the stereo and sound equipment store data in Table 3.7. Refer to Figure 3.13 as we present the steps involved.

CD file
Stereo

Excel's covariance function, COVAR, can be used to compute the population covariance by entering the following formula into cell F1:

=COVAR(B2:B11,C2:C11)

Similarly, the formula =CORREL(B2:B11,C2:C11) is entered into cell F2 to compute the sample correlation coefficient. The worksheet shown in the foreground shows the values computed using the Excel functions. Note that the value of the sample correlation coefficient (.93) is the same as we computed using equation (3.12). However, the result provided by the Excel COVAR function, 9.9, was obtained by treating the data as if it were a population. Thus, we must adjust the Excel result of 9.9 to obtain the sample covariance. The adjustment is rather simple. First, note that the formula for the population covariance, equation (3.11), requires dividing by the total number of observations in the data set. But, the formula for the sample covariance, equation (3.10), requires dividing by the total number of observations minus 1. So, to use the Excel result of 9.9 to compute the sample covariance, we simply multiply 9.9 by $n/(n-1)$. Because $n = 10$, we obtain

$$s_{xy} = \left(\frac{10}{9}\right)9.9 = 11$$

Thus, the sample covariance for the stereo and sound equipment data is 11.

FIGURE 3.13 USING EXCEL FUNCTIONS FOR COMPUTING COVARIANCE AND CORRELATION

	A	B	C	D	E	F	G
1	Week	Commercials	Sales		Covariance	=COVAR(B2:B11,C2:C11)	
2	1	2	50		Correlation	=CORREL(B2:B11,C2:C11)	
3	2	5	57				
4	3	1	41				
5	4	3	54				
6	5	4	54				
7	6	1	38				
8	7	5	63				
9	8	3	48				
10	9	4	59				
11	10	2	46				
12							

	A	B	C	D	E	F	G
1	Week	Commercials	Sales		Covariance	9.9	
2	1	2	50		Correlation	0.93	
3	2	5	57				
4	3	1	41				
5	4	3	54				
6	5	4	54				
7	6	1	38				
8	7	5	63				
9	8	3	48				
10	9	4	59				
11	10	2	46				
12							
13							

Using Excel's Descriptive Statistics Tool

As we already demonstrated, Excel provides statistical functions to compute descriptive statistics for a data set. These functions can be used to compute one statistic at a time (e.g., mean, variance, etc.). Excel also provides a variety of Data Analysis Tools. One of these, called Descriptive Statistics, allows the user to compute a variety of descriptive statistics at once. We show here how it can be used to compute descriptive statistics for the starting salary data in Table 3.1. Refer to Figure 3.14 as we describe the steps involved.

Step 1. Select the **Tools** pull-down menu
Step 2. Choose **Data Analysis**
Step 3. Choose **Descriptive Statistics** from the list of Analysis Tools
Step 4. When the Descriptive Statistics dialog box appears:
Enter B1:B13 in the **Input Range** box
Select **Grouped By Columns**
Select **Labels in First Row**
Select **Output Range**
Enter D1 in the **Output Range** box (to identify the upper left-hand corner of the section of the worksheet where the descriptive statistics will appear)
Click **OK**

Cells D1:E15 of Figure 3.14 show the descriptive statistics provided by Excel. The boldface entries are the descriptive statistics we have covered in this chapter. The descriptive statistics that are not boldface are either covered subsequently in the text or discussed in more advanced texts.

FIGURE 3.14 USING EXCEL'S DESCRIPTIVE STATISTICS TOOL

	A	B	C	D	E	F
1	**Graduate**	**Starting Salary**		*Starting Salary*		
2	1	2850				
3	2	2950		**Mean**	2940	
4	3	3050		Standard Error	47.8199	
5	4	2880		**Median**	2905	
6	5	2755		**Mode**	2880	
7	6	2710		**Standard Deviatio**	165.653	
8	7	2890		**Sample Variance**	27440.91	
9	8	3130		Kurtosis	1.718884	
10	9	2940		Skewness	1.091109	
11	10	3325		**Range**	615	
12	11	2920		**Minimum**	2710	
13	12	2880		**Maximum**	3325	
14				**Sum**	35280	
15				**Count**	12	
16						

Chapter 4

INTRODUCTION TO PROBABILITY

CONTENTS

STATISTICS IN PRACTICE

MORTON INTERNATIONAL*
Chicago, Illinois

Morton International is a company with businesses in salt, household products, rocket motors, and specialty chemicals. Carstab Corporation, a subsidiary of Morton International, produces specialty chemicals and offers a variety of chemicals designed to meet the unique specifications of its customers. For one particular customer, Carstab produced an expensive catalyst used in chemical processing. Some, but not all, of the lots produced by Carstab met the customer's specifications for the product.

Carstab's customer agreed to test each lot after receiving it and determine whether the catalyst would perform the desired function. Lots that did not pass the customer's test would be returned to Carstab. Over time, Carstab found that the customer was accepting 60% of the lots and returning 40%. In probability terms, each Carstab shipment to the customer had a .60 probability of being accepted and a .40 probability of being returned.

Neither Carstab nor its customer was pleased with these results. In an effort to improve service, Carstab explored the possibility of duplicating the customer's test prior to shipment. However, the high cost of the special testing equipment made that alternative infeasible. Carstab's chemists then proposed a new, relatively low-cost test designed to indicate whether a lot would pass the customer's test. The probability question of interest was: What is the probability that a lot will pass the customer's test if it has passed the new Carstab test?

*The authors are indebted to Michael Haskell of Morton International for providing this Statistics in Practice.

Morton Salt: "When It Rains It Pours." © Joe Higgins/ South-Western.

A sample of lots was produced and subjected to the new Carstab test. Only lots that passed the new test were sent to the customer. Probability analysis of the data indicated that if a lot passed the Carstab test, it had a .909 probability of passing the customer's test and being accepted. Alternatively, if a lot passed the Carstab test, it had only a .091 probability of being returned. The probability analysis provided key supporting evidence for the adoption and implementation of the new testing procedure at Carstab. The new test resulted in an immediate improvement in customer service and a substantial reduction in shipping and handling costs for returned lots.

The probability of a lot being accepted by the customer after passing the new Carstab test is called a conditional probability. In this chapter, you will learn how to compute this and other probabilities that are helpful in decision making.

Business decisions are often based on an analysis of uncertainties such as the following:

1. What are the chances that sales will decrease if we increase prices?
2. What is the likelihood a new assembly method will increase productivity?
3. How likely is it that the project will be finished on time?
4. What are the odds in favor of a new investment being profitable?

Some of the earliest work on probability originated in a series of letters between Pierre de Fermat and Blaise Pascal in the 1650s.

Probability is a numerical measure of the likelihood that an event will occur. Thus, probabilities could be used as measures of the degree of uncertainty associated with the four events previously listed. If probabilities were available, we could determine the likelihood of each event occurring.

Probability values are always assigned on a scale from 0 to 1. A probability near zero indicates an event is unlikely to occur; a probability near 1 indicates an event is almost certain to occur. Other probabilities between 0 and 1 represent degrees of likelihood that an event will occur. For example, if we consider the event "rain tomorrow," we understand that when the weather report indicates "a near-zero probability of rain," it means almost no chance of rain. However, if a .90 probability of rain is reported, we know that rain is likely to occur. A .50 probability indicates that rain is just as likely to occur as not. Figure 4.1 depicts the view of probability as a numerical measure of the likelihood of an event occurring.

Pierre-Simon de Laplace's book, Theorie Analytique des Probabilities, *was published in 1812.*

FIGURE 4.1 PROBABILITY AS A NUMERICAL MEASURE OF THE LIKELIHOOD OF OCCURRENCE

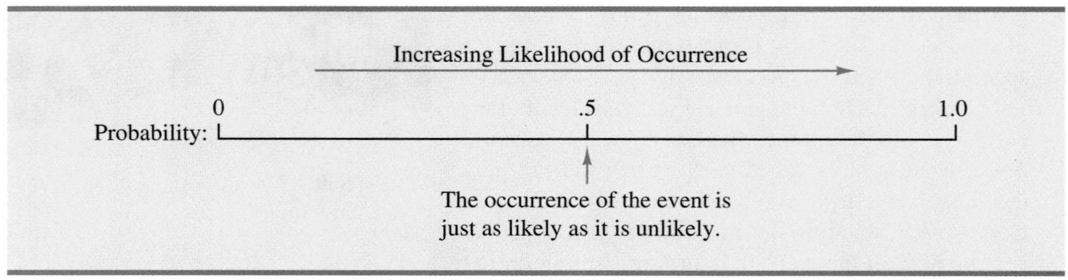

4.1 EXPERIMENTS, COUNTING RULES, AND ASSIGNING PROBABILITIES

In discussing probability, we define an **experiment** to be a process that generates well-defined outcomes. On any single repetition of an experiment, one and only one of the possible experimental outcomes will occur. Several examples of experiments and their associated outcomes follow.

Experiment	Experimental Outcomes
Toss a coin	Head, tail
Select a part for inspection	Defective, nondefective
Conduct a sales call	Purchase, no purchase
Roll a die	1, 2, 3, 4, 5, 6
Play a football game	Win, lose, tie

When we have specified all possible experimental outcomes, we have identified the **sample space** for an experiment.

Sample Space

The sample space for an experiment is the set of all experimental outcomes.

Experimental outcomes are also called sample points.

An experimental outcome is also called a **sample point** to identify it as an element of the sample space.

Consider the first experiment in the preceding table—tossing a coin. The experimental outcomes (sample points) are determined by the upward face of the coin—a head or a tail. If we let S denote the sample space, we can use the following notation to describe the sample space.

$$S = \{\text{Head, Tail}\}$$

The sample space for the second experiment in the table—selecting a part for inspection—has the following sample space and sample points.

$$S = \{\text{Defective, Nondefective}\}$$

Both of the experiments just described have two experimental outcomes (sample points). However, suppose we consider the fourth experiment listed in the table—rolling a die. The possible experimental outcomes defined as the number of dots appearing on the upward face of the die are the six points in the sample space for this experiment.

$$S = \{1, 2, 3, 4, 5, 6\}$$

Counting Rules, Combinations, and Permutations

Being able to identify and count the experimental outcomes is a necessary step in assigning probabilities. We now discuss three counting rules that are useful.

Multiple-Step Experiments. The first counting rule is for multiple-step experiments. Consider the experiment of tossing two coins. Let the experimental outcomes be defined in terms of the sequence of heads and tails appearing on the upward faces of the two coins. How many experimental outcomes are possible for this experiment? The experiment of tossing two coins can be thought of as a two-step experiment in which step 1 is the tossing of the first coin and step 2 is the tossing of the second coin. If we use H to denote a head and T to denote a tail, (H, H) indicates the experimental outcome with a head on the first coin and a head on the second coin. Continuing this notation, we can describe the sample space (S) for this coin-tossing experiment as follows:

$$S = \{(H, H), (H, T), (T, H), (T, T)\}$$

Thus, we see that four experimental outcomes are possible. In this case, it is not difficult to list all of the experimental outcomes.

The counting rule for multiple-step experiments makes it possible to determine the number of experimental outcomes without listing them.

A Counting Rule for Multiple-Step Experiments

If an experiment can be described as a sequence of k steps with n_1 possible outcomes on the first step, n_2 possible outcomes on the second step, and so on, then the total number of experimental outcomes is given by $(n_1)(n_2) \ldots (n_k)$.

Viewing the experiment of tossing two coins as a sequence of first tossing one coin ($n_1 = 2$) and then tossing the other coin ($n_2 = 2$), we can see from the counting rule that there are $(2)(2) = 4$ distinct experimental outcomes. As shown, they are $S = \{(H, H), (H, T), (T, H), (T, T)\}$. The number of experimental outcomes in an experiment involving tossing six coins is $(2)(2)(2)(2)(2)(2) = 64$.

Without the tree diagram, one might think only three experimental outcomes are possible for two tosses of a coin: 0 heads, 1 head, and 2 heads.

A graphical representation that is helpful in visualizing an experiment and enumerating outcomes in a multiple-step experiment is a **tree diagram.** Figure 4.2 shows a tree diagram for the experiment of tossing two coins. The sequence of steps is depicted by moving from left to right through the tree. Step 1 corresponds to tossing the first coin and has two branches corresponding to the two possible outcomes. Step 2 corresponds to tossing the second coin and, for each possible outcome at step 1, has two branches corresponding to the two possible outcomes at step 2. Finally, each of the points on the right end of the tree corresponds to an experimental outcome. Each path through the tree from the leftmost node to one of the nodes at the right side of the tree corresponds to a unique sequence of outcomes.

Let us now see how the counting rule for multiple-step experiments can be used in the analysis of a capacity expansion project faced by the Kentucky Power & Light Company (KP&L). KP&L is starting a project designed to increase the generating capacity of one of its plants in northern Kentucky. The project is divided into two sequential stages or steps: stage 1 (design) and stage 2 (construction). While each stage will be scheduled and controlled as closely as possible, management cannot predict beforehand the exact time required to complete each stage of the project. An analysis of similar construction projects has shown completion times for the design stage of 2, 3, or 4 months and completion times for the construction stage of 6, 7, or 8 months. In addition, because of the critical need for additional electrical power, management has set a goal of 10 months for the completion of the entire project.

Because this project has three possible completion times for the design stage (step 1) and three possible completion times for the construction stage (step 2), the counting rule for multiple-step experiments can be applied here to determine a total of $(3)(3) = 9$ experimental outcomes. To describe the experimental outcomes, we will use a two-number notation; for instance, (2, 6) will indicate that the design stage is completed in 2 months and the construction stage is completed in 6 months. This experimental outcome results in a total of $2 + 6 = 8$ months to complete the entire project. Table 4.1 summarizes the nine experimental outcomes for the KP&L problem. The tree diagram in Figure 4.3 shows how the nine outcomes (sample points) occur.

The counting rule and tree diagram have been used to help the project manager identify the experimental outcomes and determine the possible project completion times. From the information in Figure 4.3, we see that the project will be completed in 8 to 12 months, with

FIGURE 4.2 TREE DIAGRAM FOR THE EXPERIMENT OF TOSSING TWO COINS

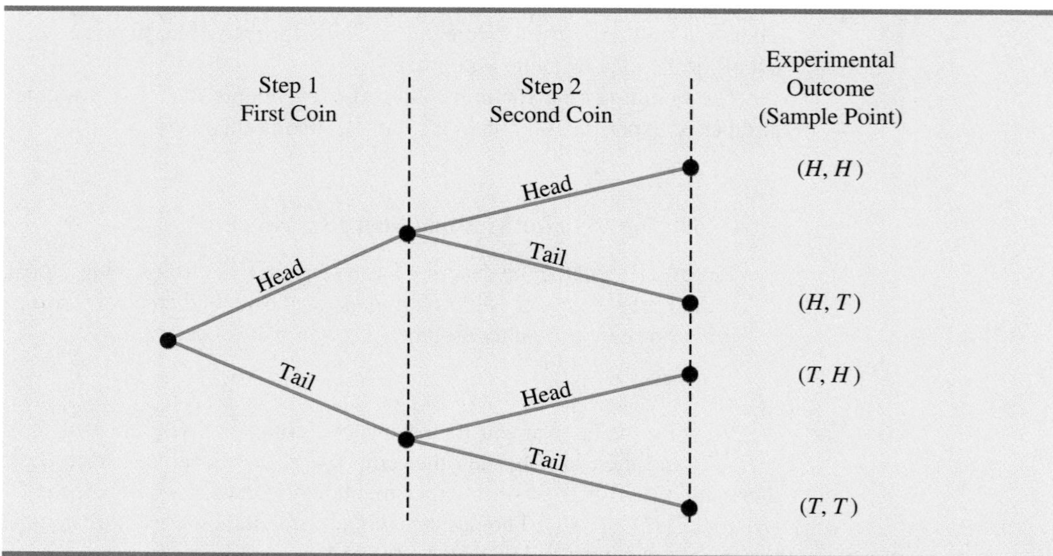

TABLE 4.1 EXPERIMENTAL OUTCOMES (SAMPLE POINTS) FOR THE KP&L PROBLEM

Completion Time (months)			
Stage 1 (Design)	**Stage 2 (Construction)**	**Notation for Experimental Outcome**	**Total Project Completion Time (months)**
2	6	(2, 6)	8
2	7	(2, 7)	9
2	8	(2, 8)	10
3	6	(3, 6)	9
3	7	(3, 7)	10
3	8	(3, 8)	11
4	6	(4, 6)	10
4	7	(4, 7)	11
4	8	(4, 8)	12

FIGURE 4.3 TREE DIAGRAM FOR THE KP&L PROJECT

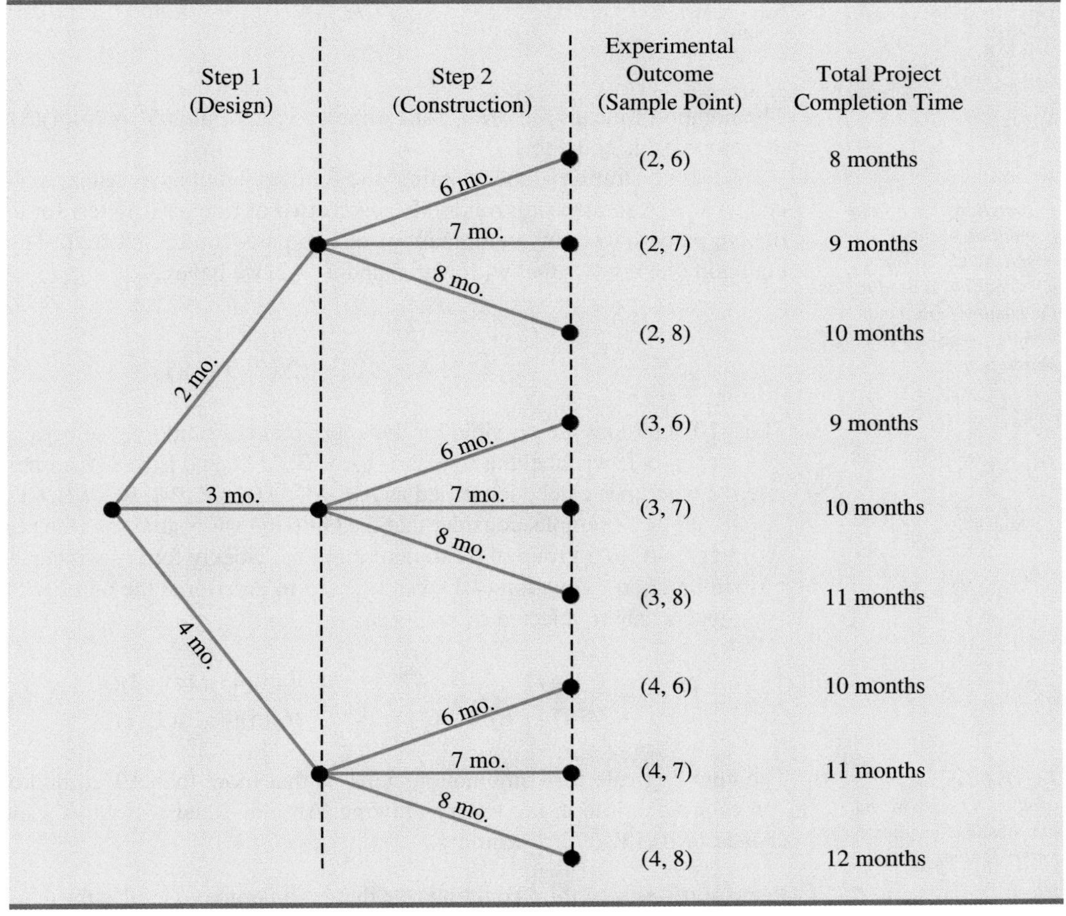

six of the nine experimental outcomes providing the desired completion time of 10 months or less. While it has been helpful to identify the experimental outcomes, we will need to consider how probability values can be assigned to the experimental outcomes before making an assessment of the probability that the project will be completed within the desired 10 months.

Combinations. A second counting rule that is often useful allows one to count the number of experimental outcomes when the experiment involves selecting n objects from a (usually larger) set of N objects. It is called the counting rule for combinations.

Counting Rule for Combinations

The number of combinations of N objects taken n at a time is

$$C_n^N = \binom{N}{n} = \frac{N!}{n!(N-n)!} \tag{4.1}$$

where

$$N! = N(N-1)(N-2)\cdots(2)(1)$$
$$n! = n(n-1)(n-2)\cdots(2)(1)$$

and

$$0! = 1$$

The notation ! means *factorial;* for example, 5 factorial is $5! = (5)(4)(3)(2)(1) = 120$. By definition, $0!$ is equal to 1.

In sampling from a finite population of size N, the counting rule for combinations is used to find the number of different samples of size n that can be selected.

As an illustration of the counting rule for combinations, consider a quality control procedure where an inspector randomly selects two of five parts to test for defects. In a group of five parts, how many combinations of two parts can be selected? The counting rule in equation (4.1) shows that with $N = 5$ and $n = 2$, we have

$$C_2^5 = \binom{5}{2} = \frac{5!}{2!(5-2)!} = \frac{(5)(4)(3)(2)(1)}{(2)(1)(3)(2)(1)} = \frac{120}{12} = 10$$

Thus, 10 outcomes are possible for the experiment of randomly selecting two parts from a group of five. If we label the five parts as A, B, C, D, and E, the 10 combinations or experimental outcomes can be identified as AB, AC, AD, AE, BC, BD, BE, CD, CE, and DE.

As another example, consider that the Ohio lottery system uses the random selection of six integers from a group of 47 to determine the weekly lottery winner. The counting rule for combinations, equation (4.1), can be used to determine the number of ways six different integers can be selected from a group of 47.

$$\binom{47}{6} = \frac{47!}{6!(47-6)!} = \frac{47!}{6!41!} = \frac{(47)(46)(45)(44)(43)(42)}{(6)(5)(4)(3)(2)(1)} = 10{,}737{,}573$$

The counting rule for combinations shows that the chance of winning the lottery is slim.

The counting rule for combinations tells us that more than 10 million experimental outcomes are possible in the lottery drawing. An individual who buys a lottery ticket has 1 chance in 10,737,573 of winning.

Permutations. A third counting rule that is sometimes useful is the counting rule for permutations. It allows one to compute the number of experimental outcomes when n objects are to be selected from a set of N objects where the order of selection is important. The same n objects selected in a different order is considered a different experimental outcome.

Counting Rule for Permutations

The number of permutations of N objects taken n at a time is given by

$$P_n^N = n!\binom{N}{n} = \frac{N!}{(N-n)!} \tag{4.2}$$

The counting rule for permutations is closely related to the one for combinations; however, an experiment will have more permutations than combinations for the same number of objects because every selection of n objects has $n!$ different ways to order them.

As an example, consider again the quality control process in which an inspector selects two of five parts to inspect for defects. How many permutations may be selected? The counting rule in equation (4.2) shows that with $N = 5$ and $n = 2$, we have

$$P_2^5 = \frac{5!}{(5-2)!} = \frac{5!}{3!} = \frac{(5)(4)(3)(2)(1)}{(3)(2)(1)} = (5)(4) = 20$$

Thus, 20 outcomes are possible for the experiment of randomly selecting two parts from a group of five when the order of selection must be taken into account. If we label the parts A, B, C, D, and E, the 20 permutations are AB, BA, AC, CA, AD, DA, AE, EA, BC, CB, BD, DB, BE, EB, CD, DC, CE, EC, DE, and ED.

Assigning Probabilities

Now let us see how probabilities can be assigned to experimental outcomes. The three approaches most frequently used are the classical, relative frequency, and subjective methods. Regardless of the method used, the probabilities assigned must satisfy two *basic requirements of probability.*

Basic Requirements for Assigning Probabilities

1. The probability assigned to each experimental outcome must be between 0 and 1, inclusively. If we let E_i denote the ith experimental outcome and $P(E_i)$ its probability, then this requirement can be written as

$$0 \leq P(E_i) \leq 1 \text{ for all } i \tag{4.3}$$

2. The sum of the probabilities for all the experimental outcomes must equal 1. For n experimental outcomes, this requirement can be written as

$$P(E_1) + P(E_2) + \cdots + P(E_n) = 1 \tag{4.4}$$

The **classical method** of assigning probabilities is appropriate when all the experimental outcomes are equally likely. If n experimental outcomes are possible, a probability of $1/n$ is assigned to each experimental outcome. When using this approach, the two basic requirements for assigning probabilities are automatically satisfied.

For an example, consider the experiment of tossing a fair coin. The two experimental outcomes—head and tail—are equally likely. Because one of the two equally likely outcomes is a head, the probability of observing a head is 1/2, or .50. Similarly, the probability of observing a tail is also 1/2, or .50.

As another example, consider the experiment of rolling a die. It would seem reasonable to conclude that the six possible outcomes are equally likely, and hence each outcome is assigned a probability of 1/6. If $P(1)$ denotes the probability that one dot appears on the upward face of the die, then $P(1) = 1/6$. Similarly, $P(2) = 1/6$. $P(3) = 1/6$, $P(4) = 1/6$, $P(5) = 1/6$, and $P(6) = 1/6$. Note that the requirements of equations (4.3) and (4.4) are both satisfied because each of the probabilities is greater than or equal to zero and they sum to one.

The **relative frequency method** of assigning probabilities is appropriate when data are available to estimate the proportion of the time the experimental outcome will occur if the experiment is repeated a large number of times. As an example consider a study of waiting times in the X-ray department for a local hospital. The number of patients waiting for service at 9:00 A.M. was recorded for 20 successive days. The following results were obtained.

Number Waiting	Number of Days Outcome Occurred
0	2
1	5
2	6
3	4
4	3
Total	20

These data show that on 2 of the 20 days, zero patients were waiting for service; on 5 of the days, one patient was waiting for service; and so on. Using the relative frequency method, we would assign a probability of $2/20 = .10$ to the experimental outcome of zero patients waiting for service, $5/20 = .25$ to the experimental outcome of one patient waiting, $6/20 = .30$ to two patients waiting, $4/20 = .20$ to three patients waiting, and $3/20 = .15$ to four patients waiting.

As with the classical method, the two basic requirements of equations (4.3) and (4.4) are automatically satisfied when the relative frequency method is used.

The **subjective method** of assigning probabilities is most appropriate when it is unrealistic to assume that the experimental outcomes are equally likely and when little relevant data are available. When the subjective method is used to assign probabilities to the experimental outcomes, we may use any information available, such as our experience or intuition. After considering all available information, a probability value that expresses our *degree of belief* (on a scale from 0 to 1) that the experimental outcome will occur is specified. Because subjective probability expresses a person's degree of belief, it is personal. Using the subjective method, different people can be expected to assign different probabilities to the same experimental outcome.

When using the subjective probability assignment method, extra care must be taken to ensure that the requirements of equations (4.3) and (4.4) are satisfied. Regardless of a person's degree of belief, the probability value assigned to each experimental outcome must be between 0 and 1, inclusive, and the sum of all the experimental outcome probabilities must equal one.

Consider the case in which Tom and Judy Elsbernd have just made an offer to purchase a house. Two outcomes are possible:

$$E_1 = \text{their offer is accepted}$$
$$E_2 = \text{their offer is rejected}$$

Judy believes that the probability their offer will be accepted is .8; thus, Judy would set $P(E_1) = .8$ and $P(E_2) = .2$. Tom, however, believes that the probability that their offer will be accepted is .6; hence, Tom would set $P(E_1) = .6$ and $P(E_2) = .4$. Note that Tom's probability estimate for E_1 reflects the fact that he is a bit more pessimistic than Judy about their offer being accepted.

Bayes' theorem (see Section 4.5) provides a means for combining subjectively determined prior probabilities with probabilities obtained by other means to obtain revised, or posterior, probabilities.

Both Judy and Tom have assigned probabilities that satisfy the two basic requirements. The fact that their probability estimates are different emphasizes the personal nature of the subjective method.

Even in business situations where either the classical or the relative frequency approach can be applied, managers may want to provide subjective probability estimates. In such cases, the best probability estimates often are obtained by combining the estimates from the classical or relative frequency approach with subjective probability estimates.

Probabilities for the KP&L Project

To perform further analysis on the KP&L project, we must develop probabilities for each of the nine experimental outcomes listed in Table 4.1. On the basis of experience and judgment, management concluded that the experimental outcomes were not equally likely. Hence, the classical method of assigning probabilities could not be used. Management then decided to conduct a study of the completion times for similar projects undertaken by KP&L over the past three years. The results of a study of 40 similar projects are summarized in Table 4.2.

After reviewing the results of the study, management decided to employ the relative frequency method of assigning probabilities. Management could have provided subjective probability estimates, but felt that the current project was quite similar to the 40 previous projects. Thus, the relative frequency method was judged best.

In using the data in Table 4.2 to compute probabilities, we note that outcome (2, 6)—stage 1 completed in 2 months and stage 2 completed in 6 months—occurred six times in the 40 projects. We can use the relative frequency method to assign a probability of 6/40 = .15 to this outcome. Similarly, outcome (2, 7) also occurred in six of the 40 projects, providing a 6/40 = .15 probability. Continuing in this manner, we obtain the probability assignments for the sample points of the KP&L project shown in Table 4.3. Note that $P(2, 6)$ represents the probability of the sample point (2, 6), $P(2, 7)$ represents the probability of the sample point (2, 7), and so on.

TABLE 4.2 COMPLETION RESULTS FOR 40 KP&L PROJECTS

Completion Time (months)			
Stage 1	**Stage 2**	**Sample Point**	**Number of Past Projects Having These Completion Times**
2	6	(2, 6)	6
2	7	(2, 7)	6
2	8	(2, 8)	2
3	6	(3, 6)	4
3	7	(3, 7)	8
3	8	(3, 8)	2
4	6	(4, 6)	2
4	7	(4, 7)	4
4	8	(4, 8)	6
		Total	40

TABLE 4.3 PROBABILITY ASSIGNMENTS FOR THE KP&L PROBLEM BASED ON THE RELATIVE FREQUENCY METHOD

Sample Point	Project Completion Time	Probability of Sample Point
(2, 6)	8 months	$P(2, 6) = 6/40 = .15$
(2, 7)	9 months	$P(2, 7) = 6/40 = .15$
(2, 8)	10 months	$P(2, 8) = 2/40 = .05$
(3, 6)	9 months	$P(3, 6) = 4/40 = .10$
(3, 7)	10 months	$P(3, 7) = 8/40 = .20$
(3, 8)	11 months	$P(3, 8) = 2/40 = .05$
(4, 6)	10 months	$P(4, 6) = 2/40 = .05$
(4, 7)	11 months	$P(4, 7) = 4/40 = .10$
(4, 8)	12 months	$P(4, 8) = 6/40 = .15$
		Total 1.00

NOTES AND COMMENTS

1. In statistics, the notion of an experiment is somewhat different from the notion of an experiment in the physical sciences. In the physical sciences, an experiment is usually conducted in a laboratory or a controlled environment in order to learn about cause and effect. In statistical experiments, the outcomes are determined by probability. Even though the experiment is repeated in exactly the same way, an entirely different outcome may occur. Because of this influence of probability on the outcome, the experiments of statistics are sometimes called *random experiments.*
2. When drawing a random sample without replacement from a population of size N, the counting rule for combinations is used to find the number of different samples of size n that can be selected.

EXERCISES

Methods

1. An experiment has three steps with three outcomes possible for the first step, two outcomes possible for the second step, and four outcomes possible for the third step. How many experimental outcomes exist for the entire experiment?

2. How many ways can three items be selected from a group of six items? Use the letters A, B, C, D, E, and F to identify the items, and list each of the different combinations of three items.
3. How many permutations of three items can be selected from a group of six? Use the letters A, B, C, D, E, and F to identify the items, and list each of the permutations of the three items (B, D, F).
4. Consider the experiment of tossing a coin three times.
   a. Develop a tree diagram for the experiment.
   b. List the experimental outcomes.
   c. What is the probability for each experimental outcome?
5. Suppose an experiment has five equally likely outcomes: E_1, E_2, E_3, E_4, E_5. Assign probabilities to each outcome and show that the conditions in equations (4.3) and (4.4) are satisfied. What method did you use?

6. An experiment with three outcomes has been repeated 50 times, and it was learned that E_1 occurred 20 times. E_2 occurred 13 times, and E_3 occurred 17 times. Assign probabilities to the outcomes. What method did you use?

7. A decision maker has subjectively assigned the following probabilities to the four outcomes of an experiment: $P(E_1) = .10$, $P(E_2) = .15$, $P(E_3) = .40$, and $P(E_4) = .20$. Are these valid probability assignments? Check to see whether equations (4.3) and (4.4) are satisfied.

Applications

8. In the city of Milford, applications for zoning changes go through a two-step process: a review by the planning commission and a final decision by the city council. At step 1 the planning commission will review the zoning change request and make a positive or negative recommendation concerning the change. At step 2 the city council will review the planning commission's recommendation and then vote to approve or to disapprove the zoning change. An application for a zoning change has just been submitted by the developer of an apartment complex. Consider the application process as an experiment.
   a. How many sample points are there for this experiment? List the sample points.
   b. Construct a tree diagram for the experiment.

SELF test

9. Simple random sampling uses a sample of size n from a population of size N to obtain data that can be used to make inferences about the characteristics of a population. Suppose we have a population of 50 bank accounts and want to take a random sample of four accounts in order to learn about the population. How many different random samples of four accounts are possible?

10. The availability of venture capital has provided a big boost in funds available to companies in recent years. According to Venture Economics (*Investor's Business Daily,* April 28, 2000), 2,374 venture capital disbursements were made in 1999. Of these 1,434 were to companies in California, 390 were to companies in Massachusetts, 217 were to companies in New York, and 112 were to companies in Colorado. Twenty-two percent of the companies receiving funds were in the early stages of development and 55% of the companies were in an expansion stage.

    Suppose you want to randomly choose one of these companies to learn about how they used the funds.
    a. What is the probability the company chosen will be from California?
    b. What is the probability the company chosen will not be from one of the four states mentioned?
    c. What is the probability the company will not be in the early stages of development?
    d. Assuming the companies in the early stages of development were evenly distributed across the country, how many Massachusetts companies receiving venture capital funds were in their early stages of development?
    e. The total amount of funds invested was $32.4 billion. Estimate the amount that went to Colorado.

11. Strom Construction has made bids on two contracts. The owner has identified the possible outcomes and subjectively assigned the following probabilities.

Experimental Outcome	Obtain Contract 1	Obtain Contract 2	Probability
1	Yes	Yes	.15
2	Yes	No	.15
3	No	Yes	.30
4	No	No	.25

   a. Are these valid probability assignments? Why or why not?
   b. What, if anything, needs to be done to make the probability assignments valid?

12. State lotteries have become popular means of fund raising. The Powerball lottery is open to participants across several states. When entering the Powerball lottery, a participant selects five numbers from the digits 1 through 49 and then selects a powerball number from the digits 1 through 42. One week in 1998 the Powerball lottery jackpot reached $150,000,000, in addition to a payoff of $100,000 for anybody selecting the first five numbers correctly.

a. Compute the number of ways the first five numbers can be selected.

b. What is the probability of winning the $100,000 prize by selecting the first five numbers correctly?

c. What is the probability of winning the Powerball jackpot?

13. A company that manufactures toothpaste is studying five different package designs. Assuming that one design is just as likely to be selected by a consumer as any other design, what selection probability would you assign to each of the package designs? In an actual experiment, 100 consumers were asked to pick the design they preferred. The following data were obtained. Do the data appear to confirm the belief that one design is just as likely to be selected as another? Explain.

Design	Number of Times Preferred
1	5
2	15
3	30
4	40
5	10

4.2 EVENTS AND THEIR PROBABILITIES

Until now we have used the term *event* much as it would be used in everyday language. We must now introduce the formal definition of an event as it relates to probability.

Event

An *event* is a collection of sample points.

For an example, let us return to the KP&L problem and assume that the project manager is interested in the event that the entire project can be completed in 10 months or less. Referring to Table 4.3, we see that six sample points—(2, 6), (2, 7), (2, 8), (3, 6), (3, 7), and (4, 6)—provide a project completion time of 10 months or less. Let C denote the event that the project is completed in 10 months or less; we write

$$C = \{(2, 6), (2, 7), (2, 8), (3, 6), (3, 7), (4, 6)\}$$

Event C is said to occur if *any one* of these six sample points appears as the experimental outcome.

Other events that might be of interest to KP&L management include the following.

L = The event that the project is completed in *less* than 10 months
M = The event that the project is completed in *more* than 10 months

Using the information in Table 4.3, we see that these events consist of the following sample points.

$$L = \{(2, 6), (2, 7), (3, 6)\}$$
$$M = \{(3, 8), (4, 7), (4, 8)\}$$

A variety of additional events can be defined for the KP&L problem, but in each case the event must be identified as a collection of sample points for the experiment.

Given the probabilities of the sample points shown in Table 4.3, we can use the following definition to compute the probability of any event that KP&L management might want to consider.

Probability of an Event

The probability of any event is equal to the sum of the probabilities of the sample points in the event.

Using this definition, we calculate the probability of a particular event by adding the probabilities of the sample points (experimental outcomes) that make up the event. We can now compute the probability that the project will take 10 months or less to complete. Because this event is given by $C = \{(2, 6), (2, 7), (2, 8), (3, 6), (3, 7), (4, 6)\}$, the probability of event C, denoted $P(C)$, is given by

$$P(C) = P(2, 6) + P(2, 7) + P(2, 8) + P(3, 6) + P(3, 7) + P(4, 6)$$

Refer to the sample point probabilities in Table 4.3; we have

$$P(C) = .15 + .15 + .05 + .10 + .20 + .05 = .70$$

Similarly, because the event that the project is completed in less than 10 months is given by $L = \{(2, 6), (2, 7), (3, 6)\}$, the probability of this event is given by

$$\begin{aligned} P(L) &= P(2, 6) + P(2, 7) + P(3, 6) \\ &= .15 + .15 + .10 = .40 \end{aligned}$$

Finally, for the event that the project is completed in more than 10 months, we have $M = \{(3, 8), (4, 7), (4, 8)\}$ and thus

$$\begin{aligned} P(M) &= P(3, 8) + P(4, 7) + P(4, 8) \\ &= .05 + .10 + .15 = .30 \end{aligned}$$

Using these probability results, we can now tell KP&L management that there is a .70 probability that the project will be completed in 10 months or less, a .40 probability that the project will be completed in less than 10 months, and a .30 probability that the project will be completed in more than 10 months. This procedure of computing event probabilities can be repeated for any event of interest to the KP&L management.

Any time that we can identify all the sample points of an experiment and assign probabilities to each, we can compute the probability of an event using the definition. However, in many experiments the number of sample points is large and the identification of the sample points, as well as the determination of their associated probabilities, is extremely cumbersome, if not impossible. In the remaining sections of this chapter, we present some basic probability relationships that can be used to compute the probability of an event without knowledge of all the sample point probabilities.

NOTES AND COMMENTS

1. The sample space, S, is an event. Because it contains all the experimental outcomes, it has a probability of 1; that is, $P(S) = 1$.
2. When the classical method is used to assign probabilities, the assumption is that the experimental outcomes are equally likely. In such cases, the probability of an event can be computed by counting the number of experimental outcomes in the event and dividing the result by the total number of experimental outcomes.

EXERCISES

Methods

14. An experiment has four equally likely outcomes: E_1, E_2, E_3, and E_4.

- **a.** What is the probability that E_2 occurs?
- **b.** What is the probability that any two of the outcomes occur (e.g., E_1 or E_3)?
- **c.** What is the probability that any three of the outcomes occur (e.g., E_1 or E_2 or E_4)?

15. Consider the experiment of selecting a card from a deck of 52 cards. Each card corresponds to a sample point with a 1/52 probability.

- **a.** List the sample points in the event an ace is selected.
- **b.** List the sample points in the event a club is selected.
- **c.** List the sample points in the event a face card (jack, queen, or king) is selected.
- **d.** Find the probabilities associated with each of the events in (a), (b), and (c).

16. Consider the experiment of rolling a pair of dice. Suppose that we are interested in the sum of the face values showing on the dice.

- **a.** How many sample points are possible? (Hint: Use the counting rule for multiple-step experiments.)
- **b.** List the sample points.
- **c.** What is the probability of obtaining a value of 7?
- **d.** What is the probability of obtaining a value of 9 or greater?
- **e.** Because each roll has six possible even values (2, 4, 6, 8, 10, and 12) and only five possible odd values (3, 5, 7, 9, and 11), the dice should show even values more often than odd values. Do you agree with this statement? Explain.
- **f.** What method did you use to assign the probabilities requested?

Applications

17. Refer to the KP&L sample points and sample point probabilities in Tables 4.2 and 4.3.

- **a.** The design stage (stage 1) will run over budget if it takes 4 months to complete. List the sample points in the event the design stage is over budget.
- **b.** What is the probability that the design stage is over budget?
- **c.** The construction stage (stage 2) will run over budget if it takes 8 months to complete. List the sample points in the event the construction stage is over budget.
- **d.** What is the probability that the construction stage is over budget?
- **e.** What is the probability that both stages are over budget?

18. Suppose that a manager of a large apartment complex provides the following subjective probability estimates about the number of vacancies that will exist next month.

Vacancies	Probability
0	.05
1	.15
2	.35
3	.25
4	.10
5	.10

List the sample points in each of the following events and provide the probability of the event.

a. No vacancies
b. At least four vacancies
c. Two or fewer vacancies

19. The manager of a furniture store sells from 0 to 4 china hutches each week. On the basis of past experience, the following probabilities are assigned to sales of 0, 1, 2, 3, or 4 hutches: $P(0) = .08$; $P(1) = .18$; $P(2) = .32$; $P(3) = .30$; and $P(4) = .12$.

a. Are these valid probability assignments? Why or why not?
b. Let A be the event that 2 or fewer are sold in one week. Find $P(A)$.
c. Let B be the event that 4 or more are sold in one week. Find $P(B)$.

20. *Fortune* magazine publishes an annual issue containing information on *Fortune* 500 companies. The following data show the six states with the largest number of *Fortune* 500 companies as well as the number of companies headquartered in those states (*Fortune,* April 17, 2000).

State	Number of Companies
New York	56
California	53
Texas	43
Illinois	37
Ohio	28
Pennsylvania	28

Suppose a *Fortune* 500 company is chosen for a follow-up questionnaire. What are the probabilities of the following events?

a. Let N be the event the company is headquartered in New York. Find $P(N)$.
b. Let T be the event the company is headquartered in Texas. Find $P(T)$.
c. Let B be the event the company is headquartered in one of these six states. Find $P(B)$.

21. A survey of 50 students at Tarpon Springs College about the number of extracurricular activities resulted in the data shown.

Number of Activities	Frequency
0	8
1	20
2	12
3	6
4	3
5	1

a. Let A be the event that a student participates in at least 1 activity. Find $P(A)$.
b. Let B be the event that a student participates in 3 or more activities. Find $P(B)$.
c. What is the probability that a student participates in exactly 2 activities?

4.3 SOME BASIC RELATIONSHIPS OF PROBABILITY

Complement of an Event

Given an event A, the **complement of A** is defined to be the event consisting of all sample points that are *not* in A. The complement of A is denoted by A^c. Figure 4.4 is a diagram, known as a **Venn diagram,** which illustrates the concept of a complement. The rectangular area represents the sample space for the experiment and as such contains all possible sample points. The circle represents event A and contains only the sample points that belong to A. The shaded region of the rectangle contains all sample points not in event A, and is by definition the complement of A.

In any probability application, either event A or its complement A^c must occur. Therefore, we have

$$P(A) + P(A^c) = 1$$

Solving for $P(A)$, we obtain the following result.

Computing Probability Using the Complement

$$P(A) = 1 - P(A^c) \tag{4.5}$$

Equation (4.5) shows that the probability of an event A can be computed easily if the probability of its complement, $P(A^c)$, is known.

As an example, consider the case of a sales manager who, after reviewing sales reports, states that 80% of new customer contacts result in no sale. By allowing A to denote the event of a sale and A^c to denote the event of no sale, the manager is stating that $P(A^c) = .80$. Using equation (4.5), we see that

$$P(A) = 1 - P(A^c) = 1 - .80 = .20$$

We can conclude that a new customer contact has a .20 probability of resulting in a sale.

In another example, a purchasing agent states a .90 probability that a supplier will send a shipment that is free of defective parts. Using the complement, we can conclude that there is a $1 - .90 = .10$ probability that the shipment will contain defective parts.

FIGURE 4.4 COMPLEMENT OF EVENT A

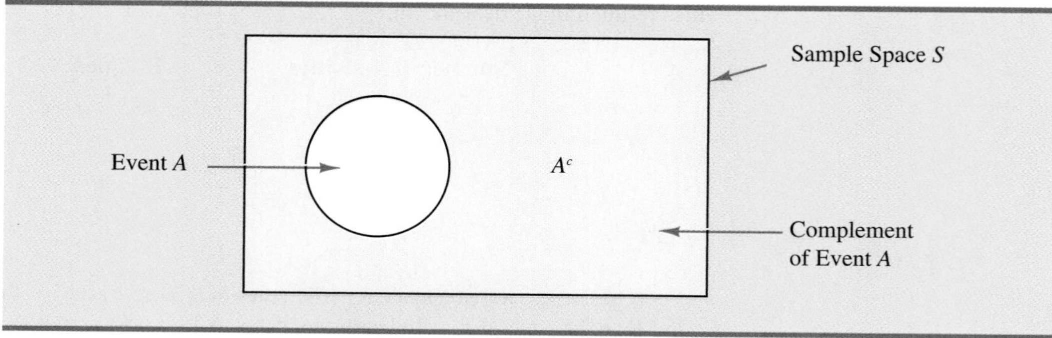

Addition Law

The addition law is helpful when we have two events and are interested in knowing the probability that at least one of the events occurs. That is, with events A and B we are interested in knowing the probability that event A or event B or both occur.

Before we present the addition law, we need to discuss two concepts related to the combination of events: the *union* of events and the *intersection* of events. Given two events A and B, the **union of A and B** is defined as follows.

Union of Two Events

The *union* of A and B is the event containing *all* sample points belonging to A *or* B *or both.* The union is denoted by $A \cup B$.

The union of an event and its complement is the entire sample space.

The Venn diagram in Figure 4.5 depicts the union of events A and B. Note that the two circles contain all the sample points in event A as well as all the sample points in event B. The fact that the circles overlap indicates that some sample points are contained in both A and B.

The definition of the **intersection of two events A and B** follows.

Intersection of Two Events

Given two events A and B, the *intersection* of A and B is the event containing the sample points belonging to *both A and B*. The intersection is denoted by $A \cap B$.

The Venn diagram depicting the intersection of the two events is shown in Figure 4.6. The area where the two circles overlap is the intersection; it contains the sample points that are in both A and B.

Let us now continue with a discussion of the addition law. The **addition law** provides a way to compute the probability of event A or B or both A and B occurring. In other words, the addition law is used to compute the probability of the union of two events, $A \cup B$. The addition law is written as follows.

Addition Law

$$P(A \cup B) = P(A) + P(B) - P(A \cap B) \tag{4.6}$$

FIGURE 4.5 UNION OF EVENTS A AND B

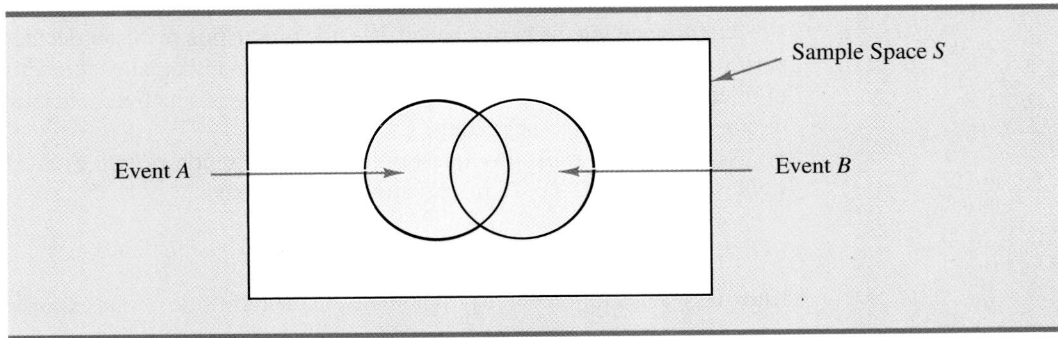

FIGURE 4.6 INTERSECTION OF EVENTS A AND B

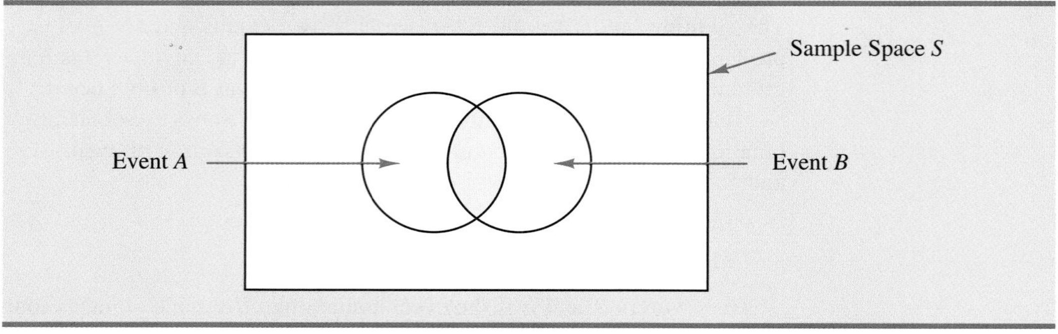

To understand the addition law intuitively, note that the first two terms in the addition law, $P(A) + P(B)$, account for all the sample points in $A \cup B$. However, because the sample points in the intersection $A \cap B$ are in both A and B, when we compute $P(A) + P(B)$, we are in effect counting each of the sample points in $A \cap B$ twice. We correct for this overcounting by subtracting $P(A \cap B)$.

As an example of an application of the addition law, let us consider the case of a small assembly plant with 50 employees. Each worker is expected to complete work assignments on time and in such a way that the assembled product will pass a final inspection. On occasion, some of the workers fail to meet the performance standards by completing work late and/or assembling a defective product. At the end of a performance evaluation period, the production manager found that 5 of the 50 workers had completed work late, 6 of the 50 workers had assembled a defective product, and 2 of the 50 workers had both completed work late *and* assembled a defective product.

Let

$$L = \text{the event that the work is completed late}$$
$$D = \text{the event that the assembled product is defective}$$

The relative frequency information leads to the following probabilities.

$$P(L) = \frac{5}{50} = .10$$

$$P(D) = \frac{6}{50} = .12$$

$$P(L \cap D) = \frac{2}{50} = .04$$

After reviewing the performance data, the production manager decided to assign a poor performance rating to any employee whose work was either late or defective; thus the event of interest is $L \cup D$. What is the probability that the production manager assigned an employee a poor performance rating?

Note that the probability question is about the union of two events. Specifically, we want to know $P(L \cup D)$. Using equation (4.6), we have

$$P(L \cup D) = P(L) + P(D) - P(L \cap D)$$

Knowing values for the three probabilities on the right side of this expression, we can write

$$P(L \cup D) = .10 + .12 - .04 = .18$$

This calculation indicates a .18 probability that a randomly selected employee received a poor performance rating.

As another example of the addition law, consider a recent study conducted by the personnel manager of a major computer software company. It was found that 30% of the employees who left the firm within two years did so primarily because they were dissatisfied with their salary, 20% left because they were dissatisfied with their work assignments, and 12% of the former employees indicated dissatisfaction with *both* their salary and their work assignments. What is the probability that an employee who leaves within two years does so because of dissatisfaction with salary, dissatisfaction with the work assignment, or both?

Let

$$S = \text{the event that the employee leaves because of salary}$$
$$W = \text{the event that the employee leaves because of work assignment}$$

We have $P(S) = .30$, $P(W) = .20$, and $P(S \cap W) = .12$. Using equation (4.6), the addition law, we have

$$P(S \cup W) = P(S) + P(W) - P(S \cap W) = .30 + .20 - .12 = .38.$$

We find a .38 probability that an employee leaves for salary or work assignment reasons.

Before we conclude our discussion of the addition law, let us consider a special case that arises for **mutually exclusive events.**

Mutually Exclusive Events

Two events are said to be *mutually exclusive* if the events have no sample points in common.

Events A and B are mutually exclusive if, when one event occurs, the other cannot occur. Thus, a requirement for A and B to be mutually exclusive is that their intersection must contain no sample points. The Venn diagram depicting two mutually exclusive events A and B is shown in Figure 4.7. In this case $P(A \cap B) = 0$ and the addition law can be written as follows.

Addition Law for Mutually Exclusive Events

$$P(A \cup B) = P(A) + P(B)$$

FIGURE 4.7 MUTUALLY EXCLUSIVE EVENTS

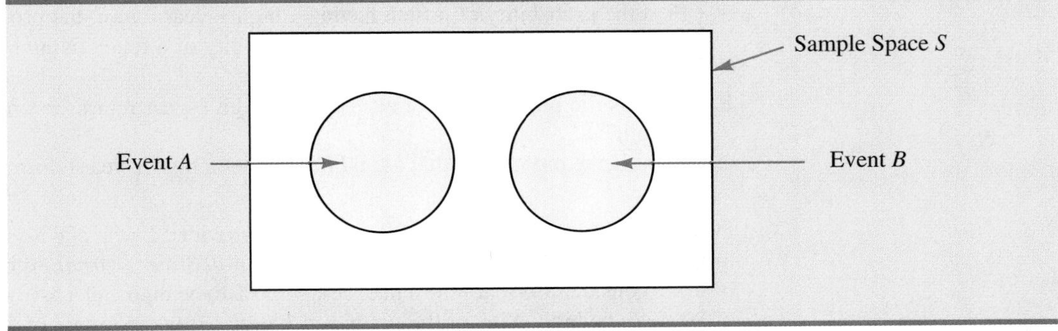

EXERCISES

Methods

22. Suppose that we have a sample space with five equally likely experimental outcomes: E_1, E_2, E_3, E_4, E_5. Let

$$A = \{E_1, E_2\}$$
$$B = \{E_3, E_4\}$$
$$C = \{E_2, E_3, E_5\}$$

   a. Find $P(A)$, $P(B)$, and $P(C)$.
   b. Find $P(A \cup B)$. Are A and B mutually exclusive?
   c. Find A^c, C^c, $P(A^c)$, and $P(C^c)$.
   d. Find $A \cup B^c$ and $P(A \cup B^c)$.
   e. Find $P(B \cup C)$.

23. Suppose that we have a sample space $S = \{E_1, E_2, E_3, E_4, E_5, E_6, E_7\}$, where $E_1, E_2, \ldots, E_7$ denote the sample points. The following probability assignments apply: $P(E_1) = .05$, $P(E_2) = .20$, $P(E_3) = .20$, $P(E_4) = .25$, $P(E_5) = .15$, $P(E_6) = .10$, and $P(E_7) = .05$. Let

$$A = \{E_1, E_4, E_6\}$$
$$B = \{E_2, E_4, E_7\}$$
$$C = \{E_2, E_3, E_5, E_7\}$$

   a. Find $P(A)$, $P(B)$, and $P(C)$.
   b. Find $A \cup B$ and $P(A \cup B)$.
   c. Find $A \cap B$ and $P(A \cap B)$.
   d. Are events A and C mutually exclusive?
   e. Find B^c and $P(B^c)$.

Applications

24. According to a *Business Week*/Harris poll, 14% of adults felt the chance of a big crash in the stock market during 1998 was very likely, and 43% felt the chance of a big crash during 1998 was somewhat likely (*Business Week,* December 29, 1997). If we were to choose an adult at random, what is the probability that she or he would indicate a crash was not likely?

25. Data on the 30 largest bond funds provided 1-year and 5-year percentage returns for the period ending March 31, 2000 (*The Wall Street Journal,* April 10, 2000). Suppose we consider a 1-year return in excess of 2% to be high and a 5-year return in excess of 44% to be high. One half of the funds had a 1-year return in excess of 2%, 12 of the funds had a 5-year return in excess of 44%, and six of the funds had a 1-year return in excess of 2% and a 5-year return in excess of 44%.
   a. Find the probability of a fund having a high 1-year return, the probability of a fund having a high 5-year return, and the probability of a fund having both a high 1-year return and a high 5-year return.
   b. What is the probability that a fund has a high 1-year return or a high 5-year return or both?
   c. What is the probability that a fund has neither a high 1-year return nor a high 5-year return?

26. Data on the 30 largest stock and balanced funds provided 1-year and 5-year percentage returns for the period ending March 31, 2000 (*The Wall Street Journal,* April 10, 2000). Suppose we consider a 1-year return in excess of 50% to be high and a 5-year return in excess of 300% to be high. Nine of the funds had 1-year returns in excess of 50%, seven of the

funds had 5-year returns in excess of 300%, and five of the funds had both 1-year returns in excess of 50% and 5-year returns in excess of 300%.

a. What is the probability of a high 1-year return and what is the probability of a high 5-year return?

b. What is the probability of both a high 1-year return and a high 5-year return?

c. What is the probability of neither a high 1-year return nor a high 5-year return?

27. *Business Week* surveyed its subscribers concerning whether they consumed or served wine during an average week. The findings were that 57% consumed or served domestic wine, 33% consumed or served imported wine, and 63% consumed or served wine (*Business Week* 1996 Worldwide Subscriber Study). What is the probability that a *Business Week* subscriber consumes or serves both domestic and imported wine in an average week?

28. A survey of magazine subscribers showed that 45.8% rented a car during the past 12 months for business reasons, 54% rented a car during the past 12 months for personal reasons, and 30% rented a car during the past 12 months for both business and personal reasons.

a. What is the probability that a subscriber rented a car during the past 12 months for business or personal reasons?

b. What is the probability that a subscriber did not rent a car during the past 12 months for either business or personal reasons?

29. According to the Census Bureau, deaths in the United States occur at a rate of 2,425,000 per year. The National Center for Health Statistics reported that the three leading causes of death during 1997 were heart disease (725,790), cancer (537,390), and stroke (159,877). Let H, C, and S represent the events that a person dies of heart disease, cancer, and stroke, respectively.

a. Use the data to estimate $P(H)$, $P(C)$, and $P(S)$.

b. Are events H and C mutually exclusive? Find $P(H \cap C)$.

c. What is the probability that a person dies from heart disease or cancer?

d. What is the probability that a person dies from cancer or a stroke?

e. Find the probability that someone dies from a cause other than one of these three.

4.4 CONDITIONAL PROBABILITY

Often, the probability of an event is influenced by whether a related event has occurred. Suppose we have an event A with probability $P(A)$. If we obtain new information and learn that a related event, denoted by B, has occurred, we will want to take advantage of this information by calculating a new probability for event A. This new probability of event A is called a **conditional probability** and is written $P(A \mid B)$. The notation $\mid$ is used to denote the fact that we are considering the probability of event A *given* the condition that event B has occurred. Hence, the notation $P(A \mid B)$ is read "the probability of A given B."

As an illustration of the application of conditional probability, consider the situation of the promotion status of male and female officers of a major metropolitan police force in the eastern United States. The police force consists of 1200 officers, 960 men and 240 women. Over the past two years, 324 officers on the police force have been awarded promotions. The specific breakdown of promotions for male and female officers is shown in Table 4.4.

TABLE 4.4 PROMOTION STATUS OF POLICE OFFICERS OVER THE PAST TWO YEARS

	Man	Woman	Totals
Promoted	288	36	324
Not Promoted	672	204	876
Totals	960	240	1200

After reviewing the promotion record, a committee of female officers raised a discrimination case on the basis that 288 male officers had received promotions but only 36 female officers had received promotions. The police administration argued that the relatively low number of promotions for female officers was due not to discrimination, but to the fact that there are relatively few female officers on the police force. Let us show how conditional probability could be used to analyze the discrimination charge.

Let

$$M = \text{event an officer is a man}$$
$$W = \text{event an officer is a woman}$$
$$A = \text{event an officer is promoted}$$
$$A^c = \text{event an officer is not promoted}$$

Dividing the data values in Table 4.4 by the total of 1200 officers enables us to summarize the available information with the following probability values.

$P(M \cap A) = 288/1200 = .24 =$ probability that a randomly selected officer is a man *and* is promoted

$P(M \cap A^c) = 672/1200 = .56 =$ probability that a randomly selected officer is a man *and* is not promoted

$P(W \cap A) = 36/1200 = .03 =$ probability that a randomly selected officer is a woman *and* is promoted

$P(W \cap A^c) = 204/1200 = .17 =$ probability that a randomly selected officer is a woman *and* is not promoted

Because each of these values gives the probability of the intersection of two events, the probabilities are called **joint probabilities.** Table 4.5, which provides a summary of the probability information for the police officer promotion situation, is referred to as a *joint probability table.*

The values in the margins of the joint probability table provide the probabilities of each event separately. That is, $P(M) = .80$, $P(W) = .20$, $P(A) = .27$, and $P(A^c) = .73$. These probabilities are referred to as *marginal probabilities* because of their location in the margins of the joint probability table. We note that the marginal probabilities are found by summing the joint probabilities in the corresponding row or column of the joint probability table. For instance, the marginal probability of being promoted is $P(A) = P(M \cap A) + P(W \cap A) = .24 + .03 = .27$. From the marginal probabilities, we see that 80% of the force is male, 20% of the force is female, 27% of all officers received promotions, and 73% were not promoted.

TABLE 4.5 JOINT PROBABILITY TABLE FOR PROMOTIONS

Joint probabilities appear in the body of the table.

	Man (M)	Woman (W)	Totals
Promoted (A)	.24	.03	.27
Not Promoted (A^c)	.56	.17	.73
Totals	.80	.20	1.00

Marginal probabilities appear in the margins of the table.

Let us begin the conditional probability analysis by computing the probability that an officer is promoted given that the officer is a man. In conditional probability notation, we are attempting to determine $P(A \mid M)$. To calculate $P(A \mid M)$, we first realize that this notation simply means that we are considering the probability of the event A (promotion) given that the condition designated as event M (the officer is a man) is known to exist. Thus $P(A \mid M)$ tells us that we are now concerned only with the promotion status of the 960 male officers. Because 288 of the 960 male officers received promotions, the probability of being promoted given that the officer is a man is 288/960 = .30. In other words, given that an officer is a man, that officer had a 30% chance of receiving a promotion over the past two years.

This procedure was easy to apply because the values in Table 4.4 show the number of officers in each category. We now want to demonstrate how conditional probabilities such as $P(A \mid M)$ can be computed directly from related event probabilities rather than the frequency data of Table 4.4.

We have shown that $P(A \mid M) = 288/960 = .30$. Let us now divide both the numerator and denominator of this fraction by 1200, the total number of officers in the study.

$$P(A \mid M) = \frac{288}{960} = \frac{288/1200}{960/1200} = \frac{.24}{.80} = .30$$

We now see that the conditional probability $P(A \mid M)$ can be computed as .24/.80. Refer to the joint probability table (Table 4.5). Note in particular that .24 is the joint probability of A and M; that is, $P(A \cap M) = .24$. Also note that .80 is the marginal probability that a randomly selected officer is a man; that is, $P(M) = .80$. Thus, the conditional probability $P(A \mid M)$ can be computed as the ratio of the joint probability $P(A \cap M)$ to the marginal probability $P(M)$.

$$P(A \mid M) = \frac{P(A \cap M)}{P(M)} = \frac{.24}{.80} = .30$$

The fact that conditional probabilities can be computed as the ratio of a joint probability to a marginal probability provides the following general formula for conditional probability calculations for two events A and B.

Conditional Probability

$$P(A \mid B) = \frac{P(A \cap B)}{P(B)} \tag{4.7}$$

or

$$P(B \mid A) = \frac{P(A \cap B)}{P(A)} \tag{4.8}$$

The Venn diagram in Figure 4.8 is helpful in obtaining an intuitive understanding of conditional probability. The circle on the right shows that event B has occurred; the portion of the circle that overlaps with event A denotes the event $(A \cap B)$. We know that once event B has occurred, the only way that we can also observe event A is for the event $(A \cap B)$ to occur. Thus, the ratio $P(A \cap B)/P(B)$ provides the conditional probability that we will observe event A given event B has already occurred.

FIGURE 4.8 CONDITIONAL PROBABILITY

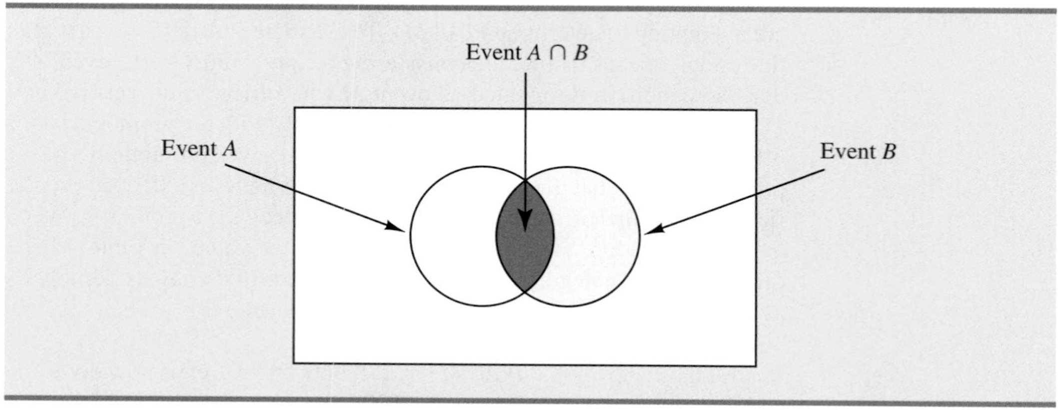

Let us return to the issue of discrimination against the female officers. The marginal probability in row 1 of Table 4.5 shows that the probability of promotion of an officer is $P(A) = .27$ (regardless of whether that officer is male or female). However, the critical issue in the discrimination case involves the two conditional probabilities $P(A \mid M)$ and $P(A \mid W)$. That is, what is the probability of a promotion *given* that the officer is a man, and what is the probability of a promotion given that the officer is a woman? If these two probabilities are equal, a discrimination argument has no basis because the chances of a promotion are the same for male and female officers. However, a difference in the two conditional probabilities will support the position that male and female officers are treated differently in promotion decisions.

We have already determined that $P(A \mid M) = .30$. Let us now use the probability values in Table 4.5 and the basic relationship of conditional probability in equation (4.7) to compute the probability that an officer is promoted given that the officer is a woman; that is, $P(A \mid W)$. Using equation (4.7), we obtain

$$P(A \mid W) = \frac{P(A \cap W)}{P(W)} = \frac{.03}{.20} = .15$$

What conclusion do you draw? The probability of a promotion given that the officer is a man is .30, twice the .15 probability of a promotion given that the officer is a woman. While the use of conditional probability does not in itself prove that discrimination exists in this case, the conditional probability values support the argument presented by the female officers.

Independent Events

In the preceding illustration, $P(A) = .27$, $P(A \mid M) = .30$, and $P(A \mid W) = .15$. We see that the probability of a promotion (event A) is affected or influenced by whether the officer is a man or a woman. Particularly, because $P(A \mid M) \neq P(A)$, we would say that events A and M are dependent events. That is, the probability of event A (promotion) is altered or affected by knowing event M (the officer is a man) has occurred. Similarly, with $P(A \mid W) \neq P(A)$, we would say that events A and W are *dependent events*. However, if the probability of event A is not changed by the occurrence of event M—that is, $P(A \mid M) = P(A)$—we would say that events A and M are **independent events**. This situation leads to the following definition of the independence of two events.

Independent Events

Two events A and B are independent if

$$P(A \mid B) = P(A) \tag{4.9}$$

or

$$P(B \mid A) = P(B) \tag{4.10}$$

Otherwise, the events are dependent.

Multiplication Law

Whereas the addition law of probability is used to compute the probability of a union of two events, the multiplication law is used to compute the probability of an intersection of two events. The multiplication law is based on the definition of conditional probability. Using equations (4.7) and (4.8) and solving for $P(A \cap B)$, we obtain the **multiplication law.**

Multiplication Law

$$P(A \cap B) = P(B)P(A \mid B) \tag{4.11}$$

or

$$P(A \cap B) = P(A)P(B \mid A) \tag{4.12}$$

To illustrate the use of the multiplication law, consider a newspaper circulation department where it is known that 84% of the households in a particular neighborhood subscribe to the daily edition of the paper. If we let D denote the event that a household subscribes to the daily edition, $P(D) = .84$. In addition, it is known that the probability that a household who already holds a daily subscription also subscribes to the Sunday edition (event S) is .75; that is, $P(S \mid D) = .75$. What is the probability that a household subscribes to both the Sunday and daily editions of the newspaper? Using the multiplication law, we compute the desired $P(S \cap D)$ as

$$P(S \cap D) = P(D)P(S \mid D) = .84(.75) = .63$$

We now know that 63% of the households subscribe to both the Sunday and daily editions.

Before concluding this section, let us consider the special case of the multiplication law when the events involved are independent. Recall that events A and B are independent whenever $P(A \mid B) = P(A)$ or $P(B \mid A) = P(B)$. Hence, using equations (4.11) and (4.12) for the special case of independent events, we obtain the following multiplication law.

Multiplication Law for Independent Events

$$P(A \cap B) = P(A)P(B) \tag{4.13}$$

To compute the probability of the intersection of two independent events, we simply multiply the corresponding probabilities. Note that the multiplication law for independent events provides another way to determine whether A and B are independent. That is, if $P(A \cap B) = P(A)P(B)$, then A and B are independent; if $P(A \cap B) \neq P(A)P(B)$, then A and B are dependent.

As an application of the multiplication law for independent events, consider the situation of a service station manager who knows from past experience that 80% of the customers use a credit card when they purchase gasoline. What is the probability that the next two customers purchasing gasoline will each use a credit card? If we let

A = the event that the first customer uses a credit card
B = the event that the second customer uses a credit card

then the event of interest is $A \cap B$. Given no other information, we can reasonably assume that A and B are independent events. Thus,

$$P(A \cap B) = P(A)P(B) = (.80)(.80) = .64$$

To summarize this section, we note that our interest in conditional probability is motivated by the fact that events are often related. In such cases, we say the events are dependent and the conditional probability formulas in equations (4.7) and (4.8) must be used to compute the event probabilities. If two events are not related they are independent and neither event's probability is affected by whether the other event occurred.

NOTES AND COMMENTS

Do not confuse the notion of mutually exclusive events with that of independent events. Two events with nonzero probabilities cannot be both mutually exclusive and independent. If one mutually exclusive event is known to occur, the other cannot occur; thus, the probability of the other event occurring is reduced to zero. They are therefore dependent.

EXERCISES

Methods

30. Suppose that we have two events, A and B, with $P(A) = .50$, $P(B) = .60$, and $P(A \cap B) = .40$.

a. Find $P(A \mid B)$.
b. Find $P(B \mid A)$.
c. Are A and B independent? Why or why not?

31. Assume that we have two events, A and B, that are mutually exclusive. Assume further that we know $P(A) = .30$ and $P(B) = .40$.

a. What is $P(A \cap B)$?
b. What is $P(A \mid B)$?
c. A student in statistics argues that the concepts of mutually exclusive events and independent events are really the same, and that if events are mutually exclusive they must be independent. Do you agree with this statement? Use the probability information in this problem to justify your answer.
d. What general conclusion would you make about mutually exclusive and independent events given the results of this problem?

Applications

32. A Daytona Beach nightclub has the following data on the age and marital status of 140 customers.

		Marital Status	
		Single	Married
Age	Under 30	77	14
	30 or Over	28	21

a. Develop a joint probability table for these data.
b. Use the marginal probabilities to comment on the age of customers attending the club.
c. Use the marginal probabilities to comment on the marital status of customers attending the club.
d. What is the probability of finding a customer who is single and under the age of 30?
e. If a customer is under 30, what is the probability that he or she is single?
f. Is marital status independent of age? Explain, using probabilities.

SELF test

33. In a survey of MBA students, the following data were obtained on "students' first reason for application to the school in which they matriculated."

		Reason for Application			
		School Quality	School Cost or Convenience	Other	Totals
Enrollment Status	Full Time	421	393	76	890
	Part Time	400	593	46	1039
	Totals	821	986	122	1929

a. Develop a joint probability table for these data.
b. Use the marginal probabilities of school quality, cost/convenience, and other to comment on the most important reason for choosing a school.
c. If a student goes full time, what is the probability that school quality is the first reason for choosing a school?
d. If a student goes part time, what is the probability that school quality is the first reason for choosing a school?
e. Let *A* denote the event that a student is full time and let *B* denote the event that the student lists school quality as the first reason for applying. Are events *A* and *B* independent? Justify your answer.

34. The following table shows the distribution of blood types in the general population (Hoxworth Blood Center, Cincinnati, Ohio).

	A	B	AB	O
Rh+	34%	9%	4%	38%
Rh−	6%	2%	1%	6%

a. What is the probability a person will have type O blood?
b. What is the probability a person will be Rh−?
c. What is the probability a married couple will both be Rh−?
d. What is the probability a married couple will both have type AB blood?
e. What is the probability a person will be Rh− given she or he has type O blood?
f. What is the probability a person will have type B blood given he or she is Rh+?

35. "Since 1950, the January Barometer has predicted the annual course of the stock market with amazing accuracy" (*1998 Stock Trader's Almanac*). Over the 48 years from 1950 through 1997, the stock market has been up in January 31 times; it has been up for the year 36 times; and it has been up for the year and up for January 29 times.

a. Estimate the probability the stock market will be up in January.
b. Estimate the probability the stock market will be up for the year.
c. What is the probability the stock market will be up for the year given it is up in January?
d. Do the probabilities suggest that the stock market's January performance and its annual performance are independent events? Explain.

36. A study of job satisfaction was conducted for four occupations: cabinetmaker, lawyer, physical therapist, and systems analyst. Job satisfaction was measured on a scale of 0–100. The data obtained are summarized in the following crosstabulation.

	Satisfaction Score				
Occupation	**Under 50**	**50–59**	**60–69**	**70–79**	**80–89**
Cabinetmaker	0	2	4	3	1
Lawyer	6	2	1	1	0
Physical Therapist	0	5	2	1	2
Systems Analyst	2	1	4	3	0

a. Develop a joint probability table.
b. What is the probability one of the participants studied had a satisfaction score in the 80s?
c. What is the probability of a satisfaction score in the 80s given the study participant was a physical therapist?
d. What is the probability one of the participants studied was a lawyer?
e. What is the probability one of the participants was a lawyer and received a score under 50?
f. What is the probability of a satisfaction score under 50 given a person is a lawyer?
g. What is the probability of a satisfaction score of 70 or higher?

37. A purchasing agent has placed rush orders for a particular raw material with two different suppliers, A and B. If neither order arrives in 4 days, the production process must be shut down until at least one of the orders arrives. The probability that supplier A can deliver the material in 4 days is .55. The probability that supplier B can deliver the material in 4 days is .35.

a. What is the probability that both suppliers will deliver the material in 4 days? Because two separate suppliers are involved, we are willing to assume independence.
b. What is the probability that at least one supplier will deliver the material in 4 days?
c. What is the probability that the production process will be shut down in 4 days because of a shortage of raw material (that is, both orders are late)?

38. A survey of 1035 workers by the Institute for the Future and the Gallup Organization found that workers are being inundated by messages (*The Cincinnati Enquirer,* November 2, 1998). The study indicated that each worker gets an average of 190 messages per day. The following table shows the breakdown by type of message.

Source	Daily Messages	Source	Daily Messages
Telephone	52	E-mail	30
Voice mail	22	Interoffice mail	18
U.S. mail	18	Fax	15
Post-it note	11	Phone msg. slip	10
Pager	4	Overnight courier	4
Cellular phone	3	U.S. Express mail	3

a. For any particular worker, what is the probability the next message will be on the telephone?

b. For the next two messages, what is the probability the first will be via e-mail and the second will be via fax?

c. What is the probability the next message received will be via telephone call or inter-office mail?

4.5 BAYES' THEOREM

In the discussion of conditional probability, we indicated that revising probabilities when new information is obtained is an important phase of probability analysis. Often, we begin the analysis with initial or **prior probability** estimates for specific events of interest. Then, from sources such as a sample, a special report, or a product test, we obtain additional information about the events. Given this new information, we update the prior probability values by calculating revised probabilities, referred to as **posterior probabilities. Bayes' theorem** provides a means for making these probability calculations. The steps in this probability revision process are shown in Figure 4.9.

As an application of Bayes' theorem, consider a manufacturing firm that receives shipments of parts from two different suppliers. Let A_1 denote the event that a part is from supplier 1 and A_2 denote the event that a part is from supplier 2. Currently, 65% of the parts purchased by the company are from supplier 1 and the remaining 35% are from supplier 2. Hence, if a part is selected at random, we would assign the prior probabilities $P(A_1) = .65$ and $P(A_2) = .35$.

The quality of the purchased parts varies with the source of supply. Historical data suggest that the quality ratings of the two suppliers are as shown in Table 4.6. If we let G denote the event that a part is good and B denote the event that a part is bad, the information in Table 4.6 provides the following conditional probability values.

$$P(G \mid A_1) = .98 \quad P(B \mid A_1) = .02$$
$$P(G \mid A_2) = .95 \quad P(B \mid A_2) = .05$$

The tree diagram in Figure 4.10 depicts the process of the firm receiving a part from one of the two suppliers and then discovering that the part is good or bad as a two-step

FIGURE 4.9 PROBABILITY REVISION USING BAYES' THEOREM

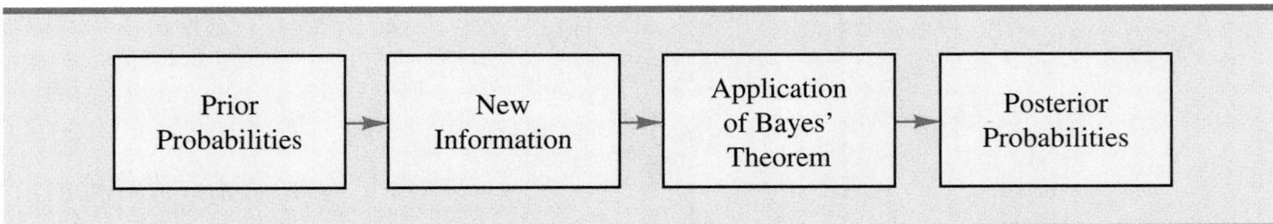

TABLE 4.6 HISTORICAL QUALITY LEVELS OF TWO SUPPLIERS

	Percentage Good Parts	Percentage Bad Parts
Supplier 1	98	2
Supplier 2	95	5

FIGURE 4.10 TWO-STEP TREE DIAGRAM

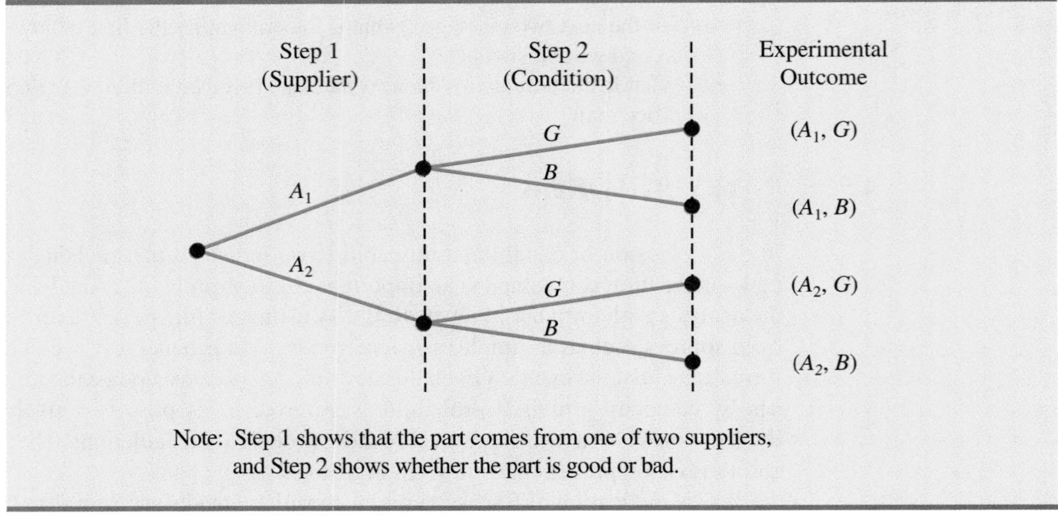

Note: Step 1 shows that the part comes from one of two suppliers, and Step 2 shows whether the part is good or bad.

experiment. We see that four experimental outcomes are possible; two correspond to the part being good and two correspond to the part being bad.

Each of the experimental outcomes is the intersection of two events, so we can use the multiplication rule to compute the probabilities. For instance,

$$P(A_1, G) = P(A_1 \cap G) = P(A_1)P(G \mid A_1)$$

The process of computing these joint probabilities can be depicted in what is called a probability tree (see Figure 4.11). From left to right through the tree, the probabilities for each branch at step 1 are prior probabilities and the probabilities for each branch at step 2 are conditional probabilities. To find the probabilities of each experimental outcome, we

FIGURE 4.11 PROBABILITY TREE FOR TWO-SUPPLIER EXAMPLE

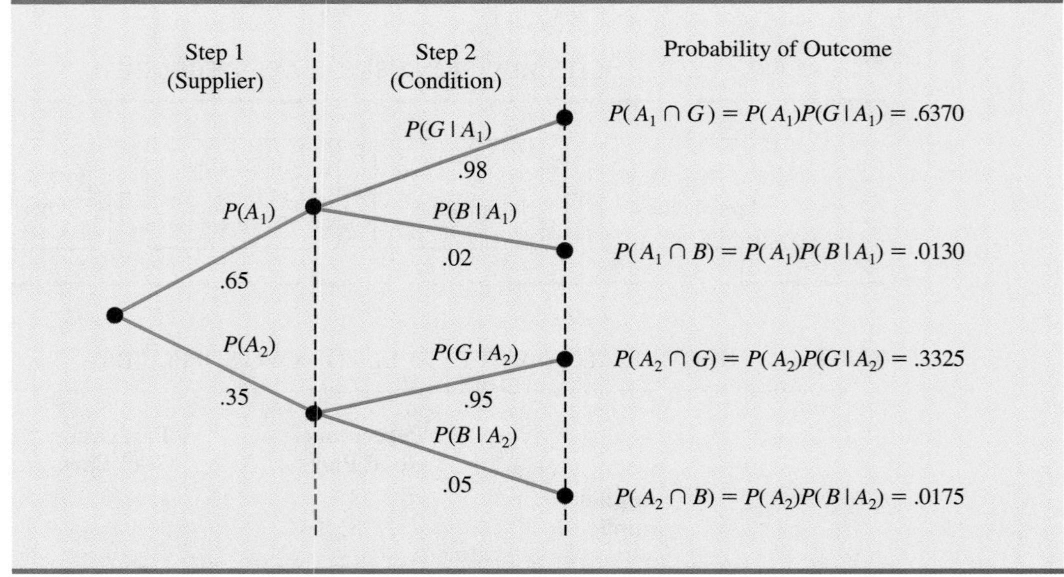

simply multiply the probabilities on the branches leading to the outcome. Each of these joint probabilities is shown in Figure 4.11 along with the known probabilities for each branch.

Suppose now that the parts from the two suppliers are used in the firm's manufacturing process and that a machine breaks down because it attempts to process a bad part. Given the information that the part is bad, what is the probability that it came from supplier 1 and what is the probability that it came from supplier 2? With the information in the probability tree (Figure 4.11), Bayes' theorem can be used to answer these questions.

Letting B denote the event that the part is bad, we are looking for the posterior probabilities $P(A_1 \mid B)$ and $P(A_2 \mid B)$. From the law of conditional probability, we know that

$$P(A_1 \mid B) = \frac{P(A_1 \cap B)}{P(B)} \tag{4.14}$$

Referring to the probability tree, we see that

$$P(A_1 \cap B) = P(A_1)P(B \mid A_1) \tag{4.15}$$

To find $P(B)$, we note that event B can occur in only two ways: $(A_1 \cap B)$ and $(A_2 \cap B)$. Therefore, we have

$$\begin{aligned} P(B) &= P(A_1 \cap B) + P(A_2 \cap B) \\ &= P(A_1)P(B \mid A_1) + P(A_2)P(B \mid A_2) \end{aligned} \tag{4.16}$$

Substituting from equations (4.15) and (4.16) into equation (4.14) and writing a similar result for $P(A_2 \mid B)$, we obtain Bayes' theorem for the case of two events.

The Reverend Thomas Bayes (1702–1761), a Presbyterian minister, is credited with the original work leading to the version of Bayes' theorem in use today.

Bayes' Theorem (Two-Event Case)

$$P(A_1 \mid B) = \frac{P(A_1)P(B \mid A_1)}{P(A_1)P(B \mid A_1) + P(A_2)P(B \mid A_2)} \tag{4.17}$$

$$P(A_2 \mid B) = \frac{P(A_2)P(B \mid A_2)}{P(A_1)P(B \mid A_1) + P(A_2)P(B \mid A_2)} \tag{4.18}$$

Using equation (4.17) and the probability values provided in the example, we have

$$\begin{aligned} P(A_1 \mid B) &= \frac{P(A_1)P(B \mid A_1)}{P(A_1)P(B \mid A_1) + P(A_2)P(B \mid A_2)} \\ &= \frac{(.65)(.02)}{(.65)(.02) + (.35)(.05)} = \frac{.0130}{.0130 + .0175} \\ &= \frac{.0130}{.0305} = .4262 \end{aligned}$$

In addition, using equation (4.18), we find $P(A_2 \mid B)$.

$$\begin{aligned} P(A_2 \mid B) &= \frac{(.35)(.05)}{(.65)(.02) + (.35)(.05)} \\ &= \frac{.0175}{.0130 + .0175} = \frac{.0175}{.0305} = .5738 \end{aligned}$$

Note that in this application we started with a probability of .65 that a part selected at random was from supplier 1. However, given information that the part is bad, the probability that the part is from supplier 1 drops to .4262. In fact, if the part is bad, it has better than a 50–50 chance that it came from supplier 2; that is, $P(A_2 \mid B) = .5738$.

Bayes' theorem is applicable when the events for which we want to compute posterior probabilities are mutually exclusive and their union is the entire sample space.* Bayes' theorem can be extended to the case of n mutually exclusive events $A_1, A_2, \ldots, A_n$, whose union is the entire sample space. In such a case, Bayes' theorem for the computation of any posterior probability $P(A_i \mid B)$ has the following form.

Bayes' Theorem

$$P(A_i \mid B) = \frac{P(A_i)P(B \mid A_i)}{P(A_1)P(B \mid A_1) + P(A_2)P(B \mid A_2) + \cdots + P(A_n)P(B \mid A_n)} \qquad \textbf{(4.19)}$$

With prior probabilities $P(A_1), P(A_2), \ldots, P(A_n)$ and the appropriate conditional probabilities $P(B \mid A_1), P(B \mid A_2), \ldots, P(B \mid A_n)$, equation (4.19) can be used to compute the posterior probability of the events $A_1, A_2, \ldots, A_n$.

Tabular Approach

A tabular approach is helpful in conducting the Bayes' theorem calculations. Such an approach is shown in Table 4.7 for the parts supplier problem. The computations shown there are done in the following steps.

Step 1. Prepare the following three columns:
Column 1—The mutually exclusive events for which posterior probabilities are desired.
Column 2—The prior probabilities for the events.
Column 3—The conditional probabilities of the new information *given* each event.

Step 2. In column 4, compute the joint probabilities for each event and the new information B by using the multiplication law. These joint probabilities are found by multiplying the prior probabilities in column 2 by the corresponding conditional probabilities in column 3; that is, $P(A_i \cap B) = P(A_i)P(B \mid A_i)$.

Step 3. Sum the joint probabilities in column 4. The sum is the probability of the new information, $P(B)$. Thus we see in Table 4.7 that there is a .0130 probability of a bad part and supplier 1 and a .0175 probability of a bad part and supplier 2. Because these are the only two ways in which a bad part can be obtained, the sum .0130 + .0175 shows an overall probability of .0305 of finding a bad part from the combined shipments of the two suppliers.

Step 4. In column 5, compute the posterior probabilities using the basic relationship of conditional probability.

$$P(A_i \mid B) = \frac{P(A_i \cap B)}{P(B)}$$

Note that the joint probabilities $P(A_i \cap B)$ are in column 4 and the probability $P(B)$ is the sum of column 4.

*If the union of events is the entire sample space, the events are said to be *collectively exhaustive*.

TABLE 4.7 SUMMARY OF BAYES' THEOREM CALCULATIONS FOR THE TWO-SUPPLIER PROBLEM

(1) Events A_i	(2) Prior Probabilities $P(A_i)$	(3) Conditional Probabilities $P(B \mid A_i)$	(4) Joint Probabilities $P(A_i \cap B)$	(5) Posterior Probabilities $P(A_i \mid B)$
A_1	.65	.02	.0130	.0130/.0305 = .4262
A_2	.35	.05	.0175	.0175/.0305 = .5738
	1.00		$P(B)$ = .0305	1.0000

NOTES AND COMMENTS

1. Bayes' theorem is used extensively in decision analysis. The prior probabilities are often subjective estimates provided by a decision maker. Sample information is obtained and posterior probabilities are computed for use in choosing the best decision.
2. An event and its complement are mutually exclusive, and their union is the entire sample space. Thus, Bayes' theorem is always applicable for computing posterior probabilities of an event and its complement.

EXERCISES

Methods

39. The prior probabilities for events A_1 and A_2 are $P(A_1) = .40$ and $P(A_2) = .60$. It is also known that $P(A_1 \cap A_2) = 0$. Suppose $P(B \mid A_1) = .20$ and $P(B \mid A_2) = .05$.
    a. Are A_1 and A_2 mutually exclusive? Explain.
    b. Compute $P(A_1 \cap B)$ and $P(A_2 \cap B)$.
    c. Compute $P(B)$.
    d. Apply Bayes' theorem to compute $P(A_1 \mid B)$ and $P(A_2 \mid B)$.
40. The prior probabilities for events A_1, A_2, and A_3 are $P(A_1) = .20$, $P(A_2) = .50$, and $P(A_3) = .30$. The conditional probabilities of event B given A_1, A_2, and A_3 are $P(B \mid A_1) = .50$, $P(B \mid A_2) = .40$, and $P(B \mid A_3) = .30$.
    a. Compute $P(B \cap A_1)$, $P(B \cap A_2)$, and $P(B \cap A_3)$.
    b. Apply Bayes' theorem, equation (4.19), to compute the posterior probability $P(A_2 \mid B)$.
    c. Use the tabular approach to applying Bayes' theorem to compute $P(A_1 \mid B)$, $P(A_2 \mid B)$, and $P(A_3 \mid B)$.

Applications

41. A consulting firm has submitted a bid for a large research project. The firm's management initially felt they had a 50–50 chance of getting the project. However, the agency to which the bid was submitted has subsequently requested additional information on the bid. Past experience indicates that on 75% of the successful bids and 40% of the unsuccessful bids the agency requested additional information.
    a. What is the prior probability of the bid being successful (that is, prior to the request for additional information)?
    b. What is the conditional probability of a request for additional information given that the bid will ultimately be successful?
    c. Compute a posterior probability that the bid will be successful given that a request for additional information has been received.

42. A local bank is reviewing its credit card policy with a view toward recalling some of its credit cards. In the past approximately 5% of cardholders have defaulted and the bank has been unable to collect the outstanding balance. Hence, management has established a prior probability of .05 that any particular cardholder will default. The bank has further found that the probability of missing one or more monthly payments is .20 for customers who do not default. Of course, the probability of missing one or more payments for those who default is 1.

a. Given that a customer has missed a monthly payment, compute the posterior probability that the customer will default.

b. The bank would like to recall its card if the probability that a customer will default is greater than .20. Should the bank recall its card if the customer misses a monthly payment? Why or why not?

43. Small cars get better gas mileage, but they are not as safe as bigger cars. Small cars accounted for 18% of the vehicles on the road, but accidents involving small cars led to 11,898 fatalities during a recent year (*Reader's Digest,* May, 2000). The probability of an accident involving a small car leading to a fatality is .128 and the probability of an accident not involving a small car leading to a fatality is .05. Suppose you learn of an accident involving a fatality. What is the probability a small car was involved?

44. A city has a professional basketball team playing at home and a professional hockey team playing away on the same night. A professional basketball team has a .641 probability of winning a home game and a professional hockey team has a .462 probability of winning an away game. Historically, when both teams play on the same night, the chance that the next morning's leading sports story will be about the basketball game is 60% and the chance that it will be about the hockey game is 40%. Suppose that on the morning after these games the newspaper's leading sports story begins with the headline "We Win!!" What is the probability that the story is about the basketball team?

45. *M. D. Computing* (May 1991) describes the use of Bayes' theorem and the use of conditional probability in medical diagnosis. Prior probabilities of diseases are based on the physician's assessment of such things as geographical location, seasonal influence, occurrence of epidemics, and so forth. Assume that a patient is believed to have one of two diseases, denoted D_1 and D_2 with $P(D_1) = .60$ and $P(D_2) = .40$ and that medical research has determined the probability associated with each symptom that may accompany the diseases. Suppose that, given diseases D_1 and D_2, the probabilities that the patient will have symptoms S_1, S_2, or S_3 are as follows.

		Symptoms		
		S_1	S_2	S_3
Disease	D_1	.15	.10	.15 ← $P(S_3 \mid D_1)$
	D_2	.80	.15	.03

After a certain symptom is found to be present, the medical diagnosis may be aided by finding the revised probabilities of each particular disease. Compute the posterior probabilities of each disease given the following medical findings.

a. The patient has symptom S_1.

b. The patient has symptom S_2.

c. The patient has symptom S_3.

d. For the patient with symptom S_1 in part (*a*), suppose we also find symptom S_2. What are the revised probabilities of D_1 and D_2?

SUMMARY

In this chapter we introduced basic probability concepts and illustrated how probability analysis can be used to provide helpful information for decision making. We described how probability can be interpreted as a numerical measure of the likelihood that an event will occur. In addition, we saw that the probability of an event can be computed either by summing the probabilities of the experimental outcomes (sample points) comprising the event or by using the relationships established by the addition, conditional probability, and multiplication laws of probability. For cases in which additional information is available, we showed how Bayes' theorem can be used to obtain revised or posterior probabilities.

GLOSSARY

Probability A numerical measure of the likelihood that an event will occur.

Experiment A process that generates well-defined outcomes.

Sample space The set of all experimental outcomes.

Sample point An element of the sample space. A sample point represents an experimental outcome.

Tree diagram A graphical representation helpful in identifying the sample points of an experiment involving multiple steps.

Basic requirements of probability Two requirements that restrict the manner in which probability assignments can be made: (1) for each experimental outcome E_i we must have $0 \leq P(E_i) \leq 1$; (2) considering all experimental outcomes, we must have $P(E_1) + P(E_2) + \cdots + P(E_n) = 1$.

Classical method A method of assigning probabilities that assumes the experimental outcomes are equally likely.

Relative frequency method A method of assigning probabilities on the basis of experimentation or historical data.

Subjective method A method of assigning probabilities on the basis of judgment.

Event A collection of sample points.

Complement of event *A* The event consisting of all sample points that are not in A.

Venn diagram A graphical representation for showing symbolically the sample space and operations involving events. The sample space is represented by a rectangle and events are represented as circles within the sample space.

Union of events *A* and *B* The event containing all sample points that are in A, in B, or in both. The union is denoted $A \cup B$.

Intersection of *A* and *B* The event containing all sample points that are in both A and B. The intersection is denoted $A \cap B$.

Addition law A probability law used to compute the probability of a union of two events, denoted A and B. It is $P(A \cup B) = P(A) + P(B) - P(A \cap B)$. For mutually exclusive events, because $P(A \cap B) = 0$, it reduces to $P(A \cup B) = P(A) + P(B)$.

Mutually exclusive events Events that have no sample points in common; that is, $A \cap B$ is empty and $P(A \cap B) = 0$.

Conditional probability The probability of an event given that another event has occurred. The conditional probability of A given B is $P(A \mid B) = P(A \cap B)/P(B)$.

Joint probability The probability of two events both occurring; that is, the probability of the intersection of two events.

Independent events Two events A and B where $P(A \mid B) = P(A)$ or $P(B \mid A) = P(B)$; that is, the events have no influence on each other.

Multiplication law A probability law used to compute the probability of an intersection of two events, denoted A and B. It is $P(A \cap B) = P(A)P(B \mid A)$ or $P(A \cap B) = P(B)P(A \mid B)$. For independent events it reduces to $P(A \cap B) = P(A)P(B)$.

Prior probabilities Initial estimates of the probabilities of events.

Posterior probabilities Revised probabilities of events based on additional information.

Bayes' theorem A method used to compute posterior probabilities.

KEY FORMULAS

Counting Rule for Combinations

$$C_n^N = \binom{N}{n} = \frac{N!}{n!(N-n)!} \tag{4.1}$$

Counting Rule for Permutations

$$P_n^N = n!\binom{N}{n} = \frac{N!}{(N-n)!} \tag{4.2}$$

Computing Probability Using the Complement

$$P(A) = 1 - P(A^c) \tag{4.5}$$

Addition Law

$$P(A \cup B) = P(A) + P(B) - P(A \cap B) \tag{4.6}$$

Conditional Probability

$$P(A \mid B) = \frac{P(A \cap B)}{P(B)} \tag{4.7}$$

$$P(B \mid A) = \frac{P(A \cap B)}{P(A)} \tag{4.8}$$

Multiplication Law

$$P(A \cap B) = P(B)P(A \mid B) \tag{4.11}$$

$$P(A \cap B) = P(A)P(B \mid A) \tag{4.12}$$

Multiplication Law for Independent Events

$$P(A \cap B) = P(A)P(B) \tag{4.13}$$

Bayes' Theorem

$$P(A_i \mid B) = \frac{P(A_i)P(B \mid A_i)}{P(A_1)P(B \mid A_1) + P(A_2)P(B \mid A_2) + \cdots + P(A_n)P(B \mid A_n)} \tag{4.19}$$

SUPPLEMENTARY EXERCISES

46. In a *Business Week*/Harris Poll, 1035 adults were asked about their attitudes toward business (*Business Week,* September 11, 2000). One question asked: "How would you rate large U.S. companies on making good products and competing in a global environment?" The responses were: excellent—18%, pretty good—50%, only fair—26%, poor—5%, and don't know/no answer—1%.

a. What is the probability that a person rated U.S. companies pretty good or excellent?
b. How many respondents rated U.S. companies poor?
c. How many respondents did not know or didn't answer?

47. A financial manager has just made two new investments—one in the oil industry and one in municipal bonds. After a 1-year period, each of the investments will be classified as either successful or unsuccessful. Consider the making of the two investments as an experiment.

a. How many sample points exist for this experiment?
b. Show a tree diagram and list the sample points.
c. Let O = the event that the oil investment is successful and M = the event that the municipal bond investment is successful. List the sample points in O and in M.
d. List the sample points in the union of the events ($O \cup M$).
e. List the sample points in the intersection of the events ($O \cap M$).
f. Are events O and M mutually exclusive? Explain.

48. A survey of American opinion asked: Are you satisfied or dissatisfied with the state of the U.S. economy today? (*The Wall Street Journal,* June 27, 1997). The following table provides the responses for all adults and the distribution by age groups.

	Satisfied (%)	Dissatisfied (%)	Other (%)
All Adults	61	37	2
18–34	64	35	1
35–49	58	41	1
50–64	57	40	3
65+	70	26	4

a. What is the probability a randomly selected adult is satisfied?
b. Which age groups report a higher level of satisfaction than the average for all adults?
c. What is the probability an adult 65 or over did not indicate they were satisfied?

49. A study of 31,000 hospital admissions in New York State was conducted by a Harvard research team led by Paul Weiler (*Business Week,* March 27, 1995). They found that 4% of the admissions led to treatment-caused injuries. One-seventh of these treatment-caused injuries resulted in death, and one-fourth were caused by negligence. Malpractice claims were filed in one out of 7.5 cases involving negligence, and payments were made in one out of every two claims.

a. What is the probability a person admitted to the hospital will suffer a treatment-caused injury due to negligence?
b. What is the probability a person admitted to the hospital will die from a treatment-caused injury?
c. In the case of a negligent treatment-caused injury, what is the probability a malpractice claim will be paid?

50. A telephone survey was used to determine viewer response to a new television show. The following data were obtained.

Rating	Frequency
Poor	4
Below average	8
Average	11
Above average	14
Excellent	13

a. What is the probability that a randomly selected viewer will rate the new show as average or better?

b. What is the probability that a randomly selected viewer will rate the new show below average or worse?

51. *Business Week* surveyed its subscribers concerning the number of cars they owned or leased. Responses were obtained to three questions: How many cars do you own? How many cars do you lease? How many cars do you have (owned or leased)? Table 4.8 contains the results obtained from 932 households (*Business Week* 1996 Worldwide Subscriber Study). In interpreting the table note that the third row shows that 401 households own 2 cars, 47 households lease 2 cars, and 447 households have 2 cars (owned and/or leased).

a. What is the probability a household leases 1 car?

b. What is the probability a household owns 2 or fewer cars?

c. What is the probability a household has 3 or more cars?

d. What is the probability a household does not own or lease a car?

52. A GMAC MBA new-matriculants survey provided the following data for 2018 students.

		Applied to More Than One School	
		Yes	**No**
Age Group	**23 and under**	207	201
	24–26	299	379
	27–30	185	268
	31–35	66	193
	36 and over	51	169

a. For a randomly selected MBA student, prepare a joint probability table for the experiment consisting of observing the student's age and whether the student applied to one or more schools.

b. What is the probability that a randomly selected applicant is 23 or under?

c. What is the probability that a randomly selected applicant is older than 26?

d. What is the probability that a randomly selected applicant applied to more than one school?

53. Refer again to the data from the GMAC new-matriculants survey in Exercise 52.

a. Given that a person applied to more than one school, what is the probability that the person is 24–26 years old?

b. Given that a person is in the 36-and-over age group, what is the probability that the person applied to more than one school?

c. What is the probability that a person is 24–26 years old *or* applied to more than one school?

d. Suppose a person is known to have applied to only one school. What is the probability that the person is 31 or more years old?

e. Is the number of schools applied to independent of age? Explain.

TABLE 4.8 NUMBER OF CARS OWNED OR LEASED, AND TOTAL NUMBER OF CARS PER HOUSEHOLD

	Number of Households		
Cars	Own	Lease	Have
0	65	708	19
1	242	168	168
2	401	47	447
3	149	9	186
4 or more	75	0	112
Total	932	932	932

54. An IBD/TIPP poll was conducted to learn about attitudes toward investment and retirement (*Investor's Business Daily,* May 5, 2000). One question asked male and female respondents how important they felt level of risk was in choosing a retirement investment. The following joint probability table was constructed from the data provided. Important means the respondent said level of risk was either important or very important.

	Male	**Female**	**Total**
Important	.22	.27	.49
Not Important	.28	.23	.51
Total	.50	.50	1.00

a. What is the probability a survey respondent will say level of risk is important?
b. What is the probability a male respondent will say level of risk is important?
c. What is the probability a female respondent will say level of risk is important?
d. Is level of risk independent of the gender of the respondent? Why or why not?
e. Do male and female attitudes toward risk differ?

55. A large consumer goods company has been running a television advertisement for one of its soap products. A survey was conducted. On the basis of this survey, probabilities were assigned to the following events.

B = individual purchased the product
S = individual recalls seeing the advertisement
$B \cap S$ = individual purchased the product and recalls seeing the advertisement

The probabilities assigned were $P(B) = .20$, $P(S) = .40$, and $P(B \cap S) = .12$.

a. What is the probability of an individual's purchasing the product given that the individual recalls seeing the advertisement? Does seeing the advertisement increase the probability that the individual will purchase the product? As a decision maker, would you recommend continuing the advertisement (assuming that the cost is reasonable)?
b. Assume that individuals who do not purchase the company's soap product buy from its competitors. What would be your estimate of the company's market share? Would you expect that continuing the advertisement will increase the company's market share? Why or why not?
c. The company has also tested another advertisement and assigned it values of $P(S) = .30$ and $P(B \cap S) = .10$. What is $P(B \mid S)$ for this other advertisement? Which advertisement seems to have had the bigger effect on customer purchases?

56. Cooper Realty is a small real estate company located in Albany, New York, specializing primarily in residential listings. They have recently become interested in determining the likelihood of one of their listings being sold within a certain number of days. An analysis of company sales of 800 homes in previous years produced the following data.

		Days Listed Until Sold			
		Under 30	**31–90**	**Over 90**	Total
Initial Asking Price	**Under $50,000**	50	40	10	100
	$50,000–$99,999	20	150	80	250
	$100,000–$150,000	20	280	100	400
	Over $150,000	10	30	10	50
	Total	100	500	200	800

a. If A is defined as the event that a home is listed for over 90 days before being sold, estimate the probability of A.

b. If B is defined as the event that the initial asking price is under $50,000, estimate the probability of B.

c. What is the probability of $A \cap B$?

d. Assuming that a contract has just been signed to list a home that has an initial asking price of less than $50,000, what is the probability that the home will take Cooper Realty more than 90 days to sell?

e. Are events A and B independent?

57. A company has studied the number of lost-time accidents occurring at its Brownsville, Texas, plant. Historical records show that 6% of the employees had lost-time accidents last year. Management believes that a special safety program will reduce such accidents to 5% during the current year. In addition, it estimates that 15% of employees who had lost-time accidents last year will have a lost-time accident during the current year.

a. What percentage of the employees will have lost-time accidents in both years?

b. What percentage of the employees will have at least one lost-time accident over the two-year period?

58. The Dallas IRS auditing staff is concerned with identifying potentially fraudulent tax returns. From past experience they believe that the probability of finding a fraudulent return given that the return contains deductions for contributions exceeding the IRS standard is .20. Given that the deductions for contributions do not exceed the IRS standard, the probability of a fraudulent return decreases to .02. If 8% of all returns exceed the IRS standard for deductions due to contributions, what is the best estimate of the percentage of fraudulent returns?

59. An oil company has purchased an option on land in Alaska. Preliminary geologic studies have assigned the following prior probabilities.

$$P(\text{high-quality oil}) = .50$$
$$P(\text{medium-quality oil}) = .20$$
$$P(\text{no oil}) = .30$$

a. What is the probability of finding oil?

b. After 200 feet of drilling on the first well, a soil test is taken. The probabilities of finding the particular type of soil identified by the test follow.

$$P(\text{soil} \mid \text{high-quality oil}) = .20$$
$$P(\text{soil} \mid \text{medium-quality oil}) = .80$$
$$P(\text{soil} \mid \text{no oil}) = .20$$

How should the firm interpret the soil test? What are the revised probabilities, and what is the new probability of finding oil?

60. A Bayesian approach can be used to revise probabilities that a prospect field will produce oil (*Oil & Gas Journal,* January 11, 1988). In one case, geological assessment indicates a 25% chance that the field will produce oil. Further, there is an 80% chance that a particular well will strike oil given that oil is present in the prospect field.

a. Suppose that one well is drilled on the field and it comes up dry. What is the probability that the prospect field will produce oil?

b. If two wells come up dry, what is the probability that the field will produce oil?

c. The oil company would like to keep looking as long as the chances of finding oil are greater than 1%. How many dry wells must be drilled before the field will be abandoned?

Case Problem HAMILTON COUNTY JUDGES

Hamilton County judges try thousands of cases per year. In an overwhelming majority of the cases disposed, the verdict stands as rendered. However, some cases are appealed, and of those appealed, some of the cases are reversed. Kristen DelGuzzi of *The Cincinnati Enquirer* conducted a study of cases handled by Hamilton County judges over the years 1994 through 1996 (*The Cincinnati Enquirer,* January 11, 1998). Shown in Table 4.9 are the results for 182,908 cases handled (disposed) by 38 judges in Common Pleas Court, Domestic Relations Court, and Municipal Court. Two of the judges (Dinkelacker and Hogan) did not serve in the same court for the entire three-year period.

The purpose of the newspaper's study was to evaluate the performance of the judges. Appeals are often the result of mistakes made by judges, and the newspaper wanted to know which judges were doing a good job and which were making too many mistakes. You have been called in to assist in the data analysis. Use your knowledge of probability and conditional probability to help with the ranking of the judges. You also may be able to analyze the likelihood of cases handled by different courts being appealed and reversed.

Managerial Report

Prepare a report with your rankings of the judges. Also, include an analysis of the likelihood of appeal and case reversal in the three courts. At a minimum, your report should include the following:

1. The probability of cases being appealed and reversed in the three different courts.
2. The probability of a case being appealed for each judge.
3. The probability of a case being reversed for each judge.
4. The probability of reversal given an appeal for each judge.
5. Rank the judges within each court. State the criteria you used and provide a rationale for your choice.

TABLE 4.9 TOTAL CASES DISPOSED, APPEALED, AND REVERSED IN HAMILTON COUNTY COURTS DURING 1994 THROUGH 1996

Common Pleas Court

Judge	Total Cases Disposed	Appealed Cases	Reversed Cases
Fred Cartolano	3037	137	12
Thomas Crush	3372	119	10
Patrick Dinkelacker	1258	44	8
Timothy Hogan	1954	60	7
Robert Kraft	3138	127	7
William Mathews	2264	91	18
William Morrissey	3032	121	22
Norbert Nadel	2959	131	20
Arthur Ney, Jr.	3219	125	14
Richard Niehaus	3353	137	16
Thomas Nurre	3000	121	6
John O'Connor	2969	129	12
Robert Ruehlman	3205	145	18
J. Howard Sundermann	955	60	10
Ann Marie Tracey	3141	127	13
Ralph Winkler	3089	88	6
Total	43945	1762	199

CD file

Judge

Domestic Relations Court

Judge	Total Cases Disposed	Appealed Cases	Reversed Cases
Penelope Cunningham	2729	7	1
Patrick Dinkelacker	6001	19	4
Deborah Gaines	8799	48	9
Ronald Panioto	12970	32	3
Total	30499	106	17

Municipal Court

Judge	Total Cases Disposed	Appealed Cases	Reversed Cases
Mike Allen	6149	43	4
Nadine Allen	7812	34	6
Timothy Black	7954	41	6
David Davis	7736	43	5
Leslie Isaiah Gaines	5282	35	13
Karla Grady	5253	6	0
Deidra Hair	2532	5	0
Dennis Helmick	7900	29	5
Timothy Hogan	2308	13	2
James Patrick Kenney	2798	6	1
Joseph Luebbers	4698	25	8
William Mallory	8277	38	9
Melba Marsh	8219	34	7
Beth Mattingly	2971	13	1
Albert Mestemaker	4975	28	9
Mark Painter	2239	7	3
Jack Rosen	7790	41	13
Mark Schweikert	5403	33	6
David Stockdale	5371	22	4
John A. West	2797	4	2
Total	108464	500	104

DISCRETE PROBABILITY DISTRIBUTIONS

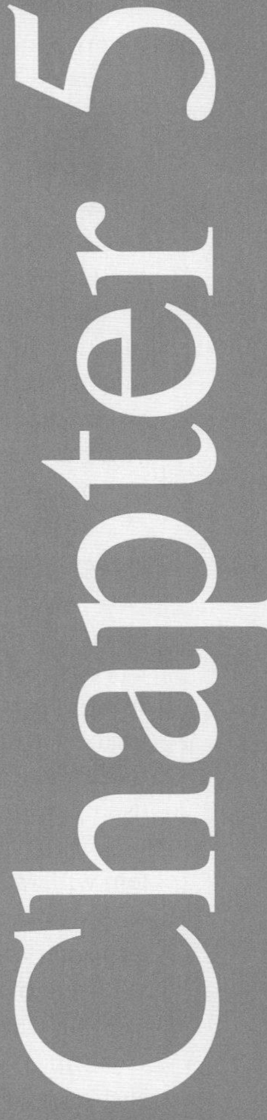

CONTENTS

STATISTICS IN PRACTICE

CITIBANK*
Long Island City, New York

Citibank, a division of Citigroup, makes available a wide range of financial services, including checking and savings accounts, loans and mortgages, insurance, and investment services, within the framework of a unique strategy for delivering those services called Citibanking. Citibanking entails a consistent brand identity all over the world, consistent product offerings, and high-level customer service. Citibanking lets you manage your money anytime, anywhere, anyway you choose. Whether you need to save for the future or borrow for today, you can do it all at Citibank.

Citibanking's state-of-the-art automatic teller machines (ATMs) located in Citicard Banking Centers (CBCs), let customers do all their banking in one place with the touch of a finger, 24 hours a day, 7 days a week. More than 150 different banking functions from deposits to managing investments can be performed with ease. Citibanking ATMs are so much more than just cash machines that customers today use them for 80% of their transactions.

Each Citibank CBC operates as a waiting line system with randomly arriving customers seeking service at one of the ATMs. If all ATMs are busy, the arriving customers wait in line. Periodic CBC capacity studies are used to analyze customer waiting times and to determine whether additional ATMs are needed.

Data collected by Citibank showed that the random customer arrivals followed a probability distribution known as the Poisson probability distribution. Using the Poisson probability distribution, Citibank can compute probabilities for the number of customers arriving at a CBC during any time period and make decisions concerning the number of ATMs needed. For example, let x = the number of customers arriving during a one-minute period. Assuming that a particular CBC has a mean arrival rate of 2 customers per minute, the following table shows the probabilities for the number of customers arriving during a one-minute period.

A Citibank ATM in Manhattan. © PhotoDisc, Inc.

x	**Probability**
0	.1353
1	.2707
2	.2707
3	.1804
4	.0902
5 or more	.0527

Discrete probability distributions, such as the one used by Citibank, are the topic of this chapter. In addition to the Poisson probability distribution, you will learn about the binomial and hypergeometric probability distributions and how they can be used to provide helpful probability information.

*The authors are indebted to Ms. Stacey Karter, Citibank, for providing this Statistics in Practice.

In this chapter we continue the study of probability by introducing the concepts of random variables and probability distributions. The focus of this chapter is discrete probability distributions. Three special discrete probability distributions—the binomial, Poisson, and hypergeometric—are covered.

5.1 RANDOM VARIABLES

In Chapter 4 we defined the concept of an experiment and its associated experimental outcomes. A random variable provides a means for describing experimental outcomes using numerical values.

Random variables must assume numerical values.

Random Variable

A **random variable** is a numerical description of the outcome of an experiment.

In effect, a random variable associates a numerical value with each possible experimental outcome. The particular numerical value of the random variable depends on the outcome of the experiment. A random variable can be classified as being either *discrete* or *continuous* depending on the numerical values it assumes.

Discrete Random Variables

A random variable that may assume either a finite number of values or an infinite sequence of values such as 0, 1, 2, . . . is referred to as a **discrete random variable.** For example, consider the experiment of an accountant taking the certified public accountant (CPA) examination. The examination has four parts. We can define a random variable as x = the number of parts of the CPA examination passed. It is a discrete random variable because it may assume the finite number of values 0, 1, 2, 3, or 4.

As another example of a discrete random variable, consider the experiment of cars arriving at a tollbooth. The random variable of interest is x = the number of cars arriving during a one-day period. The possible values for x come from the sequence of integers 0, 1, 2, and so on. Hence, x is a discrete random variable assuming one of the values in this infinite sequence.

Although many experiments have outcomes that are naturally described by numerical values, others do not. For example, a survey question might ask an individual to recall the message in a recent television commercial. This experiment would have two possible outcomes: the individual cannot recall the message and the individual can recall the message. We can still describe these experimental outcomes numerically by defining the discrete random variable x as follows: let $x = 0$ if the individual cannot recall the message and $x = 1$ if the individual can recall the message. The numerical values for this random variable are arbitrary (we could have used 5 and 10), but they are acceptable in terms of the definition of a random variable—namely, x is a random variable because it provides a numerical description of the outcome of the experiment.

Table 5.1 provides some additional examples of discrete random variables. Note that in each example the discrete random variable assumes a finite number of values or an infinite sequence of values such as 0, 1, 2, ... Discrete random variables such as these are discussed in detail in this chapter.

Continuous Random Variables

A random variable that may assume any numerical value in an interval or collection of intervals is called a **continuous random variable.** Experimental outcomes that are based on measurement scales such as time, weight, distance, and temperature can be described by continuous random variables. For example, consider an experiment of monitoring incoming telephone calls to the claims office of a major insurance company. Suppose the random variable of interest is x = the time between consecutive incoming calls in minutes. This random variable may assume any value in the interval $x \geq 0$. Actually, an infinite number of values are possible for x, including values such as 1.26 minutes, 2.751 minutes, 4.3333 minutes, and so on. As another example, consider a 90-mile section of interstate highway I-75 north of Atlanta, Georgia. For an emergency ambulance service located in Atlanta, we might define the random variable as x = number of miles to the location of the next traffic accident along this section of I-75. In this case, x would be a continuous random

TABLE 5.1 EXAMPLES OF DISCRETE RANDOM VARIABLES

Experiment	Random Variable (x)	Possible Values for the Random Variable
Contact five customers	Number of customers who place an order	0, 1, 2, 3, 4, 5
Inspect a shipment of 50 radios	Number of defective radios	0, 1, 2, · · · , 49, 50
Operate a restaurant for one day	Number of customers	0, 1, 2, 3, · · ·
Sell an automobile	Gender of the customer	0 if male; 1 if female

variable assuming any value in the interval $0 \le x \le 90$. Additional examples of continuous random variables are listed in Table 5.2. Note that each example describes a random variable that may assume any value in an interval of values. Continuous random variables and their probability distributions will be the topic of Chapter 6.

TABLE 5.2 EXAMPLES OF CONTINUOUS RANDOM VARIABLES

Experiment	Random Variable (x)	Possible Values for the Random Variable
Operate a bank	Time between customer arrivals in minutes	$x \ge 0$
Fill a soft drink can (max = 12.1 ounces)	Number of ounces	$0 \le x \le 12.1$
Work on a project to construct a new library	Percentage of project complete after 6 months	$0 \le x \le 100$
Test a new chemical process	Temperature when the desired reaction takes place (min 150° F; max 212° F)	$150 \le x \le 212$

NOTES AND COMMENTS

One way to determine whether a random variable is discrete or continuous is to think of the values of the random variable as points on a line segment. Choose two points representing values of the random variable. If the entire line segment between the two points also represents possible values for the random variable, then the random variable is continuous.

EXERCISES

Methods

1. Consider the experiment of tossing a coin twice.
   a. List the experimental outcomes.
   b. Define a random variable that represents the number of heads occurring on the two tosses.
   c. Show what value the random variable would assume for each of the experimental outcomes.
   d. Is this random variable discrete or continuous?

2. Consider the experiment of a worker assembling a product and recording how long it takes.
   a. Define a random variable that represents the time in minutes required to assemble the product.
   b. What values may the random variable assume?
   c. Is the random variable discrete or continuous?

Applications

3. Three students have interviews scheduled for summer employment at the Brookwood Institute. In each case the result of the interview will be that a position is either offered or not offered. Experimental outcomes are defined in terms of the results of the three interviews.
   a. List the experimental outcomes.
   b. Define a random variable that represents the number of offers made. Is it a discrete or continuous random variable?
   c. Show the value of the random variable for each of the experimental outcomes.
4. Suppose we know home mortgage rates for 12 Florida lending institutions. Assume that the random variable of interest is the number of lending institutions in this group that offers a 30-year fixed rate of 8.5% or less. What values may this random variable assume?
5. To perform a certain type of blood analysis, lab technicians must perform two procedures. The first procedure requires either 1 or 2 separate steps, and the second procedure requires either 1, 2, or 3 steps.
   a. List the experimental outcomes associated with performing the blood analysis.
   b. If the random variable of interest is the total number of steps required to do the complete analysis (both procedures), show what value the random variable will assume for each of the experimental outcomes.
6. Listed is a series of experiments and associated random variables. In each case, identify the values that the random variable can assume and state whether the random variable is discrete or continuous.

Experiment	**Random Variable (x)**
a. Take a 20-question examination	Number of questions answered correctly
b. Observe cars arriving at a tollbooth for 1 hour	Number of cars arriving at tollbooth
c. Audit 50 tax returns	Number of returns containing errors
d. Observe an employee's work	Number of nonproductive hours in an 8-hour workday
e. Weigh a shipment of goods	Number of pounds

5.2 DISCRETE PROBABILITY DISTRIBUTIONS

The **probability distribution** for a random variable describes how probabilities are distributed over the values of the random variable. For a discrete random variable x, the probability distribution is defined by a **probability function,** denoted by $f(x)$. The probability function provides the probability for each value of the random variable.

As an illustration of a discrete random variable and its probability distribution, consider the sales of automobiles at DiCarlo Motors in Saratoga, New York. Over the past 300 days of operation, sales data show 54 days with no automobiles sold, 117 days with 1 automobile sold, 72 days with 2 automobiles sold, 42 days with 3 automobiles sold, 12 days with 4 automobiles sold, and 3 days with 5 automobiles sold. Suppose we consider the experiment of selecting a day of operation at DiCarlo Motors and define the random variable of interest as $x =$ the number of automobiles sold during a day. From historical data, we know x is a discrete random variable that can assume the values 0, 1, 2, 3, 4, or 5. In probability function notation, $f(0)$ provides the probability of 0 automobiles sold, $f(1)$ provides

TABLE 5.3 PROBABILITY DISTRIBUTION FOR THE NUMBER OF AUTOMOBILES SOLD DURING A DAY AT DICARLO MOTORS

x	$f(x)$
0	.18
1	.39
2	.24
3	.14
4	.04
5	.01
Total	1.00

the probability of 1 automobile sold, and so on. Because historical data show 54 of 300 days with 0 automobiles sold, we assign the value $54/300 = .18$ to $f(0)$, indicating that the probability of 0 automobiles being sold during a day is .18. Similarly, because 117 of 300 days had 1 automobile sold, we assign the value $117/300 = .39$ to $f(1)$, indicating that the probability of exactly 1 automobile being sold during a day is .39. Continuing in this way for the other values of the random variable, we compute the values for $f(2)$, $f(3)$, $f(4)$, and $f(5)$ as shown in Table 5.3, the probability distribution for the number of automobiles sold during a day at DiCarlo Motors.

A primary advantage of defining a random variable and its probability distribution is that once the probability distribution is known, it is relatively easy to determine the probability of a variety of events that may be of interest to a decision maker. For example, using the probability distribution for DiCarlo Motors as shown in Table 5.3, we see that the most probable number of automobiles sold during a day is 1 with a probability of $f(1) = .39$. In addition, there is an $f(3) + f(4) + f(5) = .14 + .04 + .01 = .19$ probability of selling 3 or more automobiles during a day. These probabilities, plus others the decision maker may ask about, provide information that can help the decision maker understand the process of selling automobiles at DiCarlo Motors.

In the development of a probability function for any discrete random variable, the following two conditions must be satisfied.

These conditions are the analogs to the two basic requirements for assigning probabilities to experimental outcomes presented in Chapter 4.

Required Conditions for a Discrete Probability Function

$$f(x) \geq 0 \tag{5.1}$$

$$\Sigma f(x) = 1 \tag{5.2}$$

Table 5.3 shows that the probabilities for the random variable x satisfy equation (5.1); $f(x)$ is greater than or equal to 0 for all values of x. In addition, the probabilities sum to 1 so equation (5.2) is satisfied. Thus, the DiCarlo Motors probability function is a valid discrete probability function.

We can also present probability distributions graphically. In Figure 5.1 the values of the random variable x for DiCarlo Motors are shown on the horizontal axis and the probability associated with these values is shown on the vertical axis.

In addition to tables and graphs, a formula that gives the probability function, $f(x)$, for every value of x is often used to describe probability distributions. The simplest example of a discrete probability distribution given by a formula is the **discrete uniform probability distribution.** Its probability function is given.

Discrete Uniform Probability Function

$$f(x) = 1/n \tag{5.3}$$

where

$n =$ the number of values the random variable may assume

For example, consider the experiment of rolling a die and define the random variable x to be the number coming up. There are $n = 6$ possible values for the random variable; $x = 1, 2, 3, 4, 5, 6$. Thus, the probability function for this random variable is

$$f(x) = 1/6 \qquad x = 1, 2, 3, 4, 5, 6$$

The possible values of the random variable and the associated probabilities are shown.

x	$f(x)$
1	1/6
2	1/6
3	1/6
4	1/6
5	1/6
6	1/6

Note that the values of the random variable are equally likely.

FIGURE 5.1 GRAPHICAL REPRESENTATION OF THE PROBABILITY DISTRIBUTION FOR THE NUMBER OF AUTOMOBILES SOLD DURING A DAY AT DICARLO MOTORS

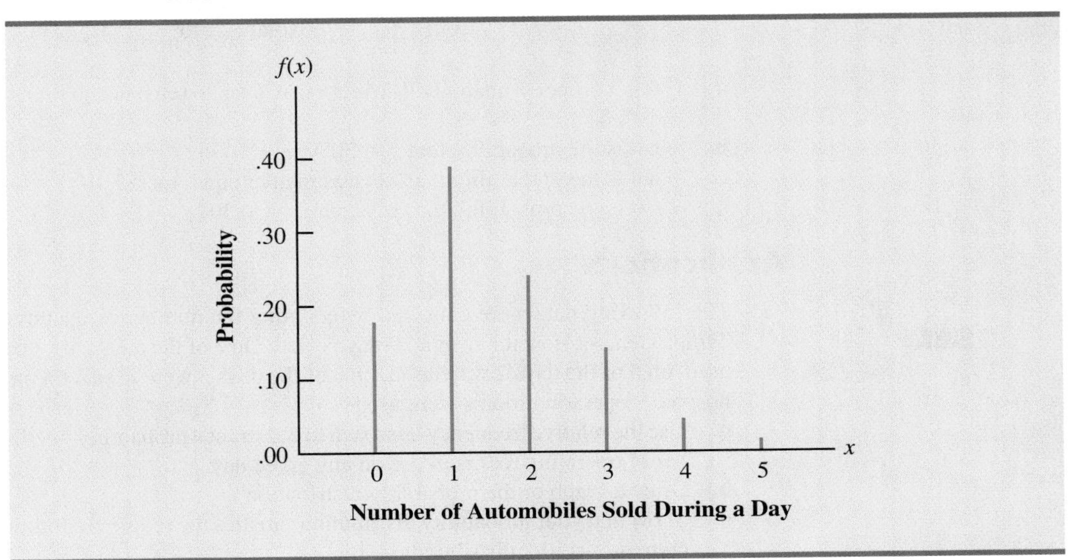

As another example, consider the random variable x with the following discrete probability distribution.

x	$f(x)$
1	1/10
2	2/10
3	3/10
4	4/10

This probability distribution can be defined by the formula

$$f(x) = \frac{x}{10} \qquad \text{for } x = 1, 2, 3, \text{ or } 4$$

Evaluating $f(x)$ for a given value of the random variable will provide the associated probability. For example, using the preceding probability function, we see that $f(2) = 2/10$ provides the probability that the random variable assumes a value of 2.

The more widely used discrete probability distributions generally are specified by formulas. Three important cases are the binomial, Poisson, and hypergeometric probability distributions; they are discussed later in the chapter.

EXERCISES

Methods

7. The probability distribution of the random variable x is shown as follows.

x		$f(x)$
20		.20
25		.15
30		.25
35		.40
	Total	1.00

   a. Is this a proper probability distribution? Check to see that equations (5.1) and (5.2) are satisfied.
   b. What is the probability that $x = 30$?
   c. What is the probability that x is less than or equal to 25?
   d. What is the probability that x is greater than 30?

Applications

8. The following data were collected by counting the number of operating rooms in use at Tampa General Hospital over a 20-day period: On 3 of the days only 1 operating room was used, on 5 of the days 2 were used, on 8 of the days 3 were used, and on 4 days all 4 of the hospital's operating rooms were used.
   a. Use the relative frequency approach to construct a probability distribution for the number of operating rooms in use on any given day.
   b. Draw a graph of the probability distribution.
   c. Show that your probability distribution satisfies the required conditions for a valid discrete probability distribution.

9. The following data show the number of employees at each of the five executive levels in the federal government (U.S. Office of Personnel Management, *Pay Structure of the Federal Civil Service,* 1996).

Executive Level	Number of Employees
1	15
2	32
3	84
4	300
5	31
Total	462

Suppose we want to select a sample of executive-level employees for a survey about working conditions. Let x be a random variable indicating the executive level of one employee randomly selected.

a. Use the data to develop a probability distribution for x. Specify the values for the random variable and the corresponding values for the probability function $f(x)$.
b. Draw a graph of the probability distribution.
c. Show that the probability distribution satisfies equations (5.1) and (5.2).

10. Table 5.4 shows the percent frequency distributions of job satisfaction scores for a sample of information systems (IS) senior executives and IS middle managers (*Computerworld,* May 26, 1997). The scores range from a low of 1 (very dissatisfied) to a high of 5 (very satisfied).

a. Develop a probability distribution for the job satisfaction score of a senior executive.
b. Develop a probability distribution for the job satisfaction score of a middle manager.
c. What is the probability a senior executive will report a job satisfaction score of 4 or 5?
d. What is the probability a middle manager is very satisfied?
e. Compare the overall job satisfaction of senior executives and middle managers.

11. A technician services mailing machines at companies in the Phoenix area. Depending on the type of malfunction, the service call can take 1, 2, 3, or 4 hours. The different types of malfunctions occur at about the same frequency.

a. Develop a probability distribution for the duration of a service call.
b. Draw a graph of the probability distribution.
c. Show that your probability distribution satisfies the conditions required for a discrete probability function.
d. What is the probability a service call will take 3 hours?
e. A service call has just come in, but the type of malfunction is unknown. It is 3:00 P.M.; service technicians usually get off at 5:00 P.M. What is the probability the service technician will have to work overtime to fix the machine today?

TABLE 5.4 PERCENT FREQUENCY DISTRIBUTION OF JOB SATISFACTION SCORES FOR INFORMATION SYSTEMS EXECUTIVES AND MIDDLE MANAGERS

Job Satisfaction Score	IS Senior Executives (%)	IS Middle Managers (%)
1	5	4
2	9	10
3	3	12
4	42	46
5	41	28
Total	100	100

12. The director of admissions at Lakeville Community College has subjectively assessed a probability distribution for x, the number of entering students, as follows.

x	$f(x)$
1000	.15
1100	.20
1200	.30
1300	.25
1400	.10

a. Is this a valid probability distribution?
b. What is the probability of 1200 or fewer entering students?

13. A psychologist has determined that the number of hours required to obtain the trust of a new patient is either 1, 2, or 3. Let x be a random variable indicating the time in hours required to gain the patient's trust. The following probability function has been proposed.

$$f(x) = \frac{x}{6} \qquad \text{for } x = 1, 2, \text{ or } 3$$

a. Is this a valid probability function? Explain.
b. What is the probability that it takes exactly 2 hours to gain the patient's trust?
c. What is the probability that it takes at least 2 hours to gain the patient's trust?

14. The following table is a partial probability distribution for the MRA Company's projected profits (x = profit in thousands of dollars) for the first year of operation (the negative value denotes a loss).

x	$f(x)$
−100	.10
0	.20
50	.30
100	.25
150	.10
200	

a. What is the proper value for $f(200)$? What is your interpretation of this value?
b. What is the probability that MRA will be profitable?
c. What is the probability that MRA will make at least $100,000?

5.3 EXPECTED VALUE AND VARIANCE

Expected Value

The **expected value,** or mean, of a random variable is a measure of the central location for the random variable. The mathematical expression for the expected value of a discrete random variable x follows.

Expected Value of a Discrete Random Variable

$$E(x) = \mu = \Sigma x f(x) \tag{5.4}$$

Both the notations $E(x)$ and μ are used to denote the expected value of a random variable.

Equation (5.4) shows that to compute the expected value of a discrete random variable, we must multiply each value of the random variable by the corresponding probability $f(x)$ and

The expected value is a weighted average of the values the random variable may assume. The weights are the probabilities.

then add the resulting products. Using the DiCarlo Motors automobile sales example from Section 5.2, we show the calculation of the expected value for the number of automobiles sold during a day in Table 5.5. The sum of the entries in the $xf(x)$ column shows that the expected value is 1.50 automobiles per day. We therefore know that although sales of 0, 1, 2, 3, 4, or 5 automobiles are possible on any one day, over time DiCarlo can anticipate selling an average of 1.50 automobiles per day. Assuming 30 days of operation during a month, we can use the expected value of 1.50 to anticipate average monthly sales of 30(1.50) = 45 automobiles.

Variance

While the expected value provides the mean value for the random variable, we often need a measure of variability, or dispersion. Just as we used the variance in Chapter 3 to summarize the variability in data, we now use **variance** to summarize the variability in the values of a random variable. The mathematical expression for the variance of a discrete random variable follows.

Variance of a Discrete Random Variable

$$\text{Var}(x) = \sigma^2 = \Sigma(x - \mu)^2 f(x) \qquad \textbf{(5.5)}$$

The variance is a weighted average of the squared deviations of a random variable from its mean. The weights are the probabilities.

As equation (5.5) shows, an essential part of the variance formula is the deviation, $x - \mu$, which measures how far a particular value of the random variable is from the expected value, or mean, μ. In computing the variance of a random variable, the deviations are squared and then weighted by the corresponding value of the probability function. The sum of these weighted squared deviations for all values of the random variable is referred to as the *variance*. The notations Var(x) and σ^2 are both used to denote the variance of a random variable.

The calculation of the variance for the probability distribution of the number of automobiles sold during a day at DiCarlo Motors is summarized in Table 5.6. We see that the variance is 1.25. The **standard deviation**, σ, is defined as the positive square root of the variance. Thus, the standard deviation for the number of automobiles sold during a day is

$$\sigma = \sqrt{1.25} = 1.118$$

The standard deviation is measured in the same units as the random variable ($\sigma = 1.118$ automobiles) and therefore is often preferred in describing the variability of a random variable. The variance σ^2 is measured in squared units and is thus more difficult to interpret.

TABLE 5.5 CALCULATION OF THE EXPECTED VALUE FOR THE NUMBER OF AUTOMOBILES SOLD DURING A DAY AT DICARLO MOTORS

x	$f(x)$	$xf(x)$
0	.18	0(.18) = .00
1	.39	1(.39) = .39
2	.24	2(.24) = .48
3	.14	3(.14) = .42
4	.04	4(.04) = .16
5	.01	5(.01) = .05
		1.50

$E(x) = \mu = \Sigma xf(x)$

TABLE 5.6 CALCULATION OF THE VARIANCE FOR THE NUMBER OF AUTOMOBILES SOLD DURING A DAY AT DICARLO MOTORS

x	$x - \mu$	$(x - \mu)^2$	$f(x)$	$(x - \mu)^2 f(x)$
0	$0 - 1.50 = -1.50$	2.25	.18	2.25(.18) = .4050
1	$1 - 1.50 = -.50$	.25	.39	.25(.39) = .0975
2	$2 - 1.50 = .50$	.25	.24	.25(.24) = .0600
3	$3 - 1.50 = 1.50$	2.25	.14	2.25(.14) = .3150
4	$4 - 1.50 = 2.50$	6.25	.04	6.25(.04) = .2500
5	$5 - 1.50 = 3.50$	12.25	.01	12.25(.01) = .1225
				1.2500

$\sigma^2 = \Sigma(x - \mu)^2 f(x)$

EXERCISES

Methods

15. The following table is a probability distribution for the random variable x.

x	$f(x)$
3	.25
6	.50
9	.25
Total	1.00

a. Compute $E(x)$, the expected value of x.
b. Compute σ^2, the variance of x.
c. Compute σ, the standard deviation of x.

16. The following table is a probability distribution for the random variable y.

y	$f(y)$
2	.20
4	.30
7	.40
8	.10
Total	1.00

a. Compute $E(y)$.
b. Compute Var(y) and σ.

Applications

17. A volunteer ambulance service handles 0 to 5 service calls on any given day. The probability distribution for the number of service calls is as follows.

Number of Service Calls	Probability
0	.10
1	.15
2	.30
3	.20
4	.15
5	.10

a. What is the expected number of service calls?
b. What is the variance in the number of service calls? What is the standard deviation?

18. *The Statistical Abstract of the United States,* 1997, shows that the average number of television sets per household is 2.3. Assume that the probability distribution for the number of television sets per household in New Orleans is as shown in the following table.

x	$f(x)$
0	.01
1	.23
2	.41
3	.20
4	.10
5	.05

a. Compute the expected value of the number of television sets per household and compare it with the average reported in the *Statistical Abstract.*
b. What are the variance and standard deviation of the number of television sets per household?

19. The actual shooting records of four basketball teams showed the probability of making a 2-point basket was .50 and the probability of making a 3-point basket was .39.
a. What is the expected value of a 2-point shot for these teams?
b. What is the expected value of a 3-point shot for these teams?
c. If the probability of making a 2-point basket is greater than the probability of making a 3-point basket, why do coaches allow some players to shoot the 3-point shot if they have the opportunity? Use expected value to explain your answer.

20. The probability distribution for damage claims paid by the Newton Automobile Insurance Company on collision insurance is shown as follows.

Payment ($)	Probability
0	.90
400	.04
1000	.03
2000	.01
4000	.01
6000	.01

a. Use the expected collision payment to determine the collision insurance premium that would enable the company to break even.
b. The insurance company charges an annual rate of $260 for the collision coverage. What is the expected value of the collision policy for a policyholder? (Hint: It is the expected payments from the company minus the cost of coverage.) Why does the policyholder purchase a collision policy with this expected value?

21. Shown are the probability distributions of job satisfaction scores for a sample of information systems (IS) senior executives and IS middle managers (*Computerworld,* May 26, 1997). The scores range from a low of 1 (very dissatisfied) to a high of 5 (very satisfied).

	Probability	
Job Satisfaction Score	IS Senior Executives	IS Middle Managers
1	.05	.04
2	.09	.10
3	.03	.12
4	.42	.46
5	.41	.28
Total	1.00	1.00

a. What is the expected value of the job satisfaction score for senior executives?
b. What is the expected value of the job satisfaction score for middle managers?
c. Compute the variance of job satisfaction scores for executives and middle managers.
d. Compute the standard deviation of job satisfaction scores for both probability distributions.
e. Compare the overall job satisfaction of senior executives and middle managers.

22. The demand for a product of Carolina Industries varies greatly from month to month. The probability distribution in the following table, based on the past 2 years of data, shows the company's monthly demand.

Unit Demand	Probability
300	.20
400	.30
500	.35
600	.15

a. If the company bases monthly orders on the expected value of the monthly demand, what should Carolina's monthly order quantity be for this product?
b. Assume that each unit demanded generates $70 in revenue and that each unit ordered costs $50. How much will the company gain or lose in a month if it places an order based on your answer to part (a) and the actual demand for the item is 300 units?

23. According to a survey, 95% of subscribers to *The Wall Street Journal Interactive Edition* have a computer at home. For those households, the probability distributions for the number of laptop and desktop computers are given (*The Wall Street Journal Interactive Edition Subscriber Study,* 1999).

	Probability	
Number of Computers	Laptop	Desktop
0	.47	.06
1	.45	.56
2	.06	.28
3	.02	.10

a. What is the expected value of the number of computers per household for each type?
b. What is the variance in the number of computers per household for each computer type?
c. Make some comparisons between the number of laptops and the number of desktops owned by the *Journal's* subscribers.

24. The J. R. Ryland Computer Company is considering a plant expansion that will enable the company to begin production of a new computer product. The company's president must determine whether to make the expansion a medium- or large-scale project. An uncertainty is the demand for the new product, which for planning purposes may be low demand, medium demand, or high demand. The probability estimates for demand are .20, .50, and .30, respectively. Letting x and y indicate the annual profit in $1000s, the firm's planners have developed the following profit forecasts for the medium- and large-scale expansion projects.

		Medium-Scale Expansion Profit		Large-Scale Expansion Profit	
		x	$f(x)$	y	$f(y)$
Demand	**Low**	50	.20	0	.20
	Medium	150	.50	100	.50
	High	200	.30	300	.30

a. Compute the expected value for the profit associated with the two expansion alternatives. Which decision is preferred for the objective of maximizing the expected profit?
b. Compute the variance for the profit associated with the two expansion alternatives. Which decision is preferred for the objective of minimizing the risk or uncertainty?

5.4 BINOMIAL PROBABILITY DISTRIBUTION

The binomial probability distribution is a discrete probability distribution that has many applications. It is associated with a multiple-step experiment that we call the binomial experiment.

A Binomial Experiment

A **binomial experiment** has the following four properties.

Properties of a Binomial Experiment

1. The experiment consists of a sequence of n identical trials.
2. Two outcomes are possible on each trial. We refer to one as a *success* and the other as a *failure.*
3. The probability of a success, denoted by p, does not change from trial to trial. Consequently, the probability of a failure, denoted by $1 - p$, does not change from trial to trial.
4. The trials are independent.

Jakob Bernoulli (1654–1705), the first of the Bernoulli family of Swiss mathematicians, published a treatise on probability that contained the theory of permutations and combinations, as well as the Binomial Theorem.

If properties 2, 3, and 4 are present, we say the trials are generated by a Bernoulli process. If, in addition, property 1 is present, we say we have a **binomial experiment.** Figure 5.2 depicts one possible sequence of outcomes of a binomial experiment involving eight trials. This case presents the possibility of five successes and three failures.

In a binomial experiment, our interest is in the *number of successes occurring in the n trials.* If we let x denote the number of successes occurring in the n trials, we see that x can assume the values of $0, 1, 2, 3, \ldots, n$. Because the number of values is finite, x is a *discrete* random variable. The probability distribution associated with this random variable is called the **binomial probability distribution.** For example, consider the experiment of tossing a coin five times and on each toss observing whether the coin lands with a head or a tail on

FIGURE 5.2 DIAGRAM OF AN EIGHT-TRIAL BINOMIAL EXPERIMENT

Property 1: The experiment consists of $n = 8$ identical trials.

Property 2: Each trial results in either success (S) or failure (F).

Trials →	1	2	3	4	5	6	7	8
Outcomes →	S	F	F	S	S	F	S	S

its upward face. Suppose we are interested in counting the number of heads appearing over the five tosses. Does this experiment have the properties of a binomial experiment? What is the random variable of interest? Note that:

1. The experiment consists of five identical trials; each trial involves the tossing of one coin.
2. Two outcomes are possible for each trial: a head or a tail. We can designate head a success and tail a failure.
3. The probability of a head and the probability of a tail are the same for each trial, with $p = .5$ and $1 - p = .5$.
4. The trials or tosses are independent because the outcome on any one trial is not affected by what happens on other trials or tosses.

Thus, the properties of a binomial experiment are satisfied. The random variable of interest is x = the number of heads appearing in the five trials. In this case, x can assume the values of 0, 1, 2, 3, 4, or 5.

As another example, consider an insurance salesperson who visits 10 randomly selected families. The outcome associated with each visit is classified as a success if the family purchases an insurance policy and a failure if the family does not. From past experience, the salesperson knows the probability that a randomly selected family will purchase an insurance policy is .10. Checking the properties of a binomial experiment, we observe that:

1. The experiment consists of 10 identical trials; each trial involves contacting one family.
2. Two outcomes are possible on each trial: the family purchases a policy (success) or the family does not purchase a policy (failure).
3. The probabilities of a purchase and a nonpurchase are assumed to be the same for each sales call, with $p = .10$ and $1 - p = .90$.
4. The trials are independent because the families are randomly selected.

Because the four assumptions are satisfied, this example is a binomial experiment. The random variable of interest is the number of sales obtained in contacting the 10 families. In this case, x can assume the values of 0, 1, 2, 3, 4, 5, 6, 7, 8, 9, and 10.

Property 3 of the binomial experiment is called the *stationarity assumption* and is sometimes confused with property 4, independence of trials. To see how they differ, consider again the case of the salesperson calling on families to sell insurance policies. If, as the day wore on, the salesperson got tired and lost enthusiasm, the probability of success (selling a policy) might drop to .05, for example, by the tenth call. In such a case, property 3 (stationarity) would not be satisfied, and we would not have a binomial experiment. Even if property 4 held—that is, the purchase decisions of each family were made independently—it would not be a binomial experiment if property 3 was not satisfied.

In applications involving binomial experiments, a special mathematical formula, called the **binomial probability function,** can be used to compute the probability of x successes in the n trials. Using probability concepts introduced in Chapter 4, we will show in the context of an illustrative problem how the formula can be developed.

Martin Clothing Store Problem

Let us consider the purchase decisions of the next three customers who enter the Martin Clothing Store. On the basis of past experience, the store manager estimates the probability that any one customer will make a purchase is .30. What is the probability that two of the next three customers will make a purchase?

Using a tree diagram (Figure 5.3), we can see that the experiment of observing the three customers each making a purchase decision has eight possible outcomes. Using S to denote success (a purchase) and F to denote failure (no purchase), we are interested in experimental outcomes involving two successes in the three trials (purchase decisions). Next, let us verify that the experiment involving the sequence of three purchase decisions can be viewed as a binomial experiment. Checking the four requirements for a binomial experiment, we note that:

1. The experiment can be described as a sequence of three identical trials, one trial for each of the three customers who will enter the store.
2. Two outcomes—the customer makes a purchase (success) or the customer does not make a purchase (failure)—are possible for each trial.
3. The probability that the customer will make a purchase (.30) or will not make a purchase (.70) is assumed to be the same for all customers.
4. The purchase decision of each customer is independent of the decisions of the other customers.

Hence, the properties of a binomial experiment are present.

The number of experimental outcomes resulting in exactly x successes in n trials can be computed from the following formula.*

Number of Experimental Outcomes Providing Exactly x Successes in n Trials

$$\binom{n}{x} = \frac{n!}{x!(n-x)!} \tag{5.6}$$

where

$$n! = n(n-1)(n-2)\cdots(2)(1)$$

and

$$0! = 1$$

Now let us return to the Martin Clothing Store experiment involving three customer purchase decisions. Equation (5.6) can be used to determine the number of experimental outcomes involving two purchases; that is, the number of ways of obtaining $x = 2$ successes in the $n = 3$ trials. From equation (5.6) we have

$$\binom{n}{x} = \binom{3}{2} = \frac{3!}{2!(3-2)!} = \frac{(3)(2)(1)}{(2)(1)(1)} = \frac{6}{2} = 3$$

Equation (5.6) shows that three of the outcomes yield two successes. From Figure 5.3 we see these three outcomes are denoted by *SSF*, *SFS*, and *FSS*.

Using equation (5.6) to determine how many experimental outcomes have three successes (purchases) in the three trials, we obtain

$$\binom{n}{x} = \binom{3}{3} = \frac{3!}{3!(3-3)!} = \frac{3!}{3!0!} = \frac{(3)(2)(1)}{3(2)(1)(1)} = \frac{6}{6} = 1$$

*This formula was introduced in Chapter 4 to determine the number of combinations of n objects selected x at a time. For the binomial experiment, this combinatorial formula provides the number of experimental outcomes (sequences of n trials) resulting in x successes.

FIGURE 5.3 TREE DIAGRAM FOR THE MARTIN CLOTHING STORE PROBLEM

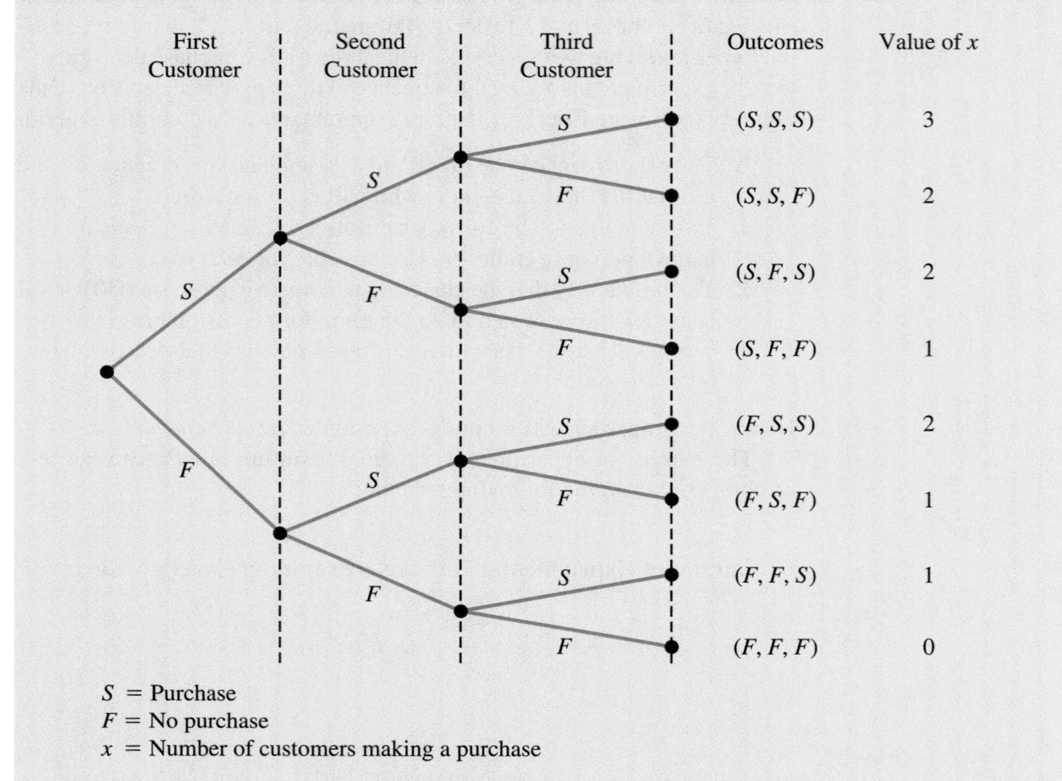

From Figure 5.3 we see that the one experimental outcome with three successes is identified by *SSS*.

We know that equation (5.6) can be used to determine the number of experimental outcomes that result in x successes. But, if we are to determine the probability of x successes in n trials, we must also know the probability associated with each of these experimental outcomes. Because the trials of a binomial experiment are independent, we can simply multiply the probabilities associated with each trial outcome to find the probability of a particular sequence of successes and failures.

The probability of purchases by the first two customers and no purchase by the third customer is given by

$$pp(1 - p)$$

With a .30 probability of a purchase on any one trial, the probability of a purchase on the first two trials and no purchase on the third is given by

$$(.30)(.30)(.70) = (.30)^2(.70) = .063$$

Two other sequences of outcomes result in two successes and one failure. The probabilities for all three sequences involving two successes are shown as follows.

Trial Outcomes			Success-Failure Notation	Probability of Experimental Outcome
1st Customer	2nd Customer	3rd Customer		
Purchase	Purchase	No purchase	*SSF*	$pp(1-p) = p^2(1-p)$ $= (.30)^2(.70) = .063$
Purchase	No purchase	Purchase	*SFS*	$p(1-p)p = p^2(1-p)$ $= (.30)^2(.70) = .063$
No purchase	Purchase	Purchase	*FSS*	$(1-p)pp = p^2(1-p)$ $= (.30)^2(.70) = .063$

Observe that all three outcomes with two successes have exactly the same probability. This observation holds in general. In any binomial experiment, all sequences of trial outcomes yielding x successes in n trials have the *same probability* of occurrence. The probability of each sequence of trials yielding x successes in n trials follows.

$$\text{Probability of a Particular Sequence of Trial Outcomes with } x \text{ Successes in } n \text{ Trials} = p^x(1-p)^{(n-x)} \quad \textbf{(5.7)}$$

For the Martin Clothing Store, this formula shows that any outcome with two successes has a probability of $p^2(1-p)^{(3-2)} = p^2(1-p)^1 = (.30)^2(.70)^1 = .063$.

Because equation (5.6) shows the number of outcomes in a binomial experiment with x successes and equation (5.7) gives the probability for each sequence involving x successes, we combine equations (5.6) and (5.7) to obtain the following binomial probability function.

Binomial Probability Function

$$f(x) = \binom{n}{x} p^x(1-p)^{(n-x)} \quad \textbf{(5.8)}$$

where

$f(x)$ = the probability of x successes in n trials

n = the number of trials

$$\binom{n}{x} = \frac{n!}{x!(n-x)!}$$

p = the probability of a success on any one trial

$(1-p)$ = the probability of a failure on any one trial

In the Martin Clothing Store example, let us compute the probability that no customer makes a purchase, exactly one customer makes a purchase, exactly two customers make a purchase, and all three customers make a purchase. The calculations are summarized in Table 5.7, which gives the probability distribution of the number of customers making a purchase. Figure 5.4 is a graph of this probability distribution.

The binomial probability function can be applied to *any* binomial experiment. If we are satisfied that a situation has the properties of a binomial experiment and if we know the values of n, p, and $(1-p)$, we can use equation (5.8) to compute the probability of x successes in the n trials.

TABLE 5.7 PROBABILITY DISTRIBUTION FOR THE NUMBER OF CUSTOMERS MAKING A PURCHASE

x	$f(x)$
0	$\frac{3!}{0!3!}(.30)^0(.70)^3 = .343$
1	$\frac{3!}{1!2!}(.30)^1(.70)^2 = .441$
2	$\frac{3!}{2!1!}(.30)^2(.70)^1 = .189$
3	$\frac{3!}{3!0!}(.30)^3(.70)^0 = .027$
	1.000

FIGURE 5.4 GRAPHICAL REPRESENTATION OF THE PROBABILITY DISTRIBUTION FOR THE MARTIN CLOTHING STORE PROBLEM

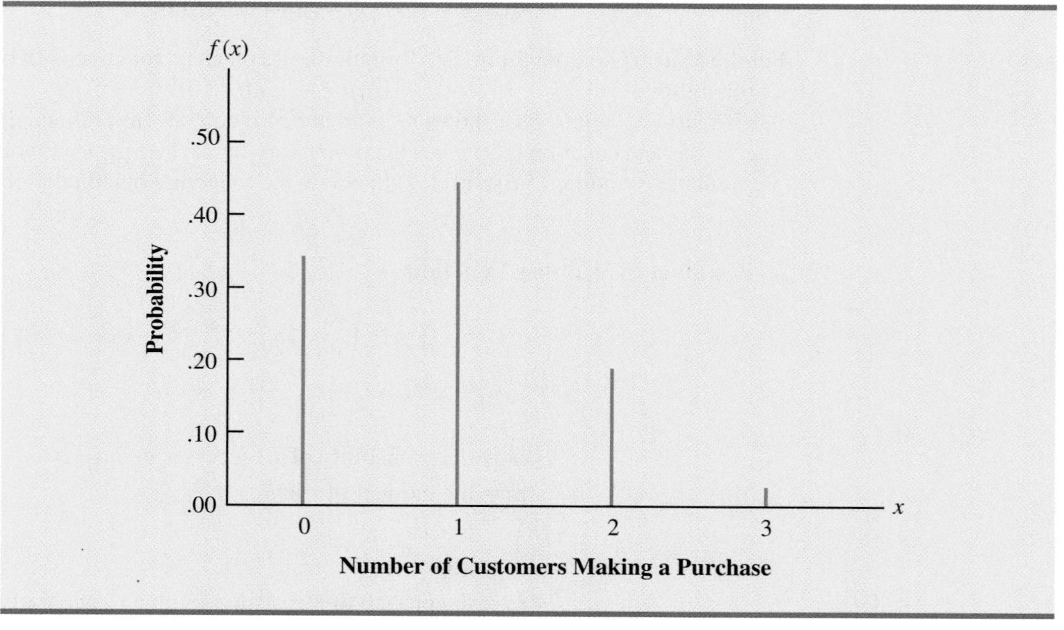

If we consider variations of the Martin experiment, such as 10 customers rather than three entering the store, the binomial probability function given by equation (5.8) is still applicable. Suppose we have a binomial experiment with $n = 10$, $x = 4$, and $p = .30$. The probability of making exactly four sales to 10 potential customers entering the store is

$$f(4) = \frac{10!}{4!6!}(.30)^4(.70)^6 = .2001$$

Using Tables of Binomial Probabilities

Tables have been developed that give the probability of x successes in n trials for a binomial experiment. The tables are generally easy to use and quicker than (5.8). A table of binomial probabilities is provided as Table 5 of Appendix B. A portion of this table is given in Table 5.8. To use this table, we must specify the values of n, p, and x for the binomial experiment of interest. In the example at the top of Table 5.8, we see that the probability of $x = 3$ successes in a binomial experiment with $n = 10$ and $p = .40$ is .2150. You can use equation (5.8) to verify that you would obtain the same answer if you used the binomial probability function directly.

Now let us use Table 5.8 to verify the probability of four successes in 10 trials for the Martin Clothing Store problem. Note that the value of $f(4) = .2001$ can be read directly from the table of binomial probabilities, with $n = 10$, $x = 4$, and $p = .30$.

With modern calculators, these tables are almost unnecessary. It is easy to evaluate (5.8) directly.

While the tables of binomial probabilities are relatively easy to use, it is impossible to have tables that show all possible values of n and p that might be encountered in a binomial experiment. However, with today's calculators, using (5.8) to calculate the desired probability is not difficult, especially if the number of trials is not large. In the exercises, you should practice using (5.8) to compute the binomial probabilities unless the problem specifically requests that you use the binomial probability table.

Statistical software packages such as Minitab and spreadsheet packages such as Excel also provide a capability for computing binomial probabilities. Consider the Martin Clothing Store example with $n = 10$ and $p = .30$. Figure 5.5 shows the binomial probabilities generated by Minitab for all possible values of x. Note that these values are the same as those found in the $p = .30$ column of Table 5.8. Appendix 5.1 gives the step-by-step procedure for using Minitab to generate the output in Figure 5.5. Appendix 5.2 describes how Excel can be used to compute binomial probabilities.

TABLE 5.8 SELECTED VALUES FROM THE BINOMIAL PROBABILITY TABLE EXAMPLE: $n = 10$, $x = 3$, $p = .40$; $f(3) = .2150$

						p					
n	x	**.05**	**.10**	**.15**	**.20**	**.25**	**.30**	**.35**	**.40**	**.45**	**.50**
9	0	.6302	.3874	.2316	.1342	.0751	.0404	.0207	.0101	.0046	.0020
	1	.2985	.3874	.3679	.3020	.2253	.1556	.1004	.0605	.0339	.0176
	2	.0629	.1722	.2597	.3020	.3003	.2668	.2162	.1612	.1110	.0703
	3	.0077	.0446	.1069	.1762	.2336	.2668	.2716	.2508	.2119	.1641
	4	.0006	.0074	.0283	.0661	.1168	.1715	.2194	.2508	.2600	.2461
	5	.0000	.0008	.0050	.0165	.0389	.0735	.1181	.1672	.2128	.2461
	6	.0000	.0001	.0006	.0028	.0087	.0210	.0424	.0743	.1160	.1641
	7	.0000	.0000	.0000	.0003	.0012	.0039	.0098	.0212	.0407	.0703
	8	.0000	.0000	.0000	.0000	.0001	.0004	.0013	.0035	.0083	.0176
	9	.0000	.0000	.0000	.0000	.0000	.0000	.0001	.0003	.0008	.0020
10	0	.5987	.3487	.1969	.1074	.0563	.0282	.0135	.0060	.0025	.0010
	1	.3151	.3874	.3474	.2684	.1877	.1211	.0725	.0403	.0207	.0098
	2	.0746	.1937	.2759	.3020	.2816	.2335	.1757	.1209	.0763	.0439
	3	.0105	.0574	.1298	.2013	.2503	.2668	.2522	**.2150**	.1665	.1172
	4	.0010	.0112	.0401	.0881	.1460	.2001	.2377	.2508	.2384	.2051
	5	.0001	.0015	.0085	.0264	.0584	.1029	.1536	.2007	.2340	.2461
	6	.0000	.0001	.0012	.0055	.0162	.0368	.0689	.1115	.1596	.2051
	7	.0000	.0000	.0001	.0008	.0031	.0090	.0212	.0425	.0746	.1172
	8	.0000	.0000	.0000	.0001	.0004	.0014	.0043	.0106	.0229	.0439
	9	.0000	.0000	.0000	.0000	.0000	.0001	.0005	.0016	.0042	.0098
	10	.0000	.0000	.0000	.0000	.0000	.0000	.0000	.0001	.0003	.0010

FIGURE 5.5 MINITAB OUTPUT SHOWING BINOMIAL PROBABILITIES FOR THE MARTIN CLOTHING STORE PROBLEM

x	P(X = x)
0.00	0.0282
1.00	0.1211
2.00	0.2335
3.00	0.2668
4.00	0.2001
5.00	0.1029
6.00	0.0368
7.00	0.0090
8.00	0.0014
9.00	0.0001
10.00	0.0000

Expected Value and Variance for the Binomial Probability Distribution

In Section 5.3 we provided formulas for computing the expected value and variance of a discrete random variable. In the special case where the random variable has a binomial probability distribution with a known number of trials n and a known probability of success p, the general formulas for the expected value and variance can be simplified. The results follow.

Expected Value and Variance for the Binomial Probability Distribution

$$E(x) = \mu = np \tag{5.9}$$

$$\text{Var}(x) = \sigma^2 = np(1 - p) \tag{5.10}$$

For the Martin Clothing Store problem with three customers, we can use equation (5.9) to compute the expected number of customers making a purchase.

$$E(x) = np = 3(.30) = .9$$

Suppose that for the next month the Martin Clothing Store forecasts 1000 customers will enter the store. What is the expected number of customers who will make a purchase? The answer is $\mu = np = (1000)(.3) = 300$. Thus, to increase the expected number of sales, Martin's must induce more customers to enter the store and/or somehow increase the probability that any individual customer will make a purchase after entering.

For the Martin Clothing Store problem with three customers, we see that the variance and standard deviation for the number of customers making a purchase are

The np chart and lot acceptance sampling are quality control procedures that are excellent examples of the use of the binomial distribution. These procedures are discussed in Chapter 20, Statistical Methods for Quality Control.

$$\sigma^2 = np(1 - p) = 3(.3)(.7) = .63$$
$$\sigma = \sqrt{.63} = .79$$

For the next 1000 customers entering the store, the variance and standard deviation for the number of customers making a purchase are

$$\sigma^2 = np(1 - p) = 1000(.3)(.7) = 210$$
$$\sigma = \sqrt{210} = 14.49$$

NOTES AND COMMENTS

1. The binomial tables in Appendix B show values of p only up to and including $p = .50$. It would appear that such tables cannot be used when the probability of success exceeds $p = .50$. However, they can be used by noting that the probability of $n - x$ failures is also the probability of x successes. When the probability of success is greater than $p = .50$, one can compute the probability of $n - x$ failures instead. The probability of failure, $1 - p$, will be less than .50 when $p > .50$.
2. Some sources present binomial tables in a cumulative form. In using such tables, one must subtract to find the probability of x successes in n trials. For example, $f(2) = P(x \leq 2) - P(x \leq 1)$. Our tables provide these probabilities directly. To compute cumulative probabilities using our tables, one simply sums the individual probabilities. For example, to compute $P(x \leq 2)$ using our tables, we sum $f(0) + f(1) + f(2)$.

EXERCISES

Methods

25. Consider a binomial experiment with two trials and $p = .4$.

a. Draw a tree diagram showing it as a two-trial experiment (see Figure 5.3).
b. Compute the probability of one success, $f(1)$.
c. Compute $f(0)$.
d. Compute $f(2)$.
e. Find the probability of at least one success.
f. Find the expected value, variance, and standard deviation.

26. Consider a binomial experiment with $n = 10$ and $p = .10$.

a. Find $f(0)$.
b. Find $f(2)$.
c. Find $P(x \leq 2)$.
d. Find $P(x \geq 1)$.
e. Find $E(x)$.
f. Find Var(x) and σ.

27. Consider a binomial experiment with $n = 20$ and $p = .70$. Use the binomial tables (Table 5 of Appendix B) to answer parts (a) through (d).

a. Find $f(12)$.
b. Find $f(16)$.
c. Find $P(x \geq 16)$.
d. Find $P(x \leq 15)$.
e. Find $E(x)$.
f. Find Var(x) and σ.

Applications

28. The 1999 Youth and Money Survey, sponsored by the American Savings Education Council, the Employee Benefit Research Institute, and Mathew Greenwald & Associates, talked to a thousand students ages 16–22 about personal finance. The survey found that 33% of the students have their own credit card.

a. In a sample of six students, what is the probability that two will have their own credit card?
b. In a sample of six students, what is the probability that at least two will have their own credit card?
c. In a sample of 10 students, what is the probability that none will have their own credit card?

29. According to a *Business Week*/Harris Poll of 1035 adults, 40% of those surveyed agreed strongly with the proposition that business has too much power over American life (*Business Week,* September 11, 2000). Assume this percentage is representative of the American population and that a sample of 20 individuals has been taken from a cross-section of the American population to learn about the role of business in their lives. What is the probability that at least five of these individuals will feel that business has too much power over American life?

SELF test

30. When a new machine is functioning properly, only 3% of the items produced are defective. Assume that we will randomly select two parts produced on the machine and that we are interested in the number of defective parts found.

a. Describe the conditions under which this situation would be a binomial experiment.
b. Draw a tree diagram similar to Figure 5.3 showing this problem as a two-trial experiment.
c. How many experimental outcomes result in exactly one defect being found?
d. Compute the probabilities associated with finding no defects, exactly one defect, and two defects.

31. Five percent of American truck drivers are women (*Statistical Abstract of the United States,* 1997). Suppose 10 truck drivers are selected randomly to be interviewed about quality of work conditions.

a. Is the selection of the 10 drivers a binomial experiment? Explain.
b. What is the probability that two of the drivers will be women?
c. What is the probability that none will be women?
d. What is the probability that at least one will be a woman?

32. Military radar and missile detection systems are designed to warn a country against enemy attacks. A reliability question is whether a detection system will be able to identify an attack and issue a warning. Assume that a particular detection system has a .90 probability of detecting a missile attack. Use the binomial probability distribution to answer the following questions.

a. What is the probability that a single detection system will detect an attack?
b. If two detection systems are installed in the same area and operate independently, what is the probability that at least one of the systems will detect the attack?
c. If three systems are installed, what is the probability that at least one of the systems will detect the attack?
d. Would you recommend that multiple detection systems be used? Explain.

33. Fifty percent of midsize manufacturers planned to have management representatives visit Canada and Mexico to take advantage of opportunities created by the North American Free Trade Agreement (*Grant Thornton Survey of American Manufacturers,* 1995). An export-import group in Toronto, Canada, has invited 20 U.S. midsize manufacturers to participate in a conference to explore trade opportunities.

a. What is the probability that 12 or more of these manufacturers will send representatives?
b. What is the probability that no more than 5 of these manufacturers will send representatives?
c. How many of these manufacturers would you expect to send representatives?
d. What are the variance and standard deviation of the number of manufacturers sending representatives?

34. Forty percent of business travelers carry either a cell phone or a laptop (*USA Today,* September 12, 2000). In a sample of 15 business travelers,

a. What is the probability three have a cell phone or laptop?
b. What is the probability that 12 of the travelers have neither a cell phone nor a laptop?
c. What is the probability that at least three of the travelers have a cell phone or a laptop?

35. A university found that 20% of its students withdraw without completing the introductory statistics course. Assume that 20 students have registered for the course this quarter.

a. What is the probability that two or fewer will withdraw?
b. What is the probability that exactly four will withdraw?

c. What is the probability that more than three will withdraw?
d. What is the expected number of withdrawals?

36. For the special case of a binomial random variable, we stated that the variance measure could be computed from the formula $\sigma^2 = np(1 - p)$. For the Martin Clothing Store problem data in Table 5.7, we found $\sigma^2 = np(1 - p) = .63$. Use the general definition of variance for a discrete random variable, equation (5.5), and the data in Table 5.7 to verify that the variance is in fact .63.

37. Twenty-nine percent of lawyers and judges are women (*Statistical Abstract of the United States,* 1997). In a jurisdiction with 30 lawyers and judges, what is the expected number of women? What are the variance and standard deviation?

5.5 POISSON PROBABILITY DISTRIBUTION

The Poisson probability distribution is often used to model arrival rates in waiting line situations.

In this section we consider a discrete random variable that is often useful in estimating the number of occurrences over a specified interval of time or space. For example, the random variable of interest might be the number of arrivals at a car wash in one hour, the number of repairs needed in 10 miles of highway, or the number of leaks in 100 miles of pipeline. If the following two properties are satisfied, the number of occurrences is a random variable described by the **Poisson probability distribution.**

Properties of a Poisson Experiment

1. The probability of an occurrence is the same for any two intervals of equal length.
2. The occurrence or nonoccurrence in any interval is independent of the occurrence or nonoccurrence in any other interval.

The **Poisson probability function** is given by equation (5.11).

Poisson Probability Function

Siméon Poisson taught mathematics at the Ecole Polytechnique in Paris from 1802 to 1808. In 1837, he published a work entitled, "Researches on the probability of criminal and civil verdicts," which includes a discussion of what later became known as the Poisson distribution.

$$f(x) = \frac{\mu^x e^{-\mu}}{x!} \quad \textbf{(5.11)}$$

where

$f(x)$ = the probability of x occurrences in an interval
μ = expected value or mean number of occurrences in an interval
e = 2.71828

Before we consider a specific example to see how the Poisson distribution can be applied, note that the number of occurrences, x, has no upper limit. It is a discrete random variable that may assume an infinite sequence of values ($x = 0, 1, 2, \ldots$).

An Example Involving Time Intervals

Suppose that we are interested in the number of arrivals at the drive-up teller window of a bank during a 15-minute period on weekday mornings. If we can assume that the probability of a car arriving is the same for any two time periods of equal length and that the

Bell Labs has made substantial use of the Poisson distribution in modeling the arrival of phone calls.

arrival or nonarrival of a car in any time period is independent of the arrival or nonarrival in any other time period, the Poisson probability function is applicable. Suppose these assumptions are satisfied and an analysis of historical data shows that the average number of cars arriving in a 15-minute period of time is 10; in this case, the following probability function applies.

$$f(x) = \frac{10^x e^{-10}}{x!}$$

The random variable here is x = number of cars arriving in any 15-minute period.

If management wanted to know the probability of exactly five arrivals in 15 minutes, we would set $x = 5$ and thus obtain

$$\begin{array}{l}\text{Probability of Exactly} \\ \text{5 Arrivals in 15 Minutes}\end{array} = f(5) = \frac{10^5 e^{-10}}{5!} = .0378$$

Although this probability was determined by evaluating the probability function with $\mu = 10$ and $x = 5$, it is often easier to refer to tables for the Poisson probability distribution. These tables provide probabilities for specific values of x and μ. We have included such a table as Table 7 of Appendix B. For convenience, we have reproduced a portion of this table as Table 5.9. Note that to use the table of Poisson probabilities, we need know only the values of x and μ. From Table 5.9 we see that the probability of five arrivals in a 15-minute period is found by locating the value in the row of the table corresponding to $x = 5$ and the column of the table corresponding to $\mu = 10$. Hence, we obtain $f(5) = .0378$.

Our illustration involves a 15-minute period, but other time periods can be used. Suppose we want to compute the probability of one arrival in a 3-minute period. Because 10 is the expected number of arrivals in a 15-minute period, we see that $10/15 = 2/3$ is the expected number of arrivals in a 1-minute period and that (2/3)(3 minutes) = 2 is the expected number of arrivals in a 3-minute period. Thus, the probability of x arrivals in a 3-minute time period with $\mu = 2$ is given by the following Poisson probability function.

$$f(x) = \frac{2^x e^{-2}}{x!}$$

Suppose now we want to compute the probability of one arrival in a 3-minute period. We can use this function to find

$$\begin{array}{l}\text{Probability of Exactly} \\ \text{1 Arrival in 3 Minutes}\end{array} = f(1) = \frac{2^1 e^{-2}}{1!} = .2707$$

An Example Involving Length or Distance Intervals

Let us illustrate an application not involving time intervals in which the Poisson probability distribution is useful. Suppose we are concerned with the occurrence of major defects in a highway 1 month after resurfacing. We will assume that the probability of a defect is the same for any two highway intervals of equal length and that the occurrence or nonoccurrence of a defect in any one interval is independent of the occurrence or nonoccurrence of a defect in any other interval. Hence, the Poisson probability distribution can be applied.

TABLE 5.9 SELECTED VALUES FROM THE POISSON PROBABILITY TABLES
EXAMPLE: $\mu = 10, x = 5; f(5) = .0378$

	μ									
x	**9.1**	**9.2**	**9.3**	**9.4**	**9.5**	**9.6**	**9.7**	**9.8**	**9.9**	**10**
0	.0001	.0001	.0001	.0001	.0001	.0001	.0001	.0001	.0001	.0000
1	.0010	.0009	.0009	.0008	.0007	.0007	.0006	.0005	.0005	.0005
2	.0046	.0043	.0040	.0037	.0034	.0031	.0029	.0027	.0025	.0023
3	.0140	.0131	.0123	.0115	.0107	.0100	.0093	.0087	.0081	.0076
4	.0319	.0302	.0285	.0269	.0254	.0240	.0226	.0213	.0201	.0189
5	.0581	.0555	.0530	.0506	.0483	.0460	.0439	.0418	.0398	**.0378**
6	.0881	.0851	.0822	.0793	.0764	.0736	.0709	.0682	.0656	.0631
7	.1145	.1118	.1091	.1064	.1037	.1010	.0982	.0955	.0928	.0901
8	.1302	.1286	.1269	.1251	.1232	.1212	.1191	.1170	.1148	.1126
9	.1317	.1315	.1311	.1306	.1300	.1293	.1284	.1274	.1263	.1251
10	.1198	.1210	.1219	.1228	.1235	.1241	.1245	.1249	.1250	.1251
11	.0991	.1012	.1031	.1049	.1067	.1083	.1098	.1112	.1125	.1137
12	.0752	.0776	.0799	.0822	.0844	.0866	.0888	.0908	.0928	.0948
13	.0526	.0549	.0572	.0594	.0617	.0640	.0662	.0685	.0707	.0729
14	.0342	.0361	.0380	.0399	.0419	.0439	.0459	.0479	.0500	.0521
15	.0208	.0221	.0235	.0250	.0265	.0281	.0297	.0313	.0330	.0347
16	.0118	.0127	.0137	.0147	.0157	.0168	.0180	.0192	.0204	.0217
17	.0063	.0069	.0075	.0081	.0088	.0095	.0103	.0111	.0119	.0128
18	.0032	.0035	.0039	.0042	.0046	.0051	.0055	.0060	.0065	.0071
19	.0015	.0017	.0019	.0021	.0023	.0026	.0028	.0031	.0034	.0037
20	.0007	.0008	.0009	.0010	.0011	.0012	.0014	.0015	.0017	.0019
21	.0003	.0003	.0004	.0004	.0005	.0006	.0006	.0007	.0008	.0009
22	.0001	.0001	.0002	.0002	.0002	.0002	.0003	.0003	.0004	.0004
23	.0000	.0001	.0001	.0001	.0001	.0001	.0001	.0001	.0002	.0002
24	.0000	.0000	.0000	.0000	.0000	.0000	.0000	.0001	.0001	.0001

Suppose we learn that major defects 1 month after resurfacing occur at the average rate of two per mile. Let us find the probability of no major defects in a particular 3-mile section of the highway. Because we are interested in an interval with a length of 3 miles, $\mu =$ (2 defects/mile)(3 miles) = 6 represents the expected number of major defects over the 3-mile section of highway. Using equation (5.11), the probability of no major defects is $f(0) = 6^0 e^{-6}/0! = .0025$. Thus, it is unlikely that no major defects will occur in the 3-mile section. In fact, this example indicates a $1 - .0025 = .9975$ probability of at least one major defect in the 3-mile highway section.

EXERCISES

Methods

38. Consider a Poisson probability distribution with $\mu = 3$.

a. Write the appropriate Poisson probability function.

b. Find $f(2)$.

c. Find $f(1)$.

d. Find $P(x \geq 2)$.

39. Consider a Poisson probability distribution with an average number of occurrences per time period of two.
- **a.** Write the appropriate Poisson probability function.
- **b.** What is the average number of occurrences in three time periods?
- **c.** Write the appropriate Poisson probability function to determine the probability of x occurrences in three time periods.
- **d.** Find the probability of two occurrences in one time period.
- **e.** Find the probability of six occurrences in three time periods.
- **f.** Find the probability of five occurrences in two time periods.

Applications

40. Phone calls arrive at the rate of 48 per hour at the reservation desk for Regional Airways.
- **a.** Find the probability of receiving three calls in a 5-minute interval of time.
- **b.** Find the probability of receiving exactly 10 calls in 15 minutes.
- **c.** Suppose no calls are currently on hold. If the agent takes 5 minutes to complete the current call, how many callers do you expect to be waiting by that time? What is the probability that none will be waiting?
- **d.** If no calls are currently being processed, what is the probability that the agent can take 3 minutes for personal time without being interrupted by a call?

41. During the period of time phone-in registrations are being taken at a local university, calls come in at the rate of one every 2 minutes.
- **a.** What is the expected number of calls in 1 hour?
- **b.** What is the probability of three calls in 5 minutes?
- **c.** What is the probability of no calls in a 5-minute period?

42. The mean number of times per year that Barron's subscribers take domestic flights for personal reasons is four (*Barron's* 1995 Primary Reader Survey).
- **a.** What is the probability a subscriber takes two domestic flights for personal reasons in a year?
- **b.** What is the mean number of domestic flights for personal reasons in a 3-month period?
- **c.** What is the probability a subscriber takes one or more domestic flights for personal reasons in a 6-month period?

43. Airline passengers arrive randomly and independently at the passenger-screening facility at a major international airport. The mean arrival rate is 10 passengers per minute.
- **a.** What is the probability of no arrivals in a 1-minute period?
- **b.** What is the probability that three or fewer passengers arrive in a 1-minute period?
- **c.** What is the probability of no arrivals in a 15-second period?
- **d.** What is the probability of at least one arrival in a 15-second period?

44. Investment activities for subscribers to *The Wall Street Journal Interactive Edition* show that the average number of security transactions per year is approximately 15 (*The Wall Street Journal* Subscriber Study, 1999). Assume that a particular investor makes transactions at this rate. Furthermore, assume that the probability of a transaction for this investor is the same for any two months and transactions in one month are independent of transactions in any other month. Answer the following questions.
- **a.** What is the mean number of transactions per month?
- **b.** What is the probability of no stock transactions during a month?
- **c.** What is the probability of exactly one stock transaction during a month?
- **d.** What is the probability of more than one stock transaction during a month?

45. Each year, 450 accidental deaths due to firearms occur in the 15–24 age group (National Safety Council, *Accident Facts,* 1996).
- **a.** What is the average number of accidental deaths due to firearms per week?
- **b.** What is the probability of no accidental deaths due to firearms in a typical week?
- **c.** What is the probability of two or more accidental deaths due to firearms in a typical day?

5.6 HYPERGEOMETRIC PROBABILITY DISTRIBUTION

The hypergeometric probability distribution is closely related to the binomial probability distribution. The key difference between the two probability distributions is that with the hypergeometric distribution, the trials are not independent; and the probability of success changes from trial to trial.

The usual notation in applications of the hypergeometric probability distribution is to let r denote the number of elements in the population of size N that are labeled success and $N - r$ denote the number of elements in the population that are labeled failure. The **hypergeometric probability function** is used to compute the probability that in a random sample of n elements, selected without replacement, we will obtain x elements labeled success and $n - x$ elements labeled failure. For this result to occur, we must obtain x successes from the r successes in the population and $n - x$ failures from the $N - r$ failures. The following hypergeometric probability function provides $f(x)$, the probability of obtaining x successes in a sample of size n.

Hypergeometric Probability Function

$$f(x) = \frac{\binom{r}{x}\binom{N-r}{n-x}}{\binom{N}{n}} \quad \text{for } 0 \le x \le r \tag{5.12}$$

where

$f(x)$ = probability of x successes in n trials
n = number of trials
N = number of elements in the population
r = number of elements in the population labeled success

Note that $\binom{N}{n}$ represents the number of ways a sample of size n can be selected from a population of size N; $\binom{r}{x}$ represents the number of ways that x successes can be selected from a total of r successes in the population; and $\binom{N-r}{n-x}$ represents the number of ways that $n - x$ failures can be selected from a total of $N - r$ failures in the population.

To illustrate the computations involved in using equation (5.12), let us consider the problem of selecting two people from a five-member committee to send to a Las Vegas convention. Assume that the five-member committee consists of three women and two men. To determine the probability of randomly selecting two women, we can use equation (5.12) with $n = 2$, $N = 5$, $r = 3$, and $x = 2$.

$$f(2) = \frac{\binom{3}{2}\binom{2}{0}}{\binom{5}{2}} = \frac{\left(\frac{3!}{2!1!}\right)\left(\frac{2!}{2!0!}\right)}{\left(\frac{5!}{2!3!}\right)} = \frac{3}{10} = .30$$

Suppose we learn later that three committee members will be allowed to make the trip. If we use $n = 3$, $N = 5$, $r = 3$, and $x = 2$, the probability that exactly two of the three members will be women is

$$f(2) = \frac{\binom{3}{2}\binom{2}{1}}{\binom{5}{2}} = \frac{\left(\frac{3!}{2!1!}\right)\left(\frac{2!}{1!1!}\right)}{\left(\frac{5!}{2!3!}\right)} = \frac{6}{10} = .60$$

As another illustration, suppose a population consists of 10 items, four of which are classified as defective and six of which are classified as nondefective. What is the probability that a random sample of size three will contain two defective items? For this problem we can think of obtaining a defective item as a "success." Using equation (5.12) with $n = 3$, $N = 10$, $r = 4$, and $x = 2$, we can compute $f(2)$ as follows.

$$f(2) = \frac{\binom{4}{2}\binom{6}{1}}{\binom{10}{3}} = \frac{\left(\frac{4!}{2!2!}\right)\left(\frac{6!}{1!5!}\right)}{\left(\frac{10!}{3!7!}\right)} = \frac{36}{120} = .30$$

EXERCISES

Methods

46. Suppose $N = 10$ and $r = 3$. Compute the hypergeometric probabilities for the following values of n and x.
    a. $n = 4, x = 1$
    b. $n = 2, x = 2$
    c. $n = 2, x = 0$
    d. $n = 4, x = 2$
47. Suppose $N = 15$ and $r = 4$. What is the probability of $x = 3$ for $n = 10$?

Applications

48. According to *Beverage Digest,* Coke Classic and Pepsi ranked number one and number two in sales (*The Wall Street Journal Almanac,* 1998). Assume that in a group of 10 individuals, six preferred Coke Classic and four preferred Pepsi. A random sample of three of these individuals is selected.
    a. What is the probability that exactly two preferred Coke Classic?
    b. What is the probability that the majority (either two or three) preferred Pepsi?
49. Blackjack, or "twenty-one" as it is frequently called, is a popular gambling game played in Las Vegas casinos. A player is dealt two cards. Face cards (jacks, queens, and kings) and tens have a point value of 10. Aces have a point value of 1 or 11. A 52-card deck has 16 cards with a point value of 10 (jacks, queens, kings, and tens) and four aces.
    a. What is the probability that both cards dealt are aces or 10-point cards?
    b. What is the probability that both of the cards are aces?
    c. What is the probability that both of the cards have a point value of 10?
    d. A blackjack is a 10-point card and an ace for a value of 21. Use your answers to parts (a), (b), and (c) to determine the probability that a player is dealt blackjack. (Hint: Part (d) is not a hypergeometric problem. Develop your own logical relationship as to how the hypergeometric probabilities from parts (a), (b), and (c) can be combined to answer this question.)

50. Axline Computers manufactures personal computers at two plants, one in Texas and the other in Hawaii. The Texas plant has 40 employees; the Hawaii plant has 20. A random sample of 10 employees is to be asked to fill out a benefits questionnaire.

a. What is the probability that none of the employees are at the plant in Hawaii?
b. What is the probability that one of the employees is at the plant in Hawaii?
c. What is the probability that two or more of the employees are at the plant in Hawaii?
d. What is the probability that nine of the employees are at the plant in Texas?

51. Of the 25 students (14 boys and 11 girls) in the sixth-grade class at St. Andrew School, five students were absent Thursday.

a. What is the probability that two of the absent students were girls?
b. What is the probability that two of the absent students were boys?
c. What is the probability that all of the absent students were boys?
d. What is the probability that none of the absent students were boys?

52. A shipment of 10 items has two defective and eight nondefective units. In the inspection of the shipment, a sample of units will be selected and tested. If a defective unit is found, the shipment of 10 units will be rejected.

a. If a sample of three items is selected, what is the probability that the shipment will be rejected?
b. If a sample of four items is selected, what is the probability that the shipment will be rejected?
c. If a sample of five items is selected, what is the probability that the shipment will be rejected?
d. If management would like a .90 probability of rejecting a shipment with two defective and eight nondefective units, how large a sample would you recommend?

SUMMARY

The concept of a random variable was introduced to provide a numerical description of the outcome of an experiment. We saw that the probability distribution for a random variable describes how the probabilities are distributed over the values the random variable can assume. For any discrete random variable x, the probability distribution is defined by a probability function, denoted by $f(x)$, which provides the probability associated with each value of the random variable. Once the probability function has been defined, we can compute the expected value and the variance for the random variable.

The binomial probability distribution can be used to determine the probability of x successes in n trials whenever the experiment has the following properties:

1. The experiment consists of a sequence of n identical trials.
2. Two outcomes are possible on each trial, one called success and the other failure.
3. The probability of a success p does not change from trial to trial. Consequently, the probability of failure, $1 - p$, does not change from trial to trial.
4. The trials are independent.

When the four conditions hold, a binomial probability function can be used to determine the probability of x successes in n trials. Formulas were also presented for the mean and variance of the binomial probability distribution.

The Poisson probability distribution is used when it is desirable to determine the probability of x occurrences over an interval of time or space. The following assumptions are necessary for the Poisson distribution to be applicable.

1. The probability of an occurrence of the event is the same for any two intervals of equal length.
2. The occurrence or nonoccurrence of the event in any interval is independent of the occurrence or nonoccurrence of the event in any other interval.

A third discrete probability distribution, the hypergeometric, was introduced in Section 5.6. Like the binomial, it is used to compute the probability of x successes in n trials. But, in contrast to the binomial, the probability of success changes from trial to trial.

GLOSSARY

Random variable A numerical description of the outcome of an experiment.

Discrete random variable A random variable that may assume either a finite number of values or an infinite sequence of values.

Continuous random variable A random variable that may assume any value in an interval or collection of intervals.

Probability distribution A description of how the probabilities are distributed over the values the random variable can assume.

Probability function A function, denoted by $f(x)$, that provides the probability that x assumes a particular value for a discrete random variable.

Discrete uniform probability distribution A probability distribution for which each possible value of the random variable has the same probability.

Expected value A measure of the mean, or central location, of a random variable.

Variance A measure of the variability, or dispersion, of a random variable.

Standard deviation The positive square root of the variance.

Binomial experiment A probability experiment having the four properties stated at the beginning of Section 5.4.

Binomial probability distribution A probability distribution showing the probability of x successes in n trials of a binomial experiment.

Binomial probability function The function used to compute probabilities in a binomial experiment.

Poisson probability distribution A probability distribution showing the probability of x occurrences of an event over a specified interval of time or space.

Poisson probability function The function used to compute Poisson probabilities.

Hypergeometric probability function The function used to compute the probability of x successes in n trials when the trials are dependent.

KEY FORMULAS

Discrete Uniform Probability Function

$$f(x) = 1/n \tag{5.3}$$

where $n =$ the number of values the random variable may assume

Expected Value of a Discrete Random Variable

$$E(x) = \mu = \Sigma x f(x) \tag{5.4}$$

Variance of a Discrete Random Variable

$$\text{Var}(x) = \sigma^2 = \Sigma(x - \mu)^2 f(x) \tag{5.5}$$

Number of Experimental Outcomes Providing Exactly x Successes in n Trials

$$\binom{n}{x} = \frac{n!}{x!(n-x)!} \tag{5.6}$$

Binomial Probability Function

$$f(x) = \binom{n}{x} p^x (1-p)^{(n-x)} \tag{5.8}$$

Expected Value for the Binomial Probability Distribution

$$E(x) = \mu = np \tag{5.9}$$

Variance for the Binomial Probability Distribution

$$\text{Var}(x) = \sigma^2 = np(1-p) \tag{5.10}$$

Poisson Probability Function

$$f(x) = \frac{\mu^x e^{-\mu}}{x!} \tag{5.11}$$

Hypergeometric Probability Function

$$f(x) = \frac{\binom{r}{x}\binom{N-r}{n-x}}{\binom{N}{n}} \quad \text{for } 0 \le x \le r \tag{5.12}$$

SUPPLEMENTARY EXERCISES

53. A *Wall Street Journal*/NBC News poll asked 2004 adults how they felt about their opportunities for career advancement (*The Wall Street Journal,* September 19, 1997). Twenty-three percent indicated they were very satisfied; 38% indicated they were somewhat satisfied; 3% indicated they were not sure; 18% indicated they were somewhat dissatisfied; and 18% indicated they were very dissatisfied. Let x be a random variable representing level of satisfaction. We let $x = 5$ for very satisfied, $x = 4$ for somewhat satisfied, $x = 3$ for not sure, $x = 2$ for somewhat dissatisfied, and $x = 1$ for very dissatisfied.

a. Use the preceding numerical values to develop a probability distribution for level of satisfaction with opportunities for career advancement.

b. Compute the expected level of satisfaction. Does a higher score indicate more or less satisfaction?

c. Compute the variance and standard deviation for the satisfaction scores.

54. An opinion survey of 1009 adults was conducted by Louis Harris & Associates, Inc. (*Business Week,* December 29, 1997). Table 5.10 shows the probability distribution of responses to a question concerning the stock market's valuation.

a. Show that the probability distribution in Table 5.10 satisfies the properties of all probability distributions.

b. What are the expected value and variance of the probability distribution of opinions?

c. Comment on whether people think the stock market is overvalued.

TABLE 5.10 PROBABILITY DISTRIBUTION OF OPINIONS CONCERNING STOCK MARKET'S VALUATION

Market Valuation	Random Variable (x)	Probability $f(x)$
Very undervalued	1	0.02
Somewhat undervalued	2	0.06
Fairly valued	3	0.28
Somewhat overvalued	4	0.54
Very overvalued	5	0.10

55. The budgeting process for a midwestern college resulted in expense forecasts for the coming year (in millions of dollars) of \$9, \$10, \$11, \$12, and \$13. Because the actual expenses are unknown, the following respective probabilities are assigned: .3, .2, .25, .05, and .2.

a. Show the probability distribution for the expense forecast.

b. What is the expected value of the expense forecast for the coming year?

c. What is the variance in the expense forecast for the coming year?

d. If income projections for the year are estimated at \$12 million, comment on the financial position of the college.

56. Kristen DelGuzzi conducted a study of Hamilton County judges and courts over a 3-year period (*The Cincinnati Enquirer,* January 11, 1998). One finding was that 4% of the cases decided in Common Pleas Court were appealed.

a. If 20 cases are heard on a particular day, what is the probability that 3 will be appealed?

b. If 20 cases are heard on a particular day, what is the probability none will be appealed?

c. It is not unusual for 1200 cases to be heard in a month. What is the expected number of appeals?

d. If 1200 cases are heard in a month, what are the variance and standard deviation in the number of appeals?

57. A company is planning to interview Internet users to learn how its proposed web site will be received by different age groups. According to the Bureau of the Census 40% of individuals aged 18 to 54, and 12% of individuals aged 55 and over use the Internet (*Statistical Abstract of the United States,* 2000).

a. How many people from the 18–54 age group must be contacted to find an expected number of at least 10 Internet users?

b. How many people from the 55 and over age group must be contacted to find an expected number of at least 10 Internet users?

c. If you contact the number of 18–54-year-old people suggested in part (a), what is the standard deviation of the number who will be Internet users?

d. If you contact the number of 55-and-over-year-old people suggested in part (b), what is the standard deviation of the number who will be Internet users?

58. Many companies use a quality control technique called *acceptance sampling* to monitor incoming shipments of parts, raw materials, and so on. In the electronics industry, component parts are commonly shipped from suppliers in large lots. Inspection of a sample of n components can be viewed as the n trials of a binomial experiment. The outcome for each component tested (trial) will be that the component is classified as good or defective. Reynolds Electronics accepts a lot from a particular supplier if the defective components in the lot do not exceed 1%. Suppose a random sample of five items from a recent shipment has been tested.

a. Assume that 1% of the shipment is defective. Compute the probability that no items in the sample are defective.

b. Assume that 1% of the shipment is defective. Compute the probability that exactly one item in the sample is defective.

c. What is the probability of observing one or more defective items in the sample if 1% of the shipment is defective?

d. Would you feel comfortable accepting the shipment if one item were found to be defective? Why or why not?

59. The unemployment rate is 4.1% (*Barron's,* September 4, 2000). Assume that 100 employable people are selected randomly.

a. What is the expected number who are unemployed?

b. What is the variance and standard deviation of the number who are unemployed?

60. In December 1997, the U.S. Department of Justice issued an injunction against Microsoft Corporation for bundling its Internet Explorer® Web browser with its Windows 95® operating system (*Fortune,* February 2, 1998). Public opinion was divided on whether Microsoft was a monopoly. In a *Fortune* poll, 41% of respondents agreed with the statement, "Microsoft is a monopoly." Suppose a sample of 800 people has been taken.

a. How many would you expect to agree that Microsoft is a monopoly?

b. What is the standard deviation of the number of respondents who believe Microsoft is a monopoly?

c. What is the standard deviation of the number of respondents who do not believe Microsoft is a monopoly?

61. Cars arrive at a car wash randomly and independently; the probability of an arrival is the same for any two time intervals of equal length. The mean arrival rate is 15 cars per hour. What is the probability that 20 or more cars will arrive during any given hour of operation?

62. A new automated production process has had an average of 1.5 breakdowns per day. Because of the cost associated with a breakdown, management is concerned about the possibility of having three or more breakdowns during a day. Assume that breakdowns occur randomly, that the probability of a breakdown is the same for any two time intervals of equal length, and that breakdowns in one period are independent of breakdowns in other periods. What is the probability of having three or more breakdowns during a day?

63. A regional director responsible for business development in the state of Pennsylvania is concerned about the number of small business failures. If the mean number of small business failures per month is 10, what is the probability that exactly four small businesses will fail during a given month? Assume that the probability of a failure is the same for any two months and that the occurrence or nonoccurrence of a failure in any month is independent of failures in any other month.

64. Customer arrivals at a bank are random and independent; the probability of an arrival in any 1-minute period is the same as the probability of an arrival in any other 1-minute period. Answer the following questions, assuming a mean arrival rate of three customers per minute.

a. What is the probability of exactly three arrivals in a 1-minute period?

b. What is the probability of at least three arrivals in a 1-minute period?

65. Most people are familiar with the game of five-card draw poker. With 52 cards including four aces, what is the probability that the deal of the five cards provides:

a. A pair of aces?

b. Exactly one ace?

c. No aces?

d. At least one ace?

66. According to the WTA Tour and ATP Tour, four of the top 10 women tennis players use Wilson rackets (*USA Today,* March 2, 1995). Suppose two of these players have reached the finals of a tournament.

a. What is the probability that exactly one uses a Wilson racket?

b. What is the probability that both use Wilson rackets?

c. What is the probability that neither uses a Wilson racket?

Appendix 5.1 DISCRETE PROBABILITY DISTRIBUTIONS WITH MINITAB

Statistical packages such as Minitab offer a relatively easy and efficient procedure for computing binomial probabilities. In this appendix, we show the step-by-step procedure for determining the binomial probabilities for the Martin Clothing Store problem in Section 5.4. Recall that the desired binomial probabilities are based on $n = 10$ and $p = .30$. Before beginning the Minitab routine, the user must enter the desired values of the random variable x into a column of the worksheet. We entered the values 0, 1, 2, . . . , 10 in column 1 (see Figure 5.5) to generate the entire binomial probability distribution. The Minitab steps to obtain the desired binomial probabilities follow.

Step 1. Select the **Calc** pull-down menu
Step 2. Choose **Probability Distributions**
Step 3. Choose **Binomial**
Step 4. When the Binomial Distribution dialog box appears:
Select **Probability**
Enter 10 in the **Number of trials** box
Enter .3 in the **Probability of success** box
Enter C1 in the **Input column** box
Click **OK**

The Minitab output with the binomial probabilities will appear as shown in Figure 5.5.

Minitab provides Poisson probabilities in a similar manner. The only differences are in step 3, where the Poisson option would be selected, and step 4, where the **Mean** would be entered rather than the number of trials and the probability of success.

Appendix 5.2 DISCRETE PROBABILITY DISTRIBUTIONS WITH EXCEL

Excel provides functions for computing probabilities for the binomial, Poisson, and hypergeometric distributions introduced in this chapter. The Excel function for computing binomial probabilities is BINOMDIST. It has four arguments: x (the number of successes), n (the number of trials), p (the probability of success), and cumulative. FALSE is used for the fourth argument (cumulative) if we want the probability of x successes, and TRUE is used for the fourth argument if we want the cumulative probability of x or fewer successes. Here we show how to compute the probabilities of 0 through 10 successes in the Martin Clothing Store problem of Section 5.4 (see Figure 5.5).

As we describe the worksheet development refer to Figure 5.6; the formula worksheet is in the background and the value worksheet is in the foreground. The following steps will generate the desired probabilities:

Step 1. Use the BINOMDIST function to compute the probability of $x = 0$ by entering the following formula into cell B2:

=BINOMDIST(A2,10,0.3,FALSE)

Step 2. Copy the formula in cell B2 into cells B3:B12.

The value worksheet in the foreground of Figure 5.6 shows that the probabilities obtained are the same as in Figure 5.5. Poisson and hypergeometric probabilities can be computed in a similar fashion. The POISSON and HYPERGEOMDIST functions are used. Excel's function wizard can help the user in entering the proper arguments for these functions (see Appendix 2.2).

FIGURE 5.6 EXCEL WORKSHEET FOR COMPUTING BINOMIAL PROBABILITIES

	A	B	C	D
1	*x*	*f(x)*		
2	0	=BINOMDIST(A2,10,0.3,FALSE)		
3	1	=BINOMDIST(A3,10,0.3,FALSE)		
4	2	=BINOMDIST(A4,10,0.3,FALSE)		
5	3	=BINOMDIST(A5,10,0.3,FALSE)		
6	4	=BINOMDIST(A6,10,0.3,FALSE)		
7	5	=BINOMDIST(A7,10,0.3,FALSE)		
8	6	=BINOMDIST(A8,10,0.3,FALSE)		
9	7	=BINOMDIST(A9,10,0.3,FALSE)		
10	8	=BINOMDIST(A10,10,0.3,FALSE)		
11	9	=BINOMDIST(A11,10,0.3,FALSE)		
12	10	=BINOMDIST(A12,10,0.3,FALSE)		
13				
14				

	A	B	C	D
1	*x*	*f(x)*		
2	0	0.0282		
3	1	0.1211		
4	2	0.2335		
5	3	0.2668		
6	4	0.2001		
7	5	0.1029		
8	6	0.0368		
9	7	0.0090		
10	8	0.0014		
11	9	0.0001		
12	10	0.0000		
13				

Chapter 6

CONTINUOUS PROBABILITY DISTRIBUTIONS

CONTENTS

STATISTICS IN PRACTICE

PROCTER & GAMBLE*

Cincinnati, Ohio

Procter & Gamble (P&G) is in the consumer products business worldwide. P&G produces and markets such products as detergents, disposable diapers, over-the-counter pharmaceuticals, dentifrices, bar soaps, mouthwashes, and paper towels. It has the leading brand in more categories than any other consumer products company.

As a leader in the application of statistical methods in decision making, P&G employs people with diverse academic backgrounds: engineering, statistics, operations research, and business. The major quantitative technologies for which these people provide support are probabilistic decision and risk analysis, advanced simulation, quality improvement, and quantitative methods (e.g., linear programming, regression analysis, probability analysis).

Some of Procter & Gamble's many well-known products. © Joe Higgins/South-Western.

The Industrial Chemicals Division of P&G is a major supplier of fatty alcohols derived from natural substances such as coconut oil and from petroleum-based derivatives. The division wanted to know the economic risks and opportunities of expanding its fatty-alcohol production facilities, and P&G's experts in probabilistic decision and risk analysis were called in to help. After structuring and modeling the problem, they determined that the key to profitability was the cost difference between the petroleum- and coconut-based raw materials. Future costs were unknown, but the analysts were able to represent them with the following continuous random variables.

x = the coconut oil price per pound of fatty alcohol

and

y = the petroleum raw material price per pound of fatty alcohol

Because the key to profitability was the difference between these two random variables, a third random variable, $d = x - y$, was used in the analysis. Experts were interviewed to determine the probability distribution for x and y. In turn, this information was used to develop a probability distribution for the difference in prices d. This continuous probability distribution showed a .90 probability that the price difference would be \$.0655 or less and a .50 probability that the price difference would be \$.035 or less. In addition, there was only a .10 probability that the price difference would be \$.0045 or less.[†]

The Industrial Chemicals Division thought that being able to quantify the impact of raw material price differences was key to reaching a consensus. The probabilities obtained were used in a sensitivity analysis of the raw material price difference. The analysis yielded sufficient insight to form the basis for a recommendation to management.

The use of continuous random variables and their probability distributions was helpful to P&G in analyzing the economic risks associated with its fatty-alcohol production. In this chapter, you will gain an understanding of continuous random variables and their probability distributions including one of the most important probability distributions in statistics, the normal distribution.

*The authors are indebted to Joel Kahn of Procter & Gamble for providing this Statistics in Practice.

[†]The price differences stated here have been modified to protect proprietary data.

In the preceding chapter we discussed discrete random variables and their probability distributions. In this chapter we turn to the study of continuous random variables. Specifically, we discuss three continuous probability distributions: the uniform, the normal, and the exponential.

A fundamental difference separates discrete and continuous random variables in terms of how probabilities are computed. For a discrete random variable, the probability function $f(x)$ provides the probability that the random variable assumes a particular value. With continuous random variables the counterpart of the probability function is the **probability density function,** also denoted by $f(x)$. The difference is that the probability density function does not directly provide probabilities. However, the area under the graph of $f(x)$ corresponding to a given interval does provide the probability that the continuous random variable x assumes a value in that interval. So when we compute probabilities for continuous random variables we are computing the probability that the random variable assumes any value in an interval.

One of the implications of the definition of probability for continuous random variables is that the probability of any particular value of the random variable is zero, because the area under the graph of $f(x)$ at any particular point is zero. In Section 6.1 we demonstrate these concepts for a continuous random variable that has a uniform probability distribution.

Much of the chapter is devoted to describing and showing applications of the normal probability distribution. The normal probability distribution is of major importance; it is used extensively in statistical inference. The chapter closes with a discussion of the exponential probability distribution.

6.1 UNIFORM PROBABILITY DISTRIBUTION

Whenever the probability is proportional to the length of the interval, the random variable is uniformly distributed.

Consider the random variable x that represents the flight time of an airplane traveling from Chicago to New York. Suppose the flight time can be any value in the interval from 120 minutes to 140 minutes. Because the random variable x can assume any value in that interval, x is a continuous rather than a discrete random variable. Let us assume that sufficient actual flight data are available to conclude that the probability of a flight time within any 1-minute interval is the same as the probability of a flight time within any other 1-minute interval contained in the larger interval from 120 to 140 minutes. With every 1-minute interval being equally likely, the random variable x is said to have a **uniform probability distribution.** The probability density function, which defines the uniform probability distribution for the flight time random variable, is

$$f(x) = \begin{cases} 1/20 & \text{for } 120 \le x \le 140 \\ 0 & \text{elsewhere} \end{cases}$$

Figure 6.1 is a graph of this probability density function. In general, the uniform probability density function for a random variable x is found by using the following formula.

Uniform Probability Density Function

$$f(x) = \begin{cases} \dfrac{1}{b - a} & \text{for } a \le x \le b \\ 0 & \text{elsewhere} \end{cases} \qquad \textbf{(6.1)}$$

In the flight-time example, $a = 120$ and $b = 140$.

FIGURE 6.1 UNIFORM PROBABILITY DENSITY FUNCTION FOR FLIGHT TIME

As noted in the introduction, for a continuous random variable, we consider probability only in terms of the likelihood that a random variable has a value within a specified interval. In the flight-time example, an acceptable probability question is: What is the probability that the flight time is between 120 and 130 minutes? That is, what is $P(120 \leq x \leq 130)$? Because the flight time must be between 120 and 140 minutes and because the probability is described as being uniform over this interval, we feel comfortable saying $P(120 \leq x \leq 130) = .50$. In the following subsection we show that this probability can be computed as the area under the graph of $f(x)$ from 120 to 130.

Area as a Measure of Probability

Let us make an observation about the graph in Figure 6.2. Consider the area under the graph of $f(x)$ in the interval from 120 to 130. The area is rectangular, and the area of a rectangle is simply the width multiplied by the height. With the width of the interval equal to $130 - 120 = 10$ and the height equal to the value of the probability density function $f(x) = 1/20$, we have area = width $\times$ height $= 10(1/20) = 10/20 = .50$.

FIGURE 6.2 AREA PROVIDES PROBABILITY OF FLIGHT TIME BETWEEN 120 AND 130 MINUTES

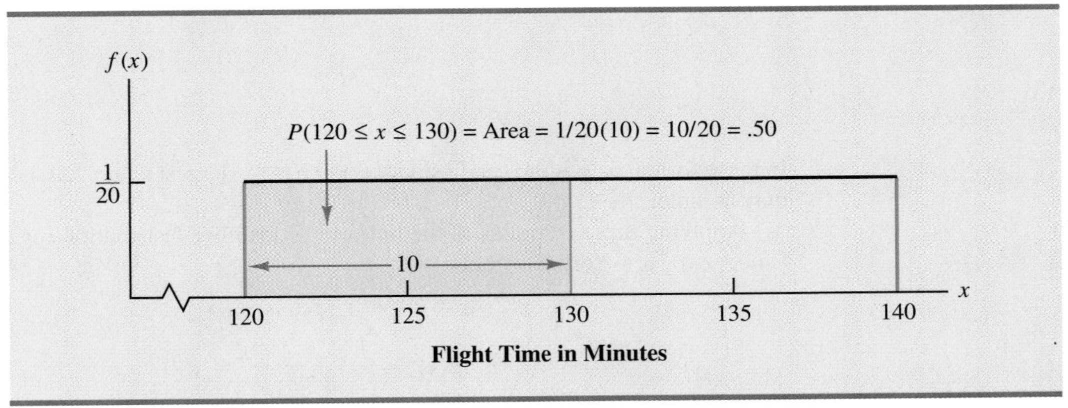

What observation can you make about the area under the graph of $f(x)$ and probability? They are identical! Indeed, it is true for all continuous random variables. Once a probability density function $f(x)$ has been identified, the probability that x takes a value between some lower value x_1 and some higher value x_2 can be found by computing the area under the graph of $f(x)$ over the interval x_1 to x_2.

Given the uniform probability distribution for flight time and using the interpretation of area as probability, we can answer any number of probability questions about flight times. For example, what is the probability of a flight time between 128 and 136 minutes? The width of the interval is $136 - 128 = 8$. With the uniform height of $f(x) = 1/20$, we see that $P(128 \leq x \leq 136) = 8(1/20) = .40$.

Note that $P(120 \leq x \leq 140) = 20(1/20) = 1$. That is, the total area under the graph of $f(x)$ is equal to 1. This property holds for all continuous probability distributions and is the analog of the condition that the sum of the probabilities must equal 1 for a discrete probability function. For a continuous probability density function, we must also require that $f(x) \geq 0$ for all values of x. This requirement is the analog of the requirement that $f(x) \geq 0$ for discrete probability functions.

Two major differences stand out between the treatment of continuous random variables and the treatment of their discrete counterparts.

To see that the probability of any single point is 0, refer to Figure 6.2 and compute the probability of a single point, say, $x = 125$. $P(x = 125) = P(125 \leq x \leq 125) = 0(1/20) = 0$.

1. We no longer talk about the probability of the random variable assuming a particular value. Instead, we talk about the probability of the random variable assuming a value within some given interval.
2. The probability of the random variable assuming a value within some given interval from x_1 to x_2 is defined to be the area under the graph of the probability density function between x_1 and x_2. It implies that the probability of a continuous random variable assuming any particular value exactly is zero, because the area under the graph of $f(x)$ at a single point is zero.

The calculation of the expected value and variance for a continuous random variable is analogous to that for a discrete random variable. However, because the computational procedure involves integral calculus, we leave the derivation of the appropriate formulas to more advanced texts.

For the uniform continuous probability distribution introduced in this section, the formulas for the expected value and variance are

$$E(x) = \frac{a + b}{2}$$

$$\text{Var}(x) = \frac{(b - a)^2}{12}$$

In these formulas, a is the smallest value and b is the largest value that the random variable may assume.

Applying these formulas to the uniform probability distribution for flight times from Chicago to New York, we obtain

$$E(x) = \frac{(120 + 140)}{2} = 130$$

$$\text{Var}(x) = \frac{(140 - 120)^2}{12} = 33.33$$

The standard deviation of flight times can be found by taking the square root of the variance. Thus, $\sigma = 5.77$ minutes.

NOTES AND COMMENTS

1. For a continuous random variable the probability of any particular value is zero; thus, $P(a \le x \le b) = P(a < x < b)$. This result shows that the probability of a random variable assuming a value in any interval is the same whether or not the endpoints are included.
2. To see more clearly why the height of a probability density function is not a probability, think about a random variable with the following uniform probability distribution.

$$f(x) = \begin{cases} 2 & \text{for } 0 \le x \le .5 \\ 0 & \text{elsewhere} \end{cases}$$

The height of the probability density function, $f(x)$, is 2 for values of x between 0 and .5. However, we know probabilities can never be greater than 1. Thus, we see that $f(x)$ cannot be interpreted as the probability of x.

EXERCISES

Methods

1. The random variable x is known to be uniformly distributed between 1.0 and 1.5.
   a. Show the graph of the probability density function.
   b. Find $P(x = 1.25)$.
   c. Find $P(1.0 \le x \le 1.25)$.
   d. Find $P(1.20 < x < 1.5)$.
2. The random variable x is known to be uniformly distributed between 10 and 20.
   a. Show the graph of the probability density function.
   b. Find $P(x < 15)$.
   c. Find $P(12 \le x \le 18)$.
   d. Find $E(x)$.
   e. Find $\text{Var}(x)$.

Applications

3. Delta Airlines quotes a flight time of 2 hours, 5 minutes for its flights from Cincinnati to Tampa. Suppose we believe that actual flight times are uniformly distributed between 2 hours and 2 hours, 20 minutes.
   a. Show the graph of the probability density function for flight times.
   b. What is the probability that the flight will be no more than five minutes late?
   c. What is the probability that the flight will be more than 10 minutes late?
   d. What is the expected flight time?

4. Most computer languages have a function that can be used to generate random numbers. In Excel, the RAND function can be used to generate random numbers between 0 and 1. If we let x denote a random number generated, then x is a continuous random variable with the following probability density function.

$$f(x) = \begin{cases} 1 & \text{for } 0 \le x \le 1 \\ 0 & \text{elsewhere} \end{cases}$$

a. Graph the probability density function.
b. What is the probability of generating a random number between .25 and .75?
c. What is the probability of generating a random number with a value less than or equal to .30?
d. What is the probability of generating a random number with a value greater than .60?

5. The driving distance for the top 60 women golfers on the LPGA tour is between 238.9 and 261.2 yards (*Golfweek,* December 6, 1997). Assume that the driving distance for these women is uniformly distributed over this interval.
   a. Give a mathematical expression for the probability density function of driving distance.
   b. What is the probability the driving distance for one of these women is less than 250 yards?
   c. What is the probability the driving distance for one of these women is at least 255 yards?
   d. What is the probability the driving distance for one of these women is between 245 and 260 yards?
   e. How many of these women drive the ball at least 250 yards?
6. The label on a bottle of liquid detergent shows contents to be 12 ounces per bottle. The production operation fills the bottle uniformly according to the following probability density function.

$$f(x) = \begin{cases} 8 & \text{for } 11.975 \le x \le 12.10 \\ 0 & \text{elsewhere} \end{cases}$$

   a. What is the probability that a bottle will be filled with between 12 and 12.05 ounces?
   b. What is the probability that a bottle will be filled with 12.02 or more ounces?
   c. Quality control accepts a bottle that is filled to within .02 ounces of the number of ounces shown on the container label. What is the probability that a bottle of this liquid detergent will fail to meet the quality control standard?
7. Suppose we are interested in bidding on a piece of land and we know there is one other bidder.* The seller has announced that the highest bid in excess of $10,000 will be accepted. Assume that the competitor's bid x is a random variable that is uniformly distributed between $10,000 and $15,000.
   a. Suppose you bid $12,000. What is the probability that your bid will be accepted?
   b. Suppose you bid $14,000. What is the probability that your bid will be accepted?
   c. What amount should you bid to maximize the probability that you get the property?
   d. Suppose you know someone who is willing to pay you $16,000 for the property. Would you consider bidding less than the amount in part (c)? Why or why not?

6.2 NORMAL PROBABILITY DISTRIBUTION

Abraham de Moivre, a French mathematician, published The Doctrine of Chances *in 1733. He derived the normal probability distribution.*

The most important probability distribution for describing a continuous random variable is the **normal probability distribution.** The normal probability distribution has been used in a wide variety of practical applications in which the random variables are heights and weights of people, test scores, scientific measurements, amounts of rainfall, and so on. It is also widely used in statistical inference, which is the major topic of the remainder of this book. In such applications, the normal probability distribution provides a description of the likely results obtained through sampling.

---

*This exercise is based on a problem suggested to us by Professor Roger Myerson of Northwestern University.

Normal Curve

The form, or shape, of the normal probability distribution is illustrated by the bell-shaped curve in Figure 6.3. The probability density function that defines the bell-shaped curve of the normal probability distribution follows.

Normal Probability Density Function

$$f(x) = \frac{1}{\sigma\sqrt{2\pi}} e^{-(x-\mu)^2/2\sigma^2} \tag{6.2}$$

where

$$\begin{aligned} \mu &= \text{mean} \\ \sigma &= \text{standard deviation} \\ \pi &= 3.14159 \\ e &= 2.71828 \end{aligned}$$

We make several observations about the characteristics of the normal probability distribution.

The normal curve has two parameters, μ and σ. They determine the location and shape of the normal probability distribution.

1. The entire family of normal probability distributions is differentiated by its mean μ and its standard deviation σ.
2. The highest point on the normal curve is at the mean, which is also the median and mode of the distribution.
3. The mean of the distribution can be any numerical value: negative, zero, or positive. Three normal distributions with the same standard deviation but three different means (−10, 0, and 20) are shown here.

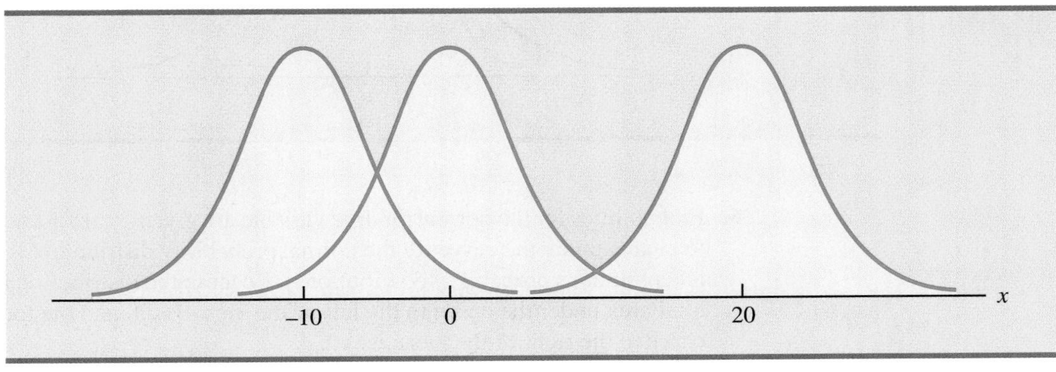

4. The normal probability distribution is symmetric, with the shape of the curve to the left of the mean a mirror image of the shape of the curve to the right of the mean. The tails of the curve extend to infinity in both directions and theoretically never touch the horizontal axis.

FIGURE 6.3 BELL-SHAPED CURVE FOR THE NORMAL PROBABILITY DISTRIBUTION

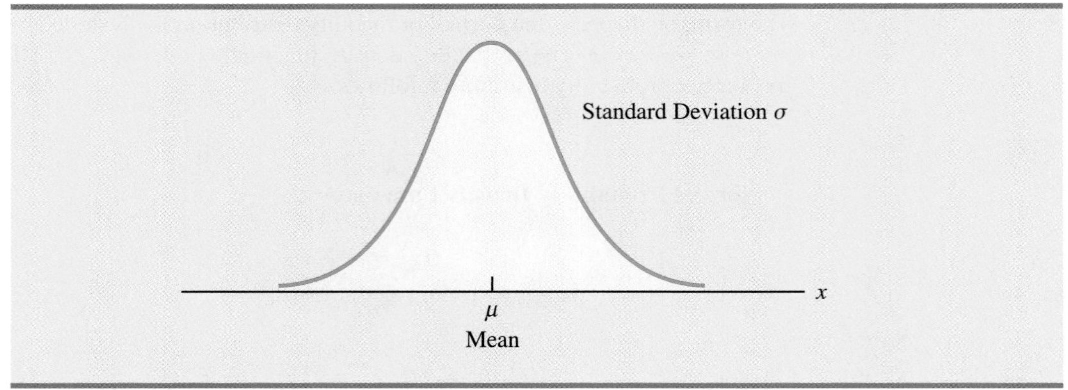

5. The standard deviation determines the width of the curve. Larger values of the standard deviation result in wider, flatter curves, showing more variability in the data. Two normal distributions with the same mean but with different standard deviations are shown here.

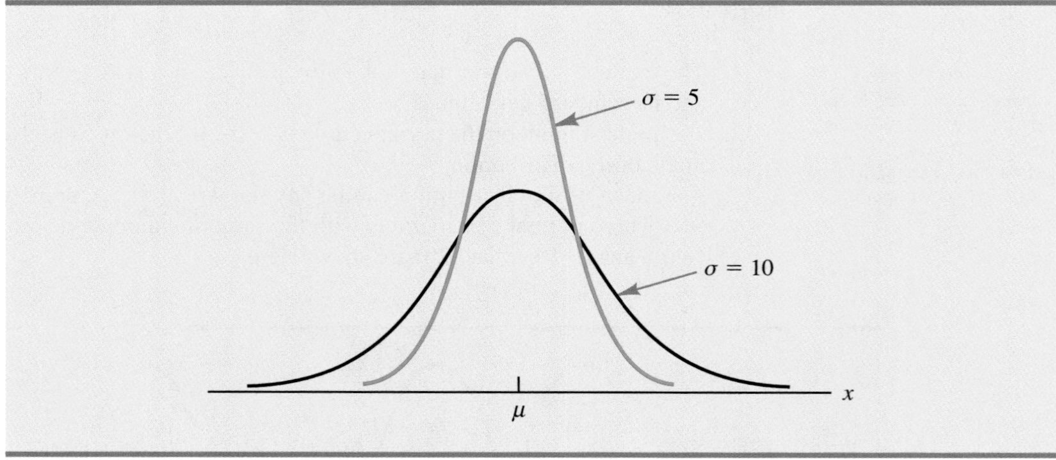

6. Probabilities for the normal random variable are given by areas under the curve. The total area under the curve for the normal probability distribution is 1 (this is true for all continuous probability distributions). Because the distribution is symmetric, the total area under the curve to the left of the mean is .50 and the total area under the curve to the right of the mean is .50.
7. The percentage of values in some commonly used intervals are:
   a. 68.26% of the values of a normal random variable are within plus or minus one standard deviation of its mean.
   b. 95.44% of the values of a normal random variable are within plus or minus two standard deviations of its mean.
   c. 99.72% of the values of a normal random variable are within plus or minus three standard deviations of its mean.

   Figure 6.4 shows properties (a), (b), and (c) graphically.

These percentages are the basis for the empirical rule introduced in Section 3.3.

FIGURE 6.4 AREAS UNDER THE CURVE FOR ANY NORMAL PROBABILITY DISTRIBUTION

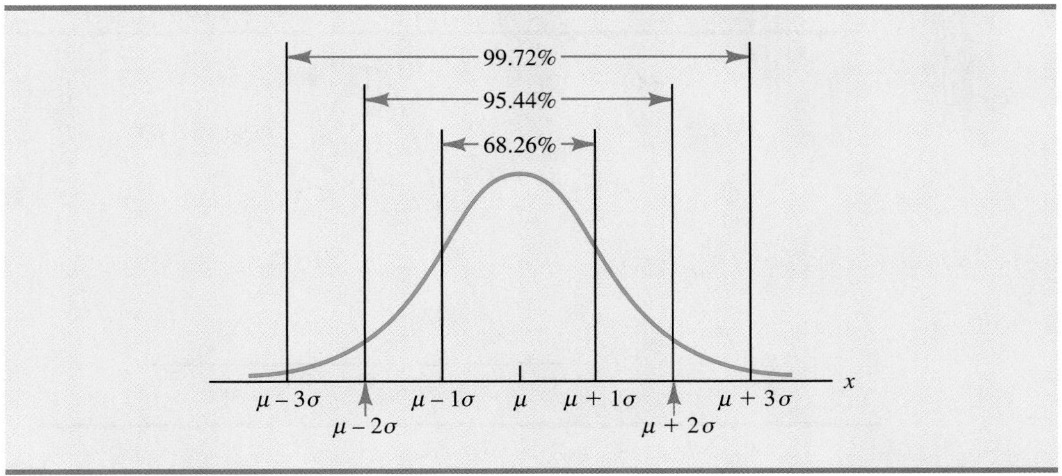

Standard Normal Probability Distribution

For the normal probability density function, the height of the curve varies and calculus is required to compute the areas that represent probability.

Given a z value, we use the standard normal table to find the appropriate probability (an area under the curve).

A random variable that has a normal distribution with a mean of zero and a standard deviation of one is said to have a **standard normal probability distribution.** The letter z is commonly used to designate this particular normal random variable. Figure 6.5 is the graph of the standard normal probability distribution. It has the same general appearance as other normal distributions, but with the special properties of $\mu = 0$ and $\sigma = 1$.

As with other continuous random variables, probability calculations with any normal probability distribution are made by computing areas under the graph of the probability density function. Thus, to find the probability that a normal random variable is within any specific interval, we must compute the area under the normal curve over that interval. For the standard normal probability distribution, areas under the normal curve have been computed and are available in tables that can be used in computing probabilities. Table 6.1 is such a table; it is also available as Table 1 of Appendix B and inside the front cover of this text.

To see how the table of areas under the curve for the standard normal probability distribution (Table 6.1) can be used to find probabilities, let us consider some examples. Later, we will see how this same table can be used to compute probabilities for any normal distribution. To begin, let us see how we can compute the probability that the z value for the

FIGURE 6.5 THE STANDARD NORMAL PROBABILITY DISTRIBUTION

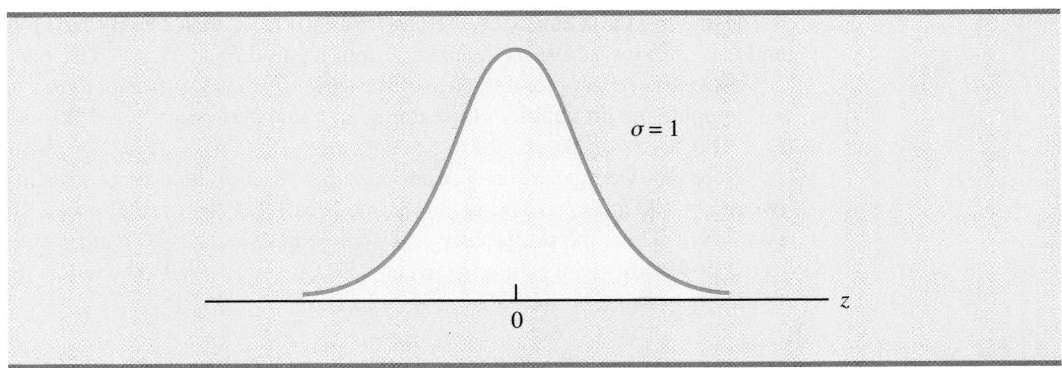

standard normal random variable will be between .00 and 1.00; that is, $P(.00 \leq z \leq 1.00)$. The shaded region in the following graph shows this probability.

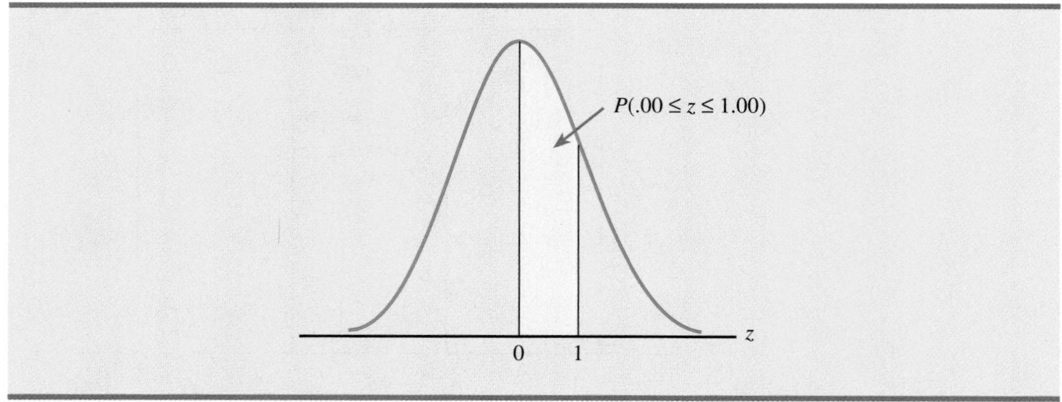

The entries in Table 6.1 give the area under the standard normal curve between the mean, $z = 0$, and a specified value of z (see the graph at the top of the table). In this case, we are interested in the area between $z = 0$ and $z = 1.00$. Thus, we must find the entry in the table corresponding to $z = 1.00$. First, we find 1.0 in the left column of the table and then find .00 in the top row of the table. By looking in the body of the table, we find that the 1.0 row and the .00 column intersect at the value of .3413. We have found the desired probability: $P(.00 \leq z \leq 1.00) = .3413$. A portion of Table 6.1 showing these steps follows.

z	**.00**	**.01**	**.02**
.			
.			
.			
.9	.3159	.3186	.3212
1.0	.3413	.3438	.3461
1.1	.3643	.3665	.3686
1.2	.3849	.3869	.3888
.			
.			
.			

$P(.00 \leq z \leq 1.00)$

Using the same approach, we can find $P(.00 \leq z \leq 1.25)$. By first locating the 1.2 row and then moving across to the .05 column, we find $P(.00 \leq z \leq 1.25) = .3944$.

As another example of the use of the table of areas for the standard normal distribution, we compute the probability of obtaining a z value between $z = -1.00$ and $z = 1.00$; that is, $P(-1.00 \leq z \leq 1.00)$.

Note that we have already used Table 6.1 to show that the probability of a z value between $z = .00$ and $z = 1.00$ is .3413, and recall that the normal probability distribution is *symmetric*. Thus, the probability of a z value between $z = .00$ and $z = -1.00$ is the same as the probability of a z value between $z = .00$ and $z = +1.00$. Hence, the probability of a z value between $z = -1.00$ and $z = +1.00$ is

$$P(-1.00 \leq z \leq .00) + P(.00 \leq z \leq 1.00) = .3413 + .3413 = .6826$$

TABLE 6.1 AREAS, OR PROBABILITIES, FOR THE STANDARD NORMAL DISTRIBUTION

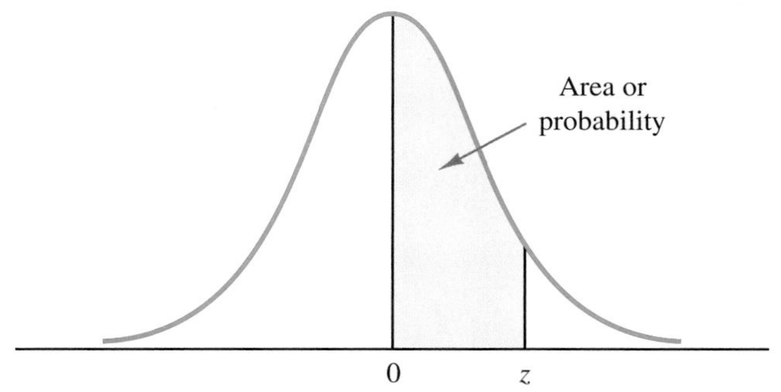

z	.00	.01	.02	.03	.04	.05	.06	.07	.08	.09
.0	.0000	.0040	.0080	.0120	.0160	.0199	.0239	.0279	.0319	.0359
.1	.0398	.0438	.0478	.0517	.0557	.0596	.0636	.0675	.0714	.0753
.2	.0793	.0832	.0871	.0910	.0948	.0987	.1026	.1064	.1103	.1141
.3	.1179	.1217	.1255	.1293	.1331	.1368	.1406	.1443	.1480	.1517
.4	.1554	.1591	.1628	.1664	.1700	.1736	.1772	.1808	.1844	.1879
.5	.1915	.1950	.1985	.2019	.2054	.2088	.2123	.2157	.2190	.2224
.6	.2257	.2291	.2324	.2357	.2389	.2422	.2454	.2486	.2518	.2549
.7	.2580	.2612	.2642	.2673	.2704	.2734	.2764	.2794	.2823	.2852
.8	.2881	.2910	.2939	.2967	.2995	.3023	.3051	.3078	.3106	.3133
.9	.3159	.3186	.3212	.3238	.3264	.3289	.3315	.3340	.3365	.3389
1.0	.3413	.3438	.3461	.3485	.3508	.3531	.3554	.3577	.3599	.3621
1.1	.3643	.3665	.3686	.3708	.3729	.3749	.3770	.3790	.3810	.3830
1.2	.3849	.3869	.3888	.3907	.3925	.3944	.3962	.3980	.3997	.4015
1.3	.4032	.4049	.4066	.4082	.4099	.4115	.4131	.4147	.4162	.4177
1.4	.4192	.4207	.4222	.4236	.4251	.4265	.4279	.4292	.4306	.4319
1.5	.4332	.4345	.4357	.4370	.4382	.4394	.4406	.4418	.4429	.4441
1.6	.4452	.4463	.4474	.4484	.4495	.4505	.4515	.4525	.4535	.4545
1.7	.4554	.4564	.4573	.4582	.4591	.4599	.4608	.4616	.4625	.4633
1.8	.4641	.4649	.4656	.4664	.4671	.4678	.4686	.4693	.4699	.4706
1.9	.4713	.4719	.4726	.4732	.4738	.4744	.4750	.4756	.4761	.4767
2.0	.4772	.4778	.4783	.4788	.4793	.4798	.4803	.4808	.4812	.4817
2.1	.4821	.4826	.4830	.4834	.4838	.4842	.4846	.4850	.4854	.4857
2.2	.4861	.4864	.4868	.4871	.4875	.4878	.4881	.4884	.4887	.4890
2.3	.4893	.4896	.4898	.4901	.4904	.4906	.4909	.4911	.4913	.4916
2.4	.4918	.4920	.4922	.4925	.4927	.4929	.4931	.4932	.4934	.4936
2.5	.4938	.4940	.4941	.4943	.4945	.4946	.4948	.4949	.4951	.4952
2.6	.4953	.4955	.4956	.4957	.4959	.4960	.4961	.4962	.4963	.4964
2.7	.4965	.4966	.4967	.4968	.4969	.4970	.4971	.4972	.4973	.4974
2.8	.4974	.4975	.4976	.4977	.4977	.4978	.4979	.4979	.4980	.4981
2.9	.4981	.4982	.4982	.4983	.4984	.4984	.4985	.4985	.4986	.4986
3.0	.4986	.4987	.4987	.4988	.4988	.4989	.4989	.4989	.4990	.4990

This probability is shown graphically in the following figure.

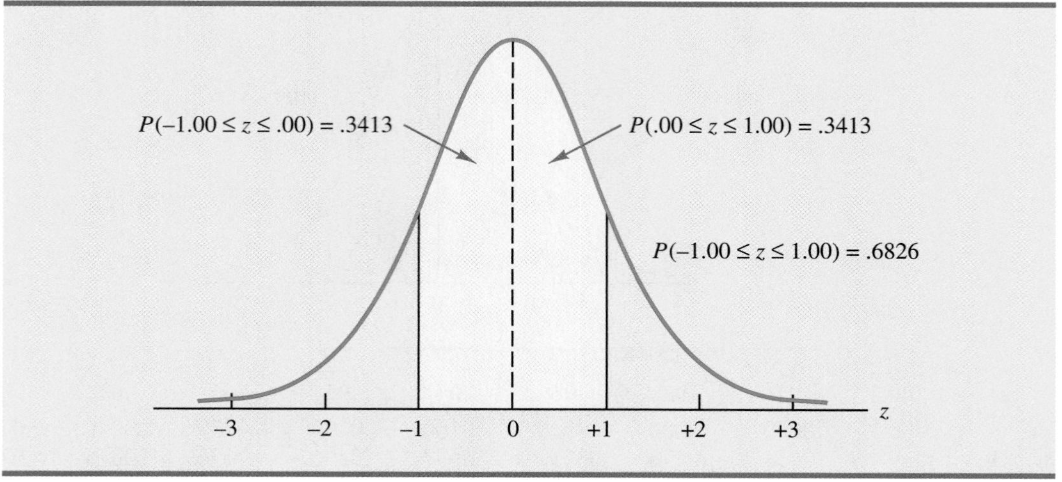

These probability calculations are the basis for observation 7 on page 220.

In a similar manner, we can use the values in Table 6.1 to show that the probability of a z value between -2.00 and $+2.00$ is $.4772 + .4772 = .9544$ and that the probability of a z value between -3.00 and $+3.00$ is $.4986 + .4986 = .9972$. Because we know that the total probability or total area under the curve for any continuous random variable must be 1.0000, the probability .9972 tells us that the value of z will almost always be between -3.00 and $+3.00$.

Next, we compute the probability of obtaining a z value of at least 1.58; that is, $P(z \geq 1.58)$. First, we use the $z = 1.5$ row and the .08 column of Table 6.1 to find that $P(.00 \leq z \leq 1.58) = .4429$. Now, because the normal probability distribution is symmetric, we know that 50% of the area under the curve must be above the mean (i.e., $z = 0$) and 50% of the area under the curve must be below the mean. If .4429 is the area between the mean and $z = 1.58$, then the area or probability corresponding to $z \geq 1.58$ must be $.5000 - .4429 = .0571$. This probability is shown in the following figure.

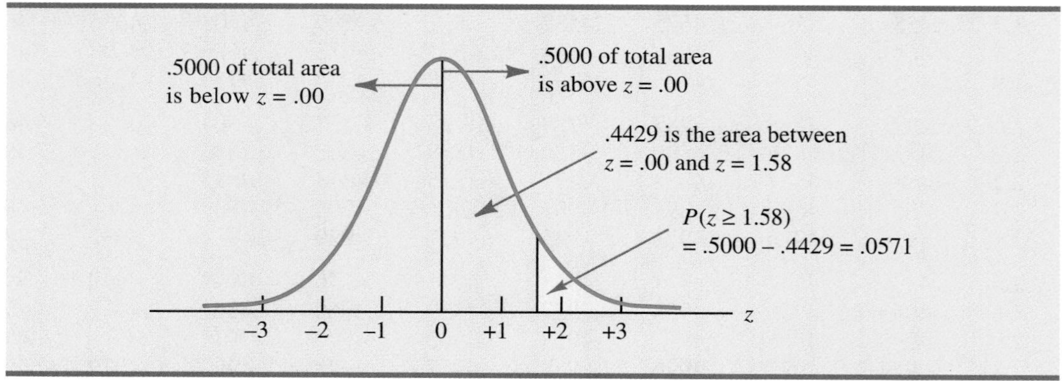

As another illustration, consider the probability that the random variable z assumes a value of $-.50$ or larger; that is, $P(z \geq -.50)$. To make this computation, we note that the probability we are seeking can be written as the sum of two probabilities: $P(z \geq -.50) = P(-.50 \leq z \leq .00) + P(z \geq 0.00)$. We have previously seen that $P(z \geq .00) = .50$. Also, we know that since the normal distribution is symmetric, $P(-.50 \leq z \leq .00) = P(.00 \leq z \leq .50)$. Referring to Table 6.1, we find that $P(.00 \leq z \leq .50) = .1915$. Therefore $P(z \geq -.50) = .1915 + .5000 = .6915$. The following graph shows this probability.

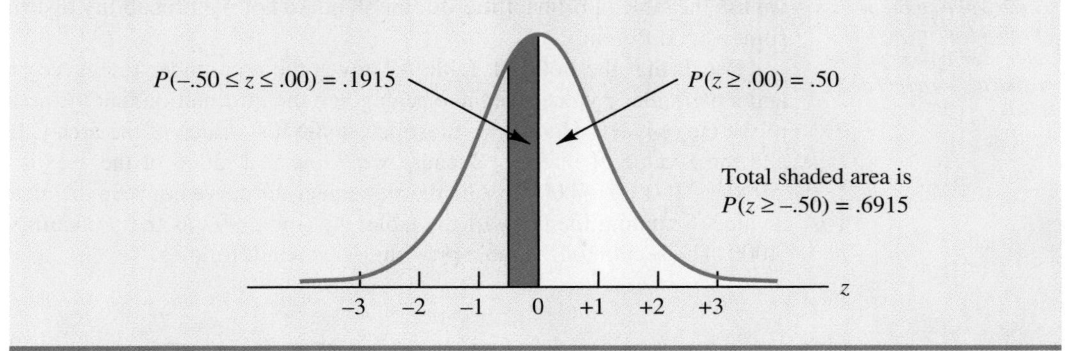

Next, we compute the probability of obtaining a z value between 1.00 and 1.58; that is, $P(1.00 \le z \le 1.58)$. From our previous examples, we know that there is a .3413 probability of a z value between $z = 0.00$ and $z = 1.00$ and that there is a .4429 probability of a z value between $z = 0.00$ and $z = 1.58$. Hence, there must be a $.4429 - .3413 = .1016$ probability of a z value between $z = 1.00$ and $z = 1.58$. Thus, $P(1.00 \le z \le 1.58) = .1016$. This situation is shown graphically in the following figure.

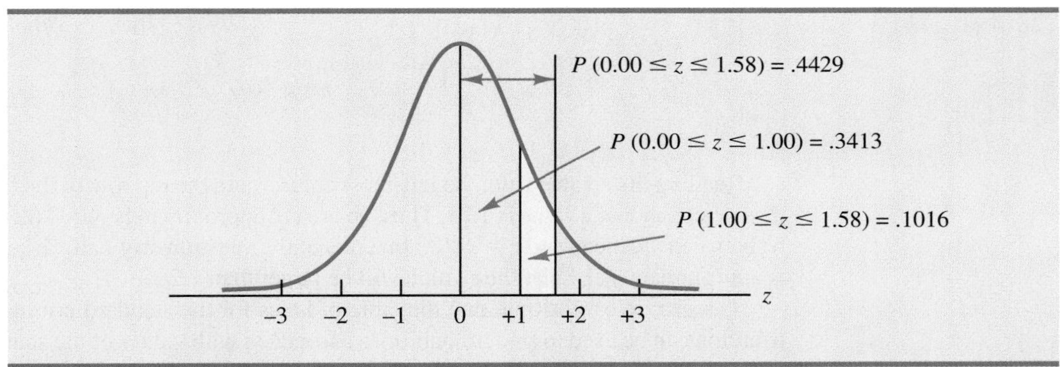

As a final illustration, let us find a z value such that the probability of obtaining a larger z value is .10. The following figure shows this situation graphically.

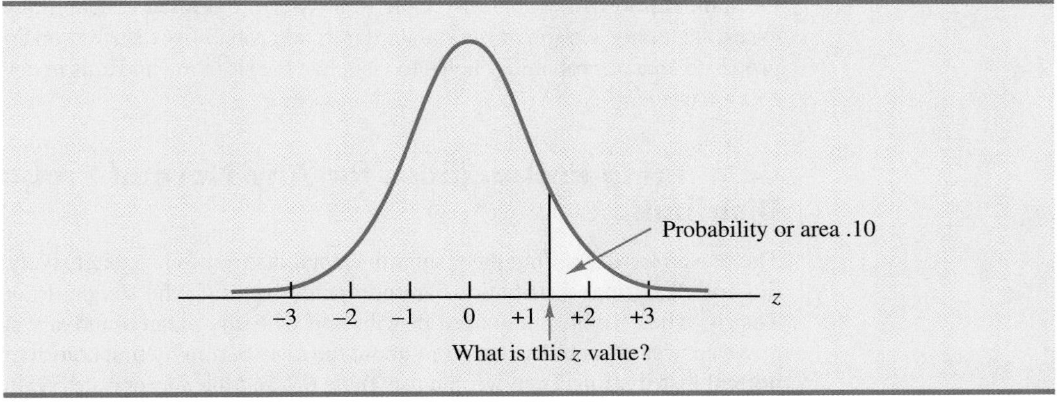

This problem is the inverse of those in the preceding examples. Previously, we specified the z value of interest and then found the corresponding probability, or area. In this example, we are given the probability and asked to find the corresponding z value. To do so,

Given a probability, we can use the standard normal table in an inverse fashion to find the corresponding z value.

we use the table of probabilities for the standard normal probability distribution (Table 6.1) somewhat differently.

Recall that the body of Table 6.1 gives the area under the curve between the mean and a particular z value. We have been given the information that the area in the upper tail of the curve is .10. Hence, we must determine how much of the area is between the mean and the z value of interest. Because we know that .5000 of the area is above the mean, $.5000 - .1000 = .4000$ must be the area under the curve between the mean and the desired z value. Scanning the body of the table, we find .3997 as the probability value closest to .4000. The section of the table providing this result follows.

z	.06	.07	.08	.09
.				
.				
.				
1.0	.3554	.3577	.3599	.3621
1.1	.3770	.3790	.3810	.3830
1.2	.3962	.3980	.3997	.4015
1.3	.4131	.4147	.4162	.4177
1.4	.4279	.4292	.4306	.4319
.				
.				
.				

Area value in body of table closest to .9000

Reading the z value from the left-most column and the top row of the table, we find that the corresponding z value is 1.28. Thus, an area of approximately .4000 (actually .3997) will be between the mean and $z = 1.28$.* In terms of the question originally asked, the probability is approximately .10 that the z value will be larger than 1.28.

The examples illustrate that the table of areas for the standard normal probability distribution can be used to find probabilities associated with values of the standard normal random variable z. Two types of questions can be asked. The first type of question specifies a value, or values, for z and asks us to use the table to determine the corresponding areas, or probabilities. The second type of question provides an area, or probability, and asks us to use the table to determine the corresponding z value. Thus, we need to be flexible in using the standard normal probability table to answer the desired probability question. In most cases, sketching a graph of the standard normal probability distribution and shading the appropriate area or probability helps to visualize the situation and aids in determining the correct answer.

Computing Probabilities for Any Normal Probability Distribution

The reason for discussing the standard normal distribution so extensively is that probabilities for all normal distributions are computed by using the standard normal distribution. That is, when we have a normal distribution with any mean μ and any standard deviation σ, we answer probability questions about the distribution by first converting to the standard normal distribution. Then we can use Table 6.1 and the appropriate z values to find the de-

*We could use interpolation in the body of the table to get a better approximation of the z value that corresponds to an area of .4000. Doing so to provide one more decimal place of accuracy would yield a z value of 1.282. However, in most practical situations, sufficient accuracy is obtained by simply using the table value closest to the desired probability.

sired probabilities. The formula used to convert any normal random variable x with mean μ and standard deviation σ to the standard normal distribution follows.

The formula for the standard normal random variable is similar to the formula we introduced in Chapter 3 for computing z scores for a data set.

Converting to the Standard Normal Distribution

$$z = \frac{x - \mu}{\sigma} \tag{6.3}$$

A value of x equal to its mean μ results in $z = (\mu - \mu)/\sigma = 0$. Thus, we see that a value of x equal to its mean μ corresponds to a value of z at its mean 0. Now suppose that x is one standard deviation above its mean; that is, $x = \mu + \sigma$. Applying equation (6.3), we see that the corresponding z value is $z = [(\mu + \sigma) - \mu]/\sigma = \sigma/\sigma = 1$. Thus, a value of x that is one standard deviation above its mean corresponds to $z = 1$. In other words, we can interpret z as the number of standard deviations that the normal random variable x is from its mean μ.

To see how this conversion enables us to compute probabilities for any normal distribution, suppose we have a normal distribution with $\mu = 10$ and $\sigma = 2$. What is the probability that the random variable x is between 10 and 14? Using equation (6.3) we see that at $x = 10$, $z = (x - \mu)/\sigma = (10 - 10)/2 = 0$ and that at $x = 14$, $z = (14 - 10)/2 = 4/2 = 2$. Thus, the answer to our question about the probability of x being between 10 and 14 is given by the equivalent probability that z is between 0 and 2 for the standard normal distribution. In other words, the probability that we are seeking is the probability that the random variable x is between its mean and two standard deviations above the mean. Using $z = 2.00$ and Table 6.1, we see that the probability is .4772. Hence the probability that x is between 10 and 14 is .4772.

Grear Tire Company Problem

We turn now to an application of the normal probability distribution. Suppose the Grear Tire Company has just developed a new steel-belted radial tire that will be sold through a national chain of discount stores. Because the tire is a new product, Grear's managers believe that the mileage guarantee offered with the tire will be an important factor in the acceptance of the product. Before finalizing the tire mileage guarantee policy, Grear's managers want probability information about the number of miles the tires will last.

From actual road tests with the tires, Grear's engineering group has estimated the mean tire mileage is $\mu = 36{,}500$ miles and that the standard deviation is $\sigma = 5000$. In addition, the data collected indicate a normal distribution is a reasonable assumption. What percentage of the tires can be expected to last more than 40,000 miles? In other words, what is the probability that the tire mileage will exceed 40,000? This question can be answered by finding the area of the shaded region in Figure 6.6.

At $x = 40{,}000$, we have

$$z = \frac{x - \mu}{\sigma} = \frac{40{,}000 - 36{,}500}{5000} = \frac{3500}{5000} = .70$$

Refer now to the bottom of Figure 6.6. We see that a value of $x = 40{,}000$ on the Grear Tire normal distribution corresponds to a value of $z = .70$ on the standard normal distribution. Using Table 6.1, we see that the area between the mean and $z = .70$ is .2580. Referring again to Figure 6.6, we see that the area between $x = 36{,}500$ and $x = 40{,}000$ on the Grear Tire normal distribution is the same. Thus, $.5000 - .2580 = .2420$ is the probability that x will exceed 40,000. We can conclude that about 24.2% of the tires will exceed 40,000 in mileage.

FIGURE 6.6 GREAR TIRE COMPANY MILEAGE DISTRIBUTION

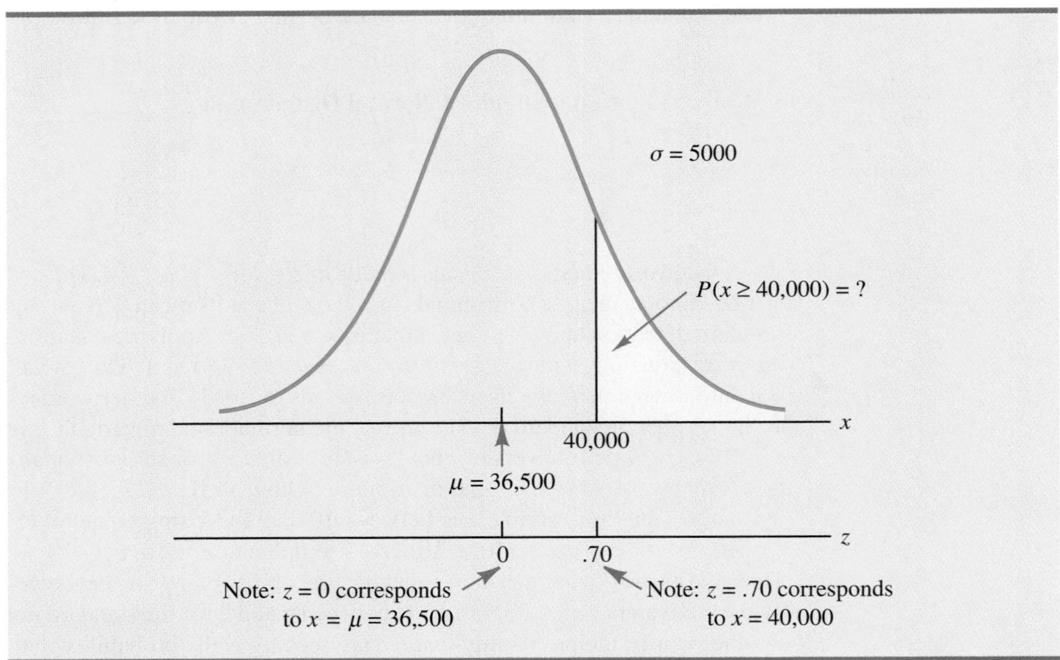

Let us now assume that Grear is considering a guarantee that will provide a discount on replacement tires if the original tires do not exceed the mileage stated in the guarantee. What should the guarantee mileage be if Grear wants no more than 10% of the tires to be eligible for the discount guarantee? This question is interpreted graphically in Figure 6.7.

According to Figure 6.7, 40% of the area must be between the mean and the unknown guarantee mileage. We look up .4000 in the body of Table 6.1 and see that this area is at approximately 1.28 standard deviations below the mean. That is, $z = -1.28$ is the value of the standard normal random variable corresponding to the desired mileage guarantee on the Grear Tire normal distribution. To find the mileage x corresponding to $z = -1.28$, we have

$$z = \frac{x - \mu}{\sigma} = -1.28$$
$$x - \mu = -1.28\sigma$$
$$x = \mu - 1.28\sigma$$

With $\mu = 36{,}500$ and $\sigma = 5000$,

$$x = 36{,}500 - 1.28(5000) = 30{,}100$$

With the guarantee set at 30,000 miles, the actual percentage eligible for the guarantee will be 9.68%.

Thus, a guarantee of 30,100 miles will meet the requirement that approximately 10% of the tires will be eligible for the guarantee. Perhaps, with this information, the firm will set its tire mileage guarantee at 30,000 miles.

Again, we see the important role that probability distributions play in providing decision-making information. Namely, once a probability distribution is established for a particular application, it can be used quickly and easily to obtain probability information about the problem. Probability does not make a decision recommendation directly, but it provides information that helps the decision maker better understand the risks and uncer-

FIGURE 6.7 GREAR'S DISCOUNT GUARANTEE

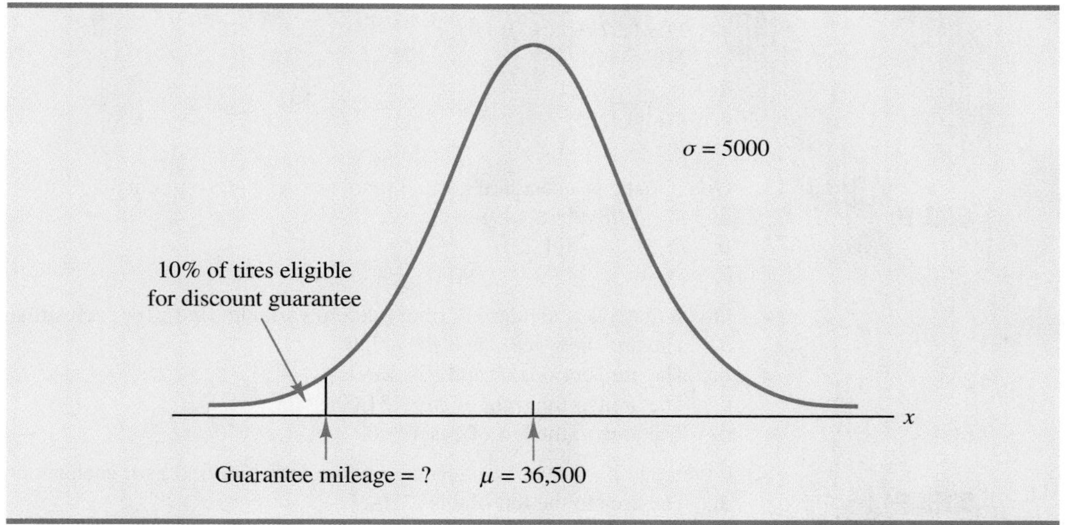

tainties associated with the problem. Ultimately, this information may assist the decision maker in reaching a good decision.

EXERCISES

Methods

8. Using Figure 6.4 as a guide, sketch a normal curve for a random variable x that has a mean of $\mu = 100$ and a standard deviation of $\sigma = 10$. Label the horizontal axis with values of 70, 80, 90, 100, 110, 120, and 130.

9. A random variable is normally distributed with a mean of $\mu = 50$ and a standard deviation of $\sigma = 5$.

a. Sketch a normal curve for the probability density function. Label the horizontal axis with values of 35, 40, 45, 50, 55, 60, and 65. Figure 6.4 shows that the normal curve almost touches the horizontal axis at three standard deviations below and at three standard deviations above the mean (in this case at 35 and 65).

b. What is the probability the random variable will assume a value between 45 and 55?

c. What is the probability the random variable will assume a value between 40 and 60?

10. Draw a graph for the standard normal distribution. Label the horizontal axis at values of $-3, -2, -1, 0, 1, 2$, and 3. Then use the table of probabilities for the standard normal distribution to compute the following probabilities.

a. $P(0 \leq z \leq 1)$

b. $P(0 \leq z \leq 1.5)$

c. $P(0 < z < 2)$

d. $P(0 < z < 2.5)$

11. Given that z is a standard normal random variable, compute the following probabilities.

a. $P(-1 \leq z \leq 0)$

b. $P(-1.5 \leq z \leq 0)$

c. $P(-2 < z < 0)$

d. $P(-2.5 \leq z \leq 0)$

e. $P(-3 \leq z \leq 0)$

12. Given that z is a standard normal random variable, compute the following probabilities.
   a. $P(0 \leq z \leq .83)$
   b. $P(-1.57 \leq z \leq 0)$
   c. $P(z > 44)$
   d. $P(z \geq -.23)$
   e. $P(z < 1.20)$
   f. $P(z \leq -.71)$

13. Given that z is a standard normal random variable, compute the following probabilities.
   a. $P(-1.98 \leq z \leq .49)$
   b. $P(.52 \leq z \leq 1.22)$
   c. $P(-1.75 \leq z \leq -1.04)$

14. Given that z is a standard normal random variable, find z for each situation.
   a. The area between 0 and z is .4750.
   b. The area between 0 and z is .2291.
   c. The area to the right of z is .1314.
   d. The area to the left of z is .6700.

15. Given that z is a standard normal random variable, find z for each situation.
   a. The area to the left of z is .2119.
   b. The area between $-z$ and z is .9030.
   c. The area between $-z$ and z is .2052.
   d. The area to the left of z is .9948.
   e. The area to the right of z is .6915.

16. Given that z is a standard normal random variable, find z for each situation.
   a. The area to the right of z is .01.
   b. The area to the right of z is .025.
   c. The area to the right of z is .05.
   d. The area to the right of z is .10.

Applications

17. The average American male adult is 5 feet 9 inches tall (*Astounding Averages,* 1995). Assume the standard deviation is 3 inches in answering the following questions.
   a. What is the probability an adult male is taller than 6 feet?
   b. What is the probability an adult male is shorter than 5 feet?
   c. What is the probability an adult male is between 5 feet 6 inches and 5 feet 10 inches?
   d. What is the probability an adult male is no more than 6 feet tall?

18. The average time a subscriber spends reading *The Wall Street Journal* is 49 minutes (*The Wall Street Journal* Subscriber Study, 1996). Assume the standard deviation is 16 minutes and that the times are normally distributed.
   a. What is the probability a subscriber will spend at least 1 hour reading the *Journal*?
   b. What is the probability a subscriber will spend no more than 30 minutes reading the *Journal*?
   c. For the 10% who spend the most time reading the *Journal,* how much time do they spend?

19. The average amount of precipitation in Dallas, Texas, during the month of April is 3.5 inches (*The World Almanac,* 2000). Assume that a normal distribution applies and that the standard deviation is .8 inches.
   a. What percentage of the time does the amount of rainfall in April exceed 5 inches?
   b. What percentage of the time is the amount of rainfall in April less than 3 inches?
   c. A month is classified as extremely wet if the amount of rainfall is in the upper 10% for that month. How much precipitation must fall before a month of April is classified as extremely wet?

20. According to a survey, subscribers to *The Wall Street Journal Interactive Edition* spend an average of 27 hours per week using the computer at work (*WSJ.com* Subscriber Study, 1999). Assume the normal distribution applies and that the standard deviation is 8 hours.

a. What is the probability a randomly selected subscriber spends less than 11 hours using the computer at work?

b. What percentage of the subscribers spends more than 40 hours per week using the computer at work?

c. A person is classified as a heavy user if he or she is in the upper 20% in terms of hours of usage. How many hours must a subscriber use the computer in order to be classified as a heavy user?

21. A person must score in the upper 2% of the population on an IQ test to qualify for membership in Mensa, the international high-IQ society (*US Airways Attache,* September, 2000). If IQ scores are normally distributed with a mean of 100 and a standard deviation of 15, what score must a person get to qualify for Mensa?

22. According to the Bureau of Labor Statistics, the average weekly pay for a U.S. production worker was $441.84 in 1998 (*The World Almanac,* 2000). Assume that available data indicate that wages are normally distributed with a standard deviation of $90.

a. What is the probability that a worker earns between $400 and $500?

b. How much does a production worker have to make to be in the top 20% of wage earners?

c. For a randomly selected production worker, what is the probability the worker earns less than $250 per week?

23. The time needed to complete a final examination in a particular college course is normally distributed with a mean of 80 minutes and a standard deviation of 10 minutes. Answer the following questions.

a. What is the probability of completing the exam in one hour or less?

b. What is the probability that a student will complete the exam in more than 60 minutes but less than 75 minutes?

c. Assume that the class has 60 students and that the examination period is 90 minutes in length. How many students do you expect will be unable to complete the exam in the allotted time?

24. The daily trading volumes (millions of shares) for stocks traded on the New York Stock Exchange for 12 days in August and September are shown here (*Barron's,* August 7, 2000, September 4, 2000, and September 11, 2000).

917	983	1,046
944	723	783
813	1,057	766
836	992	973

The probability distribution of trading volume is approximately normal.

a. Compute the mean and standard deviation for the daily trading volume to use as estimates of the population mean and standard deviation.

b. What is the probability that on a particular day the trading volume will be less than 800 million shares?

c. What is the probability that trading volume will exceed 1 billion shares?

d. If the exchange wants to issue a press release on the top 5% of trading days, what volume will trigger a release?

25. The average ticket price for a major league baseball game was $11.98 in 1998 (*USA Today,* November 11, 1998). Adding the cost of food, parking, and souvenirs, the average cost for a family of four to attend a game was approximately $110.00. Assume the normal distribution applies and that the standard deviation is $20.00.

a. What is the probability the cost will exceed $100.00?

b. What is the probability a family will spend $90.00 or less?

c. What is the probability a family will spend between $80.00 and $130.00?

6.3 EXPONENTIAL PROBABILITY DISTRIBUTION

A continuous probability distribution that is useful in describing the time it takes to complete a task is the **exponential probability distribution.** The exponential random variable can be used to describe such things as the time between arrivals at a car wash, the time required to load a truck, the distance between major defects in a highway, and so on. The exponential probability density function follows.

Exponential Probability Density Function

$$f(x) = \frac{1}{\mu} e^{-x/\mu} \qquad \text{for } x \geq 0, \mu > 0 \tag{6.4}$$

As an example of the exponential probability distribution, suppose that x = the time it takes to load a truck at the Schips loading dock follows such a distribution. If the mean, or average, time to load a truck is 15 minutes ($\mu = 15$), the appropriate probability density function is

$$f(x) = \frac{1}{15} e^{-x/15}$$

Figure 6.8 is the graph of this probability density function.

Computing Probabilities for the Exponential Distribution

In waiting line applications, the exponential distribution is often used for service time.

As with any continuous probability distribution, the area under the curve corresponding to an interval provides the probability that the random variable assumes a value in that interval. In the Schips loading dock example, the probability that loading a truck will take 6 minutes or less ($x \leq 6$) is defined to be the area under the curve in Figure 6.8 from $x = 0$ to

FIGURE 6.8 EXPONENTIAL PROBABILITY DISTRIBUTION FOR THE SCHIPS LOADING DOCK EXAMPLE

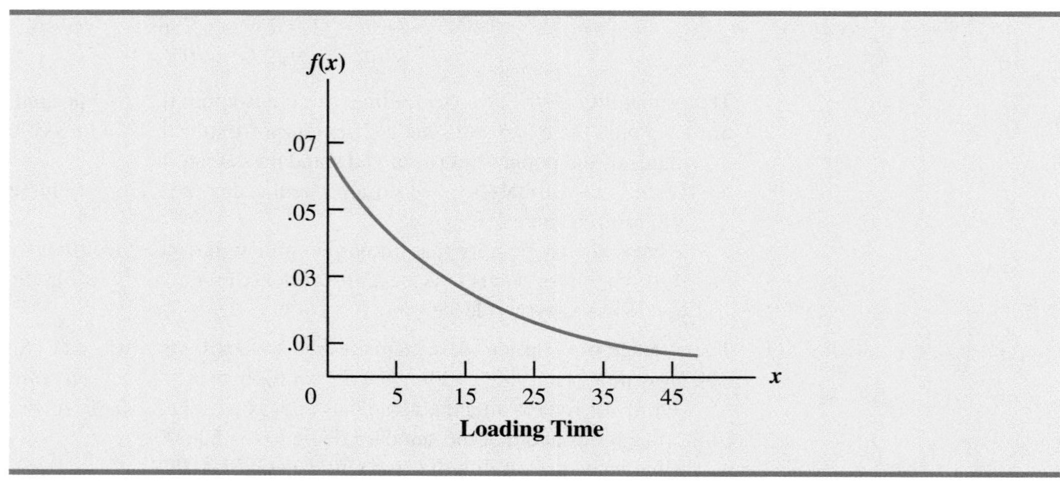

$x = 6$. Similarly, the probability that loading a truck will take 18 minutes or less ($x \leq 18$) is the area under the curve from $x = 0$ to $x = 18$. Note also that the probability that loading a truck will take between 6 minutes and 18 minutes ($6 \leq x \leq 18$) is given by the area under the curve from $x = 6$ to $x = 18$.

To compute exponential probabilities such as those just described, we use the following formula. It provides the cumulative probability of obtaining a value for the exponential random variable of less than or equal to some specific value of x, denoted by x_0.

Exponential Distribution: Cumulative Probabilities

$$P(x \leq x_0) = 1 - e^{-x_0/\mu} \tag{6.5}$$

For the Schips loading dock example, x = loading time and equation (6.5) can be written as

$$P(x \leq x_0) = 1 - e^{-x_0/15}$$

Hence, the probability that loading a truck will take 6 minutes or less $P(x \leq 6)$ is

$$P(x \leq 6) = 1 - e^{-6/15} = .3297$$

Figure 6.9 shows the area or probability for a loading time of 6 minutes or less. Note also that the probability of loading a truck in 18 minutes or less $P(x \leq 18)$ is

$$P(x \leq 18) = 1 - e^{-18/15} = .6988$$

Thus, the probability that loading a truck will take between 6 minutes and 18 minutes is equal to $.6988 - .3297 = .3691$. Probabilities for any other interval can be computed similarly.

FIGURE 6.9 PROBABILITY OF A LOADING TIME OF SIX MINUTES OR LESS

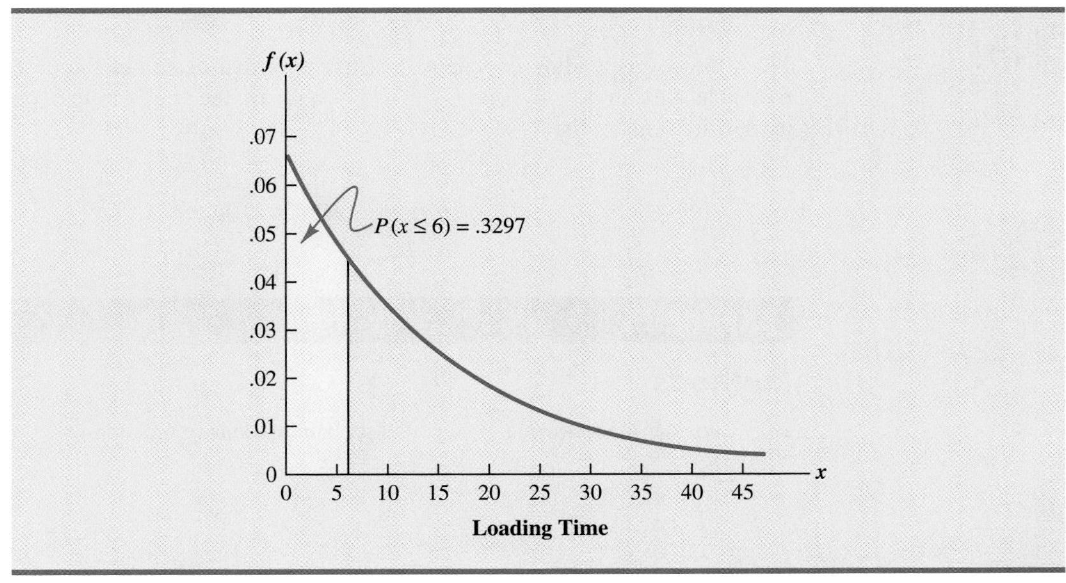

Relationship Between the Poisson and Exponential Distributions

In Section 5.5 we introduced the Poisson distribution as a discrete probability distribution that is often useful in examining the number of occurrences of an event over a specified interval of time or space. Recall that the Poisson probability function is

$$f(x) = \frac{\mu^x e^{-\mu}}{x!}$$

where

$$\mu = \text{expected value or mean number of occurrences in an interval}$$

The continuous exponential probability distribution is related to the discrete Poisson distribution. If the Poisson distribution provides an appropriate description of the number of occurrences per interval, the exponential distribution provides a description of the length of the interval between occurrences.

If arrivals follow a Poisson distribution, the time between arrivals must follow an exponential distribution.

To illustrate this relationship, suppose the number of cars that arrive at a car wash during 1 hour is described by a Poisson probability distribution with a mean of 10 cars per hour. The Poisson probability function that gives the probability of x arrivals per hour is

$$f(x) = \frac{10^x e^{-10}}{x!}$$

Because the average number of arrivals is 10 cars per hour, the average time between cars arriving is

$$\frac{1 \text{ hour}}{10 \text{ cars}} = .1 \text{ hour/car}$$

Thus, the corresponding exponential distribution that describes the time between the arrivals has a mean of $\mu = .1$ hour per car; as a result, the appropriate exponential probability density function is

$$f(x) = \frac{1}{.1} e^{-x/.1} = 10e^{-10x}$$

EXERCISES

Methods

26. Consider the following exponential probability density function.

$$f(x) = \frac{1}{8} e^{-x/8} \qquad \text{for } x \geq 0$$

a. Find $P(x \leq 6)$.
b. Find $P(x \leq 4)$.

c. Find $P(x \geq 6)$.
d. Find $P(4 \leq x \leq 6)$.

27. Consider the following exponential probability density function.

$$f(x) = \frac{1}{3} e^{-x/3} \qquad \text{for } x \geq 0$$

a. Write the formula for $P(x \leq x_0)$.
b. Find $P(x \leq 2)$.
c. Find $P(x \geq 3)$.
d. Find $P(x \leq 5)$.
e. Find $P(2 \leq x \leq 5)$.

Applications

28. *Internet Magazine* monitors Internet service providers (ISPs) and provides statistics on their performance. The average time to download a web page for free ISPs is approximately 20 seconds for European web pages (*Internet Magazine,* January, 2000). Assume the time to download a web page follows an exponential distribution.
a. What is the probability it will take less than 10 seconds to download a web page?
b. What is the probability it will take more than 30 seconds to download a web page?
c. What is the probability it will take between 10 and 30 seconds to download a web page?

29. The time between arrivals of vehicles at a particular intersection follows an exponential probability distribution with a mean of 12 seconds.
a. Sketch this exponential probability distribution.
b. What is the probability that the arrival time between vehicles is 12 seconds or less?
c. What is the probability that the arrival time between vehicles is 6 seconds or less?
d. What is the probability of 30 or more seconds between vehicle arrivals?

30. The lifetime (hours) of an electronic device is a random variable with the following exponential probability density function.

$$f(x) = \frac{1}{50} e^{-x/50} \qquad \text{for } x \geq 0$$

a. What is the mean lifetime of the device?
b. What is the probability that the device will fail in the first 25 hours of operation?
c. What is the probability that the device will operate 100 or more hours before failure?

31. Sparagowski & Associates conducted a study of service times at the drive-up window of fast-food restaurants. The average time between placing an order and receiving the order at McDonald's restaurants was 2.78 minutes (*The Cincinnati Enquirer,* July 9, 2000). Waiting times, such as these, frequently follow an exponential distribution.
a. What is the probability that a customer's service time is less than 2 minutes?
b. What is the probability that a customer's service time is more than 5 minutes?
c. What is the probability that a customer's service time is more than 2.78 minutes?

32. According to *Barron's 1998 Primary Reader Survey,* the average annual number of investment transactions for a subscriber is 30 (*http://www.barronsmag.com,* July 28, 2000). Suppose the number of transactions in a year follows the Poisson probability distribution.
a. Show the probability distribution for the time between investment transactions.
b. What is the probability of no transactions during the month of January for a particular subscriber?
c. What is the probability that the next transaction will occur within the next half month for a particular subscriber?

SUMMARY

This chapter extended the discussion of probability distributions to the case of continuous random variables. The major conceptual difference between discrete and continuous probability distributions involves the method of computing probabilities. With discrete distributions, the probability function $f(x)$ provides the probability that the random variable x assumes various values. With continuous probability distributions, we associate a probability density function, denoted by $f(x)$. The probability density function does not provide probability values for a continuous random variable directly. Probabilities are given by areas under the curve or graph of the probability density function $f(x)$. Because the area under the curve above a single point is zero, we observe that the probability of any particular value is zero for a continuous random variable.

Three continuous probability distributions—the uniform, normal, and exponential distributions—were treated in detail. The normal probability distribution is used widely in statistical inference and will be used extensively throughout the remainder of the text.

GLOSSARY

Probability density function A function used to compute probabilities for a continuous random variable. The area under the graph of a probability density function over an interval represents probability.

Uniform probability distribution A continuous probability distribution for which the probability that the random variable will assume a value in any interval is the same for each interval of equal length.

Normal probability distribution A continuous probability distribution. Its probability density function is bell shaped and determined by its mean μ and standard deviation σ.

Standard normal probability distribution A normal distribution with a mean of zero and a standard deviation of one.

Exponential probability distribution A continuous probability distribution that is useful in computing probabilities for the time it takes to complete a task.

KEY FORMULAS

Uniform Probability Density Function

$$f(x) = \begin{cases} \dfrac{1}{b-a} & \text{for } a \le x \le b \\ 0 & \text{elsewhere} \end{cases} \tag{6.1}$$

Normal Probability Density Function

$$f(x) = \frac{1}{\sigma\sqrt{2\pi}} e^{-(x-\mu)^2/2\sigma^2} \tag{6.2}$$

Converting to the Standard Normal Distribution

$$z = \frac{x-\mu}{\sigma} \tag{6.3}$$

Exponential Probability Density Function

$$f(x) = \frac{1}{\mu} e^{-x/\mu} \qquad \text{for } x \geq 0, \mu > 0 \tag{6.4}$$

Exponential Distribution: Cumulative Probabilities

$$P(x \leq x_0) = 1 - e^{-x_0/\mu} \tag{6.5}$$

SUPPLEMENTARY EXERCISES

33. A business executive has been transferred from Chicago to Atlanta and needs to sell her house in Chicago quickly. The executive's employer has offered to buy the house for \$210,000, but the offer expires at the end of the week. The executive does not currently have a better offer, but can afford to leave the house on the market for another month. From conversations with her realtor, the executive believes the price she will get by leaving the house on the market for another month is uniformly distributed between \$200,000 and \$225,000.

a. If she leaves the house on the market for another month, what is the mathematical expression for the probability density function of the sales price?

b. If she leaves it on the market for another month, what is the probability she will get at least \$215,000 for the house?

c. If she leaves it on the market for another month, what is the probability she will get less than \$210,000?

d. Should the executive leave the house on the market for another month? Why or why not?

34. Sixty-eight percent of the debt owed by American families is home mortgage or equity credit (*Federal Reserve Bulletin,* January 1997). The median amount of mortgage debt for families with the head of the household under 35 years old is \$63,000. Assume the amount of mortgage debt for this group is normally distributed and the standard deviation is \$15,000.

a. What is the mean amount of mortgage debt for this group?

b. How much mortgage debt do the 10% with the smallest debt have?

c. What percent of these families have mortgage debt in excess of \$80,000?

d. The upper 5% of mortgage debt is in excess of what amount?

35. Motorola used the normal distribution to determine the probability of defects and the number of defects expected in a production process (*APICS—The Performance Advantage,* July 1991). Assume a production process is designed to produce items with a weight of 10 ounces and that the process mean is 10. Calculate the probability of a defect and the expected number of defects for a 1000-unit production run in the following situations.

a. The process standard deviation is .15, and the process control is set at plus or minus one standard deviation. Units with weights less than 9.85 or greater than 10.15 ounces will be classified as defects.

b. Through process design improvements, the process standard deviation can be reduced to .05. Assume the process control remains the same, with weights less than 9.85 or greater than 10.15 ounces being classified as defects.

c. What is the advantage of reducing process variation and setting process control limits at a greater number of standard deviations from the mean?

36. The mean hourly operating cost of a USAir 737 airplane is \$2071 (*The Tampa Tribune,* February 17, 1995). Assume that the hourly operating cost for the airplane is normally distributed.

a. If 11% of the hourly operating costs are \$1800 or less, what is the standard deviation of hourly operating cost?

b. What is the probability that the hourly operating cost of a USAir 737 airplane is between \$2000 and \$2500?

c. What is the hourly operating cost of the 3% of the airplanes that have the lowest operating cost?

37. The sales of High-Brite Toothpaste are believed to be approximately normally distributed, with a mean of 10,000 tubes per week and a standard deviation of 1500 tubes per week.

- **a.** What is the probability that more than 12,000 tubes will be sold in any given week?
- **b.** To have a .95 probability that the company will have sufficient stock to cover the weekly demand, how many tubes should be produced?

38. Ward Doering Auto Sales is considering offering a special service contract that will cover the total cost of any service work required on leased vehicles. From experience, the company manager estimates that yearly service costs are approximately normally distributed, with a mean of $150 and a standard deviation of $25.

- **a.** If the company offers the service contract to customers for a yearly charge of $200, what is the probability that any one customer's service costs will exceed the contract price of $200?
- **b.** What is Ward's expected profit per service contract?

39. The Asian currency crisis of late 1997 and early 1998 was expected to lead to substantial job losses in the United States as inexpensive imports flooded markets. California was expected to be especially hard hit. The Economic Policy Institute estimated that the mean number of job losses in California would be 126,681 (*St. Petersburg Times,* January 24, 1998). Assume the number of jobs lost in California is normally distributed with a standard deviation of 30,000.

- **a.** What is the probability the number of lost jobs is between 80,000 and 150,000?
- **b.** What is the probability the number of lost jobs will be less than 50,000?
- **c.** What cutoff value will provide a .95 probability that the number of lost jobs will not exceed the value?

40. Assume that the test scores from a college admissions test are normally distributed, with a mean of 450 and a standard deviation of 100.

- **a.** What percentage of the people taking the test score between 400 and 500?
- **b.** Suppose someone receives a score of 630. What percentage of the people taking the test score better? What percentage score worse?
- **c.** If a particular university will not admit anyone scoring below 480, what percentage of the persons taking the test would be acceptable to the university?

41. According to *Advertising Age,* the average base salary for women working as copywriters in advertising firms is higher than the average base salary for men. The average base salary for women is $67,000 and the average base salary for men is $65,500 (*Working Woman,* July/August, 2000). Assume salaries are normally distributed and that the standard deviation is $7,000 for both men and women.

- **a.** What is the probability of a woman receiving a salary in excess of $75,000?
- **b.** What is the probability of a man receiving a salary in excess of $75,000?
- **c.** What is the probability of a woman receiving a salary below $50,000?
- **d.** How much would a woman have to make to have a higher salary than 99% of her male counterparts?

42. A machine fills containers with a particular product. The standard deviation of filling weights is known from past data to be .6 ounce. If only 2% of the containers hold less than 18 ounces, what is the mean filling weight for the machine? That is, what must μ equal? Assume the filling weights have a normal distribution.

43. The time in minutes for which a student uses a computer terminal at the computer center of a major university follows an exponential probability distribution with a mean of 36 minutes. Assume a student arrives at the terminal just as another student is beginning to work on the terminal.

- **a.** What is the probability that the wait for the second student will be 15 minutes or less?
- **b.** What is the probability that the wait for the second student will be between 15 and 45 minutes?
- **c.** What is the probability that the second student will have to wait an hour or more?

44. A new automated production process has been averaging two breakdowns per day, and the number of breakdowns per day follows a Poisson probability distribution.

a. What is the mean time between breakdowns, assuming eight hours of operation per day?

b. Show the exponential probability density function that can be used for the time between breakdowns.

c. What is the probability that the process will run 1 hour or more before another breakdown?

d. What is the probability that the process can run a full 8-hour shift without a breakdown?

45. The time (in minutes) a checkout lane is idle between customers at a supermarket follows an exponential probability distribution with a mean of 1.2 minutes.

a. Show the probability density function for this distribution.

b. What is the probability that the next customer will arrive between .5 and 1.0 minute after a customer is served?

c. What is the probability of the checkout lane being idle for more than a minute between customers?

46. The time (in minutes) between telephone calls at an insurance claims office has the following exponential probability distribution.

$$f(x) = .50e^{-.50x} \qquad \text{for } x \geq 0$$

a. What is the mean time between telephone calls?

b. What is the probability of having 30 seconds or less between telephone calls?

c. What is the probability of having 1 minute or less between telephone calls?

d. What is the probability of having 5 or more minutes without a telephone call?

Appendix 6.1 CONTINUOUS PROBABILITY DISTRIBUTIONS WITH MINITAB

Let us demonstrate the Minitab procedure for computing continuous probabilities by referring to the Grear Tire Company problem where tire mileage was described by a normal probability distribution with $\mu = 36{,}500$ and $\sigma = 5000$. One question asked was: What is the probability that the tire mileage will exceed 40,000 miles?

For continuous probability distributions, Minitab gives a cumulative probability; that is, Minitab gives the probability that the random variable will assume a value less than or equal to a specified constant. For the Grear tire mileage question, Minitab can be used to determine the cumulative probability that the tire mileage will be less than or equal to 40,000 miles. (The specified constant in this case is 40,000.) After obtaining the cumulative probability from Minitab, we must subtract it from one to determine the probability that the tire mileage will exceed 40,000 miles.

Prior to using Minitab to compute a probability, one must enter the specified constant into a column of the worksheet. For the Grear tire mileage question we entered the specified constant of 40,000 into column 1 of the Minitab worksheet. The steps in using Minitab to compute the cumulative probability of the normal random variable assuming a value less than or equal to 40,000 follow.

Step 1. Select the **Calc** pull-down menu
Step 2. Choose **Probability Distributions**
Step 3. Choose **Normal**
Step 4. When the Normal Distribution dialog box appears:
Select **Cumulative probability**
Enter 36500 in the **Mean** box
Enter 5000 in the **Standard deviation** box
Enter C1 in the **Input column** box (the column containing 40,000)
Click **OK**

After the user clicks **OK,** Minitab will print the cumulative probability that the normal random variable assumes a value less than or equal to 40,000. Minitab will show that this probability is .7580. Because we are interested in the probability that the tire mileage will be greater than 40,000, the desired probability is 1 − .7580 = .2420.

A second question in the Grear Tire Company problem was: What mileage guarantee should Grear set to ensure that no more than 10% of the tires qualify for the guarantee? Here we are given a probability and want to find the corresponding value for the random variable. Minitab uses an inverse calculation routine to find the value of the random variable associated with a given cumulative probability. First, we must enter the cumulative probability into a column of the Minitab worksheet (say C1). In this case, the desired cumulative probability is .10. Then, the first three steps of the Minitab procedure are as already listed. In step 4, we select **Inverse cumulative probability** instead of **Cumulative probability** and complete the remaining parts of the step. Minitab then displays the mileage guarantee of 30,100 miles.

Minitab is capable of computing probabilities for other continuous probability distributions, including the exponential probability distribution. To compute exponential probabilities, follow the procedure shown previously for the normal probability distribution and select the **Exponential** option in step 3. Step 4 is as shown, with the exception that entering the standard deviation is not required. Output for cumulative probabilities and inverse cumulative probabilities is identical to that described for the normal probability distribution.

Appendix 6.2 CONTINUOUS PROBABILITY DISTRIBUTIONS WITH EXCEL

Excel has the capability of computing probabilities for several continuous probability distributions, including the normal and exponential probability distributions. In this appendix, we describe how Excel can be used to compute probabilities for any normal probability distribution. The procedures for the exponential and other continuous probability distributions are similar to the one we describe for the normal probability distribution.

Let us return to the Grear Tire Company problem where the tire mileage was described by a normal probability distribution with $\mu = 36{,}500$ and $\sigma = 5000$. Assume we are interested in the probability that tire mileage will exceed 40,000 miles.

Excel's NORMDIST function provides cumulative probabilities for a normal probability distribution. The general form of the function is NORMDIST (x, μ, σ, cumulative). For the fourth argument, TRUE is specified if a cumulative probability is desired. Thus, to compute the cumulative probability that the tire mileage will be less than or equal to 40,000 miles we would enter the following formula into any cell of an Excel worksheet:

=NORMDIST(40000,36500,5000,TRUE)

At this point, .7580 will appear in the cell where the formula was entered, indicating that the probability of tire mileage being less than or equal to 40,000 miles is .7580. Therefore, the probability that tire mileage will exceed 40,000 miles is 1 − .7580 = .2420.

Excel's NORMINV function uses an inverse computation to find the x value corresponding to a given cumulative probability. For instance, suppose we want to find the guaranteed mileage Grear should offer so that no more than 10% of the tires will be eligible for the guarantee. We would enter the following formula into any cell of an Excel worksheet:

=NORMINV(.1,36500,5000)

At this point, 30092 will appear in the cell where the formula was entered indicating that the probability of a tire not lasting 30,092 miles is .10.

The Excel function for computing exponential probabilities is EXPONDIST. Using it is straightforward. But, if one needs help specifying the proper values for the arguments, Excel's function wizard can be used (see Appendix 2.2).

Chapter 7

SAMPLING AND SAMPLING DISTRIBUTIONS

CONTENTS

STATISTICS IN PRACTICE

MEAD CORPORATION*
Dayton, Ohio

Mead Corporation is a diversified paper and forest products company that has more than 16,000 employees. It operates worldwide in 32 countries, with customers located in 98 countries. Mead holds a leading position in paper production, with an annual capacity of 1.8 million tons. The company's products include textbook paper, glossy magazine paper, beverage packaging systems, and office products. Mead's internal consulting group uses sampling to provide a variety of information that enables the company to obtain significant productivity benefits and remain competitive.

For example, Mead maintains large woodland holdings, which provide the trees that are the raw material for many of the company's products. Managers need reliable and accurate information about the timberlands and forests to evaluate the company's ability to meet its future raw material needs. What is the present volume in the forests? What is the past growth of the forests? What is the projected future growth of the forests? With answers to these important questions, Mead's managers can develop plans for the future including long-term planting and harvesting schedules for the trees.

How does Mead obtain the information it needs about its vast forest holdings? Data collected from sample plots throughout the forests are the basis for learning about the population of trees owned by the company. To identify the sample plots, the timberland holdings are first divided into three sections based on location and types of trees. Using maps and random numbers, Mead analysts identify random samples of $\frac{1}{5}$ to $\frac{1}{7}$ acre plots in each section of the forest. The sample plots are where Mead foresters collect data and learn about the forest population.

*The authors are indebted to Dr. Edward P. Winkofsky, Mead Corporation, for providing this Statistics in Practice.

Random sampling of its forest holdings enables Mead Corporation to meet future raw material needs. © Larry Goldstein/Tony Stone.

Foresters throughout the organization participate in the field data collection process. Periodically, two-person teams gather information on each tree in every sample plot. The sample data are entered into the company's continuous forest inventory (CFI) computer system. Reports from the CFI system include a number of frequency distribution summaries containing statistics on types of trees, present forest volume, past forest growth rates, and projected future forest growth and volume. Sampling and the associated statistical summaries of the sample data provide the reports that are essential for the effective management of Mead's forests and timberlands.

In this chapter you will learn about simple random sampling and the sample selection process. In addition, you will learn how statistics such as the sample mean and sample proportion are used to estimate the population mean and population proportion. The important concept of a sampling distribution is also introduced.

In Chapter 1, we defined a *population* and a *sample* as two important aspects of a statistical study. The definitions are restated here.

1. A *population* is the set of all the elements of interest in a study.
2. A *sample* is a subset of the population.

The purpose of *statistical inference* is to develop estimates and test hypotheses about the characteristics of a population using information contained in a sample. Let us begin by citing two situations in which sampling is conducted to give a manager or decision maker information about a population.

1. A tire manufacturer developed a new tire designed to provide an increase in mileage over the firm's current line of tires. To estimate the mean number of miles provided by the new tires, the manufacturer selected a sample of 120 new tires for testing. The test results provided a sample mean of 36,500 miles. Hence, an estimate of the mean tire mileage for the population of new tires was 36,500 miles.
2. Members of a political party were considering supporting a particular candidate for election to the U.S. Senate, and party leaders wanted an estimate of the proportion of registered voters favoring the candidate. The time and cost associated with contacting every individual in the population of registered voters were prohibitive. Hence, a sample of 400 registered voters was selected and 160 of the 400 voters indicated a preference for the candidate. An estimate of the proportion of the population of registered voters favoring the candidate was 160/400 = .40.

The preceding examples show how sampling and the sample results can be used to develop estimates of population characteristics. Note that in the tire mileage example, collecting the data on tire life involves wearing out each tire tested. Clearly it is not feasible to test every tire in the population; a sample is the only realistic way to obtain the desired tire mileage data. In the example involving the election, contacting every registered voter in the population is theoretically possible, but the time and cost in doing so are prohibitive; thus, a sample of registered voters is preferred.

A sample mean provides an estimate of a population mean, and a sample proportion provides an estimate of a population proportion. With estimates such as these, some sampling error or ± value can be expected. A key point in this chapter is that statistical methods can be used to make probability statements about the size of the sampling error.

The examples illustrate some of the reasons for using samples. However, it is important to realize that sample results provide only *estimates* of the values of the population characteristics. That is, we do not expect the sample mean of 36,500 miles to exactly equal the mean mileage for all tires in the population, nor do we expect exactly 40% of the population of registered voters to favor the candidate. The reason is simply that the sample contains only a portion of the population. With proper sampling methods, the sample results will provide "good" estimates of the population characteristics. But how good can we expect the sample results to be? Fortunately, statistical procedures are available for answering this question.

In this chapter we show how simple random sampling can be used to select a sample from a population. We then show how data obtained from a simple random sample can be used to compute estimates of a population mean, a population standard deviation, and a population proportion. In addition, we introduce the important concept of a sampling distribution. As we show, knowledge of the appropriate sampling distribution is what enables us to make statements about the goodness of the sample results. The last section discusses some alternatives to simple random sampling that are often employed in practice.

7.1 THE ELECTRONICS ASSOCIATES SAMPLING PROBLEM

The director of personnel for Electronics Associates, Inc. (EAI), has been assigned the task of developing a profile of the company's 2500 managers. The characteristics to be identified include the mean annual salary for the managers and the proportion of managers having completed the company's management training program.

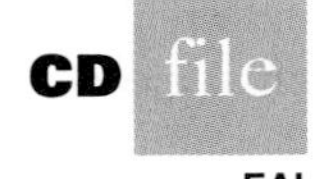

EAI

Using the 2500 managers as the population for this study, we can find the annual salary and the training program status for each individual by referring to the firm's personnel records. The data file containing this information for all 2500 managers in the population is on the disk at the back of the book.

Using the formulas presented in Chapter 3, we can compute the population mean and the population standard deviation for the annual salary data.

$$\text{Population mean: } \mu = \$51{,}800$$
$$\text{Population standard deviation: } \sigma = \$4000$$

Furthermore, the data for the training program status show that 1500 of the 2500 managers have completed the training program. Letting p denote the proportion of the population having completed the training program, we see that $p = 1500/2500 = .60$.

A **parameter** is a numerical characteristic of a population. The population mean annual salary ($\mu = \$51,800$), the population standard deviation of annual salary ($\sigma = \$4000$), and the population proportion having completed the training program ($p = .60$) are parameters of the population of EAI managers.

Now, suppose that the necessary information on all the EAI managers was not readily available in the company's database. Employee records for recent hires might not be on the computer. Perhaps only the salary information (and not the training program status) is in the employee's record. Or, maybe the employee records at some locations are on the computer, but records at other locations are not. For whatever reason, it is often the case that it is not possible, or perhaps too costly, to collect and process information on all elements of a population. In such cases, information from a sample can often be used to develop estimates of the population parameters of interest.

Often the cost of collecting information from a sample is substantially less than from a population, especially when personal interviews must be conducted to collect the information.

The question we now consider is how the firm's director of personnel can obtain estimates of the population parameters of interest by using a sample of managers rather than all 2500 managers in the population. Suppose that a sample of 30 managers will be used. Clearly, the time and the cost of developing a profile would be substantially less for 30 managers than for the entire population. If the personnel director could be assured that a sample of 30 managers would provide adequate information about the population of 2500 managers, working with a sample would be preferable to working with the entire population. Let us explore the possibility of using a sample for the EAI study by first considering how we can identify a sample of 30 managers.

7.2 SIMPLE RANDOM SAMPLING

Several methods can be used to select a sample from a population; one of the most common is **simple random sampling.** The definition of a simple random sample and the process of selecting a simple random sample depend on whether the population is *finite* or *infinite.* Because the EAI sampling problem involves a finite population of 2500 managers, we first consider sampling from a finite population.

Sampling from a Finite Population

A simple random sample of size n from a finite population of size N is defined as follows.

Simple Random Sample (Finite Population)

A simple random sample of size n from a finite population of size N is a sample selected such that each possible sample of size n has the same probability of being selected.

Computer-generated random numbers can also be used to implement the random sample selection process. Excel provides a function for generating random numbers in its worksheets.

One procedure for selecting a simple random sample from a finite population is to choose the elements for the sample one at a time in such a way that each of the elements remaining in the population has the same probability of being selected. Sampling n elements in this way will satisfy the definition of a simple random sample from a finite population.

To select a simple random sample from the finite population of EAI managers, we first assign each manager a number. For example, we can assign the managers the numbers 1 to 2500 in the order that their names appear in the EAI personnel file. Next, we refer to the table of random numbers shown in Table 7.1. Using the first row of the table, each digit, 6, 3, 2, . . . , is a random digit having an equal chance of occurring. Because the largest

TABLE 7.1 RANDOM NUMBERS

63271	59986	71744	51102	15141	80714	58683	93108	13554	79945
88547	09896	95436	79115	08303	01041	20030	63754	08459	28364
55957	57243	83865	09911	19761	66535	40102	26646	60147	15702
46276	87453	44790	67122	45573	84358	21625	16999	13385	22782
55363	07449	34835	15290	76616	67191	12777	21861	68689	03263
69393	92785	49902	58447	42048	30378	87618	26933	40640	16281
13186	29431	88190	04588	38733	81290	89541	70290	40113	08243
17726	28652	56836	78351	47327	18518	92222	55201	27340	10493
36520	64465	05550	30157	82242	29520	69753	72602	23756	54935
81628	36100	39254	56835	37636	02421	98063	89641	64953	99337
84649	48968	75215	75498	49539	74240	03466	49292	36401	45525
63291	11618	12613	75055	43915	26488	41116	64531	56827	30825
70502	53225	03655	05915	37140	57051	48393	91322	25653	06543
06426	24771	59935	49801	11082	66762	94477	02494	88215	27191
20711	55609	29430	70165	45406	78484	31639	52009	18873	96927
41990	70538	77191	25860	55204	73417	83920	69468	74972	38712
72452	36618	76298	26678	89334	33938	95567	29380	75906	91807
37042	40318	57099	10528	09925	89773	41335	96244	29002	46453
53766	52875	15987	46962	67342	77592	57651	95508	80033	69828
90585	58955	53122	16025	84299	53310	67380	84249	25348	04332
32001	96293	37203	64516	51530	37069	40261	61374	05815	06714
62606	64324	46354	72157	67248	20135	49804	09226	64419	29457
10078	28073	85389	50324	14500	15562	64165	06125	71353	77669
91561	46145	24177	15294	10061	98124	75732	00815	83452	97355
13091	98112	53959	79607	52244	63303	10413	63839	74762	50289

number in the population list of EAI managers, 2500, has four digits, we will select random numbers from the table in sets or groups of four digits. Even though we may start the selection of random numbers anywhere in the table and move systematically in a direction of our choice, we will use the first row of Table 7.1 and move from left to right. The four-digit random numbers are

6327 1599 8671 7445 1102 1514 1807

Because the numbers in the table are random, these four-digit numbers are equally probable or equally likely.

We can now use these four-digit random numbers to give each manager in the population an equal chance of being included in the random sample. The first number, 6327, is greater than 2500. It does not correspond to one of the numbered managers in the population, and hence is discarded. The second number, 1599, is between 1 and 2500. Thus the first manager selected for the random sample is number 1599 on the list of EAI managers. Continuing this process, we ignore the numbers 8671 and 7445 before identifying managers number 1102, 1514, and 1807 to be included in the random sample. This process continues until the simple random sample of 30 EAI managers has been obtained.

In implementing the simple random sample selection process, it is possible that a random number used previously may appear again in the table before the sample of 30 EAI managers has been selected. Because we do not want to select a manager more than one time, any previously used random numbers are ignored because the corresponding manager is already included in the sample. Selecting a sample in this manner is referred to as

sampling without replacement. If we had selected the sample such that previously used random numbers were acceptable and specific managers could be included in the sample two or more times, we would be **sampling with replacement.** Sampling with replacement is a valid way of identifying a simple random sample. However, sampling without replacement is the sampling procedure used most often. When we refer to simple random sampling, we will assume that the sampling is without replacement.

Sampling from an Infinite Population

In practice, a population being studied is usually considered infinite if it involves an ongoing process that makes listing or counting every element in the population impossible.

Many sampling situations in business and economics involve finite populations, but in some situations the population is either infinite or so large that for practical purposes it must be treated as infinite. In sampling from an infinite population, we must use a new definition of a simple random sample. In addition, because the elements in an infinite population cannot be listed and numbered, we must use a different process for selecting the sample.

Suppose we want to estimate the average time between placing an order and receiving food for customers at a fast-food restaurant during the 11:30 A.M. to 1:30 P.M. lunch period. If we consider the population as being all possible customer visits, we see that it would not be feasible to develop a list of all possible visits. In fact, if we define the population as being all customer visits that could conceivably occur during the lunch period, we can consider the population as being infinite. Our task is to select a simple random sample of n customers from this population. The definition of a simple random sample from an infinite population follows.

Simple Random Sample (Infinite Population)

A simple random sample from an infinite population is a sample selected such that the following conditions are satisfied.

1. Each element selected comes from the same population.
2. Each element is selected independently.

A random number selection procedure cannot be used for an infinite population because a listing of the population is impossible. In this case, a sample selection procedure must be specially devised to select the items independently and thus avoid a selection bias that gives higher selection probabilities to certain types of elements.

For the problem of selecting a simple random sample of customer visits at a fast-food restaurant, we find that the first condition as defined is satisfied by any customer visit occurring during the 11:30 A.M. to 1:30 P.M. lunch period while the restaurant is operating with its regular staff under "normal" operating conditions. The second condition is satisfied by ensuring that the selection of a particular customer does not influence the selection of any other customer. That is, the customers are selected independently.

A well-known fast-food restaurant has implemented a simple random sampling procedure for just such a situation. The sampling procedure is based on the fact that some customers present discount coupons for special prices on sandwiches, drinks, french fries, and so on. Whenever a customer presents a discount coupon, the next customer served is selected for the sample. Because the customers present discount coupons randomly and independently, the firm is satisfied that the sampling plan satisfies the two conditions for a simple random sample from an infinite population.

NOTES AND COMMENTS

1. Finite populations are often defined by lists such as organization membership rosters, student enrollment records, credit card account lists, inventory product numbers, and so on. Infinite populations are often defined by an ongoing process whereby the elements of the population consist of items generated as though the process would operate indefinitely under the same conditions. In such cases, it is impossible to obtain a list of all elements in the population. For example, populations consisting of all possible parts to be manufactured, all possible customer

visits, all possible bank transactions, and so on can be classified as infinite populations.

2. The number of different simple random samples of size n that can be selected from a finite population of size N is

$$\frac{N!}{n!(N-n)!}$$

In this formula, $N!$ and $n!$ are the factorial computations discussed in Chapter 4. For the EAI problem with $N = 2500$ and $n = 30$, this expression can be used to show that there are approximately 2.75×10^{69} different simple random samples of 30 EAI managers.

3. Computers can be used to generate random numbers for selecting random samples. For example, given the EAI population of 2500 managers, Excel's function =RANDBETWEEN (1,2500) can be used to generate random numbers between 1 and 2500. These computer-generated random numbers can then be used to identify the random sample of 30 managers the same way we used the random numbers from Table 7.1.

EXERCISES

Methods

1. Consider a finite population with five elements labeled A, B, C, D, and E. Ten possible simple random samples of size two can be selected.
   a. List the 10 samples beginning with AB, AC, and so on.
   b. Using simple random sampling, what is the probability that each sample of size two is selected?
   c. Assume random number 1 corresponds to A, random number 2 corresponds to B, and so on. List the simple random sample of size two that will be selected by using the random digits 8 0 5 7 5 3 2.

2. Assume a finite population has 350 elements. Using the last three digits of each of the following five-digit random numbers (601, 022, 448, . . .), determine the first four elements that will be selected for the simple random sample.

   98601 73022 83448 02147 34229 27553 84147 93289 14209

Applications

3. *Fortune* publishes data on sales, profits, assets, stockholders' equity, market value, and earnings per share for the 500 largest U.S. industrial corporations (The *Fortune* 500, 2000). Assume that you want to select a simple random sample of 10 corporations from the *Fortune* 500 list. Use the last three digits in column 9 of Table 7.1, beginning with 554. Read down the column and identify the numbers of the 10 corporations that would be selected.

4. Ten companies with widely held stocks are shown here (*USA Today,* September 6, 2000).

AT&T	IBM
America Online	Johnson & Johnson
Cisco Systems	Microsoft
General Electric	Motorola
Intel	Pfizer

   a. Assume that a random sample of five of these companies will be selected for an in-depth study of common business practices in large corporations. Beginning with the first random digit in Table 7.1 and reading down the column, use the single-digit random numbers to select a simple random sample of five companies to be used in the study.
   b. According to the second point in Notes and Comments, how many different simple random samples of size five can be selected from the list of 10 companies?

5. A student government organization is interested in estimating the proportion of students who favor a mandatory "pass-fail" grading policy for elective courses. A list of names and addresses of the 645 students enrolled during the current quarter is available from the registrar's office. Using three-digit random numbers in row 10 of Table 7.1 and moving across the row from left to right, identify the first 10 students who would be selected using simple random sampling. The three-digit random numbers begin with 816, 283, and 610.

6. The *County and City Data Book,* published by the Bureau of the Census, lists information on 3139 counties throughout the United States. Assume that a national study will collect data from 30 randomly selected counties. Use four-digit random numbers from the last column of Table 7.1 to identify the numbers corresponding to the first five counties selected for the sample. Ignore the first digits and begin with the four-digit random numbers 9945, 8364, 5702, and so on.

7. Assume that we want to identify a simple random sample of 12 of the 372 doctors practicing in a particular city. The doctors' names are available from a local medical organization. Use the eighth column of five-digit random numbers in Table 7.1 to identify the 12 doctors for the sample. Ignore the first two random digits in each five-digit grouping of the random numbers. This process begins with random number 108 and proceeds down the column of random numbers.

8. The following table provides the NCAA football top 25 teams at the beginning of the 2000 season (*Sports Illustrated,* August 14, 2000). Use the ninth column of the random numbers in Table 7.1 beginning with 13554, to select a simple random sample of six football teams. Begin with team 13 and use the first two digits in each row of the ninth column for your selection process. Which six football teams are selected for the simple random sample?

1. Nebraska	14. Tennessee
2. Florida State	15. TCU
3. Alabama	16. Purdue
4. Michigan	17. Mississippi
5. Wisconsin	18. USC
6. Kansas State	19. Penn State
7. Georgia	20. Southern Miss
8. Clemson	21. Illinois
9. Texas	22. Ohio State
10. Miami	23. Oklahoma
11. Florida	24. Colorado State
12. Virginia Tech	25. Colorado
13. Washington	

9. *Business Week* provided performance data and annual ratings for 895 mutual funds (*Business Week,* February 3, 1997). Assume that a simple random sample of 12 of the 895 mutual funds will be selected for a follow-up study on the performance of mutual funds. Use the fourth column of the random numbers in Table 7.1, beginning with 51102, to select the simple random sample of 12 mutual funds. Begin with mutual fund 511 and use the first three digits in each row of the fourth column for your selection process. What are the numbers of the 12 mutual funds in the simple random sample?

10. Indicate whether the following populations should be considered finite or infinite.
   a. All registered voters in the state of California.
   b. All television sets that could be produced by the Allentown, Pennsylvania, plant of the TV-M Company.

c. All orders that could be processed by a mail-order firm.
d. All emergency telephone calls that could come into a local police station.
e. All components that Fibercon, Inc., produced on the second shift on May 17.

7.3 POINT ESTIMATION

Now that we have described how to select a simple random sample, let us return to the EAI problem. Assume that a simple random sample of 30 managers has been selected and that the corresponding data on annual salary and management training program participation are as shown in Table 7.2. The notation x_1, x_2, and so on is used to denote the annual salary of the first manager in the sample, the annual salary of the second manager in the sample, and so on. Participation in the management training program is indicated by Yes in the management training program column.

To estimate the value of a population parameter, we compute a corresponding characteristic of the sample, referred to as a **sample statistic.** For example, to estimate the population mean μ and the population standard deviation σ for the annual salary of EAI managers, we simply use the data in Table 7.2 to calculate the corresponding sample statistics: the sample mean $\bar{x}$ and the sample standard deviation s. Using the formulas for a sample mean and a sample standard deviation presented in Chapter 3, the sample mean is

$$\bar{x} = \frac{\Sigma x_i}{n} = \frac{1{,}554{,}420}{30} = \$51{,}814.00$$

and the sample standard deviation is

$$s = \sqrt{\frac{\Sigma(x_i - \bar{x})^2}{n - 1}} = \sqrt{\frac{325{,}009{,}260}{29}} = \$3347.72$$

TABLE 7.2 ANNUAL SALARY AND TRAINING PROGRAM STATUS FOR A SIMPLE RANDOM SAMPLE OF 30 EAI MANAGERS

Annual Salary ($)	Management Training Program?	Annual Salary ($)	Management Training Program?
x_1 = 49,094.30	Yes	x_{16} = 51,766.00	Yes
x_2 = 53,263.90	Yes	x_{17} = 52,541.30	No
x_3 = 49,643.50	Yes	x_{18} = 44,980.00	Yes
x_4 = 49,894.90	Yes	x_{19} = 51,932.60	Yes
x_5 = 47,621.60	No	x_{20} = 52,973.00	Yes
x_6 = 55,924.00	Yes	x_{21} = 45,120.90	Yes
x_7 = 49,092.30	Yes	x_{22} = 51,753.00	Yes
x_8 = 51,404.40	Yes	x_{23} = 54,391.80	No
x_9 = 50,957.70	Yes	x_{24} = 50,164.20	No
x_{10} = 55,109.70	Yes	x_{25} = 52,973.60	No
x_{11} = 45,922.60	Yes	x_{26} = 50,241.30	No
x_{12} = 57,268.40	No	x_{27} = 52,793.90	No
x_{13} = 55,688.80	Yes	x_{28} = 50,979.40	Yes
x_{14} = 51,564.70	No	x_{29} = 55,860.90	Yes
x_{15} = 56,188.20	No	x_{30} = 57,309.10	No

In addition, by computing the proportion of managers in the sample who responded Yes, we can estimate the proportion of managers in the population who have completed the management training program. Table 7.2 shows that 19 of the 30 managers in the sample have completed the training program. Thus, the sample proportion, denoted by $\bar{p}$, is given by

$$\bar{p} = \frac{19}{30} = .63$$

This value is used as an estimate of the population proportion p.

By making the preceding computations, we have performed the statistical procedure called *point estimation.* In point estimation we use the data from the sample to compute a value of a sample statistic that serves as an estimate of a population parameter. Using the terminology of point estimation, we refer to $\bar{x}$ as the **point estimator** of the population mean μ, s as the *point estimator* of the population standard deviation σ, and $\bar{p}$ as the *point estimator* of the population proportion p. The actual numerical value obtained for $\bar{x}$, s, or $\bar{p}$ in a particular sample is called the **point estimate** of the parameter. Thus, for the sample of 30 EAI managers, \$51,814.00 is the point estimate of μ, \$3,347.72 is the point estimate of σ, and .63 is the point estimate of p. Table 7.3 summarizes the sample results and compares the point estimates to the actual values of the population parameters.

Sampling error is the result of using a subset of the population (the sample), and not the entire population.

As Table 7.3 shows, none of the point estimates are exactly equal to the corresponding population parameters. This variation is to be expected because only a sample and not a census of the entire population is being used to develop the estimate. The absolute value of the difference between an unbiased point estimate and the corresponding population parameter is called the **sampling error.*** For a sample mean, sample standard deviation, and sample proportion, the sampling errors are $|\bar{x} - \mu|$, $|s - \sigma|$, and $|\bar{p} - p|$, respectively. Thus, for the EAI sample, the sampling errors are $|\$51{,}814 - \$51{,}800| = \$14$ for the sample mean, $|\$3{,}347.72 - \$4000.00| = \$652.28$ for the sample standard deviation, and $|.63 - .60| = .03$ for the sample proportion.

We were able to compute the sampling errors here because the population parameters were known. However, in an actual sampling application we will not be able to calculate the sampling error exactly because the value of the population parameter will not be known.

TABLE 7.3 SUMMARY OF POINT ESTIMATES OBTAINED FROM A SIMPLE RANDOM SAMPLE OF 30 EAI MANAGERS

Population Parameter	Parameter Value	Point Estimator	Point Estimate
μ = Population mean annual salary	\$51,800.00	$\bar{x}$ = Sample mean annual salary	\$51,814.00
σ = Population standard deviation for annual salary	\$4,000.00	s = Sample standard deviation for annual salary	\$3,347.72
p = Population proportion having completed the management training program	.60	$\bar{p}$ = Sample proportion having completed the management training program	.63

*Technically the difference between the point estimate and the population parameter is called sampling error only when the point estimator is unbiased. The point estimators ($\bar{x}$, s, and $\bar{p}$) used here are unbiased.

We will show how statisticians analyze the sample data in order to make probability statements about the size of the sampling error.

NOTES AND COMMENTS

In our discussion of point estimators, we use $\bar{x}$ to denote a sample mean and $\bar{p}$ to denote a sample proportion. Our use of $\bar{p}$ is based on the fact that the sample proportion is also a *sample mean.* For instance, suppose that in a sample of size n with data values $x_1, x_2, \ldots, x_n$, we let $x_i = 1$ when a characteristic of interest is present for the ith observation and $x_i = 0$ when the characteristic is not present. Then the sample proportion is computed by $\Sigma x_i/n$, which is the formula for a sample mean. We also like the consistency of using the bar over the letter to remind the reader that the sample proportion $\bar{p}$ estimates the population proportion just as the sample mean $\bar{x}$ estimates the population mean. Some texts use $\hat{p}$ instead of $\bar{p}$ to denote the sample proportion.

EXERCISES

Methods

11. The following data are from a simple random sample.

5 8 10 7 10 14

    a. What is the point estimate of the population mean?
    b. What is the point estimate of the population standard deviation?

12. A survey question for a sample of 150 individuals yielded 75 Yes responses, 55 No responses, and 20 No Opinions.
    a. What is the point estimate of the proportion in the population who respond Yes?
    b. What is the point estimate of the proportion in the population who respond No?

Applications

13. A simple random sample of 5 months of sales data provided the following information:

Month:	1	2	3	4	5
Units Sold:	94	100	85	94	92

    a. What is the point estimate of the population mean number of units sold per month?
    b. What is the point estimate of the population standard deviation?

14. A sample of 784 children ages 9 to 14 were asked about cash provided by their parents (*Consumer Reports,* January 1997). The responses were as follows.

Source of Cash	**Frequency**
Allowance only	149
Handouts & chores plus an allowance	219
Handouts & chores but no allowance	251
Nothing	165
Total	784

    a. What proportion of children receive an allowance as their only source of cash?
    b. What proportion of children receive cash as handouts and for chores, but nothing in the form of an allowance?
    c. Considering all sources, what proportion of children receive at least some cash from their parents?

15. *Appliance Magazine* provided estimates of the life expectancy of household appliances (*USA Today,* September 5, 2000). A simple random sample of 10 VCRs shows the following useful life in years.

6.5 8.0 6.2 7.4 7.0 8.4 9.5 4.6 5.0 7.4

a. What is the point estimate of the population mean life expectancy for VCRs?
b. What is the point estimate of the population standard deviation for life expectancy of VCRs?

16. The U.S. Department of Transportation reports statistics on how frequently major airline flights arrive at or before their scheduled arrival times (*Associated Press,* September 8, 2000). Assume that the estimated proportion of flights arriving on time for an airline is based on a sample of 1400 flights. If 1117 of the flights arrive on time, what is the point estimate of the proportion of all flights that arrive on time?

17. A Louis Harris poll used a survey of 1008 adults to learn about how people feel about the economy (*Business Week,* August 7, 2000). Responses were as follow:

595 adults	The economy is growing.
332 adults	The economy is staying about the same.
81 adults	The economy is shrinking.

Develop the point estimate of the following population parameters.
a. The proportion of all adults who feel the economy is growing.
b. The proportion of all adults who feel the economy is staying about the same.
c. The proportion of all adults who feel the economy is shrinking.

7.4 INTRODUCTION TO SAMPLING DISTRIBUTIONS

In the preceding section we used a simple random sample of 30 EAI managers to develop point estimates of the mean and standard deviation of annual salary for the population of all EAI managers as well as the proportion of the managers in the population who have completed the company's management training program. Suppose we select another simple random sample of 30 EAI managers, and an analysis of the data from the second sample provides the following information.

$$\text{Sample Mean } \bar{x} = \$52{,}669.70$$
$$\text{Sample Standard Deviation } s = \$4{,}239.07$$
$$\text{Sample Proportion } \bar{p} = .70$$

These results show that different values of $\bar{x}$, s, and $\bar{p}$ have been obtained with the second sample. In general, a second simple random sample will not contain the same elements as the first. Let us imagine carrying out the same process of selecting a new simple random sample of 30 managers over and over again, each time computing values of $\bar{x}$, s, and $\bar{p}$. In this way we could begin to identify the variety of values that these point estimators can have. To illustrate, we repeated the simple random sampling process for the EAI problem until we obtained 500 samples of 30 managers each and the corresponding $\bar{x}$, s, and $\bar{p}$ values. A portion of the results is shown in Table 7.4. Table 7.5 gives the frequency and relative frequency distributions for the 500 $\bar{x}$ values. Figure 7.1 is the relative frequency histogram for the $\bar{x}$ values.

The concept of a sampling distribution is one of the most important topics in this chapter. The ability to understand the material in subsequent chapters depends heavily on the ability to understand and use the sampling distributions presented in this chapter.

Recall that in Chapter 5 we defined a random variable as a numerical description of the outcome of an experiment. If we consider the process of selecting a simple random sample as an experiment, the sample mean $\bar{x}$ is the numerical description of the outcome of the experiment. Thus, the sample mean $\bar{x}$ is a random variable. As a result, just like other random variables, $\bar{x}$ has a mean or expected value, a variance, and a probability distribution. Because the various possible values of $\bar{x}$ are the result of different simple random samples, the probability distribution of $\bar{x}$ is called the **sampling distribution** of $\bar{x}$. Knowledge of this

TABLE 7.4 VALUES OF $\bar{x}$, s, and $\bar{p}$ FROM 500 SIMPLE RANDOM SAMPLES OF 30 EAI MANAGERS

Sample Number	Sample Mean ($\bar{x}$)	Sample Standard Deviation (s)	Sample Proportion ($\bar{p}$)
1	51,814.00	3,347.72	.63
2	52,669.70	4,239.07	.70
3	51,780.30	4,433.43	.67
4	51,587.90	3,985.32	.53
.	.	.	.
.	.	.	.
.	.	.	.
500	51,752.00	3,857.82	.50

sampling distribution and its properties will enable us to make probability statements about how close the sample mean $\bar{x}$ is to the population mean μ.

Let us return to Figure 7.1. We would need to enumerate every possible sample of 30 managers and compute each sample mean to completely determine the sampling distribution of $\bar{x}$. However, the histogram of 500 $\bar{x}$ values gives an approximation of this sampling distribution. From the approximation we observe the bell-shaped appearance of the distribution. We note that the estimate of the population mean provided by $\bar{x}$ could be off by more than $2000. But we also note that the largest concentration of the $\bar{x}$ values and the mean of the 500 $\bar{x}$ values is near the population mean $\mu = \$51,800$. We will describe the properties of the sampling distribution of $\bar{x}$ more fully in the next section.

The 500 values of the sample standard deviation s and the 500 values of the sample proportion $\bar{p}$ are summarized by the relative frequency histograms in Figures 7.2 and 7.3. As in the case of $\bar{x}$, both s and $\bar{p}$ are random variables that provide numerical descriptions of the outcome of a simple random sample. If every possible sample of size 30 were selected from the population and if a value of s and a value of $\bar{p}$ were computed for each sample, the resulting probability distributions would be called the sampling distribution of s and the sampling distribution of $\bar{p}$, respectively. The relative frequency histograms of the 500 sample values, Figures 7.2 and 7.3, provide a general idea of the appearance of these two sampling distributions.

TABLE 7.5 FREQUENCY DISTRIBUTION OF $\bar{x}$ FROM 500 SIMPLE RANDOM SAMPLES OF 300 EAI MANAGERS

Mean Annual Salary ($)	Frequency	Relative Frequency
49,500.00–49,999.99	2	.004
50,000.00–50,499.99	16	.032
50,500.00–50,999.99	52	.104
51,000.00–51,499.99	101	.202
51,500.00–51,999.99	133	.266
52,000.00–52,499.99	110	.220
52,500.00–52,999.99	54	.108
53,000.00–53,499.99	26	.052
53,500.00–53,999.99	6	.012
Totals	500	1.000

FIGURE 7.1 RELATIVE FREQUENCY HISTOGRAM OF $\bar{x}$ VALUES FROM 500 SIMPLE RANDOM SAMPLES OF SIZE 30 EACH

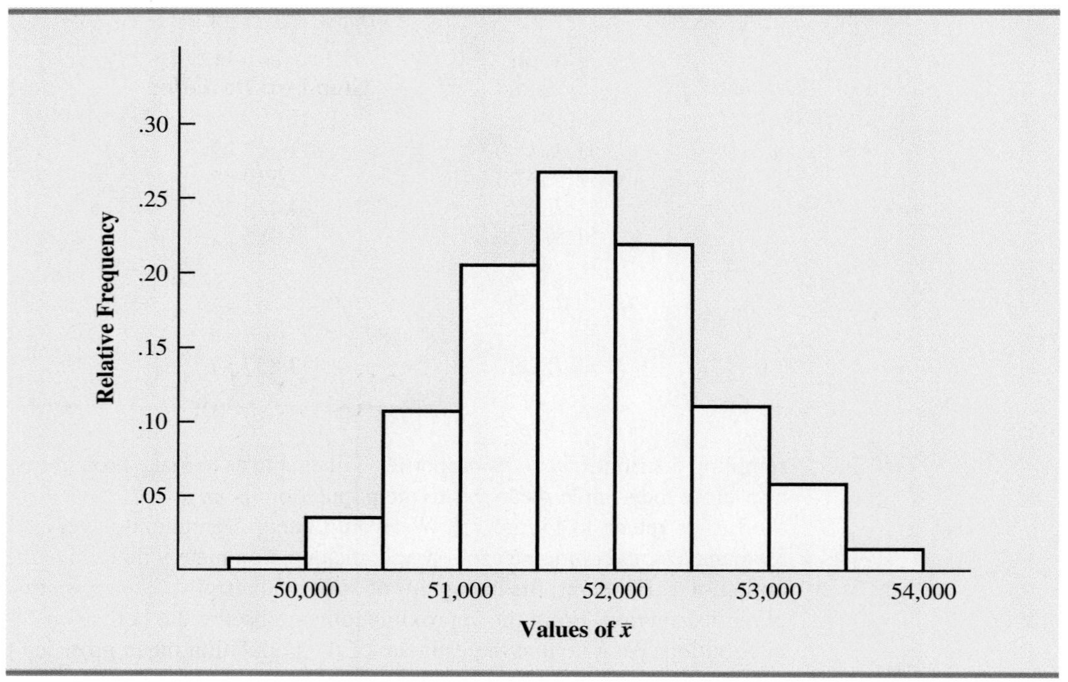

FIGURE 7.2 RELATIVE FREQUENCY HISTOGRAM OF s VALUES FROM 500 SIMPLE RANDOM SAMPLES OF SIZE 30 EACH

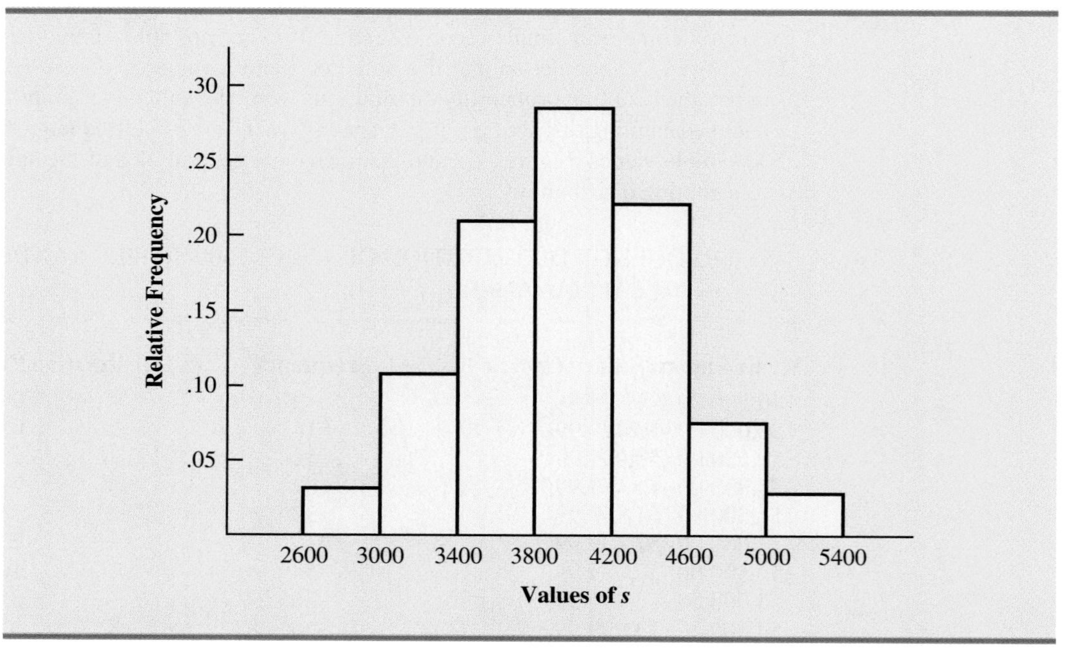

FIGURE 7.3 RELATIVE FREQUENCY HISTOGRAM OF $\bar{p}$ VALUES FROM 500 SIMPLE RANDOM SAMPLES OF SIZE 30 EACH

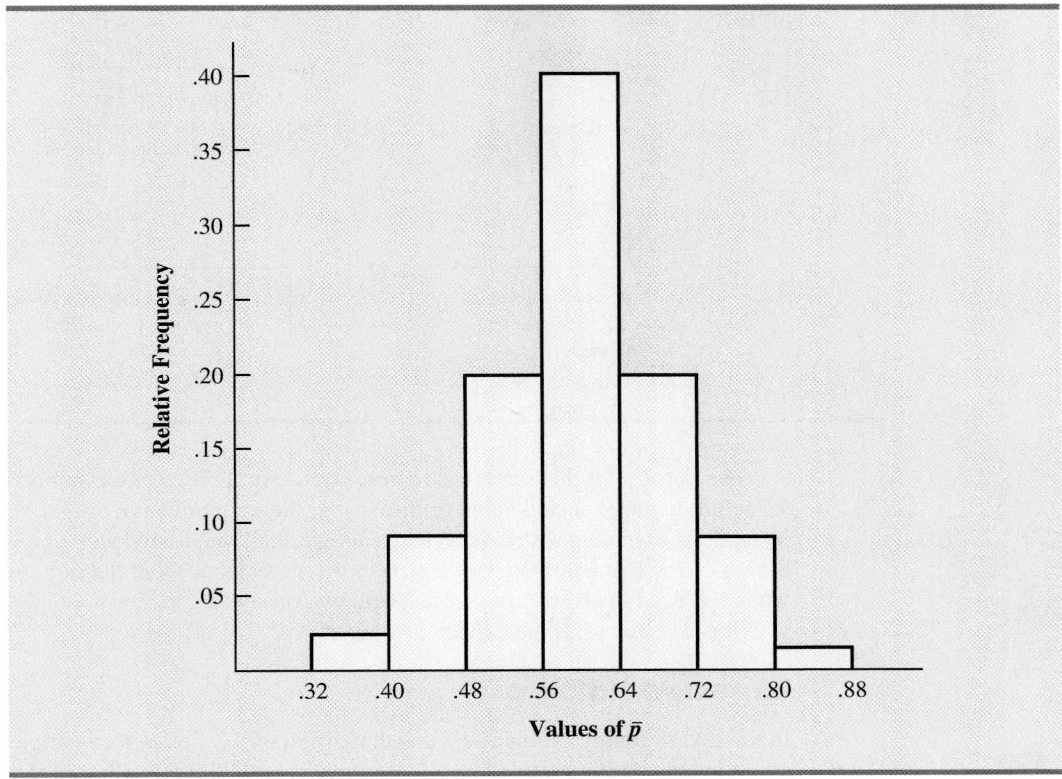

In practice, we select only one simple random sample from the population. We repeated the sampling process 500 times in this section simply to illustrate that many different samples are possible and that the different samples generate a variety of values for the sample statistics $\bar{x}$, s, and $\bar{p}$. The probability distribution of any particular sample statistic is called the sampling distribution of the statistic. In Section 7.5 we show the characteristics of the sampling distribution of $\bar{x}$. In Section 7.6 we show the characteristics of the sampling distribution of $\bar{p}$.

7.5 SAMPLING DISTRIBUTION OF $\bar{x}$

One of the most common statistical procedures is the use of a sample mean $\bar{x}$ to make inferences about a population mean μ. This process is shown in Figure 7.4. On each repetition of the process, we can anticipate obtaining a different value for the sample mean $\bar{x}$. The probability distribution for all possible values of the sample mean $\bar{x}$ is called the sampling distribution of the sample mean $\bar{x}$.

Sampling Distribution of $\bar{x}$

The sampling distribution of $\bar{x}$ is the probability distribution of all possible values of the sample mean, $\bar{x}$.

FIGURE 7.4 THE STATISTICAL PROCESS OF USING A SAMPLE MEAN TO MAKE INFERENCES ABOUT A POPULATION MEAN

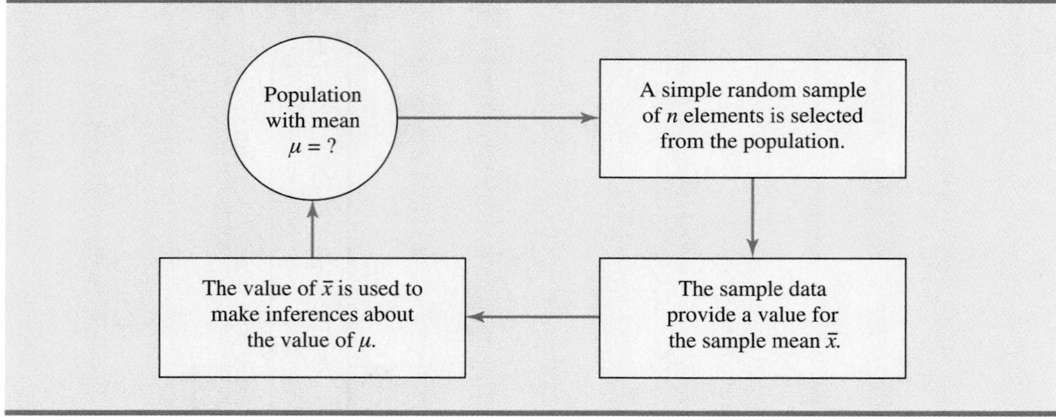

The purpose of this section is to describe the properties of the sampling distribution of $\bar{x}$, including the expected value or mean of $\bar{x}$, the standard deviation of $\bar{x}$, and the shape or form of the sampling distribution itself. As we shall see, knowledge of the sampling distribution of $\bar{x}$ will enable us to make probability statements about the sampling error involved when $\bar{x}$ is used to estimate μ. Let us begin by considering the mean of all possible $\bar{x}$ values, which is referred to as the expected value of $\bar{x}$.

Expected Value of $\bar{x}$

In the EAI sampling problem we saw that different simple random samples result in a variety of values for the sample mean $\bar{x}$. Because many different values of the random variable $\bar{x}$ are possible, we are often interested in the mean of all possible values of $\bar{x}$ that can be generated by the various simple random samples. The mean of the $\bar{x}$ random variable is the expected value of $\bar{x}$. Let $E(\bar{x})$ represent the expected value of $\bar{x}$ and μ represent the mean of the population from which we are sampling. It can be shown that with simple random sampling, $E(\bar{x})$ and μ are equal.

The expected value of $\bar{x}$ equals the mean of the population from which the sample is drawn.

Expected Value of $\bar{x}$

$$E(\bar{x}) = \mu \tag{7.1}$$

where

$$E(\bar{x}) = \text{the expected value of } \bar{x}$$
$$\mu = \text{the population mean}$$

This result shows that with simple random sampling, the expected value or mean for $\bar{x}$ is equal to the mean of the population. In Section 7.1 we saw that the mean annual salary for the population of EAI managers is μ = \$51,800. Thus, according to (7.1), the mean of all possible sample means for the EAI study is also \$51,800.

Standard Deviation of $\bar{x}$

Let us define the standard deviation of the sampling distribution of $\bar{x}$. We will use the following notation.

$\sigma_{\bar{x}}$ = the standard deviation of the sampling distribution of $\bar{x}$
σ = the standard deviation of the population
n = the sample size
N = the population size

It can be shown that with simple random sampling, the standard deviation of $\bar{x}$ depends on whether the population is finite or infinite. The two expressions for the standard deviation of $\bar{x}$ follow.

Standard Deviation of $\bar{x}$

Finite Population | *Infinite Population*

$$\sigma_{\bar{x}} = \sqrt{\frac{N-n}{N-1}}\left(\frac{\sigma}{\sqrt{n}}\right) \qquad \sigma_{\bar{x}} = \frac{\sigma}{\sqrt{n}} \tag{7.2}$$

In comparing the two expressions in (7.2), we see that the factor $\sqrt{(N-n)/(N-1)}$ is required for the finite population case but not for the infinite population case. This factor is commonly referred to as the **finite population correction factor.** In many practical sampling situations, we find that the population involved, although finite, is "large," whereas the sample size is relatively "small." In such cases the finite population correction factor $\sqrt{(N-n)/(N-1)}$ is close to 1. As a result, the difference between the values of the standard deviation of $\bar{x}$ for the finite and infinite population cases becomes negligible. Then, $\sigma_{\bar{x}} = \sigma/\sqrt{n}$ becomes a good approximation to the standard deviation of $\bar{x}$ even though the population is finite. This observation leads to the following general guideline or rule of thumb for computing the standard deviation of $\bar{x}$.

Use the Following Expression to Calculate the Standard Deviation of $\bar{x}$

$$\sigma_{\bar{x}} = \frac{\sigma}{\sqrt{n}} \tag{7.3}$$

whenever

1. The population is infinite; or
2. The population is finite *and* the sample size is less than or equal to 5% of the population size; that is, $n/N \leq .05$.

Problem 21 shows that when $n/N \leq .05$, the finite population correction factor has little effect on the value of $\sigma_{\bar{x}}$.

In cases where $n/N > .05$, the finite population version of (7.2) should be used in the computation of $\sigma_{\bar{x}}$. Unless otherwise noted, throughout the text we will assume that the population size is "large," $n/N \leq .05$, and (7.3) can be used to compute $\sigma_{\bar{x}}$.

Now let us return to the EAI study and determine the standard deviation of all possible sample means that can be generated with samples of 30 EAI managers. In Section 7.1 we saw that the population standard deviation for the annual salary data is $\sigma = 4000$. In this case the population is finite, with $N = 2500$. However, with a sample size of 30, we have $n/N = 30/2500 = .012$. Following the rule of thumb given in (7.3), we can ignore the finite population correction factor and use (7.3) to compute the standard deviation of $\bar{x}$.

$$\sigma_{\bar{x}} = \frac{\sigma}{\sqrt{n}} = \frac{4000}{\sqrt{30}} = 730.30$$

Later we will see that the value of $\sigma_{\bar{x}}$ is helpful in determining how far the sample mean may be from the population mean. Because of the role that $\sigma_{\bar{x}}$ plays in computing possible sampling errors, $\sigma_{\bar{x}}$ is referred to as the **standard error of the mean.**

Central Limit Theorem

The final step in identifying the characteristics of the sampling distribution of $\bar{x}$ is to determine the form of the probability distribution of $\bar{x}$. We consider two cases: one in which the population distribution is unknown and one in which the population distribution is known to be normally distributed.

The central limit theorem applies to any population. Thus, it can be used to describe the sampling distribution of $\bar{x}$ even when the population distribution is unknown.

When the population distribution is unknown, we rely on one of the most important theorems in statistics—the **central limit theorem.** A statement of the central limit theorem as it applies to the sampling distribution of $\bar{x}$ follows.

Central Limit Theorem

In selecting simple random samples of size n from a population, the sampling distribution of the sample mean $\bar{x}$ can be approximated by a *normal probability distribution* as the sample size becomes large.

Figure 7.5 shows how the central limit theorem works for three different populations. The top panel of the figure shows that each population is not normal. However, note what begins to happen to the sampling distribution of $\bar{x}$ as the sample size is increased. When the samples are of size two, we see that the sampling distribution of $\bar{x}$ begins to take on an appearance different from that of the population distribution. For samples of size five, we see all three sampling distributions beginning to take on a bell-shaped appearance. Finally, the samples of size 30 show all three sampling distributions to be approximately normal. Thus, for sufficiently large samples, the sampling distribution of $\bar{x}$ can be approximated by a normal probability distribution. However, how large must the sample size be before we can assume that the central limit theorem applies? Statistical researchers have investigated this question by studying the sampling distribution of $\bar{x}$ for a variety of populations and a variety of sample sizes. General statistical practice is to assume that for most applications, the sampling distribution of $\bar{x}$ can be approximated by a normal probability distribution whenever the sample size is 30 or more. In effect, a sample size of 30 or more is assumed to satisfy the large-sample condition of the central limit theorem. This observation is so important that we restate it.

The sampling distribution of $\bar{x}$ can be approximated by a normal probability distribution whenever the sample size is large. The large-sample condition can be assumed for simple random samples of size 30 or more.

The central limit theorem is the key to identifying the approximate form of the sampling distribution of $\bar{x}$ whenever the population distribution is unknown. However, we may encounter some sampling situations in which the population is assumed or believed to have a normal probability distribution. When this condition occurs, the following result identifies the form of the sampling distribution of $\bar{x}$.

Whenever the population has a normal probability distribution, the sampling distribution of $\bar{x}$ has a normal probability distribution for any sample size.

FIGURE 7.5 ILLUSTRATION OF THE CENTRAL LIMIT THEOREM FOR THREE POPULATIONS

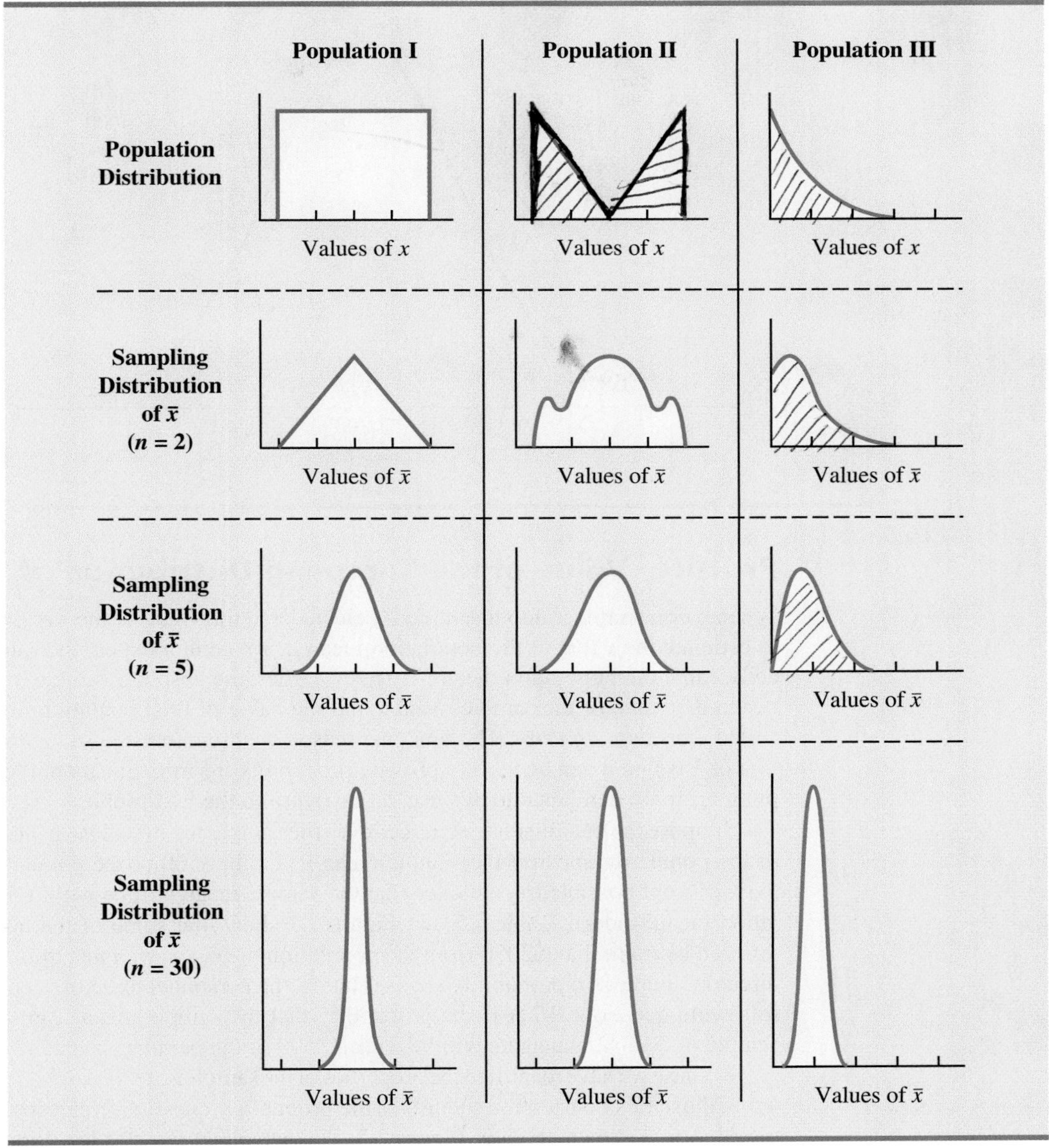

In summary, if we use a large ($n \geq 30$) simple random sample, the central limit theorem enables us to conclude that the sampling distribution of $\bar{x}$ can be approximated by a normal probability distribution. When the simple random sample is small ($n < 30$), the sampling distribution of $\bar{x}$ can be considered normal only if we are willing to assume that the population has a normal probability distribution.

Sampling Distribution of $\bar{x}$ for the EAI Problem

For the EAI study we have shown that $E(\bar{x}) = 51{,}800$ and $\sigma_{\bar{x}} = 730.30$. Because we are using a simple random sample of 30 managers, the central limit theorem enables us to conclude that the sampling distribution of $\bar{x}$ is approximately normal as shown in Figure 7.6.

FIGURE 7.6 SAMPLING DISTRIBUTION OF $\bar{x}$ FOR THE MEAN ANNUAL SALARY OF A SIMPLE RANDOM SAMPLE OF 30 EAI MANAGERS

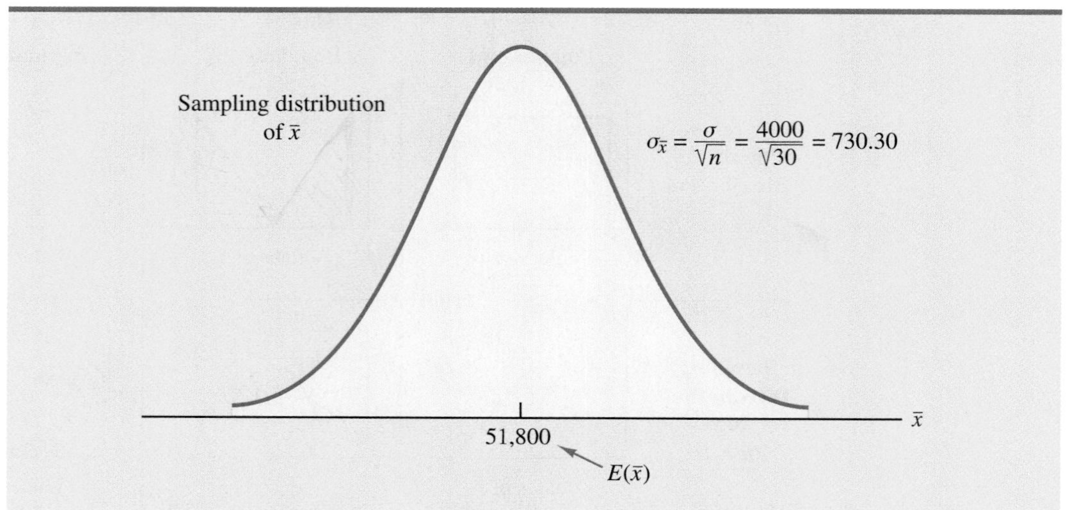

Practical Value of the Sampling Distribution of $\bar{x}$

Whenever a simple random sample is selected and the value of the sample mean $\bar{x}$ is used to estimate the value of the population mean μ, we cannot expect the sample mean to exactly equal the population mean. As stated earlier, the absolute value of the difference between the value of the sample mean $\bar{x}$ and the value of the population mean μ, $|\bar{x} - \mu|$, is called the *sampling error*. The practical reason we are interested in the sampling distribution of $\bar{x}$ is that it can be used to provide probability information about the size of the sampling error. To demonstrate this use, let us return to the EAI problem.

Suppose the personnel director believes the sample mean will be an acceptable estimate of the population mean if the sample mean is within \$500 of the population mean. However, it is not possible to guarantee that the sample mean will be within \$500 of the population mean. Indeed, Table 7.5 and Figure 7.1 show that some of the 500 sample means differed by more than \$2000 from the population mean. So we must think of the personnel director's request in probability terms. That is, the personnel director is concerned with the following question: What is the probability that the sample mean from a simple random sample of 30 EAI managers will be within \$500 of the population mean?

Because we have identified the properties of the sampling distribution of $\bar{x}$ (see Figure 7.6), we will use this distribution to answer the probability question. Refer to the sampling distribution of $\bar{x}$ shown again in Figure 7.7. The personnel director is asking about the probability that the sample mean is between \$51,300 and \$52,300. If the value of the sample mean $\bar{x}$ is in this interval, the value of $\bar{x}$ will be within \$500 of the population mean. The appropriate probability is given by the shaded area of the sampling distribution shown in Figure 7.7. Because the sampling distribution is normal, with mean 51,800 and standard deviation 730.30, we can use the standard normal probability distribution table to find the area or probability. At $\bar{x} = 51{,}300$, we have

$$z = \frac{51{,}300 - 51{,}800}{730.30} = -.68$$

Referring to the standard normal probability distribution table, we find an area between $z = 0$ and $z = -.68$ of .2518. Similar calculations for $\bar{x} = 52{,}300$ show an area between

FIGURE 7.7 THE PROBABILITY OF A SAMPLE MEAN BEING WITHIN $500 OF THE POPULATION MEAN

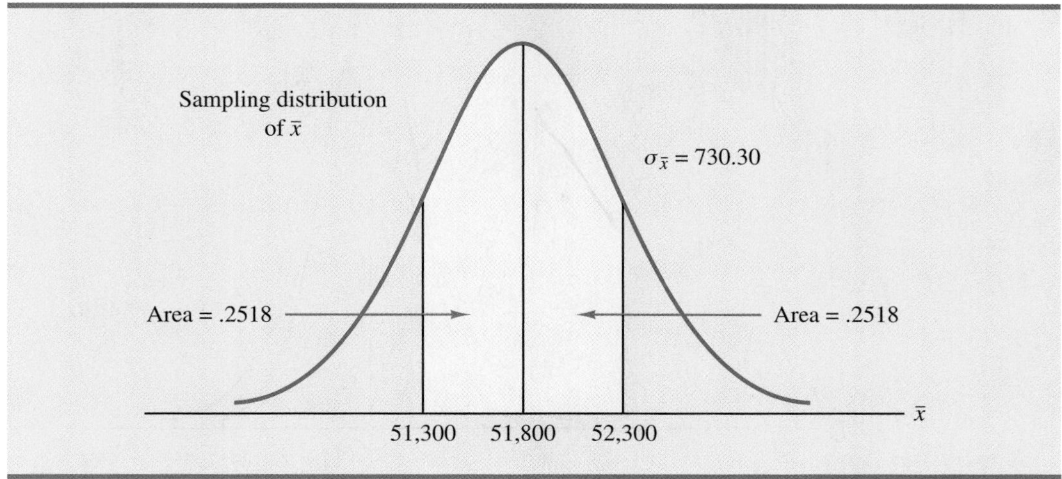

$z = 0$ and $z = +.68$ of .2518. Thus, the probability of the value of the sample mean being between 51,300 and 52,300 is $.2518 + .2518 = .5036$.

The preceding computations show that a simple random sample of 30 EAI managers has a .5036 probability of providing a sample mean $\bar{x}$ that is within $500 of the population mean. Thus, there is a $1 - .5036 = .4964$ probability that the sample mean will miss the population mean by more than $500. In other words, a simple random sample of 30 EAI managers has roughly a 50–50 chance of providing a sample mean within the allowable $500. Perhaps a larger sample size should be considered. Let us explore this possibility by considering the relationship between the sample size and the sampling distribution of $\bar{x}$.

The sampling distribution of $\bar{x}$ can be used to provide probability information about how close the sample mean $\bar{x}$ is to the population mean μ.

Relationship Between the Sample Size and the Sampling Distribution of $\bar{x}$

Suppose that in the EAI sampling problem we select a simple random sample of 100 EAI managers instead of the 30 originally considered. Intuitively, it would seem that with more data provided by the larger sample size, the sample mean based on $n = 100$ should provide a better estimate of the population mean than the sample mean based on $n = 30$. To see how much better, let us consider the relationship between the sample size and the sampling distribution of $\bar{x}$.

First note that $E(\bar{x}) = \mu$ regardless of the sample size. Thus, the mean of all possible values of $\bar{x}$ is equal to the population mean μ regardless of the sample size n. However, note that the standard error of the mean, $\sigma_{\bar{x}} = \sigma/\sqrt{n}$, is related to the square root of the sample size. Specifically, whenever the sample size is increased, the standard error of the mean $\sigma_{\bar{x}}$ is decreased. With $n = 30$, the standard error of the mean for the EAI problem is 730.30. However, with the increase in the sample size to $n = 100$, the standard error of the mean is decreased to

$$\sigma_{\bar{x}} = \frac{\sigma}{\sqrt{n}} = \frac{4000}{\sqrt{100}} = 400$$

The sampling distributions of $\bar{x}$ with $n = 30$ and $n = 100$ are shown in Figure 7.8. Because the sampling distribution with $n = 100$ has a smaller standard error, the values of $\bar{x}$ have less variation and tend to be closer to the population mean than the values of $\bar{x}$ with $n = 30$.

FIGURE 7.8 A COMPARISON OF THE SAMPLING DISTRIBUTIONS OF $\bar{x}$ FOR SIMPLE RANDOM SAMPLES OF $n = 30$ AND $n = 100$ EAI MANAGERS

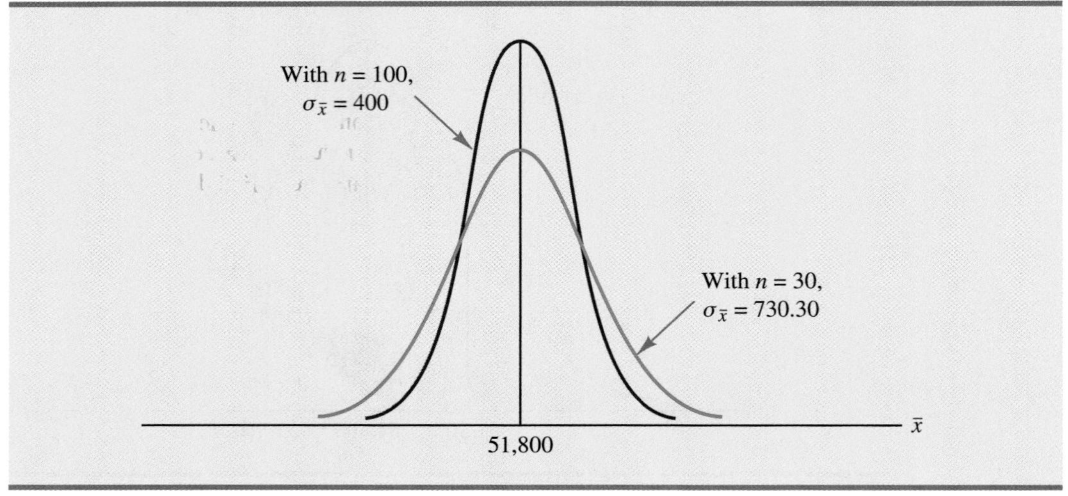

We can use the sampling distribution of $\bar{x}$ for the case with $n = 100$ to compute the probability that a simple random sample of 100 EAI managers will provide a sample mean that is within \$500 of the population mean. Because the sampling distribution is normal, with mean 51,800 and standard deviation 400, we can use the standard normal probability distribution table to find the area or probability. At $\bar{x} = 51{,}300$ (Figure 7.9), we have

$$z = \frac{51{,}300 - 51{,}800}{400} = -1.25$$

FIGURE 7.9 THE PROBABILITY OF A SAMPLE MEAN BEING WITHIN \$500 OF THE POPULATION MEAN WHEN A SIMPLE RANDOM SAMPLE OF 100 EAI MANAGERS IS USED

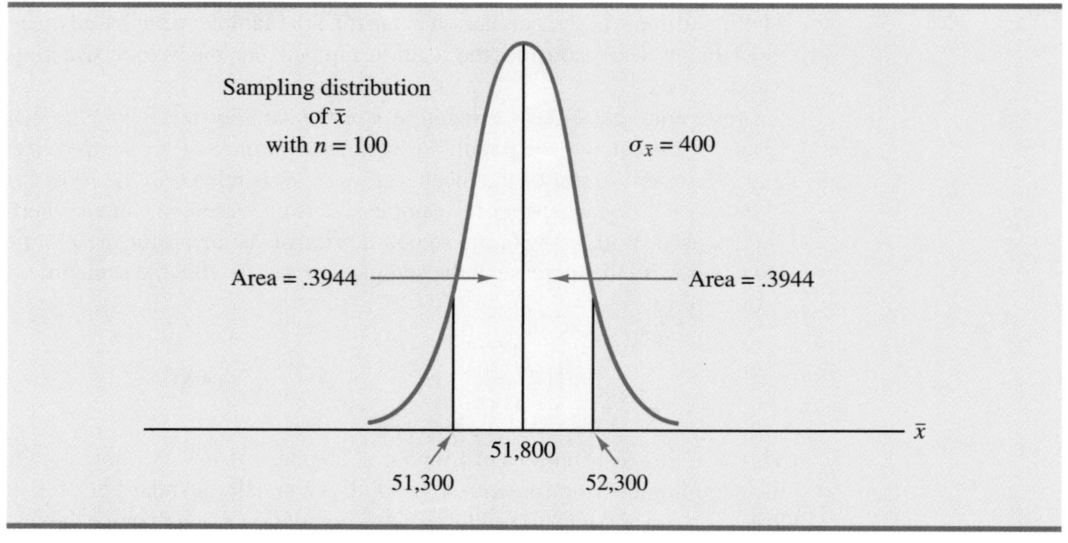

Referring to the standard normal probability distribution table, we find an area between $z = 0$ and $z = -1.25$ of .3944. With a similar calculation for $\bar{x} = 52{,}300$, we see that the probability of the value of the sample mean being between 51,300 and 52,300 is .3944 + .3944 = .7888. Thus, by increasing the sample size from 30 to 100 EAI managers, we have increased the probability of obtaining a sample mean within \$500 of the population mean from .5036 to .7888.

The important point in this discussion is that as the sample size is increased, the standard error of the mean is decreased. As a result, the larger sample size will provide a higher probability that the sample mean is within a specified distance of the population mean.

NOTES AND COMMENTS

1. In presenting the sampling distribution of $\bar{x}$ for the EAI problem, we took advantage of the fact that the population mean $\mu = 51{,}800$ and the population standard deviation $\sigma = 4000$ were known. However, in general, the values of the population mean μ and the population standard deviation σ that are needed to determine the sampling distribution of $\bar{x}$ will be unknown. In Chapter 8 we will show how the sample mean $\bar{x}$ and the sample standard deviation s are used when μ and σ are unknown.
2. The theoretical proof of the central limit theorem requires independent observations in the sample. This condition is met for infinite populations and for finite populations where sampling is done with replacement. Although the central limit theorem does not directly address sampling without replacement from finite populations, general statistical practice has been to apply the findings of the central limit theorem in this situation when the population size is large.

EXERCISES

Methods

18. A population has a mean of 200 and a standard deviation of 50. A simple random sample of size 100 will be taken and the sample mean $\bar{x}$ will be used to estimate the population mean.
    a. What is the expected value of $\bar{x}$?
    b. What is the standard deviation of $\bar{x}$?
    c. Show the sampling distribution of $\bar{x}$.
    d. What does the sampling distribution of $\bar{x}$ show?

19. A population has a mean of 200 and a standard deviation of 50. Suppose a simple random sample of size 100 is selected and $\bar{x}$ is used to estimate μ.
    a. What is the probability that the sample mean will be within ± 5 of the population mean?
    b. What is the probability that the sample mean will be within ± 10 of the population mean?
20. Assume the population standard deviation is $\sigma = 25$. Compute the standard error of the mean, $\sigma_{\bar{x}}$, for sample sizes of 50, 100, 150, and 200. What can you say about the size of the standard error of the mean as the sample size is increased?
21. Suppose a simple random sample of size 50 is selected from a population with $\sigma = 10$. Find the value of the standard error of the mean in each of the following cases (use the finite population correction factor if appropriate).
    a. The population size is infinite.
    b. The population size is $N = 50{,}000$.
    c. The population size is $N = 5000$.
    d. The population size is $N = 500$.

22. A population has a mean of 400 and a standard deviation of 50. The probability distribution of the population is unknown.

a. A researcher will use simple random samples of either 10, 20, 30, or 40 items to collect data about the population. With which of these sample-size alternatives will we be able to use a normal probability distribution to describe the sampling distribution of $\bar{x}$? Explain.

b. Show the sampling distribution of $\bar{x}$ for the instances in which the normal probability distribution is appropriate.

23. A population has a mean of 100 and a standard deviation of 16. What is the probability that a sample mean will be within ± 2 of the population mean for each of the following sample sizes?

a. $n = 50$

b. $n = 100$

c. $n = 200$

d. $n = 400$

e. What is the advantage of a larger sample size?

Applications

24. Refer to the EAI sampling problem. Suppose the simple random sample had contained 60 managers.

a. Sketch the sampling distribution of $\bar{x}$ when simple random samples of size 60 are used.

b. What happens to the sampling distribution of $\bar{x}$ if simple random samples of size 120 are used?

c. What general statement can you make about what happens to the sampling distribution of $\bar{x}$ as the sample size is increased? Does this generalization seem logical? Explain.

25. In the EAI sampling problem (see Figure 7.7), we showed that for $n = 30$, there was .5036 probability of obtaining a sample mean within $\pm$\$500 of the population mean.

a. What is the probability that $\bar{x}$ is within \$500 of the population mean if a sample of size 60 is used?

b. Answer part (a) for a sample of size 120.

26. The mean price per gallon of regular gasoline sold in the United States is \$1.20 (*The Energy Information Administration,* March 3, 1997). Assume the population mean price per gallon is $\mu = 1.20$, and the population standard deviation is $\sigma = .10$. Suppose that a random sample of 50 gasoline stations will be selected, and a sample mean price per gallon will be computed for data collected from the 50 gasoline stations.

a. Show the sampling distribution of the sample mean $\bar{x}$ where $\bar{x}$ is the sample mean price per gallon for the 50 gasoline stations.

b. What is the probability that the simple random sample will provide a sample mean within 2 cents, .02, of the population mean?

c. What is the probability that the simple random sample will provide a sample mean within 1 cent, .01, of the population mean?

27. The College Board American College Testing Program reported a population mean SAT score of $\mu = 1017$ (*The New York Times,* 2000 Almanac). Assume that the population standard deviation is $\sigma = 100$.

a. What is the probability that a random sample of 75 students will provide a sample mean SAT score within 10 of the population mean?

b. What is the probability a random sample of 75 students will provide a sample mean SAT score within 20 of the population mean?

28. The mean annual starting salary for marketing majors is \$34,000 (*Time,* May 8, 2000). Assume that for the population of graduates with a marketing major, the mean annual starting salary is $\mu = 34{,}000$, and the standard deviation is $\sigma = 2000$.

a. What is the probability that a simple random sample of marketing majors will have a sample mean within $\pm\$250$ of the population mean for each of the following sample sizes: 30, 50, 100, 200 and 400?

b. What is the advantage of a larger sample size when attempting to estimate the population mean?

29. The mean monthly rental for a two-bedroom apartment in Atlanta is \$982 (*Elle,* September 1998). Assume that the population mean is \$982 and the population standard deviation is \$210.

a. What is the probability that a simple random sample of 40 two-bedroom apartments will provide a sample mean monthly rental within $\pm\$100$ of the population mean?

b. What is the probability that a simple random sample of 40 two-bedroom apartments will provide a sample mean monthly rental within $\pm\$25$ of the population mean?

c. Discuss the results in parts (a) and (b).

30. The population mean for the price of a new one-family home is \$166,500 (U.S. Bureau of the Census, *New One-Family Houses Sold,* 1997). Assume that the population standard deviation is \$42,000 and that a sample of 100 new one-family homes will be selected.

a. Show the sampling distribution of the sample mean price for new homes based on the sample of 100.

b. What is the probability that the sample mean for the 100 new homes will be within \$10,000 of the population mean?

c. Repeat part (b) for values of \$5000, \$2500, and \$1000.

d. To estimate the population mean price to within $\pm\$2500$ or $\pm\$1000$, what would you recommend?

31. *Business Week* reports that its subscribers who plan to purchase a new vehicle plan to spend a mean of \$27,100 (*Business Week,* Subscriber Profile, 1996). Assume that the new vehicle price for the population of *Business Week* subscribers has a mean of $\mu = \$27{,}100$ and a standard deviation of $\sigma = \$5200$.

a. What is the probability that the sample mean new vehicle price for a sample of 30 subscribers is within \$1000 of the population mean?

b. What is the probability that the sample mean new vehicle price for a sample of 50 subscribers is within \$1000 of the population mean?

c. What is the probability that the sample mean new vehicle price for a sample of 100 subscribers is within \$1000 of the population mean?

d. Would you recommend a sample size of 30, 50, or 100 if it is desired to have at least a .90 probability that the sample mean is within \$1000 of the population mean?

32. To estimate the mean age for a population of 4000 employees, a simple random sample of 40 employees is selected.

a. Would you use the finite population correction factor in calculating the standard error of the mean? Explain.

b. If the population standard deviation is $\sigma = 8.2$ years, compute the standard error both with and without using the finite population correction factor. What is the rationale for ignoring the finite population correction factor whenever $n/N \leq .05$?

c. What is the probability that the sample mean age of the employees will be within ± 2 years of the population mean age?

7.6 SAMPLING DISTRIBUTION OF $\bar{p}$

In many situations in business and economics, we use the sample proportion $\bar{p}$ to make statistical inferences about the population proportion p. This process is depicted in Figure 7.10. On each repetition of the process, we can anticipate obtaining a different value for the

FIGURE 7.10 THE STATISTICAL PROCESS OF USING A SAMPLE PROPORTION TO MAKE INFERENCES ABOUT A POPULATION PROPORTION

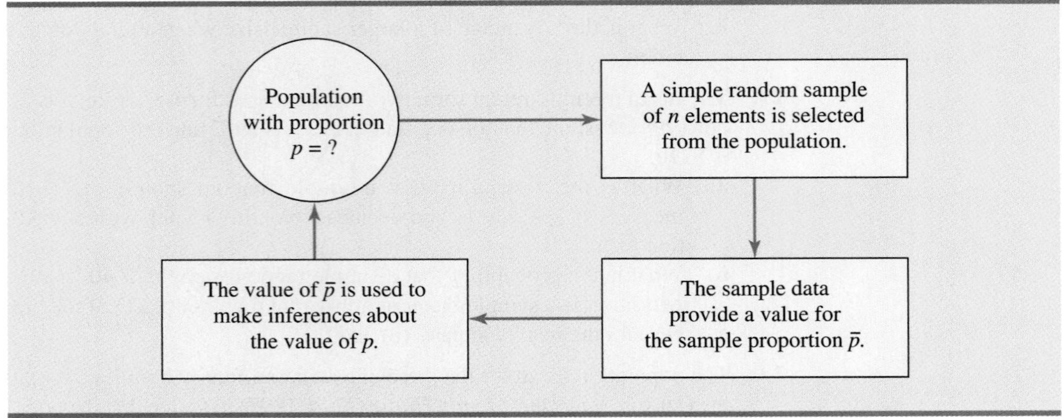

sample proportion $\bar{p}$. The probability distribution for all possible values of the sample proportion $\bar{p}$ is called the sampling distribution of the sample proportion $\bar{p}$.

Sampling Distribution of $\bar{p}$

The sampling distribution of $\bar{p}$ is the probability distribution of all possible values of the sample proportion $\bar{p}$.

To determine how close the sample proportion $\bar{p}$ is to the population proportion p, we need to understand the properties of the sampling distribution of $\bar{p}$: the expected value of $\bar{p}$, the standard deviation of $\bar{p}$, and the shape of the sampling distribution of $\bar{p}$.

Expected Value of $\bar{p}$

The expected value of $\bar{p}$, the mean of all possible values of $\bar{p}$, can be expressed as follows.

Expected Value of $\bar{p}$

$$E(\bar{p}) = p \tag{7.4}$$

where

$$E(\bar{p}) = \text{the expected value of } \bar{p}$$
$$p = \text{the population proportion}$$

Equation (7.4) shows that the mean of all possible $\bar{p}$ values is equal to the population proportion p. Recall that in Section 7.1 we noted that $p = .60$ for the EAI population, where p is the proportion of the population of managers who had participated in the company's management training program. Thus, the expected value of $\bar{p}$ for the EAI sampling problem is .60.

Standard Deviation of $\bar{p}$

The standard deviation of $\bar{p}$ is referred to as the **standard error of the proportion.** Just as we found for the sample mean $\bar{x}$, the standard deviation of $\bar{p}$ depends on whether the population is finite or infinite. The two expressions for computing the standard deviation of $\bar{p}$ follow.

Standard Deviation of $\bar{p}$

Finite Population	*Infinite Population*	
$\sigma_{\bar{p}} = \sqrt{\frac{N-n}{N-1}}\sqrt{\frac{p(1-p)}{n}}$	$\sigma_{\bar{p}} = \sqrt{\frac{p(1-p)}{n}}$	**(7.5)**

Comparing the two expressions in (7.5), we see that the only difference is the use of the finite population correction factor $\sqrt{(N-n)/(N-1)}$.

As was the case with the sample mean $\bar{x}$, the difference between the expressions for the finite population and the infinite population becomes negligible if the size of the finite population is large in comparison to the sample size. We follow the same rule of thumb that we recommended for the sample mean. That is, if the population is finite with $n/N \leq .05$, we will use $\sigma_{\bar{p}} = \sqrt{p(1-p)/n}$. However, if the population is finite and if $n/N > .05$, the finite population correction factor should be used, as shown in (7.5). Again, unless specifically noted, throughout the text we will assume that the population size is large in relation to the sample size and thus the finite population correction factor is unnecessary.

For the EAI study we know that the population proportion of managers who have participated in the management training program is $p = .60$. With $n/N = 30/2500 = .012$, we can ignore the finite population correction factor when we compute the standard deviation of $\bar{p}$. For the simple random sample of 30 managers, $\sigma_{\bar{p}}$ is

$$\sigma_{\bar{p}} = \sqrt{\frac{p(1-p)}{n}} = \sqrt{\frac{.60(1-.60)}{30}} = .0894$$

Form of the Sampling Distribution of $\bar{p}$

Now that we know the mean and standard deviation of $\bar{p}$, we want to consider the form of the sampling distribution of $\bar{p}$. Applying the central limit theorem as it relates to $\bar{p}$ produces the following result.

The sampling distribution of $\bar{p}$ can be approximated by a normal probability distribution whenever the sample size is large.

With $\bar{p}$, the sample size can be considered large whenever the following two conditions are satisfied.

$$np \geq 5$$
$$n(1-p) \geq 5$$

Recall that for the EAI sampling problem we know that the population proportion of managers who have participated in the training program is $p = .60$. With a simple random sample of size 30, we have $np = 30(.60) = 18$ and $n(1-p) = 30(.40) = 12$. Thus, the sampling distribution of $\bar{p}$ can be approximated by a normal probability distribution as shown in Figure 7.11.

Practical Value of the Sampling Distribution of $\bar{p}$

Whenever a simple random sample is selected and the value of the sample proportion $\bar{p}$ is used to estimate the value of the population proportion p, we anticipate some sampling error. In this case, the sampling error is the absolute value of the difference between the value of the sample proportion $\bar{p}$ and the value of the population proportion p. The practical value of the sampling distribution of $\bar{p}$ is that it can be used to provide probability information about the sampling error.

Suppose, in the EAI problem, the personnel director wants to know the probability of obtaining a value of $\bar{p}$ that is within .05 of the population proportion of EAI managers who have participated in the training program. That is, what is the probability of obtaining a sample with a sample proportion $\bar{p}$ between .55 and .65? The shaded area in Figure 7.12 shows this probability. Using the fact that the sampling distribution of $\bar{p}$ can be approximated by a normal probability distribution with a mean of .60 and a standard deviation of $\sigma_{\bar{p}} = .0894$, we find that the standard normal random variable corresponding to $\bar{p} = .55$ has a value of $z = (.55 - .60)/.0894 = -.56$. Referring to the standard normal probability distribution table, we see that the area between $z = -.56$ and $z = 0$ is .2123. Similarly, at $\bar{p} = .65$ we find an area between $z = 0$ and $z = .56$ of .2123. Thus, the probability of selecting a sample that provides a sample proportion $\bar{p}$ within .05 of the population proportion p is $.2123 + .2123 = .4246$.

If we consider increasing the sample size to $n = 100$, the standard error of the proportion becomes

$$\sigma_{\bar{p}} = \sqrt{\frac{.60(1 - .60)}{100}} = .0490$$

With a sample size of 100 EAI managers, the probability of the sample proportion having a value within .05 of the population proportion can now be computed. Because the sampling

FIGURE 7.11 SAMPLING DISTRIBUTION OF $\bar{p}$ FOR THE PROPORTION OF EAI MANAGERS WHO HAVE PARTICIPATED IN THE MANAGEMENT TRAINING PROGRAM

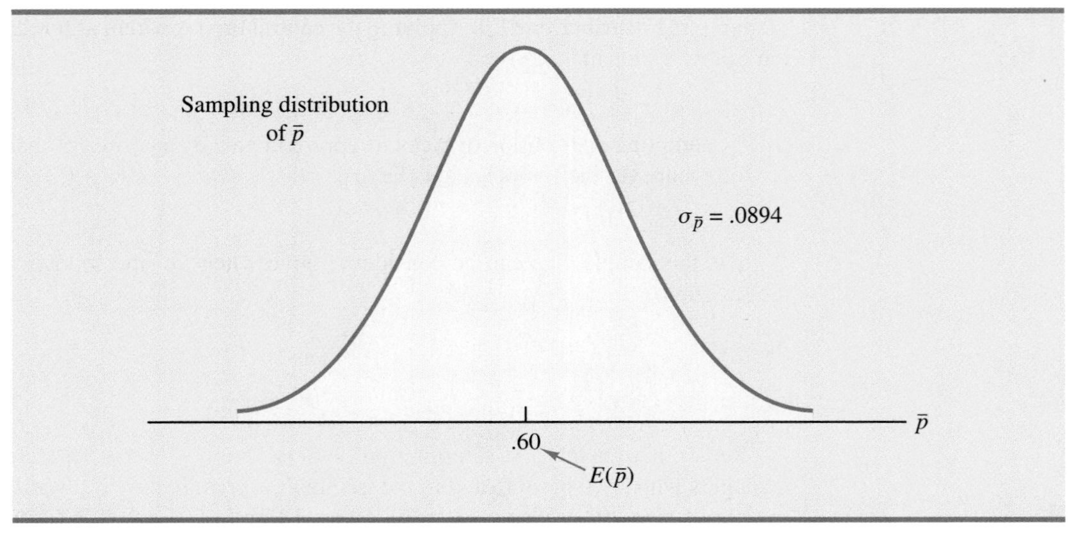

FIGURE 7.12 SAMPLING DISTRIBUTION OF $\bar{p}$ FOR THE EAI SAMPLING PROBLEM

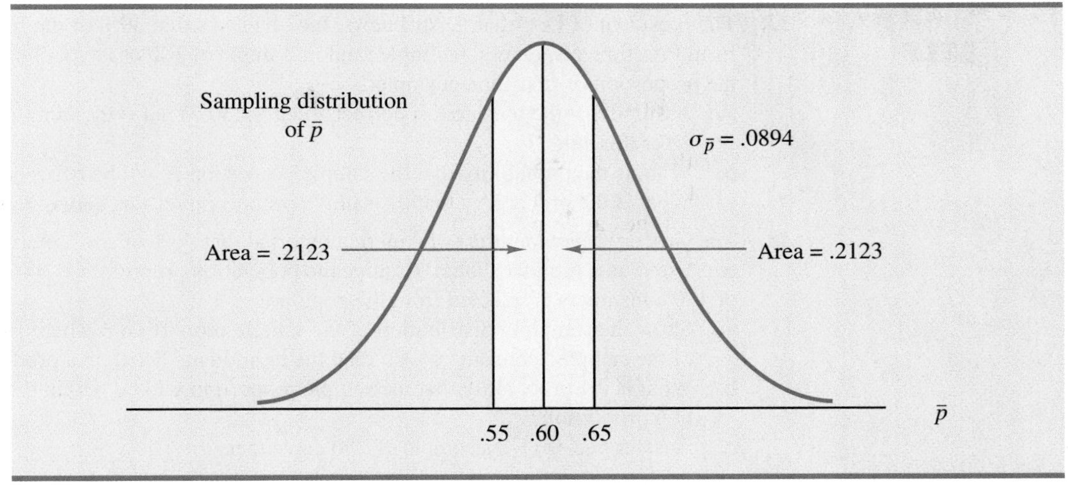

distribution is approximately normal, with mean .60 and standard deviation .0490, we can use the standard normal probability distribution table to find the area or probability. At $\bar{p} = .55$, we have $z = (.55 - .60)/.0490 = -1.02$. Referring to the standard normal probability distribution table, we see that the area between $z = -1.02$ and $z = 0$ is .3461. Similarly, at .65 the area between $z = 0$ and $z = 1.02$ is .3461. Thus, if the sample size is increased from 30 to 100, the probability that the sample proportion $\bar{p}$ is within .05 of the population proportion p will increase to $.3461 + .3461 = .6922$.

EXERCISES

Methods

33. A simple random sample of size 100 is selected from a population with $p = .40$.

a. What is the expected value of $\bar{p}$?

b. What is the standard deviation of $\bar{p}$?

c. Show the sampling distribution of $\bar{p}$.

d. What does the sampling distribution of $\bar{p}$ show?

34. A population proportion is .40. A simple random sample of size 200 will be taken and the sample proportion $\bar{p}$ will be used to estimate the population proportion.

a. What is the probability that the sample proportion will be within $\pm .03$ of the population proportion?

b. What is the probability that the sample proportion will be within $\pm .05$ of the population proportion?

35. Assume that the population proportion is .55. Compute the standard error of the proportion, $\sigma_{\bar{p}}$, for sample sizes of 100, 200, 500, and 1000. What can you say about the size of the standard error of the proportion as the sample size is increased?

36. The population proportion is .30. What is the probability that a sample proportion will be within $\pm .04$ of the population proportion for each of the following sample sizes?

a. $n = 100$

b. $n = 200$

c. $n = 500$

d. $n = 1000$

e. What is the advantage of a larger sample size?

Applications

37. The president of Doerman Distributors, Inc., believes that 30% of the firm's orders come from first-time customers. A simple random sample of 100 orders will be used to estimate the proportion of first-time customers.
    a. Assume that the president is correct and $p = .30$. What is the sampling distribution of $\bar{p}$ for this study?
    b. What is the probability that the sample proportion $\bar{p}$ will be between .20 and .40?
    c. What is the probability that the sample proportion will be between .25 and .35?
38. The Grocery Manufacturers of America reported that 76% of consumers read the ingredients listed on a product's label. Assume the population proportion is $p = .76$ and a sample of 400 consumers is selected from the population.
    a. Show the sampling distribution of the sample proportion $\bar{p}$ where $\bar{p}$ is the proportion of the sampled consumers who read the ingredients listed on a product's label.
    b. What is the probability that the sample proportion will be within $\pm.03$ of the population proportion?
    c. Answer part (b) for a sample of 750 consumers.
39. *Time*/CNN voter polls monitored public opinion for the presidential candidates during the 2000 presidential election campaign. One *Time*/CNN poll conducted by Yankelovich Partners, Inc., used a sample of 589 likely voters (*Time,* June 26, 2000). Assume the population proportion for a presidential candidate is $p = .50$. Let $\bar{p}$ be the sample proportion of likely voters favoring the presidential candidate.
    a. Show the sampling distribution of $\bar{p}$.
    b. What is the probability the *Time*/CNN poll will provide a sample proportion within $\pm.04$ of the population proportion?
    c. What is the probability the *Time*/CNN poll will provide a sample proportion within $\pm.03$ of the population proportion?
    d. What is the probability the *Time*/CNN poll will provide a sample proportion within $\pm.02$ of the population proportion?
40. Although most people believe breakfast is the most important meal of the day, 25% of adults skip breakfast (*U.S. News & World Report,* November 10, 1997). Assume the population proportion is $p = .25$, and $\bar{p}$ is the sample proportion of adults who skip breakfast based on a sample of 200 adults.
    a. Show the sampling distribution of $\bar{p}$.
    b. What is the probability that the sample proportion will be within $\pm.03$ of the population proportion?
    c. What is the probability that the sample proportion will be within $\pm.05$ of the population proportion?
41. The Institute for Women's Policy Research reported that women constitute 37% of all union members. Suppose the population proportion of women who are union members is $p = .37$, and a simple random sample of 1000 union members is selected.
    a. Show the sampling distribution of $\bar{p}$, the proportion of women in the sample.
    b. What is the probability that the sample proportion will be within $\pm.03$ of the population proportion?
    c. Answer part (b) for a simple random sample of 500.
42. Assume that 15% of the items produced in an assembly line operation are defective, but that the firm's production manager is not aware of this situation. Assume further that 50 parts are tested by the quality assurance department to determine the quality of the assembly operation. Let $\bar{p}$ be the sample proportion found defective by the quality assurance test.
    a. Show the sampling distribution for $\bar{p}$.
    b. What is the probability that the sample proportion will be within $\pm.03$ of the population proportion that is defective?

c. If the test shows $\bar{p} = .10$ or more, the assembly line operation will be shut down to check for the cause of the defects. What is the probability that the sample of 50 parts will lead to the conclusion that the assembly line should be shut down?

43. The Food Marketing Institute shows that 17% of households spend more than \$100 per week on groceries. Assume the population proportion is $p = .17$ and a simple random sample of 800 households will be selected from the population.

a. Show the sampling distribution of $\bar{p}$, the sample proportion of households spending more than \$100 per week on groceries.

b. What is the probability that the sample proportion will be within $\pm.02$ of the population proportion?

c. Answer part (b) for a sample of 1600 households.

7.7 PROPERTIES OF POINT ESTIMATORS

In this chapter we have shown how sample statistics such as a sample mean $\bar{x}$, a sample standard deviation s, and a sample proportion $\bar{p}$ can be used as point estimators of their corresponding population parameters μ, σ, and p. It is intuitively appealing that each of these sample statistics is the point estimator of its corresponding population parameter. However, before using a sample statistic as a point estimator, statisticians check to see whether the sample statistic has certain properties associated with good point estimators. In this section we discuss the properties of good point estimators: unbiasedness, efficiency, and consistency.

Because several different sample statistics can be used as point estimators of different population parameters, we will use the following general notation in this section.

$$\theta = \text{the population parameter of interest}$$
$$\hat{\theta} = \text{the sample statistic or point estimator of } \theta$$

The notation θ is the Greek letter theta, and the notation $\hat{\theta}$ is pronounced "theta-hat." In general, θ represents any population parameter such as a population mean, population standard deviation, population proportion, and so on; $\hat{\theta}$ represents the corresponding sample statistic such as the sample mean, sample standard deviation, and sample proportion.

Unbiasedness

If the expected value of the sample statistic is equal to the population parameter being estimated, the sample statistic is said to be an *unbiased estimator* of the population parameter. The property of **unbiasedness** is defined as follows.

Unbiasedness

The sample statistic $\hat{\theta}$ is an unbiased estimator of the population parameter θ if

$$E(\hat{\theta}) = \theta \tag{7.6}$$

where

$$E(\hat{\theta}) = \text{the expected value of the sample statistic } \hat{\theta}$$

Hence, the expected value, or mean, of all possible values of an unbiased sample statistic is equal to the population parameter being estimated.

Figure 7.13 shows the cases of unbiased and biased point estimators. In the illustration showing the unbiased estimator, the mean of the sampling distribution is equal to the value of the population parameter. The sampling errors balance out in this case, because sometimes the value of the point estimator $\hat{\theta}$ may be less than θ and other times it may be greater than θ. In the case of a biased estimator, the mean of the sampling distribution is less than or greater than the value of the population parameter. In the illustration in Figure 7.13(b), $E(\hat{\theta})$ is greater than θ; thus, the sample statistic has a high probability of overestimating the value of the population parameter. The amount of the bias is shown in the figure.

In discussing the sampling distributions of the sample mean and the sample proportion, we stated that $E(\bar{x}) = \mu$ and $E(\bar{p}) = p$. Thus, both $\bar{x}$ and $\bar{p}$ are unbiased estimators of their corresponding population parameters μ and p.

We defer a more detailed disucssion of the sampling distribution of the sample standard deviation s and the sample variance s^2 until Chapter 11. However, it can be shown that $E(s^2) = \sigma^2$. Thus, we conclude that the sample variance s^2 is an unbiased estimator of the population variance σ^2. In fact, when we first presented the formulas for the sample variance and the sample standard deviation in Chapter 3, $n - 1$ rather than n was used in the denominator. The reason for using $n - 1$ rather than n is to make the sample variance an unbiased estimator of the population variance. If we had used n in the denominator, the sample variance would have been a biased estimator, tending to slightly underestimate the population variance.

Efficiency

Assume that a simple random sample of n elements can be used to provide two unbiased point estimators of the same population parameter. In this situation, we would prefer to use the point estimator with the smaller standard deviation, because it tends to provide estimates closer to the population parameter. The point estimator with the smaller standard deviation is said to have greater **relative efficiency** than the other.

Figure 7.14 shows the sampling distributions of two unbiased point estimators, $\hat{\theta}_1$ and $\hat{\theta}_2$. Note that the standard deviation of $\hat{\theta}_1$ is less than the standard deviation of $\hat{\theta}_2$; thus, values of $\hat{\theta}_1$ have a greater chance of being close to the parameter θ than do values of $\hat{\theta}_2$. Be-

FIGURE 7.13 EXAMPLES OF UNBIASED AND BIASED POINT ESTIMATORS

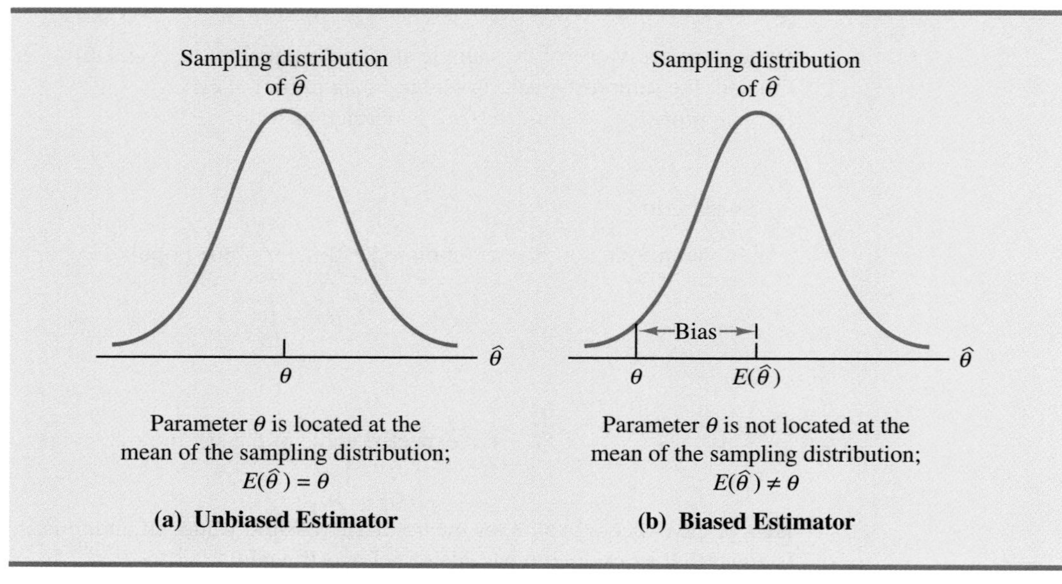

FIGURE 7.14 SAMPLING DISTRIBUTIONS OF TWO UNBIASED POINT ESTIMATORS

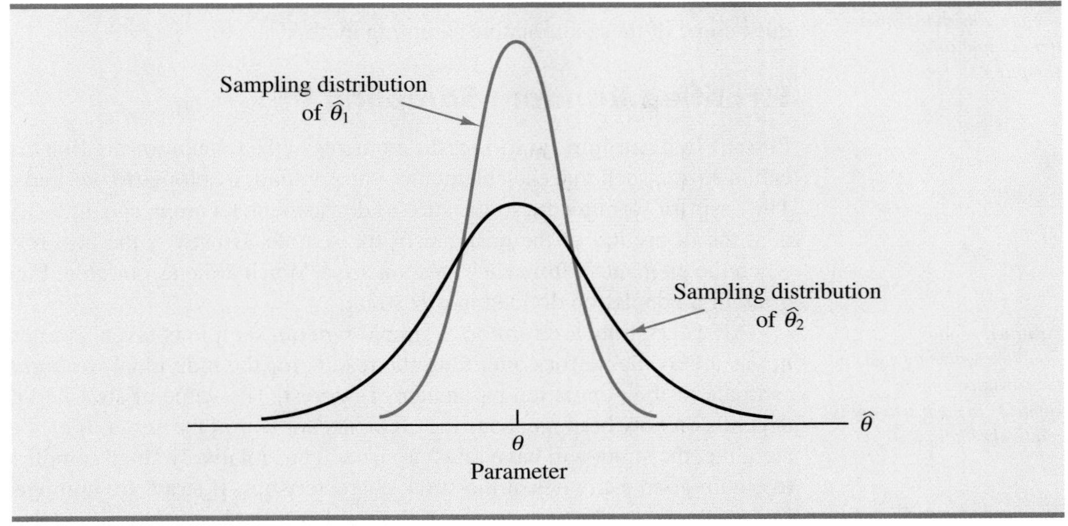

When sampling from a normal population, the standard deviation of the sample mean is less than the standard deviation of the sample median. Thus, the sample mean is more efficient than the sample median.

cause the standard deviation of point estimator $\hat{\theta}_1$ is less than the standard deviation of point estimator $\hat{\theta}_2$, $\hat{\theta}_1$ is relatively more efficient than $\hat{\theta}_2$ and is the preferred point estimator.

Consistency

A third property associated with good point estimators is **consistency.** Loosely speaking, a point estimator is consistent if the values of the point estimator tend to become closer to the population parameter as the sample size becomes larger. In other words, a large sample size tends to provide a better point estimate than a small sample size. Note that for the sample mean $\bar{x}$, we showed that the standard deviation of $\bar{x}$ is given by $\sigma_{\bar{x}} = \sigma/\sqrt{n}$. Because $\sigma_{\bar{x}}$ is related to the sample size such that larger sample sizes provide smaller values for $\sigma_{\bar{x}}$, we conclude that a larger sample size tends to provide point estimates closer to the population mean μ. In this sense, we can say that the sample mean $\bar{x}$ is a consistent estimator of the population mean μ. Using a similar rationale, we can also conclude that the sample proportion $\bar{p}$ is a consistent estimator of the population proportion p.

NOTES AND COMMENTS

In Chapter 3 we stated that the mean and the median are two measures of central location. In this chapter we discussed only the mean. The reason is that in sampling from a normal population, where the population mean and population median are identical, the standard error of the median is approximately 25% larger than the standard error of the mean. Recall that in the EAI problem where $n = 30$, the standard error of the mean is $\sigma_{\bar{x}} = 730.30$. The standard error of the median for this problem would be approximately $1.25 \times (730.30) = 913$. As a result, the sample mean is more efficient and will have a higher probability of being within a specified distance of the population mean.

7.8 OTHER SAMPLING METHODS

We have described the simple random sampling procedure and discussed the properties of the sampling distributions of $\bar{x}$ and $\bar{p}$ when simple random sampling is used. However, simple random sampling is not the only sampling method available. Such methods as stratified

This section provides a brief introduction to sampling methods other than simple random sampling.

random sampling, cluster sampling, and systematic sampling are alternatives that in some situations have advantages over simple random sampling. In this section we briefly introduce some of these alternative sampling methods.

Stratified Random Sampling

In **stratified random sampling,** the elements in the population are first divided into groups called *strata,* such that each element in the population belongs to one and only one stratum. The basis for forming the strata, such as department, location, age, industry type, and so on, is at the discretion of the designer of the sample. However, the best results are obtained when the elements within each stratum are as much alike as possible. Figure 7.15 is a diagram of a population divided into H strata.

Stratified random sampling works best when the variance among elements in each stratum is relatively small.

After the strata are formed, a simple random sample is taken from each stratum. Formulas are available for combining the results for the individual stratum samples into one estimate of the population parameter of interest. The value of stratified random sampling depends on how homogeneous the elements are within the strata. If elements within strata are alike, the strata will have low variances. Thus relatively small sample sizes can be used to obtain good estimates of the strata characteristics. If strata are homogeneous, the stratified random sampling procedure will provide results just as precise as those of simple random sampling by using a smaller total sample size.

Cluster Sampling

Cluster sampling works best when each cluster provides a small-scale representation of the population.

In **cluster sampling,** the elements in the population are first divided into separate groups called *clusters.* Each element of the population belongs to one and only one cluster (see Figure 7.16). A simple random sample of the clusters is then taken. All elements within each sampled cluster form the sample. Cluster sampling tends to provide the best results when the elements within the clusters are not alike. In the ideal case, each cluster is a representative small-scale version of the entire population. The value of cluster sampling depends on how representative each cluster is of the entire population. If all clusters are alike in this regard, sampling a small number of clusters will provide good estimates of the population parameters.

One of the primary applications of cluster sampling is area sampling, where clusters are city blocks or other well-defined areas. Cluster sampling generally requires a larger total sample size than either simple random sampling or stratified random sampling. However, it can result in cost savings because of the fact that when an interviewer is sent to a sampled cluster (e.g., a city-block location), many sample observations can be obtained in a rela-

FIGURE 7.15 DIAGRAM FOR STRATIFIED RANDOM SAMPLING

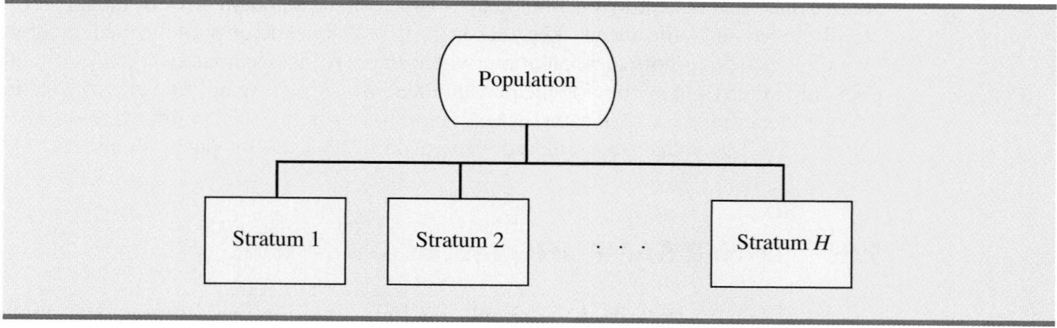

FIGURE 7.16 DIAGRAM FOR CLUSTER SAMPLING

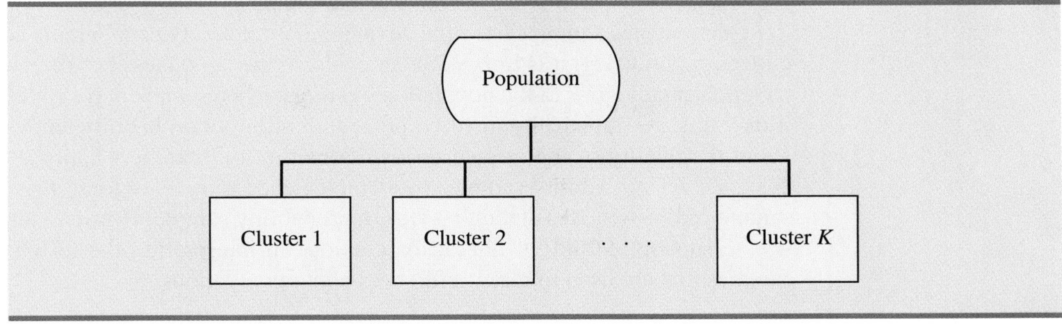

tively short time. Hence, a larger sample size may be obtainable with a significantly lower total cost.

Systematic Sampling

In some sampling situations, especially those with large populations, it is time-consuming to select a simple random sample by first finding a random number and then counting or searching through the list of the population until the corresponding element is found. An alternative to simple random sampling is **systematic sampling.** For example, if a sample size of 50 is desired from a population containing 5000 elements, we will sample one element for every 5000/50 = 100 elements in the population. A systematic sample for this case involves selecting randomly one of the first 100 elements from the population list. Other sample elements are identified by starting with the first sampled element and then selecting every 100th element that follows in the population list. In effect, the sample of 50 is identified by moving systematically through the population and identifying every 100th element after the first randomly selected element. The sample of 50 usually will be easier to identify in this way than it would be if simple random sampling were used. Because the first element selected is a random choice, a systematic sample is usually assumed to have the properties of a simple random sample. This assumption is especially applicable when the list of elements in the population is a random ordering of the elements.

Convenience Sampling

The sampling methods discussed thus far are referred to as *probability sampling* techniques. Elements selected from the population have a known probability of being included in the sample. The advantage of probability sampling is that the sampling distribution of the appropriate sample statistic generally can be identified. Formulas such as the ones for simple random sampling presented in this chapter can be used to determine the properties of the sampling distribution. Then the sampling distribution can be used to make probability statements about the sampling error associated with the results.

Convenience sampling is a *nonprobability sampling* technique. As the name implies, the sample is identified primarily by convenience. Elements are included in the sample without prespecified or known probabilities of being selected. For example, a professor conducting research at a university may use student volunteers to constitute a sample simply because they are readily available and will participate as subjects for little or no cost. Similarly, an inspector may sample a shipment of oranges by selecting oranges haphazardly from among several crates. Labeling each orange and using a probability method of sampling would be impractical. Samples such as wildlife captures and volunteer panels for consumer research are also convenience samples.

Convenience samples have the advantage of relatively easy sample selection and data collection; however, it is impossible to evaluate the "goodness" of the sample in terms of its representativeness of the population. A convenience sample may provide good results or it may not; no statistically justified procedure allows a probability analysis and inference about the quality of the sample results. Sometimes researchers apply statistical methods designed for probability samples to a convenience sample, arguing that the convenience sample can be treated as though it were a probability sample. However, this argument cannot be supported, and we should be cautious in interpreting the results of convenience samples that are used to make inferences about populations.

Judgment Sampling

One additional nonprobability sampling technique is **judgment sampling.** In this approach, the person most knowledgeable on the subject of the study selects elements of the population that he or she feels are most representative of the population. Often this method is a relatively easy way of selecting a sample. For example, a reporter may sample two or three senators, judging that those senators reflect the general opinion of all senators. However, the quality of the sample results depends on the judgment of the person selecting the sample. Again, great caution is warranted in drawing conclusions based on judgment samples used to make inferences about populations.

NOTES AND COMMENTS

We recommend using probability sampling methods: simple random sampling, stratified random sampling, cluster sampling, or systematic sampling. For these methods, formulas are available for evaluating the "goodness" of the sample results in terms of the closeness of the results to the population characteristics being estimated. An evaluation of the goodness cannot be made with convenience or judgment sampling. Thus, great care should be used in interpreting the results when nonprobability sampling methods are used.

SUMMARY

In this chapter we presented the concepts of simple random sampling and sampling distributions. We demonstrated how a simple random sample can be selected and how the data collected for the sample can be used to develop point estimates of population parameters. Because different simple random samples provide a variety of different values for the point estimators, point estimators such as $\bar{x}$ and $\bar{p}$ are random variables. The probability distribution of such a random variable is called a sampling distribution. In particular, we described the sampling distributions of the sample mean $\bar{x}$ and the sample proportion $\bar{p}$.

In considering the characteristics of the sampling distributions of $\bar{x}$ and $\bar{p}$, we stated that $E(\bar{x}) = \mu$ and $E(\bar{p}) = p$. After developing the standard deviation or standard error formulas for these estimators, we showed how the central limit theorem provided the basis for using a normal probability distribution to approximate these sampling distributions in the large-sample case. Rules of thumb were given for determining when large-sample conditions were satisfied. Other sampling methods including stratified random sampling, cluster sampling, systematic sampling, convenience sampling, and judgment sampling were discussed.

GLOSSARY

Parameter A numerical characteristic of a population, such as a population mean μ, a population standard deviation σ, a population proportion p, and so on.

Simple random sampling Finite population: a sample selected such that each possible sample of size n has the same probability of being selected. Infinite population: a sample selected such that each element comes from the same population and the elements are selected independently.

Sampling without replacement Once an element has been included in the sample, it is removed from the population and cannot be selected a second time.

Sampling with replacement Once an element has been included in the sample, it is returned to the population. A previously selected element can be selected again and therefore may appear in the sample more than once.

Sample statistic A sample characteristic, such as a sample mean $\bar{x}$, a sample standard deviation s, a sample proportion $\bar{p}$, and so on. The value of the sample statistic is used to estimate the value of the population parameter.

Point estimate A single numerical value used as an estimate of a population parameter.

Point estimator The sample statistic, such as $\bar{x}$, s, or $\bar{p}$, that provides the point estimate of the population parameter.

Sampling error The absolute value of the difference between an unbiased point estimator and the corresponding population parameter. For a sample mean, sample standard deviation, and sample proportion, the sampling errors are $|\bar{x} - \mu|$, $|s - \sigma|$, and $|\bar{p} - p|$, respectively.

Sampling distribution A probability distribution consisting of all possible values of a sample statistic.

Finite population correction factor The term $\sqrt{(N - n)/(N - 1)}$ that is used in the formulas for $\sigma_{\bar{x}}$ and $\sigma_{\bar{p}}$ whenever a finite population, rather than an infinite population, is being sampled. The generally accepted rule of thumb is to ignore the finite population correction factor whenever $n/N \leq .05$.

Standard error The standard deviation of a point estimator.

Central limit theorem A theorem that enables one to use the normal probability distribution to approximate the sampling distribution of $\bar{x}$ and $\bar{p}$ whenever the sample size is large.

Unbiasedness A property of a point estimator when the expected value of the point estimator is equal to the population parameter it estimates.

Relative efficiency Given two unbiased point estimators of the same population parameter, the point estimator with the smaller standard deviation is more efficient.

Consistency A property of a point estimator that is present whenever larger sample sizes tend to provide point estimates closer to the population parameter.

Stratified random sampling A probability sampling method in which the population is first divided into strata and a simple random sample is then taken from each stratum.

Cluster sampling A probability sampling method in which the population is first divided into clusters and then a simple random sample of the clusters is taken.

Systematic sampling A probability sampling method in which we randomly select one of the first k elements and then select every kth element thereafter.

Convenience sampling A nonprobability method of sampling whereby elements are selected for the sample on the basis of convenience.

Judgment sampling A nonprobability method of sampling whereby elements are selected for the sample based on the judgment of the person doing the study.

KEY FORMULAS

Expected Value of $\bar{x}$

$$E(\bar{x}) = \mu \tag{7.1}$$

Standard Deviation of $\bar{x}$

Finite Population *Infinite Population*

$$\sigma_{\bar{x}} = \sqrt{\frac{N-n}{N-1}}\left(\frac{\sigma}{\sqrt{n}}\right) \qquad \sigma_{\bar{x}} = \frac{\sigma}{\sqrt{n}} \tag{7.2}$$

Expected Value of $\bar{p}$

$$E(\bar{p}) = p \tag{7.4}$$

Standard Deviation of $\bar{p}$

Finite Population *Infinite Population*

$$\sigma_{\bar{p}} = \sqrt{\frac{N-n}{N-1}}\sqrt{\frac{p(1-p)}{n}} \qquad \sigma_{\bar{p}} = \sqrt{\frac{p(1-p)}{n}} \tag{7.5}$$

SUPPLEMENTARY EXERCISES

44. *Business Week's* Corporate Scoreboard provides quarterly data on sales, profits, net income, return on equity, price/earnings ratio, and earnings per share for 899 companies (*Business Week,* August 14, 2000). The companies can be numbered 1 to 899 in the order they appear on the Corporate Scoreboard list. Begin at the bottom of the second column of random digits in Table 7.1. Ignoring the first two digits in each group and using three-digit random numbers beginning with 112, read *up* the column to identify the first eight companies to be included in a simple random sample.

45. The mean television viewing time for teens is 3 hours per day (*Barron's,* November 8, 1999). Assume the population mean is $\mu = 3$ and the population standard deviation is $\sigma = 1.2$ hours. Suppose a sample of 50 teens will be used to monitor television viewing time. Let $\bar{x}$ denote the sample mean viewing time.

a. Show the sample distribution of $\bar{x}$.

b. What is the probability the sample mean will be within $\pm.25$ hours of the population mean?

46. The mean travel time to work for individuals in Chicago is 31.5 minutes (*1998 Information Please Almanac*). Assume the population mean is $\mu = 31.5$ minutes and the population standard deviation is $\sigma = 12$ minutes. A sample of 50 Chicago residents is selected.

a. Show the sampling distribution of $\bar{x}$ where $\bar{x}$ is the sample mean travel time to work for the 50 Chicago residents.

b. What is the probability that the sample mean will be within ± 1 minute of the population mean?

c. What is the probability that the sample mean will be within ± 3 minutes of the population mean?

47. The U.S. Bureau of Labor Statistics reported the mean hourly wage rate for individuals in executive, administrative, and managerial occupations is \$24.07 (*The Wall Street Journal Almanac 1998*). Assume the population mean is $\mu = \$24.07$ and the population standard deviation is $\sigma = \$4.80$. A sample of 120 individuals in executive, administrative, and managerial occupations will be selected.

a. What is the probability that the sample mean will be within $\pm\$0.50$ of the population mean?

b. What is the probability that the sample mean will be within $\pm\$1.00$ of the population mean?

48. According to *USA Today* (April 11, 1995), the mean number of days per year that business travelers are on the road for business is 115. The standard deviation is 60 days per year. Assume that these results apply to the population of business travelers and that a sample of 50 business travelers will be selected from the population.

a. What is the value of the standard error of the mean?

b. What is the probability that the sample mean will be more than 115 days per year?

c. What is the probability that the sample mean will be within ± 5 days of the population mean?

d. How would the probability change in part (c) if the sample size were increased to 100?

49. Three firms have inventories that differ in size. Firm A has a population of 2000 items, firm B has a population of 5000 items, and firm C has a population of 10,000 items. The population standard deviation for the cost of the items is $\sigma = 144$. A statistical consultant recommends that each firm take a sample of 50 items from its population to provide statistically valid estimates of the average cost per item. Managers of the small firm state that because it has the smallest population, it should be able to obtain the data from a much smaller sample than that required by the larger firms. However, the consultant states that to obtain the same standard error and thus the same precision in the sample results, all firms should use the same sample size regardless of population size.

a. Using the finite population correction factor, compute the standard error for each of the three firms given a sample of size 50.

b. What is the probability that for each firm the sample mean $\bar{x}$ will be within ± 25 of the population mean μ?

50. A researcher reports survey results by stating that the standard error of the mean is 20. The population standard deviation is 500.

a. How large was the sample used in this survey?

b. What is the probability that the estimate would be within ± 25 of the population mean?

51. A production process is checked periodically by a quality control inspector. The inspector selects simple random samples of 30 finished products and computes the sample mean product weights $\bar{x}$. If test results over a long period of time show that 5% of the $\bar{x}$ values are over 2.1 pounds and 5% are under 1.9 pounds, what are the mean and the standard deviation for the population of products produced with this process?

52. What is the most important factor for business travelers when they are staying in a hotel? According to *USA Today,* 74% of business travelers state that having a smoke-free room is the most important factor (*USA Today,* April 11, 1995). Assume that the population proportion is $p = .74$ and that a sample of 200 business travelers will be selected.

a. Show the sampling distribution of $\bar{p}$, the sample proportion of business travelers stating that a smoke-free room is the most important factor when staying in a hotel.

b. What is the probability that the sample proportion will be within $\pm.04$ of the population proportion?

c. What is the probability that the sample proportion will be within $\pm.02$ of the population proportion?

53. A market research firm conducts telephone surveys with a 40% historical response rate. What is the probability that in a new sample of 400 telephone numbers, at least 150 individuals will cooperate and respond to the questions? In other words, what is the probability that the sample proportion will be at least 150/400 = .375?

54. According to ORC International, 71% of Internet users connect their computers to the Internet by normal telephone lines (*USA Today,* January 18, 2000). Assume a population proportion $p = .71$.

a. What is the probability that a sample proportion from a simple random sample of 350 Internet users will be within $\pm.05$ of the population proportion?

b. What is the probability that a sample proportion from a simple random sample of 350 Internet users will be .75 or greater?

55. The proportion of individuals insured by the All-Driver Automobile Insurance Company who have received at least one traffic ticket during a 5-year period is .15.

a. Show the sampling distribution of $\bar{p}$ if a random sample of 150 insured individuals is used to estimate the proportion having received at least one ticket.

b. What is the probability that the sample proportion will be within $\pm.03$ of the population proportion?

56. Lori Jeffrey is a successful sales representative for a major publisher of college textbooks. Historically, Lori obtains a book adoption on 25% of her sales calls. Viewing her sales calls for one month as a sample of all possible sales calls, assume that a statistical analysis of the data yields a standard error of the proportion of .0625.

a. How large was the sample used in this analysis? That is, how many sales calls did Lori make during the month?

b. Let $\bar{p}$ indicate the sample proportion of book adoptions obtained during the month. Show the sampling distribution $\bar{p}$.

c. Using the sampling distribution of $\bar{p}$, compute the probability that Lori will obtain book adoptions on 30% or more of her sales calls during a one-month period.

Appendix 7.1 THE EXPECTED VALUE AND STANDARD DEVIATION OF $\bar{x}$

In this appendix we present the mathematical basis for the expressions for $E(\bar{x})$, the expected value of $\bar{x}$ as given by equation (7.1), and $\sigma_{\bar{x}}$, the standard deviation of $\bar{x}$ as given by equation (7.2).

Expected Value of $\bar{x}$

Assume a population with mean μ and variance σ^2. A simple random sample of size n is selected with individual observations denoted $x_1, x_2, \ldots, x_n$. A sample mean $\bar{x}$ is computed as follows.

$$\bar{x} = \frac{\Sigma x_i}{n}$$

With repeated simple random samples of size n, $\bar{x}$ is a random variable that takes different numerical values depending on the specific n items selected. The expected value of the random variable $\bar{x}$ is the mean of all possible $\bar{x}$ values.

$$\begin{aligned}\text{Mean of } \bar{x} = E(\bar{x}) &= E\left(\frac{\Sigma x_i}{n}\right)\\ &= \frac{1}{n}[E(x_1 + x_2 + \cdots + x_n)]\\ &= \frac{1}{n}[E(x_1) + E(x_2) + \cdots + E(x_n)]\end{aligned}$$

For any x_i we have $E(x_i) = \mu$, therefore we can write

$$\begin{aligned} E(\bar{x}) &= \frac{1}{n}(\mu + \mu + \cdots + \mu) \\ &= \frac{1}{n}(n\mu) = \mu \end{aligned}$$

This result shows that the mean of all possible $\bar{x}$ values is the same as the population mean μ. That is, $E(\bar{x}) = \mu$.

Standard Deviation of $\bar{x}$

Again assume a population with mean μ, variance σ^2, and a sample mean given by

$$\bar{x} = \frac{\Sigma x_i}{n}$$

With repeated simple random samples of size n, we know that $\bar{x}$ is a random variable that takes different numerical values depending on the specific n items selected. What follows is the derivation of the expression for the standard deviation of the $\bar{x}$ values, $\sigma_{\bar{x}}$, for the case of an infinite population. The derivation of the expression for $\sigma_{\bar{x}}$ for a finite population when sampling is done without replacement is more difficult and is beyond the scope of this text.

Returning to the infinite population case, recall that a simple random sample from an infinite population consists of observations $x_1, x_2, \ldots, x_n$ that are independent. The following two expressions are general formulas for the variance of random variables.

$$\text{Var}(ax) = a^2\,\text{Var}(x) \tag{7.7}$$

where a is a constant and x is a random variable, and

$$\text{Var}(x + y) = \text{Var}(x) + \text{Var}(y) \tag{7.8}$$

where x and y are *independent* random variables. Using equations (7.7) and (7.8), we can develop the expression for the variance of the random variable $\bar{x}$ as follows.

$$\text{Var}(\bar{x}) = \text{Var}\left(\frac{\Sigma x_i}{n}\right) = \text{Var}\left(\frac{1}{n}\Sigma x_i\right)$$

Using equation (7.7) with $1/n$ as the constant, we have

$$\begin{aligned} \text{Var}(\bar{x}) &= \left(\frac{1}{n}\right)^2 \text{Var}(\Sigma x_i) \\ &= \left(\frac{1}{n}\right)^2 \text{Var}(x_1 + x_2 + \cdots + x_n) \end{aligned}$$

In the infinite population case, the random variables $x_1, x_2, \ldots, x_n$ are independent. Thus, equation (7.8) enables us to write

$$\text{Var}(\bar{x}) = \left(\frac{1}{n}\right)^2 [\text{Var}(x_1) + \text{Var}(x_2) + \cdots + \text{Var}(x_n)]$$

For any x_i, we have $\text{Var}(x_i) = \sigma^2$, therefore we have

$$\text{Var}(\bar{x}) = \left(\frac{1}{n}\right)^2 (\sigma^2 + \sigma^2 + \cdots + \sigma^2)$$

With n values of σ^2 in this expression, we have

$$\text{Var}(\bar{x}) = \left(\frac{1}{n}\right)^2 (n\sigma^2) = \frac{\sigma^2}{n}$$

Taking the square root provides the formula for the standard deviation of $\bar{x}$.

$$\sigma_{\bar{x}} = \sqrt{\text{Var}(\bar{x})} = \frac{\sigma}{\sqrt{n}}$$

Appendix 7.2 RANDOM SAMPLING WITH MINITAB

If a population list is available in a Minitab file, Minitab can be used to select a simple random sample from the population. For example, a list of the top 100 metropolitan areas in the United States and Canada are provided in column 1 of the data set MetAreas (*Places Rated Almanac—The Millennium Edition 2000*). Column 2 contains the overall rating of each metropolitan area. The first 10 metropolitan areas in the data set and their corresponding ratings are shown in Table 7.6.

Suppose that you would like to select a simple random sample of 30 metropolitan areas in order to do an in-depth study of the cost of living in the United States and Canada. The following steps can be used to select the sample.

Step 1. Select the **Calc** pull-down menu
Step 2. Choose **Random Data**
Step 3. Choose **Samples From Columns**
Step 4. When the Sample From Columns dialog box appears:
Enter 30 in the **Sample** __ box
Enter C1 C2 in the **Sample 30 rows from column(s)** box
Enter C3 C4 in the **Store samples in** box
Step 5. Click **OK**

The random sample of 30 metropolitan areas appears in columns C3 and C4.

TABLE 7.6 OVERALL RATING FOR THE FIRST 10 METROPOLITAN AREAS IN THE DATA SET METAREAS

MetAreas

Metropolitan Area	Rating
Albany, NY	64.18
Albuquerque, NM	66.16
Appleton, WI	60.56
Atlanta, GA	69.97
Austin, TX	71.48
Baltimore, MD	69.75
Birmingham, AL	69.59
Boise City, ID	68.36
Boston, MA	68.99
Buffalo, NY	66.10

Appendix 7.3 RANDOM SAMPLING WITH EXCEL

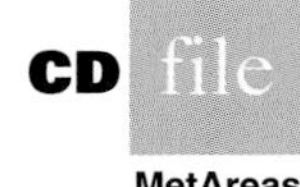

MetAreas

If a population list is available in an Excel file, Excel can be used to select a simple random sample from the population. For example, a list of the top 100 metropolitan areas in the United States and Canada are provided in column A of the data set MetAreas (*Places Rated Almanac—The Millennium Edition 2000*). Column B contains the overall rating of each metropolitan area. The first 10 metropolitan areas in the data set, and their corresponding ratings are shown in Table 7.6. Assume that you would like to select a simple random sample of 30 metropolitan areas in order to do an in-depth study of the cost of living in the United States and Canada.

The rows of any Excel data set can be placed in a random order by adding an extra column to the data set and filling the column with random numbers using the =RAND() function. Then using Excel's sort ascending capability on the random number column, the rows of the data set will be reordered randomly. The random sample of size n appears in the first n rows of the reordered data set.

In the MetAreas data set, labels are in row 1 and the 100 metropolitan areas are in rows 2 to 101. The following steps can be used to select a simple random sample of 30 metropolitan areas.

Step 1. Enter =RAND() in cell C2
Step 2. Copy cell C2 to cells C3:C101
Step 3. Select any cell in Column C
Step 4. Click the **Sort Ascending** button on the tool bar

The random sample of 30 metropolitan areas appears in rows 2 to 31 of the reordered data set. The random numbers in column C are no longer necessary and can be deleted if desired.

Chapter 8

INTERVAL ESTIMATION

CONTENTS

STATISTICS IN PRACTICE

DOLLAR GENERAL CORPORATION*
Goodlettsville, Tennessee

Dollar General Corporation was founded in 1939 as a dry goods wholesale company. After World War II, the company began opening retail locations in rural south-central Kentucky. Today Dollar General Corporation operates more than 4300 stores across the middle and southeastern United States. Emphasizing small-store convenience, Dollar General markets health and beauty aids, cleaning supplies, housewares, stationery, apparel, shoes, and domestic items at everyday low prices.

Being in an inventory-intense business with more than 20,000 products, Dollar General made the decision to adopt the LIFO (last-in-first-out) method of inventory valuation. This method matches current costs against current revenues, which minimizes the effect of radical price changes on profit and loss results. In addition, the LIFO method reduces net income and thereby income taxes during periods of inflation. LIFO also brings disposable cash generated from sales in line with income and allows for the replacement of inventory at current costs.

Accounting practices require that a LIFO index be established for inventory under the LIFO method of valuation. For example, a LIFO index of 1.028 indicates that the company's inventory value at current costs reflects a 2.8% increase due to inflation over the most recent one-year period.

The establishment of a LIFO index requires that the year-end inventory count for each product be valued at the current year-end cost and at the preceding year-end cost. To avoid counting the inventory of every product in more than 4300 retail locations, a random sample of 800 products is selected from 100 retail locations and three warehouses. Physical inventories for the sampled products are taken at the end of the year. Accounting personnel then provide the current-year and preceding-year costs needed to construct the LIFO index.

This store in the Cleveland area is one of more than 4300 Dollar General Stores. © Jim Baron/The Image Finders.

For a recent year, the LIFO index was 1.030. However, because this index is a sample estimate of the population's LIFO index, a statement about the precision of the estimate was required. On the basis of the sample results and a 95% confidence level, the margin of error was computed to be .006. Thus, the interval from 1.024 to 1.036 provides the 95% confidence interval estimate of the population LIFO index. This precision was judged to be good.

In this chapter you will learn how to compute the margin of error associated with a sample mean and a sample proportion. Then, you will learn how to use this information to construct and interpret confidence interval estimates of a population mean and a population proportion. You will also learn how to determine the sample size needed to ensure that the margin of error will be within acceptable limits.

*The authors are indebted to Robert S. Knaul, Controller, Dollar General Corporation, for providing this Statistics in Practice.

In Chapter 7 we stated that a point estimator is a sample statistic used to estimate a lation parameter. For instance, the sample mean $\bar{x}$ is a point estimator of the mean μ, and the sample proportion $\bar{p}$ is a point estimator of the population cause a point estimator does not provide information about how clos population parameter, statisticians prefer to use an **interval estima** formation about the precision of the estimate.

An interval estimate of a population parameter is constructed by subtracting and adding a value, called the **margin of error,** to a point estimate. All of the interval estimates that we develop in this chapter will be of the form

$$\text{Point Estimate} \pm \text{Margin of Error}$$

Specifically, we will show how to develop an interval estimate for a population mean μ and a population proportion p. An interval estimate of the population mean will take the following form

$$\bar{x} \pm \text{Margin of Error}$$

and an interval estimate of a population proportion will take the following form

$$\bar{p} \pm \text{Margin of Error}$$

The inclusion of the margin of error provides the precision information about the estimate. As we will show, the sampling distribution of $\bar{x}$ and the sampling distribution of $\bar{p}$ presented in Chapter 7 play important roles in the development of interval estimates of a population mean μ and a population proportion p.

8.1 INTERVAL ESTIMATION OF A POPULATION MEAN: LARGE-SAMPLE CASE

In this section we show how to use a simple random sample to develop an interval estimate of a population mean. We focus on the large-sample case where the sample size is at least 30. We begin with a situation where the population standard deviation σ is assumed known. We then consider the case where the population standard deviation σ is unknown and, as a result, is estimated by the sample standard deviation s.

CJW Estimation Problem

The development of an interval estimate of a population mean can be illustrated by considering the monthly customer service survey conducted by CJW, Inc. CJW provides an Internet ordering service, which offers its customers a specialty line of sporting equipment and accessory products. Customer service in the form of easy on-line ordering, correctly filled orders, timely delivery of orders, and prompt response to customer inquiries is critical to the ongoing success of the company.

CJW's quality assurance team uses a customer service survey to measure customer satisfaction with its on-line ordering system. Each month, a questionnaire is sent to a random sample of customers who have placed an order during the previous month. The sampled customers are asked to complete a questionnaire that rates the service provided by CJW. Each customer's response is converted into a satisfaction score that ranges from 0 (worst possible rating) to 100 (best possible rating). Based on each month's survey, a sample mean customer satisfaction score is computed. The sample mean satisfaction score is used as a point estimate of the population mean satisfaction score for all CJW customers. The sample mean provides CJW management with a timely measure of the quality of its on-line ordering service. Corrective action can be taken promptly if low customer satisfaction scores begin to emerge.

Prior monthly customer service surveys have shown that the standard deviation of the satisfaction scores has stabilized at a value of 20. Hence, the historical data provide a basis for assuming that the population standard deviation is known with $\sigma = 20$. The most recent CJW customer satisfaction survey provided data on the satisfaction scores for a sample of

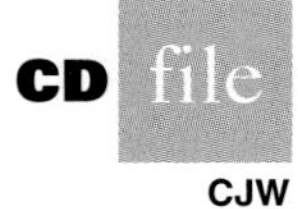

CJW

100 customers ($n = 100$). The sample mean satisfaction score, $\bar{x} = 82$, provides a point estimate of the population mean μ. In the discussion that follows, we will compute the margin of error associated with $\bar{x} = 82$, and then develop an interval estimate of the population mean satisfaction score for all CJW customers.

Sampling Error

Any time a sample mean is used to provide a point estimate of a population mean, someone may ask: How good is the estimate? The "how good" question is a way of asking about the error involved when the value of $\bar{x}$ is used as a point estimate of μ. In general, the absolute value of the difference between an unbiased point estimator and the population parameter is called the **sampling error.** For the case of a sample mean estimating a population mean, the sampling error is defined as follows:

$$\text{Sampling Error} = |\bar{x} - \mu| \tag{8.1}$$

In practice, the value of the sampling error cannot be determined exactly because the population mean μ is unknown. However, the sampling distribution of $\bar{x}$ can be used to make probability statements about the sampling error, which we will illustrate using the CJW example.

The central limit theorem applies to any population. Thus, the methodology presented in this section can be used even if the population distribution is unknown.

With a sample size $n = 100$ and a population standard deviation $\sigma = 20$, the central limit theorem introduced in Chapter 7 enables us to conclude that the sampling distribution of $\bar{x}$ can be approximated by a normal distribution with a mean μ and a standard deviation $\sigma_{\bar{x}} = \sigma/\sqrt{n} = 20/\sqrt{100} = 2$. This sampling distribution is shown in Figure 8.1. Because the sampling distribution shows how values of $\bar{x}$ are distributed around μ, the sampling distribution provides information about the possible differences between $\bar{x}$ and μ. This information provides the basis for a probability statement about the sampling error.

Using the tables of areas for the standard normal probability distribution, we find that 95% of the values of a normally distributed random variable are within ± 1.96 standard deviations of the mean. Because the sampling distribution of $\bar{x}$ can be approximated by a normal distribution, 95% of all $\bar{x}$ values must be within $\pm 1.96\sigma_{\bar{x}}$ of the mean μ. In the CJW example, $1.96\sigma_{\bar{x}} = 1.96(2) = 3.92$. Thus, we can conclude that 95% of the sample means that can be obtained using a sample size $n = 100$ will be within ± 3.92 of the population mean μ.

FIGURE 8.1 SAMPLING DISTRIBUTION OF THE SAMPLE MEAN SATISFACTION SCORE FROM SIMPLE RANDOM SAMPLES OF 100 CUSTOMERS

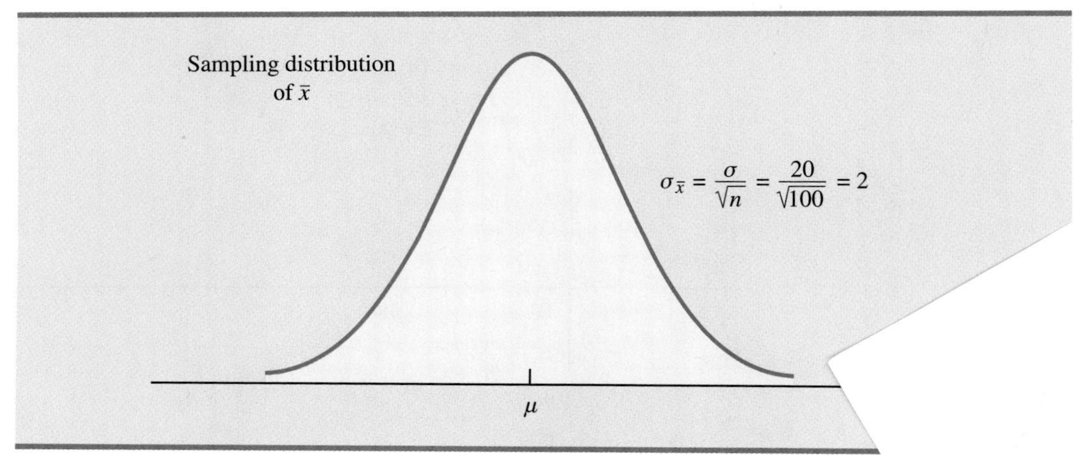

The location of the sample means that differ from μ by 3.92 or less is shown in Figure 8.2. Note that if a sample mean is in the region denoted by "95% of all $\bar{x}$ values," it provides a sampling error $|\bar{x} - \mu|$ of 3.92 or less. However, if a sample mean is in either the lower tail or the upper tail of the sampling distribution, the sampling error $|\bar{x} - \mu|$ will be greater than 3.92. Therefore we can make the following probability statement about the sampling error for the CJW problem.

> There is a .95 probability that the sample mean will provide a sampling error of 3.92 or less.

This probability statement is a **precision statement** that tells CJW about the sampling error that can exist if the sample mean from a simple random sample of 100 customers is used to estimate the population mean. The value 3.92, which provides an upper limit on the sampling error, is referred to as the *margin of error.*

Large-Sample Case With σ Assumed Known

The CJW sample of $n = 100$ customers meets the large-sample condition of $n \geq 30$. In addition, prior customer surveys provided data that enabled CJW to assume that the population standard deviation is known with $\sigma = 20$. Let us show how we can use the precision statement to develop a confidence interval estimate of a population mean.

As we pointed out at the beginning of the chapter, an interval estimate of a population mean takes the following form:

$$\bar{x} \pm \text{Margin of Error}$$

For the most recent CJW survey of $n = 100$ customers, the sample mean is $\bar{x} = 82$. We have shown that using the .95 probability precision statement, the margin of error is 3.92. Thus, 82 ± 3.92 provides an interval estimate of the population mean. Thus, we obtain $82 - 3.92 = 78.08$ as the lower limit and $82 + 3.92 = 85.92$ as the upper limit of the interval estimate. Hence, 78.08 to 85.92 is an interval estimate of the population mean.

FIGURE 8.2 SAMPLING DISTRIBUTION OF $\bar{x}$ SHOWING THE LOCATION OF SAMPLE MEANS THAT PROVIDE A SAMPLING ERROR OF 3.92 OR LESS

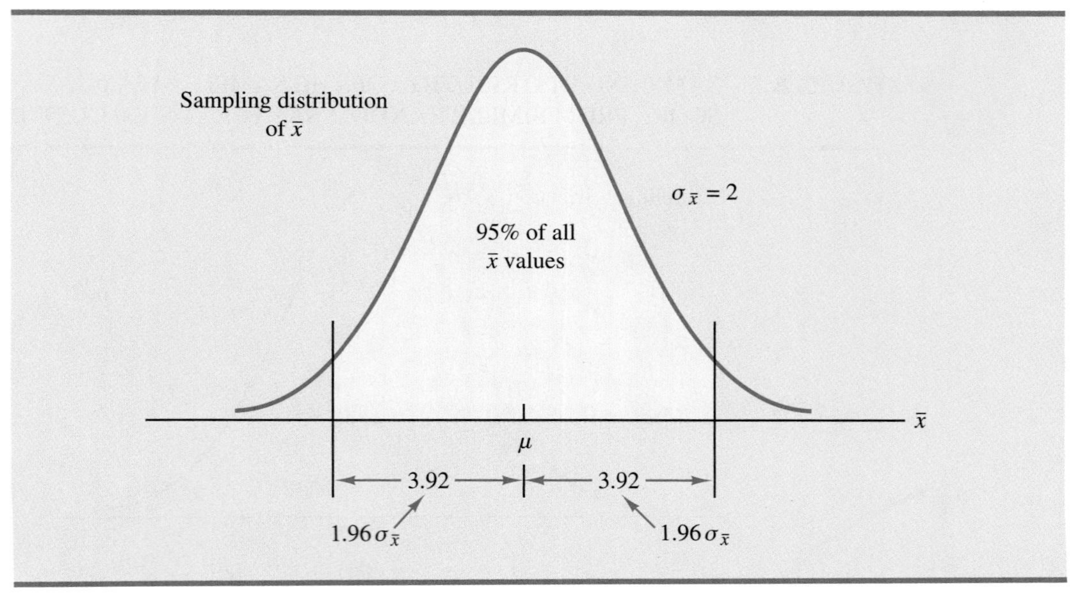

To explain how to interpret an interval estimate of a population mean μ, let us consider possible values of the sample mean $\bar{x}$ that could be obtained from three *different* random samples, each consisting of 100 CJW customers. Suppose the first sample mean turns out to have the value shown as $\bar{x}_1$ in Figure 8.3. In this case, Figure 8.3 shows that the interval formed by subtracting 3.92 from $\bar{x}_1$ and adding 3.92 to $\bar{x}_1$ includes the population mean μ. Now consider what happens if the sample mean turns out to have the value shown as $\bar{x}_2$ in Figure 8.3. Although this sample mean is different from the first sample mean, we see that the interval formed by subtracting 3.92 from $\bar{x}_2$ and adding 3.92 to $\bar{x}_2$ also includes the population mean μ. However, consider what happens if the sample mean turns out to have the value shown as $\bar{x}_3$ in Figure 8.3. In this case, the interval formed by subtracting 3.92 from $\bar{x}_3$ and adding 3.92 to $\bar{x}_3$ does not include the population mean μ. The reason is that $\bar{x}_3$ is in the upper tail of the distribution and is farther than 3.92 from μ. Hence, subtracting and adding 3.92 to $\bar{x}_3$ forms an interval that does not include μ.

This discussion provides insight as to why the interval is called a 95% confidence interval.

Any sample mean $\bar{x}$ that is within the shaded region of Figure 8.3 will provide an interval that contains the population mean μ. Because 95% of all possible sample means are in the shaded region, 95% of all intervals formed by subtracting 3.92 from $\bar{x}$ and adding 3.92 to $\bar{x}$ will include the population mean μ. Thus we are *95% confident* that the interval constructed from $\bar{x} - 3.92$ to $\bar{x} + 3.92$ will include the population mean μ. Because 95% of all possible sample means will provide an interval that includes the population mean, the interval is established at the 95% **confidence level.** The value .95 is referred to as the **confidence coefficient,** and the interval estimate $\bar{x} \pm 3.92$ is called a 95% confidence interval.

FIGURE 8.3 INTERVALS FORMED FROM SELECTED SAMPLE MEANS AT LOCATIONS $\bar{x}_1$, $\bar{x}_2$, AND $\bar{x}_3$

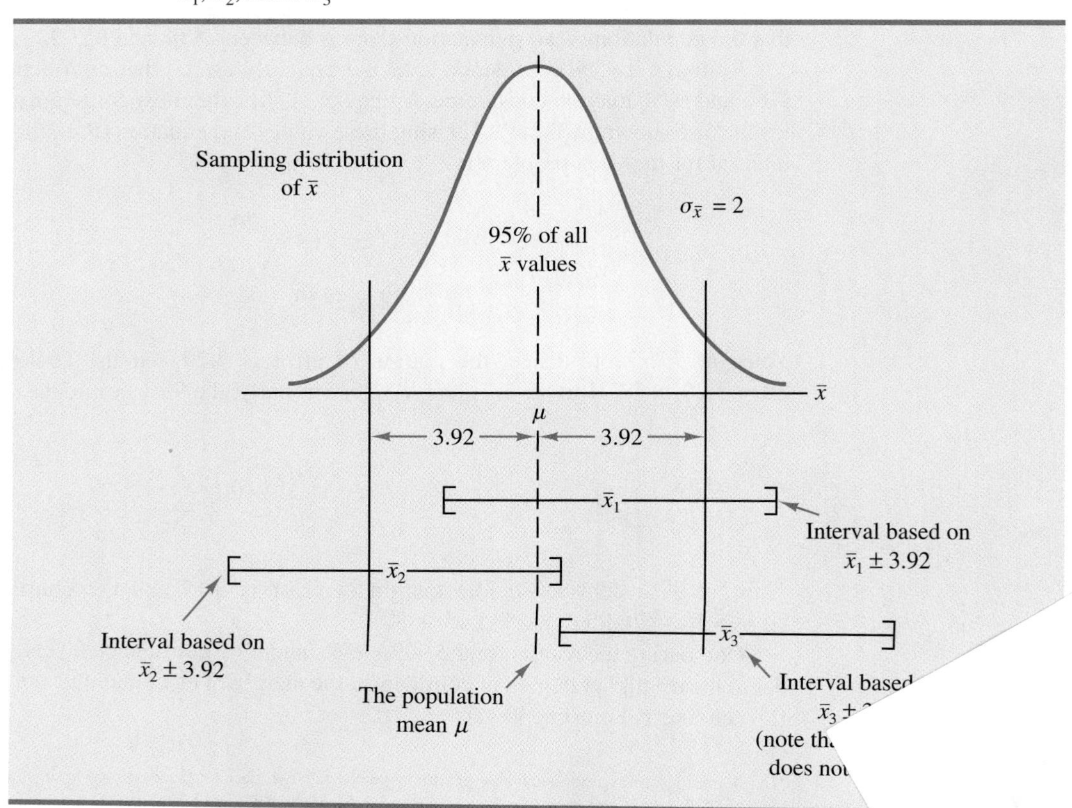

The general procedure for developing an interval estimate of a population mean whenever we have a large sample and the population standard deviation is assumed known follows.

Interval Estimate of a Population Mean: Large-Sample Case ($n \geq 30$) With σ Assumed Known

$$\bar{x} \pm z_{\alpha/2}\frac{\sigma}{\sqrt{n}} \tag{8.2}$$

where $(1 - \alpha)$ is the confidence coefficient and $z_{\alpha/2}$ is the z value providing an area $\alpha/2$ in the upper tail of the standard normal probability distribution.

Let us use equation (8.2) to construct a 95% confidence interval for the CJW problem. For a 95% confidence interval, the confidence coefficient is $(1 - \alpha) = .95$ and thus, $\alpha = .05$. Using the tables of areas for the standard normal distribution, an area of $\alpha/2 = .05/2 = .025$ in the upper tail provides $z_{.025} = 1.96$. With the CJW sample mean $\bar{x} = 82$, $\sigma = 20$ assumed known, and a sample size $n = 100$, we obtain

$$82 \pm 1.96\frac{20}{\sqrt{100}}$$

$$82 \pm 3.92$$

Thus, using equation (8.2), the margin of error is 3.92 and the confidence interval is $82 - 3.92 = 78.08$ to $82 + 3.92 = 85.92$. Thus, as shown previously, CJW can be 95% confident that the population mean satisfaction score is between 78.08 and 85.92.

Although a 95% confidence level is frequently used, other confidence levels such as 90% and 99% may be considered. Values of $z_{\alpha/2}$ for the most commonly used confidence levels* are shown in Table 8.1. Using these values and equation (8.2), the 90% confidence interval for the CJW problem is

$$82 \pm 1.645\frac{20}{\sqrt{100}}$$

$$82 \pm 3.29$$

Thus, at 90% confidence, the margin of error is 3.29 and the confidence interval is $82 - 3.29 = 78.71$ to $82 + 3.29 = 85.29$. Similarly, the 99% confidence interval is

$$82 \pm 2.576\frac{20}{\sqrt{100}}$$

$$82 \pm 5.15$$

Thus, at 99% confidence, the margin of error is 5.15 and the confidence interval is $82 - 5.15 = 76.85$ to $82 + 5.15 = 87.15$.

Comparing the results for the 90%, 95%, and 99% confidence levels, we see that in order to have a higher degree of confidence, the margin of error and thus the width of the confidence interval must be larger.

*The standard normal probability table provides z values with two decimal places. However, it is general practice to use z values with three decimal places for the confidence levels of 90%, 95%, and 99%.

TABLE 8.1 VALUES OF $z_{\alpha/2}$ FOR THE MOST COMMONLY USED CONFIDENCE LEVELS

Confidence Level	α	$\alpha/2$	$z_{\alpha/2}$
90%	.10	.05	1.645
95%	.05	.025	1.960
99%	.01	.005	2.576

Large-Sample Case With σ Estimated by s

A difficulty in using equation (8.2) to compute an interval estimate of a population mean is that in many applications there is no basis for assuming that the population standard deviation is known. In such cases, the sample standard deviation s is used to estimate σ. In the large-sample case ($n \geq 30$), the central limit theorem and the fact that the sample standard deviation s provides a good estimate of σ as the sample size becomes large,* enables us to use the following procedure to develop an interval estimate of a population mean.

With the population standard deviation σ unknown, we use the sample standard deviation s to estimate σ.

Interval Estimate of a Population Mean: Large-Sample Case ($n \geq 30$) With σ Estimated by s

$$\bar{x} \pm z_{\alpha/2} \frac{s}{\sqrt{n}} \tag{8.3}$$

where s is the sample standard deviation, $(1 - \alpha)$ is the confidence coefficient, and $z_{\alpha/2}$ is the z-value providing an area of $\alpha/2$ in the upper tail of the standard normal probability distribution.

To illustrate this interval estimation procedure, let us consider a sampling study designed to estimate the credit card debt of U.S. households. A sample of 85 households provided the credit card balances shown in Table 8.2. With $n = 85$, we have the large-sample case. In addition, because no historical data are available about credit card balances per household, the population standard deviation σ will be estimated by the sample standard deviation s. Let us develop a 95% confidence interval estimate of the population mean credit card balance per household.

First, we use the data in Table 8.2 to compute the sample mean $\bar{x} = \$5900$ and the sample standard deviation $s = \$3058$. At 95% confidence, $z_{\alpha/2} = z_{.025} = 1.96$. With a sample size of $n = 85$, equation (8.3) shows

1.96

$$5900 \pm 1.96 \frac{3058}{\sqrt{85}}$$

$$5900 \pm 650$$

Hence, the margin of error is \$650 and the 95% confidence interval estimate of the population mean is $5900 - 650 = \$5250$ to $5900 + 650 = \$6550$. Thus, we can be 95% confident that the population mean credit card balance for all households is between \$5250 and \$6…

Minitab, Excel, and other software packages provide easy computations of c… intervals for the large-sample case when s is used to estimate σ. The proc…

*Large-sample theory shows that as the sample size increases the sample variance s^2 converge… lation variance σ^2. This convergence property enables us to use s to estimate σ and to use equati… terval estimate of a population mean.

TABLE 8.2 CREDIT CARD BALANCES FOR A SAMPLE OF 85 HOUSEHOLDS

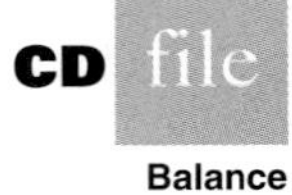

Balance

9619	5994	3344	7888	7581	9980
5364	4652	13627	3091	12545	8718
8348	5376	968	943	7959	8452
7348	5998	4714	8762	2563	4935
381	7530	4334	1407	6787	5938
2998	3678	4911	6644	5071	5266
1686	3581	1920	7644	9536	10658
1962	5625	3780	11169	4459	3910
4920	5619	3478	7979	8047	7503
5047	9032	6185	3258	8083	1582
6921	13236	1141	8660	2153	
5759	4447	7577	7511	8003	
8047	609	4667	14442	6795	
3924	414	5219	4447	5915	
3470	7636	6416	6550	7164	

In Chapter 7, we pointed out that $\sigma_{\bar{x}} = \sigma/\sqrt{n}$ is the standard error of the mean. Minitab provides an estimate of the standard error of the mean by using s to estimate σ.

Minitab and Excel are described in Appendixes 8.1 and 8.2. The results of the Minitab interval estimation procedure are shown in Figure 8.4. The sample of 85 households provides a sample mean credit card balance of \$5900, a sample standard deviation of \$3058, a standard error of the mean of \$332, and a 95% confidence interval of \$5250 to \$6550.

FIGURE 8.4 THE MINITAB CONFIDENCE INTERVAL FOR THE CREDIT CARD BALANCE SURVEY

```
Variable    N    Mean    StDev    SE Mean         95.0% CI
Balance    85    5900     3058        332    (5250, 6550)
```

NOTES AND COMMENTS

1. In developing an interval estimate of the population mean, we specify the desired confidence coefficient $(1 - \alpha)$ before selecting the sample. Thus, prior to selecting the sample, we conclude that there is a $1 - \alpha$ probability that the confidence interval we eventually compute will contain the population mean μ. However, once the sample is taken, the sample mean $\bar{x}$ is computed, and the particular interval estimate is determined, the resulting interval may or may not contain μ. If $1 - \alpha$ is reasonably large, we can be confident that the resulting interval will contain μ because we know that if we use this procedure repeatedly, $100(1 - \alpha)$ percent of all possible intervals developed in this way will contain μ.
2. In practical applications, the population standard deviation σ is rarely known. However, historical data, theoretical, or other reasons may provide the user with very good information about the value of σ. If this value of σ is believed to be better than an estimate of σ that can be obtained by using a sample standard deviation s, equation (8.2) should be used for the large-sample case with σ assumed known.
3. Note that the sample size n appears in the denominator of the interval estimation expressions (8.2) and (8.3). Thus, if a particular sample size provides too wide an interval to be of any practical use, we may want to consider increasing the sample size. With n in the denominator, a larger sample size will provide a smaller margin of error, a narrower interval, and a greater precision. The procedure for determining the size of a simple random sample necessary to obtain a desired precision is discussed in Section 8.3.

EXERCISES

Methods

1. A simple random sample of 40 items resulted in a sample mean of 25. The population standard deviation is $\sigma = 5$.
   a. What is the standard error of the mean, $\sigma_{\bar{x}}$?
   b. At a 95% probability, what is the margin of error?

SELF test

2. A simple random sample of 50 items resulted in a sample mean of 32 and a sample standard deviation of 6.
   a. Provide a 90% confidence interval for the population mean.
   b. Provide a 95% confidence interval for the population mean.
   c. Provide a 99% confidence interval for the population mean.
3. A sample of 60 items resulted in a sample mean of 80 and a sample standard deviation of 15.
   a. Compute the 95% confidence interval for the population mean.
   b. Assume that the same sample mean and sample standard deviation were obtained from a sample of 120 items. Provide a 95% confidence interval for the population mean.
   c. What is the effect of a larger sample size on the interval estimate of a population mean?
4. A 95% confidence interval for a population mean was reported to be 122 to 130. If the sample mean is 126 and the sample standard deviation is 16.07, what sample size was used in this study?

Applications

5. In an effort to estimate the mean amount spent per customer for dinner at a major Atlanta restaurant, data were collected for a sample of 49 customers. Assume a population standard deviation of $5.
   a. At 95% confidence, what is the margin of error?
   b. If the sample mean is $24.80, what is the 95% confidence interval for the population mean?
6. Mean weekly earnings of individuals working in various industries were reported in *The New York Times 1998 Almanac.* The mean weekly earnings for individuals in the service industry were $369. Assume that this result was based on a sample of 250 service individuals and that the sample standard deviation was $50. Compute the 95% confidence interval for the population mean weekly earnings for individuals working in the service industry.
7. The undergraduate grade point average (GPA) for students admitted to the top graduate business schools was 3.37 (*Best Graduate Schools, U.S. News and World Report,* 2001 Edition). Assume this estimate was based on a sample of 120 students admitted to the top schools. Using past years' data, the population standard deviation can be assumed known with $\sigma = .28$. What is the 95% confidence interval estimate of the mean undergraduate GPA for students admitted to the top graduate business schools?
8. The National Association of Independent Colleges and Universities reported that students graduating from public colleges and universities have an average debt of $12,000 upon graduation (*Kiplinger's Personal Finance Magazine,* November 1998). Assume that this average or mean amount is based on a sample of 245 students and that, based on past ies, the population standard deviation for the debt upon graduation is $2200.
   a. Develop a 90% confidence interval estimate of the population mean
   b. Develop a 95% confidence interval estimate of the population m
   c. Develop a 99% confidence interval estimate of the population n
   d. Discuss what happens to the width of the confidence interval as t is increased. Does this result seem reasonable? Explain.

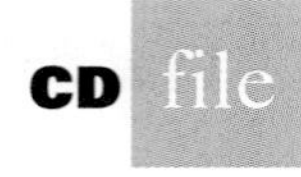
Workers

9. A summary of the mean hourly pay for production workers was reported in *The Wall Street Journal,* November 25, 1997. A sample of 60 hourly pay rates for production workers is contained in the data set Workers.
   a. Use Workers to develop a point estimate of the mean hourly pay rate for the population of production workers.
   b. What is the sample standard deviation?
   c. What is the 95% confidence interval for the mean hourly pay rate for the population of production workers?

10. Nielsen Media Research reports the household mean television viewing time during the 8 P.M. to 11 P.M. time period is 7.75 hours per week (*The World Almanac 2000*). Assuming a sample size of 180 households and a sample standard deviation of 3.45 hours, what is the 95% confidence interval estimate of the mean television viewing time per week during the 8 P.M. to 11 P.M. time period?

11. The International Air Transport Association surveys business travelers to develop quality ratings for transatlantic gateway airports. The maximum possible rating is 10. Suppose a simple random sample of 50 business travelers is selected and each traveler is asked to provide a rating for the Miami International Airport. The ratings obtained from the sample of 50 business travelers follow.

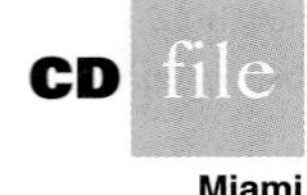
Miami

6	4	6	8	7	7	6	3	3	8	10	4	8
7	8	7	5	9	5	8	4	3	8	5	5	4
4	4	8	4	5	6	2	5	9	9	8	4	8
9	9	5	9	7	8	3	10	8	9	6		

Develop a 95% confidence interval estimate of the population mean rating for Miami.

12. Thirty fast-food restaurants including Wendy's, McDonald's, and Burger King were visited during the summer of 2000 (*The Cincinnati Enquirer,* July 9, 2000). During each visit, the customer went to the drive-through and ordered a basic meal such as a "combo" meal or a sandwich, fries, and shake. The time between pulling up to the message board and the receiving of the filled order was recorded. The times in minutes for the 30 visits are as follows:

CD file
FastFood

0.9	1.0	1.2	2.2	1.9	3.6	2.8	5.2	1.8	2.1
6.8	1.3	3.0	4.5	2.8	2.3	2.7	5.7	4.8	3.5
2.6	3.3	5.0	4.0	7.2	9.1	2.8	3.6	7.3	9.0

   a. What is the estimate of the population mean drive-through time at fast-food restaurants?
   b. At 95% confidence, what is the margin of error?
   c. What is the 95% confidence interval estimate of the population mean?

8.2 INTERVAL ESTIMATION OF A POPULATION MEAN: SMALL-SAMPLE CASE

The interval-estimation procedures developed in Section 8.1 were based on large-sample theory and the use of the central limit theorem. In such cases, the sampling distribution of $\bar{x}$ can be approximated by a normal probability distribution regardless of the population. As a result, the large-sample procedures do not require any assumption about the distribution of the population.

The methodology presented in this section is based on he assumption that the ɔulation has a normal ibution.

In this section we show how to develop an interval estimate of a population mean for the small-sample case ($n < 30$). With a small sample, the sampling distribution of $\bar{x}$ depends on the distribution of the population. The interval estimation procedures that follow are based on the assumption that the population has a normal distribution. If this assumption is appropriate, the methodology presented in this section can be used to compute an interval estimate of a population mean. However, if this assumption is not appropriate, we recommend increasing the sample size to $n \geq 30$ in order to use the large-sample procedures presented in Section 8.1.

In the small-sample case, we begin with the situation where the population standard deviation σ is assumed known. We then consider the case where the population standard deviation σ is estimated by the sample standard deviation s.

Small-Sample Case With σ Assumed Known

We begin by making the assumption that the population has a normal distribution and the population standard deviation σ is assumed known. Under these conditions, the sampling distribution of $\bar{x}$ is normally distributed with mean μ and standard deviation $\sigma_{\bar{x}} = \sigma/\sqrt{n}$ for any sample size. Thus, given that the sampling distribution of $\bar{x}$ is a normal distribution, the small-sample interval estimation procedure with σ assumed known is identical to the large-sample interval estimation presented in Section 8.1. This procedure is stated as follows:

Interval Estimate of a Population Mean: Small-Sample Case ($n < 30$) With σ Assumed Known

Assumption: The population has a normal probability distribution.

$$\bar{x} \pm z_{\alpha/2}\frac{\sigma}{\sqrt{n}} \tag{8.4}$$

where $(1 - \alpha)$ is the confidence coefficient and $z_{\alpha/2}$ is the z-value providing an area $\alpha/2$ in the upper tail of the standard normal probability distribution

With σ assumed known, $z_{\alpha/2}$ based on the desired confidence level, the sample size n, and the sample mean $\bar{x}$, equation (8.4) can be used to compute an interval estimate of the population mean. Because this computation is identical to the computation made in Section 8.1, we shall not present a new numerical example here.

Small-Sample Case With σ Estimated by s

We begin by making the assumption that the population has a normal distribution. If there is no basis for assuming that the population standard deviation σ is known, the sample standard deviation s is used to estimate σ. Under these conditions, the interval estimation procedure is based on a probability distribution known as the ***t* distribution.**

William Sealy Gosset, writing under the name "Student," is the founder of the t distribution. Gosset, an Oxford graduate in mathematics, worked for the Guinness Brewery in Dublin, Ireland. He developed a new small-sample theory of statistics while working on small-scale materials and temperature experiments in the brewery.

The t distribution is a family of similar probability distributions, with a specific t distribution depending on a parameter known as the **degrees of freedom.** The t distribution with one degree of freedom is unique, as is the t distribution with two degrees of freedom. with three degrees of freedom, and so on. As the number of degrees of freedom creases, the difference between the t distribution and the standard normal pr distribution becomes smaller and smaller. Figure 8.5 shows t distributio 20 degrees of freedom and their relationship to the standard normal tion. Note that a t distribution with more degrees of freedom has less closely resembles the standard normal probability distribution. Note als the t distribution is zero.

FIGURE 8.5 COMPARISON OF THE STANDARD NORMAL DISTRIBUTION WITH t DISTRIBUTIONS HAVING 10 AND 20 DEGREES OF FREEDOM

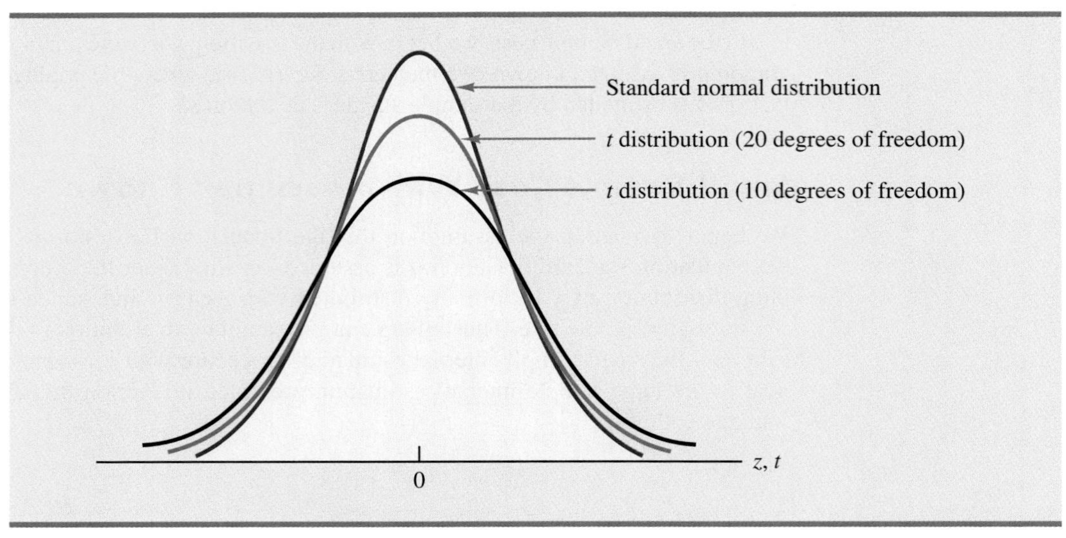

As the degrees of freedom increase, the t distribution approaches the standard normal distribution. In fact, standard normal distribution z values can be found in the infinite degrees of freedom row of the t distribution table.

We will place a subscript on t to indicate the area in the upper tail of the t distribution. For example, just as we used $z_{.025}$ to indicate the z value providing a .025 area in the upper tail of a standard normal probability distribution, we will use $t_{.025}$ to indicate a .025 area in the upper tail of a t distribution. In general, we will use the notation $t_{\alpha/2}$ to represent a t value with an area of $\alpha/2$ in the upper tail of the t distribution. See Figure 8.6.

Table 8.3 is a table for the t distribution. This table is also shown inside the front cover of the text. Note, for example, that for a t distribution with 10 degrees of freedom, $t_{.025} = 2.228$. Similarly, for a t distribution with 20 degrees of freedom, $t_{.025} = 2.086$. As the degrees of freedom continue to increase, $t_{.025}$ approaches $z_{.025} = 1.96$.

Now that we have an idea of what the t distribution is, let us see how it is used to develop an interval estimate of a population mean. Assume that the population has a normal

FIGURE 8.6 t DISTRIBUTION WITH $\alpha/2$ AREA OR PROBABILITY IN THE UPPER TAIL

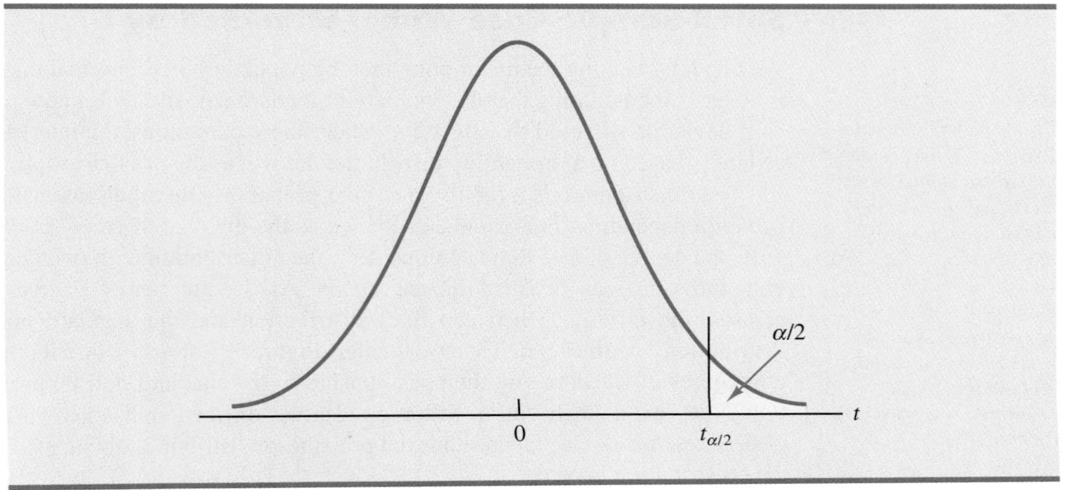

TABLE 8.3 *t* DISTRIBUTION TABLE FOR AREAS IN THE UPPER TAIL. EXAMPLE: WITH 10 DEGREES OF FREEDOM, $t_{.025} = 2.228$

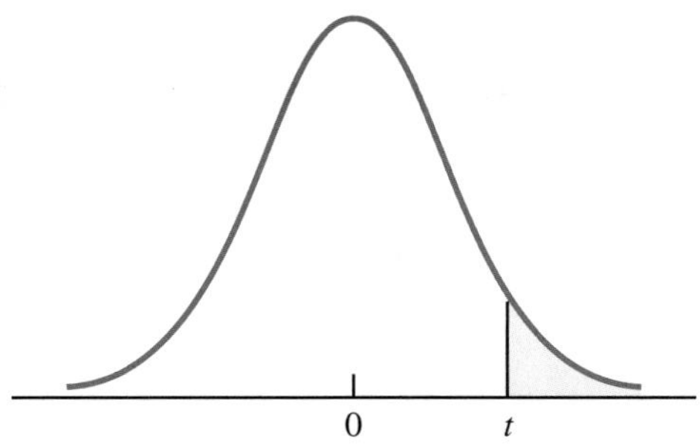

Degrees of Freedom	Upper-Tail Area (Shaded)				
	.10	**.05**	**.025**	**.01**	**.005**
1	3.078	6.314	12.706	31.821	63.657
2	1.886	2.920	4.303	6.965	9.925
3	1.638	2.353	3.182	4.541	5.841
4	1.533	2.132	2.776	3.747	4.604
5	1.476	2.015	2.571	3.365	4.032
6	1.440	1.943	2.447	3.143	3.707
7	1.415	1.895	2.365	2.998	3.499
8	1.397	1.860	2.306	2.896	3.355
9	1.383	1.833	2.262	2.821	3.250
10	1.372	1.812	2.228	2.764	3.169
11	1.363	1.796	2.201	2.718	3.106
12	1.356	1.782	2.179	2.681	3.055
13	1.350	1.771	2.160	2.650	3.012
14	1.345	1.761	2.145	2.624	2.977
15	1.341	1.753	2.131	2.602	2.947
16	1.337	1.746	2.120	2.583	2.921
17	1.333	1.740	2.110	2.567	2.898
18	1.330	1.734	2.101	2.552	2.878
19	1.328	1.729	2.093	2.539	2.861
20	1.325	1.725	2.086	2.528	2.845
21	1.323	1.721	2.080	2.518	2.831
22	1.321	1.717	2.074	2.508	2.819
23	1.319	1.714	2.069	2.500	2.807
24	1.318	1.711	2.064	2.492	2.797
25	1.316	1.708	2.060	2.485	2.787
26	1.315	1.706	2.056	2.479	2.779
27	1.314	1.703	2.052	2.473	2.771
28	1.313	1.701	2.048	2.467	2.763
29	1.311	1.699	2.045	2.462	2.756
30	1.310	1.697	2.042	2.457	2.75[illegible]
40	1.303	1.684	2.021	2.423	[illegible]
60	1.296	1.671	2.000	2.390	[illegible]
120	1.289	1.658	1.980	2.358	[illegible]
∞	1.282	1.645	1.960	2.326	[illegible]

probability distribution and that the sample standard deviation s is used to estimate the population standard deviation σ. The following interval estimation procedure is applicable.

Interval Estimate of a Population Mean: Small-Sample Case ($n < 30$) With σ Estimated by s

Assumption: The population has a normal probability distribution.

$$\bar{x} \pm t_{\alpha/2} \frac{s}{\sqrt{n}} \quad \textbf{(8.5)}$$

where s is the sample standard deviation, $(1 - \alpha)$ is the confidence coefficient, and $t_{\alpha/2}$ is the t value providing an area $\alpha/2$ in the upper tail of the t distribution with $n - 1$ degrees of freedom.

The reason the number of degrees of freedom associated with the t value in equation (8.5) is $n - 1$ has to do with the use of s as an estimate of the population standard deviation σ. The expression for the sample standard deviation is

$$s = \sqrt{\frac{\Sigma(x_i - \bar{x})^2}{n - 1}}$$

Degrees of freedom refer to the number of independent pieces of information that go into the computation of $\Sigma(x_i - \bar{x})^2$. The n pieces of information involved in computing $\Sigma(x_i - \bar{x})^2$ are as follows: $x_1 - \bar{x}, x_2 - \bar{x}, \ldots, x_n - \bar{x}$. In Section 3.2 we indicated that $\Sigma(x_i - \bar{x}) = 0$ for any data set. Thus, only $n - 1$ of the $x_i - \bar{x}$ values are independent; that is, if we know $n - 1$ of the values, the remaining value can be determined exactly by using the condition that the sum of the $x_i - \bar{x}$ values must be 0. Thus, $n - 1$ is the number of degrees of freedom associated with $\Sigma(x_i - \bar{x})^2$ and hence the number of degrees of freedom for the t distribution in expression (8.5).

Let us demonstrate the small-sample interval estimation procedure by considering the training program evaluation conducted by Scheer Industries. Scheer's director of manufacturing is interested in a computer-assisted program that can be used to train the firm's maintenance employees for machine repair operations. The expectation is that the computer-assisted method will reduce the time needed to train employees. To evaluate the training method, the director of manufacturing has requested an estimate of the population mean training time for the computer-assisted program.

Suppose management has agreed to train 15 employees using the computer-assisted program. The data on training days required for each employee in the sample are listed in Table 8.4. The sample mean and sample standard deviation for these data follow.

$$\bar{x} = \frac{\Sigma x_i}{n} = \frac{808}{15} = 53.87 \text{ days}$$

$$s = \sqrt{\frac{\Sigma(x_i - \bar{x})^2}{n - 1}} = \sqrt{\frac{651.73}{14}} = 6.82 \text{ days}$$

The point estimate of the population mean training time is 53.87 days. We can obtain information about the precision of this estimate by developing an interval estimate of the population mean. Because the population standard deviation is unknown, we will use the sample standard deviation $s = 6.82$ days to estimate σ. With the small sample size, $n = 15$,

TABLE 8.4 TRAINING TIME IN DAYS FOR THE COMPUTER-ASSISTED TRAINING PROGRAM AT SCHEER INDUSTRIES

Training

Employee	Time	Employee	Time	Employee	Time
1	52	6	59	11	54
2	44	7	50	12	58
3	55	8	54	13	60
4	44	9	62	14	62
5	45	10	46	15	63

The t distribution is based on the assumption that the population has a normal probability distribution. However, confidence intervals based on the t distribution can be used as long as the population distribution does not differ extensively from a normal probability distribution.

we will use (8.5) to develop a 95% confidence interval estimate of the population mean. If we assume that the population of training times has a normal probability distribution, the t distribution with $n - 1 = 14$ degrees of freedom is the appropriate probability distribution for the interval estimation procedure. We see from Table 8.3 that with 14 degrees of freedom, $t_{\alpha/2} = t_{.025} = 2.145$. Using (8.5), we have

$$\bar{x} \pm t_{.025} \frac{s}{\sqrt{n}}$$

$$53.87 \pm 2.145\left(\frac{6.82}{\sqrt{15}}\right)$$

$$53.87 \pm 3.78$$

Thus, the margin of error is 3.78 days and the 95% confidence interval estimate of the population mean is 50.09 days to 57.65 days.

Minitab can be used to provide the confidence interval for the population mean in the Scheer Industries example. The output is shown in Figure 8.7. The sample of 15 employees provides a sample mean of 53.87 days, a sample standard deviation of 6.82 days, a standard error of the mean of 1.76 days, and a 95% confidence interval from 50.09 to 57.65 days.

The Role of the Population Distribution

As we close this section, let us point out that one of the considerations in developing an interval estimate of a population mean is determining what is known about the population distribution. The large-sample interval estimation procedures in Section 8.1 make no assumption about the population distribution and can be applied to a population having any distribution. The interval estimation procedures in Section 8.2 are based on the assumption that the population has a normal distribution and can be applied to a population having a normal, or near-normal distribution.

In practice, if the sample size is large ($n \geq 30$) and the population distribution is unknown, we rely on the procedures in Section 8.1. In this case, an interval estimate of population mean can be constructed as follows:

σ assumed known

$$\bar{x} \pm z_{\alpha/2} \frac{\sigma}{\sqrt{n}}$$

σ estimated by s

$$\bar{x} \pm z_{\alpha/2} \frac{s}{\sqrt{n}}$$

FIGURE 8.7 THE MINITAB CONFIDENCE INTERVAL FOR THE SCHEER INDUSTRIES PROBLEM

```
Variable     N      Mean     StDev     SE Mean          95.0 % CI
Time        15     53.87      6.82        1.76     (50.09, 57.65)
```

Most importantly, the large-sample theory supports the use of expressions (8.2) and (8.3) for a population having any distribution.

If the sample size is small ($n < 30$) and the population has a normal distribution, the procedures in Section 8.2 apply. In this case we can develop an interval estimate of a population mean as follows:

σ assumed known

$$\bar{x} \pm z_{\alpha/2} \frac{\sigma}{\sqrt{n}} \tag{8.4}$$

σ estimated by s

$$\bar{x} \pm t_{\alpha/2} \frac{s}{\sqrt{n}} \tag{8.5}$$

In fact, if the population is known to have a normal distribution, expressions (8.4) and (8.5) can be used to develop an interval estimate of a population mean in both the large-sample and small-sample cases.

In most practical applications, the population distribution is unknown. Because expressions (8.2) and (8.3) apply to a population with any distribution, we make no assumption about the population distribution and use (8.2) or (8.3) provided the sample size is large ($n \geq 30$). Thus, the only time it is necessary to address the assumption that the population has a normal distribution and consider using expressions (8.4) and (8.5) is when the sample size is small ($n < 30$). Figure 8.8 summarizes the interval-estimation procedures we have covered and provides a practical guide for computing an interval estimate of a population mean.

EXERCISES

Methods

13. For a t distribution with 12 degrees of freedom, find the area, or probability, that is in each region.

- **a.** To the left of 1.782
- **b.** To the right of -1.356
- **c.** To the right of 2.681
- **d.** To the left of -1.782
- **e.** Between -2.179 and $+2.179$
- **f.** Between -1.356 and $+1.782$

14. Find the t value(s) for each of the following examples.

- **a.** Upper tail area of .05 with 18 degrees of freedom.
- **b.** Lower tail area of .10 with 22 degrees of freedom.
- **c.** Upper tail area of .01 with 5 degrees of freedom.
- **d.** 90% of the area is between these two t values with 14 degrees of freedom.
- **e.** 95% of the area is between these two t valu.es with 28 degrees of freedom.

FIGURE 8.8 SUMMARY OF INTERVAL ESTIMATION PROCEDURES FOR A POPULATION MEAN

Is n large ($n \geq 30$)?

Yes → Can σ be assumed known?

- Yes → Use $\bar{x} \pm z_{\alpha/2} \frac{\sigma}{\sqrt{n}}$
- No → Use the sample standard deviation s to estimate σ → Use $\bar{x} \pm z_{\alpha/2} \frac{s}{\sqrt{n}}$

No → Is the population approximately normal?

- Yes → Can σ be assumed known?
  - Yes → Use $\bar{x} \pm z_{\alpha/2} \frac{\sigma}{\sqrt{n}}$
  - No → Use the sample standard deviation s to estimate σ → Use $\bar{x} \pm t_{\alpha/2} \frac{s}{\sqrt{n}}$
- No → Increase the sample size to $n \geq 30$ in order to develop an interval estimate

15. The following data have been collected for a sample from a normal population: 10, 8, 12, 15, 13, 11, 6, 5.
    a. What is the point estimate of the population mean?
    b. What is the point estimate of the population standard deviation?
    c. What is the 95% confidence interval for the population mean?

16. A simple random sample of 20 observations from a normal population resulted in a sample mean of 17.25 and a sample standard deviation of 3.3.
    a. Develop a 90% confidence interval for the population mean.
    b. Develop a 95% confidence interval for the population mean.
    c. Develop a 99% confidence interval for the population mean.

Applications

17. In the testing of a new production method, 18 employees were selected randomly and asked to try the new method. The sample mean production rate for the 18 employees was 80 parts per hour and the sample standard deviation was 10 parts per hour. Provide 90% and 95% confidence intervals for the population mean production rate for the new method, assuming the population has a normal probability distribution.

18. A *USA Today* study of rental car gasoline prices found the following prices per gallon at 12 major airports (*USA Today,* April 4, 2000).

1.58	1.53	1.60	1.55	1.80	1.75	1.58	1.62	1.69	1.21	1.50	1.55

a. What is the point estimate of the population mean price per gallon?
b. What is the point estimate of the population standard deviation?
c. Assuming a normal population, what is the 95% confidence interval estimate of the population mean rental car price per gallon?

CD file
TVtime

19. The American Association of Advertising Agencies records data on nonprogram minutes on half-hour, prime-time television shows. Representative data in minutes for a sample of 20 prime-time shows on major networks at 8:30 P.M. follow.

6.0	6.6	5.8
7.0	6.3	6.2
7.2	5.7	6.4
7.0	6.5	6.2
6.0	6.5	7.2
7.3	7.6	6.8
6.0	6.2	

Assuming a normal population, provide a point estimate and a 95% confidence interval for the mean number of nonprogram minutes on half-hour, prime-time television shows at 8:30 P.M.

20. Sales personnel for Skillings Distributors are required to submit weekly reports listing the customer contacts made during the week. A sample of 61 weekly contact reports showed a mean of 22.4 customer contacts per week for the sales personnel. The sample standard deviation was five contacts.

a. Use the large-sample case (8.3) to develop a 95% confidence interval for the mean number of weekly customer contacts for the population of sales personnel.
b. Assume that the population of weekly contact data has a normal distribution. Use the t distribution with 60 degrees of freedom to develop a 95% confidence interval for the mean number of weekly customer contacts.
c. Compare your answers for parts (a) and (b). Comment on why in the large-sample case it is permissible to base interval estimates on the procedure used in part (a) even though the t distribution may also be applicable too.

21. Rising prescription drug prices caused the U.S. Congress to consider laws that would force pharmaceutical companies to offer prescription discounts to senior citizens without drug benefits. The House Government Reform Committee provided data on the prescription cost for some of the most widely used drugs (*Newsweek,* May 8, 2000). Assume the following data show a sample of the prescription cost in dollars for Zocor, a drug used to lower cholesterol.

110	112	115	99	100	98	104	126

Assuming a normal population, what is the 95% confidence interval estimate of the population mean cost for a prescription of Zocor?

22. The number of hours Americans sleep each night varies considerably with 12% of the population sleeping less than six hours to 3% sleeping more than eight hours (*The Macmillan Visual Almanac,* 1996). A following sample of 25 individuals reports the hours of sleep per night.

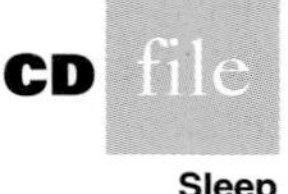
Sleep

6.9	7.6	6.5	6.2	5.3
7.8	7.0	5.5	7.6	6.7
7.3	6.6	7.1	6.9	6.0
6.8	6.5	7.2	5.8	8.6
7.6	7.1	6.0	7.2	7.7

a. What is the point estimate of the population mean number of hours of sleep each night?
b. Assuming that the population has a normal distribution, develop a 95% confidence interval for the population mean number of hours of sleep each night.

8.3 DETERMINING THE SAMPLE SIZE

In the large-sample case with σ assumed known, the interval estimate of the population mean is given by

$$\bar{x} \pm z_{\alpha/2} \frac{\sigma}{\sqrt{n}} \tag{8.2}$$

The quantity $z_{\alpha/2}(\sigma/\sqrt{n})$ is the margin of error. Thus, we see that $z_{\alpha/2}$, the population standard deviation σ, and the sample size n combine to determine the margin of error. Once we select a confidence coefficient $1 - \alpha$, $z_{\alpha/2}$ can be determined. Then, if we have a value for σ, we can determine the sample size n needed to provide any desired margin of error. Development of the formula used to compute the required sample size n follows.

Let E = the desired margin of error

$$E = z_{\alpha/2} \frac{\sigma}{\sqrt{n}}$$

If a desired margin of error is selected prior to sampling, the procedures in this section can be used to determine the sample size necessary to satisfy the margin of error requirement.

Solving for $\sqrt{n}$, we have

$$\sqrt{n} = \frac{z_{\alpha/2}\sigma}{E}$$

Squaring both sides of this equation, we obtain the following expression for the sample size.

Sample Size for an Interval Estimate of a Population Mean

$$n = \frac{(z_{\alpha/2})^2\sigma^2}{E^2} \tag{8.6}$$

This sample size will provide the desired margin of error at the chosen confidence level.

In equation (8.6) E is the margin of error that the user is willing to accept and the value of $z_{\alpha/2}$ follows directly from the confidence level to be used in developing the interval estimate. Although user preference must be considered, 95% confidence is the most frequently chosen value ($z_{.025} = 1.96$).

Finally, use of (8.6) requires a value for the population standard deviation σ. In most cases, σ will be unknown. However, we can still use (8.6) if we have a preliminary or *planning value* for σ. In practice, one of the following procedures can be chosen.

A planning value for the population standard deviation σ must be specified before the sample size can be determined. Three methods of obtaining a planning value for σ are discussed here.

1. Use the sample standard deviation from a previous sample of the same or similar units.
2. Use a pilot study to select a preliminary sample. The sample standard deviation from the preliminary sample can be used as the planning value for σ.
3. Use judgment or a "best guess" for the value of σ. For example, we might begin by estimating the largest and smallest data values in the population. The difference between the largest and smallest values provides an estimate of the range for the data. Finally, the range divided by four is often suggested as a rough approximation of the standard deviation and thus an acceptable planning value for σ.

Let us demonstrate the use of equation (8.6) to determine the sample size by considering the following example. A previous study that investigated the cost of renting automobiles in the United States found that the mean cost of renting a medium-size automobile was approximately \$55 per day. Suppose that the organization that conducted this study would like to conduct a new study in order to estimate the population mean daily rental cost for a medium-size automobile in the United States. In designing the new study, the project director has specified that the population mean daily rental cost should be estimated with a margin of error of \$2 and a 95% level of confidence.

Using equation (8.6), we see that the project director has specified a margin of error $E = 2$. The 95% level of confidence indicates $z_{.025} = 1.96$. Thus, we only need a planning value for the population standard deviation σ in order to compute the required sample size. At this point, an analyst reviewed the sample data from the original study and found that the sample standard deviation for the daily rental cost was \$9.65. Using this value as the planning value for σ, we have

Equation (8.6) provides the minimum sample size needed to satisfy the desired margin of error requirement. If the computed sample size is not an integer, round up to the next integer value to determine the recommended sample size.

$$n = \frac{(z_{\alpha/2})^2\sigma^2}{E^2} = \frac{(1.96)^2(9.65)^2}{2^2} = 89.43$$

Thus, the sample size for the new study needs to be at least 89.43 medium-size automobile rentals in order to satisfy the project director's \$2 margin-of-error requirement. In cases where the computed n is not an integer, we *round up* to the next integer value; hence, the recommended sample size is 90 medium-size automobile rentals.

EXERCISES

Methods

23. How large a sample should be selected to provide a 95% confidence interval with a margin of error of 5? Assume that the population standard deviation is 25.

SELF test

24. The range for a set of data is estimated to be 36.
    a. What is the planning value for the population standard deviation?
    b. Using 95% confidence, how large a sample should be used to provide a margin of error of 3?
    c. Using 95% confidence, how large a sample should be used to provide a margin of error of 2?

Applications

25. Refer to the Scheer Industries example in Section 8.2. Use $\sigma = 6.82$ days as a planning value for the population standard deviation.
    a. Assuming 95% confidence, what sample size would be required to obtain a margin of error of 1.5 days?
    b. If the precision statement was made with 90% confidence, what sample size would be required to obtain a margin of error of 2 days?
26. *Bride's* magazine reported that the mean cost of a wedding is \$19,000 (*USA Today,* April 17, 2000). Assume that the population standard deviation is \$9,400. *Bride's* plans to use an annual survey to monitor the cost of a wedding. Use 95% confidence.
    a. What is the recommended sample size if the desired margin of error is \$1000?
    b. What is the recommended sample size if the desired margin of error is \$500?
    c. What is the recommended sample size if the desired margin of error is \$200?

27. Annual starting salaries for college graduates with business administration degrees are believed to have a standard deviation of approximately \$2000. Assume that a 95% confidence interval estimate of the mean annual starting salary is desired. How large a sample should be taken if the desired margin of error is

a. \$500?

b. \$200?

c. \$100?

28. RealFacts, a real estate research firm, provides monthly mean apartment rental costs for Los Angeles County (*Los Angeles Times,* August 1, 1999). Assume the population standard deviation is \$220 and that the desired margin of error is \$50.

a. What is the recommended sample size for a 90% confidence interval estimate of the population mean rental cost?

b. What is the recommended sample size for a 95% confidence interval?

c. What is the recommended sample size for a 99% confidence interval?

d. If the desired margin of error is fixed, what happens to the sample size as the confidence level is increased?

29. The travel-to-work time for residents of the 15 largest cities in the United States is reported in the *1998 Information Please Almanac.* Suppose that a preliminary simple random sample of residents of San Francisco is used to develop a planning value of 6.25 minutes for the population standard deviation.

a. If we want to estimate the population mean travel-to-work time for San Francisco residents with a margin of error of 2 minutes, what sample size should be used? Assume a 95% confidence.

b. If we want to estimate the population mean travel-to-work time for San Francisco residents with a margin of error of 1 minute, what sample size should be used? Assume 95% confidence.

30. The sample standard deviation of P/E ratios for stocks listed on the New York Stock Exchange is $s = 7.8$ (*The Wall Street Journal,* March 19, 1998). Assume that we are interested in estimating the population mean P/E ratio for all stocks listed on the New York Stock Exchange. How many stocks should be included in the sample if we want a margin of error of 2? Use 95% confidence.

8.4 INTERVAL ESTIMATION OF A POPULATION PROPORTION

In Chapter 7 we showed that a sample proportion $\bar{p}$ is an unbiased estimator of a population proportion p and that for large samples the sampling distribution of $\bar{p}$ can be approximated by a normal probability distribution, as shown in Figure 8.9. Recall that the use of the normal distribution as an approximation of the sampling distribution of $\bar{p}$ is based on the large-sample condition that both np and $n(1 - p)$ are 5 or more. We will be using the sampling distribution of $\bar{p}$ to make probability statements about the sampling error whenever a sample proportion $\bar{p}$ is used to estimate a population proportion p. In this case, the sampling error is defined as the absolute value of the difference between $\bar{p}$ and p, written $|\bar{p} - p|$.

When the sample size is large, the following precision statement can be made about the sampling error.

> There is a $1 - \alpha$ probability that the value of the sample proportion will provide a sampling error of $z_{\alpha/2}\sigma_{\bar{p}}$ or less.

Hence, for a proportion, the quantity $z_{\alpha/2}\sigma_{\bar{p}}$ is the margin of error.

Once we know the margin of error, $z_{\alpha/2}\sigma_{\bar{p}}$, we can subtract and add this value to $\bar{p}$ to obtain an interval estimate of the population proportion. Such an interval estimate is given by

$$\bar{p} \pm z_{\alpha/2}\sigma_{\bar{p}} \qquad \textbf{(8.7)}$$

FIGURE 8.9 NORMAL APPROXIMATION OF THE SAMPLING DISTRIBUTION OF $\bar{p}$ WHEN $np \geq 5$ AND $n(1 - p) \geq 5$

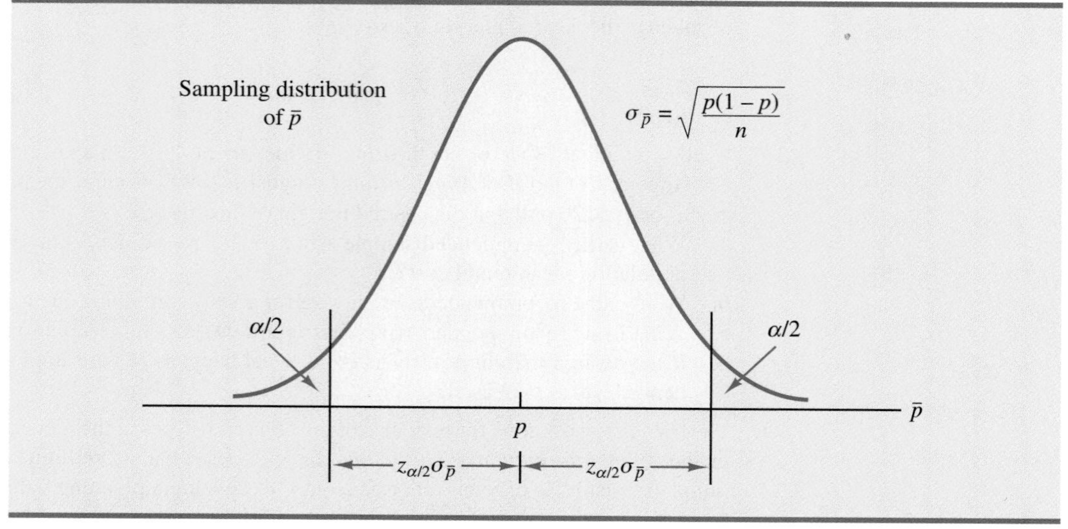

where $1 - \alpha$ is the confidence coefficient. Because $\sigma_{\bar{p}} = \sqrt{p(1 - p)/n}$, we can rewrite (8.7) as

$$\bar{p} \pm z_{\alpha/2} \sqrt{\frac{p(1 - p)}{n}} \tag{8.8}$$

When developing confidence intervals for proportions, $\sigma_{\bar{p}}$ is referred to as the standard error of the proportion and the quantity $z_{\alpha/2}\sqrt{\bar{p}(1 - \bar{p})/n}$ provides the margin of error.

To use expression (8.8) to develop an interval estimate of a population proportion p, the value of p would have to be known. Because the value of p is what we are trying to estimate, we simply substitute the sample proportion $\bar{p}$ for p. The resulting general expression for a confidence interval estimate of a population proportion follows.*

Interval Estimate of a Population Proportion

$$\bar{p} \pm z_{\alpha/2} \sqrt{\frac{\bar{p}(1 - \bar{p})}{n}} \tag{8.9}$$

where $1 - \alpha$ is the confidence coefficient and $z_{\alpha/2}$ is the z value providing an area of $\alpha/2$ in the upper tail of the standard normal probability distribution.

Let us use the following example to illustrate the computation of the margin of error and interval estimate for a population proportion. Ferrell Calvillo Communications conducted a national survey of 902 women golfers to learn how women golfers view themselves as being treated at golf courses in the United States. The survey found that 397 women golfers were satisfied with the availability of tee times. Thus, the point estimate of

*An unbiased estimate of $\sigma_{\bar{p}}^2$ is $\bar{p}(1 - \bar{p})/(n - 1)$, which suggests that $\sqrt{\bar{p}(1 - \bar{p})/(n - 1)}$ should be used in place of $\sqrt{\bar{p}(1 - \bar{p})/n}$ in (8.9). However, the bias introduced by using n in the denominator does not cause any difficulty because a large sample is generally used to estimate a population proportion. In such cases the numerical difference between the results obtained by using n and those obtained by using $n - 1$ is negligible.

the proportion of the population of women golfers who are satisfied with the availability of tee times is 397/902 = .44. Using (8.9) and a 95% confidence level, we have

$$\bar{p} \pm z_{\alpha/2}\sqrt{\frac{\bar{p}(1-\bar{p})}{n}}$$

$$.44 \pm 1.96\sqrt{\frac{.44(1-.44)}{902}}$$

$$.44 \pm .0324$$

Thus, the margin of error is .0324 and the 95% confidence interval estimate of the population proportion is .4076 to .4724. Using percentages, the survey results enable us to state that with 95% confidence between 40.76% and 47.24% of all women golfers are satisfied with the availability of tee times.

Determining the Sample Size

Let us consider the question of how large the sample size should be to obtain an estimate of a population proportion at a specified level of precision. The rationale for the sample size determination in developing interval estimates of p is similar to the rationale used in Section 8.3 to determine the sample size for estimating a population mean.

Previously in this section we pointed out that the margin of error associated with an estimate of a population proportion is $z_{\alpha/2}\sigma_{\bar{p}}$. With $\sigma_{\bar{p}} = \sqrt{p(1-p)/n}$, the margin of error is based on the values of $z_{\alpha/2}$, the population proportion p, and the sample size n. For a confidence coefficient $1-\alpha$, $z_{\alpha/2}$ can be determined. Then, for a specified value of the population proportion p, the margin of error is determined by the sample size n. Larger sample sizes provide a smaller margin of error and better precision.

Let E = the desired margin of error

$$E = z_{\alpha/2}\sqrt{\frac{p(1-p)}{n}}$$

Solving this equation for n provides the following formula for the sample size.

Sample Size for an Interval Estimate of a Population Proportion

$$n = \frac{(z_{\alpha/2})^2 p(1-p)}{E^2} \qquad \textbf{(8.10)}$$

Because the population proportion p is what we are attempting to estimate from the sample, a planning value for p must be available in order to use (8.10). Four ways to obtain this planning value are listed here.

In equation (8.10), the desired margin of error E must be specified by the user; in most cases, E = .10 or less. User preference also specifies the confidence level and thus the corresponding value of $z_{\alpha/2}$. Finally, because the population proportion p is unknown, the use of (8.10) requires a planning value for p. In practice, the planning value can be chosen by one of the following procedures.

1. Use the sample proportion from a previous sample of the same or similar units.
2. Use a pilot study to select a preliminary sample. The sample proportion from this sample can be used as the planning value for p.
3. Use judgment or a "best guess" for the value of p.
4. If none of the preceding alternatives apply, use a planning value of p = .50.

Let us return to the survey of women golfers and assume that the company is interested in conducting a new survey to estimate the current proportion of the population of women golfers who are satisfied with the availability of tee times. How large should the sample be if the survey director wants to estimate the population proportion with a margin of error of .025 at a 95% confidence? With $E = .025$ and $z_{\alpha/2} = 1.96$, we need a planning value of p to answer the sample size question. Using the previous survey result of $\bar{p} = .44$ as the planning value for p, (8.10) shows that

$$n = \frac{(z_{\alpha/2})^2 p(1 - p)}{E^2} = \frac{(1.96)^2(.44)(1 - .44)}{(.025)^2} = 1514.51$$

Thus, the sample size must be at least 1514.51 women golfers to satisfy the margin of error requirement. Rounding up to the next integer value indicates that a sample of 1515 women golfers is recommended.

The fourth alternative suggested for selecting a planning value for p is to use $p = .50$. This value of p is frequently used when no other information is available. To understand why, note that the numerator of equation (8.10) shows that the sample size is proportional to the quantity $p(1 - p)$. A larger value for the quantity $p(1 - p)$ will result in a larger sample size. Table 8.5 gives some possible values of $p(1 - p)$. Note that the largest value of $p(1 - p)$ occurs when $p = .50$. Thus, in case of any uncertainty about an appropriate planning value for p, we know that $p = .50$ will provide the largest sample size recommendation. In effect, we are being on the safe or conservative side in recommending the largest possible sample size. If the proportion turns out to be different from the .50 planning value, the precision statement will be better than anticipated. Thus, in using $p = .50$, we are guaranteeing that the sample size will be sufficient to obtain the desired margin of error.

Surveys that ask many questions about proportions are often designed by using a planning value $p = .50$ for the sample size computation.

In the survey of women golfers example, a planning value of $p = .50$ would have provided the sample size

$$n = \frac{(z_{\alpha/2})^2 p(1 - p)}{E^2} = \frac{(1.96)^2(.50)(1 - .50)}{(.025)^2} = 1536.64$$

Thus a slightly larger sample size of 1537 women golfers would be recommended.

TABLE 8.5 SOME POSSIBLE VALUES FOR $p(1 - p)$

p	$p(1 - p)$	
.10	(.10)(.90) = .09	
.30	(.30)(.70) = .21	
.40	(.40)(.60) = .24	
.50	(.50)(.50) = .25	← Largest value for $p(1 - p)$
.60	(.60)(.40) = .24	
.70	(.70)(.30) = .21	
.90	(.90)(.10) = .09	

NOTES AND COMMENTS

The margin of error for estimating a population proportion is almost always .10 or less. In national public opinion polls conducted by organizations such as Gallup and Harris, a .03 or .04 margin of error is generally reported. With such margins of error, equation (8.10) will almost always provide a sample size that is large enough to satisfy the large-sample requirements of $np \geq 5$ and $n(1 - p) \geq 5$.

EXERCISES

Methods

31. A simple random sample of 400 items provides 100 Yes responses.
    a. What is the point estimate of the proportion of the population that would provide Yes responses?
    b. What is the standard error of the proportion, $\sigma_{\bar{p}}$?
    c. Compute the 95% confidence interval for the population proportion.
32. A simple random sample of 800 units generates a sample proportion $\bar{p} = .70$.
    a. Provide a 90% confidence interval for the population proportion.
    b. Provide a 95% confidence interval for the population proportion.
33. In a survey, the planning value for the population proportion p is .35. How large a sample should be taken to provide a 95% confidence interval with a margin of error of .05?
34. Using 95% confidence, how large a sample should be taken to obtain a margin of error for the estimation of a population proportion of .03? Assume that past data are not available for developing a planning value for p.

Applications

35. A *Time*/CNN poll asked 814 adults to respond to a series of questions about their feelings toward the state of affairs within the United States. A total of 562 adults responded Yes to the question: Do you feel things are going well in the United States these days? (*Time,* August 11, 1997).
    a. What is the point estimate of the proportion of the adult population that feels things are going well in the United States?
    b. At 90% confidence, what is the margin of error?
    c. What is the 90% confidence interval for the proportion of the adult population that feels things are going well in the United States?
36. A survey by the Society for Human Resource Management asked 346 job seekers why employees change jobs so frequently (*The Wall Street Journal,* March 28, 2000). The answer selected most (152 times) was "higher compensation elsewhere."
    a. What is the point estimate of the proportion of job seekers who would select "higher compensation elsewhere" as the reason for changing jobs?
    b. What is the 95% confidence interval estimate of the population proportion?
37. A survey by Wirthlin Worldwide collected data on attitudes toward the quality of customer service in retail stores. The survey found that 28% of Americans feel customer service is better today than it was two years ago (*USA Today,* January 20, 1998). If 650 adults were included in the sample, develop a 95% confidence interval for the proportion of the population of adults who feel customer service is better today than it was 2 years ago.
38. Audience profile data collected at the ESPN SportsZone web site showed that 26% of the users were women (*USA Today,* January 21, 1998). Assume that this percentage was based on a sample of 400 users.
    a. Using 95% confidence, what is the margin of error associated with the estimated proportion of users who are women?
    b. What is the 95% confidence interval for the population proportion of ESPN Sportszone web site users who are women?
    c. How large a sample should be taken if the desired margin of error is 3%?

39. An Employee Benefit Research Institute survey explored the reasons small employers offer a retirement plan to their employees (*USA Today,* April 4, 2000). The reason "competitive advantage in recruitment/retention" was anticipated 33% of the time.

a. What sample size is recommended if a survey goal is to estimate the proportion of small employers who offer a retirement plan primarily for "competitive advantage in recruitment/retention" with a margin of error of 3%. Use 95% confidence.

b. Repeat part (a) using 99% confidence.

40. An Associated Press poll of 1018 adults found 255 adults planned to spend less money on gifts during the 1998 holiday season compared to the previous year (ICR Media survey, November 13–17, 1998).

a. What is the point estimate of the proportion of all adults who planned to spend less money on gifts during the 1998 holiday season?

b. Using 95% confidence, what is the margin of error associated with this estimate?

41. An American Express retail survey found that 16% of U.S. consumers had used the Internet to buy gifts during the 1999 holiday season (*USA Today,* January 18, 2000). If 1285 customers participated in the survey, what is the margin of error and what is the interval estimate of the population proportion of customers using the Internet to buy gifts? Use 95% confidence.

42. A *USA Today*/CNN/Gallup poll for the 2000 presidential campaign sampled 491 potential voters in June (*USA Today,* June 9, 2000). A primary purpose of the poll was to obtain an estimate of the proportion of potential voters who favor each candidate. Assume a planning value for the population proportion of $p = .50$ and a 95% confidence level.

a. Using $p = .50$, what was the planned margin of error for the June poll?

b. As the November election gets closer, better precision and smaller margins of error are desired. Assume the following margins of error are requested for surveys to be conducted during the presidential campaign. Compute the recommended sample size for each survey.

Survey	Margin of Error
September	.04
October	.03
Early November	.02
Pre-Election Day	.01

43. The League of American Theatres and Producers uses an ongoing audience tracking survey that provides up-to-date information about Broadway theater audiences (*Playbill,* Winter 1997). Every week, the League distributes a one-page survey on random theater seats at a rotation roster of Broadway shows. The survey questionnaire takes only 5 minutes to complete and enables audiences to communicate their views about the theater industry.

a. How large a sample should be taken if the desired margin of error on any population proportion is .04? Use 95% confidence and a planning value of $p = .50$.

b. Assume that the sample size recommended in part (a) is implemented and that during one week 445 theatergoers indicated that they do not live in New York City. What is the point estimate of the proportion of Broadway audiences that do not live in New York City?

c. Using the data in part (b), what is the 95% confidence interval for the population proportion of Broadway audiences that do not live in New York City?

SUMMARY

In this chapter we presented methods for developing a confidence interval for a population mean μ and a population proportion p. The purpose of developing a confidence interval is to provide information about the precision of the estimate. A wide confidence interval indicates poor precision; in such cases, the sample size can be increased to reduce the width of the confidence interval and improve the precision of the estimate.

The expression used to compute an interval estimate of a population mean depends on whether the sample size is large ($n \geq 30$) or small ($n < 30$), whether the population standard deviation can be assumed known or is estimated by the sample standard deviation s, and in some cases whether the population has a normal or approximately normal probability distribution. If the sample size is large, no assumption is required about the distribution of the population, and $z_{\alpha/2}$ is used in the computation of the interval estimate. If the sample size is small, the population must have a normal or approximately normal probability distribution in order to develop an interval estimate of μ. If the population has a normal probability distribution, $z_{\alpha/2}$ is used in the computation of the interval estimate when σ is assumed known, whereas $t_{\alpha/2}$ is used when σ is estimated by the sample standard deviation s. Finally, if the sample size is small and the assumption of a normally distributed population is inappropriate, we recommend increasing the sample size to $n \geq 30$ in order to use the large-sample interval estimation procedure.

In addition, we showed how to determine the sample size so that interval estimates of μ and p would have a specified margin of error. In practice, the sample sizes required for interval estimates of a population proportion are generally large. Hence, we provided the large-sample interval estimation formulas for a population proportion where both $np \geq 5$ and $n(1 - p) \geq 5$.

GLOSSARY

Interval estimate An estimate of a population parameter that provides an interval believed to contain the value of the parameter. It has the form point estimate ± margin of error.

Margin of error The ± value added to and subtracted from a point estimate in order to develop an interval estimate of a population parameter.

Sampling error The absolute value of the difference between the value of an unbiased point estimator, such as the sample mean $\bar{x}$, and the value of the population parameter it estimates, such as the population mean μ. In the case of a population mean, the sampling error is $|\bar{x} - \mu|$. In the case of a population proportion, the sampling error is $|\bar{p} - p|$.

Precision statement A probability statement about the sampling error.

Confidence level The confidence associated with an interval estimate. For example, if an interval estimation procedure provides intervals such that 95% of the intervals formed using the procedure will include the population parameter, the interval estimate is said to be constructed at the 95% confidence level.

Confidence Coefficient The confidence level expressed as a decimal value. For example, .95 is the confidence coefficient for a 95% confidence level.

***t* Distribution** A family of probability distributions that can be used to develop an interval estimate of a population mean whenever the population standard deviation σ is estimated by the sample standard deviation s and the population has a normal or near-normal probability distribution.

Degrees of freedom A parameter of the t distribution. When the t distribution is used in the computation of an interval estimate of a population mean, the appropriate t distribution has $n - 1$ degrees of freedom, where n is the size of the simple random sample.

KEY FORMULAS

Sampling Error When Estimating μ

$$|\bar{x} - \mu| \tag{8.1}$$

Interval Estimate of a Population Mean: Large-Sample Case ($n \geq 30$) With σ Assumed Known

$$\bar{x} \pm z_{\alpha/2} \frac{\sigma}{\sqrt{n}} \tag{8.2}$$

Interval Estimate of a Population Mean: Large-Sample Case ($n \geq 30$) With σ Estimated by s

$$\bar{x} \pm z_{\alpha/2} \frac{s}{\sqrt{n}} \tag{8.3}$$

Interval Estimate of a Population Mean: Small-Sample Case ($n < 30$) With σ Assumed Known

$$\bar{x} \pm z_{\alpha/2} \frac{\sigma}{\sqrt{n}} \tag{8.4}$$

Interval Estimate of a Population Mean: Small-Sample Case ($n < 30$) With σ Estimated by s

$$\bar{x} \pm t_{\alpha/2} \frac{s}{\sqrt{n}} \tag{8.5}$$

Sample Size for an Interval Estimate of a Population Mean

$$n = \frac{(z_{\alpha/2})^2\sigma^2}{E^2} \tag{8.6}$$

Interval Estimate of a Population Proportion

$$\bar{p} \pm z_{\alpha/2}\sqrt{\frac{\bar{p}(1 - \bar{p})}{n}} \tag{8.9}$$

Sample Size for an Interval Estimate of a Population Proportion

$$n = \frac{(z_{\alpha/2})^2 p(1 - p)}{E^2} \tag{8.10}$$

SUPPLEMENTARY EXERCISES

44. A survey of first-time home buyers found that the mean of annual household income was \$50,000 (*CNBC.com,* July 11, 2000). Assume the survey used a sample of 400 first-time home buyers and that the sample standard deviation in income was \$20,500.

a. Using 95% confidence, what is the margin of error for this study?

b. What is the 95% confidence interval for the population mean annual household income for first-time home buyers?

45. A survey conducted by the American Automobile Association showed that a family of four spends an average of \$215.60 per day while on vacation. Suppose a sample of 64 families of four vacationing at Niagara Falls resulted in a sample mean of \$252.45 per day and a sample standard deviation of \$74.50.
   a. Develop a 95% confidence interval estimate of the mean amount spent per day by a family of four visiting Niagara Falls.
   b. Using the confidence interval from part (a), does it appear that the population mean amount spent per day by families visiting Niagara Falls is different from the mean reported by the American Automobile Association? Explain.

ActTemps

46. A survey by Accountemps asked a sample of 200 executives to provide data on the number of minutes per day office workers waste trying to locate mislabeled, misfiled, or misplaced items (*Astounding Averages,* by D. D. Dauphinais and K. Droste, 1995). Data consistent with this survey are contained in the data set ActTemps.
   a. Use ActTemps to develop a point estimate of the number of minutes per day office workers waste trying to locate mislabeled, misfiled, or misplaced items.
   b. What is the sample standard deviation?
   c. What is the 95% confidence interval for the mean number of minutes wasted per day?

47. Arthur D. Little, Inc., estimates that approximately 70% of mail received by a household is advertisements (*Time,* July 14, 1997). A sample of 20 households shows the following data for the number of advertisements received and the total number of pieces of mail received during one week.

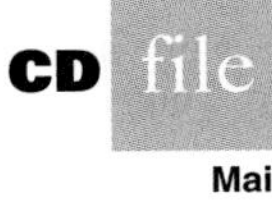
Mail

Household	Advertisements	Total Mail	Household	Advertisements	Total Mail
1	24	35	11	13	19
2	9	14	12	16	28
3	18	30	13	20	27
4	9	12	14	17	22
5	15	28	15	21	24
6	23	33	16	21	33
7	13	20	17	15	25
8	17	20	18	15	24
9	20	23	19	18	24
10	20	25	20	12	16

   a. What is the point estimate of the mean number of advertisements received per week? What is the 95% confidence interval for the population mean?
   b. What is the point estimate of the mean number of pieces of mail received per week? What is the 95% confidence interval for the population mean?
   c. Use the point estimates in parts (a) and (b). Are these estimates in agreement with the statement that approximately 70% of mailings are advertisements?

48. The Money & Investing section of the *The Wall Street Journal* contains a summary of daily investment performance for stock exchanges, overseas markets, options, commodities, futures, and so on. In the New York Stock Exchange section, information is provided on each stock's 52-week high price per share, 52-week low price per share, dividend rate, yield, P/E ratio, daily volume, daily high price per share, daily low price per share, closing price per share, and daily net change. The P/E (price/earnings) ratio for each stock is determined by dividing the price of a share of stock by the earnings per share reported by the company for the most recent four quarters. A sample of 10 stocks taken from *The Wall Street Journal* (September 29, 2000) provided the following data on P/E ratios: 5, 7, 9, 10, 14, 23, 20, 15, 3, 26.
   a. What is the point estimate of the mean P/E ratio for the population of all stocks listed on the New York Stock Exchange?
   b. What is the point estimate of the standard deviation of the P/E ratios for the population of all stocks listed on the New York Stock Exchange?

c. At 95% confidence, what is the interval estimate of the mean P/E ratio for the population of all stocks listed on the New York Stock Exchange? Assume that the population has a normal distribution.

d. Comment on the precision of the results.

49. Many Americans who work in large offices also work at home or in the office on weekends (*USA Today,* June 18, 1997). How large a sample should be taken to estimate the population mean amount of time worked on weekends with a margin of error of 10 minutes? Use 95% confidence and assume that the planning value for the population standard deviation is 45 minutes.

50. Mileage tests are conducted for a particular model of automobile. If the desired precision is a 98% confidence interval with a margin of error of 1 mile per gallon, how many automobiles should be used in the test? Assume that preliminary mileage tests indicate the standard deviation to be 2.6 miles per gallon.

51. In developing patient appointment schedules, a medical center wants an estimate of the mean time that a staff member spends with each patient. How large a sample should be taken if the desired margin of error is 2 minutes at a 95% level of confidence? How large a sample should be taken for a 99% level of confidence? Use a planning value for the population standard deviation of 8 minutes.

52. Annual salary plus bonus data for chief executive officers are presented in the *Business Week* 47th Annual Pay Survey (*Business Week,* April 21, 1997). A preliminary sample showed that the standard deviation is $675 with data provided in thousands of dollars. How many chief executive officers should be in a sample if we want to estimate the population mean annual salary plus bonus with a margin of error of $100,000. (Note: The desired margin of error would be $E = 100$ if the data are in thousands of dollars.) Use 95% confidence.

53. The National Center for Education Statistics reported that 47% of college students work to pay for tuition and living expenses (*The Tampa Tribune,* January 22, 1997). Assume that a sample of 450 college students was used in the study.

a. Provide a 95% confidence interval for the population proportion of college students who work to pay for tuition and living expenses.

b. Provide a 99% confidence interval for the population proportion of college students who work to pay for tuition and living expenses.

c. What happens to the margin of error as the confidence is increased from 95% to 99%?

54. A *USA Today*/CNN/Gallup survey of 369 working parents found 200 who said they spend too little time with their children because of work commitments (*USA Today,* April 10, 1995).

a. What is the point estimate of the proportion of the population of working parents who feel they spend too little time with their children because of work commitments?

b. At 95% confidence, what is the margin of error?

c. What is the 95% confidence interval estimate of the population proportion of working parents who feel they spend too little time with their children because of work commitments?

55. A *Time*/CNN poll of 1400 American adults asked, "Where would you rather go in your spare time?" The top response by 504 adults was a shopping mall.

a. What is the point estimate of the proportion of adults who would prefer going to a shopping mall in their spare time?

b. At 95% confidence, what is the margin of error associated with this estimate?

56. A well-known bank credit card firm is interested in estimating the proportion of credit cardholders who carry a nonzero balance at the end of the month and incur an interest charge. Assume that the desired margin of error is .03 at 98% confidence.

a. How large a sample should be selected if it is anticipated that roughly 70% of the firm's cardholders carry a nonzero balance at the end of the month?

b. How large a sample should be selected if no planning value for the population proportion could be specified?

57. In a survey, 200 people were asked to identify their major source of news information; 110 stated that their major source was television news.

a. Construct a 95% confidence interval for the proportion of people in the population who consider television their major source of news information.

b. How large a sample would be necessary to estimate the population proportion with a margin of error of .05 at 95% confidence?

58. A Roper Starch survey asked employees ages 18 to 29 if they would prefer better health insurance or a raise in salary (*USA Today,* September 5, 2000). Answer the following questions if 340 of 500 employees said they would prefer better health insurance over a raise.

a. What is the point estimate of the proportion of employees ages 18 to 29 who would prefer better health insurance?

b. What is the 95% confidence interval estimate of the population proportion?

59. The *1997 Statistical Abstract of the United States* reports the percentage of people 18 years of age and older who smoke. Assume that a study is being designed to collect new data on smokers and nonsmokers. The best preliminary estimate of the population proportion who smoke is 30%.

a. How large a sample should be taken to estimate the proportion of smokers in the population with a margin of error of .02? Use 95% confidence.

b. Assume that the study uses your sample size recommendation in part (a) and finds 520 smokers. What is the point estimate of the proportion of smokers in the population?

c. What is the 95% confidence interval for the proportion of smokers in the population?

60. Although airline schedules and cost are important factors for business travelers when choosing an airline carrier, a *USA Today* survey found that business travelers list an airline's frequent flyer program as the most important factor (*USA Today,* April 11, 1995). From a sample of $n = 1993$ business travelers who responded to the survey, 618 listed a frequent flyer program as the most important factor.

a. What is the point estimate of the proportion of the population of business travelers who believe a frequent flyer program is the most important factor when choosing an airline carrier?

b. Develop a 95% confidence interval estimate of the population proportion.

c. How large a sample would be required to report the margin of error of .01 at 95% confidence? Would you recommend that *USA Today* attempt to provide this degree of precision? Why or why not?

Case Problem 1 BOCK INVESTMENT SERVICES

Lisa Rae Bock started Bock Investment Services (BIS) in 1994 with the goal of making BIS the leading money market advisory service in South Carolina. To provide better service for her present clients and to attract new clients, she has developed a weekly newsletter. Lisa has been considering adding a new feature to the newsletter that will report the results of a weekly telephone survey of fund managers. To investigate the feasibility of offering this service, and to determine what type of information to include in the newsletter, Lisa selected a simple random sample of 45 money market funds. A portion of the data obtained is shown in Table 8.6, which reports fund assets and yields for the past seven and 30 days. Before calling the money market fund managers to obtain additional data, Lisa decided to do some preliminary analysis of the data already collected.

TABLE 8.6 DATA FOR BOCK INVESTMENT SERVICES

CD file

Bock

Money Market Fund	Assets ($ millions)	7-day Yield (%)	30-day Yield (%)
Amcore	103.9	4.10	4.08
Alger	156.7	4.79	4.73
Arch MM/Trust	496.5	4.17	4.13
BT Instit Treas	197.8	4.37	4.32
Benchmark Div	2755.4	4.54	4.47
Bradford	707.6	3.88	3.83
Capital Cash	1.7	4.29	4.22
Cash Mgt Trust	2707.8	4.14	4.04
Composite	122.8	4.03	3.91
Cowen Standby	694.7	4.25	4.19
Cortland	217.3	3.57	3.51
Declaration	38.4	2.67	2.61
Dreyfus	4832.8	4.01	3.89
Elfun	81.7	4.51	4.41
FFB Cash	506.2	4.17	4.11
Federated Master	738.7	4.41	4.34
Fidelity Cash	13272.8	4.51	4.42
Flex-fund	172.8	4.60	4.48
Fortis	105.6	3.87	3.85
Franklin Money	996.8	3.97	3.92
Freedom Cash	1079.0	4.07	4.01
Galaxy Money	801.4	4.11	3.96
Government Cash	409.4	3.83	3.82
Hanover Cash	794.3	4.32	4.23
Heritage Cash	1008.3	4.08	4.00
Infinity/Alpha	53.6	3.99	3.91
John Hancock	226.4	3.93	3.87
Landmark Funds	481.3	4.28	4.26
Liquid Cash	388.9	4.61	4.64
MarketWatch	10.6	4.13	4.05
Merrill Lynch Money	27005.6	4.24	4.18
NCC Funds	113.4	4.22	4.20
Nationwide	517.3	4.22	4.14
Overland	291.5	4.26	4.17
Pierpont Money	1991.7	4.50	4.40
Portico Money	161.6	4.28	4.20
Prudential MoneyMart	6835.1	4.20	4.16
Reserve Primary	1408.8	3.91	3.86
Schwab Money	10531.0	4.16	4.07
Smith Barney Cash	2947.6	4.16	4.12
Stagecoach	1502.2	4.18	4.13
Strong Money	470.2	4.37	4.29
Transamerica Cash	175.5	4.20	4.19
United Cash	323.7	3.96	3.89
Woodward Money	1330.0	4.24	4.21

Source: Barron's, October 3, 1994.

Managerial Report

1. Use appropriate descriptive statistics to summarize the data on assets and yields for the money market funds.
2. Develop a 95% confidence interval estimate of the mean assets, mean 7-day yield, and mean 30-day yield for the population of money market funds. Provide a managerial interpretation of each interval estimate.
3. Discuss the implication of your findings in terms of how Lisa could use this type of information in preparing her weekly newsletter.
4. What other information would you recommend that Lisa gather to provide the most useful information to her clients?

Case Problem 2 GULF REAL ESTATE PROPERTIES

Gulf Real Estate Properties, Inc., is a real estate firm located in Southwest Florida. The company, which advertises itself as "expert in the real estate market," monitors condominium sales by collecting data on location, list price, sale price, and number of days it takes to sell each unit. Each condominium is classified as *Gulf View* if it is located directly on the Gulf of Mexico or *No Gulf View* if it is located on the bay or golf course, near but not on the Gulf. Sample data from the multiple listing service in Naples, Florida, provided recent sales data for 40 Gulf View condominiums and 18 No Gulf View condominiums.* Prices are in thousands of dollars. The data are shown in Table 8.7.

Managerial Report

1. Use appropriate descriptive statistics to summarize each of the three variables for the 40 Gulf View condominiums.
2. Use appropriate descriptive statistics to summarize each of the three variables for the 18 No Gulf View condominiums.
3. Compare your summary results. Discuss any specific statistical results that would help a real estate agent understand the condominium market.
4. Develop a 95% confidence interval estimate of the population mean sales price and population mean number of days to sell for Gulf View condominiums. Interpret your results.
5. Develop a 95% confidence interval estimate of the population mean sales price and population mean number of days to sell for No Gulf View condominiums. Interpret your results.
6. Assume the branch manager has requested estimates of the mean selling price of Gulf View condominiums with a margin of error of $40,000 and the mean selling price of No Gulf View condominiums with a margin of error of $15,000. Using 95% confidence, how large should the sample sizes be?
7. Gulf Real Estate Properties has just signed contracts for two new listings: a Gulf View condominium with a list price of $589,000 and a No Gulf View condominium with a list price of $285,000. What is your estimate of the final selling price and number of days required to sell each of these units?

Case Problem 3 METROPOLITAN RESEARCH, INC.

Metropolitan Research, Inc., is a consumer research organization that takes surveys designed to evaluate a wide variety of products and services available to consumers. In one particular study, Metropolitan was interested in learning about consumer satisfaction with

*Data based on condominium sales reported in the Naples MLS (Coldwell Banker, June 2000).

TABLE 8.7 SALES DATA FOR GULF REAL ESTATE PROPERTIES

CD file GulfProp

Gulf View Condominiums			No Gulf View Condominiums		
List Price	Sale Price	Days to Sell	List Price	Sale Price	Days to Sell
495.0	475.0	130	217.0	217.0	182
379.0	350.0	71	148.0	135.5	338
529.0	519.0	85	186.5	179.0	122
552.5	534.5	95	239.0	230.0	150
334.9	334.9	119	279.0	267.5	169
550.0	505.0	92	215.0	214.0	58
169.9	165.0	197	279.0	259.0	110
210.0	210.0	56	179.9	176.5	130
975.0	945.0	73	149.9	144.9	149
314.0	314.0	126	235.0	230.0	114
315.0	305.0	88	199.8	192.0	120
885.0	800.0	282	210.0	195.0	61
975.0	975.0	100	226.0	212.0	146
469.0	445.0	56	149.9	146.5	137
329.0	305.0	49	160.0	160.0	281
365.0	330.0	48	322.0	292.5	63
332.0	312.0	88	187.5	179.0	48
520.0	495.0	161	247.0	227.0	52
425.0	405.0	149			
675.0	669.0	142			
409.0	400.0	28			
649.0	649.0	29			
319.0	305.0	140			
425.0	410.0	85			
359.0	340.0	107			
469.0	449.0	72			
895.0	875.0	129			
439.0	430.0	160			
435.0	400.0	206			
235.0	227.0	91			
638.0	618.0	100			
629.0	600.0	97			
329.0	309.0	114			
595.0	555.0	45			
339.0	315.0	150			
215.0	200.0	48			
395.0	375.0	135			
449.0	425.0	53			
499.0	465.0	86			
439.0	428.5	158			

the performance of automobiles produced by a major Detroit manufacturer. A questionnaire sent to owners of one of the manufacturer's full-sized cars revealed several complaints about early transmission problems. To learn more about the transmission failures, Metropolitan used a sample of actual transmission repairs provided by a transmission repair firm in the Detroit area. The following data show the actual number of miles that 50 vehicles had been driven at the time of transmission failure.

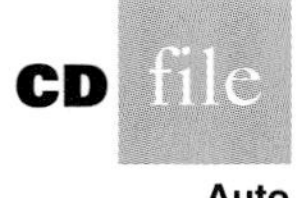
Auto

85,092	32,609	59,465	77,437	32,534	64,090	32,464	59,902
39,323	89,641	94,219	116,803	92,857	63,436	65,605	85,861
64,342	61,978	67,998	59,817	101,769	95,774	121,352	69,568
74,276	66,998	40,001	72,069	25,066	77,098	69,922	35,662
74,425	67,202	118,444	53,500	79,294	64,544	86,813	116,269
37,831	89,341	73,341	85,288	138,114	53,402	85,586	82,256
77,539	88,798						

Managerial Report

1. Use appropriate descriptive statistics to summarize the transmission failure data.
2. Develop a 95% confidence interval for the mean number of miles driven until transmission failure for the population of automobiles that have had transmission failure. Provide a managerial interpretation of the interval estimate.
3. Discuss the implication of your statistical finding in terms of the belief that some owners of the automobiles have experienced early transmission failures.
4. How many repair records should be sampled if the research firm wants the population mean number of miles driven until transmission failure to be estimated with a margin of error of 5000 miles. Use 95% confidence.
5. What other information would you like to gather to evaluate the transmission failure problem more fully?

Appendix 8.1 INTERVAL ESTIMATION OF A POPULATION MEAN WITH MINITAB

We describe the use of Minitab in constructing confidence intervals for a population mean in four different cases.

Large-Sample Case With σ Assumed Known

CJW

We illustrate using the CJW example in Section 8.1. The population standard deviation $\sigma = 20$ is assumed known. The satisfaction scores for the sample of 100 CJW customers have been entered into column C1 of a Minitab worksheet. The following steps can be used to compute a 95% confidence interval estimate of the population mean.

Step 1. Select the **Stat** pull-down menu
Step 2. Choose **Basic Statistics**
Step 3. Choose **1-Sample Z**
Step 4. When the 1-Sample Z dialog box appears:
Enter C1 in the **Variables** box
Enter 20 in the **Sigma** box
Step 5. Click **OK**

The Minitab default is a 95% confidence level. In order to specify a different confidence level such as 90% add the following to Step 4.

Select **Options**
When the 1-Sample Z Options dialog box appears:
Enter 90 in the **Confidence level** box
Click **OK**

Large-Sample Case With σ Estimated by s

Balance

We illustrate using the data in Table 8.2 showing the credit card balances for a sample of 85 households. The data have been entered into column C1 of a Minitab worksheet. In this case the population standard deviation σ is unknown and will be estimated by the sample standard deviation s. The steps used to compute the sample standard deviation are as follow:

Step 1. Select the **Calc** pull-down menu
Step 2. Choose **Column Statistics**
Step 3. When the Column Statistics dialog box appears:
Select **Standard deviation**
Enter C1 in the **Input variable** box
Enter stdev in the **Store result in** box
Click **OK**

The output shows the sample standard deviation is 3058. This value is stored in Minitab as stdev.

The interval estimate of the population mean can now be obtained by following the steps shown for the large-sample case with σ assumed known. The only change is in step 4 where the sample standard deviation s is used to estimate σ. This is accomplished by entering stdev in the **Sigma** box.

Small-Sample Case With σ Assumed Known

The small-sample case with σ assumed known uses the same step-by-step procedure previously described for the large-sample case with σ assumed known.

Small-Sample Case With σ Estimated by s

Training

We illustrate using the data in Table 8.4 showing the training program time in days for a sample of 15 employees. The data have been entered in column C1 of a Minitab worksheet. In this small-sample case the population standard deviation σ is unknown and will be estimated by the sample standard deviation s. The following steps can be used to compute a 95% confidence interval estimate of the population mean.

Step 1. Select the **Stat** pull-down menu
Step 2. Choose **Basic Statistics**
Step 3. Choose **1-Sample t**
Step 4. When the 1-Sample t dialog box appears:
Enter C1 in the **Variables** box
Step 5. Click **OK**

The Minitab default is a 95% confidence level. In order to specify a different confidence level such as 90% add the following to Step 4.

Select **Options**
When the 1-Sample t dialog box appears:
Enter 90 in the **Confidence level** box
Click **OK**

Appendix 8.2 INTERVAL ESTIMATION OF A POPULATION MEAN WITH EXCEL

We describe the use of Excel in constructing confidence intervals for a population mean in four different cases.

Large-Sample Case With σ Assumed Known

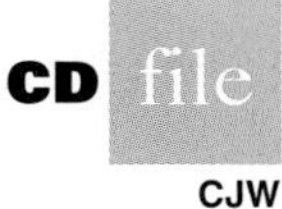

We illustrate using the CJW example in Section 8.1. The population standard deviation $\sigma = 20$ is assumed known. The data have been entered into an Excel worksheet with the label Score in cell A1 and the satisfaction scores for the sample of 100 CJW customers in cells A2:A101. The following steps can be used to compute a 95% confidence interval estimate of the population mean.

Step 1. Compute the sample mean in cell C2

=AVERAGE(A2:A101)

Step 2. Compute the margin of error* in cell C3

=CONFIDENCE(.05,20,100)

Step 3. Compute the lower limit of the interval estimate in cell C4

=C2−C3

Step 4. Compute the upper limit of the interval estimate in cell C5

=C2+C3

The formatted output with labels added appears in Figure 8.10.

Large-Sample Case With σ Estimated by s

We illustrate using the credit card balance data in Table 8.2. The data have been entered into an Excel worksheet with the label Balance in cell A1 and the credit card balances for the sample of 85 households in cells A2:A86. The population standard deviation σ is unknown

FIGURE 8.10 EXCEL OUTPUT FOR THE CJW INTERVAL ESTIMATE OF A POPULATION MEAN

Note: Rows 10–95 are hidden.

	A	B	C	D
1	**Score**			
2	98	**Sample Mean**	82	
3	79	**Margin of Error**	3.92	
4	100	**Lower Limit**	78.08	
5	86	**Upper Limit**	85.92	
6	89			
7	55			
8	68			
9	37			
96	44			
97	61			
98	96			
99	96			
100	97			
101	66			
102				

*The general form of the margin of error function is =CONFIDENCE(alpha, standard deviation, sample size), where alpha = 1 − Confidence Coefficient = 1 − .95 = .05.

and will be estimated by the sample standard deviation s. Compute the sample standard deviation in cell C1.

=STDEV(A2:A86)

Then develop the interval estimate using the same steps described for the large-sample case with σ assumed known.

Step 1. Compute the sample mean in cell C2

=AVERAGE(A2:A86)

The confidence function in step 2 uses C1 because the sample standard deviation used to estimate σ appears in cell C1.

Step 2. Compute the margin of error in cell C3

=CONFIDENCE(.05,C1,85)

Step 3. Compute the lower limit of the interval estimate in cell C4

=C2−C3

Step 4. Compute the upper limit of the interval estimate in cell C5

=C2+C3

The output with the sample standard deviation appearing in cell C1 will be similar to the output shown in Figure 8.10.

Small-Sample Case With σ Assumed Known

If the assumption that the population has a normal distribution is appropriate, the small-sample case with σ assumed known uses the same step-by-step procedure previously described under the large-sample case with σ assumed known.

Small-Sample Case With σ Estimated by s

Training

The data in Table 8.4 show the training program times for a sample of 15 employees. The data have been entered into an Excel worksheet with the label Time in cell A1 and the training program times in days in cells A2:A16. In this case the population standard deviation σ is unknown and will be estimated by the sample standard deviation s. The following steps can be used to compute a 95% confidence interval estimate of the population mean.

Step 1. Select the **Tools** pull-down menu
Step 2. Choose **Data Analysis**
Step 3. When the Data Analysis dialog box appears:
  Choose **Descriptive Statistics**
  Click **OK**
Step 4. When the Descriptive Statistics dialog box appears:
  Enter A1:A16 in the **Input Range** box
  Select **Labels in First Row**
  Select **Summary Statistics**
  Select **Confidence Level for Mean** and enter 95 in the box
  Select **Output Range** and enter B1 in the box
  Click **OK**

The sample mean, 53.87, appears in cell C3 and the margin of error, 3.78, appears in cell C16. Note that the Excel label for the margin of error is Confidence Level (95%). The confidence interval can be obtained by subtracting and then adding the margin of error to the sample mean. The formula =C3−C16 can be used to place the lower limit in cell C17 and the formula =C3+C16 can be used to place the upper limit in cell C18.

HYPOTHESIS TESTING

Chapter 9

CONTENTS

STATISTICS IN PRACTICE

HARRIS CORPORATION*
Melbourne, Florida

Harris Corporation is a communications equipment company that provides products and services for worldwide markets including wireless, broadcast, government, and network support systems. The company has sales and service facilities in more than 80 countries. Many of the Harris products require medium- to high-volume production operations, including printed circuit assembly, final product assembly, and testing.

One of the company's high-volume products has an assembly called an RF deck. Each RF deck consists of 16 electronic components soldered to a machined casting that forms the plated surface of the deck. During a manufacturing run, a problem developed in the soldering process; the flow of solder onto the deck did not meet the quality criteria established for the product. After considering a variety of factors that might affect the soldering process, an engineer made the preliminary determination that the soldering problem was most likely due to defective platings.

The engineer wondered whether the proportion of defective platings in the Harris inventory exceeded that set by the supplier's design specifications. With p indicating the proportion of defective platings in the Harris inventory and p_0 indicating the proportion of defective platings set by the supplier's design specifications, the following hypotheses were formulated.

$$H_0: p \leq p_0$$
$$H_a: p > p_0$$

H_0 indicates that the Harris inventory has a defective plating proportion less than or equal to that set by the design specifications. Such a proportion would be judged acceptable, and the engineer would need to look for other causes of the soldering problem. However, H_a indicates that the Harris inventory has a defective plating proportion greater than that set by the design specifications. In that case, excessive defective platings may well be the cause of the soldering problem and action should be taken to determine why the defective proportion in inventory is so high.

Harris Corporation used hypothesis testing to solve a problem in the soldering process for circuit boards.

Tests made on a sample of platings from the Harris inventory resulted in the rejection of H_0. The conclusion was that H_a was true and that the proportion of defective platings in inventory exceeded that set by the supplier's design specifications. Further investigation of the inventory area led to the conclusion that the underlying problem was shelf contamination during storage. By altering the storage environment, the engineer was able to solve the problem.

In this chapter you will learn how to formulate hypotheses about a population mean and a population proportion. Through the analysis of sample data, you will be able to determine whether a hypothesis should or should not be rejected. Appropriate conclusions and actions will be demonstrated for testing research hypotheses, testing the validity of a claim, and decision making.

*The authors are indebted to Richard A. Marshall of the Harris Corporation for providing this Statistics in Practice.

In Chapters 7 and 8 we showed how a sample could be used to develop point and interval estimates of population parameters. In this chapter we continue the discussion of statistical inference by showing how hypothesis testing can be used to determine whether a statement about the value of a population parameter should or should not be rejected.

In hypothesis testing we begin by making a tentative assumption about a population parameter. This tentative assumption is called the **null hypothesis** and is denoted by H_0. We then define another hypothesis, called the **alternative hypothesis,** which is the opposite of what is stated in the null hypothesis. The alternative hypothesis is denoted by H_a. The hypothesis testing procedure involves using data from a sample to test the two competing statements indicated by H_0 and H_a.

The purpose of this chapter is to show how hypothesis tests can be conducted about a population mean and a population proportion. We begin by providing examples that illustrate approaches to developing null and alternative hypotheses.

9.1 DEVELOPING NULL AND ALTERNATIVE HYPOTHESES

Learning to formulate hypotheses correctly will take practice. Expect some initial confusion over the proper choice for H_0 and H_a. The examples in this section show a variety of forms for H_0 and H_a depending upon the application.

In some applications it may not be obvious how the null and alternative hypotheses should be formulated. Care must be taken to be sure the hypotheses are structured appropriately and that the hypothesis testing conclusion provides the information the researcher or decision maker wants. Guidelines for establishing the null and alternative hypotheses will be given for three types of situations in which hypothesis testing procedures are commonly employed.

Testing Research Hypotheses

Consider a particular automobile model that currently attains an average fuel efficiency of 24 miles per gallon. A product research group has developed a new fuel injection system specifically designed to increase the miles-per-gallon rating. To evaluate the new system, several will be manufactured, installed in automobiles, and subjected to research-controlled driving tests. Note that the product research group is looking for evidence to conclude that the new system *increases* the mean miles-per-gallon rating. In this case, the research hypothesis is that the new system will provide a mean miles-per-gallon rating exceeding 24; that is, $\mu > 24$. As a general guideline, a research hypothesis should be stated as the *alternative hypothesis.* Hence, the appropriate null and alternative hypotheses for the study are:

The research hypothesis is generally stated as the alternative hypothesis. The conclusion that the research hypothesis is true can be made if the null hypothesis is rejected.

$$H_0: \mu \leq 24$$
$$H_a: \mu > 24$$

If the sample results indicate that H_0 cannot be rejected, researchers cannot conclude that the new fuel injection system is better. Perhaps more research and subsequent testing should be conducted. However, if the sample results indicate that H_0 can be rejected, researchers can make the inference that H_a: $\mu > 24$ is true. With this conclusion, the researchers have the statistical support necessary to state that the new system increases the mean number of miles per gallon. Action to begin production with the new system may be undertaken. In research studies such as these, the null and alternative hypotheses should be formulated so that the rejection of H_0 supports the conclusion and action being sought. The research hypothesis therefore should be expressed as the alternative hypothesis.

Testing the Validity of a Claim

As an illustration of testing the validity of a claim, consider the situation of a manufacturer of soft drinks who states that two-liter containers of its products have an average of at least 67.6 fluid ounces. A sample of two-liter containers will be selected, and the contents will be measured to test the manufacturer's claim. In this type of hypothesis testing situation, we generally assume that the manufacturer's claim is true unless the sample evidence proves otherwise. Using this approach for the soft-drink example, we would state the null and alternative hypotheses as follows.

$$H_0: \mu \geq 67.6$$
$$H_a: \mu < 67.6$$

If the sample results indicate H_0 cannot be rejected, the manufacturer's claim cannot be challenged. However, if the sample results indicate H_0 can be rejected, the inference will be made that H_a: $\mu < 67.6$ is true. With this conclusion, statistical evidence indicates that the manufacturer's claim is incorrect and that the soft drink containers are being filled with a mean less than the claimed 67.6 ounces. Appropriate action against the manufacturer may be considered.

A manufacturer's claim is usually given the benefit of the doubt and stated as the null hypothesis. The conclusion that the claim is false can be made if the null hypothesis is rejected.

In any situation that involves testing the validity of a claim, the null hypothesis is generally based on the assumption that the claim is true. The alternative hypothesis is then formulated so that rejection of H_0 will provide statistical evidence that the stated assumption is incorrect. Action to correct the claim should be considered whenever H_0 is rejected.

Testing in Decision-Making Situations

In testing research hypotheses or testing the validity of a claim, action is taken if H_0 is rejected. In many instances, however, action must be taken both when H_0 cannot be rejected and when H_0 can be rejected. In general, this type of situation occurs when a decision maker must choose between two courses of action, one associated with the null hypothesis and another associated with the alternative hypothesis. For example, on the basis of a sample of parts from a shipment that has just been received, a quality control inspector must decide whether to accept the shipment or to return the shipment to the supplier because it does not meet specifications. Assume that specifications for a particular part require a mean length of two inches per part. If the mean length is greater or less than the two-inch standard, the parts will cause quality problems in the assembly operation. In this case, the null and alternative hypotheses would be formulated as follows.

$$H_0\colon \mu = 2$$
$$H_a\colon \mu \neq 2$$

If the sample results indicate H_0 cannot be rejected, the quality control inspector will have no reason to doubt that the shipment meets specifications, and the shipment will be accepted. However, if the sample results indicate H_0 should be rejected, the conclusion will be that the parts do not meet specifications. In this case, the quality control inspector will have sufficient evidence to return the shipment to the supplier. Thus, we see that for these types of situations, action is taken both when H_0 cannot be rejected and when H_0 can be rejected.

Summary of Forms for Null and Alternative Hypotheses

Let μ_0 denote the specific numerical value being considered in the null and alternative hypotheses. In general, a hypothesis test about the values of a population mean μ take one of the following three forms.

The three possible forms of hypotheses H_0 and H_a are shown here. Note that the equality always *appears in the null hypothesis H_0. Focusing on what you want to prove and placing it in the alternative hypothesis H_a will help select the proper form of the hypotheses for a particular application.*

$$H_0\colon \mu \geq \mu_0 \qquad H_0\colon \mu \leq \mu_0 \qquad H_0\colon \mu = \mu_0$$
$$H_a\colon \mu < \mu_0 \qquad H_a\colon \mu > \mu_0 \qquad H_a\colon \mu \neq \mu_0$$

In many situations, the choice of H_0 and H_a is not obvious and judgment is necessary to select the proper form. However, as the preceding forms show, the equality part of the expression (either $\geq$, $\leq$, or $=$) *always* appears in the null hypothesis. In selecting the proper form of H_0 and H_a, keep in mind that the alternative hypothesis is what the test is attempting to establish. Hence, asking whether the user is looking for evidence to support $\mu < \mu_0$, $\mu > \mu_0$, or $\mu \neq \mu_0$ will help determine H_a. The following exercises are designed to provide practice in choosing the proper form for a hypothesis test.

EXERCISES

1. The manager of the Danvers-Hilton Resort Hotel has stated that the mean guest bill for a weekend is \$600 or less. A member of the hotel's accounting staff has noticed that the total charges for guest bills have been increasing in recent months. The accountant will use a sample of weekend guest bills to test the manager's claim.
   a. Which form of the hypotheses should be used to test the manager's claim? Explain.

$$H_0: \mu \geq 600 \qquad H_0: \mu \leq 600 \qquad H_0: \mu = 600$$
$$H_a: \mu < 600 \qquad H_a: \mu > 600 \qquad H_a: \mu \neq 600$$

   b. What conclusion is appropriate when H_0 cannot be rejected?
   c. What conclusion is appropriate when H_0 can be rejected?

2. The manager of an automobile dealership is considering a new bonus plan designed to increase sales volume. Currently, the mean sales volume is 14 automobiles per month. The manager wants to conduct a research study to see whether the new bonus plan increases sales volume. To collect data on the plan, a sample of sales personnel will be allowed to sell under the new bonus plan for a 1-month period.
   a. Develop the null and alternative hypotheses that are most appropriate for this research situation.
   b. Comment on the conclusion when H_0 cannot be rejected.
   c. Comment on the conclusion when H_0 can be rejected.
3. A production line operation is designed to fill cartons with laundry detergent to a mean weight of 32 ounces. A sample of cartons is periodically selected and weighed to determine whether underfilling or overfilling is occurring. If the sample data lead to a conclusion of underfilling or overfilling, the production line will be shut down and adjusted to obtain proper filling.
   a. Formulate the null and alternative hypotheses that will help in deciding whether to shut down and adjust the production line.
   b. Comment on the conclusion and the decision when H_0 cannot be rejected.
   c. Comment on the conclusion and the decision when H_0 can be rejected.
4. Because of high production-changeover time and costs, a director of manufacturing must convince management that a proposed manufacturing method reduces costs before the new method can be implemented. The current production method operates with a mean cost of \$220 per hour. A research study is to be conducted in which the cost of the new method will be measured over a sample production period.
   a. Develop the null and alternative hypotheses that are most appropriate for this study.
   b. Comment on the conclusion when H_0 cannot be rejected.
   c. Comment on the conclusion when H_0 can be rejected.

9.2 TYPE I AND TYPE II ERRORS

The null and alternative hypotheses are competing statements about the population. Either the null hypothesis H_0 is true or the alternative hypothesis H_a is true, but not both. Ideally the hypothesis testing procedure should lead to the acceptance of H_0 when H_0 is true and the rejection of H_0 when H_a is true. Unfortunately, the correct conclusions are not always possible. Because hypothesis tests are based on sample information, we must allow for the possibility of errors. Table 9.1 illustrates the two kinds of errors that can be made in hypothesis testing.

The first row of Table 9.1 shows what can happen when the conclusion is to accept H_0. If H_0 is true, this conclusion is correct. However, if H_a is true, we have made a **Type II error**;

TABLE 9.1 ERRORS AND CORRECT CONCLUSIONS IN HYPOTHESIS TESTING

		Population Condition	
		H_0 True	**H_a True**
Conclusion	**Accept H_0**	Correct Conclusion	Type II Error
	Reject H_0	Type I Error	Correct Conclusion

that is, we have accepted H_0 when it is false. The second row of Table 9.1 shows what can happen when the conclusion is to reject H_0. If H_0 is true, we have made a **Type I error;** that is, we have rejected H_0 when it is true. However, if H_a is true, rejecting H_0 is correct.

Recall the hypothesis testing illustration discussed in Section 9.1 in which an automobile product research group had developed a new fuel injection system designed to increase the miles-per-gallon rating of a particular automobile. With the current model obtaining an average of 24 miles per gallon, the hypothesis test was formulated as follows.

$$H_0: \mu \leq 24$$
$$H_a: \mu > 24$$

The alternative hypothesis, $H_a: \mu > 24$, indicates that the researchers are looking for sample evidence that will support the conclusion that the population mean miles per gallon is greater than 24.

In this application, the Type I error of rejecting H_0 when it is true corresponds to the researchers claiming that the new system improves the miles-per-gallon rating ($\mu > 24$) when in fact the new system is not any better than the current system. In contrast, the Type II error of accepting H_0 when it is false corresponds to the researchers concluding that the new system is not any better than the current system ($\mu \leq 24$) when in fact the new system improves miles-per-gallon performance.

In practice, the person conducting the hypothesis test specifies the maximum allowable probability of making a Type I error, called the **level of significance** for the test. Common choices for the level of significance are .05 and .01. Referring to the second row of Table 9.1, note that the conclusion to *reject* H_0 indicates that either a Type I error or a correct conclusion has been made. Thus, if the probability of making a Type I error is controlled for by selecting a small value for the level of significance, we have a high degree of confidence that the conclusion to reject H_0 is correct. In such cases, we have statistical support for concluding that H_0 is false and H_a is true. Any action suggested by the alternative hypothesis H_a is appropriate.

Although most applications of hypothesis testing control for the probability of making a Type I error, they do not always control for the probability of making a Type II error. Hence, if we decide to accept H_0, we cannot determine how confident we can be with that decision. Because of the uncertainty associated with making a Type II error, statisticians often recommend that we use the statement "do not reject H_0" instead of "accept H_0." Using the statement "do not reject H_0" carries the recommendation to withhold both judgment and action. In effect, by never directly accepting H_0, the statistician avoids the risk of making a Type II error. Whenever the probability of making a Type II error has not been determined and controlled, we will not make the conclusion to accept H_0. In such cases, only two conclusions are possible: *do not reject* H_0 or *reject* H_0.

If the sample data are consistent with the null hypothesis H_0, we will follow the practice of concluding "do not reject H_0." This conclusion is preferred over "accept H_0," because the conclusion to accept H_0 puts us at risk of making a Type II error.

Although controlling for a Type II error in hypothesis testing is not common, it can be done. In fact, in Sections 9.7 and 9.8, we will illustrate procedures for determining and controlling the probability of making a Type II error. If proper controls have been established for this error, action based on the "do not reject H_0" conclusion can be appropriate.

NOTES AND COMMENTS

Many applications of hypothesis testing have a decision-making goal. The conclusion *reject* H_0 provides the statistical support to conclude that H_a is true and take whatever action is appropriate. The statement "do not reject H_0", although inconclusive, often forces managers to behave as though H_0 is true. In this case, managers need to be aware of the fact that such action may be the result of a Type II error.

EXERCISES

5. Americans spend an average of 8.6 minutes per day reading newspapers (*USA Today*, April 10, 1995). A researcher believes that individuals in management positions spend more than the national average time per day reading newspapers. A sample of individuals in management positions will be selected by the researcher. Data on newspaper-reading times will be used to test the following null and alternative hypotheses.

$$H_0: \mu \leq 8.6$$
$$H_a: \mu > 8.6$$

a. What is the Type I error in this situation? What are the consequences of making this error?
b. What is the Type II error in this situation? What are the consequences of making this error?

6. The label on a three-quart container of orange juice claims that the orange juice contains an average of one gram of fat or less. Answer the following questions for a hypothesis test that could be used to test the claim on the label.
a. Develop the appropriate null and alternative hypotheses.
b. What is the Type I error in this situation? What are the consequences of making this error?
c. What is the Type II error in this situation? What are the consequences of making this error?

7. Carpetland salespersons have had sales averaging \$8000 per week. Steve Contois, the firm's vice president, has proposed a compensation plan with new selling incentives. Steve hopes that the results of a trial selling period will enable him to conclude that the compensation plan increases the average sales per salesperson.
a. Develop the appropriate null and alternative hypotheses.
b. What is the Type I error in this situation? What are the consequences of making this error?
c. What is the Type II error in this situation? What are the consequences of making this error?

8. Suppose a new production method will be implemented if a hypothesis test supports the conclusion that the new method reduces the mean operating cost per hour.
a. State the appropriate null and alternative hypotheses if the mean cost for the current production method is \$220 per hour.
b. What is the Type I error in this situation? What are the consequences of making this error?
c. What is the Type II error in this situation? What are the consequences of making this error?

9.3 ONE-TAILED TESTS ABOUT A POPULATION MEAN: LARGE-SAMPLE CASE

The Federal Trade Commission (FTC) periodically conducts studies designed to test the claims manufacturers make about their products. For example, the label on a large can of Hilltop Coffee states that the can contains three pounds of coffee. Suppose we want to check this claim by using hypothesis testing.

The first step of hypothesis testing is to develop the appropriate null and alternative hypotheses. In the Hilltop Coffee example, the FTC is concerned about detecting underfilled products where customers are receiving less than the manufacturer claims. Thus, if cans of Hilltop Coffee have a population mean weight of 3 *or more* pounds, the label is correct. However, if cans of Hilltop Coffee have a population mean weight of *less than* 3 pounds, the label is incorrect. In this case, a FTC charge of underfilling is warranted and appropriate follow-up action against Hilltop is justified.

We begin by tentatively assuming that Hilltop's label claim is true. With μ denoting the population mean weight per can, the null and alternative hypotheses are as follows:

$$H_0\colon \mu \geq 3$$
$$H_a\colon \mu < 3$$

Thus, if sample data indicate that H_0 cannot be rejected, no action will be taken against Hilltop. However, if sample data indicate that H_0 can be rejected, the FTC will have statistical evidence to conclude that the alternative hypothesis, $H_a\colon \mu < 3$, is true. In this case, a FTC charge of underfilling would be appropriate.

Suppose that a random sample of 36 coffee cans will be selected and that the sample mean weight per can will be used to conduct the hypothesis test. A sample mean that is less than 3 pounds provides support for the conclusion that Hilltop's label claim is not true. However, how much less than 3 pounds would the sample mean have to be before the FTC would be willing to reject H_0 and charge Hilltop with a label violation?

To answer this question, let us tentatively assume that the null hypothesis is true with $\mu = 3$. In addition, data from earlier tests with Hilltop coffee cans show that the population standard deviation may be assumed known with $\sigma = .18$ pounds. Using the sampling distribution of $\bar{x}$ as discussed in Chapters 7 and 8, the sampling distribution of $\bar{x}$ for the Hilltop study can be approximated by a normal probability distribution. Figure 9.1 shows the sampling distribution of $\bar{x}$ when the null hypothesis is true at $\mu = 3$. We will now show how the value of the sample mean and the sampling distribution in Figure 9.1 can be used to test the hypotheses $H_0\colon \mu \geq 3$ and $H_a\colon \mu < 3$.

The methodology of hypothesis testing requires that we specify the maximum allowable probability of making a Type I error; this probability is called the **level of significance**

FIGURE 9.1 SAMPLING DISTRIBUTION OF $\bar{x}$ FOR THE HILLTOP COFFEE STUDY WHEN THE NULL HYPOTHESIS IS TRUE ($\mu = 3$)

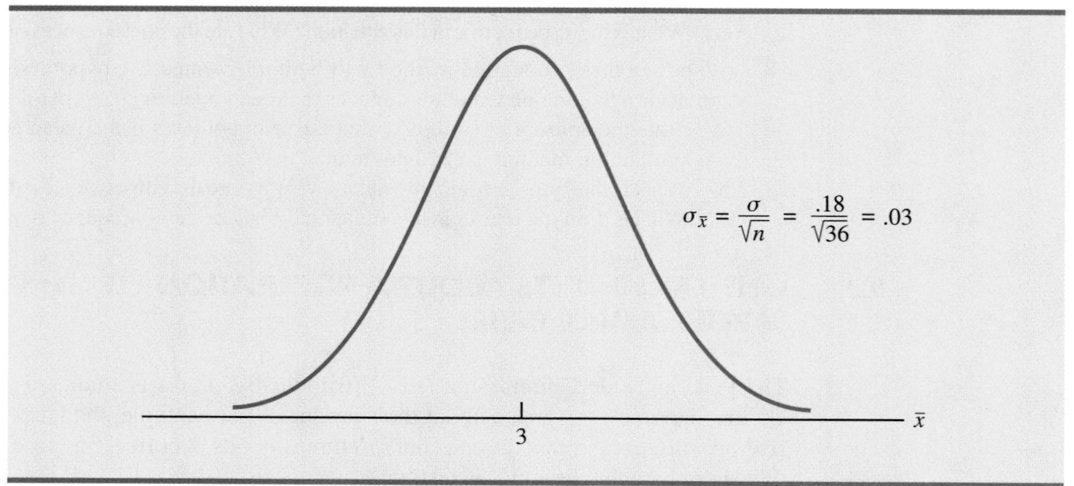

for the test. The level of significance is denoted by α and represents the probability of making a Type I error when the null hypothesis is true as an equality. In the Hilltop Coffee study, the director of the FTC testing program made the following statement: "If Hilltop is meeting its weight specifications exactly ($\mu = 3$), I would like a 99% chance of not taking any action against the company. Even though I do not want to accuse Hilltop of wrongly of underfilling its product, I am willing to have a 1% chance of making this error." Thus, the FTC director is specifying $\alpha = .01$ as the maximum allowable probability of making a Type I error. Figure 9.2 shows the sampling distribution of $\bar{x}$ and the **rejection region** corresponding to a level of significance of $\alpha = .01$. Note that when H_0 is true with $\mu = 3$, there is a .01 probability of $\bar{x}$ being in the rejection region; thus, there is a .01 probability of making a Type I error.

The question of whether an observed sample mean $\bar{x}$ is in the rejection region can be answered by computing a **test statistic.** In the large-sample case with the population standard deviation σ assumed known, the test statistic is given by

$$z = \frac{\bar{x} - \mu_0}{\sigma/\sqrt{n}} \tag{9.1}$$

In this expression, μ_0 is the value of the population mean appearing in H_0 and n is the sample size. The sampling distribution of the test statistic z is a standard normal probability distribution; thus, the computed value of z can be used to determine whether the sample mean $\bar{x}$ is in the rejection region. In the Hilltop coffee example, $\mu_0 = 3$, $\sigma = .18$ and $n = 36$. Thus, once the sample mean $\bar{x}$ is identified, the value of the test statistic is computed as follows.

$$z = \frac{\bar{x} - \mu_0}{\sigma/\sqrt{n}} = \frac{\bar{x} - 3}{.18/\sqrt{36}}$$

We are now ready to proceed by using the sample results to draw the hypothesis testing conclusion. We will show that two methods can be used: one based on the observed value of the test statistic and one based on the p-value criterion. We begin with the test statistic method.

FIGURE 9.2 REJECTION REGION FOR HILLTOP COFFEE WITH $\alpha = .01$ LEVEL OF SIGNIFICANCE

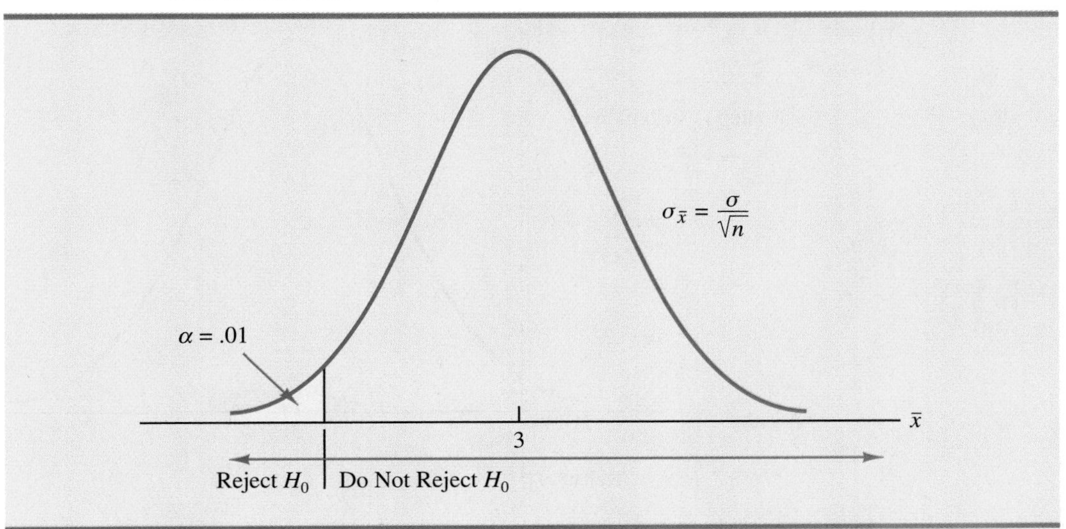

Using the Test Statistic

The sampling distribution of $\bar{x}$ and the sampling distribution of the test statistic z are shown in Figure 9.3. Each value of $\bar{x}$ has a corresponding z-value. Previously, we stated that the Hilltop Coffee hypothesis test would be made at the $\alpha = .01$ level of significance. Thus the rejection region for H_0 is as shown in Figure 9.3.

Because the test statistic z has a standard normal probability distribution, we can use the standard normal probability distribution table to find the z-value with an area of $\alpha = .01$ in the lower tail of the distribution. Using the table, we find that this z-value is $z = -2.33$, which is located as shown in Figure 9.3. The value of the test statistic that defines the rejection region is called the **critical value** for the test. Using the critical value, $z = -2.33$, we can develop the following rejection rule for the Hilltop hypothesis test:

$$\text{Reject } H_0 \text{ if } z < -2.33$$

The rejection rule tells us that if the sample mean $\bar{x}$ provides a value of the test statistic z that is less than -2.33, we reject the null hypothesis H_0: $\mu \geq 3$ and conclude that the alter-

FIGURE 9.3 SAMPLING DISTRIBUTION OF $\bar{x}$ AND DISTRIBUTION OF z FOR HILLTOP COFFEE WITH $\alpha = .01$

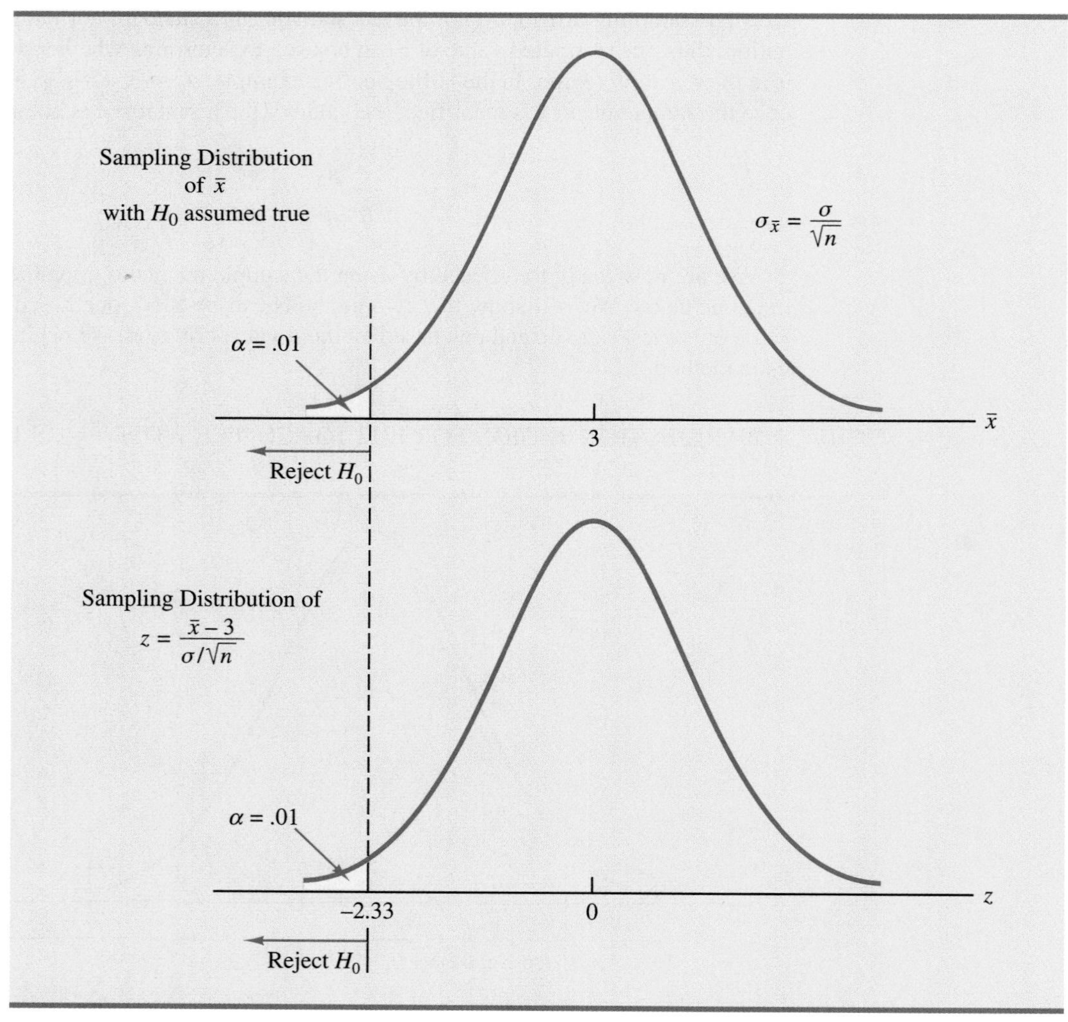

native hypothesis H_a: $\mu < 3$ is true. However, if the sample mean $\bar{x}$ provides a value of the test statistic z that is not in the rejection region, $z < -2.33$, we cannot reject H_0: $\mu \geq 3$. Note that the rejection region in Figure 9.3 is in one-tail of the sampling distribution. In such cases, we say the test is a **one-tailed** test.

Suppose the sample of 36 cans provides a sample mean of $\bar{x} = 2.92$ pounds. Is $\bar{x} = 2.92$ in the rejection region? To answer this question, we compute the test statistic z.

$$z = \frac{\bar{x} - \mu_0}{\sigma/\sqrt{n}} = \frac{2.92 - 3}{.18/\sqrt{36}} = -2.67$$

Comparing $z = -2.67$ to the critical value for the test, $z = -2.33$, we see that $z = -2.67$ is less than -2.33. Thus, $z = -2.67$ is in the rejection region (see Figure 9.4) and hence we reject H_0: $\mu \geq 3$. In this case, the FTC is justified in concluding that H_a: $\mu < 3$ is true, and FTC action against Hilltop for underfilling its product is warranted.

Using the p-Value

Another approach that can be used to draw a hypothesis testing conclusion is based on a probability called a **p-value.** Assuming the null hypothesis is true, the p-value is the probability of obtaining a sample result that is at least as unlikely as what is observed. In the Hilltop Coffee example, the rejection region is in the lower tail; therefore the p-value is the probability of obtaining a value for the sample mean that is less than or equal to the value of the observed sample mean. The p-value is also called the *observed level of significance.*

The p-value for a one-tailed test is always the area from the test statistic toward the rejection tail of the distribution.

In the Hilltop Coffee example, the sample mean $\bar{x} = 2.92$ pounds provided a value for the test statistic of $z = -2.67$. Because the rejection region is in the lower tail of the distribution, the p-value is the probability of z being less than -2.67. Using the standard normal probability distribution table, we find that the area between the mean and $z = -2.67$ is .4962. Thus the p-value is $.5000 - .4962 = .0038$, which is shown in Figure 9.5. This p-value indicates a small probability of obtaining a sample mean as small as $\bar{x} = 2.92$ when H_0 is true.

FIGURE 9.4 VALUE OF THE TEST STATISTIC FOR $\bar{x} = 2.92$ IS IN THE REJECTION REGION

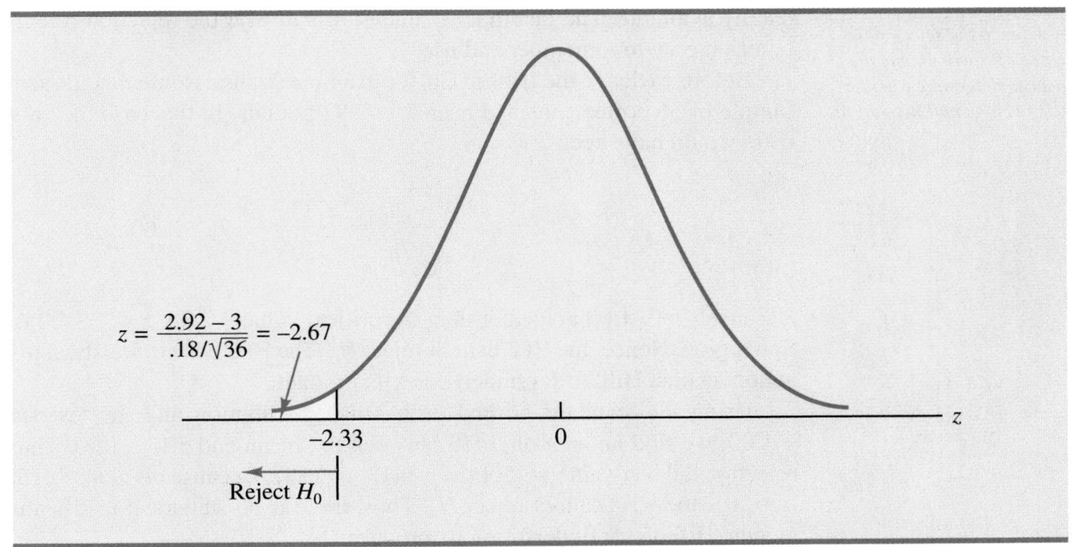

FIGURE 9.5 p-VALUE FOR THE HILLTOP COFFEE STUDY WHEN $\bar{x} = 2.92$ AND $z = -2.67$

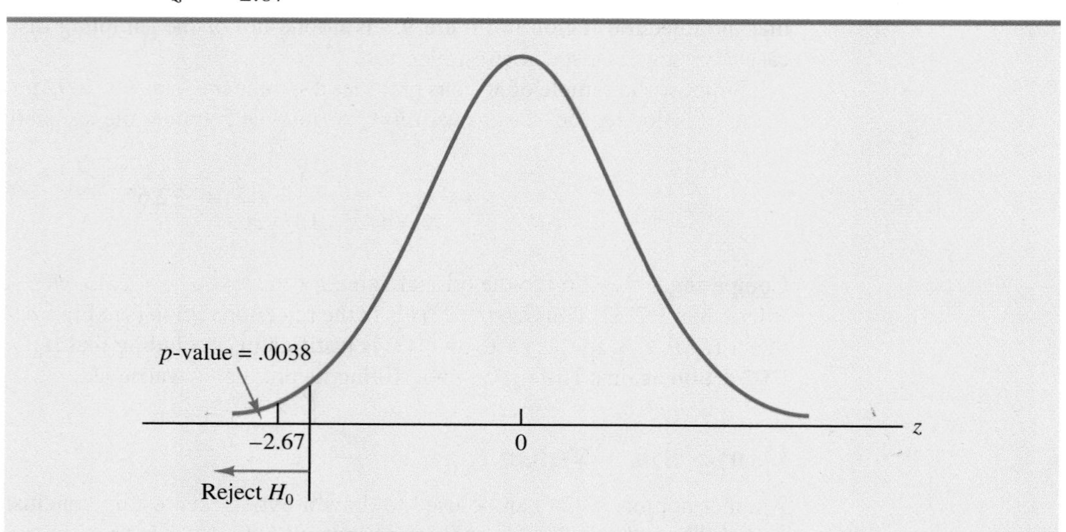

The rejection rule using the p-value follows.

Regardless of the form of the hypothesis test, always reject H_0 if p-value $< \alpha$.

$$\text{Reject } H_0 \text{ if } p\text{-value} < \alpha$$

Sometimes a researcher may simply report the p-value for a particular sample result. The user of the research findings can select any level of significance, α, he or she is willing to tolerate. Comparing the reported p-value to the selected value of α will enable the user to determine whether to reject the null hypothesis.

Because the p-value is the area in the rejection tail, if p-value $< \alpha$, the sample result must be in the rejection region. For the Hilltop Coffee example, the p-value $= .0038$ is less than $\alpha = .01$; thus, we can reject the null hypothesis H_0: $\mu \geq 3$. The FTC is justified in concluding that H_a: $\mu < 3$ is true. FTC action against Hilltop for underfilling its product is warranted.

Note that the test statistic method and the p-value method will always provide the same hypothesis testing conclusion. Choice of one method over the other is largely personal preference. Either or both methods may be used. Computer printouts often make the p-value readily available. The fact that p-value $< \alpha$ is *always* the rejection rule makes the p-value criterion easy to remember and use.

Before we leave the Hilltop Coffee example, let us assume that the sample mean for the sample of 36 coffee cans had been $\bar{x} = 2.97$ pounds. In this case the value of the test statistic would have been

$$z = \frac{\bar{x} - \mu_0}{\sigma/\sqrt{n}} = \frac{2.97 - 3}{.18/\sqrt{36}} = -1.00$$

Because $z = -1.00$ is greater than the critical value -2.33, $z = -1.00$ is not in the rejection region. Hence, the FTC cannot reject H_0. The FTC has no statistical justification to take action against Hilltop for underfilling its product.

Using the standard normal probability distribution and the test statistic value $z = -1.00$, we find an area of .3413 between the mean and $z = -1.00$. Thus, the area in the rejection tail is p-value $= .5000 - .3413 = .1587$. Because p-value $= .1587$ is greater than $\alpha = .01$, the FTC cannot reject H_0. Thus, there is no statistical justification to take action against Hilltop for underfilling its product.

Summary: One-Tailed Tests About a Population Mean

Let us generalize the hypothesis testing procedure for one-tailed tests about a population mean. We consider the large-sample case ($n \geq 30$) in which the central limit theorem enables us to assume that the sampling distribution of $\bar{x}$ can be approximated by a normal probability distribution. In the large-sample case, we may either use historical data as a basis for assuming the population standard deviation σ is known or estimate σ by using the sample standard deviation s. The general form of a lower-tail test, where μ_0 is the stated value for the population mean, follows.

In many applications the sample standard deviation s is used in the computation of the test statistic because the population standard deviation σ is unknown.

Large-Sample ($n \geq 30$) Hypothesis Test About a Population Mean for a One-Tailed Test of the Form

$$H_0: \mu \geq \mu_0$$
$$H_a: \mu < \mu_0$$

Test Statistic: σ Assumed Known

$$z = \frac{\bar{x} - \mu_0}{\sigma/\sqrt{n}} \tag{9.1}$$

Test Statistic: σ Estimated by s

$$z = \frac{\bar{x} - \mu_0}{s/\sqrt{n}} \tag{9.2}$$

Rejection Rule

Using test statistic: Reject H_0 if $z < -z_\alpha$
Using p-value: Reject H_0 if p-value $< \alpha$

A second form of the one-tailed test rejects the null hypothesis when the test statistic is in the upper tail of the sampling distribution. This one-tailed test and rejection rule are summarized next (see Figure 9.6). Again, we are considering the large-sample case.

Large-Sample ($n \geq 30$) Hypothesis Test About a Population Mean for a One-Tailed Test of the Form

$$H_0: \mu \leq \mu_0$$
$$H_a: \mu > \mu_0$$

Test Statistic: σ Assumed Known

$$z = \frac{\bar{x} - \mu_0}{\sigma/\sqrt{n}} \tag{9.1}$$

Test Statistic: σ Estimated by s

$$z = \frac{\bar{x} - \mu_0}{s/\sqrt{n}} \tag{9.2}$$

Rejection Rule

Using test statistic: Reject H_0 if $z > z_\alpha$
Using p-value: Reject H_0 if p-value $< \alpha$

FIGURE 9.6 REJECTION REGION FOR AN UPPER-TAIL HYPOTHESIS TEST ABOUT A POPULATION MEAN

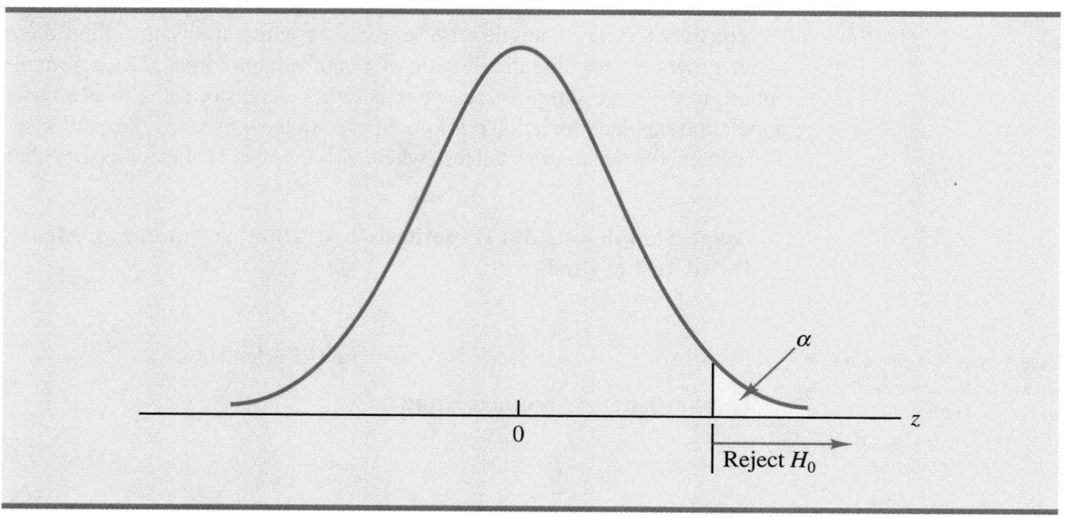

Steps of Hypothesis Testing

In conducting the Hilltop Coffee hypothesis test, we carried out the steps that can be used for any hypothesis test. A summary of the steps follows. Note that either the test statistic or the p-value may be used to draw the hypothesis testing conclusion.

Steps of Hypothesis Testing

1. Develop the null and alternative hypotheses.
2. Specify the level of significance α.
3. Select the test statistic that will be used to test the hypothesis.

Using the Test Statistic

4. Use the level of significance to determine the critical value for the test statistic and state the rejection rule for H_0.
5. Collect the sample data and compute the value of the test statistic.
6. Use the value of the test statistic and the rejection rule to determine whether to reject H_0.

Using the p-Value

4. Collect the sample data and compute the value of the test statistic.
5. Use the value of the test statistic to compute the p-value.
6. Reject H_0 if p-value $< \alpha$.

NOTES AND COMMENTS

The p-value is a measure of the likelihood of the sample results when the null hypothesis is assumed to be true. The smaller the p-value, the less likely it is that the sample results came from a situation where the null hypothesis is true. Most statistical software packages provide the p-value associated with a hypothesis test. The user can then compare the p-value to α and draw a hypothesis test conclusion without referring to a statistical table.

EXERCISES

Note to Student: Some of the hypothesis testing exercises that follow will ask you to use the test statistic and other exercises will ask you to use the *p*-value. Both methods will provide the same hypothesis testing conclusion. We have provided exercises with both methods so that you will have practice using both. In an actual application, you may select either based on personal preference.

Methods

9. Consider the following hypothesis test.

$$H_0: \mu \geq 10$$
$$H_a: \mu < 10$$

A sample of 50 provided a sample mean of 9.46 and sample standard deviation of 2.

a. At $\alpha = .05$, what is the critical value for z? What is the rejection rule?

b. Compute the value of the test statistic z. What is your conclusion?

SELF test

10. Consider the following hypothesis test.

$$H_0: \mu \leq 15$$
$$H_a: \mu > 15$$

A sample of 40 provided a sample mean of 16.5 and sample standard deviation of 7.

a. At $\alpha = .02$, what is the critical value for z, and what is the rejection rule?

b. Compute the value of the test statistic z.

c. What is the *p*-value?

d. What is your conclusion?

11. Consider the following hypothesis test.

$$H_0: \mu \geq 25$$
$$H_a: \mu < 25$$

A sample of 100 is used and the population standard deviation is assumed to be 12. Use $\alpha = .05$. Provide the value of the test statistic z and your conclusion for each of the following sample results.

a. $\bar{x} = 22.0$ **b.** $\bar{x} = 24.0$ **c.** $\bar{x} = 23.5$ **d.** $\bar{x} = 22.8$

12. Consider the following hypothesis test.

$$H_0: \mu \leq 5$$
$$H_a: \mu > 5$$

Assume the following test statistics. Compute the corresponding *p*-values and make the appropriate conclusions based on $\alpha = .05$.

a. $z = 1.82$ **b.** $z = .45$ **c.** $z = 1.50$ **d.** $z = 3.30$ **e.** $z = -1.00$

Applications

13. Individuals filing federal income tax returns prior to March 31 had an average refund of $1056. Consider the population of "last-minute" filers who mail their returns during the last 5 days of the income tax period (typically April 10 to April 15).

a. A researcher suggests that one of the reasons individuals wait until the last 5 days to file their returns is that on average those individuals have a lower refund than early filers. Develop appropriate hypotheses such that rejection of H_0 will support the researcher's contention.

b. Using $\alpha = .05$, what is the critical value for the test statistic and what is the rejection rule?

c. For a sample of 400 individuals who filed a return between April 10 and April 15, the sample mean refund was \$910 and the sample standard deviation was \$1600. Compute the value of the test statistic.
d. What is your conclusion?
e. What is the p-value for the test?

14. A Nielsen survey provided the estimate that the mean number of hours of television viewing per household is 7.25 hours per day (*New York Daily News,* November 2, 1997). Assume that the Nielsen survey involved 200 households and that the sample standard deviation was 2.5 hours per day. Ten years ago the population mean number of hours of television viewing per household was reported to be 6.70 hours. Letting μ = the population mean number of hours of television viewing per household in 1997, test the hypotheses H_0: $\mu \leq 6.70$ and H_a: $\mu > 6.70$. Use $\alpha = .01$.
   a. What is the critical value of the test statistic and what is the rejection rule?
   b. Compute the value of the test statistic.
   c. What is your conclusion?

15. According to the National Automobile Dealers Association, the mean price for used cars is \$10,192. A manager of a Kansas City used car dealership reviewed a sample of 100 recent used car sales at the dealership. The sample mean price was \$9300 and the sample standard deviation was \$4500. Letting μ denote the population mean price for used cars at the Kansas City dealership, test H_0: $\mu \geq 10{,}192$ and H_a: $\mu < 10{,}192$ at a .05 level of significance.
   a. What is the critical value of the test statistic and what is the rejection rule?
   b. Compute the value of the test statistic.
   c. What is your conclusion?

16. Media Metrix, Inc., tracks Internet users in seven countries: Australia, Britain, Canada, France, Germany, Japan, and the United States. According to recent measurement figures, American home users rank first in Internet usage with a mean of 13 hours per month (*The Washington Post,* August 4, 2000). Assume that in a follow-up study involving a sample of 145 Canadian Internet users, the sample mean was 10.8 hours per month and the sample standard deviation was 9.2 hours.
   a. Formulate the null and alternative hypotheses that can be used to determine whether the sample data support the conclusion that Canadian Internet users have a population mean less than the U.S. mean of 13 hours per month.
   b. Using $\alpha = .01$, what is the critical value for the test statistic? State the rejection rule.
   c. What is the value of the test statistic?
   d. What is your conclusion?

17. Fowle Marketing Research, Inc., bases charges to a client on the assumption that telephone surveys can be completed in a mean time of 15 minutes or less. If a longer mean survey time is necessary, a premium rate is charged. Suppose a sample of 35 surveys shows a sample mean of 17 minutes and a sample standard deviation of 4 minutes. Is the premium rate justified?
   a. Formulate the null and alternative hypotheses for this application.
   b. Compute the value of the test statistic.
   c. What is the p-value?
   d. Using $\alpha = .01$, what is your conclusion?

18. A Channel One Network study reported that teens spend a mean of \$5.72 per visit to fast-food restaurants such as McDonald's, Burger King, and Wendy's (*USA Today,* October 5, 1998). In a follow-up study, a sample of 102 teen visits to fast-food restaurants in Chicago found a sample mean of \$5.98 and a sample standard deviation of \$1.24.
   a. Formulate the null and alternative hypotheses that can be used to determine whether the sample data support the conclusion that teens in Chicago have a population mean expenditure of more than \$5.72 per visit to fast-food restaurants.

b. What is the value of the test statistic?
c. What is the p-value?
d. Using $\alpha = .05$, what is your conclusion?

19. The national mean sales price for new one-family homes is \$181,900 (*The New York Times Almanac 2000*). A sample of 40 one-family home sales in the South showed a sample mean of \$166,400 and a sample standard deviation of \$33,500.

a. Formulate the null and alternative hypotheses that can be used to determine whether the sample data support the conclusion that the population mean sales price for new one-family homes in the South is less than the national mean of \$181,900.
b. What is the value of the test statistic?
c. What is the p-value?
d. Using $\alpha = .01$, what is your conclusion?

20. According to the National Association of Colleges and Employers, the 2000 mean annual salary of business degree graduates in accounting was \$37,000 (*Time,* May 8, 2000). In a follow-up study in June 2001, a sample of 48 graduating accounting majors found a sample mean of \$38,100 and a sample standard deviation of \$5,200.

a. Formulate the null and alternative hypotheses that can be used to determine whether the sample data support the conclusion that June 2001 graduates in accounting have a mean salary greater than the 2000 mean annual salary of \$37,000.
b. What is the value of the test statistic?
c. What is the p-value?
d. Using $\alpha = .05$, what is your conclusion?

9.4 TWO-TAILED TESTS ABOUT A POPULATION MEAN: LARGE-SAMPLE CASE

Two-tailed tests differ from one-tailed hypothesis tests in that two-tailed tests have rejection regions in both the lower and the upper tails of the sampling distribution. Let us introduce an example to show how and why two-tailed tests are conducted.

The United States Golf Association (USGA) has established rules that manufacturers of golf equipment must meet if their products are to be acceptable for use in USGA events. One of the rules for the manufacture of golf balls states: "A brand of golf ball, when tested on apparatus approved by the USGA on the outdoor range at the USGA Headquarters . . . shall not cover an average distance in carry and roll exceeding 280 yards. . . ." Suppose Superflight, Inc., has recently developed a high-technology manufacturing method that can produce golf balls having an average distance in carry and roll of 280 yards.

Superflight realizes, however, that if the new manufacturing process goes out of adjustment, the process may produce golf balls for which the average distance is either less than 280 yards or greater than 280 yards. In the first case, sales may decline as a result of marketing an inferior product, and in the second case, the golf balls may be rejected by the USGA for exceeding the 280-yard distance rule.

Superflight's management has instituted a quality control program to monitor the manufacturing process. Periodically, a hypothesis test will be used to help determine whether the manufacturing process is operating incorrectly. We will begin by assuming that the manufacturing process is functioning correctly, with the golf balls having the desired mean distance of 280 yards. The null and alternative hypotheses are as follow:

$$H_0: \mu = 280$$
$$H_a: \mu \neq 280$$

If the sample data indicate that H_0 cannot be rejected, no action will be taken to adjust the manufacturing process. However, if the sample data indicate that H_0 can be rejected, we can conclude that the golf balls do not have the desired mean distance of 280 yards; in this case corrective action should be taken to return the manufacturing process to its desired state.

The quality control team has selected $\alpha = .05$ as the level of significance for the hypothesis test. With a sample size of 36 golf balls used each time the quality control test is performed, we have a large-sample case. As a result the test statistic is

The test statistic is the same as used in Section 9.3.

$$z = \frac{\bar{x} - \mu_0}{\sigma/\sqrt{n}}$$

As usual, we make the tentative assumption that the null hypothesis is true. We want to reject H_0: $\mu = 280$ when the test statistic z indicates that the sample mean distance $\bar{x}$ is significantly less than 280 yards or is significantly greater than 280 yards. Thus, H_0 may be rejected for values of z in either the lower tail or upper tail of the sampling distribution. Therefore, the hypothesis test is called a **two-tailed** test.

Figure 9.7 shows the sampling distribution of z with the two-tailed rejection region for $\alpha = .05$. With two-tailed hypothesis tests, we will always determine the rejection region by placing an area or probability of $\alpha/2$ in each tail of the distribution. The values of z that provide an area of .025 in each tail can be found from the standard normal probability distribution table. We see in Figure 9.7 that $-z_{.025} = -1.96$ identifies an area of .025 in the lower tail and $z_{.025} = +1.96$ identifies an area of .025 in the upper tail. Thus, $z = -1.96$ and $z = 1.96$ are the critical values for the test and the rejection rule is as follows.

$$\text{Reject } H_0 \text{ if } z < -1.96 \text{ or if } z > 1.96$$

Suppose that a simple random sample of 36 golf balls provided the data in Table 9.2. For these data we obtained a sample mean of $\bar{x} = 278.5$ yards and a sample standard deviation of $s = 12$ yards. With the value $\mu_0 = 280$ from the null hypothesis and the sample

FIGURE 9.7 REJECTION REGION FOR THE TWO-TAILED HYPOTHESIS TEST FOR SUPERFLIGHT, INC.

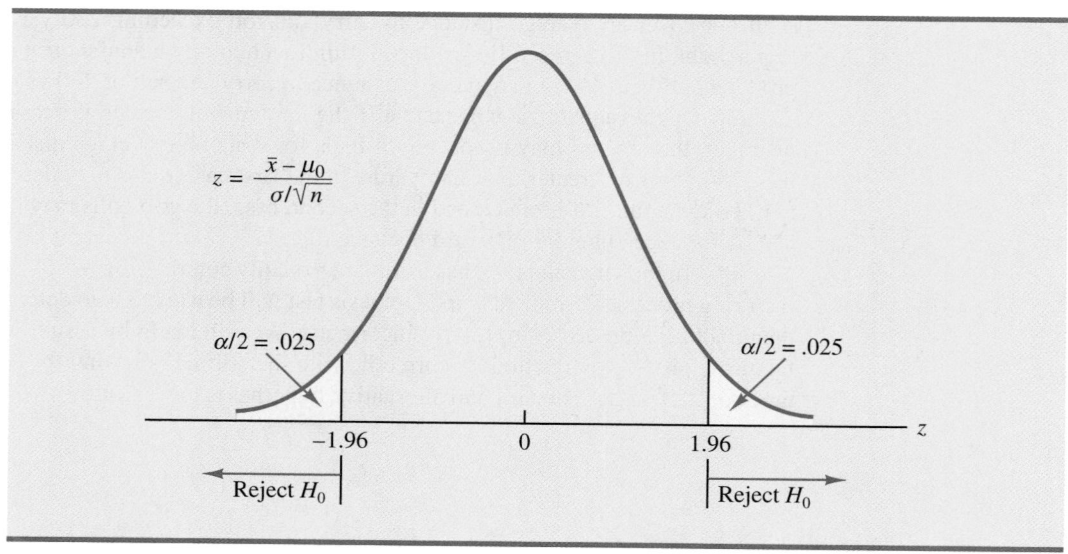

TABLE 9.2 DISTANCE DATA FOR A SIMPLE RANDOM SAMPLE OF 36 SUPERFLIGHT GOLF BALLS

CD file

Distance

Ball	Yards	Ball	Yards	Ball	Yards
1	269	13	296	25	272
2	300	14	265	26	285
3	268	15	271	27	293
4	278	16	279	28	281
5	282	17	284	29	269
6	263	18	260	30	299
7	301	19	275	31	263
8	295	20	282	32	264
9	288	21	260	33	273
10	278	22	266	34	291
11	276	23	270	35	274
12	286	24	293	36	277

standard deviation $s = 12$ used to estimate the population standard deviation σ, the value of the test statistic is

$$z = \frac{\bar{x} - \mu_0}{s/\sqrt{n}} = \frac{278.5 - 280}{12/\sqrt{36}} = -.75$$

According to the rejection rule, H_0 cannot be rejected. The sample results indicate no reason to doubt the assumption that the manufacturing process is producing golf balls with a mean distance of 280 yards.

p-Values for Two-Tailed Tests

Let us compute the p-value for the Superflight golf ball example. The sample mean of $\bar{x} = 278.5$ has a corresponding $z = -.75$. The table for the standard normal probability distribution shows that the area between the mean and $z = -.75$ is .2734. Thus, the area in the lower tail is $.5000 - .2734 = .2266$. Looking at Figure 9.7, we see that the lower-tail portion of the rejection region has an area or probability of $\alpha/2 = .05/2 = .025$. Thus, with $.2266 > .025$, the test statistic is not in the rejection region, and the null hypothesis cannot be rejected.

In a two-tailed test, the p-value is found by doubling the area in the tail. The doubling of the area is done so that the p-value can be compared directly to α. As a result, the same rejection rule can be used for both one-tailed and two-tailed tests.

One question remains: What value should we report as the p-value for the two-tailed test? At first glance, you may be inclined to say the p-value is .2266. If that is your choice, you will have to remember two different rules: one for the one-tailed test, which is to reject H_0 if p-value $< \alpha$, and another for the two-tailed test, which is to reject H_0 if p-value $< \alpha/2$. Alternatively, suppose we define the p-value for a two-tailed test as *double* the area found in the tail of the distribution. Thus, for the Superflight example, we would define the p-value to be $2(.2266) = .4532$. The advantage of this definition of the p-value for a two-tailed test is that the p-value can be compared directly to the level of significance α. Hence, with $.4532 > .05$, we see that the null hypothesis cannot be rejected. By remembering that the p-value for a two-tailed test is simply double the area found in the tail of the distribution, the previous rule to reject H_0 if p-value $< \alpha$ can be used for all hypothesis tests.

Minitab can be used to provide hypothesis test information for a population mean. The computer output for the Superflight hypothesis test is shown in Figure 9.8. The sample mean $\bar{x} = 278.5$, the test statistic $z = -.75$ and the p-value $= .453$ are computed automatically. The step-by-step procedure used to obtain the Minitab output is described in

FIGURE 9.8 MINITAB OUTPUT FOR THE SUPERFLIGHT GOLF BALL HYPOTHESIS TEST

```
Test of mu = 280.00 vs mu not = 280.00
Variable     N       Mean    StDev    SE Mean        Z        P
Yards       36     278.50    12.00       2.00    -0.75    0.453
```

Appendix 9.1. Note that the Minitab output will also provide a 95% confidence interval for the population mean.

Summary: Two-Tailed Tests About a Population Mean

Let μ_0 represent the value of the mean in the hypotheses. The general form of the two-tailed hypothesis test about a population mean follows.

Large-Sample ($n \geq 30$) Hypothesis Test About a Population Mean for a Two-Tailed Test of the Form

$$H_0: \mu = \mu_0$$
$$H_a: \mu \neq \mu_0$$

Test Statistic: σ Assumed Known

$$z = \frac{\bar{x} - \mu_0}{\sigma/\sqrt{n}} \tag{9.1}$$

Test Statistic: σ Estimated by s

$$z = \frac{\bar{x} - \mu_0}{s/\sqrt{n}} \tag{9.2}$$

Rejection Rule

Using test statistic: Reject H_0 if $z < -z_{\alpha/2}$ or if $z > z_{\alpha/2}$
Using p-value: Reject H_0 if p-value $< \alpha$

Relationship Between Interval Estimation and Hypothesis Testing

In Chapter 8 we showed how to develop a confidence interval estimate of a population mean. In the large-sample case, the confidence interval estimate of a population mean corresponding to a $1 - \alpha$ confidence coefficient is given by

$$\bar{x} \pm z_{\alpha/2}\frac{\sigma}{\sqrt{n}} \tag{9.3}$$

when σ is assumed known and

$$\bar{x} \pm z_{\alpha/2}\frac{s}{\sqrt{n}} \tag{9.4}$$

when σ is estimated by s.

Conducting a hypothesis test requires us first to make an assumption about the value of a population parameter. In the case of the population mean, the two-tailed hypothesis test has the form

$$H_0: \mu = \mu_0$$
$$H_a: \mu \neq \mu_0$$

where μ_0 is the hypothesized value for the population mean. Using the two-tailed rejection rule, we see that the region over which we do not reject H_0 includes all values of the sample mean $\bar{x}$ that are within $-z_{\alpha/2}$ and $+z_{\alpha/2}$ standard errors of μ_0. Thus, the do-not-reject region for the sample mean $\bar{x}$ in a two-tailed hypothesis test with a level of significance of α is given by

$$\mu_0 \pm z_{\alpha/2} \frac{\sigma}{\sqrt{n}} \tag{9.5}$$

when σ is assumed known and

$$\mu_0 \pm z_{\alpha/2} \frac{s}{\sqrt{n}} \tag{9.6}$$

when σ is estimated by s.

A close look at (9.3) and (9.5) provides insight about the relationship between the estimation and hypothesis testing approaches to statistical inference. Note in particular that both procedures require the computation of the values $z_{\alpha/2}$ and $\sigma/\sqrt{n}$. Focusing on α, we see that a confidence coefficient of $(1 - \alpha)$ for interval estimation corresponds to a level of significance of α in hypothesis testing. For example, a 95% confidence interval corresponds to a .05 level of significance for hypothesis testing. Furthermore, (9.3) and (9.5) show that because $z_{\alpha/2}\,(\sigma/\sqrt{n})$ is the plus or minus value for both expressions, if $\bar{x}$ is in the do-not-reject region defined by (9.5), the hypothesized value μ_0 will be in the confidence interval defined by (9.3). Conversely, if the hypothesized value μ_0 is in the confidence interval defined by (9.3), the sample mean $\bar{x}$ will be in the do-not-reject region for the hypothesis H_0: $\mu = \mu_0$ as defined by (9.5). These observations lead to the following procedure for using a confidence interval to conduct a two-tailed hypothesis test.

A Confidence Interval Approach to Testing a Hypothesis of the Form

$$H_0: \mu = \mu_0$$
$$H_a: \mu \neq \mu_0$$

1. Select a simple random sample from the population and use the value of the sample mean $\bar{x}$ to develop the confidence interval for the population mean μ. If σ is assumed known, compute the interval estimate by using

$$\bar{x} \pm z_{\alpha/2} \frac{\sigma}{\sqrt{n}}$$

If σ is estimated by s, compute the interval estimate by using

$$\bar{x} \pm z_{\alpha/2} \frac{s}{\sqrt{n}}$$

For a two-tailed hypothesis test, the null hypothesis can be rejected if the confidence interval does not include μ_0.

2. If the confidence interval contains the hypothesized value μ_0, do not reject H_0. Otherwise, reject H_0.

Let us return to the Superflight golf ball study, which resulted in the following two-tailed test.

$$H_0: \mu = 280$$
$$H_a: \mu \neq 280$$

To test this hypothesis with a level of significance of $\alpha = .05$, we sampled 36 golf balls and found a sample mean distance of $\bar{x} = 278.5$ yards and a sample standard deviation of $s = 12$ yards. Using these results with $z_{.025} = 1.96$, we find that the 95% confidence interval estimate of the population mean is

$$\bar{x} \pm z_{.025} \frac{s}{\sqrt{n}}$$
$$278.5 \pm 1.96 \frac{12}{\sqrt{36}}$$
$$278.5 \pm 3.92$$

or

$$274.58 \text{ to } 282.42$$

This finding enables the quality control manager to conclude with 95% confidence that the mean distance for the population of golf balls is between 274.58 and 282.42 yards. Because the hypothesized value for the population mean, $\mu_0 = 280$, is in this interval, the hypothesis testing conclusion is that the null hypothesis, H_0: $\mu = 280$, cannot be rejected.

Note that this discussion and example pertain to two-tailed hypothesis tests about a population mean. However, the same confidence interval and hypothesis testing relationship exist for other population parameters. In addition, the relationship can be extended to one-tailed tests about population parameters. Doing so, however, requires the development of one-sided confidence intervals.

NOTES AND COMMENTS

1. The p-value depends only on the sample outcome. However, it is necessary to know whether the hypothesis test being investigated is one-tailed or two-tailed. Given the value of $\bar{x}$ in a sample, the p-value for a two-tailed test will always be *twice* the area in the tail of the sampling distribution at the value of $\bar{x}$.
2. The interval estimation approach to hypothesis testing helps to highlight the role of the sample size. From (9.3) we can see that larger sample sizes n lead to narrower confidence intervals. Thus, for a given level of significance α, a larger sample is less likely to lead to an interval containing μ_0 when the null hypothesis is false. That is, the larger sample size will provide a higher probability of rejecting H_0 when H_0 is false.

EXERCISES

Methods

21. Consider the following hypothesis test.

$$H_0: \mu = 10$$
$$H_a: \mu \neq 10$$

A sample of 36 provided a sample mean of 11 and sample standard deviation of 2.5.

a. At $\alpha = .05$, what is the rejection rule?

b. Compute the value of the test statistic z. What is your conclusion?

22. Consider the following hypothesis test.

$$H_0: \mu = 15$$
$$H_a: \mu \neq 15$$

A sample of 50 provided a sample mean of 14.2 and sample standard deviation of 5.

a. At $\alpha = .02$, what is the rejection rule?

b. Compute the value of the test statistic z.

c. What is the p-value?

d. What is your conclusion?

23. Consider the following hypothesis test.

$$H_0: \mu = 25$$
$$H_a: \mu \neq 25$$

A sample of 80 is used and the population standard deviation is 10. Use $\alpha = .05$. Compute the value of the test statistic z and specify your conclusion for each of the following sample results.

a. $\bar{x} = 22.0$ **b.** $\bar{x} = 27.0$ **c.** $\bar{x} = 23.5$ **d.** $\bar{x} = 28.0$

24. Consider the following hypothesis test.

$$H_0: \mu = 5$$
$$H_a: \mu \neq 5$$

Assume the following test statistics. Compute the corresponding p-values and specify your conclusions based on $\alpha = .05$.

a. $z = 1.80$ **b.** $z = -.45$ **c.** $z = 2.05$ **d.** $z = -3.50$ **e.** $z = -1.00$

Applications

25. The mean length of a work week for the population of workers is 39.2 hours (*Investor's Business Daily,* September 11, 2000). Test the hypotheses: $H_0: \mu = 39.2$ and $H_a: \mu \neq 39.2$ with $\alpha = .05$.

a. What are the critical values for the test statistic and what is the rejection rule for H_0?

b. Assume a follow-up sample of 112 workers showed a sample mean of 38.5 hours and a sample standard deviation 4.8 hours. What is the value of the test statistic?

c. Can the null hypothesis be rejected? What is your conclusion?

d. What is the p-value?

26. CNN and ActMedia provided a television channel that showed news, features, and ads, and was targeted to individuals waiting in grocery checkout lines. The television programs were designed with an 8-minute cycle on the assumption that the population mean time a shopper stands in line at a grocery store checkout is 8 minutes (*Astounding Averages,* 1995). A sample of 120 shoppers at a major grocery store showed a sample mean waiting time of 7.5 minutes with a sample standard deviation of 3.2 minutes. Test H_0: $\mu = 8$ and H_a: $\mu \neq 8$ with $\alpha = .05$.
   a. What are the critical values of the test statistic and what is the rejection rule?
   b. Compute the value of the test statistic?
   c. What is your conclusion?

27. A production line operates with a mean filling weight of 16 ounces per container. Overfilling or underfilling is a serious problem, and the production line should be shut down if either occurs. From past data, σ is assumed to be .8 ounces. A quality control inspector samples 30 items every 2 hours and at that time makes the decision of whether to shut the line down for adjustment.
   a. With a .05 level of significance, what is the rejection rule for the hypothesis testing procedure?
   b. If a sample mean of $\bar{x} = 16.32$ ounces were found, what action would you recommend?
   c. If $\bar{x} = 15.82$ ounces, what action would you recommend?
   d. What is the p-value for parts (b) and (c)?

28. A Gallup survey found the mean charitable contribution on federal tax returns was $1075 (*USA Today,* April 10, 2000). Assume a sample of April 2001 tax returns will be used to conduct a hypothesis test designed to determine whether any change has occurred in the mean charitable contributions.
   a. Formulate the null and alternative hypotheses.
   b. Assume that a sample of 200 tax returns has a sample mean of $1160 and a sample standard deviation of $840. What is the value of the test statistic?
   c. What is the p-value?
   d. Using $\alpha = .05$, what is your conclusion?

29. Historically, evening long-distance phone calls from a particular city have averaged 15.2 minutes per call. In a random sample of 35 calls, the sample mean time was 14.3 minutes per call, with a sample standard deviation of 5 minutes. Use this sample information to test whether the mean duration of long-distance phone calls has changed.
   a. Compute the value of the test statistic.
   b. What is the p-value?
   c. Using $\alpha = .05$, what is your conclusion?

30. The Florida Department of Labor and Employment Security reported the state mean annual wage of $26,133 (*The Naples Daily News,* February 13, 1999). A test of wages by county can be conducted to see which counties have a mean annual wage that differs from the state mean.
   a. Formulate the null and alternative hypotheses that can be used to determine whether the sample data from a particular county support the conclusion that the population mean annual wage for the county differs from the state mean of $26,133.
   b. A sample of 550 people in Collier County showed a sample mean annual wage of $25,457 and a sample standard deviation of $7,600. Compute the value of the test statistic.
   c. What is the p-value?
   d. Using $\alpha = .05$, what is your conclusion?

31. At Western University the historical mean scholarship examination score of entering students has been 900, with a standard deviation of 180. Each year a sample of applications is taken to see whether the examination scores are at the same level as in previous years.

The null hypothesis tested is H_0: $\mu = 900$. A sample of 200 students in this year's class provided a sample mean score of 935. Use a .05 level of significance.

a. Use a confidence interval approach to conduct a hypothesis test.

b. Use a test statistic to test this hypothesis.

c. What is the p-value?

32. An industry pays an average wage rate of \$15.00 per hour. A sample of 36 workers from one company showed a mean wage of $\bar{x} = \$14.50$ and a sample standard deviation of $s = \$0.60$.

a. A one-sided confidence interval uses the sample results to establish either an upper limit or a lower limit for the value of the population parameter. For this exercise establish an upper 95% confidence limit for the hourly wage rate paid by the company. The form of this one-sided confidence interval requires that we be 95% confident that the population mean is this value or less. What is the 95% confidence statement for this one-sided confidence interval?

b. Use the one-sided confidence interval result to test H_0: $\mu \geq 15$. What is your conclusion? Explain.

9.5 TESTS ABOUT A POPULATION MEAN: SMALL-SAMPLE CASE

The methodology presented in this section is based on the assumption that the population has a normal distribution.

The procedures for hypothesis tests about a population mean discussed in Sections 9.3 and 9.4 were based on the central limit theorem and large-sample theory. In Section 9.5, we consider tests about a population mean using a small sample ($n < 30$). In this case the sampling distribution of $\bar{x}$ depends on the distribution of the population. In fact, the hypothesis testing procedures for the small-sample case require the assumption that the population has a normal probability distribution. If this assumption is appropriate, the methodology present in this section can be used. However, if this assumption is not appropriate, the best alternative is to increase the sample size to $n \geq 30$ and rely on the large-sample hypothesis testing procedures presented in Sections 9.3 and 9.4.

Let us first consider the situation where the sample size is small ($n < 30$), the population is assumed to have a normal distribution and based on historical data, theoretical or other reasons, the population standard deviation σ is assumed known. Under these conditions the sampling distribution of $\bar{x}$ is normally distributed with mean μ and standard deviation $\sigma_{\bar{x}} = \sigma/\sqrt{n}$, for any sample size. Because the sampling distribution of $\bar{x}$ is a normal distribution, the small-sample hypothesis testing procedures are identical to the large-sample hypothesis testing procedures presented in Sections 9.3 and 9.4. The test statistic is again

$$z = \frac{\bar{x} - \mu_0}{\sigma/\sqrt{n}} \tag{9.7}$$

Because the hypothesis testing computations are identical to the computations we have made previously, we shall not present a new numerical example at this time.

The small-sample hypothesis testing procedure with σ estimated by s is based on the same theory used for the small-sample interval estimation procedure presented in Section 8.2. Both use a normal population assumption and the t distribution.

Next consider the situation where the sample size is small ($n < 30$), the population is assumed to have a normal distribution and the population standard deviation σ is estimated by the sample standard deviation s. In this case, the t distribution can be used to make inferences about the value of the population mean. In using the t distribution for hypothesis tests, the test statistic is

$$t = \frac{\bar{x} - \mu_0}{s/\sqrt{n}} \tag{9.8}$$

This test statistic has a t distribution with $n - 1$ degrees of freedom.

Let us consider an example of a one-tailed hypothesis test about a population mean for the small-sample case. The International Air Transport Association surveys business travelers to develop ratings of transatlantic gateway airports. The maximum possible rating is 10. A magazine devoted to business travel has decided to classify airports according to the rating they receive. Airports that have a population mean rating of 7 or more will be designated as providing superior service. Suppose a simple random sample of 12 business travelers have been asked to rate London's Heathrow airport, and that the 12 ratings obtained are 7, 8, 10, 8, 6, 9, 6, 7, 7, 8, 9, and 8. The sample mean for these data is $\bar{x} = 7.75$ and the sample standard deviation is $s = 1.215$. Assuming that the population of ratings can be approximated by a normal probability distribution, should Heathrow be designated as providing superior service?

Heathrow

Using a .05 level of significance, we want a test to determine whether the population mean rating for the Heathrow airport is greater than 7. The null and alternative hypotheses follow.

$$H_0: \mu \leq 7$$
$$H_a: \mu > 7$$

The Heathrow airport will be designated as providing superior service if H_0 can be rejected.

The rejection region is in the upper tail of the sampling distribution. With $n - 1 = 12 - 1 = 11$ degrees of freedom, the t distribution table inside the front cover of the text shows that the critical value is $t_{.05} = 1.796$. Thus, the rejection rule is

$$\text{Reject } H_0 \text{ if } t > 1.796$$

Using (9.8) with $\bar{x} = 7.75$ and $s = 1.215$, we have the following value for the test statistic.

$$t = \frac{\bar{x} - \mu_0}{s/\sqrt{n}} = \frac{7.75 - 7}{1.215/\sqrt{12}} = 2.14$$

Because 2.14 is greater than 1.796, the null hypothesis is rejected. At the .05 level of significance, we can conclude that the population mean rating for the Heathrow airport is greater than 7. Thus, Heathrow can be designated as providing superior service. Figure 9.9 shows that the value of the test statistic is in the rejection region.

p-Values and the t Distribution

Let us consider the p-value for the Heathrow airport hypothesis test. As discussed in Sections 9.3 and 9.4, the p-value can be interpreted as the observed level of significance for the test. The usual rule applies: If the p-value is less than the level of significance α, the null hypothesis can be rejected. Unfortunately, the format of the t distribution table provided in most statistics textbooks does not have sufficient detail to determine the exact p-value for the test. However, we can still use the t distribution table to identify a range for the p-value. For example, the t distribution used in the Heathrow airport hypothesis test has 11 degrees of freedom. Referring to the t distribution table, we see that the row for 11 degrees of freedom provides the following information.

Area in Upper Tail	.10	.05	.025	.01	.005
***t* Value**	1.363	1.796	2.201	2.718	3.106

The computed value for the test statistic was $t = 2.14$. The p-value is the area in the tail corresponding to $t = 2.14$. From the above information, we see 2.14 is between 1.796 and 2.201. Thus, although we cannot determine the exact p-value, we know that the p-value

FIGURE 9.9 VALUE OF THE TEST STATISTIC ($t = 2.14$) FOR THE HEATHROW AIRPORT HYPOTHESIS TEST

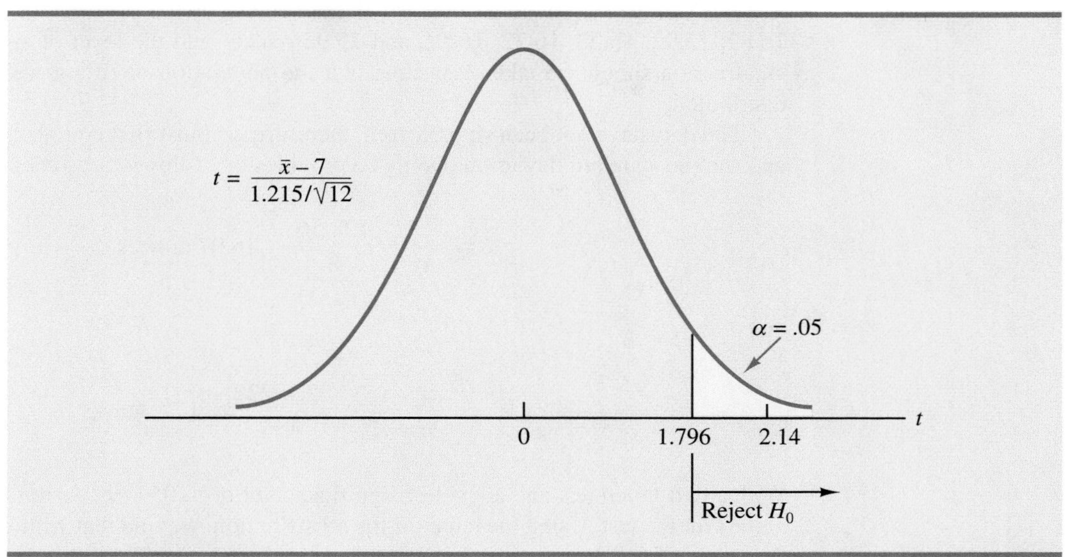

The p-values are difficult to obtain directly from a table of the t distribution. Generally, the best that can be done with the table is to specify a range for the p-value such as between .05 and .025.

must be between .05 and .025. With a level of significance of $\alpha = .05$, we know that the p-value must be less than .05; thus, the null hypothesis is rejected.

Minitab can be used to provide small-sample hypothesis test information for a population mean. The computer output for the Heathrow Airport hypothesis test is shown in Figure 9.10. The sample mean $\bar{x} = 7.75$, the test statistic $t = 2.14$, and the p-value $= .028$ are computed automatically. The step-by-step procedure used to obtain the Minitab output is described in Appendix 9.1.

Two-Tailed Test

As an example of a two-tailed hypothesis test about a population mean using a small sample, consider the following production problem. A production process is designed to fill containers with a mean filling weight of $\mu = 16$ ounces. It is undesirable for the process to underfill containers because the consumer will not receive the amount of product indicated on the container label. It is equally undesirable for the process to overfill containers; in that case, the firm loses money because the process is placing more product in the container than is required. Quality assurance personnel periodically select a simple random sample of eight containers in order to conduct the following two-tailed hypothesis test.

$$H_0: \mu = 16$$
$$H_a: \mu \neq 16$$

FIGURE 9.10 MINITAB OUTPUT FOR THE HEATHROW AIRPORT RATING HYPOTHESIS TEST

```
Test of mu = 7.000 vs mu > 7.000
Variable     N      Mean     StDev     SE Mean       T        P
Rating      12     7.750     1.215       0.351    2.14    0.028
```

If H_0 is rejected the production manager will request that the production process be stopped and that the mechanism for regulating filling weights be readjusted to ensure a mean filling weight of 16 ounces. If the sample yields data values of 16.02, 16.22, 15.82, 15.92, 16.22, 16.32, 16.12, and 15.92 ounces and the level of significance is .05, what action should be taken? Assume that the population of filling weights is normally distributed.

The data have not been summarized, therefore we must first compute the sample mean and sample standard deviation. Doing so provides the following results.

$$\bar{x} = \frac{\Sigma x_i}{n} = \frac{128.56}{8} = 16.07 \text{ ounces}$$

and

$$s = \sqrt{\frac{\Sigma(x_i - \bar{x})^2}{n - 1}} = \sqrt{\frac{.22}{7}} = .177 \text{ ounces}$$

With a two-tailed test and a level of significance of $\alpha = .05$, $-t_{.025}$ and $t_{.025}$ are the critical values for the test. Using the table for the t distribution, we find that with $n - 1 = 8 - 1 = 7$ degrees of freedom, $-t_{.025} = -2.365$ and $t_{.025} = +2.365$. Thus, the rejection rule is

$$\text{Reject } H_0 \text{ if } t < -2.365 \text{ or if } t > 2.365$$

Using $\bar{x} = 16.07$ and $s = .18$, we have

$$t = \frac{\bar{x} - \mu_0}{s/\sqrt{n}} = \frac{16.07 - 16.00}{.177/\sqrt{8}} = 1.12$$

Because $t = 1.12$ is not in the rejection region, the null hypothesis cannot be rejected. The evidence is not sufficient to adjust the production process.

Using the t distribution table and the row for seven degrees of freedom, we see that the computed t value of 1.12 has an upper-tail area of *more than* .10. Although the format of the t distribution table prevents us from being more specific, we can at least conclude that the two-tailed p-value is greater than $2(.10) = .20$. Because this value is greater than the .05 level of significance, we see that the p-value leads to the same conclusion; that is, do not reject H_0. The computer solution for this hypothesis test shows the exact p-value is .301.

EXERCISES

Methods

33. Consider the following hypothesis test.

$$H_0: \mu \leq 10$$
$$H_a: \mu > 10$$

A sample of 16 provides a sample mean of 11 and sample standard deviation of 3.

a. With $\alpha = .05$, what is the rejection rule?

b. Compute the value of the test statistic t. What is your conclusion?

34. Consider the following hypothesis test.

$$H_0: \mu = 20$$
$$H_a: \mu \neq 20$$

Data from a sample of six items are: 18, 20, 16, 19, 17, 18.

a. Compute the sample mean.
b. Compute the sample standard deviation.
c. With $\alpha = .05$, what is the rejection rule?
d. Compute the value of the test statistic t.
e. What is your conclusion?

35. Consider the following hypothesis test.

$$H_0: \mu \geq 15$$
$$H_a: \mu < 15$$

A sample of 22 provided the sample standard deviation $s = 8$. Use $\alpha = .05$. Compute the value of the test statistic t and your conclusion for each of the following sample results.

a. $\bar{x} = 13.0$ **b.** $\bar{x} = 11.5$ **c.** $\bar{x} = 15.0$ **d.** $\bar{x} = 19.0$

36. Consider the following hypothesis test.

$$H_0: \mu \leq 50$$
$$H_a: \mu > 50$$

Assume a sample of 16 items provided the following test statistics. What can you say about the p-values in each case? What are your conclusions based on $\alpha = .05$?

a. $t = 2.602$ **b.** $t = 1.341$ **c.** $t = 1.960$ **d.** $t = 1.055$ **e.** $t = 3.261$

Applications

37. The population mean earnings per share for financial services corporations including American Express, E*Trade Group, Goldman Sachs, and Merrill Lynch was \$3 (*Business Week,* August 14, 2000). In 2001, a sample of 10 financial services corporations provided the following earnings per share data:

1.92	2.16	3.63	3.16	4.02	3.14	2.20	2.34	3.05	2.38

a. Formulate the null and alternative hypotheses that can be used to determine whether the population mean earnings per share in 2001 differ from \$3 reported in 2000.
b. Using $\alpha = .05$, what are the critical values for the test statistic, and what is the rejection rule?
c. Compute the sample mean.
d. Compute the sample standard deviation.
e. Compute the value of the test statistic.
f. What is your conclusion?
g. What can you say about the p-value?

38. The average U.S. household spends \$90 per day (*American Demographics,* August 1997). Assume a sample of 25 households in Corning, New York, showed a sample mean daily expenditure of \$84.50 with a sample standard deviation of \$14.50. Test $H_0: \mu = 90$ and $H_a: \mu \neq 90$ with $\alpha = .05$.

a. What are the critical values of the test statistic, and what is the rejection rule?
b. Compute the value of the test statistic.
c. What is your conclusion?
d. What can you say about the p-value?

39. On the average, a housewife with a husband and two children is estimated to work 55 hours or less per week on household-related activities. The hours worked during a week for a sample of eight housewives are: 58, 52, 64, 63, 59, 62, 62, and 55. Test H_0: $\mu \le 55$ and H_a: $\mu > 55$ with $\alpha = .05$.

a. What is the critical value for the test, and what is the rejection rule?
b. Compute the sample mean.
c. Compute the sample standard deviation.
d. Compute the value of the test statistic.
e. What is your conclusion?
f. What can you say about the *p*-value?

40. The cost of a one-carat VS2 clarity, H color diamond from the Diamond Source USA is \$4000 (*www.diasource.com,* July 2000). A Midwestern jeweler makes calls to contacts in the diamond district of New York City to see whether the mean price of similar diamonds differs from \$4000. The jeweler agrees to collect cost data from 14 New York City contacts.

a. Formulate the null and alternative hypotheses that can be used to determine if the mean price in New York City differs from \$4000.
b. Using $\alpha = .05$, what are the critical values for the test, and what is the rejection rule?
c. Assume the sample of 14 provides a sample mean price of \$4120 and a sample standard deviation of \$275. Compute the value of the test statistic.
d. What is your conclusion?
e. What can you say about the *p*-value?

41. Callaway Golf Company's new forged titanium ERC driver has been described as "illegal" because it promised driving distances that exceeded the USGA's standard. *Golf Digest* compared actual driving distances with the ERC driver and a USGA approved driver with a population mean driving distance of 280 yards. Using nine test drives, the mean driving distance by the ERC driver was 286.9 yards (*Golf World,* May 12, 2000). Answer the following questions assuming a sample standard deviation driving distance of 10 yards.

a. Formulate the null and alternative hypotheses that can be used to determine whether the new ERC driver has a population mean driving distance greater than 280 yards.
b. On average, how many yards farther did the golf ball travel with the ERC driver?
c. Using $\alpha = .05$, what is the critical value for the test, and what is the rejection rule?
d. Compute the value of the test statistic.
e. What is your conclusion?
f. What can you say about the *p*-value?

42. Joan's Nursery specializes in custom-designed landscaping for residential areas. The estimated labor cost associated with a particular landscaping proposal is based on the number of plantings of trees, shrubs, and so on to be used for the project. For cost-estimating purposes, managers use 2 hours of labor time for the planting of a medium-size tree. Actual times from a sample of 10 plantings during the past month follow (times in hours).

1.9 1.7 2.8 2.4 2.6 2.5 2.8 3.2 1.6 2.5

Using a .05 level of significance, test to see whether the mean tree-planting time exceeds 2 hours.

a. State the null and alternative hypotheses.
b. What is the critical value for the test, and what is the rejection rule?
c. Compute the sample mean.
d. Compute the sample standard deviation.
e. Compute the value of the test statistic.
f. What is your conclusion?
g. What can you say about the *p*-value?

9.6 TESTS ABOUT A POPULATION PROPORTION

With the symbol p denoting the population proportion and p_0 denoting a particular hypothesized value for the population proportion, the three forms for a hypothesis test about a population proportion are as follows.

$$H_0\colon p \geq p_0 \qquad H_0\colon p \leq p_0 \qquad H_0\colon p = p_0$$
$$H_a\colon p < p_0 \qquad H_a\colon p > p_0 \qquad H_a\colon p \neq p_0$$

The first two forms are one-tailed tests, whereas the third form is a two-tailed test. The specific form used depends on the application.

Hypothesis tests about a population proportion are based on the difference between the sample proportion $\bar{p}$ and the hypothesized population proportion p_0. The methods used to conduct the tests are similar to the procedures used for hypothesis tests about a population mean. The only difference is that we use the sample proportion $\bar{p}$ and its standard deviation $\sigma_{\bar{p}}$ in developing the test statistic. We begin by formulating null and alternative hypotheses about the value of the population proportion. Then, using the value of the sample proportion $\bar{p}$ and its standard deviation $\sigma_{\bar{p}}$, we compute a value for the test statistic z. Comparing the value of the test statistic to the critical value or comparing the p-value to α enables us to determine whether the null hypothesis should be rejected.

Either the test statistic or the p-value can be used to make the hypothesis testing conclusion.

Let us illustrate hypothesis testing for a population proportion by considering the situation faced by Pine Creek golf course. Over the past few months, 20% of the players at Pine Creek have been women. In an effort to increase the proportion of women playing, Pine Creek used a special promotion to attract women golfers. After one week, a random sample of 400 players showed 300 men and 100 women. Course managers would like to determine whether the data support the conclusion that the proportion of women playing at Pine Creek has increased.

To determine whether the effect of the promotion has been to increase the proportion of women golfers, we state the following null and alternative hypotheses.

$$H_0\colon p \leq .20$$
$$H_a\colon p > .20$$

As usual, we begin the hypothesis testing procedure by assuming that H_0 is true with $p = .20$. Using the sample proportion $\bar{p}$ to estimate p, we next consider the sampling distribution of $\bar{p}$. Because $\bar{p}$ is an unbiased estimator of p, we know that if $p = .20$, the mean of the sampling distribution of $\bar{p}$ is .20. In addition, we know from Chapter 7 that the standard deviation of $\bar{p}$ is given by

$$\sigma_{\bar{p}} = \sqrt{\frac{p(1 - p)}{n}}$$

With the assumed value of $p = .20$ and a sample size of $n = 400$, the standard deviation of $\bar{p}$ is

$$\sigma_{\bar{p}} = \sqrt{\frac{.20(1 - .20)}{400}} = .02$$

In Chapter 7 we saw that the sampling distribution of $\bar{p}$ can be approximated by a normal probability distribution if both np and $n(1 - p)$ are greater than or equal to 5. In the Pine Creek case, $np = 400(.20) = 80$ and $n(1 - p) = 400(.80) = 320$; thus, the normal probability distribution approximation is appropriate. The sampling distribution of $\bar{p}$ is shown in Figure 9.11.

Because the sampling distribution of $\bar{p}$ is approximately normal, the following test statistic can be used.

In hypothesis testing, a value of the population proportion, p_0, is assumed in H_0. Therefore, when H_0 is true, $\sigma_{\bar{p}}$ can be computed by using p_0 as shown in (9.10). The sample proportion, $\bar{p}$, is not used in the computation of $\sigma_{\bar{p}}$.

Test Statistic for Tests About a Population Proportion

$$z = \frac{\bar{p} - p_0}{\sigma_{\bar{p}}} \tag{9.9}$$

where

$$\sigma_{\bar{p}} = \sqrt{\frac{p_0(1 - p_0)}{n}} \tag{9.10}$$

Let us assume that $\alpha = .05$ has been selected as the level of significance for the test. With the critical value $z_{.05} = 1.645$, the upper-tail rejection region for the hypothesis test (see Figure 9.12) provides the following rejection rule.

$$\text{Reject } H_0 \text{ if } z > 1.645$$

Once the rejection rule has been determined, we collect the data, compute the value of the point estimate $\bar{p}$, and compute the corresponding value of the test statistic z. By comparing the value of z to the critical value ($z_{.05} = 1.645$), we can decide whether to reject the null hypothesis.

FIGURE 9.11 SAMPLING DISTRIBUTION OF $\bar{p}$ FOR THE PROPORTION OF WOMEN GOLFERS AT PINE CREEK GOLF COURSE

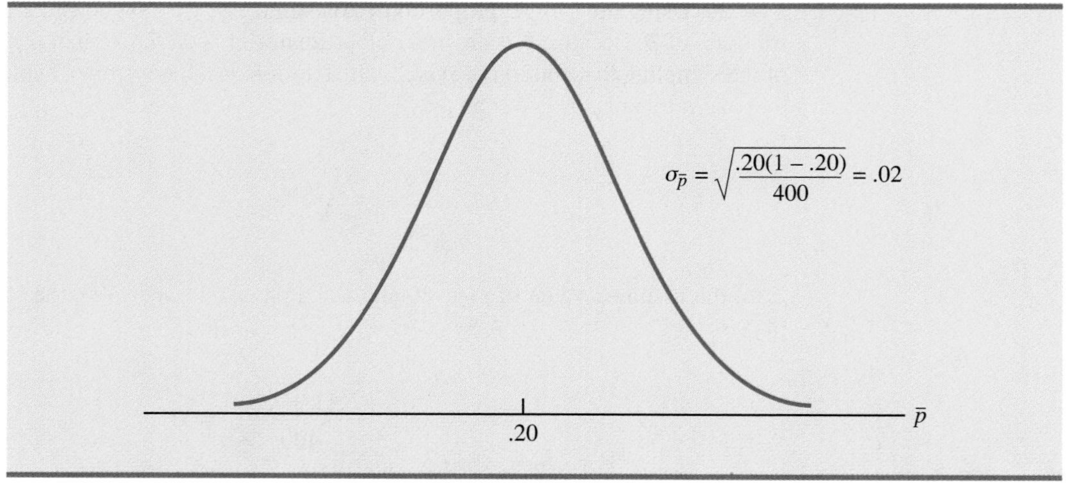

FIGURE 9.12 REJECTION REGION FOR THE PINE CREEK GOLF COURSE HYPOTHESIS TEST

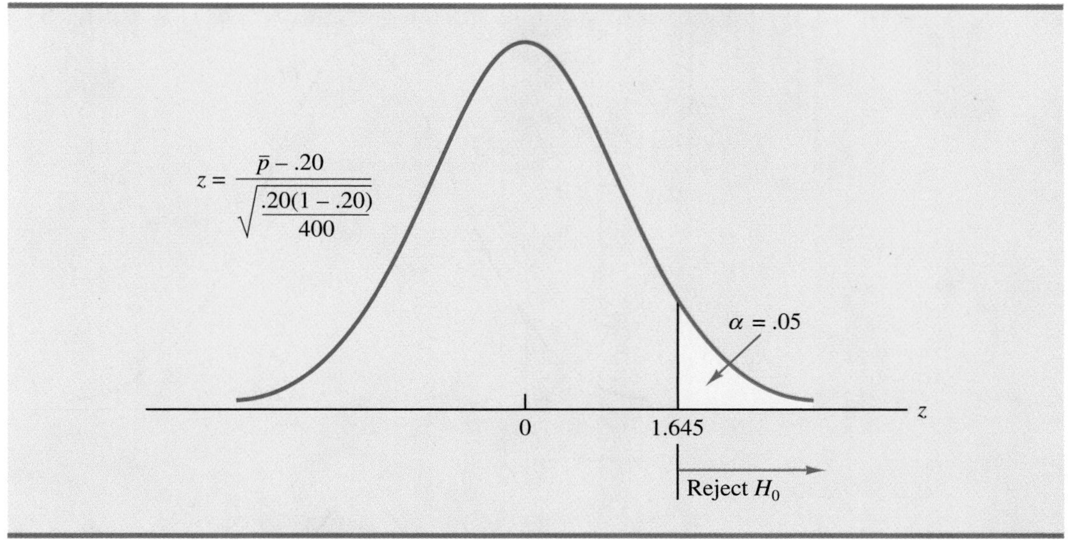

A random sample of 400 players taken during the promotion showed 100 players were women. Thus, $\bar{p} = 100/400 = .25$. With $\sigma_{\bar{p}} = .02$, the value of the test statistic is

$$z = \frac{\bar{p} - p_0}{\sigma_{\bar{p}}} = \frac{.25 - .20}{.02} = 2.5$$

Because $z = 2.5 > 1.645$, we can reject H_0. The Pine Creek managers can conclude the proportion of women players increased during the promotion.

Using the table of areas for the standard normal probability distribution, we find that the p-value for the test can also be computed. For example, given the test statistic $z = 2.50$, the table of areas shows a .4938 area or probability between the mean and $z = 2.50$. Thus, the p-value for the test is $.5000 - .4938 = .0062$. With a p-value that is less than $\alpha = .05$, the null hypothesis can be rejected.

We see that hypothesis tests about a population proportion and a population mean are similar; the primary difference is that the test statistic is based on the sampling distribution of $\bar{x}$ when the hypothesis test involves a population mean and on the sampling distribution of $\bar{p}$ when the hypothesis test involves a population proportion. Comparing the test statistic to the critical value or comparing the p-value to α may be used to reach the hypothesis-testing conclusion. Figure 9.13 summarizes the decision rules for hypothesis tests about a population proportion. We assume the large-sample case of $np \geq 5$ and $n(1 - p) \geq 5$ where the normal probability distribution can be used to approximate the sampling distribution of $\bar{p}$.

NOTES AND COMMENTS

We have not shown the procedure for small-sample hypothesis tests involving population proportions. In the small-sample case, the sampling distribution of $\bar{p}$ follows the binomial distribution and hence the normal approximation is not applicable. More advanced texts show how hypothesis tests are conducted for this situation. However, in practice, small-sample tests are rarely conducted for a population proportion.

FIGURE 9.13 SUMMARY OF REJECTION RULES FOR HYPOTHESIS TESTS ABOUT A POPULATION PROPORTION

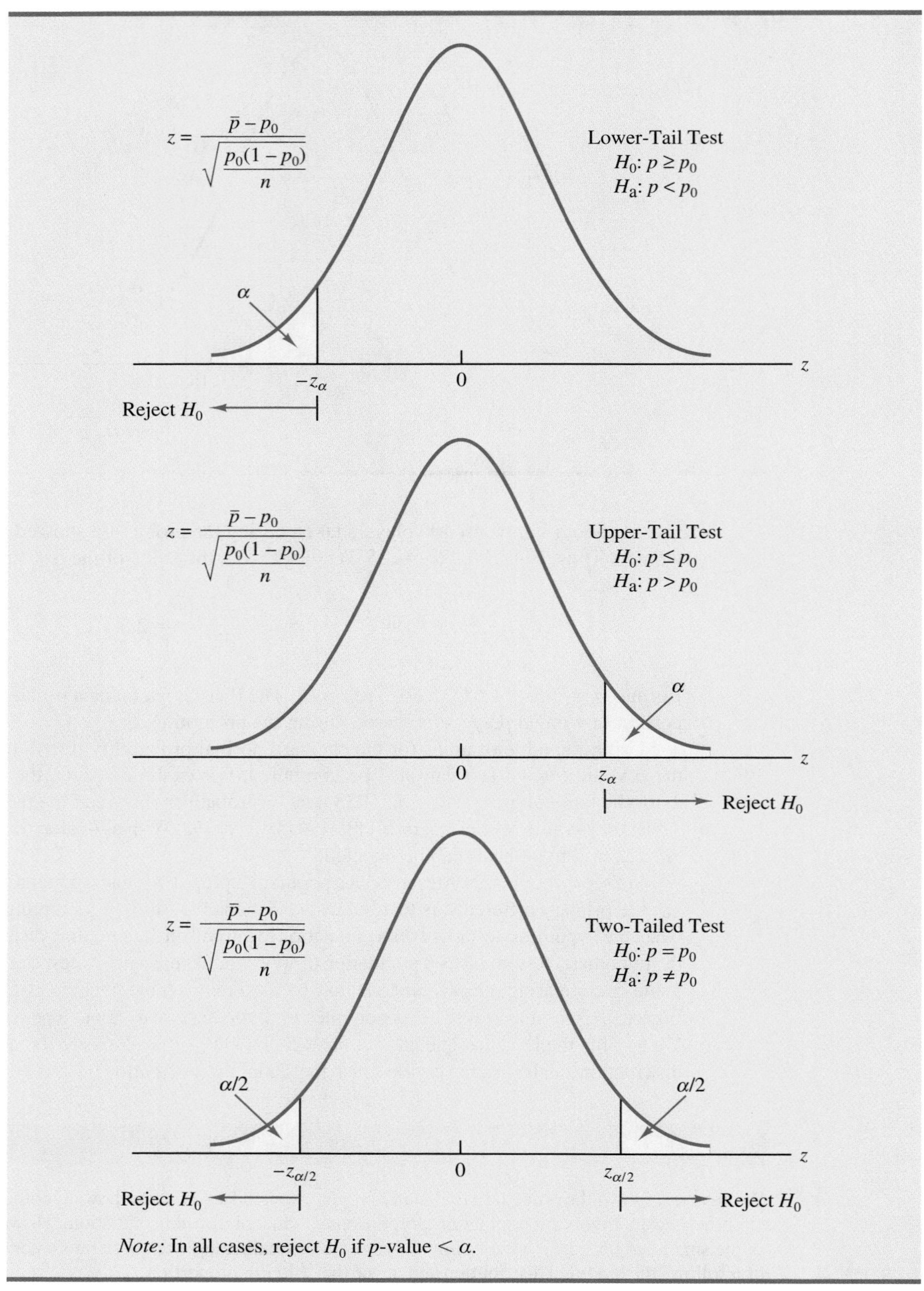

Note: In all cases, reject H_0 if p-value $< \alpha$.

EXERCISES

Methods

43. Consider the following hypothesis test.

$$H_0: p \leq .50$$
$$H_a: p > .50$$

A sample of 200 provided a sample proportion $\bar{p} = .57$.

a. At $\alpha = .05$, what is the rejection rule?

b. Compute the value of the test statistic z. What is your conclusion?

44. Consider the following hypothesis test.

$$H_0: p = .20$$
$$H_a: p \neq .20$$

A sample of 400 provided a sample proportion of $\bar{p} = .175$.

a. At $\alpha = .05$, what is the rejection rule?

b. Compute the value of the test statistic z.

c. What is the p-value?

d. What is your conclusion?

45. Consider the following hypothesis test.

$$H_0: p \geq .75$$
$$H_a: p < .75$$

A sample of 300 was selected. Using $\alpha = .05$, provide the value of the test statistic z, the p-value, and your conclusion for each of the following sample results.

a. $\bar{p} = .68$ **b.** $\bar{p} = .72$ **c.** $\bar{p} = .70$ **d.** $\bar{p} = .77$

Applications

46. The Heldrich Center for Workforce Development found that 40% of Internet users received more than 10 e-mail messages per day (*USA Today,* May 7, 2000). In 2001, a similar study on the use of e-mail was repeated. The purpose of the study was see whether use of e-mail has increased.

a. Formulate the null and alternative hypotheses to determine whether an increase has occurred in the proportion of Internet users receiving more than 10 e-mails message per day.

b. Using $\alpha = .05$, what is the critical value for the test, and what is the rejection rule?

c. If a sample of 420 Internet users found 188 receiving more than 10 e-mail messages per day, what is the sample proportion and what is the value of the test statistic?

d. What is your conclusion?

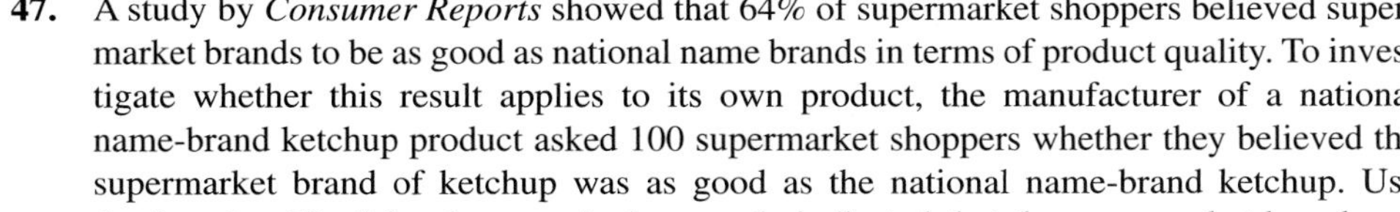

47. A study by *Consumer Reports* showed that 64% of supermarket shoppers believed supermarket brands to be as good as national name brands in terms of product quality. To investigate whether this result applies to its own product, the manufacturer of a national name-brand ketchup product asked 100 supermarket shoppers whether they believed the supermarket brand of ketchup was as good as the national name-brand ketchup. Use the fact that 52 of the shoppers in the sample indicated that the supermarket brand was as good as the national name brand to test H_0: $p \geq .64$ and H_a: $p < .64$. Use a .05 level of significance.

a. What is the critical value, and what is the rejection rule?
b. Compute the value of the test statistic.
c. What is your conclusion?
d. What is the p-value?

48. Burger King sponsored what it believed to be the biggest product giveaway in fast-food history when it initiated "FreeFryDay" on Friday, January 2, 1998. Anyone who stopped at one of Burger King's 7400 restaurants received a small order of the new Burger King fries free of charge. A national taste test comparing the new Burger King fries to McDonald's fries was the basis for the Burger King promotion. Of the 500 customers who participated in the taste test, 285 expressed a taste preference for the Burger King fries (*USA Today,* December 10, 1997). Let p = the population proportion who favors the Burger King fries. Consider the hypothesis following test:

$$H_0: p \leq .50$$
$$H_a: p > .50$$

a. What is the point estimate of the population proportion that favors the Burger King fries?
b. Compute the value of the test statistic.
c. What is the p-value?
d. Using $\alpha = .01$, what is your conclusion?
e. Do you believe that the national taste test statistics helped Burger King's management decide to implement the "FreeFryDay" campaign? Discuss.

49. In a study of driving practices in Strathcona County, Alberta, Canada, it was found that 48% of drivers did not stop at stop sign intersections on county roads (*Edmonton Journal,* July 19, 2000). Assume that a follow-up study two months later found the 360 of 800 drivers did not stop at stop sign intersections on county roads.

a. Formulate the null and alternative hypotheses assuming the purpose of the follow-up study is to see whether the proportion of drivers who did not stop has changed.
b. Using $\alpha = .05$, what are the critical values for the test, and what is the rejection rule?
c. What is the sample proportion of drivers who do not stop?
d. Compute the value of the test statistic.
e. What is your conclusion?

50. Shell Oil office workers were asked which work schedule appealed most: working five 8-hour days or four 10-hour days (*USA Today,* September 11, 2000). Let p equal the proportion of office workers preferring the four 10-hour day alternative. Test the hypotheses H_0: $p = .50$ and H_a: $p \neq .50$ at $\alpha = .01$. A sample of 105 office workers showed that 67 workers preferred the four 10-hour day schedule.

a. What is the sample proportion preferring the four 10-hour day schedule?
b. What is the value of the test statistic?
c. What is the p-value?
d. What is your conclusion? Is a statistically significant preference indicated between the two alternatives?

51. Drugstore.com was the first e-Commerce company to offer Internet drugstore retailing. Drugstore.com customers were provided the opportunity to buy health, beauty, personal care, wellness and pharmaceutical replenishment products over the Internet. At the end of 10 months of operation, the company reported 44% of orders were from repeat customers (*Drugstore.com Annual Report,* January 2, 2000). Assume that Drugstore.com will use a sample of customer orders each quarter to determine if the proportion of orders from repeat customers has changed from the initial $p = .44$.

a. Formulate the null and alternative hypotheses.
b. During the first quarter a sample of 500 orders showed 205 repeat customers. What is the *p*-value? What is your conclusion using $\alpha = .05$?
c. During the second quarter a sample of 500 orders showed 245 repeat customers. What is the *p*-value? What is your conclusion using $\alpha = .05$?

52. Microsoft Outlook is believed to be the most widely used e-mail manager. A Microsoft executive claims that Microsoft Outlook is used by at least 75% of the Internet users. A Merrill Lynch study reported 72% use Microsoft Outlook (CNBC, June 2000). Test the following hypotheses: H_0: $p \geq .75$, H_a: $p < .75$ and $\alpha = .05$.
a. Assuming the Merrill Lynch sample size was 300 users, compute the value of the test statistic.
b. What is the *p*-value?
c. Should the executive's claim of at least 75% be rejected?

53. A fast-food restaurant plans a special offer that will enable customers to purchase specially designed drink glasses featuring well-known cartoon characters. If more than 15% of the customers will purchase the glasses, the special offer will be implemented. A preliminary test has been set up at several locations, and 88 of 500 customers have purchased the glasses. Should the special glass offer be implemented? Conduct a hypothesis test that will support your decision. Use a .01 level of significance. What is your recommendation?

54. Based on rates of errors, third base is the most difficult position to play in professional baseball. Considering all third basemen, an error occurs on 4.7% of the chances (*ESPN The Magazine,* July 12, 1999). A sample of games played by all-star third baseman Brooks Robinson showed that he had 35 errors in 1182 chances. Does the sample support the conclusion that Brooks Robinson makes fewer errors than the typical third baseman? Use $\alpha = .01$.
a. What are the appropriate null and alternative hypotheses?
b. What is the sample proportion error percentage for Brooks Robinson?
c. Compute the value of the test statistic.
d. What is the *p*-value?
e. What is your conclusion?

55. At least 20% of all workers are believed to be willing to work fewer hours for less pay to obtain more time for personal and leisure activities. A *USA Today*/CNN/Gallup poll with a sample of 596 respondents found 83 willing to work fewer hours for less pay to obtain more personal and leisure time (*USA Today,* April 10, 1995). Test H_0: $p \geq .20$ and H_a: $p < .20$ using a .05 level of significance. What is your conclusion?

9.7 HYPOTHESIS TESTING AND DECISION MAKING

In Section 9.1 we noted three types of situations in which hypothesis testing is used:

1. Testing research hypotheses.
2. Testing the validity of a claim.
3. Testing in decision-making situations.

In the first two situations, action is taken only when the null hypothesis H_0 is rejected and hence the alternative hypothesis H_a is concluded to be true. In the third situation—decision making—it is necessary to take action when the null hypothesis is not rejected as well as when it is rejected.

The hypothesis testing procedures presented thus far have limited applicability in a decision-making situation because it is not considered appropriate to accept H_0 and take action based on the conclusion that H_0 is true. The reason for not taking action when the test results indicate *do not reject* H_0 is that the decision to accept H_0 exposes the decision maker to the risk of making a Type II error; that is, accepting H_0 when it is false. With the hypothesis testing procedures described in the preceding sections, the probability of a Type I error is controlled by establishing a level of significance for the test. However, the probability of making the Type II error is not controlled.

Clearly, in certain decision-making situations the decision maker may want—and in some cases may be forced—to take action with both the conclusion *do not reject* H_0 and the conclusion *reject* H_0. A good illustration of this situation is lot-acceptance sampling, a topic we will discuss in more depth in Chapter 20. For example, a quality control manager must decide to accept a shipment of batteries from a supplier or to return the shipment because of poor quality. Assume that design specifications require batteries from the supplier to have a mean useful life of at least 120 hours. To evaluate the quality of an incoming shipment, a sample of 36 batteries will be selected and tested. On the basis of the sample, a decision must be made to accept the shipment of batteries or to return it to the supplier because of poor quality. Let μ denote the mean number of hours of useful life for batteries in the shipment. The null and alternative hypotheses about the population mean follow.

$$H_0: \mu \geq 120$$
$$H_a: \mu < 120$$

If H_0 is rejected, the alternative hypothesis is concluded to be true. This conclusion indicates that the appropriate decision is to return the shipment to the supplier. However, if H_0 is not rejected, the decision maker must still determine what action should be taken. Thus, without directly concluding that H_0 is true, but merely by not rejecting it, the decision maker will have made the decision to accept the shipment as being of satisfactory quality.

In such decision-making situations, it is recommended that the hypothesis testing procedure be extended to include consideration of the probability of making a Type II error. Because a decision will be made and action taken when we do not reject H_0, knowledge of the probability of making a Type II error will be helpful. In Sections 9.8 and 9.9 we explain how to compute the probability of making a Type II error and how the sample size can be adjusted to help control the probability of making a Type II error.

9.8 CALCULATING THE PROBABILITY OF TYPE II ERRORS

In this section we show how to calculate the probability of making a Type II error for a hypothesis test about a population mean. We illustrate the procedure by using the lot-acceptance example described in Section 9.7. The null and alternative hypotheses about the mean number of hours of useful life for a shipment of batteries are H_0: $\mu \geq 120$ and H_a: $\mu < 120$. If H_0 is rejected, the decision will be to return the shipment to the supplier because the mean hours of useful life are less than the specified 120 hours. If H_0 is not rejected, the decision will be to accept the shipment.

Suppose a level of significance of $\alpha = .05$ is used to conduct the hypothesis test. The test statistic is

$$z = \frac{\bar{x} - \mu}{\sigma/\sqrt{n}} = \frac{\bar{x} - 120}{\sigma/\sqrt{n}}$$

With $z_{.05} = 1.645$, the rejection rule for the lower-tail test becomes

$$\text{Reject } H_0 \text{ if } z < -1.645$$

Suppose a sample of 36 batteries will be selected and based upon previous testing that the standard deviation for the population is assumed known with $\sigma = 12$ hours. The rejection rule indicates that we will reject H_0 if

$$z = \frac{\bar{x} - 120}{12/\sqrt{36}} < -1.645$$

Solving for $\bar{x}$ in the preceding expression indicates that we will reject H_0 if

$$\bar{x} < 120 - 1.645\left(\frac{12}{\sqrt{36}}\right) = 116.71$$

Rejecting H_0 when $\bar{x} < 116.71$ means that we will make the decision to accept the shipment whenever

$$\bar{x} \geq 116.71$$

With this information, we are ready to compute probabilities associated with making a Type II error. First, recall that we make a Type II error whenever the true shipment mean is less than 120 hours and we make the decision to accept H_0: $\mu \geq 120$. Hence, to compute the probability of making a Type II error, we must select a value of μ less than 120 hours. For example, suppose the shipment is considered to be of poor quality if the batteries have a mean life of $\mu = 112$ hours. If $\mu = 112$ is really true, what is the probability of accepting H_0: $\mu \geq 120$ and hence committing a Type II error? Note that this probability is the probability that the sample mean $\bar{x}$ is greater than or equal to 116.71 when $\mu = 112$.

Figure 9.14 shows the sampling distribution of $\bar{x}$ when the mean is $\mu = 112$. The shaded area in the upper tail gives the probability of obtaining $\bar{x} \geq 116.71$. Using the standard normal probability distribution calculation based on Figure 9.14, we see that

$$z = \frac{\bar{x} - \mu}{\sigma/\sqrt{n}} = \frac{116.71 - 112}{12/\sqrt{36}} = 2.36$$

FIGURE 9.14 PROBABILITY OF A TYPE II ERROR WHEN $\mu = 112$

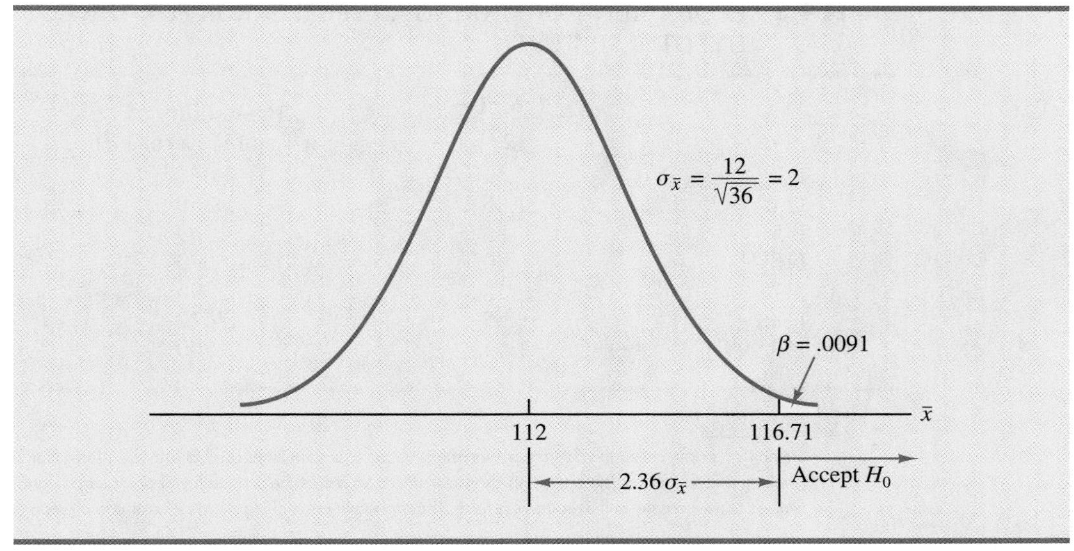

The standard normal probability distribution table shows that with $z = 2.36$, the area in the upper tail is $.5000 - .4909 = .0091$. Thus, .0091 is the probability of making a Type II error when $\mu = 112$. Denoting the probability of making a Type II error as β, we see that when $\mu = 112$, $\beta = .0091$. Therefore, we can conclude that if the mean of the population is 112 hours, the probability of making a Type II error is only .0091.

We can repeat these calculations for other values of μ less than 120. Doing so will show a different probability of making a Type II error for each value of μ. For example, suppose the shipment of batteries has a mean useful life of $\mu = 115$ hours. Because we will accept H_0 whenever $\bar{x} \geq 116.71$, the z value for $\mu = 115$ is given by

$$z = \frac{\bar{x} - \mu}{\sigma/\sqrt{n}} = \frac{116.71 - 115}{12/\sqrt{36}} = .86$$

From the standard normal probability distribution table, we find that the area in the upper tail of the standard normal probability distribution for $z = .86$ is $.5000 - .3051 = .1949$. Thus, the probability of making a Type II error is $\beta = .1949$ when the true mean is $\mu = 115$.

As Table 9.3 shows, the probability of a Type II error depends on the value of the population mean μ. For values of μ near μ_0, the probability of making the Type II error can be high. Because the value of μ is never known, the probability of making the Type II error cannot be stated.

In Table 9.3 we show the probability of making a Type II error for a variety of values of μ less than 120. Note that as μ increases toward 120, the probability of making a Type II error increases toward an upper bound of .95. However, as μ decreases to values farther below 120, the probability of making a Type II error diminishes. This pattern is what we should expect. When the true population mean μ is close to the null hypothesis value of $\mu = 120$, the probability is high that we will make a Type II error. However, when the true population mean μ is far below the null hypothesis value of $\mu = 120$, the probability is low that we will make a Type II error.

The probability of correctly rejecting H_0 when it is false is called the **power** of the test. For any particular value of μ, the power is $1 - \beta$. That is, the probability of correctly rejecting the null hypothesis is one minus the probability of making a Type II error. Values of power are listed in Table 9.3. On the basis of these values, the power associated with each value of μ is shown graphically in Figure 9.15. Such a graph is called a **power curve.** Note that the power curve extends over the values of μ for which the null hypothesis is false. The height of the power curve at any value of μ indicates the probability of correctly rejecting H_0 when H_0 is false.*

TABLE 9.3 PROBABILITY OF MAKING A TYPE II ERROR FOR THE LOT-ACCEPTANCE HYPOTHESIS TEST

Value of μ	$z = \frac{116.71 - \mu}{12/\sqrt{36}}$	Probability of a Type II Error (β)	Power ($1 - \beta$)
112	2.36	.0091	.9909
114	1.36	.0869	.9131
115	.86	.1949	.8051
116.71	.00	.5000	.5000
117	−.15	.5596	.4404
118	−.65	.7422	.2578
119.999	−1.645	.9500	.0500

*Another graph, called the *operating characteristic curve,* is sometimes used to provide information about the probability of making a Type II error. The operating characteristic curve shows the probability of accepting H_0 and thus provides β for the values of μ where the null hypothesis is false. The probability of making a Type II error can be read directly from this graph.

FIGURE 9.15 POWER CURVE FOR THE LOT-ACCEPTANCE HYPOTHESIS TEST

In summary, the following step-by-step procedure can be used to compute the probability of making a Type II error in hypothesis tests about a population mean.

1. Formulate the null and alternative hypotheses.
2. Use the level of significance α to establish a rejection rule based on the test statistic.
3. Using the rejection rule, solve for the value of the sample mean that identifies the rejection region for the test.
4. Use the results from step 3 to state the values of the sample mean that lead to the acceptance of H_0. It also defines the acceptance region for the test.
5. Using the sampling distribution of $\bar{x}$ for any value of μ from the alternative hypothesis, and the acceptance region from step 4, compute the probability that the sample mean will be in the acceptance region. This probability is the probability of making a Type II error at the chosen value of μ.

EXERCISES

Methods

56. Consider the following hypothesis test.

$$H_0: \mu \geq 10$$
$$H_a: \mu < 10$$

The sample size is 120 and the population standard deviation is
Use $\alpha = .05$.

a. If the population mean is 9, what is the probability
conclusion *do not reject* H_0?

b. What type of error would be made if the actu
that H_0: $\mu \geq 10$ is true?

c. What is the probability of making a Ty

57. Consider the following hypothesis test.

$$H_0: \mu = 20$$
$$H_a: \mu \neq 20$$

A sample of 200 items will be taken and the population standard deviation is $\sigma = 10$. Use $\alpha = .05$. Compute the probability of making a Type II error if the population mean is:

a. $\mu = 18.0$ **b.** $\mu = 22.5$ **c.** $\mu = 21.0$

Applications

58. Fowle Marketing Research, Inc., bases charges to a client on the assumption that telephone surveys can be completed within 15 minutes or less. If more time is required, a premium rate is charged. With a sample of 35 surveys, a standard deviation of 4 minutes, and a level of significance of .01, the sample mean will be used to test the null hypothesis $H_0: \mu \leq 15$.

a. What is your interpretation of the Type II error for this problem? What is its impact on the firm?

b. What is the probability of making a Type II error when the actual mean time is $\mu = 17$ minutes?

c. What is the probability of making a Type II error when the actual mean time is $\mu = 18$ minutes?

d. Sketch the general shape of the power curve for this test.

59. A consumer research group is interested in testing an automobile manufacturer's claim that a new economy model will travel at least 25 miles per gallon of gasoline ($H_0: \mu \geq 25$).

a. With a .02 level of significance and a sample of 30 cars, what is the rejection rule based on the value of $\bar{x}$ for the test to determine whether the manufacturer's claim should be rejected? Assume that σ is three miles per gallon.

b. What is the probability of committing a Type II error if the actual mileage is 23 miles per gallon?

c. What is the probability of committing a Type II error if the actual mileage is 24 miles per gallon?

d. What is the probability of committing a Type II error if the actual mileage is 25.5 miles per gallon?

60. *Young Adult* magazine states the following hypotheses about the mean age of its subscribers.

$$H_0: \mu = 28$$
$$H_a: \mu \neq 28$$

a. What would it mean to make a Type II error in this situation?

b. The population standard deviation is assumed known at $\sigma = 6$ years and the sample size is 100. With $\alpha = .05$, what is the probability of accepting H_0 for μ equal to 26, 27, 29, and 30?

c. What is the power at $\mu = 26$? What does this tell you?

61. A production line operation is tested for filling-weight accuracy using the following hypotheses.

Hypothesis	Conclusion and Action
H_0: $\mu = 16$	Filling okay; keep running
H_a: $\mu \neq 16$	Filling off standard; stop and adjust machine

The sample size is 30 and the population standard deviation is $\sigma = .8$. Use $\alpha = .05$.

What would a Type II error mean in this situation?

Vhat is the probability of making a Type II error when the machine is overfilling by nces?

c. What is the power of the statistical test when the machine is overfilling by .5 ounces?
d. Show the power curve for this hypothesis test. What information does it contain for the production manager?

62. Refer to Exercise 58. Assume the firm selects a sample of 50 surveys and repeat parts (b) and (c). What observation can you make about how increasing the sample size affects the probability of making a Type II error?

63. Sparr Investments, Inc., specializes in tax-deferred investment opportunities for its clients. Recently Sparr has offered a payroll deduction investment program for the employees of a particular company. Sparr estimates that the employees are currently averaging \$100 or less per month in tax-deferred investments. A sample of 40 employees will be used to test Sparr's hypothesis about the current level of investment activity among the population of employees. Assume the employee monthly tax-deferred investment amounts have a standard deviation of \$75 and that a .05 level of significance will be used in the hypothesis test.
a. What is the Type II error in this situation?
b. What is the probability of the Type II error if the actual mean employee monthly investment is \$120?
c. What is the probability of the Type II error if the actual mean employee monthly investment is \$130?
d. Assume a sample size of 80 employees is used and repeat parts (b) and (c).

9.9 DETERMINING THE SAMPLE SIZE FOR A HYPOTHESIS TEST ABOUT A POPULATION MEAN

Assume that a hypothesis test is to be conducted about the value of a population mean. The level of significance specified by the user determines the probability of making a Type I error for the test. By controlling the sample size, the user can also control the probability of making a Type II error. Let us show how a sample size can be determined for the following one-tailed test about a population mean.

$$H_0: \mu \geq \mu_0$$
$$H_a: \mu < \mu_0$$

The upper part of Figure 9.16 is the sampling distribution of $\bar{x}$ when H_0 is true and $\mu = \mu_0$. Note that the user's specified level of significance α determines the rejection region for the test. Let c denote the critical value such that $\bar{x} < c$ determines the rejection region for the test. With z_α indicating the z value corresponding to an area of α in the tail of the standard normal probability distribution, we compute c by using the following formula.

$$c = \mu_0 - z_\alpha \frac{\sigma}{\sqrt{n}}$$

...cal

Now consider the sampling distribution in the lower part of Figur... we have selected a value of the population mean, denoted by μ_a, t... case when H_0 is false and H_a is true with $\mu_a < \mu_0$. Let us assume... probability of a Type II error that can be tolerated if the true... probability is shown as β in Figure 9.16. With z_β indicating... area of β in the tail of the standard normal probability d... value c by the following formula.

(9.12)

$$c = \mu_a$$

FIGURE 9.16 DETERMINING THE SAMPLE SIZE FOR SPECIFIED LEVELS OF THE TYPE I (α) AND TYPE II (β) ERRORS

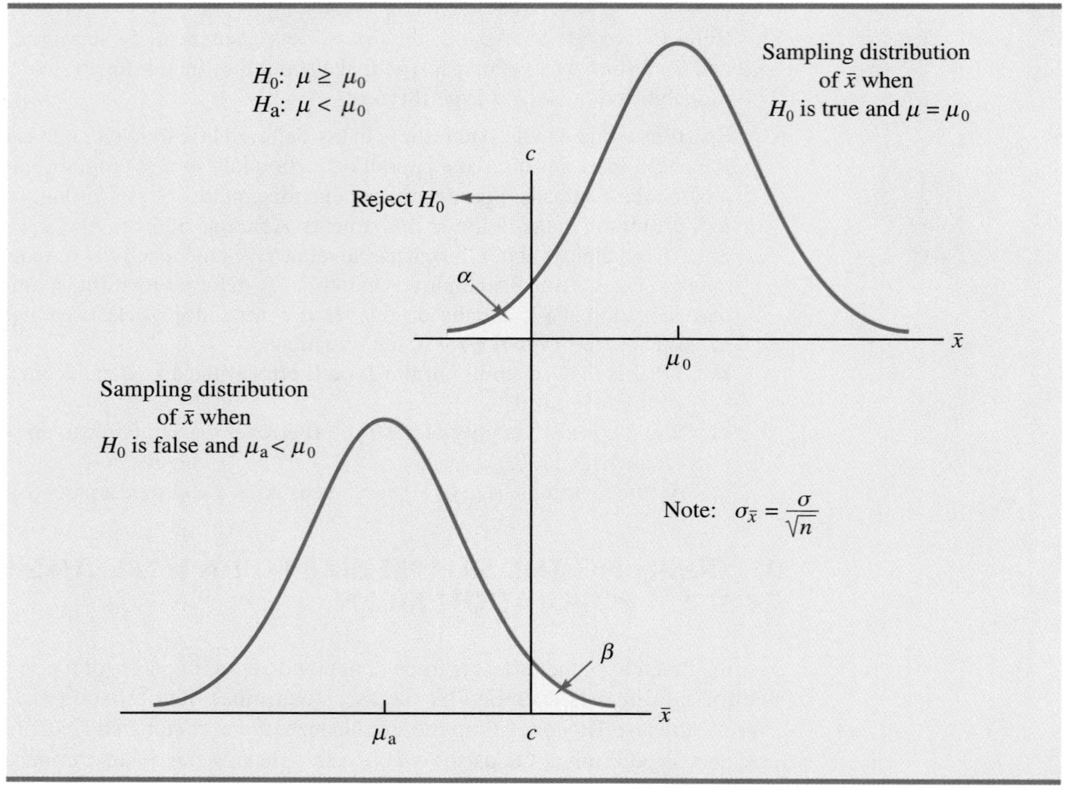

Because (9.11) and (9.12) are both expressions for c, we know they must be equal and thus the following expression must be true.

$$\mu_0 - z_\alpha \frac{\sigma}{\sqrt{n}} = \mu_a + z_\beta \frac{\sigma}{\sqrt{n}}$$

To determine the required sample size, we first solve for the $\sqrt{n}$ as follows.

$$\mu_0 - \mu_a = z_\alpha \frac{\sigma}{\sqrt{n}} + z_\beta \frac{\sigma}{\sqrt{n}}$$

$$\mu_0 - \mu_a = \frac{(z_\alpha + z_\beta)\sigma}{\sqrt{n}}$$

and

$$\sqrt{n} = \frac{(z_\alpha + z_\beta)\sigma}{(\mu_0 - \mu_a)}$$

uaring both sides of the expression provides the following sample size formula for a one-
hypothesis test about a population mean.

Recommended Sample Size for a One-Tailed Hypothesis Test About a Population Mean

$$n = \frac{(z_\alpha + z_\beta)^2\sigma^2}{(\mu_0 - \mu_a)^2} \tag{9.13}$$

where

z_α = z value providing an area of α in the tail of a standard normal distribution
z_β = z value providing an area of β in the tail of a standard normal distribution
σ = the population standard deviation
μ_0 = the value of the population mean in the null hypothesis
μ_a = the value of the population mean used for the Type II error

Note: In a two-tailed hypothesis test, use (9.13) with $z_{\alpha/2}$ replacing z_α.

Although the logic of equation (9.13) was developed for the hypothesis test shown in Figure 9.16, it holds for any one-tailed test about a population mean. Note that in a two-tailed hypothesis test about a population mean, $z_{\alpha/2}$ is used instead of z_α in (9.13).

Let us return to the lot acceptance example from Sections 9.7 and 9.8. The design specification for the shipment of batteries indicated a mean useful life of at least 120 hours for the batteries. Shipments were rejected if H_0: $\mu \geq 120$ was rejected. Let us assume that the quality control manager makes the following statements about the allowable probabilities for the Type I and Type II errors.

Type I error statement: If the mean life of the batteries in the shipment is $\mu = 120$, I am willing to risk an $\alpha = .05$ probability of rejecting the shipment.

Type II error statement: If the mean life of the batteries in the shipment is five hours under the specification (i.e., $\mu = 115$), I am willing to risk a $\beta = .10$ probability of accepting the shipment.

These statements are based on the judgment of the manager. Someone else might specify different restrictions on the probabilities. However, statements about the allowable probabilities of both errors must be made before the sample size can be determined.

In the example, $\alpha = .05$ and $\beta = .10$. Using the standard normal probability distribution, we have $z_{.05} = 1.645$ and $z_{.10} = 1.28$. From the statements about the error probabili- ties, we note that $\mu_0 = 120$ and $\mu_a = 115$. Finally, the population standard deviat assumed known at $\sigma = 12$. By using (9.13), we find that the recommended sam the lot acceptance example is

$$n = \frac{(1.645 + 1.28)^2(12)^2}{(120 - 115)^2} = 49.3$$

Rounding up, the recommended sample size is 50.

Because both the Type I and Type II error probabili able levels with $n = 50$, the quality control manager and reject H_0 statements for the hypothesis test. with allowable probabilities of making Type I

We can make three observations about the relationship among α, β, and the sample size n.

1. Once two of the three values are known, the other can be computed.
2. For a given level of significance α, increasing the sample size will reduce β.
3. For a given sample size, decreasing α will increase β, whereas increasing α will decrease β.

The third observation should be kept in mind when the probability of a Type II error is not being controlled. It suggests that one should not choose unnecessarily small values for the level of significance α. For a given sample size, choosing a smaller level of significance means more exposure to a Type II error. Inexperienced users of hypothesis testing often think that smaller values of α are always better. They are better if we are concerned only about making a Type I error. However, smaller values of α have the disadvantage of increasing the probability of making a Type II error.

EXERCISES

Methods

64. Consider the following hypothesis test.

$$H_0: \mu \geq 10$$
$$H_a: \mu < 10$$

The sample size is 120 and the population standard deviation is 5. Use $\alpha = .05$. If the actual population mean is 9, the probability of a Type II error is .2912. Suppose the researcher wants to reduce the probability of a Type II error to .10 when the actual population mean is 9. What sample size is recommended?

65. Consider the following hypothesis test.

$$H_0: \mu = 20$$
$$H_a: \mu \neq 20$$

The population standard deviation is 10. Use $\alpha = .05$. How large a sample should be taken if the researcher is willing to accept a .05 probability of making a Type II error when the actual population mean is 22?

Applications

66. Suppose the project director for the Hilltop Coffee study (see Section 9.3) had asked for a .10 probability of claiming that Hilltop was not in violation when it really was underfilling by one ounce ($\mu_a = 2.9375$ pounds). What sample size would have been recommended?

SELF test

67. A special industrial battery must have a life of at least 400 hours. A hypothesis test is to be conducted with a .02 level of significance. If the batteries from a particular production run have an actual mean use life of 385 hours, the production manager wants a sampling procedure that only 10% of the time would show erroneously that the batch is acceptable. What sample size is recommended for the hypothesis test? Use 30 hours as an estimate of the population standard deviation.

68. *Young Adult* magazine states the following hypotheses about the mean age of its subscribers.

$$H_0: \mu = 28$$
$$H_a: \mu \neq 28$$

If the manager conducting the test will permit a .15 probability of making a Type II error when the true mean age is 29, what sample size should be selected? Assume $\sigma = 6$ and a .05 level of significance.

69. An automobile mileage study tested the following hypotheses.

Hypothesis	Conclusion
H_0: $\mu \geq 25$ mpg	Manufacturer's claim supported
H_a: $\mu < 25$ mpg	Manufacturer's claim rejected; average mileage per gallon less than stated

For $\sigma = 3$ and a .02 level of significance, what sample size would be recommended if the researcher wants an 80% chance of detecting that μ is less than 25 miles per gallon when it is actually 24?

SUMMARY

Hypothesis testing is a statistical procedure that uses sample data to determine whether a statement about the value of a population parameter should or should not be rejected. The hypotheses, which come from a variety of sources, must be two competing statements: a null hypothesis, H_0, and an alternative hypothesis, H_a. In some applications it is not obvious how the null and alternative hypotheses should be formulated. In Section 9.1, we suggested guidelines for developing hypotheses in three types of situations most frequently encountered.

Figure 9.17* summarizes the test statistics used in hypothesis tests about a population mean and provides a practical guide for selecting the hypothesis testing procedure. The figure shows that the test statistic depends on whether the sample size is large, whether the population standard deviation is assumed known, and in some cases whether the population has a normal or approximately normal probability distribution. If the sample size is large ($n \geq 30$), the z test statistic is used to conduct the hypothesis test. If the sample size is small ($n < 30$), the population must have a normal or approximately normal distribution to conduct a hypothesis test about the value of μ. The z test statistic is used if σ is assumed known, whereas the t test statistic is used if σ is estimated by the sample standard deviation s. Finally, note that if the sample size is small and the assumption of a normally distributed population is inappropriate, we recommend increasing the sample size to $n \geq 30$.

The rejection rule for hypothesis testing procedures involves comparing the value of the test statistic with a critical value. For lower-tail tests, the null hypothesis is rejected if the value of the test statistic is less than the critical value. For upper-tail tests, the null hypothesis is rejected if the test statistic is greater than the critical value. For two-tailed tests, the null hypothesis is rejected for values of the test statistic in either tail of the sampling distribution.

We also saw that p-values could be used for hypothesis testing. The p-value yields the probability, when the null hypothesis is true, of obtaining a sample result that is at least as unlikely as what is observed. When p-values are used to conduct a hypothesis test, the rejection rule calls for rejecting the null hypothesis whenever the p-value is less than α.

Extensions of hypothesis testing procedures to include an analysis of the Type II error were also presented. In Section 9.8, we showed how to compute the probability of making a Type II error. In Section 9.9 we showed how to determine a sample size that would enable us to control the probability of making both a Type I error and a Type II error.

*In some cases, nonparametric statistical methods can be used to conduct a hypothesis test. These methods are discussed in Chapter 19.

FIGURE 9.17 SUMMARY OF THE TEST STATISTICS TO BE USED IN A HYPOTHESIS TEST ABOUT A POPULATION MEAN

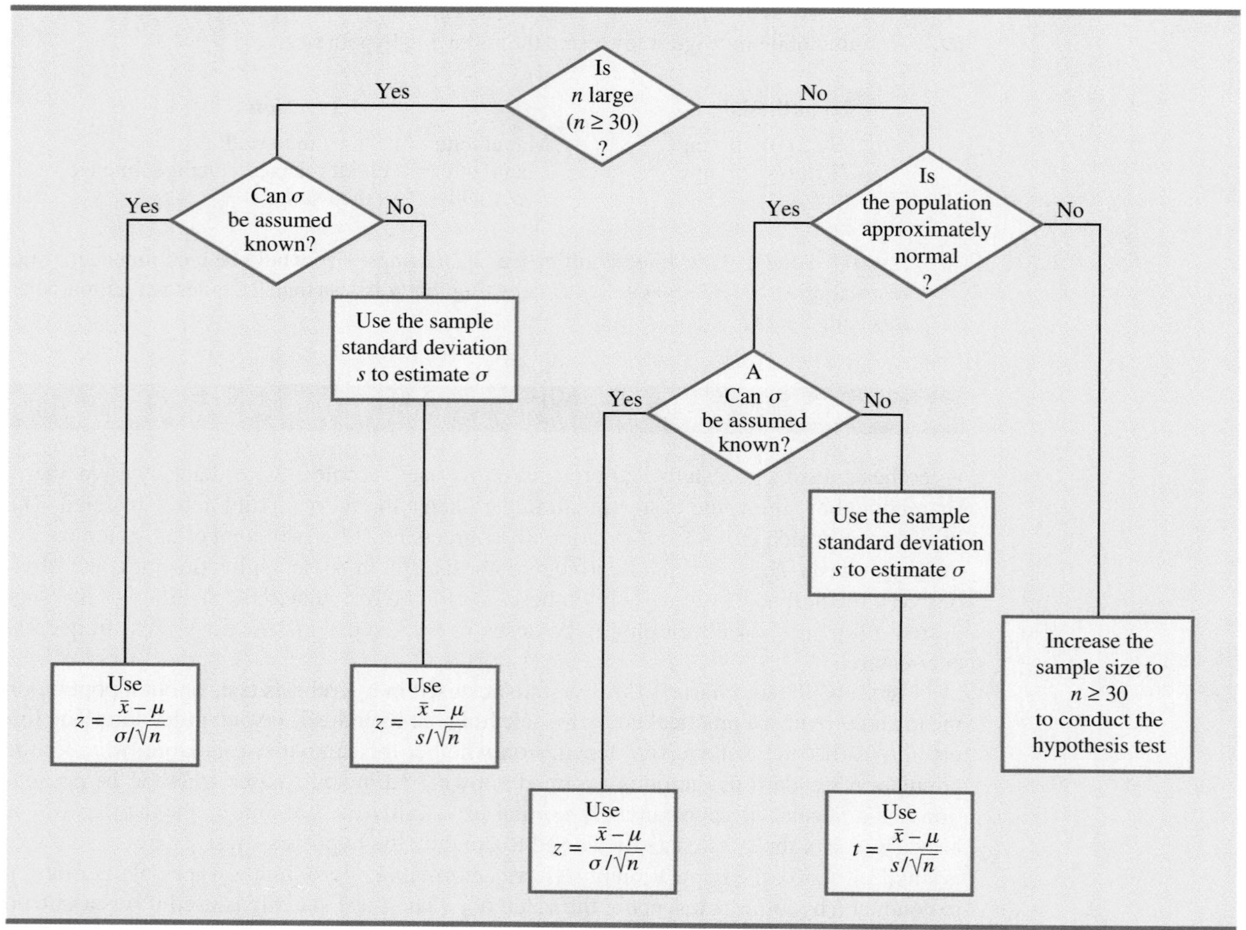

GLOSSARY

Null hypothesis The hypothesis tentatively assumed true in the hypothesis testing procedure.

Alternative hypothesis The hypothesis concluded to be true if the null hypothesis is rejected.

Type I error The error of rejecting H_0 when it is true.

Type II error The error of accepting H_0 when it is false.

Level of significance The maximum allowable probability of making a Type I error.

Rejection region The range of values that will lead to the rejection of a null hypothesis.

Test statistic A statistic whose value is used to determine whether a null hypothesis can be rejected.

Critical value A value that is compared with the test statistic to determine whether H_0 should be rejected.

One-tailed test A hypothesis test in which rejection of the null hypothesis occurs for values of the test statistic in one tail of the sampling distribution.

Two-tailed test A hypothesis test in which rejection of the null hypothesis occurs for values of the test statistic in either tail of the sampling distribution.

***p*-value** The probability, when the null hypothesis is true, of obtaining a sample result that is at least as unlikely as what is observed. It is often called the observed level of significance.

Power The probability of correctly rejecting H_0 when it is false.

Power curve A graph of the probability of rejecting H_0 for all possible values of the population parameter not satisfying the null hypothesis. The power curve provides the probability of correctly rejecting the null hypothesis.

KEY FORMULAS

Test Statistic for a Large-Sample ($n \geq 30$) Hypothesis Test About a Population Mean

σ assumed known
$$z = \frac{\bar{x} - \mu_0}{\sigma/\sqrt{n}} \tag{9.1}$$

σ estimated by s
$$z = \frac{\bar{x} - \mu_0}{s/\sqrt{n}} \tag{9.2}$$

Test Statistic for a Small-Sample ($n < 30$) Hypothesis Test About a Population Mean

σ assumed known
$$z = \frac{\bar{x} - \mu_0}{\sigma/\sqrt{n}} \tag{9.7}$$

σ estimated by s
$$t = \frac{\bar{x} - \mu_0}{s/\sqrt{n}} \tag{9.8}$$

Test Statistic for Hypothesis Test About a Population Proportion

$$z = \frac{\bar{p} - p_0}{\sigma_{\bar{p}}} \tag{9.9}$$

where

$$\sigma_{\bar{p}} = \sqrt{\frac{p_0(1 - p_0)}{n}} \tag{9.10}$$

Sample Size for a One-Tailed Hypothesis Test About a Population Mean

$$n = \frac{(z_\alpha + z_\beta)^2\sigma^2}{(\mu_0 - \mu_a)^2} \tag{9.13}$$

In a two-tailed test, replace z_α with $z_{\alpha/2}$.

SUPPLEMENTARY EXERCISES

70. The population mean annual salary for public school teachers in the state of New York is \$45,250. A sample mean annual salary of public school teachers in New York City is \$47,000 (*Time,* April 3, 2000). Assume the New York City results are based on a sample of 95 teachers. The sample standard deviation was \$6,300.

a. Formulate the null and alternative hypotheses that can be used to determine whether the sample data support the conclusion that public school teachers in New York City have a higher mean salary than the public school teachers in the state of New York.

b. What is the value of the test statistic?

c. What is the *p*-value?

d. Using $\alpha = .01$, what is your conclusion?

71. The Ford Taurus is listed as having a highway fuel efficiency average of 30 miles per gallon (1995 *Motor Trend* New Car Buyer's Guide). A consumer interest group conducts automobile mileage tests seeking statistical evidence to show that automobile manufacturers overstate the miles-per-gallon ratings for particular models. In the case of the Ford Taurus, hypotheses for the test would be stated H_0: $\mu \geq 30$ and H_a: $\mu < 30$. In a sample of 50 mileage tests with the Ford Taurus, the consumer interest group obtained a sample mean highway mileage rating of 29.5 miles per gallon and a sample standard deviation of 1.8 miles per gallon. What conclusion should be drawn from the sample results? Use a .01 level of significance.

72. The chamber of commerce of a Florida Gulf Coast community advertises that area residential property is available at a mean cost of $25,000 or less per lot. Using a .05 level of significance, test the validity of this claim. Suppose a sample of 32 properties provided a sample mean of $26,000 per lot and a sample standard deviation of $2500. What is the *p*-value?

73. A bath soap manufacturing process is designed to produce a mean of 120 bars of soap per batch. Quantities over or under the standard are undesirable. A sample of 10 batches shows the following numbers of bars of soap. The population is assumed to be normal.

108	118	120	122	119	113	124	122	120	123

Using a .05 level of significance, test to see whether the sample results indicate that the manufacturing process is functioning properly.

74. The monthly rent for a two-bedroom apartment in a particular city is reported to average $550. Suppose we want to test H_0: $\mu = 550$ versus H_a: $\mu \neq 550$. A sample of 36 two-bedroom apartments is selected. The sample mean turns out to be $\bar{x} = \$562$, with a sample standard deviation of $s = \$40$.

a. Use the value of the test statistic to draw a conclusion with $\alpha = .05$.

b. Compute the *p*-value.

75. Stout Electric Company operates a fleet of trucks that provide electrical service to the construction industry. Monthly mean maintenance cost has been $75 per truck. A random sample of 40 trucks provided a sample mean maintenance cost of $82.50 per month, with a sample standard deviation of $30. Managers want a test to determine whether the mean monthly maintenance cost has increased.

a. Using a .05 level of significance, what is the rejection rule for this test?

b. What is your conclusion based on the sample mean of $82.50?

c. What is the *p*-value associated with this sample result? What is your conclusion based on the *p*-value?

76. In making bids on building projects, Sonneborn Builders, Inc., assumes construction workers are idle no more than 15% of the time. Hence, for a normal 8-hour shift, the mean idle time per worker should be 72 minutes or less per day. A sample of 30 construction workers had a mean idle time of 80 minutes per day. The sample standard deviation was 20 minutes. Suppose a hypothesis test is to be designed to test the validity of the company's assumption.

a. What is the *p*-value associated with the sample result?

b. What is your conclusion?

77. Sixty percent of Americans believe that business profits are distributed unfairly (General Social Surveys, National Opinion Research Center, University of Chicago). Suppose a sample of 40 Midwesterners showed that 27 believe business profits are distributed unfairly.

a. Do these results justify the inference that a larger proportion of Midwesterners believe that business profits are distributed unfairly? Use $\alpha = .05$.

b. What is the *p*-value?

78. An airline promotion to business travelers was based on the assumption that approximately two-thirds of business travelers use a laptop computer on overnight business trips. Test this assumption by testing the hypotheses H_0: $p = .67$ and H_a: $p \neq .67$ at $\alpha = .05$. Use the sample results from an American Express on-line line survey which found 355 of 546 business travelers use a laptop computer on overnight business trips (*The Cincinnati Enquirer,* August 31, 1998).

a. What is the sample proportion?

b. Compute the value of the test statistic.

c. What is the p-value?

d. Can the claim of two-thirds be rejected? Discuss.

79. In 1993 the Immigration and Naturalization service reported that 79% of foreign travelers visiting the United States stated that the primary purpose of their visit was to enjoy a vacation. In a follow-up study conducted in 2001, a sample of 500 foreign visitors showed 360 said that their primary reason for visiting the United States was to enjoy a vacation. Was the population proportion of foreign travelers vacationing in the United States in 2001 less than the proportion reported previously? Support your conclusion with a statistical test using a .05 level of significance.

80. The Gallup Organization conducted a survey of 1350 people for the National Occupational Information Coordinating Committee, a panel Congress created to improve the use of job information. A research question related to the study was: Do individuals hold jobs that they planned to hold or do they hold jobs for such reasons as chance or lack of choice? Let p indicate the population proportion of individuals who hold jobs that they planned to hold.

a. If the hypotheses are stated H_0: $p \geq .50$ and H_a: $p < .50$, discuss the research hypothesis H_a in terms of what the researcher is investigating.

b. The Gallup poll found that 41% of the respondents hold jobs they planned to hold. What is your conclusion at a .01 level of significance? Discuss.

81. A well-known doctor hypothesized that 75% of women wear shoes that are too small. A study of 356 women by the American Orthopedic Foot and Ankle Society found 313 women wore shoes that were at least one size too small. Test H_0: $p = .75$ and H_a: $p \neq .75$ at $\alpha = .01$. What is your conclusion?

82. The Department of Transportation reported Amtrack trains had a 78% on-time arrival record over the previous 12 months (*USA Today,* November 23, 1998). Assume that in 2001, a study found 330 of 400 Amtrack trains arrived on time. Does the sample indicate the Amtrack arrival rate has changed? Test H_0: $p = .78$ and H_a: $p \neq .78$ at $\alpha = .05$.

a. What is the sample proportion of Amtrack trains arriving on time?

b. Compute the value of the test statistic.

c. What is the p-value?

d. What is your conclusion?

83. A radio station in Myrtle Beach announced that at least 90% of the hotels and motels would be full for the Memorial Day weekend. The station advised listeners to make reservations in advance if they planned to be in the resort over the weekend. On Saturday night a sample of 58 hotels and motels showed 49 with a no-vacancy sign and nine with vacancies. What is your reaction to the radio station's claim after seeing the sample evidence? Use $\alpha = .05$ in making the statistical test. What is the p-value?

84. Environmental health indicators include air quality, water quality, and food quality. Twenty-five years ago, 47% of U.S. food samples contained pesticide residues (*U.S. News & World Report,* April 17, 2000). In a recent study, 44 of 125 food samples contained pesticide residues. Test the hypotheses: H_0: $p \geq .47$ and H_a: $p < .47$.

a. What is the sample proportion?

b. Compute the value of the test statistic.

c. What is the p-value?

d. Using $\alpha = .01$, what is your conclusion?

85. Refer again to Exercise 76.

a. What is the probability of making a Type II error when the population mean idle time is 80 minutes?

b. What is the probability of making a Type II error when the population mean idle time is 75 minutes?

c. What is the probability of making a Type II error when the population mean idle time is 70 minutes?

d. Sketch the power curve for this problem.

86. A federal funding program is available to low-income neighborhoods. To qualify for the funding a neighborhood must have a mean household income of less than \$15,000 per year. Neighborhoods with mean annual household income of \$15,000 or more do not qualify. Funding decisions are based on a sample of residents in the neighborhood. A hypothesis test with a .02 level of significance is conducted. If the funding guidelines call for a maximum probability of .05 of not funding a neighborhood with a mean annual household income of \$14,000, what sample size should be used in the funding decision study? Use $\sigma = \$4000$ as a planning value.

87. H_0: $\mu = 120$ and H_a: $\mu \neq 120$ are used to test whether a bath soap production process is meeting the standard output of 120 bars per batch. Use a .05 level of significance for the test and a planning value of 5 for the standard deviation.

a. If the mean output drops to 117 bars per batch, the firm wants to have a 98% chance of concluding that the standard production output is not being met. How large a sample should be selected?

b. With your sample size from part (a), what is the probability of concluding that the process is operating satisfactorily for each of the following actual mean outputs: 117, 118, 119, 121, 122, and 123 bars per batch? That is, what is the probability of a Type II error in each case?

Case Problem 1 QUALITY ASSOCIATES, INC.

Quality Associates, Inc., is a consulting firm that advises its clients about sampling and statistical procedures that can be used to control their manufacturing processes. In one particular application, a client gave Quality Associates a sample of 800 observations taken during a time in which that client's process was operating satisfactorily. The sample standard deviation for these data was .21; hence, the population standard deviation was assumed to be .21. Quality Associates then suggested that random samples of size 30 be taken periodically to monitor the process on an ongoing basis. By analyzing the new samples, the client could quickly learn whether the process was operating satisfactorily. When the process was not operating satisfactorily, corrective action could be taken to eliminate the problem. The design specification indicated the mean for the process should be 12. The hypotheses test suggested by Quality Associates follows.

$$H_0: \mu = 12$$
$$H_a: \mu \neq 12$$

Corrective action will be taken any time H_0 is rejected.

The following samples were collected at hourly intervals during the first day of operation of the new statistical process control procedure. These data are available in the data set Quality.

Quality

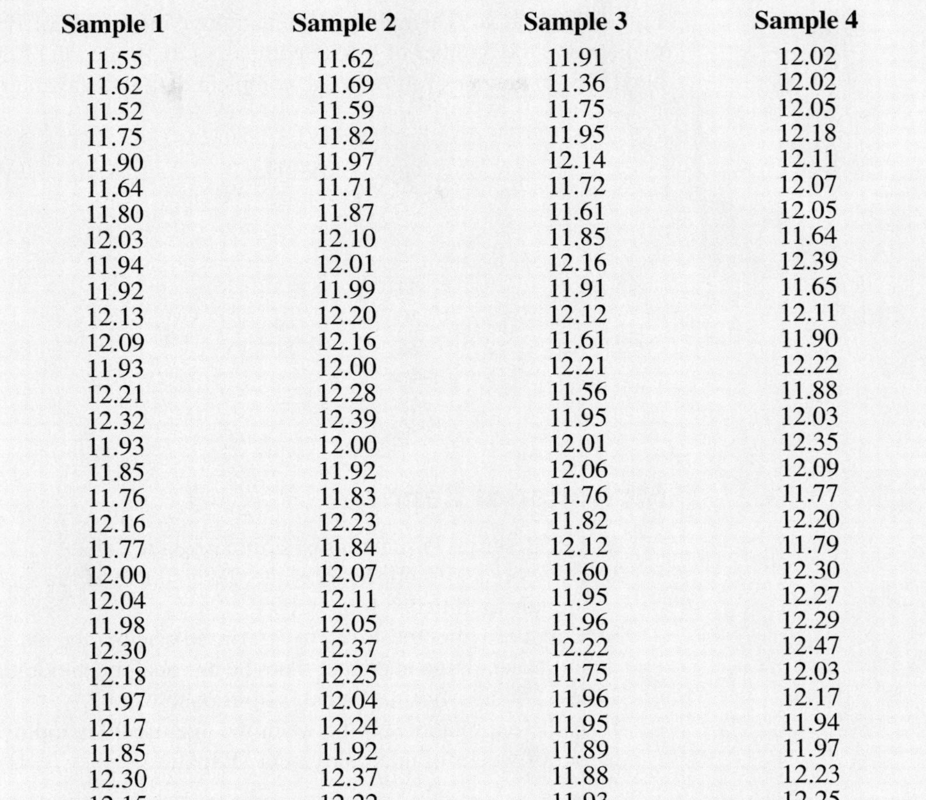

Sample 1	Sample 2	Sample 3	Sample 4
11.55	11.62	11.91	12.02
11.62	11.69	11.36	12.02
11.52	11.59	11.75	12.05
11.75	11.82	11.95	12.18
11.90	11.97	12.14	12.11
11.64	11.71	11.72	12.07
11.80	11.87	11.61	12.05
12.03	12.10	11.85	11.64
11.94	12.01	12.16	12.39
11.92	11.99	11.91	11.65
12.13	12.20	12.12	12.11
12.09	12.16	11.61	11.90
11.93	12.00	12.21	12.22
12.21	12.28	11.56	11.88
12.32	12.39	11.95	12.03
11.93	12.00	12.01	12.35
11.85	11.92	12.06	12.09
11.76	11.83	11.76	11.77
12.16	12.23	11.82	12.20
11.77	11.84	12.12	11.79
12.00	12.07	11.60	12.30
12.04	12.11	11.95	12.27
11.98	12.05	11.96	12.29
12.30	12.37	12.22	12.47
12.18	12.25	11.75	12.03
11.97	12.04	11.96	12.17
12.17	12.24	11.95	11.94
11.85	11.92	11.89	11.97
12.30	12.37	11.88	12.23
12.15	12.22	11.93	12.25

Managerial Report

1. Conduct the hypothesis test for each sample at the .01 level of significance and determine what action, if any, should be taken. Provide the test statistic and p-value for each test.
2. Compute the standard deviation for each of the four samples. Does the assumption of .21 for the population standard deviation appear reasonable?
3. Compute limits for the sample mean $\bar{x}$ around $\mu = 12$ such that, as long as a new sample mean is within those limits, the process will be considered to be operating satisfactorily. If $\bar{x}$ exceeds the upper limit or if $\bar{x}$ is below the lower limit, corrective action will be taken. These limits are referred to as upper and lower control limits for quality control purposes.
4. Discuss the implications of changing the level of significance to a larger value. What mistake or error could increase if that were done?

Case Problem 2 UNEMPLOYMENT STUDY

Each month the U.S. Bureau of Labor Statistics publishes a variety of unemployment statistics, including the number of individuals who are unemployed and the mean length of time the individuals have been unemployed. For November 1998, the Bureau of Labor Statistics reported that the national mean length of time of unemployment was 14.6 weeks.

The mayor of Philadelphia has requested a study on the status of unemployment in the Philadelphia area. A sample of 50 unemployed residents of Philadelphia included data on the age and the number of weeks without a job. A portion of the data collected in November 1998 is shown as follows. The complete data set is available in the data file BLS.

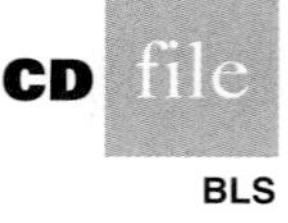
BLS

Age	Weeks	Age	Weeks
56	22	25	5
35	19	40	20
22	7	25	12
57	37	25	1
40	18	59	33
22	11	49	26
48	6	33	13
48	22		

Managerial Report

1. Use descriptive statistics to summarize the data.
2. Develop a 95% confidence interval estimate of the mean age of unemployed individuals in Philadelphia.
3. Conduct a hypothesis test to determine whether the mean duration of unemployment in Phiiladelphia is greater than the national mean duration of 14.6 weeks. Use a .01 level of significance. What is your conclusion?
4. Is there a relationship between the age of an unemployed individual and the number of weeks of unemployment? Explain.

Appendix 9.1 HYPOTHESIS TESTING WITH MINITAB

In this appendix, we describe how to conduct hypothesis tests about a population mean in both the large- and small-sample cases.

Large-Sample Case

Distance

We illustrate the large-sample case by using the data on golf ball distances in Table 9.2. The data have been entered into column C1 of a Minitab worksheet. The level of significance for the hypothesis test is $\alpha = .05$ and the population standard deviation σ is estimated by the sample standard deviation s. The following steps can be used to test the hypothesis H_0: $\mu = 280$ versus H_a: $\mu \neq 280$.

Step 1. Select the **Calc** pull-down menu
Step 2. Choose **Column Statistics**
Step 3. When the Column Statistics dialog box appears:
Select **Standard Deviation**
Enter C1 in the **Input variable** box
Enter stdev in the **Store result in** box
Click **OK**
Step 4. Select the **Stat** pull-down menu
Step 5. Choose **Basic Statistics**
Step 6. Choose **1-Sample Z**
Step 7. When the 1-Sample Z dialog box appears:
Enter C1 in the **Variables** box
Enter stdev in the **Sigma** box

Enter 280 in the **Test mean** box
Select **Options**

Step 8. Enter 95 in the **Confidence level** box*
Select not equal in the **Alternative** box
Click **OK**

Step 9. Click **OK**

In addition to the hypothesis testing results shown in Figure 9.8, the Minitab procedure provides a 95% confidence interval for the population mean.

The Superflight golf ball study used the sample standard deviation s to estimate the population standard deviation σ. In a large-sample hypothesis test where σ is assumed known, steps 1 to 3 of the preceding procedure may be skipped. In this case, simply start with step 4 and enter the appropriate value of σ in the **Sigma** box during step 7. Finally, the Superflight golf ball study involved a two-tailed hypothesis test. However, the procedure can be easily modified for a one-tailed hypothesis test by selecting the less than or greater than option in the **Alternative** box in step 8.

Small-Sample Case

Heathrow

The ratings that 12 business travelers gave for Heathrow airport were listed in Section 9.5. The data have been entered in column C1 of a Minitab worksheet. The level of significance for the test is $\alpha = .05$, and the population standard deviation σ will be estimated by the sample standard deviation s. The following steps can be used to test the hypothesis H_0: $\mu \leq 7$ against H_a: $\mu > 7$.

Step 1. Select the **Stat** pull-down menu

Step 2. Choose **Basic Statistics**

Step 3. Choose **1-Sample t**

Step 4. When the 1-Sample t dialog box appears:
Enter C1 in the **Variables** box
Enter 7 in the **Test mean** box
Select **Options**

Step 5. Enter 95 in the **Confidence level** box
Select greater than in the **Alternative** box
Click **OK**

Step 6. Click **OK**

This procedure uses the sample standard deviation s to estimate the population standard deviation σ. In a small-sample hypothesis test where σ is assumed known, steps 4 through 9 of the large-sample 1-Sample Z procedure can be used with the appropriate value of σ entered in the **Sigma** box during step 7. Finally, the Heathrow airport rating study involved a one-tailed hypothesis test. The preceding steps can be easily modified for other hypothesis tests by selecting the less than or not equal option in the **Alternative** box in step 5.

Appendix 9.2 HYPOTHESIS TESTING WITH EXCEL

Large-Sample Case

Distance

We illustrate the large-sample case by using the data on golf ball distances in Table 9.2. The label Yards appears in cell A1 and the 36 distance values appear in cells A2 to A37 of an Excel worksheet. Note that the level of significance for the hypothesis test is $\alpha = .05$, and the population standard deviation σ is estimated by the sample standard deviation s.

*Minitab provides both hypothesis testing and interval estimation results simultaneously. The user may select any confidence level for the interval estimate of the population mean: 95% confidence is suggested here.

A relatively easy way to do hypothesis testing with Excel is to develop your own spreadsheet and use the p-value criterion to draw the conclusion. The worksheet we developed is shown in Figure 9.18. The steps that can be used to test H_0: $\mu = 280$ and H_a: $\mu \neq 280$ are as follows:

Step 1. Enter 280 in cell D2
Step 2. Compute the sample size in cell D6

=COUNT(A2:A37)

Step 3. Compute the sample mean in cell D7

=AVERAGE(A2:A37)

Step 4. Compute the sample standard deviation in cell D8

=STDEV(A2:A37)

Step 5. Compute the test statistic in cell D9

=(D7-D2)/(D8/SQRT(D6))

Step 6. Compute the p-value in cell D10

=2*(1-NORMSDIST(ABS(D9)))

Cell D9 contains the formula for the test statistic

$$z = \frac{\bar{x} - \mu_0}{s/\sqrt{n}}$$

Cell D10 contains the formula for computing the p-value, which is the normal distribution two-tailed area associated with the test statistic z in cell D9.

The Superflight golf ball study used the sample standard deviation s to estimate the population standard deviation σ. In a large-sample hypothesis test where σ is assumed

FIGURE 9.18 EXCEL SPREADSHEET FOR THE SUPERFLIGHT GOLF BALL HYPOTHESIS TEST

	A	B	C	D	E
1	**Yards**		**Hypothesis Test**		
2	269		Test Mean	280	
3	300				
4	268				
5	278		**Sample Results**		
6	282		Sample Size	36	
7	263		Mean	278.5	
8	301		Standard Deviation	12	
9	295		Test Statistic	-0.75	
10	288		p-value	0.453	
11	278				
12	276				
13	286				
36	274				
37	277				

Note: Rows 14–35 are hidden.

known, the value of σ can be entered directly into cell D8 because no computation is necessary. Finally, the Superflight golf ball study involved a two-tailed hypothesis test. However, this procedure can be easily modified for a one-tailed hypothesis test. The cell D10 formula =NORMSDIST(D9) provides the p-value when the reject region is in the lower tail and the cell D10 formula =1-NORMSDIST(D9) provides the p-value when the reject region is in the upper tail.

Small-Sample Case

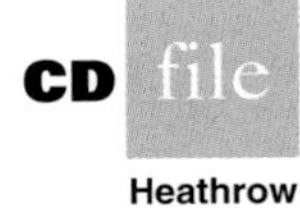

Heathrow

The ratings that 12 business travelers gave for Heathrow airport were listed in Section 9.5. The data have been entered in an Excel worksheet with the label Rating in cell A1 and the 12 ratings in cells A2:A13. The level of significance for the test is $\alpha = .05$, and the population standard deviation σ will be estimated by the sample standard deviation s. With the spreadsheet layout as shown in Figure 9.18, the following steps can be used to test the hypotheses H_0: $\mu \leq 7$ and H_a: $\mu > 7$.

Step 1. Enter 7 in cell D2

Step 2. Compute the sample size in cell D6

=COUNT(A2:A13)

Step 3. Compute the sample mean in cell D7

=AVERAGE(A2:A13)

Step 4. Compute the sample standard deviation in cell D8

=STDEV(A2:A13)

Step 5. Compute the test statistic in cell D9

=(D7-D2)/(D8/SQRT(D6))

Step 6. Compute the p-value in cell D10

=IF(D9>0,TDIST(D9,D6-1,1),1-TDIST(ABS(D9),D6-1,1))

Cell D9 contains the formula for the test statistic

$$t = \frac{\bar{x} - \mu_0}{s/\sqrt{n}}$$

Cell D10 contains the formula for computing the p-value, which is the t distribution upper-tail area associated with the test statistic t. This formula is relatively complex because the expression for computing the p-value depends on whether the test statistic in cell D9 is positive or negative.

The Heathrow airport study used the sample standard deviation s to estimate the population standard deviation σ. In a small-sample hypothesis test where σ is assumed known, the value of σ can be entered directly into cell D8 because no computation is necessary. In this case, the test statistic is z and its associated p-value is computed using the NORMSDIST function exactly as described for the large-sample case.

Finally, the Heathrow airport small-sample study involved a one-tailed hypothesis test with an upper-tail rejection region. However, this procedure can be easily modified for other hypothesis tests. For example, the cell D10 formula =IF(D9>0,1-TDIST(D9,D6-1,1), TDIST(ABS(D9),D6-1,1)) provides the p-value when the reject region is in the lower tail and the cell D10 formula =TDIST(ABS(D9),D6-1,2) provides the p-value for a two-tailed test.

Chapter 10

STATISTICAL INFERENCE ABOUT MEANS AND PROPORTIONS WITH TWO POPULATIONS

CONTENTS

STATISTICS IN PRACTICE

FISONS CORPORATION
Rochester, New York

Fisons Corporation, Rochester, New York, is a unit of Fisons Plc., UK. Fisons opened its United States operations in 1966.

Fisons' Pharmaceutical Division uses extensive statistical procedures to test and develop new drugs. The testing process in the pharmaceutical industry usually consists of three stages: (1) preclinical testing, (2) testing for long-term usage and safety, and (3) clinical efficacy testing. At each successive stage, the chance that a drug will pass the rigorous tests decreases; however, the cost of further testing increases dramatically. Industry surveys indicate that on average the research and development for one new drug costs $250 million and takes 12 years. Hence, it is important to eliminate unsuccessful new drugs in the early stages of the testing process, as well as identify promising ones for further testing.

Statistics plays a major role in pharmaceutical research, where government regulations are stringent and rigorously enforced. In preclinical testing, a two- or three-population statistical study typically is used to determine whether a new drug should continue to be studied in the long-term usage and safety program. The populations may consist of the new drug, a control, and a standard drug. The preclinical testing process begins when a new drug is sent to the pharmacology group for evaluation of efficacy—the capacity of the drug to produce the desired effects. As part of the process, a statistician is asked to design an experiment that can be used to test the new drug. The design must specify the sample size and the statistical methods of analysis. In a two-population study, one sample is used to obtain data on the efficacy of the new drug (population 1) and a second sample is used to obtain data on the efficacy of a standard drug (population 2). Depending on the intended use, the new and standard drugs are tested in such disciplines as neurology, cardiology, and immunology. In most studies, the statistical method involves hypothesis testing for the difference between the means of the new drug population and the standard drug population. If a new drug lacks efficacy or produces undesirable effects in comparison with the standard drug, the new drug is rejected and withdrawn from further testing. Only new drugs that show promising comparisons with the standard drugs are forwarded to the long-term usage and safety testing program.

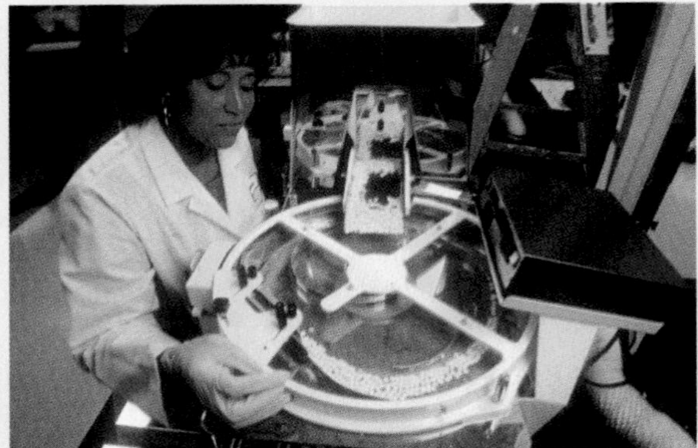

Statistical methods are used to test and develop new drugs. © Mark Richards/PhotoEdit.

Further data collection and multipopulation studies are conducted in the long-term usage and safety testing program and in the clinical testing programs. The Food and Drug Administration (FDA) requires that statistical methods be defined prior to such testing to avoid data-related biases. In addition, to avoid human biases, some of the clinical trials are double or triple blind. That is, neither the subject nor the investigator knows what drug is administered to whom. If the new drug meets all requirements in relation to the standard drug, a new drug application (NDA) is filed with the FDA. The application is rigorously scrutinized by statisticians and scientists at the agency.

In this chapter you will learn how to construct interval estimates and make hypothesis tests about means and proportions with two populations. Techniques will be presented for analyzing independent random samples as well as matched samples.

In Chapters 8 and 9 we showed how to develop interval estimates and conduct hypothesis tests for situations involving one population mean and one population proportion. In this chapter we continue our discussion of statistical inference by showing how interval estimates and hypothesis tests can be developed for situations involving two populations, when the difference between the two population means or the two population proportions is of prime importance. For example, we may want to develop an interval estimate of the difference between the mean starting salary for a population of men and the mean starting salary for a population of women or conduct a hypothesis test to determine whether any difference is present between the proportion of defective parts in a population of parts produced by supplier A and the proportion of defective parts in a population of parts produced by supplier B. We will begin our discussion of statistical inference about means and proportions with two populations by showing how an interval estimate of the difference between the means of two populations can be developed for a sampling study conducted by Greystone Department Stores, Inc.

10.1 ESTIMATION OF THE DIFFERENCE BETWEEN THE MEANS OF TWO POPULATIONS: INDEPENDENT SAMPLES

Greystone Department Stores, Inc., operates two stores in Buffalo, New York; one is in the inner city and the other is in a suburban shopping center. The regional manager has noticed that products that sell well in one store do not always sell well in the other. The manager believes this situation may be attributable to differences in customer demographics at the two locations. Customers may differ in age, education, income, and so on. Suppose the manager has asked us to investigate the difference between the mean ages of the customers who shop at the two stores.

Let us define population 1 as all customers who shop at the inner-city store and population 2 as all customers who shop at the suburban store.

μ_1 = mean of population 1 (i.e., the mean age of all customers who shop at the inner-city store)

μ_2 = mean of population 2 (i.e., the mean age of all customers who shop at the suburban store)

The difference between the two population means is $\mu_1 - \mu_2$.

To estimate $\mu_1 - \mu_2$, we will select a simple random sample of n_1 customers from population 1 and a simple random sample of n_2 customers from population 2. Since the simple random sample of n_1 customers is selected independently of the simple random sample of n_2 customers, we have the case of **independent simple random samples.**

$\bar{x}_1$ = sample mean age for the simple random sample of n_1 inner-city customers

$\bar{x}_2$ = sample mean age for the simple random sample of n_2 suburban customers

Because $\bar{x}_1$ is a point estimator of μ_1 and $\bar{x}_2$ is a point estimator of μ_2, the point estimator of the difference between the two population means is expressed as follows.

Point Estimator of the Difference Between the Means of Two Populations

$$\bar{x}_1 - \bar{x}_2 \tag{10.1}$$

Thus, we see that the point estimator of the difference between the two population means is the difference between the sample means of the two independent simple random samples. Figure 10.1 provides an overview of the process used to estimate the difference between two population means based on two independent simple random samples.

FIGURE 10.1 ESTIMATING THE DIFFERENCE BETWEEN THE MEANS OF TWO POPULATIONS

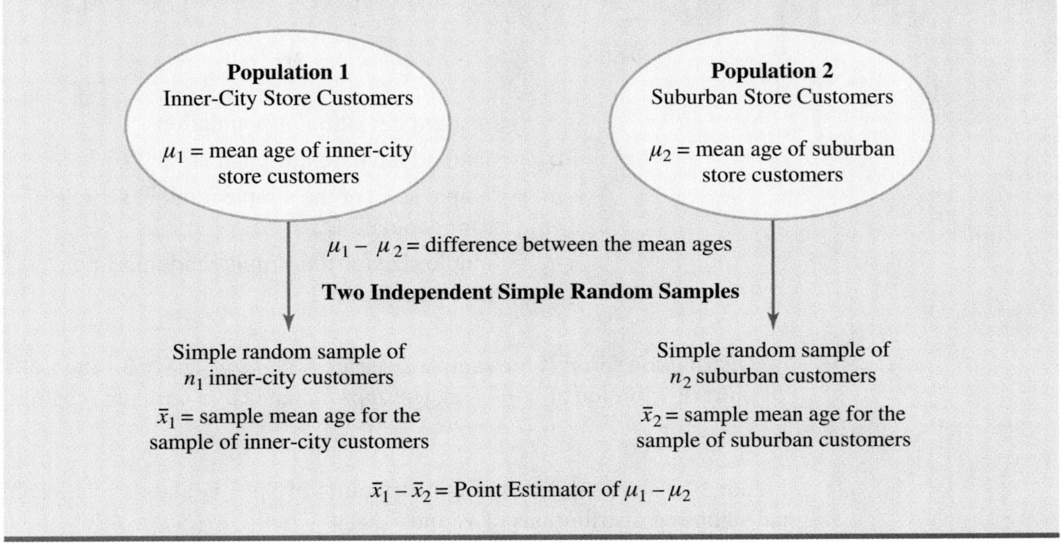

Assume the customer age data collected from the two independent simple random samples of Greystone customers provide the following results.

Store	Number of Customers Sampled	Sample Mean Age	Sample Standard Deviation
Inner City	36	$\bar{x}_1 = 40$ years	$s_1 = 9$ years
Suburban	49	$\bar{x}_2 = 35$ years	$s_2 = 10$ years

Using (10.1), we find that a point estimate of the difference between the mean ages of the two populations is $\bar{x}_1 - \bar{x}_2 = 40 - 35 = 5$ years. Thus, we are led to believe that the customers at the inner-city store have a mean age five years greater than the mean age of the suburban store customers. However, as with all point estimates, we know that five years is only one of many possible estimates of the difference between the mean ages of the two populations. If Greystone selected another simple random sample of 36 inner-city customers and another simple random sample of 49 suburban customers, the difference between the two new sample means would probably not equal five years. The sampling distribution of $\bar{x}_1 - \bar{x}_2$ is the probability distribution of the difference in sample means for all possible sets of two independent simple random samples.

Sampling Distribution of $\bar{x}_1 - \bar{x}_2$

We can use the sampling distribution of $\bar{x}_1 - \bar{x}_2$ to develop an interval estimate of the difference between the two population means in much the same way as we used the sampling distribution of $\bar{x}$ for interval estimation with a single population mean. The sampling distribution of $\bar{x}_1 - \bar{x}_2$ has the following properties.

Sampling Distribution of $\bar{x}_1 - \bar{x}_2$

$$\text{Expected Value: } E(\bar{x}_1 - \bar{x}_2) = \mu_1 - \mu_2 \qquad \textbf{(10.2)}$$

$$\text{Standard Deviation:} \quad \sigma_{\bar{x}_1-\bar{x}_2} = \sqrt{\frac{\sigma_1^2}{n_1} + \frac{\sigma_2^2}{n_2}} \tag{10.3}$$

where

σ_1 = standard deviation of population 1
σ_2 = standard deviation of population 2
n_1 = sample size for the simple random sample from population 1
n_2 = sample size for the simple random sample from population 2

Distribution form: If the sample sizes are both *large* ($n_1 \geq 30$ and $n_2 \geq 30$), the sampling distribution of $\bar{x}_1 - \bar{x}_2$ can be approximated by a normal probability distribution.

Figure 10.2 shows the sampling distribution of $\bar{x}_1 - \bar{x}_2$ and its relationship to the individual sampling distributions of $\bar{x}_1$ and $\bar{x}_2$.

Let us now develop an interval estimate of the difference between the means of two populations. We consider two cases, one in which the sample sizes are large ($n_1 \geq 30$ and $n_2 \geq 30$) and another in which one or both sample sizes are small ($n_1 < 30$ and/or $n_2 < 30$). We consider the large-sample case first.

Interval Estimate of $\mu_1 - \mu_2$: Large-Sample Case

In the large-sample case, the sampling distribution of $\bar{x}_1 - \bar{x}_2$ can be approximated by a normal probability distribution. With this approximation we can use the following expression to develop an interval estimate of the difference between the means of the two populations.

Interval Estimate of the Difference Between the Means of Two Populations: Large-Sample Case ($n_1 \geq 30$ and $n_2 \geq 30$) with σ_1 and σ_2 Assumed Known

$$\bar{x}_1 - \bar{x}_2 \pm z_{\alpha/2}\sigma_{\bar{x}_1-\bar{x}_2} \tag{10.4}$$

where $1 - \alpha$ is the confidence coefficient

Note that to use (10.4) to develop an interval estimate of the difference between the means of two populations, we must know the value of $\sigma_{\bar{x}_1-\bar{x}_2}$, the standard deviation of the sampling distribution of $\bar{x}_1 - \bar{x}_2$. However, (10.3) shows that the value of $\sigma_{\bar{x}_1-\bar{x}_2}$ depends on the values of σ_1 and σ_2, the standard deviations of each of the populations. When the population standard deviations are unknown, we can use the sample standard deviations as estimates of the population standard deviations and estimate $\sigma_{\bar{x}_1-\bar{x}_2}$ as follows.

Point Estimator of $\sigma_{\bar{x}_1-\bar{x}_2}$

$$s_{\bar{x}_1-\bar{x}_2} = \sqrt{\frac{s_1^2}{n_1} + \frac{s_2^2}{n_2}} \tag{10.5}$$

FIGURE 10.2 SAMPLING DISTRIBUTION OF $\bar{x}_1 - \bar{x}_2$ AND ITS RELATIONSHIP TO THE INDIVIDUAL SAMPLING DISTRIBUTION OF $\bar{x}_1$ AND $\bar{x}_2$

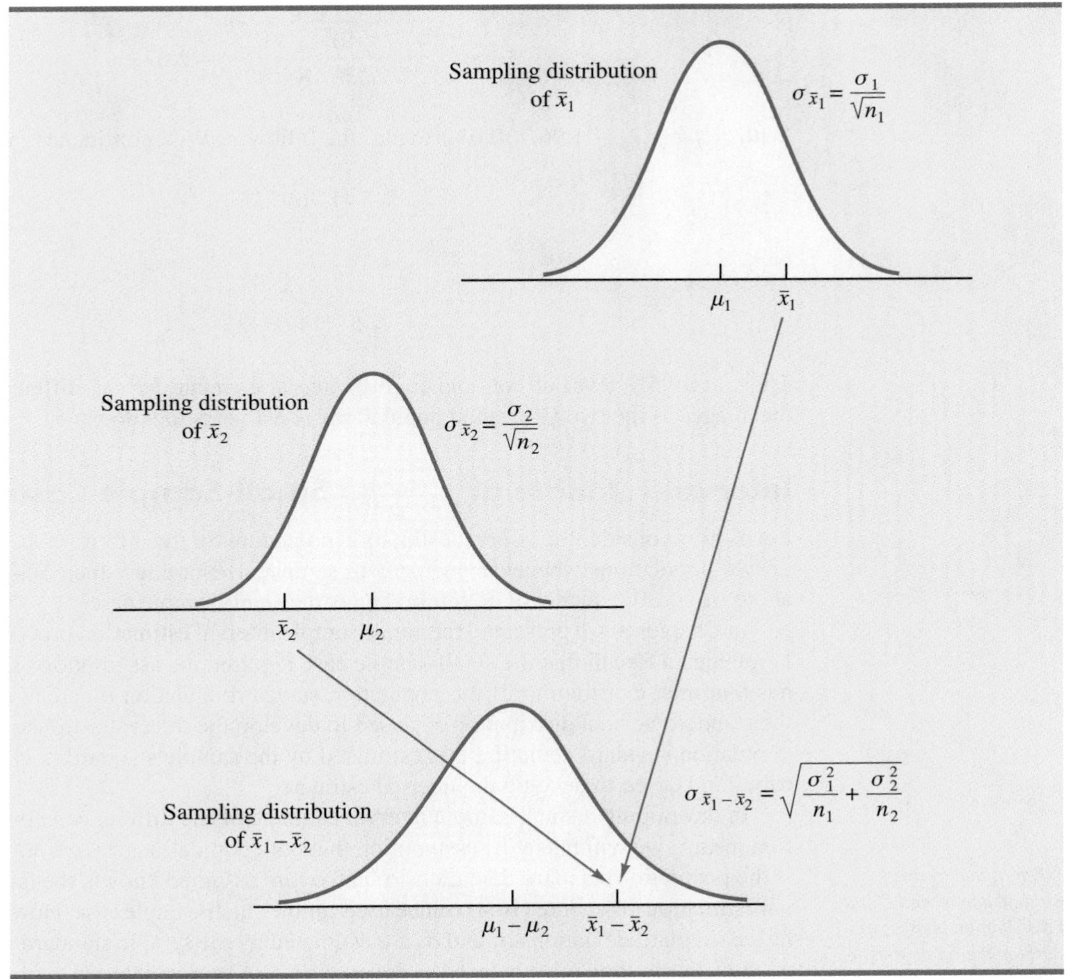

Thus, in the large-sample case, we can use the point estimator of $\sigma_{\bar{x}_1-\bar{x}_2}$ to develop the following confidence interval estimate of the difference between the two population means.

Interval Estimate of the Difference Between the Means of Two Populations: Large-Sample Case ($n_1 \geq 30$ and $n_2 \geq 30$) With σ_1 and σ_2 Estimated by s_1 and s_2

$$\bar{x}_1 - \bar{x}_2 \pm z_{\alpha/2} s_{\bar{x}_1-\bar{x}_2} \quad \textbf{(10.6)}$$

where $1 - \alpha$ is the confidence coefficient.

Let us use expression (10.6) to develop a confidence interval estimate of the difference between the mean ages of the two customer populations in the Greystone Department Store study. Recall that the sample mean age and sample standard deviation for the simple random sample of 36 inner-city customers are $\bar{x}_1 = 40$ years and $s_1 = 9$ years, respectively; the sample mean and sample standard deviation for the simple random sample of 49 suburban customers are $\bar{x}_2 = 35$ years and $s_2 = 10$ years, respectively. Using (10.5) to estimate $\sigma_{\bar{x}_1 - \bar{x}_2}$, we have

$$s_{\bar{x}_1 - \bar{x}_2} = \sqrt{\frac{(9)^2}{36} + \frac{(10)^2}{49}} = 2.07$$

With $z_{\alpha/2} = z_{.025} = 1.96$, (10.6) provides the following 95% confidence interval.

$$5 \pm (1.96)(2.07)$$

or

$$5 \pm 4.06$$

Thus, at a 95% level of confidence, the interval estimate for the difference between the mean ages of the two Greystone populations is .94 years to 9.06 years.

Interval Estimate of $\mu_1 - \mu_2$: Small-Sample Case

Let us now consider the interval estimation procedure for the difference between the means of two populations whenever one or both sample sizes are less than 30—that is, $n_1 < 30$ and/or $n_2 < 30$, which will be referred to as the small-sample case.

In Chapter 8, we presented the small-sample interval estimation procedures for a population mean. Recall that the small-sample case required the assumption that the population has a normal distribution. If the population standard deviation σ can be assumed known, the standard normal distribution z is used to develop the interval estimate. However, if the population standard deviation σ is estimated by the sample standard deviation s, the t distribution is used to develop the interval estimate.

When σ_1 and σ_2 are estimated by s_1 and s_2, the t distribution is used to develop the small-sample interval estimate of the difference between two population means.

In developing a small-sample interval estimate of the difference between two population means, we will make the assumption that both populations have a normal distribution. If the population standard deviations σ_1 and σ_2 are assumed known, the large-sample interval estimation procedure (10.4) can be used for the small-sample case. However, if the population standard deviations σ_1 and σ_2 are estimated by the sample standard deviations s_1 and s_2, the t distribution must be used to develop a small-sample interval estimate of the difference between two population means. The details of this procedure and an example follow. We begin by making the following assumptions:

If the sample sizes are equal, the procedure in this section provides acceptable results even if the population variances are not equal. Thus, whenever possible, a researcher should consider equal sample sizes with $n_1 = n_2$.

1. Both populations have normal distributions.
2. The variances of the populations are equal ($\sigma_1^2 = \sigma_2^2 = \sigma^2$).

Given these assumptions, the sampling distribution of $\bar{x}_1 - \bar{x}_2$ is normally distributed regardless of the sample sizes. The expected value of $\bar{x}_1 - \bar{x}_2$ is $\mu_1 - \mu_2$. Because of the equal variances assumption, (10.3) can be written

$$\sigma_{\bar{x}_1 - \bar{x}_2} = \sqrt{\frac{\sigma^2}{n_1} + \frac{\sigma^2}{n_2}} = \sqrt{\sigma^2\left(\frac{1}{n_1} + \frac{1}{n_2}\right)} \quad \textbf{(10.7)}$$

The sampling distribution of $\bar{x}_1 - \bar{x}_2$ is shown in Figure 10.3.

Because equation (10.7) is based on the assumption that $\sigma_1^2 = \sigma_2^2 = \sigma^2$, we do not need separate estimates of σ_1^2 and σ_2^2. In fact, we can combine the data from the two samples to provide the best single estimate of σ^2. The process of combining the results of two independent simple random samples to provide one estimate of σ^2 is referred to as *pooling*. The

FIGURE 10.3 SAMPLING DISTRIBUTION OF $\bar{x}_1 - \bar{x}_2$ WHEN THE POPULATIONS HAVE NORMAL DISTRIBUTIONS WITH EQUAL VARIANCES

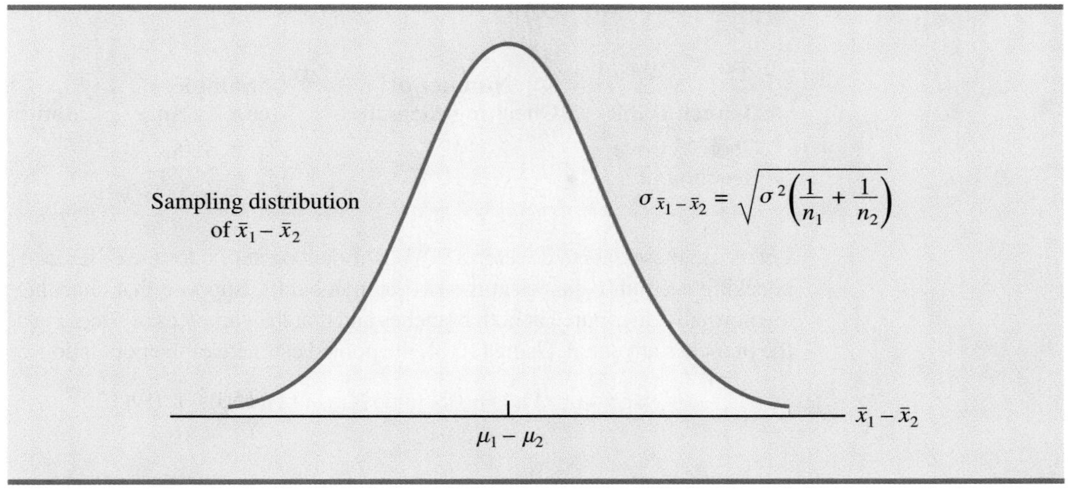

pooled variance estimator of σ^2, denoted by s^2, is a weighted average of the two sample variances s_1^2 and s_2^2. The formula for the pooled estimator of σ^2 follows.

A pooled variance estimator of σ^2 is a weighted average of the two sample variances and will always be a value between s_1^2 and s_2^2.

Pooled Estimator of σ^2

$$s^2 = \frac{(n_1 - 1)s_1^2 + (n_2 - 1)s_2^2}{n_1 + n_2 - 2} \tag{10.8}$$

With s^2 as the pooled estimator of σ^2, we can use (10.7) to obtain the following estimator of $\sigma_{\bar{x}_1 - \bar{x}_2}$.

Point Estimator of $\sigma_{\bar{x}_1 - \bar{x}_2}$ When $\sigma_1^2 = \sigma_2^2 = \sigma^2$

$$s_{\bar{x}_1 - \bar{x}_2} = \sqrt{s^2\left(\frac{1}{n_1} + \frac{1}{n_2}\right)} \tag{10.9}$$

The t distribution can now be used to compute an interval estimate of the difference between the means of the two populations. Because $n_1 - 1$ degrees of freedom are associated with the sample from population 1 and $n_2 - 1$ degrees of freedom are associated with the sample from population 2, the t distribution will have $n_1 + n_2 - 2$ degrees of freedom. The interval estimation procedure follows.

Interval Estimate of the Difference Between the Means of Two Populations: Small-Sample Case ($n_1 < 30$ and/or $n_2 < 30$) With σ_1 and σ_2 Estimated by s_1 and s_2

$$\bar{x}_1 - \bar{x}_2 \pm t_{\alpha/2}\, s_{\bar{x}_1 - \bar{x}_2} \tag{10.10}$$

where the t value is based on a t distribution with $n_1 + n_2 - 2$ degrees of freedom and where $1 - \alpha$ is the confidence coefficient.

Let us demonstrate the interval estimation procedure for a sampling study conducted by the Clearview National Bank. Independent random samples of checking account balances for customers at two Clearview branch banks yielded the following results.

Branch Bank	Number of Checking Accounts	Sample Mean Balance	Sample Standard Deviation
Cherry Grove	12	$\bar{x}_1 = \$1000$	$s_1 = \$150$
Beechmont	10	$\bar{x}_2 = \$\ 920$	$s_2 = \$120$

Let us use these data to develop a 90% confidence interval for the difference between the mean checking account balances at the two branch banks. Suppose that checking account balances are normally distributed at both branches and that the variances of checking account balances at the branches are equal. Using (10.8), the pooled estimate of the population variance becomes

$$s^2 = \frac{(n_1 - 1)s_1^2 + (n_2 - 1)s_2^2}{n_1 + n_2 - 2} = \frac{(11)(150)^2 + (9)(120)^2}{12 + 10 - 2} = 18{,}855$$

Using (10.9), the corresponding estimate of $\sigma_{\bar{x}_1 - \bar{x}_2}$ is

$$s_{\bar{x}_1 - \bar{x}_2} = \sqrt{s^2\left(\frac{1}{n_1} + \frac{1}{n_2}\right)} = \sqrt{18{,}855\left(\frac{1}{12} + \frac{1}{10}\right)} = 58.79$$

The appropriate t distribution for the interval estimation procedure has $n_1 + n_2 - 2 = 12 + 10 - 2 = 20$ degrees of freedom. With $\alpha = .10$, $t_{\alpha/2} = t_{.05} = 1.725$. Thus, using expression (10.10), we see that the interval estimate becomes

$$\bar{x}_1 - \bar{x}_2 \pm t_{.05}\, s_{\bar{x}_1 - \bar{x}_2}$$
$$1000 - 920 \pm (1.725)(58.79)$$
$$80 \pm 101.41$$

At a 90% confidence level, the interval estimate of the difference between the mean account balances at the two branch banks is −\$21.41 to \$181.41. The fact that the interval includes a negative range of values indicates that the actual difference between the two means, $\mu_1 - \mu_2$, may be negative. Thus, μ_2 could actually be larger than μ_1, indicating that the population mean balance could be greater for the Beechmont branch even though the results show a greater sample mean balance at the Cherry Grove branch. The fact that the confidence interval contains the value 0 can be interpreted as indicating that we do not have sufficient evidence to conclude that the population mean account balances differ between the two branches.

NOTES AND COMMENTS

1. The use of the t distribution in the small-sample procedure presented in this section is based on the assumptions that both populations have a normal distribution and that $\sigma_1^2 = \sigma_2^2$. Fortunately, this procedure is a *robust* statistical procedure, meaning that it is relatively insensitive to these assumptions. For instance, if $\sigma_1^2 \neq \sigma_2^2$, the procedure provides acceptable results if n_1 and n_2 are approximately equal.
2. The t distribution is not restricted to the small-sample situation; it is applicable whenever both populations are normally distributed and the variances of the populations are equal. However, (10.4) and (10.6) show how to determine an interval estimate of the difference between the means of two populations when the sample sizes are large. Thus, in the large-sample case, use of the t distribution and its corresponding assumptions are not required. We therefore do not refer to the t distribution until we have a small-sample case.

EXERCISES

Methods

1. Consider the following results for two independent random samples taken from two populations.

Sample 1	Sample 2
$n_1 = 50$	$n_2 = 35$
$\bar{x}_1 = 13.6$	$\bar{x}_2 = 11.6$
$s_1 = 2.2$	$s_2 = 3.0$

   a. What is the point estimate of the difference between the two population means?
   b. Provide a 90% confidence interval for the difference between the two population means.
   c. Provide a 95% confidence interval for the difference between the two population means.

2. Consider the following results for two independent random samples taken from two populations.

Sample 1	Sample 2
$n_1 = 10$	$n_2 = 8$
$\bar{x}_1 = 22.5$	$\bar{x}_2 = 20.1$
$s_1 = 2.5$	$s_2 = 2.0$

   a. What is the point estimate of the difference between the two population means?
   b. What is the pooled estimate of the population variance?
   c. Develop a 95% confidence interval for the difference between the two population means.

3. Consider the following data for two independent random samples taken from two populations.

Sample 1		Sample 2	
10	7	8	7
12	7	8	4
9	9	6	9

   a. Compute the two sample means.
   b. Compute the two sample standard deviations.
   c. What is the point estimate of the difference between the two population means?
   d. What is the pooled estimate of the population variance?
   e. Develop a 95% confidence interval for the difference between the two population means.

Applications

4. Gasoline prices increased substantially from 1999 to 2000. The American Automobile Association provided information on the mean cost per gallon for self-serve regular unleaded gasoline over the 2 years (*AAA Going Places,* May/June, 2000). Assume the following results were obtained from independent samples of locations throughout the country.

2000 Cost	1999 Cost
$\bar{x}_1 = \$1.58$	$\bar{x}_2 = \$0.98$
$s_1 = \$0.12$	$s_2 = \$0.08$
$n_1 = 50$	$n_2 = 42$

   a. What is the point estimate of the increase in the mean cost per gallon from 1999 to 2000?
   b. What is the 95% confidence interval estimate of the increase in the mean cost per gallon from 1999 to 2000?

5. The U.S. Department of Transportation provides the number of miles that residents of the 75 largest metropolitan areas travel per day in a car. Suppose that for a simple random sample of 50 Buffalo residents the mean is 22.5 miles a day and the standard deviation is 8.4 miles a day, and for an independent simple random sample of 100 Boston residents the mean is 18.6 miles a day and the standard deviation is 7.4 miles a day.
   a. What is the point estimate of the difference between the mean number of miles that Buffalo residents travel per day and the mean number of miles that Boston residents travel per day?
   b. What is the 95% confidence interval for the difference between the two population means?

6. The International Air Transport Association surveyed business travelers to determine ratings of various international airports. The maximum possible score was 10. Suppose 50 business travelers were asked to rate the Miami airport and 50 other business travelers were asked to rate the Los Angeles airport. The rating scores follow.

Miami

6	4	6	8	7	7	6	3	3	8	10	4	8
7	8	7	5	9	5	8	4	3	8	5	5	4
4	4	8	4	5	6	2	5	9	9	8	4	8
9	9	5	9	7	8	3	10	8	9	6		

Los Angeles

10	9	6	7	8	7	9	8	10	7	6	5	7
3	5	6	8	7	10	8	4	7	8	6	9	9
5	3	1	8	9	6	8	5	4	6	10	9	8
3	2	7	9	5	3	10	3	5	10	8		

Develop a 95% confidence interval estimate of the difference between the mean ratings of the Miami and Los Angeles airports.

7. A Cornell University study of wage differentials between men and women reported that one of the reasons wages for men are higher than wages for women is that men tend to have more years of work experience than women (*Business Week,* August 28, 2000). Assume the following sample summaries show the years of experience for each group.

Men	**Women**
$\bar{x}_1 = 14.9$ years	$\bar{x}_2 = 10.3$ years
$s_1 = 5.2$ years	$s_2 = 3.8$ years
$n_1 = 100$	$n_2 = 85$

   a. What is the point estimate of the difference between the two population means?
   b. As 95% confidence, what is the margin of error?
   c. What is the 95% confidence interval estimate of the difference between the two population means?

8. An urban planning group is interested in estimating the difference between the mean household incomes for two neighborhoods in a large metropolitan area. Independent random samples of households in the neighborhoods provided the following results.

Neighborhood 1	**Neighborhood 2**
$n_1 = 8$	$n_2 = 12$
$\bar{x}_1 = \$15{,}700$	$\bar{x}_2 = \$14{,}500$
$s_1 = \$700$	$s_2 = \$850$

   a. Develop a point estimate of the difference between the mean incomes in the two neighborhoods.
   b. Develop a 95% confidence interval for the difference between the mean incomes in the two neighborhoods.
   c. What assumptions were made to compute the interval estimates in part (b)?

9. The National Association of Home Builders provided data on the cost of the most popular home remodeling projects (*USA Today*, June 17, 1997). Sample data on cost in thousands of dollars for two types of remodeling projects are as follows.

Kitchen	Master Bedroom
25.2	18.0
17.4	22.9
22.8	26.4
21.9	24.8
19.7	26.9
23.0	17.8
19.7	24.6
16.9	21.0
21.8	
23.6	

a. Develop a point estimate of the difference between the population mean remodeling costs for the two types of projects.
b. Develop a 90% confidence interval for the difference between the two population means.

10. Suppose independent random samples of 15 unionized women and 20 nonunionized women in manufacturing have been selected and the following hourly wage rates are found.

Union

Union Workers

22.40	18.90	16.70	14.05	16.20	20.00	16.10	16.30	19.10
16.50	18.50	19.80	17.00	14.30	17.20			

Nonunion Workers

17.60	14.40	16.60	15.00	17.65	15.00	17.55	13.30	11.20
15.90	19.20	11.85	16.65	15.20	15.30	17.00	15.10	14.30
13.90	14.50							

a. What is the point estimate of the difference between hourly wages for the two populations?
b. What is the pooled estimate of the population variance?
c. Develop a 95% confidence interval estimate of the difference between the two population means.
d. Does there appear to be any difference in the mean wage rate for these two groups? Explain.

10.2 HYPOTHESIS TESTS ABOUT THE DIFFERENCE BETWEEN THE MEANS OF TWO POPULATIONS: INDEPENDENT SAMPLES

In this section we present procedures that can be used to test hypotheses about the difference between the means of two populations. The methodology is again divided into large-sample ($n_1 \geq 30$, $n_2 \geq 30$) and small-sample ($n_1 < 30$ and/or $n_2 < 30$) cases.

Large-Sample Case

As part of a study to evaluate differences in educational quality between two training centers, a standardized examination is given to individuals who are trained at the two centers.

The examination scores are a major factor in assessing any quality differences between the centers. The means for the two centers are as follows.

μ_1 = the mean examination score for the population of individuals trained at center A

μ_2 = the mean examination score for the population of individuals trained at center B

We begin with the tentative assumption that no difference exists in the training quality provided at the two centers. Hence, in terms of the mean examination scores, the null hypothesis is that $\mu_1 - \mu_2 = 0$. If sample evidence leads to the rejection of this hypothesis, we will conclude that the mean examination scores differ for the two populations. This conclusion indicates a quality differential between the two centers and that a follow-up study investigating the reasons for the differential may be warranted. The null and alternative hypotheses are written as follows.

$$H_0\colon \mu_1 - \mu_2 = 0$$
$$H_a\colon \mu_1 - \mu_2 \neq 0$$

Following the hypothesis testing procedure from Chapter 9, we make the tentative assumption that H_0 is true. Using the difference between the sample means as the point estimator of the difference between the population means, we consider the sampling distribution of $\bar{x}_1 - \bar{x}_2$ when H_0 is true. For the large-sample case, this distribution is as shown in Figure 10.4. Because the sampling distribution of $\bar{x}_1 - \bar{x}_2$ can be approximated by a normal probability distribution, the following test statistic is used.

$$z = \frac{(\bar{x}_1 - \bar{x}_2) - (\mu_1 - \mu_2)}{\sqrt{\sigma_1^2/n_1 + \sigma_2^2/n_2}} \tag{10.11}$$

If σ_1 and σ_2 are unknown, we can use the sample standard deviations s_1 and s_2 to compute the test statistic.

FIGURE 10.4 SAMPLING DISTRIBUTION OF $\bar{x}_1 - \bar{x}_2$ WITH $H_0\colon \mu_1 - \mu_2 = 0$

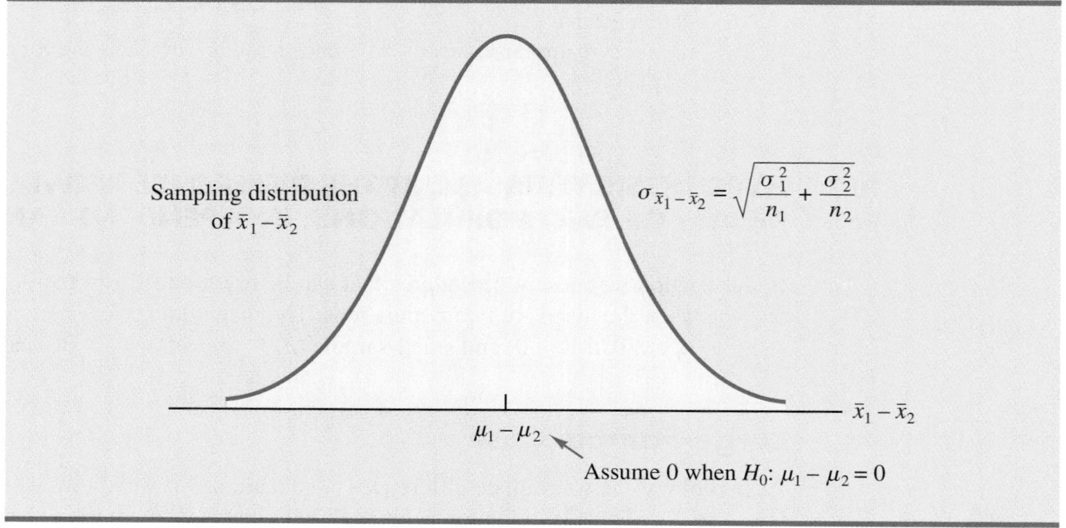

The value of z given by equation (10.11) can be interpreted as the number of standard deviations $\bar{x}_1 - \bar{x}_2$ is from the value of $\mu_1 - \mu_2$ specified in H_0. For $\alpha = .05$ and thus $z_{\alpha/2} = z_{.025} = 1.96$, the rejection region for the two-tailed hypothesis test is shown in Figure 10.5. Using the test statistic, the rejection rule is

$$\text{Reject } H_0 \text{ if } z < -1.96 \text{ or if } z > +1.96$$

Let us assume that independent random samples of individuals trained at the two centers provide the examination scores in Table 10.1; summary statistics are given in Table 10.2. Using s_1 and s_2 to estimate σ_1 and σ_2, we find that the test statistic z given by (10.11) for the null hypothesis H_0: $\mu_1 - \mu_2 = 0$ becomes

$$z = \frac{(82.5 - 78) - 0}{\sqrt{(8)^2/30 + (10)^2/40}} = 2.09$$

With $z = 2.09 > 1.96$, the conclusion is to reject H_0. Thus, we can conclude that μ_1 and μ_2 are not equal and that the two centers differ in terms of educational quality.

FIGURE 10.5 REJECTION REGION FOR THE TWO-TAILED HYPOTHESIS TEST WITH $\alpha = .05$

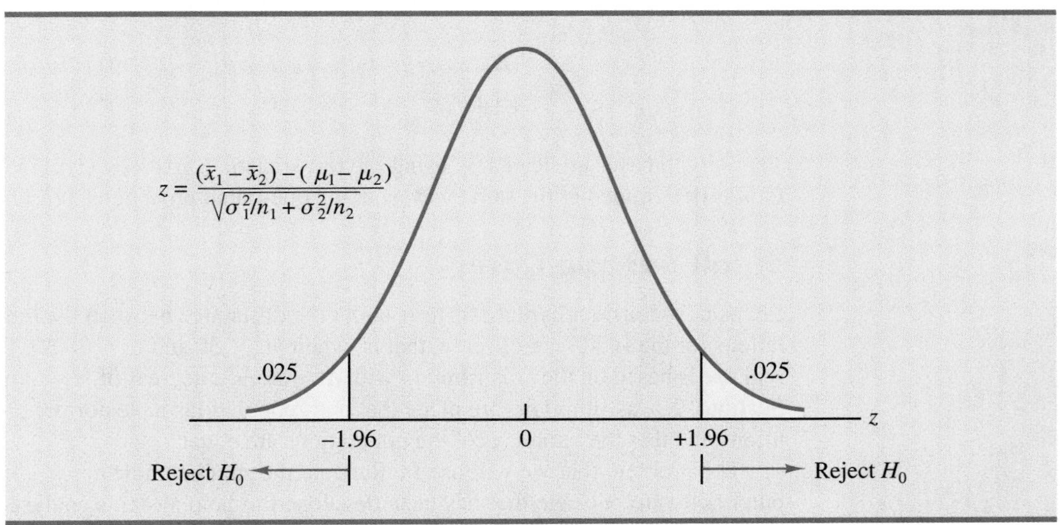

TABLE 10.1 EXAMINATION SCORE DATA

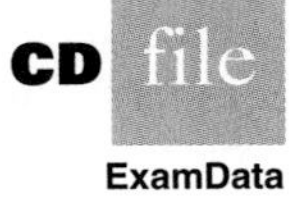

Training Center A			Training Center B			
97	83	91	64	66	91	84
90	84	87	85	83	78	85
94	76	73	72	74	87	85
79	82	92	64	70	93	84
78	85	64	74	82	89	59
87	85	74	93	82	79	62
83	91	88	70	75	84	91
89	72	88	79	78	65	83
76	86	74	79	99	78	80
84	70	73	75	57	66	76

TABLE 10.2 EXAMINATION SCORE RESULTS

Training Center A	Training Center B
$n_1 = 30$	$n_2 = 40$
$\bar{x}_1 = 82.5$	$\bar{x}_2 = 78$
$s_1 = 8$	$s_2 = 10$

The p-value can also be used for this hypothesis test.

With $z = 2.09$, the standard normal probability distribution table can be used to compute the p-value for this two-tailed test. With an area of .4817 between the mean and $z = 2.09$, the p-value is $2(.5000 - .4817) = .0366$; because the p-value is less than $\alpha = .05$, the p-value approach also results in the rejection of H_0.

In this hypothesis test, we were interested in determining whether the means of the two populations differ. We did not have a prior belief that one mean might be greater than or less than the other, therefore the hypotheses $H_0\colon \mu_1 - \mu_2 = 0$ and $H_a\colon \mu_1 - \mu_2 \neq 0$ were appropriate. In other hypothesis tests about the difference between the means of two populations, we may want to find out whether one mean is greater than or perhaps less than the other mean. In these cases, a one-tailed hypothesis test would be appropriate. The two forms of a one-tailed test about the difference between two population means follow.

$$H_0\colon \mu_1 - \mu_2 \leq 0 \qquad H_0\colon \mu_1 - \mu_2 \geq 0$$
$$H_a\colon \mu_1 - \mu_2 > 0 \qquad H_a\colon \mu_1 - \mu_2 < 0$$

These hypotheses are tested by using the test statistic z given by (10.11). The rejection region is determined in the same way as in the one-tailed procedures presented in Chapter 9.

Small-Sample Case

Let us now consider hypothesis tests about the difference between the means of two populations for the small-sample case, that is, where $n_1 < 30$ and/or $n_2 < 30$. The procedure we will use is based on the t distribution with $n_1 + n_2 - 2$ degrees of freedom. As discussed in Section 10.1, assumptions are made that both populations have normal probability distributions and that the variances of the populations are equal.

The problem that we will use to illustrate the small-sample case involves a new computer software package that has been developed to help systems analysts reduce the time required to design, develop, and implement an information system. To evaluate the benefits of the new software package, a random sample of 24 systems analysts is selected. Each analyst is given specifications for a hypothetical information system. Then 12 of the analysts are instructed to produce the information system by using current technology. The other 12 analysts are trained in the use of the new software package and then instructed to use it to produce the information system.

This study has two populations: a population of systems analysts using the current technology and a population of systems analysts using the new software package. In terms of the time required to complete the information system design project, the population means are as follow.

μ_1 = the mean project completion time for systems analysts using the current technology

μ_2 = the mean project completion time for systems analysts using the new software package

The researcher in charge of the new software evaluation project hopes to show that the new software package will provide a shorter mean project completion time. Thus, the researcher is looking for evidence to conclude that μ_2 is less than μ_1; in this case, the difference between the two population means, $\mu_1 - \mu_2$, will be greater than zero. The research hypothesis $\mu_1 - \mu_2 > 0$ is stated as the alternative hypothesis.

$$H_0: \mu_1 - \mu_2 \leq 0$$
$$H_a: \mu_1 - \mu_2 > 0$$

The researcher is looking for evidence to reject H_0 and conclude that the new software package provides a smaller mean completion time.

Suppose that the 24 analysts complete the study with the results shown in Table 10.3. Under the assumption that the variances of the populations are equal, (10.8) is used to compute the pooled estimate of σ^2.

$$s^2 = \frac{(n_1 - 1)s_1^2 + (n_2 - 1)s_2^2}{n_1 + n_2 - 2} = \frac{11(40)^2 + 11(44)^2}{12 + 12 - 2} = 1768$$

The test statistic for the small-sample case is

$$t = \frac{(\bar{x}_1 - \bar{x}_2) - (\mu_1 - \mu_2)}{\sqrt{s^2\left(\frac{1}{n_1} + \frac{1}{n_2}\right)}} \qquad \textbf{(10.12)}$$

In the case of two independent random samples of sizes n_1 and n_2, the t distribution will have $n_1 + n_2 - 2$ degrees of freedom. For $\alpha = .05$, the t distribution table shows that with $12 + 12 - 2 = 22$ degrees of freedom, $t_{.05} = 1.717$. Thus, using the test statistic, the rejection region for the one-tailed test is

$$\text{Reject } H_0 \text{ if } t > 1.717$$

TABLE 10.3 COMPLETION TIME DATA AND SUMMARY STATISTICS FOR THE SOFTWARE TESTING STUDY

Software

	Current Technology	New Software
	300	276
	280	222
	344	310
	385	338
	372	200
	360	302
	288	317
	321	260
	376	320
	290	312
	301	334
	283	265
Summary Statistics		
Sample size	$n_1 = 12$	$n_2 = 12$
Sample mean	$\bar{x}_1 = 325$	$\bar{x}_2 = 288$
Sample standard deviation	$s_1 = 40$	$s_2 = 44$

The sample data and (10.12) provide the following value for the test statistic.

$$t = \frac{(325 - 288) - 0}{\sqrt{1768\left(\frac{1}{12} + \frac{1}{12}\right)}} = 2.16$$

Checking the rejection region, we see that $t = 2.16$ allows the rejection of H_0 at the .05 level of significance. Thus, the sample results enable the researcher to conclude that $\mu_1 - \mu_2 > 0$ and that the new software package provides a smaller mean completion time.

Minitab can be used for testing hypotheses about the difference between the means of two populations. The output comparing the current and new software technology is shown in Figure 10.6. The last line of the output shows $t = 2.16$ and a p-value $= .021$. With $\alpha = .05$, the null hypothesis can be rejected. Thus, the mean completion time for the new software is less than the mean completion time with the current technology.

FIGURE 10.6 MINITAB OUTPUT FOR THE HYPOTHESIS TEST ABOUT THE CURRENT AND NEW SOFTWARE TECHNOLOGY

```
Two sample T for Current vs New
              N        Mean        StDev        SE Mean
Current      12       325.0         40.0             12
New          12       288.0         44.0             13
Difference = mu Current - mu New
Estimate for difference:  37.0
T-Test of difference = 0 (vs >): T = 2.16  P-value = 0.021  DF = 22
Both use Pooled StDev = 42.0
```

NOTES AND COMMENTS

1. In the preceding section, we stated that using the t distribution for inferences about the means of two populations is fairly insensitive to the assumptions of normal populations and equal variances. However, if a user feels strongly that these assumptions are not appropriate for a particular application, one of the following actions should be taken:
   a. Consider the nonparametric Wilcoxon rank sum test presented in Chapter 19.
   b. If the populations are approximately normally distributed but the variances are not equal ($\sigma_1^2 \neq \sigma_2^2$), use (10.5) to estimate $\sigma_{\bar{x}_1 - \bar{x}_2}$. The t distribution can still be used, with the degrees of freedom given by

$$\frac{[1/n_1 + (s_2^2/s_1^2)/n_2]^2}{\left[\frac{1}{n_1^2(n_1 - 1)}\right] + \left[\frac{(s_2^2/s_1^2)^2}{n_2^2(n_2 - 1)}\right]}$$

   c. Increase both sample sizes to the large-sample case with $n_1 \geq 30$ and $n_2 \geq 30$.
2. In hypothesis tests about the difference between the means of two populations, the null hypothesis almost always contains the assumption that there is no difference between the means. Hence, the following null hypotheses are possible choices.

$$H_0: \mu_1 - \mu_2 = 0$$
$$H_0: \mu_1 - \mu_2 \leq 0$$
$$H_0: \mu_1 - \mu_2 \geq 0$$

In some instances, we may want to determine whether there is a nonzero difference D_0 between the population means. The specific value chosen for D_0 depends on the application under study. However, in this case, the null hypothesis may be of the following forms.

$$H_0: \mu_1 - \mu_2 = D_0$$
$$H_0: \mu_1 - \mu_2 \leq D_0$$
$$H_0: \mu_1 - \mu_2 \geq D_0$$

The hypothesis testing computations remain the same with the exception that D_0 is used for the value of $\mu_1 - \mu_2$ in equations (10.11) and (10.12).

EXERCISES

Methods

11. Consider the following hypothesis test.

$$H_0: \mu_1 - \mu_2 \leq 0$$
$$H_a: \mu_1 - \mu_2 > 0$$

The following results are for two independent samples taken from the two populations.

Sample 1	Sample 2
$n_1 = 40$	$n_2 = 50$
$\bar{x}_1 = 25.2$	$\bar{x}_2 = 22.8$
$s_1 = 5.2$	$s_2 = 6.0$

a. With $\alpha = .05$, what is your hypothesis testing conclusion?
b. What is the p-value?

12. Consider the following hypothesis test.

$$H_0: \mu_1 - \mu_2 = 0$$
$$H_a: \mu_1 - \mu_2 \neq 0$$

The following results are for two independent samples taken from the two populations.

Sample 1	Sample 2
$n_1 = 80$	$n_2 = 70$
$\bar{x}_1 = 104$	$\bar{x}_2 = 106$
$s_1 = 8.4$	$s_2 = 7.6$

a. With $\alpha = .05$, what is your hypothesis testing conclusion?
b. What is the p-value?

13. Consider the following hypothesis test.

$$H_0: \mu_1 - \mu_2 = 0$$
$$H_a: \mu_1 - \mu_2 \neq 0$$

The following results are for two independent samples taken from the two populations. With $\alpha = .05$, what is your hypothesis testing conclusion?

Sample 1	Sample 2
$n_1 = 8$	$n_2 = 7$
$\bar{x}_1 = 1.4$	$\bar{x}_2 = 1.0$
$s_1 = 0.4$	$s_2 = 0.6$

Applications

14. Coastal areas of the United States including Cape Cod, the Outer Banks, the Carolinas, and the Gulf Coast had relatively high population growth rates during the 1990s. Data were collected on residents living in the coastal communities as well as on residents living in noncoastal areas throughout the United States (*USA Today,* July 21, 2000).

Assume that the following sample results were obtained on the ages of individuals in the two populations:

Coastal Areas	Noncoastal Areas
$\bar{x}_1 = 39.3$ years	$\bar{x}_2 = 35.4$ years
$s_1 = 16.8$ years	$s_2 = 15.2$ years
$n_1 = 150$	$n_2 = 175$

Test the hypothesis of no difference between the mean ages of the two populations. Use $\alpha = .05$.

a. Formulate the null and alternative hypotheses.
b. What is the rejection rule?
c. What is the value of the test statistic?
d. What is your conclusion?
e. What is the p-value?

15. The Greystone Department Store study in Section 10.1 supplied the following data on customer ages from independent random samples taken at two store locations.

Inner-City Store	Suburban Store
$n_1 = 36$	$n_2 = 49$
$\bar{x}_1 = 40$ years	$\bar{x}_2 = 35$ years
$s_1 = 9$ years	$s_2 = 10$ years

For $\alpha = .05$, test H_0: $\mu_1 - \mu_2 = 0$ against the alternative H_a: $\mu_1 - \mu_2 \neq 0$. What is your conclusion about the mean ages of the populations of customers at the two stores? What is the p-value?

16. The Educational Testing Service conducted a study to investigate differences between the scores of male and female students on the Scholastic Aptitude Test. The study identified a random sample of 562 female and 852 male students who had achieved the same high score on the mathematics portion of the test. That is, the female and male students were viewed as having similarly high abilities in mathematics. The SAT verbal scores for the two samples are as follows.

Female Students	Male Students
$\bar{x}_1 = 547$	$\bar{x}_2 = 525$
$s_1 = 83$	$s_2 = 78$

Do the data support the conclusion that given a population of female students and a population of male students with similarly high mathematical abilities, the female students will have a significantly higher verbal ability? Test at a .02 level of significance. What is your conclusion?

17. A firm is studying the delivery times of two raw material suppliers. The firm is basically satisfied with supplier A and is prepared to stay with that supplier if the mean delivery time is the same as or less than that of supplier B. However, if the firm finds that the mean delivery time of supplier B is less than that of supplier A, it will begin making raw material purchases from supplier B.

a. What are the null and alternative hypotheses for this situation?
b. Assume that independent samples show the following delivery time characteristics for the two suppliers.

Supplier A	Supplier B
$n_1 = 50$	$n_2 = 30$
$\bar{x}_1 = 14$ days	$\bar{x}_2 = 12.5$ days
$s_1 = 3$ days	$s_2 = 2$ days

With $\alpha = .05$, what is your conclusion for the hypotheses from part (a)? What action do you recommend in terms of supplier selection?

18. Arnold Palmer and Tiger Woods are two of the best golfers to ever play the game. The question was raised as to how these two golfers would have compared if both were playing at the top of their game. The following sample data show the results of 18-hole scores during a PGA tournament competition. Palmer's scores are from his 1960 season, while Wood's scores are from his 1999 season (*Golf Magazine,* February 2000).

Palmer, 1960	**Woods, 1999**
$\bar{x}_1 = 69.95$	$\bar{x}_2 = 69.56$
$n_1 = 112$	$n_2 = 84$

Use the sample results to test the hypothesis of no difference between the population mean 18-hole scores for the two golfers.

a. Assuming a population standard deviation of 2.5 for both golfers, what is the value of the test statistic?
b. What is the p-value?
c. Using $\alpha = .01$, what is your conclusion?

19. The mean time to locate flight information on Internet web sites of the major airline companies is generally 2 to 3 minutes (*USA Today,* September 11, 2000). Sample results representative of the times for Delta Airlines and Northwest Airlines are as follows:

Delta	**Northwest**
$\bar{x}_1 = 2.5$ minutes	$\bar{x}_2 = 2.1$ minutes
$s_1 = .8$ minutes	$s_2 = 1.1$ minutes
$n_1 = 22$	$n_2 = 20$

a. Formulate the hypotheses if the purpose is to test for a significant difference between the mean times for the two airlines.
b. Using $\alpha = .05$, what is the rejection rule?
c. Compute the value of the test statistic.
d. What is your conclusion?
e. What can you say about the p-value?

20. Periodically, Merrill Lynch customers are asked to evaluate Merrill Lynch financial consultants and services (2000 Merril Lynch Client Satisfaction Survey). Higher ratings on the client satisfaction survey indicate better service with 7 the maximum service rating. Independent samples of service ratings for two financial consultants are summarized below. Consultant A has 10 years of experience while consultant B has 1 year of experience. Using $\alpha = .05$, test to see whether the consultant with the more experience has the higher population mean service rating.

Consultant A	**Consultant B**
$\bar{x}_1 = 6.82$	$\bar{x}_2 = 6.25$
$s_1 = .64$	$s_2 = .75$
$n_1 = 16$	$n_2 = 10$

a. Formulate the null and alternative hypotheses.
b. What is the rejection rule?
c. Compute the value of the test statistic.
d. What is your conclusion?
e. What can you say about the p-value?

10.3 INFERENCES ABOUT THE DIFFERENCE BETWEEN THE MEANS OF TWO POPULATIONS: MATCHED SAMPLES

Suppose a manufacturing company has two methods by which employees can perform a production task. To maximize production output, the company wants to identify the method with the shortest mean completion time. Let μ_1 denote the mean completion time for production method 1 and μ_2 denote the mean completion time for production method 2. With no preliminary indication of the preferred production method, we begin by tentatively assuming that the two production methods have the same mean completion time. Thus, the null hypothesis is H_0: $\mu_1 - \mu_2 = 0$. If this hypothesis is rejected, we can conclude that the mean completion times differ. In this case, the method providing the shorter mean completion time would be recommended. The null and alternative hypotheses are written as follows.

$$H_0: \mu_1 - \mu_2 = 0$$
$$H_a: \mu_1 - \mu_2 \neq 0$$

In choosing the sampling procedure that will be used to collect production time data and test the hypotheses, we consider two alternative designs. One is based on **independent samples** and the other is based on **matched samples.**

1. *Independent sample design:* A simple random sample of workers is selected and each worker in the sample uses method 1. A second independent simple random sample of workers is selected and each worker in this sample uses method 2. The test of the difference between means is based on the procedures in Section 10.2.
2. *Matched sample design:* One simple random sample of workers is selected. Each worker first uses one method and then uses the other method. The order of the two methods is assigned randomly to the workers, with some workers performing method 1 first and others performing method 2 first. Each worker provides a pair of data values, one value for method 1 and another value for method 2.

In the matched sample design the two production methods are tested under similar conditions (i.e., with the same workers); hence this design often leads to a smaller sampling error than the independent sample design. The primary reason is that in a matched sample design, variation between workers is eliminated as a source of sampling error.

Let us demonstrate the analysis of a matched sample design by assuming it is the method used to test the difference between the two production methods. A random sample of six workers is used. The data on completion times for the six workers are given in Table 10.4. Note that each worker provides a pair of data values, one for each production method. Also note that the last column contains the difference in completion times d_i for each worker in the sample.

The key to the analysis of the matched sample design is to realize that we consider only the column of differences. We therefore have six data values (.6, −.2, .5, .3, .0, and .6) that will be used to analyze the difference between the means of the two production methods.

Let μ_d = the mean of the *difference* values for the population of workers. With this notation, the null and alternative hypotheses are rewritten as follows.

$$H_0: \mu_d = 0$$
$$H_a: \mu_d \neq 0$$

The formulas for the sample mean and sample standard deviation use the notation d to remind the user that the data show the difference between the matched sample data values. Other than the use of the d notation, the formulas for the sample mean and sample standard deviation are the same ones used previously in the text.

If H_0 can be rejected, we can conclude that the mean completion times differ.

The d notation is a reminder that the matched sample provides *difference* data. The sample mean and sample standard deviation for the six difference values in Table 10.4 follow.

$$\bar{d} = \frac{\Sigma d_i}{n} = \frac{1.8}{6} = .30$$

$$s_d = \sqrt{\frac{\Sigma(d_i - \bar{d})^2}{n - 1}} = \sqrt{\frac{.56}{5}} = .335$$

TABLE 10.4 TASK COMPLETION TIMES FOR A MATCHED SAMPLE DESIGN

CD file
Matched

Worker	Completion Time for Method 1 (minutes)	Completion Time for Method 2 (minutes)	Difference in Completion Times (d_i)
1	6.0	5.4	.6
2	5.0	5.2	−.2
3	7.0	6.5	.5
4	6.2	5.9	.3
5	6.0	6.0	.0
6	6.4	5.8	.6

In Chapter 9 we stated that if the population can be assumed to be normally distributed, the t distribution with $n - 1$ degrees of freedom can be used to test the null hypothesis about a population mean. Assuming difference data are normally distributed, the test statistic becomes

$$t = \frac{\bar{d} - \mu_d}{s_d/\sqrt{n}} \tag{10.13}$$

With $\alpha = .05$ and $n - 1 = 5$ degrees of freedom ($t_{.025} = 2.571$), the rejection rule for the two-tailed test becomes

$$\text{Reject } H_0 \text{ if } t < -2.571 \text{ or if } t > 2.571$$

With $\bar{d} = .30$, $s_d = .335$, and $n = 6$, the value of the test statistic for the null hypothesis H_0: $\mu_d = 0$ is

$$t = \frac{\bar{d} - \mu_d}{s_d/\sqrt{n}} = \frac{.30 - 0}{.335/\sqrt{6}} = 2.20$$

Once the difference data have been computed, the t distribution procedure for matched samples is the same as the one-population procedure described in Chapters 8 and 9.

Since $t = 2.20$ is not in the rejection region, the sample data do not provide sufficient evidence to reject H_0.

Using the 5 degrees of freedom, the t distribution table shows $t = 2.20$ is between $t_{.025} = 2.571$ and $t_{.025} = 2.015$. Thus, the one-tailed area associated with $t = 2.20$ is between .025 and .05. For a two-tailed test, we know the p-value is between .05 and .10. Because this p-value is greater than $\alpha = .05$, H_0 is not rejected. Using Minitab and the data in Table 10.4 shows p-value $= .08$.

In addition we can obtain an interval estimate of the difference between the two population means by using the single population methodology of Chapter 8. The calculation follows.

$$\bar{d} \pm t_{\alpha/2}\frac{s_d}{\sqrt{n}}$$

$$0.3 \pm 2.571\left(\frac{.335}{\sqrt{6}}\right)$$

$$0.3 \pm .35$$

Thus, the 95% confidence interval for the difference between the means of the two production methods is −.05 minutes to .65 minutes. Note that because the confidence interval includes the value of zero, the sample data do not provide sufficient evidence to reject H_0.

NOTES AND COMMENTS

1. In the example presented in this section, workers performed the production task with first one method and then the other method. This example illustrates a matched sample design in which each sampled item (worker) provides a pair of data values. It is also possible to use different but "similar" items to provide the pair of data values. For example, a worker at one location could be matched with a similar worker at another location (similarity based on age, education, sex, experience, etc.). The pairs of workers would provide the difference data that could be used in the matched sample analysis.
2. A matched sample procedure for inferences about two population means generally provides better precision than the independent sample approach, therefore it is the recommended design. However, in some applications the matching cannot be achieved, or perhaps the time and cost associated with matching are excessive. In such cases, the independent sample design should be used.
3. The example presented in this section had a sample size of six workers and thus was a small-sample case. The t distribution was used in both the hypothesis test and interval estimation computations. If the sample size is large ($n \geq 30$), use of the t distribution is unnecessary; in such cases, statistical inferences can be based on the z values of the standard normal probability distribution.

EXERCISES

Methods

21. Consider the following hypothesis test.

$$H_0: \mu_d \leq 0$$
$$H_a: \mu_d > 0$$

The following data are from matched samples taken from two populations.

	Population	
Element	1	2
1	21	20
2	28	26
3	18	18
4	20	20
5	26	24

a. Compute the difference value for each element.
b. Compute $\bar{d}$.
c. Compute the standard deviation s_d.
d. Test the hypothesis using $\alpha = .05$. What is your conclusion?

22. The following data are from matched samples taken from two populations.

	Population	
Element	1	2
1	11	8
2	7	8
3	9	6
4	12	7
5	13	10
6	15	15
7	15	14

a. Compute the difference value for each element.
b. Compute $\bar{d}$.
c. Compute the standard deviation s_d.
d. What is the point estimate of the difference between the two population means?
e. Provide a 95% confidence interval for the difference between the two population means.

Applications

23. A market research firm used a sample of individuals to rate the purchase potential of a particular product before and after the individuals saw a new television commercial about the product. The purchase potential ratings were based on a 0 to 10 scale, with higher values indicating a higher purchase potential. The null hypothesis stated that the mean rating "after" would be less than or equal to the mean rating "before." Rejection of this hypothesis would show that the commercial improved the mean purchase potential rating. Use $\alpha = .05$ and the following data to test the hypothesis and comment on the value of the commercial.

	Purchase Rating			Purchase Rating	
Individual	After	Before	Individual	After	Before
1	6	5	5	3	5
2	6	4	6	9	8
3	7	7	7	7	5
4	4	3	8	6	6

24. A sample of 10 international telephone calls provided the Sprint and WorldCom calling rates per minute for calls from the United States (*World Traveler,* July 2000).

Country	Sprint	WorldCom
Australia	.46	.26
Belgium	.69	.40
Brazil	.92	.53
Colombia	.55	.53
Denmark	.50	.26
France	.46	.26
Germany	.46	.26
Hong Kong	.92	.40
Japan	.69	.40
United Kingdom	.46	.26

Provide a 95% confidence interval estimate of the difference between the two population means.

25. The cost of transportation from the airport to the downtown area depends on the method of transportation. One-way costs for taxi and shuttle bus transportation for a sample of 10 major cities follow. Provide a 95% confidence interval for the mean cost increase associated with taxi transportation.

City	Taxi ($)	Shuttle Bus ($)	City	Taxi ($)	Shuttle Bus ($)
Atlanta	15.00	7.00	Minneapolis	16.50	7.50
Chicago	22.00	12.50	New Orleans	18.00	7.00
Denver	11.00	5.00	New York (LaGuardia)	16.00	8.50
Houston	15.00	4.50	Philadelphia	20.00	8.00
Los Angeles	26.00	11.00	Washington, D.C.	10.00	5.00

26. Rental car gasoline prices per gallon were sampled at eight major airports. Data for Hertz and National car rental companies follow (*USA Today,* April 4, 2000).

Airport	Hertz	National
Boston Logan	1.55	1.56
Chicago O'Hare	1.62	1.59
Los Angeles	1.72	1.78
Miami	1.65	1.49
New York (JFK)	1.72	1.51
New York (LaGuardia)	1.67	1.50
Orange County, CA	1.68	1.77
Washington (Dulles)	1.52	1.41

Use $\alpha = .05$ to test the hypothesis of no difference between the population mean prices per gallon for the two companies.

27. A survey was made of Book-of-the-Month-Club members to ascertain whether members spend more time watching television than they do reading. Assume a sample of 15 respondents provided the following data on weekly hours of television watching and weekly hours of reading. Using a .05 level of significance, can you conclude that Book-of-the-Month-Club members spend more hours per week watching television than reading?

CD file TVRead

Respondent	Television	Reading	Respondent	Television	Reading
1	10	6	9	4	7
2	14	16	10	8	8
3	16	8	11	16	5
4	18	10	12	5	10
5	15	10	13	8	3
6	14	8	14	19	10
7	10	14	15	11	6
8	12	14			

28. The 1997 price/earnings ratios and estimated 1998 price/earnings ratios for a sample of 12 stocks follow (*Kiplinger's Personal Finance Magazine,* November 1997).

CD file PERatio

Stock	1997 P/E Ratio	Est. 1998 P/E Ratio
Coca-Cola	40	32
Walt Disney	33	23
Du Pont	24	16
Eastman Kodak	21	13
General Electric	30	23
General Mills	25	19
IBM	19	14
Merck	29	21
McDonald's	20	17
Motorola	35	20
Philip Morris	17	13
Xerox	20	17

a. Use $\alpha = .05$ and test to see whether there is any change in the population mean P/E ratios for the 2-year period.

b. Provide a 95% confidence interval for the change in the population mean price/earnings ratios for the 2-year period. What is your conclusion?

29. A manufacturer produces both a deluxe and a standard model of an automatic sander designed for home use. Selling prices obtained from a sample of retail outlets follow.

Retail Outlet	Model Price ($) Deluxe	Standard	Retail Outlet	Model Price ($) Deluxe	Standard
1	39	27	5	40	30
2	39	28	6	39	34
3	45	35	7	35	29
4	38	30			

a. The manufacturer's suggested retail prices for the two models show a $10 price differential. Using a .05 level of significance, test that the mean difference between the prices of the two models is $10.

b. What is the 95% confidence interval for the difference between the mean prices of the two models?

10.4 INFERENCES ABOUT THE DIFFERENCE BETWEEN THE PROPORTIONS OF TWO POPULATIONS

A tax preparation firm is interested in comparing the quality of work at two of its regional offices. By randomly selecting samples of tax returns prepared at each office and having the sample returns verified for accuracy, the firm will be able to estimate the proportion of erroneous returns prepared at each office. Of particular interest is the difference between these proportions.

p_1 = proportion of erroneous returns for population 1 (office 1)
p_2 = proportion of erroneous returns for population 2 (office 2)
$\bar{p}_1$ = sample proportion for a simple random sample from population 1
$\bar{p}_2$ = sample proportion for a simple random sample from population 2

The difference between the two population proportions is given by $p_1 - p_2$. The point estimator of $p_1 - p_2$ follows.

Point Estimator of the Difference Between the Proportions of Two Populations

$$\bar{p}_1 - \bar{p}_2 \tag{10.14}$$

Thus, the point estimator of the difference between two population proportions is the difference between the sample proportions of two independent simple random samples.

Sampling Distribution of $\bar{p}_1 - \bar{p}_2$

In the study of the difference between two population proportions, $\bar{p}_1 - \bar{p}_2$ is the point estimator of interest. As we have seen in several preceding cases, the sampling distribution of the point estimator is a key factor in developing interval estimates and in testing hypotheses about the parameter of interest. The properties of the sampling distribution of $\bar{p}_1 - \bar{p}_2$ follow.

Sampling Distribution of $\bar{p}_1 - \bar{p}_2$

$$\text{Expected Value: } E(\bar{p}_1 - \bar{p}_2) = p_1 - p_2 \tag{10.15}$$

$$\text{Standard deviation: } \sigma_{\bar{p}_1 - \bar{p}_2} = \sqrt{\frac{p_1(1 - p_1)}{n_1} + \frac{p_2(1 - p_2)}{n_2}} \tag{10.16}$$

where

n_1 = sample size for the simple random sample from population 1
n_2 = sample size for the simple random sample from population 2

Distribution form: If the sample sizes are large [i.e., n_1p_1, $n_1(1 - p_1)$, n_2p_2, and $n_2(1 - p_2)$ are all greater than or equal to 5], the sampling distribution of $\bar{p}_1 - \bar{p}_2$ can be approximated by a normal probability distribution.

Figure 10.7 shows the sampling distribution of $\bar{p}_1 - \bar{p}_2$.

Interval Estimation of $p_1 - p_2$

Let us assume that independent simple random samples of tax returns from the two offices provide the following information.

Office 1	**Office 2**
$n_1 = 250$	$n_2 = 300$
Number of returns with errors = 35	Number of returns with errors = 27

The sample proportions for the two offices follow.

$$\bar{p}_1 = \frac{35}{250} = .14$$

$$\bar{p}_2 = \frac{27}{300} = .09$$

The point estimate of the difference between the proportions of erroneous tax returns for the two populations is $\bar{p}_1 - \bar{p}_2 = .14 - .09 = .05$. Thus, we estimate that office 1 has a .05, or 5%, greater error rate than office 2.

FIGURE 10.7 SAMPLING DISTRIBUTION OF $\bar{p}_1 - \bar{p}_2$

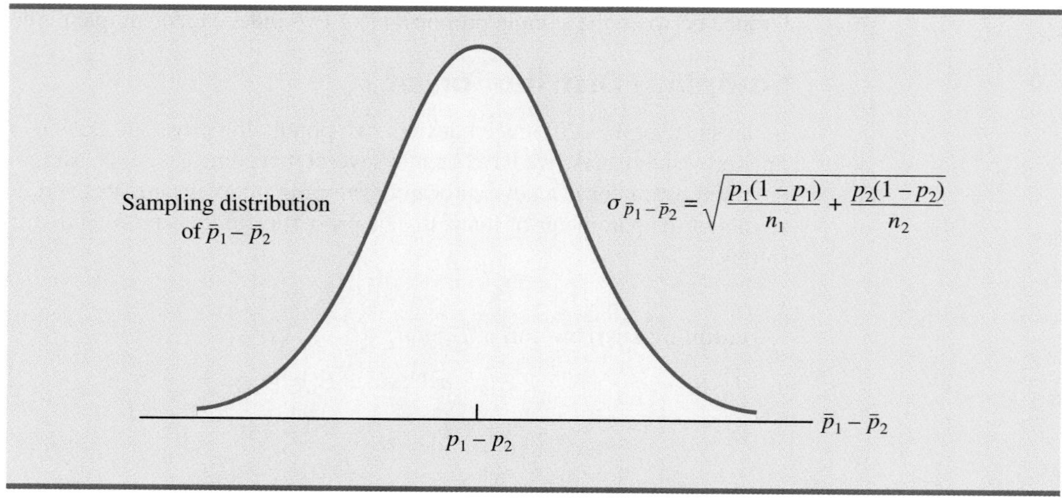

In order to develop a confidence interval for the difference between two populations, we need an estimate of the $\sigma_{\bar{p}_1-\bar{p}_2}$ shown in (10.16). However, equation (10.16) cannot be used directly because the two population proportions, p_1 and p_2, are unknown. In this case, we use the sample proportion $\bar{p}_1$ to estimate p_1 and the sample proportion $\bar{p}_2$ to estimate p_2. Doing so provides the following point estimator of $\sigma_{\bar{p}_1-\bar{p}_2}$.

Point Estimator of $\sigma_{\bar{p}_1-\bar{p}_2}$

$$s_{\bar{p}_1-\bar{p}_2} = \sqrt{\frac{\bar{p}_1(1-\bar{p}_1)}{n_1} + \frac{\bar{p}_2(1-\bar{p}_2)}{n_2}} \tag{10.17}$$

Using equation (10.17), the following expression provides a confidence interval estimate of the difference between the proportions of two populations. We use $z_{\alpha/2}$ because the sampling distribution of $\bar{p}_1 - \bar{p}_2$ can be approximated by a normal distribution when the sample sizes are large.

Interval Estimate of the Difference Between the Proportions of Two Populations: Large-Sample Case With n_1p_1, $n_1(1-p_1)$, n_2p_2, and $n_2(1-p_2) \geq 5$

$$\bar{p}_1 - \bar{p}_2 \pm z_{\alpha/2}s_{\bar{p}_1-\bar{p}_2} \tag{10.18}$$

where $1 - \alpha$ is the confidence coefficient.

Using equation (10.17), we have

$$s_{\bar{p}_1-\bar{p}_2} = \sqrt{\frac{.14(.86)}{250} + \frac{.09(.91)}{300}} = .0275$$

With a 90% confidence interval $z_{\alpha/2} = z_{.05} = 1.645$, (10.18) provides the following interval estimate.

$$(.14 - .09) \pm 1.645(.0275)$$
$$.05 \pm .045$$

Thus, the 90% confidence interval for the difference between the two population error rates is .005 to .095.

Hypothesis Tests About $p_1 - p_2$

As an example of hypothesis tests about the difference between the proportions of two populations, consider the data collected in the preceding example and assume that the firm is attempting to determine whether the error proportions differ between the two offices. Let us illustrate the procedure by testing the following.

$$H_0: p_1 - p_2 = 0$$
$$H_a: p_1 - p_2 \neq 0$$

FIGURE 10.8 SAMPLING DISTRIBUTION OF $\bar{p}_1 - \bar{p}_2$ WITH H_0: $p_1 - p_2 = 0$

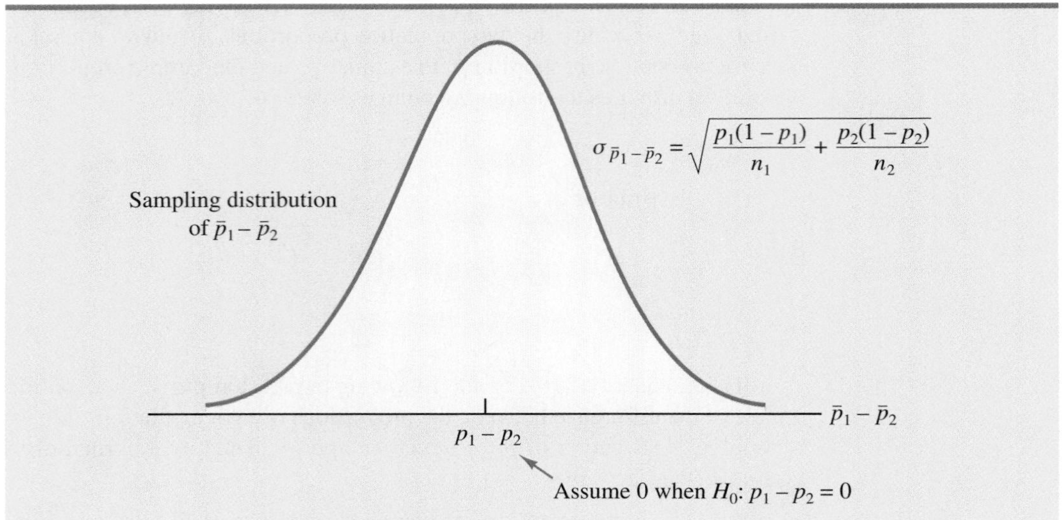

Figure 10.8 shows the sampling distribution of $\bar{p}_1 - \bar{p}_2$ based on the assumption of no difference between the two population proportions; that is, $p_1 - p_2 = 0$. With the sampling distribution approximately normal, the test statistic for the difference between two population proportions can be written

$$z = \frac{(\bar{p}_1 - \bar{p}_2) - (p_1 - p_2)}{\sigma_{\bar{p}_1 - \bar{p}_2}} \tag{10.19}$$

With $\alpha = .10$, $z_{\alpha/2} = z_{.05} = 1.645$. Using the test statistic, the rejection rule is

$$\text{Reject } H_0 \text{ if } z < -1.645 \text{ or if } z > 1.645$$

The computation of z in equation (10.19) requires a value for the standard error of the difference between proportions $\sigma_{\bar{p}_1 - \bar{p}_2}$. Because this standard error is unknown, it must be estimated from the sample data. We may be tempted to use $\bar{p}_1$ and $\bar{p}_2$ in (10.18) as we did with the interval estimation procedure, but in hypothesis testing we often adjust (10.18) to a slightly different form. For the *special case* in which the hypotheses involve no difference between the population proportions (i.e., either H_0: $p_1 - p_2 = 0$, H_0: $p_1 - p_2 \leq 0$, or H_0: $p_1 - p_2 \geq 0$), (10.18) is modified to reflect the fact that when we assume H_0 to be true at the equality, we are assuming $p_1 = p_2$. When making this assumption, we combine or *pool* the two sample proportions to provide one estimate. This pooled estimator, denoted by $\bar{p}$, follows.

In hypothesis tests where H_0 is $p_1 - p_2 = 0$, $p_1 - p_2 \leq 0$, or $p_1 - p_2 \geq 0$, the 0 indicates no difference between the two population proportions p_1 and p_2. In this case, the two sample proportions, $\bar{p}_1$ and $\bar{p}_2$, should be combined to provide one estimate of the population proportion; see (10.20). Then $s_{\bar{p}_1 - \bar{p}_2}$ can be computed as shown in (10.21).

$$\bar{p} = \frac{n_1\bar{p}_1 + n_2\bar{p}_2}{n_1 + n_2} \tag{10.20}$$

With $\bar{p}$ used in place of both $\bar{p}_1$ and $\bar{p}_2$, (10.17) is revised to

$$s_{\bar{p}_1 - \bar{p}_2} = \sqrt{\bar{p}(1 - \bar{p})\left(\frac{1}{n_1} + \frac{1}{n_2}\right)} \tag{10.21}$$

Using equations (10.20) and (10.21), we can now proceed with the calculations.

$$\bar{p} = \frac{250(.14) + 300(.09)}{550} = \frac{62}{550} = .113$$

$$s_{\bar{p}_1 - \bar{p}_2} = \sqrt{(.113)(.887)\left(\frac{1}{250} + \frac{1}{300}\right)} = .0271$$

Using (10.19), the value of the test statistic becomes

$$z = \frac{(\bar{p}_1 - \bar{p}_2) - (p_1 - p_2)}{s_{\bar{p}_1 - \bar{p}_2}} = \frac{(.14 - .09) - 0}{.0271} = 1.85$$

Because $1.85 > 1.645$, the null hypothesis is rejected. The area in the tail associated with $z = 1.85$ is $.5000 - .4678 = .0322$. Thus, the two-tailed p-value is $2(.0322) = .0644$, which is less than $\alpha = .10$. The sample evidence indicates a significant difference between the two population error rates.

As we saw with the hypothesis tests about differences between two population means, one-tailed tests can be developed for the difference between two population proportions. The one-tailed rejection regions are established as in the one-tailed hypothesis testing procedures for a single population proportion.

EXERCISES

Methods

30. Consider the following results for two independent samples taken from two populations.

Sample 1	Sample 2
$n_1 = 400$	$n_2 = 300$
$\bar{p}_1 = .48$	$\bar{p}_2 = .36$

a. What is the point estimate of the difference between the two population proportions?
b. Develop a 90% confidence interval for the difference between the two population proportions.
c. Develop a 95% confidence interval for the difference between the two population proportions.

31. Consider the hypothesis test

$$H_0: p_1 - p_2 \leq 0$$
$$H_a: p_1 - p_2 > 0$$

The following results are for two independent samples taken from the two populations.

Sample 1	Sample 2
$n_1 = 200$	$n_2 = 300$
$\bar{p}_1 = .22$	$\bar{p}_2 = .16$

a. With $\alpha = .05$, what is your hypothesis testing conclusion?
b. What is the p-value?

Applications

32. A *Business Week*/Harris survey asked senior executives at large corporations their opinions about the economic outlook for the future (*Business Week,* June 16, 1997). One question was "Do you think that there will be an increase in the number of full-time employees at your company over the next 12 months?" In May 1997, 220 of 400 executives answered yes, while in December 1996, 192 of 400 executives had answered yes. Provide a 95% confidence interval estimate for the difference between the proportions at the two points in time. What is your interpretation of the interval estimate?

33. A Gallup poll found that 16% of 505 men and 25% of 496 women surveyed favored a law forbidding the sale of all beer, wine, and liquor throughout the nation. Develop a 95% confidence interval for the difference between the proportion of women who favor such a ban and the proportion of men who favor such a ban.

34. A CNN/*USA Today* poll evaluated the popularity of major league baseball after the Mark McGwire and Sammy Sosa record-setting home-run chase during the 1998 season. A sample of 1082 found 682 people responding yes to the question "Are you a fan of major league baseball?" (*USA Today,* September 17, 1998). During the baseball strike of April 1995, a sample of 1008 found only 413 people responding yes to the same question.

a. Use the sample proportions to estimate the fan support for major league baseball in 1998 and in 1995. What is the point estimate of the increase in major league baseball fan support from the strike in 1995 to the McGwire/Sosa home-run chase of 1998?

b. Develop a 95% confidence interval estimate of the increase in fan support.

35. *Fortune* enlisted Yankelovich Partners to poll 600 adults on their ideas about marriage, divorce, and the contribution of the corporate wife to her executive husband (*Fortune,* February 2, 1998). In response to the statement, "In a divorce in a long-term marriage where the husband works outside the home and the wife is not employed for pay, the wife should be entitled to half the assets accumulated during the marriage," 279 of 300 women agreed and 255 of 300 men agreed.

a. What are the sample proportions for the two populations, women and men?

b. Test the hypothesis H_0: $p_1 - p_2 = 0$ using $\alpha = .05$. What is your conclusion?

c. Provide a 95% confidence interval for the difference between the population proportions.

36. A sample of 1545 men and a sample of 1691 women were used to compare the amount of housework done by women and men in dual-earner marriages. The study showed that 67.5% of the men felt the division of housework was fair and 60.8% of the women felt the division of housework was fair (*American Journal of Sociology,* September 1994). Is the proportion of men who felt the division of housework was fair greater than the proportion of women who felt the division of housework was fair? Support your conclusion with a statistical test using a .05 level of significance.

37. In a test of the quality of two television commercials, each commercial was shown in a separate test area six times over a 1-week period. The following week a telephone survey was conducted to identify individuals who had seen the commercials. The individuals who had seen the commercials were asked to state the primary message in the commercials. The following results were recorded.

Commercial	Number Who Saw Commercial	Number Who Recalled Primary Message
A	150	63
B	200	60

a. Using $\alpha = .05$, test the hypothesis that there is no difference in the recall proportions for the two commercials.

b. Compute a 95% confidence interval for the difference between the recall proportions for the two populations.

38. The stock market declined in early 2000 with many stocks dropping below their 1997 highs. In fact, 81.5% of New York Stock Exchange (NYSE) stocks and 72.4% of NASDAQ stocks were trading below their 1997 highs (*Barron's,* May 22, 2000). If the data were based on 232 NYSE stocks and 210 NASDAQ stocks, test the hypothesis H_0: $p_1 - p_2 = 0$ with $\alpha = .05$. What is the p-value? What is your conclusion?

39. *Yahoo Internet Life* sponsored surveys in several metropolitan areas to estimate the proportion of adults using the Internet at work (*USA Today,* May 7, 2000). Results showed 40% of Washington, D.C., adults use the Internet at work, while 32% of San Francisco adults use the Internet at work. If the sample sizes are 240 and 250 respectively, do the sample results indicate that the population proportion of adults using the Internet at work in Washington, D.C., is greater than the population proportion in San Francisco? What is the p-value? Using $\alpha = .05$, what is the conclusion?

SUMMARY

In this chapter we discussed procedures for developing interval estimates and conducting hypothesis tests involving two populations. First, we showed how to make inferences about the difference between the means of two populations when independent simple random samples are selected. We considered both the large- and small-sample cases. The z values from the standard normal probability distribution are used for inferences about the difference between two population means when the sample sizes are large. In the small-sample case, if the populations are normally distributed with equal variances, the t distribution is used for inferences.

Inferences about the difference between the means of two populations were then discussed for the matched sample design. In the matched sample design each element provides a pair of data values, one from each population. The difference between the paired data values is then used in the statistical analysis. The matched sample design is generally preferred to the independent sample design because the matched-sample procedure often reduces the sampling error and improves the precision of the estimate.

Finally, interval estimation and hypothesis testing about the difference between two population proportions were discussed. Statistical procedures for analyzing the difference between proportions for two populations are similar to the procedures for analyzing the difference between two population means.

GLOSSARY

Pooled variance An estimate of the variance of a population based on the combination of two sample variances. The pooled variance estimate is appropriate whenever the variances of two populations are assumed equal.

Independent samples Samples selected from two populations in such a way that the elements making up one sample are chosen independently of the elements making up the other sample.

Matched samples Samples in which each data value of one sample is matched with a corresponding data value of the other sample.

KEY FORMULAS

Point Estimator of the Difference Between the Means of Two Populations

$$\bar{x}_1 - \bar{x}_2 \tag{10.1}$$

Expected Value of $\bar{x}_1 - \bar{x}_2$

$$E(\bar{x}_1 - \bar{x}_2) = \mu_1 - \mu_2 \tag{10.2}$$

Standard Deviation of $\bar{x}_1 - \bar{x}_2$

$$\sigma_{\bar{x}_1 - \bar{x}_2} = \sqrt{\frac{\sigma_1^2}{n_1} + \frac{\sigma_2^2}{n_2}} \tag{10.3}$$

Interval Estimate of the Difference Between the Means of Two Populations: Large-Sample Case ($n_1 \geq 30$ and $n_2 \geq 30$) With σ_1 and σ_2 Assumed Known

$$\bar{x}_1 - \bar{x}_2 \pm z_{\alpha/2}\sigma_{\bar{x}_1 - \bar{x}_2} \tag{10.4}$$

Point Estimator of $\sigma_{\bar{x}_1 - \bar{x}_2}$

$$s_{\bar{x}_1 - \bar{x}_2} = \sqrt{\frac{s_1^2}{n_1} + \frac{s_2^2}{n_2}} \tag{10.5}$$

Interval Estimate of the Difference Between the Means of Two Populations: Large-Sample Case ($n_1 \geq 30$ and $n_2 \geq 30$) With σ_1 and σ_2 Estimated by s_1 and s_2

$$\bar{x}_1 - \bar{x}_2 \pm z_{\alpha/2}s_{\bar{x}_1 - \bar{x}_2} \tag{10.6}$$

Standard Deviation of $\bar{x}_1 - \bar{x}_2$ When $\sigma_1^2 = \sigma_2^2$

$$\sigma_{\bar{x}_1 - \bar{x}_2} = \sqrt{\frac{\sigma^2}{n_1} + \frac{\sigma^2}{n_2}} = \sqrt{\sigma^2\left(\frac{1}{n_1} + \frac{1}{n_2}\right)} \tag{10.7}$$

Pooled Estimator of σ^2

$$s^2 = \frac{(n_1 - 1)s_1^2 + (n_2 - 1)s_2^2}{n_1 + n_2 - 2} \tag{10.8}$$

Point Estimator of $\sigma_{\bar{x}_1 - \bar{x}_2}$ when $\sigma_1^2 = \sigma_2^2 = \sigma^2$

$$s_{\bar{x}_1 - \bar{x}_2} = \sqrt{s^2\left(\frac{1}{n_1} + \frac{1}{n_2}\right)} \tag{10.9}$$

Interval Estimate of the Difference Between the Means of Two Populations: Small-Sample Case ($n_1 < 30$ and/or $n_2 < 30$) With σ_1 and σ_2 Estimated by s_1 and s_2

$$\bar{x}_1 - \bar{x}_2 \pm t_{\alpha/2}s_{\bar{x}_1 - \bar{x}_2} \tag{10.10}$$

Test Statistic for Hypothesis Tests About the Difference Between the Means of Two Populations (Large-Sample Case)

$$z = \frac{(\bar{x}_1 - \bar{x}_2) - (\mu_1 - \mu_2)}{\sqrt{\sigma_1^2/n_1 + \sigma_2^2/n_2}} \tag{10.11}$$

Test Statistic for Hypothesis Tests About the Difference Between the Means of Two Populations (Small-Sample Case)

$$t = \frac{(\bar{x}_1 - \bar{x}_2) - (\mu_1 - \mu_2)}{\sqrt{s^2\left(\frac{1}{n_1} + \frac{1}{n_2}\right)}} \tag{10.12}$$

Sample Mean for Matched Samples

$$\bar{d} = \frac{\Sigma d_i}{n}$$

Sample Standard Deviation for Matched Samples

$$s_d = \sqrt{\frac{\Sigma(d_i - \bar{d})^2}{n - 1}}$$

Test Statistic for Matched Samples (Small-Sample Case)

$$t = \frac{\bar{d} - \mu_d}{s_d/\sqrt{n}} \tag{10.13}$$

Point Estimator of the Difference Between the Proportions of Two Populations

$$\bar{p}_1 - \bar{p}_2 \tag{10.14}$$

Expected Value of $\bar{p}_1 - \bar{p}_2$

$$E(\bar{p}_1 - \bar{p}_2) = p_1 - p_2 \tag{10.15}$$

Standard Deviation of $\bar{p}_1 - \bar{p}_2$

$$\sigma_{\bar{p}_1 - \bar{p}_2} = \sqrt{\frac{p_1(1 - p_1)}{n_1} + \frac{p_2(1 - p_2)}{n_2}} \tag{10.16}$$

Point Estimator of $\sigma_{\bar{p}_1 - \bar{p}_2}$

$$s_{\bar{p}_1 - \bar{p}_2} = \sqrt{\frac{\bar{p}_1(1 - \bar{p}_1)}{n_1} + \frac{\bar{p}_2(1 - \bar{p}_2)}{n_2}} \tag{10.17}$$

Interval Estimate of the Difference Between the Proportions of Two Populations: Large-Sample Case With n_1p_1, $n_1(1 - p_1)$, n_2p_2, and $n_2(1 - p_2) \geq 5$

$$\bar{p}_1 - \bar{p}_2 \pm z_{\alpha/2} s_{\bar{p}_1 - \bar{p}_2} \tag{10.18}$$

Test Statistic for Hypothesis Tests About the Difference Between the Proportions of Two Populations

$$z = \frac{(\bar{p}_1 - \bar{p}_2) - (p_1 - p_2)}{\sigma_{\bar{p}_1 - \bar{p}_2}} \tag{10.19}$$

Pooled Estimator of the Population Proportion

$$\bar{p} = \frac{n_1\bar{p}_1 + n_2\bar{p}_2}{n_1 + n_2} \tag{10.20}$$

Point Estimator of $\sigma_{\bar{p}_1 - \bar{p}_2}$ When $p_1 = p_2$

$$s_{\bar{p}_1 - \bar{p}_2} = \sqrt{\bar{p}(1 - \bar{p})\left(\frac{1}{n_1} + \frac{1}{n_2}\right)} \tag{10.21}$$

SUPPLEMENTARY EXERCISES

40. Starting annual salaries for individuals with master's and bachelor's degrees in business were collected in two independent random samples. Use the following data to develop a 90% confidence interval estimate of the increase in starting salary that can be expected upon completion of a master's program.

Master's Degree	Bachelor's Degree
$n_1 = 60$	$n_2 = 80$
$\bar{x}_1 = \$40{,}000$	$\bar{x}_2 = \$35{,}000$
$s_1 = \$2{,}500$	$s_2 = \$2{,}000$

41. Safegate Foods, Inc., is redesigning the checkout lanes in its supermarkets throughout the country. Two designs have been suggested. Tests on customer checkout times have been conducted at two stores where the two new systems have been installed. A summary of the sample data follows.

System A	System B
$n_1 = 120$	$n_2 = 100$
$\bar{x}_1 = 4.1$ minutes	$\bar{x}_2 = 3.3$ minutes
$s_1 = 2.2$ minutes	$s_2 = 1.5$ minutes

Test at the .05 level of significance to determine whether there is a difference between the mean checkout times of the two systems. Which system is preferred?

42. Mutual funds are classified as *load* or *no-load* funds. Load funds require an investor to pay an initial fee based on a percentage of the amount invested in the fund. The no-load funds do not require this initial fee. Some financial advisors argue that the load mutual funds may be worth the extra fee because these funds provide a higher mean rate of return than the no-load mutual funds. A sample of 30 load mutual funds and a sample of 30 no-load mutual funds were selected from *Barron's Lipper Mutual Funds Quarterly,* January 12, 1998. Data were collected on the annual return for the funds over a 5-year period. The data are contained in the data set Mutual. The data for the first five load and first five no-load mutual funds are as follows.

CD 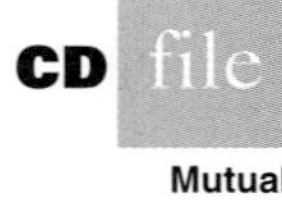
Mutual

Mutual Funds—Load	Return	Mutual Funds—No Load	Return
American National Growth	15.51	Amana Income Fund	13.24
Arch Small Cap Equity	14.57	Berger One Hundred	12.13
Bartlett Cap Basic	17.73	Columbia International Stock	12.17
Calvert World International	10.31	Dodge & Cox Balanced	16.06
Colonial Fund A	16.23	Evergreen Fund	17.61

a. Formulate H_0 and H_a such that rejection of H_0 leads to the conclusion that the load mutual funds have a higher mean annual return over the 5-year period.

b. Use the 60 mutual funds in the data set Mutual to conduct the hypothesis test. Using $\alpha = .05$, what is your conclusion?

c. What is the p-value?

43. Samples of final examination scores for two statistics classes with different instructors provided the following results.

Instructor A	Instructor B
$n_1 = 12$	$n_2 = 15$
$\bar{x}_1 = 72$	$\bar{x}_2 = 78$
$s_1 = 8$	$s_2 = 10$

With $\alpha = .05$, test whether these data are sufficient to conclude that the mean grades for the two classes differ.

44. Figure Perfect, Inc., is a women's figure salon that specializes in weight reduction programs. Weights for a sample of clients before and after a six-week introductory program follow.

	Weight	
Client	Before	After
1	140	132
2	160	158
3	210	195
4	148	152
5	190	180
6	170	164

Using $\alpha = .05$, test to determine whether the introductory program provides a statistically significant weight loss. What is your conclusion?

45. The Asian economy faltered during the last few months of 1997. Investors anticipated that the downturn in the Asian economy would have a negative effect on the earnings of companies in the United States during the fourth quarter of 1997. The following sample data show the earnings per share for the fourth quarter of 1996 and the fourth quarter of 1997 (*The Wall Street Journal,* January 28, 1998).

Asia

Company	Earnings 1996	Earnings 1997
Atlantic Richfield	1.16	1.17
Balchem Corp.	0.16	0.13
Black & Decker Corp.	0.97	1.02
Dial Corp.	0.18	0.23
DSC Communications	0.15	−0.32
Eastman Chemical	0.77	0.36
Excel Communications	0.28	−0.14
Federal Signal	0.40	0.29
Ford Motor Company	0.97	1.45
GTE Corp	0.81	0.73
ITT Industries	0.59	0.60
Kimberly-Clark	0.61	−0.27
Minnesota Mining & Mfr.	0.91	0.89
Procter & Gamble	0.63	0.71

a. Formulate H_0 and H_a such that rejection of H_0 leads to the conclusion that the mean earnings per share for the fourth quarter of 1997 are less than for the fourth quarter of 1996.
b. Use the data in the file Asia to conduct the hypothesis test. Using $\alpha = .05$, what is your conclusion?

46. A *Business Week*/Harris poll asked 1035 adults how well U.S. companies competed in the global economy, 704 respondents answered good/excellent (*Business Week,* September 11, 2000). In a similar poll of 1004 adults in 1996, 582 respondents answered good/excellent. Can the sample results be used to conclude that the proportion of adults responding good/excellent has *increased* over the 4 years from 1996 to 2000?
a. State the null and alternative hypotheses.
b. Compute the p-value.
c. Using $\alpha = .01$, what is your conclusion?

47. A large automobile insurance company selected samples of single and married male policyholders and recorded the number who had made an insurance claim over the preceding three-year period.

Single Policyholders	Married Policyholders
$n_1 = 400$	$n_2 = 900$
Number making claims = 76	Number making claims = 90

a. Using $\alpha = .05$, test to determine whether the claim rates differ between single and married male policyholders.

b. Provide a 95% confidence interval for the difference between the proportions for the two populations.

48. Medical tests were conducted to learn about drug-resistant tuberculosis. Of 142 cases tested in New Jersey, nine were found to be drug-resistant. Of 268 cases tested in Texas, five were found to be drug-resistant. Do these data suggest a statistically significant difference between the proportions of drug-resistant cases in the two states? Test H_0: $p_1 - p_2 = 0$ at the .02 level of significance. What is the p-value and what is your conclusion?

Case Problem PAR, INC.

Par, Inc., is a major manufacturer of golf equipment. Management believes that Par's market share could be increased with the introduction of a cut-resistant, longer-lasting golf ball. Therefore, the research group at Par has been investigating a new golf ball coating designed to resist cuts and provide a more durable ball. The tests with the coating have been promising.

One of the researchers voiced concern about the effect of the new coating on driving distances. Par would like the new cut-resistant ball to offer driving distances comparable to those of the current-model golf ball. To compare the driving distances for the two balls, 40 balls of both the new and current models were subjected to distance tests. The testing was performed with a mechanical hitting machine so that any difference between the mean distances for the two models could be attributed to a difference in the design. The results of the tests, with distances measured to the nearest yard, follow. These data are available on the data disk in the data set Golf.

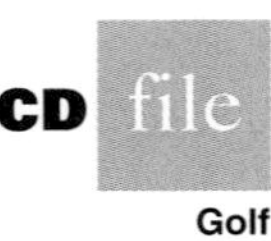

Golf

Model		Model		Model		Model	
Current	New	Current	New	Current	New	Current	New
264	277	270	272	263	274	281	283
261	269	287	259	264	266	274	250
267	263	289	264	284	262	273	253
272	266	280	280	263	271	263	260
258	262	272	274	260	260	275	270
283	251	275	281	283	281	267	263
258	262	265	276	255	250	279	261
266	289	260	269	272	263	274	255
259	286	278	268	266	278	276	263
270	264	275	262	268	264	262	279

Managerial Report

1. Formulate and present the rationale for a hypothesis test that Par could use to compare the driving distances of the current and new golf balls.
2. Analyze the data to provide the hypothesis testing conclusion. What is the p-value for your test? What is your recommendation for Par, Inc.?

3. Provide descriptive statistical summaries of the data for each model.
4. What is the 95% confidence interval for the population mean of each model, and what is the 95% confidence interval for the difference between the means of the two populations?
5. Do you see a need for larger sample sizes and more testing with the golf balls? Discuss.

Appendix 10.1 TWO POPULATION MEANS WITH MINITAB

We describe how to use Minitab for three hypothesis tests about the difference between the means of two populations.

Large-Sample Case

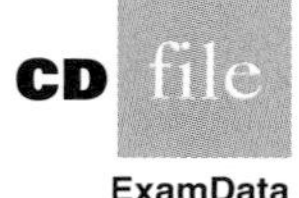

ExamData

We will use the data on examination scores at two training centers shown in Table 10.1. The Center A scores appear in column C1 and the Center B scores appear in column C2. Minitab does not have a two-sample hypothesis testing procedure that uses the z test statistic. However, the Minitab 2-Sample t test procedure provides a good approximation in the large-sample case. The steps necessary to use the Minitab 2-Sample t test for the training center data follow.

Step 1. Select the **Stat** pull-down menu
Step 2. Choose **Basic Statistics**
Step 3. Choose **2-Sample t**
Step 4. When the 2-Sample t dialog box appears:
Select **Samples in different columns**
Enter C1 in the **First** box
Enter C2 in the **Second** box
Make sure **Assume equal variances** is not selected
Select **Options**
Step 5. Enter 95 in the **Confidence level** box*
Enter 0 in the **Test Mean** box
Enter not equal in the **Alternative** box
Click **OK**
Step 6. Click **OK**

The Minitab output shows p-value = .04. With a .05 level of significance, we reject H_0.

In this large-sample case, we have stated that the z test statistic should be used when population distribution is not known. However, Minitab does not provide a two-sample z test. As a result we must adapt the 2-Sample t test to our needs. For the large-sample case with s used to estimate σ, the t-value provided by Minitab is the same as the z value of 2.09. Thus, the only difference with the 2-Sample t test is that the p-value = .04 is slightly larger than the p-value = .0366 that would have been obtained using the standard normal probability distribution with $z = 2.09$. In most cases, this slight difference in the p-value does not affect the hypothesis testing conclusion.

Small-Sample Case

The Minitab 2-Sample t procedure described for the large-sample case can be used for a small-sample hypothesis test about the difference between two population means. The only difference is that **Assume equal variances** is *selected* in step 4. Finally, the **Alternative** box in step 5 may be less than, not equal, or greater than depending upon the alternative hypothesis H_a.

*Minitab provides both hypothesis testing and interval estimation results simultaneously. The user may select any confidence level for the interval estimate of the population mean: 95% is suggested here.

Matched Samples

Matched

We use the data on production times in Table 10.4 to illustrate the matched-sample procedure. The completion times for method 1 are entered into column C1 and the completion times for method 2 are entered into column C2. The Minitab steps for the matched-sample hypothesis test are as follows:

Step 1. Select the **Stat** pull-down menu
Step 2. Choose **Basic Statistics**
Step 3. Choose **Paired t**
Step 4. When the Paired t dialog box appears:
  Enter C1 in the **First sample** box
  Enter C2 in the **Second sample** box
  Select **Options**
Step 5. Enter 95 in the **Confidence level**
  Enter 0 in the **Test Mean** box
  Enter not equal in the **Alternative** box
  Click **OK**
Step 6. Click **OK**

Appendix 10.2 TWO POPULATION MEANS WITH EXCEL

We describe how to use Excel for three hypothesis tests about the difference between the means of two populations.

Large-Sample Case

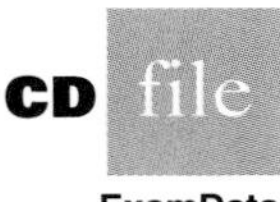
ExamData

We use the examination scores in Table 10.1. The label Center A is in cell A1 and the label Center B is in cell B1. The Center A scores are in cells A2:A31 and the Center B scores are in cells B2:B41. We will assume that the VAR function has been used to compute the variances of the two populations. Thus, we will use $s_1^2 = (8)^2 = 64$ and $s_2^2 = (10)^2 = 100$. The steps to conduct the test follow:

Step 1. Select the **Tools** pull-down menu
Step 2. Choose **Data Analysis**
Step 3. When the Data Analysis dialog box appears:
  Choose **z-Test: Two Sample for Means**
  Click **OK**
Step 4. When the z-Test: Two Sample for Means dialog box appears:
  Enter A1:A31 in the **Variable 1 Range** box
  Enter B1:B41 in the **Variable 2 Range** box
  Enter 0 in the **Hypothesized Mean Difference** box
  Enter 64 in the **Variable 1 Variance** box
  Enter 100 in the **Variable 2 Variance** box
  Select **Labels**
  Enter .05 in the **Alpha** box
  Select **Output Range** and enter C1 in the box
  Click **OK**

The value of the test statistic $z = 2.09$ appears in cell D8, and the p-value $= .0366$ appears in cell D11.

Small-Sample Case

Software

We use the data for the software testing study in Table 10.3. The data have been entered into an Excel worksheet with the label Current in cell A1 and the label New in cell B1. The completion times for the current technology are in cells A2:A13, and the completion times for the new software are in cells B2:B13. The following steps can be used to conduct a small-sample hypothesis test about the difference between two population means.

Step 1. Select **Tools** pull-down menu
Step 2. Choose **Data Analysis**
Step 3. When the Data Analysis dialog box appears:
Choose **t-Test: Two Sample Assuming Equal Variances**
Click **OK**
Step 4. When the t-Test: Two Sample Assuming Equal Variances dialog box appears:
Enter A1:A13 in the **Variable 1 Range** box
Enter B1:B13 in the **Variable 2 Range** box
Enter 0 in the **Hypothesized Mean Difference** box
Select **Labels**
Enter .05 in the **Alpha** box
Select **Output Range** and enter C1 in the box
Click **OK**

The value of the test statistic $t = 2.16$ appears in cell D10, and the one-tailed p-value = .021 appears in cell D11.

Matched Samples

Matched

We use the matched-sample completion times in Table 10.4 to illustrate. The data are entered into a worksheet with the label Method 1 in cell A1 and the label Method 2 in cell B1. The completion times for method 1 are in cells A2:A7 and the completion times for method 2 are in cells B2:B7. The Excel procedure uses the steps previously described for the small-sample case except the user chooses the **t-Test: Paired Two Sample for Means** data analysis tool in Step 3. The variable 1 range is A1:A7 and the variable 2 range is B1:B7. The value of the test statistic $t = 2.196$ appears in cell D10, and the two-tailed p-value = .08 appears in cell D13.

Chapter 11

INFERENCES ABOUT POPULATION VARIANCES

CONTENTS

STATISTICS IN PRACTICE

U.S. GENERAL ACCOUNTING OFFICE*
Washington, D.C.

The U.S. General Accounting Office (GAO) is an independent, nonpolitical audit organization in the legislative branch of the federal government. The GAO evaluators determine the effectiveness of current and proposed federal programs. To carry out their duties, evaluators must be proficient in records review, legislative research, and statistical analysis techniques.

In one case, the GAO evaluators studied a Department of Interior program established to help clean up the nation's rivers and lakes. As part of this program, federal grants were made to small cities throughout the United States. Congress asked the GAO to determine how effectively the program was operating. To do so, the GAO examined records and visited the sites of several waste treatment plants.

One objective of the GAO audit was to ensure that the effluent (treated sewage) at the plants met certain standards. Among other things the audits reviewed sample data on the oxygen content, the pH level, and the amount of suspended solids in the effluent. A requirement of the program was that a variety of tests be taken daily at each plant and that the collected data be sent periodically to the state engineering department. The GAO's investigation of the data showed whether various characteristics of the effluent were within acceptable limits.

For example, the mean or average pH level of the effluent was examined carefully. In addition, the variance in the reported pH levels was reviewed. The following hypothesis test was conducted about the variance in pH level for the population of effluent.

$$H_0: \sigma^2 = \sigma_0^2$$
$$H_a: \sigma^2 \neq \sigma_0^2$$

In this test, σ_0^2 is the population variance in pH level expected at a properly functioning plant. In one particular plant, the null hypothesis was rejected. Further analysis showed that this plant had a variance in pH level that was significantly less than normal.

Effluent at this water treatment plant must fall within a statistically determined pH range. © John Boykin/The Stock Market.

The auditors visited the plant to examine the measuring equipment and to discuss their statistical findings with the plant manager. The auditors found that the measuring equipment was not being used because the operator did not know how to work it. Instead, the operator had been told by an engineer what level of pH was acceptable and had simply recorded similar values without actually conducting the test. The unusually low variance in this plant's data had resulted in rejection of H_0. The GAO suspected that other plants might have similar problems and recommended an operator training program to improve the data collection aspect of the pollution control program.

In this chapter you will learn how to conduct statistical inferences about the variances of one and two populations. Two new distributions, the chi-square distribution and the F distribution, will be introduced and used to make interval estimates and hypothesis tests about population variances.

*The authors thank Mr. Art Foreman and Mr. Dale Ledman of the U.S. General Accounting Office for providing this Statistics in Practice.

In the preceding four chapters we examined methods of statistical inference involving population means and population proportions. In this chapter we expand the discussion to situations involving inferences about population variances. As an example of a case in which a variance can provide important decision-making information, consider the production process of filling containers with a liquid detergent product. The filling mechanism for the process is adjusted so that the mean filling quantity is 16 ounces per container. Although a mean of 16 ounces is desired, the variance of the fillings is also critical. That is, even with the filling mechanism properly adjusted for the mean of 16 ounces, we cannot expect every container to have exactly 16 ounces. By selecting a sample of containers, we can compute a sample variance for the number of ounces placed in a container. This value will serve as an estimate of the variance for the population of containers being filled by the production process. If the sample variance is modest, the production process will be continued. However, if the sample variance is excessive, overfilling and underfilling may be occurring even though the mean may be correct at 16 ounces. In this case, the filling mechanism will be readjusted in an attempt to reduce the filling variance for the containers.

In many manufacturing applications, controlling the process variance is extremely important in maintaining quality.

In the first section we consider inferences about the variance of a single population. Subsequently, we will discuss procedures that can be used to make inferences about the variances of two populations.

11.1 INFERENCES ABOUT A POPULATION VARIANCE

In preceding chapters we used the sample variance

$$s^2 = \frac{\Sigma(x_i - \bar{x})^2}{n - 1} \tag{11.1}$$

as the point estimator of the population variance σ^2. In using the sample variance as a basis for making inferences about a population variance, the sampling distribution of the quantity $(n - 1)s^2/\sigma^2$ is helpful. This sampling distribution is described as follows.

Sampling Distribution of $(n - 1)s^2/\sigma^2$

Whenever a simple random sample of size n is selected from a normal population, the sampling distribution of

$$\frac{(n - 1)s^2}{\sigma^2} \tag{11.2}$$

has a chi-square distribution with $n - 1$ degrees of freedom.

The chi-square distribution is based on sampling from a normal population.

Figure 11.1 shows some possible forms of the sampling distribution of $(n - 1)s^2/\sigma^2$.

Tables of areas or probabilities are readily available for the chi-square distribution. Since the sampling distribution of $(n - 1)s^2/\sigma^2$ is known to have a chi-square distribution whenever a simple random sample of size n is selected from a normal population, we can use the chi-square distribution to develop interval estimates and conduct hypothesis tests about a population variance.

Interval Estimation of σ^2

To show how the chi-square distribution can be used to develop a confidence interval estimate of a population variance σ^2, suppose that we are interested in estimating the population variance for the production filling process mentioned at the beginning of this chapter. A sample of 20 containers is taken, and the sample variance for the filling quantities is found to be $s^2 = .0025$. However, we know we cannot expect the variance of a sample of 20 containers to provide the exact value of the variance for the population of containers filled by

FIGURE 11.1 EXAMPLES OF THE SAMPLING DISTRIBUTION OF $(n - 1)s^2/\sigma^2$ (A CHI-SQUARE DISTRIBUTION)

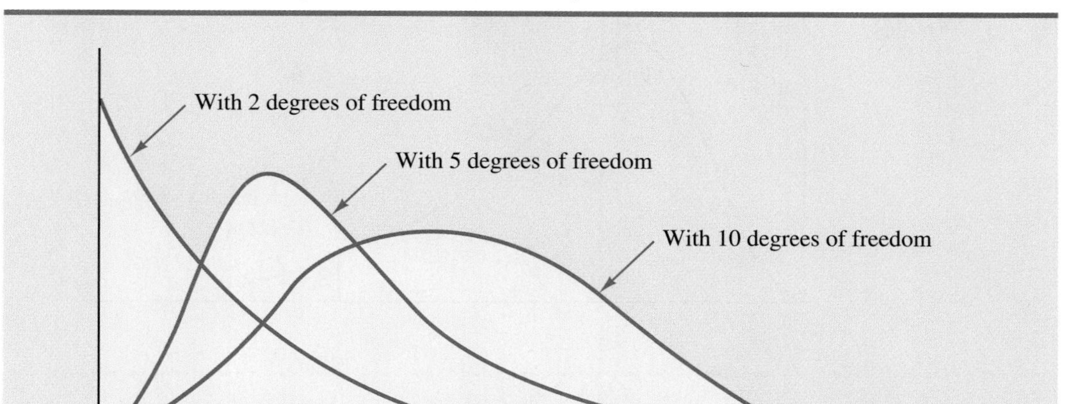

the production process. Hence, our interest will be in developing an interval estimate for the population variance.

We will use the notation χ^2_α to denote the value for the chi-square distribution that provides an area or probability of α to the *right* of the stated χ^2_α value. For example, in Figure 11.2 the chi-square distribution with 19 degrees of freedom is shown with $\chi^2_{.025} = 32.8523$ indicating that 2.5% of the chi-square values are to the right of 32.8523, and $\chi^2_{.975} = 8.90655$ indicating that 97.5% of the chi-square values are to the right of 8.90655. Refer to Table 11.1 and verify that these chi-square values with 19 degrees of freedom (19th row of the table) are correct. Table 3 of Appendix B provides a more extensive table of chi-square distribution values.

From the graph in Figure 11.2 we see that .95, or 95%, of the chi-square values are between $\chi^2_{.975}$ and $\chi^2_{.025}$. That is, there is a .95 probability of obtaining a χ^2 value such that

$$\chi^2_{.975} \leq \chi^2 \leq \chi^2_{.025}$$

However, we stated in (11.2) that $(n - 1)s^2/\sigma^2$ follows a chi-square distribution, therefore we can substitute $(n - 1)s^2/\sigma^2$ for the χ^2 and write

$$\chi^2_{.975} \leq \frac{(n - 1)s^2}{\sigma^2} \leq \chi^2_{.025} \tag{11.3}$$

In effect, (11.3) provides an interval estimate in that .95, or 95%, of all possible values for $(n - 1)s^2/\sigma^2$ will be in the interval $\chi^2_{.975}$ to $\chi^2_{.025}$. We now need to do some algebraic manipulations with (11.3) to develop an interval estimate for the population variance σ^2. Working with the leftmost inequality in (11.3), we have

$$\chi^2_{.975} \leq \frac{(n - 1)s^2}{\sigma^2}$$

Thus

$$\sigma^2\chi^2_{.975} \leq (n - 1)s^2$$

TABLE 11.1 SELECTED VALUES FROM THE CHI-SQUARE DISTRIBUTION TABLE*

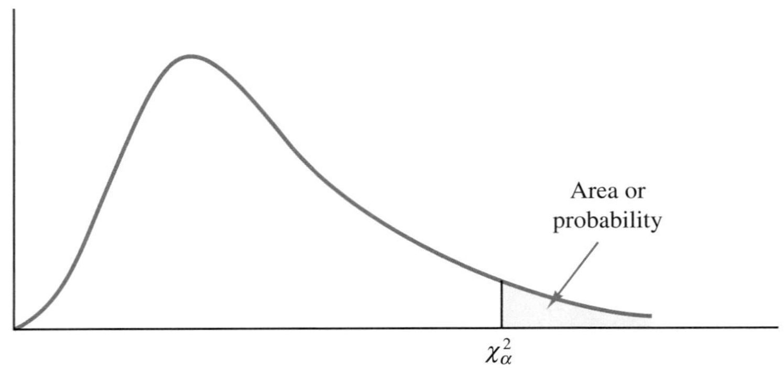

Degrees of Freedom	Area in Upper Tail					
	.99	.975	.95	.05	.025	.01
1	$157{,}088 \times 10^{-9}$	$982{,}069 \times 10^{-9}$	$393{,}214 \times 10^{-8}$	3.84146	5.02389	6.63490
2	.0201007	.0506356	.102587	5.99147	7.37776	9.21034
3	.114832	.215795	.351846	7.81473	9.34840	11.3449
4	.297110	.484419	.710721	9.48773	11.1433	13.2767
5	.554300	.831211	1.145476	11.0705	12.8325	15.0863
6	.872085	1.237347	1.63539	12.5916	14.4494	16.8119
7	1.239043	1.68987	2.16735	14.0671	16.0128	18.4753
8	1.646482	2.17973	2.73264	15.5073	17.5346	20.0902
9	2.087912	2.70039	3.32511	16.9190	19.0228	21.6660
10	2.55821	3.24697	3.94030	18.3070	20.4831	23.2093
11	3.05347	3.81575	4.57481	19.6751	21.9200	24.7250
12	3.57056	4.40379	5.22603	21.0261	23.3367	26.2170
13	4.10691	5.00874	5.89186	22.3621	24.7356	27.6883
14	4.66043	5.62872	6.57063	23.6848	26.1190	29.1413
15	5.22935	6.26214	7.26094	24.9958	27.4884	30.5779
16	5.81221	6.90766	7.96164	26.2962	28.8454	31.9999
17	6.40776	7.56418	8.67176	27.5871	30.1910	33.4087
18	7.01491	8.23075	9.39046	28.8693	31.5264	34.8053
19	7.63273	8.90655	10.1170	30.1435	32.8523	36.1908
20	8.26040	9.59083	10.8508	31.4104	34.1696	37.5662
21	8.89720	10.28293	11.5913	32.6705	35.4789	38.9321
22	9.54249	10.9823	12.3380	33.9244	36.7807	40.2894
23	10.19567	11.6885	13.0905	35.1725	38.0757	41.6384
24	10.8564	12.4011	13.8484	36.4151	39.3641	42.9798
25	11.5240	13.1197	14.6114	37.6525	40.6465	44.3141
26	12.1981	13.8439	15.3791	38.8852	41.9232	45.6417
27	12.8786	14.5733	16.1513	40.1133	43.1944	46.9630
28	13.5648	15.3079	16.9279	41.3372	44.4607	48.2782
29	14.2565	16.0471	17.7083	42.5569	45.7222	49.5879
30	14.9535	16.7908	18.4926	43.7729	46.9792	50.8922
0	22.1643	24.4331	26.5093	55.7585	59.3417	63.6907
[illegible]	29.7067	32.3574	34.7642	67.5048	71.4202	76.1539
[illegible]	37.4848	40.4817	43.1879	79.0819	83.2976	88.3794

*For a[illegible]ional χ^2 distribution values, see Table 3 of Appendix B.

FIGURE 11.2 A CHI-SQUARE DISTRIBUTION WITH 19 DEGREES OF FREEDOM

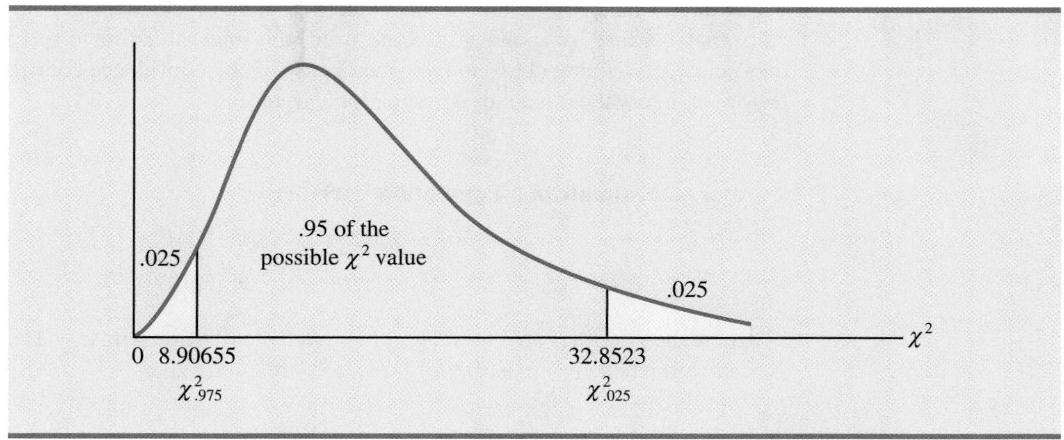

or

$$\sigma^2 \leq \frac{(n-1)s^2}{\chi^2_{.975}} \tag{11.4}$$

Performing similar algebraic manipulations with the rightmost inequality in (11.3) gives

$$\frac{(n-1)s^2}{\chi^2_{.025}} \leq \sigma^2 \tag{11.5}$$

The results of (11.4) and (11.5) can be combined to provide

$$\frac{(n-1)s^2}{\chi^2_{.025}} \leq \sigma^2 \leq \frac{(n-1)s^2}{\chi^2_{.975}} \tag{11.6}$$

Because (11.3) is true for 95% of the $(n-1)s^2/\sigma^2$ values, (11.6) provides a 95% confidence interval estimate of the population variance σ^2.

Let us return to the problem of providing an interval estimate of the population variance of filling quantities. Recall that the sample of 20 containers provided a sample variance of $s^2 = .0025$. With a sample size of 20, we have 19 degrees of freedom. As shown in Figure 11.2, we have already determined that $\chi^2_{.925} = 8.90655$ and $\chi^2_{.025} = 32.8523$. Using these values in (11.6) provides the following interval estimate for the population variance.

$$\frac{(19)(.0025)}{32.8523} \leq \sigma^2 \leq \frac{(19)(.0025)}{8.90655}$$

or

$$.0014 \leq \sigma^2 \leq .0053$$

A confidence interval for a population standard deviation can be found by computing the square roots of the lower limit and upper limit of the confidence interval for the population variance.

Taking the square root of these values provides the following 95% confidence interval for the population standard deviation.

$$.0374 \leq \sigma \leq .0728$$

Thus, we have illustrated the process of using the chi-square distribution to establish interval estimates of a population variance and a population standard deviation. Note specifically that because $\chi^2_{.975}$ and $\chi^2_{.025}$ were used, the interval estimate has a .95 confidence coefficient. Extending (11.6) to the general case of any confidence coefficient, we have the following interval estimate of a population variance.

Interval Estimate of a Population Variance

$$\frac{(n-1)s^2}{\chi^2_{\alpha/2}} \le \sigma^2 \le \frac{(n-1)s^2}{\chi^2_{(1-\alpha/2)}} \tag{11.7}$$

where the χ^2 values are based on a chi-square distribution with $n - 1$ degrees of freedom and where $1 - \alpha$ is the confidence coefficient.

Hypothesis Testing

Let us now consider an example and the statistical methodology necessary to test hypotheses about the value of a population variance. The St. Louis Metro Bus Company has recently made a concerted effort to promote an image of reliability by encouraging its drivers to maintain consistent schedules. As a standard policy the company expects arrival times at various bus stops to have low variability. In terms of the variance of arrival times, the company standard specifies an arrival-time variance of 4 or less with arrival times measured in minutes. Periodically, the company collects arrival-time data at various bus stops to determine whether the variability guideline is being maintained. The sample results are used to test the following hypotheses.

$$H_0: \sigma^2 \le 4$$
$$H_a: \sigma^2 > 4$$

In tentatively assuming H_0 to be true, we are assuming that the variance of arrival times is within the company guidelines. We will reject H_0 only if the sample evidence indicates that the guidelines are not being maintained. In this sense, rejection of H_0 suggests that follow-up steps are necessary to reduce the arrival-time variance.

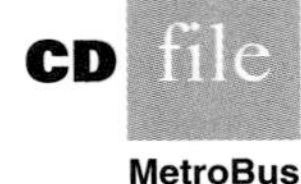

MetroBus

Assume that a random sample of 10 bus arrivals will be taken at a particular downtown intersection. If the population of arrival times has a normal probability distribution, we know from (11.2) that the quantity $(n - 1)s^2/\sigma^2$ has a chi-square distribution with $n - 1$ degrees of freedom. Thus, the test statistic

$$\chi^2 = \frac{(n-1)s^2}{\sigma^2} \tag{11.8}$$

has a chi-square distribution with $n - 1 = 9$ degrees of freedom. With the null hypothesis $\sigma^2 = 4$, a sample size $n = 10$, and a sample variance s^2, (11.8) provides the following observed χ^2 value.

$$\chi^2 = \frac{9s^2}{4}$$

Figure 11.3 is the chi-square distribution showing the rejection region for this one-tailed test. Note that we will reject H_0 only if the sample variance s^2 leads to a large χ^2 value. With

FIGURE 11.3 REJECTION REGION FOR THE ST. LOUIS METRO BUS TEST WITH $\alpha = .05$

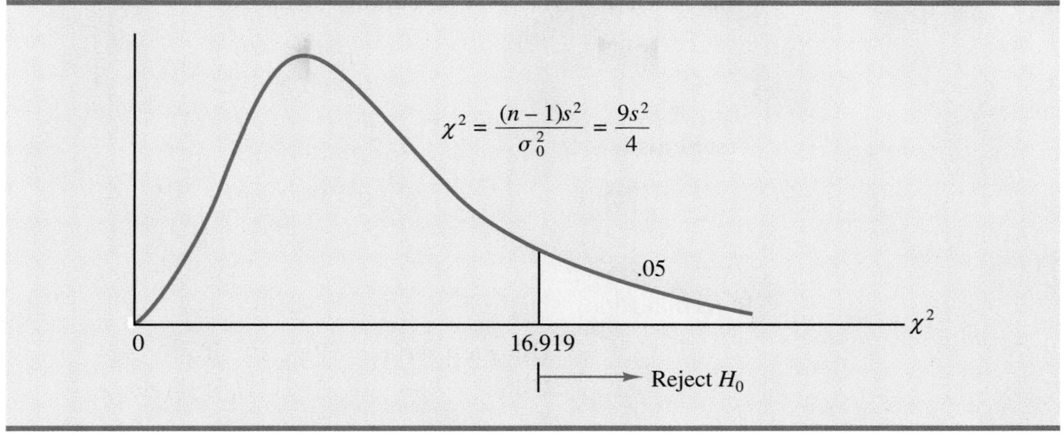

$\alpha = .05$, Table 11.1 shows that with 9 degrees of freedom $\chi^2_{.05} = 16.919$. With this critical value for the test, the rejection rule is:

$$\text{Reject } H_0 \text{ if } \chi^2 > 16.919$$

Let us assume that the sample of arrival times for 10 buses shows a sample variance of $s^2 = 4.8$. Is this sample evidence sufficient to reject H_0 and conclude that the buses are not meeting the company's arrival-time variance guideline? With $s^2 = 4.8$ we obtain the following.

$$\chi^2 = \frac{9(4.8)}{4} = 10.8$$

The rejection rule Reject H_0 if p-value $< \alpha$ applies to this hypothesis test. Appendixes 11.1 and 11.2 show how Minitab and Excel can be used to compute the p-value.

Because $\chi^2 = 10.8$ is less than 16.919, we cannot reject H_0. Hence, the sample variance of $s^2 = 4.8$ is insufficient evidence to conclude that the arrival-time variance is not meeting the company standard.

The *p*-value criterion can also be used for this hypothesis test. The usual rejection rule applies: Reject H_0 if *p*-value $< \alpha$. However, because it is difficult to determine the *p*-value directly from the tables of a chi-square distribution, a computer software package such as Minitab or Excel is required. Appendixes 11.1 and 11.2 show the procedures that can be used. In the preceding example, the *p*-value associated with the test statistic $\chi^2 = 10.8$ is .29. With a *p*-value $= .29 > \alpha = .05$, the null hypothesis cannot be rejected.

In practice, one-tailed tests are the most frequently encountered tests about population variances. That is, in situations involving arrival times, production times, filling weights, part dimensions, and so on, low variances are generally desired, whereas large variances tend to be unacceptable. With a statement about the maximum allowable population variance, we can test the null hypothesis that the population variance is less than or equal to the maximum allowable value against the alternative hypothesis that the population variance is greater than the maximum allowable value.

However, just as we saw with population means and proportions, other forms of the hypotheses can be developed. The one-tailed test for H_0: $\sigma^2 \geq \sigma_0^2$ is similar to the preceding test except that the rejection region is in the lower tail of the chi-square distribution at the critical value of $\chi^2_{(1-\alpha)}$. We now summarize the procedures for one-tailed tests about a population variance.

One-Tailed Test About a Population Variance

$$H_0: \sigma^2 \le \sigma_0^2$$
$$H_a: \sigma^2 > \sigma_0^2$$

Test Statistic

$$\chi^2 = \frac{(n-1)s^2}{\sigma_0^2}$$

Rejection Rule

Using test statistic: Reject H_0 if $\chi^2 > \chi^2_\alpha$
Using p-value: Reject H_0 if p-value $< \alpha$

where σ_0^2 is the hypothesized value for the population variance and χ^2_α is based on a chi-square distribution with $n - 1$ degrees of freedom.

One-Tailed Test About a Population Variance

$$H_0: \sigma^2 \ge \sigma_0^2$$
$$H_a: \sigma^2 < \sigma_0^2$$

Test Statistic

$$\chi^2 = \frac{(n-1)s^2}{\sigma_0^2}$$

Rejection Rule

Using test statistic: Reject H_0 if $\chi^2 < \chi^2_{(1-\alpha)}$
Using p-value: Reject H_0 if p-value $< \alpha$

where σ_0^2 is the hypothesized value for the population variance and $\chi^2_{(1-\alpha)}$ is based on a chi-square distribution with $n - 1$ degrees of freedom.

The two-tailed test with H_0: $\sigma^2 = \sigma_0^2$, like other two-tailed tests, places an area of $\alpha/2$ in each tail of the distribution to establish the two critical values for the test. The decision rule for conducting a two-tailed test about a population variance follows.

Two-Tailed Test About a Population Variance

$$H_0: \sigma^2 = \sigma_0^2$$
$$H_a: \sigma^2 \ne \sigma_0^2$$

Test Statistic

$$\chi^2 = \frac{(n-1)s^2}{\sigma_0^2}$$

Rejection Rule

Using test statistic: Reject H_0 if $\chi^2 < \chi^2_{(1-\alpha/2)}$ or if $\chi^2 > \chi^2_{\alpha/2}$
Using p-value: Reject H_0 if p-value $< \alpha$

where σ_0^2 is the hypothesized value for the population variance and $\chi^2_{(1-\alpha/2)}$ and $\chi^2_{\alpha/2}$ are based on a chi-square distribution with $n - 1$ degrees of freedom.

Let us demonstrate the use of the chi-square distribution in conducting a two-tailed test about a population variance by considering a situation faced by a bureau of motor vehicles. Historically, the variance in test scores for individuals applying for driver's licenses has been $\sigma^2 = 100$. A new examination with new test questions has been developed. Administrators of the bureau of motor vehicles would like the variance in the test scores for the new examination to remain at the historical level. To evaluate the variance in the new examination test scores, the following two-tailed hypothesis test has been proposed.

$$H_0\colon \sigma^2 = 100$$
$$H_a\colon \sigma^2 \neq 100$$

Rejection of H_0 will indicate that a change in the variance has occurred and suggest that some questions in the new examination may need revision to make the variance of the new test scores similar to the variance of the old test scores. A sample of 30 applicants for driver's licenses will be given the new version of the examination.

The chi-square distribution can be used to conduct this two-tailed test. With a .05 level of significance, the critical values will be $\chi^2_{.975}$ and $\chi^2_{.025}$. With $n - 1 = 29$ degrees of freedom, Table 11.1 shows that $\chi^2_{.925} = 16.0471$ and $\chi^2_{.025} = 45.7222$. The rejection rule for the two-tailed test becomes

$$\text{Reject } H_0 \text{ if } \chi^2 < 16.0471 \text{ or if } \chi^2 > 45.7222$$

What is the appropriate conclusion if the sample of 30 new examination scores provides a sample variance of $s^2 = 64$? With $H_0\colon \sigma^2 = 100$, the value of the χ^2 statistic is computed to be

$$\chi^2 = \frac{(n-1)s^2}{\sigma_0^2} = \frac{29(64)}{100} = 18.56$$

Because 18.56 is not in the rejection region, we are not able to reject H_0. There is no statistical evidence that the variance in the new examination scores differs from the historical variance in examination scores. The procedures in Appendixes 11.1 and 11.2 can be used to show that the p-value for this hypothesis test is .136.

EXERCISES

Methods

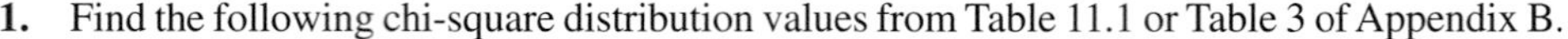

1. Find the following chi-square distribution values from Table 11.1 or Table 3 of Appendix B.
   a. $\chi^2_{.05}$ with df = 5 **d.** $\chi^2_{.01}$ with df = 10
   b. $\chi^2_{.025}$ with df = 15 **e.** $\chi^2_{.95}$ with df = 18
   c. $\chi^2_{.975}$ with df = 20

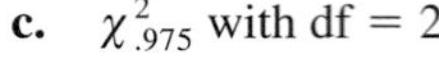

SELF test

2. A sample of 20 items provides a sample standard deviation of 5.
   a. Compute the 90% confidence interval estimate of the population variance.
   b. Compute the 95% confidence interval estimate of the population variance.
   c. Compute the 95% confidence interval estimate of the population standard deviation.

3. A sample of 16 items provides a sample standard deviation of 8. Test the following hypotheses using $\alpha = .05$. What is your conclusion?

$$H_0\colon \sigma^2 \leq 50$$
$$H_a\colon \sigma^2 > 50$$

Applications

4. The variance in drug weights is critical in the pharmaceutical industry. For a specific drug, with weights measured in grams, a sample of 18 units provided a sample variance of $s^2 = .36$.
   a. Construct a 90% confidence interval estimate of the population variance for the weights of this drug.
   b. Construct a 90% confidence interval estimate of the population standard deviation.

5. The daily car rental rates for a sample of eight cities follow (*The Wall Street Journal,* December 12, 1997).

City	Daily Car Rental Rate
Atlanta	$47
Chicago	50
Dallas	53
New Orleans	45
Phoenix	40
Pittsburgh	43
San Francisco	39
Seattle	37

   a. Compute the variance and the standard deviation for these data.
   b. What is the 95% confidence interval estimate of the variance of car rental rates for the population?
   c. What is the 95% confidence interval estimate of the standard deviation for the population?

6. The Oppenheimer Capital Appreciation mutual fund provided a 17.6% average annual return over a 5-year period (*Fortune,* August 18, 1997). Assume that the following data show the annual returns for each of the 5 years: 10.8, 34.2, 4.2, 9.4, and 29.4. The standard deviation of the annual returns can be used as a measure of risk, with a larger standard deviation indicating more variation and therefore more uncertainty in the annual returns. A standard deviation of 18.2% was reported for the 10 aggressive growth mutual funds similar to the Oppenheimer Capital Appreciation mutual fund.
   a. Using the data as a sample of five annual returns, what is the sample standard deviation measure of risk for the Oppenheimer Capital Appreciation mutual fund?
   b. What is the 95% confidence interval for the population standard deviation of annual returns for the Oppenheimer Capital Appreciation mutual fund?

7. A sample of eight earnings per share estimates for 1998 is shown here (*Barron's,* December 8, 1997).

Company	Estimated Earnings per Share
AT&T	2.92
Caterpillar	4.65
Eastman Kodak	4.27
Exxon	3.09
Hewlett-Packard	3.57
IBM	7.04
McDonald's	2.64
Wal-Mart	1.74

a. Compute the sample variance and sample standard deviation for these data.
b. What is the 95% confidence interval for the population variance?
c. What is the 95% confidence interval for the population standard deviation?

8. A group of 12 security analysts provided estimates of the year 2001 earnings per share for Qualcomm, Inc. (*Zacks.com,* June 13, 2000). The data are as follows:

1.40 1.40 1.45 1.49 1.37 1.27 1.40 1.55 1.40 1.42 1.48 1.63

a. Compute the sample variance for the earnings per share estimate.
b. Compute the sample standard deviation for the earnings per share estimate.
c. Provide 95% confidence interval estimates of the population variance and the population standard deviation.

9. An automotive part must be machined to close tolerances to be acceptable to customers. Production specifications call for a maximum variance in the lengths of the parts of .0004. Suppose the sample variance for 30 parts turns out to be $s^2 = .0005$. Using $\alpha = .05$, test to see whether the population variance specification is being violated.

10. The average useful life of a VCR is 6 years with a standard deviation of .75 years (*Consumer Reports 1995 Buying Guide*). A sample of the useful life of 30 television sets provided a sample standard deviation of 2 years. Construct a hypothesis test that can be used to determine whether the standard deviation of the useful life of television sets is significantly greater than the standard deviation of the useful life of VCRs. With a .05 level of significance, what is your conclusion?

11. Home mortgage interest rates for 30-year fixed-rate loans vary throughout the country. During the summer of 2000, data available from various parts of the country suggested that the standard deviation of the interest rates was .096 (*The Wall Street Journal,* September 8, 2000). The corresponding variance in interest rates would be $(.096)^2 = .009216$. Consider a follow-up study in the summer of 2001. The interest rates for 30-year fixed rate loans at a sample of 20 lending institutions had a sample standard deviation of .114. Conduct a hypothesis test H_0: $\sigma^2 = .009216$ to see whether the sample data indicate that the variability in interest rates changed. Use $\alpha = .05$. What is your conclusion?

12. A *Fortune* study found that the variance in the number of vehicles owned or leased by subscribers to *Fortune* magazine is .94. Assume a sample of 12 subscribers to another magazine provided the following data on the number of vehicles owned or leased: 2, 1, 2, 0, 3, 2, 2, 1, 2, 1, 0, and 1.
a. Compute the sample variance in the number of vehicles owned or leased by the 12 subscribers.
b. Test the hypothesis H_0: $\sigma^2 = .94$ to determine if the variance in the number of vehicles owned or leased by subscribers of the other magazine differ from $\sigma^2 = .94$ for *Fortune.* Using a .05 level of significance, what is your conclusion?

11.2 INFERENCES ABOUT THE VARIANCES OF TWO POPULATIONS

In some statistical applications we might want to compare the variances in product quality resulting from two different production processes, the variances in assembly times for two assembly methods, or the variances in temperatures for two heating devices. In making comparisons about the variances of two populations, we will be using data collected from two independent random samples, one from population 1 and another from population 2. The two sample variances s_1^2 and s_2^2 will be the basis for making inferences about the two population variances σ_1^2 and σ_2^2. Whenever the variances of two normal populations are equal ($\sigma_1^2 = \sigma_2^2$), the sampling distribution of the ratio of the two sample variances s_1^2/s_2^2 is as follows.

Sampling Distribution of s_1^2/s_2^2 When $\sigma_1^2 = \sigma_2^2$

Whenever independent simple random samples of sizes n_1 and n_2 are selected from two normal populations with equal variances, the sampling distribution of

$$\frac{s_1^2}{s_2^2} \tag{11.9}$$

The F distribution is based on sampling from two normal populations.

has an F distribution with $n_1 - 1$ degrees of freedom for the numerator and $n_2 - 1$ degrees of freedom for the denominator; s_1^2 is the sample variance for the random sample of n_1 items from population 1, and s_2^2 is the sample variance for the random sample of n_2 items from population 2.

Figure 11.4 is a graph of the F distribution with 20 degrees of freedom for both the numerator and denominator. As can be seen from this graph, the F distribution is not symmetric, and the F values can never be negative. The shape of any particular F distribution depends on its numerator and denominator degrees of freedom.

We will use F_α to denote the value for the F distribution that provides an area or probability of α to the *right* of the stated F_α value. For example, as noted in Figure 11.4, $F_{.05}$ denotes the upper 5% of the F values for an F distribution with 20 degrees of freedom for the numerator and 20 degrees of freedom for the denominator. Table 11.2 shows that for this

FIGURE 11.4 *F* DISTRIBUTION WITH 20 DEGREES OF FREEDOM FOR THE NUMERATOR AND 20 DEGREES OF FREEDOM FOR THE DENOMINATOR

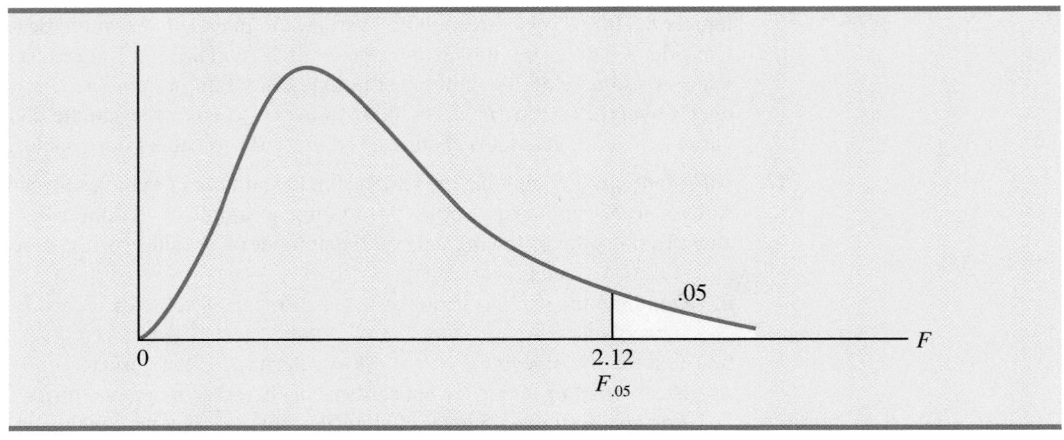

particular F distribution, $F_{.05} = 2.12$. Table 4 of Appendix B is a more extensive table of F distribution values. Let us show how the F distribution can be used for a hypothesis test about the variances of two populations.

Dullus County Schools is renewing its school bus service contract for the coming year and must select one of two bus companies, the Milbank Company or the Gulf Park Company. We will use the variance of the arrival or pickup/delivery times as a primary measure of the quality of the bus service. Low variance values indicate the more consistent and higher

TABLE 11.2 SELECTED VALUES FROM THE F DISTRIBUTION TABLE*

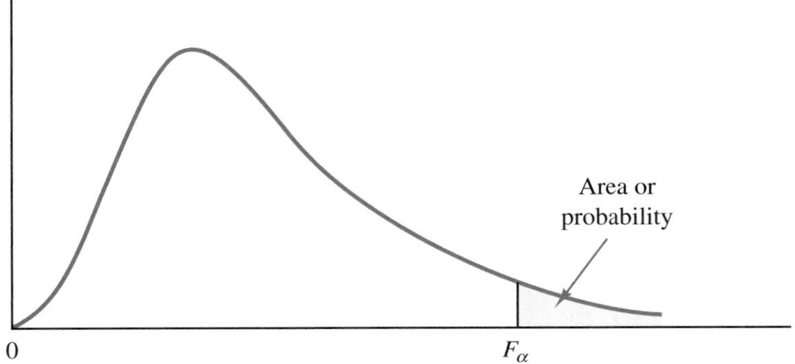

Table of $F_{.05}$ Values

Denominator Degrees of Freedom	Numerator Degrees of Freedom							
	6	7	8	9	10	12	15	20
1	234.0	236.8	238.9	240.5	241.9	243.9	245.9	248.0
2	19.33	19.35	19.37	19.38	19.40	19.41	19.43	19.45
3	8.94	8.89	8.85	8.81	8.79	8.74	8.70	8.66
4	6.16	6.09	6.04	6.00	5.96	5.91	5.86	5.80
5	4.95	4.88	4.82	4.77	4.74	4.68	4.62	4.56
6	4.28	4.21	4.15	4.10	4.06	4.00	3.94	3.87
7	3.87	3.79	3.73	3.68	3.64	3.57	3.51	3.44
8	3.58	3.50	3.44	3.39	3.35	3.28	3.22	3.15
9	3.37	3.29	3.23	3.18	3.14	3.07	3.01	2.94
10	3.22	3.14	3.07	3.02	2.98	2.91	2.85	2.77
11	3.09	3.01	2.95	2.90	2.85	2.79	2.72	2.65
12	3.00	2.91	2.85	2.80	2.75	2.69	2.62	2.54
13	2.92	2.83	2.77	2.71	2.67	2.60	2.53	2.46
14	2.85	2.76	2.70	2.65	2.60	2.53	2.46	2.39
15	2.79	2.71	2.64	2.59	2.54	2.48	2.40	2.33
16	2.74	2.66	2.59	2.54	2.49	2.42	2.35	2.28
17	2.70	2.61	2.55	2.49	2.45	2.38	2.31	2.23
18	2.66	2.58	2.51	2.46	2.41	2.34	2.27	2.19
19	2.63	2.54	2.48	2.42	2.38	2.31	2.23	2.16
20	2.60	2.51	2.45	2.39	2.35	2.28	2.20	2.12
21	2.57	2.49	2.42	2.37	2.32	2.25	2.18	2.10
22	2.55	2.46	2.40	2.34	2.30	2.23	2.15	2.07
23	2.53	2.44	2.37	2.32	2.27	2.20	2.13	2.05
24	2.51	2.42	2.36	2.30	2.25	2.18	2.11	2.03

*For additional F distribution values, see Table 4 of Appendix B.

quality service. If the variances of arrival times associated with the two services are equal, Dullus School administrators will select the company offering the better financial terms. However, if the sample data on bus arrival times for the two companies indicate a significant difference between the variances, the administrators may want to give special consideration to the company with the better or lower variance service. The appropriate hypotheses follow.

$$H_0: \sigma_1^2 = \sigma_2^2$$
$$H_a: \sigma_1^2 \neq \sigma_2^2$$

If H_0 can be rejected, the conclusion of unequal service quality is appropriate. In that case, the company with the lower sample variance would be preferred.

Assume that the hypothesis test will be conducted with $\alpha = .10$. Furthermore, assume that we obtain samples of arrival times from school systems currently using the two school bus services. A sample of 25 arrival times is available for the Milbank service (population 1) and a sample of 16 arrival times is available for the Gulf Park service (population 2). Figure 11.5 is the graph of the F distribution with $n_1 - 1 = 24$ degrees of freedom for the numerator and $n_2 - 1 = 15$ degrees of freedom for the denominator. Note that the two-tailed rejection region is indicated by the critical values at $F_{.95}$ and $F_{.05}$.

CD file
Dullus

Let us assume that the two samples of bus arrival times resulted in sample variances of $s_1^2 = 48$ for the Milbank service and $s_2^2 = 20$ for the Gulf Park service. What conclusion is appropriate? Assume that the two populations of arrival times have normal probability distributions, and assume that H_0 is true with $\sigma_1^2 = \sigma_2^2$. The F distribution can be used to reach a conclusion. Specifically, we compute $F = s_1^2/s_2^2$ and use the rejection region shown in Figure 11.5. Thus, we find

$$F = \frac{s_1^2}{s_2^2} = \frac{48}{20} = 2.40$$

Using Table 4 of Appendix B, we find that the upper-tail critical value with 24 numerator degrees of freedom and 15 denominator degrees of freedom is $F_{.05} = 2.29$. Although Table 4 does not provide $F_{.95}$ values, note that the determination of this lower-tail critical value is not necessary. We can already observe that $F = 2.40$ exceeds $F_{.05} = 2.29$. Thus, at the .10

FIGURE 11.5 REJECTION REGION FOR THE DULLUS COUNTY SCHOOL BUS EXAMPLE WITH $\alpha = .10$

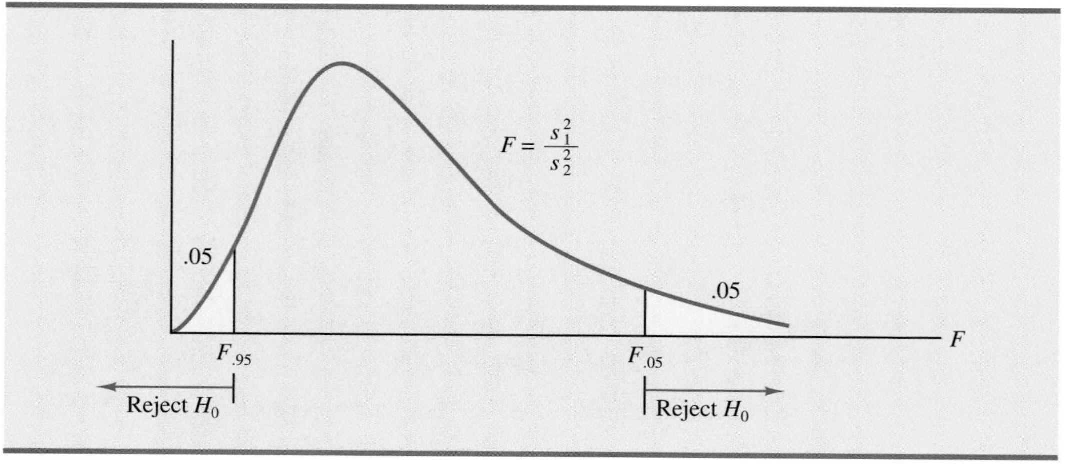

level of significance, H_0 is rejected. This finding leads us to the conclusion that the two bus services differ in terms of pickup/delivery time variances. The recommendation is that the Dullus School administrators give special consideration to the better or lower variance service offered by the Gulf Park Company.

The rejection rule Reject H_0 if p-value $< \alpha$ also applies to this test.

The p-value criterion can also be used for a hypothesis test about two population variances. The usual rejection rule applies: Reject H_0 if p-value $< \alpha$. However, like the chi-square distribution, it is difficult to determine the p-value directly from the tables of the F distribution. Again, a computer software package such as Minitab or Excel is required. Appendixes 11.1 and 11.2 show the procedures that can be used. In the school bus study, $F = 2.40$ provided a p-value $= .082$. With a p-value $= .082 < \alpha = .10$, the null hypothesis of equal variances is rejected.

In the Dullus School example, you might feel that we were lucky in carrying out the two-tailed test because the lower-tail critical value $F_{.95}$, which is not provided in the F distribution table, was not necessary. We were able to draw the appropriate hypothesis-testing conclusion without knowing the value of $F_{.95}$. If you ever need to know a lower-tail F value, $F_{(1-\alpha)}$, it can be determined by using an upper-tail F_α value and the following relationship.

$$F_{(1-\alpha),\text{df}_1,\text{df}_2} = \frac{1}{F_{\alpha,\text{df}_2,\text{df}_1}} \quad \textbf{(11.10)}$$

Equation (11.10) shows how lower-tail F values can be computed from the corresponding upper-tail F values. With the upper-tail F values shown in Table 11.3 and in Table 4 of Appendix B, the corresponding lower-tail F values can be computed. Tables with lower-tail F values are not necessary.

Thus, with 24 degrees of freedom in the numerator ($\text{df}_1 = 24$) and 15 degrees of freedom in the denominator ($\text{df}_2 = 15$), $F_{.95}$ is the reciprocal of the $F_{.05}$ value with 15 degrees of freedom in the numerator and 24 degrees of freedom in the denominator. In the F distribution table, we find that $F_{.05}$ with 15 numerator and 24 denominator degrees of freedom is 2.11. Hence, the value of $F_{.95}$ with 24 degrees of freedom in the numerator and 15 degrees of freedom in the denominator is

$$F_{.95} = \frac{1}{2.11} = .47$$

Although (11.10) can be used to compute a lower-tail F value, common practice is to conduct the hypothesis test computations so that only upper-tail F values are needed. In hypothesis tests with H_0: $\sigma_1^2 = \sigma_2^2$, we simply denote the population with the *larger sample variance* as population 1. That is, deciding which population to denote as either population 1 or 2 is arbitrary. By labeling the population with the larger sample variance as population 1, we guarantee that a rejection of H_0 can occur only in the *upper tail.* Although the lower-tail critical value still exists, we do not need to know its value because the convention of using the population with the largest sample variance as population 1 always places the ratio s_1^2/s_2^2 in the upper-tail direction. In the Dullus School example, population 1, the Milbank bus service, had the largest sample variance. Hence, we proceeded directly with the test. If the Gulf Park bus service had provided the largest sample variance, we would have denoted Gulf Park as population 1 and followed the same statistical testing procedure. A summary of the two-tailed test for the equality of two population variances with this procedure follows.

Two-Tailed Test About the Variances of Two Populations

$$H_0: \sigma_1^2 = \sigma_2^2$$
$$H_a: \sigma_1^2 \neq \sigma_2^2$$

Denote the population providing the *largest sample variance* as population 1.

Test Statistic

$$F = \frac{s_1^2}{s_2^2}$$

Rejection Rule

Using test statistic: Reject H_0 if $F > F_{\alpha/2}$
Using p-value: Reject H_0 if p-value $< \alpha$

where the value of $F_{\alpha/2}$ is based on an F distribution with $n_1 - 1$ degrees of freedom for the numerator and $n_2 - 1$ degrees of freedom for the denominator.

A test hypothesis about two population variances can be formulated so that the rejection region is always in the upper tail of the distribution. This approach eliminates the need for computing lower-tail F distribution values.

One-tailed tests involving two population variances are also possible. Again the F distribution is used, with the one-tailed rejection region enabling us to conclude whether one population variance is significantly greater or significantly less than the other. Only upper-tail F values are needed. For any one-tailed test we set up the null hypothesis so that the rejection region is in the upper tail, which can be accomplished by labeling the population with the larger variance in H_a as population 1. The general procedure follows.

One-Tailed Test About the Variances of Two Populations

$$H_0\colon \sigma_1^2 \le \sigma_2^2$$
$$H_a\colon \sigma_1^2 > \sigma_2^2$$

Test Statistic

$$F = \frac{s_1^2}{s_2^2}$$

Rejection Rule

Using test statistic: Reject H_0 if $F > F_\alpha$
Using p-value: Reject H_0 if p-value $< \alpha$

where the value of F_α is based on an F distribution with $n_1 - 1$ degrees of freedom for the numerator and $n_2 - 1$ degrees of freedom for the denominator.

Let us demonstrate the use of the F distribution to conduct a one-tailed test about the variances of two populations by considering a public opinion survey. Samples of 31 men and 41 women will be used to study attitudes about current political issues. The researcher conducting the study wants to test to see whether the sample data indicate that women have a greater variation in attitude on political issues than men. In the form of the one-tailed hypothesis test given previously, women will be denoted as population 1 and men will be denoted as population 2. The hypothesis test will be stated as follows.

$$H_0\colon \sigma_{\text{women}}^2 \le \sigma_{\text{men}}^2$$
$$H_a\colon \sigma_{\text{women}}^2 > \sigma_{\text{men}}^2$$

If H_0 is rejected, the researcher will have the statistical support necessary to conclude that women have a greater variation in attitude on political issues.

With the sample variance for women in the numerator and the sample variance for men in the denominator, the F distribution with $41 - 1 = 40$ degrees of freedom in the numerator and $31 - 1 = 30$ degrees of freedom in the denominator will be used to conduct the one-tailed test. With a .05 level of significance, the rejection region is based on $F_{.05}$. Using Table 4 of Appendix B, we find that $F_{.05} = 1.79$. Thus, the rejection rule becomes

$$\text{Reject } H_0 \text{ if } F > 1.79$$

where F is computed from the ratio of the two sample variances s_1^2/s_2^2.

Assume that the survey shows a sample variance of $s_1^2 = 120$ for the 41 women and a sample variance of $s_2^2 = 80$ for the 31 men. What is the appropriate statistical conclusion? The F statistic becomes

$$F = \frac{s_1^2}{s_2^2} = \frac{120}{80} = 1.50$$

With 1.50 less than 1.79, H_0 cannot be rejected. Hence, the sample results do not support the conclusion that women have a greater variation than men in their attitudes on political issues. With $F = 1.50$, the p-value for the test is .126.

NOTES AND COMMENTS

1. Researchers have confirmed the fact that the F distribution is sensitive to the assumption of normal populations. The F distribution should not be used unless it is reasonable to assume that both populations are at least approximately normally distributed.
2. The t test for the equality of two population means discussed in Chapter 10 requires the assumption that $\sigma_1^2 = \sigma_2^2$. It is sometimes suggested that a hypothesis test H_0: $\sigma_1^2 = \sigma_2^2$ based on the F distribution be used prior to the t test to indicate whether the t test assumption of $\sigma_1^2 = \sigma_2^2$ appears valid. Unfortunately, researchers have reported that the F test has a small probability of detecting cases in which the t test performs poorly, and therefore have concluded that the F test is not an effective preliminary test for the t test (*The American Statistician,* November 1990).

EXERCISES

Methods

13. Find the following F distribution values from Table 4 of Appendix B.

a. $F_{.05}$ with degrees of freedom 12 and 10
b. $F_{.025}$ with degrees of freedom 20 and 15
c. $F_{.01}$ with degrees of freedom 8 and 12
d. $F_{.975}$ with degrees of freedom 10 and 20

14. A sample of 16 items from population 1 has a sample variance $s_1^2 = 5.8$ and a sample of 20 items from population 2 has a sample variance $s_2^2 = 2.4$. Test the following hypotheses at the .05 level of significance.

$$H_0: \sigma_1^2 \le \sigma_2^2$$
$$H_a: \sigma_1^2 > \sigma_2^2$$

What is your conclusion?

15. Consider the following hypothesis test.

$$H_0: \sigma_1^2 = \sigma_2^2$$
$$H_a: \sigma_1^2 \neq \sigma_2^2$$

What is your conclusion if $n_1 = 25$, $s_1^2 = 4.0$, $n_2 = 21$, and $s_2^2 = 8.2$? Use $\alpha = .05$.

Applications

16. Media Metrix and Jupiter Communications gathered data on the time adults and the time teens spend on-line during a month (*USA Today,* September 14, 2000). The study concluded that on average, adults spend more time on-line than teens. Assume that a follow-up study sampled 25 adults and 30 teens. The standard deviations of the time on-line during a month were 94 minutes and 58 minutes, respectively. Do the sample results support the conclusion that adults have a greater variance in on-line time than teens? Use $\alpha = .01$.

17. Most individuals are aware of the fact that the average annual repair cost for an automobile depends on the age of the automobile. A researcher is interested in finding out whether the variance of the annual repair costs also increases with the age of the automobile. A sample of 25 automobiles 4 years old showed a sample standard deviation for annual repair costs of \$170 and a sample of 25 automobiles 2 years old showed a sample standard deviation for annual repair costs of \$100.
    a. State the null and alternative versions of the research hypothesis that the variance in annual repair costs is larger for the older automobiles.
    b. For a .01 level of significance, what is your conclusion? Discuss the reasonableness of your findings.

18. The standard deviation in the 12-month earnings per share for 10 companies in the airline industry was 4.27 and the standard deviation in the 12-month earnings per share for 7 companies in the automotive industry was 2.27 (*Business Week,* August 14, 2000). Conduct a test for equal variances at $\alpha = .05$. What is your conclusion about the variability in earnings per share for the airline industry and the automotive industry?

19. The variance in a production process is an important measure of the quality of the process. A large variance often signals an opportunity for improvement in the process by finding ways to reduce the process variance. Data showing the weight of tea bags in grams for two machines were presented in *Quality Progress* (February 1995). Conduct a statistical test to determine whether there is a significant difference between the variances in the bag weights for the two machines. Use a .05 level of significance. What is your conclusion? Which machine, if either, provides the greater opportunity for quality improvements?

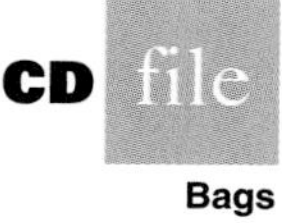

Bags

Machine 1	2.95	3.45	3.50	3.75	3.48	3.26	3.33	3.20
	3.16	3.20	3.22	3.38	3.90	3.36	3.25	3.28
	3.20	3.22	2.98	3.45	3.70	3.34	3.18	3.35
	3.12							
Machine 2	3.22	3.30	3.34	3.28	3.29	3.25	3.30	3.27
	3.38	3.34	3.35	3.19	3.35	3.05	3.36	3.28
	3.30	3.28	3.30	3.20	3.16	3.33		

20. On the basis of data provided by a Romac salary survey, the variance in annual salaries for seniors in public accounting firms is approximately 2.1 and the variance in annual salaries for managers in public accounting firms is approximately 11.1. The salary data were provided in thousands of dollars. Assuming that the salary data were based on samples of 25 seniors and 25 managers, test the hypothesis that the population variances in the salaries are equal. With a .05 level of significance, what is your conclusion?

21. The Dow Jones Industrial Average varies as investors buy and sell shares of the 30 stocks that make up the average. Samples of the Dow Jones Industrial Average taken at different times during the first 5 days of November 1997 and the first 5 days of December 1997 are as follow (*Barron's,* December 8, 1997).

DowJones

November	December
7493	8066
7525	8209
7760	7842
7499	7943
7555	7846
7690	8071
7668	8055
7600	8159
7516	7828
7711	8109

a. Compute the variances of the Dow Jones Industrial Average for the two time periods.
b. Using a .05 level of significance, test to determine whether the population variances for the two time periods are equal. What is your conclusion?

22. A research hypothesis is that the variance of stopping distances of automobiles on wet pavement is substantially greater than the variance of stopping distances of automobiles on dry pavement. In the research study, 16 automobiles traveling at the same speeds are tested for stopping distances on wet pavement and then tested for stopping distances on dry pavement. On wet pavement, the standard deviation of stopping distances is 32 feet. On dry pavement, the standard deviation is 16 feet.
a. At a .05 level of significance, do the sample data justify the conclusion that the variance in stopping distances on wet pavement is greater than the variance in stopping distances on dry pavement?
b. What are the implications of your statistical conclusions in terms of driving safety recommendations?

SUMMARY

In this chapter we have presented statistical procedures that can be used to make inferences about population variances. In the process we have introduced two new probability distributions: the chi-square distribution and the F distribution. The chi-square distribution can be used as the basis for interval estimation and hypothesis tests about the variance of a normal population.

We illustrated the use of the F distribution in hypothesis tests about the variances of two normal populations. In particular, we showed that with independent simple random samples of sizes n_1 and n_2 selected from two normal populations with equal variances $\sigma_1^2 = \sigma_2^2$, the sampling distribution of the ratio of the two sample variances s_1^2/s_2^2 has an F distribution with $n_1 - 1$ degrees of freedom for the numerator and $n_2 - 1$ degrees of freedom for the denominator.

KEY FORMULAS

Interval Estimate of a Population Variance

$$\frac{(n-1)s^2}{\chi^2_{\alpha/2}} \le \sigma^2 \le \frac{(n-1)s^2}{\chi^2_{(1-\alpha/2)}} \qquad \textbf{(11.7)}$$

Test Statistic for One Population Variance

$$\chi^2 = \frac{(n-1)s^2}{\sigma_0^2} \tag{11.8}$$

Test Statistic for Two Population Variances

$$F = \frac{s_1^2}{s_2^2} \tag{11.9}$$

SUPPLEMENTARY EXERCISES

23. Because of staffing decisions, managers of the Gibson-Marimont Hotel are interested in the variability in the number of rooms occupied per day during a particular season of the year. A sample of 20 days of operation shows a sample mean of 290 rooms occupied per day and a sample standard deviation of 30 rooms.
    a. What is the point estimate of the population variance?
    b. Provide a 90% confidence interval estimate of the population variance.
    c. Provide a 90% confidence interval estimate of the population standard deviation.
24. Initial public offerings (IPOs) of stocks are on average underpriced. The standard deviation measures the dispersion, or variation, in the underpricing-overpricing indicator. A sample of 13 Canadian IPOs that were subsequently traded on the Toronto Stock Exchange had a standard deviation of 14.95. Develop a 95% confidence interval estimate of the population standard deviation for the underpricing-overpricing indicator.
25. The estimated daily living costs for an executive traveling to various major cities follow (*Business Traveler,* June 1995). The estimates include a single room at a four-star hotel, beverages, breakfast, taxi fares, and incidental costs.

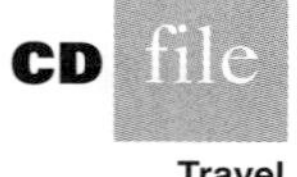

Travel

City	Daily Living Cost	City	Daily Living Cost
Bangkok	$242.87	Madrid	$283.56
Bogota	260.93	Mexico City	212.00
Bombay	139.16	Milan	284.08
Cairo	194.19	Paris	436.72
Dublin	260.76	Rio de Janeiro	240.87
Frankfurt	355.36	Seoul	310.41
Hong Kong	346.32	Tel Aviv	223.73
Johannesburg	165.37	Toronto	181.25
Lima	250.08	Warsaw	238.20
London	326.76	Washington, D.C.	250.61

   a. Compute the sample mean.
   b. Compute the sample standard deviation.
   c. Compute a 95% confidence interval for the population standard deviation.
26. Part variability is critical in the manufacturing of ball bearings. Large variances in the size of the ball bearings cause bearing failure and rapid wearout. Production standards call for a maximum variance of .0001 when the bearing sizes are measured in inches. A sample of 15 bearings shows a sample standard deviation of .014 inches.
    a. Using $\alpha = .10$, determine whether the sample indicates that the maximum acceptable variance is being exceeded.
    b. Compute the 90% confidence interval estimate of the variance of the ball bearings in the population.

27. The filling variance for boxes of cereal is designed to be .02 or less. A sample of 41 boxes of cereal shows a sample standard deviation of .16 ounces. Using $\alpha = .05$, determine whether the variance in the cereal box fillings is exceeding the design specification.

28. City Trucking, Inc., claims consistent delivery times for its routine customer deliveries. A sample of 22 truck deliveries shows a sample variance of 1.5. Test to determine whether H_0: $\sigma^2 \leq 1$ can be rejected. Use $\alpha = .10$.

29. Using a sample of 9 days over the past 6 months, a dentist has seen the following numbers of patients: 22, 25, 20, 18, 15, 22, 24, 19, and 26. If the number of patients seen per day is normally distributed, would an analysis of these sample data reject the hypothesis that the variance in the number of patients seen per day is equal to 10? Use a .10 level of significance. What is your conclusion?

30. A sample standard deviation for the number of passengers taking a particular airline flight is 8. A 95% confidence interval estimate of the population standard deviation is 5.86 passengers to 12.62 passengers.

a. Was a sample size of 10 or 15 used in the statistical analysis?

b. If the sample standard deviation of $s = 8$ had been based on a sample of 25 flights, what change would you expect in the confidence interval for the population standard deviation? Compute a 95% confidence interval estimate of σ with a sample of size of 25.

31. Each day the major stock markets have a group of leading gainers in price (stocks that go up the most). On one day the standard deviation in the percent change for a sample of 10 NASDAQ leading gainers was 15.8. On the same day, the standard deviation in the percent change for a sample of 10 NYSE leading gainers was 7.9 (*USA Today,* September 14, 2000). Conduct a test for equal population variances to see whether it can be concluded that there is a difference in the volatility of the leading gainers on the two exchanges. Use $\alpha = .10$. What is your conclusion?

32. The grade point averages of 352 students who completed a college course in financial accounting have a standard deviation of .940. The grade point averages of 73 students who dropped out of the same course have a standard deviation of .797. Do the data indicate a difference between the variances of grade point averages for students who completed a financial accounting course and students who dropped out? Use a .05 level of significance. Note: $F_{.025}$ with 351 and 72 degrees of freedom is 1.466.

33. The accounting department analyzes the variance of the weekly unit costs reported by two production departments. A sample of 16 cost reports for each of the two departments shows cost variances of 2.3 and 5.4, respectively. Is this sample sufficient to conclude that the two production departments differ in terms of unit cost variance? Use $\alpha = .10$.

34. Two new assembly methods are tested and the variances in assembly times are reported. Using $\alpha = .10$, test for equality of the two population variances.

Method	Sample Size	Sample Variance
A	31	25
B	25	12

Case Problem AIR FORCE TRAINING PROGRAM

An Air Force introductory course in electronics uses a personalized system of instruction whereby each student views a videotaped lecture and then is given a programmed instruction text. The students work independently with the text until they have completed the training and passed a test. Of concern is the varying pace at which the students complete this portion of their training program. Some students are able to cover the programmed instruction text relatively quickly, whereas other students work much longer with the text and require additional time to complete the course. The fast students wait until the slow students complete the introductory course before the entire group proceeds together with other aspects of their training.

A proposed alternative system involves use of computer-assisted instruction. In this method, all students view the same videotaped lecture and then each is assigned to a computer terminal for further instruction. The computer guides the student, working independently, through the self-training portion of the course.

To compare the proposed and current methods of instruction, an entering class of 122 students was assigned randomly to one of the two methods. One group of 61 students used the current programmed-text method and the other group of 61 students used the proposed computer-assisted method. The time in hours was recorded for each student in the study. The following data are provided on the data disk in the data set Training.

Course Completion Times (hours) for Current Training Method

76	76	77	74	76	74	74	77	72	78	73
78	75	80	79	72	69	79	72	70	70	81
76	78	72	82	72	73	71	70	77	78	73
79	82	65	77	79	73	76	81	69	75	75
77	79	76	78	76	76	73	77	84	74	74
69	79	66	70	74	72					

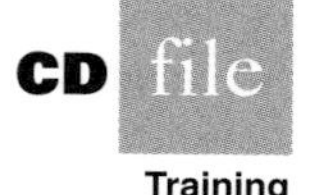

Training

Course Completion Times (hours) for Proposed Computer-Assisted Method

74	75	77	78	74	80	73	73	78	76	76
74	77	69	76	75	72	75	72	76	72	77
73	77	69	77	75	76	74	77	75	78	72
77	78	78	76	75	76	76	75	76	80	77
76	75	73	77	77	77	79	75	75	72	82
76	76	74	72	78	71					

Managerial Report

1. Use appropriate descriptive statistics to summarize the training time data for each method. What similarities and/or differences do you observe from the sample data?
2. Use the methods of Chapter 10 to comment on any difference between the population means for the two methods. Discuss your findings.
3. Compute the standard deviation and variance for each training method. Conduct a hypothesis test about the equality of population variances for the two training methods. Discuss your findings.
4. What conclusion can you reach about any differences between the two methods? What is your recommendation? Explain.
5. Can you suggest other data or testing that might be desirable before making a final decision on the training program to be used in the future?

Appendix 11.1 POPULATION VARIANCES WITH MINITAB

We describe how to use Minitab to compute the p-values for the chi-square test statistic and F test statistic. These p-values can be used in hypothesis tests about the variance of one population or the variances of two populations.

One Population

MetroBus

We will use the data for the St. Louis Metro Bus example in Section 11.1. The arrival times appear in column C1. In order to compute the p-value for a hypothesis test about a population variance, the user must first compute the sample variance and the corresponding chi-square test statistic* $\chi^2 = 10.8$. With the hypotheses H_0: $\sigma^2 \leq 4$ and H_a: $\sigma^2 > 4$, Minitab can be used to compute the following upper-tail p-value.

Step 1. Select the **Calc** pull-down menu
Step 2. Choose **Probability Distributions**
Step 3. Choose **Chi-Square**
Step 4. When the Chi-Square Distribution dialog box appears:
Select **Cumulative probability**
Enter 9 in the **Degrees of freedom** box
Select **Input constant** and enter 10.8 in the box
Enter CumProb in the **Optional storage** box
Click **OK**
Step 5. Select the **Calc** pull-down menu
Step 6. Choose **Calculator**
When the Calculator dialog box appears:
Enter p-value in the **Store results in variable** box
Enter 1-CumProb in the **Expression** box
Click **OK**

This procedure applies when the rejection region is in the upper tail of the chi-square distribution. If the hypothesis test is a one-tailed test with the rejection region in the lower tail, enter CumProb in the **Expression** box in step 6. If the hypothesis test is a two-tailed test, enter 2*(1-CumProb) in the **Expression** box in step 6 if the χ^2 test statistic falls in the upper tail of the distribution. Enter 2*CumProb in the **Expression** box in step 6 if the χ^2 test statistic falls in the lower tail of the distribution.

Two Populations

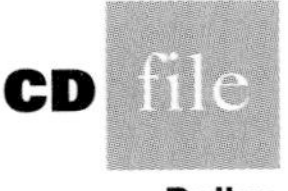
Dullus

We will use the data for the Dullus School bus study in Section 11.2. The arrival times for Milbank appear in column C1, and the arrival times for Gulf Park appear in column C2. The following Minitab procedure can be used to conduct hypothesis test H_0: $\sigma_1^2 = \sigma_2^2$ and H_a: $\sigma_1^2 \neq \sigma_2^2$.

Step 1. Select the **Stat** pull-down menu
Step 2. Choose **Basic Statistics**
Step 3. Choose **2-Variances**
Step 4. When the 2-Variances dialog box appears:
Select **Samples in different columns**
Enter C1 in the **First** box
Enter C2 in the **Second** box
Click **OK**

Information about the test will be displayed in the section entitled F-Test (normal distribution) showing the test statistic $F = 2.40$ and the p-value $= .082$. This Minitab procedure specifically performs the two-tailed test for the equality of population variances. Thus, if this Minitab routine is used for a one-tailed test, remembering that the area in one tail is

*Using Minitab to compute a sample variance s^2 was discussed in Chapter 3. The corresponding chi-square test statistic given by equation (11.8) can be obtained with a relatively easy hand calculation.

one-half of the area for the two-tailed p-value should make it relatively easy to compute the p-value for the one-tailed test.

Appendix 11.2 POPULATION VARIANCES WITH EXCEL

We describe how to use Excel to compute the p-values for the chi-square test statistic and F test statistic. These p-values can be used in hypothesis tests about the variance of one population or the variances of two populations.

One Population

MetroBus

We will use the data for the St. Louis Metro Bus example in Section 11.1. The Excel worksheet has the label Time in cell A1 and the 10 arrival-times in cells A2 to A11. The hypothesis test is H_0: $\sigma^2 \leq 4$ and H_a: $\sigma^2 > 4$. The easiest way to use Excel for this hypothesis test is to develop your own spreadsheet and use the p-value to draw the conclusion. The spreadsheet we developed is shown in Figure 11.6. The cell entries are as follow:

Step 1. Enter 4 in cell D2
Step 2. Compute the sample size in cell D6

=COUNT(A2:A11)

Step 3. Compute the sample variance in cell D7

=VAR(A2:A11)

Step 4. Compute the test statistic in cell D8

=(D6-1)*D7/D2

Step 5. Compute the p-value in cell D9

=CHIDIST(D8,D6-1)

FIGURE 11.6 EXCEL SPREADSHEET FOR THE ST. LOUIS METRO BUS HYPOTHESIS TEST

	A	B	C	D	E
1	**Time**		**Hypothesis Test**		
2	15.2		Test Variance	4	
3	17.5				
4	19.6				
5	16.6		**Sample Results**		
6	21.3		Sample Size	10	
7	17.1		Sample Variance	4.8	
8	15.0		Test Statistic	10.8	
9	15.5		p-Value	0.290	
10	20.0				
11	16.2				
12					

Cell D8 contains the formula for computing the test statistic

$$\chi^2 = \frac{(n - 1)s^2}{\sigma_0^2}$$

Cell D9 contains the formula for computing the *p*-value, which is the upper-tail area based on the value of the test statistic.

If the hypothesis test had been a one-tailed test with the rejection region in the lower tail, =1-CHIDIST(D8, D6-1) would have been used in cell D9. A two-tailed test requires the *p*-value cell formula =IF(CHIDIST(D8,D6-1)<.5,2*(CHIDIST(D8,D6-1)), 2*(1-CHIDIST(D8,D6-1))).

Two Populations

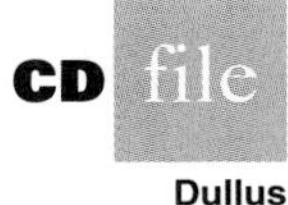

Dullus

We will use the data for the Dullus School bus study in Section 11.2. The Excel worksheet has the label Milbank in cell A1 and the label Gulf Park in cell B1. The times for the Milbank sample are in cells A2:A26 and the times for the Gulf Park sample are in cells B2:B17. The steps to conduct the hypothesis test H_0: $\sigma_1^2 = \sigma_2^2$ and H_a: $\sigma_1^2 \neq \sigma_2^2$ are as follow:

Step 1. Select the **Tools** pull-down menu
Step 2. Choose **Data Analysis**
Step 3. When the Data Analysis dialog box appears:
 Choose **F-Test Two-Sample for Variances**
 Click **OK**
Step 4. When the F-Test Two Sample for Variances dialog box appears:
 Enter A1:A26 in the **Variable 1 Range** box
 Enter B1:A17 in the **Variable 2 Range** box
 Select **Labels**
 Enter .05 in the **Alpha** box
  (Note: This Excel procedure uses alpha as the area in one tail.)
 Select **Output Range** and enter C1 in the box
 Click **OK**

The output P(F ≤ f) = .041 is the one-tail area associated with the test statistic $F = 2.40$. Thus, the two-tailed *p*-value is 2(.041) = .082. If the hypothesis test had been a one-tailed test, the one-tail area in the cell labeled P(F ≤ f) provides the information necessary to determine the *p*-value for the test.

Chapter 12

TESTS OF GOODNESS OF FIT AND INDEPENDENCE

CONTENTS

STATISTICS IN PRACTICE

UNITED WAY*
Rochester, New York

United Way of Greater Rochester is a nonprofit organization dedicated to improving the quality of life for all people in the seven counties it serves by meeting the community's most important human care needs.

The annual United Way/Red Cross fund-raising campaign, conducted each spring, funds hundreds of programs offered by more than 200 service providers. These providers meet a wide variety of human needs—physical, mental, and social—and serve people of all ages, backgrounds, and economic means.

Because of enormous volunteer involvement, United Way of Greater Rochester is able to hold its operating costs at just eight cents of every dollar raised.

The United Way of Greater Rochester decided to conduct a survey to learn more about community perceptions of charities. Focus-group interviews were held with professional, service, and general worker groups to get preliminary information on perceptions. The information obtained was then used to help develop the questionnaire for the survey. The questionnaire was pretested, modified, and distributed to 440 individuals; 323 completed questionnaires were obtained.

A variety of descriptive statistics, including frequency distributions and crosstabulations, were provided from the data collected. An important part of the analysis involved the use of contingency tables and chi-square tests of independence. One use of such statistical tests was to determine whether perceptions of administrative expenses were independent of occupation.

The hypotheses for the test of independence were:

H_0: Perception of United Way administrative expenses is independent of the occupation of the respondent.

H_a: Perception of United Way administrative expenses is not independent of the occupation of the respondent.

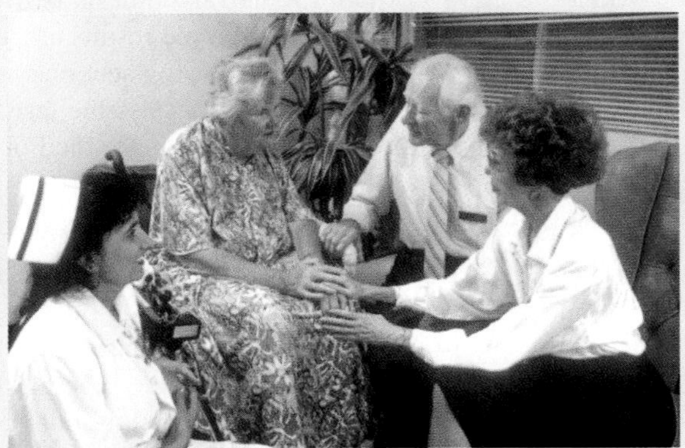

Statistical surveys help United Way adjust its program to better meet the needs of the people it serves. © Tony Freeman/PhotoEdit.

Two questions in the survey provided the data for the statistical test. One question obtained data on perceptions of the percentage of funds going to administrative expenses (up to 10%, 11–20%, and 21% or more). The other question asked for the occupation of the respondent.

The chi-square test at a 5% level of significance led to rejection of the null hypothesis of independence and to the conclusion that perceptions of United Way's administrative expenses did vary by occupation. Actual administrative expenses were less than 9%, but 35% of the respondents perceived that administrative expenses were 21% or more. Hence, many had inaccurate perceptions of administrative costs. In this group, production-line, clerical, sales, and professional-technical employees had more inaccurate perceptions than other groups.

The community perceptions study helped United Way of Rochester to develop adjustments to its program and fund-raising activities. In this chapter, you will learn how a statistical test of independence, such as that described here, is conducted.

*The authors are indebted to Dr. Philip R. Tyler, Marketing Consultant to the United Way, for providing this Statistics in Practice.

In Chapter 11 we showed how the chi-square distribution could be used in estimation and in hypothesis tests about a population variance. Here, we introduce two additional hypothesis testing procedures, both based on the use of the chi-square distribution. Like other hypothesis testing procedures, these tests compare sample results with those that are expected when the null hypothesis is true. The conclusion of the hypothesis test is based on how "close" the sample results are to the expected results.

In the following section we introduce a goodness of fit test for a multinomial population. Later we discuss the test for independence using contingency tables and then show goodness of fit tests for the Poisson and normal probability distributions.

12.1 GOODNESS OF FIT TEST: A MULTINOMIAL POPULATION

The assumptions for the multinomial experiment parallel those for the binomial experiment with the exception that the multinomial has three or more outcomes per trial.

In this section we consider the case in which each element of a population is assigned to one and only one of several classes or categories. Such a population is a **multinomial population.** The multinomial probability distribution can be thought of as an extension of the binomial distribution to the case of three or more categories of outcomes. On each trial of a multinomial experiment, one and only one of the outcomes occurs. Each trial of the experiment is assumed to be independent, and the probabilities of the outcomes stay the same for each trial.

As an example, consider the market share study being conducted by Scott Marketing Research. Over the past year market shares have stabilized at 30% for company A, 50% for company B, and 20% for company C. Recently company C has developed a "new and improved" product that has replaced its current entry in the market. Scott Marketing Research has been retained by company C to determine whether the new product will alter market shares.

In this case, the population of interest is a multinomial population; each customer is classified as buying from company A, company B, or company C. Thus, we have a multinomial population with three classifications or categories. Let us use the following notation for the proportions.

$$\begin{aligned} p_{\mathrm{A}} &= \text{market share for company A} \\ p_{\mathrm{B}} &= \text{market share for company B} \\ p_{\mathrm{C}} &= \text{market share for company C} \end{aligned}$$

Scott Marketing Research will conduct a sample survey and compute the proportion preferring each company's product. A hypothesis test will then be conducted to see whether the new product has caused a change in market shares. Assuming that company C's new product will not alter the market shares, the null and alternative hypotheses are stated as follows.

$$\begin{aligned} &H_0\text{: } p_{\mathrm{A}} = .30, p_{\mathrm{B}} = .50, \text{ and } p_{\mathrm{C}} = .20 \\ &H_{\mathrm{a}}\text{: The population proportions are not} \\ &\qquad p_{\mathrm{A}} = .30, p_{\mathrm{B}} = .50, \text{ and } p_{\mathrm{C}} = .20 \end{aligned}$$

If the sample results lead to the rejection of H_0, Scott Marketing Research will have evidence that the introduction of the new product has had an impact on the market shares.

Let us assume that the market research firm has used a consumer panel of 200 customers for the study. Each individual has been asked to specify a purchase preference among the three alternatives: company A's product, company B's product, and company C's new product. The 200 responses are summarized here.

The consumer panel of 200 customers in which each individual is asked to select one of three alternatives is equivalent to a multinomial experiment consisting of 200 trials.

Observed Frequency		
Company A's Product	**Company B's Product**	**Company C's New Product**
48	98	54

We now can perform a **goodness of fit test** that will determine whether the sample of 200 customer purchase preferences is consistent with the null hypothesis. The goodness of fit test is based on a comparison of the sample of *observed* results with the *expected* results under the assumption that the null hypothesis is true. Hence, the next step is to compute expected purchase preferences for the 200 customers under the assumption that $p_A = .30$, $p_B = .50$, and $p_C = .20$. Doing so provides the expected results.

Expected Frequency		
Company A's Product	**Company B's Product**	**Company C's New Product**
200(.30) = 60	200(.50) = 100	200(.20) = 40

Thus, we see that the expected frequency for each category is found by multiplying the sample size of 200 by the hypothesized proportion for the category.

The goodness of fit test now focuses on the differences between the observed frequencies and the expected frequencies. Large differences between observed and expected frequencies cast doubt on the assumption that the hypothesized proportions or market shares are correct. Whether the differences between the observed and expected frequencies are "large" or "small" is a question answered with the aid of the following test statistic.

Test Statistic for Goodness of Fit

$$\chi^2 = \sum_{i=1}^{k} \frac{(f_i - e_i)^2}{e_i} \tag{12.1}$$

where

f_i = observed frequency for category i
e_i = expected frequency for category i
k = the number of categories

Note: The test statistic has a chi-square distribution with $k - 1$ degrees of freedom provided that the expected frequencies are 5 *or more* for all categories.

Let us return to the market share data for the three companies. Since the expected frequencies are all 5 or more, we can proceed with the computation of the chi-square test statistic. The calculations necessary to compute the chi-square test statistic for the Scott Marketing Research market share study are shown in Table 12.1. We see that the value of the test statistic is $\chi^2 = 7.34$.

TABLE 12.1 COMPUTATION OF THE CHI-SQUARE TEST STATISTIC FOR THE SCOTT MARKETING RESEARCH MARKET SHARE STUDY

Category	Hypothesized Proportion	Observed Frequency (f_i)	Expected Frequency (e_i)	Difference $(f_i - e_i)$	Squared Difference $(f_i - e_i)^2$	Squared Difference Divided by Expected Frequency $(f_i - e_i)^2/e_i$
Company A	.30	48	60	−12	144	2.40
Company B	.50	98	100	−2	4	0.04
Company C	.20	54	40	14	196	4.90
Total		200				7.34

The test for goodness of fit is always a one-tailed test with the rejection region in the upper tail of the chi-square distribution.

Suppose we test the null hypothesis that the multinomial population has the proportions of $p_A = .30$, $p_B = .50$, and $p_C = .20$ at the $\alpha = .05$ level of significance. We will reject the null hypothesis if the differences between the observed and expected frequencies are *large,* therefore we will place a rejection area of .05 in the upper tail of the chi-square distribution. Checking the chi-square distribution table (Table 3 of Appendix B), we find that with $k - 1 = 3 - 1 = 2$ degrees of freedom, $\chi^2_{.05} = 5.99$. With $7.34 > 5.99$, we reject H_0. In rejecting H_0 we are concluding that the introduction of the new product by company C will alter the current market share structure.

The p-value criterion can be used for the goodness of fit test.

Computer software packages such as Minitab and Excel enable us to make the goodness of fit calculations and conduct the goodness of fit test using the *p*-value criterion. The steps for these procedures are described in Appendixes 12.1 and 12.2. Using Minitab, we find that the *p*-value for the Scott Marketing Research hypothesis test is .025. Using the rejection rule, Reject H_0 if *p*-value $< \alpha$, we see that .025 is less than $\alpha = .05$. Thus, we reject H_0 and conclude that the introduction of the new product will alter the current market share structure.

Although no further conclusions can be made as a result of the test, we can compare the observed and expected frequencies informally to obtain an idea of how the market share structure may change. Considering company C, we find that the observed frequency of 54 is larger than the expected frequency of 40. Because the expected frequency was based on current market shares, the larger observed frequency suggests that the new product will have a positive effect on company C's market share. Comparisons of the observed and expected frequencies for the other two companies indicate that company C's gain in market share will hurt company A more than company B.

Let us summarize the general steps that can be used to conduct a goodness of fit test for any hypothesized multinomial population distribution.

Multinomial Distribution Goodness of Fit Test: A Summary

1. State the null and alternative hypotheses.

   H_0: The population follows a multinomial probability distribution with specified probabilities for each of the *k* categories

   H_a: The population does not follow a multinomial probability distribution with the specified probabilities for each of the *k* categories

2. Select a random sample and record the observed frequencies f_i for each category.
3. Assuming the null hypothesis is true, determine the expected frequency e_i in each category by multiplying the category probability by the sample size.
4. Compute the value of the test statistic.

$$\chi^2 = \sum_{i=1}^{k} \frac{(f_i - e_i)^2}{e_i}$$

5. Rejection rule:

Using test statistic: Reject H_0 if $\chi^2 > \chi^2_\alpha$
Using p-value: Reject H_0 if p-value $< \alpha$

where α is the level of significance for the test and there are $k - 1$ degrees of freedom.

EXERCISES

Methods

1. Test the following hypotheses by using the χ^2 goodness of fit test.

$$H_0: p_A = .40,\ p_B = .50,\ \text{and}\ p_C = .20$$
$$H_a: \text{The population proportions are not}\ p_A = .30,\ p_B = .50,\ \text{and}\ p_C = .20$$

A sample of size 200 yielded 60 in category A, 120 in category B, and 20 in category C. Use $\alpha = .01$ and test to see whether the proportions are as stated in H_0.

2. Suppose we have a multinomial population with four categories: A, B, C, and D. The null hypothesis is that the proportion of items is the same in every category. The null hypothesis is

$$H_0: p_A = p_B = p_C = p_D = .25$$

A sample of size 300 yielded the following results.

A: 85 B: 95 C: 50 D: 70

Use $\alpha = .05$ to determine whether H_0 should be rejected.

Applications

3. During the first 13 weeks of the television season, the Saturday evening 8:00 P.M. to 9:00 P.M. audience proportions were recorded as ABC 29%, CBS 28%, NBC 25%, and independents 18%. A sample of 300 homes two weeks after a Saturday night schedule revision yielded the following viewing audience data: ABC 95 homes, CBS 70 homes, NBC 89 homes, and independents 46 homes. Test with $\alpha = .05$ to determine whether the viewing audience proportions have changed.

4. M&M/MARS, makers of M&M® Chocolate Candies, conducted a national poll in which more than 10 million people indicated their preference for a new color. The tally of this poll resulted in the replacement of tan-colored M&Ms with a new blue color. In the brochure "Colors," made available by M&M/MARS Consumer Affairs, the distribution of colors for the plain candies is as follows:

Brown	Yellow	Red	Orange	Green	Blue
30%	20%	20%	10%	10%	10%

In a study reported in *Chance* (no. 4, 1996), samples of 1-pound bags were used to determine whether the reported percentages were indeed valid. The following results were obtained for one sample of 506 plain candies.

Brown	Yellow	Red	Orange	Green	Blue
177	135	79	41	36	38

Use $\alpha = .05$ to determine whether these data support the percentages reported by the company.

5. One of the questions on the *Business Week* 1996 Subscriber Study was, "When making investment purchases, do you use full service or discount brokerage firms?" Survey results showed that 264 respondents use full service brokerage firms only, 255 use discount brokerage firms only, and 229 use both full service and discount firms. Use $\alpha = .10$ to determine whether there are any differences in preference among the three service choices.

6. Negative appeals have been recognized as an effective method of persuasion in advertising. A study in *The Journal of Advertising* (Summer 1997) reported the results of a content analysis of guilt advertisements in 24 magazines. The number of ads with guilt appeals that appeared in selected magazine types follow.

Magazine Type	Number of Ads With Guilt Appeals
News and opinion	20
General editorial	15
Family-oriented	30
Business/financial	22
Female-oriented	16
African-American	12

Using $\alpha = .10$ test to see whether there is a difference in the proportion of ads with guilt appeals among the six types of magazines.

7. Consumer panel preferences for three proposed store displays follow.

Display A	Display B	Display C
43	53	39

Use $\alpha = .05$ and test to see whether there is a difference in preference among the three display designs.

8. How well do airline companies serve their customers? A study showed the following customer ratings: 3% excellent, 28% good, 45% fair, and 24% poor (*Business Week,* September 11, 2000). In a follow-up study of service by telephone companies, assume that a sample of 400 adults found the following customer ratings: 24 excellent, 124 good, 172 fair, and 80 poor. Is the distribution of the customer ratings for telephone companies different from the distribution of customer ratings for airline companies? Test with $\alpha = .01$. What is your conclusion?

12.2 TEST OF INDEPENDENCE

Another important application of the chi-square distribution involves using sample data to test for the independence of two variables. Let us illustrate the test of independence by considering the study conducted by the Alber's Brewery of Tucson, Arizona. Alber's manufactures and distributes three types of beer: light, regular, and dark. In an analysis of the market segments for the three beers, the firm's market research group has raised the question of whether preferences for the three beers differ among male and female beer drinkers. If beer preference is independent of the gender of the beer drinker, one advertising campaign will be initiated for all of Alber's beers. However, if beer preference depends on the gender of the beer drinker, the firm will tailor its promotions to different target markets.

A test of independence addresses the question of whether the beer preference (light, regular, or dark) is independent of the gender of the beer drinker (male, female). The hypotheses for this test of independence are:

H_0: Beer preference is independent of the gender of the beer drinker

H_a: Beer preference is not independent of the gender of the beer drinker

To test whether two variables are independent, one sample is selected and crosstabulation is used to summarize the data for the two variables simultaneously.

Table 12.2 can be used to describe the situation being studied. After identification of the population as all male and female beer drinkers, a sample can be selected and each individual asked to state his or her preference for the three Alber's beers. Every individual in the sample will be classified in one of the six cells in the table. For example, an individual may be a male preferring regular beer (cell (1,2)), a female preferring light beer (cell (2,1)), a female preferring dark beer (cell (2,3)), and so on. Because we have listed all possible combinations of beer preference and gender or, in other words, listed all possible contingencies, Table 12.2 is called a **contingency table.** The test of independence uses the contingency table format and for that reason is sometimes referred to as a *contingency table test.*

Suppose a simple random sample of 150 beer drinkers has been selected. After tasting each beer, the individuals in the sample are asked to state their preference or first choice. The crosstabulation in Table 12.3 summarizes the responses for the study. As we see, the data for the test of independence are collected in terms of counts or frequencies for each cell or category. Of the 150 individuals in the sample, 20 were men who favored light beer, 40 were men who favored regular beer, 20 were men who favored dark beer, and so on.

The data in Table 12.3 are the observed frequencies for the six classes or categories. If we can determine the expected frequencies under the assumption of independence between beer preference and gender of the beer drinker, we can use the chi-square distribution to determine whether there is a significant difference between observed and expected frequencies.

Expected frequencies for the cells of the contingency table are based on the following rationale. First we assume that the null hypothesis of independence between beer

TABLE 12.2 CONTINGENCY TABLE FOR BEER PREFERENCE AND GENDER OF BEER DRINKER

		Beer Preference		
		Light	**Regular**	**Dark**
Gender	**Male**	cell(1,1)	cell(1,2)	cell(1,3)
	Female	cell(2,1)	cell(2,2)	cell(2,3)

TABLE 12.3 SAMPLE RESULTS FOR BEER PREFERENCES OF MALE AND FEMALE BEER DRINKERS (OBSERVED FREQUENCIES)

		Beer Preference			
		Light	**Regular**	**Dark**	**Total**
Gender	**Male**	20	40	20	80
	Female	30	30	10	70
	Total	50	70	30	150

preference and gender of the beer drinker is true. Then we note that in the entire sample of 150 beer drinkers, a total of 50 prefer light beer, 70 prefer regular beer, and 30 prefer dark beer. In terms of fractions we conclude that $^{50}/_{150} = {}^{1}/_{3}$ of the beer drinkers prefer light beer, $^{70}/_{150} = {}^{7}/_{15}$ prefer regular beer, and $^{30}/_{150} = {}^{1}/_{5}$ prefer dark beer. If the *independence* assumption is valid, we argue that these fractions must be applicable to both male and female beer drinkers. Thus, under the assumption of independence, we would expect the sample of 80 male beer drinkers to show that $(^{1}/_{3})80 = 26.67$ prefer light beer, $(^{7}/_{15})80 = 37.33$ prefer regular beer, and $(^{1}/_{5})80 = 16$ prefer dark beer. Application of the same fractions to the 70 female beer drinkers provides the expected frequencies shown in Table 12.4.

Let e_{ij} denote the expected frequency for the contingency table category in row i and column j. With this notation, let us reconsider the expected frequency calculation for males (row $i = 1$) who prefer regular beer (column $j = 2$); that is, expected frequency e_{12}. Following the preceding argument for the computation of expected frequencies, we can show that

$$e_{12} = (^{7}/_{15})80 = 37.33$$

This expression can be written slightly differently as

$$e_{12} = (^{7}/_{15})80 = (^{70}/_{150})80 = \frac{(80)(70)}{150} = 37.33$$

Note that 80 in the expression is the total number of males (row 1 total), 70 is the total number of individuals preferring regular beer (column 2 total), and 150 is the total sample size. Hence, we see that

$$e_{12} = \frac{\text{(Row 1 Total)(Column 2 Total)}}{\text{Sample Size}}$$

TABLE 12.4 EXPECTED FREQUENCIES IF BEER PREFERENCE IS INDEPENDENT OF THE GENDER OF THE BEER DRINKER

		Beer Preference			
		Light	**Regular**	**Dark**	**Total**
Gender	**Male**	26.67	37.33	16.00	80
	Female	23.33	32.67	14.00	70
	Total	50.00	70.00	30.00	150

Generalization of the expression shows that the following formula provides the expected frequencies for a contingency table in the test of independence.

Expected Frequencies for Contingency Tables Under the Assumption of Independence

$$e_{ij} = \frac{(\text{Row } i \text{ Total})(\text{Column } j \text{ Total})}{\text{Sample Size}} \tag{12.2}$$

Using the formula for male beer drinkers who prefer dark beer, we find an expected frequency of $e_{13} = (80)(30)/150 = 16.00$, as shown in Table 12.4. Use equation (12.2) to verify the other expected frequencies shown in Table 12.4.

The test procedure for comparing the observed frequencies of Table 12.3 with the expected frequencies of Table 12.4 is similar to the goodness of fit calculations made in Section 12.1. Specifically, the χ^2 value based on the observed and expected frequencies is computed as follows.

Test Statistic for Independence

$$\chi^2 = \sum_i \sum_j \frac{(f_{ij} - e_{ij})^2}{e_{ij}} \tag{12.3}$$

where

f_{ij} = observed frequency for contingency table category in row i and column j

e_{ij} = expected frequency for contingency table category in row i and column j based on the assumption of independence

Note: With n rows and m columns in the contingency table, the test statistic has a chi-square distribution with $(n - 1)(m - 1)$ degrees of freedom provided that the expected frequencies are five or more for all categories.

The double summation in equation (12.3) is used to indicate that the calculation must be made for all the cells in the contingency table.

By reviewing the expected frequencies in Table 12.4, we see that the expected frequencies are five or more for each category. We therefore proceed with the computation of the chi-square test statistic. The calculations necessary to compute the chi-square test statistic for determining whether beer preference is independent of the gender of the beer drinker are shown in Table 12.5. We see that the value of the test statistic is $\chi^2 = 6.13$.

The test for independence is always a one-tailed test with the rejection region in the upper tail of the chi-square distribution.

The number of degrees of freedom for the appropriate chi-square distribution is computed by multiplying the number of rows minus 1 by the number of columns minus 1. With two rows and three columns, we have $(2 - 1)(3 - 1) = (1)(2) = 2$ degrees of freedom for the test of independence of beer preference and gender of the beer drinker. With $\alpha = .05$ for the level of significance of the test, Table 3 of Appendix B shows an upper-tail χ^2 value of $\chi^2_{.05} = 5.99$. Note that we are again using the upper-tail value because we will reject the null hypothesis only if the differences between observed and expected frequencies provide a large χ^2 value. In our example, $\chi^2 = 6.13$ is greater than the critical value of $\chi^2_{.05} = 5.99$. Thus, we reject the null hypothesis of independence and conclude that beer preference is not independent of the gender of the beer drinker.

Computer software packages such as Minitab and Excel can simplify the computations for a test of independence and provide the p-value for the test. The steps used to obtain

TABLE 12.5 COMPUTATION OF THE CHI-SQUARE TEST STATISTIC FOR DETERMINING WHETHER BEER PREFERENCE IS INDEPENDENT OF THE GENDER OF THE BEER DRINKER

Gender	Beer Preference	Observed Frequency (f_{ij})	Expected Frequency (e_{ij})	Difference ($f_{ij} - e_{ij}$)	Squared Difference $(f_{ij} - e_{ij})^2$	Squared Difference Divided by Expected Frequency $(f_{ij} - e_{ij})^2/e_{ij}$
Male	Light	20	26.67	−6.67	44.49	1.67
Male	Regular	40	37.33	2.67	7.13	0.19
Male	Dark	20	16.00	4.00	16.00	1.00
Female	Light	30	23.33	6.67	44.49	1.91
Female	Regular	30	32.67	−2.67	7.13	0.22
Female	Dark	10	14.00	−4.00	16.00	1.14
	Total	150				6.13

FIGURE 12.1 MINITAB OUTPUT FOR THE ALBER'S BREWERY TEST OF INDEPENDENCE

```
Expected counts are printed below observed counts

         Light     Regular       Dark     Total
    1       20          40         20        80
         26.67       37.33      16.00

    2       30          30         10        70
         23.33       32.67      14.00

Total       50          70         30       150
DF = 2, P-Value = 0.047
```

computer results for a test of independence are presented in Appendixes 12.1 and 12.2. The Minitab output for the Alber's Brewery test of independence is shown in Figure 12.1. The p-value $= .047$ is less than $\alpha = .05$. Thus, we reject H_0 and conclude that beer preference is not independent of the gender of the beer drinker.

Although no further conclusions can be made as a result of the test, we can compare the observed and expected frequencies informally to obtain an idea about the dependence between beer preference and gender. Refer to Tables 12.3 and 12.4. We see that male beer drinkers have higher observed than expected frequencies for both regular and dark beers, whereas female beer drinkers have a higher observed than expected frequency only for light beer. These observations give us insight about the beer preference differences between male and female beer drinkers.

Let us summarize the steps in a contingency table test of independence.

Test of Independence: A Summary

1. State the null and alternative hypotheses.

   H_0: The column variable is independent of the row variable

   H_a: The column variable is not independent of the row variable

2. Select a random sample and record the observed frequencies for each cell of the contingency table.

3. Use equation (12.2) to compute the expected frequency for each cell.
4. Use equation (12.3) to compute the value of the test statistic.
5. Rejection rule:

Using test statistic: Reject H_0 if $\chi^2 > \chi^2_\alpha$
Using p-value: Reject H_0 if p-value $< \alpha$

where α is the level of significance, with n rows and m columns providing $(n - 1)(m - 1)$ degrees of freedom.

NOTES AND COMMENTS

The test statistic for the chi-square tests in this chapter requires an expected frequency of five for each category. When a category has fewer than five, it is often appropriate to combine two adjacent categories to obtain an expected frequency of five or more in each category.

EXERCISES

Methods

9. The following 2×3 contingency table contains observed frequencies for a sample of 200. Test for independence of the row and column variables using the χ^2 test with $\alpha = .025$.

	Column Variable		
Row Variable	**A**	**B**	**C**
P	20	44	50
Q	30	26	30

10. The following 3×3 contingency table contains observed frequencies for a sample of 240. Test for independence of the row and column variables using the χ^2 test with $\alpha = .05$.

	Column Variable		
Row Variable	**A**	**B**	**C**
P	20	30	20
Q	30	60	25
R	10	15	30

Applications

11. One of the questions on the *Business Week* 1996 Subscriber Study was, "In the past 12 months, when traveling for business, what type of airline ticket did you purchase most often?" The data obtained are shown in the following contingency table.

	Type of Flight	
Type of Ticket	**Domestic Flights**	**International Flights**
First class	29	22
Business/executive class	95	121
Full fare economy/coach class	518	135

Using $\alpha = .05$, test for the independence of type of flight and type of ticket. What is your conclusion?

12. In a study of brand loyalty in the automotive industry, new-car customers were asked whether the make of their new car was the same as the make of their previous car (*Business Week,* May 8, 2000). The breakdown of 600 responses shows the brand loyalty for domestic, European, and Asian cars.

	Manufacturer		
Purchased	**Domestic**	**European**	**Asian**
Same Make	125	55	68
Different Make	140	105	107

a. Test a hypothesis to determine whether brand loyalty is independent of the manufacturer. Use $\alpha = .05$. What is your conclusion?

b. If a significant difference is found, which manufacturer appears to have the greatest brand loyalty?

13. Starting positions for business and engineering graduates are classified by industry as shown in the following table.

	Industry			
Degree Major	**Oil**	**Chemical**	**Electrical**	**Computer**
Business	30	15	15	40
Engineering	30	30	20	20

Use $\alpha = .01$ and test for independence of degree major and industry type.

14. The results of a study conducted by Marist Institute for Public Opinion showed who men and women say is the most difficult person for them to buy holiday gifts for (*USA Today,* December 15, 1997). Suppose that the following data were obtained in a follow-up study consisting of 100 men and 100 women.

	Gender	
Most Difficult to Buy For	**Men**	**Women**
Spouse	37	25
Parents	28	31
Children	7	19
Siblings	8	3
In-laws	4	10
Other relatives	16	12

Use $\alpha = .05$ and test for independence of gender and the most difficult person to buy for. What is your conclusion?

15. Negative appeals have been recognized as an effective method of persuasion in advertising. A study in *The Journal of Advertising* (Summer 1997) reported the results of a content analysis of guilt and fear advertisements in 24 magazines. The number of ads with guilt and fear appeals that appeared in selected magazine types follows.

	Type of Appeal	
Magazine Type	**Number of Ads With Guilt Appeals**	**Number of Ads With Fear Appeals**
News and opinion	20	10
General editorial	15	11
Family-oriented	30	19
Business/financial	22	17
Female-oriented	16	14
African-American	12	15

Use the chi-square test of independence with a .01 level of significance to analyze the data. What is your conclusion?

16. Businesses are increasingly placing orders on-line. The Performance Measurement Group collected data on the rates of correctly filled electronic orders by industry (*Investor's Business Daily,* May 8, 2000). Assume a sample of 700 electronic orders provided the following results.

	Industry			
Order	**Pharmaceutical**	**Consumer**	**Computers**	**Telecommunications**
Correct	207	136	151	178
Incorrect	3	4	9	12

a. Test a hypothesis to determine whether order fulfillment is independent of industry. Use $\alpha = .05$. What is your conclusion?

b. Which industry has the highest percentage of correctly filled orders?

17. Three suppliers provide the following data on defective parts.

	Part Quality		
Supplier	**Good**	**Minor Defect**	**Major Defect**
A	90	3	7
B	170	18	7
C	135	6	9

Use $\alpha = .05$ and test for independence between supplier and part quality. What does the result of your analysis tell the purchasing department?

18. A study of educational levels of voters and their political party affiliations yielded the following results.

	Party Affiliation		
Educational Level	**Democratic**	**Republican**	**Independent**
Did not complete high school	40	20	10
High school degree	30	35	15
College degree	30	45	25

Use $\alpha = .01$ and determine whether party affiliation is independent of the educational level of the voters.

19. On the syndicated Siskel and Ebert television show the hosts often created the impression that they strongly disagreed about which movies were best. An article in *Chance* (no. 2, 1997) reported the results of ratings of 160 movies by Siskel and Ebert. Each review is categorized as Pro ("thumbs up"), Con ("thumbs down"), or Mixed.

	Ebert Rating		
Siskel Rating	**Con**	**Mixed**	**Pro**
Con	24	8	13
Mixed	8	13	11
Pro	10	9	64

Use the chi-square test of independence with a .01 level of significance to analyze the data. What is your conclusion?

12.3 GOODNESS OF FIT TEST: POISSON AND NORMAL DISTRIBUTIONS

In Section 12.1 we introduced the goodness of fit test for a multinomial population. In general, the goodness of fit test can be used with any hypothesized probability distribution. In this section we illustrate the goodness of fit test procedure for cases in which the population is hypothesized to have a Poisson or a normal probability distribution. As we shall see, the goodness of fit test and the use of the chi-square distribution for the test follow the same general procedure used for the goodness of fit test in Section 12.1.

Poisson Distribution

Let us illustrate the goodness of fit test for the case in which the hypothesized population distribution is a Poisson distribution. As an example, consider the arrival of customers at Dubek's Food Market in Tallahassee, Florida. Because of some recent staffing problems, Dubek's managers have asked a local consulting firm to assist with the scheduling of clerks for the checkout lanes. After reviewing the checkout lane operation, the consulting firm makes a recommendation for a clerk-scheduling procedure. The procedure, based on a mathematical analysis of waiting lines, is applicable only if the number of customers arriving during a specified time period follows the Poisson probability distribution. Therefore, before the scheduling process is implemented, data on customer arrivals must be collected and a statistical test conducted to see whether an assumption of a Poisson distribution for arrivals is reasonable.

We define the arrivals at the store in terms of the *number of customers* entering the store during 5-minute intervals. Hence, the following null and alternative hypotheses are appropriate for the Dubek's Food Market study.

H_0: The number of customers entering the store during 5-minute intervals has a Poisson probability distribution

H_a: The number of customers entering the store during 5-minute intervals does not have a Poisson distribution

TABLE 12.6 OBSERVED FREQUENCY OF DUBEK'S CUSTOMER ARRIVALS FOR A SAMPLE OF 128 5-MINUTE TIME PERIODS

Number of Customers Arriving	Observed Frequency
0	2
1	8
2	10
3	12
4	18
5	22
6	22
7	16
8	12
9	6
Total	128

If a sample of customer arrivals indicates H_0 cannot be rejected, Dubek's will proceed with the implementation of the consulting firm's scheduling procedure. However, if the sample leads to the rejection of H_0, the assumption of the Poisson distribution for the arrivals cannot be made, and other scheduling procedures will have to be considered.

To test the assumption of a Poisson distribution for the number of arrivals during weekday morning hours, a store employee randomly selects a sample of 128 5-minute intervals during weekday mornings over a 3-week period. For each 5-minute interval in the sample, the store employee records the number of customer arrivals. In summarizing the data, the employee determines the number of 5-minute intervals having no arrivals, the number of 5-minute intervals having one arrival, the number of 5-minute intervals having two arrivals, and so on. These data are summarized in Table 12.6.

Table 12.6 gives the observed frequencies for the 10 categories. We now want to use a goodness of fit test to determine whether the sample of 128 time periods supports the hypothesized Poisson probability distribution. To conduct the goodness of fit test, we need to consider the expected frequency for each of the 10 categories under the assumption that the Poisson distribution of arrivals is true. That is, we need to compute the expected number of time periods in which no customers, one customer, two customers, and so on would arrive if, in fact, the customer arrivals have a Poisson distribution.

The Poisson probability function, which was first introduced in Chapter 5, is

$$f(x) = \frac{\mu^x e^{-\mu}}{x!} \tag{12.4}$$

In this function, μ represents the mean or expected number of customers arriving per 5-minute period, x is the random variable indicating the number of customers arriving during a 5-minute period, and $f(x)$ is the probability that x customers will arrive in a 5-minute interval.

Before we use equation (12.4) to compute Poisson probabilities, we must obtain an estimate of μ, the mean number of customer arrivals during a 5-minute time period. The sample mean for the data in Table 12.6 provides this estimate. With no customers arriving in two 5-minute time periods, one customer arriving in eight 5-minute time periods, and so on, the total number of customers who arrived during the sample of 128 5-minute time periods is given by $0(2) + 1(8) + 2(10) + \cdots + 9(6) = 640$. The 640 customer arrivals over the sample of 128 periods provide a mean arrival rate of $\mu = 640/128 = 5$ customers per 5-minute period. With this value for the mean of the Poisson probability distribution, an estimate of the Poisson probability function for Dubek's Food Market is

$$f(x) = \frac{5^x e^{-5}}{x!} \tag{12.5}$$

The above probability function can be evaluated for different values of x to determine the probability associated with each category of arrivals. These probabilities, which can also be found in Table 7 of Appendix B, are given in Table 12.7. For example, the probability of zero customers arriving during a 5-minute interval is $f(0) = .0067$, the probability of one customer arriving during a 5-minute interval is $f(1) = .0337$, and so on. As we saw in Section 12.1, the expected frequencies for the categories are found by multiplying the probabilities by the sample size. For example, the expected number of periods with zero arrivals is given by $(.0067)(128) = .8576$, the expected number of periods with one arrival is given by $(.0337)(128) = 4.3136$, and so on.

TABLE 12.7 EXPECTED FREQUENCY OF DUBEK'S CUSTOMER ARRIVALS, ASSUMING A POISSON PROBABILITY DISTRIBUTION WITH $\mu = 5$

Number of Customers Arriving (x)	Poisson Probability $f(x)$	Expected Number of 5-Minute Time Periods With x Arrivals, $128f(x)$
0	.0067	0.8576
1	.0337	4.3136
2	.0842	10.7776
3	.1404	17.9712
4	.1755	22.4640
5	.1755	22.4640
6	.1462	18.7136
7	.1044	13.3632
8	.0653	8.3584
9	.0363	4.6464
10 or more	.0318	4.0704
	Total	128.0000

When the expected number in some category is less than five, the assumptions for the χ^2 test are not satisfied. When this happens, adjacent categories can be combined to increase the expected number to five.

Before we make the usual chi-square calculations to compare the observed and expected frequencies, note that in Table 12.7, four of the categories have an expected frequency less than five. This condition violates the requirements for use of the chi-square distribution. However, expected category frequencies less than five cause no difficulty, because adjacent categories can be combined to satisfy the "at least five" expected frequency requirement. In particular, we will combine 0 and 1 into a single category and then combine 9 with "10 or more" into another single category. Thus, the rule of a minimum expected frequency of five in each category is satisfied. Table 12.8 shows the observed and expected frequencies after combining categories.

As in Section 12.1, the goodness of fit test focuses on the differences between observed and expected frequencies, $f_i - e_i$. Thus, we will use the observed and expected frequencies shown in Table 12.8, to compute the chi-square test statistic.

$$\chi^2 = \sum_{i=1}^{k} \frac{(f_i - e_i)^2}{e_i}$$

The calculations necessary to compute the chi-square test statistic are shown in Table 12.9. The value of the test statistic is $\chi^2 = 10.9766$.

The Poisson distribution has a single parameter μ that is estimated from the sample data. This causes a loss of one degree of freedom.

In general, the chi-square distribution for a goodness of fit test has $k - p - 1$ degrees of freedom, where k is the number of categories and p is the number of population parameters estimated from the sample data. For the Poisson distribution goodness of fit test we are considering, Table 12.9 shows $k = 9$ categories. Because the sample data were used to estimate the mean of the Poisson distribution, $p = 1$. Thus, there are $k - p - 1 = k - 2 = 9 - 2 = 7$ degrees of freedom for the chi-square distribution in the Dubek's Food Market study.

Suppose we test the null hypothesis that the probability distribution for the customer arrivals is a Poisson distribution with a .05 level of significance. Table 3 of Appendix B shows that with seven degrees of freedom, $\chi^2_{.05} = 14.07$. As in similar one-tailed tests, we will reject H_0 if $\chi^2 > \chi^2_\alpha$.

Checking the preceding calculations, we find a computed $\chi^2 = 10.9766$. This value is less than the critical value of 14.07, therefore we cannot reject the null hypothesis. Hence, for this analysis, the assumption of a Poisson probability distribution for weekday morning customer arrivals cannot be rejected. With this statistical finding, Dubek's managers will proceed with the consulting firm's scheduling procedure for weekday mornings.

TABLE 12.8 OBSERVED AND EXPECTED FREQUENCIES FOR DUBEK'S CUSTOMER ARRIVALS AFTER COMBINING CATEGORIES

Number of Customers Arriving		Observed Frequency (f_i)	Expected Frequency (e_i)
0 or 1		10	5.1712
2		10	10.7776
3		12	17.9712
4		18	22.4640
5		22	22.4640
6		22	18.7136
7		16	13.3632
8		12	8.3584
9 or more		6	8.7168
	Total	128	128.0000

TABLE 12.9 COMPUTATION OF THE CHI-SQUARE TEST STATISTIC FOR THE DUBEK'S FOOD MARKET STUDY

Number of Customers Arriving (x)	Observed Frequency (f_i)	Expected Frequency (e_i)	Difference ($f_i - e_i$)	Squared Difference $(f_i - e_i)^2$	Squared Difference Divided by Expected Frequency $(f_i - e_i)^2/e_i$
0 or 1	10	5.1712	4.8288	23.3173	4.5091
2	10	10.7776	−0.7776	0.6047	0.0561
3	12	17.9712	−5.9712	35.6552	1.9840
4	18	22.4640	−4.4640	19.9273	0.8871
5	22	22.4640	−0.4640	0.2153	0.0096
6	22	18.7136	3.2864	10.8004	0.5771
7	16	13.3632	2.6368	6.9527	0.5203
8	12	8.3584	3.6416	13.2613	1.5866
9 or more	6	8.7168	−2.7168	7.3810	0.8468
Total	128	128.0000			10.9766

Appendixes 12.1 and 12.2 provide the Minitab and Excel steps that can be used to conduct this goodness of fit test. Using Minitab, the p-value $= .14$. With the p-value $> \alpha = .05$, we cannot reject the null hypothesis that the population has a Poisson distribution. A summary of the steps involved in conducting a Poisson distribution goodness of fit test follows:

Poisson Distribution Goodness of Fit Test: A Summary

1. State the null and alternative hypotheses.

   H_0: The population has a Poisson probability distribution

   H_a: The population does not have a Poisson probability distribution

2. Select a random sample and
   a. Record the observed frequency f_i for each value of the Poisson random variable.
   b. Compute the mean number of occurrences μ.
3. Compute the expected frequency of occurrences e_i for each value of the Poisson random variable. Multiply the sample size by the Poisson probability of occurrence for each value of the Poisson random variable. If there are fewer than five expected occurrences for some values, combine adjacent values and reduce the number of categories as necessary.
4. Compute the value of the test statistic.

$$\chi^2 = \sum_{i=1}^{k} \frac{(f_i - e_i)^2}{e_i}$$

5. Rejection rule:

   Using test statistic: Reject H_0 if $\chi^2 > \chi^2_\alpha$

   Using p-value: Reject if p-value $< \alpha$

   where α is the level of significance and there are $k - 2$ degrees of freedom.

Normal Distribution

The goodness of fit test for a normal probability distribution is also based on the use of the chi-square distribution. It is similar to the procedure we discussed for the Poisson distribution. In particular, observed frequencies for several categories of sample data are compared to expected frequencies under the assumption that the population has a normal distribution. Because the normal probability distribution is continuous, we must modify the way the categories are defined and how the expected frequencies are computed. Let us demonstrate the goodness of fit test for a normal probability distribution by considering the job applicant test data for Chemline, Inc., listed in Table 12.10.

TABLE 12.10

CHEMLINE EMPLOYEE APTITUTDE TEST SCORES FOR 50 RANDOMLY CHOSEN JOB APPLICANTS

71	66	61	65	54	93
60	86	70	70	73	73
55	63	56	62	76	54
82	79	76	68	53	58
85	80	56	61	61	64
65	62	90	69	76	79
77	54	64	74	65	65
61	56	63	80	56	71
79	84				

Chemline hires approximately 400 new employees annually for its four plants located throughout the United States. The personnel director has asked whether a normal distribution could be applied to the population of test scores. If such a distribution can be used, the distribution would be helpful in evaluating specific test scores. That is, scores in the upper 20%, lower 40%, and so on, could be identified quickly. Hence, we want to test the null hypothesis that the population of test scores follows a normal probability distribution.

Let us first use the data in Table 12.10 to develop estimates of the mean and standard deviation of the normal distribution that will be considered in the null hypothesis. We use the sample mean $\bar{x}$ and the sample standard deviation s as point estimators of the mean and standard deviation of the normal distribution. The calculations follow.

$$\bar{x} = \frac{\Sigma x_i}{n} = \frac{3421}{50} = 68.42$$

$$s = \sqrt{\frac{\Sigma(x_i - \bar{x})^2}{n - 1}} = \sqrt{\frac{5310.0369}{49}} = 10.41$$

Using these values, we state the following hypotheses about the distribution of the job applicant test scores.

H_0: The population of test scores has a normal distribution with mean 68.42 and standard deviation 10.41

H_a: The population of test scores does not have a normal distribution with mean 68.42 and standard deviation 10.41

The hypothesized normal distribution is shown in Figure 12.2.

Now let us consider a way of defining the categories for a goodness of fit test involving a normal distribution. For the discrete probability distribution in the Poisson distribution test, the categories were readily defined in terms of the number of customers arriving, such as 0, 1, 2, and so on. However, with the continuous normal probability distribution, we must use a different procedure for defining the categories. We need to define the categories in terms of *intervals* of test scores.

Recall the rule of thumb for an expected frequency of at least five in each interval or category. We have to define the categories of test scores such that the expected frequencies will be at least five for each category. With a sample size of 50, one way of doing this is to divide the normal distribution into 10 equal-probability intervals (see Figure 12.3). With a sample size of 50, we would expect five outcomes in each interval or category, and the rule of thumb for expected frequencies would be satisfied.

With a continuous probability distribution, establish intervals such that each interval has an [e]xpected frequency of five [or mo]re.

Let us look more closely at the procedure for calculating the category boundaries. When the normal probability distribution is assumed, the standard normal probability tables can be used to determine these boundaries. First consider the test score cutting off the lowest 10% of the test scores. From Table 1 of Appendix B we find that the z value for this test score is -1.28. Therefore, the test score of $x = 68.42 - 1.28(10.41) = 55.10$ provides this

FIGURE 12.2 HYPOTHESIZED NORMAL DISTRIBUTION OF TEST SCORES FOR THE CHEMLINE JOB APPLICANTS

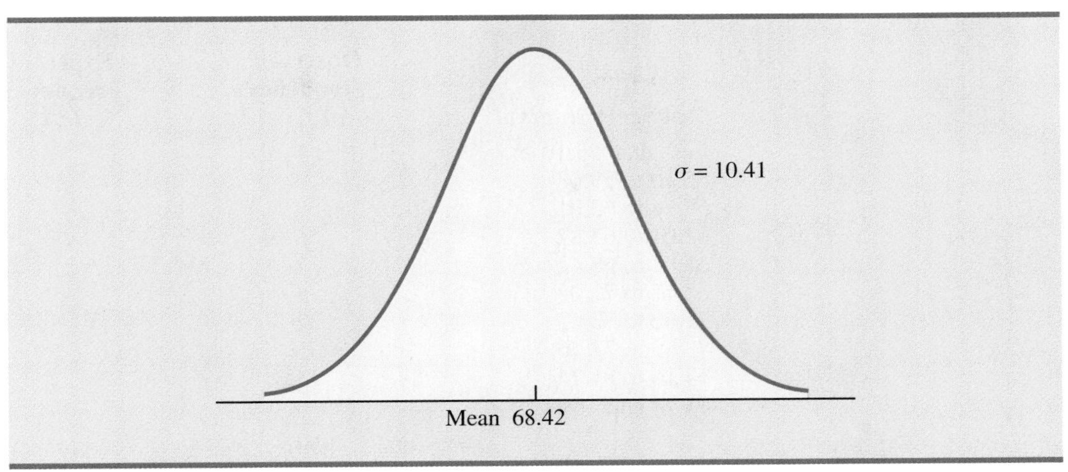

cutoff value for the lowest 10% of the scores. For the lowest 20%, we find $z = -.84$, and thus $x = 68.42 - .84(10.41) = 59.68$. Working through the normal distribution in that way provides the following test score values.

Lower 10%:	$68.42 - 1.28(10.41) = 55.10$
Lower 20%:	$68.42 - .84(10.41) = 59.68$
Lower 30%:	$68.42 - .52(10.41) = 63.01$
Lower 40%:	$68.42 - .25(10.41) = 65.82$
Mid-score:	$68.42 + 0(10.41) = 68.42$
Upper 40%:	$68.42 + .25(10.41) = 71.02$
Upper 30%:	$68.42 + .52(10.41) = 73.83$
Upper 20%:	$68.42 + .84(10.41) = 77.16$
Upper 10%:	$68.42 + 1.28(10.41) = 81.74$

These cutoff or interval boundary points are identified on the graph in Figure 12.3.

FIGURE 12.3 NORMAL PROBABILITY DISTRIBUTION FOR THE CHEMLINE EXAMPLE WITH 10 EQUAL-PROBABILITY INTERVALS

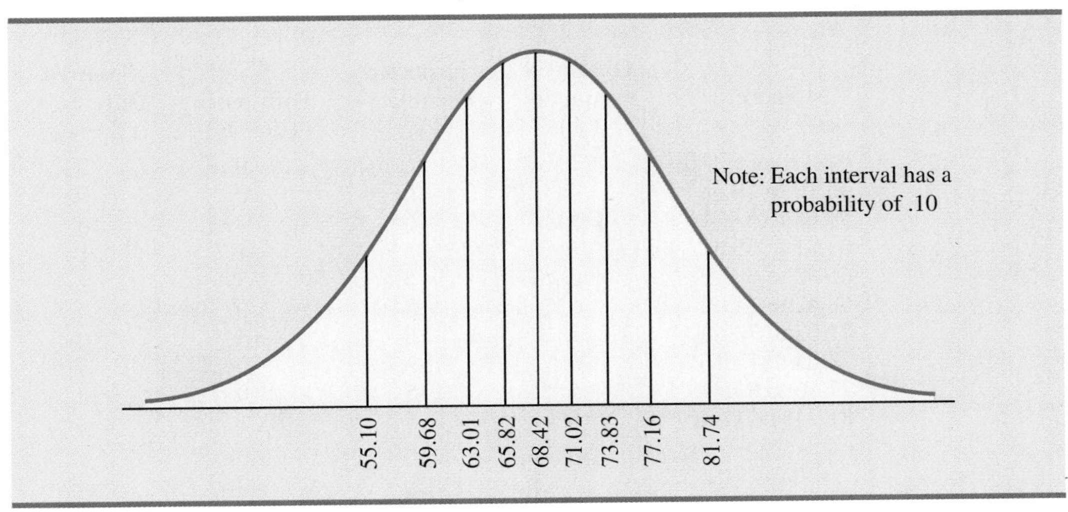

TABLE 12.11 OBSERVED AND EXPECTED FREQUENCIES FOR CHEMLINE JOB-APPLICANT TEST SCORES

Test Score Interval	Observed Frequency (f_i)	Expected Frequency (e_i)
Less than 55.10	5	5
55.10 to 59.68	5	5
59.68 to 63.01	9	5
63.01 to 65.82	6	5
65.82 to 68.42	2	5
68.42 to 71.02	5	5
71.02 to 73.83	2	5
73.83 to 77.16	5	5
77.16 to 81.74	5	5
81.74 and Over	6	5
Total	50	50

With the categories or intervals of test scores now defined and with the known expected frequency of five per category, we can return to the sample data of Table 12.10 and determine the observed frequencies for the categories. Doing so provides the results in Table 12.11.

With the results in Table 12.11, the goodness of fit calculations proceed exactly as before. Namely, we compare the observed and expected results by computing a χ^2 value. The computations necessary to compute the chi-square test statistic are shown in Table 12.12. We see that the value of the test statistic is $\chi^2 = 7.2$.

To determine whether the computed χ^2 value of 7.2 is large enough to reject H_0, we need to refer to the appropriate chi-square probability distribution tables. Using the rule for computing the number of degrees of freedom for the goodness of fit test, we have $k - p - 1 = 10 - 2 - 1 = 7$ degrees of freedom, where there are $k = 10$ categories and

TABLE 12.12 COMPUTATION OF THE CHI-SQUARE TEST STATISTIC FOR THE CHEMLINE JOB-APPLICANT EXAMPLE

Test Score Interval	Observed Frequency (f_i)	Expected Frequency (e_i)	Difference ($f_i - e_i$)	Squared Difference $(f_i - e_i)^2$	Squared Difference Divided by Expected Frequency $(f_i - e_i)^2/e_i$
Less than 55.10	5	5	0	0	0.0
55.10 to 59.68	5	5	0	0	0.0
59.68 to 63.01	9	5	4	16	3.2
63.01 to 65.82	6	5	1	1	0.2
65.82 to 68.42	2	5	−3	9	1.8
68.42 to 71.02	5	5	0	0	0.0
71.02 to 73.83	2	5	−3	9	1.8
73.83 to 77.16	5	5	0	0	0.0
77.16 to 81.74	5	5	0	0	0.0
81.74 and Over	6	5	1	1	0.2
Total	50	50			7.2

Estimating the two parameters of the normal distribution will cause a loss of two degrees of freedom in the χ^2 test.

$p = 2$ parameters (mean and standard deviation) estimated from the sample data. Using a .10 level of significance for this hypothesis test, we have $\chi^2_{.10} = 12.017$ for the upper-tail rejection region. With $7.2 < 12.017$, we conclude that the null hypothesis cannot be rejected. Hence, the hypothesis that the probability distribution for the Chemline job applicant test scores is a normal probability distribution cannot be rejected.

The p-value can also be used to test the hypothesis that the population of Chemline test scores is normally distributed with a mean of 68.42 and a standard deviation of 10.41. Using Minitab, the p-value is 0.408. Thus, p-value $> \alpha = .05$, and we cannot reject H_0. A summary of the normal distribution goodness of fit test follows.

Normal Distribution Goodness of Fit Test: A Summary

1. State the null and alternative hypotheses.

   H_0: The population has a normal probability distribution
   H_a: The population does not have a normal probability distribution

2. Select a random sample and
   a. Compute the sample mean and sample standard deviation.
   b. Define intervals of values so that the expected frequency is at least five for each interval. Using equal probability intervals is a good approach.
   c. Record the observed frequency of data values f_i in each interval defined.
3. Compute the expected number of occurrences e_i for each interval of values defined in step 2(b). Multiply the sample size by the probability of a normal random variable being in the interval.
4. Compute the value of the test statistic.

$$\chi^2 = \sum_{i=1}^{k} \frac{(f_i - e_i)^2}{e_i}$$

5. Rejection rule:

   Using test statistic: Reject H_0 if $\chi^2 > \chi^2_\alpha$
   Using p-value: Reject H_0 if p-value $< \alpha$

   where α is the level of significance and there are $k - 3$ degrees of freedom.

EXERCISES

Methods

20. Data on the number of occurrences per time period and observed frequencies follow. Use $\alpha = .05$ and the goodness of fit test to see whether the data fit a Poisson distribution.

Number of Occurrences	Observed Frequency
0	39
1	30
2	30
3	18
4	3

21. The following data are believed to have come from a normal probability distribution. Use the goodness of fit test and $\alpha = .025$ to test this claim.

17	23	22	24	19	23	18	22	20	13	11	21	18	20	21
21	18	15	24	23	23	43	29	27	26	30	28	33	23	29

Applications

22. The number of automobile accidents per day in a particular city is believed to have a Poisson distribution. A sample of 80 days during the past year gives the following data. Do these data support the belief that the number of accidents per day has a Poisson distribution? Use $\alpha = .05$.

Number of Accidents	Observed Frequency (days)
0	34
1	25
2	11
3	7
4	3

23. The number of incoming phone calls at a company switchboard during 1-minute intervals is believed to have a Poisson distribution. Use $\alpha = .10$ and the following data to test the assumption that the incoming phone calls have a Poisson distribution.

Number of Incoming Phone Calls During a 1-Minute Interval	Observed Frequency
0	15
1	31
2	20
3	15
4	13
5	4
6	2
Total	100

24. The weekly demand for a product is believed to be normally distributed. Use a goodness of fit test and the following data to test this assumption. Use $\alpha = .10$. The sample mean is 24.5 and the sample standard deviation is 3.

18	20	22	27	22
25	22	27	25	24
26	23	20	24	26
27	25	19	21	25
26	25	31	29	25
25	28	26	28	24

25. Use $\alpha = .01$ and conduct a goodness of fit test to see whether the following sample appears to have been selected from a normal probability distribution.

55	86	94	58	55	95	55	52	69	95	90	65	87	50	56
55	57	98	58	79	92	62	59	88	65					

After you complete the goodness of fit calculations, construct a histogram of the data. Does the histogram representation support the conclusion reached with the goodness of fit test? (*Note:* $\bar{x} = 71$ and $s = 17$.)

SUMMARY

In this chapter we introduced the goodness of fit test and the test of independence, both of which are based on the use of the chi-square distribution. The purpose of the goodness of fit test is to determine whether a hypothesized probability distribution can be used as a model for a particular population of interest. The computations for conducting the goodness of fit test involve comparing observed frequencies from a sample with expected frequencies when the hypothesized probability distribution is assumed true. A chi-square distribution is used to determine whether the differences between observed and expected frequencies are large enough to reject the hypothesized probability distribution. We illustrated the goodness of fit test for multinomial, Poisson, and normal probability distributions.

A test of independence for two variables is an extension of the methodology employed in the goodness of fit test for a multinomial population. A contingency table is used to determine the observed and expected frequencies. Then a chi-square value is computed. Large chi-square values, caused by large differences between observed and expected frequencies, lead to the rejection of the null hypothesis of independence.

GLOSSARY

Multinomial population A population in which each element is assigned to one and only one of several categories. The multinomial probability distribution extends the binomial probability distribution from two to three or more categories.

Goodness of fit test A statistical test conducted to determine whether to reject a hypothesized probability distribution for a population.

Contingency table A table used to summarize observed and expected frequencies for a test of independence.

KEY FORMULAS

Test Statistic for Goodness of Fit

$$\chi^2 = \sum_{i=1}^{k} \frac{(f_i - e_i)^2}{e_i} \tag{12.1}$$

Expected Frequencies for Contingency Tables Under the Assumption of Independence

$$e_{ij} = \frac{(\text{Row } i \text{ Total})(\text{Column } j \text{ Total})}{\text{Sample Size}} \tag{12.2}$$

Test Statistic for Independence

$$\chi^2 = \sum_i \sum_j \frac{(f_{ij} - e_{ij})^2}{e_{ij}} \tag{12.3}$$

SUPPLEMENTARY EXERCISES

26. In setting sales quotas, the marketing manager makes the assumption that order potentials are the same for each of four sales territories. A sample of 200 sales follows. Should the manager's assumption be rejected? Use $\alpha = .05$.

Sales Territories			
I	II	III	IV
60	45	59	36

27. Seven percent of mutual fund investors rate corporate stocks "very safe," 58% rate them "somewhat safe," 24% rate them "not very safe," 4% rate them "not at all safe," and 7% are "not sure." A *Business Week*/Harris poll asked 529 mutual fund investors how they would rate corporate bonds on safety. The responses are as follows.

Safety Rating	Frequency
Very safe	48
Somewhat safe	323
Not very safe	79
Not at all safe	16
Not sure	63
Total	529

Do mutual fund investors' attitudes toward corporate bonds differ from their attitudes toward corporate stocks? Support your conclusion with a statistical test using $\alpha = .01$.

28. A community park is to be opened soon. A sample of 140 individuals has been asked to state their preference for when they would most like to visit the park. The sample results follow.

Weekday	Saturday	Sunday	Holiday
20	20	40	60

In developing a staffing plan, should the park manager plan on the same number of individuals visiting the park each day? Support your conclusion with a statistical test using $\alpha = .05$.

29. A regional transit authority is concerned about the number of riders on one of its bus routes. In setting up the route, the assumption is that the number of riders is the same on every day from Monday through Friday. Using the following data, test with $\alpha = .05$ to determine whether the transit authority's assumption is correct.

Day	Number of Riders
Monday	13
Tuesday	16
Wednesday	28
Thursday	17
Friday	16

30. The results of *Computerworld's* Annual Job Satisfaction Survey showed that 28% of Information Systems (IS) managers are very satisfied with their job, 46% are somewhat satisfied, 12% are neither satisfied or dissatisfied, 10% are somewhat dissatisfied, and 4% are very dissatisfied (*Computerworld,* May 26, 1997). Suppose that a sample of 500 computer programmers yielded the following results.

Category	Number of Respondents
Very satisfied	105
Somewhat satisfied	235
Neither	55
Somewhat dissatisfied	90
Very dissatisfied	15

Test using $\alpha = .05$ to determine whether the job satisfaction for computer programmers is different from the job satisfaction for IS managers.

31. A sample of parts provided the following contingency table data on part quality by production shift.

Shift	Number Good	Number Defective
First	368	32
Second	285	15
Third	176	24

Use $\alpha = .05$ and test the hypothesis that part quality is independent of the production shift. What is your conclusion?

32. *The Wall Street Journal* 1996 Subscriber Study showed data on the employment status of subscribers. Sample results corresponding to subscribers of the eastern and western editions are shown here.

	Region	
Employment Status	**Eastern Edition**	**Western Edition**
Full-time	1105	574
Part-time	31	15
Self-employed/consultant	229	186
Not employed	485	344

Using $\alpha = .05$, test the hypothesis that employment status is independent of the region. What is your conclusion?

33. A lending institution supplied the following data on loan approvals by four loan officers. Use $\alpha = .05$ and test to determine whether the loan approval decision is independent of the loan officer reviewing the loan application.

	Loan Approval Decision	
Loan Officer	**Approved**	**Rejected**
Miller	24	16
McMahon	17	13
Games	35	15
Runk	11	9

34. Data on the marital status of men and women ages 20 to 29 were obtained as part of a national survey. The results from a sample of 350 men and 400 women follow. These data are representative of results published in the *U.S. Current Population Report* (*The Statistical Abstract of the United States,* 1999).

	Marital Status		
Gender	**Never Married**	**Married**	**Divorced**
Men	234	106	10
Women	216	168	16

a. Using $\alpha = .01$, test for independence between marital status and gender. What is your conclusion?

b. Summarize the percent in each marital status category for men and for women.

35. The following crosstabulation shows industry type and P/E ratio for 100 companies in the consumer products and banking industries.

		P/E Ratio					
		5–9	**10–14**	**15–19**	**20–24**	**25–29**	**Total**
Industry	**Consumer**	4	10	18	10	8	50
	Banking	14	14	12	6	4	50
	Total	18	24	30	16	12	100

Does there appear to be a relationship between industry type and P/E ratio? Support your conclusion with a statistical test using $\alpha = .05$.

36. The following data were collected on the number of emergency ambulance calls for an urban county and a rural county in Virginia.

		Day of Week							
		Sun	**Mon**	**Tue**	**Wed**	**Thur**	**Fri**	**Sat**	**Total**
County	**Urban**	61	48	50	55	63	73	43	393
	Rural	7	9	16	13	9	14	10	78
	Total	68	57	66	68	72	87	53	471

Conduct a test for independence using $\alpha = .05$. What is your conclusion?

37. A random sample of final examination grades for a college course follows.

55	85	72	99	48	71	88	70	59	98	80	74	93	85	74
82	90	71	83	60	95	77	84	73	63	72	95	79	51	85
76	81	78	65	75	87	86	70	80	64					

Use $\alpha = .05$ and test to determine whether a normal probability distribution should be rejected as being representative of the population's distribution of grades.

38. The office occupancy rates were reported for four California metropolitan areas. Do the following data suggest that the office vacancies were independent of metropolitan area? Use a .05 level of significance. What is your conclusion?

Occupancy Status	**Los Angeles**	**San Diego**	**San Francisco**	**San Jose**
Occupied	160	116	192	174
Vacant	40	34	33	26

39. A salesperson makes four calls per day. A sample of 100 days gives the following frequencies of sales volumes.

Number of Sales	Observed Frequency (days)
0	30
1	32
2	25
3	10
4	3
Total	100

Records show sales are made to 30% of all sales calls. Assuming independent sales calls, the number of sales per day should follow a binomial probability distribution. The binomial probability function presented in Chapter 5 is

$$f(x) = \frac{n!}{x!(n - x)!} p^x(1 - p)^{n-x}$$

For this exercise, assume that the population has a binomial distribution with $n = 4$, $p = .30$, and $x = 0, 1, 2, 3$, and 4.

a. Compute the expected frequencies for $x = 0, 1, 2, 3$, and 4 by using the binomial probability function. Combine categories if necessary to satisfy the requirement that the expected frequency is five or more for all categories.

b. Use the goodness of fit test to determine whether the assumption of a binomial distribution should be rejected. Use $\alpha = .05$. Because no parameters of the binomial distribution were estimated from the sample data, the degrees of freedom are $k - 1$ when k is the number of categories.

Case Problem A BIPARTISAN AGENDA FOR CHANGE

In a study conducted by Zogby International for the *Democrat and Chronicle,* more than 700 New Yorkers were polled to determine whether the New York state government works. Respondents surveyed were asked questions involving pay cuts for state legislators, restrictions on lobbyists, terms limits for legislators, and whether state citizens should be able to put matters directly on the state ballot for a vote (*Democrat and Chronicle,* December 7, 1997). The results regarding several proposed reforms had broad support, crossing all demographic and political lines.

Suppose that a follow-up survey of 100 individuals who live in the western region of New York was conducted. The party affiliation (Democrat, Independent, Republican) of each individual surveyed was recorded, as well as their responses to the following three questions.

1. Should legislative pay be cut for every day the state budget is late?
   Yes ____ No ____
2. Should there be more restrictions on lobbyists?
   Yes ____ No ____
3. Should there be term limits requiring that legislators serve a fixed number of years?
   Yes ____ No ____

NYReform

The responses were coded using 1 for a *yes* response and 2 for a *no* response. The complete data set is available on the data disk in the data set named NYReform.

Managerial Report

1. Use descriptive statistics to summarize the data from this study. What are your preliminary conclusions about the independence of the response (yes or no) and party affiliation for each of the three questions in the survey?
2. With regard to question 1, test for the independence of the response (yes and no) and party affiliation. Use $\alpha = .05$.
3. With regard to question 2, test for the independence of the response (yes and no) and party affiliation. Use $\alpha = .05$.
4. With regard to question 3, test for the independence of the response (yes and no) and party affiliation. Use $\alpha = .05$.
5. Does it appear that there is broad support for change across all political lines? Explain.

Appendix 12.1 TESTS OF GOODNESS OF FIT AND INDEPENDENCE WITH MINITAB

We describe how to use Minitab for a goodness of fit test and a test of independence.

Goodness of Fit Test

This Minitab procedure can be used for goodness of fit tests for the multinomial distribution in Section 12.1 and the Poisson and normal distributions in Section 12.3. The user must obtain the observed frequencies, calculate the expected frequencies, and enter both the observed and expected frequencies in a Minitab worksheet. Column C1 is labeled Observed and contains the observed frequencies. Column C2 is labeled Expected and contains the expected frequencies. Using the Scott Marketing Research example presented in Section 12.1, open a Minitab worksheet and enter the observed frequencies 48, 98, and 54 in column C1 and enter the expected frequencies 60, 100, and 40 in column C2. The Minitab steps for the goodness of fit test follow.

Step 1. Select the **Calc** pull-down menu
Step 2. Choose **Calculator**
Step 3. When the Calculator dialog box appears:
 Enter ChiSquare in the **Store result in variable** box
 Enter Sum((Observed-Expected)**2/Expected) in the **Expression** box
 Click **OK**
Step 4. Select the **Calc** pull-down menu
Step 5. Choose **Probability Distributions**
Step 6. Choose **Chi-Square**
Step 7. When the Chi-Square Distribution dialog box appears:
 Select **Cumulative probability**
 Enter 2 in the **Degrees of freedom** box
 Select **Input column** and enter ChiSquare in the box
 Enter CumProb in the **Optional storage** box
 Click **OK**
Step 8. Select the **Calc** pull-down menu
Step 9. Choose **Calculator**
 When the Calculator dialog box appears:
 Enter p-value in the **Store results in variable** box
 Enter 1-CumProb in the **Expression** box
 Click **OK**

Test of Independence

We begin with a new Minitab worksheet and enter the observed frequency data for the Alber's Brewery example from Section 12.2 into columns 1, 2, and 3, respectively. Thus, we entered the observed frequencies corresponding to a light beer preference (20 and 30) in C1, the observed frequencies corresponding to a regular beer preference (40 and 30) in C2, and the observed frequencies corresponding to a dark beer preference (20 and 10) in C3. The Minitab steps for the test of independence are as follows.

Step 1. Select the **Stat** pull-down menu
Step 2. Select the **Tables** pull-down menu
Step 3. Choose **Chi-Square Test**
Step 4. When the Chi-Square Test dialog box appears:
Enter C1-C3 in the **Columns containing the table** box
Click **OK**

The p-value is 0.047. With p-value $< \alpha = 0.05$, we reject H_0.

Appendix 12.2 TESTS OF GOODNESS OF FIT AND INDEPENDENCE WITH EXCEL

We describe how to use Excel for a goodness of fit test and a test of independence.

Goodness of Fit Test

This Excel procedure can be used for goodness of fit tests for the multinomial distribution in Section 12.1 and the Poisson and normal distributions in Section 12.3. The user must obtain the observed frequencies, calculate the expected frequencies, and enter both the observed and expected frequencies in an Excel worksheet. Using the Scott Marketing Research example presented in Section 12.1, open an Excel worksheet and enter the label Observed in cell A1 and the label Expected in cell B1. Then enter the observed frequencies 48, 98, and 54 in cells A2:A4 and enter the expected frequencies 60, 100, and 40 in cells B2:B4.

The χ^2 test statistic and the p-value can be computed as follows:

Step 1. Enter the following formula in cell C2

=(A2-B2)^2/B2

Step 2. Copy cell C2 to cell C3:C4
Step 3. Compute the value of χ^2 in cell C5

=SUM(C2:C4)

Step 4. Compute the p-value in cell C6

=CHIDIST(C5,2)

In step 4, the entry 2 is the degrees of freedom for the test. In the Scott Marketing Research example, the multinomial distribution goodness of fit test with $k = 3$ categories has $k - 1 = 2$ degrees of freedom. This procedure can be extended to the goodness of fit tests for the Poisson and normal distributions in Section 12.3. The CHIDIST function in step 4 contains the degrees of freedom appropriate for the test.

FIGURE 12.4 EXCEL SPREADSHEET FOR THE ALBER'S BREWERY TEST OF INDEPENDENCE

	A	B	C	D	E	F
1	**Observed Frequencies**					
2			**Beer Preference**			
3	**Gender**	Light	Regular	Dark	**Total**	
4	Male	20	40	20	80	
5	Female	30	30	10	70	
6	**Total**	50	70	30	150	
7						
8	**Expected Frequencies**					
9			**Beer Preference**			
10	**Gender**	Light	Regular	Dark	**Total**	
11	Male	26.67	37.33	16	80	
12	Female	23.33	32.67	14	70	
13	**Total**	50	70	30	150	
14						
15				**p-value=**	0.047	
16						

Test of Independence

The Excel procedure for the test of independence requires the user to obtain the observed frequencies and then calculate the expected frequencies. Two tables, one with the observed frequencies and one with the expected frequencies must be entered into an Excel worksheet. Figure 12.4 shows the Excel worksheet that we developed for the Alber's Brewery example in Section 12.2. The observed frequencies appear in cells B4:D5 and the expected frequencies appear in cells B11:D12. The *p*-value for the test of independence is obtained by entering the following function in cell E15:

=CHITEST(B4:D5,B11:D12)

ANALYSIS OF VARIANCE AND EXPERIMENTAL DESIGN

Chapter 13

CONTENTS

STATISTICS IN PRACTICE

BURKE MARKETING SERVICES, INC.*

Cincinnati, Ohio

Burke Marketing Services, Inc., is one of the most experienced market research firms in the industry. Burke writes more proposals, on more projects, every day than any other market research company in the world. Supported by state-of-the-art technology, Burke offers a wide variety of research capabilities, providing answers to nearly any marketing question.

In one study, Burke was retained by a firm to evaluate potential new versions of a children's dry cereal. To maintain confidentiality, we refer to the cereal manufacturer as the Anon Company. The four key factors that Anon's product developers thought would enhance the taste of the cereal were:

1. Ratio of wheat to corn in the cereal flake.
2. Type of sweetener: sugar, honey, or artificial.
3. Presence or absence of flavor bits with a fruit taste.
4. Short or long cooking time.

An experiment was designed to determine what effects these four factors had on cereal taste. For example, one test cereal was made with a specified ratio of wheat to corn, sugar as the sweetener, flavor bits, and a short cooking time; another test cereal was made with a different ratio of wheat to corn and the other three factors the same, and so on. Groups of children then taste-tested the cereals and stated what they thought about the taste of each.

*The authors are indebted to Dr. Ronald Tatham of Burke Marketing Services for providing this Statistics in Practice.

Burke's research provides valuable statistical information on what customers want from a product. © Mary Kate Denny/PhotoEdit.

Analysis of variance was the statistical method used to study the data obtained from the taste tests. The results of the analysis showed that:

- The flake composition and sweetener type were very influential in taste evaluation.
- The flavor bits actually detracted from the taste of the cereal.
- The cooking time had no effect on the taste.

This information helped Anon identify the factors that would lead to the best-tasting cereal.

The experimental design employed by Burke and the subsequent analysis of variance were helpful in making a product design recommendation. In this chapter, we will see how such procedures are carried out.

Sir Ronald Alymer Fisher (1890–1962) invented the branch of statistics known as experimental design. In addition to being accomplished in statistics, he was a noted scientist in the field of genetics.

In this chapter we introduce a statistical procedure called *analysis of variance* (ANOVA). First, we show how ANOVA can be used to test for the equality of three or more population means using data obtained from an observational study. Then, we discuss the use of ANOVA for analyzing data obtained from three types of experimental studies: a completely randomized design, a randomized block design, and a factorial experiment. In the following chapters we will see that ANOVA plays a key role in analyzing the results of regression analysis involving both experimental and observational data.

13.1 AN INTRODUCTION TO ANALYSIS OF VARIANCE

National Computer Products, Inc. (NCP), manufactures printers and fax machines at plants located in Atlanta, Dallas, and Seattle. To measure how much employees at these plants know about total quality management, a random sample of six employees was selected from each plant and given a quality awareness examination. The examination scores obtained for these 18 employees are listed in Table 13.1. The sample means, sample variances, and sample standard deviations for each group are also provided. Managers want to use these data to test the hypothesis that the mean examination score is the same for all three plants.

We will define population 1 as all employees at the Atlanta plant, population 2 as all employees at the Dallas plant, and population 3 as all employees at the Seattle plant. Let

$$\mu_1 = \text{mean population score for population 1}$$
$$\mu_2 = \text{mean population score for population 2}$$
$$\mu_3 = \text{mean population score for population 3}$$

Although we will never know the actual values of μ_1, μ_2, and μ_3, we want to use the sample results to test the following hypotheses.

If H_0 is rejected, we cannot conclude that all population means are different. Rejecting H_0 means that at least two population means have different values.

$$H_0: \mu_1 = \mu_2 = \mu_3$$
$$H_a: \text{Not all population means are equal}$$

As we will demonstrate shortly, analysis of variance is a statistical procedure that can be used to determine whether the observed differences in the three sample means are large enough to reject H_0.

In the introduction to this chapter we stated that analysis of variance can be used to analyze data obtained from both an observational study and an experimental study. In order to provide a common set of terminology for discussing the use of analysis of variance in both types of studies, we need to introduce the concepts of a response variable, a factor, and a treatment.

The two variables in the NCP example are plant location and score on the quality awareness examination. Because the objective is to determine whether the mean examination score is the same for plants located in Atlanta, Dallas, and Seattle, examination score is referred to as the dependent or *response variable* and plant location as the independent variable or *factor.* In general, the values of a factor selected for investigation are referred to as levels of the factor or *treatments.* Thus, in the NCP example the three treatments are Atlanta, Dallas, and Seattle. These three treatments define the populations of interest in the NCP example. For each treatment or population, the response variable is the examination score.

TABLE 13.1 EXAMINATION SCORES FOR 18 EMPLOYEES

CD 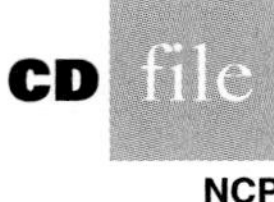

NCP

Observation	Plant 1 Atlanta	Plant 2 Dallas	Plant 3 Seattle
1	85	71	59
2	75	75	64
3	82	73	62
4	76	74	69
5	71	69	75
6	85	82	67
Sample mean	79	74	66
Sample variance	34	20	32
Sample standard deviation	5.83	4.47	5.66

Assumptions for Analysis of Variance

Three assumptions are required to use analysis of variance.

If the sample sizes are equal, analysis of variance is not sensitive to departures from the assumption of normally distributed populations.

1. **For each population, the response variable is normally distributed.** Implication: In the NCP example, the examination scores (response variable) must be normally distributed at each plant.
2. **The variance of the response variable, denoted σ^2, is the same for all of the populations.** Implication: In the NCP example, the variance of examination scores must be the same for all three plants.
3. **The observations must be independent.** Implication: In the NCP example, the examination score for each employee must be independent of the examination score for any other employee.

A Conceptual Overview

If the means for the three populations are equal, we would expect the three sample means to be close together. In fact, the closer the three sample means are to one another, the more evidence we have for the conclusion that the population means are equal. Alternatively, the more the sample means differ, the more evidence we have for the conclusion that the population means are not equal. In other words, if the variability among the sample means is "small," it supports H_0; if the variability among the sample means is "large," it supports H_a.

If the null hypothesis, H_0: $\mu_1 = \mu_2 = \mu_3$, is true, we can use the variability among the sample means to develop an estimate of σ^2. First, note that if the assumptions for analysis of variance are satisfied, each sample will have come from the same normal probability distribution with mean μ and variance σ^2. Recall from Chapter 7 that the sampling distribution of the sample mean $\bar{x}$ for a simple random sample of size n from a normal population will be normally distributed with mean μ and variance σ^2/n. Figure 13.1 illustrates such a sampling distribution.

Thus, if the null hypothesis is true, we can think of each of the three sample means, $\bar{x}_1 = 79$, $\bar{x}_2 = 74$, and $\bar{x}_3 = 66$, from Table 13.1 as values drawn at random from the sampling distribution shown in Figure 13.1. In this case, the mean and variance of the three $\bar{x}$ values can be used to estimate the mean and variance of the sampling distribution. When the sample sizes are equal, as in the NCP example, the best estimate of the mean of the sam-

FIGURE 13.1 SAMPLING DISTRIBUTION OF $\bar{x}$ GIVEN H_0 IS TRUE

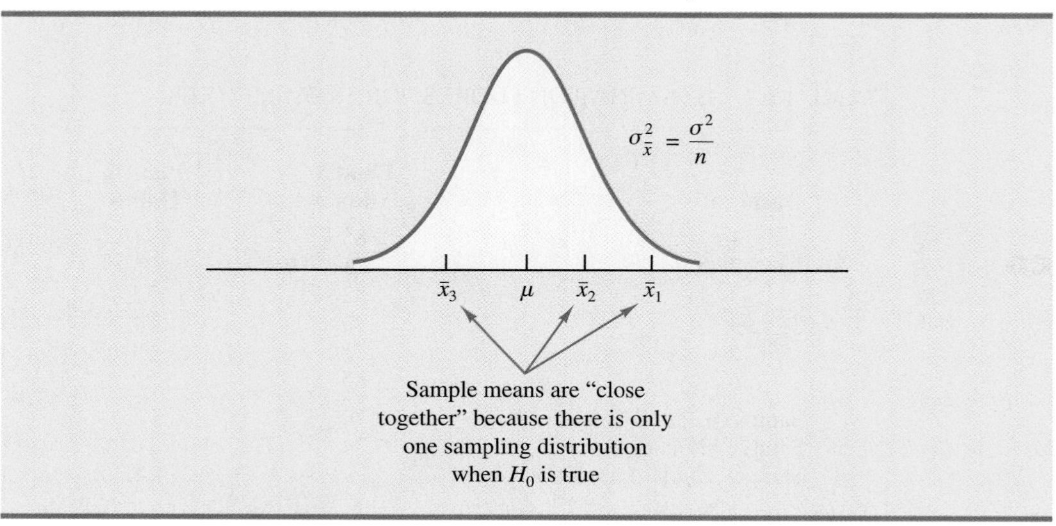

pling distribution of $\bar{x}$ is the mean or average of the sample means. Thus, in the NCP example, an estimate of the mean of the sampling distribution of $\bar{x}$ is $(79 + 74 + 66)/3 = 73$. We refer to this estimate as the *overall sample mean*. An estimate of the variance of the sampling distribution of $\bar{x}$, $\sigma^2_{\bar{x}}$, is provided by the variance of the three sample means.

$$s^2_{\bar{x}} = \frac{(79 - 73)^2 + (74 - 73)^2 + (66 - 73)^2}{3 - 1} = \frac{86}{2} = 43$$

Because $\sigma^2_{\bar{x}} = \sigma^2/n$, solving for σ^2 gives

$$\sigma^2 = n\sigma^2_{\bar{x}}$$

Hence,

$$\text{Estimate of } \sigma^2 = n \text{ (Estimate of } \sigma^2_{\bar{x}}) = ns^2_{\bar{x}} = 6(43) = 258$$

The result, $ns^2_{\bar{x}} = 258$, is referred to as the *between-treatments* estimate of σ^2.

The between-treatments estimate of σ^2 is based on the assumption that the null hypothesis is true. In this case, each sample comes from the same population, and there is only one sampling distribution of $\bar{x}$. To illustrate what happens when H_0 is false, suppose the population means all differ. Note that because the three samples are from normal populations with different means, they will result in three different sampling distributions. Figure 13.2 shows that in this case, the sample means are not as close together as they were when H_0 was true. Thus, $s^2_{\bar{x}}$ will be larger, causing the between-treatments estimate of σ^2 to be larger. In general, when the population means are not equal, the between-treatments estimate will overestimate the population variance σ^2.

The variation within each of the samples also has an effect on the conclusion we reach in analysis of variance. When a simple random sample is selected from each population, each of the sample variances provides an unbiased estimate of σ^2. Hence, we can combine or pool the individual estimates of σ^2 into one overall estimate. The estimate of σ^2 obtained in this way is called the *pooled* or *within-treatments* estimate of σ^2. Because each sample variance provides an estimate of σ^2 based only on the variation within each sample, the

FIGURE 13.2 SAMPLING DISTRIBUTION FOR $\bar{x}$ GIVEN H_0 IS FALSE

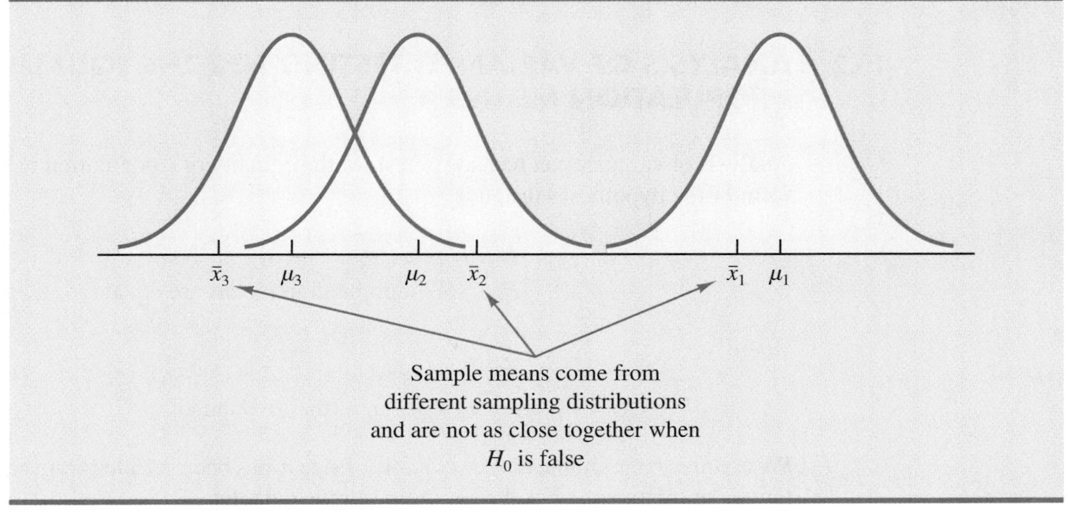

within-treatments estimate of σ^2 is not affected by whether the population means are equal. When the sample sizes are equal, the within-treatments estimate of σ^2 can be obtained by computing the average of the individual sample variances. For the NCP example we obtain

$$\text{Within-Treatments Estimate of } \sigma^2 = \frac{34 + 20 + 32}{3} = \frac{86}{3} = 28.67$$

In the NCP example, the between-treatments estimate of σ^2 (258) is much larger than the within-treatments estimate of σ^2 (28.67). In fact, the ratio of these two estimates is $258/28.67 = 9.00$. Recall, however, that the between-treatments approach provides a good estimate of σ^2 only if the null hypothesis is true; if the null hypothesis is false, the between-treatments approach overestimates σ^2. The within-treatments approach provides a good estimate of σ^2 in either case. Thus, if the null hypothesis is true, the two estimates will be similar and their ratio will be close to 1. If the null hypothesis is false, the between-treatments estimate will be larger than the within-treatments estimate, and their ratio will be large. In the next section we will show how large this ratio must be to reject H_0.

In summary, the logic behind ANOVA is based on the development of two independent estimates of the common population variance σ^2. One estimate of σ^2 is based on the variability among the sample means themselves, and the other estimate of σ^2 is based on the variability of the data within each sample. By comparing these two estimates of σ^2, we will be able to determine whether the population means are equal.

NOTES AND COMMENTS

1. In Sections 10.1 and 10.2 we presented statistical methods for testing the hypothesis that the means of two populations are equal. ANOVA can also be used to test the hypothesis that the means of two populations are equal. In practice, however, analysis of variance is usually not used except when dealing with three or more population means.
2. In Section 10.2 we discussed how to test for the equality of two population means whenever one or both sample sizes are less than 30. As part of that discussion we illustrated the process of combining the results of two independent random samples to provide one estimate of σ^2; that process was referred to as pooling, and the resulting sample variance was referred to as the pooled estimator of σ^2. In analysis of variance, the within-treatments estimate of σ^2 is simply the generalization of that concept to the case of more than two samples; that is why we also referred to the within-treatments estimator as the pooled estimator of σ^2.

13.2 ANALYSIS OF VARIANCE: TESTING FOR THE EQUALITY OF *k* POPULATION MEANS

Analysis of variance can be used to test for the equality of k population means. The general form of the hypotheses tested is

$$H_0\colon \mu_1 = \mu_2 = \cdots = \mu_k$$
$$H_a\colon \text{Not all population means are equal}$$

where

$$\mu_j = \text{mean of the } j\text{th population}$$

We assume that a simple random sample of size n_j has been selected from each of the k populations or treatments. For the resulting sample data, let

x_{ij} = value of observation i for treatment j
n_j = number of observations for treatment j
$\bar{x}_j$ = sample mean for treatment j
s_j^2 = sample variance for treatment j
s_j = sample standard deviation for treatment j

The formulas for the sample mean and sample variance for treatment j are as follow.

$$\bar{x}_j = \frac{\sum_{i=1}^{n_j} x_{ij}}{n_j} \tag{13.1}$$

$$s_j^2 = \frac{\sum_{i=1}^{n_j} (x_{ij} - \bar{x}_j)^2}{n_j - 1} \tag{13.2}$$

The overall sample mean, denoted $\bar{\bar{x}}$, is the sum of all the observations divided by the total number of observations. That is,

$$\bar{\bar{x}} = \frac{\sum_{j=1}^{k} \sum_{i=1}^{n_j} x_{ij}}{n_T} \tag{13.3}$$

where

$$n_T = n_1 + n_2 + \cdots + n_k \tag{13.4}$$

If the size of each sample is n, $n_T = kn$; in this case (13.3) reduces to

$$\bar{\bar{x}} = \frac{\sum_{j=1}^{k} \sum_{i=1}^{n_j} x_{ij}}{nk} = \frac{\sum_{j=1}^{k} \sum_{i=1}^{n_j} x_{ij}/n}{k} = \frac{\sum_{j=1}^{k} \bar{x}_j}{k} \tag{13.5}$$

In other words, whenever the sample sizes are the same, the overall sample mean is just the average of the k sample means.

Because each sample in the NCP example consists of $n = 6$ observations, the overall sample mean can be computed by using (13.5). For the data in Table 13.1 we obtained the following result.

$$\bar{\bar{x}} = \frac{79 + 74 + 66}{3} = 73$$

If the null hypothesis is true ($\mu_1 = \mu_2 = \mu_3 = \mu$), the overall sample mean of 73 is the best estimate of the population mean μ.

Between-Treatments Estimate of Population Variance

In the preceding section, we introduced the concept of a between-treatments estimate of σ^2 and showed how to compute it when the sample sizes were equal. This estimate of σ^2 is called the *mean square due to treatments* and is denoted MSTR. The general formula for computing MSTR is

$$\text{MSTR} = \frac{\sum_{j=1}^{k} n_j(\bar{x}_j - \bar{\bar{x}})^2}{k - 1} \tag{13.6}$$

The numerator in (13.6) is called the *sum of squares due to treatments* and is denoted SSTR. The denominator, $k - 1$, represents the degrees of freedom associated with SSTR. Hence, the mean square due to treatments can be computed by the following formula.

Mean Square Due to Treatments

$$\text{MSTR} = \frac{\text{SSTR}}{k - 1} \tag{13.7}$$

where

$$\text{SSTR} = \sum_{j=1}^{k} n_j(\bar{x}_j - \bar{\bar{x}})^2 \tag{13.8}$$

If H_0 is true, MSTR provides an unbiased estimate of σ^2. However, if the means of the k populations are not equal, MSTR is not an unbiased estimate of σ^2; in fact, in that case, MSTR should overestimate σ^2.

For the NCP data in Table 13.1, we obtain the following results.

$$\text{SSTR} = \sum_{j=1}^{k} n_j(\bar{x}_j - \bar{\bar{x}})^2 = 6(79 - 73)^2 + 6(74 - 73)^2 + 6(66 - 73)^2 = 516$$

$$\text{MSTR} = \frac{\text{SSTR}}{k - 1} = \frac{516}{2} = 258$$

Within-Treatments Estimate of Population Variance

In the preceding section we introduced the concept of a within-treatments estimate of σ^2 and showed how to compute it when the sample sizes were equal. This estimate of σ^2 is called the *mean square due to error* and is denoted MSE. The general formula for computing MSE is

$$\text{MSE} = \frac{\sum_{j=1}^{k} (n_j - 1)s_j^2}{n_T - k} \tag{13.9}$$

The numerator in (13.9) is called the *sum of squares due to error* and is denoted SSE. The denominator of MSE is referred to as the degrees of freedom associated with SSE. Hence, the formula for MSE can also be stated as follows.

Mean Square Due to Error

$$\text{MSE} = \frac{\text{SSE}}{n_T - k} \tag{13.10}$$

where

$$\text{SSE} = \sum_{j=1}^{k} (n_j - 1)s_j^2 \tag{13.11}$$

Note that MSE is based on the variation within each of the treatments; it is not influenced by whether the null hypothesis is true. Thus, MSE always provides an unbiased estimate of σ^2.

For the NCP data in Table 13.1 we obtain the following results.

$$\text{SSE} = \sum_{j=1}^{k} (n_j - 1)s_j^2 = (6 - 1)34 + (6 - 1)20 + (6 - 1)32 = 430$$

$$\text{MSE} = \frac{\text{SSE}}{n_T - k} = \frac{430}{18 - 3} = \frac{430}{15} = 28.67$$

Comparing the Variance Estimates: The F Test

Let us assume that the null hypothesis is true. In that case, MSTR and MSE provide two independent, unbiased estimates of σ^2. Recall from Chapter 11 that for normal populations, the sampling distribution of the ratio of two independent estimates of σ^2 follows an F distribution. Hence, if the null hypothesis is true and the ANOVA assumptions are valid, the sampling distribution of MSTR/MSE is an F distribution with numerator degrees of freedom equal to $k - 1$ and denominator degrees of freedom equal to $n_T - k$.

If the means of the k populations are not equal, the value of MSTR/MSE will be inflated because MSTR overestimates σ^2. Hence, we will reject H_0 if the resulting value of MSTR/MSE appears to be too large to have been selected at random from an F distribution with degrees of freedom $k - 1$ in the numerator and $n_T - k$ in the denominator. The value of MSTR/MSE that will cause us to reject H_0 depends on α, the level of significance. Once α is selected, a critical value can be determined. Figure 13.3 shows the sampling distribution of MSTR/MSE and the rejection region associated with a level of significance equal to α where F_α denotes the critical value.

Suppose the manager responsible for making the decision in the National Computer Products example was willing to accept a probability of making a Type I error of $\alpha = .05$. From Table 4 of Appendix B we can determine the critical F value by locating the value corresponding to numerator degrees of freedom equal to $k - 1 = 3 - 1 = 2$ and denominator degrees of freedom equal to $n_T - k = 18 - 3 = 15$. Thus, we obtain the value $F_{.05} = 3.68$. Note that this value tells us that if we were to select a value at random from an F distribution with two numerator degrees of freedom and 15 denominator degrees of freedom, only 5% of the time would we observe a value greater than 3.68. Moreover, the theory behind the analysis of variance tells us that if the null hypothesis is true, the ratio of MSTR/MSE would be a value from this F distribution. Hence, the appropriate rejection rule for the NCP example is written

$$\text{Reject } H_0 \text{ if } F = \text{MSTR/MSE} > 3.68$$

FIGURE 13.3 SAMPLING DISTRIBUTION OF MSTR/MSE; THE CRITICAL VALUE FOR REJECTING THE NULL HYPOTHESIS OF EQUALITY OF MEANS IS F_α

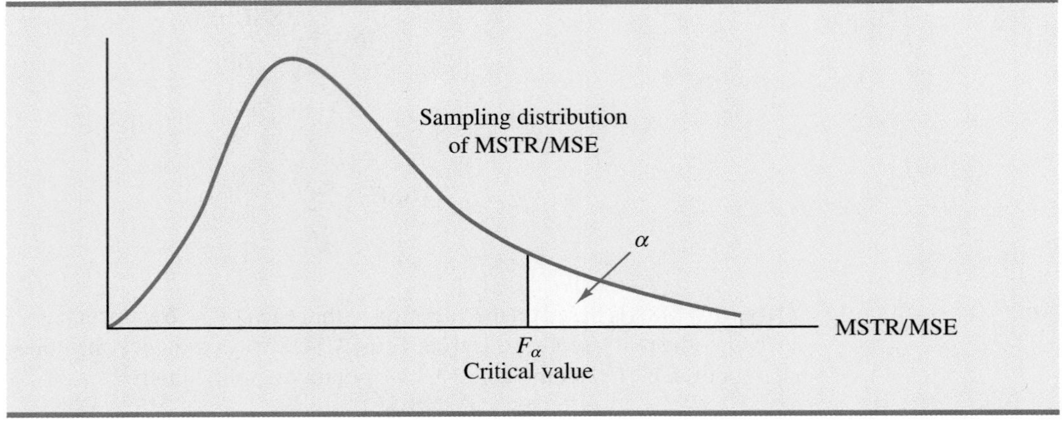

Recall that MSTR = 258 and MSE = 28.67. Because F = MSTR/MSE = 258/28.67 = 9.00 is greater than the critical value, $F_{.05} = 3.68$, we have sufficient evidence to reject the null hypothesis that the means of the three populations are equal. In other words, analysis of variance supports the conclusion that the population mean examination scores at the three NCP plants are not equal.

The p-value criterion can also be used for this hypothesis test. The usual rejection rule applies: Reject H_0 if p-value $< \alpha$. However, because it is difficult to determine the p-value directly from tables of the F probability distribution, a computer software package such as Minitab or Excel is required. Appendixes 13.1 and 13.2 show the procedures that can be used, and a discussion of the computer results for analysis of variance is provided in a subsection that follows. For the NCP example, the p-value associated with the test statistic $F = 9.00$ is .003. With a p-value $= .003 < \alpha = .05$, the null hypothesis that the means of the three populations are equal can be rejected. A summary of the overall procedure for testing for the equality of k population means follows.

Test for the Equality of *k* Population Means

$$H_0: \mu_1 = \mu_2 = \cdots = \mu_k$$
$$H_a: \text{Not all population means are equal}$$

Test Statistic

$$F = \frac{\text{MSTR}}{\text{MSE}} \quad \textbf{(13.12)}$$

Rejection Rule

Using test statistic: Reject H_0 if $F > F_\alpha$
Using p-value: Reject H_0 if p-value $< \alpha$

where the value of F_α is based on an F distribution with $k - 1$ numerator degrees of freedom and $n_T - k$ denominator degrees of freedom.

ANOVA Table

The results of the preceding calculations can be displayed conveniently in a table referred to as the analysis of variance or **ANOVA table.** Table 13.2 is the analysis of variance table for the National Computer Products example. The sum of squares associated with the source of variation referred to as "Total" is called the total sum of squares (SST). Note that the results for the NCP example suggest that SST = SSTR + SSE, and that the degrees of freedom associated with this total sum of squares is the sum of the degrees of freedom associated with the between-treatments estimate of σ^2 and the within-treatments estimate of σ^2.

We point out that SST divided by its degrees of freedom $n_T - 1$ is nothing more than the overall sample variance that would be obtained if we treated the entire set of 18 observations as one data set. With the entire data set as one sample, the formula for computing the total sum of squares, SST, is

$$\text{SST} = \sum_{j=1}^{k}\sum_{i=1}^{n_j}(x_{ij} - \bar{\bar{x}})^2 \tag{13.13}$$

It can be shown that the results we observed for the analysis of variance table for the NCP example also apply to other problems. That is,

Analysis of variance can be thought of as a statistical procedure for partitioning the total sum of squares into separate components.

$$\text{SST} = \text{SSTR} + \text{SSE} \tag{13.14}$$

In other words, SST can be partitioned into two sums of squares: the sum of squares due to treatments and the sum of squares due to error. Note also that the degrees of freedom corresponding to SST, $n_T - 1$, can be partitioned into the degrees of freedom corresponding to SSTR, $k - 1$, and the degrees of freedom corresponding to SSE, $n_T - k$. The analysis of variance can be viewed as the process of **partitioning** the total sum of squares and the degrees of freedom into their corresponding sources: treatments and error. Dividing the sum of squares by the appropriate degrees of freedom provides the variance estimates and the F value used to test the hypothesis of equal population means.

Computer Results for Analysis of Variance

Because of the widespread availability of statistical computer packages, analysis of variance computations with large sample sizes and/or a large number of populations can be performed easily. In Figure 13.4 we show output for the NCP example obtained from the Minitab computer package. The first part of the computer output contains the familiar ANOVA table format. Comparing Figure 13.4 with Table 13.2, we see that the same information is available, although some of the headings are slightly different. The heading Source is used for the source of variation column, and Factor identifies the treatments row.

TABLE 13.2 ANALYSIS OF VARIANCE TABLE FOR THE NCP EXAMPLE

Source of Variation	Sum of Squares	Degrees of Freedom	Mean Square	F
Treatments	516	2	258.00	9.00
Error	430	15	28.67	
Total	946	17		

FIGURE 13.4 MINITAB OUTPUT FOR THE NCP ANALYSIS OF VARIANCE

```
Analysis of Variance
Source      DF        SS        MS        F        p
Factor       2     516.0     258.0     9.00    0.003
Error       15     430.0      28.7
Total       17     946.0
                                    Individual 95% CIs For Mean
                                    Based on Pooled StDev
 Level       N      Mean     StDev  ---+---------+---------+---------+---
Atlanta      6    79.000     5.831                     (------*------)
Dallas       6    74.000     4.472               (------*-----)
Seattle      6    66.000     5.657   (-----*------)
                                    ---+---------+---------+---------+---
Pooled StDev =     5.354            63.0      70.0      77.0      84.0
```

The sum of squares and degrees of freedom columns are interchanged, and a p-value is provided for the F test. Thus, at the $\alpha = .05$ level of significance, we reject H_0 because the p value $= 0.003 < \alpha = .05$.

Note that following the ANOVA table the computer output contains the respective sample sizes, the sample means, and the standard deviations. In addition, Minitab provides a figure that shows individual 95% confidence interval estimates of each population mean. In developing these confidence interval estimates, Minitab uses MSE as the estimate of σ^2. Thus, the square root of MSE provides the best estimate of the population standard deviation σ. This estimate of σ on the computer output is Pooled StDev; it is equal to 5.354. To provide an illustration of how these interval estimates are developed, we will compute a 95% confidence interval estimate of the population mean for the Atlanta plant.

From our study of interval estimation in Chapter 8, we know that the general form of an interval estimate of a population mean is

$$\bar{x} \pm t_{\alpha/2} \frac{s}{\sqrt{n}} \tag{13.15}$$

where s is the estimate of the population standard deviation σ. In the analysis of variance the best estimate of σ is provided by the square root of MSE or the Pooled StDev, therefore we use a value of 5.354 for s in (13.15). The degrees of freedom for the t value is 15, the degrees of freedom associated with the within-treatments estimate of σ^2. Hence, with $t_{.025} = 2.131$ we obtain

$$79 \pm 2.131 \frac{5.354}{\sqrt{6}} = 79 \pm 4.66$$

Thus, the individual 95% confidence interval for the Atlanta plant goes from $79 - 4.66 = 74.34$ to $79 + 4.6 = 83.66$. Because the sample sizes are equal for the NCP example, the individual confidence intervals for the Dallas and Seattle plants are also constructed by adding and subtracting 4.66 from each sample mean. Thus, in the figure provided by Minitab we see that the widths of the confidence intervals are the same.

NOTES AND COMMENTS

1. The overall sample mean can also be computed as a weighted average of the k sample means.

$$\bar{\bar{x}} = \frac{n_1\bar{x}_1 + n_2\bar{x}_2 + \cdots + n_k\bar{x}_k}{n_T}$$

In problems where the sample means are provided, this formula is simpler than (13.3) for computing the overall mean.

2. If each sample consists of n observations, (13.6) can be written as

$$\text{MSTR} = \frac{n\sum_{j=1}^{k}(\bar{x}_j - \bar{\bar{x}})^2}{k-1} = n\left[\frac{\sum_{j=1}^{k}(\bar{x}_j - \bar{\bar{x}})^2}{k-1}\right] = ns_{\bar{x}}^2$$

Note that this result is the same as presented in Section 13.1 when we introduced the concept of the between-treatments estimate of σ^2. Equation (13.6) is simply a generalization of this result to the unequal sample-size case.

3. If each sample has n observations, $n_T = kn$; thus, $n_T - k = k(n-1)$, and (13.9) can be rewritten as

$$\text{MSE} = \frac{\sum_{j=1}^{k}(n-1)s_j^2}{k(n-1)} = \frac{(n-1)\sum_{j=1}^{k}s_j^2}{k(n-1)} = \frac{\sum_{j=1}^{k}s_j^2}{k}$$

In other words, if the sample sizes are the same, MSE is just the average of the k sample variances. Note that it is the same result we used in Section 13.1 when we introduced the concept of the within-treatments estimate of σ^2.

4. Confidence interval estimates for each of the k population means can be developed using

$$\bar{x} \pm t_{\alpha/2}\frac{\sqrt{\text{MSE}}}{\sqrt{n}}$$

The degrees of freedom for the t value are the degrees of freedom associated with the within-treatments estimate of σ^2.

EXERCISES

Methods

1. Samples of five observations were selected from each of three populations. The data obtained follow.

Observation	Sample 1	Sample 2	Sample 3
1	32	44	33
2	30	43	36
3	30	44	35
4	26	46	36
5	32	48	40
Sample mean	30	45	36
Sample variance	6.00	4.00	6.50

   a. Compute the between-treatments estimate of σ^2.
   b. Compute the within-treatments estimate of σ^2.
   c. At the $\alpha = .05$ level of significance, can we reject the null hypothesis that the means of the three populations are equal?
   d. Set up the ANOVA table for this problem.

2. Four observations were selected from each of three populations. The data obtained follow.

Observation	Sample 1	Sample 2	Sample 3
1	165	174	169
2	149	164	154
3	156	180	161
4	142	158	148
Sample mean	153	169	158
Sample variance	96.67	97.33	82.00

a. Compute the between-treatments estimate of σ^2.
b. Compute the within-treatments estimate of σ^2.
c. At the $\alpha = .05$ level of significance, can we reject the null hypothesis that the three population means are equal? Explain.
d. Set up the ANOVA table for this problem.

3. Samples were selected from three populations. The data obtained follow.

	Sample 1	Sample 2	Sample 3
	93	77	88
	98	87	75
	107	84	73
	102	95	84
		85	75
		82	
$\bar{x}_j$	100	85	79
s_j^2	35.33	35.60	43.50

a. Compute the between-treatments estimate of σ^2.
b. Compute the within-treatments estimate of σ^2.
c. At the $\alpha = .05$ level of significance, can we reject the null hypothesis that the three population means are equal? Explain.
d. Set up the ANOVA table for this problem.

4. A random sample of 16 observations was selected from each of four populations. A portion of the ANOVA table follows.

Source of Variation	Sum of Squares	Degrees of Freedom	Mean Square	F
Treatments			400	
Error				
Total	1500			

a. Provide the missing entries for the ANOVA table.
b. At the $\alpha = .05$ level of significance, can we reject the null hypothesis that the means of the four populations are equal?

5. Random samples of 25 observations were selected from each of three populations. For these data, SSTR = 120 and SSE = 216.
a. Set up the ANOVA table for this problem.
b. At the $\alpha = .05$ level of significance, what is the critical F value?
c. At the $\alpha = .05$ level of significance, can we reject the null hypothesis that the three population means are equal?

Applications

6. To test whether the mean time needed to mix a batch of material is the same for machines produced by three manufacturers, the Jacobs Chemical Company obtained the following data on the time (in minutes) needed to mix the material. Use these data to test whether the population mean times for mixing a batch of material differ for the three manufacturers. Use $\alpha = .05$.

Manufacturer		
1	2	3
20	28	20
26	26	19
24	31	23
22	27	22

7. Managers at all levels of an organization need adequate information to perform their respective tasks. One study investigated the effect the source has on the dissemination of information. In this particular study the sources of information were a superior, a peer, and a subordinate. In each case, a measure of dissemination was obtained, with higher values indicating greater dissemination of information. Using $\alpha = .05$ and the following data, test whether the source of information significantly affects dissemination. What is your conclusion, and what does it suggest about the use and dissemination of information?

Superior	Peer	Subordinate
8	6	6
5	6	5
4	7	7
6	5	4
6	3	3
7	4	5
5	7	7
5	6	5

8. A study investigated the perception of corporate ethical values among individuals specializing in marketing. Using $\alpha = .05$ and the following data (higher scores indicate higher ethical values), test for significant differences in perception among the three groups.

Marketing Managers	Marketing Research	Advertising
6	5	6
5	5	7
4	4	6
5	4	5
6	5	6
4	4	6

9. A study reported in the *Journal of Small Business Management* concluded that self-employed individuals experience higher job stress than individuals who are not self-employed. In this study job stress was assessed with a 15-item scale designed to measure various aspects of ambiguity and role conflict. Ratings for each of the 15 items were made using a scale with 1–5 response options ranging from strong agreement to strong disagreement. The sum of the ratings for the 15 items for each individual surveyed is between 15 and 75, with higher values indicating a higher degree of job stress (*Journal of Small Business Management,* October 1997). Suppose that a similar approach, using a 20-item scale with 1–5 response options, was used to measure the job stress of individuals for 15 randomly selected real estate agents, 15 architects, and 15 stockbrokers. The results obtained follow.

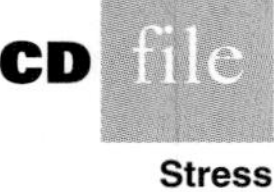

Stress

Real Estate Agent	Architect	Stockbroker
81	43	65
48	63	48
68	60	57
69	52	91
54	54	70
62	77	67
76	68	83
56	57	75
61	61	53
65	80	71
64	50	54
69	37	72
83	73	65
85	84	58
75	58	58

Using $\alpha = .05$, test for any significant difference in job stress among the three professions.

10. The *Business Week* Global 1000 ranks companies on the basis of their market value (*Business Week,* July 7, 1997). The following table shows the price/earnings ratios for 29 companies classified as being in the finance economic sector. An industry code of 1 indicates a banking firm, a code of 2 a financial services firm, and a code of 3 an insurance firm. At the .05 level of significance, test whether the mean price/earnings ratio is the same for these three groups of financial firms.

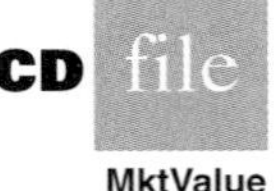
MktValue

Company	Industry Code	P/E	Company	Industry Code	P/E
Citicorp	1	15	MBNA	2	24
NationsBank	1	14	Cincinnati Financial	2	19
Wells Fargo	1	25	Franklin Resources	2	22
First Union	1	13	Fannie Mae	2	17
KeyCorp	1	14	American International Group	3	21
Chase Manhattan	1	12	Allstate	3	14
Fifth Third Bancorp	1	23	Marsh & McLennan	3	20
Bank of New York	1	17	American General	3	16
First Chicago NBD	1	13	Cigna	3	12
Mellon Bank	1	16	Lincoln National	3	13
Fleet Financial Group	1	15	AFLAC	3	21
First Bank System	1	16	Equitable	3	11
American Express	2	19	Chubb	3	20
Travelers	2	15	General Re	3	15
Merrill Lynch	2	12			

13.3 MULTIPLE COMPARISON PROCEDURES

When we use analysis of variance to test whether the means of k populations are equal, rejection of the null hypothesis allows us to conclude only that the population means are *not all equal.* In some cases we will want to go a step further and determine where the differences among means occur. The purpose of this section is to introduce two **multiple comparison procedures** that can be used to conduct statistical comparisons between pairs of population means.

Fisher's LSD

Suppose that analysis of variance has provided statistical evidence to reject the null hypothesis of equal population means. In this case, Fisher's least significant difference (LSD) procedure can be used to determine where the differences occur. To illustrate the use of Fisher's LSD procedure in making pairwise comparisons of population means, recall the NCP example introduced in Section 13.1. Using analysis of variance, we concluded that the population mean examination scores are not the same at the three plants. In this case, the follow-up question is: We believe the plants differ, but where do the differences occur? That is, do the means of populations 1 and 2 differ? Or those of populations 1 and 3? Or those of populations 2 and 3?

In Chapter 10 we presented a statistical procedure for testing the hypothesis that the means of two populations are equal. With a slight modification in how we estimate the population variance, Fisher's LSD procedure is based on the t test statistic presented for the two-population case. The following table provides a summary of Fisher's LSD procedure.

Fisher's LSD Procedure

$$H_0: \mu_i = \mu_j$$
$$H_a: \mu_i \neq \mu_j$$

Test Statistic

$$t = \frac{\bar{x}_i - \bar{x}_j}{\sqrt{\text{MSE}\left(\frac{1}{n_i} + \frac{1}{n_j}\right)}} \tag{13.16}$$

Rejection Rule

$$\text{Reject } H_0 \text{ if } t < -t_{\alpha/2} \text{ or } t > t_{\alpha/2}$$

where the value of $t_{\alpha/2}$ is based on a t distribution with $n_T - k$ degrees of freedom.

Let us now apply this procedure to determine whether there is a significant difference between the means of population 1 (Atlanta) and population 2 (Dallas). Table 13.1 shows that the sample mean is 79 for the Atlanta plant and 74 for the Dallas plant. Table 13.2 shows that the value of MSE is 28.67; it is the estimate of σ^2 and is based on 15 degrees of freedom. At the .05 level of significance, the t distribution table shows that with $n_T - k = 18 - 3 = 15$ degrees of freedom, $t_{.025} = 2.131$. Thus, if $t < -2.131$ or $t > 2.131$, we reject H_0: $\mu_1 = \mu_2$. For the NCP data we obtain the following t value.

$$t = \frac{79 - 74}{\sqrt{28.67\left(\frac{1}{6} + \frac{1}{6}\right)}} = 1.62$$

Because $t = 1.62$, we do not have sufficient statistical evidence to reject the null hypothesis; hence, we cannot conclude that the population mean score at the Atlanta plant is different from the population mean score at the Dallas plant.

Many practitioners find it easier to determine how large the difference between the sample means must be to reject H_0. In this case the test statistic is $\bar{x}_i - \bar{x}_j$, and the test is conducted by the following procedure.

Fisher's LSD Procedure Based on the Test Statistic $\bar{x}_i - \bar{x}_j$

$$H_0: \mu_i = \mu_j$$
$$H_a: \mu_i \neq \mu_j$$

Test Statistic

$$\bar{x}_i - \bar{x}_j$$

Rejection Rule at a Level of Significance α

$$\text{Reject } H_0 \text{ if } |\bar{x}_i - \bar{x}_j| > \text{LSD}$$

where

$$\text{LSD} = t_{\alpha/2}\sqrt{\text{MSE}\left(\frac{1}{n_i} + \frac{1}{n_j}\right)} \qquad \textbf{(13.17)}$$

For the NCP example the value of LSD is

$$\text{LSD} = 2.131\sqrt{28.67\left(\frac{1}{6} + \frac{1}{6}\right)} = 6.59$$

Note that when the sample sizes are equal, only one value for LSD is computed. In such cases we can simply compare the magnitude of the difference between any two sample means with the value of LSD. For example, the difference between the sample means for population 1 (Atlanta) and population 3 (Seattle) is $79 - 66 = 13$. This difference is greater than 6.59, which means we can reject the null hypothesis that the population mean examination score for the Atlanta plant is equal to the population mean score for the Seattle plant. Similarly, with the difference between the sample means for populations 2 and 3 of $74 - 66 = 8 > 6.59$, we can also reject the hypothesis that the population mean examination score for the Dallas plant is equal to the population mean examination score for the Seattle plant. In effect, our conclusion is that the Atlanta and Dallas plants both differ from the Seattle plant.

Fisher's LSD can also be used to develop a confidence interval estimate of the difference between the means of two populations. The general procedure follows.

Confidence Interval Estimate of the Difference Between Two Population Means Using Fisher's LSD Procedure

$$\bar{x}_i - \bar{x}_j \pm \text{LSD} \qquad \textbf{(13.18)}$$

where

$$\text{LSD} = t_{\alpha/2}\sqrt{\text{MSE}\left(\frac{1}{n_i} + \frac{1}{n_j}\right)} \qquad \textbf{(13.19)}$$

and $t_{\alpha/2}$ is based on a t distribution with $n_T - k$ degrees of freedom.

If the confidence interval in (13.18) includes the value zero, we cannot reject the hypothesis that the two population means are equal. However, if the confidence interval does not include the value zero, we conclude that there is a difference between the population means. For the NCP example, recall that LSD $= 6.59$ (corresponding to $t_{.025} = 2.131$). Thus, a 95% confidence interval estimate of the difference between the means of populations 1 and 2 is $79 - 74 \pm 6.59 = 5 \pm 6.59 = -1.59$ to 11.59; because this interval includes zero, we cannot reject the hypothesis that the two population means are equal.

Type I Error Rates

We began the discussion of Fisher's LSD procedure with the premise that analysis of variance had given us statistical evidence to reject the null hypothesis of equal population means. We showed how Fisher's LSD procedure can be used in such cases to determine where the differences occur. Technically, it is referred to as a *protected* or *restricted* LSD test because it is employed only if we first find a significant F value by using analysis of variance. To see why this distinction is important in multiple comparison tests, we need to explain the difference between a *comparisonwise* Type I error rate and an *experimentwise* Type I error rate.

In the NCP example we used Fisher's LSD procedure to make three pairwise comparisons.

Test 1	Test 2	Test 3
$H_0\colon \mu_1 = \mu_2$	$H_0\colon \mu_1 = \mu_3$	$H_0\colon \mu_2 = \mu_3$
$H_a\colon \mu_1 \neq \mu_2$	$H_a\colon \mu_1 \neq \mu_3$	$H_a\colon \mu_2 \neq \mu_3$

In each case, we used a level of significance of $\alpha = .05$. Therefore, for each test, if the null hypothesis is true, the probability that we will make a Type I error is $\alpha = .05$; hence, the probability that we will not make a Type I error on each test is $1 - .05 = .95$. In discussing multiple comparison procedures we refer to this probability of a Type I error ($\alpha = .05$) as the **comparisonwise Type I error rate;** comparisonwise Type I error rates indicate the level of significance associated with a single pairwise comparison.

Let us now consider a slightly different question. What is the probability that in making three pairwise comparisons, we will commit a Type I error on at least one of the three tests? To answer this question, note that the probability that we will not make a Type I error on any of the three tests is $(.95)(.95)(.95) = .8574$.* Therefore, the probability of making at least one Type I error is $1 - .8574 = .1426$. Thus, when we use Fisher's LSD procedure to make all three pairwise comparisons, the Type I error rate associated with this approach is not .05, but actually .1426; we refer to this error rate as the *overall* or **experimentwise Type I error rate.** To avoid confusion, we denote the experimentwise Type I error rate as α_{EW}.

The experimentwise Type I error rate gets larger for problems with more populations. For example, a problem with five populations has 10 possible pairwise comparisons. If we tested all possible pairwise comparisons by using Fisher's LSD with a comparisonwise error rate of $\alpha = .05$, the experimentwise Type I error rate would be $1 - (1 - .05)^{10} = .40$. In such cases, practitioners look to alternatives that provide better control over the experimentwise error rate.

One alternative for controlling the overall experimentwise error rate, referred to as the Bonferroni adjustment, involves using a smaller comparisonwise error rate for each test. For example, if we want to test C pairwise comparisons and want the maximum probability of making a Type I error for the overall experiment to be α_{EW}, we simply use a comparisonwise error rate equal to α_{EW}/C. In the NCP example, if we want to use Fisher's LSD procedure to test all three pairwise comparisons with a maximum experimentwise error rate of $\alpha_{EW} = .05$, we set the comparisonwise error rate to be $\alpha = .05/3 = .017$. For a problem with five populations and 10 possible pairwise comparisons, the Bonferroni adjustment would suggest a comparisonwise error rate of $.05/10 = .005$. Recall from our discussion of hypothesis testing in Chapter 9 that for a fixed sample size, any decrease in the probability

*The assumption is that the three tests are independent, and hence the joint probability of the three events can be obtained by simply multiplying the individual probabilities. In fact, the three tests are not independent because MSE is used in each test; therefore, the error involved is even greater than that shown.

of making a Type I error will result in an increase in the probability of making a Type II error, which corresponds to accepting the hypothesis that the two population means are equal when in fact they are not equal. As a result, many practitioners are reluctant to perform individual tests with a low comparisonwise Type I error rate because of the increased risk of making a Type II error.

Several other procedures, such as Tukey's procedure and Duncan's multiple range test, have been developed to help in such situations. However, there is considerable controversy in the statistical community as to which procedure is "best." The truth is that no one procedure is best for all types of problems.

EXERCISES

Methods

11. In Exercise 1, five observations were selected from each of three populations. For these data, $\bar{x}_1 = 30$, $\bar{x}_2 = 45$, $\bar{x}_3 = 36$, and MSE = 5.5. At the $\alpha = .05$ level of significance, the null hypothesis of equal population means was rejected. In the following calculations, use $\alpha = .05$.
    a. Using Fisher's LSD procedure, test to see whether there is a significant difference between the means of populations 1 and 2, populations 1 and 3, and populations 2 and 3.
    b. Use Fisher's LSD procedure to develop a 95% confidence interval estimate of the difference between the means of populations 1 and 2.
12. Four observations were selected from each of three populations. The data obtained are shown. In the following calculations, use $\alpha = .05$.

	Sample 1	Sample 2	Sample 3
	63	82	69
	47	72	54
	54	88	61
	40	66	48
$\bar{x}_j$	51	77	58
s_j^2	96.67	97.34	81.99

    a. Using analysis of variance, test for a significant difference among the means of the three populations.
    b. Use Fisher's LSD procedure to see which means are different.

Applications

13. Refer to Exercise 6. At the $\alpha = .05$ level of significance, use Fisher's LSD procedure to test for the equality of the means for manufacturers 1 and 3. What conclusion can you draw after carrying out this test?

14. Refer to Exercise 13. Use Fisher's LSD procedure to develop a 95% confidence interval estimate of the difference between the means of population 1 and population 2.
15. Refer to Exercise 8. At the $\alpha = .05$ level of significance, we can conclude that there are differences in the perceptions for marketing managers, marketing research specialists, and advertising specialists. Use the procedures in this section to determine where the differences occur. Use $\alpha = .05$.

16. To test for any significant difference in the number of hours between breakdowns for four machines, the following data were obtained.

Machine 1	Machine 2	Machine 3	Machine 4
6.4	8.7	11.1	9.9
7.8	7.4	10.3	12.8
5.3	9.4	9.7	12.1
7.4	10.1	10.3	10.8
8.4	9.2	9.2	11.3
7.3	9.8	8.8	11.5

a. At the $\alpha = .05$ level of significance, what is the difference, if any, in the population mean times among the four machines?

b. Use Fisher's LSD procedure to test for the equality of the means for machines 2 and 4. Use a .05 level of significance.

17. Refer to Exercise 16. Use the Bonferroni adjustment to test for a significant difference between all pairs of means. Assume that a maximum overall experimentwise error rate of .05 is desired.

18. Refer to Exercise 10. At the .05 level of significance, we can conclude that there are differences between the mean price/earnings ratios of banking firms, financial services firms, and insurance firms. Use the procedures in this section to determine where the differences occur. Use $\alpha = .05$.

13.4 AN INTRODUCTION TO EXPERIMENTAL DESIGN

Cause-and-effect relationships can be difficult to establish in observational studies; such relationships are easier to establish in experimental studies.

Statistical studies can be classified as being either experimental or observational. In an *experimental study,* variables of interest are identified. Then, one or more factors in the study are controlled so that data can be obtained about how the factors influence the variables. In *observational* or *nonexperimental* studies, no attempt is made to control the factors. A survey (see Chapter 21) is perhaps the most common type of observational study.

The NCP example that we used to introduce analysis of variance is an illustration of an observational statistical study. To measure how much NCP employees knew about total quality management, a random sample of six employees was selected from each of NCP's three plants and given a quality-awareness examination. The examination scores for these employees were then analyzed by analysis of variance to test the hypothesis that the population mean examination scores were equal for the three plants.

As an example of an experimental statistical study, let us consider the problem facing Chemitech, Inc. Chemitech has developed a new filtration system for municipal water supplies. The components for the new filtration system will be purchased from several suppliers, and Chemitech will assemble the components at its plant in Columbia, South Carolina. The industrial engineering group has been given the responsibility of determining the best assembly method for the new filtration system. After considering a variety of possible approaches, the group has narrowed the alternatives to three: method A, method B, and method C. These methods differ in the sequence of steps used to assemble the product. Managers at Chemitech want to determine which assembly method can produce the greatest number of filtration systems per week.

In the Chemitech experiment, assembly method is the independent variable or **factor**. Because there are three assembly methods corresponding to this factor, we say that there are three treatments associated with this experiment; each **treatment** corresponds to each of the three assembly methods. The Chemitech problem is an example of a **single-factor experiment** involving a qualitative factor (method of assembly). Other experiments may consist of multiple factors; some factors may be qualitative and some may be quantitative.

The three assembly methods or treatments define the three populations of interest for the Chemitech experiment. One population is all Chemitech employees who use assembly method A, another is those who use method B, and the third is those who use method C. Note that for each population the dependent or response variable is the number of filtration systems assembled per week, and the primary statistical objective of the experiment is to determine whether the mean number of units produced per week is the same for all three populations.

Randomization is the process of assigning the treatments to the experimental units at random. Prior to the work of Sir R. A. Fisher, treatments were assigned on a systematic or subjective basis.

Suppose a random sample of three employees is selected from all assembly workers at the Chemitech production facility. In experimental design terminology, the three randomly selected workers are the **experimental units.** The experimental design that we will use for the Chemitech problem is called a **completely randomized design.** This type of design requires that each of the three assembly methods or treatments be assigned randomly to one of the experimental units or workers. For example, method A might be randomly assigned to the second worker, method B to the first worker, and method C to the third worker. The concept of *randomization,* as illustrated in this example, is an important principle of all experimental designs.

Note that the experiment would result in only one measurement or number of units assembled for each treatment. In other words, we have a sample size of one corresponding to each treatment. To obtain additional data for each assembly method, we must repeat or replicate the basic experimental process. Suppose, for example, that instead of selecting just three workers at random we had selected 15 workers and then randomly assigned each of the three treatments to five of the workers. Because each method of assembly is assigned to five workers, we say that five replicates have been obtained. The process of *replication* is another important principle of experimental design. Figure 13.5 shows the completely randomized design for the Chemitech experiment.

FIGURE 13.5 COMPLETELY RANDOMIZED DESIGN FOR EVALUATING THE CHEMITECH ASSEMBLY METHOD EXPERIMENT

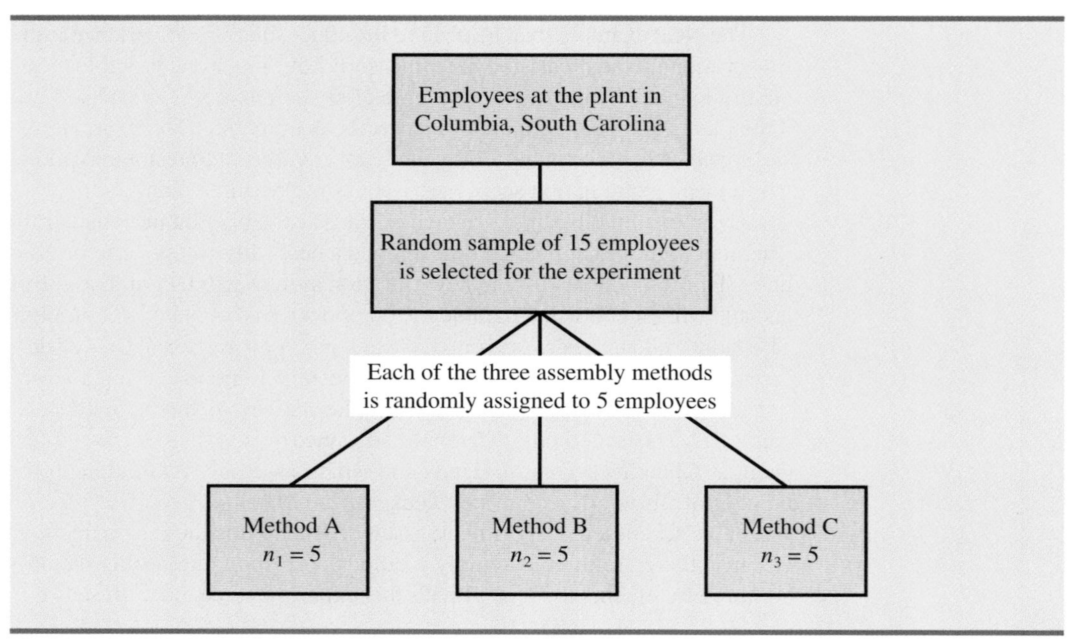

Data Collection

Once we are satisfied with the experimental design, we proceed by collecting and analyzing the data. In the Chemitech case, the employees would be instructed in how to perform the assembly method they have been assigned and then would begin assembling the new filtration systems using that method. Suppose assignment and training has been done, and the number of units assembled by each employee during one week is as shown in Table 13.3. The sample mean number of units produced with each of the three assembly methods is reported in the following table.

Assembly Method	Mean Number Produced
A	62
B	66
C	52

From these data, method B appears to result in higher production rates than either of the other methods.

The real issue is whether the three sample means observed are different enough for us to conclude that the means of the populations corresponding to the three methods of assembly are different. To write this question in statistical terms, we introduce the following notation.

$$\mu_1 = \text{mean number of units produced per week for method A}$$
$$\mu_2 = \text{mean number of units produced per week for method B}$$
$$\mu_3 = \text{mean number of units produced per week for method C}$$

Although we will never know the actual values of μ_1, μ_2, and μ_3, we want to use the sample means to test the following hypotheses.

$$H_0\colon \mu_1 = \mu_2 = \mu_3$$
H_a: Not all population means are equal

The problem we face in analyzing data from a completely randomized experimental design is the same problem we faced when we first introduced analysis of variance as a method for testing whether the means of more than two populations are equal. In the next section we

TABLE 13.3 NUMBER OF UNITS PRODUCED BY 15 WORKERS

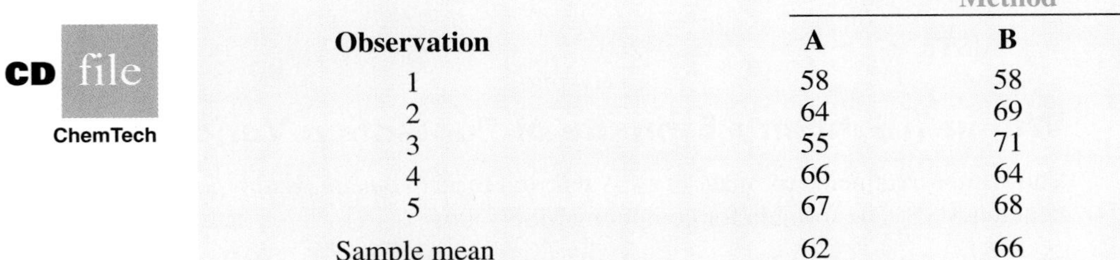

	Method		
Observation	A	B	C
1	58	58	48
2	64	69	57
3	55	71	59
4	66	64	47
5	67	68	49
Sample mean	62	66	52
Sample variance	27.5	26.5	31.0
Sample standard deviation	5.24	5.15	5.57

ChemTech

will show how analysis of variance is applied in problem situations such as the Chemitech assembly method.

NOTES AND COMMENTS

1. Randomization in experimental design is the analog of probability sampling in an observational study.
2. In many medical experiments, potential bias is eliminated by using a double-blind study. In such studies neither the physician applying the treatment nor the subject knows which treatment is being applied. Many other types of experiments could benefit from this type of study.

13.5 COMPLETELY RANDOMIZED DESIGNS

The hypotheses we want to test when analyzing the data from a completely randomized design are exactly the same as the general form of the hypotheses we presented in Section 13.2.

$$H_0: \mu_1 = \mu_2 = \cdots = \mu_k$$
$$H_a: \text{Not all population means are equal}$$

Hence, to test for the equality of means in situations where the data have been collected in a completely randomized experimental design, we can use analysis of variance as introduced in Sections 13.1 and 13.2. Recall that analysis of variance requires the calculation of two independent estimates of the population variance σ^2.

Between-Treatments Estimate of Population Variance

The between-treatments estimate of σ^2 is referred to as the *mean square due to treatments* and is denoted MSTR. The formula for computing MSTR follows:

$$\text{MSTR} = \frac{\sum_{j=1}^{k} n_j(\bar{x}_j - \bar{\bar{x}})^2}{k - 1} \tag{13.20}$$

The numerator in (13.20) is called the *sum of squares between* or *sum of squares due to treatments* and is denoted SSTR. The denominator $k - 1$ represents the degrees of freedom associated with SSTR.

For the Chemitech data in Table 13.3, we obtain the following results (note: $\bar{\bar{x}} = 60$).

$$\text{SSTR} = \sum_{j=1}^{k} n_j(\bar{x}_j - \bar{\bar{x}})^2 = 5(62 - 60)^2 + 5(66 - 60)^2 + 5(52 - 60)^2 = 520$$

$$\text{MSTR} = \frac{\text{SSTR}}{k - 1} = \frac{520}{3 - 1} = 260$$

Within-Treatments Estimate of Population Variance

The within-treatments estimate of σ^2 is referred to as the *mean square due to error* and is denoted MSE. The formula for computing MSE follows.

$$\text{MSE} = \frac{\sum_{j=1}^{k} (n_j - 1)s_j^2}{n_T - k} \tag{13.21}$$

The numerator in (13.21) is called the *sum of squares within* or *sum of squares due to error* and is denoted SSE. The denominator of MSE is referred to as the degrees of freedom associated with the within-treatment variance estimate.

For the Chemitech data in Table 13.3, we obtain the following results.

$$\text{SSE} = \sum_{j=1}^{k}(n_j - 1)s_j^2 = 4(27.5) + 4(26.5) + 4(31) = 340$$

$$\text{MSE} = \frac{\text{SSE}}{n_T - k} = \frac{340}{15 - 3} = 28.33$$

Comparing the Variance Estimates: The F Test

If the null hypothesis is true and the ANOVA assumptions are valid, the sampling distribution of MSTR/MSE is an F distribution with numerator degrees of freedom equal to $k - 1$ and denominator degrees of freedom equal to $n_T - k$. Recall also that if the means of the k populations are not equal, the value of MSTR/MSE will be inflated because MSTR overestimates σ^2. Hence we will reject H_0 if the resulting value of MSTR/MSE appears to be too large to have been selected at random from an F distribution with degrees of freedom $k - 1$ in the numerator and $n_T - k$ in the denominator.

For the Chemitech problem the value of $F = \text{MSTR/MSE} = 260/28.33 = 9.18$. The critical F value is based on two numerator degrees of freedom and 12 denominator degrees of freedom. For a .05 level of significance, Table 4 of Appendix B shows a value of $F_{.05} = 3.89$. When the observed value of F is greater than the critical value, we reject the null hypothesis and conclude that not all of the population means are equal.

ANOVA Table

We can now write the result that shows how the total sum of squares, SST, is partitioned.

$$\text{SST} = \text{SSTR} + \text{SSE} \qquad \textbf{(13.22)}$$

This result also holds true for the degrees of freedom associated with each of these sums of squares; that is, the total degrees of freedom is the sum of the degrees of freedom associated with SSTR and SSE. The general form of the ANOVA table for a completely randomized design is shown in Table 13.4; Table 13.5 is the corresponding ANOVA table for the Chemitech problem.

Pairwise Comparisons

We can use Fisher's LSD procedure to test all possible pairwise comparisons for the Chemitech problem. At the 5% level of significance, the t distribution table shows that with

TABLE 13.4 ANOVA TABLE FOR A COMPLETELY RANDOMIZED DESIGN

Source of Variation	Sum of Squares	Degrees of Freedom	Mean Square	F
Treatments	SSTR	$k - 1$	$\text{MSTR} = \frac{\text{SSTR}}{k - 1}$	$\frac{\text{MSTR}}{\text{MSE}}$
Error	SSE	$n_T - k$	$\text{MSE} = \frac{\text{SSE}}{n_T - k}$	
Total	SST	$n_T - 1$		

TABLE 13.5 ANOVA TABLE FOR THE CHEMITECH PROBLEM

Source of Variation	Sum of Squares	Degrees of Freedom	Mean Square	F
Treatments	520	2	260.00	9.18
Error	340	12	28.33	
Total	860	14		

$n_T - k = 15 - 3 = 12$ degrees of freedom, $t_{.025} = 2.179$. Using MSE = 28.33 in (13.17), we obtain Fisher's least significant difference.

$$\text{LSD} = t_{\alpha/2}\sqrt{\text{MSE}\left(\frac{1}{n_i} + \frac{1}{n_j}\right)} = 2.179\sqrt{28.33\left(\frac{1}{5} + \frac{1}{5}\right)} = 7.34$$

If the magnitude of the difference between any two sample means exceeds 7.34, we can reject the hypothesis that the corresponding population means are equal. For the Chemitech data in Table 13.3, we obtain the following results.

Sample Differences	Significant?
Method A − Method B = 62 − 66 = −4	No
Method A − Method C = 62 − 52 = 10	Yes
Method B − Method C = 66 − 52 = 14	Yes

Thus, the difference in the population means is attributable to the difference between the means for method A and method C and the difference between the means for method B and method C. Methods A and B therefore are preferred to method C. However, more testing should be done to compare method A with method B. The current study does not provide sufficient evidence to conclude that these two methods differ.

EXERCISES

Methods

19. The following data are from a completely randomized design.

Observation	Treatment A	Treatment B	Treatment C
1	162	142	126
2	142	156	122
3	165	124	138
4	145	142	140
5	148	136	150
6	174	152	128
$\bar{x}_j$	156	142	134
s_j^2	164.4	131.2	110.4

a. Compute the sum of squares between treatments.
b. Compute the mean square between treatments.

c. Compute the sum of squares due to error.
d. Compute the mean square due to error.
e. At the $\alpha = .05$ level of significance, test whether the means for the three treatments are equal.

20. Refer to Exercise 19.
a. Set up the ANOVA table.
b. At the $\alpha = .05$ level of significance, use Fisher's least significant difference procedure to test all possible pairwise comparisons. What conclusion can you draw after carrying out this procedure?

21. In a completely randomized experimental design, seven experimental units were used for each of the five levels of the factor. Complete the following ANOVA table.

Source of Variation	Sum of Squares	Degrees of Freedom	Mean Square	*F*
Treatments	300			
Error				
Total	460			

22. Refer to Exercise 21.
a. What hypotheses are implied in this problem?
b. At the $\alpha = .05$ level of significance, can we reject the null hypothesis in part (a)? Explain.

23. In an experiment designed to test the output levels of three different treatments, the following results were obtained: SST = 400, SSTR = 150, $n_T = 19$. Set up the ANOVA table and test for any significant difference between the mean output levels of the three treatments. Use $\alpha = .05$.

24. In a completely randomized experimental design, 12 experimental units were used for the first treatment, 15 for the second treatment, and 20 for the third treatment. Complete the following analysis of variance. At a .05 level of significance, is there a significant difference between the treatments?

Source of Variation	Sum of Squares	Degrees of Freedom	Mean Square	*F*
Treatments	1200			
Error				
Total	1800			

25. Develop the analysis of variance computations for the experimental design below. At $\alpha = .05$, is there a significant difference between the treatment means?

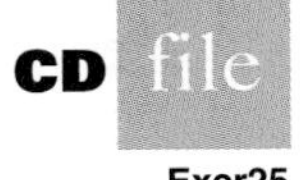

Exer25

	Treatment		
	A	B	C
	136	107	92
	120	114	82
	113	125	85
	107	104	101
	131	107	89
	114	109	117
	129	97	110
	102	114	120
		104	98
		89	106
$\bar{x}_j$	119	107	100
s_j^2	146.86	96.44	173.78

Applications

26. Three different methods for assembling a product were proposed by an industrial engineer. To investigate the number of units assembled correctly with each method, 30 employees were randomly selected and randomly assigned to the three proposed methods in such a way that each method was used by 10 workers. The number of units assembled correctly was recorded, and the analysis of variance procedure was applied to the resulting data set. The following results were obtained: SST = 10,800; SSTR = 4560.
    a. Set up the ANOVA table for this problem.
    b. Using $\alpha = .05$, test for any significant difference in the means for the three assembly methods.

27. In an experiment designed to test the breaking strength of four types of cables, the following results were obtained: SST = 85.05, SSTR = 61.64, $n_T = 24$. Set up the ANOVA table and test for any significant difference in the mean breaking strength of the four cables. Use $\alpha = .05$.

28. To study the effect of temperature on yield in a chemical process, five batches were produced at each of three temperature levels. The results follow. Construct an analysis of variance table. Using a .05 level of significance, test to see whether the temperature level appears to have an effect on the mean yield of the process.

Temperature		
50°C	60°C	70°C
34	30	23
24	31	28
36	34	28
39	23	30
32	27	31

29. Auditors must make judgments about various aspects of an audit on the basis of their own direct experience, indirect experience, or a combination of the two. In a study, auditors were asked to make judgments about the frequency of errors to be found in an audit. The judgments by the auditors were then compared to the actual results. Suppose the following data were obtained from a similar study; lower scores indicate better judgments.

CD file
AudJudg

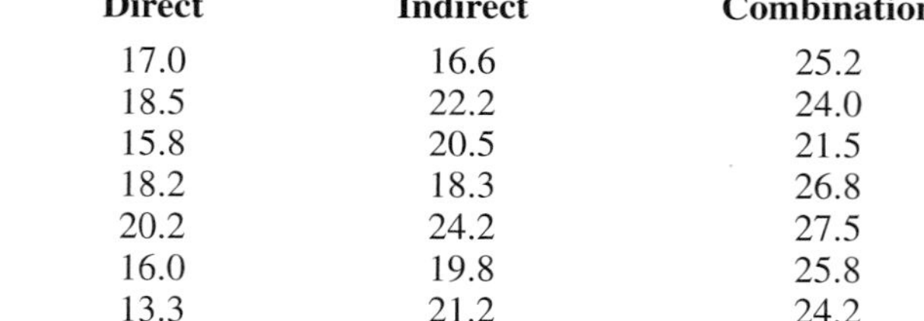

Direct	Indirect	Combination
17.0	16.6	25.2
18.5	22.2	24.0
15.8	20.5	21.5
18.2	18.3	26.8
20.2	24.2	27.5
16.0	19.8	25.8
13.3	21.2	24.2

Using $\alpha = .05$, test to see whether the basis for the judgment affects the quality of the judgment. What is your conclusion?

30. Four different paints are advertised as having the same drying time. To check the manufacturer's claims, five samples were tested for each of the paints. The time in minutes until the paint was dry enough for a second coat to be applied was recorded. The following data were obtained.

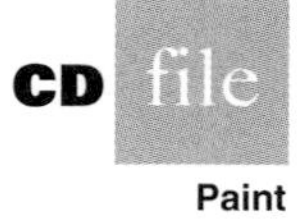

Paint

Paint 1	Paint 2	Paint 3	Paint 4
128	144	133	150
137	133	143	142
135	142	137	135
124	146	136	140
141	130	131	153

At the $\alpha = .05$ level of significance, test to see whether the mean drying time is the same for each type of paint.

31. Three top-of-the-line mid-size automobiles manufactured in the United States have been test-driven and compared on a variety of criteria by a well-known automotive magazine. In the area of gasoline mileage performance, five automobiles of each brand were each test-driven 500 miles; the miles per gallon data obtained follow. Use the analysis of variance procedure with $\alpha = .05$ to determine whether there is a significant difference in the mean number of miles per gallon for the three types of automobiles.

Automobile		
A	**B**	**C**
19	19	24
21	20	26
20	22	23
19	21	25
21	23	27

32. Refer to Exercise 29. Use Fisher's least significant difference procedure to test all possible pairwise comparisons. What conclusion can you draw after carrying out this procedure? Use $\alpha = .05$.

33. Refer to Exercise 31. Use Fisher's least significant difference procedure to test all possible pairwise comparisons. What conclusion can you draw after carrying out this procedure? Use $\alpha = .05$.

13.6 RANDOMIZED BLOCK DESIGN

Thus far we have considered the completely randomized experimental design. Recall that to test for a difference among treatment means, we computed an F value by using the ratio

$$F = \frac{\text{MSTR}}{\text{MSE}} \tag{13.23}$$

A completely randomized design is useful when the experimental units are homogeneous. If the experimental units are heterogeneous, blocking is often used to form homogeneous groups.

A problem can arise whenever differences due to extraneous factors (ones not considered in the experiment) cause the MSE term in this ratio to become large. In such cases, the F value in (13.23) can become small, signaling no difference among treatment means when in fact such a difference exists.

In this section we present an experimental design known as a **randomized block design**. Its purpose is to control some of the extraneous sources of variation by removing such variation from the MSE term. This design tends to provide a better estimate of the true error variance and leads to a more powerful hypothesis test in terms of the ability to detect differences among treatment means. To illustrate, let us consider a stress study for air traffic controllers.

Air Traffic Controller Stress Test

A study measuring the fatigue and stress of air traffic controllers has resulted in proposals for modification and redesign of the controller's work station. After consideration of several designs for the work station, three specific alternatives have been selected as having the best potential for reducing controller stress. The key question is: To what extent do the three alternatives differ in terms of their effect on controller stress? To answer this question, we need to design an experiment that will provide measurements of air traffic controller stress under each alternative.

Experimental studies in business often involve experimental units that are highly heterogeneous; as a result, randomized block designs are often employed.

Blocking in experimental design is similar to stratification in sampling.

In a completely randomized design, a random sample of controllers would be assigned to each work station alternative. However, controllers are believed to differ substantially in their ability to handle stressful situations. What is high stress to one controller might be only moderate or even low stress to another. Hence, when considering the within-group source of variation (MSE), we must realize that this variation includes both random error and error due to individual controller differences. In fact, managers expected controller variability to be a major contributor to the MSE term.

One way to separate the effect of the individual differences is to use a randomized block design. Such a design will identify the variability stemming from individual controller differences and remove it from the MSE term. The randomized block design calls for a single sample of controllers. Each controller in the sample is tested with each of the three work station alternatives. In experimental design terminology, the work station is the *factor of interest* and the controllers are the *blocks.* The three treatments or populations associated with the work station factor correspond to the three work station alternatives. For simplicity, we refer to the work station alternatives as system A, system B, and system C.

The *randomized* aspect of the randomized block design is the random order in which the treatments (systems) are assigned to the controllers. If every controller were to test the three systems in the same order, any observed difference in systems might be due to the order of the test rather than to true differences in the systems.

To provide the necessary data, the three work station alternatives were installed at the Cleveland Control Center in Oberlin, Ohio. Six controllers were selected at random and assigned to operate each of the systems. A follow-up interview and a medical examination of each controller participating in the study provided a measure of the stress for each controller on each system. The data are reported in Table 13.6.

Table 13.7 is a summary of the stress data collected. In this table we have included column totals (treatments) and row totals (blocks) as well as some sample means that will be helpful in making the sum of squares computations for the ANOVA procedure. Because lower stress values are viewed as better, the sample data seem to favor system B with its mean stress rating of 13. However, the usual question remains: Do the sample results justify the conclusion that the population mean stress levels for the three systems differ? That is, are the differences statistically significant? An analysis of variance computation similar to the one performed for the completely randomized design can be used to answer this statistical question.

ANOVA Procedure

The ANOVA procedure for the randomized block design requires us to partition the sum of squares total (SST) into three groups: sum of squares due to treatments, sum of

TABLE 13.6 A RANDOMIZED BLOCK DESIGN FOR THE AIR TRAFFIC CONTROLLER STRESS TEST

		Treatments		
		System A	**System B**	**System C**
Blocks	**Controller 1**	15	15	18
	Controller 2	14	14	14
	Controller 3	10	11	15
	Controller 4	13	12	17
	Controller 5	16	13	16
	Controller 6	13	13	13

TABLE 13.7 SUMMARY OF STRESS DATA FOR THE AIR TRAFFIC CONTROLLER STRESS TEST

		Treatments			Row or Block Totals	Block Means
		System A	System B	System C		
Blocks	Controller 1	15	15	18	48	$\bar{x}_{1\cdot} = 48/3 = 16.0$
	Controller 2	14	14	14	42	$\bar{x}_{2\cdot} = 42/3 = 14.0$
	Controller 3	10	11	15	36	$\bar{x}_{3\cdot} = 36/3 = 12.0$
	Controller 4	13	12	17	42	$\bar{x}_{4\cdot} = 42/3 = 14.0$
	Controller 5	16	13	16	45	$\bar{x}_{5\cdot} = 45/3 = 15.0$
	Controller 6	13	13	13	39	$\bar{x}_{6\cdot} = 39/3 = 13.0$
Column or Treatment Totals		81	78	93	252	$\bar{\bar{x}} = \frac{252}{18} = 14.0$
Treatment Means		$\bar{x}_{\cdot 1} = \frac{81}{6} = 13.5$	$\bar{x}_{\cdot 2} = \frac{78}{6} = 13.0$	$\bar{x}_{\cdot 3} = \frac{93}{6} = 15.5$		

squares due to blocks, and sum of squares due to error. The formula for this partitioning follows.

$$\text{SST} = \text{SSTR} + \text{SSBL} + \text{SSE} \tag{13.24}$$

This sum of squares partition is summarized in the ANOVA table for the randomized block design as shown in Table 13.8. The notation used in the table is

$$k = \text{the number of treatments}$$
$$b = \text{the number of blocks}$$
$$n_T = \text{the total sample size } (n_T = kb)$$

Note that the ANOVA table also shows how the $n_T - 1$ total degrees of freedom are partitioned such that $k - 1$ degrees of freedom go to treatments, $b - 1$ go to blocks, and $(k - 1)(b - 1)$ go to the error term. The mean square column shows the sum of squares

TABLE 13.8 ANOVA TABLE FOR THE RANDOMIZED BLOCK DESIGN WITH k TREATMENTS AND b BLOCKS

Source of Variation	Sum of Squares	Degrees of Freedom	Mean Square	F
Treatments	SSTR	$k - 1$	$\text{MSTR} = \frac{\text{SSTR}}{k - 1}$	$\frac{\text{MSTR}}{\text{MSE}}$
Blocks	SSBL	$b - 1$	$\text{MSBL} = \frac{\text{SSBL}}{b - 1}$	
Error	SSE	$(k - 1)(b - 1)$	$\text{MSE} = \frac{\text{SSE}}{(k - 1)(b - 1)}$	
Total	SST	$n_T - 1$		

divided by the degrees of freedom, and F = MSTR/MSE is the F ratio used to test for a significant difference among the treatment means. The primary contribution of the randomized block design is that, by including blocks, we have removed the individual controller differences from the MSE term and obtained a more powerful test for the stress differences in the three work station alternatives.

Computations and Conclusions

To compute the F statistic needed to test for a difference among treatment means with a randomized block design, we need to compute MSTR and MSE. To calculate these two mean squares, we must first compute SSTR and SSE; in doing so, we will also compute SSBL and SST. To simplify the presentation, we perform the calculations in four steps. In addition to k, b, and n_T as previously defined, the following notation is used.

$$
\begin{aligned}
x_{ij} &= \text{value of the observation corresponding to treatment } j \text{ in block } i \\
\bar{x}_{\cdot j} &= \text{sample mean of the } j\text{th treatment} \\
\bar{x}_{i\cdot} &= \text{sample mean for the } i\text{th block} \\
\bar{\bar{x}} &= \text{overall sample mean}
\end{aligned}
$$

Step 1. Compute the total sum of squares (SST).

$$\text{SST} = \sum_{i=1}^{b}\sum_{j=1}^{k}(x_{ij} - \bar{\bar{x}})^2 \tag{13.25}$$

Step 2. Compute the sum of squares due to treatments (SSTR).

$$\text{SSTR} = b\sum_{j=1}^{k}(\bar{x}_{\cdot j} - \bar{\bar{x}})^2 \tag{13.26}$$

Step 3. Compute the sum of squares due to blocks (SSBL).

$$\text{SSBL} = k\sum_{i=1}^{b}(\bar{x}_{i\cdot} - \bar{\bar{x}})^2 \tag{13.27}$$

Step 4. Compute the sum of squares due to error (SSE).

$$\text{SSE} = \text{SST} - \text{SSTR} - \text{SSBL} \tag{13.28}$$

For the air traffic controller data in Table 13.7, these steps lead to the following sums of squares.

Step 1. $\text{SST} = (15 - 14)^2 + (15 - 14)^2 + (18 - 14)^2 + \cdots + (13 - 14)^2 = 70$
Step 2. $\text{SSTR} = 6[(13.5 - 14)^2 + (13.0 - 14)^2 + (15.5 - 14)^2] = 21$
Step 3. $\text{SSBL} = 3[(16 - 14)^2 + (14 - 14)^2 + (12 - 14)^2 + (14 - 14)^2 + (15 - 14)^2 + (13 - 14)^2] = 30$
Step 4. $\text{SSE} = 70 - 21 - 30 = 19$

These sums of squares divided by their degrees of freedom provide the corresponding mean square values shown in Table 13.9. The F ratio used to test for differences between treatment means is MSTR/MSE = 10.5/1.9 = 5.53. Checking the F values in Table 4 of

TABLE 13.9 ANOVA TABLE FOR THE AIR TRAFFIC CONTROLLER STRESS TEST

Source of Variation	Sum of Squares	Degrees of Freedom	Mean Square	F
Treatments	21	2	10.5	10.5/1.9 = 5.53
Blocks	30	5	6.0	
Error	19	10	1.9	
Total	70	17		

Appendix B, we find that the critical F value at $\alpha = .05$ (two numerator degrees of freedom and 10 denominator degrees of freedom) is 4.10. With $F = 5.53$, we reject the null hypothesis H_0: $\mu_1 = \mu_2 = \mu_3$ and conclude that the population mean stress levels differ for the three work station alternatives.

Some general comments can be made about the randomized block design. The experimental design described in this section is a *complete* block design; the word "complete" indicates that each block is subjected to all k treatments. That is, all controllers (blocks) were tested with all three systems (treatments). Experimental designs in which some but not all treatments are applied to each block are referred to as *incomplete* block designs. A discussion of incomplete block designs is beyond the scope of this text.

Because each controller in the air traffic controller stress test was required to use all three systems, this approach guarantees a complete block design. In some cases, however, blocking is carried out with "similar" experimental units in each block. For example, assume that in a pretest of air traffic controllers, the population of controllers was divided into groups ranging from extremely high-stress individuals to extremely low-stress individuals. The blocking could still be accomplished by having three controllers from each of the stress classifications participate in the study. Each block would then consist of three controllers in the same stress group. The randomized aspect of the block design would be the random assignment of the three controllers in each block to the three systems.

Finally, note that the ANOVA table shown in Table 13.8 provides an F value to test for treatment effects but *not* for blocks. The reason is that the experiment was designed to test a single factor—work station design. The blocking based on individual stress differences was conducted to remove such variation from the MSE term. However, the study was not designed to test specifically for individual differences in stress.

Some analysts compute F = MSB/MSE and use that statistic to test for significance of the blocks. Then they use the result as a guide to whether the same type of blocking would be desired in future experiments. However, if individual stress difference is to be a factor in the study, a different experimental design should be used. A test of significance on blocks should not be performed as a basis for a conclusion about a second factor.

NOTES AND COMMENTS

1. The matched-samples t test introduced in Chapter 10 is an example of a randomized block design with two blocks.
2. The error degrees of freedom are less for a randomized block design than for a completely randomized design because $b - 1$ degrees of freedom are lost for the b blocks. If n is small, the potential effects due to blocks can be masked because of the loss of error degrees of freedom; for large n, the effects are minimized.

EXERCISES

Methods

34. Consider the experimental results for the following randomized block design. Make the calculations necessary to set up the analysis of variance table.

		Treatments		
		A	**B**	**C**
Blocks	**1**	10	9	8
	2	12	6	5
	3	18	15	14
	4	20	18	18
	5	8	7	8

Using $\alpha = .05$, test for any significant differences.

35. The following data were obtained for a randomized block design involving five treatments and three blocks: SST = 430, SSTR = 310, SSBL = 85. Set up the ANOVA table and test for any significant differences. Use $\alpha = .05$.

36. An experiment has been conducted for four treatments with eight blocks. Complete the following analysis of variance table.

Source of Variation	Sum of Squares	Degrees of Freedom	Mean Square	F
Treatments	900			
Blocks	400			
Error				
Total	1800			

Using $\alpha = .05$, test for any significant differences.

Applications

37. An automobile dealer conducted a test to determine if the time in minutes needed to complete a minor engine tune-up depends on whether a computerized engine analyzer or an electronic analyzer is used. Because tune-up time varies among compact, intermediate, and full-size cars, the three types of cars were used as blocks in the experiment. The data obtained follow.

		Analyzer	
		Computerized	**Electronic**
Car	**Compact**	50	42
	Intermediate	55	44
	Full-size	63	46

Using $\alpha = .05$, test for any significant differences.

38. Five different auditing procedures were compared in terms of total audit time. To control for possible variation due to the person conducting the audit, four accountants were selected randomly and treated as blocks in the experiment. The following values were obtained by the ANOVA procedure: SST = 100, SSTR = 45, SSBL = 36. Using $\alpha = .05$, test for any significant difference in the mean total audit time for the five auditing procedures.

39. An important factor in selecting software for word-processing and database management systems is the time required to learn how to use the system. To evaluate three file management systems, a firm designed a test involving five word-processing operators. Since operator variability was believed to be a significant factor, each of the five operators was trained on each of the three file management systems. The data obtained follow.

		System		
		A	B	C
Operator	1	16	16	24
	2	19	17	22
	3	14	13	19
	4	13	12	18
	5	18	17	22

Using $\alpha = .05$, test for any difference in the mean training time (in hours) for the three systems.

40. A study reported in the *Journal of the American Medical Association* investigated the cardiac demands of heavy snow shoveling. Ten healthy men underwent exercise testing with a treadmill and a cycle ergometer modified for arm cranking. The men then cleared two tracts of heavy, wet snow by using a lightweight plastic snow shovel and an electric snow thrower. Each subject's heart rate, blood pressure, oxygen uptake, and perceived exertion during snow removal were compared with the values obtained during treadmill and arm-crank ergometer testing. Suppose the following table gives the heart rates in beats per minute for each of the 10 subjects.

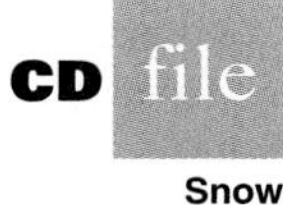

Snow

Subject	Treadmill	Arm-Crank Ergometer	Snow Shovel	Snow Thrower
1	177	205	180	98
2	151	177	164	120
3	184	166	167	111
4	161	152	173	122
5	192	142	179	151
6	193	172	205	158
7	164	191	156	117
8	207	170	160	123
9	177	181	175	127
10	174	154	191	109

At the .05 level of significance, test for any significant differences.

13.7 FACTORIAL EXPERIMENTS

The experimental designs we have considered thus far enable us to draw statistical conclusions about one factor. However, in some experiments we want to draw conclusions about more than one variable or factor. **Factorial experiments** and their corresponding ANOVA computations are valuable designs when simultaneous conclusions about two or more factors are required. The term *factorial* is used because the experimental conditions include all possible combinations of the factors. For example, for a levels of factor A and b levels of factor B, the experiment will involve collecting data on ab treatment combinations. In this section we will show the analysis for a two-factor factorial experiment. The basic approach can be extended to experiments involving more than two factors.

As an illustration of a two-factor factorial experiment, we will consider a study involving the Graduate Management Admissions Test (GMAT), a standardized test used by graduate schools of business to evaluate an applicant's ability to pursue a graduate program in that field. Scores on the GMAT range from 200 to 800, with higher scores implying higher aptitude.

In an attempt to improve students' performance on the GMAT exam, a major Texas university is considering offering the following three GMAT preparation programs.

1. A 3-hour review session covering the types of questions generally asked on the GMAT.
2. A 1-day program covering relevant exam material, along with the taking and grading of a sample exam.
3. An intensive 10-week course involving the identification of each student's weaknesses and the setting up of individualized programs for improvement.

Hence, one factor in this study is the GMAT preparation program, which has three treatments: 3-hour review, 1-day program, and 10-week course. Before selecting the preparation program to adopt, further study will be conducted to determine how the proposed programs affect GMAT scores.

The GMAT is usually taken by students from three colleges: the College of Business, the College of Engineering, and the College of Arts and Sciences. Therefore, a second factor of interest in the experiment is whether a student's undergraduate college affects the GMAT score. This second factor, undergraduate college, also has three treatments: business, engineering, and arts and sciences. The factorial design for this experiment with three treatments corresponding to factor A, the preparation program, and three treatments corresponding to factor B, the undergraduate college, will have a total of $3 \times 3 = 9$ treatment combinations. These treatment combinations or experimental conditions are summarized in Table 13.10.

Assume that a sample of two students will be selected corresponding to each of the nine treatment combinations shown in Table 13.10: two business students will take the 3-hour review, two will take the 1-day program, and two will take the 10-week course. In addition, two engineering students and two arts and sciences students will take each of the three preparation programs. In experimental design terminology, the sample size of two for each treatment combination indicates that we have two **replications.** Additional replications and a larger sample size could easily have been used, but we elected to minimize the computational aspects for this illustration.

This experimental design requires that six students who plan to attend graduate school be randomly selected from *each* of the three undergraduate colleges. Then two students from each college should be assigned randomly to each preparation program, resulting in a total of 18 students being used in the study.

Let us assume that the students have been randomly selected, have participated in the preparation program, and have taken the GMAT. The scores obtained are reported in Table 13.11.

The analysis of variance computations with the data in Table 13.11 will provide answers to the following questions.

- **Main effect (factor A):** Do the preparation programs differ in terms of effect on GMAT scores?
- **Main effect (factor B):** Do the undergraduate colleges differ in terms of effect on GMAT scores?
- **Interaction effect (factors A and B):** Do students in some colleges do better on one type of preparation program whereas others do better on a different type of preparation program?

The term **interaction** refers to a new effect that we can now study because we have used a factorial experiment. If the interaction effect has a significant impact on the GMAT scores, we can conclude that the effect of the type of preparation program depends on the undergraduate college.

TABLE 13.10 NINE TREATMENT COMBINATIONS FOR THE TWO-FACTOR GMAT EXPERIMENT

		Factor B: College		
		Business	**Engineering**	**Arts and Sciences**
Factor A:	3-hour review	1	2	3
Preparation	1-day program	4	5	6
Program	10-week course	7	8	9

TABLE 13.11 GMAT SCORES FOR THE TWO-FACTOR EXPERIMENT

CD file
GMAT

		Factor B: College		
		Business	**Engineering**	**Arts and Sciences**
Factor A: Preparation Program	**3-hour review**	500 580	540 460	480 400
	1-day program	460 540	560 620	420 480
	10-week course	560 600	600 580	480 410

ANOVA Procedure

The ANOVA procedure for the two-factor factorial experiment is similar to the completely randomized experiment and the randomized block experiment in that we again partition the sum of squares and the degrees of freedom into their respective sources. The formula for partitioning the sum of squares for the two-factor factorial experiments follows.

$$\text{SST} = \text{SSA} + \text{SSB} + \text{SSAB} + \text{SSE} \tag{13.29}$$

The partitioning of the sum of squares and degrees of freedom is summarized in Table 13.12. The following notation is used.

a = number of levels of factor A
b = number of levels of factor B
r = number of replications
n_T = total number of observations taken in the experiment; $n_T = abr$

Computations and Conclusions

To compute the F statistics needed to test for the significance of factor A, factor B, and the interaction, we need to compute MSA, MSB, MSAB, and MSE. To calculate these four mean

TABLE 13.12 ANOVA TABLE FOR THE TWO-FACTOR FACTORIAL EXPERIMENT WITH r REPLICATIONS

Source of Variation	Sum of Squares	Degrees of Freedom	Mean Square	F
Factor A	SSA	$a-1$	$\text{MSA} = \dfrac{\text{SSA}}{a-1}$	$\dfrac{\text{MSA}}{\text{MSE}}$
Factor B	SSB	$b-1$	$\text{MSB} = \dfrac{\text{SSB}}{b-1}$	$\dfrac{\text{MSB}}{\text{MSE}}$
Interaction	SSAB	$(a-1)(b-1)$	$\text{MSAB} = \dfrac{\text{SSAB}}{(a-1)(b-1)}$	$\dfrac{\text{MSAB}}{\text{MSE}}$
Error	SSE	$ab(r-1)$	$\text{MSE} = \dfrac{\text{SSE}}{ab(r-1)}$	
Total	SST	n_T-1		

squares, we must first compute SSA, SSB, SSAB, and SSE; in doing so we will also compute SST. To simplify the presentation, we perform the calculations in five steps. In addition to a, b, r, and n_T as previously defined, the following notation is used.

x_{ijk} = observation corresponding to the kth replicate taken from treatment i of factor A and treatment j of factor B

$\bar{x}_{i\cdot}$ = sample mean for the observations in treatment i (factor A)

$\bar{x}_{\cdot j}$ = sample mean for the observations in treatment j (factor B)

$\bar{x}_{ij}$ = sample mean for the observations corresponding to the combination of treatment i (factor A) and treatment j (factor B)

$\bar{\bar{x}}$ = overall sample mean of all n_T observations

Step 1. Compute the total sum of squares.

$$\text{SST} = \sum_{i=1}^{a}\sum_{j=1}^{b}\sum_{k=1}^{r}(x_{ijk} - \bar{\bar{x}})^2 \tag{13.30}$$

Step 2. Compute the sum of squares for factor A.

$$\text{SSA} = br\sum_{i=1}^{a}(\bar{x}_{i\cdot} - \bar{\bar{x}})^2 \tag{13.31}$$

Step 3. Compute the sum of squares for factor B.

$$\text{SSB} = ar\sum_{j=1}^{b}(\bar{x}_{\cdot j} - \bar{\bar{x}})^2 \tag{13.32}$$

Step 4. Compute the sum of squares for interaction.

$$\text{SSAB} = r\sum_{i=1}^{a}\sum_{j=1}^{b}(\bar{x}_{ij} - \bar{x}_{i\cdot} - \bar{x}_{\cdot j} + \bar{\bar{x}})^2 \tag{13.33}$$

Step 5. Compute the sum of squares due to error.

$$\text{SSE} = \text{SST} - \text{SSA} - \text{SSB} - \text{SSAB} \tag{13.34}$$

Table 13.13 reports the data collected in the experiment and the various sums that will help us with the sum of squares computations. Using (13.30) to (13.34), we have the following sums of squares for the GMAT two-factor factorial experiment.

Step 1. $\text{SST} = (500 - 515)^2 + (580 - 515)^2 + (540 - 515)^2 + \cdots + (410 - 515)^2 = 82{,}450$

Step 2. $\text{SSA} = (3)(2)[(493.33 - 515)^2 + (513.33 - 515)^2 + (538.33 - 515)^2] = 6100$

Step 3. $\text{SSB} = (3)(2)[(540 - 515)^2 + (560 - 515)^2 + (445 - 515)^2] = 45{,}300$

Step 4. $\text{SSAB} = 2[(540 - 493.33 - 540 + 515)^2 + (500 - 493.33 - 560 + 515)^2 + \cdots + (445 - 538.33 - 445 + 515)^2] = 11{,}200$

Step 5. $\text{SSE} = 82{,}450 - 6100 - 45{,}300 - 11{,}200 = 19{,}850$

TABLE 13.13 GMAT SUMMARY DATA FOR THE TWO-FACTOR EXPERIMENT

		Factor B: College				
Treatment combination totals (→ 1080)		Business	Engineering	Arts and Sciences	Row Totals	Factor A Means
Factor A: Preparation Program	3-hour review	500 580 1080 $\bar{x}_{11} = \frac{1080}{2} = 540$	540 460 1000 $\bar{x}_{12} = \frac{1000}{2} = 500$	480 400 880 $\bar{x}_{13} = \frac{880}{2} = 440$	2960	$\bar{x}_{1\cdot} = \frac{2960}{6} = 493.33$
	1-day program	460 540 1000 $\bar{x}_{21} = \frac{1000}{2} = 500$	560 620 1180 $\bar{x}_{22} = \frac{1180}{2} = 590$	420 480 900 $\bar{x}_{23} = \frac{900}{2} = 450$	3080	$\bar{x}_{2\cdot} = \frac{3080}{6} = 513.33$
	10-week course	560 600 1160 $\bar{x}_{31} = \frac{1160}{2} = 580$	600 580 1180 $\bar{x}_{32} = \frac{1180}{2} = 590$	480 410 890 $\bar{x}_{33} = \frac{890}{2} = 445$	3230	$\bar{x}_{3\cdot} = \frac{3230}{6} = 538.33$
Column Totals		3240	3360	2670	9270 ← Overall total	
Factor B Means		$\bar{x}_{\cdot 1} = \frac{3240}{6} = 540$	$\bar{x}_{\cdot 2} = \frac{3360}{6} = 560$	$\bar{x}_{\cdot 3} = \frac{2670}{6} = 445$	$\bar{\bar{x}} = \frac{9270}{18} = 515$	

TABLE 13.14 ANOVA TABLE FOR THE TWO-FACTOR GMAT STUDY

Source of Variation	Sum of Squares	Degrees of Freedom	Mean Square	F
Factor A	6,100	2	3,050	3,050/2206 = 1.38
Factor B	45,300	2	22,650	22,650/2206 = 10.27
Interaction	11,200	4	2,800	2,800/2206 = 1.27
Error	19,850	9	2,206	
Total	82,450	17		

These sums of squares divided by their corresponding degrees of freedom, as shown in Table 13.14, provide the appropriate mean square values for testing the two main effects (preparation program and undergraduate college) and the interaction effect. The F ratio used to test for differences among preparation programs is 1.38. The critical F value at $\alpha = .05$ (with two numerator degrees of freedom and nine denominator degrees of freedom) is 4.26. With $F = 1.38$, we cannot reject the null hypothesis and must conclude that there is no significant difference among the three preparation programs. However, for the undergraduate college effect, $F = 10.27$ exceeds the critical F value of 4.26. Hence, the analysis of variance results enables us to conclude that there is a difference in GMAT test scores among the three undergraduate colleges; that is, the three undergraduate colleges do not provide the same preparation for performance on the GMAT. Finally, the interaction F value of $F = 1.27$ (critical F value $= 3.63$ at $\alpha = .05$) means that we cannot identify a significant interaction effect. Therefore, we have no reason to believe that the three preparation programs differ in their ability to prepare students from the different colleges for the GMAT.

Undergraduate college was found to be a significant factor. Checking the calculations in Table 13.13, we see that the sample means are: business students $\bar{x}_{\cdot 1} = 540$, engineering students $\bar{x}_{\cdot 2} = 560$, and arts and sciences students $\bar{x}_{\cdot 3} = 445$. Tests on individual treatment means can be conducted; yet after reviewing the three sample means, we would anticipate no difference in preparation for business and engineering graduates. However, the arts and sciences students appear to be significantly less prepared for the GMAT than students in the other colleges. Perhaps this observation will lead the university to consider other options for assisting these students in preparing for graduate management admission tests.

Because of the computational effort involved in any modest- to large-size factorial experiment, the computer usually plays an important role in performing and summarizing the analysis of variance computations. Figure 13.6 is the Minitab computer printout for the analysis of variance of the GMAT two-factor factorial experiment.

FIGURE 13.6 MINITAB OUTPUT FOR THE GMAT TWO-FACTOR DESIGN

```
SOURCE          DF       SS       MS       F        P
Factor A         2     6100     3050    1.38    0.299
Factor B         2    45300    22650   10.27    0.005
Interaction      4    11200     2800    1.27    0.350
Error            9    19850     2206
Total           17    82450
```

EXERCISES

Methods

41. A factorial experiment involving two levels of factor A and three levels of factor B resulted in the following data.

		Factor B		
		Level 1	Level 2	Level 3
Factor A	Level 1	135 165	90 66	75 93
	Level 2	125 95	127 105	120 136

Test for any significant main effect and any interaction. Use $\alpha = .05$.

42. The calculations for a factorial experiment involving four levels of factor A, three levels of factor B, and three replications resulted in the following data: SST = 280, SSA = 26, SSB = 23, SSAB = 175. Set up the ANOVA table and test for any significant main effects and any interaction effect. Use $\alpha = .05$.

Applications

43. A mail-order catalog firm designed a factorial experiment to test the effect of the size of a magazine advertisement and the advertisement design on the number of catalog requests received (1000s). Three advertising designs and two different-size advertisements were considered. The data obtained follow. Use the ANOVA procedure for factorial designs to test for any significant effects due to type of design, size of advertisement, or interaction. Use $\alpha = .05$.

		Size of Advertisement	
		Small	Large
Design	A	8 12	12 8
	B	22 14	26 30
	C	10 18	18 14

44. An amusement park has been studying methods for decreasing the waiting time (minutes) for rides by loading and unloading riders more efficiently. Two alternative loading/unloading methods have been proposed. To account for potential differences due to the type of ride and the possible interaction between the method of loading and unloading and the type of ride, a factorial experiment was designed. Using the following data, test for any significant effect due to the loading and unloading method, the type of ride, and interaction. Use $\alpha = .05$.

	Type of Ride		
	Roller Coaster	Screaming Demon	Log Flume
Method 1	41 43	52 44	50 46
Method 2	49 51	50 46	48 44

45. The U.S. Bureau of Labor Statistics collects information on the earnings of men and women for different occupations. Suppose that a reporter for *The Tampa Tribune* wanted to investigate whether there were any differences between the weekly salaries of men and women employed as financial managers, computer programmers, and pharmacists. A sample of five men and women was selected from each of the three occupations, and the weekly salary for each individual in the sample was recorded. The data obtained follow.

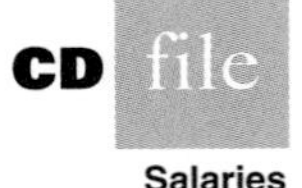
Salaries

Weekly Salary ($)	Occupation	Gender
872	Financial Manager	Male
859	Financial Manager	Male
1028	Financial Manager	Male
1117	Financial Manager	Male
1019	Financial Manager	Male
519	Financial Manager	Female
702	Financial Manager	Female
805	Financial Manager	Female
558	Financial Manager	Female
591	Financial Manager	Female
747	Computer Programmer	Male
766	Computer Programmer	Male
901	Computer Programmer	Male
690	Computer Programmer	Male
881	Computer Programmer	Male
884	Computer Programmer	Female
765	Computer Programmer	Female
685	Computer Programmer	Female
700	Computer Programmer	Female
671	Computer Programmer	Female
1105	Pharmacist	Male
1144	Pharmacist	Male
1085	Pharmacist	Male
903	Pharmacist	Male
998	Pharmacist	Male
813	Pharmacist	Female
985	Pharmacist	Female
1006	Pharmacist	Female
1034	Pharmacist	Female
817	Pharmacist	Female

At the $\alpha = .05$ level of significance, test for any significant effect due to occupation, gender, and interaction.

46. A study reported in *The Accounting Review* examined the separate and joint effects of two levels of time pressure (low and moderate) and three levels of knowledge (naive, declarative, and procedural) on key word selection behavior in tax research. Subjects were given a tax case containing a set of facts, a tax issue, and a key word index consisting of 1336 key words. They were asked to select the key words they believed would refer them to a tax authority relevant to resolving the tax case. Prior to the experiment, a group of tax experts had determined that there were 19 relevant key words in the index. Subjects in the naive group had little or no declarative or procedural knowledge, subjects in the declarative group had significant declarative knowledge but little or no procedural knowledge, and subjects in the procedural group had significant declarative knowledge and procedural knowledge. Declarative knowledge consists of knowledge of both the applicable tax rules and the technical terms used to describe such rules. Procedural knowledge is knowledge of the rules that guide the tax researcher's search for relevant key words. Subjects in the low time pressure situation were told they had 25 minutes to complete the problem, an amount of time which should be "more than adequate" to complete the case; subjects in

the moderate time pressure situation were told they would have "only" 11 minutes to complete the case (*The Accounting Review,* January 1995). Suppose 25 subjects were selected for each of the six treatment combinations and the sample means for each treatment combination are as follows (standard deviations are in parentheses).

		Knowledge		
		Naive	**Declarative**	**Procedural**
Time Pressure	**Low**	1.13 (1.12)	1.56 (1.33)	2.00 (1.54)
	Moderate	0.48 (0.80)	1.68 (1.36)	2.86 (1.80)

Use the ANOVA procedure to test for any significant differences due to time pressure, knowledge, and interaction. Use a .05 level of significance. Assume that the total sum of squares for this experiment is 327.50.

SUMMARY

In this chapter we have shown how analysis of variance can be used to test for differences among means of several populations or treatments. We introduced the completely randomized design, the randomized block design, and the two-factor factorial experiment. The completely randomized design and the randomized block design are used to draw conclusions about differences in the means of a single factor. The primary purpose of blocking in the randomized block design is to remove extraneous sources of variation from the error term. Such blocking provides a better estimate of the true error variance and a better test to determine whether the population or treatment means of the factor differ significantly.

We showed that the basis for the statistical tests used in analysis of variance and experimental design is the development of two independent estimates of the population variance σ^2. In the single-factor case, one estimator is based on the variation between the treatments; this estimator provides an unbiased estimate of σ^2 only if the means $\mu_1, \mu_2, \ldots, \mu_k$ are all equal. A second estimator of σ^2 is based on the variation of the observations within each sample; this estimator will always provide an unbiased estimate of σ^2. By computing the ratio of these two estimators (the F statistic) we developed a rejection rule for determining whether to reject the null hypothesis that the population or treatment means are equal. In all the experimental designs considered, the partitioning of the sum of squares and degrees of freedom into their various sources enabled us to compute the appropriate values for the analysis of variance calculations and tests. We also showed how Fisher's LSD procedure and the Bonferroni adjustment can be used to perform pairwise comparisons to determine which means are different.

GLOSSARY

ANOVA table A table used to summarize the analysis of variance computations and results. It contains columns showing the source of variation, the sum of squares, the degrees of freedom, the mean square, and the F value(s).

Partitioning The process of allocating the total sum of squares and degrees of freedom to the various components.

Multiple comparison procedures Statistical procedures that can be used to conduct statistical comparisons between pairs of population means.

Comparisonwise Type I error rate The probability of a Type I error associated with a single pairwise comparison.

Experimentwise Type I error rate The probability of making a Type I error on at least one of several pairwise comparisons.

Factor Another word for the independent variable of interest.

Treatments Different levels of a factor.

Single-factor experiment An experiment involving only one factor with k populations or treatments.

Experimental units The objects of interest in the experiment.

Completely randomized design An experimental design in which the treatments are randomly assigned to the experimental units.

Blocking The process of using the same or similar experimental units for all treatments. The purpose of blocking is to remove a source of variation from the error term and hence provide a more powerful test for a difference in population or treatment means.

Randomized block design An experimental design employing blocking.

Factorial experiment An experimental design that allows statistical conclusions about two or more factors.

Replication The number of times each experimental condition is repeated in an experiment.

Interaction The effect produced when the levels of one factor interact with the levels of another factor in influencing the response variable.

KEY FORMULAS

Testing for the Equality of k Population Means

Sample Mean for Treatment j

$$\bar{x}_j = \frac{\sum_{i=1}^{n_j} x_{ij}}{n_j} \tag{13.1}$$

Sample Variance for Treatment j

$$s_j^2 = \frac{\sum_{i=1}^{n_j} (x_{ij} - \bar{x}_j)^2}{n_j - 1} \tag{13.2}$$

Overall Sample Mean

$$\bar{\bar{x}} = \frac{\sum_{j=1}^{k} \sum_{i=1}^{n_j} x_{ij}}{n_T} \tag{13.3}$$

$$n_T = n_1 + n_2 + \cdots + n_k \tag{13.4}$$

Mean Square Due to Treatments

$$\text{MSTR} = \frac{\text{SSTR}}{k - 1} \tag{13.7}$$

Sum of Squares Due to Treatments

$$\text{SSTR} = \sum_{j=1}^{k} n_j(\bar{x}_j - \bar{\bar{x}})^2 \tag{13.8}$$

Mean Square Due to Error

$$\text{MSE} = \frac{\text{SSE}}{n_T - k} \tag{13.10}$$

Sum of Squares Due to Error

$$\text{SSE} = \sum_{j=1}^{k} (n_j - 1)s_j^2 \tag{13.11}$$

Test Statistic for the Equality of *k* Population Means

$$F = \frac{\text{MSTR}}{\text{MSE}} \tag{13.12}$$

Total Sum of Squares

$$\text{SST} = \sum_{j=1}^{k} \sum_{i=1}^{n_j} (x_{ij} - \bar{\bar{x}})^2 \tag{13.13}$$

Partition of Sum of Squares

$$\text{SST} = \text{SSTR} + \text{SSE} \tag{13.14}$$

Multiple Comparison Procedures

Test Statistic for Fisher's LSD

$$t = \frac{\bar{x}_i - \bar{x}_j}{\sqrt{\text{MSE}\left(\frac{1}{n_i} + \frac{1}{n_j}\right)}} \tag{13.16}$$

Fisher's LSD

$$\text{LSD} = t_{\alpha/2}\sqrt{\text{MSE}\left(\frac{1}{n_i} + \frac{1}{n_j}\right)} \tag{13.17}$$

Completely Randomized Designs

Mean Square Due to Treatments

$$\text{MSTR} = \frac{\sum_{j=1}^{k} n_j(\bar{x}_j - \bar{\bar{x}})^2}{k - 1} \tag{13.20}$$

Mean Square Due to Error

$$\text{MSE} = \frac{\sum_{j=1}^{k} (n_j - 1)s_j^2}{n_T - k} \tag{13.21}$$

F **Test Statistic**

$$F = \frac{\text{MSTR}}{\text{MSE}} \tag{13.23}$$

Randomized Block Designs

Total Sum of Squares

$$\text{SST} = \sum_{i=1}^{b}\sum_{j=1}^{k}(x_{ij} - \bar{\bar{x}})^2 \tag{13.25}$$

Sum of Squares Due to Treatments

$$\text{SSTR} = b\sum_{j=1}^{k}(\bar{x}_{\cdot j} - \bar{\bar{x}})^2 \tag{13.26}$$

Sum of Squares Due to Blocks

$$\text{SSBL} = k\sum_{i=1}^{b}(\bar{x}_{i\cdot} - \bar{\bar{x}})^2 \tag{13.27}$$

Sum of Squares Due to Error

$$\text{SSE} = \text{SST} - \text{SSTR} - \text{SSBL} \tag{13.28}$$

Factorial Experiments

Total Sum of Squares

$$\text{SST} = \sum_{i=1}^{a}\sum_{j=1}^{b}\sum_{k=1}^{r}(x_{ijk} - \bar{\bar{x}})^2 \tag{13.30}$$

Sum of Squares for Factor A

$$\text{SSA} = br\sum_{i=1}^{a}(\bar{x}_{i\cdot} - \bar{\bar{x}})^2 \tag{13.31}$$

Sum of Squares for Factor B

$$\text{SSB} = ar\sum_{j=1}^{b}(\bar{x}_{\cdot j} - \bar{\bar{x}})^2 \tag{13.32}$$

Sum of Squares for Interaction

$$\text{SSAB} = r\sum_{i=1}^{a}\sum_{j=1}^{b}(\bar{x}_{ij} - \bar{x}_{i\cdot} - \bar{x}_{\cdot j} + \bar{\bar{x}})^2 \tag{13.33}$$

Sum of Squares for Error

$$\text{SSE} = \text{SST} - \text{SSA} - \text{SSB} - \text{SSAB} \tag{13.34}$$

SUPPLEMENTARY EXERCISES

47. A simple random sample of the asking prices ($1000s) of four houses currently for sale in each of two residential areas resulted in the following data.

Area 1	Area 2
92	90
89	102
98	96
105	88

a. Use the procedure developed in Chapter 10 to test whether the mean asking price is the same in both areas. Use $\alpha = .05$.

b. Use the ANOVA procedure to test whether the mean asking price is the same. Compare your analysis with part (a). Use $\alpha = .05$.

c. Suppose that data were collected for another residential area. The asking prices for the simple random sample from the third area were $81,000, $86,000, $75,000, and $90,000. Is the mean asking price the same for all three areas? Use $\alpha = .05$.

48. Buyers of sport utility vehicles (SUVs) and pickup trucks find a wide choice in today's marketplace. One of the factors that is important to many buyers is the resale value of the vehicle. The following table shows the resale value (%) after 2 years for 10 SUVs, 10 small pickup trucks, and 10 large pickup trucks (*Kiplinger's New Cars & Trucks 2000 Buyer's Guide*).

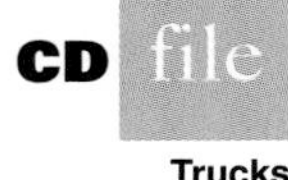
Trucks

Sport Utility	Resale Value	Small Pickup	Resale Value
Chevrolet Blazer LS	55	Chevrolet S-10 Extended Cab	46
Ford Explorer Sport	57	Dodge Dakota Club Cab Sport	53
GMC Yukon XL 1500	67	Ford Ranger XLT Regular Cab	48
Honda CR-V	65	Ford Ranger XLT Supercab	55
Isuzu VehiCross	62	GMC Sonoma Regular Cab	44
Jeep Cherokee Limited	57	Isuzu Hombre Spacecab	41
Mercury Mountaineer	59	Mazda B4000 SE Cab Plus	51
Nissan Pathfinder XE	54	Nissan Frontier XE Regular Cab	51
Toyota 4Runner	55	Toyota Tacoma Xtracab	49
Toyota RAV4	55	Toyota Tacoma Xtracab V6	50

Full-Size Pickup	Resale Value
Chevrolet K2500	60
Chevrolet Silverado 2500 Ext	64
Dodge Ram 1500	54
Dodge Ram Quad Cab 2500	63
Dodge Ram Regular Cab 2500	59
Ford F150 XL	58
Ford F-350 Super Duty Crew Cab XL	64
GMC New Sierra 1500 Ext Cab	68
Toyota Tundra Access Cab Limited	53
Toyota Tundra Regular Cab	58

At the $\alpha = .05$ level of significance, test for any significant difference in the mean resale value for the three types of vehicles.

49. The following data show the age of 12 top executives for major marketers of food, retail, and personal care products (*Advertising Age,* December 1, 1997).

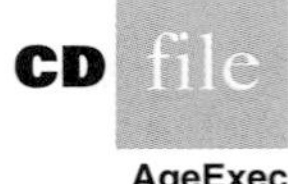

AgeExec

Company	Category	Executive	Age
Campbell Soup Co.	Food	Dale F. Morrison	48
General Mills	Food	Stephen W. Sanger	51
Kellogg Co.	Food	Arnold G. Langbo	59
RJR Nabisco	Food	Stephen F. Goldstone	51
Estee Lauder Cos.	Personal Care	Leonard A. Lauder	64
Gillette Co.	Personal Care	Alfred M. Zeien	67
Procter & Gamble Co.	Personal Care	John E. Pepper	59
Unilever	Personal Care	Morris Tabaksblat	59
Federated Department Stores	Retail	James W. Zimmerman	53
J.C. Penny Co.	Retail	James E. Oesterreicher	55
Sears Roebuck & Co.	Retail	Arthur C. Martinez	57
Kmart Corp.	Retail	Floyd Hall	58

At the $\alpha = .05$ level of significance, is there a significant difference in the mean age of executives for the three categories of companies?

50. A study reported in the *Journal of Small Business Management* concluded that self-employed individuals do not experience higher job satisfaction than individuals who are not self-employed. In this study, job satisfaction is measured using 18 items, each of which is rated using a Likert-type scale with 1–5 response options ranging from strong agreement to strong disagreement. A higher score on this scale indicates a higher degree of job satisfaction. The sum of the ratings for the 18 items, ranging from 18–90, is used as the measure of job satisfaction (*Journal of Small Business Management,* October 1997). Suppose that this approach was used to measure the job satisfaction for lawyers, physical therapists, cabinetmakers, and systems analysts. The results obtained for a sample of 40 individuals from each profession follow.

SatisJob

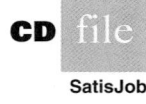

Lawyer	Physical Therapist	Cabinetmaker	Systems Analyst
44	55	54	44
42	78	65	73
74	80	79	71
42	86	69	60
53	60	79	64
50	59	64	66
45	62	59	41
48	52	78	55
64	55	84	76
38	50	60	62

At the $\alpha = .05$ level of significance, test for any difference in the job satisfaction among the four professions.

51. Crown Plaza Hotels and Resorts offered special weekend rates at hotels at resorts nationwide. A sample of 30 properties from three regions of the country provided the following room rates (*USA Today,* April 14, 2000).

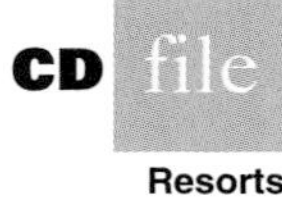

Resorts

West	Rate ($)	South	Rate ($)	Northeast	Rate ($)
Albuquerque	89	Atlanta	105	Albany	89
Irvine	79	Dallas	80	Boston	139
Las Vegas	119	Greenville	79	Hartford	85
Los Angeles	99	Houston	79	New York	159
Palo Alto	109	Jackson	69	Philadelphia	99
Phoenix	149	Macon	69	Pittsfield	99
Portland	79	Miami	89	Providence	149
San Francisco	139	Orlando	119	Washington	159
San Jose	99	Richmond	109	White Plains	109
Seattle	119	Tampa	119	Worchester	124

At the $\alpha = .05$ level of significance, test whether the mean rates are the same for the three regions.

52. According to the sixth annual survey of ad agency employees conducted by the accounting firm Altschuler, Melvoin & Glasser, ad agency employees can expect another banner year in compensation (*Advertising Age,* December 1, 1997). To investigate whether there is any difference in the annual compensation for art directors, suppose that a sample of 10 art directors was selected from each of four regions: West, South, North Central, and Northeast. The base salary ($1000s) for each of the individuals sampled follows.

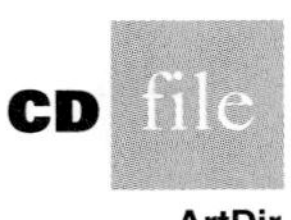

ArtDir

West	South	North Central	Northeast
60.9	50.8	49.5	65.9
45.9	39.6	42.3	58.6
62.1	44.2	35.5	49.3
66.6	40.0	49.1	52.9
68.0	53.9	56.7	48.5
65.0	45.4	41.4	52.9
49.4	61.1	51.3	52.4
62.3	42.3	49.4	48.1
62.6	38.4	42.1	46.5
57.2	38.3	55.7	45.9

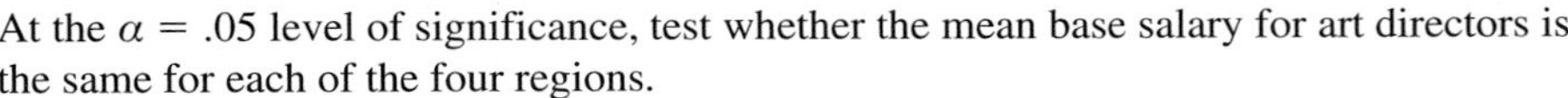

At the $\alpha = .05$ level of significance, test whether the mean base salary for art directors is the same for each of the four regions.

53. The National Football League rates prospects position by position on a scale that ranges from 5 to 9. The ratings are interpreted as follows: 8–9 should start the first year; 7.0–7.9 should start; 6.0–6.9 will make the team as backup; and 5.0–5.9, can make the club and contribute. The following table shows the ratings for three positions for 40 NFL prospects (*USA Today,* April 14, 2000). Does there appear to be any significant effect on the rating due to the player's position?

CD file

NFL

Wide Receiver		Guard		Offensive Tackle	
Name	**Rating**	**Name**	**Rating**	**Name**	**Rating**
Peter Warrick	9.0	Cosey Coleman	7.4	Chris Samuels	8.5
Plaxico Burress	8.8	Travis Claridge	7.0	Stockar McDougle	8.0
Sylvester Morris	8.3	Kaulana Noa	6.8	Chris McIngosh	7.8
Travis Taylor	8.1	Leander Jordan	6.7	Adrian Klemm	7.6
Laveranues Coles	8.0	Chad Clifton	6.3	Todd Wade	7.3
Dez White	7.9	Manula Savea	6.1	Marvel Smith	7.1
Jerry Porter	7.4	Ryan Johanningmei	6.0	Michael Thompson	6.8
Ron Dugans	7.1	Mark Tauscher	6.0	Bobby Williams	6.8
Todd Pinkston	7.0	Blaine Saipaia	6.0	Darnell Alford	6.4
Dennis Northcutt	7.0	Richard Mercier	5.8	Terrance Beadles	6.3
Anthony Lucas	6.9	Damion McIntosh	5.3	Tutan Reyes	6.1
Darrell Jackson	6.6	Jeno James	5.5	Greg Robinson-Ran	6.0
Danny Farmer	6.5	Al Jackson	5.5		
Sherrod Gideon	6.4				
Trevor Gaylor	6.2				

54. In a completely randomized experimental design, three brands of paper towels were tested for their ability to absorb water. Equal-size towels were used, with four sections of towels tested per brand. The absorbency rating data follow. At a .05 level of significance, does there appear to be a difference in the ability of the brands to absorb water?

Brand		
x	y	z
91	99	83
100	96	88
88	94	89
89	99	76

55. Following are the percentage changes in the Dow Jones Industrial Average in each of the four years of six presidential terms. Does there appear to be any significant effect due to the year of the presidential term on stock market performance? Use $\alpha = .05$.

CD file
MktPerf

First Year	Second Year	Third Year	Fourth Year
10.9	−18.9	15.2	4.3
−15.2	4.8	6.1	14.6
−16.7	−27.6	38.3	17.9
−17.3	−3.1	4.2	14.9
−9.2	19.6	20.3	−3.7
27.7	22.6	2.3	11.8
27.0	−4.3	20.3	4.2
13.7	2.1	33.5	26.0

56. Three different assembly methods have been proposed for a new product. A completely randomized experimental design was chosen to determine which assembly method results in the greatest number of parts produced per hour, and 30 workers were randomly selected and assigned to use one of the proposed methods. The number of units produced by each worker follows.

Assembly

Method		
A	B	C
97	93	99
73	100	94
93	93	87
100	55	66
73	77	59
91	91	75
100	85	84
86	73	72
92	90	88
95	83	86

Use these data and test to see whether the mean number of parts produced is the same with each method. Use $\alpha = .05$.

57. Hargreaves Automotive Parts, Inc., wanted to compare the mileage for four different types of brake linings. Thirty linings of each type were produced and placed on a fleet of rental cars. The number of miles that each brake lining lasted until it no longer met the required federal safety standard was recorded, and an average value was computed for each type of lining. The following data were obtained.

Type	Sample Size	Sample Mean	Standard Deviation
A	30	32,000	1450
B	30	27,500	1525
C	30	34,200	1650
D	30	30,300	1400

Test to see whether the corresponding population means are equal. Use $\alpha = .05$.

58. A manufacturer of batteries for electronic toys and calculators is considering three new battery designs. An attempt was made to determine whether the mean lifetime in hours is the same for each of the three designs.

Design A	Design B	Design C
78	112	115
98	99	101
88	101	100
96	116	120

Test to see whether the population means are equal. Use $\alpha = .05$.

59. A study was conducted to investigate browsing activity by shoppers. Each shopper was initially classified as a nonbrowser, light browser, or heavy browser. For each shopper in the study a measure was obtained to determine how comfortable the shopper was in a store. Higher scores indicated greater comfort. Suppose the following data are from a related study.

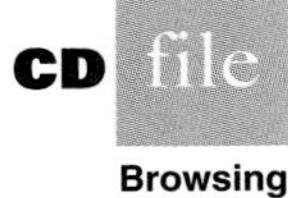

Browsing

Nonbrowser	Light Browser	Heavy Browser
4	5	5
5	6	7
6	5	5
3	4	7
3	7	4
4	4	6
5	6	5
4	5	7

a. Using $\alpha = .05$, test for differences among comfort levels for the three types of browsers.
b. Use Fisher's LSD procedure to compare the comfort levels of nonbrowsers and light browsers. Use $\alpha = .05$. What is your conclusion?

60. A research firm tests the miles-per-gallon characteristics of three brands of gasoline. Because of different gasoline performance characteristics in different brands of automobiles, five brands of automobiles are selected and treated as blocks in the experiment. That is, each brand of automobile is tested with each type of gasoline. The results of the experiment (in miles per gallon) follow.

		Gasoline Brands		
		I	II	III
Automobiles	A	18	21	20
	B	24	26	27
	C	30	29	34
	D	22	25	24
	E	20	23	24

At $\alpha = .05$, is there a significant difference in the mean miles-per-gallon characteristics of the three brands of gasoline?

61. Analyze the experimental data provided in Exercise 60 using the ANOVA procedure for completely randomized designs. Compare your findings with those obtained in Exercise 60. What is the advantage of attempting to remove the block effect?

62. Each month *Internet Magazine* accesses more than 100 Internet service providers (ISPs) in order to check the availability of the ISP and test the speed of the connection by measuring the time (seconds) it takes to download a number of popular web pages. The following data show the download time for 22 free ISPs for web sites located in the United Kingdom, United States, and Europe (*Internet Magazine,* January 2000).

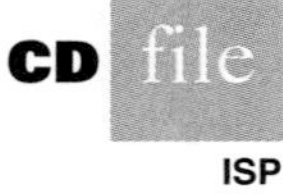

ISP

ISP Name	U.K.	U.S.	Europe
Abel Gratis	10.62	14.64	17.08
Breathe	11.67	14.14	19.86
btclick.com	12.12	16.43	21.30
Bun	11.13	14.09	15.83
Cable & Wireless Life	9.99	13.07	18.43
conX	12.63	15.97	22.12
Freebeeb	11.71	15.52	19.57
Free-Online	13.77	13.98	23.35
Freeserve	10.65	13.62	25.56
FreeUK	12.20	14.96	18.95
Icom-Web	9.62	11.66	15.91
IPNet	13.82	16.70	22.86
I-way Soho	14.86	12.86	19.32
LineOne	12.01	17.82	21.88
Madasafish	13.38	15.59	19.61
NetDirect Online	11.71	15.52	19.57
Netscape Online	10.84	12.66	16.52
Screaming.net (BT Line)	13.23	15.91	23.08
Telinco Internet Services	12.83	15.34	18.76
UK Online	10.39	13.28	21.04
UKPeople	13.79	19.82	19.76
Virgin Net	12.17	15.47	21.94

At $\alpha = .05$, is there a significant difference in the mean download time for web sites located in the three countries?

63. A factorial experiment was designed to test for any significant differences in the time needed to perform English to foreign language translations with two computerized language translators. Because the type of language translated was also considered a significant factor, translations were made with both systems for three different languages: Spanish, French, and German. Use the following data for translation time in hours.

	Language		
	Spanish	**French**	**German**
System 1	8 12	10 14	12 16
System 2	6 10	14 16	16 22

Test for any significant differences due to language translator, type of language, and interaction. Use $\alpha = .05$.

64. A manufacturing company designed a factorial experiment to determine whether the number of defective parts produced by two machines differed and if the number of defective parts produced also depended on whether the raw material needed by each machine was loaded manually or by an automatic feed system. The following data give the numbers of defec-

tive parts produced. Using $\alpha = .05$, test for any significant effect due to machine, loading system, and interaction.

	Loading System	
	Manual	**Automatic**
Machine 1	30	30
	34	26
Machine 2	20	24
	22	28

Case Problem 1 WENTWORTH MEDICAL CENTER

As part of a long-term study of individuals 65 years of age or older, sociologists and physicians at the Wentworth Medical Center in upstate New York investigated the relationship between geographic location and depression. A sample of 60 individuals, all in reasonably good health, was selected; 20 individuals were residents of Florida, 20 were residents of New York, and 20 were residents of North Carolina. Each of the individuals sampled was given a standardized test to measure depression. The data collected follow; higher test scores indicate higher levels of depression. These data are available on the data disk in the file Medical1.

A second part of the study considered the relationship between geographic location and depression for individuals 65 years of age or older who had a chronic health condition such as arthritis, hypertension, and/or heart ailment. A sample of 60 individuals with such conditions was identified. Again, 20 were residents of Florida, 20 were residents of New York, and 20 were residents of North Carolina. The levels of depression recorded for this study follow. These data are available on the data disk in the file Medical2.

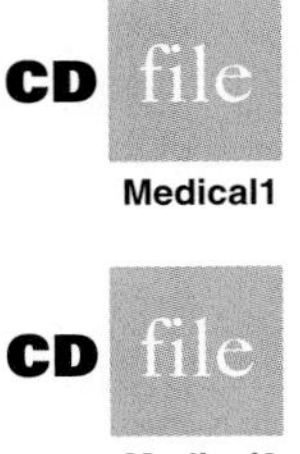

Data from Medical1			Data from Medical2		
Florida	**New York**	**North Carolina**	**Florida**	**New York**	**North Carolina**
3	8	10	13	14	10
7	11	7	12	9	12
7	9	3	17	15	15
3	7	5	17	12	18
8	8	11	20	16	12
8	7	8	21	24	14
8	8	4	16	18	17
5	4	3	14	14	8
5	13	7	13	15	14
2	10	8	17	17	16
6	6	8	12	20	18
2	8	7	9	11	17
6	12	3	12	23	19
6	8	9	15	19	15
9	6	8	16	17	13
7	8	12	15	14	14
5	5	6	13	9	11
4	7	3	10	14	12
7	7	8	11	13	13
3	8	11	17	11	11

Managerial Report

1. Use descriptive statistics to summarize the data from the two studies. What are your preliminary observations about the depression scores?
2. Use analysis of variance on both data sets. State the hypotheses being tested in each case. What are your conclusions?
3. Use inferences about individual treatment means where appropriate. What are your conclusions?
4. Discuss extensions of this study or other analyses that you feel might be helpful.

Case Problem 2 COMPENSATION FOR ID PROFESSIONALS

For the last 10 years *Industrial Distribution* has been tracking compensation of industrial distribution (ID) professionals. Results for the 358 respondents in the 1997 Annual Salary Survey showed that 27% of the respondents work for companies with sales over $40 million, with the typical ID professional working for a $12 million firm. Those who work for small to mid-sized companies (between $6 million and $20 million) report higher earnings than those in larger firms. The lowest paid employees work for firms with sales of less than $1 million. The typical outside salesperson made $50,000 in 1996, and the typical inside salesperson earned just $30,000 (*Industrial Distribution,* November 1997). Suppose that a local chapter of ID professionals in the greater San Francisco area conducted a survey of its membership to study the relationship, if any, between the years of experience and salary for individuals employed in outside and inside sales positions. On the survey, respondents were asked to specify one of three levels of years of experience: low (1–10 years); medium (11–20 years); and high (21 or more years). A portion of the data obtained follow. The complete data set, consisting of 120 observations, is available on the data disk in the file IDSalary.

CD file
IDSalary

Observation	Salary $	Position	Experience
1	28938	Inside	Medium
2	27694	Inside	Medium
3	45515	Outside	Low
4	27031	Inside	Medium
5	37283	Outside	Low
6	32718	Inside	Low
7	54081	Outside	High
8	23621	Inside	Low
9	47835	Outside	High
10	29768	Inside	Medium
.	.	.	.
.	.	.	.
.	.	.	.
115	33080	Inside	High
116	53702	Outside	Medium
117	58131	Outside	Medium
118	32788	Inside	High
119	28070	Inside	Medium
120	35259	Outside	Low

Managerial Report

1. Use descriptive statistics to summarize the data.
2. Develop a 95% confidence interval estimate of the mean annual salary for all salespersons, regardless of years of experience and type of position.
3. Develop a 95% confidence interval estimate of the mean salary for outside salespersons. Compare your results with the national value reported by *Industrial Distribution.*
4. Develop a 95% confidence interval estimate of the mean salary for inside salespersons. Compare your results with the national value reported by *Industrial Distribution.*
5. Ignoring the years of experience, develop a 95% confidence interval estimate of the mean difference between the annual salary for outside salespersons and the mean annual salary for inside salespersons. What is your conclusion?
6. Use analysis of variance to test for any significant differences due to position. Use a .05 level of significance, and for now, ignore the effect of years of experience.
7. Use analysis of variance to test for any significant differences due to years of experience. Use a .05 level of significance, and for now, ignore the effect of position.
8. At the .05 level of significance test for any significant differences due to position, years of experience, and interaction. Use inferences about individual treatment means where appropriate.

Appendix 13.1 ANALYSIS OF VARIANCE AND EXPERIMENTAL DESIGN WITH MINITAB

Single-Factor Observational Studies and Completely Randomized Designs

In Section 13.2 we showed how analysis of variance could be used to test for the equality of k population means using data from an observational study. In Section 13.5 we showed how the same approach could be used to test for the equality of k population means in situations where the data have been collected in a completely randomized design. To illustrate how Minitab can be used to test for the equality of k population means for both of these cases, we show how to test whether the mean examination score is the same at each plant in the National Computer Products example introduced in Section 13.1. The examination score data has been entered into the first three columns of a Minitab worksheet; column 1 is labeled Atlanta, column 2 is labeled Dallas, and column 3 is labeled Seattle. The following steps produce the Minitab output in Figure 13.4

NCP

Step 1. Select the **Stat** pull-down menu
Step 2. Choose **ANOVA**
Step 3. Choose **Oneway (Unstacked)**
Step 4. When the Oneway Analysis of Variance dialog box appears:
Enter C1-C3 in the **Responses (in separate columns)** box
Click **OK**

Randomized Block Designs

In Section 13.6 we showed how analysis of variance could be used to test for the equality of k population means using data from a randomized block design. To illustrate how Minitab can be used for this type of experimental design, we show how to test whether the mean stress levels for air traffic controllers is the same for three work stations. The stress level scores shown in Table 13.6 have been entered into column 1 of a Minitab worksheet. Coding the treatments as 1 for System A, 2 for System B, and 3 for System C, the coded values for the treatments are entered into column 2 of the worksheet. Finally, the corresponding number of each controller (1, 2, 3, 4, 5, 6) is entered into column 3. Thus, the values in the first row of the worksheet are 15, 1, 1; the values in row 2 are 15, 2, 1; the values in row 3 are 18, 3, 1; the values in row 4 are 14, 1, 2; and so on. The following steps produce the Minitab output corresponding to the ANOVA table shown in Table 13.9.

Step 1. Select the **Stat** pull-down menu
Step 2. Choose **ANOVA**
Step 3. Choose **Two-way**
Step 4. When the Two-way Analysis of Variance dialog box appears:
Enter C1 in the **Response** box
Enter C2 in the **Row factor** box
Enter C3 in the **Column factor** box
Select **Fit additive model**
Click **OK**

Factorial Experiments

In Section 13.7 we showed how analysis of variance could be used to test for the equality of k population means using data from a factorial experiment. To illustrate how Minitab can be used for this type of experimental design, we show how to analyze the data for the two-factor GMAT experiment introduced in that section. The GMAT scores shown in Table 13.11 have been entered into column 1 of a Minitab worksheet; column 1 is labeled Score, column 2 is labeled Factor A, and column 3 is labeled Factor B. Coding the Factor A preparation programs as 1 for the 3-hour review, 2 for the one-day program, and 3 for the 10-week course, the coded values for Factor A are entered into column 2 of the worksheet. Coding the Factor B Colleges as 1 for Business, 2 for Engineering, and 3 for Arts and Sciences, the coded values for Factor B are entered into column 3. Thus, the values in the first row of the worksheet are 500, 1, 1; the values in row 2 are 580, 1, 1; the values in row 3 are 540, 1, 2; the values in row 4 are 460, 1, 2; and so on. The following steps produce the Minitab output corresponding to the ANOVA table shown in Figure 13.6.

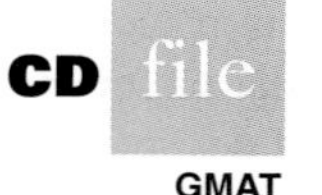

GMAT

Step 1. Select the **Stat** pull-down menu
Step 2. Choose **ANOVA**
Step 3. Choose **Two-way**
Step 4. When the Two-way Analysis of Variance dialog box appears:
Enter C1 in the **Response** box
Enter C2 in the **Row factor** box
Enter C3 in the **Column factor** box
Click **OK**

Appendix 13.2 ANALYSIS OF VARIANCE AND EXPERIMENTAL DESIGN WITH EXCEL

Single-Factor Observational Studies and Completely Randomized Designs

In Section 13.2 we showed how analysis of variance could be used to test for the equality of k population means using data from an observational study. In Section 13.5 we showed how the same approach could be used to test for equality of k population means in situations where the data have been collected in a completely randomized design. To illustrate how Excel can be used to test for the equality of k population means for both of these cases, we show how to test whether the mean examination score is the same at each plant in the National Computer Products example introduced in Section 13.1. The examination score data has been entered into worksheet rows 2 to 7 of columns A, B, and C as shown in Figure 13.7; note that the cells in row 1 for columns A, B, and C are labeled Atlanta, Dallas, and Seattle. The following steps are used to obtain the output shown in cells A10:G24; the ANOVA portion of this output corresponds to the ANOVA table shown in Table 13.2.

NCP

Step 1. Select the **Tools** pull-down menu
Step 2. Choose **Data Analysis**
Step 3. Choose **Anova: Single-Factor** from the list of Analysis Tools
Click **OK**
Step 4. When the Anova: Single-Factor dialog box appears:
Enter A1:C7 in **Input Range** box
Select **Columns**
Select **Labels in First Row**
Select **Output Range** and enter A10 in the box
Click **OK**

Randomized Block Designs

In Section 13.6 we showed how analysis of variance could be used to test for the equality of k population means using data from a randomized block design. To illustrate how Excel

FIGURE 13.7 EXCEL SOLUTION FOR THE NCP ANALYSIS OF VARIANCE EXAMPLE

	A	B	C	D	E	F	G	H
1	**Atlanta**	**Dallas**	**Seattle**					
2	85	71	59					
3	75	75	64					
4	82	73	62					
5	76	74	69					
6	71	69	75					
7	85	82	67					
8								
9								
10	Anova: Single Factor							
11								
12	**SUMMARY**							
13	*Groups*	*Count*	*Sum*	*Average*	*Variance*			
14	Atlanta	6	474	79	34			
15	Dallas	6	444	74	20			
16	Seattle	6	396	66	32			
17								
18								
19	ANOVA							
20	Source of Variateion	SS	df	MS	F	P-Value	F crit	
21	Between Groups	516	2	258	9	0.002703	3.682317	
22	Within Groups	430	15	28.66667				
23								
24	Total	946	17					
25								

AirTraf

can be used for this type of experimental design, we show how to test whether the mean stress levels for air traffic controllers are the same for three work stations. The stress level scores shown in Table 13.6 have been entered into worksheet rows 2 to 7 of columns B, C, and D as shown in Figure 13.8. The cells in rows 2 to 7 of column A contain the number of each controller (1, 2, 3, 4, 5, 6). The following steps produce the Excel output corresponding to the ANOVA table shown in Table 13.9.

Step 1. Select the **Tools** pull-down menu
Step 2. Choose **Data Analysis**
Step 3. Choose **Anova: Two-Factor Without Replication** from the list of Analysis Tools
Click **OK**
Step 4. When the Anova: Two-Factor Without Replication dialog box appears:
Enter A1:D7 in **Input Range** box
Select **Labels**
Select **Output Range** and enter A10 in the box
Click **OK**

FIGURE 13.8 EXCEL SOLUTION FOR THE AIR TRAFFIC CONTROLLER STRESS TEST

	A	B	C	D	E	F	G	H
1	**Controller**	**System A**	**System B**	**System C**				
2	1	15	15	18				
3	2	14	14	14				
4	3	10	11	15				
5	4	13	12	17				
6	5	16	13	16				
7	6	13	13	13				
8								
9								
10	Anova: Two-Factor Without Replication							
11								
12	*Summary*	*Count*	*Sum*	*Average*	*Variance*			
13	1	3	48	16	3			
14	2	3	42	14	0			
15	3	3	36	12	7			
16	4	3	42	14	7			
17	5	3	45	15	3			
18	6	3	39	13	0			
19								
20	System A	6	81	13.5	4.3			
21	System B	6	78	13	2			
22	System C	6	93	15.5	3.5			
23								
24								
25	*ANOVA*							
26	*Source of Variation*	*SS*	*df*	*MS*	*F*	*P-Value*	*F crit*	
27	Rows	30	5	6	3.157895	0.057399	3.325837	
28	Columns	21	2	10.5	5.526316	0.024181	4.102816	
29	Error	19	10	1.9				
30								
31	*Total*	70	17					
32								

Factorial Experiments

In Section 13.7 we showed how analysis of variance could be used to test for the equality of k population means using data from a factorial experiment. To illustrate how Excel can be used for this type of experimental design, we show how to analyze the data for the two-factor GMAT experiment introduced in that section. The GMAT scores shown in Table 13.11 have been entered into worksheet rows 2 to 7 of columns B, C, and D as shown in Figure 13.9.

FIGURE 13.9 EXCEL SOLUTION FOR THE TWO-FACTOR GMAT EXPERIMENT

	A	B	C	D	E	F	G	H
1		**Business**	**Engineering**	**Arts and Sciences**				
2	**3-hour review**	500	540	480				
3		580	460	400				
4	**1-day program**	460	560	420				
5		540	620	480				
6	**10-week course**	560	600	480				
7		600	580	410				
8								
9								
10	Anova: Two-Factor With Replication							
11								
12	SUMMARY	Business	Engineering	Arts and Sciences	Total			
13	*3-hour review*							
14	Count	2	2	2	6			
15	Sum	1080	1000	880	2960			
16	Average	540	500	440	493.3333			
17	Variance	3200	3200	3200	3946.667			
18								
19	*1-day program*							
20	Count	2	2	2	6			
21	Sum	1000	1180	900	3080			
22	Average	500	590	450	513.3333			
23	Variance	3200	1800	1800	5386.667			
24								
25	*10-week course*							
26	Count	2	2	2	6			
27	Sum	1160	1180	890	3230			
28	Average	580	590	445	538.3333			
29	Variance	800	200	2450	5936.667			
30								
31	*Total*							
32	Count	6	6	6				
33	Sum	3240	3360	2670				
34	Average	540	560	445				
35	Variance	2720	3200	1510				
36								
37								
38	ANOVA							
39	*Source of Variation*	*SS*	*df*	*MS*	*F*	*P-Value*	*F crit*	
40	Sample	6100	2	3050	1.382872	0.299436	4.256492	
41	Columns	45300	2	22650	10.26952	0.004757	4.256492	
42	Interaction	11200	4	2800	1.269521	0.350328	3.63309	
43	Within	19850	9	2205.555556				
44								
45	Total	82450	17					
46								

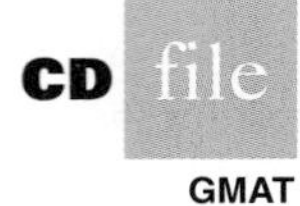

GMAT

The following steps are used to obtain the output shown in cells A10:G45; the ANOVA portion of this output corresponds to the ANOVA table shown in Table 13.14.

Step 1. Select the **Tools** pull-down menu
Step 2. Choose **Data Analysis**
Step 3. Choose **Anova: Two-Factor With Replication** from the list of Analysis Tools
  Click **OK**
Step 4. When the Anova: Two-Factor With Replication dialog box appears:
  Enter A1:D7 in **Input Range** box
  Enter 2 in **Rows per sample** box
  Select **Output Range** and enter A10 in the box
  Click **OK**

SIMPLE LINEAR REGRESSION

Chapter 14

CONTENTS

STATISTICS IN PRACTICE

POLAROID CORPORATION*
Cambridge, Massachusetts

Polaroid's consumer photography business began in 1947 when the company's founder, Dr. Edwin Land, announced a one-step dry process for producing a finished photograph within 1 minute after taking the picture. The first Polaroid Land camera and Polaroid Land film went on sale in 1949. Since then, Polaroid's continuous experimentation and development in chemistry, optics, and electronics have produced photographic systems of ever higher quality, reliability, and convenience.

Polaroid's other major business segment, technical and industrial photography, focuses on making Polaroid's instant photography a key component of the growing number of imaging systems used in today's visual communications environment. To this end, Polaroid markets a wide variety of instant photographic systems, cameras, components, and films for professional, industrial, scientific, and medical uses. Other businesses include magnetics, sunglasses, industrial polarizers, chemicals, custom coating, and holography.

Sensitometry, the measurement of the sensitivity of photographic materials, provides information on many characteristics of film, such as its useful exposure range. Within Polaroid's central sensitometry laboratory, scientists systematically sample and analyze instant films that have been stored at temperature and humidity levels approximating those to which the films will be subjected after they have been purchased by consumers. To investigate the relationship between film speed and the age of a Polaroid extended range, color professional print film, Polaroid's central sensitometry lab selected film samples ranging in age (time since manufacture) from 1 to 13 months. The data showed that film speed decreases with age, and that a straight-line or linear relationship could be used to approximate the relationship between change in film speed and age of the film.

*The authors are indebted to Lawrence Friedman, Manager, Photographic Quality, for providing this Statistics in Practice.

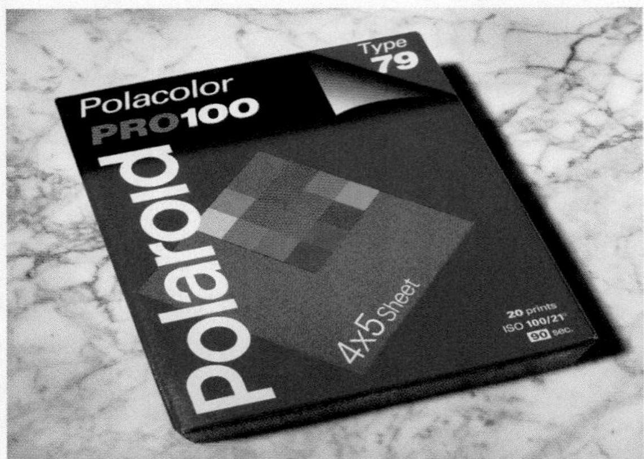

Regression analysis enables Polaroid to produce films with the performance levels its customers require. © Joe Higgins/South-Western.

Using regression analysis, Polaroid was able to develop the following equation relating the change in the film speed to the film's age.

$$\hat{y} = -19.8 - 7.6x$$

where

$$\hat{y} = \text{change in film speed}$$
$$x = \text{film age in months}$$

This equation shows that the average decrease in film speed is 7.6 units per month. The information provided by this analysis, when coupled with consumer purchase and use patterns, enables Polaroid to make manufacturing adjustments that help the company produce films with the performance levels its customers require.

In this chapter you will learn how regression analysis can be used to develop an equation relating two variables, such as the change in film speed to the age of the film in the Polaroid example. In subsequent chapters we will extend this concept to cases involving more than two variables.

Managerial decisions often are based on the relationship between two or more variables. For example, after considering the relationship between advertising expenditures and sales, a marketing manager might attempt to predict sales for a given level of advertising expenditures. In another case, a public utility might use the relationship between the daily high temperature and the demand for electricity to predict electricity usage on the basis of next month's anticipated daily high temperatures. Sometimes a manager will rely on intuition to judge how two variables are related. However, if data can be obtained, a statistical procedure called *regression analysis* can be used to develop an equation showing how the variables are related.

The statistical methods used in studying the relationship between two variables were first employed by Sir Francis Galton (1822–1911). Galton was interested in studying the relationship between a father's height and the son's height. Galton's disciple, Karl Pearson (1857–1936), analyzed the relationship between the father's height and the son's height for 1078 pairs of subjects.

In regression terminology, the variable being predicted is called the **dependent variable.** The variable or variables being used to predict the value of the dependent variable are called the **independent variables.** For example, in analyzing the effect of advertising expenditures on sales, a marketing manager's desire to predict sales would suggest making sales the dependent variable. Advertising expenditure would be the independent variable used to help predict sales. In statistical notation, y denotes the dependent variable and x denotes the independent variable.

In this chapter we consider the simplest type of regression analysis involving one independent variable and one dependent variable in which the relationship between the variables is approximated by a straight line. It is called **simple linear regression.** Regression analysis involving two or more independent variables is called multiple regression analysis; multiple regression and cases involving curvilinear relationships are covered in Chapters 15 and 16.

14.1 SIMPLE LINEAR REGRESSION MODEL

Armand's Pizza Parlors is a chain of Italian-food restaurants located in a five-state area. The most successful locations for Armand's have been near college campuses. The managers believe that quarterly sales for these restaurants (denoted by y) are related positively to the size of the student population (denoted by x); that is, restaurants near campuses with a large population tend to generate more sales than those located near campuses with a small population. Using regression analysis, we can develop an equation showing how the dependent variable y is related to the independent variable x.

Regression Model and Regression Equation

In the Armand's Pizza Parlors example, every restaurant has associated with it a value of x (student population) and a corresponding value of y (quarterly sales). The equation that describes how y is related to x and an error term is called the **regression model.** The regression model used in simple linear regression follows.

Simple Linear Regression Model

$$y = \beta_0 + \beta_1 x + \epsilon \tag{14.1}$$

In the simple linear regression model, y is a linear function of x (the $\beta_0 + \beta_1 x$ part) plus ϵ. β_0 and β_1 are referred to as the parameters of the model, and ϵ (the Greek letter epsilon) is a random variable referred to as the error term. The error term accounts for the variability in y that cannot be explained by the linear relationship between x and y.

In Section 14.4 we will discuss the assumptions for the simple linear regression model and ϵ. One of the assumptions is that the mean or expected value of ϵ is zero. A consequence of this assumption is that the mean or expected value of y, denoted $E(y)$, is equal to $\beta_0 + \beta_1 x$; in other words, the mean value of y is a linear function of x. The equation that describes how the mean value of y is related to x is called the **regression equation.** The regression equation for simple linear regression follows.

Simple Linear Regression Equation

$$E(y) = \beta_0 + \beta_1 x \tag{14.2}$$

The graph of the simple linear regression equation is a straight line; β_0 is the y intercept of the regression line, β_1 is the slope, and $E(y)$ is the mean or expected value of y for a given value of x. Examples of possible regression lines for simple linear regression are shown in Figure 14.1. The regression line in panel A of the figure shows that the mean value of y is related positively to x, with larger values of $E(y)$ associated with larger values of x. The regression line in panel B shows that the mean value of y is related negatively to x, with smaller values of $E(y)$ associated with larger values of x. The regression line in panel C shows the case in which the mean value of y is not related to x; that is, the mean value of y is the same for every value of x.

Estimated Regression Equation

If the values of the parameters β_0 and β_1 were known, we could use equation (14.2) to compute the mean value of y for a known value of x. Unfortunately, the parameter values are not known in practice and must be estimated by using sample data. Sample statistics (denoted b_0 and b_1) are computed as estimates of the parameters β_0 and β_1. Substituting the values of the sample statistics b_0 and b_1 for β_0 and β_1 in the regression equation, we obtain the **estimated regression equation.** The estimated regression equation for simple linear regression follows.

Estimated Simple Linear Regression Equation

$$\hat{y} = b_0 + b_1 x \tag{14.3}$$

The graph of the estimated simple linear regression equation is called the *estimated regression line;* b_0 is the y intercept, b_1 is the slope, and $\hat{y}$ *is the estimated value of* y for a given value of x. In the next section, we show how the least squares method can be used to compute the values of b_0 and b_1 in the estimated regression equation. Figure 14.2 is a summary of the estimation process for simple linear regression.

FIGURE 14.1 POSSIBLE REGRESSION LINES IN SIMPLE LINEAR REGRESSION

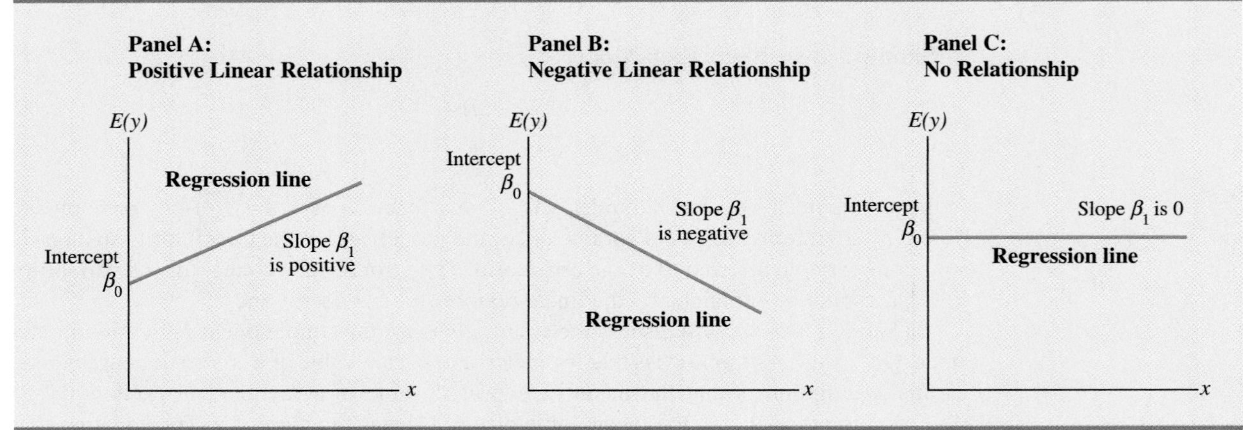

FIGURE 14.2 THE ESTIMATION PROCESS IN SIMPLE LINEAR REGRESSION

Regression Model
$y = \beta_0 + \beta_1 x + \epsilon$
Regression Equation
$E(y) = \beta_0 + \beta_1 x$
Unknown Parameters
β_0, β_1

Sample Data:

x	y
x_1	y_1
x_2	y_2
.	.
.	.
.	.
x_n	y_n

Estimated Regression Equation
$\hat{y} = b_0 + b_1 x$
Sample Statistics
b_0, b_1

b_0 and b_1
provide estimates of
β_0 and β_1

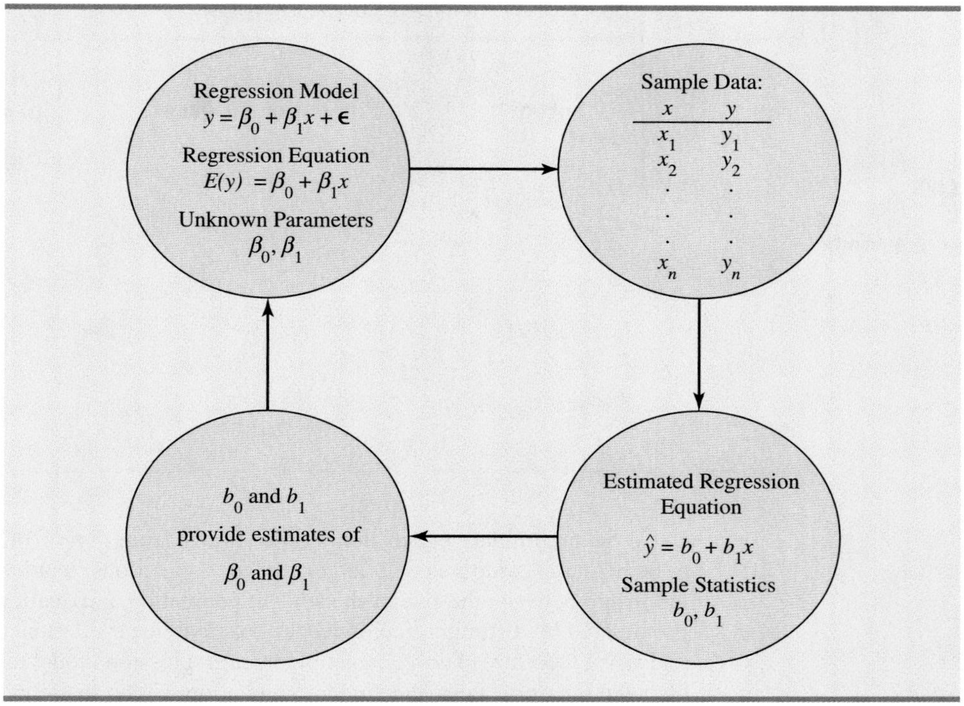

The estimation of β_0 and β_1 is a statistical process much like the estimation of μ discussed in Chapter 7. β_0 and β_1 are the unknown parameters of interest, and β_0 and β_1 are the sample statistics used to estimate the parameters.

NOTES AND COMMENTS

Regression analysis cannot be interpreted as a procedure for establishing a cause-and-effect relationship between variables. It can only indicate how or to what extent variables are associated with each other. Any conclusions about cause and effect must be based on the judgment of the individual or individuals most knowledgeable about the application.

14.2 LEAST SQUARES METHOD

In simple linear regression, each observation consists of two values: one for the independent variable and one for the dependent variable.

The **least squares method** is a procedure for using sample data to find the estimated regression equation. To illustrate the least squares method, suppose data were collected from a sample of 10 Armand's Pizza Parlor restaurants located near college campuses. For the ith observation or restaurant in the sample, x_i is the size of the student population (in thousands) and y_i is the quarterly sales (in thousands of dollars). The values of x_i and y_i for the 10 restaurants in the sample are summarized in Table 14.1. We see that restaurant 1, with $x_1 = 2$ and $y_1 = 58$, is near a campus with 2000 students and has quarterly sales of \$58,000. Restaurant 2, with $x_2 = 6$ and $y_2 = 105$, is near a campus with 6000 students and has quarterly sales of \$105,000. The largest sales value is for restaurant 10, which is near a campus with 26,000 students and has quarterly sales of \$202,000.

Figure 14.3 is a scatter diagram of the data in Table 14.1. The size of the student population is shown on the horizontal axis and the value of quarterly sales is shown on the vertical axis. **Scatter diagrams** for regression analysis are constructed with values of the independent variable x on the horizontal axis and values of the dependent variable y on the vertical axis. The scatter diagram enables us to observe the data graphically and to draw preliminary conclusions about the possible relationship between the variables.

TABLE 14.1 STUDENT POPULATION AND QUARTERLY SALES DATA FOR 10 ARMAND'S PIZZA PARLORS

Armand's

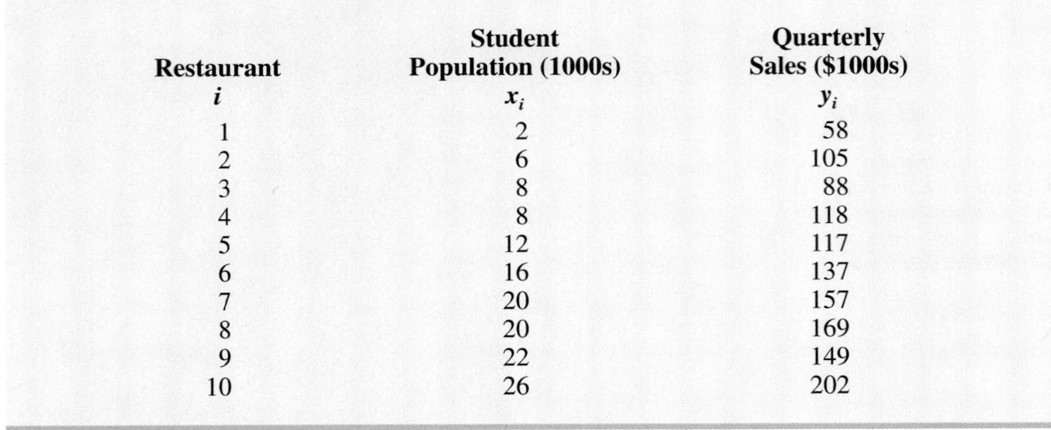

Restaurant i	Student Population (1000s) x_i	Quarterly Sales ($1000s) y_i
1	2	58
2	6	105
3	8	88
4	8	118
5	12	117
6	16	137
7	20	157
8	20	169
9	22	149
10	26	202

What preliminary conclusions can be drawn from Figure 14.3? Quarterly sales appear to be higher at campuses with larger student populations. In addition, for these data the relationship between the size of the student population and quarterly sales appears to be approximated by a straight line; indeed, a positive linear relationship is indicated between x and y. We therefore choose the simple linear regression model to represent the relationship between quarterly sales and student population. Given that choice, our next task is to use

FIGURE 14.3 SCATTER DIAGRAM OF STUDENT POPULATION AND QUARTERLY SALES FOR ARMAND'S PIZZA PARLORS

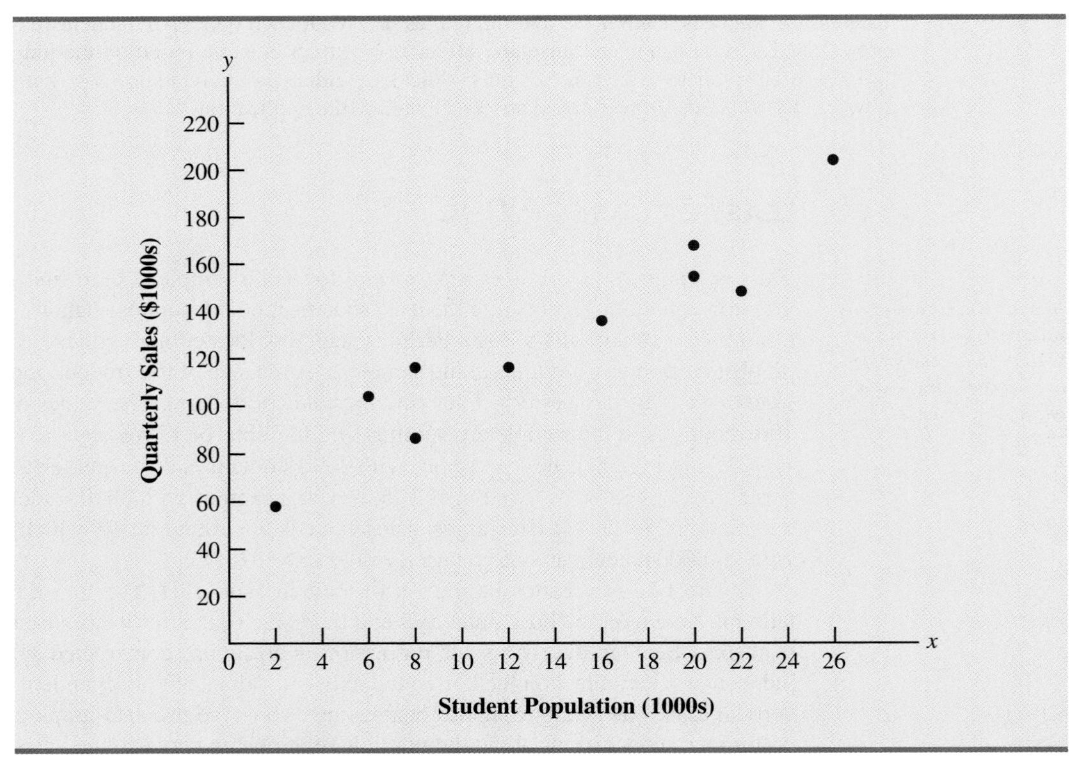

the sample data in Table 14.1 to determine the values of b_0 and b_1 in the estimated simple linear regression equation. For the ith restaurant, the estimated regression equation provides

$$\hat{y}_i = b_0 + b_1 x_i \tag{14.4}$$

where

$\hat{y}_i$ = estimated value of quarterly sales ($1000s) for the ith restaurant
b_0 = the y intercept of the estimated regression line
b_1 = the slope of the estimated regression line
x_i = size of the student population (1000s) for the ith restaurant

With y_i denoting the observed (actual) sales for restaurant i and $\hat{y}_i$ in (14.4) representing the estimated value of sales for restaurant i, every restaurant in the sample will have an observed value of sales y_i and an estimated value of sales $\hat{y}_i$. For the estimated regression line to provide a good fit to the data, we want the differences between the observed sales values and the estimated sales values to be small.

The least squares method uses the sample data to provide the values of b_0 and b_1 that minimize the *sum of the squares of the deviations* between the observed values of the dependent variable y_i and the estimated values of the dependent variable $\hat{y}_i$. The criterion for the least squares method is given by (14.5).

Carl Friedrich Gauss (1777–1855) proposed the least squares method.

Least Squares Criterion

$$\min \Sigma(y_i - \hat{y}_i)^2 \tag{14.5}$$

where

y_i = observed value of the dependent variable for the ith observation
$\hat{y}_i$ = estimated value of the dependent variable for the ith observation

Differential calculus can be used to show (see Appendix 14.1) that the values of b_0 and b_1 that minimize (14.5) can be found by using (14.6) and (14.7).

In computing b_1 with a calculator, carry as many significant digits as possible in the intermediate calculations. We recommend carrying at least four significant digits.

Slope and y-Intercept for the Estimated Regression Equation*

$$b_1 = \frac{\Sigma(x_i - \bar{x})(y_i - \bar{y})}{\Sigma(x_i - \bar{x})^2} \tag{14.6}$$

$$b_0 = \bar{y} - b_1\bar{x} \tag{14.7}$$

where

x_i = value of the independent variable for the ith observation
y_i = value of the dependent variable for the ith observation
$\bar{x}$ = mean value for the independent variable
$\bar{y}$ = mean value for the dependent variable
n = total number of observations

*An alternate formula for b_1 is

$$b_1 = \frac{\Sigma x_i y_i - (\Sigma x_i \Sigma y_i)/n}{\Sigma x_i^2 - (\Sigma x_i)^2/n}$$

This form of equation (14.6) is often recommended when using a calculator to compute b_1.

Some of the calculations necessary to develop the least squares estimated regression equation for Armand's Pizza Parlors are shown in Table 14.2. With the sample of 10 restaurants, we have $n = 10$ observations. Because equations (14.6) and (14.7) require $\bar{x}$ and $\bar{y}$, we begin the calculations by computing $\bar{x}$ and $\bar{y}$.

$$\bar{x} = \frac{\Sigma x_i}{n} = \frac{140}{10} = 14$$

$$\bar{y} = \frac{\Sigma y_i}{n} = \frac{1300}{10} = 130$$

Using equations (14.6), (14.7), and the information in Table 14.2, we can compute the slope and intercept of the estimated regression equation for Armand's Pizza Parlors. The calculation of the slope (b_1) proceeds as follows.

$$\begin{aligned} b_1 &= \frac{\Sigma(x_i - \bar{x})(y_i - \bar{y})}{\Sigma(x_i - \bar{x})^2} \\ &= \frac{2840}{568} \\ &= 5 \end{aligned}$$

The calculation of the y intercept (b_0) follows.

$$\begin{aligned} b_0 &= \bar{y} - b_1\bar{x} \\ &= 130 - 5(14) \\ &= 60 \end{aligned}$$

Thus, the estimated regression equation is

$$\hat{y} = 60 + 5x$$

Figure 14.4 shows the graph of this equation on the scatter diagram.

TABLE 14.2 CALCULATIONS FOR THE LEAST SQUARES ESTIMATED REGRESSION EQUATION FOR ARMAND PIZZA PARLORS

Restaurant i	x_i	y_i	$x_i - \bar{x}$	$y_i - \bar{y}$	$(x_i - \bar{x})(y_i - \bar{y})$	$(x_i - \bar{x})^2$
1	2	58	−12	−72	864	144
2	6	105	−8	−25	200	64
3	8	88	−6	−42	252	36
4	8	118	−6	−12	72	36
5	12	117	−2	−13	26	4
6	16	137	2	7	14	4
7	20	157	6	27	162	36
8	20	169	6	39	234	36
9	22	149	8	19	152	64
10	26	202	12	72	864	144
Totals	140	1300			2840	568
	Σx_i	Σy_i			$\Sigma(x_i - \bar{x})(y_i - \bar{y})$	$\Sigma(x_i - \bar{x})^2$

FIGURE 14.4 GRAPH OF THE ESTIMATED REGRESSION EQUATION FOR ARMAND'S PIZZA PARLORS: $\hat{y} = 60 + 5x$

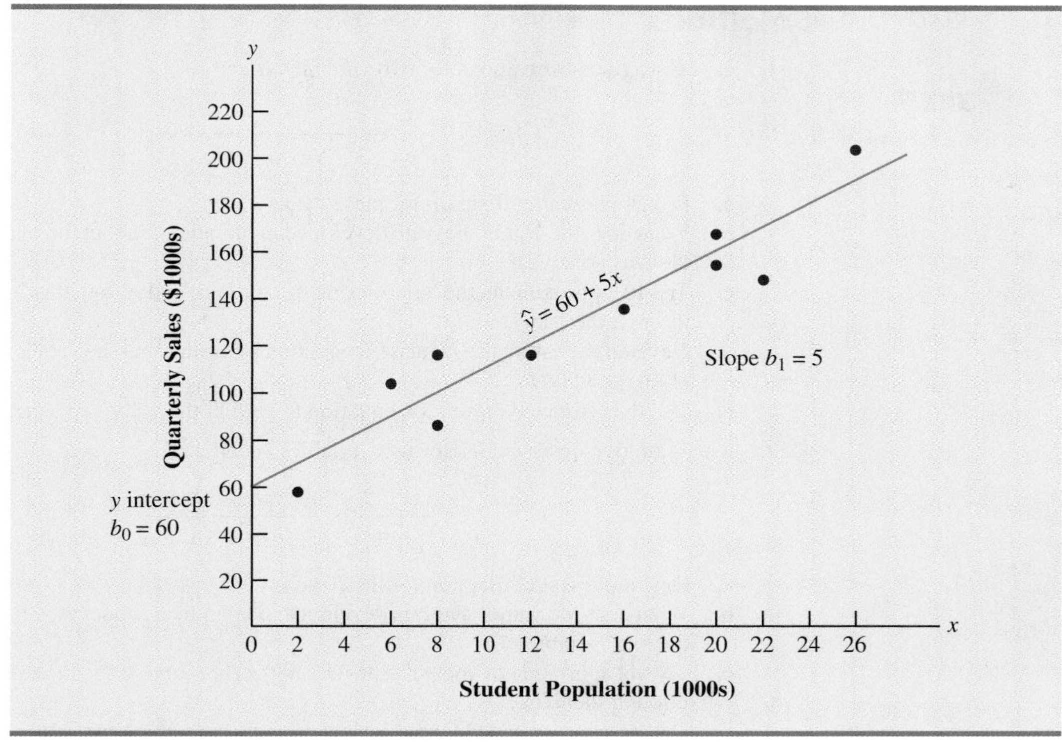

The slope of the estimated regression equation ($b_1 = 5$) is positive, implying that as student population increases, sales increase. In fact, we can conclude (based on sales measured in $1000s and student population in 1000s) that an increase in the student population of 1000 is associated with an increase of $5000 in expected sales; that is, sales are expected to increase by $5 per student.

Using the estimated regression equation to make predictions outside the range of the values of the independent variable should be done with caution because outside that range we cannot be sure that the same relationship is valid.

If we believe the least squares estimated regression equation adequately describes the relationship between x and y, it would seem reasonable to use the estimated regression equation to predict the value of y for a given value of x. For example, if we wanted to predict sales for a restaurant to be located near a campus with 16,000 students, we would compute

$$\hat{y} = 60 + 5(16) = 140$$

Hence, we would predict quarterly sales of $140,000 for this restaurant. In the following sections we will discuss methods for assessing the appropriateness of using the estimated regression equation for estimation and prediction.

NOTES AND COMMENTS

The least squares method provides an estimated regression equation that minimizes the sum of squared deviations between the observed values of the dependent variable y_i and the estimated values of the dependent variable $\hat{y}_i$. This is the least squares criterion for choosing the equation that provides the best fit. If some other criterion were used, such as minimizing the sum of the absolute deviations between y_i and $\hat{y}_i$, a different equation would be obtained. In practice, the least squares method is the most widely used.

EXERCISES

Methods

1. Given are five observations for two variables, x and y.

x_i	1	2	3	4	5
y_i	3	7	5	11	14

   a. Develop a scatter diagram for these data.
   b. What does the scatter diagram developed in (a) indicate about the relationship between the two variables?
   c. Try to approximate the relationship between x and y by drawing a straight line through the data.
   d. Develop the estimated regression equation by computing the values of b_0 and b_1 using (14.6) and (14.7).
   e. Use the estimated regression equation to predict the value of y when $x = 4$.

2. Given are five observations for two variables, x and y.

x_i	2	3	5	1	8
y_i	25	25	20	30	16

   a. Develop a scatter diagram for these data.
   b. What does the scatter diagram developed in (a) indicate about the relationship between the two variables?
   c. Try to approximate the relationship between x and y by drawing a straight line through the data.
   d. Develop the estimated regression equation by computing the values of b_0 and b_1 using (14.6) and (14.7).
   e. Use the estimated regression equation to predict the value of y when $x = 6$.

3. Given are five observations collected in a regression study on two variables.

x_i	2	4	5	7	8
y_i	2	3	2	6	4

   a. Develop a scatter diagram for these data.
   b. Develop the estimated regression equation for these data.
   c. Use the estimated regression equation to predict the value of y when $x = 4$.

Applications

4. The following data were collected on the height (inches) and weight (pounds) of women swimmers.

Height	68	64	62	65	66
Weight	132	108	102	115	128

   a. Develop a scatter diagram for these data with height as the independent variable.
   b. What does the scatter diagram developed in (a) indicate about the relationship between the two variables?
   c. Try to approximate the relationship between height and weight by drawing a straight line through the data.
   d. Develop the estimated regression equation by computing the values of b_0 and b_1 using (14.6) and (14.7).
   e. If a swimmer's height is 63 inches, what would you estimate her weight to be?

5. The following data show the case sales (millions) and the media expenditures (millions of dollars) for seven major brands of soft drinks (*Superbrands '98,* October 20, 1997).

Brand	Media Expenditures ($)	Case Sales
Coca-Cola Classic	131.3	1929.2
Pepsi-Cola	92.4	1384.6
Diet Coke	60.4	811.4
Sprite	55.7	541.5
Dr. Pepper	40.2	536.9
Mountain Dew	29.0	535.6
7-Up	11.6	219.5

a. Develop a scatter diagram for these data with media expenditures as the independent variable.
b. What does the scatter diagram developed in (a) indicate about the relationship between the two variables?
c. Draw a straight line through the data to approximate a linear relationship between media expenditures and case sales.
d. Use the least squares method to develop the estimated regression equation.
e. Provide an interpretation for the slope of the estimated regression equation.
f. Predict the case sales for a brand with a media expenditure of $70 million.

6. Airline performance data for U.S. airlines were reported in *The Wall Street Journal Almanac 1998.* Data on the percentage of flights arriving on time and the number of complaints per 100,000 passengers follow.

Airline	Percentage on Time	Complaints
Southwest	81.8	0.21
Continental	76.6	0.58
Northwest	76.6	0.85
US Airways	75.7	0.68
United	73.8	0.74
American	72.2	0.93
Delta	71.2	0.72
America West	70.8	1.22
TWA	68.5	1.25

a. Develop a scatter diagram for these data with percentage on time as the independent variable.
b. What does the scatter diagram developed in (a) indicate about the relationship between the two variables?
c. Develop the estimated regression equation showing how the number of complaints per 100,000 passengers is related to the percentage of flights arriving on time.
d. Provide an interpretation for the slope of the estimated regression equation.
e. What is the estimated number of complaints per 100,000 passengers if the percentage of flights arriving on time is 80%?

7. The Dow Jones Industrial Average (DJIA) and the Standard & Poor's 500 (S&P) indexes are both used as measures of overall movement in the stock market. The DJIA is based on the price movements of 30 large companies; the S&P 500 is an index composed of 500 stocks. Some say the S&P 500 is a better measure of stock market performance because it is broader based. The closing prices for the DJIA and the S&P 500 for 10 weeks, beginning with February 11, 2000, follow (*Barron's,* April 17, 2000).

CD file
DowS&P

Date	DJIA	S&P
February 11	10425	1387
February 18	10220	1346
February 25	9862	1333

Continued

Date	DJIA	S&P
March 3	10367	1409
March 10	9929	1395
March 17	10595	1464
March 24	11113	1527
March 31	10922	1499
April 7	11111	1516
April 14	10306	1357

a. Develop a scatter diagram for these data with DJIA as the independent variable.
b. Develop the least squares estimated regression equation.
c. Suppose the closing price for the DJIA is 11,000. Estimate the closing price for the S&P 500.

8. Nielsen Media Research collects data showing which advertisers get the most exposure during prime-time TV on ABC, CBS, NBC, Fox, UPN, and WB networks. Data showing the number of household exposures in millions and the number of times the ad was aired for the week of April 28–May 4, 1997, follow (*USA Today,* May 5, 1997).

Advertised Brand	Times Ad Aired	Household Exposures
Wendy's	28	191.7
Ford Escort	20	174.6
Austin Powers movie	14	161.3
Nissan	16	161.1
Pizza Hut	16	147.7
Saturn	16	146.3
Father's Day movie	11	138.2

a. Develop the estimated regression equation showing how the number of times an ad is aired is related to the number of household exposures.
b. Provide an interpretation for the slope of the estimated regression equation.
c. What is the estimated number of household exposures if an ad is aired 15 times?

9. A sales manager has collected the following data on annual sales and years of experience.

Salesperson	Years of Experience	Annual Sales ($1000s)
1	1	80
2	3	97
3	4	92
4	4	102
5	6	103
6	8	111
7	10	119
8	10	123
9	11	117
10	13	136

a. Develop a scatter diagram for these data with years of experience as the independent variable.
b. Develop an estimated regression equation that can be used to predict annual sales given the years of experience.
c. Use the estimated regression equation to predict annual sales for a salesperson with 9 years of experience.

10. *PC World* provided ratings for the top 15 notebook PCs (*PC World,* February 2000). The performance score is a measure of how fast a PC can run a mix of common business applications as compared to how fast a baseline machine can run them. For example, a PC with a performance score of 200 is twice as fast as the baseline machine. A 100-point scale was used to provide an overall rating for each notebook tested in the study. A score in the 90s is exceptional, while one in the 70s is above average. The performance scores and the overall ratings for the 15 notebooks are shown below.

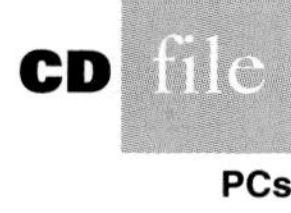

PCs

	Performance Score	Overall Rating
AMS Tech Roadster 15CTA380	115	67
Compaq Armada M700	191	78
Compaq Prosignia Notebook 150	153	79
Dell Inspiron 3700 C466GT	194	80
Dell Inspiron 7500 R500VT	236	84
Dell Latitude Cpi A366XT	184	76
Enpower ENP-313 Pro	184	77
Gateway Solo 9300LS	216	92
HP Pavilion Notebook PC	185	83
IBM ThinkPad I Series 1480	183	78
Micro Express NP7400	189	77
Micron TransPort NX PII-400	202	78
NEC Versa SX	192	78
Sceptre Soundx 5200	141	73
Sony VAIO PCG-F340	187	77

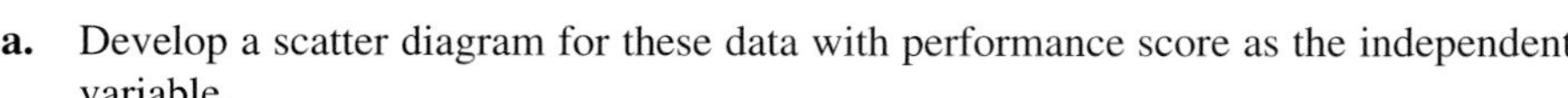

a. Develop a scatter diagram for these data with performance score as the independent variable.
b. Develop the least squares estimated regression equation.
c. Estimate the overall rating for a new PC that has a performance score of 225.

11. The following data show the gaming revenue and the hotel revenue, in millions of dollars, for 10 Las Vegas casino hotels (*Cornell Hotel And Restaurant Administration Quarterly,* October 1997).

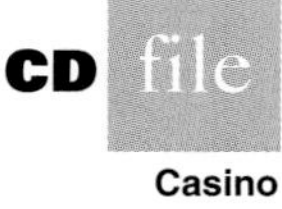

Casino

Company	Hotel Revenue	Gaming Revenue
Boyd Gaming	$303.5	$548.2
Circus Circus Enterprises	664.8	664.8
Grand Casinos	121.0	270.7
Hilton Corp. Gaming Div.	429.6	511.0
MGM Grand, Inc.	373.1	404.7
Mirage Resorts	670.9	782.8
Primadonna Resorts	66.4	130.7
Rio Hotel & Casino	105.8	105.5
Sahara Gaming	102.4	148.7
Station Casinos	135.8	358.5

a. Develop a scatter diagram for these data with hotel revenue as the independent variable.
b. Does there appear to be a linear relationship between the two variables?
c. Develop the estimated regression equation relating the gaming revenue to the hotel revenue.
d. Suppose that the hotel revenue was $500 million. What is an estimate of the gaming revenue?

12. The following table gives the number of employees and the revenue (millions of dollars) for 20 companies (*Fortune,* April 17, 2000).

CD file
EmpRev

	Employees	Revenue ($millions)
Sprint	77,600	19,930
Chase Manhattan	74,801	33,710
Computer Sciences	50,000	7,660
Wells Fargo	89,355	21,795
Sunbeam	12,200	2,398
CBS	29,000	7,510
Time Warner	69,722	27,333
Steelcase	16,200	2,743
Georgia-Pacific	57,000	17,796
Toro	1,275	4,673
American Financial	9,400	3,334
Fluor	53,561	12,417
Phillips Petroleum	15,900	13,852
Cardinal Health	36,000	25,034
Borders Group	23,500	2,999
MCI Worldcom	77,000	37,120
Consolidated Edison	14,269	7,491
IBP	45,000	14,075
Super Value	50,000	17,421
H&R Block	4,200	1,669

a. Develop a scatter diagram for these data with number of employees as the independent variable.
b. What does the scatter diagram developed in (a) indicate about the relationship between number of employees and revenue?
c. Develop the estimated regression equation for these data.
d. Use the estimated regression equation to predict the revenue for a firm that has 75,000 employees.

13. To the Internal Revenue Service, the reasonableness of total itemized deductions depends on the taxpayer's adjusted gross income. Large deductions, which include charity and medical deductions, are more reasonable for taxpayers with large adjusted gross incomes. If a taxpayer claims larger than average itemized deductions for a given level of income, the chances of an IRS audit are increased. Data (in $1000s) on adjusted gross income and the average or reasonable amount of itemized deductions follow.

Adjusted Gross Income ($1000s)	Total Itemized Deductions ($1000s)
22	9.6
27	9.6
32	10.1
48	11.1
65	13.5
85	17.7
120	25.5

a. Develop a scatter diagram for these data with adjusted gross income as the independent variable.
b. Use the least squares method to develop the estimated regression equation.
c. Estimate a reasonable level of total itemized deductions for a taxpayer with an adjusted gross income of $52,500. If this taxpayer has claimed total itemized deductions of $20,400, would the IRS agent's request for an audit appear justified? Explain.

14. The following data report the occupancy rates (%) and room rates ($) for the largest U.S. hotel markets (*The Wall Street Journal Almanac 1998*).

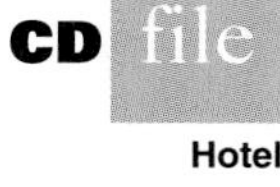

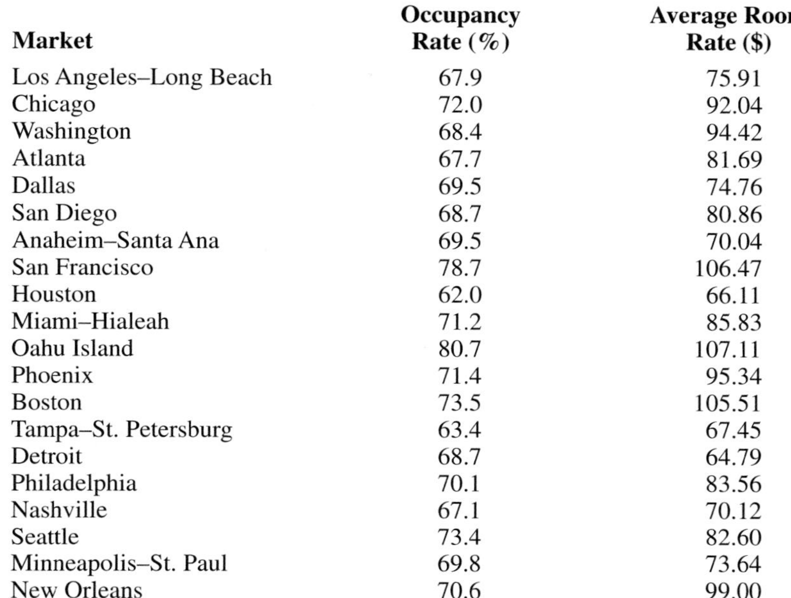

Market	Occupancy Rate (%)	Average Room Rate ($)
Los Angeles–Long Beach	67.9	75.91
Chicago	72.0	92.04
Washington	68.4	94.42
Atlanta	67.7	81.69
Dallas	69.5	74.76
San Diego	68.7	80.86
Anaheim–Santa Ana	69.5	70.04
San Francisco	78.7	106.47
Houston	62.0	66.11
Miami–Hialeah	71.2	85.83
Oahu Island	80.7	107.11
Phoenix	71.4	95.34
Boston	73.5	105.51
Tampa–St. Petersburg	63.4	67.45
Detroit	68.7	64.79
Philadelphia	70.1	83.56
Nashville	67.1	70.12
Seattle	73.4	82.60
Minneapolis–St. Paul	69.8	73.64
New Orleans	70.6	99.00

a. Develop a scatter diagram for these data with average room rate as the independent variable.

b. Develop the estimated regression equation relating the occupancy rate to the average room rate.

c. Estimate the occupancy rate for a hotel with an average room rate of $80.

14.3 COEFFICIENT OF DETERMINATION

For the Armand's Pizza Parlors example, we developed the estimated regression equation $\hat{y} = 60 + 5x$ to approximate the linear relationship between the size of the student population x and quarterly sales y. A question now is: How well does the estimated regression equation fit the data? In this section, we show that the **coefficient of determination** provides a measure of the goodness of fit for the estimated regression equation.

For the ith observation, the difference between the observed value of the dependent variable, y_i, and the estimated value of the dependent variable, $\hat{y}_i$, is called the **ith residual.** The ith residual represents the error in using $\hat{y}_i$ to estimate y_i. Thus, for the ith observation, the residual is $y_i - \hat{y}_i$. The sum of squares of these residuals or errors is the quantity that is minimized by the least squares method. This quantity, also known as the *sum of squares due to error,* is denoted by SSE.

Sum of Squares Due to Error

$$\text{SSE} = \Sigma(y_i - \hat{y}_i)^2 \qquad \textbf{(14.8)}$$

The value of SSE is a measure of the error in using the estimated regression equation to estimate the values of the dependent variable in the sample.

In Table 14.3 we show the calculations required to compute the sum of squares due to error for the Armand's Pizza Parlors example. For instance, for restaurant 1 the values of the independent and dependent variables are $x_1 = 2$ and $y_1 = 58$. Using the estimated regression equation, we find that the estimated value of quarterly sales for restaurant 1 is $\hat{y}_1 = 60 + 5(2) = 70$. Thus, the error in using $\hat{y}_1$ to estimate y_1 for restaurant 1 is $y_1 - \hat{y}_1 = 58 - 70 = -12$. The squared error, $(-12)^2 = 144$, is shown in the last column of Table 14.3. After computing and squaring the residuals for each restaurant in the sample, we sum them to obtain SSE = 1530. Thus, SSE = 1530 measures the error in using the estimated regression equation $\hat{y} = 60 + 5x$ to predict sales.

Now suppose we are asked to develop an estimate of quarterly sales without knowledge of the size of the student population. Without knowledge of any related variables, we would use the sample mean as an estimate of quarterly sales at any given restaurant. Table 14.2 shows that for the sales data, $\Sigma y_i = 1300$. Hence, the mean value of quarterly sales for the sample of 10 Armand's restaurants is $\bar{y} = \Sigma y_i/n = 1300/10 = 130$. In Table 14.4 we show the sum of squared deviations obtained by using the sample mean $\bar{y} = 130$ to estimate the value of quarterly sales for each restaurant in the sample. For the ith restaurant in the sample, the difference $y_i - \bar{y}$ provides a measure of the error involved in using $\bar{y}$ to estimate sales. The corresponding sum of squares, called the *total sum of squares,* is denoted SST.

Total Sum of Squares

$$\text{SST} = \Sigma(y_i - \bar{y})^2 \tag{14.9}$$

The sum at the bottom of the last column in Table 14.4 is the total sum of squares for Armand's Pizza Parlors; it is SST = 15,730.

In Figure 14.5 we show the estimated regression line $\hat{y} = 60 + 5x$ and the line corresponding to $\bar{y} = 130$. Note that the points cluster more closely around the estimated regression line than they do about the line $\bar{y} = 130$. For example, for the 10th restaurant in the sample we see that the error is much larger when $\bar{y} = 130$ is used as an estimate of y_{10} than when $\hat{y}_{10} = 60 + 5(26) = 190$ is used. We can think of SST as a measure of how well the observations cluster about the $\bar{y}$ line and SSE as a measure of how well the observations cluster about the $\hat{y}$ line.

TABLE 14.3 CALCULATION OF SSE FOR ARMAND'S PIZZA PARLORS

Restaurant i	x_i = Student Population (1000s)	y_i = Quarterly Sales ($1000s)	$\hat{y}_i = 60 + 5x_i$	$y_i - \hat{y}_i$	$(y_i - \hat{y}_i)^2$
1	2	58	70	−12	144
2	6	105	90	15	225
3	8	88	100	−12	144
4	8	118	100	18	324
5	12	117	120	−3	9
6	16	137	140	−3	9
7	20	157	160	−3	9
8	20	169	160	9	81
9	22	149	170	−21	441
10	26	202	190	12	144
					SSE = 1530

TABLE 14.4 COMPUTATION OF THE TOTAL SUM OF SQUARES FOR ARMAND'S PIZZA PARLORS

Restaurant i	x_i = Student Population (1000s)	y_i = Quarterly Sales ($1000s)	$y_i - \bar{y}$	$(y_i - \bar{y})^2$
1	2	58	−72	5,184
2	6	105	−25	625
3	8	88	−42	1,764
4	8	118	−12	144
5	12	117	−13	169
6	16	137	7	49
7	20	157	27	729
8	20	169	39	1,521
9	22	149	19	361
10	26	202	72	5,184
				SST = 15,730

FIGURE 14.5 DEVIATIONS ABOUT THE ESTIMATED REGRESSION LINE AND THE LINE $y = \bar{y}$ FOR ARMAND'S PIZZA PARLORS

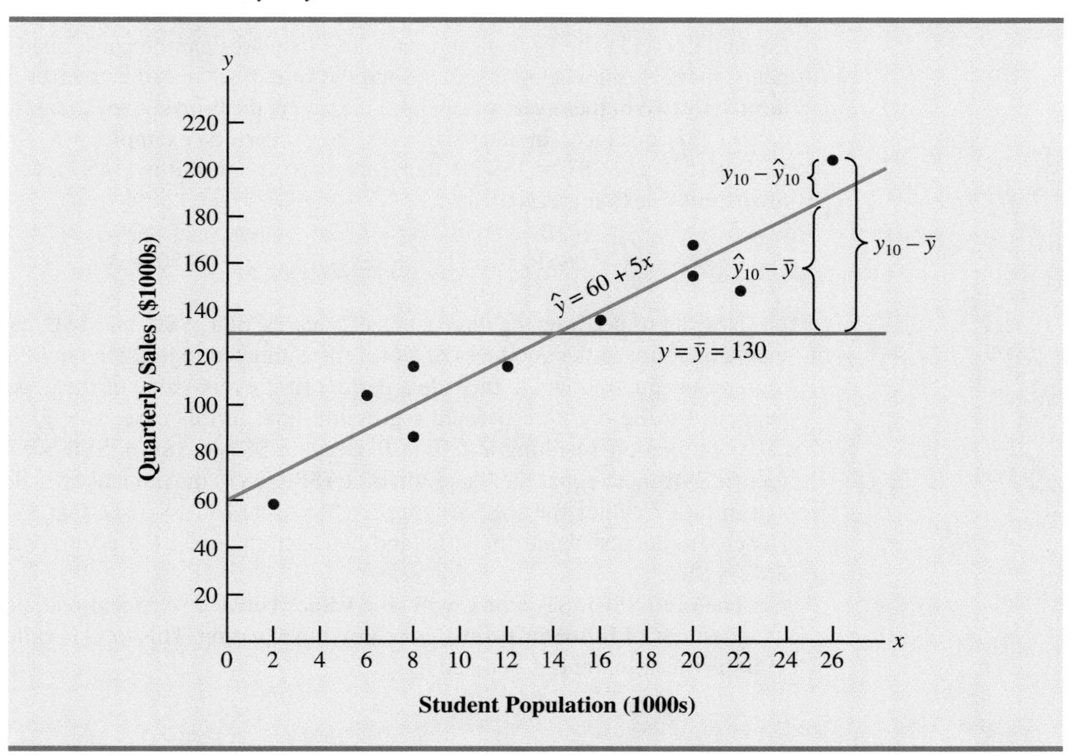

To measure how much the $\hat{y}$ values on the estimated regression line deviate from $\bar{y}$, another sum of squares is computed. This sum of squares, called the *sum of squares due to regression,* is denoted SSR.

Sum of Squares Due to Regression

$$\text{SSR} = \Sigma(\hat{y}_i - \bar{y})^2 \tag{14.10}$$

From the preceding discussion, we should expect that SST, SSR, and SSE are related. Indeed, the relationship among these three sums of squares provides one of the most important results in statistics.

SSR can be thought of as the explained portion of SST, and SSE can be thought of as the unexplained portion of SST.

Relationship Among SST, SSR, and SSE

$$\text{SST} = \text{SSR} + \text{SSE} \tag{14.11}$$

where

SST = total sum of squares
SSR = sum of squares due to regression
SSE = sum of squares due to error

Equation (14.11) shows that the total sum of squares can be partitioned into two components, the regression sum of squares and the sum of squares due to error. Hence, if the values of any two of these sum of squares are known, the third sum of squares can be computed easily. For instance, in the Armand's Pizza Parlors example, we already know that SSE = 1530 and SST = 15,730; therefore, solving for SSR in (14.11), we find that the sum of squares due to regression is

$$\text{SSR} = \text{SST} - \text{SSE} = 15{,}730 - 1530 = 14{,}200$$

Now let us see how the three sums of squares, SST, SSR, and SSE, can be used to provide a measure of the goodness of fit for the estimated regression equation. The estimated regression equation would provide a perfect fit if every value of the dependent variable y_i happened to lie on the estimated regression line. In this case, $y_i - \hat{y}_i$ would be zero for each observation, resulting in SSE = 0. Because SST = SSR + SSE, we see that for a perfect fit SSR must equal SST, and the ratio (SSR/SST) must equal one. Poorer fits will result in larger values for SSE. Solving for SSE in (14.11), we see that SSE = SST − SSR. Hence, the largest value for SSE (and hence the poorest fit) occurs when SSR = 0 and SSE = SST.

The ratio SSR/SST, which will take values between zero and one, is used to evaluate the goodness of fit for the estimated regression equation. This ratio is called the **coefficient of determination** and is denoted by r^2.

Coefficient of Determination

$$r^2 = \frac{\text{SSR}}{\text{SST}} \tag{14.12}$$

For the Armand's Pizza Parlors example, the value of the coefficient of determination is

$$r^2 = \frac{\text{SSR}}{\text{SST}} = \frac{14{,}200}{15{,}730} = .9027$$

When we express the coefficient of determination as a percentage, r^2 can be interpreted as the percentage of the total sum of squares that can be explained by using the estimated regression equation. For Armand's Pizza Parlors, we can conclude that 90.27% of the total sum of squares can be explained by using the estimated regression equation $\hat{y} = 60 + 5x$ to predict sales. In other words, 90.27% of the variability in sales can be explained by the linear relationship between the size of the student population and sales. We should be pleased to find such a good fit for the estimated regression equation.

Correlation Coefficient

In Chapter 3 we introduced the **correlation coefficient** as a descriptive measure of the strength of linear association between two variables, x and y. Values of the correlation coefficient are always between -1 and $+1$. A value of $+1$ indicates that the two variables x and y are perfectly related in a positive linear sense. That is, all data points are on a straight line that has a positive slope. A value of -1 indicates that x and y are perfectly related in a negative linear sense, with all data points on a straight line that has a negative slope. Values of the correlation coefficient close to zero indicate that x and y are not linearly related.

In Section 3.5 we presented the equation for computing the sample correlation coefficient. If a regression analysis has already been performed and the coefficient of determination r^2 has been computed, the sample correlation coefficient can be computed as follows.

Sample Correlation Coefficient

$$\begin{aligned} r_{xy} &= (\text{sign of } b_1)\sqrt{\text{Coefficient of Determination}} \\ &= (\text{sign of } b_1)\sqrt{r^2} \end{aligned} \tag{14.13}$$

where

$$b_1 = \text{the slope of the estimated regression equation } \hat{y} = b_0 + b_1x$$

The sign for the sample correlation coefficient is positive if the estimated regression equation has a positive slope ($b_1 > 0$) and negative if the estimated regression equation has a negative slope ($b_1 < 0$).

For the Armand's Pizza Parlor example, the value of the coefficient of determination corresponding to the estimated regression equation $\hat{y} = 60 + 5x$ is .9027. Because the slope of the estimated regression equation is positive, (14.13) shows that the sample correlation coefficient is $+\sqrt{.9027} = +.9501$. With a sample correlation coefficient of $r_{xy} = +.9501$, we would conclude that there is a strong positive linear association between x and y.

In the case of a linear relationship between two variables, both the coefficient of determination and the sample correlation coefficient provide measures of the strength of the relationship. The coefficient of determination provides a measure between zero and one whereas the sample correlation coefficient provides a measure between -1 and $+1$. Although the sample correlation coefficient is restricted to a linear relationship between two variables, the coefficient of determination can be used for nonlinear relationships and for relationships that have two or more independent variables. In that sense, the coefficient of determination has a wider range of applicability.

NOTES AND COMMENTS

1. In developing the least squares estimated regression equation and computing the coefficient of determination, we made no probabilistic assumptions and no statistical tests for significance of the relationship between x and y. Larger values of r^2 imply that the least squares line provides a better fit to the data; that is, the observations are more closely grouped about the least squares line. But, using only r^2, we can draw no conclusion about whether the relationship between x and y is statistically significant. Such a conclusion must be based on considerations that involve the sample size and the properties of the appropriate sampling distributions of the least squares estimators.
2. As a practical matter, for typical data found in the social sciences, values of r^2 as low as .25 are often considered useful. For data in the physical and life sciences, r^2 values of .60 or greater are often found; in fact, in some cases, r^2 values greater than .90 can be found. In business applications, r^2 values vary greatly, depending on the unique characteristics of each application.

EXERCISES

Methods

15. The data from Exercise 1 follow.

x_i	1	2	3	4	5
y_i	3	7	5	11	14

The estimated regression equation for these data is $\hat{y} = .20 + 2.60x$.

a. Compute SSE, SST, and SSR using (14.8), (14.9), and (14.10).

b. Compute the coefficient of determination r^2. Comment on the goodness of fit.

c. Compute the sample correlation coefficient.

16. The data from Exercise 2 follow.

x_i	2	3	5	1	8
y_i	25	25	20	30	16

The estimated regression equation for these data is $\hat{y} = 30.33 - 1.88x$.

a. Compute SSE, SST, and SSR.

b. Compute the coefficient of determination r^2. Comment on the goodness of fit.

c. Compute the sample correlation coefficient.

17. The data from Exercise 3 follow.

x_i	2	4	5	7	8
y_i	2	3	2	6	4

The estimated regression equation for these data is $\hat{y} = .75 + .51x$. What percentage of the total sum of squares can be accounted for by the estimated regression equation? What is the value of the sample correlation coefficient?

Applications

18. The following data are the monthly salaries y and the grade point averages x for students who had obtained a bachelor's degree in business administration with a major in information systems. The estimated regression equation for these data is $\hat{y} = 1790.5 + 581.1x$.

GPA	Monthly Salary ($)
2.6	3300
3.4	3600
3.6	4000
3.2	3500
3.5	3900
2.9	3600

a. Compute SST, SSR, and SSE.
b. Compute the coefficient of determination r^2. Comment on the goodness of fit.
c. What is the value of the sample correlation coefficient?

19. The data from Exercise 7 follow.

CD file
DowS&P

	x = DJIA	y = S&P
February 11	10425	1387
February 18	10220	1346
February 25	9862	1333
March 3	10367	1409
March 10	9929	1395
March 17	10595	1464
March 24	11113	1527
March 31	10922	1499
April 7	11111	1516
April 14	10306	1357

The estimated regression equation for these data is $\hat{y} = -137.63 + .1489x$. What percentage of the total sum of squares can be accounted for by the estimated regression equation? Comment on the goodness of fit. What is the sample correlation coefficient?

20. The typical household income and typical home price for a sample of 18 cities follow (*Places Rated Almanac,* 2000). Data are in thousands of dollars.

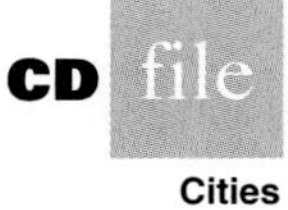
Cities

	Income	Home Price
Akron, OH	74.1	114.9
Atlanta, GA	82.4	126.9
Birmingham, AL	71.2	130.9
Bismarck, ND	62.8	92.8
Cleveland, OH	79.2	135.8
Columbia, SC	66.8	116.7
Denver, CO	82.6	161.9
Detroit, MI	85.3	145.0
Fort Lauderdale, FL	75.8	145.3
Hartford, CT	89.1	162.1
Lancaster, PA	75.2	125.9
Madison, WI	78.8	145.2
Naples, FL	100.0	173.6
Nashville, TN	77.3	125.9
Philadelphia, PA	87.0	151.5
Savannah, GA	67.8	108.1
Toledo, OH	71.2	101.1
Washington, DC	97.4	191.9

a. Use these data to develop an estimated regression equation that could be used to estimate the typical home price for a city given the typical household income.
b. Compute r^2. Would you feel comfortable using this estimated regression equation to estimate the typical home price for a city?
c. Estimate the typical home price for a city that has a typical household income of $95,000.

21. An important application of regression analysis in accounting is in the estimation of cost. By collecting data on volume and cost and using the least squares method to develop an estimated regression equation relating volume and cost, an accountant can estimate the cost associated with a particular manufacturing operation Consider the following sample of production volumes and total cost data for a manufacturing operation.

Production Volume (units)	Total Cost ($)
400	4000
450	5000
550	5400
600	5900
700	6400
750	7000

a. Use these data to develop an estimated regression equation that could be used to predict the total cost for a given production volume.
b. What is the variable or additional cost per unit produced?
c. Compute the coefficient of determination. What percentage of the variation in total cost can be explained by production volume?
d. The company's production schedule shows 500 units must be produced next month. What is the estimated total cost for this operation?

22. Are company presidents and chief executive officers paid according to the profit performance of the company? The following table lists corporate data on percentage change in return on equity over a 2-year period and percentage change in the pay of presidents and chief executive officers immediately after the 2-year period (*Business Week,* April 21, 1997).

Company	Two-Year Change in Return on Equity (%)	Change in Executive Compensation (%)
Dow Chemical	201.3	18
Rohm & Haas	146.5	28
Morton International	76.7	10
Union Carbide	158.2	28
Praxair	−34.9	15
Air Products & Chemicals	73.2	−9
Eastman Chemical	−7.9	−20

a. Develop the estimated regression equation with the 2-year percentage change in return on equity as the independent variable.
b. Compute r^2. Would you feel comfortable using the percentage change in return on equity over a 2-year period to predict the percentage change in the pay of presidents and chief executive officers? Discuss.
c. What is the sample correlation coefficient? Does it reflect a strong or weak relationship between return on equity and executive compensation?

14.4 MODEL ASSUMPTIONS

In conducting a regression analysis, we begin by making an assumption about the appropriate model for the relationship between the dependent and independent variable(s). For the case of simple linear regression, the assumed regression model is

$$y = \beta_0 + \beta_1 x + \epsilon$$

Then, the least squares method is used to develop values for b_0 and b_1, the estimates of the model parameters β_0 and β_1, respectively. The resulting estimated regression equation is

$$\hat{y} = b_0 + b_1 x$$

We saw that the value of the coefficient of determination (r^2) is a measure of the goodness of fit of the estimated regression equation. However, even with a large value of r^2, the estimated regression equation should not be used until further analysis of the appropriateness of the assumed model has been conducted. An important step in determining whether the assumed model is appropriate involves testing for the significance of the relationship. The tests of significance in regression analysis are based on the following assumptions about the error term ϵ.

Assumptions About the Error Term ϵ in the Regression Model

$$y = \beta_0 + \beta_1 x + \epsilon$$

1. The error term ϵ is a random variable with a mean or expected value of zero; that is, $E(\epsilon) = 0$.
   Implication: β_0 and β_1 are constants, therefore $E(\beta_0) = \beta_0$ and $E(\beta_1) = \beta_1$; thus, for a given value of x, the expected value of y is

   $$E(y) = \beta_0 + \beta_1 x \qquad \textbf{(14.14)}$$

   As we indicated previously, equation (14.14) is referred to as the regression equation.
2. The variance of ϵ, denoted by σ^2, is the same for all values of x.
   Implication: The variance of y equals σ^2 and is the same for all values of x.
3. The values of ϵ are independent.
   Implication: The value of ϵ for a particular value of x is not related to the value of ϵ for any other value of x; thus, the value of y for a particular value of x is not related to the value of y for any other value of x.
4. The error term ϵ is a normally distributed random variable.
   Implication: Because y is a linear function of ϵ, y is also a normally distributed random variable.

Figure 14.6 is an illustration of the model assumptions and their implications; note that in this graphical interpretation, the value of $E(y)$ changes according to the specific value of x considered. However, regardless of the x value, the probability distribution of ϵ and hence the probability distributions of y are normally distributed, each with the same variance. The specific value of the error ϵ at any particular point depends on whether the actual value of y is greater than or less than $E(y)$.

At this point, we must keep in mind that we are also making an assumption or hypothesis about the form of the relationship between x and y. That is, we have assumed that a straight line represented by $\beta_0 + \beta_1 x$ is the basis for the relationship between the variables. We must not lose sight of the fact that some other model, for instance $y = \beta_0 + \beta_1 x^2 + \epsilon$, may turn out to be a better model for the underlying relationship.

FIGURE 14.6 ASSUMPTIONS FOR THE REGRESSION MODEL

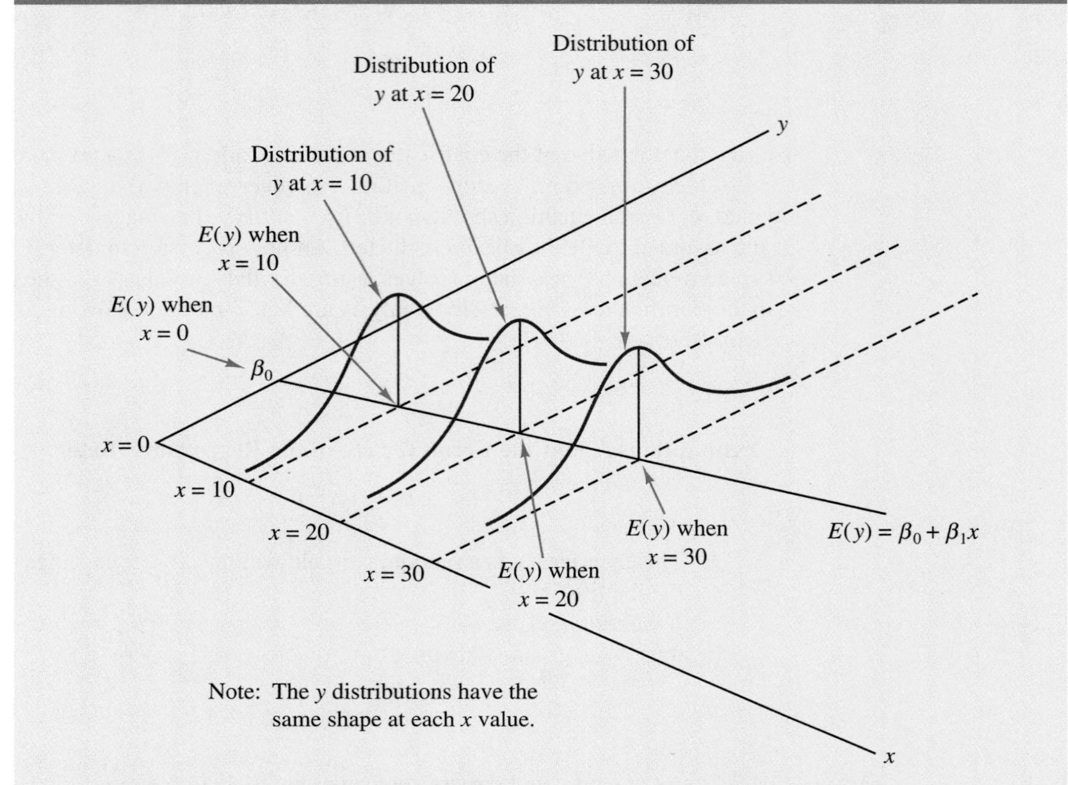

14.5 TESTING FOR SIGNIFICANCE

In a simple linear regression equation the mean or expected value of y is a linear function of x: $E(y) = \beta_0 + \beta_1 x$. If the value of β_1 is zero, $E(y) = \beta_0 + (0)x = \beta_0$. In this case, the mean value of y does not depend on the value of x and hence we would conclude that x and y are not linearly related. Alternatively, if the value of β_1 is not equal to zero, we would conclude that the two variables are related. Thus, to test for a significant regression relationship, we must conduct a hypothesis test to determine whether the value of β_1 is zero. Two tests are commonly used. Both require an estimate of σ^2, the variance of ϵ in the regression model.

Estimate of σ^2

From the regression model and its assumptions we can conclude that σ^2, the variance of ϵ, also represents the variance of the y values about the regression line. Recall that the deviations of the y values about the estimated regression line are called residuals. Thus, SSE, the sum of squared residuals, is a measure of the variability of the actual observations about the estimated regression line. The **mean square error** (MSE) provides the estimate of σ^2; it is SSE divided by its degrees of freedom.

With $\hat{y}_i = b_0 + b_1 x_i$, SSE can be written as

$$\text{SSE} = \Sigma(y_i - \hat{y}_i)^2 = \Sigma(y_i - b_0 - b_1 x_i)^2$$

Every sum of squares has associated with it a number called its degrees of freedom. Statisticians have shown that SSE has $n - 2$ degrees of freedom because two parameters (β_0 and β_1) must be estimated to compute SSE. Thus, the mean square is computed by dividing SSE by $n - 2$. MSE provides an unbiased estimator of σ^2. Because the value of MSE provides an estimate of σ^2, the notation s^2 is also used.

Mean Square Error (Estimate of σ^2)

$$s^2 = \text{MSE} = \frac{\text{SSE}}{n - 2} \tag{14.15}$$

In Section 14.3 we showed that for the Armand's Pizza Parlors example, SSE = 1530; hence,

$$s^2 = \text{MSE} = \frac{1530}{8} = 191.25$$

provides an unbiased estimate of σ^2.

To estimate σ we take the square root of s^2. The resulting value, s, is referred to as the **standard error of the estimate.**

Standard Error of the Estimate

$$s = \sqrt{\text{MSE}} = \sqrt{\frac{\text{SSE}}{n - 2}} \tag{14.16}$$

For the Armand's Pizza Parlors example, $s = \sqrt{\text{MSE}} = \sqrt{191.25} = 13.829$. In the following discussion, we use the standard error of the estimate in the tests for a significant relationship between x and y.

t Test

The simple linear regression model is $y = \beta_0 + \beta_1 x + \epsilon$. If x and y are linearly related, we must have $\beta_1 \neq 0$. The purpose of the t test is to see whether we can conclude that $\beta_1 \neq 0$. We will use the sample data to test the following hypotheses about the parameter β_1.

$$H_0: \beta_1 = 0$$
$$H_a: \beta_1 \neq 0$$

If H_0 is rejected, we will conclude that $\beta_1 \neq 0$ and that a statistically significant relationship exists between the two variables. However, if H_0 cannot be rejected, we will have insufficient evidence to conclude that a significant relationship exists. The properties of the sampling distribution of b_1, the least squares estimator of β_1, provide the basis for the hypothesis test.

First, let us consider what would have happened if we had used a different random sample for the same regression study. For example, suppose that Armand's Pizza Parlors had used the sales records of a different sample of 10 restaurants. A regression analysis of this new sample might result in an estimated regression equation similar to our previous estimated regression equation $\hat{y} = 60 + 5x$. However, it is doubtful that we would obtain exactly the same equation (with an intercept of exactly 60 and a slope of exactly 5). Indeed,

b_0 and b_1, the least squares estimators, are sample statistics that have their own sampling distributions. The properties of the sampling distribution of b_1 follow.

Sampling Distribution of b_1

Expected Value

$$E(b_1) = \beta_1$$

Standard Deviation

$$\sigma_{b_1} = \frac{\sigma}{\sqrt{\Sigma(x_i - \bar{x})^2}} \tag{14.17}$$

Distribution Form

Normal

Note that the expected value of b_1 is equal to β_1, so b_1 is an unbiased estimator of β_1.

Because we do not know the value of σ, we develop an estimate of σ_{b_1}, denoted s_{b_1}, by estimating σ with s in equation (14.17). Thus, we obtain the following estimate of σ_{b_1}.

The standard deviation of b_1 is also referred to as the standard error of b_1. Thus, s_{b_1} provides an estimate of the standard error of b_1.

Estimated Standard Deviation of b_1

$$s_{b_1} = \frac{s}{\sqrt{\Sigma(x_i - \bar{x})^2}} \tag{14.18}$$

For Armand's Pizza Parlors, $s = 13.829$. Hence, using $\Sigma(x_i - \bar{x})^2 = 568$ as shown in Table 14.2, we have

$$s_{b_1} = \frac{13.829}{\sqrt{568}} = .5803$$

as the estimated standard deviation of b_1.

The t test for a significant relationship is based on the fact that the test statistic

$$\frac{b_1 - \beta_1}{s_{b_1}}$$

follows a t distribution with $n - 2$ degrees of freedom. If the null hypothesis is true, then $\beta_1 = 0$ and $t = b_1/s_{b_1}$.

Let us conduct this test of significance for Armand's Pizza Parlors at the $\alpha = .01$ level of significance. The test statistic is

$$t = \frac{b_1}{s_{b_1}} = \frac{5}{.5803} = 8.62$$

From Table 2 of Appendix B we find that the two-tailed t value corresponding to $\alpha = .01$ and $n - 2 = 10 - 2 = 8$ degrees of freedom is $t_{.005} = 3.355$. With $8.62 > 3.355$, we reject H_0 and conclude at the .01 level of significance that β_1 is not equal to zero. The statistical evidence is sufficient to conclude that we have a significant relationship between student population and sales.

The p-value criterion can also be used to test for a significant relationship. The usual rejection rule applies: Reject H_0 if p-value $< \alpha$. However, because it is difficult to determine the p-value directly from tables of the t probability distribution, a computer software package such as Minitab or Excel is used. Appendixes 14.3 and 14.4 show the procedures that can be used; a discussion of the computer results for regression analysis is provided in Section 14.7. For Armand's Pizza Parlors, the p-value associated with the test statistic $t = 8.62$ is .000 (to three decimal places). With a p-value $= .000 < \alpha = .05$, we reject H_0 and conclude that we have a significant relationship between student population and sales. A summary of the t test for a significant relationship follows.

t Test for Significance in Simple Linear Regression

$$H_0\colon \beta_1 = 0$$
$$H_a\colon \beta_1 \neq 0$$

Test Statistic

$$t = \frac{b_1}{s_{b_1}} \tag{14.19}$$

Rejection Rule

Using test statistic: Reject H_0 if $t < -t_{\alpha/2}$ or if $t > t_{\alpha/2}$
Using p-value: Reject H_0 if p-value $< \alpha$

where $t_{\alpha/2}$ is based on a t distribution with $n - 2$ degrees of freedom.

Confidence Interval for β_1

The form of a confidence interval for β_1 is as follows:

$$b_1 \pm t_{\alpha/2} s_{b_1}$$

The point estimator is b_1 and the margin of error is $t_{\alpha/2} s_{b_1}$. The confidence coefficient associated with this interval is $1 - \alpha$, and $t_{\alpha/2}$ is the t value providing an area of $\alpha/2$ in the upper tail of a t distribution with $n - 2$ degrees of freedom. For example, suppose that we wanted to develop a 99% confidence interval estimate of β_1 for Armand's Pizza Parlors. From Table 2 of Appendix B we find that the t value corresponding to $\alpha = .01$ and $n - 2 = 10 - 2 = 8$ degrees of freedom is $t_{.005} = 3.355$. Thus, the 99% confidence interval estimate of β_1 is

$$b_1 \pm t_{\alpha/2} s_{b_1} = 5 \pm 3.355(.5803) = 5 \pm 1.95$$

or 3.05 to 6.95.

In using the t test for significance, the hypotheses tested were

$$H_0\colon \beta_1 = 0$$
$$H_a\colon \beta_1 \neq 0$$

At the $\alpha = .01$ level of significance, we can use the 99% confidence interval as an alternative for drawing the hypothesis testing conclusion for the Armand's data. Because 0, the hypothesized value of β_1, is not included in the confidence interval (3.05 to 6.95), we can reject H_0 and conclude that a significant statistical relationship exists between population and sales. In general, a confidence interval can be used to test any two-sided hypothesis about β_1. If the hypothesized value of β_1 is contained in the confidence interval, do not reject H_0. Otherwise, reject H_0.

F Test

An F test, based on the F probability distribution, can also be used to test for significance in regression. With only one independent variable, the F test will provide the same conclusion as the t test; that is, if the t test indicates $\beta_1 \neq 0$ and hence a significant relationship, the F test will also indicate a significant relationship. But with more than one independent variable, only the F test can be used to test for an overall significant relationship.

The logic behind the use of the F test for determining whether the regression relationship is statistically significant is based on the development of two independent estimates of σ^2. We have just seen that MSE provides an estimate of σ^2. If the null hypothesis H_0: $\beta_1 = 0$ is true, the sum of squares due to regression, SSR, divided by its degrees of freedom provides another independent estimate of σ^2. This estimate is called the *mean square due to regression,* or simply the *mean square regression,* and is denoted MSR. In general,

$$\text{MSR} = \frac{\text{SSR}}{\text{regression degrees of freedom}}$$

For the models we consider in this text, the regression degrees of freedom is always equal to the number of independent variables; thus,

$$\text{MSR} = \frac{\text{SSR}}{\text{number of independent variables}} \tag{14.20}$$

Because we consider only regression models with one independent variable in this chapter, we have MSR = SSR/1 = SSR. Hence, for Armand's Pizza Parlors, MSR = SSR = 14,200.

If the null hypothesis (H_0: $\beta_1 = 0$) is true, MSR and MSE are two independent estimates of σ^2 and the sampling distribution of MSR/MSE follows an F distribution with numerator degrees of freedom equal to one and denominator degrees of freedom equal to $n - 2$. Therefore, when $\beta_1 = 0$, the value of MSR/MSE should be close to one. However, if the null hypothesis is false ($\beta_1 \neq 0$), MSR will overestimate σ^2 and the value of MSR/MSE will be inflated; thus, large values of MSR/MSE lead to the rejection of H_0 and the conclusion that the relationship between x and y is statistically significant.

Let us conduct the F test for the Armand's Pizza Parlors example. The test statistic is

$$F = \frac{\text{MSR}}{\text{MSE}} = \frac{14{,}200}{191.25} = 74.25$$

The F test and the t test provide identical results for simple linear regression.

From Table 4 of Appendix B we find that the F value corresponding to $\alpha = .01$ with one degree of freedom in the numerator and $n - 2 = 10 - 2 = 8$ degrees of freedom in the denominator is $F_{.01} = 11.26$. With $74.25 > 11.26$, we reject H_0 and conclude at the .01 level of significance that β_1 is not equal to zero. The F test has provided the statistical evidence necessary to conclude that we have a significant relationship between student population and sales.

The p-value criterion can also be used with the F test. The usual rejection rule applies: Reject H_0 if p-value $< \alpha$. However, because it is difficult to determine the p-value directly from tables of the F probability distribution, a computer software package such as Minitab or Excel is used. Appendixes 14.3 and 14.4 show the procedures that can be used; a discussion of the computer results for regression analysis is provided in Section 14.7. For Armand's Pizza Parlors, the p-value associated with the test statistic $F = 74.25$ is .000 (to three decimal places). With a p-value $= .000 < \alpha = .05$, we reject H_0 and conclude that we have a significant relationship between student population and sales. A summary of the F test for a significant relationship follows.

If H_0 is false, MSE *still provides an unbiased estimate of σ^2 and* MSR *overestimates σ^2. If H_0 is true, both* MSE *and* MSR *provide unbiased estimates of σ^2; in this case the value of* MSR/MSE *should be close to 1.*

F Test for Significance in Simple Linear Regression

$$H_0\colon \beta_1 = 0$$
$$H_a\colon \beta_1 \neq 0$$

Test Statistic

$$F = \frac{\text{MSR}}{\text{MSE}} \tag{14.21}$$

Rejection Rule

Using test statistic: Reject H_0 if $F > F_\alpha$
Using p-value: Reject H_0 if p-value $< \alpha$

where F_α is based on an F distribution with 1 degree of freedom in the numerator and $n - 2$ degrees of freedom in the denominator.

In Chapter 13 we covered analysis of variance (ANOVA) and showed how an **ANOVA table** could be used to provide a convenient summary of the computational aspects of analysis of variance. A similar ANOVA table can be used to summarize the results of the F test for significance in regression. Table 14.5 is the general form of the ANOVA table for simple linear regression. Table 14.6 is the ANOVA table with the F test computations performed for

In every analysis of variance table the total sum of squares is the sum of the regression sum of squares and the error sum of squares; in addition, the total degrees of freedom is the sum of the regression degrees of freedom and the error degrees of freedom.

TABLE 14.5 GENERAL FORM OF THE ANOVA TABLE FOR SIMPLE LINEAR REGRESSION

Source of Variation	Sum of Squares	Degrees of Freedom	Mean Square	F
Regression	SSR	1	$\text{MSR} = \frac{\text{SSR}}{1}$	$F = \frac{\text{MSR}}{\text{MSE}}$
Error	SSE	$n - 2$	$\text{MSE} = \frac{\text{SSE}}{n-2}$	
Total	SST	$n - 1$		

TABLE 14.6 ANOVA TABLE FOR THE ARMAND'S PIZZA PARLORS PROBLEM

Source of Variation	Sum of Squares	Degrees of Freedom	Mean Square	F
Regression	14,200	1	$\frac{14{,}200}{1} = 14{,}200$	$\frac{14{,}200}{191.25} = 74.25$
Error	1,530	8	$\frac{1530}{8} = 191.25$	
Total	15,730	9		

Armand's Pizza Parlors. Regression, Error, and Total are the labels for the three sources of variation, with SSR, SSE, and SST appearing as the corresponding sum of squares in column two. The degrees of freedom, 1 for SSR, $n - 2$ for SSE, and $n - 1$ for SST, are shown in column 3. Column 4 contains the values of MSR and MSE and column 5 contains the value of $F = \text{MSR/MSE}$. Almost all computer printouts of regression analysis include an ANOVA table summary of the test for significance.

Some Cautions About the Interpretation of Significance Tests

Regression analysis, which can be used to identify how variables are associated with one another, cannot be used as evidence of a cause-and-effect relationship.

Rejecting the null hypothesis $H_0\colon \beta_1 = 0$ and concluding that the relationship between x and y is significant does not enable us to conclude that a cause-and-effect relationship is present between x and y. Concluding a cause-and-effect relationship is warranted only if the analyst has some type of theoretical justification that the relationship is in fact causal. In the Armand's Pizza Parlors example, we can conclude that there is a significant relationship between the size of the student population x and quarterly sales y; moreover, the estimated regression equation $\hat{y} = 60 + 5x$ provides the least squares estimate of the relationship. We cannot, however, conclude that changes in student population x *cause* changes in quarterly sales y just because we have identified a statistically significant relationship. The appropriateness of such a cause-and-effect conclusion is left to supporting theoretical justification and to good judgment on the part of the analyst. Armand's managers felt that increases in the student population were a likely cause of increased quarterly sales. Thus, the result of the significance test enabled them to conclude that a cause-and-effect relationship was present.

In addition, just because we are able to reject $H_0\colon \beta_1 = 0$ and demonstrate statistical significance does not enable us to conclude that the relationship between x and y is linear. We can state only that x and y are related and that a linear relationship explains a significant portion of the variability in y over the range of values for x observed in the sample. Figure 14.7 illustrates this situation. The test for significance calls for the rejection of the

FIGURE 14.7 EXAMPLE OF A LINEAR APPROXIMATION OF A NONLINEAR RELATIONSHIP

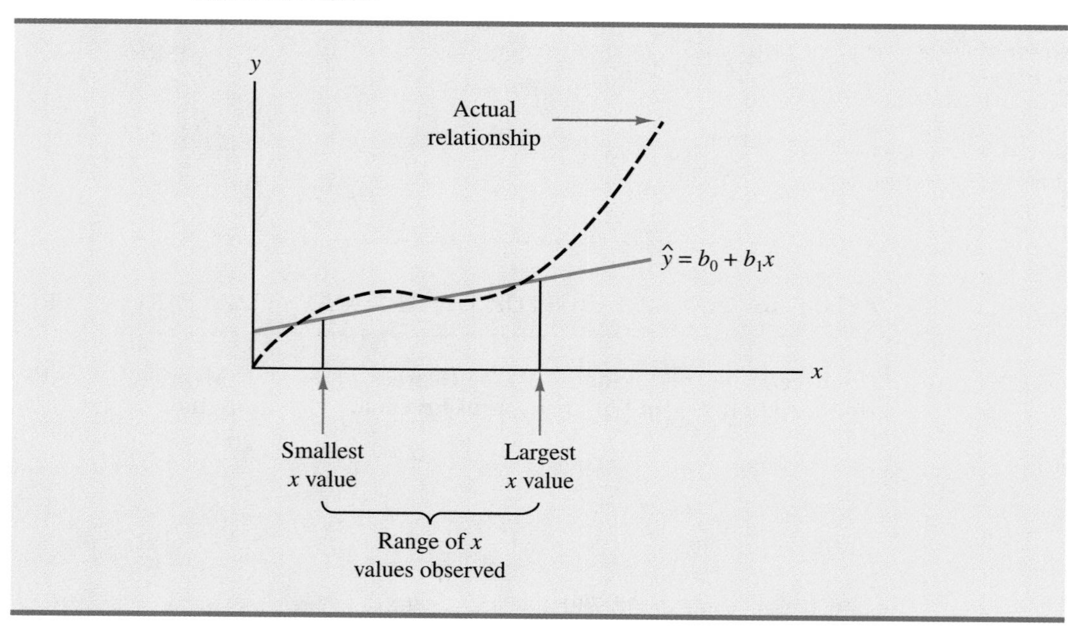

null hypothesis H_0: $\beta_1 = 0$ and leads to the conclusion that x and y are significantly related, but the figure shows that the actual relationship between x and y is not linear. Although the linear approximation provided by $\hat{y} = b_0 + b_1x$ is good over the range of x values observed in the sample, it becomes poor for x values outside that range.

Given a significant relationship, we should feel confident in using the estimated regression equation for predictions corresponding to x values within the range of the x values observed in the sample. For Armand's Pizza Parlors, this range corresponds to values of x between 2 and 26. But unless other reasons indicate that the model is valid beyond this range, predictions outside the range of the independent variable should be made with caution. For Armand's Pizza Parlors, because the regression relationship has been found significant at the .01 level, we should feel confident using it to predict sales for restaurants where the associated student population is between 2000 and 26,000.

NOTES AND COMMENTS

1. The assumptions made about the error term (Section 14.4) are what allow the tests of statistical significance in this section. The properties of the sampling distribution of b_1 and the subsequent t and F tests follow directly from these assumptions.
2. Do not confuse statistical significance with practical significance. With very large sample sizes, statistically significant results can be obtained for small values of b_1; in such cases, one must exercise care in concluding that the relationship has practical significance.
3. A test of significance for a linear relationship between x and y can also be performed by using the sample correlation coefficient r_{xy}. With ρ_{xy} denoting the population correlation coefficient, the hypotheses are as follows.

$$H_0: \rho_{xy} = 0$$
$$H_a: \rho_{xy} \neq 0$$

A significant relationship can be concluded if H_0 is rejected. The details of this test are provided in Appendix 14.2. However, the t and F tests presented previously in this section give the *same result* as the test for significance using the correlation coefficient. Conducting a test for significance using the correlation coefficient therefore is not necessary if a t or F test has already been conducted.

EXERCISES

Methods

23. The data from Exercise 1 follow.

x_i	1	2	3	4	5
y_i	3	7	5	11	14

a. Compute the mean square error using (14.15).
b. Compute the standard error of the estimate using (14.16).
c. Compute the estimated standard deviation of b_1 using (14.18).
d. Use the t test to test the following hypotheses ($\alpha = .05$):

$$H_0: \beta_1 = 0$$
$$H_a: \beta_1 \neq 0$$

e. Use the F test to test the hypotheses in (d) at a .05 level of significance. Present the results in the analysis of variance table format.

24. The data from Exercise 2 follow.

x_i	2	3	5	1	8
y_i	25	25	20	30	16

a. Compute the mean square error using (14.15).
b. Compute the standard error of the estimate using (14.16).
c. Compute the estimated standard deviation of b_1 using (14.18).
d. Use the t test to test the following hypotheses ($\alpha = .05$):

$$H_0: \beta_1 = 0$$
$$H_a: \beta_1 \neq 0$$

e. Use the F test to test the hypotheses in (d) at a .05 level of significance. Present the results in the analysis of variance table format.

25. The data from Exercise 3 follow.

x_i	2	4	5	7	8
y_i	2	3	2	6	4

a. What is the value of the standard error of the estimate?
b. Test for a significant relationship by using the t test. Use $\alpha = .05$.
c. Use the F test to test for a significant relationship. Use $\alpha = .05$. What is your conclusion?

Applications

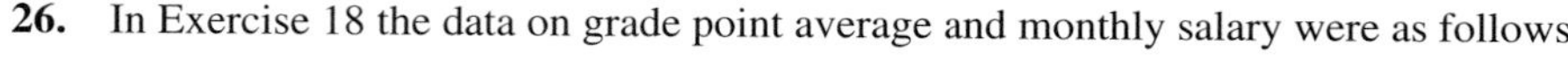

26. In Exercise 18 the data on grade point average and monthly salary were as follows.

GPA	Monthly Salary ($)	GPA	Monthly Salary ($)
2.6	3300	3.2	3500
3.4	3600	3.5	3900
3.6	4000	2.9	3600

a. Does the t test indicate a significant relationship between grade point average and monthly salary? What is your conclusion? Use $\alpha = .05$.
b. Test for a significant relationship using the F test. What is your conclusion? Use $\alpha = .05$.
c. Show the ANOVA table.

27. Refer to Exercise 9, where an estimated regression equation relating years of experience and annual sales was developed. At a .05 level of significance, determine whether years of experience and annual sales are related.

28. Refer to Exercise 10, where an estimated regression equation relating the performance score and the overall rating for notebook PCs was developed. At the .05 level of significance, test whether these two variables are related. Show the ANOVA table. What is your conclusion?

29. Refer to Exercise 21, where data on production volume and cost were used to develop an estimated regression equation relating production volume and cost for a particular manufacturing operation. Using $\alpha = .05$, test whether the production volume is significantly related to the total cost. Show the ANOVA table. What is your conclusion?

30. Refer to Exercise 22, where the following data were used to determine whether company presidents and chief executive officers are paid on the basis of company profit performance (*Business Week,* April 21, 1997).

Company	Two-Year Change in Return on Equity (%)	Change in Executive Compensation (%)
Dow Chemical	201.3	18
Rohm & Haas	146.5	28
Morton International	76.7	10
Union Carbide	158.2	28
Praxair	−34.9	15
Air Products & Chemicals	73.2	−9
Eastman Chemical	−7.9	−20

Does the evidence indicate a significant relationship between the two variables? Conduct the appropriate statistical test and state your conclusion. Use $\alpha = .05$.

31. Refer to Exercise 20, where an estimated regression equation was developed relating typical household income and typical home price. Test whether the typical household income for a city and the typical home price are related at the .01 level of significance.

14.6 USING THE ESTIMATED REGRESSION EQUATION FOR ESTIMATION AND PREDICTION

The simple linear regression model is an assumption about the relationship between x and y. Using the least squares method, we obtained the estimated simple linear regression equation. If the results show a statistically significant relationship between x and y, and the fit provided by the estimated regression equation appears to be good, the estimated regression equation should be useful for estimation and prediction.

Point Estimation

In the Armand's Pizza Parlors example, the estimated regression equation $\hat{y} = 60 + 5x$ provides an estimate of the relationship between the size of the student population x and quarterly sales y. We can use the estimated regression equation to develop a point estimate of the mean value of y for a particular value of x or to predict an individual value of y corresponding to a given value of x. For instance, suppose Armand's managers want a point estimate of the mean quarterly sales for all restaurants located near college campuses with 10,000 students. Using the estimated regression equation $\hat{y} = 60 + 5x$, we see that for $x = 10$ (or 10,000 students), $\hat{y} = 60 + 5(10) = 110$. Thus, a point estimate of the mean quarterly sales for all restaurants located near campuses with 10,000 students is \$110,000.

Now suppose Armand's managers want to predict sales for an individual restaurant located near Talbot College, a school with 10,000 students. In this case we are not interested in the mean value for all restaurants located near campuses with 10,000 students; we are just interested in predicting quarterly sales for one individual restaurant. As it turns out, the point estimate is the same as the point estimate for the mean value of y. Hence, we would predict quarterly sales of $\hat{y} = 60 + 5(10) = 110$ or \$110,000 for this one restaurant.

Interval Estimation

Confidence intervals and prediction intervals indicate the precision of the regression results. Narrower intervals indicate a higher degree of precision.

Point estimates do not provide any information about the precision associated with an estimate. For that we must develop interval estimates much like those in Chapters 8, 10, and 11. The first type of interval estimate, a **confidence interval estimate,** is an interval estimate of the *mean value of y* for a given value of x. The second type of interval estimate, a

prediction interval estimate, is used whenever we want an interval estimate of an *individual value of y* corresponding to a given value of x. The point estimate of the mean value of y is the same as the point estimate of an individual value of y. But, the interval estimates we obtain for the two cases are different.

Confidence Interval Estimate of the Mean Value of y

The estimated regression equation provides a point estimate of the mean value of y for a given value of x. In describing the confidence interval estimation procedure, we will use the following notation.

$$x_p = \text{the particular or given value of the independent variable } x$$
$$E(y_p) = \text{the mean or expected value of the dependent variable } y \text{ corresponding to the given } x_p$$
$$\hat{y}_p = b_0 + b_1 x_p = \text{the point estimate of } E(y_p) \text{ when } x = x_p$$

Using this notation to estimate the mean sales for all Armand's restaurants located near a campus with 10,000 students, we have $x_p = 10$, and $E(y_p)$ denotes the unknown mean value of sales for all restaurants where $x_p = 10$. The estimate of $E(y_p)$ is provided by $\hat{y}_p = 60 + 5(10) = 110$.

In general, we cannot expect $\hat{y}_p$ to equal $E(y_p)$ exactly. If we want to make an inference about how close $\hat{y}_p$ is to the true mean value $E(y_p)$, we will have to estimate the variance of $\hat{y}_p$. The formula for estimating the variance of $\hat{y}_p$ given x_p, denoted by $s^2_{\hat{y}_p}$, is

$$s^2_{\hat{y}_p} = s^2\left[\frac{1}{n} + \frac{(x_p - \bar{x})^2}{\Sigma(x_i - \bar{x})^2}\right] \tag{14.22}$$

The estimate of the standard deviation of $\hat{y}_p$ is given by the square root of (14.22).

$$s_{\hat{y}_p} = s\sqrt{\frac{1}{n} + \frac{(x_p - \bar{x})^2}{\Sigma(x_i - \bar{x})^2}} \tag{14.23}$$

The computational results for Armand's Pizza Parlors in Section 14.5 provided $s = 13.829$. With $x_p = 10$, $\bar{x} = 14$, and $\Sigma(x_i - \bar{x})^2 = 568$, we can use (14.23) to obtain

$$s_{\hat{y}_p} = 13.829\sqrt{\frac{1}{10} + \frac{(10 - 14)^2}{568}}$$
$$= 13.829\sqrt{.1282} = 4.95$$

The general expression for a confidence interval estimate of $E(y_p)$ follows.

The margin of error associated with this internal estimate is $t_{\alpha/2}s_{\hat{y}_p}$

Confidence Interval Estimate of $E(y_p)$

$$\hat{y}_p \pm t_{\alpha/2}s_{\hat{y}_p} \tag{14.24}$$

where the confidence coefficient is $1 - \alpha$ and $t_{\alpha/2}$ is based on a t distribution with $n - 2$ degrees of freedom.

Using (14.24) to develop a 95% confidence interval estimate of the mean quarterly sales for all Armand's restaurants located near campuses with 10,000 students, we need the value of t for $\alpha/2 = .025$ and $n - 2 = 10 - 2 = 8$ degrees of freedom. Using Table 2 of Appendix B, we have $t_{.025} = 2.306$. Thus, with $\hat{y}_p = 110$ and a margin of error of $t_{\alpha/2}s_{\hat{y}_p} = 2.306(4.95) = 11.415$, the 95% confidence interval estimate is

$$110 \pm 11.415$$

In dollars, the 95% confidence interval for the mean quarterly sales of all restaurants near campuses with 10,000 students is \$110,000 ± \$11,415. Therefore, the confidence interval estimate for the mean quarterly sales when the student population is 10,000 is \$98,585 to \$121,415.

Note that the estimated standard deviation of $\hat{y}_p$ given by equation (14.23) is smallest when $x_p = \bar{x}$ and the quantity $x_p - \bar{x} = 0$. In this case, the estimated standard deviation of $\hat{y}_p$ becomes

$$s_{\hat{y}_p} = s\sqrt{\frac{1}{n} + \frac{(\bar{x} - \bar{x})^2}{\Sigma(x_i - \bar{x})^2}} = s\sqrt{\frac{1}{n}}$$

This result implies that we can make the best or most precise estimate of the mean value of y whenever $x_p = \bar{x}$. In fact, the further x_p is from $\bar{x}$, the larger $x_p - \bar{x}$ becomes. As a result, confidence intervals for the mean value of y will become wider as x_p deviates more from $\bar{x}$. This pattern is shown graphically in Figure 14.8.

FIGURE 14.8 CONFIDENCE INTERVALS FOR THE MEAN SALES y AT GIVEN VALUES OF STUDENT POPULATION x

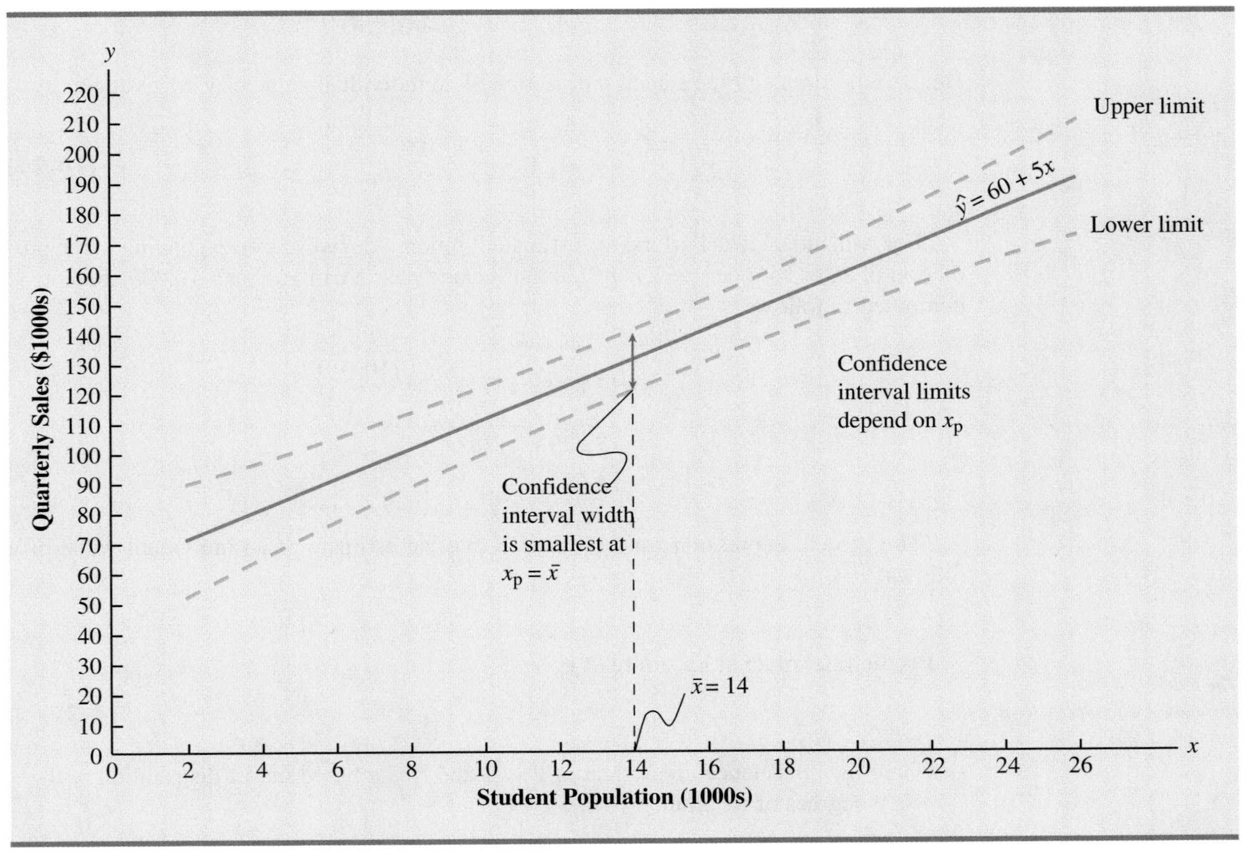

Prediction Interval Estimate of an Individual Value of y

Suppose that instead of estimating the mean value of sales for all Armand's restaurants located near campuses with 10,000 students, we want to estimate the sales for an individual restaurant located near Talbot College, a school with 10,000 students. As noted previously, the point estimate of an individual value of y given $x = x_p$ is provided by the estimated regression equation with $\hat{y}_p = b_0 + b_1 x_p$. For the restaurant at Talbot College, we have $x_p = 10$ and a corresponding estimated quarterly sales of $\hat{y}_p = 60 + 5(10) = 110$, or \$110,000. Note that this value is the same as the point estimate of the mean sales for all restaurants located near campuses with 10,000 students.

To develop a prediction interval estimate, we must first determine the variance associated with using $\hat{y}_p$ as an estimate of an individual value of y when $x = x_p$. This variance is made up of the sum of the following two components.

1. The variance of individual y values about the mean $E(y_p)$, an estimate of which is given by s^2.
2. The variance associated with using $\hat{y}_p$ to estimate $E(y_p)$, an estimate of which is given by $s^2_{\hat{y}_p}$.

The formula for estimating the variance of an individual value of y_p, denoted by s^2_{ind}, is

$$\begin{aligned} s^2_{\text{ind}} &= s^2 + s^2_{\hat{y}_p} \\ &= s^2 + s^2\left[\frac{1}{n} + \frac{(x_p - \bar{x})^2}{\Sigma(x_i - \bar{x})^2}\right] \\ &= s^2\left[1 + \frac{1}{n} + \frac{(x_p - \bar{x})^2}{\Sigma(x_i - \bar{x})^2}\right] \end{aligned} \quad \textbf{(14.25)}$$

Hence, an estimate of the standard deviation of an individual value of y_p is given by

$$s_{\text{ind}} = s\sqrt{1 + \frac{1}{n} + \frac{(x_p - \bar{x})^2}{\Sigma(x_i - \bar{x})^2}} \quad \textbf{(14.26)}$$

For Armand's Pizza Parlors, the estimated standard deviation corresponding to the prediction of sales for one specific restaurant located near a campus with 10,000 students is computed as follows.

$$\begin{aligned} s_{\text{ind}} &= 13.829\sqrt{1 + \frac{1}{10} + \frac{(10 - 14)^2}{568}} \\ &= 13.829\sqrt{1.1282} \\ &= 14.69 \end{aligned}$$

The general expression for a prediction interval estimate of an individual value of y follows.

The margin of error associated with this interval estimate is $t_{\alpha/2}s_{\text{ind}}$.

Prediction Interval Estimate of y_p

$$\hat{y}_p \pm t_{\alpha/2}s_{\text{ind}} \quad \textbf{(14.27)}$$

where the confidence coefficient is $1 - \alpha$ and $t_{\alpha/2}$ is based on a t distribution with $n - 2$ degrees of freedom.

The 95% prediction interval for quarterly sales at Armand's Talbot College restaurant can be found by using $t_{.025} = 2.306$ and $s_{\text{ind}} = 14.69$. Thus, with $\hat{y}_{\text{p}} = 110$ and a margin of error of $t_{\alpha/2}s_{\text{ind}} = 2.306(14.69) = 33.875$, the 95% prediction interval estimate is

$$110 \pm 33.875$$

In dollars, this prediction interval is \$110,000 ± \$33,875 or \$76,125 to \$143,875. Note that the prediction interval for an individual restaurant is wider than the confidence interval for the mean sales of all restaurants located near campuses with 10,000 students (\$98,585 to \$121,415). The difference reflects the fact that we are able to estimate the mean value of y more precisely than we can an individual value of y.

Both confidence interval estimates and prediction interval estimates are most precise when the value of the independent variable is $x_{\text{p}} = \bar{x}$. The general shapes of confidence intervals and the wider prediction intervals are shown together in Figure 14.9.

FIGURE 14.9 CONFIDENCE AND PREDICTION INTERVALS FOR SALES y AT GIVEN VALUES OF STUDENT POPULATION x

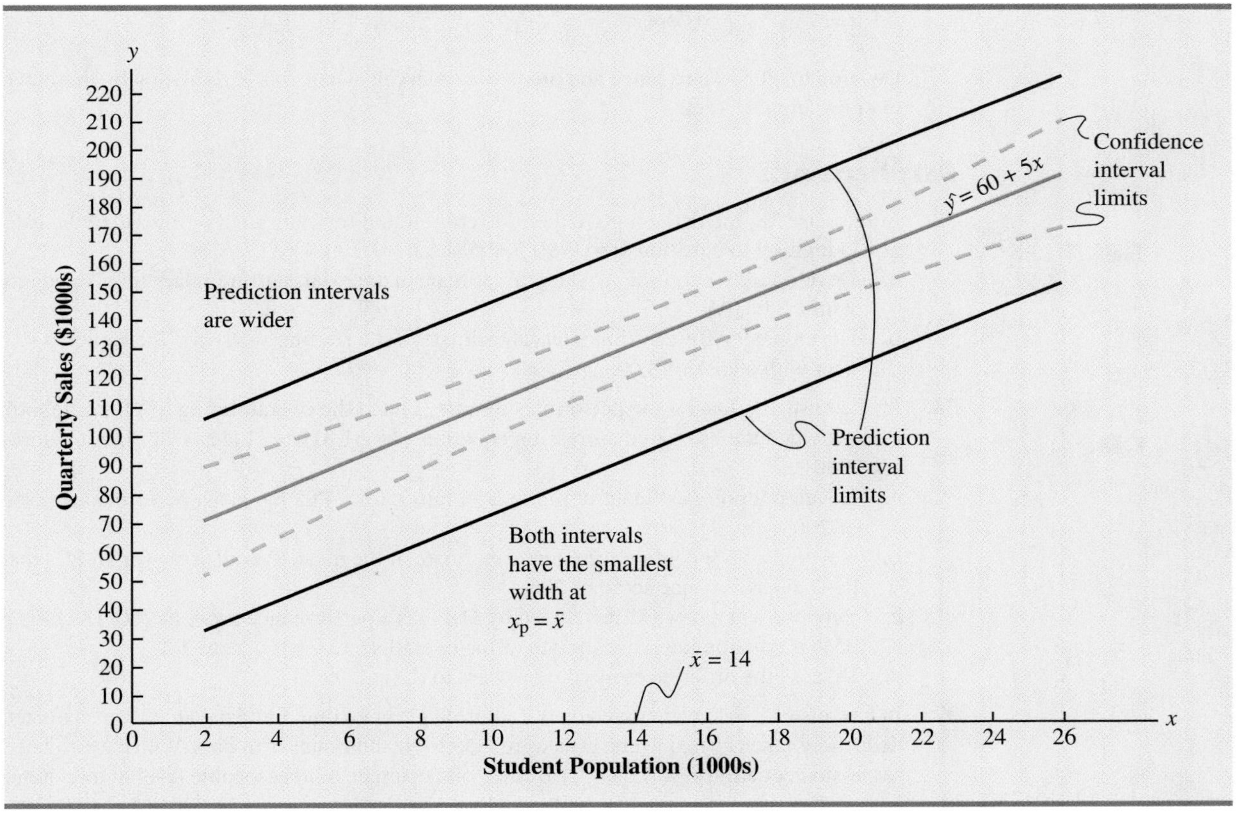

EXERCISES

Methods

32. The data from Exercise 1 follow.

x_i	1	2	3	4	5
y_i	3	7	5	11	14

a. Use (14.23) to estimate the standard deviation of $\hat{y}_p$ when $x = 4$.
b. Use (14.24) to develop a 95% confidence interval estimate of the expected value of y when $x = 4$.
c. Use (14.26) to estimate the standard deviation of an individual value of y when $x = 4$.
d. Use (14.27) to develop a 95% prediction interval for y when $x = 4$.

33. The data from Exercise 2 follow.

x_i	2	3	5	1	8
y_i	25	25	20	30	16

a. Estimate the standard deviation of $\hat{y}_p$ when $x = 3$.
b. Develop a 95% confidence interval estimate of the expected value of y when $x = 3$.
c. Estimate the standard deviation of an individual value of y when $x = 3$.
d. Develop a 95% prediction interval for y when $x = 3$.

34. The data from Exercise 3 follow.

x_i	2	4	5	7	8
y_i	2	3	2	6	4

Develop the 95% confidence and prediction intervals when $x = 3$. Explain why these two intervals are different.

Applications

35. In Exercise 18, the data on grade point average x and monthly salary y provided the estimated regression equation $\hat{y} = 1790.5 + 581.1x$.
a. Develop a 95% confidence interval estimate of the mean starting salary for all students with a 3.0 GPA.
b. Develop a 95% prediction interval estimate of the starting salary for Joe Heller, a student with a GPA of 3.0.

PCs

36. In Exercise 10, data on the performance score (x) and the overall rating (y) for notebook PCs provided the estimated regression equation $\hat{y} = 51.819 + .1452x$ (*PC World,* February 2000).
a. Develop a point estimate of the overall rating for a PC that has a performance score of 200.
b. Develop a 95% confidence interval estimate of the mean overall score for all PCs that have a performance score of 200.
c. Suppose that a new PC developed by Dell has a performance score of 200. Develop a 95% prediction interval estimate of the overall score for this new PC.
d. Discuss the differences in your answers to (b) and (c).

37. In Exercise 13, data were given on the adjusted gross income x and the amount of itemized deductions taken by taxpayers. Data were reported in thousands of dollars. With the estimated regression equation $\hat{y} = 4.68 + .16x$, the point estimate of a reasonable level of total itemized deductions for a taxpayer with an adjusted gross income of \$52,500 is \$13,080.
a. Develop a 95% confidence interval estimate of the mean amount of total itemized deductions for all taxpayers with an adjusted gross income of \$52,500.
b. Develop a 95% prediction interval estimate for the amount of total itemized deductions for a particular taxpayer with an adjusted gross income of \$52,500.
c. If the particular taxpayer referred to in (b) has claimed total itemized deductions of \$20,400, would the IRS agent's request for an audit appear to be justified?
d. Using your answer to (b), give the IRS agent a guideline as to the amount of total itemized deductions a taxpayer with an adjusted gross income of \$52,500 should have before an audit is recommended.

38. Refer to Exercise 21, where data on the production volume x and total cost y for a particular manufacturing operation were used to develop the estimated regression equation $\hat{y} = 1246.67 + 7.6x$.

a. The company's production schedule shows that 500 units must be produced next month. What is the point estimate of the total cost for next month?

b. Develop a 99% prediction interval estimate of the total cost for next month.

c. If an accounting cost report at the end of next month shows that the actual production cost during the month was $6000, should managers be concerned about incurring such a high total cost for the month? Discuss.

39. Nielsen Media Research collects data showing the number of households tuned in to shows that carry a particular advertisement. This information is useful to advertisers because it tells them how many consumers they are reaching. The following data show the number of household exposures in millions and the number of times the ad was aired for the week of October 27–November 2, 1997 (*USA Today,* November 17, 1997).

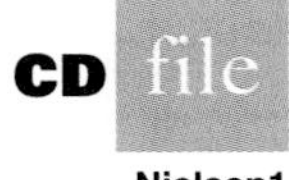

Nielsen1

Advertised Brand	Times Ad Aired	Household Exposures
McDonald's	49	359.6
Burger King	42	296.1
HBO	30	271.6
Red Corner movie	26	251.1
Pizza Hut	31	229.3
Sears	20	186.9
Isuzu Rodeo	21	186.3
MCI	24	172.7
Sprint	15	166.0
JC Penney	19	162.1

a. Use these data to develop an estimated regression equation that could be used to predict the number of household exposures given the number of times the ad is aired.

b. Did the estimated regression equation provide a good fit? Explain.

c. Develop a 95% confidence interval estimate for the number of household exposures for all ads that are aired 35 times.

d. Suppose that Wendy's was considering airing a particular ad 35 times. Develop a 95% prediction interval estimate for the number of household exposures for this particular ad.

14.7 COMPUTER SOLUTION

Performing the regression analysis computations without the help of a computer can be quite time consuming. In this section we discuss how the computational burden can be minimized by using a computer software package such as Minitab.

We entered Armand's student population and sales data into a Minitab worksheet. The independent variable was named Pop and the dependent variable was named Sales to assist with interpretation of the computer output. Using Minitab, we obtained the printout for Armand's Pizza Parlors shown in Figure 14.10.* The interpretation of this printout follows.

*The Minitab steps necessary to generate the output are given in Appendix 14.3.

FIGURE 14.10 MINITAB OUTPUT FOR THE ARMAND'S PIZZA PARLORS PROBLEM

```
The regression equation is
Sales = 60.0 + 5.00 Pop

Predictor        Coef        Stdev     t-ratio        p
Constant       60.000        9.226        6.50    0.000
Pop            5.0000       0.5803        8.62    0.000

s = 13.83      R-sq = 90.3%     R-sq(adj) = 89.1%

Analysis of Variance

SOURCE        DF         SS          MS         F         p
Regression     1      14200       14200     74.25     0.000
Error          8       1530         191
Total          9      15730

      Fit  Stdev.Fit          95% C.I.           95% P.I.
   110.00       4.95   ( 98.58, 121.42)   ( 76.12, 143.88)
```


1. Minitab prints the estimated regression equation as Sales = 60.0 + 5.00 Pop.
2. A table is printed that shows the values of the coefficients b_0 and b_1, the standard deviation of each coefficient, the t value obtained by dividing each coefficient value by its standard deviation, and the p-value associated with the t test. Thus, to test H_0: $\beta_1 = 0$ versus H_a: $\beta_1 \neq 0$, we could compare 8.62 (located in the t-ratio column) to the appropriate critical value. This is the procedure described for the t test in Section 14.5. Alternatively, we could use the p-value provided by Minitab to perform the same test. Because the p-value in this case is zero (to three decimal places), the sample results indicate that the null hypothesis (H_0: $\beta_1 = 0$) should be rejected.
3. Minitab prints the standard error of the estimate, $s = 13.83$, as well as information about the goodness of fit. Note that "R-sq = 90.3%" is the coefficient of determination expressed as a percentage. The output "R-Sq (adj) = 89.1%" is discussed in Chapter 15.
4. The ANOVA table is printed below the heading Analysis of Variance. Note that DF is an abbreviation for degrees of freedom and that MSR is given as 14,200 and MSE as 191. The ratio of these two values provides the F value of 74.25; in Section 14.5, we showed how the F value can be used to determine whether there is a significant relationship between Sales and Pop. Minitab also prints the p-value associated with this F test. Because the p-value is zero (to three decimal places), the relationship is judged statistically significant.
5. The 95% confidence interval estimate of the expected sales and the 95% prediction interval estimate of sales for an individual restaurant located near a campus with 10,000 students are printed below the ANOVA table. The confidence interval is (98.58, 121.42) and the prediction interval is (76.12, 143.88) as we showed in Section 14.6.

EXERCISES

Applications

40. The commercial division of a real estate firm is conducting a regression analysis of the relationship between x, annual gross rents (\$1000s), and y, selling price (\$1000s) for apartment buildings. Data have been collected on several properties recently sold, and the following output has been obtained in a computer run.

```
The regression equation is
Y = 20.0 + 7.21 X

Predictor        Coef      Stdev    t-ratio
Constant       20.000     3.2213       6.21
X               7.210     1.3626       5.29

Analysis of Variance

SOURCE          DF          SS
Regression       1     41587.3
Error            7
Total            8     51984.1
```

a. How many apartment buildings were in the sample?
b. Write the estimated regression equation.
c. What is the value of s_{b_1}?
d. Use the F statistic to test the significance of the relationship at a .05 level of significance.
e. Estimate the selling price of an apartment building with gross annual rents of \$50,000.

41. Following is a portion of the computer output for a regression analysis relating y = maintenance expense (dollars per month) to x = usage (hours per week) of a particular brand of computer terminal.

```
The regression equation is

Y = 6.1092 + .8951 X

Predictor          Coef        Stdev
Constant         6.1092       0.9361
X                0.8951       0.1490

Analysis of Variance

SOURCE            DF          SS            MS

Regression         1     1575.76       1575.76
Error              8      349.14         43.64
Total              9     1924.90
```

a. Write the estimated regression equation.
b. Use a t test to determine whether monthly maintenance expense is related to usage at the .05 level of significance.
c. Use the estimated regression equation to predict monthly maintenance expense for any terminal that is used 25 hours per week.

42. A regression model relating x, number of salespersons at a branch office, to y, annual sales at the office ($1000s), has been developed. The computer output from a regression analysis of the data follows.

```
The regression equation is
Y = 80.0 + 50.00 X

Predictor        Coef        Stdev    t-ratio
Constant         80.0       11.333       7.06
X                50.0        5.482       9.12

Analysis of Variance

SOURCE          DF            SS                MS
Regression       1        6828.6            6828.6
Error           28        2298.8              82.1
Total           29        9127.4
```

a. Write the estimated regression equation.
b. How many branch offices were involved in the study?
c. Compute the F statistic and test the significance of the relationship at a .05 level of significance.
d. Predict the annual sales at the Memphis branch office. This branch has 12 salespersons.

43. The National Association of Home Builders ranks the most and least affordable housing markets, based on the proportion of homes that a family earning the median income in that market could afford to buy. Data showing the median income ($1000s) and the median sale price ($1000s) for a sample of 12 housing markets that appeared on the list of most affordable markets follow (*The Wall Street Journal Almanac 1998*).

CD file
HomeBldg

Market	**Income ($1000s)**	**Price ($1000s)**
Syracuse, NY	41.8	76
Springfield, IL	47.7	91
Lima, OH	40.0	65
Dayton, OH	44.3	88
Beaumont, TX	37.3	70
Lakeland, FL	35.9	73
Baton Rouge, LA	39.3	85
Nashua, NH	56.9	118
Racine, WI	46.7	81
Des Moines, IA	48.3	89
Minneapolis–St. Paul, MN	54.6	110
Wilmington, DE–MD	55.5	110

a. Use these data to develop an estimated regression equation that could be used to predict the median price for a market given the median income for the market.
b. Did the estimated regression equation provide a good fit? Explain.
c. Develop a 95% confidence interval estimate of the median price for all homes sold in a market in which the median income is $40,100.
d. The median income in Binghamton, New York, was reported to be $40,100. Develop a 95% prediction interval estimate of the median sales price for homes sold in the Binghamton market.

44. Cushman & Wakefield, Inc., collects data showing the office building vacancy rates and rental rates for markets in the United States. The following data show the overall vacancy rates (%) and the average rental rates (per square foot) for the central business district for 18 selected markets (*The Wall Street Journal Almanac 1998*).

CD file
OffRates

Market	Vacancy Rate (%)	Average Rental Rate ($)
Atlanta	21.9	18.54
Boston	6.0	33.70
Hartford	22.8	19.67
Baltimore	18.1	21.01
Washington	12.7	35.09
Philadelphia	14.5	19.41
Miami	20.0	25.28
Tampa	19.2	17.02
Chicago	16.0	24.04
San Francisco	6.6	31.42
Phoenix	15.9	18.74
San Jose	9.2	26.76
West Palm Beach	19.7	27.72
Detroit	20.0	18.20
Brooklyn	8.3	25.00
Downtown, NY	17.1	29.78
Midtown, NY	10.8	37.03
Midtown South, NY	11.1	28.64

a. Develop a scatter diagram for these data; plot the vacancy rate on the horizontal axis.
b. Does there appear to be any relationship between these two variables?
c. Develop the estimated regression equation that could be used to predict the average rental rate given the overall vacancy rate.
d. Test the significance of the relationship at the .05 level of significance.
e. Did the estimated regression equation provide a good fit? Explain.
f. Predict the expected rental rate for markets with a 25% vacancy rate in the central business district.
g. The overall vacancy rate in the central business district in Ft. Lauderdale is 11.3%. Predict the expected rental rate for Ft. Lauderdale.

14.8 RESIDUAL ANALYSIS: VALIDATING MODEL ASSUMPTIONS

Residual analysis is the primary tool for determining whether the assumed regression model is appropriate.

As we have previously noted, the *residual* for observation i is the difference between the observed value of the dependent variable (y_i) and the estimated value of the dependent variable ($\hat{y}_i$).

Residual for Observation i

$$y_i - \hat{y}_i \quad (14.28)$$

where

y_i is the observed value of the dependent variable
$\hat{y}_i$ is the estimated value of the dependent variable

In other words, the ith residual is the error resulting from using the estimated regression equation to predict the value of y_i. The residuals for the Armand's Pizza Parlors example are computed in Table 14.7. The observed values of the dependent variable are in the second column and the estimated values of the dependent variable, obtained using the estimated

TABLE 14.7 RESIDUALS FOR ARMAND'S PIZZA PARLORS

Student Population x_i	Sales y_i	Estimated Sales $\hat{y}_i = 60 + 5x_i$	Residuals $y_i - \hat{y}_i$
2	58	70	−12
6	105	90	15
8	88	100	−12
8	118	100	18
12	117	120	−3
16	137	140	−3
20	157	160	−3
20	169	160	9
22	149	170	−21
26	202	190	12

regression equation $\hat{y} = 60 + 5x$, are in the third column. The corresponding residuals are in the fourth column. An analysis of these residuals will help determine whether the assumptions that have been made about the regression model are appropriate.

Let us now review the regression assumptions for the Armand's Pizza Parlors example. A simple linear regression model was assumed.

$$y = \beta_0 + \beta_1 x + \epsilon \qquad \textbf{(14.29)}$$

This model indicates that we assumed quarterly sales (y) to be a linear function of the size of the student population (x) plus an error term ϵ. In Section 14.4 we made the following assumptions about the error term ϵ.

1. $E(\epsilon) = 0$.
2. The variance of ϵ, denoted by σ^2, is the same for all values of x.
3. The values of ϵ are independent.
4. The error term ϵ has a normal probability distribution.

These assumptions provide the theoretical basis for the t test and the F test used to determine whether the relationship between x and y is significant, and for the confidence and prediction interval estimates presented in Section 14.6. If the assumptions about the error term ϵ appear questionable, the hypothesis tests about the significance of the regression relationship and the interval estimation results may not be valid.

The residuals provide the best information about ϵ; hence an analysis of the residuals is an important step in determining whether the assumptions for ϵ are appropriate. Much of residual analysis is based on an examination of graphical plots. In this section, we discuss the following residual plots.

1. A plot of the residuals against values of the independent variable x.
2. A plot of residuals against the predicted values of the dependent variable $\hat{y}$.
3. A standardized residual plot.
4. A normal probability plot.

Residual Plot Against x

A residual plot against the independent variable x is a graph in which the values of the independent variable are represented by the horizontal axis and the corresponding residual values are represented by the vertical axis. A point is plotted for each residual. The first coordinate for each point is given by the value of x_i and the second coordinate is given by the corresponding value of the residual $y_i - \hat{y}_i$. For a residual plot against x with the Armand's

Pizza Parlors data from Table 14.7, the coordinates of the first point are (2, −12), corresponding to $x_1 = 2$ and $y_1 - \hat{y}_1 = -12$; the coordinates of the second point are (6, 15), corresponding to $x_2 = 6$ and $y_2 - \hat{y}_2 = 15$, and so on. Figure 14.11 is the resulting residual plot.

Before interpreting the results for this residual plot, let us consider some general patterns that might be observed in any residual plot. Three examples are shown in Figure 14.12. If the assumption that the variance of ϵ is the same for all values of x and the assumed regression model is an adequate representation of the relationship between the variables, the residual plot should give an overall impression of a horizontal band of points such as the one in panel A of Figure 14.12. However, if the variance of ϵ is not the same for all values of x—for example, if variability about the regression line is greater for larger values of x—a pattern such as the one in panel B of Figure 14.12 could be observed. In this case, the assumption of a constant variance of ϵ is violated. Another possible residual plot is shown in panel C. In this case, we would conclude that the assumed regression model is not an adequate representation of the relationship between the variables. A curvilinear regression model or multiple regression model should be considered.

Now let us return to the residual plot for Armand's Pizza Parlors shown in Figure 14.11. The residuals appear to approximate the horizontal pattern in panel A of Figure 14.12. Hence, we conclude that the residual plot does not provide evidence that the assumptions made for Armand's regression model should be challenged. At this point, we are confident in the conclusion that Armand's simple linear regression model is valid.

Experience and good judgment are always factors in the effective interpretation of residual plots. Seldom does a residual plot conform precisely to one of the patterns shown in Figure 14.12. Yet analysts who frequently conduct regression studies and frequently review residual plots become adept at understanding the differences between patterns that are reasonable and patterns that indicate the assumptions of the model should be questioned. A residual plot as shown here is one of the techniques that is used to assess the validity of the assumptions for a regression model.

FIGURE 14.11 PLOT OF THE RESIDUALS AGAINST THE INDEPENDENT VARIABLE x FOR ARMAND'S PIZZA PARLORS

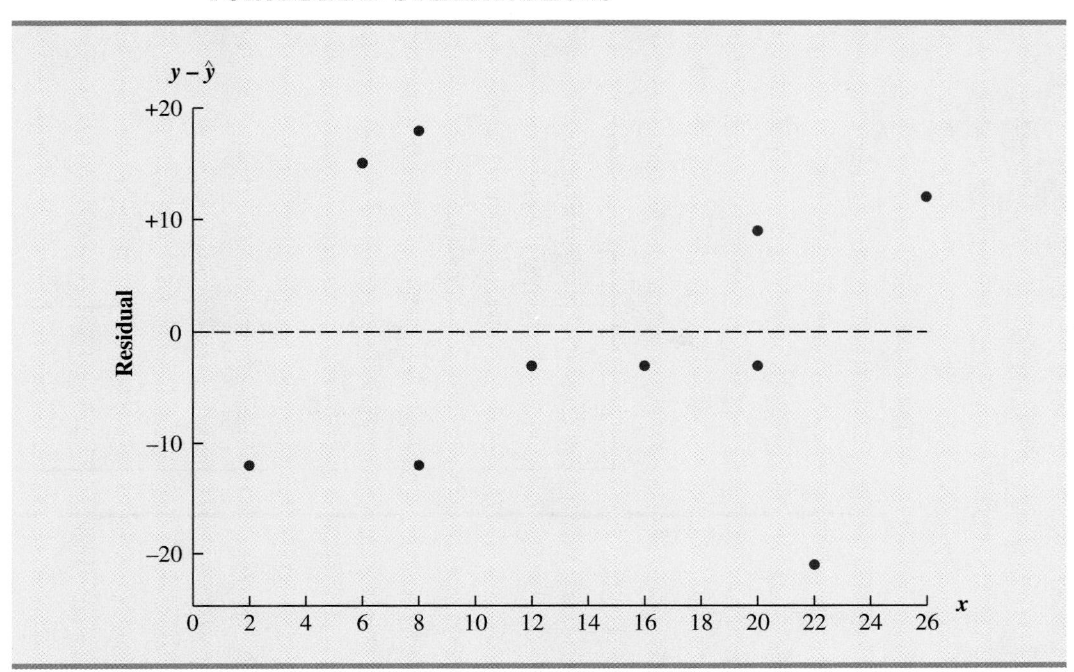

FIGURE 14.12 RESIDUAL PLOTS FROM THREE REGRESSION STUDIES

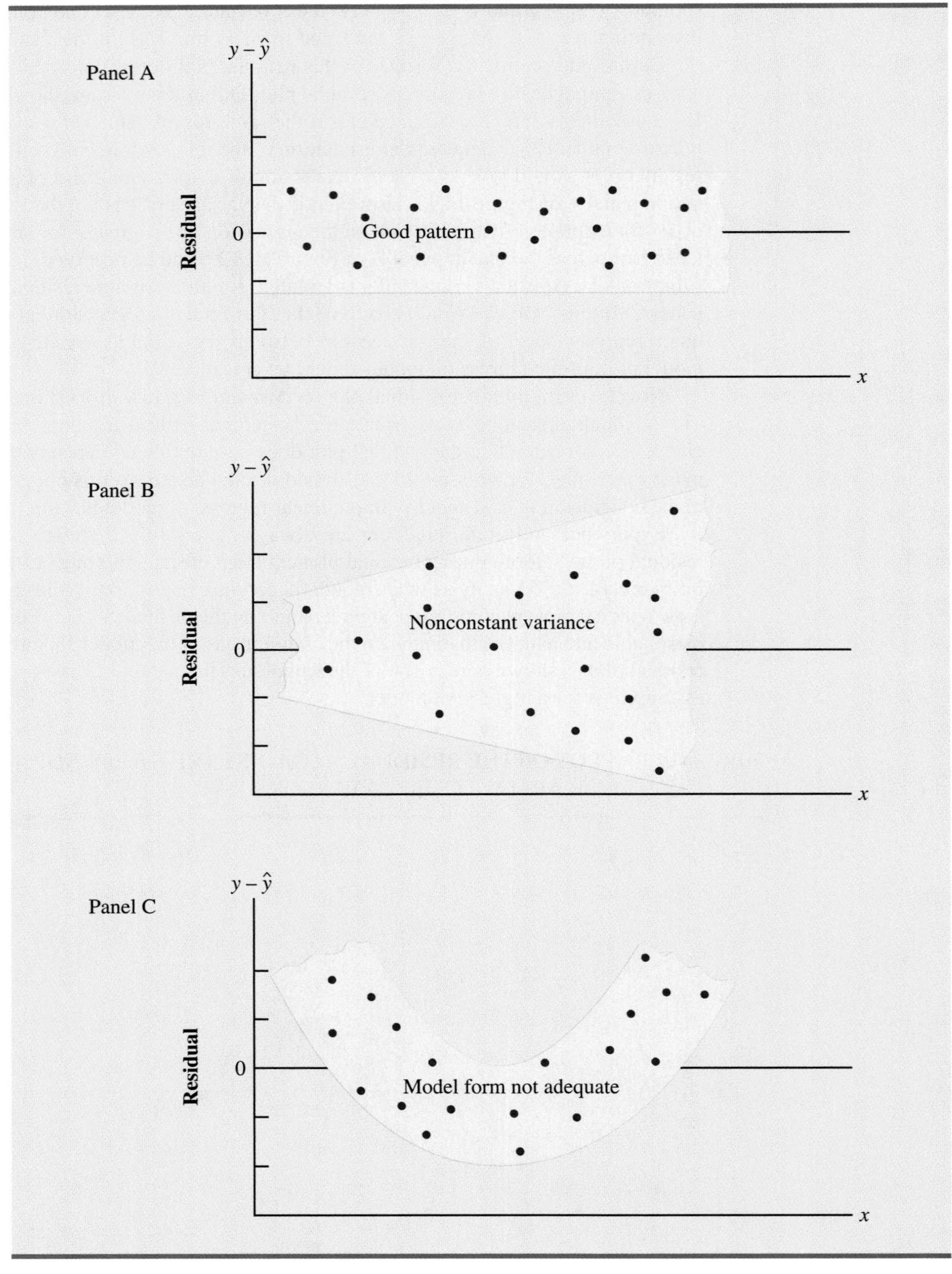

Residual Plot Against $\hat{y}$

Another residual plot represents the predicted value of the dependent variable $\hat{y}$ on the horizontal axis and the residual values on the vertical axis. A point is plotted for each residual. The first coordinate for each point is given by $\hat{y}_i$ and the second coordinate is given by the corresponding value of the ith residual $y_i - \hat{y}_i$. With the Armand's data from Table 14.7, the coordinates of the first point are (70, −12), corresponding to $\hat{y}_1 = 70$ and $y_1 - \hat{y}_1 = -12$; the coordinates of the second point are (90, 15), and so on. Figure 14.13 is the residual plot. Note that the pattern of this residual plot is the same as the pattern of the residual plot against the independent variable x. It is not a pattern that would lead us to question the model assumptions. For simple linear regression, both the residual plot against x and the residual plot against $\hat{y}$ provide the same pattern. For multiple regression analysis, the residual plot against $\hat{y}$ is more widely used because of the presence of more than one independent variable.

Standardized Residuals

Many of the residual plots provided by computer software packages use a standardized version of the residuals. As we have seen in preceding chapters, a random variable is standardized by subtracting its mean and dividing the result by its standard deviation. With the least squares method, the mean of the residuals is zero. Thus, simply dividing each residual by its standard deviation provides the **standardized residual.**

It can be shown that the standard deviation of residual i depends on the standard error of the estimate s and the corresponding value of the independent variable x_i.

Standard Deviation of the *i*th Residual*

$$s_{y_i - \hat{y}_i} = s\sqrt{1 - h_i} \tag{14.30}$$

where

$$s_{y_i - \hat{y}_i} = \text{the standard deviation of residual } i$$
$$s = \text{the standard error of the estimate}$$
$$h_i = \frac{1}{n} + \frac{(x_i - \bar{x})^2}{\Sigma(x_i - \bar{x})^2} \tag{14.31}$$

Note that (14.30) shows that the ith residual standard deviation depends on x_i because of the presence of h_i in the formula.† Once the standard deviation of each residual has been calculated, we can compute the standardized residual by dividing each residual by its corresponding standard deviation.

Standardized Residual for Observation *i*

$$\frac{y_i - \hat{y}_i}{s_{y_i - \hat{y}_i}} \tag{14.32}$$

Table 14.8 shows the calculation of the standardized residuals for Armand's Pizza Parlors. Recall that previous calculations showed $s = 13.829$. Figure 14.14 is the plot of the standardized residuals against the independent variable x.

*This equation actually provides an estimate of the standard deviation of the ith residual, because s is used instead of σ.

†h_i is referred to as the *leverage* of observation i. Leverage will be discussed further when we consider influential observations in Section 14.9.

TABLE 14.8 COMPUTATION OF STANDARDIZED RESIDUALS FOR ARMAND'S PIZZA PARLORS

Restaurant i	x_i	$x_i - \bar{x}$	$(x_i - \bar{x})^2$	$\frac{(x_i - \bar{x})^2}{\Sigma(x_i - \bar{x})^2}$	h_i	$s_{y_i - \hat{y}_i}$	$y_i - \hat{y}_i$	Standardized Residual
1	2	−12	144	.2535	.3535	11.1193	−12	−1.0792
2	6	−8	64	.1127	.2127	12.2709	15	1.2224
3	8	−6	36	.0634	.1634	12.6493	−12	−.9487
4	8	−6	36	.0634	.1634	12.6493	18	1.4230
5	12	−2	4	.0070	.1070	13.0682	−3	−.2296
6	16	2	4	.0070	.1070	13.0682	−3	−.2296
7	20	6	36	.0634	.1634	12.6493	−3	−.2372
8	20	6	36	.0634	.1634	12.6493	9	.7115
9	22	8	64	.1127	.2127	12.2709	−21	−1.7114
10	26	12	144	.2535	.3535	11.1193	12	1.0792
		Total	568					

Note: The values of the residuals were computed in Table 14.7.

FIGURE 14.13 PLOT OF RESIDUALS AGAINST THE PREDICTED VALUES $\hat{y}$ FOR ARMAND'S PIZZA PARLORS

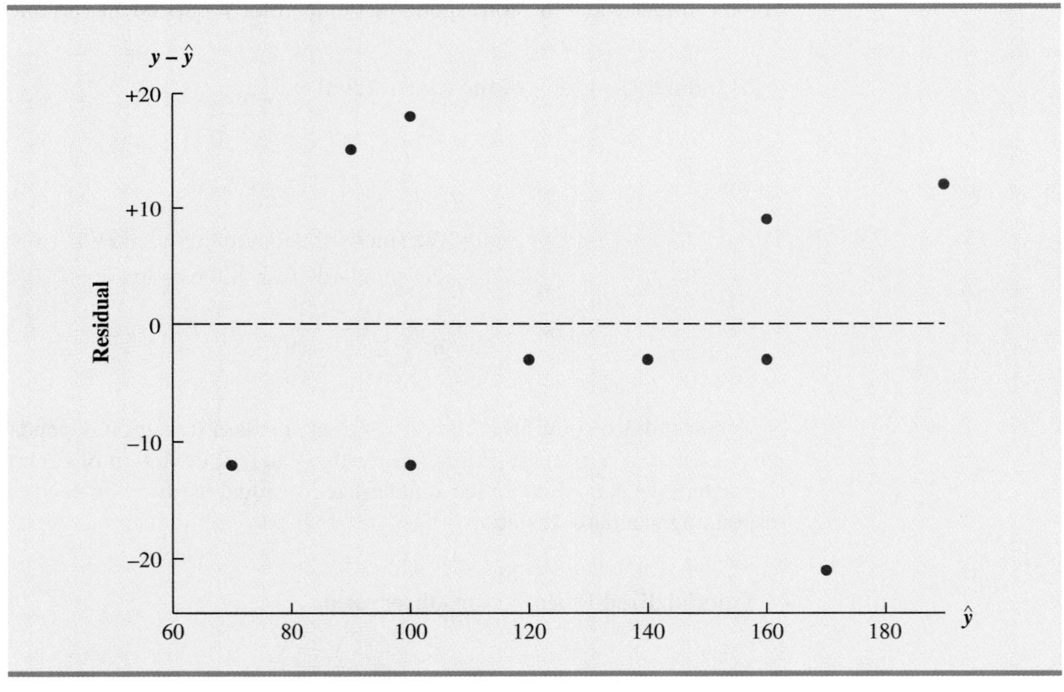

Small departures from normality do not have a great effect on the statistical tests used in regression analysis.

The standardized residual plot can provide insight about the assumption that the error term ϵ has a normal distribution. If this assumption is satisfied, the distribution of the standardized residuals should appear to come from a standard normal probability distribution.* Thus, when looking at a standardized residual plot, we should expect to see approximately

*Because s is used instead of σ in (14.30), the probability distribution of the standardized residuals is not technically normal. However, in most regression studies, the sample size is large enough that a normal approximation is very good.

FIGURE 14.14 PLOT OF THE STANDARDIZED RESIDUALS AGAINST THE INDEPENDENT VARIABLE x FOR ARMAND'S PIZZA PARLORS

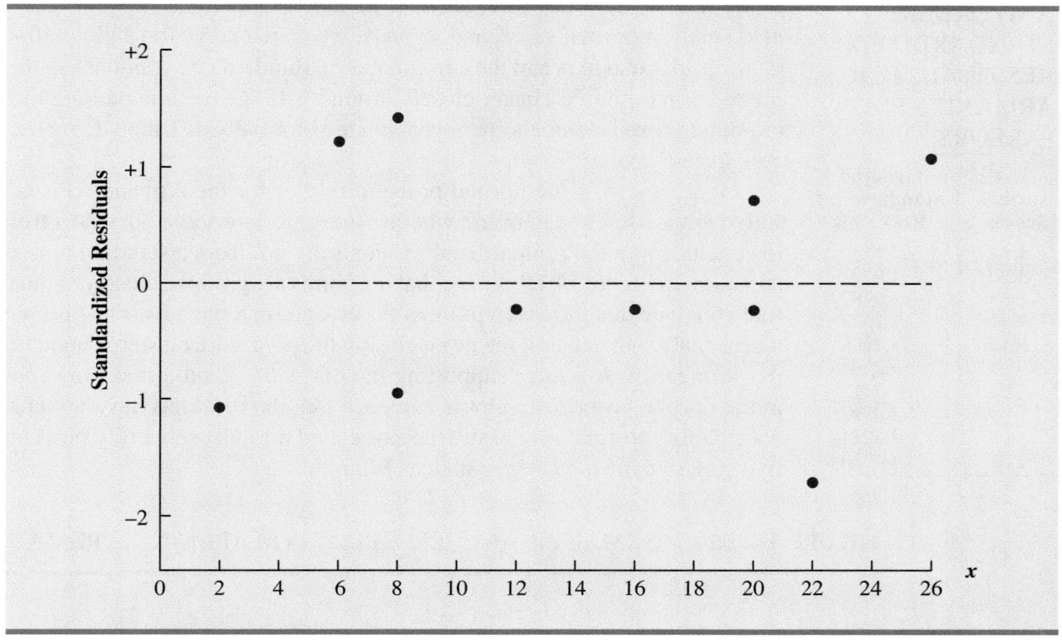

95% of the standardized residuals between −2 and +2. We see in Figure 14.14 that for the Armand's example all standardized residuals are between −2 and +2. Therefore, on the basis of the standardized residuals, we have no reason to question the assumption that ϵ has a normal distribution.

Because of the effort required to compute the estimated values of $\hat{y}$, the residuals, and the standardized residuals, most statistical packages provide these values as optional regression output. Hence, residual plots can be easily obtained. For large problems computer packages are the only practical means for developing the residual plots we have discussed in this section.

Normal Probability Plot

Another approach for determining the validity of the assumption that the error term has a normal distribution is the **normal probability plot.** To show how a normal probability plot is developed, we introduce the concept of *normal scores.*

TABLE 14.9

NORMAL SCORES FOR $n = 10$

Order Statistic	Normal Score
1	−1.55
2	−1.00
3	−.65
4	−.37
5	−.12
6	.12
7	.37
8	.65
9	1.00
10	1.55

Suppose 10 values are selected randomly from a normal probability distribution with a mean of zero and a standard deviation of one, and that the sampling process is repeated over and over with the values in each sample of 10 ordered from smallest to largest. For now, let us consider only the smallest value in each sample. The random variable representing the smallest value obtained in repeated sampling is called the first-order statistic.

Statisticians have shown that for samples of size 10 from a standard normal probability distribution, the expected value of the first-order statistic is −1.55. This expected value is called a normal score. For the case with a sample of size $n = 10$, there are 10 order statistics and 10 normal scores (see Table 14.9). In general, if we have a data set consisting of n observations, there are n order statistics and hence n normal scores.

Let us now show how the 10 normal scores can be used to determine whether the standardized residuals for Armand's Pizza Parlors appear to come from a standard normal probability distribution. We begin by ordering the 10 standardized residuals from Table 14.8. The

TABLE 14.10 NORMAL SCORES AND ORDERED STANDARDIZED RESIDUALS FOR ARMAND'S PIZZA PARLORS

Normal Scores	Ordered Standardized Residuals
−1.55	−1.7114
−1.00	−1.0792
−.65	−.9487
−.37	−.2372
−.12	−.2296
.12	−.2296
.37	.7115
.65	1.0792
1.00	1.2224
1.55	1.4230

10 normal scores and the ordered standardized residuals are shown together in Table 14.10. If the normality assumption is satisfied, the smallest standardized residual should be close to the smallest normal score, the next smallest standardized residual should be close to the next smallest normal score, and so on. If we were to develop a plot with the normal scores on the horizontal axis and the corresponding standardized residuals on the vertical axis, the plotted points should cluster closely around a 45-degree line passing through the origin if the standardized residuals are approximately normally distributed. Such a plot is referred to as a *normal probability plot.*

Figure 14.15 is the normal probability plot for the Armand's Pizza Parlors example. Judgment is used to determine whether the pattern observed deviates from the line enough to conclude that the standardized residuals are not from a standard normal probability distribution. In Figure 14.15, we see that the points are grouped closely about the line. We therefore conclude that the assumption of the error term having a normal probability distribution is reasonable. In general, the more closely the points are clustered about the 45-degree line, the stronger the evidence supporting the normality assumption. Any substantial curvature in the normal probability plot is evidence that the residuals have not come from a normal distribution. Normal scores and the associated normal probability plot can be obtained easily using statistical packages such as Minitab.

FIGURE 14.15 NORMAL PROBABILITY PLOT FOR ARMAND'S PIZZA PARLORS

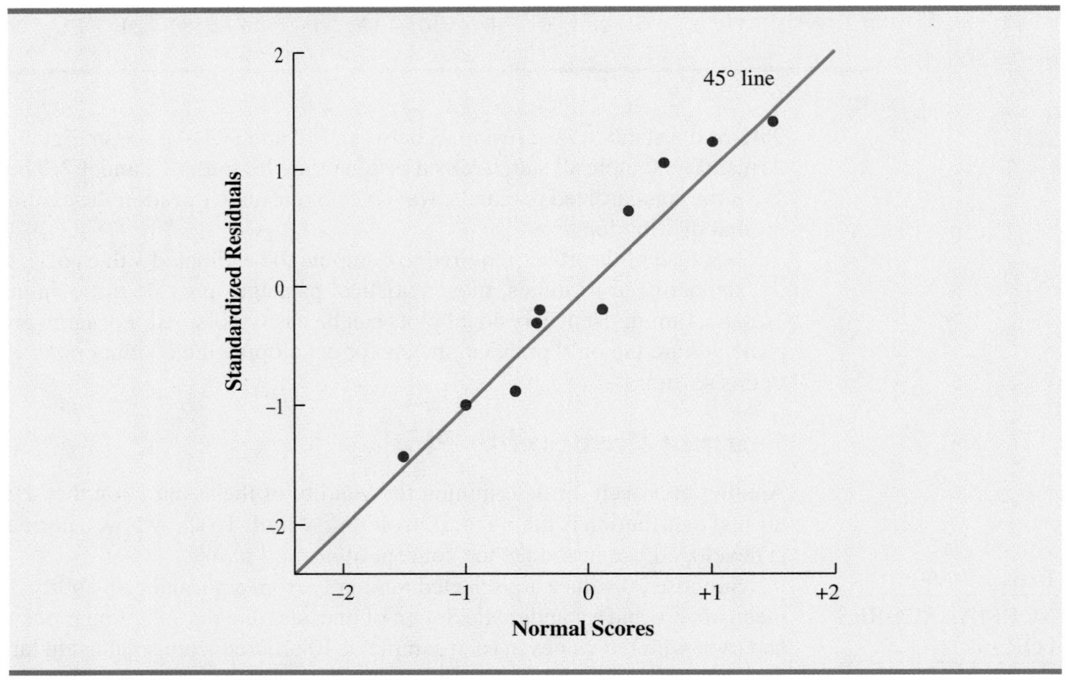

NOTES AND COMMENTS

1. We use residual and normal probability plots to validate the assumptions of a regression model. If our review indicates that one or more assumptions are questionable, a different regression model and/or a transformation of the data should be considered. The appropriate corrective action when the assumptions are violated must be based on good judgment; recommendations from an experienced statistician can be valuable.
2. Analysis of residuals is the primary method statisticians use to verify that the assumptions as-

sociated with a regression model are valid. Even if no violations are found, it does not necessarily follow that the model will yield good predictions. However, if additional statistical tests support the conclusion of significance and the coefficient of determination is large, we should be able to develop good estimates and predictions using the estimated regression equation.

EXERCISES

Methods

45. Given are data for two variables, x and y.

x_i	6	11	15	18	20
y_i	6	8	12	20	30

a. Develop an estimated regression equation for these data.
b. Compute the residuals.
c. Develop a plot of the residuals against the independent variable x. Do the assumptions about the error terms seem to be satisfied?
d. Compute the standardized residuals.
e. Develop a plot of the standardized residuals against $\hat{y}$. What conclusions can you draw from this plot?

46. The following data were used in a regression study.

Observation	x_i	y_i	Observation	x_i	y_i
1	2	4	6	7	6
2	3	5	7	7	9
3	4	4	8	8	5
4	5	6	9	9	11
5	7	4			

a. Develop an estimated regression equation for these data.
b. Construct a plot of the residuals. Do the assumptions about the error term seem to be satisfied?

Applications

47. Data on advertising expenditures ($1000s) and revenue ($1000s) for the Four Seasons Restaurant follow.

Advertising Expenditures	Revenue
1	19
2	32
4	44
6	40
10	52
14	53
20	54

a. Let x equal advertising expenditures ($1000s) and y equal revenue ($1000s). Use the method of least squares to develop a straight line approximation of the relationship between the two variables.
b. Test whether revenue and advertising expenditures are related at a .05 level of significance.

c. Prepare a residual plot of $y - \hat{y}$ versus $\hat{y}$. Use the result of (a) to obtain the values of $\hat{y}$.
d. What conclusions can you draw from residual analysis? Should this model be used, or should we look for a better one?

48. Refer to Exercise 9, where an estimated regression equation relating years of experience and annual sales was developed.
a. Compute the residuals and construct a residual plot for this problem.
b. Do the assumptions about the error terms seem reasonable in light of the residual plot?

49. American Depository Receipts (ADRs) are certificates traded on the NYSE representing shares of a foreign company held on deposit in a bank in its home country. The following table shows the price/earnings (P/E) ratio and the percentage return on investment (ROE) for 10 Indian companies that are likely new ADRs (*Bloomberg Personal Finance,* April 2000).

CD file ADRs

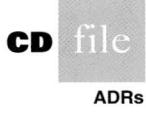

	ROE	P/E
Bharti Televentures	6.43	36.88
Gujarat Ambuja Cements	13.49	27.03
Hindalco Industries	14.04	10.83
ICICI	20.67	5.15
Mahanagar Telephone Nigam	22.74	13.35
NIIT	46.23	95.59
Pentamedia Graphics	28.90	54.85
Satyam Computer Services	54.01	189.21
Silverline Technologies	28.02	75.86
Videsh Sanchar Nigam	27.04	13.17

a. Use a computer package to develop an estimated regression equation relating y = P/E and x = ROE.
b. Construct a residual plot of the standardized residuals against the independent variable.
c. Do the assumptions about the error terms and model form seem reasonable in light of the residual plot?

14.9 RESIDUAL ANALYSIS: OUTLIERS AND INFLUENTIAL OBSERVATIONS

In Section 14.8 we showed how residual analysis could be used to determine when violations of assumptions about the regression model have occurred. In this section, we discuss how residual analysis can be used to identify observations that can be classified as outliers or as being especially influential in determining the estimated regression equation. Some steps that should be taken when such observations have been found are discussed.

TABLE 14.11 DATA SET ILLUSTRATING THE EFFECT OF AN OUTLIER

x_i	y_i
1	45
1	55
2	50
3	75
3	40
3	45
4	30
4	35
5	25
6	15

Detecting Outliers

Figure 14.16 is a scatter diagram for a data set that has an outlier, a data point (observation) that does not fit the trend shown by the remaining data. Outliers represent observations that are suspect and warrant careful examination. They may represent erroneous data; if so, the data should be corrected. They may signal a violation of model assumptions; if so, another model should be considered. Finally, they may simply be unusual values that have occurred by chance. In this case, they should be retained.

To illustrate the process of detecting outliers, consider the data set in Table 14.11; Figure 14.17 is a scatter diagram. Except for observation 4 ($x_4 = 3$, $y_4 = 75$), a pattern suggesting a negative linear relationship is apparent. Indeed, given the pattern of the rest of the data, we would have expected y_4 to be much smaller and hence would identify the corre-

FIGURE 14.16 A DATA SET WITH AN OUTLIER

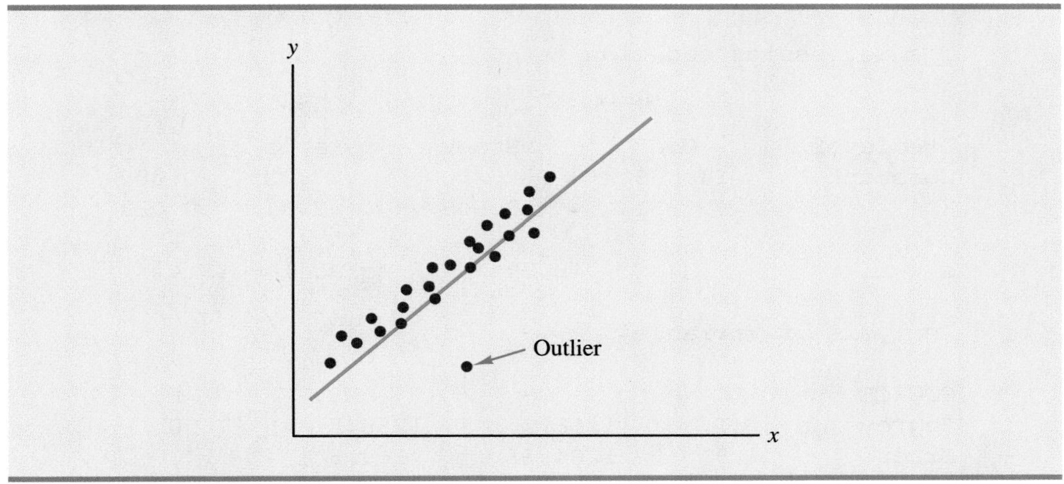

sponding observation as an outlier. For the case of simple linear regression, one can often detect outliers by simply examining the scatter diagram.

The standardized residuals can also be used to identify outliers. If an observation deviates greatly from the pattern of the rest of the data (e.g., the outlier in Figure 14.16), the corresponding standardized residual will be large in absolute value. Many computer packages automatically identify observations with standardized residuals that are large in absolute value. In Figure 14.18 we show the Minitab output from a regression analysis of the data in Table 14.11. The next to last line of the output shows that the standardized residual for observation 4 is 2.67. Minitab identifies any observation with a standardized residual of less than -2 or greater than $+2$ as an unusual observation; in such cases, the observation is printed on a separate line with an R next to the standardized residual, as shown in Figure 14.18. With normally distributed errors, standardized residuals should be outside these limits approximately 5% of the time.

FIGURE 14.17 SCATTER DIAGRAM FOR OUTLIER DATA SET

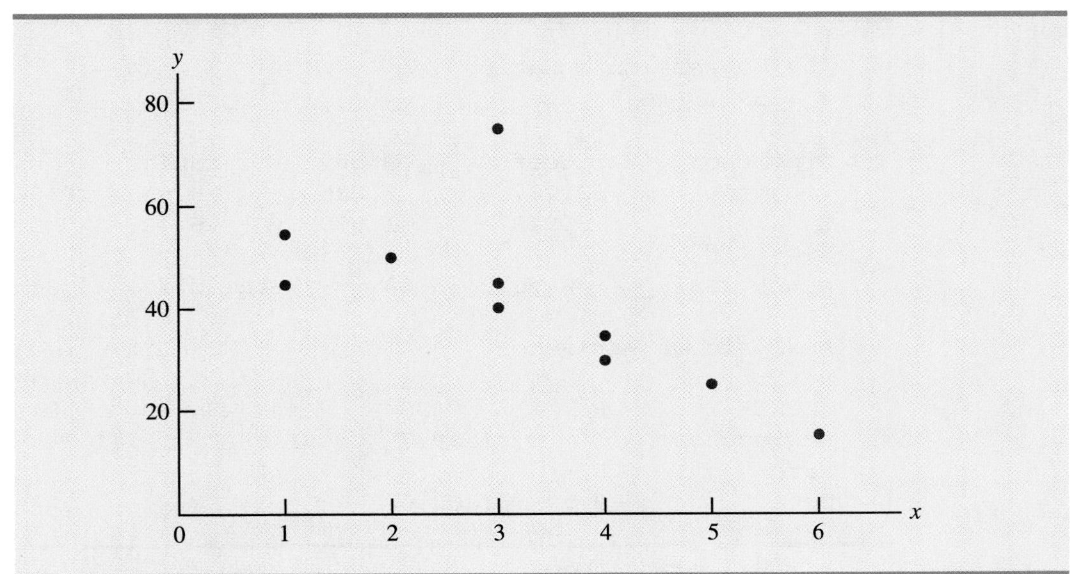

FIGURE 14.18 MINITAB OUTPUT FOR REGRESSION ANALYSIS OF THE OUTLIER DATA SET

```
The regression equation is
Y = 65.0 - 7.33 X

Predictor         Coef          Stdev     t-ratio         p
Constant        64.958          9.258        7.02     0.000
X               -7.331          2.608       -2.81     0.023

s = 12.67         R-sq = 49.7%       R-sq(adj) = 43.4%

Analysis of Variance

SOURCE          DF           SS            MS          F         p
Regression       1       1268.2        1268.2       7.90     0.023
Error            8       1284.3         160.5
Total            9       2552.5

Unusual Observations
Obs.         X            Y        Fit Stdev.Fit  Residual    St.Resid
  4       3.00        75.00      42.97      4.04     32.03       2.67R

R denotes an obs. with a large st. resid.
```

In deciding how to handle an outlier, we should first check to see whether it is a valid observation. Perhaps an error has been made in initially recording the data or in entering the data into the computer file. For example, suppose that in checking the data for the outlier in Table 14.17, we find that an error has been made and that the correct value for observation 4 is $x_4 = 3$, $y_4 = 30$. Figure 14.19 is the Minitab output obtained after correc-

FIGURE 14.19 MINITAB OUTPUT FOR THE REVISED OUTLIER DATA SET

```
The regression equation is
Y = 59.2 - 6.95 X

Predictor         Coef          Stdev     t-ratio         p
Constant        59.237          3.835       15.45     0.000
X               -6.949          1.080       -6.43     0.000

s = 5.248         R-sq = 83.8%       R-sq(adj) = 81.8%

Analysis of Variance

SOURCE          DF           SS            MS          F         p
Regression       1       1139.7        1139.7      41.38     0.000
Error            8        220.3          27.5
Total            9       1360.0
```

tion of the value of y_4. We see that using the incorrect data value had a substantial effect on the goodness of fit. With the correct data, the value of R-sq has increased from 49.7% to 83.8% and the value of b_0 has decreased from 64.958 to 59.237. The slope of the line has changed from −7.331 to −6.949. The identification of the outlier enabled us to correct the data error and improve the regression results.

Detecting Influential Observations

Sometimes one or more observations have a strong influence on the results obtained. Figure 14.20 shows an example of an **influential observation** in simple linear regression. The estimated regression line has a negative slope. However, if the influential observation were dropped from the data set, the slope of the estimated regression line would change from negative to positive and the *y*-intercept would be smaller. Clearly, this one observation is much more influential in determining the estimated regression line than any of the others; dropping one of the other observations from the data set would have little effect on the estimated regression equation.

Influential observations can be identified from a scatter diagram when only one independent variable is present. An influential observation may be an outlier (an observation with a *y* value that deviates substantially from the trend), it may correspond to an *x* value far away from its mean (e.g., see Figure 14.20), or it may be caused by a combination of the two (a somewhat off-trend *y* value and a somewhat extreme *x* value).

Because influential observations may have such a dramatic effect on the estimated regression equation, they must be examined carefully. We should first check to make sure that no error has been made in collecting or recording the data. If an error has occurred, it can be corrected and a new estimated regression equation can be developed. If the observation is valid, we might consider ourselves fortunate to have it. Such a point, if valid, can contribute to a better understanding of the appropriate model and can lead to a better estimated regression equation. The presence of the influential observation in Figure 14.20, if valid, would suggest trying to obtain data on intermediate values of *x* to understand better the relationship between *x* and *y*.

Observations with extreme values for the independent variables are called **high leverage points.** The influential observation in Figure 14.20 is a point with high leverage. The leverage of an observation is determined by how far the values of the independent variables

FIGURE 14.20 A DATA SET WITH AN INFLUENTIAL OBSERVATION

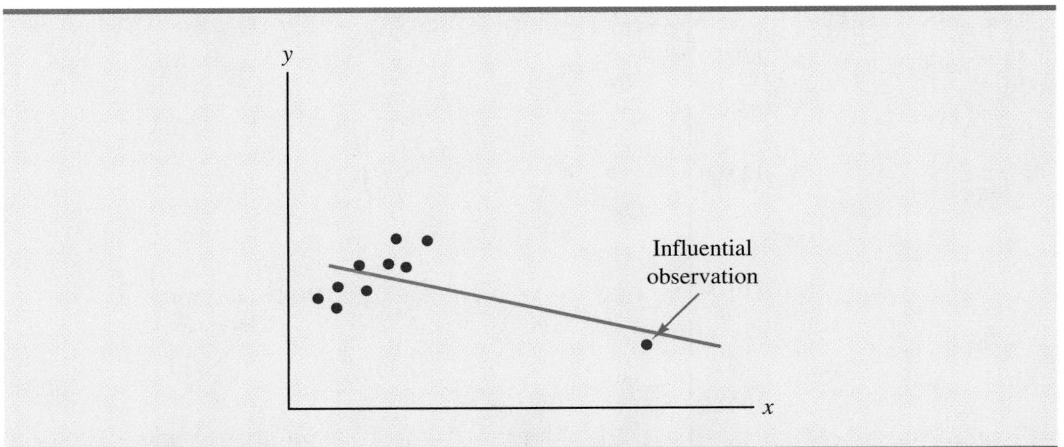

are from their mean values. For the single-independent-variable case, the leverage of the ith observation, denoted h_i, can be computed by using (14.33).

Leverage of Observation i

$$h_i = \frac{1}{n} + \frac{(x_i - \bar{x})^2}{\Sigma(x_i - \bar{x})^2} \tag{14.33}$$

TABLE 14.12 DATA SET WITH A HIGH LEVERAGE OBSERVATION

x_i	y_i
10	125
10	130
15	120
20	115
20	120
25	110
70	100

From the formula, it is clear that the farther x_i is from its mean $\bar{x}$, the higher the leverage of observation i.

Many statistical packages automatically identify observations with high leverage as part of the standard regression output. As an illustration of how the Minitab statistical package identifies points with high leverage, let us consider the data set in Table 14.12.

From Figure 14.21, a scatter diagram for the data set in Table 14.12, it is clear that observation 7 ($x = 70$, $y = 100$) is an observation with an extreme value of x. Hence, we would expect it to be identified as a point with high leverage. For this observation, the leverage is computed by using (14.33) as follows.

$$h_7 = \frac{1}{n} + \frac{(x_7 - \bar{x})^2}{\Sigma(x_i - \bar{x})^2} = \frac{1}{7} + \frac{(70 - 24.286)^2}{2621.43} = .94$$

Computer software packages are essential for performing the computations to identify influential observations. Minitab's selection rule is discussed here.

For the case of simple linear regression, Minitab identifies observations as having high leverage if $h_i > 6/n$; for the data set in Table 14.12, $6/n = 6/7 = .86$. Since $h_7 = .94 > .86$, Minitab will identify observation 7 as an observation whose x value gives it large influence. Figure 14.22 shows the Minitab output for a regression analysis of this data set. Observation 7 ($x = 70$, $y = 100$) is identified as having large influence; it is printed on a separate line at the bottom, with an X in the right margin.

Influential observations that are caused by an interaction of large residuals and high leverage can be difficult to detect. Diagnostic procedures are available that take both into account in determining when an observation is influential. One such measure, called Cook's D statistic, will be discussed in Chapter 15.

FIGURE 14.21 SCATTER DIAGRAM FOR THE DATA SET WITH A HIGH LEVERAGE OBSERVATION

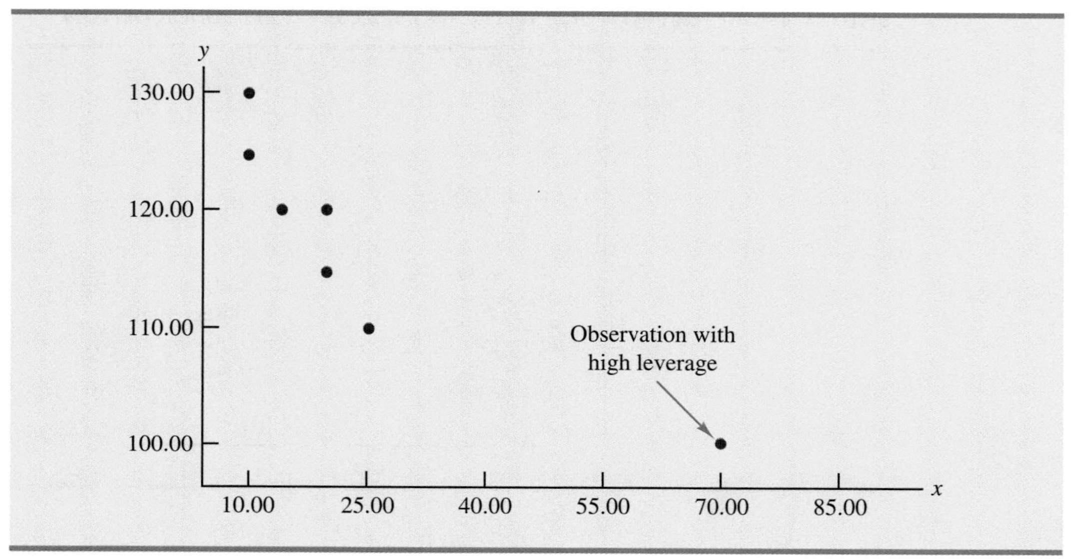

FIGURE 14.22 MINITAB OUTPUT FOR THE DATA SET WITH A HIGH LEVERAGE OBSERVATION

```
The regression equation is
Y = 127 -0.425 X

Predictor        Coef       Stdev    t-ratio        p
Constant      127.466       2.961      43.04    0.000
X            -0.42507     0.09537      -4.46    0.007

s = 4.883        R-sq = 79.9%      R-sq(adj) = 75.9%

Analysis of Variance

SOURCE        DF          SS           MS          F         p
Regression     1      473.65       473.65      19.87     0.007
Error          5      119.21        23.84
Total          6      592.86

Unusual Observations
Obs.      X          Y        Fit  Stdev.Fit   Residual    St.Resid
  7    70.0     100.00      97.71       4.73       2.29      1.91 X

X denotes an obs. whose X value gives it large influence
```

NOTES AND COMMENTS

Once an observation has been identified as potentially influential because of a large residual or high leverage, its impact on the estimated regression equation should be evaluated. More advanced texts discuss diagnostics for doing so. However, if one is not familiar with the more advanced material, a simple procedure is to run the regression analysis with and without the observation. Although time-consuming, this approach will reveal the influence of the observation on the results.

EXERCISES

Methods

50. Consider the following data for two variables, x and y.

x_i	135	110	130	145	175	160	120
y_i	145	100	120	120	130	130	110

a. Compute the standardized residuals for these data. Do there appear to be any outliers in the data? Explain.

b. Plot the standardized residuals against $\hat{y}$. Does this plot reveal any outliers?

c. Develop a scatter diagram for these data. Does the scatter diagram indicate any outliers in the data? In general, what implications does this finding have for simple linear regression?

51. Consider the following data for two variables, x and y.

x_i	4	5	7	8	10	12	12	22
y_i	12	14	16	15	18	20	24	19

a. Compute the standardized residuals for these data. Do there appear to be any outliers in the data? Explain.

b. Compute the leverage values for these data. Do there appear to be any influential observations in these data? Explain.

c. Develop a scatter diagram for these data. Does the scatter diagram indicate any influential observations? Explain.

Applications

52. The following data show the media expenditures ($ millions) and the shipments in bbls. (millions) for 10 major brands of beer (*Superbrands '98,* October 20, 1997).

Beer

Brand	Media Expenditures ($millions)	Shipments
Budweiser	120.0	36.3
Bud Light	68.7	20.7
Miller Lite	100.1	15.9
Coors Light	76.6	13.2
Busch	8.7	8.1
Natural Light	0.1	7.1
Miller Genuine Draft	21.5	5.6
Miller High Life	1.4	4.4
Busch Light	5.3	4.3
Milwaukee's Best	1.7	4.3

a. Develop the estimated regression equation for these data.

b. Uses residual analysis to determine whether any outliers and/or influential observations are present. Briefly summarize your findings and conclusions.

53. Nielsen Media Research collects data showing the number of households tuned in to shows that carry a particular advertisement. This information is useful to advertisers because it tells them how many consumers they are reaching. The following data show the number of household exposures in millions and the number of times the ad was aired for the week of November 24–30, 1997 (*USA Today,* December 15, 1997).

Nielsen2

Advertised Brand	Times Ad Aired	Household Exposures
Sears	95	758.8
JC Penney	46	323.0
Burger King	41	275.3
Polaroid One Step Express	38	241.8
Wendy's	29	219.9
McDonald's	32	198.5
Target	25	193.8
Kmart	21	189.7
Visa	21	161.9
Nissan Frontier	16	160.0

a. Develop the estimated regression equation that can be used to predict the number of household exposures given the number of times the ad was aired.

b. Use residual analysis to determine whether any outliers and/or influential observations are present. Briefly summarize your findings and conclusions.

54. The market capitalization and the salary of the chief executive officer (CEO) for 20 companies are shown in the following table (*The Wall Street Journal,* February 24, 2000, and April 6, 2000).

CD file CEO

	Market Cap. ($millions)	CEO Salary ($1000s)
Anheuser-Busch	32,977.4	1,130
AT&T	162,365.1	1,400
Charles Schwab	31,363.8	800
Chevron	56,849.0	1,350
DuPont	68,848.0	1,000
General Elec.	507,216.8	3,325
Gillette	44,180.1	978
IBM	194,455.9	2,000
Johnson & Johnson	143,131.0	1,365
Kimberly-Clark	35,377.5	950
Merrill Lynch	31,062.1	700
Motorola	92,923.7	1,275
Philip Morris	54,421.2	1,625
Procter & Gamble	144,152.9	1,318.3
Qualcomm	116,840.8	773
Schering-Plough	62,259.4	1,200
Sun Microsystems	120,966.5	116
Texaco	30,040.7	950
USWest	36,450.8	897
Walt Disney	61,288.1	750

a. Develop the estimated regression equation that can be used to predict the CEO salary given the market capitalization.

b. Use residual analysis to determine whether any outliers and/or influential observations are present. Briefly summarize your findings and conclusions.

SUMMARY

In this chapter we showed how regression analysis can be used to determine how a dependent variable y is related to an independent variable x. In simple linear regression, the regression model is $y = \beta_0 + \beta_1 x + \epsilon$. The simple linear regression equation $E(y) = \beta_0 + \beta_1 x$ describes how the mean or expected value of y is related to x. We used sample data and the least squares method to develop the estimated regression equation $\hat{y} = b_0 + b_1 x$. In effect, b_0 and b_1 are the sample statistics used to estimate the unknown model parameters β_0 and β_1.

The coefficient of determination was presented as a measure of the goodness of fit for the estimated regression equation; it can be interpreted as the proportion of the variation in the dependent variable y that can be explained by the estimated regression equation. We reviewed correlation as a descriptive measure of the strength of a linear relationship between two variables.

The assumptions about the regression model and its associated error term ϵ were discussed, and t and F tests, based on those assumptions, were presented as a means for determining whether the relationship between two variables is statistically significant. We showed how to use the estimated regression equation to develop confidence interval estimates of the mean value of y and prediction interval estimates of individual values of y.

The chapter concluded with a section on the computer solution of regression problems and two sections on the use of residual analysis to validate the model assumptions and to identify outliers and influential observations.

GLOSSARY

Dependent variable The variable that is being predicted or explained. It is denoted by y.

Independent variable The variable that is doing the predicting or explaining. It is denoted by x.

Simple linear regression Regression analysis involving one independent variable and one dependent variable in which the relationship between the variables is approximated by a straight line.

Regression model The equation describing how y is related to x and an error term; in simple linear regression, the regression model is $y = \beta_0 + \beta_1 x + \epsilon$.

Regression equation The equation that describes how the mean or expected value of the dependent variable is related to the independent variable; in simple linear regression, $E(y) = \beta_0 + \beta_1 x$.

Estimated regression equation The estimate of the regression equation developed from sample data by using the least squares method. For simple linear regression, the estimated regression equation is $\hat{y} = b_0 + b_1 x$.

Least squares method The procedure used to develop the estimated regression equation. The objective is to minimize $\Sigma(y_i - \hat{y}_i)^2$.

Scatter diagram A graph of bivariate data in which the independent variable is on the horizontal axis and the dependent variable is on the vertical axis.

Coefficient of determination A measure of the goodness of fit of the estimated regression equation. It can be interpreted as the proportion of the variability in the dependent variable y that is explained by the estimated regression equation.

***i*th Residual** The difference between the observed value of the dependent variable and the value predicted using the estimated regression equation; that is, for the ith observation the ith residual is $y_i - \hat{y}_i$.

Correlation coefficient A measure of the strength of the linear relationship between two variables (previously discussed in Chapter 3).

Mean Square Error The unbiased estimate of the variance of the error term, σ^2. It is denoted by MSE or s^2.

Standard Error of the Estimate The square root of the mean square error, denoted by s. It is the estimate of σ, the standard deviation of the error term ϵ.

ANOVA table The analysis of variance table used to summarize the computations associated with the F test for significance.

Confidence interval estimate The interval estimate of the mean value of y for a given value of x.

Prediction interval estimate The interval estimate of an individual value of y for a given value of x.

Residual analysis The analysis of the residuals used to determine whether the assumptions made about the regression model appear to be valid. Residual analysis is also used to identify outliers and influential observations.

Residual plots Graphical representations of the residuals that can be used to determine whether the assumptions made about the regression model appear to be valid.

Standardized residual The value obtained by dividing a residual by its standard deviation.

Normal probability plot A graph of the standardized residuals plotted against values of the normal scores. This plot helps determine whether the assumption that the error term has a normal probability distribution appears to be valid.

Outlier A data point or observation that does not fit the trend shown by the remaining data.

Influential observation An observation that has a strong influence or effect on the regression results.

High leverage points Observations with extreme values for the independent variables.

KEY FORMULAS

Simple Linear Regression Model

$$y = \beta_0 + \beta_1 x + \epsilon \tag{14.1}$$

Simple Linear Regression Equation

$$E(y) = \beta_0 + \beta_1 x \tag{14.2}$$

Estimated Simple Linear Regression Equation

$$\hat{y} = b_0 + b_1 x \tag{14.3}$$

Least Squares Criterion

$$\min \Sigma(y_i - \hat{y}_i)^2 \tag{14.5}$$

Slope and y-Intercept for the Estimated Regression Equation

$$b_1 = \frac{\Sigma(x_i - \bar{x})(y_i - \bar{y})}{\Sigma(x_i - \bar{x})^2} \tag{14.6}$$

$$b_0 = \bar{y} - b_1\bar{x} \tag{14.7}$$

Sum of Squares Due to Error

$$\text{SSE} = \Sigma(y_i - \hat{y}_i)^2 \tag{14.8}$$

Total Sum of Squares

$$\text{SST} = \Sigma(y_i - \bar{y})^2 \tag{14.9}$$

Sum of Squares Due to Regression

$$\text{SSR} = \Sigma(\hat{y}_i - \bar{y})^2 \tag{14.10}$$

Relationship Among SST, SSR, and SSE

$$\text{SST} = \text{SSR} + \text{SSE} \tag{14.11}$$

Coefficient of Determination

$$r^2 = \frac{\text{SSR}}{\text{SST}} \tag{14.12}$$

Sample Correlation Coefficient

$$\begin{aligned} r_{xy} &= (\text{sign of } b_1)\sqrt{\text{Coefficient of Determination}} \\ &= (\text{sign of } b_1)\sqrt{r^2} \end{aligned} \tag{14.13}$$

Mean Square Error (Estimate of σ^2)

$$s^2 = \text{MSE} = \frac{\text{SSE}}{n - 2} \tag{14.15}$$

Standard Error of the Estimate

$$s = \sqrt{\text{MSE}} = \sqrt{\frac{\text{SSE}}{n - 2}} \tag{14.16}$$

Standard Deviation of b_1

$$\sigma_{b_1} = \frac{\sigma}{\sqrt{\Sigma(x_i - \bar{x})^2}} \tag{14.17}$$

Estimated Standard Deviation of b_1

$$s_{b_1} = \frac{s}{\sqrt{\Sigma(x_i - \bar{x})^2}} \tag{14.18}$$

***t* Test Statistic**

$$t = \frac{b_1}{s_{b_1}} \tag{14.19}$$

Mean Square Regression

$$\text{MSR} = \frac{\text{SSR}}{\text{number of independent variables}} \tag{14.20}$$

The *F* Test Statistic

$$F = \frac{\text{MSR}}{\text{MSE}} \tag{14.21}$$

Estimated Standard Deviation of $\hat{y}_p$

$$s_{\hat{y}_p} = s\sqrt{\frac{1}{n} + \frac{(x_p - \bar{x})^2}{\Sigma(x_i - \bar{x})^2}} \tag{14.23}$$

Confidence Interval Estimate of $E(\hat{y}_p)$

$$\hat{y}_p \pm t_{\alpha/2} s_{\hat{y}_p} \tag{14.24}$$

Estimated Standard Deviation of an Individual Value

$$s_{\text{ind}} = s\sqrt{1 + \frac{1}{n} + \frac{(x_p - \bar{x})^2}{\Sigma(x_i - \bar{x})^2}} \tag{14.26}$$

Prediction Interval Estimate of y_p

$$\hat{y}_p \pm t_{\alpha/2} s_{\text{ind}} \tag{14.27}$$

Residual for Observation *i*

$$y_i - \hat{y}_i \tag{14.28}$$

Standard Deviation of the *i*th Residual

$$s_{y_i - \hat{y}_i} = s\sqrt{1 - h_i} \tag{14.30}$$

Standardized Residual for Observation *i*

$$\frac{y_i - \hat{y}_i}{s_{y_i - \hat{y}_i}} \tag{14.32}$$

Leverage of Observation *i*

$$h_i = \frac{1}{n} + \frac{(x_i - \bar{x})^2}{\Sigma(x_i - \bar{x})^2} \tag{14.33}$$

SUPPLEMENTARY EXERCISES

55. Does a high value of r^2 imply that two variables are causally related? Explain.

56. In your own words, explain the difference between an interval estimate of the mean value of y for a given x and an interval estimate for an individual value of y for a given x.

57. What is the purpose of testing whether $\beta_1 = 0$? If we reject $\beta_1 = 0$, does it imply a good fit?

58. The data in the following table show the number of shares selling (millions) and the expected price (average of projected low price and projected high price) for 10 selected initial public stock offerings (*USA Today,* November 17, 1997).

IPO

Company	Shares Selling	Expected Price ($)
American Physician	5.0	15
Apex Silver Mines	9.0	14
Dan River	6.7	15
Franchise Mortgage	8.75	17
Gene Logic	3.0	11
International Home Foods	13.6	19
PRT Group	4.6	13
Rayovac	6.7	14
RealNetworks	3.0	10
Software AG Systems	7.7	13

a. Develop an estimated regression equation with the number of shares selling as the independent variable and the expected price as the dependent variable.
b. At the .05 level of significance, is there a significant relationship between the two variables?
c. Did the estimated regression equation provide a good fit? Explain.
d. Use the estimated regression equation to estimate the expected price for a firm considering an initial public offering of 6 million shares.

59. Corporate share repurchase programs are often touted as a benefit for shareholders. But, Robert Gabele, director of insider research for First Call/Thomson Financial, has noted that many of these have been undertaken solely to acquire stock for a company's incentive options for top managers. Across all companies, existing stock options in 1998 represented 6.2 percent of all common shares outstanding. The following data shows the number of shares covered by option grants and the number of shares outstanding for 13 companies (*Bloomberg Personal Finance,* January/February 2000).

CD file
Options

	Shares of Option Grants Outstanding (millions)	Common Shares Outstanding (millions)
Adobe Systems	20.3	61.8
Apple Computer	52.7	160.9
Applied Materials	109.1	375.4
Autodesk	15.7	58.9
Best Buy	44.2	203.8
Fruit of the Loom	14.2	66.9
ITT Industries	18.0	87.9
Merrill Lynch	89.9	365.5
Novell	120.2	335.0
Parametric Technology	78.3	269.3
Reebok International	12.8	56.1
Silicon Graphics	52.6	188.8
Toys R Us	54.8	247.6

a. Develop the estimated regression equation that could be used to estimate the number of shares of option grants outstanding given the number of common shares outstanding.
b. Use the estimated regression equation to estimate the number of shares of option grants outstanding for a company that has 150 million shares of common stock outstanding.
c. Do you believe the estimated regression equation would provide a good prediction of the number of shares of option grants outstanding? Use r^2 to support your answer.

60. *Value Line* (February 24, 1995) reported that the market beta for Woolworth Corporation was 1.25. Market betas for individual stocks are determined by simple linear regression. For each stock, the dependent variable is its quarterly percentage return (capital appreciation plus dividends) minus the percentage return that could be obtained from a risk-free investment (the Treasury Bill rate is used as the risk-free rate). The independent variable is the quarterly percentage return (capital appreciation plus dividends) for the stock market (S&P 500) minus the percentage return from a risk-free investment. An estimated regression equation is developed with quarterly data; the market beta for the stock is the slope of the estimated regression equation (b_1). The value of the market beta is often interpreted as a measure of the risk associated with the stock. Market betas greater than 1 indicate that the stock is more volatile than the market average; market betas less than 1 indicate that the stock is less volatile than the market average. The following figures are the differences between the percentage return and the risk-free return for 10 quarters for the S&P 500 and IBM.

S&P 500	IBM
1.2	−0.7
−2.5	−2.0
−3.0	−5.5
2.0	4.7
5.0	1.8
1.2	4.1
3.0	2.6
−1.0	2.0
.5	−1.3
2.5	5.5

a. Develop an estimated regression equation that can be used to determine the market beta for IBM. What is IBM's market beta?
b. Test for a significant relationship at the .05 level of significance.
c. Did the estimated regression equation provide a good fit? Explain.
d. Use the market betas of Woolworth and IBM to compare the risk associated with the two stocks.

61. The daily high and low temperatures for 20 cities follow (*USA Today,* May 9, 2000).

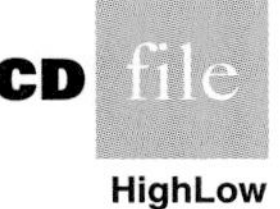
HighLow

	Low	High
Athens	54	75
Bangkok	74	92
Cairo	57	84
Copenhagen	39	64
Dublin	46	64
Havana	68	86
Hong Kong	72	81
Johannesburg	50	61
London	48	73
Manila	75	93
Melbourne	50	66
Montreal	52	64
Paris	55	77
Rio de Janeiro	61	80
Rome	54	81
Seoul	50	64
Singapore	75	90
Sydney	55	68
Tokyo	59	79
Vancouver	43	57

a. Develop a scatter diagram with low temperature on the horizontal axis and high temperature on the vertical axis.
b. What does the scatter diagram developed in (a) indicate about the relationship between the two variables?
c. Develop the estimated regression equation that could be used to predict the high temperature given the low temperature.
d. Test for a significant relationship at the .05 level of significance.
e. Did the estimated regression equation provide a good fit? Explain.
f. What is the value of the sample correlation coefficient?

62. The PJH&D Company is in the process of deciding whether to purchase a maintenance contract for its new word-processing system. Managers feel that maintenance expense should be related to usage and have collected the following information on weekly usage (hours) and annual maintenance expense ($100s).

Weekly Usage (hours)	Annual Maintenance Expense
13	17.0
10	22.0
20	30.0
28	37.0
32	47.0
17	30.5
24	32.5
31	39.0
40	51.5
38	40.0

a. Develop the estimated regression equation that relates annual maintenance expense to weekly usage.
b. Test the significance of the relationship in (a) at a .05 level of significance.
c. PJH&D expects to operate the word processor 30 hours per week. Develop a 95% prediction interval for the company's annual maintenance expense.
d. If the maintenance contract costs $3000 per year, would you recommend purchasing it? Why or why not?

63. In a manufacturing process the assembly line speed (feet per minute) was thought to affect the number of defective parts found during the inspection process. To test this theory, managers devised a situation in which the same batch of parts was inspected visually at a variety of line speeds. The table below lists the collected data.

Line Speed	Number of Defective Parts Found
20	21
20	19
40	15
30	16
60	14
40	17

a. Develop the estimated regression equation that relates line speed to the number of defective parts found.
b. At a .05 level of significance, determine whether line speed and number of defective parts found are related.
c. Did the estimated regression equation provide a good fit to the data?
d. Develop a 95% confidence interval to predict the mean number of defective parts for a line speed of 50 feet per minute.

64. A sociologist was hired by a large city hospital to investigate the relationship between the number of unauthorized days that employees are absent per year and the distance (miles) between home and work for the employees. A sample of 10 employees was chosen, and the following data were collected.

Distance to Work	Number of Days Absent
1	8
3	5
4	8
6	7
8	6
10	3
12	5
14	2
14	4
18	2

a. Develop a scatter diagram for these data. Does a linear relationship appear reasonable? Explain.
b. Develop the least squares estimated regression equation.
c. Is there a significant relationship between the two variables? Use $\alpha = .05$.
d. Did the estimated regression equation provide a good fit? Explain.
e. Use the estimated regression equation developed in (b) to develop a 95% confidence interval estimate of the expected number of days absent for employees living 5 miles from the company.

65. The regional transit authority for a major metropolitan area wants to determine whether there is any relationship between the age of a bus and the annual maintenance cost. A sample of 10 buses resulted in the following data.

Age of Bus (years)	Maintenance Cost ($)
1	350
2	370
2	480
2	520
2	590
3	550
4	750
4	800
5	790
5	950

a. Develop the least squares estimated regression equation.
b. Test to see whether the two variables are significantly related with $\alpha = .05$.
c. Did the least squares line provide a good fit to the observed data? Explain.
d. Develop a 95% prediction interval for the maintenance cost for a specific bus that is 4 years old.

66. A marketing professor at Givens College is interested in the relationship between hours spent studying and total points earned in a course. Data collected on 10 students who took the course last quarter follow.

Hours Spent Studying	Total Points Earned
45	40
30	35
90	75
60	65
105	90
65	50
90	90
80	80
55	45
75	65

a. Develop an estimated regression equation showing how total points earned is related to hours spent studying.
b. Test the significance of the model with $\alpha = .05$.
c. Predict the total points earned by Mark Sweeney. He spent 95 hours studying.
d. Develop a 95% prediction interval for the total points earned by Mark Sweeney.

67. The Transactional Records Access Clearinghouse at Syracuse University reported data showing the odds of an Internal Revenue Service audit. The following table shows the average adjusted gross income reported and the percent of the returns that were audited for 20 selected IRS districts (*The Wall Street Journal Almanac 1998*).

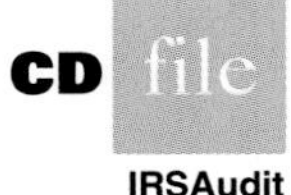

IRSAudit

District	Adjusted Gross Income	Percent Audited
Los Angeles	$36,664	1.3
Sacramento	38,845	1.1
Atlanta	34,886	1.1
Boise	32,512	1.1

Continued

District	Adjusted Gross Income	Percent Audited
Dallas	34,531	1.0
Providence	35,995	1.0
San Jose	37,799	0.9
Cheyenne	33,876	0.9
Fargo	30,513	0.9
New Orleans	30,174	0.9
Oklahoma City	30,060	0.8
Houston	37,153	0.8
Portland	34,918	0.7
Phoenix	33,291	0.7
Augusta	31,504	0.7
Albuquerque	29,199	0.6
Greensboro	33,072	0.6
Columbia	30,859	0.5
Nashville	32,566	0.5
Buffalo	34,296	0.5

a. Develop the estimated regression equation that could be used to predict the percent audited given the average adjusted gross income reported.
b. At the .05 level of significance, determine whether the adjusted gross income and the percent audited are related.
c. Did the estimated regression equation provide a good fit? Explain.
d. Use the estimated regression equation developed in (a) to calculate a 95% confidence interval estimate of the expected percent audited for districts with an average adjusted gross income of $35,000.

Case Problem 1 SPENDING AND STUDENT ACHIEVEMENT

Is the educational achievement level of students related to how much the state in which they reside spends on education? In many communities this important question is being asked by taxpayers who are being asked by their school districts to increase the amount of tax revenue spent on education. In this case, you will be asked to analyze data on spending and achievement scores in order to determine whether there is any relationship between spending and student achievement in the public schools.

The federal government's National Assessment of Educational Progress (NAEP) program is frequently used to measure the educational achievement of students. Table 14.13 shows the total current spending per pupil per year, and the composite NAEP test score for 35 states that participated in the NAEP program. These data are available on the data disk in the file named NAEP. The composite score is the sum of the math, science, and reading scores on the 1996 (1994 for reading) NAEP test. Pupils tested are in grade 8, except for reading, which is given to fourth-graders only. The maximum possible score is 1300. Table 14.14 shows the spending per pupil for 14 states that did not participate in relevant NAEP surveys. These data were reported in an article on spending and achievement level appearing in *Forbes* (November 3, 1997).

Managerial Report

1. Develop numerical and graphical summaries of the data.
2. Use regression analysis to investigate the relationship between the amount spent per pupil and the composite score on the NAEP test. Discuss your findings.

TABLE 14.13 SPENDING PER PUPIL AND COMPOSITE SCORES FOR STATES THAT PARTICIPATED IN THE NAEP PROGRAM

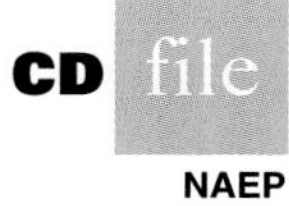
NAEP

State	Spending per Pupil ($)	Composite Score
Louisiana	4049	581
Mississippi	3423	582
California	4917	580
Hawaii	5532	580
South Carolina	4304	603
Alabama	3777	604
Georgia	4663	611
Florida	4934	611
New Mexico	4097	614
Arkansas	4060	615
Delaware	6208	615
Tennessee	3800	618
Arizona	4041	618
West Virginia	5247	625
Maryland	6100	625
Kentucky	5020	626
Texas	4520	627
New York	8162	628
North Carolina	4521	629
Rhode Island	6554	638
Washington	5338	639
Missouri	4483	641
Colorado	4772	644
Indiana	5128	649
Utah	3280	650
Wyoming	5515	657
Connecticut	7629	657
Massachusetts	6413	658
Nebraska	5410	660
Minnesota	5477	661
Iowa	5060	665
Montana	4985	667
Wisconsin	6055	667
North Dakota	4374	671
Maine	5561	675

3. Do you think that the estimated regression equation developed for these data could be used to estimate the composite scores for the states that did not participate in the NAEP program?
4. Suppose that you only considered states that spend at least $4000 per pupil but not more than $6000 per pupil. For these states, does the relationship between the two variables appear to be any different than for the complete data set? Discuss the results of your findings and whether you think deleting states with spending less than $4000 per year and more than $6000 per pupil is appropriate.
5. Develop estimates of the composite scores for the states that did not participate in the NAEP program.
6. Based upon your analyses, do you think that the educational achievement level of students is related to how much the state spends on education?

TABLE 14.14 SPENDING PER PUPIL AND FOR STATES THAT DID NOT PARTICIPATE IN THE NAEP PROGRAM

State	Spending per Pupil ($)
Idaho	3602
South Dakota	4067
Oklahoma	4265
Nevada	4658
Kansas	5164
Illinois	5297
New Hampshire	5387
Ohio	5438
Oregon	5588
Vermont	6269
Michigan	6391
Pennsylvania	6579
Alaska	7890

Case Problem 2 U.S. DEPARTMENT OF TRANSPORTATION

As part of a study on transportation safety, the U.S. Department of Transportation collected data on the number of fatal accidents per 1000 licenses and the percentage of licensed drivers under the age of 21 in a sample of 42 cities. Data collected over a 1-year period follow. These data are available on the data disk in the file named Safety.

CD file
Safety

Percent Under 21	Fatal Accidents per 1000 Licenses	Percent Under 21	Fatal Accidents per 1000 Licenses
13	2.962	17	4.100
12	0.708	8	2.190
8	0.885	16	3.623
12	1.652	15	2.623
11	2.091	9	0.835
17	2.627	8	0.820
18	3.830	14	2.890
8	0.368	8	1.267
13	1.142	15	3.224
8	0.645	10	1.014
9	1.028	10	0.493
16	2.801	14	1.443
12	1.405	18	3.614
9	1.433	10	1.926
10	0.039	14	1.643
9	0.338	16	2.943
11	1.849	12	1.913
12	2.246	15	2.814
14	2.855	13	2.634
14	2.352	9	0.926
11	1.294	17	3.256

Managerial Report

1. Develop numerical and graphical summaries of the data.
2. Use regression analysis to investigate the relationship between the number of fatal accidents and the percentage of drivers under the age of 21. Discuss your findings.
3. What conclusion and/or recommendations can you derive from your analysis?

Case Problem 3 ALUMNI GIVING

Alumni donations are an important source of revenue for colleges and universities. If administrators could determine the factors that influence increases in the percentage of alumni who make a donation, they might be able to implement policies that could lead to increased revenues. Research shows that students who are more satisfied with their contact with teachers are more likely to graduate. As a result, one might suspect that smaller class sizes and lower student-faculty ratios might lead to a higher percentage of satisfied graduates, which in turn might lead to increases in the percentage of alumni that make a donation. Table 14.15 shows data for 48 national universities (*America's Best Colleges,* Year 2000 Edition). The column labeled % of Classes Under 20 shows the percentage of classes offered with fewer than 20 students. The column labeled Student/Faculty Ratio is the number of students enrolled divided by the total number of faculty. Finally, the column labeled Alumni Giving Rate is the percentage of alumni that made a donation to the university.

Managerial Report

1. Develop numerical and graphical summaries of the data.
2. Use regression analysis to develop an estimated regression equation that could be used to predict the alumni giving rate given the percentage of classes with fewer than 20 students.
3. Use regression analysis to develop an estimated regression equation that could be used to predict the alumni giving rate given the student-faculty ratio.
4. Which of the two estimated regression equations provides the best fit? For this estimated regression equation, perform an analysis of the residuals and discuss your findings and conclusions.
5. What conclusions and/or recommendations can you derive from your analysis?

Appendix 14.1 CALCULUS-BASED DERIVATION OF LEAST SQUARES FORMULAS

As mentioned in the chapter, the least squares method is a procedure for determining the values of b_0 and b_1 that minimize the sum of squared residuals. The sum of squared residuals is given by

$$\Sigma(y_i - \hat{y}_i)^2$$

Substituting $\hat{y}_i = b_0 + b_1x_i$, we get

$$\Sigma(y_i - b_0 - b_1x_i)^2 \quad \textbf{(14.34)}$$

as the expression that must be minimized.

TABLE 14.15 DATA FOR 48 NATIONAL UNIVERSITIES

Alumni

	% of Classes Under 20	Student/Faculty Ratio	Alumni Giving Rate
Boston College	39	13	25
Brandeis University	68	8	33
Brown University	60	8	40
California Institute of Technology	65	3	46
Carnegie Mellon University	67	10	28
Case Western Reserve Univ.	52	8	31
College of William and Mary	45	12	27
Columbia University	69	7	31
Cornell University	72	13	35
Dartmouth College	61	10	53
Duke University	68	8	45
Emory University	65	7	37
Georgetown University	54	10	29
Harvard University	73	8	46
Johns Hopkins University	64	9	27
Lehigh University	55	11	40
Massachusetts Inst. of Technology	65	6	44
New York University	63	13	13
Northwestern University	66	8	30
Pennsylvania State Univ.	32	19	21
Princeton University	68	5	67
Rice University	62	8	40
Stanford University	69	7	34
Tufts University	67	9	29
Tulane University	56	12	17
U. of California–Berkeley	58	17	18
U. of California–Davis	32	19	7
U. of California–Irvine	42	20	9
U. of California–Los Angeles	41	18	13
U. of California–San Diego	48	19	8
U. of California–Santa Barbara	45	20	12
U. of Chicago	65	4	36
U. of Florida	31	23	19
U. of Illinois–Urbana Champaign	29	15	23
U. of Michigan–Ann Arbor	51	15	13
U. of North Carolina–Chapel Hill	40	16	26
U. of Notre Dame	53	13	49
U. of Pennsylvania	65	7	41
U. of Rochester	63	10	23
U. of Southern California	53	13	22
U. of Texas–Austin	39	21	13
U. of Virginia	44	13	28
U. of Washington	37	12	12
U. of Wisconsin–Madison	37	13	13
Vanderbilt University	68	9	31
Wake Forest University	59	11	38
Washington University–St. Louis	73	7	33
Yale University	77	7	50

To minimize (14.34), we must take the partial derivatives with respect to b_0 and b_1, set them equal to zero, and solve. Doing so, we get

$$\frac{\partial\Sigma(y_i - b_0 - b_1x_i)^2}{\partial b_0} = -2\Sigma(y_i - b_0 - b_1x_i) = 0 \tag{14.35}$$

$$\frac{\partial\Sigma(y_i - b_0 - b_1x_i)^2}{\partial b_1} = -2\Sigma x_i(y_i - b_0 - b_1x_i) = 0 \tag{14.36}$$

Dividing (14.35) by two and summing each term individually yields

$$-\Sigma y_i + \Sigma b_0 + \Sigma b_1x_i = 0$$

Bringing Σy_i to the other side of the equal sign and noting that $\Sigma b_0 = nb_0$, we obtain

$$nb_0 + (\Sigma x_i)b_1 = \Sigma y_i \tag{14.37}$$

Similar algebraic simplification applied to (14.36) yields

$$(\Sigma x_i)b_0 + (\Sigma x_i^2)b_1 = \Sigma x_iy_i \tag{14.38}$$

Equations (14.37) and (14.38) are known as the *normal equations*. Solving (14.37) for b_0 yields

$$b_0 = \frac{\Sigma y_i}{n} - b_1\frac{\Sigma x_i}{n} \tag{14.39}$$

Using (14.39) to substitute for b_0 in (14.38) provides

$$\frac{\Sigma x_i\Sigma y_i}{n} - \frac{(\Sigma x_i)^2}{n}b_1 + (\Sigma x_i^2)b_1 = \Sigma x_iy_i \tag{14.40}$$

Rearranging (14.40), we obtain

$$b_1 = \frac{\Sigma x_iy_i - (\Sigma x_i\Sigma y_i)/n}{\Sigma x_i^2 - (\Sigma x_i)^2/n} = \frac{\Sigma(x_i - \bar{x})(y_i - \bar{y})}{\Sigma(x_i - \bar{x})^2} \tag{14.41}$$

Because $\bar{y} = \Sigma y_i/n$ and $\bar{x} = \Sigma x_i/n$, we can rewrite (14.39) as

$$b_0 = \bar{y} - b_1\bar{x} \tag{14.42}$$

Equations (14.41) and (14.42) are the formulas (14.6) and (14.7) we used in the chapter to compute the coefficients in the estimated regression equation.

Appendix 14.2 A TEST FOR SIGNIFICANCE USING CORRELATION

Using the sample correlation coefficient r_{xy}, we can determine whether the linear relationship between x and y is significant by testing the following hypotheses about the population correlation coefficient ρ_{xy}.

$$H_0: \rho_{xy} = 0$$
$$H_a: \rho_{xy} \neq 0$$

If H_0 is rejected, we can conclude that the population correlation coefficient is not equal to zero and that the linear relationship between the two variables is significant. This test for significance follows.

A Test for Significance Using Correlation

$$H_0\colon \rho_{xy} = 0$$
$$H_a\colon \rho_{xy} \neq 0$$

Test Statistic

$$t = r_{xy}\sqrt{\frac{n-2}{1-r_{xy}^2}} \tag{14.43}$$

Rejection Rule

$$\text{Reject } H_0 \text{ if } t < -t_{\alpha/2} \text{ or if } t > t_{\alpha/2}$$

where $t_{\alpha/2}$ is based on a t distribution with $n - 2$ degrees of freedom.

In Section 14.4, we found that the sample with $n = 10$ provided the sample correlation coefficient for student population and quarterly sales of $r_{xy} = .9501$. The test statistic is

$$t = r_{xy}\sqrt{\frac{n-2}{1-r_{xy}^2}} = .9501\sqrt{\frac{10-2}{1-(.9501)^2}} = 8.61$$

From Table 2 of Appendix B, we find that the t value corresponding to $\alpha = .01$ and $n - 2 = 10 - 2 = 8$ degrees of freedom is $t_{.005} = 3.355$. With $8.61 > 3.355$, we reject H_0 and conclude at the .01 level of significance that ρ_{xy} is not equal to zero. This finding provides the statistical evidence necessary to conclude that there is a significant linear relationship between student population and quarterly sales.

Note that the test statistic t and the conclusion of a significant relationship are identical to the results obtained in Section 14.5 for the t test conducted using Armand's estimated regression equation $\hat{y} = 60 + 5x$. Performing regression analysis provides the conclusion of a significant relationship between x and y and in addition provides the equation showing how the variables are related. Most analysts therefore use modern computer packages to perform regression analysis and find that using correlation as a test of significance is unnecessary.

Appendix 14.3 REGRESSION ANALYSIS WITH MINITAB

Armand's

In Section 14.7 we discussed the computer solution of regression problems by showing Minitab's output for the Armand's Pizza Parlors problem. In this appendix, we describe the steps required to generate the Minitab computer solution. First, the data must be entered in a Minitab worksheet. Student population data were entered in column C1 and quarterly sales data were entered in column C2. The variable names Pop and Sales were entered as the column headings on the worksheet. In subsequent steps, we could refer to the data by using the variable names Pop and Sales or the column indicators C1 and C2. The following steps describe how to use Minitab to produce the regression results shown in Figure 14.10.

Step 1. Select the **Stat** pull-down menu
Step 2. Select the **Regression** pull-down menu

Step 3. Choose **Regression**
Step 4. When the Regression dialog box appears:
Enter Sales in the **Response** box
Enter Pop in the **Predictors** box
Click **OK**

The Minitab regression dialog box provides additional capabilities that can be obtained by selecting the desired options. For instance, to obtain a residual plot that shows the predicted value of the dependent variable $\hat{y}$ on the horizontal axis and the standardized residual values on the vertical axis, step 4 would be as follows:

Step 4. When the Regression dialog box appears:
Enter Sales in the **Response** box
Enter Pop in the **Predictors** box
Click the **Graphs** button
When the Regression-Graphs dialog box appears:
Select **Standardized** under Residuals for Plots
Select **Residuals versus fits** under Residual Plots
Click **OK**
When the Regression dialog box appears:
Click **OK**

Appendix 14.4 REGRESSION ANALYSIS WITH EXCEL

In this appendix we will illustrate how Excel's Regression tool can be used to perform the regression analysis computations for the Armand's Pizza Parlors problem. Refer to Figure 14.23 as we describe the steps involved. The labels Restaurant, Population, and Sales have been entered into cells A1:C1 of the worksheet. To identify each of the 10 observations, we entered the numbers 1 through 10 into cells A2:A11. The sample data have been entered into cells B2:C11. The following steps describe how to use Excel to produce the regression results.

Step 1. Select the **Tools** pull-down menu
Step 2. Choose **Data Analysis**
Step 3. Choose **Regression** from the list of Analysis Tools
Step 4. Click **OK**
Step 5. When the Regression dialog box appears:
Enter C1:C11 in the **Input Y Range** box
Enter B1:B11 in the **Input X Range** box
Select **Labels**
Select **Confidence Level**
Enter 99 in the **Confidence Level** box
Select **Output Range**
Enter A13 in the **Output Range** box
(Any upper-left-hand corner cell indicating where the output is to begin may be entered here.)
Click **OK**

The first section of the output, titled *Regression Statistics,* contains summary statistics such as the coefficient of determination (R Square). The second section of the output, titled ANOVA, contains the analysis of variance table. The last section of the output, which is not titled, contains the estimated regression coefficients and related information. We will begin our discussion of the interpretation of the regression output with the information contained in cells A28:I30.

FIGURE 14.23 EXCEL SPREADSHEET SOLUTION TO THE ARMAND'S PIZZA PARLORS PROBLEM

	A	B	C	D	E	F	G	H	I	J
1	**Restaurant**	**Population**	**Sales**							
2	1	2	58							
3	2	6	105							
4	3	8	88							
5	4	8	118							
6	5	12	117							
7	6	16	137							
8	7	20	157							
9	8	20	169							
10	9	22	149							
11	10	26	202							
12										
13	SUMMARY OUTPUT									
14										
15	*Regression Statistics*									
16	Multiple R	0.9501								
17	R Square	0.9027								
18	Adjusted R Square	0.8906								
19	Standard Error	13.8293								
20	Observations	10								
21										
22	ANOVA									
23		*df*	*SS*	*MS*	*F*	*Significance F*				
24	Regression	1	14200	14200	74.2484	2.55E-05				
25	Residual	8	1530	191.25						
26	Total	9	15730							
27										
28		*Coefficients*	*Standard Error*	*t Stat*	*P-value*	*Lower 95%*	*Upper 95%*	*Lower 99.0%*	*Upper 99.0%*	
29	Intercept	60	9.2260	6.5033	0.0002	38.7247	81.2753	29.0431	90.9569	
30	Population	5	0.5803	8.6167	2.55E-05	3.6619	6.3381	3.0530	6.9470	
31										

Interpretation of Estimated Regression Equation Output

The y intercept of the estimated regression line, $b_0 = 60$, is shown in cell B29, and the slope of the estimated regression line, $b_1 = 5$, is shown in cell B30. The label Intercept in cell A29 and the label Population in cell A30 are used to identify these two values.

In Section 14.5 we showed that the estimated standard deviation of b_1 is $s_{b_1} = .5803$. Note that the value in cell C30 is .5803. The label Standard Error in cell C28 is Excel's way of indicating that the value in cell C30 is the standard error, or standard deviation, of b_1. Recall that the t test for a significant relationship required the computation of the t statistic, $t = b_1/s_{b_1}$. For the Armand's data, the value of t that we computed was $t = 5/.5803 = 8.62$. The label in cell D28, *t Stat,* reminds us that cell D30 contains the value of the t test statistic.

In Section 14.5 we also showed that for a significance level of $\alpha = .01$ and $n - 2 = 10 - 2 = 8$ degrees of freedom, $t_{.005} = 3.355$. With $t = 8.62 > 3.355$, we were able to conclude at the .01 level of significance that β_1 is not equal to 0. In other words, the statistical evidence is sufficient to conclude that we have a significant relationship between student population and sales. The value in cell E30 is the p-value associated with the t test for significance. Excel has displayed the p-value in cell E30 using scientific notation. To obtain the decimal value, we move the decimal point 5 places to the left, obtaining a value of .0000255. Because the p-value $= .0000255 < \alpha = .01$, we can reject H_0 and conclude that we have a significant relationship between student population and quarterly sales.

The information in cells F28:I30 can be used to develop confidence interval estimates of the y intercept and slope of the estimated regression equation. Excel always provides the lower and upper limits for a 95% confidence interval. Recall that in step 4 we selected Confidence Level and entered 99 in the Confidence Level box. As a result, Excel's Regression tool also provides the lower and upper limits for a 99% confidence interval. The value in cell H30 is the lower limit for the 99% confidence interval estimate of β_1 and the value in cell I30 is the upper limit. Thus, after rounding, the 99% confidence interval estimate of β_1 is 3.05 to 6.95. The values in cells F30 and G30 provide the lower and upper limits for the 95% confidence interval. Thus, the 95% confidence interval is 3.66 to 6.34.

Interpretation of ANOVA Output

The information in cells A22:F26 is a summary of the analysis of variance computations. The three sources of variation are labeled Regression, Residual, and Total. The label *df* in cell B23 stands for degrees of freedom, the label *SS* in cell C23 stands for sum of squares, and the label *MS* in cell D23 stands for mean square.

In Section 14.5 we stated that the mean square error, obtained by dividing the error or residual sum of squares by its degrees of freedom, provides an estimate of σ^2. The value in cell D25, 191.25, is the mean square error for the Armand's regression output. In Section 14.5 we showed that an F test could also be used to test for significance in regression. The value in cell F24, .0000255, is the p-value associated with the F test for significance. Because the p-value $= .0000255 < \alpha = .01$, we can reject H_0 and conclude that we have a significant relationship between student population and quarterly sales. The label Excel uses to identify the p-value for the F test for significance, shown in cell F23, is *Significance F.*

The label Significance F may be more meaningful if you think of the value in cell F24 as the observed level of significance for the F test.

Interpretation of Regression Statistics Output

The coefficient of determination, .9027, appears in cell B17; the corresponding label, R Square, is shown in cell A17. The square root of the coefficient of determination provides the sample correlation coefficient of .9501 shown in cell B16. Note that Excel uses the label Multiple R (cell A16) to identify this value. In cell A19, the label Standard Error is used to identify the value of the standard error of the estimate shown in cell B19. Thus, the standard error of the estimate is 13.8293. We caution the reader to keep in mind that in the Excel output, the label Standard Error appears in two different places. In the Regression Statistics section of the output, the label Standard Error refers to the estimate of σ. In the Estimated Regression Equation section of the output, the label Standard Error refers to s_{b_1}, the standard deviation of the sampling distribution of b_1.

Chapter 15

MULTIPLE REGRESSION

CONTENTS

STATISTICS IN PRACTICE

CHAMPION INTERNATIONAL CORPORATION*
Stamford, Connecticut

Champion International Corporation is one of the largest forest product companies in the world, with more than 3 million acres of timberlands in the United States. It produces building materials such as lumber and plywood, white paper products such as printing and writing grades of white paper, and brown paper products such as linerboard and corrugated containers. To make these paper products, Champion's pulp mills process wood chips and chemicals to produce wood pulp. The wood pulp is then used at a paper mill to produce paper products.

In the production of white paper products, the pulp must be bleached to remove any discoloration. A key bleaching agent used in the process is chlorine dioxide, which, because of its combustible nature, is usually produced at Champion pulp mill facilities and then piped in solution form into the bleaching tower of the pulp mill. To improve one of the processes Champion uses to produce chlorine dioxide, a study was undertaken to look at process control and efficiency. One of the aspects studied was the chemical-feed rate for chlorine dioxide production.

To produce the chlorine dioxide, four chemicals flow at metered rates into the chlorine dioxide generator. The chlorine dioxide produced in the generator flows to an absorber where chilled water absorbs the chlorine dioxide gas to form a chlorine dioxide solution. The solution is then piped into the paper mill. A key part of controlling the process involves the chemical-feed rates. Historically, the chemical-feed rates were simply set by experienced operators; this approach, however, led to overcontrol by the operators. Consequently, chemical engineers at the mill requested that a set of control equations, one for each chemical feed, be developed to aid the operators in setting the rates.

Using multiple regression analysis, Champion International developed a better bleaching process for their paper products. © Lester Lefkowitz/The Stock Market.

Using multiple regression analysis, statistical analysts at Champion were able to develop an estimated multiple regression equation for each of the four chemicals used in the process. Each equation related the production of chlorine dioxide to the amount of chemical used and the concentration level of the chlorine dioxide solution. The resulting set of four equations was programmed into a microcomputer at each mill. In the new system, operators enter the concentration of the chlorine dioxide solution and the desired production rate; the computer software then calculates the chemical feed needed to achieve the desired production rate. Since the operators have begun using the control equations, the chlorine dioxide generator efficiency has increased, and the number of times the concentrations have been within acceptable ranges has increased significantly.

Champion used multiple regression analysis to develop its control equations. In this chapter we will discuss how statistical computer packages such as Minitab are used for such purposes. Most of the concepts introduced in Chapter 14 for simple linear regression can be directly extended to the multiple regression case.

*The authors are indebted to Marian Williams and Bill Griggs of Champion International Corporation for providing this Statistics in Practice.

In Chapter 14 we presented simple linear regression and demonstrated its use in developing an equation that describes the relationship between two variables. Recall that the variable being predicted or explained by the equation is called the dependent variable and the variable being used to predict or explain the dependent variable is called the independent variable. In this chapter we continue our study of regression analysis by considering situations involving two or more independent variables. This subject area is called **multiple regression analysis.** It enables us to consider more factors and thus obtain better estimates than are possible with simple linear regression.

15.1 MULTIPLE REGRESSION MODEL

Multiple regression analysis is the study of how a dependent variable y is related to two or more independent variables. In the general case, we will use p to denote the number of independent variables.

Regression Model and Regression Equation

The concepts of a regression model and a regression equation introduced in the preceding chapter are applicable in the multiple regression case. The equation that describes how the dependent variable y is related to the independent variables $x_1, x_2, \ldots x_p$ and an error term is called the **multiple regression model.** We begin with the assumption that the multiple regression model has the following form.

Multiple Regression Model

$$y = \beta_0 + \beta_1 x_1 + \beta_2 x_2 + \cdots + \beta_p x_p + \epsilon \qquad \textbf{(15.1)}$$

In the multiple regression model, $\beta_0, \beta_1, \beta_2, \ldots, \beta_p$ are the parameters and ϵ (the Greek letter epsilon) is a random variable. A close examination of this model reveals that y is a linear function of $x_1, x_2, \ldots, x_p$ (the $\beta_0 + \beta_1 x_1 + \beta_2 x_2 + \cdots + \beta_p x_p$ part) plus an error term ϵ. The error term accounts for the variability in y that cannot be explained by the linear effect of the p independent variables.

In Section 15.4 we will discuss the assumptions for the multiple regression model and ϵ. One of the assumptions is that the mean or expected value of ϵ is zero. A consequence of this assumption is that the mean or expected value of y, denoted $E(y)$, is equal to $\beta_0 + \beta_1 x_1 + \beta_2 x_2 + \cdots + \beta_p x_p$. The equation that describes how the mean value of y is related to $x_1, x_2, \ldots, x_p$ is called the **multiple regression equation.**

Multiple Regression Equation

$$E(y) = \beta_0 + \beta_1 x_1 + \beta_2 x_2 + \cdots + \beta_p x_p \qquad \textbf{(15.2)}$$

Estimated Multiple Regression Equation

If the values of $\beta_0, \beta_1, \beta_2, \ldots, \beta_p$ were known, equation (15.2) could be used to compute the mean value of y at given values of $x_1, x_2, \ldots, x_p$. Unfortunately, these parameter values will not, in general, be known and must be estimated from sample data. A simple random sample is used to compute sample statistics $b_0, b_1, b_2, \ldots, b_p$ that are used as the point estimators of the parameters $\beta_0, \beta_1, \beta_2, \ldots, \beta_p$. These sample statistics provide the following **estimated multiple regression equation.**

Estimated Multiple Regression Equation

$$\hat{y} = b_0 + b_1x_1 + b_2x_2 + \cdots + b_px_p \tag{15.3}$$

where

$b_0, b_1, b_2, \ldots, b_p$ are the estimates of $\beta_0, \beta_1, \beta_2, \ldots, \beta_p$
$\hat{y}$ = estimated value of the dependent variable

The estimation process for multiple regression is shown in Figure 15.1.

In simple linear regression, b_0 and b_1 were the sample statistics used to estimate the parameters β_0 and β_1. Multiple regression parallels this statistical inference process, with b_0, $b_1, b_2, \ldots b_p$ denoting the sample statistics used to estimate the $\beta_0, \beta_1, \beta_2, \ldots, \beta_p$.

FIGURE 15.1 THE ESTIMATION PROCESS FOR MULTIPLE REGRESSION

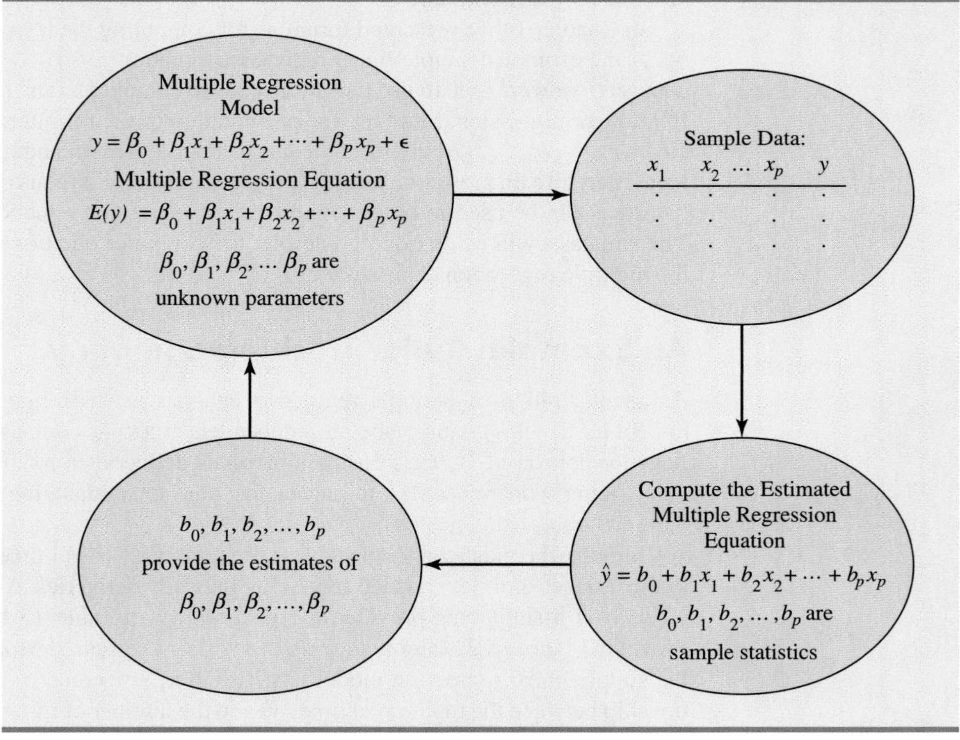

15.2 LEAST SQUARES METHOD

In Chapter 14, we used the **least squares method** to develop the estimated regression equation that best approximated the straight-line relationship between the dependent and independent variables. This same approach is used to develop the estimated multiple regression equation. The least squares criterion is restated as follows.

Least Squares Criterion

$$\min \Sigma(y_i - \hat{y}_i)^2 \tag{15.4}$$

where

y_i = observed value of the dependent variable for the ith observation
$\hat{y}_i$ = estimated value of the dependent variable for the ith observation

The estimated values of the dependent variable are computed by using the estimated multiple regression equation,

$$\hat{y} = b_0 + b_1x_1 + b_2x_2 + \cdots + b_px_p$$

As expression (15.4) shows, the least squares method uses sample data to provide the values of $b_0, b_1, b_2, \ldots, b_p$ that make the sum of squared residuals [the deviations between the observed values of the dependent variable (y_i) and the estimated values of the dependent variable ($\hat{y}_i$)] a minimum.

In Chapter 14 we presented formulas for computing the least squares estimators b_0 and b_1 for the estimated simple linear regression equation $\hat{y} = b_0 + b_1x$. With relatively small data sets, we were able to use those formulas to compute b_0 and b_1 by manual calculations. In multiple regression, however, the presentation of the formulas for the regression coefficients $b_0, b_1, b_2, \ldots, b_p$ involves the use of matrix algebra and is beyond the scope of this text. Therefore, in presenting multiple regression, we will focus on how computer software packages can be used to obtain the estimated regression equation and other information. The emphasis will be on how to interpret the computer output rather than on how to make the multiple regression computations.

An Example: Butler Trucking Company

As an illustration of multiple regression analysis, we will consider a problem faced by the Butler Trucking Company, an independent trucking company in southern California. A major portion of Butler's business involves deliveries throughout its local area. To develop better work schedules, the managers want to estimate the total daily travel time for their drivers.

Initially the managers believed that the total daily travel time would be closely related to the number of miles traveled in making the daily deliveries. A simple random sample of 10 driving assignments provided the data shown in Table 15.1 and the scatter diagram shown in Figure 15.2. After reviewing this scatter diagram, the managers hypothesized that the simple linear regression model $y = \beta_0 + \beta_1x_1 + \epsilon$ could be used to describe the relationship between the total travel time (y) and the number of miles traveled (x_1). To estimate the parameters β_0 and β_1, the least squares method was used to develop the estimated regression equation.

$$\hat{y} = b_0 + b_1x_1 \tag{15.5}$$

In Figure 15.3, we show the Minitab computer output from applying simple linear regression to the data in Table 15.1. The estimated regression equation is

$$\hat{y} = 1.27 + .0678x_1$$

TABLE 15.1 PRELIMINARY DATA FOR BUTLER TRUCKING

CD file

Butler

Driving Assignment	x_1 = Miles Traveled	y = Travel Time (hours)
1	100	9.3
2	50	4.8
3	100	8.9
4	100	6.5
5	50	4.2
6	80	6.2
7	75	7.4
8	65	6.0
9	90	7.6
10	90	6.1

At the .05 level of significance, the F value of 15.81 and its corresponding p-value of .004 indicate that the relationship is significant; that is, we can reject H_0: $\beta_1 = 0$ because the p-value is less than $\alpha = .05$. Note that the same conclusion is obtained from the t value of 3.98 and its associated p-value of .004. Thus, we can conclude that the relationship between the total travel time and the number of miles traveled is significant; longer travel times are associated with more miles traveled. With a coefficient of determination (expressed as a percentage) of R-sq = 66.4%, we see that 66.4% of the variability in travel time can be explained by the linear effect of the number of miles traveled. This finding is fairly good, but

FIGURE 15.2 SCATTER DIAGRAM OF PRELIMINARY DATA FOR BUTLER TRUCKING

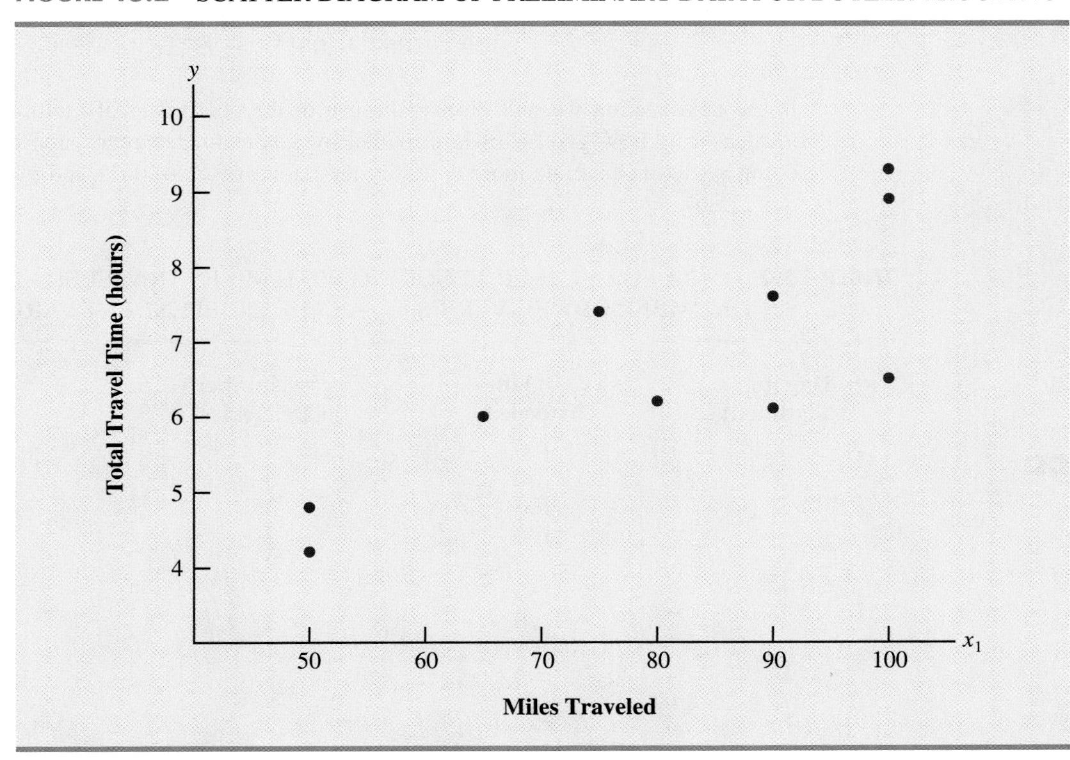

FIGURE 15.3 MINITAB OUTPUT FOR BUTLER TRUCKING WITH ONE INDEPENDENT VARIABLE

In the Minitab output the variable names Miles and Time were entered as the column headings on the worksheet; thus, x_1 = Miles and y = Time.

```
The regression equation is
Time = 1.27 + 0.0678 Miles

Predictor        Coef        Stdev     t-ratio        p
Constant        1.274        1.401        0.91    0.390
Miles         0.06783      0.01706        3.98    0.004

s = 1.002        R-sq = 66.4%      R-sq(adj) = 62.2%

Analysis of Variance

SOURCE        DF        SS          MS         F        p
Regression     1    15.871      15.871     15.81    0.004
Error          8     8.029       1.004
Total          9    23.900
```

the managers might want to consider adding a second independent variable to explain some of the remaining variability in the dependent variable.

In attempting to identify another independent variable, the managers felt that the number of deliveries could also contribute to the total travel time. The Butler Trucking data, with the number of deliveries added, are shown in Table 15.2. The Minitab computer solution with both miles traveled (x_1) and number of deliveries (x_2) as independent variables is shown in Figure 15.4. The estimated regression equation is

$$\hat{y} = -.869 + .0611x_1 + .923x_2 \tag{15.6}$$

In the next section we will discuss the use of the coefficient of multiple determination in measuring how good a fit is provided by this estimated regression equation. Before doing so, let us examine more carefully the values of $b_1 = .0611$ and $b_2 = .923$ in equation (15.6).

TABLE 15.2 DATA FOR BUTLER TRUCKING WITH MILES TRAVELED (x_1) AND NUMBER OF DELIVERIES (x_2) AS THE INDEPENDENT VARIABLES

Butler

Driving Assignment	x_1 = Miles Traveled	x_2 = Number of Deliveries	y = Travel Time (hours)
1	100	4	9.3
2	50	3	4.8
3	100	4	8.9
4	100	2	6.5
5	50	2	4.2
6	80	2	6.2
7	75	3	7.4
8	65	4	6.0
9	90	3	7.6
10	90	2	6.1

FIGURE 15.4 MINITAB OUTPUT FOR BUTLER TRUCKING WITH TWO INDEPENDENT VARIABLES

In the Minitab output the variable names Miles, Deliv, and Time were entered as the column headings on the worksheet; thus, x_1 = Miles, x_2 = Deliv, and y = Time.

```
The regression equation is
Time = - 0.869 + 0.0611 Miles + 0.923 Deliv

Predictor        Coef        Stdev     t-ratio        p
Constant      -0.8687       0.9515       -0.91    0.392
Miles        0.061135     0.009888        6.18    0.000
Deliv          0.9234       0.2211        4.18    0.004

s = 0.5731       R-sq = 90.4%       R-sq(adj) = 87.6%

Analysis of Variance

SOURCE        DF          SS          MS          F        p
Regression     2      21.601      10.800      32.88    0.000
Error          7       2.299       0.328
Total          9      23.900
```

Note on Interpretation of Coefficients

One observation can be made at this point about the relationship between the estimated regression equation with only the miles traveled as an independent variable and the equation that includes the number of deliveries as a second independent variable. The value of b_1 is not the same in both cases. In simple linear regression, we interpret b_1 as an estimate of the change in y for a one-unit change in the independent variable. In multiple regression analysis, this interpretation must be modified somewhat. That is, in multiple regression analysis, we interpret each regression coefficient as follows: b_i represents an estimate of the change in y corresponding to a one-unit change in x_i when all other independent variables are held constant. In the Butler Trucking example involving two independent variables, $b_1 = .0611$. Thus, .0611 hours is an estimate of the expected increase in travel time corresponding to an increase of one mile in the distance traveled when the number of deliveries is held constant. Similarly, because $b_2 = .923$, an estimate of the expected increase in travel time corresponding to an increase of one delivery when the number of miles traveled is held constant is .923 hours.

EXERCISES

Note to student: The exercises involving data in this and subsequent sections were designed to be solved using a computer software package.

Methods

1. The estimated regression equation for a model involving two independent variables and 10 observations follows.

$$\hat{y} = 29.1270 + .5906x_1 + .4980x_2$$

a. Interpret b_1 and b_2 in this estimated regression equation.
b. Estimate y when $x_1 = 180$ and $x_2 = 310$.

2. Consider the following data for a dependent variable y and two independent variables, x_1 and x_2.

CD file Exer2

x_1	x_2	y
30	12	94
47	10	108
25	17	112
51	16	178
40	5	94
51	19	175
74	7	170
36	12	117
59	13	142
76	16	211

a. Using these data, develop an estimated regression equation relating y to x_1. Estimate y if $x_1 = 45$.
b. Using these data, develop an estimated regression equation relating y to x_2. Estimate y if $x_2 = 15$.
c. Using these data, develop an estimated regression equation relating y to x_1 and x_2. Estimate y if $x_1 = 45$ and $x_2 = 15$.

3. In a regression analysis involving 30 observations, the following estimated regression equation was obtained.

$$\hat{y} = 17.6 + 3.8x_1 - 2.3x_2 + 7.6x_3 + 2.7x_4$$

a. Interpret b_1, b_2, b_3, and b_4 in this estimated regression equation.
b. Estimate y when $x_1 = 10$, $x_2 = 5$, $x_3 = 1$, and $x_4 = 2$.

Applications

4. A shoe store has developed the following estimated regression equation relating sales to inventory investment and advertising expenditures.

$$\hat{y} = 25 + 10x_1 + 8x_2$$

where

x_1 = inventory investment ($1000s)
x_2 = advertising expenditures ($1000s)
y = sales ($1000s)

a. Estimate sales resulting from a $15,000 investment in inventory and an advertising budget of $10,000.
b. Interpret b_1 and b_2 in this estimated regression equation.

SELF test

5. The owner of Showtime Movie Theaters, Inc., would like to estimate weekly gross revenue as a function of advertising expenditures. Historical data for a sample of 8 weeks follow.

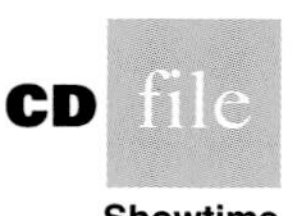

Weekly Gross Revenue ($1000s)	Television Advertising ($1000s)	Newspaper Advertising ($1000s)
96	5.0	1.5
90	2.0	2.0
95	4.0	1.5

Weekly Gross Revenue ($1000s)	Television Advertising ($1000s)	Newspaper Advertising ($1000s)
92	2.5	2.5
95	3.0	3.3
94	3.5	2.3
94	2.5	4.2
94	3.0	2.5

a. Develop an estimated regression equation with the amount of television advertising as the independent variable.
b. Develop an estimated regression equation with both television advertising and newspaper advertising as the independent variables.
c. Is the estimated regression equation coefficient for television advertising expenditures the same in part (a) and in part (b)? Interpret the coefficient in each case.
d. What is the estimate of the weekly gross revenue for a week when $3500 is spent on television advertising and $1800 is spent on newspaper advertising?

6. The following table reports the horsepower, curb weight, and the speed at ¼ mile for 16 sports and GT cars (*1998 Road & Track Sports & GT Cars*).

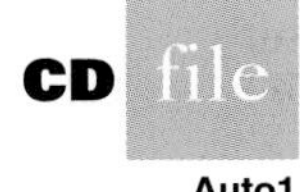

Auto1

Sports & GT Car	Curb Weight (lb.)	Horsepower	Speed at ¼ mile (mph)
Acura Integra Type R	2577	195	90.7
Acura NSX-T	3066	290	108.0
BMW Z3 2.8	2844	189	93.2
Chevrolet Camaro Z28	3439	305	103.2
Chevrolet Corvette Convertible	3246	345	102.1
Dodge Viper RT/10	3319	450	116.2
Ford Mustang GT	3227	225	91.7
Honda Prelude Type SH	3042	195	89.7
Mercedes-Benz CLK320	3240	215	93.0
Mercedes-Benz SLK230	3025	185	92.3
Mitsubishi 3000GT VR-4	3737	320	99.0
Nissan 240SX SE	2862	155	84.6
Pontiac Firebird Trans Am	3455	305	103.2
Porsche Boxster	2822	201	93.2
Toyota Supra Turbo	3505	320	105.0
Volvo C70	3285	236	97.0

a. Use curb weight as the independent variable and the speed at ¼ mile as the dependent variable. What is the estimated regression equation?
b. Use curb weight and horsepower as two independent variables and the speed at ¼ mile as the dependent variable. What is the estimated regression equation?
c. The 1999 Porsche 911 Carrera has been advertised as having a curb weight of 2910 pounds and an engine with 296 horsepower. Use the results of (b) to predict the speed at ¼ mile for the Porsche 911.

7. Heller Company manufactures lawn mowers and related lawn equipment. The managers believe the quantity of lawn mowers sold depends on the price of the mower and the price of a competitor's mower. Let

$$y = \text{quantity sold (1000s)}$$
$$x_1 = \text{price of competitor's mower (\$)}$$
$$x_2 = \text{price of Heller's mower (\$)}$$

The managers want to develop an estimated regression equation that relates quantity sold to the prices of the Heller mower and the competitor's mower. The following table lists prices in 10 cities.

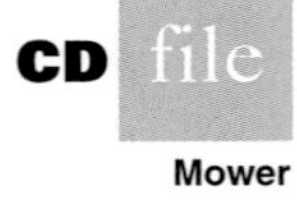
Mower

Competitor's Price (x_1)	Heller's Price (x_2)	Quantity Sold (y)
120	100	102
140	110	100
190	90	120
130	150	77
155	210	46
175	150	93
125	250	26
145	270	69
180	300	65
150	250	85

a. Determine the estimated regression equation that can be used to predict the quantity sold given the competitor's price and Heller's price.

b. Interpret b_1 and b_2.

c. Predict the quantity sold in a city where Heller prices its mower at \$160 and the competitor prices its mower at \$170.

8. The following table gives the annual return, the safety rating (0 = riskiest, 10 = safest), and the annual expense ratio for 20 foreign funds (*Mutual Funds,* March 2000).

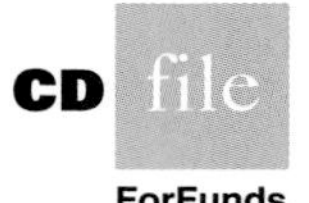
ForFunds

	Safety Rating	Annual Expense Ratio (%)	Annual Return (%)
Accessor Int'l Equity "Adv"	7.1	1.59	49
Aetna "I" International	7.2	1.35	52
Amer Century Int'l Discovery "Inv"	6.8	1.68	89
Columbia International Stock	7.1	1.56	58
Concert Inv "A" Int'l Equity	6.2	2.16	131
Dreyfus Founders Int'l Equity "F"	7.4	1.80	59
Driehaus International Growth	6.5	1.88	99
Excelsior "Inst" Int'l Equity	7.0	0.90	53
Julius Baer International Equity	6.9	1.79	77
Marshall International Stock "Y"	7.2	1.49	54
MassMutual Int'l Equity "S"	7.1	1.05	57
Morgan Grenfell Int'l Sm Cap "Inst"	7.7	1.25	61
New England "A" Int'l Equity	7.0	1.83	88
Pilgrim Int'l Small Cap "A"	7.0	1.94	122
Republic International Equity	7.2	1.09	71
Sit International Growth	6.9	1.50	51
Smith Barney "A" Int'l Equity	7.0	1.28	60
State St Research "S" Int'l Equity	7.1	1.65	50
Strong International Stock	6.5	1.61	93
Vontobel International Equity	7.0	1.50	47

a. Using these data, develop an estimated regression equation relating the annual return to the safety rating and the annual expense ratio.

b. Estimate the annual return for a firm that has a safety rating of 7.5 and annual expense ratio of 2.

9. Two experts provided subjective lists of school districts that they think are among the best in the country. For each school district the average class size, the combined SAT score, and the percentage of students who attended a 4-year college were provided.

Schools

District	Average Class Size	Combined SAT Score	% Attend 4-Year College
Blue Springs, MO	25	1083	74
Garden City, NY	18	997	77
Indianapolis, IN	30	716	40
Newport Beach, CA	26	977	51
Novi, MI	20	980	53
Piedmont, CA	28	1042	75
Pittsburg, PA	21	983	66
Scarsdale, NY	20	1110	87
Wayne, PA	22	1040	85
Weston, MA	21	1031	89
Farmingdale, NY	22	947	81
Mamaroneck, NY	20	1000	69
Mayfield, OH	24	1003	48
Morristown, NJ	22	972	64
New Rochelle, NY	23	1039	55
Newtown Square, PA	17	963	79
Omaha, NE	23	1059	81
Shaker Heights, OH	23	940	82

a. Using these data, develop an estimated regression equation relating the percentage of students who attend a 4-year college to the average class size and the combined SAT score.

b. Estimate the percentage of students who attend a 4-year college if the average class size is 20 and the combined SAT score is 1000.

10. *Auto Rental News* provided the following data, which show the number of cars in service (1000s), the number of locations, and the rental revenue ($ millions) for 15 car rental companies. (*The Wall Street Journal Almanac 1998*).

CD file

CarRent

Company	Cars	Locations	Revenue
Alamo	130	171	1180
Avis	190	1130	1500
Budget	126	1052	1500
Dollar	63.5	450	560
Enterprise	315.1	2636	2060
FRCS (Ford)	55.25	1784	312.5
Hertz	250	1200	2400
National	135	935	1200
Payless	15	100	47
PROP (Chrysler)	27	1500	160
Rent-A-Wreck	10.9	460	78
Snappy	15.5	259	85
Thrifty	34	480	340
U-Save	13.5	500	95
Value	18	45	150.1

a. Determine the estimated regression equation that can be used to predict the rental revenue given the number of cars in service.

b. Provide an interpretation for the slope of the estimated regression equation developed in part (a).

c. Determine the estimated regression equation that can be used to predict the rental revenue given the number of cars in service and the number of locations.

15.3 MULTIPLE COEFFICIENT OF DETERMINATION

In simple linear regression we showed that the total sum of squares can be partitioned into two components: the sum of squares due to regression and the sum of squares due to error. The same procedure applies to the sum of squares in multiple regression.

Relationship Among SST, SSR, and SSE

$$\text{SST} = \text{SSR} + \text{SSE} \tag{15.7}$$

where

$$\text{SST} = \text{total sum of squares} = \Sigma(y_i - \bar{y})^2$$
$$\text{SSR} = \text{sum of squares due to regression} = \Sigma(\hat{y}_i - \bar{y})^2$$
$$\text{SSE} = \text{sum of squares due to error} = \Sigma(y_i - \hat{y}_i)^2$$

Because of the computational difficulty in computing the three sums of squares, we rely on computer packages to determine those values. The analysis of variance part of the Minitab output in Figure 15.4 shows the three values for the Butler Trucking problem with two independent variables: SST $=$ 23.900, SSR $=$ 21.601, and SSE $=$ 2.299. With only one independent variable (number of miles traveled), the Minitab output in Figure 15.3 shows that SST $=$ 23.900, SSR $=$ 15.871, and SSE $=$ 8.029. The value of SST is the same in both cases because it does not depend on $\hat{y}$, but SSR increases and SSE decreases when a second independent variable (number of deliveries) is added. The implication is that the estimated multiple regression equation provides a better fit for the observed data.

In Chapter 14, we used the coefficient of determination, $r^2 = \text{SSR/SST}$, to measure the goodness of fit for the estimated regression equation. The same concept applies to multiple regression. The term **multiple coefficient of determination** indicates that we are measuring the goodness of fit for the estimated multiple regression equation. The multiple coefficient of determination, denoted R^2, is computed as follows.

Multiple Coefficient of Determination

$$R^2 = \frac{\text{SSR}}{\text{SST}} \tag{15.8}$$

The multiple coefficient of determination can be interpreted as the proportion of the variability in the dependent variable that can be explained by the estimated multiple regression equation. Hence, when multiplied by 100, it can be interpreted as the percentage of the variability in y that can be explained by the estimated regression equation.

In the two-independent-variable Butler Trucking example, with SSR $=$ 21.601 and SST $=$ 23.900, we have

$$R^2 = \frac{21.601}{23.900} = .904$$

Therefore, 90.4% of the variability in travel time y is explained by the estimated multiple regression equation with miles traveled and number of deliveries as the independent variables. In Figure 15.4, we see that the multiple coefficient of determination is also provided by the Minitab output; it is denoted by R-sq = 90.4%.

Adding independent variables causes the prediction errors to become smaller, thus reducing the sum of squares due to error, SSE. Because SSR = SST − SSE, when SSE becomes smaller, SSR becomes larger, causing R^2 = SSR/SST to increase.

Figure 15.3 shows that the R-sq value for the estimated regression equation with only one independent variable, number of miles traveled (x_1), is 66.4%. Thus, the percentage of the variability in travel times that is explained by the estimated regression equation increases from 66.4% to 90.4% when number of deliveries is added as a second independent variable. In general, R^2 always increases as independent variables are added to the model.

Many analysts prefer adjusting R^2 for the number of independent variables to avoid overestimating the impact of adding an independent variable on the amount of variability explained by the estimated regression equation. With n denoting the number of observations and p denoting the number of independent variables, the **adjusted multiple coefficient of determination** is computed as follows.

If a variable is added to the model, R^2 becomes larger even if the variable added is not statistically significant. The adjusted multiple coefficient of determination compensates for the number of independent variables in the model.

Adjusted Multiple Coefficient of Determination

$$R_a^2 = 1 - (1 - R^2)\frac{n - 1}{n - p - 1} \tag{15.9}$$

For the Butler Trucking example with $n = 10$ and $p = 2$, we have

$$R_a^2 = 1 - (1 - .904)\frac{10 - 1}{10 - 2 - 1} = .88$$

Thus, after adjusting for the two independent variables, we have an adjusted multiple coefficient of determination of .88. This value is provided by the Minitab output in Figure 15.4 as R-sq(adj) = 87.6%; the value we calculated differs because we used a rounded value of R^2 in the calculation.

NOTES AND COMMENTS

If the value of R^2 is small and the model contains a large number of independent variables, the adjusted coefficient of determination can take a negative value; in such cases, Minitab sets the adjusted coefficient of determination to zero.

EXERCISES

Methods

11. In Exercise 1, the following estimated regression equation based on 10 observations was presented.

$$\hat{y} = 29.1270 + .5906x_1 + .4980x_2$$

The values of SST and SSR are 6724.125 and 6216.375, respectively.

a. Find SSE. **b.** Compute R^2. **c.** Compute R_a^2.

d. Comment on the goodness of fit.

12. In Exercise 2, 10 observations were provided for a dependent variable y and two independent variables x_1 and x_2; for these data SST = 15,182.9, and SSR = 14,052.2.
    a. Compute R^2. **b.** Compute R_a^2.
    c. Does the estimated regression equation explain a large amount of the variability in the data? Explain.

13. In Exercise 3, the following estimated regression equation based on 30 observations was presented.

$$\hat{y} = 17.6 + 3.8x_1 - 2.3x_2 + 7.6x_3 + 2.7x_4$$

The values of SST and SSR are 1805 and 1760, respectively.
a. Compute R^2. **b.** Compute R_a^2. **c.** Comment on the goodness of fit.

Applications

14. In Exercise 4, the following estimated regression equation relating sales to inventory investment and advertising expenditures was given.

$$\hat{y} = 25 + 10x_1 + 8x_2$$

The data used to develop the model came from a survey of 10 stores; for those data, SST = 16,000 and SSR = 12,000.
a. For the estimated regression equation given, compute R^2.
b. Compute R_a^2.
c. Does the model appear to explain a large amount of variability in the data? Explain.

15. In Exercise 5, the owner of Showtime Movie Theaters, Inc., used multiple regression analysis to predict gross revenue (y) as a function of television advertising (x_1) and newspaper advertising (x_2). The estimated regression equation was

$$\hat{y} = 83.2 + 2.29x_1 + 1.30x_2$$

The computer solution provided SST = 25.5 and SSR = 23.435.
a. Compute and interpret R^2 and R_a^2.
b. When television advertising was the only independent variable, $R^2 = .653$ and $R_a^2 = .595$. Do you prefer the multiple regression results? Explain.

16. In Exercise 6, data were given on curb weight, horsepower, and speed of a sports and GT car at ¼ mile for 16 cars (*1998 Road & Track Sports & GT Cars*).
    a. Did the estimated regression equation that uses only curb weight to predict the speed at ¼ mile provide a good fit? Explain.
    b. Discuss the benefits of using both the curb weight and the horsepower to predict the speed at ¼ mile.

17. In Exercise 9, an estimated regression equation was developed relating the percentage of students who attend a 4-year college to the average class size and the combined SAT score.
    a. Compute and interpret R^2 and R_a^2.
    b. Does the estimated regression equation provide a good fit to the data? Explain.

18. Refer to Exercise 10, where data were reported on the number of cars in service (1000s), the number of locations, and the rental revenue ($ millions) for 15 car rental companies (*The Wall Street Journal Almanac 1998*).
    a. In part (c) of Exercise 10, an estimated regression equation was developed relating the rental revenue to the number of cars in service and the number of locations. What are the values of R^2 and R_a^2?
    b. Does the estimated regression equation provide a good fit to the data? Explain.

15.4 MODEL ASSUMPTIONS

In Section 15.1 we introduced the following multiple regression model.

Multiple Regression Model

$$y = \beta_0 + \beta_1 x_1 + \beta_2 x_2 + \cdots + \beta_p x_p + \epsilon \qquad \textbf{(15.10)}$$

The assumptions about the error term ϵ in the multiple regression model parallel those for the simple linear regression model.

Assumptions About the Error Term ϵ in the Multiple Regression Model $y = \beta_0 + \beta_1 x_1 + \cdots + \beta_p x_p + \epsilon$

1. The error ϵ is a random variable with mean or expected value of zero; that is, $E(\epsilon) = 0$.
   Implication: For given values of $x_1, x_2, \ldots, x_p$, the expected, or average, value of y is given by

$$E(y) = \beta_0 + \beta_1 x_1 + \beta_2 x_2 + \cdots + \beta_p x_p. \qquad \textbf{(15.11)}$$

   Equation (15.11) is the multiple regression equation we introduced in Section 15.1. In this equation, $E(y)$ represents the average of all possible values of y that might occur for the given values of $x_1, x_2, \ldots, x_p$.
2. The variance of ϵ is denoted by σ^2 and is the same for all values of the independent variables $x_1, x_2, \ldots, x_p$.
   Implication: The variance of y equals σ^2 and is the same for all values of $x_1, x_2, \ldots, x_p$.
3. The values of ϵ are independent.
   Implication: The size of the error for a particular set of values for the independent variables is not related to the size of the error for any other set of values.
4. The error ϵ is a normally distributed random variable reflecting the deviation between the y value and the expected value of y given by $\beta_0 + \beta_1 x_1 + \beta_2 x_2 + \cdots + \beta_p x_p$.
   Implication: Because $\beta_0, \beta_1, \ldots, \beta_p$ are constants for the given values of $x_1, x_2, \ldots, x_p$, the dependent variable y is also a normally distributed random variable.

To obtain more insight about the form of the relationship given by (15.11), consider the following two-independent-variable multiple regression equation.

$$E(y) = \beta_0 + \beta_1 x_1 + \beta_2 x_2$$

The graph of this equation is a plane in three-dimensional space. Figure 15.5 is such a graph. Note that ϵ is shown as the difference between the actual y value and the expected value of y, $E(y)$, when $x_1 = x_1^*$ and $x_2 = x_2^*$.

In regression analysis, the term *response variable* is often used in place of the term *dependent variable.* Furthermore, since the multiple regression equation generates a plane or surface, its graph is called a *response surface.*

FIGURE 15.5 GRAPH OF THE REGRESSION EQUATION FOR MULTIPLE REGRESSION ANALYSIS WITH TWO INDEPENDENT VARIABLES

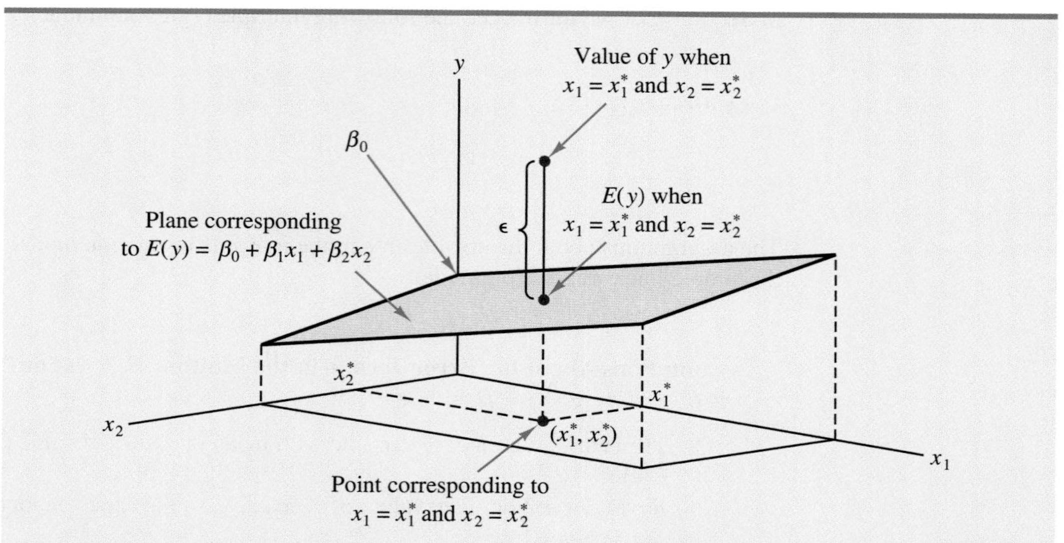

15.5 TESTING FOR SIGNIFICANCE

In this section we show how to conduct significance tests for a multiple regression relationship. The significance tests we used in simple linear regression were a t test and an F test. In simple linear regression, both tests provide the same conclusion; that is, if the null hypothesis is rejected, we conclude that $\beta_1 \neq 0$. In multiple regression, the t test and the F test have different purposes.

1. The F test is used to determine whether a significant relationship exists between the dependent variable and the set of all the independent variables; we will refer to the F test as the test for *overall significance.*
2. If the F test shows an overall significance, the t test is used to determine whether each of the individual independent variables is significant. A separate t test is conducted for each of the independent variables in the model; we refer to each of these t tests as a test for *individual significance.*

In the material that follows, we will explain the F test and the t test and apply each to the Butler Trucking Company example.

F Test

The multiple regression model as defined in Section 15.4 is

$$y = \beta_0 + \beta_1 x_1 + \beta_2 x_2 + \cdots + \beta_p x_p + \epsilon$$

The hypotheses for the F test involve the parameters of the multiple regression model.

$$H_0\colon \beta_1 = \beta_2 = \cdots = \beta_p = 0$$
H_a: One or more of the parameters is not equal to zero

If H_0 is rejected, we have sufficient statistical evidence to conclude that one or more of the parameters is not equal to zero and that the overall relationship between y and the set of independent variables $x_1, x_2, \ldots, x_p$ is significant. However, if H_0 cannot be rejected, we do not have sufficient evidence to conclude that a significant relationship is present.

Before describing the steps of the F test, we need to review the concept of *mean square.* A mean square is a sum of squares divided by its corresponding degrees of freedom. In the multiple regression case, the total sum of squares has $n - 1$ degrees of freedom, the sum of squares due to regression (SSR) has p degrees of freedom, and the sum of squares due to error has $n - p - 1$ degrees of freedom. Hence, the mean square due to regression (MSR) is SSR/p and the mean square due to error (MSE) is SSE/$(n - p - 1)$.

$$\text{MSR} = \frac{\text{SSR}}{p} \tag{15.12}$$

and

$$\text{MSE} = \frac{\text{SSE}}{n - p - 1} \tag{15.13}$$

As discussed in Chapter 14, MSE provides an unbiased estimate of σ^2, the variance of the error term ϵ. If H_0: $\beta_1 = \beta_2 = \cdots = \beta_p = 0$ is true, MSR also provides an unbiased estimate of σ^2, and the value of MSR/MSE should be close to 1. However, if H_0 is false, MSR overestimates σ^2 and the value of MSR/MSE becomes larger. To determine how large the value of MSR/MSE must be to reject H_0, we make use of the fact that if H_0 is true and the assumptions about the multiple regression model are valid, the sampling distribution of MSR/MSE is an F distribution with p degrees of freedom in the numerator and $n - p - 1$ in the denominator. A summary of the F test for significance in multiple regression follows.

F Test for Overall Significance

H_0: $\beta_1 = \beta_2 = \cdots = \beta_p = 0$

H_a: One or more of the parameters is not equal to zero

Test Statistic

$$F = \frac{\text{MSR}}{\text{MSE}} \tag{15.14}$$

Rejection Rule

Using test statistic: Reject H_0 if $F > F_\alpha$

Using p-value: Reject H_0 if p-value $< \alpha$

where F_α is based on an F distribution with p degrees of freedom in the numerator and $n - p - 1$ degrees of freedom in the denominator.

Let us apply the F test to the Butler Trucking Company multiple regression problem. With two independent variables, the hypotheses are written as follows.

H_0: $\beta_1 = \beta_2 = 0$

H_a: β_1 and/or β_2 is not equal to zero

Figure 15.6 is the Minitab output for the multiple regression model with miles traveled (x_1) and number of deliveries (x_2) as the two independent variables. In the analysis of variance part of the output, we see that MSR = 10.8 and MSE = .328. Using (15.14), we obtain the test statistic.

$$F = \frac{10.8}{.328} = 32.9$$

Note that the F value on the Minitab output is $F = 32.88$; the value we calculated differs because we used rounded values for MSR and MSE in the calculation. With a level of significance $\alpha = .01$, Table 4 of Appendix B shows that with two degrees of freedom in the numerator and seven degrees of freedom in the denominator, $F_{.01} = 9.55$. With $32.9 > 9.55$, we reject H_0: $\beta_1 = \beta_2 = 0$ and conclude that a significant relationship is present between travel time y and the two independent variables, miles traveled and number of deliveries. The p-value = 0.000 in the last column of the analysis of variance table (Figure 15.6) also indicates that we can reject H_0: $\beta_1 = \beta_2 = 0$ because the p-value is less than $\alpha = .01$.

As noted previously, the mean square error provides an unbiased estimate of σ^2, the variance of the error term ϵ. Referring to Figure 15.6, we see that the estimate of σ^2 is MSE = .328. The square root of MSE is the estimate of the standard deviation of the error term. As defined in Section 14.5, this standard deviation is called the standard error of the estimate and is denoted s. Hence, we have $s = \sqrt{\text{MSE}} = \sqrt{.328} = .573$. Note that the value of the standard error of the estimate appears in the Minitab output in Figure 15.6.

Table 15.3 is the general analysis of variance (ANOVA) table that provides the F test results for a multiple regression model. The value of the F test statistic appears in the last column and can be compared to F_α with p degrees of freedom in the numerator and $n - p - 1$ degrees of freedom in the denominator to make the hypothesis test conclusion. By reviewing the Minitab output for Butler Trucking Company in Figure 15.6, we see that Minitab's analysis of variance table contains this information. Moreover, Minitab also provides the p-value corresponding to the F test statistic.

FIGURE 15.6 MINITAB OUTPUT FOR BUTLER TRUCKING WITH TWO INDEPENDENT VARIABLES, MILES TRAVELED (x_1) AND NUMBER OF DELIVERIES (x_2)

```
The regression equation is
Time = -0.869 + 0.0611 Miles + 0.923 Deliv

Predictor        Coef       Stdev     t-ratio       p
Constant      -0.8687      0.9515       -0.91   0.392
Miles        0.061135    0.009888        6.18   0.000
Deliv          0.9234      0.2211        4.18   0.004

s = 0.5731       R-sq = 90.4%      R-sq(adj) = 87.6%

Analysis of Variance

SOURCE         DF        SS        MS        F       p
Regression      2    21.601    10.800    32.88   0.000
Error           7     2.299     0.328
Total           9    23.900
```

TABLE 15.3 ANOVA TABLE FOR A MULTIPLE REGRESSION MODEL WITH p INDEPENDENT VARIABLES

Source	Sum of Squares	Degrees of Freedom	Mean Square	F
Regression	SSR	p	$\text{MSR} = \dfrac{\text{SSR}}{p}$	$F = \dfrac{\text{MSR}}{\text{MSE}}$
Error	SSE	$n - p - 1$	$\text{MSE} = \dfrac{\text{SSE}}{n - p - 1}$	
Total	SST	$n - 1$		

t Test

If the F test shows that the multiple regression relationship is significant, a t test can be conducted to determine the significance of each of the individual parameters. The t test for individual significance follows.

t Test for Individual Significance

For any parameter β_i

$$H_0: \beta_i = 0$$
$$H_a: \beta_i \neq 0$$

Test Statistic

$$t = \frac{b_i}{s_{b_i}} \tag{15.15}$$

Rejection Rule

Using test statistic: Reject H_0 if $t < -t_{\alpha/2}$ or if $t > t_{\alpha/2}$

Using p-value: Reject H_0 if p-value $< \alpha$

where $t_{\alpha/2}$ is based on a t distribution with $n - p - 1$ degrees of freedom.

In the test statistic, s_{b_i} is the estimate of the standard deviation of b_i. The value of s_{b_i} will be provided by the computer software package.

Let us conduct the t test for the Butler Trucking regression problem. Refer to the section of Figure 15.6 that shows the Minitab output for the t-ratio calculations. Values of b_1, b_2, s_{b_1}, and s_{b_2} are as follows.

$$b_1 = .061135 \quad s_{b_1} = .009888$$
$$b_2 = .9234 \quad s_{b_2} = .2211$$

Using (15.15), we obtain the test statistic for the hypotheses involving parameters β_1 and β_2.

$$t = .061135/.009888 = 6.18$$
$$t = .9234/.2211 = 4.18$$

Note that both of these t-ratio values are provided by the Minitab output in Figure 15.6. Using $\alpha = .01$ and $n - p - 1 = 10 - 2 - 1 = 7$ degrees of freedom, we can use Table 2 of Appendix B to show $t_{.005} = 3.499$. With $6.18 > 3.499$, we reject H_0: $\beta_1 = 0$. Similarly, with $4.18 > 3.499$, we reject H_0: $\beta_2 = 0$. Note that the p-values of .000 and .004 on the Minitab output also indicate rejection of these hypotheses. Hence, both parameters are statistically significant.

Multicollinearity

We have used the term *independent variable* in regression analysis to refer to any variable being used to predict or explain the value of the dependent variable. The term does not mean, however, that the independent variables themselves are independent in any statistical sense. On the contrary, most independent variables in a multiple regression problem are correlated to some degree with one another. For example, in the Butler Trucking example involving the two independent variables x_1 (miles traveled) and x_2 (number of deliveries), we could treat the miles traveled as the dependent variable and the number of deliveries as the independent variable to determine whether those two variables are themselves related. We could then compute the sample correlation coefficient $r_{x_1x_2}$ to determine the extent to which the variables are related. Doing so yields $r_{x_1x_2} = .162$. Thus, we find some degree of linear association between the two independent variables. In multiple regression analysis, **multicollinearity** refers to the correlation among the independent variables.

To provide a better perspective of the potential problems of multicollinearity, let us consider a modification of the Butler Trucking example. Instead of x_2 being the number of deliveries, let x_2 denote the number of gallons of gasoline consumed. Clearly, x_1 (the miles traveled) and x_2 are related; that is, we know that the number of gallons of gasoline used depends on the number of miles traveled. Hence, we would conclude logically that x_1 and x_2 are highly correlated independent variables.

Assume that we obtain the equation $\hat{y} = b_0 + b_1x_1 + b_2x_2$ and find that the F test shows the relationship to be significant. Then suppose we conduct a t test on β_1 to determine whether $\beta_1 \neq 0$, and we cannot reject H_0: $\beta_1 = 0$. Does this mean that travel time is not related to miles traveled? Not necessarily. What it probably means is that with x_2 already in the model, x_1 does not make a significant contribution to determining the value of y. This interpretation makes sense in our example; if we know the amount of gasoline consumed, we do not gain much additional information useful in predicting y by knowing the miles traveled. Similarly, a t test might lead us to conclude $\beta_2 = 0$ on the grounds that, with x_1 in the model, knowledge of the amount of gasoline consumed does not add much.

A sample correlation coefficient greater than +0.70 or less than −0.70 for two independent variables is a rule of thumb warning of potential problems with multicollinearity.

To summarize, in t tests for the significance of individual parameters, the difficulty caused by multicollinearity is that it is possible to conclude that none of the individual parameters are significantly different from zero when an F test on the overall multiple regression equation indicates a significant relationship. This problem is avoided when there is little correlation among the independent variables.

Statisticians have developed several tests for determining whether multicollinearity is high enough to cause problems. According to the rule of thumb test, multicollinearity is a potential problem if the absolute value of the sample correlation coefficient exceeds .7 for any two of the independent variables. The other types of tests are more advanced and beyond the scope of this text.

When the independent variables are highly correlated, it is not possible to determine the separate effect of any particular independent variable on the dependent variable.

If possible, every attempt should be made to avoid including independent variables that are highly correlated. In practice, however, strict adherence to this policy is rarely possible. When decision makers have reason to believe substantial multicollinearity is present, they must realize that separating the effects of the individual independent variables on the dependent variable is difficult.

NOTES AND COMMENTS

Ordinarily, multicollinearity does not affect the way in which we perform our regression analysis or interpret the output from a study. However, when multicollinearity is severe—that is, when two or more of the independent variables are highly correlated with one another—we can have difficulty interpreting the results of t tests on the individual parameters. In addition to the type of problem illustrated in this section, severe cases of multicollinearity have been shown to result in least squares estimates that have the wrong sign. That is, in simulated studies where researchers created the underlying regression model and then applied the least squares technique to develop estimates of β_0, β_1, β_2, and so on, it has been shown that under conditions of high multicollinearity the least squares estimates can have a sign opposite that of the parameter being estimated. For example, β_2 might actually be $+10$ and β_2, its estimate, might turn out to be -2. Thus, little faith can be placed in the individual coefficients if multicollinearity is present to a high degree.

EXERCISES

Methods

19. In Exercise 1, the following estimated regression equation based on 10 observations was presented.

$$\hat{y} = 29.1270 + .5906x_1 + .4980x_2$$

Here SST $= 6724.125$, SSR $= 6216.375$, $s_{b_1} = .0813$, and $s_{b_2} = .0567$.

a. Compute MSR and MSE.
b. Compute F and perform the appropriate F test. Use $\alpha = .05$.
c. Perform a t test for the significance of β_1. Use $\alpha = .05$.
d. Perform a t test for the significance of β_2. Use $\alpha = .05$.

20. Refer to the data presented in Exercise 2. The estimated regression equation for these data is

$$\hat{y} = -18.4 + 2.01x_1 + 4.74x_2$$

Here SST $= 15{,}182.9$, SSR $= 14{,}052.2$, $s_{b_1} = .2471$, and $s_{b_2} = .9484$.

a. Test for a significant relationship among x_1, x_2, and y. Use $\alpha = .05$.
b. Is β_1 significant? Use $\alpha = .05$.
c. Is β_2 significant? Use $\alpha = .05$.

21. The following estimated regression equation was developed for a model involving two independent variables.

$$\hat{y} = 40.7 + 8.63x_1 + 2.71x_2$$

After x_2 was dropped from the model, the least squares method was used to obtain an estimated regression equation involving only x_1 as an independent variable.

$$\hat{y} = 42.0 + 9.01x_1$$

a. Give an interpretation of the coefficient of x_1 in both models.
b. Could multicollinearity explain why the coefficient of x_1 differs in the two models? If so, how?

Applications

22. In Exercise 4 the following estimated regression equation relating sales to inventory investment and advertising expenditures was given.

$$\hat{y} = 25 + 10x_1 + 8x_2$$

The data used to develop the model came from a survey of 10 stores; for these data SST = 16,000 and SSR = 12,000.

   a. Compute SSE, MSE, and MSR.

   b. Use an F test and a .05 level of significance to determine whether there is a relationship among the variables.

SELF test

23. Refer to Exercise 5.

   a. Use $\alpha = .01$ to test the hypotheses

$$H_0\colon \beta_1 = \beta_2 = 0$$
$$H_a\colon \beta_1 \text{ and/or } \beta_2 \text{ is not equal to zero}$$

   for the model $y = \beta_0 + \beta_1 x_1 + \beta_2 x_2 + \epsilon$, where

$$x_1 = \text{television advertising (\$1000s) and}$$
$$x_2 = \text{newspaper advertising (1000s).}$$

   b. Use $\alpha = .05$ to test the significance of β_1. Should x_1 be dropped from the model?

   c. Use $\alpha = .05$ to test the significance of β_2. Should x_2 be dropped from the model?

24. Refer to the data in Exercise 6. Use curb weight and horsepower to predict the speed of a sports and GT car at ¼ mile.

   a. Use the F test to determine the overall significance of the relationship. What is your conclusion at the .05 level of significance?

   b. Use the t test to determine the significance of each independent variable. What is your conclusion at the .05 level of significance?

25. A sample of 16 companies taken from the *Stock Investor Pro* database was used to obtain the following data on the price/earnings (P/E) ratio, the gross profit margin, and the sales growth for each company (*Stock Investor Pro*, American Association of Individual Investors, August 21, 1997).

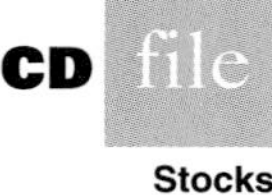

Stocks

	P/E Ratio	Gross Profit Margin (%)	Sales Growth (%)
Abbott Laboratories	22.3	23.7	10.0
American Home Products	22.6	21.1	5.3
Amoco	16.7	11.0	16.5
Bristol Meyers Squibb Co.	25.9	26.6	9.4
Chevron	18.3	11.6	18.4
Exxon	18.7	9.8	8.3
General Electric Company	13.1	13.4	13.1
Hewlett-Packard	23.3	9.7	21.9
IBM	17.3	11.5	5.6
Merck & Co., Inc.	26.2	25.6	18.9
Mobil	18.7	8.2	8.1
Pfizer	34.6	25.1	12.8
Pharmacia & Upjohn, Inc.	22.3	15.0	2.7
Procter & Gamble Co.	5.4	14.9	5.4
Texaco	12.3	7.3	23.7
Travelers Group, Inc.	28.7	17.8	28.7

a. Determine the estimated regression equation that can be used to predict the price/earnings ratio given the gross profit margin and the sales growth.
b. Use the F test to determine the overall significance of the relationship. What is your conclusion at the .05 level of significance?
c. Use the t test to determine the significance of each independent variable. What is your conclusion at the .05 level of significance?
d. Remove any independent variable that is not significant from the estimated regression equation. What is your recommended estimated regression equation? Compare the R^2 with the value of R^2 from part (a). Discuss the differences.

26. In Exercise 10 an estimated regression equation was developed relating the rental revenue to the number of cars in service and the number of locations.
a. Test for a significant relationship between the dependent variable and the two independent variables. Use $\alpha = .05$.
b. Is the number of cars in service significant? Use $\alpha = .05$.
c. Is the number of locations significant? Use $\alpha = .05$.

15.6 USING THE ESTIMATED REGRESSION EQUATION FOR ESTIMATION AND PREDICTION

The procedures for estimating the mean value of y and predicting an individual value of y in multiple regression are similar to those in regression analysis involving one independent variable. First, recall that in Chapter 14 we showed that the point estimate of the expected value of y for a given value of x was the same as the point estimate of an individual value of y. In both cases, we used $\hat{y} = b_0 + b_1x$ as the point estimate.

In multiple regression we use the same procedure. That is, we substitute the given values of $x_1, x_2, \ldots, x_p$ into the estimated regression equation and use the corresponding value of $\hat{y}$ as the point estimate. Suppose that for the Butler Trucking example we want to use the estimated regression equation involving x_1 (miles traveled) and x_2 (number of deliveries) to develop two estimates:

1. A *confidence interval estimate* of the mean travel time for all trucks that travel 100 miles and make two deliveries.
2. A *prediction interval estimate* of the travel time for *one specific* truck that travels 100 miles and makes two deliveries.

Using the estimated regression equation $\hat{y} = -.869 + .0611x_1 + .923x_2$ with $x_1 = 100$ and $x_2 = 2$, we obtain the following value of $\hat{y}$.

$$\hat{y} = -.869 + .0611(100) + .923(2) = 7.09$$

Hence, the point estimate of travel time in both cases is approximately seven hours.

To develop interval estimates for the mean value of y and for an individual value of y, we use a procedure similar to that for regression analysis involving one independent variable. The formulas required are beyond the scope of the text, but computer packages for multiple regression analysis will often provide confidence intervals once the values of $x_1, x_2, \ldots, x_p$ are specified by the user. In Table 15.4 we show the 95% confidence and prediction interval estimates for the Butler Trucking example for selected values of x_1 and x_2; these values were obtained using Minitab. Note that the interval estimate for an individual value of y is wider than the interval estimate for the expected value of y. This difference simply reflects the fact that for given values of x_1 and x_2 we can estimate the mean travel time for all trucks with more precision than we can predict the travel time for one specific truck.

TABLE 15.4 THE 95% CONFIDENCE AND PREDICTION INTERVAL ESTIMATES FOR BUTLER TRUCKING

Value of	Value of	Confidence Interval		Prediction Interval	
x_1	x_2	Lower Limit	Upper Limit	Lower Limit	Upper Limit
50	2	3.146	4.924	2.414	5.656
50	3	4.127	5.789	3.368	6.548
50	4	4.815	6.948	4.157	7.607
100	2	6.258	7.926	5.500	8.683
100	3	7.385	8.645	6.520	9.510
100	4	8.135	9.742	7.362	10.515

EXERCISES

Methods

27. In Exercise 1, the following estimated regression equation based on 10 observations was presented.

$$\hat{y} = 29.1270 + .5906x_1 + .4980x_2$$

   a. Develop a point estimate of the mean value of y when $x_1 = 180$ and $x_2 = 310$.
   b. Develop a point estimate for an individual value of y when $x_1 = 180$ and $x_2 = 310$.

SELF test

28. Refer to the data in Exercise 2. The estimated regression equation for those data is

$$\hat{y} = -18.4 + 2.01x_1 + 4.74x_2$$

   a. Develop a 95% confidence interval estimate of the mean value of y when $x_1 = 45$ and $x_2 = 15$.
   b. Develop a 95% prediction interval estimate of y when $x_1 = 45$ and $x_2 = 15$.

Applications

29. In Exercise 5, the owner of Showtime Movie Theaters, Inc., used multiple regression analysis to predict gross revenue (y) as a function of television advertising (x_1) and newspaper advertising (x_2). The estimated regression equation was

$$\hat{y} = 83.2 + 2.29x_1 + 1.30x_2$$

   a. What is the gross revenue expected for a week when $3500 is spent on television advertising ($x_1 = 3.5$) and $1800 is spent on newspaper advertising ($x_2 = 1.8$)?
   b. Provide a 95% confidence interval estimate for the mean revenue of all weeks with the expenditures listed in part (a).
   c. Provide a 95% prediction interval estimate for next week's revenue, assuming that the advertising expenditures will be allocated as in part (a).

30. In Exercise 6, data were given on curb weight, horsepower, and speed at ¼ mile for 16 sports and GT cars (*1998 Road and Track Sports & GT Cars*).
   a. Estimate the ¼-mile speed of a 1999 Porsche 911 Carrera that has a curb weight of 2910 pounds and a horsepower of 296.
   b. Provide a 95% confidence interval estimate for the ¼-mile speed of all sports and GT cars with the characteristics listed in part (a).
   c. Provide a 95% prediction interval estimate for the 1999 Porsche 911 Carrera described in part (a).

31. In Exercise 9, an estimated regression equation was developed relating the percentage of students who attend a 4-year college to the average class size and the combined SAT score.

a. Develop a 95% confidence interval estimate of the mean percentage of students who attend a 4-year college for a school district that has an average class size of 25 and whose students have a combined SAT score of 1000.

b. Suppose that a school district in Conway, South Carolina, has an average class size of 25 and a combined SAT score of 950. Develop a 95% prediction interval estimate of the percentage of students who attend a 4-year college.

15.7 QUALITATIVE INDEPENDENT VARIABLES

The independent variables may be qualitative or quantitative.

Thus far, the examples we have considered have involved quantitative independent variables such as student population, distance traveled, and number of deliveries. In many situations, however, we must work with **qualitative independent variables** such as gender (male, female), method of payment (cash, credit card, check), and so on. The purpose of this section is to show how qualitative variables are handled in regression analysis. To illustrate the use and interpretation of a qualitative independent variable, we will consider a problem facing the managers of Johnson Filtration, Inc.

An Example: Johnson Filtration, Inc.

Johnson Filtration, Inc., provides maintenance service for water-filtration systems throughout southern Florida. Customers contact Johnson with requests for maintenance service on their water-filtration systems. To estimate the service time and the service cost, Johnson's managers want to predict the repair time necessary for each maintenance request. Hence, repair time in hours is the dependent variable. Repair time is believed to be related to two factors, the number of months since the last maintenance service and the type of repair problem (mechanical or electrical). Data for a sample of 10 service calls are reported in Table 15.5.

Let y denote the repair time in hours and x_1 denote the number of months since the last maintenance service. The regression model that uses only x_1 to predict y is

$$y = \beta_0 + \beta_1 x_1 + \epsilon$$

Using Minitab to develop the estimated regression equation, we obtained the output shown in Figure 15.7. The estimated regression equation is

$$\hat{y} = 2.15 + .304x_1 \tag{15.16}$$

TABLE 15.5 DATA FOR THE JOHNSON FILTRATION EXAMPLE

Service Call	Months Since Last Service	Type of Repair	Repair Time in Hours
1	2	electrical	2.9
2	6	mechanical	3.0
3	8	electrical	4.8
4	3	mechanical	1.8
5	2	electrical	2.9
6	7	electrical	4.9
7	9	mechanical	4.2
8	8	mechanical	4.8
9	4	electrical	4.4
10	6	electrical	4.5

FIGURE 15.7 MINITAB OUTPUT FOR JOHNSON FILTRATION WITH MONTHS SINCE LAST SERVICE (x_1) AS THE INDEPENDENT VARIABLE

```
The regression equation is
Time = 2.15 + 0.304 Months

Predictor        Coef       Stdev     t-ratio        p
Constant       2.1473      0.6050        3.55    0.008
Months         0.3041      0.1004        3.03    0.016

s = 0.7810        R-sq = 53.4%      R-sq(adj) = 47.6%

Analysis of Variance

SOURCE        DF        SS          MS          F        p
Regression     1    5.5960      5.5960       9.17    0.016
Error          8    4.8800      0.6100
Total          9   10.4760
```

In the Minitab output the variable names Months and Time were entered as the column headings on the worksheet; thus, x_1 = Months and y = Time.

At the .05 level of significance, the *p*-value of .016 for the *t* (or *F*) test indicates that the number of months since the last service is significantly related to repair time. R-sq = 53.4% indicates that x_1 alone explains 53.4% of the variability in repair time.

To incorporate the type of failure into the regression model, we define the following variable.

$$x_2 = \begin{cases} 0 \text{ if the type of repair is mechanical} \\ 1 \text{ if the type of repair is electrical} \end{cases}$$

In regression analysis x_2 is called a **dummy** or *indicator* **variable.** Using this dummy variable, we can write the multiple regression model as

$$y = \beta_0 + \beta_1 x_1 + \beta_2 x_2 + \epsilon$$

Table 15.6 is the revised data set that includes the values of the dummy variable. Using Minitab and the data in Table 15.6, we can develop estimates of the model parameters. The Minitab output in Figure 15.8 shows that the estimated multiple regression equation is

$$\hat{y} = .93 + .388x_1 + 1.26x_2 \qquad \textbf{(15.17)}$$

At the .05 level of significance, the *p*-value of .001 associated with the *F* test ($F = 21.36$) indicates that the regression relationship is significant. The *t* test part of the printout in Figure 15.8 shows that both months since last service (*p*-value = .000) and type of repair (*p*-value = .005) are statistically significant. In addition, R-sq = 85.9% and R-sq(adj) = 81.9% indicate that the estimated regression equation does a good job of explaining the variability in repair times. Thus, equation (15.17) should prove helpful in estimating the repair time necessary for the various service calls.

Interpreting the Parameters

The multiple regression equation for the Johnson Filtration example is

$$E(y) = \beta_0 + \beta_1 x_1 + \beta_2 x_2 \qquad \textbf{(15.18)}$$

TABLE 15.6 DATA FOR THE JOHNSON FILTRATION EXAMPLE WITH TYPE OF REPAIR INDICATED BY A DUMMY VARIABLE ($x_2 = 0$ FOR MECHANICAL; $x_2 = 1$ FOR ELECTRICAL)

CD file
Johnson

Customer	Months Since Last Service (x_1)	Type of Repair (x_2)	Repair Time in Hours (y)
1	2	1	2.9
2	6	0	3.0
3	8	1	4.8
4	3	0	1.8
5	2	1	2.9
6	7	1	4.9
7	9	0	4.2
8	8	0	4.8
9	4	1	4.4
10	6	1	4.5

To understand how to interpret the parameters β_0, β_1, and β_2 when a qualitative variable is present, consider the case when $x_2 = 0$ (mechanical repair). Using $E(y \mid \text{mechanical})$ to denote the mean repair time *given* a mechanical repair, we have

$$E(y \mid \text{mechanical}) = \beta_0 + \beta_1 x_1 + \beta_2(0) = \beta_0 + \beta_1 x_1 \qquad \textbf{(15.19)}$$

Similarly, for an electrical repair ($x_2 = 1$), we have

$$\begin{aligned} E(y \mid \text{electrical}) &= \beta_0 + \beta_1 x_1 + \beta_2(1) = \beta_0 + \beta_1 x_1 + \beta_2 \\ &= (\beta_0 + \beta_2) + \beta_1 x_1 \end{aligned} \qquad \textbf{(15.20)}$$

FIGURE 15.8 MINITAB OUTPUT FOR JOHNSON FILTRATION WITH MONTHS SINCE LAST SERVICE (x_1) AND TYPE OF REPAIR (x_2) AS THE INDEPENDENT VARIABLES

In the Minitab output the variable names Months, Type, and Time were entered as the column headings on the worksheet; thus, x_1 = Months, x_2 = Type, and y = Time.

```
The regression equation is
Time = 0.930 + 0.388 Months + 1.26 Type

Predictor       Coef       Stdev    t-ratio        p
Constant      0.9305      0.4670       1.99    0.087
Months       0.38762     0.06257       6.20    0.000
Type          1.2627      0.3141       4.02    0.005

s = 0.4590      R-sq = 85.9%      R-sq(adj) = 81.9%

Analysis of Variance

SOURCE        DF          SS          MS          F        p
Regression     2      9.0009      4.5005      21.36    0.001
Error          7      1.4751      0.2107
Total          9     10.4760
```

Comparing equations (15.19) and (15.20), we see that the expected repair time is a linear function of x_1 for both mechanical and electrical repairs. The slope of both equations is β_1, but the y-intercept differs. The y-intercept is β_0 in (15.19) for mechanical repairs and $(\beta_0 + \beta_2)$ in (15.20) for electrical repairs. The interpretation of β_2 is that it indicates the difference between the expected repair time for an electrical repair and the expected repair time for a mechanical repair.

If β_2 is positive, the expected repair time for an electrical repair will be greater than that for a mechanical repair; if β_2 is negative, the expected repair time for an electrical repair will be less than that for a mechanical repair. Finally, if $\beta_2 = 0$, there is no difference in repair time between electrical and mechanical repairs and the type of repair is not related to the repair time.

Using the estimated multiple regression equation $\hat{y} = .93 + .388x_1 + 1.26x_2$, we see that .93 is the estimate of β_0 and 1.26 is the estimate of β_2. Thus, when $x_2 = 0$ (mechanical repair)

$$\hat{y} = .93 + .388x_1 \tag{15.21}$$

and when $x_2 = 1$ (electrical repair)

$$\begin{aligned}\hat{y} &= .93 + .388x_1 + 1.26(1) \\ &= 2.19 + .388x_1\end{aligned} \tag{15.22}$$

In effect, the use of a dummy variable for type of repair has provided two equations that can be used to predict the repair time, one corresponding to mechanical repairs and one corresponding to electrical repairs. In addition, with $b_2 = 1.26$, we have learned that, on average, electrical repairs require 1.26 hours longer than mechanical repairs.

Figure 15.9 is the plot of the Johnson data from Table 15.6. Repair time in hours (y) is represented by the vertical axis and months since last service (x_1) is represented by the horizontal axis. A data point for a mechanical repair is indicated by an M and a data point for an electrical repair is indicated by an E. Equations (15.21) and (15.22) are plotted on the graph to show graphically the two equations that can be used to predict the repair time, one corresponding to mechanical repairs and one corresponding to electrical repairs.

More Complex Qualitative Variables

A qualitative variable with k levels must be modeled using k − 1 dummy variables. Care must be taken in defining and interpreting the dummy variables.

Because the qualitative variable for the Johnson Filtration example had two levels (mechanical and electrical), defining a dummy variable with zero indicating a mechanical repair and one indicating an electrical repair was easy. However, when a qualitative variable has more than two levels, care must be taken in both defining and interpreting the dummy variables. As we will show, if a qualitative variable has k levels, $k - 1$ dummy variables are required, with each dummy variable being coded as 0 or 1.

For example, suppose a manufacturer of copy machines has organized the sales territories for a particular state into three regions: A, B, and C. The managers want to use regression analysis to help predict the number of copiers sold per week. With the number of units sold as the dependent variable, they are considering several independent variables (the number of sales personnel, advertising expenditures, and so on). Suppose the managers believe sales region is also an important factor in predicting the number of copiers sold. Because sales region is a qualitative variable with three levels, A, B and C, we will need $3 - 1 = 2$ dummy variables to represent the sales region. Each variable can be coded 0 or 1 as follows.

$$x_1 = \begin{cases} 1 \text{ if sales region B} \\ 0 \text{ otherwise} \end{cases}$$

$$x_2 = \begin{cases} 1 \text{ if sales region C} \\ 0 \text{ otherwise} \end{cases}$$

FIGURE 15.9 SCATTER DIAGRAM FOR THE JOHNSON FILTRATION REPAIR DATA FROM TABLE 15.6

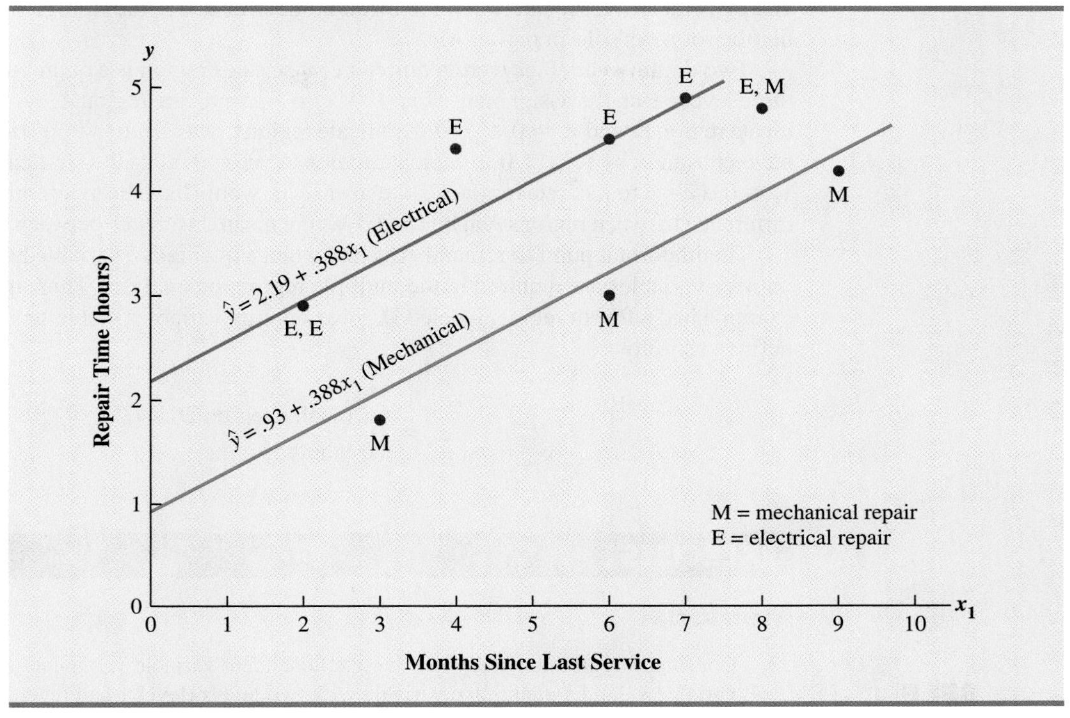

With this definition, we have the following values of x_1 and x_2.

Region	x_1	x_2
A	0	0
B	1	0
C	0	1

Observations corresponding to region A would be coded $x_1 = 0, x_2 = 0$; observations corresponding to region B would be coded $x_1 = 1, x_2 = 0$; and observations corresponding to region C would be coded $x_1 = 0, x_2 = 1$.

The regression equation relating the expected value of the number of units sold, $E(y)$, to the dummy variables would be written as

$$E(y) = \beta_0 + \beta_1 x_1 + \beta_2 x_2$$

To help us interpret the parameters β_0, β_1, and β_2, consider the following three variations of the regression equation.

$$\begin{aligned}
E(y \mid \text{region A}) &= \beta_0 + \beta_1(0) + \beta_2(0) = \beta_0 \\
E(y \mid \text{region B}) &= \beta_0 + \beta_1(1) + \beta_2(0) = \beta_0 + \beta_1 \\
E(y \mid \text{region C}) &= \beta_0 + \beta_1(0) + \beta_2(1) = \beta_0 + \beta_2
\end{aligned}$$

Thus, β_0 is the mean or expected value of sales for region A; β_1 is the difference between the mean number of units sold in region B and the mean number of units sold in region A; and β_2 is the difference between the mean number of units sold in region C and the mean number of units sold in region A.

Two dummy variables were required because sales region is a qualitative variable with three levels. But, the assignment of $x_1 = 0, x_2 = 0$ to indicate region A, $x_1 = 1, x_2 = 0$ to indicate region B, and $x_1 = 0, x_2 = 1$ to indicate region C was arbitrary. For example, we could have chosen $x_1 = 1, x_2 = 0$ to indicate region A, $x_1 = 0, x_2 = 0$ to indicate region B, and $x_1 = 0, x_2 = 1$ to indicate region C. In that case, β_1 would have been interpreted as the mean difference between regions A and B and β_2 as the mean difference between regions C and B.

The important point to remember is that when a qualitative variable has k levels, $k - 1$ dummy variables are required in the multiple regression analysis. Thus, if the sales region example had a fourth region, labeled D, three dummy variables would be necessary with x_3 defined as follows.

$$x_3 = \begin{cases} 1 \text{ if sales region D} \\ 0 \text{ otherwise} \end{cases}$$

EXERCISES

Methods

32. Consider a regression study involving a dependent variable y, a quantitative independent variable x_1, and a qualitative variable with two levels (level 1 and level 2).
    a. Write a multiple regression equation relating x_1 and the qualitative variable to y.
    b. What is the expected value of y corresponding to level 1 of the qualitative variable?
    c. What is the expected value of y corresponding to level 2 of the qualitative variable?
    d. Interpret the parameters in your regression equation.
33. Consider a regression study involving a dependent variable y, a quantitative independent variable x_1, and a qualitative independent variable with three possible levels (level 1, level 2, and level 3).
    a. How many dummy variables are required to represent the qualitative variable?
    b. Write a multiple regression equation relating x_1 and the qualitative variable to y.
    c. Interpret the parameters in your regression equation.

Applications

34. The following regression model has been proposed to predict sales at a fast-food outlet.

$$y = \beta_0 + \beta_1 x_1 + \beta_2 x_2 + \beta_3 x_3 + \epsilon$$

where

$$\begin{aligned} x_1 &= \text{number of competitors within one mile} \\ x_2 &= \text{population within one mile (1000s)} \\ x_3 &= \begin{cases} 1 \text{ if drive-up window present} \\ 0 \text{ otherwise} \end{cases} \\ y &= \text{sales (\$1000s)} \end{aligned}$$

The following estimated regression equation was developed after 20 outlets were surveyed.

$$\hat{y} = 10.1 - 4.2x_1 + 6.8x_2 + 15.3x_3$$

a. What is the expected amount of sales attributable to the drive-up window?
b. Predict sales for a store with two competitors, a population of 8000 within one mile, and no drive-up window.
c. Predict sales for a store with one competitor, a population of 3000 within one mile, and a drive-up window.

35. Refer to the Johnson Filtration problem introduced in this section. Suppose that in addition to information on the number of months since the machine was serviced and whether a mechanical or an electrical failure had occurred, the managers obtained a list showing which repairperson had performed the service. The revised data follow.

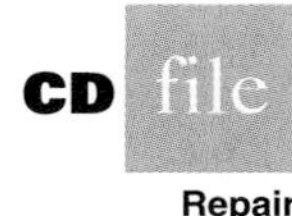

Repair

Repair Time in Hours	Months Since Last Service	Type of Repair	Repairperson
2.9	2	Electrical	Dave Newton
3.0	6	Mechanical	Dave Newton
4.8	8	Electrical	Bob Jones
1.8	3	Mechanical	Dave Newton
2.9	2	Electrical	Dave Newton
4.9	7	Electrical	Bob Jones
4.2	9	Mechanical	Bob Jones
4.8	8	Mechanical	Bob Jones
4.4	4	Electrical	Bob Jones
4.5	6	Electrical	Dave Newton

a. Ignore for now the months since the last maintenance service (x_1) and the repairperson who performed the service. Develop the estimated simple linear regression equation to predict the repair time (y) given the type of repair (x_2). Recall that $x_2 = 0$ if the type of repair is mechanical and 1 if the type of repair is electrical.
b. Does the equation that you developed in (a) provide a good fit for the observed data? Explain.
c. Ignore for now the months since the last maintenance service and the type of repair associated with the machine. Develop the estimated simple linear regression equation to predict the repair time given the repairperson who performed the service. Let $x_3 = 0$ if Bob Jones performed the service and $x_3 = 1$ if Dave Newton performed the service.
d. Does the equation that you developed in (c) provide a good fit for the observed data? Explain.

36. This problem is an extension of the situation described in Exercise 35.
a. Develop the estimated regression equation to predict the repair time given the number of months since the last maintenance service, the type of repair, and the repairperson who performed the service.
b. At the .05 level of significance, test whether the estimated regression equation developed in (a) represents a significant relationship between the independent variables and the dependent variable.
c. Is the addition of the independent variable x_3, the repairperson who performed the service, statistically significant? Use $\alpha = .05$. What explanation can you give for the results observed?

37. The National Football League rates prospects position by position on a scale that ranges from 5 to 9. The ratings are interpreted as follows: 8–9 should start the first year; 7.0–7.9 should start; 6.0–6.9 will make the team as backup; and 5.0–5.9 can make the club and contribute. The following table shows the position, weight, speed (for 40 yards), and ratings for 25 NFL prospects (*USA Today,* April 14, 2000).

CD file Football

	Position	Weight (pounds)	Speed (seconds)	Rating
Cosey Coleman	Guard	322	5.38	7.4
Travis Claridge	Guard	303	5.18	7.0
Kaulana Noa	Guard	317	5.34	6.8
Leander Jordan	Guard	330	5.46	6.7
Chad Clifton	Guard	334	5.18	6.3
Manula Savea	Guard	308	5.32	6.1
Ryan Johanningmei	Guard	310	5.28	6.0
Mark Tauscher	Guard	318	5.37	6.0
Blaine Saipaia	Guard	321	5.25	6.0
Richard Mercier	Guard	295	5.34	5.8
Damion McIntosh	Guard	328	5.31	5.3
Jeno James	Guard	320	5.64	5.0
Al Jackson	Guard	304	5.20	5.0
Chris Samuels	Offensive tackle	325	4.95	8.5
Stockar McDougle	Offensive tackle	361	5.50	8.0
Chris McIngosh	Offensive tackle	315	5.39	7.8
Adrian Klemm	Offensive tackle	307	4.98	7.6
Todd Wade	Offensive tackle	326	5.20	7.3
Marvel Smith	Offensive tackle	320	5.36	7.1
Michael Thompson	Offensive tackle	287	5.05	6.8
Bobby Williams	Offensive tackle	332	5.26	6.8
Darnell Alford	Offensive tackle	334	5.55	6.4
Terrance Beadles	Offensive tackle	312	5.15	6.3
Tutan Reyes	Offensive tackle	299	5.35	6.1
Greg Robinson-Ran	Offensive tackle	333	5.59	6.0

a. Develop a dummy variable that will account for the player's position.
b. Develop an estimated regression equation to show how rating is related to position, weight, and speed.
c. At the .05 level of significance, test whether the estimated regression equation developed in part (b) indicates a significant relationship between the independent variables and the dependent variable.
d. Does the estimated regression equation provide a good fit for the observed data? Explain.
e. Is position a significant factor in the player's rating? Use $\alpha = .05$. Explain.
f. Suppose a new offensive tackle prospect who weighs 300 pounds ran the 40 yards in 5.1 seconds. Use the estimated regression equation developed in part (b) to estimate the rating for this player.

38. A 10-year study conducted by the American Heart Association provided data on how age, blood pressure, and smoking relate to the risk of strokes. Assume that the following data are from a portion of this study. Risk is interpreted as the probability (times 100) that the patient will have a stroke over the next 10-year period. For the smoking variable, define a dummy variable with 1 indicating a smoker and 0 indicating a nonsmoker.

CD file Stroke

Risk	Age	Pressure	Smoker
12	57	152	No
24	67	163	No
13	58	155	No
56	86	177	Yes
28	59	196	No

Risk	Age	Pressure	Smoker
51	76	189	Yes
18	56	155	Yes
31	78	120	No
37	80	135	Yes
15	78	98	No
22	71	152	No
36	70	173	Yes
15	67	135	Yes
48	77	209	Yes
15	60	199	No
36	82	119	Yes
8	66	166	No
34	80	125	Yes
3	62	117	No
37	59	207	Yes

a. Using these data, develop an estimated regression equation that relates risk of a stroke to the person's age, blood pressure, and whether the person is a smoker.
b. Is smoking a significant factor in the risk of a stroke? Explain. Use $\alpha = .05$.
c. What is the probability of a stroke over the next 10 years for Art Speen, a 68-year-old smoker who has blood pressure of 175? What action might the physician recommend for this patient?

15.8 RESIDUAL ANALYSIS

In Chapter 14 we pointed out that standardized residuals were frequently used in residuals plots and in the identification of outliers. The general formula for the standardized residual for observation i follows.

Standardized Residual for Observation i

$$\frac{y_i - \hat{y}_i}{s_{y_i - \hat{y}_i}} \tag{15.23}$$

where

$$s_{y_i - \hat{y}_i} = \text{the standard deviation of residual } i$$

The general formula for the standard deviation of residual i is defined as follows.

Standard Deviation of Residual i

$$s_{y_i - \hat{y}_i} = s\sqrt{1 - h_i} \tag{15.24}$$

where

$$s = \text{standard error of the estimate}$$
$$h_i = \text{leverage of observation } i$$

As we stated in Chapter 14, the **leverage** of an observation is determined by how far the values of the independent variables are from their means. The computation of h_i, $s_{y_i - \hat{y}_i}$, and hence the standardized residual for observation i in multiple regression analysis is too complex to be done by hand. However, the standardized residuals can be easily obtained as part of the output from statistical software packages. Table 15.7 lists the predicted values, the residuals, and the standardized residuals for the Butler Trucking example presented previously in this chapter; we obtained these values by using the Minitab statistical software package. The predicted values in the table are based on the estimated regression equation $\hat{y} = -.869 + .0611x_1 + .923x_2$.

The standardized residuals and the predicted values of y from Table 15.7 are used in Figure 15.10, the standardized residual plot for the Butler Trucking multiple regression example. This standardized residual plot does not indicate any unusual abnormalities. Also, all of the standardized residuals are between -2 and $+2$; hence, we have no reason to question the assumption that the error term ϵ is normally distributed. We conclude that the model assumptions are reasonable.

A normal probability plot also can be used to determine whether the distribution of ϵ appears to be normal. The procedure and interpretation for a normal probability plot were discussed in Section 14.8. The same procedure is appropriate for multiple regression. Again, we would use a statistical software package to perform the computations and provide the normal probability plot.

Detecting Outliers

An **outlier** is an observation that is unusual in comparison with the other data; in other words, an outlier does not fit the pattern of the other data. In Chapter 14 we showed an example of an outlier and discussed how standardized residuals can be used to detect outliers. Minitab classifies an observation as an outlier if the value of its standardized residual is less than -2 or greater than $+2$. Applying this rule to the standardized residuals for the Butler Trucking example (see Table 15.7), we do not detect any outliers in the data set.

In general, the presence of one or more outliers in a data set tends to increase s, the standard error of the estimate, and hence increase $s_{y - \hat{y}_i}$, the standard deviation of residual i. Because $s_{y_i - \hat{y}_i}$ appears in the denominator of the formula for the standardized residual (15.23), the size of the standardized residual will decrease as s increases. As a result, even though a residual may be unusually large, the large denominator in (15.23) may cause the standardized

TABLE 15.7 RESIDUALS AND STANDARDIZED RESIDUALS FOR THE BUTLER TRUCKING REGRESSION ANALYSIS

Miles Traveled (x_1)	Deliveries (x_2)	Travel Time (y)	Predicted Value ($\hat{y}$)	Residual ($y - \hat{y}$)	Standardized Residual
100	4	9.3	8.93846	0.361541	0.78344
50	3	4.8	4.95830	−0.158304	−0.34962
100	4	8.9	8.93846	−0.038460	−0.08334
100	2	6.5	7.09161	−0.591609	−1.30929
50	2	4.2	4.03488	0.165121	0.38167
80	2	6.2	5.86892	0.331083	0.65431
75	3	7.4	6.48667	0.913331	1.68917
65	4	6.0	6.79875	−0.798749	−1.77372
90	3	7.6	7.40369	0.196311	0.36703
90	2	6.1	6.48026	−0.380263	−0.77639

FIGURE 15.10 STANDARDIZED RESIDUAL PLOT FOR BUTLER TRUCKING

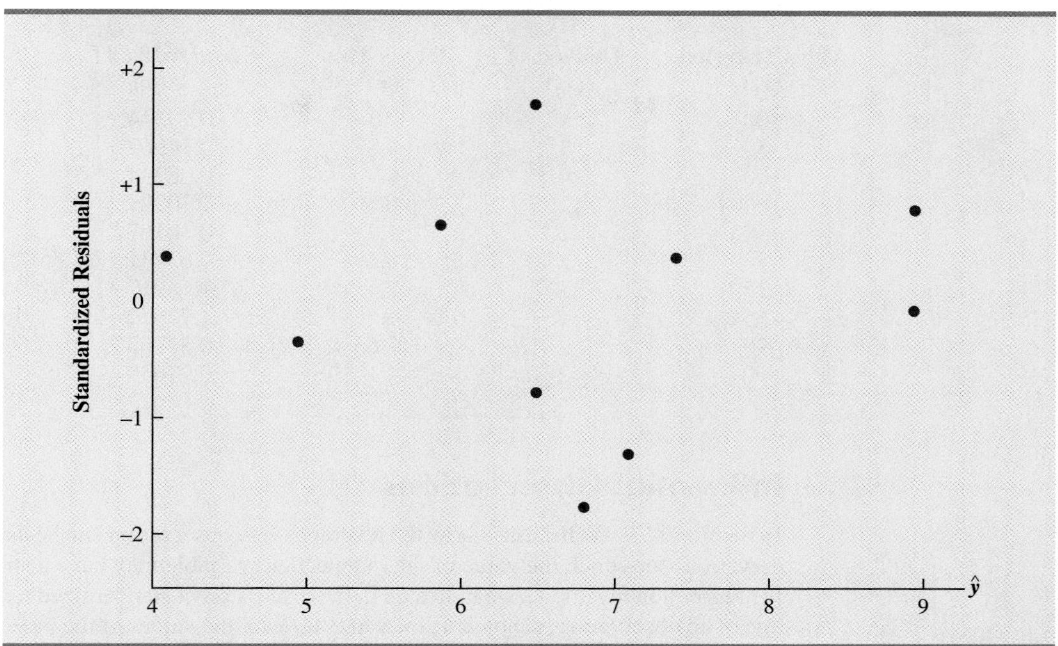

residual rule to fail to identify the observation as being an outlier. We can circumvent this difficulty by using a form of the standardized residuals called **studentized deleted residuals.**

Studentized Deleted Residuals and Outliers

Suppose the ith observation is deleted from the data set and a new estimated regression equation is developed with the remaining $n - 1$ observations. Let $s_{(i)}$ denote the standard error of the estimate based on the data set with the ith observation deleted. If we compute the standard deviation of residual i (15.24) using $s_{(i)}$ instead of s, and then compute the standardized residual for observation i (15.23) using the revised $s_{y_i - \hat{y}_i}$ value, the resulting standardized residual is called a studentized deleted residual. If the ith observation is an outlier, $s_{(i)}$ will be less than s. The absolute value of the ith studentized deleted residual therefore will be larger than the absolute value of the standardized residual. In this sense, studentized deleted residuals may detect outliers that standardized residuals do not detect.

Many statistical software packages provide an option for obtaining studentized deleted residuals. Using Minitab, we obtained the studentized deleted residuals for the Butler Trucking example; the results are reported in Table 15.8. The t distribution can be used to determine whether the studentized deleted residuals indicate the presence of outliers. Recall that p denotes the number of independent variables and n denotes the number of observations. Hence, if we delete the ith observation, the number of observations in the reduced data set is $n - 1$; in this case the error sum of squares has $(n - 1) - p - 1$ degrees of freedom. For the Butler Trucking example with $n = 10$ and $p = 2$, the degrees of freedom for the error sum of squares with the ith observation deleted is $9 - 2 - 1 = 6$. At a .05 level of significance, the t distribution (Table 2 of Appendix B) shows that with six degrees of freedom, $t_{.025} = 2.447$. If the value of the ith studentized deleted residual is less than -2.447 or greater than $+2.447$, we can conclude that the ith observation is an outlier. The studentized deleted residuals in Table 15.8 do not exceed those limits; therefore, we conclude that outliers are not present in the data set.

TABLE 15.8 STUDENTIZED DELETED RESIDUALS FOR BUTLER TRUCKING

Miles Traveled (x_1)	Deliveries (x_2)	Travel Time (y)	Standardized Residual	Studentized Deleted Residual
100	4	9.3	0.78344	0.75939
50	3	4.8	−0.34962	−0.32654
100	4	8.9	−0.08334	−0.07720
100	2	6.5	−1.30929	−1.39494
50	2	4.2	0.38167	0.35709
80	2	6.2	0.65431	0.62519
75	3	7.4	1.68917	2.03187
65	4	6.0	−1.77372	−2.21314
90	3	7.6	0.36703	0.34312
90	2	6.1	−0.77639	−0.75190

Influential Observations

In Section 14.9 we discussed how the leverage of an observation can be used to identify observations for which the value of the independent variable may have a strong influence on the regression results. As we indicated in the discussion of standardized residuals, the leverage of an observation, denoted h_i, measures how far the values of the independent variables are from their mean values. The leverage values are easily obtained as part of the output from statistical software packages. Minitab computes the leverage values and uses the rule of thumb $h_i > 3(p + 1)/n$ to identify **influential observations.** For the Butler Trucking example with $p = 2$ independent variables and $n = 10$ observations, the critical value for leverage is $3(2 + 1)/10 = .9$. The leverage values for the Butler Trucking example obtained by using Minitab are reported in Table 15.9. Because h_i does not exceed .9, we do not detect influential observations in the data set.

Using Cook's Distance Measure to Identify Influential Observations

A problem that can arise in using leverage to identify influential observations is that an observation can be identified as having high leverage and not necessarily be influential in terms of the resulting estimated regression equation. For example, Table 15.10 is a data set

TABLE 15.9 LEVERAGE AND COOK'S DISTANCE MEASURES FOR BUTLER TRUCKING

Miles Traveled (x_1)	Deliveries (x_2)	Travel Time (y)	Leverage (h_i)	Cook's D (D_i)
100	4	9.3	.351704	.110994
50	3	4.8	.375863	.024536
100	4	8.9	.351704	.001256
100	2	6.5	.378451	.347923
50	2	4.2	.430220	.036663
80	2	6.2	.220557	.040381
75	3	7.4	.110009	.117562
65	4	6.0	.382657	.650029
90	3	7.6	.129098	.006656
90	2	6.1	.269737	.074217

TABLE 15.10

DATA SET ILLUSTRATING POTENTIAL PROBLEM USING THE LEVERAGE CRITERION

x_i	y_i	Leverage h_i
1	18	.204170
1	21	.204170
2	22	.164205
3	21	.138141
4	23	.125977
4	24	.125977
5	26	.127715
15	39	.909644

consisting of eight observations and their corresponding leverage values (obtained by using Minitab). Because the leverage for the eighth observation is .91 > .75 (the critical leverage value), this observation is identified as influential. Before reaching any final conclusions, however, let us consider the situation from a different perspective.

Figure 15.11 shows the scatter diagram and the estimated regression equation corresponding to the data set in Table 15.10. We used Minitab to develop the following estimated regression equation for these data.

$$\hat{y} = 18.2 + 1.39x$$

The straight line in Figure 15.11 is the graph of this equation. Now, let us delete the observation $x = 15$, $y = 39$ from the data set and fit a new estimated regression equation to the remaining seven observations; the new estimated regression equation is

$$\hat{y} = 18.1 + 1.42x$$

We note that the y-intercept and slope of the new estimated regression equation are not significantly different from the values obtained by using all the data. Although the leverage

FIGURE 15.11 SCATTER DIAGRAM FOR THE DATA SET IN TABLE 15.10

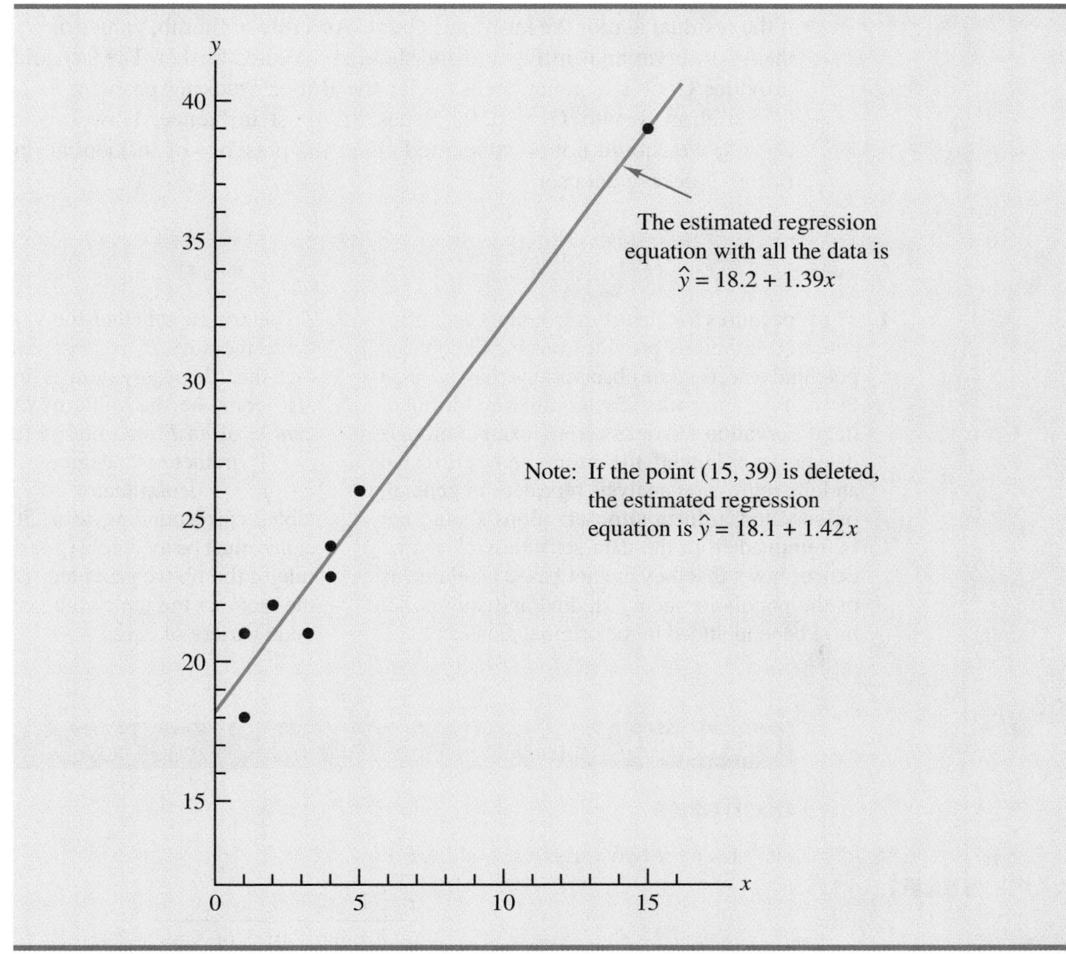

criterion identified the eighth observation as influential, this observation clearly had little influence on the results obtained. Thus, in some situations using only leverage to identify influential observations can lead to wrong conclusions.

Cook's distance measure uses both the leverage of observation i, h_i, and the residual for observation i, $(y_i - \hat{y}_i)$, to determine whether the observation is influential.

Cook's Distance Measure

$$D_i = \frac{(y_i - \hat{y}_i)^2}{(p - 1)s^2}\left[\frac{h_i}{(1 - h_i)^2}\right] \tag{15.25}$$

where

D_i = Cook's distance measure for observation i
$y_i - \hat{y}_i$ = the residual for observation i
h_i = the leverage for observation i
p = the number of independent variables
s = the standard error of the estimate

The value of Cook's distance measure will be large and indicate an influential observation if the residual and/or the leverage is large. As a rule of thumb, values of $D_i > 1$ indicate that the ith observation is influential and should be studied further. The last column of Table 15.9 provides Cook's distance measure for the Butler Trucking problem as given by Minitab. Observation 8 with $D_i = .650029$ has the most influence. However, applying the rule $D_i > 1$, we should not be concerned about the presence of influential observations in the Butler Trucking data set.

NOTES AND COMMENTS

1. The procedures for identifying outliers and influential observations provide warnings about the potential effects some observations may have on the regression results. Each outlier and influential observation warrants careful examination. If data errors are found, the errors can be corrected and the regression analysis repeated. In general, outliers and influential observations should not be removed from the data set unless clear evidence shows that they are not based on elements of the population being studied and should not have been included in the original data set.
2. To determine whether the value of Cook's distance measure D_i is large enough to conclude that the ith observation is influential, we can also compare the value of D_i to the 50th percentile of an F distribution (denoted $F_{.50}$) with $p + 1$ numerator degrees of freedom and $n - p - 1$ denominator degrees of freedom. F tables corresponding to a .50 level of significance must be available to carry out the test. The rule of thumb we provided ($D_i > 1$) is based on the fact that the table value is close to one for a wide variety of cases.

EXERCISES

Methods

39. Data for two variables, x and y, follow.

x_i	1	2	3	4	5
y_i	3	7	5	11	14

a. Develop the estimated regression equation for these data.
b. Plot the standardized residuals versus $\hat{y}$. Do there appear to be any outliers in these data? Explain.
c. Compute the studentized deleted residuals for these data. At the .05 level of significance, can any of these observations be classified as an outlier? Explain.

40. Data for two variables, x and y, follow.

x_i	22	24	26	28	40
y_i	12	21	31	35	70

a. Develop the estimated regression equation for these data.
b. Compute the studentized deleted residuals for these data. At the .05 level of significance, can any of these observations be classified as an outlier? Explain.
c. Compute the leverage values for these data. Do there appear to be any influential observations in these data? Explain.
d. Compute Cook's distance measure for these data. Are any observations influential? Explain.

Applications

41. Exercise 5 gave the following data on weekly gross revenue ($1000s), television advertising ($1000s), and newspaper advertising ($1000s) for Showtime Movie Theaters.

Showtime

Weekly Gross Revenue ($1000s)	Television Advertising ($1000s)	Newspaper Advertising ($1000s)
96	5.0	1.5
90	2.0	2.0
95	4.0	1.5
92	2.5	2.5
95	3.0	3.3
94	3.5	2.3
94	2.5	4.2
94	3.0	2.5

a. Find an estimated regression equation relating weekly gross revenue to television and newspaper advertising.
b. Plot the standardized residuals against $\hat{y}$. Does the residual plot support the assumptions about ϵ? Explain.
c. Check for any outliers in these data. What are your conclusions?
d. Are there any influential observations? Explain.

42. Exercise 6 provided data on the curb weight, horsepower, and ¼-mile speed for 16 popular sports and GT cars. Suppose that the price of each sports and GT car is also available. The complete data set is as follows:

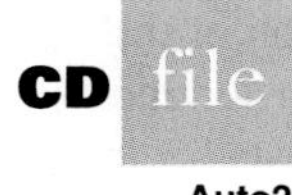

Auto2

Sports & GT Car	Price ($1000s)	Curb Weight (lb.)	Horsepower	Speed at ¼ mile (mph)
Accura Integra Type R	25.035	2577	195	90.7
Accura NSX-T	93.758	3066	290	108.0
BMW Z3 2.8	40.900	2844	189	93.2
Chevrolet Camaro Z28	24.865	3439	305	103.2
Chevrolet Corvette Convertible	50.144	3246	345	102.1
Dodge Viper RT/10	69.742	3319	450	116.2
Ford Mustang GT	23.200	3227	225	91.7

Continued

Sports & GT Car	Price ($1000s)	Curb Weight (lb.)	Horsepower	Speed at ¼ mile (mph)
Honda Prelude Type SH	26.382	3042	195	89.7
Mercedes-Benz CLK320	44.988	3240	215	93.0
Mercedes-Benz SLK230	42.762	3025	185	92.3
Mitsubishi 3000GT VR-4	47.518	3737	320	99.0
Nissan 240SX SE	25.066	2862	155	84.6
Pontiac Firebird Trans Am	27.770	3455	305	103.2
Porsche Boxster	45.560	2822	201	93.2
Toyota Supra Turbo	40.989	3505	320	105.0
Volvo C70	41.120	3285	236	97.0

a. Find the estimated regression equation, which uses price and horsepower to predict ¼-mile speed.
b. Plot the standardized residuals against $\hat{y}$. Does the residual plot support the assumption about ϵ? Explain.
c. Check for any outliers. What are your conclusions?
d. Are there any influential observations? Explain.

43. In Exercise 9, data were provided showing the percentage of students who attend a 4-year college, the average class size, and the combined SAT score.
a. Develop an estimated regression equation that can be used to predict the percentage of students who attend a 4-year college given the combined SAT score.
b. Based on the estimated regression equation developed in part (a), do these data contain any outliers and/or influential observations? Explain.
c. Develop an estimated regression equation that can be used to predict the percentage of students who attend a 4-year college given the average class size and the combined SAT score.
d. Based upon the estimated regression equation developed in part (c), do these data contain any outliers and/or influential observations? Explain.

SUMMARY

In this chapter, we introduced multiple regression analysis as an extension of simple linear regression analysis presented in Chapter 14. Multiple regression analysis enables us to understand how a dependent variable is related to two or more independent variables. The regression equation $E(y) = \beta_0 + \beta_1 x_1 + \beta_2 x_2 + \cdots + \beta_p x_p$ shows that the expected value or mean value of the dependent variable y is related to the values of the independent variables $x_1, x_2, \ldots, x_p$. Sample data and the least squares method are used to develop the estimated regression equation $\hat{y} = b_0 + b_1 x_1 + b_2 x_2 + \cdots + b_p x_p$. In effect $b_0, b_1, b_2, \ldots, b_p$ are sample statistics used to estimate the unknown model parameters $\beta_0, \beta_1, \beta_2, \ldots, \beta_p$. Computer printouts were used throughout the chapter to emphasize the fact that statistical software packages are the only realistic means of performing the numerous computations required in multiple regression analysis.

The multiple coefficient of determination was presented as a measure of the goodness of fit of the estimated regression equation. It determines the proportion of the variation of y that can be explained by the estimated regression equation. The adjusted multiple coefficient of determination is a similar measure of goodness of fit that adjusts for the number of independent variables and thus avoids overestimating the impact of adding more independent variables.

An F test and a t test were presented as ways to determine statistically whether the relationship among the variables is significant. The F test is used to determine whether there is a significant overall relationship between the dependent variable and the set of all independent variables. The t test is used to determine whether there is a significant relationship between the dependent variable and an individual independent variable given the other independent variables in the regression model. Correlation among the independent variables, known as multicollinearity, was discussed.

The section on qualitative independent variables showed how dummy variables can be used to incorporate qualitative data into multiple regression analysis. The chapter concluded with a section on how residual analysis can be used to validate the model assumptions, detect outliers, and identify influential observations. Standardized residuals, leverage, studentized deleted residuals, and Cook's distance measure were discussed.

GLOSSARY

Multiple regression analysis Regression analysis involving two or more independent variables.

Multiple regression model The mathematical equation that describes how the dependent variable y is related to the independent variables $x_1, x_2, \ldots, x_p$ and an error term ϵ.

Multiple regression equation The mathematical equation relating the expected value or mean value of the dependent variable to the values of the independent variables; that is $E(y) = \beta_0 + \beta_1 x_1 + \beta_2 x_2 + \cdots + \beta_p x_p$.

Estimated multiple regression equation The estimate of the multiple regression equation based on sample data and the least squares method; it is $\hat{y} = b_0 + b_1 x_1 + b_2 x_2 + \cdots + b_p x_p$.

Least squares method The method used to develop the estimated regression equation. It minimizes the sum of squared residuals (the deviations between the observed values of the dependent variable, y_i, and the estimated values of the dependent variable, $\hat{y}_i$).

Multiple coefficient of determination A measure of the goodness of fit of the estimated multiple regression equation. It can be interpreted as the proportion of the variability in the dependent variable that is explained by the estimated regression equation.

Adjusted multiple coefficient of determination A measure of the goodness of fit of the estimated multiple regression equation that adjusts for the number of independent variables in the model and thus avoids overestimating the impact of adding more independent variables.

Multicollinearity The term used to describe the correlation among the independent variables.

Qualitative independent variable An independent variable with qualitative data.

Dummy variable A variable used to model the effect of qualitative independent variables. A dummy variable may take only the value zero or one.

Leverage A measure of how far the values of the independent variables are from their mean values.

Outlier An observation that does not fit the pattern of the other data.

Studentized deleted residuals Standardized residuals that are based on a revised standard error of the estimate obtained by deleting observation i from the data set and then performing the regression analysis and computations.

Influential observation An observation that has a strong influence on the regression results.

Cook's distance measure A measure of the influence of an observation based on both the leverage of observation i and the residual for observation i.

KEY FORMULAS

Multiple Regression Model

$$y = \beta_0 + \beta_1 x_1 + \beta_2 x_2 + \cdots + \beta_p x_p + \epsilon \tag{15.1}$$

Multiple Regression Equation

$$E(y) = \beta_0 + \beta_1 x_1 + \beta_2 x_2 + \cdots + \beta_p x_p \tag{15.2}$$

Estimated Multiple Regression Equation

$$\hat{y} = b_0 + b_1 x_1 + b_2 x_2 + \cdots + b_p x_p \tag{15.3}$$

Least Squares Criterion

$$\min \Sigma(y_i - \hat{y}_i)^2 \tag{15.4}$$

Relationship Among SST, SSR, and SSE

$$\text{SST} = \text{SSR} + \text{SSE} \tag{15.7}$$

Multiple Coefficient of Determination

$$R^2 = \frac{\text{SSR}}{\text{SST}} \tag{15.8}$$

Adjusted Multiple Coefficient of Determination

$$R_{\text{a}}^2 = 1 - (1 - R^2)\frac{n - 1}{n - p - 1} \tag{15.9}$$

Mean Square Regression

$$\text{MSR} = \frac{\text{SSR}}{p} \tag{15.12}$$

Mean Square Error

$$\text{MSE} = \frac{\text{SSE}}{n - p - 1} \tag{15.13}$$

F Test Statistic

$$F = \frac{\text{MSR}}{\text{MSE}} \tag{15.14}$$

t Test Statistic

$$t = \frac{b_i}{s_{b_i}} \tag{15.15}$$

Standardized Residual for Observation i

$$\frac{y_i - \hat{y}_i}{s_{y_i - \hat{y}_i}} \tag{15.23}$$

Standard Deviation of Residual i

$$s_{y_i - \hat{y}_i} = s\sqrt{1 - h_i} \tag{15.24}$$

Cook's Distance Measure

$$D_i = \frac{(y_i - \hat{y}_i)^2}{(p - 1)s^2}\left[\frac{h_i}{(1 - h_i)^2}\right] \tag{15.25}$$

SUPPLEMENTARY EXERCISES

44. The admissions officer for Clearwater College developed the following estimated regression equation relating the final college GPA to the student's SAT mathematics score and high-school GPA.

$$\hat{y} = -1.41 + .0235x_1 + .00486x_2$$

where

$$\begin{aligned} x_1 &= \text{high-school grade point average} \\ x_2 &= \text{SAT mathematics score} \\ y &= \text{final college grade point average} \end{aligned}$$

a. Interpret the coefficients in this estimated regression equation.

b. Estimate the final college GPA for a student who has a high-school average of 84 and a score of 540 on the SAT mathematics test.

45. The personnel director for Electronics Associates developed the following estimated regression equation relating an employee's score on a job satisfaction test to his or her length of service and wage rate.

$$\hat{y} = 14.4 - 8.69x_1 + 13.5x_2$$

where

$$\begin{aligned} x_1 &= \text{length of service (years)} \\ x_2 &= \text{wage rate (dollars)} \\ y &= \text{job satisfaction test score (higher scores} \\ &\quad\ \text{indicate greater job satisfaction)} \end{aligned}$$

a. Interpret the coefficients in this estimated regression equation.

b. Develop an estimate of the job satisfaction test score for an employee who has 4 years of service and makes $6.50 per hour.

46. A partial computer output from a regression analysis follows.

```
The regression equation is
Y = 8.103 + 7.602 X1 + 3.111 X2

Predictor            Coef           Stdev       t-ratio
Constant           ______           2.667        _____
X1                 ______           2.105        _____
X2                 ______           0.613        _____

s = 3.35          R-sq = 92.3%      R-sq(adj) =  _____%

Analysis of Variance

SOURCE             DF              SS            MS          F
Regression        _____           1612          _____      _____
Error              12            _______        _____
Total             _____          _______
```

a. Compute the appropriate t-ratios.
b. Test for the significance of β_1 and β_2 at $\alpha = .05$.
c. Compute the entries in the DF, SS, and MS columns.
d. Compute R_a^2.

47. Recall that in Exercise 44, the admissions officer for Clearwater College developed the following estimated regression equation relating final college GPA to the student's SAT mathematics score and high-school GPA.

$$\hat{y} = -1.41 + .0235x_1 + .00486x_2$$

where

x_1 = high-school grade point average
x_2 = SAT mathematics score
y = final college grade point average

A portion of the Minitab computer output follows.

```
The regression equation is
Y = -1.41 + .0235 X1 + .00486 X2

Predictor            Coef           Stdev       t-ratio
Constant          -1.4053          0.4848        _____
X1                0.023467         0.008666      _____
X2                 ______          0.001077      _____

s = 0.1298        R-sq = ______     R-sq(adj) = ______

Analysis of Variance

SOURCE             DF              SS            MS          F
Regression        _____          1.76209        _____      _____
Error             _____          _______        _____
Total               9            1.88000
```

a. Complete the missing entries in this output.
b. Compute F and test at a .05 level of significance to see whether a significant relationship is present.
c. Did the estimated regression equation provide a good fit to the data? Explain.
d. Use the t test and $\alpha = .05$ to test $H_0: \beta_1 = 0$ and $H_0: \beta_2 = 0$.

48. Recall that in Exercise 45 the personnel director for Electronics Associates developed the following estimated regression equation relating an employee's score on a job satisfaction test to length of service and wage rate.

$$\hat{y} = 14.4 - 8.69x_1 + 13.5x_2$$

where

x_1 = length of service (years)
x_2 = wage rate (dollars)
y = job satisfaction test score (higher scores indicate greater job satisfaction)

A portion of the Minitab computer output follows.

```
The regression equation is
Y = 14.4 - 8.69 X1 + 13.52 X2

Predictor          Coef        Stdev     t-ratio
Constant         14.448        8.191        1.76
X1               ______        1.555      ______
X2               13.517        2.085      ______

s = 3.773     R-sq = ______%  R-sq(adj) = ______%

Analysis of Variance

SOURCE        DF        SS        MS        F
Regression     2    ______    ______   ______
Error     ______     71.17    ______
Total          7     720.0
```

a. Complete the missing entries in this output.
b. Compute F and test using $\alpha = .05$ to see whether a significant relationship is present.
c. Did the estimated regression equation provide a good fit to the data? Explain.
d. Use the t test and $\alpha = .05$ to test $H_0: \beta_1 = 0$ and $H_0: \beta_2 = 0$.

49. A sample of 25 computer hardware companies taken from the *Stock Investor Pro* provided the following data on the price per share, book value per share, and the return on equity per share for each (*Stock Investor Pro,* American Association of Individual Investors, August 21, 1997).

Computer

	Price per Share	Book Value per Share	Return on Equity per Share (%)
Amdahl Corporation	12.31	4.94	−49.7
Apple Computer, Inc.	21.75	9.46	−71.8
Auspex Systems, Inc.	11.00	4.95	17.2
Capital Associates	3.25	4.33	5.1
Compaq Computer Corp.	65.50	9.58	20.8
Data General Corporation	35.94	8.46	13.3

Continued

	Price per Share	Book Value per Share	Return on Equity per Share (%)
Dell Computer Corporation	82.06	2.33	74.5
Digi International	15.00	7.35	−11.9
Digital Equipment Corp.	43.00	22.40	−12.9
En Pointe Technologies	14.25	4.11	18.8
Equitrac Corporation	16.25	6.83	10.7
Franklin Electronic Pbls.	12.88	9.13	9.0
Gateway 2000, Inc.	39.13	6.07	28.8
Hewlett-Packard Company	61.50	14.14	18.7
IBM	101.38	20.12	29.9
Ingram Micro, Inc.	28.75	6.35	15.1
Maxwell Technologies, Inc.	30.50	3.78	11.8
MicroAge, Inc.	27.19	12.59	9.8
Micron Electronics, Inc.	16.31	3.64	28.3
Network Computing Devices	11.88	3.56	4.0
Pomeroy Computer Resources	33.00	10.03	16.5
Sequent Computer Systems	28.19	10.64	3.3
Silicon Graphics, Inc.	27.44	9.12	−4.3
Southern Electronics Corp.	15.13	6.15	16.1
Stratus Computer, Inc.	55.50	22.38	11.1
Sun Microsystems, Inc.	48.00	6.40	26.2
Tandem Computers, Inc.	34.24	9.49	8.7
Tech Data Corporation	38.94	10.25	14.3
Unisys Corporation	11.31	0.68	1.6
Vitech America, Inc.	14.63	3.48	24.3

a. Develop an estimated regression equation that can be used to predict the price per share given the book value per share. At the .05 level of significance, test for a significant relationship.

b. Did the estimated regression equation developed in part (a) provide a good fit to the data? Explain.

c. Develop an estimated regression equation that can be used to predict the price per share given the book value per share and the return on equity per share. At the .05 level of significance, test for overall significance.

50. Following are data on price, curb weight, horsepower, time to go from 0 to 60 miles per hour, and the speed at ¼ mile for 16 sports and GT cars (*1998 Road & Track Sports & GT Cars*).

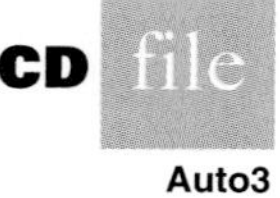

Auto3

	Price ($1000s)	Curb Weight (lb.)	Horse-power	0 to 60 (seconds)	Speed at ¼ mile (mph)
Acura Integra Type R	25.035	2577	195	7.0	90.7
Acura NSX-T	93.758	3066	290	5.0	108.0
BMW Z3 2.8	40.900	2844	189	6.6	93.2
Chevrolet Camaro Z28	24.865	3439	305	5.4	103.2
Chevrolet Corvette Convertible	50.144	3246	345	5.2	102.1
Dodge Viper RT/10	69.742	3319	450	4.4	116.2
Ford Mustang GT	23.200	3227	225	6.8	91.7
Honda Prelude Type SH	26.382	3042	195	7.7	89.7
Mercedes-Benz CLK320	44.988	3240	215	7.2	93.0
Mercedes-Benz SLK230	42.762	3025	185	6.6	92.3
Mitsubishi 3000GT VR-4	47.518	3737	320	5.7	99.0
Nissan 240SX SE	25.066	2862	155	9.1	84.6
Pontiac Firebird Trans Am	27.770	3455	305	5.4	103.2
Porsche Boxster	45.560	2822	201	6.1	93.2
Toyota Supra Turbo	40.989	3505	320	5.3	105.0
Volvo C70	41.120	3285	236	6.3	97.0

a. Develop an estimated regression equation with price, curb weight, horsepower, and time to go from 0 to 60 mph as four independent variables to predict the speed at ¼ mile.
b. Use the F test to determine the significance of the regression results. At a .05 level of significance, what is your conclusion?
c. Use the t test to determine the significance of each independent variable. At a .05 level of significance, what is your conclusion?
d. Delete any independent variable that is not significant and provide your recommended estimated regression equation.
e. Develop a standardized residual plot. Does the pattern of the residual plot appear to be reasonable?
f. Do the data contain any outliers?
g. Do the data contain any influential observations?

51. Nielsen Media Research collects data showing which advertisers get the most exposure during prime time TV on ABC, CBS, NBC, Fox, UPN, and WB networks. Data showing the number of household exposures in millions and the number of times the ad was aired for the week of April 28–May 4, 1997, follow (*USA Today,* May 5, 1997).

CD file Nielsen

Advertised Brand	Time Ad Aired	Household Exposures
Burger King	86	616.7
McDonald's	54	439.2
Sears	33	338.0
Wendy's	28	191.7
Ford Escort	20	174.6
Austin Powers movie	14	161.3
Nissan	16	161.1
Pizza Hut	16	147.7
Saturn	16	146.3
Father's Day movie	11	138.2

a. Develop a scatter diagram with number of times the ad is aired as the independent variable and household expenditures as the dependent variable. Is there anything unusual about the pattern of the points in this scatter diagram? Explain.
b. Develop the estimated regression equation showing how the number of times an ad is aired is related to the number of household exposures. Is the relationship between the two variables significant? Use a .05 level of significance.
c. Consider the addition of the independent variable BigAds, where the value of BigAds is 1 if the number of times an ad is aired is greater than 30, and 0 otherwise. Develop an estimated regression equation that can be used to predict household expenditures given the number of times an ad is aired and the dummy variable BigAds.
d. Using $\alpha = .05$, is the dummy variable added in part (c) significant?
e. What role does the dummy variable play in modeling the relationship between the number of times an ad is aired and household exposures?

52. Today's marketplace offers a wide choice to buyers of sport utility vehicles (SUVs) and pickup trucks. An important factor to many buyers is the resale value of the vehicle. The following table shows the resale value (%) after 2 years and the suggested retail price for 10 SUVs, 10 small pickup trucks, and 10 large pickup trucks (*Kiplinger's New Cars & Trucks 2000 Buyer's Guide*).

CD file Trucks

	Type of Vehicle	Suggested Retail Price ($)	Resale Value (%)
Chevrolet Blazer LS	sport utility	19,495	55
Ford Explorer Sport	sport utility	20,495	57
GMC Yukon XL 1500	sport utility	26,789	67

Continued

	Type of Vehicle	Suggested Retail Price ($)	Resale Value (%)
Honda CR-V	sport utility	18,965	65
Isuzu VehiCross	sport utility	30,186	62
Jeep Cherokee Limited	sport utility	25,745	57
Mercury Mountaineer Monterrey	sport utility	29,895	59
Nissan Pathfinder XE	sport utility	26,919	54
Toyota 4Runner	sport utility	22,418	55
Toyota RAV4	sport utility	17,148	55
Chevrolet S-10 Extended Cab	small pickup	18,847	46
Dodge Dakota Club Cab Sport	small pickup	16,870	53
Ford Ranger XLT Regular Cab	small pickup	18,510	48
Ford Ranger XLT Supercab	small pickup	20,225	55
GMC Sonoma Regular Cab	small pickup	16,938	44
Isuzu Hombre Spacecab	small pickup	18,820	41
Mazda B4000 SE Cab Plus	small pickup	23,050	51
Nissan Frontier XE Regular Cab	small pickup	12,110	51
Toyota Tacoma Xtracab	small pickup	18,228	49
Toyota Tacoma Xtracab V6	small pickup	19,318	50
Chevrolet K2500	full-size pickup	24,417	60
Chevrolet Silverado 2500 Ext	full-size pickup	24,140	64
Dodge Ram 1500	full-size pickup	17,460	54
Dodge Ram Quad Cab 2500	full-size pickup	32,770	63
Dodge Ram Regular Cab 2500	full-size pickup	23,140	59
Ford F150 XL	full-size pickup	22,875	58
Ford F-350 Super Duty Crew Cab XL	full-size pickup	34,295	64
GMC New Sierra 1500 Ext Cab	full-size pickup	27,089	68
Toyota Tundra Access Cab Limited	full-size pickup	25,605	53
Toyota Tundra Regular Cab	full-size pickup	15,835	58

a. Develop an estimated regression equation that can be used to predict the resale value given the suggested retail price. At the .05 level of significance, test for a significant relationship.

b. Did the estimated regression equation developed in part (a) provide a good fit to the data? Explain.

c. Develop an estimated regression equation that can be used to predict the resale value given the suggested retail price and the type of vehicle.

d. Use the F test to determine the significance of the regression results. At a .05 level of significance, what is your conclusion?

Case Problem 1 CONSUMER RESEARCH, INC.

Consumer Research, Inc., is an independent agency that conducts research on consumer attitudes and behaviors for a variety of firms. In one study, a client asked for an investigation of consumer characteristics that can be used to predict the amount charged by credit card users. Data were collected on annual income, household size, and annual credit card charges for a sample of 50 consumers. The data follow and are on the data disk in the data set named Consumer.

Consumer

Income ($1000s)	Household Size	Amount Charged ($)	Income ($1000s)	Household Size	Amount Charged ($)
54	3	4016	54	6	5573
30	2	3159	30	1	2583
32	4	5100	48	2	3866

Income ($1000s)	Household Size	Amount Charged ($)	Income ($1000s)	Household Size	Amount Charged ($)
50	5	4742	34	5	3586
31	2	1864	67	4	5037
55	2	4070	50	2	3605
37	1	2731	67	5	5345
40	2	3348	55	6	5370
66	4	4764	52	2	3890
51	3	4110	62	3	4705
25	3	4208	64	2	4157
48	4	4219	22	3	3579
27	1	2477	29	4	3890
33	2	2514	39	2	2972
65	3	4214	35	1	3121
63	4	4965	39	4	4183
42	6	4412	54	3	3730
21	2	2448	23	6	4127
44	1	2995	27	2	2921
37	5	4171	26	7	4603
62	6	5678	61	2	4273
21	3	3623	30	2	3067
55	7	5301	22	4	3074
42	2	3020	46	5	4820
41	7	4828	66	4	5149

Managerial Report

1. Use methods of descriptive statistics to summarize the data. Comment on the findings.
2. Develop estimated regression equations, first using annual income as the independent variable and then using household size as the independent variable. Which variable is the better predictor of annual credit card charges? Discuss your findings.
3. Develop an estimated regression equation with annual income and household size as the independent variables. Discuss your findings.
4. What is the predicted annual credit card charge for a three-person household with an annual income of $40,000?
5. Discuss the need for other independent variables that could be added to the model. What additional variables might be helpful?

Case Problem 2 NFL QUARTERBACK RATING

The National Football League (NFL) records weekly performance statistics for individuals and teams. These data can be obtained by accessing the home page for the NFL (www.nfl.com). For many fans, one of the most interesting statistics is the rating used to evaluate the passing performance of quarterbacks. Four categories are used as a basis for compiling a passing rating:

1. Percentage of touchdown passes per attempt
2. Percentage of completions per attempt
3. Percentage of interceptions per attempt
4. Average yards gained per attempt

To illustrate how to compute the rating, consider the performance for Steve Young, the highest rated quarterback in the NFL in 1997. For the 1997 season, Steve Young attempted 356 passes, completed 241 passes for a total of 3029 yards, and had 19 touchdown passes and 6 interceptions. Five steps are needed to compute the passing rating for Steve Young.

Step 1. Compute the ratio of the number of touchdowns (19) to the number of passes attempted (356); the value obtained is 19/356 = 0.0534. Divide this result by 0.05 to obtain the value for the touchdown component of the rating; the value obtained is 0.0534/0.05 = 1.0680.

Step 2. Compute the ratio of the number of passes completed (241) to the number of passes attempted (356); the value obtained is 241/356 = 0.6770. Subtract 0.3 from this result and divide by 0.2 to obtain the value for the second component of the rating; the value obtained is (0.6770 − 0.3)/0.2 = 1.8850.

Step 3. Compute the ratio of the number of interceptions (6) to the number of passes attempted (356); the value obtained is 6/356 = 0.0169. Subtract this value from 0.095 and divide the result by 0.04 to obtain the value for the interceptions component of the rating; the value obtained is (0.095 − 0.0169)/0.04 = 1.9525.

Step 4. Compute the ratio of the number of passing yards (3029) to the number of passes attempted (356); the value obtained is 3029/356 = 8.5084. Subtract 3 from this result and divide by 4 to obtain the value for the yards component of the rating; the value obtained is (8.5084 − 3)/4 = 1.3771.

Step 5. Add the sum of steps 1 through 4, multiply by 100 and divide by 6. The sum of steps 1 through 4 is 1.0680 + 1.8850 + 1.9525 + 1.3771 = 6.2826. After multiplying by 100 and dividing by 6 we obtain 104.7100 or 104.7; this figure is the passing rating reported by the NFL for Steve Young.

CD file

NFL

The passing data reported by the NFL for the 1997 season are available on the data disk in the data set named NFL. The labels for the columns are defined as follows:

Att	Number of passes attempted.
Comp	Number of passes completed.
Comp%	Number of passes completed divided by the number of passes attempted times 100.
Yds	Number of yards obtained passing.
Yds/Att	Number of yards obtained passing divided by the number of passes attempted.
TD	Number of touchdowns obtained passing.
TD%	Number of touchdowns divided by the number of passes attempted times 100.
Long	Longest pass completed.
Int	Number of interceptions thrown.
Int%	Number of interceptions divided by the number of passes attempted.
Rating	The quarterback passing rating.

Managerial Report

1. Use methods of descriptive statistics to summarize the data. Comment on the findings.
2. Develop an estimated regression equation that can be used to predict Int% given the value of Comp%. Discuss your findings.
3. Develop an estimated regression equation that can be used to predict Rating. Discuss your findings and comment on how your estimated regression equation relates to the previous discussion of computing passing ratings.
4. If the NFL hired you to compute the passing ratings for next year, what approach would you use? Explain.

Case Problem 3 PREDICTING STUDENT PROFICIENCY TEST SCORES

In order to predict how a school district would have scored when accounting for poverty and other income measures, *The Cincinnati Enquirer* gathered data from the Ohio Department of Education's Education Management Services and the Ohio Department of Taxation (*The Cincinnati Enquirer,* November 30, 1997). First, the newspaper obtained passage-rate data on the math, reading, science, writing, and citizenship proficiency exams given to fourth-, sixth-, ninth-, and 12th-graders in early 1996. By combining these data, they computed an overall percentage of students that passed the tests for each district.

The percentage of a school district's students on Aid for Dependent Children (ADC), the percentage who qualify for free or reduced-price lunches, and the district's median family income were also recorded. A portion of the data collected for the 608 school districts follows. The complete data set is available on the data disk in the data set named Enquirer.

Enquirer

Rank	School District	County	% Passed	% on ADC	% Free Lunch	Median Income ($)
1	Ottawa Hills Local	Lucas	93.85	0.11	0.00	48231
2	Wyoming City	Hamilton	93.08	2.95	4.59	42672
3	Oakwood City	Montgomery	92.92	0.20	0.38	42403
4	Madeira City	Hamilton	92.37	1.50	4.83	32889
5	Indian Hill Ex Vill	Hamilton	91.77	1.23	2.70	44135
6	Solon City	Cuyahoga	90.77	0.68	2.24	34993
7	Chagrin Falls Ex Vill	Cuyahoga	89.89	0.47	0.44	38921
8	Mariemont City	Hamilton	89.80	3.00	2.97	31823
9	Upper Arlington City	Franklin	89.77	0.24	0.92	38358
10	Granville Ex Vill	Licking	89.22	1.14	0.00	36235
⋮	⋮	⋮	⋮	⋮	⋮	⋮

The data have been ranked based on the values in the column labeled % Passed; these data are the overall percentage of students passing the tests. Data in the column labeled % on ADC are the percentage of each school district's students on ADC, and the data in the column labeled % Free Lunch are the percentage of students who qualify for free or reduced-price lunches. The column labeled Median Income shows each district's median family income. Also shown for each school district is the county in which the school district is located. Note that in some cases the value in the % Free Lunch column is 0, indicating that the district did not participate in the free lunch program.

Managerial Report

Use the methods presented in this and previous chapters to analyze this data set. Present a summary of your analysis, including key statistical results, conclusions, and recommendations, in a managerial report. Include any technical material you feel is appropriate in an appendix.

Case Problem 4 ALUMNI GIVING

Alumni donations are an important source of revenue for colleges and universities. If administrators could determine the factors that could lead to increases in the percentage of alumni who make a donation, they might be able to implement policies that could lead to increased revenues. Research shows that students who are more satisfied with their contact with teachers are more likely to graduate. As a result, one might suspect that smaller class sizes and lower student-faculty ratios might lead to a higher percentage of satisfied graduates, which in turn might lead to increases in the percentage of alumni who make a donation. In Table 15.11, the column labeled Graduation Rate is the percentage of students who initially enrolled at the university and graduated. The column labeled % of Classes Under 20 shows the percentage of classes offered with fewer than 20 students. The column labeled Student-Faculty Ratio is the number of students enrolled divided by the total number of faculty. Finally, the column labeled Alumni Giving Rate is the percentage of alumni who made a donation to the university.

Managerial Report

1. Use methods of descriptive statistics to summarize the data.
2. Develop an estimated regression equation that can be used to predict the alumni giving rate given the number of students who graduate. Discuss your findings.
3. Develop an estimated regression equation that could be used to predict the alumni giving rate using the data provided.
4. What conclusions and/or recommendations can you derive from your analysis?

TABLE 15.11 DATA FOR 48 NATIONAL UNIVERSITIES

CD file Alumni

	State	Graduation Rate	% of Classes Under 20	Student-Faculty Ratio	Alumni Giving Rate
Boston College	MA	85	39	13	25
Brandeis University	MA	79	68	8	33
Brown University	RI	93	60	8	40
California Institute of Technology	CA	85	65	3	46
Carnegie Mellon University	PA	75	67	10	28
Case Western Reserve Univ.	OH	72	52	8	31
College of William and Mary	VA	89	45	12	27
Columbia University	NY	90	69	7	31
Cornell University	NY	91	72	13	35
Dartmouth College	NH	94	61	10	53
Duke University	NC	92	68	8	45
Emory University	GA	84	65	7	37
Georgetown University	PA	91	54	10	29
Harvard University	MA	97	73	8	46
Johns Hopkins University	MD	89	64	9	27
Lehigh University	PA	81	55	11	40
Massachusetts Inst. of Technology	MA	92	65	6	44
New York University	NY	72	63	13	13
Northwestern University	IL	90	66	8	30
Pennsylvania State Univ.	PA	80	32	19	21
Princeton University	NJ	95	68	5	67
Rice University	TX	92	62	8	40
Stanford University	CA	92	69	7	34
Tufts University	MA	87	67	9	29
Tulane University	LA	72	56	12	17
U. of California–Berkeley	CA	83	58	17	18
U. of California–Davis	CA	74	32	19	7
U. of California–Irvine	CA	74	42	20	9
U. of California–Los Angeles	CA	78	41	18	13
U. of California–San Diego	CA	80	48	19	8
U. of California–Santa Barbara	CA	70	45	20	12
U. of Chicago	IL	84	65	4	36
U. of Florida	FL	67	31	23	19
U. of Illinois–Urbana Champaign	IL	77	29	15	23
U. of Michigan–Ann Arbor	MI	83	51	15	13
U. of North Carolina–Chapel Hill	NC	82	40	16	26
U. of Notre Dame	IN	94	53	13	49
U. of Pennsylvania	PA	90	65	7	41
U. of Rochester	NY	76	63	10	23
U. of Southern California	CA	70	53	13	22
U. of Texas–Austin	TX	66	39	21	13
U. of Virginia	VA	92	44	13	28
U. of Washington	WA	70	37	12	12
U. of Wisconsin–Madison	WI	73	37	13	13
Vanderbilt University	TN	82	68	9	31
Wake Forest University	NC	82	59	11	38
Washington University–St. Louis	MO	86	73	7	33
Yale University	CT	94	77	7	50

Chapter 16

REGRESSION ANALYSIS: MODEL BUILDING

CONTENTS

STATISTICS IN PRACTICE

MONSANTO COMPANY*
St. Louis, Missouri

Monsanto Company traces its roots to one entrepreneur's investment of $500 and a dusty warehouse on the Mississippi riverfront, where in 1901 John F. Queeney began manufacturing saccharin. Today, Monsanto is one of the nation's largest chemical companies, producing more than a thousand products ranging from industrial chemicals to synthetic playing surfaces used in modern sports stadiums. Monsanto is a worldwide corporation with manufacturing facilities, laboratories, technical centers, and marketing operations in 65 countries.

Monsanto's Nutrition Chemical Division manufactures and markets a methionine supplement used in poultry, swine, and cattle feed products. Because poultry growers work with high volumes and low profit margins, cost-effective poultry feed products with the best possible nutrition value are needed. Optimal feed composition will result in rapid growth and high final body weight for a given level of feed intake. The chemical industry has worked closely with poultry growers to optimize poultry feed products. Ultimately, success depends on keeping the cost of poultry low in comparison with the cost of beef and other meat products.

Monsanto used regression analysis to model the relationship between body weight y and the amount of methionine x added to the poultry feed. Initially, the following simple linear estimated regression equation was developed.

$$\hat{y} = .21 + .42x$$

This estimated regression equation proved statistically significant; however, the analysis of the residuals indicated that a curvilinear relationship would be a better model of the relationship between body weight and methionine.

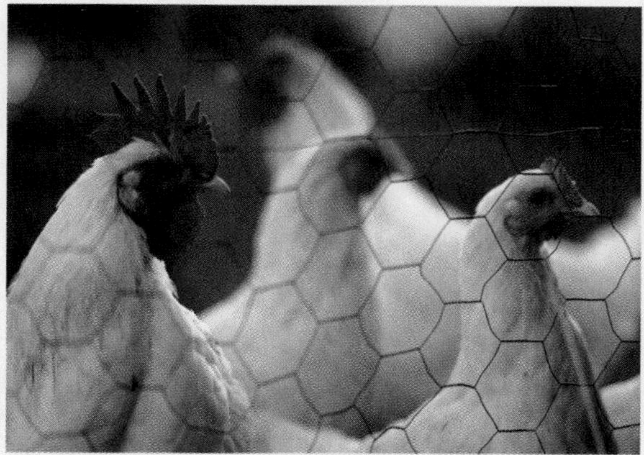

Monsanto researchers used regression analysis to develop an optimal feed composition for poultry growers.

Further research conducted by Monsanto showed that although small amounts of methionine tended to increase body weight, at some point body weight leveled off and additional amounts of the methionine were of little or no benefit. In fact, when the amount of methionine increased beyond nutritional requirements, body weight tended to decline. The following estimated multiple regression equation was used to model the curvilinear relationship between body weight and methionine.

$$\hat{y} = -1.89 + 1.32x - .506x^2$$

Use of the regression results enabled Monsanto to determine the optimal level of methionine to be used in poultry feed products.

In this chapter we will extend the discussion of regression analysis by showing how curvilinear models such as the one used by Monsanto can be developed. In addition, we will describe a variety of tools that help determine which independent variables lead to the best estimated regression equation.

*The authors are indebted to James R. Ryland and Robert M. Schisla, Senior Research Specialists, Monsanto Nutrition Chemical Division, for providing this Statistics in Practice.

Model building is the process of developing an estimated regression equation that describes the relationship between a dependent variable and one or more independent variables. The major issues in model building are finding the proper functional form of the relationship and selecting the independent variables to be included in the model. In Section 16.1 we establish the framework for model building by introducing the concept of a general linear model. Section 16.2, which provides the foundation for the more sophisticated computer-based procedures, introduces a general approach for determining when to add or delete independent variables. In Section 16.3 we consider a larger regression problem involving eight independent variables and 25 observations; this problem is used to illustrate the variable selection procedures presented in Section 16.4, including stepwise regression, the forward selection procedure, the backward elimination procedure, and best-subsets regression. In Section 16.5 we show how the Durbin-Watson test can be used to detect serial or autocorrelation, and in Section 16.6 we show how regression analysis can be used for analysis of variance and experimental design problems.

16.1 GENERAL LINEAR MODEL

Suppose we have collected data for one dependent variable y and k independent variables $x_1, x_2, \ldots, x_k$. Our objective is to use these data to develop an estimated regression equation that provides the best relationship between the dependent and independent variables. As a general framework for developing more complex relationships among the independent variables, we introduce the concept of a **general linear model** involving p independent variables.

If you can write a regression model in the form of (16.1), the standard multiple regression procedures described in Chapter 15 are applicable.

General Linear Model

$$y = \beta_0 + \beta_1 z_1 + \beta_2 z_2 + \cdots + \beta_p z_p + \epsilon \quad (16.1)$$

In equation (16.1), each of the independent variables z_j (where $j = 1, 2, \ldots, p$) is a function of $x_1, x_2, \ldots, x_k$ (the variables for which data have been collected). In some cases, each z_j may be a function of only one x variable. The simplest case is when we have collected data for just one variable x_1 and want to estimate y by using a straight-line relationship. In this case $z_1 = x_1$ and (16.1) becomes

$$y = \beta_0 + \beta_1 x_1 + \epsilon \quad (16.2)$$

Equation (16.2) is the simple linear regression model introduced in Chapter 14 with the exception that the independent variable is labeled x_1 instead of x. In the statistical modeling literature, this model is called a *simple first-order model with one predictor variable.*

Modeling Curvilinear Relationships

More complex types of relationships can be modeled with equation (16.1). To illustrate, let us consider the problem facing Reynolds, Inc., a manufacturer of industrial scales and laboratory equipment. Managers at Reynolds want to investigate the relationship between length of employment of their salespeople and the number of electronic laboratory scales sold. Table 16.1 gives the number of scales sold by 15 randomly selected salespeople for the most recent sales period and the number of months each salesperson has been employed by the firm. Figure 16.1 is the scatter diagram for these data. The scatter diagram indicates a possible curvilinear relationship between the length of time employed and the number of units sold. Before considering how to develop a curvilinear relationship for Reynolds, let us con-

Reynolds

TABLE 16.1 DATA FOR THE REYNOLDS EXAMPLE

Months Employed	Scales Sold
41	375
106	296
76	317
10	376
22	162
12	150
85	367
111	308
40	189
51	235
9	83
12	112
6	67
56	325
19	189

FIGURE 16.1 SCATTER DIAGRAM FOR THE REYNOLDS EXAMPLE

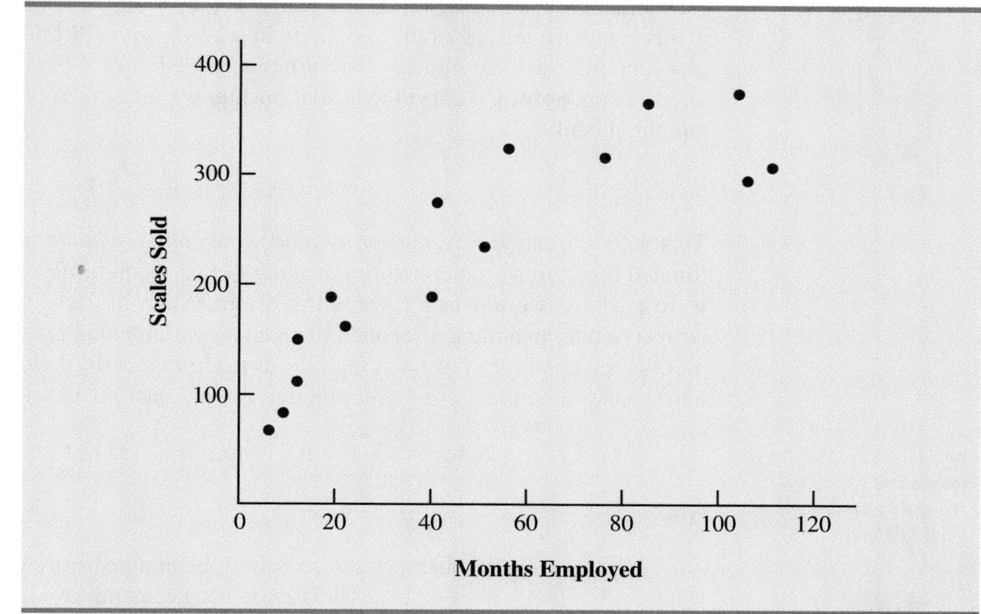

sider the Minitab output in Figure 16.2 corresponding to a simple first-order model; the estimated regression is

$$\text{Sales} = 111 + 2.38 \text{ Months}$$

where

$$\text{Sales} = \text{number of electronic laboratory scales sold}$$
$$\text{Months} = \text{the number of months the salesperson has been employed}$$

FIGURE 16.2 MINITAB OUTPUT FOR THE REYNOLDS EXAMPLE: FIRST-ORDER MODEL

```
The regression equation is
Sales = 111 + 2.38 Months

Predictor        Coef        Stdev     t-ratio        p
Constant       111.23        21.63        5.14    0.000
Months         2.3768       0.3489        6.81    0.000

s = 49.52        R-sq = 78.1%      R-sq(adj) = 76.4%

Analysis of Variance

SOURCE         DF          SS           MS          F         p
Regression      1      113783       113783      46.41     0.000
Error          13       31874         2452
Total          14      145657
```

Figure 16.3 is the corresponding standardized residual plot. Although the computer output shows that the relationship is significant (p-value = .000) and that a linear relationship explains a high percentage of the variability in sales (R-sq = 78.1%), the standardized residual plot suggests that a curvilinear relationship is needed.

To account for the curvilinear relationship, we set $z_1 = x_1$ and $z_2 = x_1^2$ in (16.1) to obtain the model

$$y = \beta_0 + \beta_1 x_1 + \beta_2 x_1^2 + \epsilon \tag{16.3}$$

This model is called a *second-order model with one predictor variable.* To develop an estimated regression equation corresponding to this second-order model, the statistical software package we are using needs the original data in Table 16.1, as well as that data corresponding to adding a second independent variable that is the square of the number of months the employee has been with the firm. In Figure 16.4 we show the Minitab output corresponding to the second-order model; the estimated regression equation is

The data for the MonthsSq independent variable is obtained by squaring the values of Months.

$$\text{Sales} = 45.3 + 6.34 \text{ Months} - .0345 \text{ MonthsSq}$$

where

MonthsSq = the square of the number of months the salesperson has been employed

Figure 16.5 is the corresponding standardized residual plot. It shows that the previous curvilinear pattern has been removed. At the .05 level of significance, the computer output shows that the overall model is significant (p-value for the F test is 0.000); note also that the p-value corresponding to the t-ratio for MonthsSq (p-value = .002) is less than .05, and hence we can conclude that adding MonthsSq to the model involving Months is significant.

FIGURE 16.3 STANDARDIZED RESIDUAL PLOT FOR THE REYNOLDS EXAMPLE: FIRST-ORDER MODEL

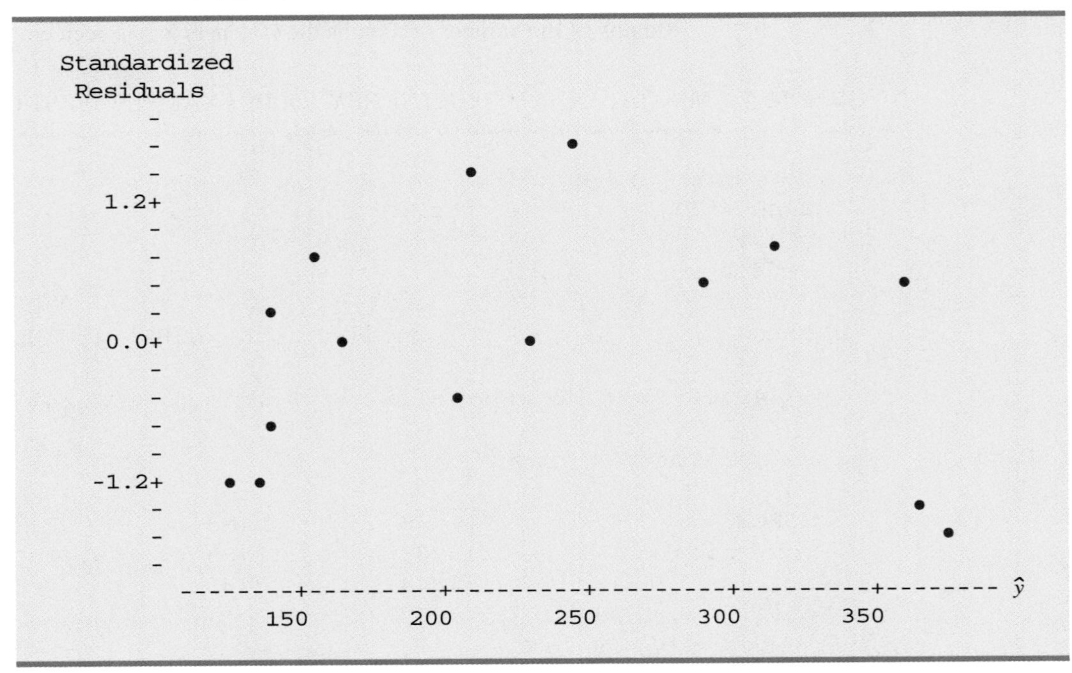

FIGURE 16.4 MINITAB OUTPUT FOR THE REYNOLDS EXAMPLE: SECOND-ORDER MODEL

```
The regression equation is
Sales = 45.3 + 6.34 Months - 0.0345 MonthsSq

Predictor         Coef        Stdev     t-ratio        p
Constant         45.35        22.77        1.99    0.070
Months           6.345        1.058        6.00    0.000
MonthsSq     -0.034486     0.008948       -3.85    0.002

s = 34.45         R-sq = 90.2%      R-sq(adj) = 88.6%

Analysis of Variance

SOURCE          DF          SS          MS          F         p
Regression       2      131413       65707      55.36     0.000
Error           12       14244        1187
Total           14      145657
```

With an R-sq(adj) value of 88.6%, we should be pleased with the fit provided by this estimated regression equation. More important, however, is seeing how easy it is to handle curvilinear relationships in regression analysis.

Clearly, many types of relationships can be modeled by using equation (16.1). The regression techniques with which we have been working are definitely not limited to linear,

FIGURE 16.5 STANDARDIZED RESIDUAL PLOT FOR THE REYNOLDS EXAMPLE: SECOND-ORDER MODEL

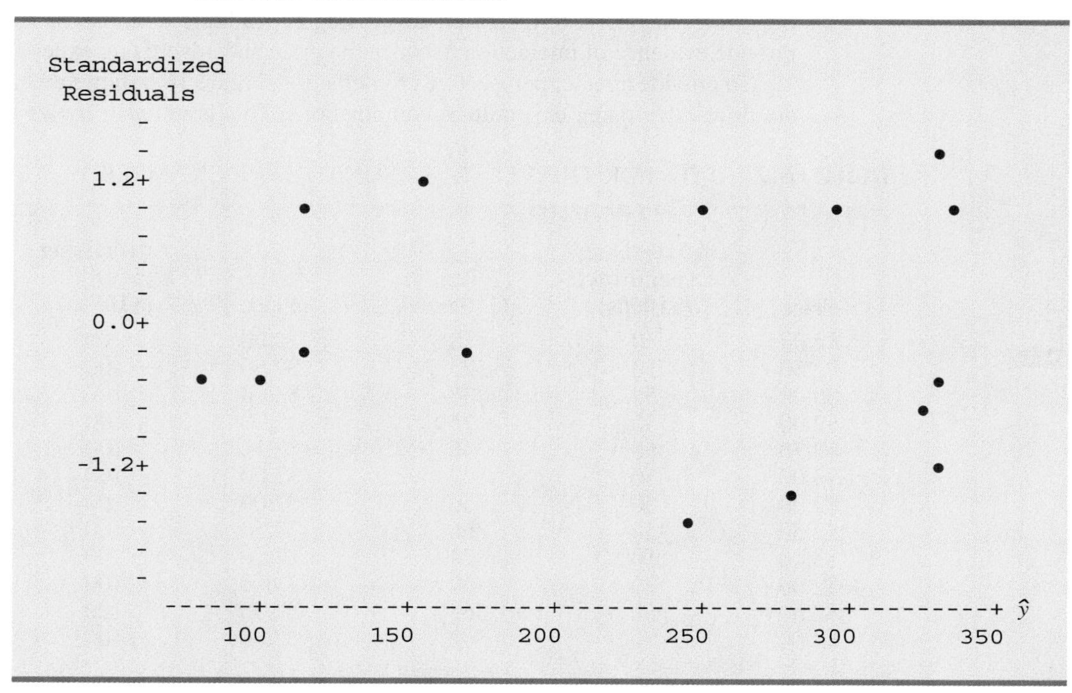

or straight-line, relationships. In multiple regression analysis the word *linear* in the term "general linear model" refers only to the fact that $\beta_0, \beta_1, \ldots, \beta_p$ all have exponents of one; it does not imply that the relationship between y and the x_i's is linear. Indeed, in this section we have seen one example of how equation (16.1) can be used to model a curvilinear relationship.

Interaction

If the original data set consists of observations for y and two independent variables x_1 and x_2, we can develop a second-order model with two predictor variables by setting $z_1 = x_1$, $z_2 = x_2$, $z_3 = x_1^2$, $z_4 = x_2^2$, and $z_5 = x_1x_2$ in the general linear model of (16.1). The model obtained is

$$y = \beta_0 + \beta_1 x_1 + \beta_2 x_2 + \beta_3 x_1^2 + \beta_4 x_2^2 + \beta_5 x_1 x_2 + \epsilon \tag{16.4}$$

In this second-order model, the variable $z_5 = x_1x_2$ is added to account for the potential effects of the two variables acting together. This type of effect is called **interaction.**

To provide an illustration of interaction and what it means, let us review the regression study conducted by Tyler Personal Care for one of its new shampoo products. Two factors believed to have the most influence on sales are unit selling price and advertising expenditure. To investigate the effects of these two variables on sales, prices of \$2.00, \$2.50, and \$3.00 were paired with advertising expenditures of \$50,000 and \$100,000 in 24 test markets. The unit sales that were observed (in 1000s) are reported in Table 16.2.

Table 16.3 is a summary of these data. Note that the mean sales corresponding to a price of \$2.00 and an advertising expenditure of \$50,000 is 461,000, and the mean sales corresponding to a price of \$2.00 and an advertising expenditure of \$100,000 is 808,000. Hence, with price held constant at \$2.00, the difference in mean sales between advertising expenditures of \$50,000 and \$100,000 is 808,000 − 461,000 = 347,000 units. When the price of the product is \$2.50, the difference in mean sales is 646,000 − 364,000 = 282,000 units. Finally, when the price is \$3.00, the difference in mean sales is 375,000 − 332,000 = 43,000 units. Clearly, the difference in mean sales between advertising expenditures of \$50,000 and \$100,000 depends on the price of the product. In other words, at higher selling prices, the effect of increased advertising expenditure diminishes. These observations provide evidence of interaction between the price and advertising expenditure variables.

To provide another perspective of interaction, Figure 16.6 shows the mean sales for the six price-advertising expenditure combinations. This graph also shows that the effect of

TABLE 16.2 DATA FOR THE TYLER PERSONAL CARE EXAMPLE

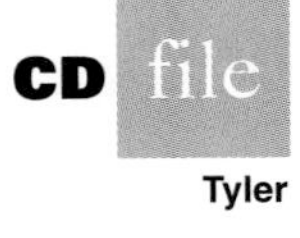
Tyler

Price	Advertising Expenditure (\$1000s)	Sales (1000s)	Price	Advertising Expenditure (\$1000s)	Sales (1000s)
\$2.00	50	478	\$2.00	100	810
\$2.50	50	373	\$2.50	100	653
\$3.00	50	335	\$3.00	100	345
\$2.00	50	473	\$2.00	100	832
\$2.50	50	358	\$2.50	100	641
\$3.00	50	329	\$3.00	100	372
\$2.00	50	456	\$2.00	100	800
\$2.50	50	360	\$2.50	100	620
\$3.00	50	322	\$3.00	100	390
\$2.00	50	437	\$2.00	100	790
\$2.50	50	365	\$2.50	100	670
\$3.00	50	342	\$3.00	100	393

TABLE 16.3 MEAN SALES (1000s) FOR THE TYLER PERSONAL CARE EXAMPLE

		Price		
		$2.00	**$2.50**	**$3.00**
Advertising Expenditure	**$50,000**	461	364	332
	$100,000	808	646	375

Mean sales of 808,000 units when price = $2.00 and advertising expenditure = $100,000

FIGURE 16.6 MEAN SALES AS A FUNCTION OF SELLING PRICE AND ADVERTISING EXPENDITURE

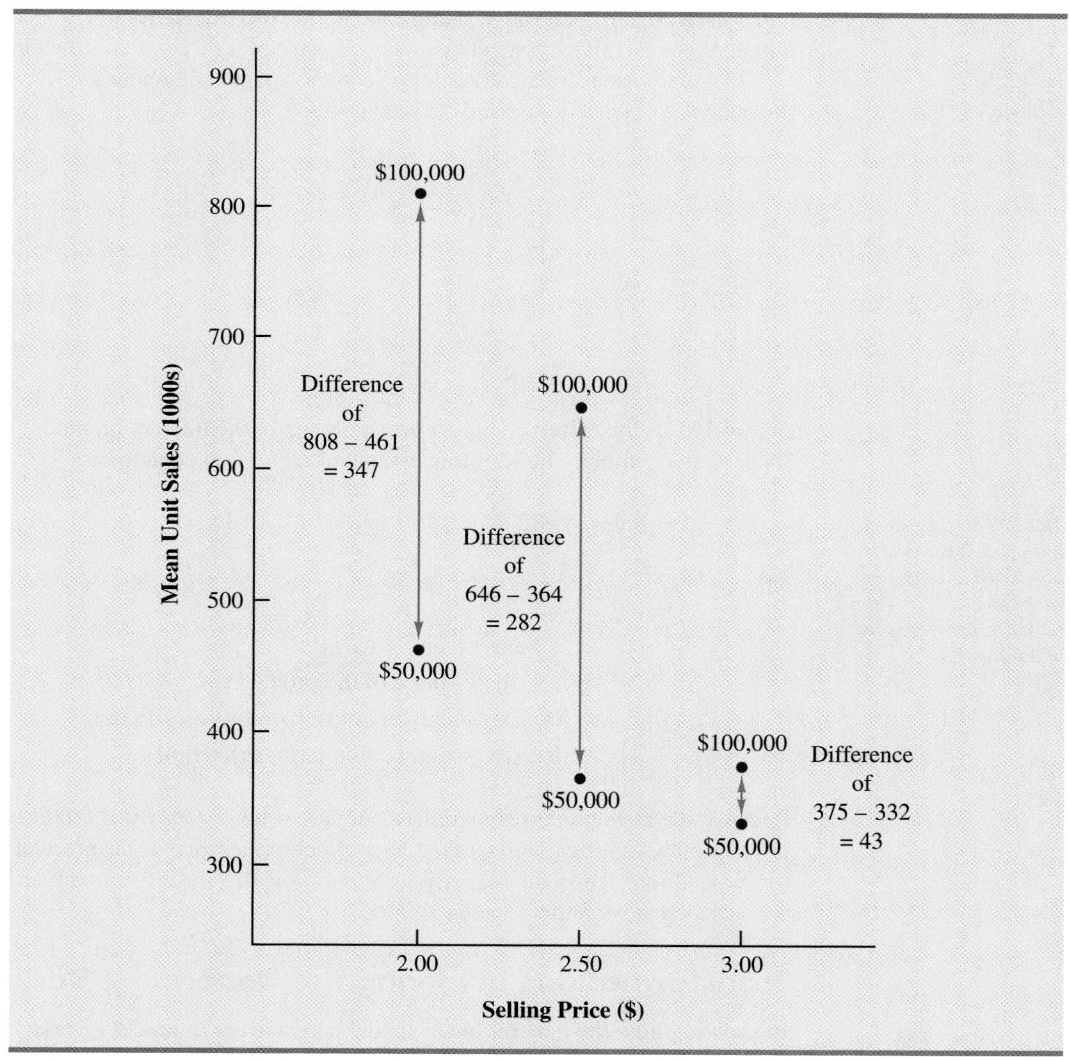

advertising expenditure on mean sales depends on the price of the product; we again see the effect of interaction. When interaction between two variables is present, we cannot study the effect of one variable on the response y independently of the other variable. In other words, meaningful conclusions can be developed only if we consider the joint effect that both variables have on the response.

To account for the effect of interaction, we will use the following regression model.

$$y = \beta_0 + \beta_1 x_1 + \beta_2 x_2 + \beta_3 x_1 x_2 + \epsilon \tag{16.5}$$

where

$$\begin{aligned} y &= \text{unit sales (1000s)} \\ x_1 &= \text{price (\$)} \\ x_2 &= \text{advertising expenditure (\$1000s)} \end{aligned}$$

Note that equation (16.5) reflects Tyler's belief that the number of units sold depends linearly on selling price and advertising expenditure (accounted for by the $\beta_1 x_1$ and $\beta_2 x_2$ terms), and that there is interaction between the two variables (accounted for by the $\beta_3 x_1 x_2$ term).

To develop an estimated regression equation, a general linear model involving three independent variables (z_1, z_2, and z_3) was used.

$$y = \beta_0 + \beta_1 z_1 + \beta_2 z_2 + \beta_3 z_3 + \epsilon \tag{16.6}$$

where

$$\begin{aligned} z_1 &= x_1 \\ z_2 &= x_2 \\ z_3 &= x_1 x_2 \end{aligned}$$

Figure 16.7 is the Minitab output corresponding to the interaction model for the Tyler Personal Care example. The resulting estimated regression equation is

The data for the PriceAdv independent variable is obtained by multiplying each value of Price times the corresponding value of AdvExp.

$$\text{Sales} = -276 + 175 \text{ Price} + 19.7 \text{ AdvExp} - 6.08 \text{ PriceAdv}$$

where

$$\begin{aligned} \text{Sales} &= \text{unit sales (1000s)} \\ \text{Price} &= \text{price of the product (\$)} \\ \text{AdvExp} &= \text{advertising expenditure (\$1000s)} \\ \text{PriceAdv} &= \text{interaction term (Price times AdvExp)} \end{aligned}$$

Because the p-value corresponding to the t test for PriceAdv is 0.000, we conclude that interaction is significant given the linear effect of the price of the product and the advertising expenditure. Thus, the regression results show that the effect of advertising expenditure on sales depends on the price.

Transformations Involving the Dependent Variable

In showing how the general linear model can be used to model a variety of possible relationships between the independent variables and the dependent variable, we have focused attention on transformations involving one or more of the independent variables. Often it

FIGURE 16.7 MINITAB OUTPUT FOR THE TYLER PERSONAL CARE EXAMPLE

```
The regression equation is
Sales = - 276 + 175 Price + 19.7 AdvExp - 6.08 PriceAdv

Predictor        Coef        Stdev     t-ratio        p
Constant       -275.8        112.8       -2.44    0.024
Price          175.00        44.55        3.93    0.001
Adver          19.680        1.427       13.79    0.000
PriceAdv      -6.0800       0.5635      -10.79    0.000

s = 28.17        R-sq = 97.8%       R-sq(adj) = 97.5%

Analysis of Variance

SOURCE        DF          SS          MS          F        p
Regression     3      709316      236439     297.87    0.000
Error         20       15875         794
Total         23      725191
```

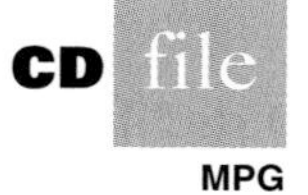

MPG

TABLE 16.4 MILES-PER-GALLON RATINGS AND WEIGHTS FOR 12 AUTOMOBILES

Weight	Miles per Gallon
2289	28.7
2113	29.2
2180	34.2
2448	27.9
2026	33.3
2702	26.4
2657	23.9
2106	30.5
3226	18.1
3213	19.5
3607	14.3
2888	20.9

is worthwhile to consider transformations involving the dependent variable y. As an illustration of when we might want to transform the dependent variable, consider the data in Table 16.4, which shows the miles-per-gallon ratings and weights for 12 automobiles. The scatter diagram in Figure 16.8 indicates a negative linear relationship between these two variables. Therefore, we use a simple first-order model to relate the two variables. The Minitab output is shown in Figure 16.9; the resulting estimated regression equation is

$$\text{MPG} = 56.1 - 0.0116 \text{ Weight}$$

where

$$\text{MPG} = \text{miles-per-gallon rating}$$
$$\text{Weight} = \text{weight of the car in pounds}$$

The model is significant (p-value for the F test is 0.000) and the fit is very good (R-sq = 93.5%). However, we note in Figure 16.9 that observation 3 is identified as having a large standardized residual.

Figure 16.10 is the standardized residual plot corresponding to the first-order model. The pattern we observe does not look like the horizontal band we should expect to find if the assumptions about the error term are valid. Instead, the variability in the residuals appears to increase as the value of $\hat{y}$ increases. In other words, we have the wedge-shaped pattern referred to in Chapters 14 and 15 as being indicative of a nonconstant variance. We are not justified in reaching any conclusions about the statistical significance of the resulting estimated regression equation when the underlying assumptions for the tests of significance do not appear to be satisfied.

Often the problem of nonconstant variance can be corrected by transforming the dependent variable to a different scale. For instance, if we work with the logarithm of the dependent variable instead of the original dependent variable, the effect will be to compress the values of the dependent variable and thus diminish the effects of nonconstant variance. Most statistical packages provide the ability to apply logarithmic transformations using

FIGURE 16.8 SCATTER DIAGRAM FOR THE MILES-PER-GALLON PROBLEM

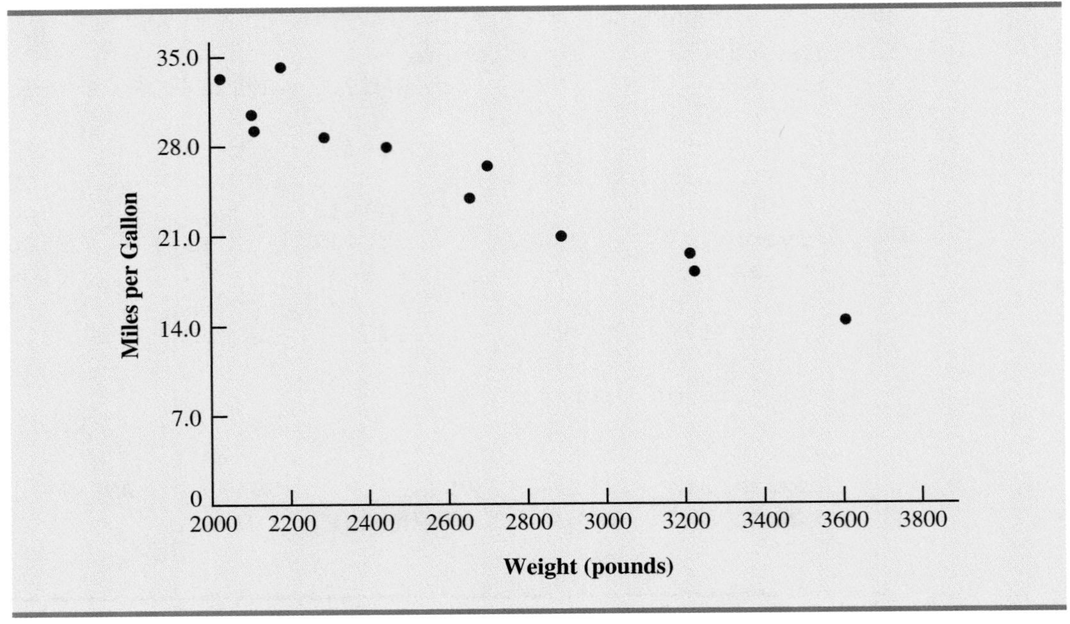

FIGURE 16.9 MINITAB OUTPUT FOR THE MILES-PER-GALLON PROBLEM

```
The regression equation is
MPG = 56.1 - 0.0116 Weight

Predictor        Coef        Stdev     t-ratio        p
Constant       56.096        2.582       21.72    0.000
Weight     -0.0116436    0.0009677      -12.03    0.000

s = 1.671       R-sq = 93.5%      R-sq(adj) = 92.9%

Analysis of Variance

SOURCE       DF         SS           MS          F         p
Regression    1     403.98       403.98     144.76     0.000
Error        10      27.91         2.79
Total        11     431.88

Unusual Observations
Obs.  Weight       MPG        Fit  Stdev.Fit  Residual   St.Resid
  3     2180    34.200     30.713      0.644     3.487      2.26R

R denotes an obs. with a large st. resid.
```

FIGURE 16.10 STANDARDIZED RESIDUAL PLOT FOR THE MILES-PER-GALLON PROBLEM

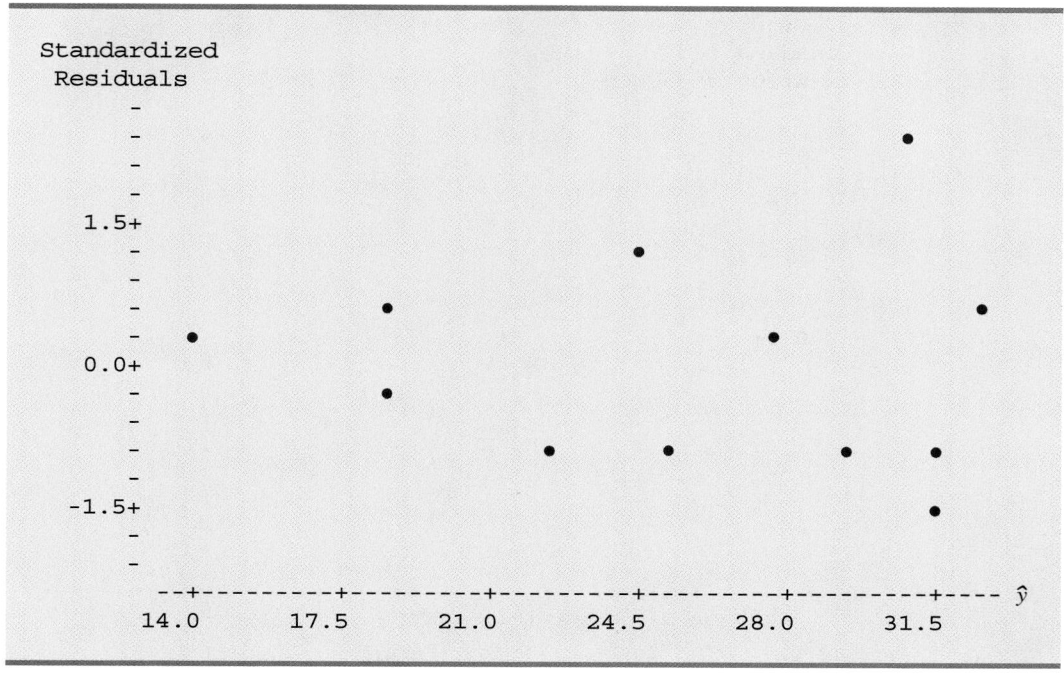

either the base 10 (common logarithm) or the base $e = 2.71828\ldots$ (natural logarithm). We applied a natural logarithmic transformation to the miles-per-gallon data and developed the estimated regression equation relating weight to the natural logarithm of miles-per-gallon. The regression results obtained by using the natural logarithm of miles-per-gallon as the dependent variable, labeled LogeMPG in the output, are shown in Figure 16.11; Figure 16.12 is the corresponding standardized residual plot.

FIGURE 16.11 MINITAB OUTPUT FOR THE MILES-PER-GALLON PROBLEM: LOGARITHMIC TRANSFORMATION

```
The regression equation is
LogeMPG = 4.52 -0.000501 Weight

Predictor        Coef        Stdev    t-ratio        p
Constant      4.52423      0.09932      45.55    0.000
Weight    -0.00050110   0.00003722     -13.46    0.000

s = 0.06425      R-sq = 94.8%      R-sq(adj) = 94.2%

Analysis of Variance

SOURCE        DF         SS          MS          F        p
Regression     1    0.74822     0.74822     181.22    0.000
Error         10    0.04129     0.00413
Total         11    0.78950
```

FIGURE 16.12 STANDARDIZED RESIDUAL PLOT FOR THE MILES-PER-GALLON PROBLEM: LOGARITHMIC TRANSFORMATION

```
Standardized
 Residuals
       -
       -                                     *              *
       -
    1.2+
       -                  *
       -
       -                                            *
       -
    0.0+                                                          *
       -                  *                  *           *
       -
       -                             *
       -                                                       *
   -1.2+     *
       -                                                       *
       -
       -
          --+---------+---------+---------+---------+---------+----  ŷ
          2.70      2.85      3.00      3.15      3.30      3.45
```

Looking at the residual plot in Figure 16.12, we see that the wedge-shaped pattern has now disappeared. Moreover, none of the observations are identified as having a large standardized residual. The model with the logarithm of miles per gallon as the dependent variable is statistically significant and provides an excellent fit to the observed data. Hence, we would recommend using the estimated regression equation

$$\text{LogeMPG} = 4.52 - 0.000501 \text{ Weight}$$

To estimate the miles-per-gallon rating for an automobile that weighs 2500 pounds, we first develop an estimate of the logarithm of the miles-per-gallon rating.

$$\text{LogeMPG} = 4.52 - 0.000501(2500) = 3.2675$$

The miles-per-gallon estimate is obtained by finding the number whose natural logarithm is 3.2675. Using a calculator with an exponential function, or raising e to the power 3.2675, we obtain 26.2 miles per gallon.

Another approach to problems of nonconstant variance is to use $1/y$ as the dependent variable instead of y. This type of transformation is called a *reciprocal transformation.* For instance, if the dependent variable is measured in miles per gallon, the reciprocal transformation would result in a new dependent variable whose units would be 1/(miles per gallon) or gallons per mile. In general, there is no way to determine whether a logarithmic transformation or a reciprocal transformation will perform best without actually trying each of them.

Nonlinear Models That Are Intrinsically Linear

Models in which the parameters ($\beta_0, \beta_1, \ldots, \beta_p$) have exponents other than one are called nonlinear models. However, for the case of the exponential model, we can perform a transformation of variables that will enable us to perform regression analysis with (16.1), the general linear model. The exponential model involves the following regression equation.

$$E(y) = \beta_0\beta_1^x \tag{16.7}$$

This model is appropriate when the dependent variable y increases or decreases by a constant percentage, instead of by a fixed amount, as x increases.

As an example, suppose sales for a product y are related to advertising expenditure x (in \$1000s) according to the following exponential model.

$$E(y) = 500(1.2)^x$$

Thus, for $x = 1$, $E(y) = 500(1.2)^1 = 600$; for $x = 2$, $E(y) = 500(1.2)^2 = 720$; and for $x = 3$, $E(y) = 500(1.2)^3 = 864$. Note that $E(y)$ is not increasing by a constant amount in this case, but by a constant percentage; the percentage increase is 20%.

We can transform this nonlinear model to a linear model by taking the logarithm of both sides of equation (16.7).

$$\log E(y) = \log \beta_0 + x \log \beta_1 \tag{16.8}$$

Now if we let $y' = \log E(y)$, $\beta_0' = \log \beta_0$, and $\beta_1' = \log \beta_1$, we can rewrite (16.8) as

$$y' = \beta_0' + \beta_1' x$$

It is clear that the formulas for simple linear regression can now be used to develop estimates of β_0' and β_1'. Denoting the estimates as b_0' and b_1' leads to the following estimated regression equation.

$$\hat{y}' = b_0' + b_1' x \tag{16.9}$$

To obtain predictions of the original dependent variable y given a value of x, we would first substitute the value of x into (16.9) and compute $\hat{y}'$. The antilog of $\hat{y}'$ would be the prediction of y, or the expected value of y.

Many nonlinear models cannot be transformed into an equivalent linear model. However, such models have had limited use in business and economic applications. Furthermore, the mathematical background needed for study of such models is beyond the scope of this text.

EXERCISES

Methods

1. Consider the following data for two variables, x and y.

x	22	24	26	30	35	40
y	12	21	33	35	40	36

a. Develop an estimated regression equation for the data of the form $\hat{y} = b_0 + b_1 x$.

b. Using the results from part (a), test for a significant relationship between x and y; use $\alpha = .05$.

c. Develop a scatter diagram for the data. Does the scatter diagram suggest an estimated regression equation of the form $\hat{y} = b_0 + b_1 x + b_2 x^2$? Explain.

d. Develop an estimated regression equation for the data of the form $\hat{y} = b_0 + b_1x + b_2x^2$.
e. Refer to part (d). Is the relationship between x, x^2, and y significant? Use $\alpha = .05$.
f. Predict the value of y when $x = 25$.

2. Consider the following data for two variables, x and y.

x	9	32	18	15	26
y	10	20	21	16	22

a. Develop an estimated regression equation for the data of the form $\hat{y} = b_0 + b_1x$. Comment on the adequacy of this equation for predicting y.
b. Develop an estimated regression equation for the data of the form $\hat{y} = b_0 + b_1x + b_2x^2$. Comment on the adequacy of this equation for predicting y.
c. Predict the value of y when $x = 20$.

3. Consider the following data for two variables, x and y.

x	2	3	4	5	7	7	7	8	9
y	4	5	4	6	4	6	9	5	11

a. Does there appear to be a linear relationship between x and y? Explain.
b. Develop the estimated regression equation relating x and y.
c. Plot the standardized residuals versus $\hat{y}$ for the estimated regression equation developed in part (b). Do the model assumptions appear to be satisfied? Explain.
d. Perform a logarithmic transformation on the dependent variable y. Develop an estimated regression equation using the transformed dependent variable. Do the model assumptions appear to be satisfied by using the transformed dependent variable? Does a reciprocal transformation work better in this case? Explain.

Applications

4. A highway department is studying the relationship between traffic flow and speed. The following model has been hypothesized.

$$y = \beta_0 + \beta_1x + \epsilon$$

where

y = traffic flow in vehicles per hour
x = vehicle speed in miles per hour

The following data were collected during rush hour for six highways leading out of the city.

Traffic Flow (y)	Vehicle Speed (x)
1256	35
1329	40
1226	30
1335	45
1349	50
1124	25

a. Develop an estimated regression equation for the data.
b. Using $\alpha = .01$, test for a significant relationship.

5. In working further with the problem of Exercise 4, statisticians suggested the use of the following curvilinear estimated regression equation.

$$\hat{y} = b_0 + b_1x + b_2x^2$$

a. Use the data of Exercise 4 to estimate the parameters of this estimated regression equation.
b. Using $\alpha = .01$, test for a significant relationship.
c. Estimate the traffic flow in vehicles per hour at a speed of 38 miles per hour.

6. A study of emergency service facilities investigated the relationship between the number of facilities and the average distance traveled to provide the emergency service (*Management Science,* July 1988). The following table gives the data collected.

Number of Facilities	Average Distance (miles)
9	1.66
11	1.12
16	.83
21	.62
27	.51
30	.47

a. Develop a scatter diagram for these data, treating average distance traveled as the dependent variable.
b. Does a simple linear model appear to be appropriate? Explain.
c. Develop an estimated regression equation for the data that you believe will best explain the relationship between these two variables.

7. The following data show the media expenditures ($ millions) and shipments in bbls. (millions) for 10 major brands of beer (*Superbrands '98,* October 20, 1997).

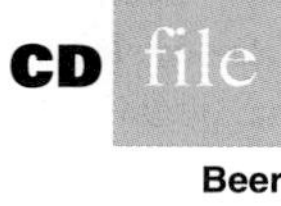

Beer

Brand	Media Expenditures	Shipments
Budweiser	$120.0	36.3
Bud Light	68.7	20.7
Miller Lite	100.1	15.9
Coors Light	76.6	13.2
Busch	8.7	8.1
Natural Light	0.1	7.1
Miller Genuine Draft	21.5	5.6
Miller High Life	1.4	4.4
Busch Light	5.3	4.3
Milwaukee's Best	1.7	4.3

a. Develop a scatter diagram for these data with media expenditures as the independent variable.
b. Does a simple linear regression model appear to be appropriate?
c. Develop an estimated regression equation for the data that you believe will best explain the relationship between these two variables.

8. In Europe the number of Internet users varies widely from country to country. In 1999, 44.3% of all Swedes used the Internet, while in France the audience is less than 10%. The disparities are expected to persist even though Internet usage is expected to grow dramatically over the next several years. The following table shows the number of Internet users in 1999 and the projected number of users in 2005 for European countries.

CD file
Internet

	1999 Internet Users (%)	2005 Projected Users (%)
Austria	12.6	53.4
Belgium	24.2	60.2
Denmark	40.4	71.2
Finland	40.9	71.4
France	9.7	53.1
Germany	15.0	59.6
Greece	3.4	15.8
Ireland	12.1	46.7
Italy	8.4	34.4
Netherlands	18.6	61.4
Norway	38.0	71.7
Portugal	4.63	36.6
Spain	7.4	39.9
Sweden	44.3	71.9
Switzerland	28.1	66.7
United Kingdom	23.6	66.8

a. Develop a scatter diagram of the data using the 1999 Internet user percentage as the independent variable. Does a simple linear regression model appear to be appropriate? Discuss.

b. Develop an estimated multiple regression equation with x = the number of 1999 Internet users and x^2 as the two independent variables.

c. Consider the nonlinear relationship shown by equation (16.7). Use logarithms to develop an estimated regression equation for this model.

d. Do you prefer the estimated regression equation developed in part (b) or part (c)? Explain.

9. In a study to test the effect of headwinds and tailwinds on typical golf ball distance, Titleist found that at a certain point tailwinds buy little extra distance; with headwinds, the deficit continues to increase at greater wind speeds (*Golf Magazine,* March 1997). Suppose that in a follow-up study the following data were obtained for Tour-level swing speeds with a driver. For these data wind speed is measured in miles per hour; negative values denote a headwind and positive values denote a tailwind. Distance is the number of yards of carry and roll.

WindGolf

Wind Speed	Distance	Wind Speed	Distance
−30	202	0	269
−30	211	10	273
−30	206	10	286
−20	221	10	287
−20	239	20	283
−20	234	20	292
−10	240	20	298
−10	245	30	292
−10	260	30	301
0	260	30	299
0	278		

a. Develop an estimated regression equation that shows how the number of yards of carry and roll is related to the wind speed.
b. Estimate the number of yards of carry and roll for a 15-mph headwind.
c. Estimate the number of yards of carry and roll for a 25-mph tailwind.
d. Do your results confirm the findings reported in *Golf Digest?* Explain.

16.2 DETERMINING WHEN TO ADD OR DELETE VARIABLES

In this section we will show how an F test can be used to determine whether it is advantageous to add one or more independent variables to a multiple regression model. This test is based on a determination of the amount of reduction in the error sum of squares resulting from adding one or more independent variables to the model. We will first illustrate how the test can be used in the context of the Butler Trucking example.

In Chapter 15, the Butler Trucking example was introduced to illustrate the use of multiple regression analysis. Recall that the managers wanted to develop an estimated regression equation to predict total daily travel time for trucks using two independent variables: miles traveled and number of deliveries. With miles traveled x_1 as the only independent variable, the least squares procedure provided the following estimated regression equation.

$$\hat{y} = 1.27 + .0678x_1$$

In Chapter 15 we showed that the error sum of squares for this model was SSE $= 8.029$. When x_2, the number of deliveries, was added as a second independent variable, we obtained the following estimated regression equation.

$$\hat{y} = -.869 + .0611x_1 + .923x_2$$

The error sum of squares for this model was SSE $= 2.299$. Clearly, adding x_2 resulted in a reduction of SSE. The question we want to answer is: Does adding the variable x_2 lead to a *significant* reduction in SSE?

We use the notation SSE(x_1) to denote the error sum of squares when x_1 is the only independent variable in the model, SSE(x_1, x_2) to denote the error sum of squares when x_1 and x_2 are both in the model, and so on. Hence, the reduction in SSE resulting from adding x_2 to the model involving just x_1 is

$$\text{SSE}(x_1) - \text{SSE}(x_1, x_2) = 8.029 - 2.299 = 5.730$$

An F test is conducted to determine whether this reduction is significant.

The numerator of the F statistic is the reduction in SSE divided by the number of independent variables added to the original model. Here only one variable, x_2, has been added; thus, the numerator of the F statistic is

$$\frac{\text{SSE}(x_1) - \text{SSE}(x_1, x_2)}{1} = 5.730$$

The result is a measure of the reduction in SSE per independent variable added to the model. The denominator of the F statistic is the mean square error for the model that includes all of the independent variables. For Butler Trucking this corresponds to the model containing both x_1 and x_2; thus, $p = 2$ and

$$\text{MSE} = \frac{\text{SSE}(x_1, x_2)}{n - p - 1} = \frac{2.299}{7} = .3284$$

The following F statistic provides the basis for testing whether the addition of x_2 is statistically significant.

$$F = \frac{\dfrac{\text{SSE}(x_1) - \text{SSE}(x_1, x_2)}{1}}{\dfrac{\text{SSE}(x_1, x_2)}{n - p - 1}} \tag{16.10}$$

The numerator degrees of freedom for this F test is equal to the number of variables added to the model, and the denominator degrees of freedom is equal to $n - p - 1$.

For the Butler Trucking problem, we obtain

$$F = \frac{\dfrac{5.730}{1}}{\dfrac{2.299}{7}} = \frac{5.730}{.3284} = 17.45$$

Refer to Table 4 of Appendix B. We find that for a level of significance of $\alpha = .05$, $F_{.05} = 5.59$. Because $F = 17.45 > F_{.05} = 5.59$, we can reject the null hypothesis that x_2 is not statistically significant; in other words, adding x_2 to the model involving only x_1 results in a significant reduction in the error sum of squares.

When we want to test for the significance of adding only one more independent variable to a model, the result found with the F test just described could also be obtained by using the t test for the significance of an individual parameter (described in Section 15.4). Indeed, the F statistic we just computed is the square of the t statistic used to test the significance of an individual parameter.

Because the t test is equivalent to the F test when only one independent variable is being added to the model, we can now further clarify the proper use of the t test for testing the significance of an individual parameter. If an individual parameter is not significant, the corresponding variable can be dropped from the model. However, if the t test shows that two or more parameters are not significant, no more than one independent variable can ever be dropped from a model on the basis of a t test; if one variable is dropped, a second variable that was not significant initially might become significant.

We now turn to a consideration of whether the addition of more than one independent variable—as a set—results in a significant reduction in the error sum of squares.

General Case

Consider the following multiple regression model involving q independent variables, where $q < p$.

$$y = \beta_0 + \beta_1 x_1 + \beta_2 x_2 + \cdots + \beta_q x_q + \epsilon \tag{16.11}$$

If we add variables $x_{q+1}, x_{q+2}, \ldots, x_p$ to this model, we obtain a model involving p independent variables.

$$\begin{aligned} y = \beta_0 &+ \beta_1 x_1 + \beta_2 x_2 + \cdots + \beta_q x_q \\ &+ \beta_{q+1} x_{q+1} + \beta_{q+2} x_{q+2} + \cdots + \beta_p x_p + \epsilon \end{aligned} \tag{16.12}$$

To test whether the addition of $x_{q+1}, x_{q+2}, \ldots, x_p$ is statistically significant, the null and alternative hypotheses can be stated as follows.

H_0: $\beta_{q+1} = \beta_{q+2} = \cdots = \beta_p = 0$
H_a: One or more of the parameters is not equal to zero

The following F statistic provides the basis for testing whether the additional independent variables are statistically significant.

$$F = \frac{\dfrac{\text{SSE}(x_1, x_2, \ldots, x_q) - \text{SSE}(x_1, x_2, \ldots, x_q, x_{q+1}, \ldots, x_p)}{p - q}}{\dfrac{\text{SSE}(x_1, x_2, \ldots, x_q, x_{q+1}, \ldots, x_p)}{n - p - 1}} \tag{16.13}$$

This computed F value is then compared with F_α, the table value with $p - q$ numerator degrees of freedom and $n - p - 1$ denominator degrees of freedom. If $F > F_\alpha$, we reject H_0 and conclude that the set of additional independent variables is statistically significant. Note that for the special case where $q = 1$ and $p = 2$, (16.13) reduces to (16.10).

Many computer packages, such as Minitab, provide extra sums of squares corresponding to the order in which each independent variable enters the model; in such cases, the computation of the F test for determining whether to add or delete a set of variables is simplified.

Many students find (16.13) somewhat complex. To provide a simpler description of this F ratio, we can refer to the model with the smaller number of independent variables as the reduced model and the model with the larger number of independent variables as the full model. If we let SSE(reduced) denote the error sum of squares for the reduced model and SSE(full) denote the error sum of squares for the full model, we can write the numerator of (16.13) as

$$\frac{\text{SSE(reduced)} - \text{SSE(full)}}{\text{number of extra terms}} \tag{16.14}$$

Note that "number of extra terms" denotes the difference between the number of independent variables in the full model and the number of independent variables in the reduced model. The denominator of (16.13) is the error sum of squares for the full model divided by the corresponding degrees of freedom; in other words, the denominator is the mean square error for the full model. Denoting the mean square error for the full model as MSE(full) enables us to write (16.13) as

$$F = \frac{\dfrac{\text{SSE(reduced)} - \text{SSE(full)}}{\text{number of extra terms}}}{\text{MSE(full)}} \tag{16.15}$$

To illustrate the use of this F statistic, suppose we have a regression problem involving 30 observations. One model with the independent variables x_1, x_2, and x_3 has an error sum of squares of 150 and a second model with the independent variables x_1, x_2, x_3, x_4, and x_5 has an error sum of squares of 100. Did the addition of the two independent variables x_4 and x_5 result in a significant reduction in the error sum of squares?

First, note that the degrees of freedom for SST is $30 - 1 = 29$ and that the degrees of freedom for the regression sum of squares for the full model is five (the number of independent variables in the full model). Thus, the degrees of freedom for the error sum of squares for the full model is $29 - 5 = 24$, and hence MSE(full) $= 100/24 = 4.17$. Therefore the F statistic is

$$F = \frac{\dfrac{150 - 100}{2}}{4.17} = 6.00$$

This computed F value is compared with the table F value with two numerator and 24 denominator degrees of freedom. At the .05 level of significance, Table 4 of Appendix B shows $F_{.05} = 3.40$. Because $F = 6.00$ is greater than 3.40, we conclude that the addition of variables x_4 and x_5 is statistically significant.

Use of *p*-Values

The *p*-value criterion can also be used to determine whether it is advantageous to add one or more independent variables to a multiple regression model. In the preceding example, we showed how to perform an F test to determine if the addition of two independent variables, x_4 and x_5, to a model with three independent variables, x_1, x_2, and x_3, was statistically significant. For this example, the computed F statistic was 6.00 and we concluded (by comparing $F = 6.00$ to the critical value $F_{.05} = 3.40$) that the addition of variables x_4 and x_5 was significant. The *p*-value associated with $F = 6.00$ (2 numerator and 24 denominator degrees of freedom) is .008. With a *p*-value $= .008 < \alpha = .05$, we also conclude that the addition of the two independent variables is statistically significant. It is difficult to determine the *p*-value directly from tables of the F distribution, but computer software packages, such as Minitab or Excel, make computing the *p*-value easy.

NOTES AND COMMENTS

Computation of the F statistic can also be based on the difference in the regression sums of squares. To show this form of the F statistic, we first note that

$$\text{SSE(reduced)} = \text{SST} - \text{SSR(reduced)}$$
$$\text{SSE(full)} = \text{SST} - \text{SSR(full)}$$

Hence

$$\begin{aligned}\text{SSE(reduced)} - \text{SSE(full)} &= [\text{SST} - \text{SSR(reduced)}] - [\text{SST} - \text{SSR(full)}] \\ &= \text{SSR (full)} - \text{SSR(reduced)}\end{aligned}$$

Thus,

$$F = \frac{\dfrac{\text{SSR(full)} - \text{SSR(reduced)}}{\text{number of extra terms}}}{\text{MSE(full)}}$$

EXERCISES

Methods

10. In a regression analysis involving 27 observations, the following estimated regression equation was developed.

$$\hat{y} = 25.2 + 5.5x_1$$

For this estimated regression equation SST = 1550 and SSE = 520.

a. At $\alpha = .05$, test whether x_1 is significant.
Suppose that variables x_2 and x_3 are added to the model and the following regression equation is obtained.

$$\hat{y} = 16.3 + 2.3x_1 + 12.1x_2 - 5.8x_3$$

For this estimated regression equation SST = 1550 and SSE = 100.

b. Use an F test and a .05 level of significance to determine whether x_2 and x_3 contribute significantly to the model.

11. In a regression analysis involving 30 observations, the following estimated regression equation was obtained.

$$\hat{y} = 17.6 + 3.8x_1 - 2.3x_2 + 7.6x_3 + 2.7x_4$$

For this estimated regression equation SST = 1805 and SSR = 1760.

a. At $\alpha = .05$, test the significance of the relationship among the variables.

Suppose variables x_1 and x_4 are dropped from the model and the following estimated regression equation is obtained.

$$\hat{y} = 11.1 - 3.6x_2 + 8.1x_3$$

For this model SST = 1805 and SSR = 1705.

b. Compute SSE(x_1, x_2, x_3, x_4).

c. Compute SSE(x_2, x_3).

d. Use an F test and a .05 level of significance to determine whether x_1 and x_4 contribute significantly to the model.

Applications

12. The table below gives some of the data available for 14 teams in the National Football League after 15 games.

a. Develop an estimated regression equation that can be used to predict the total points scored given the number of interceptions made by the team.

b. Develop an estimated regression equation that can be used to predict the total points scored given the number of interceptions made by the team, the number of rushing yards, and the number of interceptions made by the opponents.

c. At a .05 level of significance, test to see whether the addition of the number of rushing yards and the number of interceptions made by the opponents contributes significantly to the estimated regression equation developed in part (a). Explain.

Football

					Interceptions	
Team	**Won–Lost**	**Total Points**	**Rushing Yards**	**Passing Yards**	**Made by Team**	**Made by Opponent**
Atlanta	5–10	305	1907	2473	19	23
Chicago	12–3	187	2134	2718	14	24
Dallas	3–12	358	1858	3386	24	10
Detroit	4–11	292	1184	1971	15	12
Green Bay	3–12	298	1274	3046	22	20
St. Louis Rams	9–6	277	1882	3604	17	22
Minnesota	10–5	206	1744	3633	16	35
New Orleans	9–6	274	1843	2963	15	17
N.Y. Giants	10–5	277	1492	3096	14	15
Philadelphia	9–6	312	1812	3247	17	29
Phoenix	7–8	372	1909	3633	19	14
San Francisco	10–5	256	2453	3131	14	21
Tampa Bay	4–11	340	1650	3169	33	18
Washington	7–8	367	1377	3930	24	14

13. Refer to Exercise 12.

a. Develop an estimated regression equation that relates the total points scored to the number of passing yards, the number of interceptions made by the team, and the number of interceptions made by the opponents.

b. Develop an estimated regression equation using the independent variables in part (a) and the number of rushing yards.

c. At a .05 level of significance, did the number of rushing yards contribute significantly to the estimated regression equation developed in part (a)? Explain.

14. A 10-year study conducted by the American Heart Association provided data on how age, blood pressure, and smoking relate to the risk of strokes. Data from a portion of this study follow. Risk is interpreted as the probability (times 100) that a person will have a stroke over the next 10-year period. For the smoker variable, 1 indicates a smoker and 0 indicates a nonsmoker.

CD file
Stroke

Risk	Age	Blood Pressure	Smoker
12	57	152	0
24	67	163	0
13	58	155	0
56	86	177	1
28	59	196	0
51	76	189	1
18	56	155	1
31	78	120	0
37	80	135	1
15	78	98	0
22	71	152	0
36	70	173	1
15	67	135	1
48	77	209	1
15	60	199	0
36	82	119	1
8	66	166	0
34	80	125	1
3	62	117	0
37	59	207	1

a. Develop an estimated regression equation that can be used to predict the risk of stroke given the age and blood-pressure level.

b. Consider adding two independent variables to the model developed in part (a), one for the interaction between age and blood-pressure level and the other for whether the person is a smoker. Develop an estimated regression equation using these four independent variables.

c. At a .05 level of significance, test to see whether the addition of the interaction term and the smoker variable contribute significantly to the estimated regression equation developed in part (a).

15. The National Football League rates prospects position by position on a scale that ranges from 5 to 9. The ratings are interpreted as follows: 8–9 should start the first year; 7.0–7.9 should start; 6.0–6.9 will make the team as backup; and 5.0–5.9 can make the club and contribute. The following table shows the position, weight, speed (for 40 yards), and ratings for 40 NFL prospects (*USA Today,* April 14, 2000).

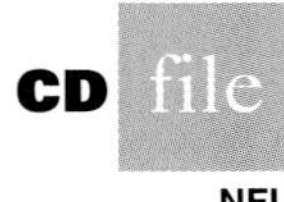

NFL

Observation	Name	Position	Weight	Speed	Rating
1	Peter Warrick	Wide receiver	194	4.53	9.0
2	Plaxico Burress	Wide receiver	231	4.52	8.8
3	Sylvester Morris	Wide receiver	216	4.59	8.3
4	Travis Taylor	Wide receiver	199	4.36	8.1
5	Laveranues Coles	Wide receiver	192	4.29	8.0
6	Dez White	Wide receiver	218	4.49	7.9
7	Jerry Porter	Wide receiver	221	4.55	7.4
8	Ron Dugans	Wide receiver	206	4.47	7.1
9	Todd Pinkston	Wide receiver	169	4.37	7.0
10	Dennis Northcutt	Wide receiver	175	4.43	7.0
11	Anthony Lucas	Wide receiver	194	4.51	6.9
12	Darrell Jackson	Wide receiver	197	4.56	6.6

Observation	Name	Position	Weight	Speed	Rating
13	Danny Farmer	Wide receiver	217	4.60	6.5
14	Sherrod Gideon	Wide receiver	173	4.57	6.4
15	Trevor Gaylor	Wide receiver	199	4.57	6.2
16	Cosey Coleman	Guard	322	5.38	7.4
17	Travis Claridge	Guard	303	5.18	7.0
18	Kaulana Noa	Guard	317	5.34	6.8
19	Leander Jordan	Guard	330	5.46	6.7
20	Chad Clifton	Guard	334	5.18	6.3
21	Manula Savea	Guard	308	5.32	6.1
22	Ryan Johanningmei	Guard	310	5.28	6.0
23	Mark Tauscher	Guard	318	5.37	6.0
24	Blaine Saipaia	Guard	321	5.25	6.0
25	Richard Mercier	Guard	295	5.34	5.8
26	Damion McIntosh	Guard	328	5.31	5.3
27	Jeno James	Guard	320	5.64	5.0
28	Al Jackson	Guard	304	5.20	5.0
29	Chris Samuels	Offensive tackle	325	4.95	8.5
30	Stockar McDougle	Offensive tackle	361	5.50	8.0
31	Chris McIngosh	Offensive tackle	315	5.39	7.8
32	Adrian Klemm	Offensive tackle	307	4.98	7.6
33	Todd Wade	Offensive tackle	326	5.20	7.3
34	Marvel Smith	Offensive tackle	320	5.36	7.1
35	Michael Thompson	Offensive tackle	287	5.05	6.8
36	Bobby Williams	Offensive tackle	332	5.26	6.8
37	Darnell Alford	Offensive tackle	334	5.55	6.4
38	Terrance Beadles	Offensive tackle	312	5.15	6.3
39	Tutan Reyes	Offensive tackle	299	5.35	6.1
40	Greg Robinson-Ran	Offensive tackle	333	5.59	6.0

a. Develop dummy variables that will account for the player's position.

b. Develop an estimated regression equation to show how rating is related to position, weight, and speed.

c. At the .05 level of significance, test whether the estimated regression equation developed in part (b) represents a significant relationship between the independent variables and the dependent variable.

d. Is position a significant factor in the player's rating? Use $\alpha = .05$. Explain.

16.3 ANALYSIS OF A LARGER PROBLEM

In introducing multiple regression analysis, we used the Butler Trucking example extensively. The small size of this problem was an advantage in exploring introductory concepts, but would make it difficult to illustrate some of the variable selection issues involved in model building. To provide an illustration of the variable selection procedures discussed in the next section, we introduce a data set consisting of 25 observations on eight independent variables. Permission to use these data was provided by Dr. David W. Cravens of the Department of Marketing at Texas Christian University. Consequently, we refer to the data set as the Cravens data.*

The Cravens data are for a company that sells products in several sales territories, each of which is assigned to a single sales representative. A regression analysis was conducted to determine whether a variety of predictor (independent) variables could explain sales in

*For details see David W. Cravens, Robert B. Woodruff, and Joe C. Stamper, "An Analytical Approach for Evaluating Sales Territory Performance," *Journal of Marketing*, 36 (January 1972): 31–37. Copyright © 1972 American Marketing Association.

TABLE 16.5 CRAVENS DATA

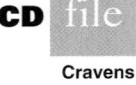

CD file Cravens

Sales	Time	Poten	AdvExp	Share	Change	Accounts	Work	Rating
3,669.88	43.10	74,065.1	4,582.9	2.51	0.34	74.86	15.05	4.9
3,473.95	108.13	58,117.3	5,539.8	5.51	0.15	107.32	19.97	5.1
2,295.10	13.82	21,118.5	2,950.4	10.91	−0.72	96.75	17.34	2.9
4,675.56	186.18	68,521.3	2,243.1	8.27	0.17	195.12	13.40	3.4
6,125.96	161.79	57,805.1	7,747.1	9.15	0.50	180.44	17.64	4.6
2,134.94	8.94	37,806.9	402.4	5.51	0.15	104.88	16.22	4.5
5,031.66	365.04	50,935.3	3,140.6	8.54	0.55	256.10	18.80	4.6
3,367.45	220.32	35,602.1	2,086.2	7.07	−0.49	126.83	19.86	2.3
6,519.45	127.64	46,176.8	8,846.2	12.54	1.24	203.25	17.42	4.9
4,876.37	105.69	42,053.2	5,673.1	8.85	0.31	119.51	21.41	2.8
2,468.27	57.72	36,829.7	2,761.8	5.38	0.37	116.26	16.32	3.1
2,533.31	23.58	33,612.7	1,991.8	5.43	−0.65	142.28	14.51	4.2
2,408.11	13.82	21,412.8	1,971.5	8.48	0.64	89.43	19.35	4.3
2,337.38	13.82	20,416.9	1,737.4	7.80	1.01	84.55	20.02	4.2
4,586.95	86.99	36,272.0	10,694.2	10.34	0.11	119.51	15.26	5.5
2,729.24	165.85	23,093.3	8,618.6	5.15	0.04	80.49	15.87	3.6
3,289.40	116.26	26,878.6	7,747.9	6.64	0.68	136.58	7.81	3.4
2,800.78	42.28	39,572.0	4,565.8	5.45	0.66	78.86	16.00	4.2
3,264.20	52.84	51,866.1	6,022.7	6.31	−0.10	136.58	17.44	3.6
3,453.62	165.04	58,749.8	3,721.1	6.35	−0.03	138.21	17.98	3.1
1,741.45	10.57	23,990.8	861.0	7.37	−1.63	75.61	20.99	1.6
2,035.75	13.82	25,694.9	3,571.5	8.39	−0.43	102.44	21.66	3.4
1,578.00	8.13	23,736.3	2,845.5	5.15	0.04	76.42	21.46	2.7
4,167.44	58.44	34,314.3	5,060.1	12.88	0.22	136.58	24.78	2.8
2,799.97	21.14	22,809.5	3,552.0	9.14	−0.74	88.62	24.96	3.9

each territory. A random sample of 25 sales territories resulted in the data in Table 16.5; the variable definitions are given in Table 16.6.

As a preliminary step, let us consider the sample correlation coefficients between each pair of variables. Figure 16.13 is the correlation matrix obtained by using the Minitab correlation command. Note that the sample correlation coefficient between Sales and Time is .623, between Sales and Poten is .598, and so on.

TABLE 16.6 VARIABLE DEFINITIONS FOR THE CRAVENS DATA

Variable	Definition
Sales	Total sales credited to the sales representative
Time	Length of time employed in months
Poten	Market potential; total industry sales in units for the sales territory*
AdvExp	Advertising expenditure in the sales territory
Share	Market share; weighted average for the past four years
Change	Change in the market share over the previous four years
Accounts	Number of accounts assigned to the sales representative*
Work	Workload; a weighted index based on annual purchases and concentrations of accounts
Rating	Sales representative overall rating on eight performance dimensions; an aggregate rating on a 1–7 scale

*These data were coded to preserve confidentiality.

FIGURE 16.13 SAMPLE CORRELATION COEFFICIENTS FOR THE CRAVENS DATA

	Sales	Time	Poten	AdvExp	Share	Change	Accounts	Work
Time	0.623							
Poten	0.598	0.454						
AdvExp	0.596	0.249	0.174					
Share	0.484	0.106	-0.211	0.264				
Change	0.489	0.251	0.268	0.377	0.085			
Accounts	0.754	0.758	0.479	0.200	0.403	0.327		
Work	-0.117	-0.179	-0.259	-0.272	0.349	-0.288	-0.199	
Rating	0.402	0.101	0.359	0.411	-0.024	0.549	0.229	-0.277

Looking at the sample correlation coefficients between the independent variables, we see that the correlation between Time and Accounts is .758; hence, if Accounts were used as an independent variable, Time would not add much more explanatory power to the model. Recall the rule-of-thumb test from the discussion of multicollinearity in Section 15.4: multicollinearity can cause problems if the absolute value of the sample correlation coefficient exceeds .7 for any two of the independent variables. If possible, then, we should avoid including both Time and Accounts in the same regression model. The sample correlation coefficient of .549 between Change and Rating is also high and may warrant further consideration.

Looking at the sample correlation coefficients between Sales and each of the independent variables can give us a quick indication of which independent variables are, by themselves, good predictors. We see that the single best predictor of Sales is Accounts, because it has the highest sample correlation coefficient (.754). Recall that for the case of one independent variable, the square of the sample correlation coefficient is the coefficient of determination. Thus, Accounts can explain $(.754)^2(100)$, or 56.85%, of the variability in Sales. The next most important independent variables are Time, Poten, and AdvExp, each with a sample correlation coefficient of approximately .6.

Although there are potential multicollinearity problems, let us consider developing an estimated regression equation using all eight independent variables. The Minitab computer package provided the results in Figure 16.14. The eight-variable multiple regression model has an adjusted coefficient of determination of 88.3%. Note, however, that the p-values for the t tests of individual parameters show that only Poten, AdvExp, and Share are significant at the $\alpha = .05$ level, given the effect of all the other variables. Hence, we might be inclined to investigate the results that would be obtained if we used just those three variables. Figure 16.15 shows the Minitab results obtained for the estimated regression equation with those three variables. We see that the estimated regression equation has an adjusted coefficient of determination of 82.7%, which, although not quite as good as that for the eight-independent-variable estimated regression equation, is high.

How can we find an estimated regression equation that will do the best job given the data available? One approach is to compute all possible regressions. That is, we could develop eight one-variable estimated regression equations (each of which corresponds to one of the independent variables), 28 two-variable estimated regression equations (the number of combinations of eight variables taken two at a time), and so on. In all, for the Cravens data, 255 different estimated regression equations involving one or more independent variables would have to be fitted to the data.

With the excellent computer packages available today, it is possible to compute all possible regressions. But doing so involves a great amount of computation and requires the

FIGURE 16.14 MINITAB OUTPUT FOR THE MODEL INVOLVING ALL EIGHT INDEPENDENT VARIABLES

```
The regression equation is
Sales = - 1508 + 2.01 Time + 0.0372 Poten + 0.151 AdvExp + 199 Share
        + 291 Change + 5.55 Accounts + 19.8 Work + 8 Rating

Predictor         Coef        Stdev    t-ratio        p
Constant        1507.8        778.6      -1.94    0.071
Time             2.010        1.931       1.04    0.313
Poten         0.037205     0.008202       4.54    0.000
AdvExp         0.15099      0.04711       3.21    0.006
Share           199.02        67.03       2.97    0.009
Change           290.9        186.8       1.56    0.139
Accounts         5.551        4.776       1.16    0.262
Work             19.79        33.68       0.59    0.565
Rating             8.2        128.5       0.06    0.950

s = 449.0       R-sq = 92.2%      R-sq(adj) = 88.3%

Analysis of Variance

SOURCE        DF          SS          MS          F        p
Regression     8    38153568     4769196      23.65    0.000
Error         16     3225984      201624
Total         24    41379552
```

FIGURE 16.15 MINITAB OUTPUT FOR THE MODEL INVOLVING Poten, AdvExp, AND Share

```
The regression equation is
Sales = - 1604 + 0.0543 Poten + 0.167 AdvExp + 283 Share

Predictor         Coef         Stdev     t-ratio        p
Constant       -1603.6         505.6       -3.17    0.005
Poten         0.054286      0.007474        7.26    0.000
AdvExp         0.16748       0.04427        3.78    0.001
Share           282.75         48.76        5.80    0.000

s = 545.5        R-sq = 84.9%      R-sq(adj) = 82.7%

Analysis of Variance

SOURCE        DF          SS           MS          F         p
Regression     3    35130240     11710080      39.35     0.000
Error         21     6249310       297586
Total         24    41379552
```

model builder to review a large volume of computer output, much of which is associated with obviously poor models. Statisticians prefer a more systematic approach to selecting the subset of independent variables that provide the best estimated regression equation. In the next section, we introduce some of the more popular approaches.

16.4 VARIABLE SELECTION PROCEDURES

Variable selection procedures are particularly useful in the early stages of building a model, but they cannot substitute for experience and judgment on the part of the analyst.

In this section we discuss four variable selection procedures: stepwise regression, forward selection, backward elimination, and best-subsets regression. Given a data set with several possible independent variables, we can use these procedures to identify which independent variables provide the best model. The first three procedures are iterative; at each step of the procedure a single independent variable is added or deleted and the new model is evaluated. The process continues until a stopping criterion indicates that the procedure cannot find a better model. The last procedure (best subsets) is not a one-variable-at-a-time procedure; it evaluates regression models involving different subsets of the independent variables.

In the stepwise regression, forward selection, and backward elimination procedures, the criterion for selecting an independent variable to add or delete from the model at each step is based on the F statistic introduced in Section 16.2. Suppose, for instance, that we are considering adding x_2 to a model involving x_1 or deleting x_2 from a model involving x_1 and x_2. To test whether the addition or deletion of x_2 is statistically significant, the null and alternative hypotheses can be stated as follows:

$$H_0\colon \beta_2 = 0$$
$$H_a\colon \beta_2 \neq 0$$

In Section 16.2 (see equation 16.10) we showed that

$$F = \frac{\dfrac{\text{SSE}(x_1) - \text{SSE}(x_1, x_2)}{1}}{\dfrac{\text{SSE}(x_1, x_2)}{n - p - 1}}$$

can be used as a criterion for determining whether the presence of x_2 in the model causes a significant reduction in the error sum of squares. The p-value corresponding to this F statistic is the criterion used to determine whether an independent variable should be added or deleted from the regression model. The usual rejection rule applies: Reject H_0 if p-value $< \alpha$.

Stepwise Regression

The stepwise regression procedure begins each step by determining whether any of the variables *already in the model* should be removed. It does so by first computing an F statistic and a corresponding p-value for each independent variable in the model. The level of significance α for determining if an independent variable should be removed from the model is referred to in Minitab as *Alpha to remove.* If the p-value for any independent variable is greater than *Alpha to remove,* the independent variable with the largest p-value is removed from the model and the stepwise regression procedure begins a new step.

If no independent variable can be removed from the model, the procedure attempts to enter another independent variable into the model. It does so by first computing an F statistic and corresponding p-value for each independent variable that is not in the model. The level of significance α for determining if an independent variable should be entered into the model is referred to in Minitab as *Alpha to enter.* The independent variable with the smallest p-value is entered into the model provided its p-value is less than *Alpha to enter.* The procedure continues in this manner until no independent variables can be deleted from or added to the model.

Figure 16.16 shows the results obtained by using the Minitab stepwise regression procedure for the Cravens data using values of .05 for *Alpha to remove* and .05 for *Alpha to enter.* The stepwise procedure terminated after four steps. The estimated regression equation identified by the Minitab stepwise regression procedure is

$$\hat{y} = -1441.93 + 9.2 \text{ Accounts} + .175 \text{ AdvExp} + .0382 \text{ Poten} + 190 \text{ Share}$$

Because the one-at-a-time procedures do not consider every possible subset for a given number of independent variables, they will not necessarily select the model with the highest R-sq value.

Note also in Figure 16.16 that $s = \sqrt{\text{MSE}}$ has been reduced from 881 with the best one-variable model (using Accounts) to 454 after four steps. The value of R-sq has been increased from 56.85% to 90.04%, and the recommended estimated regression equation has an R-Sq(adj) value of 88.05%.

In summary, at each step of the stepwise regression procedure the first consideration is to see whether any independent variable can be removed from the current model. If none of the independent variables can be removed from the model, the procedure checks to see whether any of the independent variables that are not currently in the model can be entered. Because of the nature of the stepwise regression procedure, an independent variable can enter the model at one step, be removed at a subsequent step, and then enter the model at a later step. The procedure stops when no independent variables can be removed from or entered into the model.

FIGURE 16.16 MINITAB STEPWISE REGRESSION OUTPUT FOR THE CRAVENS DATA

```
Alpha-to-Enter: 0.05       Alpha-to-Remove: 0.05

Response is Sales on 8 predictors, with N = 25

    Step          1        2         3          4
Constant     709.32    50.29   -327.24   -1441.93

Accounts       21.7     19.0      15.6        9.2
T-Value        5.50     6.41      5.19       3.22
P-Value       0.000    0.000     0.000      0.004

AdvExp                 0.227     0.216      0.175
T-Value                 4.50      4.77       4.74
P-Value                0.000     0.000      0.000

Poten                           0.0219     0.0382
T-Value                           2.53       4.79
P-Value                          0.019      0.000

Share                                         190
T-Value                                      3.82
P-Value                                     0.001

S               881      650       583        454
R-Sq          56.85    77.51     82.77      90.04
R-Sq(adj)     54.97    75.47     80.31      88.05
C-p            67.6     27.2      18.4        5.4
```

Forward Selection

The forward selection procedure starts with no independent variables. It adds variables one at a time using the same procedure as stepwise regression for determining if an independent variable should be entered into the model. However, the forward selection procedure does not permit a variable to be removed from the model once it has been entered. The procedure stops if the p-value for each of the independent variables not in the model is greater than *Alpha to enter.*

The estimated regression equation obtained using Minitab's forward selection procedure is

$$\hat{y} = -1441.93 + 9.2 \text{ Accounts} + .175 \text{ AdvExp} + .0382 \text{ Poten} + 190 \text{ Share}$$

Thus, for the Cravens data, the forward selection procedure (using .05 for *Alpha to enter*) leads to the same estimated regression equation as the stepwise procedure.

Backward Elimination

The backward elimination procedure begins with a model that includes all the independent variables. It then deletes one independent variable at a time using the same procedure as stepwise regression. However, the backward elimination procedure does not permit an independent variable to be reentered once it has been removed. The procedure stops when none of the independent variables in the model have a p-value greater than *Alpha to remove.*

The estimated regression equation obtained using Minitab's backward elimination procedure for the Cravens data (using .05 for *Alpha to remove*) is

$$\hat{y} = -1312 + 3.8 \text{ Time} + .0444 \text{ Poten} + .152 \text{ AdvExp} + 259 \text{ Share}$$

Comparing the estimated regression equation identified using the backward elimination procedure to the estimated regression equation identified using the forward selection procedure, we see that three independent variables—AdvExp, Poten, and Share—are common to both. However, the backward elimination procedure has included Time instead of Accounts.

Forward selection and backward elimination may lead to different models.

Forward selection and backward elimination are the two extremes of model building; the forward selection procedure starts with no independent variables in the model and adds independent variables one at a time, whereas the backward elimination procedure starts with all independent variables in the model and deletes variables one at a time. The two procedures may lead to the same estimated regression equation. But, as we have seen for the Cravens data, it is possible for them to lead to two different estimated regression equations. Deciding which estimated regression equation to use remains a topic for discussion. Ultimately, the analyst's judgment must be applied. The best-subsets model building procedure we discuss next provides additional model-building information to be considered before a final decision is made.

Best-Subsets Regression

Stepwise regression, forward selection, and backward elimination are approaches to choosing the regression model by adding or deleting independent variables one at a time. None of them guarantee that the best model for a given number of variables will be found. Hence, these one-variable-at-a-time methods are properly viewed as heuristics for selecting a good regression model.

Some software packages have a procedure called best-subsets regression that enables the user to find, given a specified number of independent variables, the best regression model. Minitab has such a procedure. Figure 16.17 is a portion of the computer output obtained by using the best-subsets procedure for the Cravens data set.

This output identifies the two best one-variable estimated regression equations, the two best two-variable equations, the two best three-variable equations, and so on. The criterion used in determining which estimated regression equations are best for any number

FIGURE 16.17 PORTION OF MINITAB BEST-SUBSETS REGRESSION OUTPUT

Vars	R-sq	Adj. R-sq	s	Time	Poten	AdvExp	Share	Change	Accounts	Work	Rating
1	56.8	55.0	881.09						X		
1	38.8	36.1	1049.3	X							
2	77.5	75.5	650.39			X			X		
2	74.6	72.3	691.11		X		X				
3	84.9	82.7	545.52		X	X	X				
3	82.8	80.3	582.64		X	X			X		
4	90.0	88.1	453.84		X	X	X		X		
4	89.6	87.5	463.93	X	X	X	X				
5	91.5	89.3	430.21	X	X	X	X	X			
5	91.2	88.9	436.75		X	X	X	X	X		
6	92.0	89.4	427.99	X	X	X	X	X	X		
6	91.6	88.9	438.20		X	X	X	X	X	X	
7	92.2	89.0	435.66	X	X	X	X	X	X	X	
7	92.0	88.8	440.29	X	X	X	X	X	X		X
8	92.2	88.3	449.02	X	X	X	X	X	X	X	X

of predictors is the value of the coefficient of determination (R-sq). For instance, Accounts, with an R-sq = 56.8%, provides the best estimated regression equation using only one independent variable; AdvExp and Accounts, with an R-sq = 77.5%, provides the best estimated regression equation using two independent variables; and Poten, AdvExp, and Share, with an R-sq = 84.9%, provides the best estimated regression equation with three independent variables. For the Cravens data, the adjusted coefficient of determination (Adj. R-sq = 89.4%) is largest for the model with six independent variables: Time, Poten, AdvExp, Share, Change, and Accounts. However, the best model with four independent variables (Poten, AdvExp, Share, Accounts) has an adjusted coefficient of determination almost as high (88.1%). All other things being equal, a simpler model with fewer variables is usually preferred.

Making the Final Choice

The analysis performed on the Cravens data to this point is good preparation for choosing a final model, but more analysis should be conducted before the final choice. As we have noted in Chapters 14 and 15, a careful analysis of the residuals should be done. We want the residual plot for the chosen model to resemble approximately a horizontal band. Let us assume the residuals are not a problem and that we want to use the results of the best-subsets procedure to help choose the model.

The best-subsets procedure has shown us that the best four-variable model has the independent variables Poten, AdvExp, Share, and Accounts. This also happens to be the four-variable model identified with the stepwise regression procedure. Table 16.7 is helpful in making the final choice. It shows several possible models consisting of some or all of these four independent variables.

TABLE 16.7 SELECTED MODELS INVOLVING Accounts, AdvExp, Poten, AND Share

Model	Independent Variables	Adj. R-sq
1	Accounts	55.0
2	AdvExp, Accounts	75.5
3	Poten, Share	72.3
4	Poten, AdvExp, Accounts	80.3
5	Poten, AdvExp, Share	82.7
6	Poten, AdvExp, Share, Accounts	88.1

From Table 16.7, we see that the model with just AdvExp and Accounts is good. The adjusted coefficient of determination is 75.5%, and the model with all four variables provides only a 12.6-percentage-point improvement. The simpler two-variable model might be preferred, for instance, if it is difficult to measure market potential (Poten). However, if the data are readily available and highly accurate predictions of sales are needed, the model builder would clearly prefer the model with all four variables.

NOTES AND COMMENTS

1. The stepwise procedure requires that Alpha to remove be greater than or equal to Alpha to enter. This requirement prevents the same variable from being removed and then reentered at the same step.
2. Functions of the independent variables can be used to create new independent variables for use with any of the procedures in this section. For instance, if we wanted x_1x_2 in the model to account for interaction, we would use the data for x_1 and x_2 to create the data for $z = x_1x_2$.
3. None of the procedures that add or delete variables one at a time can be guaranteed to identify the best regression model. But they are excellent approaches to finding good models—especially when little multicollinearity is present.

EXERCISES

Applications

16. Two experts provided subjective lists of school districts that they think are among the best in the country. For each school district, the following data were obtained: average class size, instructional spending per student, average teacher salary, combined SAT score, percentage of students taking the SAT, and percentage of graduates attending a 4-year college.

City	Average Class Size	Instructional Spending per Student ($)	Average Teacher Salary ($)	Combined SAT Score/ (% Taking Test)	Attend Four-Year College (%)
Blue Springs, MO	25	3,060	29,359	1083/(8)	74
Garden City, NY	18	9,700	51,000	997/(99)	77
Indianapolis, IN	30	3,222	30,482	716/(42)	40
Newport Beach, CA (Newport–Mesa)	26	4,028	37,043	977/(46)	51

Continued

City	Average Class Size	Instructional Spending per Student ($)	Average Teacher Salary ($)	Combined SAT Score/ (% Taking Test)	Attend Four-Year College (%)
Novi, MI	20	3,067	39,797	980/(15)	53
Piedmont, CA (Piedmont City)	28	4,208	37,274	1,042/(91)	75
Pittsburgh, PA (Fox Chapel area)	21	4,884	37,156	983/(80)	66
Scarsdale, NY (Edgemont)	20	9,853	31,555	1,110/(98)	87
Wayne, PA (Radnor Township)	22	5,022	40,406	1,040/(95)	85
Weston, MA	21	4,680	39,800	1,031/(99)	89
Farmingdale, NY	22	6,729	45,846	947/(75)	81
Mamaroneck, NY	20	10,405	49,625	1,000/(90)	69
Mayfield, OH	24	5,881	36,228	1,003/(25)	48
Morristown, NJ	22	6,300	37,000	972/(80)	64
New Rochelle, NY	23	8,875	41,650	1,039/(80)	55
Newtown Square, PA (Marple–Newtown)	17	5,313	38,000	963/(75)	79
Omaha, NB (Westside)	23	4,815	32,500	1,059/(31)	81
Shaker Heights, OH	23	4,370	38,639	940/(56)	82

Let the dependent variable be the percentage of graduates attending a 4-year college.

a. Develop the best one-variable estimated regression equation.
b. Use the stepwise procedure to develop the best estimated regression equation.
c. Use the backward elimination procedure to develop the best estimated regression equation.
d. Use the best-subsets regression procedure to develop the best estimated regression equation.

17. Refer to the data in Exercise 12. Let the dependent variable be the number of wins.

a. Develop the best one-variable estimated regression equation.
b. Use the stepwise procedure to develop the best estimated regression equation.
c. Use the backward elimination procedure to develop the best estimated regression equation.
d. Use the best-subsets regression procedure to develop the best estimated regression equation.

18. The Ladies Professional Golfers Association (LPGA) maintains statistics on performance and earnings for members of the LPGA Tour. Year-end LPGA Tour statistics for 1997 follow (*Golfweek,* December 6, 1997). Scoring Avg. is the average score per 18 holes of golf; Driving Distance is the average number of yards per drive; Fairways is the percentage of drives landing in the fairway, Greens is the percentage of times a player was able to hit the green in regulation; Putts is the average number of putts per round; and Sand Saves is the percentage of times a player was able to get "up and down" once in a greenside sand bunker. A green is considered hit in regulation if any part of the ball is touching the putting surface and the number of strokes taken to hit the green is not more than the value of par minus two.

CD file
LPGATour

Player	Scoring Avg.	Driving Distance	Fairways	Greens	Putts	Sand Saves
Annika Sorenstam	70.04	249.00	75.4	73.2	29.67	47.3
Karrie Webb	70.00	254.60	72.1	75.1	30.20	38.0
Kelly Robbins	70.35	256.00	69.3	78.6	29.96	44.9
Chris Johnson	70.84	249.40	66.9	70.4	30.17	39.6
Tammie Green	71.24	239.40	70.4	68.6	29.64	47.2
Juli Inkster	70.64	251.60	69.0	69.8	29.35	46.0
Liselotte Neumann	71.28	243.50	66.6	65.9	29.36	48.4
Laura Davies	70.86	258.40	56.4	68.8	29.90	40.0
Nancy Lopez	70.70	245.50	73.4	73.0	30.23	43.1
Betsy King	71.52	245.70	65.9	66.9	29.63	37.6

Player	Scoring Avg.	Driving Distance	Fairways	Greens	Putts	Sand Saves
Lorie Kane	71.47	243.30	74.9	69.4	30.22	47.8
Michelle McGann	71.52	256.20	57.8	69.6	30.05	51.0
Donna Andrews	71.01	230.60	79.1	71.7	29.94	42.3
Colleen Walkeer	72.31	230.00	73.6	63.3	29.58	40.9
Rosie Jones	71.77	227.20	77.0	66.0	29.86	41.0
Lisa Hackney	71.34	246.00	69.8	65.1	29.91	36.4
Jane Geddes	71.34	261.20	64.5	70.0	30.15	44.2
Alison Nicholas	72.31	241.70	71.8	66.4	30.43	39.4
Pat Hurst	72.14	251.50	62.4	69.1	30.58	29.9
Cindy Figg-Currier	71.90	237.20	70.1	68.0	30.40	39.3

a. If the average score per 18 holes of golf is the dependent variable, what is the best one-variable estimated regression equation? What does the estimated regression equation suggest for players on the LPGA Tour?

b. Use the methods in this section to develop the best estimated multiple regression equation for estimating a player's average score.

c. Does the estimated regression equation developed in part (b) appear reasonable in terms of interpretation?

d. Michele Redman has the following data: Driving Distance 231.6 yards, Fairways 75.7%, Greens 65.2%, Putts 30.69, and Sand Saves 50.5%. Estimate the average score for this professional golfer.

19. A sample of 16 companies taken from the *Stock Investor Pro* database was used to obtain the following data on the price/earnings (P/E) ratio, the gross profit margin, and the sales growth for each company (*Stock Investor Pro*, American Association of Individual Investors, August 21, 1997). The data in the Industry column are codes used to define the industry for each company: 1 = energy-international oil; 2 = health-drugs; and 3 = other.

StkData

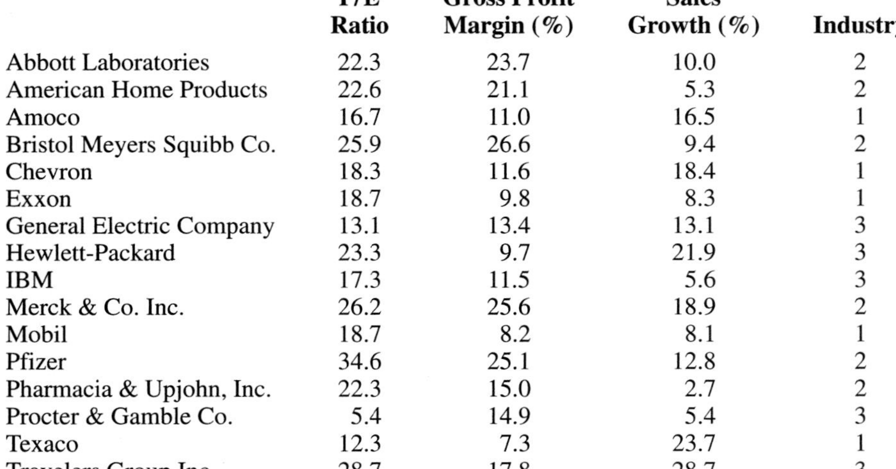

	P/E Ratio	Gross Profit Margin (%)	Sales Growth (%)	Industry
Abbott Laboratories	22.3	23.7	10.0	2
American Home Products	22.6	21.1	5.3	2
Amoco	16.7	11.0	16.5	1
Bristol Meyers Squibb Co.	25.9	26.6	9.4	2
Chevron	18.3	11.6	18.4	1
Exxon	18.7	9.8	8.3	1
General Electric Company	13.1	13.4	13.1	3
Hewlett-Packard	23.3	9.7	21.9	3
IBM	17.3	11.5	5.6	3
Merck & Co. Inc.	26.2	25.6	18.9	2
Mobil	18.7	8.2	8.1	1
Pfizer	34.6	25.1	12.8	2
Pharmacia & Upjohn, Inc.	22.3	15.0	2.7	2
Procter & Gamble Co.	5.4	14.9	5.4	3
Texaco	12.3	7.3	23.7	1
Travelers Group Inc.	28.7	17.8	28.7	3

Develop an estimated regression equation that can be used to predict price/earnings ratio. Briefly discuss the process you used to develop a recommended estimated regression equation for these data.

20. Refer to Exercise 14. Using age, blood pressure, whether a person is a smoker, and any interaction involving those variables, develop an estimated regression equation that can be used to predict risk. Briefly describe the process you used to develop an estimated regression equation for these data.

16.5 RESIDUAL ANALYSIS

In Chapters 14 and 15 we showed how residual plots can be used to detect violations of assumptions about the regression model. We looked for violations of assumptions about the error term ϵ and the assumed functional form of the model. Some of the actions that can be taken when such violations are detected have been discussed in this chapter. When a different functional form is needed, curvilinear and interaction terms can be included through the use of the general linear model. When several independent variables are considered, the variable selection procedures of the preceding section may be appropriate.

In Chapters 14 and 15 we also discussed how residual analysis can be used to identify observations that can be classified as outliers or as being influential in determining the estimated regression equation. Some steps that should be taken when such observations are found were noted. In many regression studies involving data collected over time, a special type of correlation among the error terms can cause problems; it is called **serial correlation** or **autocorrelation.** In this section we show how the **Durbin-Watson test** can be used to detect significant autocorrelation.

Autocorrelation and the Durbin-Watson Test

Often, the data used for regression studies in business and economics are collected over time. It is not uncommon for the value of y at time t, denoted by y_t, to be related to the value of y at previous time periods. In such cases, we say autocorrelation (also called serial correlation) is present in the data. If the value of y in time period t is related to its value in time period $t - 1$, first-order autocorrelation is present. If the value of y in time period t is related to the value of y in time period $t - 2$, second-order autocorrelation is present, and so on.

When autocorrelation is present, one of the assumptions of the regression model is violated: the error terms are not independent. In the case of first-order autocorrelation, the error at time t, denoted ϵ_t, will be related to the error at time period $t - 1$, denoted ϵ_{t-1}. Two cases of first-order autocorrelation are illustrated in Figure 16.18. Panel A is the case of positive autocorrelation; panel B is the case of negative autocorrelation. With positive autocorrelation we expect a positive residual in one period to be followed by a positive residual in the next period, a negative residual in one period to be followed by a negative residual in the next

FIGURE 16.18 TWO DATA SETS WITH FIRST-ORDER AUTOCORRELATION

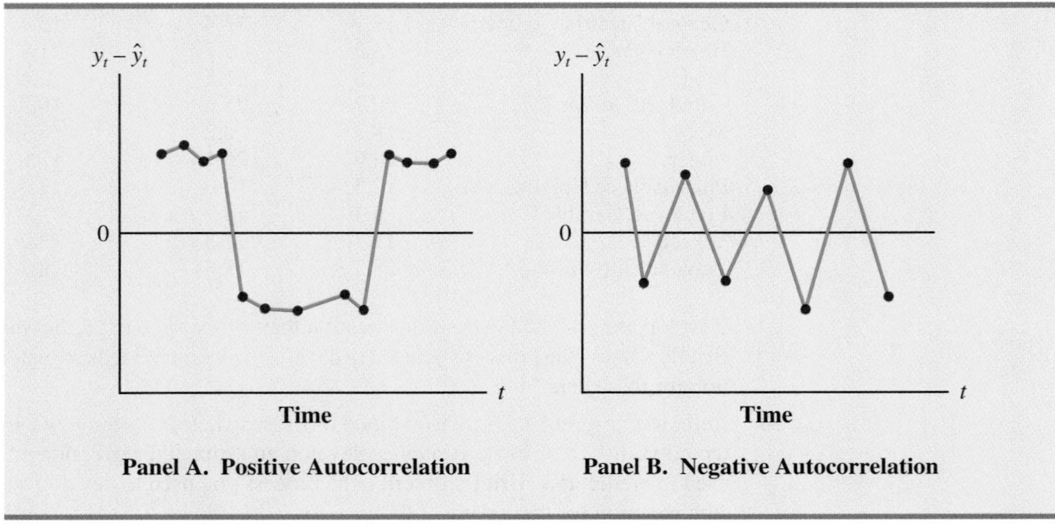

period, and so on. With negative autocorrelation, we expect a positive residual in one period to be followed by a negative residual in the next period, then a positive residual, and so on.

When autocorrelation is present, serious errors can be made in performing tests of statistical significance based upon the assumed regression model. It is therefore important to be able to detect autocorrelation and take corrective action. We will show how the Durbin-Watson statistic can be used to detect first-order autocorrelation.

Suppose the values of ϵ are not independent but are related in the following manner:

$$\epsilon_t = \rho\epsilon_{t-1} + z_t \tag{16.16}$$

where ρ is a parameter with an absolute value less than one and z_t is a normally and independently distributed random variable with a mean of zero and a variance of σ^2. From (16.16) we see that if $\rho = 0$, the error terms are not related, and each has a mean of zero and a variance of σ^2. In this case, there is no autocorrelation and the regression assumptions are satisfied. If $\rho > 0$, we have positive autocorrelation; if $\rho < 0$, we have negative autocorrelation. In either of these cases, the regression assumptions about the error term are violated.

The Durbin-Watson test for autocorrelation uses the residuals to determine whether $\rho = 0$. To simplify the notation for the Durbin-Watson statistic, we denote the ith residual by $e_i = y_i - \hat{y}_i$. The Durbin-Watson test statistic is computed as follows.

Durbin-Watson Test Statistic

$$d = \frac{\sum_{t=2}^{n}(e_t - e_{t-1})^2}{\sum_{t=1}^{n} e_t^2} \tag{16.17}$$

If successive values of the residuals are close together (positive autocorrelation), the value of the Durbin-Watson test statistic will be small. If successive values of the residuals are far apart (negative autocorrelation), the value of the Durbin-Watson statistic will be large.

The Durbin-Watson test statistic ranges in value from zero to four, with a value of two indicating no autocorrelation is present. Durbin and Watson have developed tables that can be used to determine when their test statistic indicates the presence of autocorrelation. Table 16.8 shows lower and upper bounds (d_L and d_U) for hypothesis tests using $\alpha = .05$, $\alpha = .025$, and $\alpha = .01$; n denotes the number of observations. The null hypothesis to be tested is always that there is no autocorrelation.

$$H_0\colon \rho = 0$$

The alternative hypothesis to test for positive autocorrelation is

$$H_a\colon \rho > 0$$

The alternative hypothesis to test for negative autocorrelation is

$$H_a\colon \rho < 0$$

A two-sided test is also possible. In this case the alternative hypothesis is

$$H_a\colon \rho \neq 0$$

TABLE 16.8 CRITICAL VALUES FOR THE DURBIN-WATSON TEST FOR AUTOCORRELATION

NOTE: Entries in the table are the critical values for a one-tailed Durbin-Watson test for autocorrelation. For a two-tailed test, the level of significance is doubled.

Significance Points of d_L and d_U: $\alpha = .05$

Number of Independent Variables

	1		2		3		4		5	
n	d_L	d_U	d_L	d_U	d_L	d_U	d_L	d_U	d_L	d_U
15	1.08	1.36	0.95	1.54	0.82	1.75	0.69	1.97	0.56	2.21
16	1.10	1.37	0.98	1.54	0.86	1.73	0.74	1.93	0.62	2.15
17	1.13	1.38	1.02	1.54	0.90	1.71	0.78	1.90	0.67	2.10
18	1.16	1.39	1.05	1.53	0.93	1.69	0.82	1.87	0.71	2.06
19	1.18	1.40	1.08	1.53	0.97	1.68	0.86	1.85	0.75	2.02
20	1.20	1.41	1.10	1.54	1.00	1.68	0.90	1.83	0.79	1.99
21	1.22	1.42	1.13	1.54	1.03	1.67	0.93	1.81	0.83	1.96
22	1.24	1.43	1.15	1.54	1.05	1.66	0.96	1.80	0.86	1.94
23	1.26	1.44	1.17	1.54	1.08	1.66	0.99	1.79	0.90	1.92
24	1.27	1.45	1.19	1.55	1.10	1.66	1.01	1.78	0.93	1.90
25	1.29	1.45	1.21	1.55	1.12	1.66	1.04	1.77	0.95	1.89
26	1.30	1.46	1.22	1.55	1.14	1.65	1.06	1.76	0.98	1.88
27	1.32	1.47	1.24	1.56	1.16	1.65	1.08	1.76	1.01	1.86
28	1.33	1.48	1.26	1.56	1.18	1.65	1.10	1.75	1.03	1.85
29	1.34	1.48	1.27	1.56	1.20	1.65	1.12	1.74	1.05	1.84
30	1.35	1.49	1.28	1.57	1.21	1.65	1.14	1.74	1.07	1.83
31	1.36	1.50	1.30	1.57	1.23	1.65	1.16	1.74	1.09	1.83
32	1.37	1.50	1.31	1.57	1.24	1.65	1.18	1.73	1.11	1.82
33	1.38	1.51	1.32	1.58	1.26	1.65	1.19	1.73	1.13	1.81
34	1.39	1.51	1.33	1.58	1.27	1.65	1.21	1.73	1.15	1.81
35	1.40	1.52	1.34	1.58	1.28	1.65	1.22	1.73	1.16	1.80
36	1.41	1.52	1.35	1.59	1.29	1.65	1.24	1.73	1.18	1.80
37	1.42	1.53	1.36	1.59	1.31	1.66	1.25	1.72	1.19	1.80
38	1.43	1.54	1.37	1.59	1.32	1.66	1.26	1.72	1.21	1.79
39	1.43	1.54	1.38	1.60	1.33	1.66	1.27	1.72	1.22	1.79
40	1.44	1.54	1.39	1.60	1.34	1.66	1.29	1.72	1.23	1.79
45	1.48	1.57	1.43	1.62	1.38	1.67	1.34	1.72	1.29	1.78
50	1.50	1.59	1.46	1.63	1.42	1.67	1.38	1.72	1.34	1.77
55	1.53	1.60	1.49	1.64	1.45	1.68	1.41	1.72	1.38	1.77
60	1.55	1.62	1.51	1.65	1.48	1.69	1.44	1.73	1.41	1.77
65	1.57	1.63	1.54	1.66	1.50	1.70	1.47	1.73	1.44	1.77
70	1.58	1.64	1.55	1.67	1.52	1.70	1.49	1.74	1.46	1.77
75	1.60	1.65	1.57	1.68	1.54	1.71	1.51	1.74	1.49	1.77
80	1.61	1.66	1.59	1.69	1.56	1.72	1.53	1.74	1.51	1.77
85	1.62	1.67	1.60	1.70	1.57	1.72	1.55	1.75	1.52	1.77
90	1.63	1.68	1.61	1.70	1.59	1.73	1.57	1.75	1.54	1.78
95	1.64	1.69	1.62	1.71	1.60	1.73	1.58	1.75	1.56	1.78
100	1.65	1.69	1.63	1.72	1.61	1.74	1.59	1.76	1.57	1.78

Continued

TABLE 16.8 (Continued)

Significance Points of d_L and d_U: $\alpha = .05$

Number of Independent Variables

	1		2		3		4		5	
n	d_L	d_U	d_L	d_U	d_L	d_U	d_L	d_U	d_L	d_U
15	0.95	1.23	0.83	1.40	0.71	1.61	0.59	1.84	0.48	2.09
16	0.98	1.24	0.86	1.40	0.75	1.59	0.64	1.80	0.53	2.03
17	1.01	1.25	0.90	1.40	0.79	1.58	0.68	1.77	0.57	1.98
18	1.03	1.26	0.93	1.40	0.82	1.56	0.72	1.74	0.62	1.93
19	1.06	1.28	0.96	1.41	0.86	1.55	0.76	1.72	0.66	1.90
20	1.08	1.28	0.99	1.41	0.89	1.55	0.79	1.70	0.70	1.87
21	1.10	1.30	1.01	1.41	0.92	1.54	0.83	1.69	0.73	1.84
22	1.12	1.31	1.04	1.42	0.95	1.54	0.86	1.68	0.77	1.82
23	1.14	1.32	1.06	1.42	0.97	1.54	0.89	1.67	0.80	1.80
24	1.16	1.33	1.08	1.43	1.00	1.54	0.91	1.66	0.83	1.79
25	1.18	1.34	1.10	1.43	1.02	1.54	0.94	1.65	0.86	1.77
26	1.19	1.35	1.12	1.44	1.04	1.54	0.96	1.65	0.88	1.76
27	1.21	1.36	1.13	1.44	1.06	1.54	0.99	1.64	0.91	1.75
28	1.22	1.37	1.15	1.45	1.08	1.54	1.01	1.64	0.93	1.74
29	1.24	1.38	1.17	1.45	1.10	1.54	1.03	1.63	0.96	1.73
30	1.25	1.38	1.18	1.46	1.12	1.54	1.05	1.63	0.98	1.73
31	1.26	1.39	1.20	1.47	1.13	1.55	1.07	1.63	1.00	1.72
32	1.27	1.40	1.21	1.47	1.15	1.55	1.08	1.63	1.02	1.71
33	1.28	1.41	1.22	1.48	1.16	1.55	1.10	1.63	1.04	1.71
34	1.29	1.41	1.24	1.48	1.17	1.55	1.12	1.63	1.06	1.70
35	1.30	1.42	1.25	1.48	1.19	1.55	1.13	1.63	1.07	1.70
36	1.31	1.43	1.26	1.49	1.20	1.56	1.15	1.63	1.09	1.70
37	1.32	1.43	1.27	1.49	1.21	1.56	1.16	1.62	1.10	1.70
38	1.33	1.44	1.28	1.50	1.23	1.56	1.17	1.62	1.12	1.70
39	1.34	1.44	1.29	1.50	1.24	1.56	1.19	1.63	1.13	1.69
40	1.35	1.45	1.30	1.51	1.25	1.57	1.20	1.63	1.15	1.69
45	1.39	1.48	1.34	1.53	1.30	1.58	1.25	1.63	1.21	1.69
50	1.42	1.50	1.38	1.54	1.34	1.59	1.30	1.64	1.26	1.69
55	1.45	1.52	1.41	1.56	1.37	1.60	1.33	1.64	1.30	1.69
60	1.47	1.54	1.44	1.57	1.40	1.61	1.37	1.65	1.33	1.69
65	1.49	1.55	1.46	1.59	1.43	1.62	1.40	1.66	1.36	1.69
70	1.51	1.57	1.48	1.60	1.45	1.63	1.42	1.66	1.39	1.70
75	1.53	1.58	1.50	1.61	1.47	1.64	1.45	1.67	1.42	1.70
80	1.54	1.59	1.52	1.62	1.49	1.65	1.47	1.67	1.44	1.70
85	1.56	1.60	1.53	1.63	1.51	1.65	1.49	1.68	1.46	1.71
90	1.57	1.61	1.55	1.64	1.53	1.66	1.50	1.69	1.48	1.71
95	1.58	1.62	1.56	1.65	1.54	1.67	1.52	1.69	1.50	1.71
100	1.59	1.63	1.57	1.65	1.55	1.67	1.53	1.70	1.51	1.72

Continued

TABLE 16.8 (Continued)

Significance Points of d_L and d_U: $\alpha = .05$

Number of Independent Variables

	1		2		3		4		5	
n	d_L	d_U	d_L	d_U	d_L	d_U	d_L	d_U	d_L	d_U
15	0.81	1.07	0.70	1.25	0.59	1.46	0.49	1.70	0.39	1.96
16	0.84	1.09	0.74	1.25	0.63	1.44	0.53	1.66	0.44	1.90
17	0.87	1.10	0.77	1.25	0.67	1.43	0.57	1.63	0.48	1.85
18	0.90	1.12	0.80	1.26	0.71	1.42	0.61	1.60	0.52	1.80
19	0.93	1.13	0.83	1.26	0.74	1.41	0.65	1.58	0.56	1.77
20	0.95	1.15	0.86	1.27	0.77	1.41	0.68	1.57	0.60	1.74
21	0.97	1.16	0.89	1.27	0.80	1.41	0.72	1.55	0.63	1.71
22	1.00	1.17	0.91	1.28	0.83	1.40	0.75	1.54	0.66	1.69
23	1.02	1.19	0.94	1.29	0.86	1.40	0.77	1.53	0.70	1.67
24	1.04	1.20	0.96	1.30	0.88	1.41	0.80	1.53	0.72	1.66
25	1.05	1.21	0.98	1.30	0.90	1.41	0.83	1.52	0.75	1.65
26	1.07	1.22	1.00	1.31	0.93	1.41	0.85	1.52	0.78	1.64
27	1.09	1.23	1.02	1.32	0.95	1.41	0.88	1.51	0.81	1.63
28	1.10	1.24	1.04	1.32	0.97	1.41	0.90	1.51	0.83	1.62
29	1.12	1.25	1.05	1.33	0.99	1.42	0.92	1.51	0.85	1.61
30	1.13	1.26	1.07	1.34	1.01	1.42	0.94	1.51	0.88	1.61
31	1.15	1.27	1.08	1.34	1.02	1.42	0.96	1.51	0.90	1.60
32	1.16	1.28	1.10	1.35	1.04	1.43	0.98	1.51	0.92	1.60
33	1.17	1.29	1.11	1.36	1.05	1.43	1.00	1.51	0.94	1.59
34	1.18	1.30	1.13	1.36	1.07	1.43	1.01	1.51	0.95	1.59
35	1.19	1.31	1.14	1.37	1.08	1.44	1.03	1.51	0.97	1.59
36	1.21	1.32	1.15	1.38	1.10	1.44	1.04	1.51	0.99	1.59
37	1.22	1.32	1.16	1.38	1.11	1.45	1.06	1.51	1.00	1.59
38	1.23	1.33	1.18	1.39	1.12	1.45	1.07	1.52	1.02	1.58
39	1.24	1.34	1.19	1.39	1.14	1.45	1.09	1.52	1.03	1.58
40	1.25	1.34	1.20	1.40	1.15	1.46	1.10	1.52	1.05	1.58
45	1.29	1.38	1.24	1.42	1.20	1.48	1.16	1.53	1.11	1.58
50	1.32	1.40	1.28	1.45	1.24	1.49	1.20	1.54	1.16	1.59
55	1.36	1.43	1.32	1.47	1.28	1.51	1.25	1.55	1.21	1.59
60	1.38	1.45	1.35	1.48	1.32	1.52	1.28	1.56	1.25	1.60
65	1.41	1.47	1.38	1.50	1.35	1.53	1.31	1.57	1.28	1.61
70	1.43	1.49	1.40	1.52	1.37	1.55	1.34	1.58	1.31	1.61
75	1.45	1.50	1.42	1.53	1.39	1.56	1.37	1.59	1.34	1.62
80	1.47	1.52	1.44	1.54	1.42	1.57	1.39	1.60	1.36	1.62
85	1.48	1.53	1.46	1.55	1.43	1.58	1.41	1.60	1.39	1.63
90	1.50	1.54	1.47	1.56	1.45	1.59	1.43	1.61	1.41	1.64
95	1.51	1.55	1.49	1.57	1.47	1.60	1.45	1.62	1.42	1.64
100	1.52	1.56	1.50	1.58	1.48	1.60	1.46	1.63	1.44	1.65

Source: J. Durbin and G. S. Watson, "Testing for Serial Correlation in Least Squares Regression II," *Biometrika,* 38 (1951), 159–178.

Figure 16.19 shows how the values of d_L and d_U in Table 16.8 are used to test for autocorrelation. Panel A illustrates the test for positive autocorrelation. If $d < d_L$, we conclude that positive autocorrelation is present. If $d_L \leq d \leq d_U$, we say the test is inconclusive. If $d > d_U$, we conclude that there is no evidence of positive autocorrelation.

Panel B illustrates the test for negative autocorrelation. If $d > 4 - d_L$, we conclude that negative autocorrelation is present. If $4 - d_U \leq d \leq 4 - d_L$, we say the test is inconclusive. If $d < 4 - d_U$, we conclude that there is no evidence of negative autocorrelation.

Panel C illustrates the two-sided test. If $d < d_L$ or $d > 4 - d_L$, we reject H_0 and conclude that autocorrelation is present. If $d_L \leq d \leq d_U$ or $4 - d_U \leq d \leq 4 - d_L$, we say the test is inconclusive. If $d_U < d < 4 - d_U$, we conclude that there is no evidence of autocorrelation.

If significant autocorrelation is identified, we should investigate whether we have omitted one or more key independent variables that have time-ordered effects on the dependent variable. If no such variables can be identified, including an independent variable that measures the time of the observation (for instance, the value of this variable could be one for the first observation, two for the second observation, and so on) will sometimes eliminate or reduce the autocorrelation. When these attempts to reduce or remove autocorrelation do not work, transformations on the dependent or independent variables can

FIGURE 16.19 HYPOTHESIS TEST FOR AUTOCORRELATION USING THE DURBIN-WATSON TEST

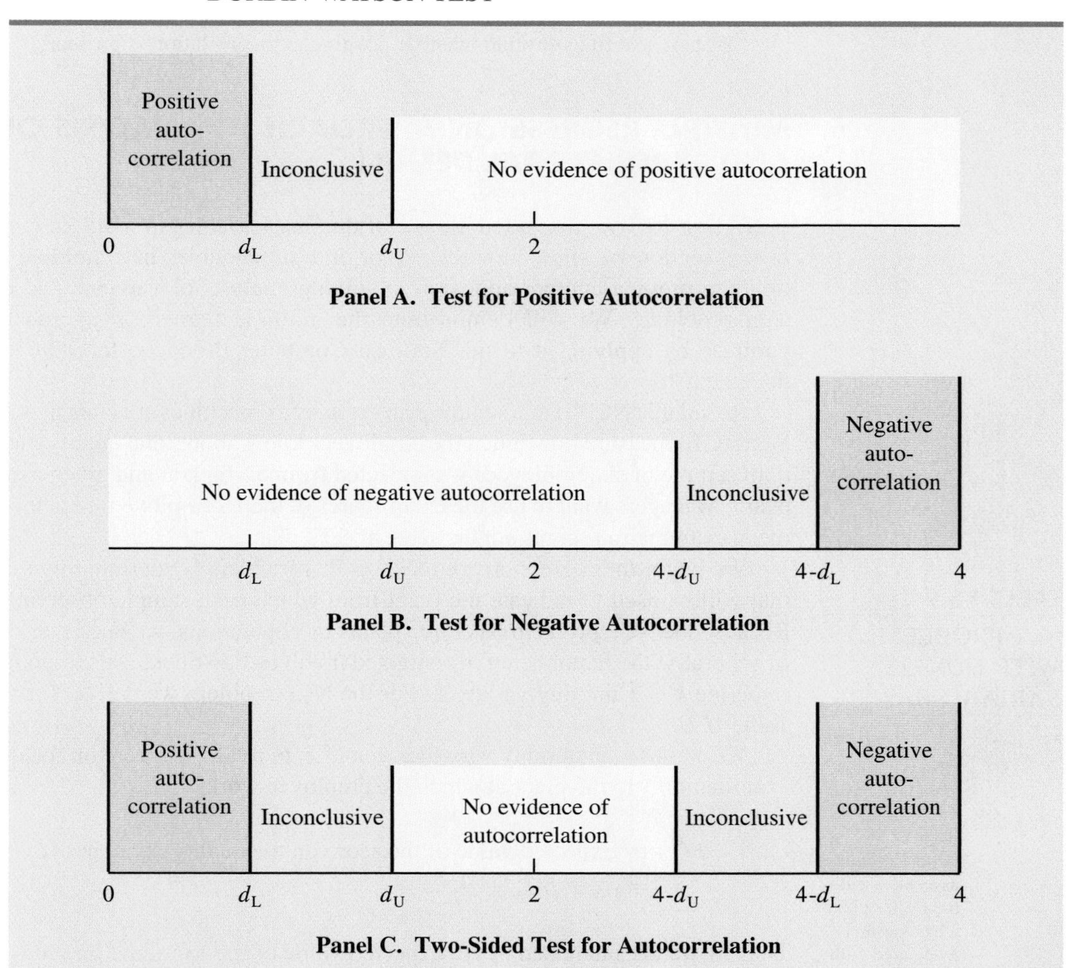

prove helpful; a discussion of such transformations can be found in more advanced texts on regression analysis.

Note that the Durbin-Watson tables list the smallest sample size as 15. The reason is that the test is generally inconclusive for smaller sample sizes; in fact, many statisticians believe the sample size should be at least 50 for the test to produce worthwhile results.

EXERCISES

Applications

21. Consider the data set in Exercise 19.
    a. Develop the estimated regression equation that can be used to predict the price/earnings ratio given the profit margin.
    b. Plot the residuals obtained from the estimated regression equation developed in part (a) as a function of the order in which the data are presented. Does any autocorrelation appear to be present in the data? Explain.
    c. At the .05 level of significance, test for any positive autocorrelation in the data.
22. Refer to the Cravens data set in Table 16.6. In Section 16.3 we showed that the estimated regression equation involving Accounts, AdvExp, Poten, and Share had an adjusted coefficient of determination of 88.1%. Use the .05 level of significance and apply the Durbin-Watson test to determine whether positive autocorrelation is present.

16.6 MULTIPLE REGRESSION APPROACH TO ANALYSIS OF VARIANCE AND EXPERIMENTAL DESIGN

In Section 15.7 we discussed the use of dummy variables in multiple regression analysis. In this section we show how the use of dummy variables in a multiple regression equation can provide another approach to solving analysis of variance and experimental design problems. We will demonstrate the multiple regression approach to analysis of variance by applying it to the National Computer Products, Inc. (NCP) problem introduced in Chapter 13.

Recall that NCP manufactures printers and fax machines at plants in Atlanta, Dallas, and Seattle. To measure how much the employees know about total quality management, a random sample of six employees was selected from each plant and given a quality-awareness exam. Managers want to use the exam scores of the 18 employees to determine whether the mean examination scores are the same at each plant.

We begin the regression approach to this problem by defining two dummy variables that will be used to indicate the plant from which each sample observation was selected. Because the NCP problem has three plants or populations, we need two dummy variables. In general, if the factor being investigated involves k distinct levels or populations, we need to define $k - 1$ dummy variables. For the NCP problem we define x_1 and x_2 as shown in Table 16.9.

TABLE 16.9 NCP PROBLEM WITH DUMMY VARIABLES

x_1	x_2	
0	0	Observation is associated with the Atlanta plant
1	0	Observation is associated with the Dallas plant
0	1	Observation is associated with the Seattle plant

We can use the dummy variables x_1 and x_2 to relate the score on the quality-awareness examination y to the plant at which the employee works.

$$E(y) = \text{Expected value of the score on the quality-awareness examination}$$
$$= \beta_0 + \beta_1 x_1 + \beta_2 x_2$$

Thus, if we are interested in the expected value of the examination score for an employee who works at the Atlanta plant, our procedure for assigning numerical values to the dummy variables x_1 and x_2 would result in setting $x_1 = x_2 = 0$. The multiple regression equation then reduces to

$$E(y) = \beta_0 + \beta_1(0) + \beta_2(0) = \beta_0$$

We can interpret β_0 as the expected value of the examination score for employees who work at the Atlanta plant.

Next let us consider the forms of the multiple regression equation for each of the other plants. For the Dallas plant, $x_1 = 1$ and $x_2 = 0$, and

$$E(y) = \beta_0 + \beta_1(1) + \beta_2(0) = \beta_0 + \beta_1$$

For the Seattle plant, $x_1 = 0$ and $x_2 = 1$, and

$$E(y) = \beta_0 + \beta_1(0) + \beta_2(1) = \beta_0 + \beta_2$$

We see that $\beta_0 + \beta_1$ represents the expected value of the examination score for employees at the Dallas plant, and $\beta_0 + \beta_2$ represents the expected value of the examination score for employees at the Seattle plant.

We now want to estimate the coefficients β_0, β_1, and β_2 and hence develop an estimate of the expected value of the examination score for each plant. The sample data consisting of 18 observations of x_1, x_2, and y were entered into Minitab. The actual input data and the output from Minitab are in Table 16.10 and Figure 16.20, respectively.

In Figure 16.20, we see that the estimates of β_0, β_1, and β_2 are $b_0 = 79$, $b_1 = -5$, and $b_2 = -13$. Thus, the best estimate of the expected value of the examination score for each plant is as follows.

Plant	**Estimate of $E(y)$**
Atlanta	$b_0 = 79$
Dallas	$b_0 + b_1 = 79 - 5 = 74$
Seattle	$b_0 + b_2 = 79 - 13 = 66$

Note that the best estimate of the expected value of the examination score for each plant obtained from the regression analysis is the same as the sample mean found previously by applying the ANOVA procedure.

TABLE 16.10 INPUT DATA FOR THE NCP PROBLEM

Atlanta			Dallas			Seattle		
x_1	x_2	y	x_1	x_2	y	x_1	x_2	y
0	0	85	1	0	71	0	1	59
0	0	75	1	0	75	0	1	64
0	0	82	1	0	73	0	1	62
0	0	76	1	0	74	0	1	69
0	0	71	1	0	69	0	1	75
0	0	85	1	0	82	0	1	67

FIGURE 16.20 MULTIPLE REGRESSION OUTPUT FOR THE NCP PROBLEM

```
The regression equation is
Y = 79.0 - 5.00 X1 - 13.0 X2

Predictor         Coef        Stdev      t-ratio        p
Constant        79.000        2.186        36.14    0.000
X1              -5.000        3.091        -1.62    0.127
X2             -13.000        3.091        -4.21    0.001

s = 5.354         R-sq = 54.5%       R-sq(adj) = 48.5%

Analysis of Variance

SOURCE           DF          SS          MS          F         p
Regression        2      516.00      258.00       9.00     0.003
Error            15      430.00       28.67
Total            17      946.00
```

Now let us see how we can use the output from the multiple regression package to perform the ANOVA test on the difference in the means for the three plants. First, we observe that if there is no difference in the means,

$$E(y) \text{ for the Dallas plant } - E(y) \text{ for the Atlanta plant } = 0$$
$$E(y) \text{ for the Seattle plant } - E(y) \text{ for the Atlanta plant } = 0$$

Because β_0 equals $E(y)$ for the Atlanta plant and $\beta_0 + \beta_1$ equals $E(y)$ for the Dallas plant, the first difference is equal to $(\beta_0 + \beta_1) - \beta_0 = \beta_1$. Moreover, since $\beta_0 + \beta_2$ equals $E(y)$ for the Seattle plant, the second difference is equal to $(\beta_0 + \beta_2) - \beta_0 = \beta_2$. We would conclude that there is no difference in the three means if $\beta_1 = 0$ and $\beta_2 = 0$. Hence, the null hypothesis for a test for difference of means can be stated as

$$H_0\colon \beta_1 = \beta_2 = 0$$

Recall that to test this type of null hypothesis about the significance of the regression relationship, we must compare the value of MSR/MSE to the critical value from an F distribution with numerator and denominator degrees of freedom equal to the degrees of freedom for the regression sum of squares and the error sum of squares, respectively. In the current problem, the regression sum of squares has two degrees of freedom and the error sum of squares has 15 degrees of freedom. Thus, we obtain the following values for MSR and MSE.

$$\text{MSR} = \frac{\text{SSR}}{2} = \frac{516}{2} = 258$$
$$\text{MSE} = \frac{\text{SSE}}{15} = \frac{430}{15} = 28.67$$

Hence, the computed F value is

$$F = \frac{\text{MSR}}{\text{MSE}} = \frac{258}{28.67} = 9.00$$

At a .05 level of significance, the critical value of F with two numerator degrees of freedom and 15 denominator degrees of freedom is 3.68. Because the observed value of F is greater than the critical value of 3.68, we reject the null hypothesis H_0: $\beta_1 = \beta_2 = 0$ and conclude that the means for the three plants are different. Alternatively, using the p-value approach we also reject H_0: $\beta_1 = \beta_2 = 0$ because p-value $= .003 < \alpha = .05$.

EXERCISES

Methods

23. Consider a completely randomized design involving four treatments: A, B, C, and D. Write a multiple regression equation that can be used to analyze these data. Define all variables.
24. Write a multiple regression equation that can be used to analyze the data for a randomized block design involving three treatments and two blocks. Define all variables.
25. Write a multiple regression equation that can be used to analyze the data for a two-factorial design with two levels for factor A and three levels for factor B. Define all variables.

Applications

26. The Jacobs Chemical Company wants to estimate the mean time (minutes) required to mix a batch of material on machines produced by three different manufacturers. To limit the cost of testing, four batches of material were mixed on machines produced by each of the three manufacturers. The times needed to mix the material follow.

Manufacturer 1	Manufacturer 2	Manufacturer 3
20	28	20
26	26	19
24	31	23
22	27	22

   a. Write a multiple regression equation that can be used to analyze the data.
   b. What are the best estimates of the coefficients in your regression equation?
   c. In terms of the regression equation coefficients, what hypotheses must we test to see whether the mean time to mix a batch of material is the same for all three manufacturers?
   d. For the $\alpha = .05$ level of significance, what conclusion should be drawn?

27. Four different paints are advertised as having the same drying time. To check the manufacturers' claims, five samples were tested for each of the paints. The time in minutes until the paint was dry enough for a second coat to be applied was recorded for each sample. The data obtained follow.

Paint 1	Paint 2	Paint 3	Paint 4
128	144	133	150
137	133	143	142
135	142	137	135
124	146	136	140
141	130	131	153

   a. Using $\alpha = .05$, test for any significant differences in mean drying time among the paints.
   b. What is your estimate of mean drying time for paint 2? How is it obtained from the computer output?

28. An automobile dealer conducted a test to determine if the time needed to complete a minor engine tune-up depends on whether a computerized engine analyzer or an electronic analyzer is used. Because tune-up time varies among compact, intermediate, and full-sized cars, the three types of cars were used as blocks in the experiment. The data (time in minutes) obtained follow.

		Car		
		Compact	**Intermediate**	**Full Size**
Analyzer	**Computerized**	50	55	63
	Electronic	42	44	46

Using $\alpha = .05$, test for any significant differences.

29. A mail-order catalog firm designed a factorial experiment to test the effect of the size of a magazine advertisement and the advertisement design on the number of catalog requests received (1000s). Three advertising designs and two sizes of advertisements were considered. The following data were obtained. Test for any significant effects due to type of design, size of advertisement, or interaction. Use $\alpha = .05$.

		Size of Advertisement	
		Small	**Large**
Design	**A**	8 12	12 8
	B	22 14	26 30
	C	10 18	18 14

SUMMARY

In this chapter we discussed several concepts used by model builders in identifying the best estimated regression equation. First, we introduced the concept of a general linear model to show how the methods discussed in Chapters 14 and 15 could be extended to handle curvilinear relationships and interaction effects. Then we discussed how transformations involving the dependent variable could be used to account for problems such as nonconstant variance in the error term.

In many applications of regression analysis, a large number of independent variables are considered. We presented a general approach based on an F statistic for adding or deleting variables from a regression model. We then introduced a larger problem involving 25 observations and eight independent variables. We saw that one issue encountered in solving larger problems is finding the best subset of the independent variables. To help in that task, we discussed several variable selection procedures: stepwise regression, forward selection, backward elimination, and best-subsets regression.

In Section 16.5, we extended the applications of residual analysis to show the Durbin-Watson test for autocorrelation. The chapter concluded with a discussion of how multiple regression models could be developed to provide another approach for solving analysis of variance and experimental design problems.

GLOSSARY

General linear model A model of the form $y = \beta_0 + \beta_1 z_1 + \beta_2 z_2 + \cdots + \beta_p z_p + \epsilon$, where each of the independent variables z_j, $j = 1, 2, \ldots, p$, is a function of $x_1, x_2, \ldots, x_k$, the variables for which data have been collected.

Interaction The effect of two independent variables acting together.

Variable selection procedures Methods for selecting a subset of the independent variables for a regression model.

Autocorrelation Correlation in the errors that arises when the error terms at successive points in time are related.

Serial correlation Same as autocorrelation.

Durbin-Watson test A test to determine whether first-order autocorrelation is present.

KEY FORMULAS

General Linear Model

$$y = \beta_0 + \beta_1 z_1 + \beta_2 z_2 + \cdots + \beta_p z_p + \epsilon \tag{16.1}$$

General F Test for Adding or Deleting $p - q$ Variables

$$F = \frac{\dfrac{\text{SSE}(x_1, x_2, \ldots, x_q) - \text{SSE}(x_1, x_2, \ldots, x_q, x_{q+1}, \ldots, x_p)}{p - q}}{\dfrac{\text{SSE}(x_1, x_2, \ldots, x_q, x_{q+1}, \ldots, x_p)}{n - p - 1}} \tag{16.13}$$

Autocorrelated Error Terms

$$\epsilon_t = \rho\epsilon_{t-1} + z_t \tag{16.16}$$

Durbin-Watson Statistic

$$d = \frac{\sum_{t=2}^{n}(e_t - e_{t-1})^2}{\sum_{t=1}^{n} e_t^2} \tag{16.17}$$

SUPPLEMENTARY EXERCISES

30. Many international funds offer more reasonable equity valuations than those found in the United States. Because international markets often move in different directions than the U.S. market, investments in foreign markets can also reduce an investor's overall risk. The following table shows the fund type (load or no-load), expense ratio (%), safety rating (0 = riskiest, 10 = safest), and the one-year performance through December 10, 1999, for 20 international funds (*Mutual Funds,* February 2000).

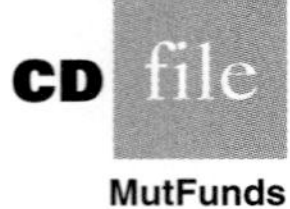
MutFunds

	Fund Type	Expense Ratio (%)	Safety Rating	Performance (%)
ABN AMRO Int'l Equity "Com"	No-load	1.38	6.9	36
Accessor Int'l Equity "Adv"	No-load	1.59	7.1	42
Artisan International	No-load	1.45	6.8	72
Columbia Int'l Stock	No-load	1.56	7.1	54
Concert Inv. "A" Int'l Equity	Load	2.16	6.3	116
Diversified Invstr Int'l Eqty	No-load	1.40	7.3	54
Driehaus Int'l Growth	No-load	1.88	6.5	92
Founders Passport	No-load	1.52	7.0	86
Guardian Baillie Fifford Int'l "A"	Load	1.62	7.1	37
Jamestown Int'l Equity	No-load	1.56	7.1	35
Julius Baer Int'l Equity	No-load	1.79	6.9	71
Aetna "I" Int'l	No-load	1.35	7.3	46
Pilgrim Int'l Value "A"	Load	1.80	7.1	42
Fidelity Diversified Int'l	No-load	1.48	7.5	42
Putnam "A" Int'l Growth	Load	1.59	6.9	55
Sit Int'l Growth	No-load	1.50	6.9	49
Touchstone Int'l Equity "A"	Load	1.60	7.5	35
United Int'l Growth "A"	Load	1.28	7.1	47
Vontobel Int'l Equity	No-load	1.50	7.0	43
Waddell & Reed Int'l Growth "B"	Load	2.46	7.0	75

a. Use the methods in this chapter to develop an estimated regression equation that can be used to estimate the performance of a fund on the basis of the data provided.

b. Did the estimated regression equation developed in part (a) provide a good fit? Explain.

c. Acorn International is a no-load fund that has an annual expense ratio of 1.12% and a safety rating of 7.6. Use the estimated regression equation developed in part (a) to estimate the one-year performance for Acorn International.

31. A study investigated the relationship between audit delay (Delay), the length of time from a company's fiscal year-end to the date of the auditor's report, and variables that describe the client and the auditor. Some of the independent variables that were included in this study follow.

Industry	A dummy variable coded 1 if the firm was an industrial company or 0 if the firm was a bank, savings and loan, or insurance company.
Public	A dummy variable coded 1 if the company was traded on an organized exchange or over the counter; otherwise coded 0.
Quality	A measure of overall quality of internal controls, as judged by the auditor, on a five-point scale ranging from "virtually none" (1) to "excellent" (5).
Finished	A measure ranging from 1 to 4, as judged by the auditor, where 1 indicates "all work performed subsequent to year-end" and 4 indicates "most work performed prior to year-end."

Suppose that in a similar study a sample of 40 companies provided the following data.

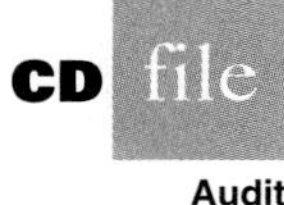
Audit

Delay	Industry	Public	Quality	Finished
62	0	0	3	1
45	0	1	3	3
54	0	0	2	2
71	0	1	1	2
91	0	0	1	1

Delay	Industry	Public	Quality	Finished
62	0	0	4	4
61	0	0	3	2
69	0	1	5	2
80	0	0	1	1
52	0	0	5	3
47	0	0	3	2
65	0	1	2	3
60	0	0	1	3
81	1	0	1	2
73	1	0	2	2
89	1	0	2	1
71	1	0	5	4
76	1	0	2	2
68	1	0	1	2
68	1	0	5	2
86	1	0	2	2
76	1	1	3	1
67	1	0	2	3
57	1	0	4	2
55	1	1	3	2
54	1	0	5	2
69	1	0	3	3
82	1	0	5	1
94	1	0	1	1
74	1	1	5	2
75	1	1	4	3
69	1	0	2	2
71	1	0	4	4
79	1	0	5	2
80	1	0	1	4
91	1	0	4	1
92	1	0	1	4
46	1	1	4	3
72	1	0	5	2
85	1	0	5	1

a. Develop the estimated regression equation using all of the independent variables.

b. Did the estimated regression equation developed in part (a) provide a good fit? Explain.

c. Develop a scatter diagram showing Delay as a function of Finished. What does this scatter diagram indicate about the relationship between Delay and Finished?

d. On the basis of your observations about the relationship between Delay and Finished, develop an alternative estimated regression equation to the one developed in (a) to explain as much of the variability in Delay as possible.

32. Refer to the data in Exercise 31. Consider a model in which only Industry is used to predict Delay. At a .01 level of significance, test for any positive autocorrelation in the data.

33. Refer to the data in Exercise 31.

a. Develop an estimated regression equation that can be used to predict Delay by using Industry and Quality.

b. Plot the residuals obtained from the estimated regression equation developed in part (a) as a function of the order in which the data are presented. Does any autocorrelation appear to be present in the data? Explain.

c. At the .05 level of significance, test for any positive autocorrelation in the data.

34. A study was conducted to investigate browsing activity by shoppers. Shoppers were classified as nonbrowsers, light browsers, and heavy browsers. For each shopper in the study, a measure was obtained to determine how comfortable the shopper was in the store. Higher scores indicated greater comfort. Assume that the following data are from this study. Use a .05 level of significance to test for differences in comfort levels among the three types of browsers.

Browsing

Nonbrowser	Light Browser	Heavy Browser
4	5	5
5	6	7
6	5	5
3	4	7
3	7	4
4	4	6
5	6	5
4	5	7

35. The following data show the percentage changes in the Dow Jones Industrial Average (DJIA) in each of the four years of eight presidential terms (*1998 Stock Trader's Almanac*). Use regression analysis to determine what effect party and year of term in office have on the change in the Dow Jones Industrial Average.

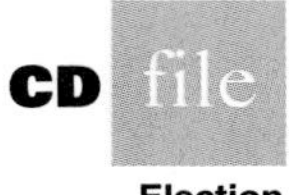

Election

President	Party	Year Elected	Year of Term	Year's Change in DJIA
Johnson	Democrat	1964	First	10.9
Johnson	Democrat		Second	−18.9
Johnson	Democrat		Third	15.2
Johnson	Democrat		Fourth	4.3
Nixon	Republican	1968	First	−15.2
Nixon	Republican		Second	4.8
Nixon	Republican		Third	6.1
Nixon	Republican		Fourth	14.6
Nixon	Republican	1972	First	−16.6
Nixon	Republican		Second	−27.6
Nixon*	Republican		Third	38.3
Nixon*	Republican		Fourth	17.9
Carter	Democrat	1976	First	−17.3
Carter	Democrat		Second	−3.1
Carter	Democrat		Third	4.2
Carter	Democrat		Fourth	14.9
Reagan	Republican	1980	First	−9.2
Reagan	Republican		Second	19.6
Reagan	Republican		Third	20.3
Reagan	Republican		Fourth	−3.7
Reagan	Republican	1984	First	−27.7
Reagan	Republican		Second	22.6
Reagan	Republican		Third	2.3
Reagan	Republican		Fourth	11.8
Bush	Republican	1988	First	27.0
Bush	Republican		Second	−4.3
Bush	Republican		Third	20.3
Bush	Republican		Fourth	4.2
Clinton	Democrat	1992	First	13.7
Clinton	Democrat		Second	2.1
Clinton	Democrat		Third	33.5
Clinton	Democrat		Fourth	26.0

*Because President Nixon resigned from office in August 1974, Gerald Ford became president and completed the remainder of Nixon's term in office.

Case Problem 1 UNEMPLOYMENT STUDY

A study provided data on variables that may be related to the number of weeks a manufacturing worker has been jobless. The dependent variable in the study (Weeks) was defined as the number of weeks a worker has been jobless due to a layoff. The following independent variables were used in the study.

Age	The age of the worker
Educ	The number of years of education
Married	A dummy variable: 1 if married, 0 otherwise
Head	A dummy variable: 1 if the head of household, 0 otherwise
Tenure	The number of years on the previous job
Manager	A dummy variable: 1 if management occupation, 0 otherwise
Sales	A dummy variable: 1 if sales occupation, 0 otherwise

Assume the data below were collected for 50 displaced workers. These data are available on the data disk in the data set named Layoffs.

Managerial Report

Use the methods presented in this and previous chapters to analyze this data set. Present a summary of your analysis, including key statistical results, conclusions, and recommendations, in a managerial report. Include any appropriate technical material (computer output, residual plots, etc.) in an appendix.

CD file
Layoffs

Weeks	Age	Educ	Married	Head	Tenure	Manager	Sales
37	30	14	1	1	1	0	0
62	27	14	1	0	6	0	0
49	32	10	0	1	11	0	0
73	44	11	1	0	2	0	0
8	21	14	1	1	2	0	0
15	26	13	1	0	7	1	0
52	26	15	1	0	6	0	0
72	33	13	0	1	6	0	0
11	27	12	1	1	8	0	0
13	33	12	0	1	2	0	0
39	20	11	1	0	1	0	0
59	35	7	1	1	6	0	0
39	36	17	0	1	9	1	0
44	26	12	1	1	8	0	0
56	36	15	0	1	8	0	0
31	38	16	1	1	11	0	1
62	34	13	0	1	13	0	0
25	27	19	1	0	8	0	0
72	44	13	1	0	22	0	0
65	45	15	1	1	6	0	0
44	28	17	0	1	3	0	1
49	25	10	1	1	1	0	0
80	31	15	1	0	12	0	0
7	23	15	1	0	2	0	0

Continued

Weeks	Age	Educ	Married	Head	Tenure	Manager	Sales
14	24	13	1	1	7	0	0
94	62	13	0	1	8	0	0
48	31	16	1	0	11	0	0
82	48	18	0	1	30	0	0
50	35	18	1	1	5	0	0
37	33	14	0	1	6	0	1
62	46	15	0	1	6	0	0
37	35	8	0	1	6	0	0
40	32	9	1	1	13	0	0
16	40	17	1	0	8	1	0
34	23	12	1	1	1	0	0
4	36	16	0	1	8	0	1
55	33	12	1	0	10	0	1
39	32	16	0	1	11	0	0
80	62	15	1	0	16	0	1
19	29	14	1	1	12	0	0
98	45	12	1	0	17	0	0
30	38	15	0	1	6	0	1
22	40	8	1	1	16	0	1
57	42	13	1	0	2	1	0
64	45	16	1	1	22	0	0
22	39	11	1	1	4	0	0
27	27	15	1	0	10	0	1
20	42	14	1	1	6	1	0
30	31	10	1	1	8	0	0
23	33	13	1	1	8	0	0

Case Problem 2 ANALYSIS OF PGA TOUR STATISTICS

The Professional Golfers Association (PGA) maintains statistics on performance and earnings for members of the PGA Tour. Year-end performance statistics for 1997 appear on the data disk in the data set named PGA Tour (*Golfweek,* November 15, 1997). Each row of the data set corresponds to a PGA Tour player. Descriptions for data, which appear on the disk, follow. In these descriptions a round refers to 18 holes of golf. PGA Tour events usually consist of four rounds of golf played over four days. Scores obtained for rounds played on Thursday and Friday of a Tour event are used to limit the size of the field for the final two rounds played on Saturday and Sunday. Players who achieve a low enough total score on Thursday and Friday to enable them to play on Saturday and Sunday, are said to have made the cut.

CD 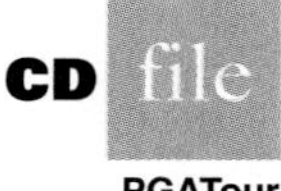

PGATour

Earnings	Total earnings in PGA Tour events
Events	Number of PGA Tour events entered
Rds	Number of rounds completed
Cuts Made	Number of events in which the player makes the cut
Rds in 60s	Number of rounds with a score in the 60s
Rds Under Par	Number of rounds with a score less than par
Birdies	Average number of birdies per round
Scoring Avg	Average score for all events

Driving Distance	Average number of yards per drive (The length of a drive is measured to the point the drive comes to rest, regardless of whether it is in the fairway. Driving distance is measured on two holes per round.)
Fairways	Average number of times a player is able to hit the fairway with his tee shot in a round
Greens	Average number of times a player is able to hit the green in regulation (A green is considered hit in regulation if any part of the ball is touching the putting surface and the difference between the value of par for the hole and the number of strokes taken to hit the green is at least 2.)
Putts	Average number of putts taken on greens hit in regulation
Sand Saves	Percent of time a player is able to get "up and down" once in a greenside sand bunker (Getting up and down means that the number of strokes taken to hit the ball out of the sand and into the hole is at most two.)

Managerial Report

Suppose that you have been hired by the commissioner of the PGA Tour to analyze the data for a presentation to be made at the annual PGA Tour dinner. The commissioner has asked whether it would be possible to use these data to determine what performance measures, such as Cuts Made, Driving Distance, Fairways, and so on, have the most effect on a player's score and amount of money earned. Use the methods presented in this and previous chapters to analyze the data. Prepare a report for the PGA Tour commissioner that summarizes your analysis, including key statistical results, conclusions, and recommendations. Include any appropriate technical material in an appendix.

Case Problem 3 PREDICTING GRADUATION RATES FOR COLLEGES AND UNIVERSITIES

The percentage of students who enroll at a college or university and actually graduate is an important statistic for university administrators. Some of the factors related to the graduation rate include the percentage of classes with fewer than 20 students, the percentage of classes with more than 50 students, the student-faculty ratio, the percentage of students who apply to the university and are admitted, the percentage of first-year students in the top 10% of their high school class, and the academic reputation of the university. To study the effect of these factors on the graduation rate, data for 48 national universities was collected (*America's Best Colleges,* Year 2000 Edition). These data are available on the data disk in the data set named GradRate. Descriptions for data, which appear on the disk, follow.

Region	The region of the country in which the university is located (North, South, Midwest, West)
Graduation Rate	The percentage of students who enroll at the university and graduate
% of Classes Under 20	The percentage of classes with fewer than 20 students
% of Classes of 50 or More	The percentage of classes with more than 50 students
Student-Faculty Ratio	The ratio of the number of students enrolled divided by the total number of faculty

Acceptance Rate	The percentage of students who apply and are accepted
1st-Year Students in Top 10% of HS Class	The percentage of students admitted who were in the top 10% of their high school class
Academic Reputation Score	A measure of the school's reputation determined by surveying administrators at other universities: measured on a scale from 1 (marginal) to 5 (distinguished)

Managerial Report

Use the methods presented in this and previous chapters to analyze this data set. Present a summary of your analysis, including key statistical results, conclusions, and recommendations, in a managerial report. Include any appropriate technical material (computer output, residual plots, etc.) in an appendix.

Chapter 17

INDEX NUMBERS

CONTENTS

STATISTICS IN PRACTICE

U.S. DEPARTMENT OF LABOR
BUREAU OF LABOR STATISTICS
Washington, D.C.

The U.S. Department of Labor, through its Bureau of Labor Statistics, compiles and distributes indexes and statistics that are indicators of business and economic activity in the United States. For instance, the Bureau compiles and publishes the Consumer Price Index, the Producer Price Index, and statistics on average hours and earnings of various groups of workers. Perhaps the most widely quoted index produced by the Bureau of Labor Statistics is the Consumer Price Index. It is often used as a measure of inflation.

In September 2000, the Bureau of Labor Statistics reported that the Consumer Price Index (CPI) decreased by .1% from the July level. This decrease was the first in 14 years. The Bureau noted, however, that the core rate of inflation was up .2% in August. The core rate excludes the volatile food and energy components of the CPI and is sometimes regarded as a better indicator of inflationary pressures. The energy index fell by 2.9% and the food index increased by .2%.

Many economists and analysts proclaimed that the CPI report confimed that the United States was in a noninflationary era. In addition to the CPI report, they pointed to very little upward pressure on wages. The Labor Department had also reported that average weekly earnings adjusted for inflation only rose .1% in August after being unchanged in each of the prior two months.

Another indicator that inflationary pressures would be low for the near future was given by the Producer Price Index. It measures price changes in wholesale markets and is often seen as a leading indicator of changes in the Consumer Price Index. The Producer Price Index fell by .2% in August. Federal Reserve policymakers had raised the overnight bank lending rate six times since June 1999 but, with this data that prices were not rising rapidly, were expected to leave interest rates unchanged for the near future.

In this chapter we will see how various indexes, such as the Consumer and Producer Price Indexes, are computed and how they should be interpreted.

The Bureau of Labor Statistics computes the Consumer Price Index from data collected on the sale of goods and services. © CORBIS.

Each month the U.S. government publishes a variety of indexes that are designed to help individuals understand current business and economic conditions. Perhaps the most widely known and cited of these indexes is the Consumer Price Index (CPI). As its name implies, the CPI is an indicator of what is happening to prices consumers are paying for items purchased. Specifically, the CPI measures changes in price over a period of time. With a given starting point or *base period* and its associated index of 100, the CPI can be used to compare current period consumer prices with those in the base period. For example, a CPI of 125 reflects the condition that consumer prices as a whole are running approximately 25% above the base period prices for the same items. Although relatively few individuals know exactly what this number means, they do know enough about the CPI to understand that an increase means higher prices.

Even though the CPI is perhaps the best-known index, many other governmental and private-sector indexes are available to help us measure and understand how economic conditions in one period compare with economic conditions in other periods. The purpose of this chapter is to describe the most widely used types of indexes. We will begin by constructing some simple index numbers to gain a better understanding of how indexes are computed.

17.1 PRICE RELATIVES

TABLE 17.1

UNLEADED GASOLINE COST

Year	Price per Gallon ($)
1984	1.21
1985	1.20
1986	.93
1987	.95
1988	.95
1989	1.02
1990	1.16
1991	1.14
1992	1.13
1993	1.11
1994	1.11
1995	1.15
1996	1.23
1997	1.23
1998	1.06

Source: U.S. Energy Administration, *Monthly Energy Review.*

The simplest form of a price index shows how the current price per unit for a given item compares to a base period price per unit for the same item. For example, Table 17.1 reports the cost of one gallon of unleaded gasoline for the years 1984 through 1998. To facilitate comparisons with other years, the actual cost-per-gallon figure can be converted to a **price relative,** which expresses the unit price in each period as a percentage of the unit price in a base period.

$$\text{Price relative in period } t = \frac{\text{Price in period } t}{\text{Base period price}}(100) \qquad \textbf{(17.1)}$$

For the gasoline prices in Table 17.1 and with 1984 as the base year, the price relatives for one gallon of unleaded gasoline in the years 1984 through 1998 can be calculated. These price relatives are listed in Table 17.2. Note how easily the price in any one year can be compared with the price in the base year by knowing the price relative. For example, the price relative of 77 in 1986 shows that the gasoline cost in 1986 was 23% below the 1984 base-year cost. Similarly, the 1998 price relative of 88 shows a 12% decrease in gasoline cost in 1998 from the 1984 base-year cost. Price relatives, such as the ones for unleaded gasoline, are extremely helpful in terms of understanding and interpreting changing economic and business conditions over time.

17.2 AGGREGATE PRICE INDEXES

TABLE 17.2

PRICE RELATIVES FOR ONE GALLON OF UNLEADED GASOLINE (1984–1998)

Year	Price Relative (Base 1984)
1984	(1.21/1.21)100 = 100.0
1985	(1.20/1.21)100 = 99.2
1986	(.93/1.21)100 = 76.9
1987	(.95/1.21)100 = 78.5
1988	(.95/1.21)100 = 78.5
1989	(1.02/1.21)100 = 84.3
1990	(1.16/1.21)100 = 95.9
1991	(1.14/1.21)100 = 94.2
1992	(1.13/1.21)100 = 93.4
1993	(1.11/1.21)100 = 91.7
1994	(1.11/1.21)100 = 91.7
1995	(1.15/1.21)100 = 95.0
1996	(1.23/1.21)100 = 101.7
1997	(1.23/1.21)100 = 101.7
1998	(1.06/1.21)100 = 87.6

Although price relatives can be used to identify price changes over time for individual items, we are often more interested in the general price change for a group of items taken as a whole. For example, if we want an index that measures the change in the overall cost of living over time, we will want the index to be based on the price changes for a variety of items, including food, housing, clothing, transportation, medical care, and so on. An **aggregate price index** is developed for the specific purpose of measuring the combined change of a group of items.

Consider the development of an aggregate price index for a group of items categorized as normal automotive operating expenses. For illustration, we limit the items included in the group to gasoline, oil, tire, and insurance expenses.

Table 17.3 gives the data for the four components of our automotive operating expense index for the years 1984 and 1998. With 1984 as the base period, an aggregate price index for the four components will give us a measure of the change in normal automotive operating expenses over the 1984–1998 period.

An unweighted aggregate index can be developed by simply summing the unit prices in the year of interest (e.g., 1998) and dividing that sum by the sum of the unit prices in the base year (1984). Let

$$P_{it} = \text{unit price for item } i \text{ in period } t$$
$$P_{i0} = \text{unit price for item } i \text{ in the base period}$$

TABLE 17.3 DATA FOR AUTOMOTIVE OPERATING EXPENSE INDEX

Item	Unit Price ($) 1984	1998
Gallon of gasoline	1.21	1.06
Quart of oil	1.50	2.20
Tires	80.00	145.00
Insurance policy	300.00	700.00

An unweighted aggregate price index in period t, denoted by I_t, is given by

$$I_t = \frac{\Sigma P_{it}}{\Sigma P_{i0}}(100) \tag{17.2}$$

where the sums are over all items in the group.

An unweighted aggregate index for normal automotive operating expenses in 1998 ($t = 1998$) is given by

$$I_{1998} = \frac{1.06 + 2.20 + 145.00 + 700.00}{1.21 + 1.50 + 80.00 + 300.00}(100) = \frac{848.26}{382.71}(100) = 222$$

If quantity of usage is the same for each item, an unweighted index gives the same value as a weighted index. In practice, however, quantities of usage are rarely the same.

From the unweighted aggregate price index, we might conclude that the price of normal automotive operating expenses increased 122% over the period from 1984 to 1998. But note that the unweighted aggregate approach to establishing a composite price index for automotive expenses is heavily influenced by the items with large per-unit prices. Consequently, items with relatively low unit prices such as gasoline and oil are dominated by the high unit-price items such as tires and insurance. The unweighted aggregate index for automotive operating expenses is too heavily influenced by price changes in tires and insurance.

Because of the sensitivity of an unweighted index to one or more high-priced items, this form of aggregate index is not widely used. A weighted aggregate price index provides a better comparison when usage quantities differ.

The philosophy behind the **weighted aggregate price index** is that each item in the group should be weighted according to its importance. In most cases, the quantity of usage is the best measure of importance. Hence, one must obtain a measure of the quantity of usage for the various items in the group. Table 17.4 gives annual usage information for each item of automotive operating expense based on the typical operation of a midsize automobile for approximately 15,000 miles per year. The quantity weights listed show the expected annual usage for this type of driving situation.

TABLE 17.4

ANNUAL USAGE INFORMATION FOR AUTOMOTIVE OPERATING EXPENSE INDEX

Item	Quantity Weights*
Gallons of gasoline	1000
Quarts of oil	15
Tires	2
Insurance policy	1

*Based on 15,000 miles per year. Tire usage is based on a 30,000-mile tire life.

Let Q_i = quantity of usage for item i. The weighted aggregate price index in period t is given by

$$I_t = \frac{\Sigma P_{it}Q_i}{\Sigma P_{i0}Q_i}(100) \tag{17.3}$$

where the sums are over all items in the group. Applied to our automotive operating expenses, the weighted aggregate price index is based on dividing total operating costs in 1998 by total operating costs in 1984.

Let $t = 1998$, and use the quantity weights in Table 17.4. We obtain the following weighted aggregate price index for automotive operating expenses in 1998.

$$I_{1998} = \frac{1.06(1000) + 2.20(15) + 145.00(2) + 700.00(1)}{1.21(1000) + 1.50(15) + 80.00(2) + 300.00(1)}(100)$$

$$= \frac{2083}{1692.5}(100) = 123$$

From this weighted aggregate price index, we would conclude that the price of automotive operating expenses has increased 23% over the period from 1984 through 1998.

Clearly, compared with the unweighted aggregate index, the weighted index provides a more accurate indication of the price change for automotive operating expenses over the 1984–1998 period. Taking the quantity of usage of gasoline into account helps to offset the large increase in insurance costs. The weighted index shows a more moderate increase in automotive operating expenses than the unweighted index. In general, the weighted aggregate index with quantities of usage as weights is the preferred method for establishing a price index for a group of items.

In the weighted aggregate price index formula (17.3), note that the quantity term Q_i does not have a second subscript to indicate the time period. The reason is that the quantities Q_i are considered fixed and do not vary with time as the prices do. The fixed weights or quantities are specified by the designer of the index at levels believed to be representative of typical usage. Once established, they are held constant or fixed for all periods of time the index is in use. Indexes for years other than 1998 require the gathering of new price data P_{it}, but the weighting quantities Q_i remain the same.

In a special case of the fixed-weight aggregate index, the quantities are determined from base-year usages. In this case we write $Q_i = Q_{i0}$, with the zero subscript indicating base-year quantity weights; (17.3) becomes

$$I_t = \frac{\Sigma P_{it}Q_{i0}}{\Sigma P_{i0}Q_{i0}}(100) \qquad \textbf{(17.4)}$$

Whenever the fixed quantity weights are determined from base-year usage, the weighted aggregate index is given the name **Laspeyres index.**

Another option for determining quantity weights is to revise the quantities each period. A quantity Q_{it} is determined for each year that the index is computed. The weighted aggregate index in period t with these quantity weights is given by

$$I_t = \frac{\Sigma P_{it}Q_{it}}{\Sigma P_{i0}Q_{it}}(100) \qquad \textbf{(17.5)}$$

Note that the same quantity weights are used for the base period (period 0) and for period t. However, the weights are based on usage in period t, not the base period. This weighted aggregate index is known as the **Paasche index.** It has the advantage of being based on current usage patterns. However, this method of computing a weighted aggregate index has two disadvantages: the normal usage quantities Q_{it} must be redetermined each year, thus adding to the time and cost of data collection, and each year the index numbers for previous years must be recomputed to reflect the effect of the new quantity weights. Because of these disadvantages, the Laspeyres index is more widely used. The automotive operating expense index was computed with base-period quantities; hence, it is a Laspeyres index. Had usage figures for 1998 been used, we would have had a Paasche index. Indeed, because of more fuel efficient cars, gasoline usage has decreased and a Paasche index would differ from a Laspeyres index.

EXERCISES

Methods

1. The following table reports prices and usage quantities for two items in 1989 and 2001.

	Quantity		Unit Price ($)	
Item	**1989**	**2001**	**1989**	**2001**
A	1500	1800	7.50	7.75
B	2	1	630.00	1500.00

   a. Compute price relatives for each item in 2001 using 1989 as the base period.
   b. Compute an unweighted aggregate price index for the two items in 2001 using 1989 as the base period.
   c. Compute a weighted aggregate price index for the two items using the Laspeyres method.
   d. Compute a weighted aggregate price index for the two items using the Paasche method.

2. An item with a price relative of 132 cost $10.75 in 2001. Its base year was 1990.
   a. What was the percentage increase or decrease in cost of the item over the 11-year period?
   b. What did the item cost in 1990?

Applications

3. A large manufacturer purchases an identical component from three independent suppliers that differ in unit price and quantity supplied. The relevant data for 1999 and 2001 are given here.

		Unit Price ($)	
Supplier	**Quantity (1999)**	**1999**	**2001**
A	150	5.45	6.00
B	200	5.60	5.95
C	120	5.50	6.20

   a. Compute the price relatives for each of the component suppliers separately. Compare the price increases by the suppliers over the 2-year period.
   b. Compute an unweighted aggregate price index for the component part in 2001.
   c. Compute a 2001 weighted aggregate price index for the component part. What is the interpretation of this index for the manufacturing firm?

4. R&B Beverages, Inc., provides a complete line of beer, wine, and soft drink products for distribution through retail outlets in central Iowa. Unit price data for 1997 and 2001 and quantities sold in cases for 1997 follow.

		Unit Price ($)	
Item	**1997 Quantity (cases)**	**1997**	**2001**
Beer	35,000	15.00	16.25
Wine	5,000	60.00	64.00
Soft drink	60,000	9.80	10.00

   Compute a weighted aggregate index for the R&B Beverage sales in 2001, with 1997 as the base period.

5. Under the LIFO inventory valuation method, a price index for inventory must be established for tax purposes. The quantity weights are based on year-ending inventory levels. Use the beginning-of-the-year price per unit as the base-period price and develop a weighted aggregate index for the total inventory value at the end of the year. What type of weighted aggregate price index must be developed for the LIFO inventory valuation?

Product	Ending Inventory	Unit Price ($) Beginning	Unit Price ($) Ending
A	500	.15	.19
B	50	1.60	1.80
C	100	4.50	4.20
D	40	12.00	13.20

17.3 COMPUTING AN AGGREGATE PRICE INDEX FROM PRICE RELATIVES

In Section 17.1 we defined the concept of a price relative and showed how a price relative can be computed with knowledge of the current-period unit price and the base-period unit price. We now want to show how aggregate price indexes like the ones developed in Section 17.2 can be computed directly from information about the price relative of each item in the group. Because of the limited use of unweighted indexes, we restrict our attention to weighted aggregate price indexes. Let us return to the automotive operating expense index of the preceding section. The necessary information for the four items is given in Table 17.5.

One must be sure prices and quantities are in the same units. For example, if prices are per case, quantity must be the number of cases and not, for instance, the number of individual units.

Let w_i be the weight applied to the price relative for item i. The general expression for a weighted average of price relatives is given by

$$I_t = \frac{\sum \frac{P_{it}}{P_{i0}}(100)w_i}{\Sigma w_i} \tag{17.6}$$

The proper choice of weights in (17.6) will enable us to compute a weighted aggregate price index from the price relatives. The proper choice of weights is given by multiplying the base-period price by the quantity of usage.

$$w_i = P_{i0}Q_i \tag{17.7}$$

TABLE 17.5 PRICE RELATIVES FOR AUTOMOTIVE OPERATING EXPENSE INDEX

Item	Unit Price ($) 1984 (P_0)	Unit Price ($) 1998 (P_t)	Price Relative $(P_t/P_0)100$	Annual Usage
Gallon of gasoline	1.21	1.06	87.6	1000
Quart of oil	1.50	2.20	146.7	15
Tires	80.00	145.00	181.3	2
Insurance policy	300.00	700.00	233.3	1

Substituting $w_i = P_{i0}Q_i$ into equation (17.6) provides the following expression for a weighted price relatives index.

$$I_t = \frac{\sum \frac{P_{it}}{P_{i0}}(100)(P_{i0}Q_i)}{\Sigma P_{i0}Q_i} \tag{17.8}$$

With the canceling of the P_{i0} terms in the numerator, an equivalent expression for the weighted price relatives index is

$$I_t = \frac{\Sigma P_{it}Q_i}{\Sigma P_{i0}Q_i}(100)$$

Thus, we see that the weighted price relatives index with $w_i = P_{i0}Q_i$ provides a price index identical to the weighted aggregate index presented in Section 17.2 by equation (17.3). Use of base-period quantities (i.e., $Q_i = Q_{i0}$) in equation (17.7) leads to a Laspeyres index. Use of current-period quantities (i.e., $Q_i = Q_{it}$) in equation (17.7) leads to a Paasche index.

Let us return to the automotive operating expense data. We can use the price relatives in Table 17.5 and equation (17.6) to compute a weighted average of price relatives. The results obtained by using the weights specified by equation (17.7) are reported in Table 17.6. The index number 123 represents a 23% increase in automotive operating expenses, which is the same as the increase identified by the weighted aggregate index computation in Section 17.2.

TABLE 17.6 AUTOMOTIVE OPERATING EXPENSE INDEX (1984–1998) BASED ON WEIGHTED PRICE RELATIVES

Item	Price Relatives $(P_{it}/P_{i0})(100)$	Base Price ($) P_{i0}	Quantity Q_i	Weight $w_i = P_{i0}Q_i$	Weighted Price Relatives $(P_{it}/P_{i0})(100)w_i$
Gasoline	87.6	1.21	1000	1,210.0	105,996.00
Oil	146.7	1.50	15	22.5	3,300.75
Tires	181.3	80.00	2	160.0	29,008.00
Insurance	233.3	300.00	1	300.0	69,990.00
			Totals	1,692.5	208,294.75

$$I_{1998} = \frac{208{,}294.75}{1{,}692.5} = 123$$

EXERCISES

Methods

6. Price relatives for three items, along with base-period prices and usage are shown in the following table. Compute a weighted aggregate price index for the current period.

Item	Price Relative	Base Period Price	Base Period Usage
A	150	22.00	20
B	90	5.00	50
C	120	14.00	40

Applications

7. The Mitchell Chemical Company produces a special industrial chemical that is a blend of three chemical ingredients. The beginning-year cost per pound, the ending-year cost per pound, and the blend proportions follow.

	Cost per Pound ($)		
Ingredient	Beginning	Ending	Quantity (pounds) per 100 Pounds of Product
A	2.50	3.95	25
B	8.75	9.90	15
C	.99	.95	60

   a. Compute the price relatives for the three ingredients.
   b. Compute a weighted average of the price relatives to develop a 1-year cost index for raw materials used in the product. What is your interpretation of this index value?

8. An investment portfolio consists of four stocks. The purchase price, current price, and number of shares are reported in the following table.

Stock	Purchase Price/Share ($)	Current Price/Share ($)	Number of Shares
Holiday Trans	15.50	17.00	500
NY Electric	18.50	20.25	200
KY Gas	26.75	26.00	500
PQ Soaps	42.25	45.50	300

   Construct a weighted average of price relatives as an index of the performance of the portfolio to date. Interpret this price index.

9. Compute the price relatives for the R&B Beverages products in Exercise 4. Use a weighted average of price relatives to show that this method provides the same index as the weighted aggregate method.

17.4 SOME IMPORTANT PRICE INDEXES

We have identified the procedures used to compute price indexes for single items or groups of items. Now let us consider some price indexes that are important measures of business and economic conditions. Specifically, we will consider the Consumer Price Index, the Producer Price Index, and the Dow Jones averages.

Consumer Price Index

The CPI includes charges for services (e.g., doctor and dentist bills) and all taxes directly associated with the purchase and use of an item.

The **Consumer Price Index (CPI),** published monthly by the U.S. Bureau of Labor Statistics, is the primary measure of the cost of living in the United States. The group of items used to develop the index consists of a *market basket* of 400 items including food, housing, clothing, transportation, and medical items. The CPI is a weighted aggregate price index with fixed weights.* The weight applied to each item in the market basket derives from a usage survey of urban families throughout the United States.

The August 2000 CPI, computed with a 1982–1984 base index of 100, was 172.7. This figure means that the cost of purchasing the market basket of goods and services had

*There are actually two Consumer Price Indexes. The Bureau of Labor Statistics publishes a Consumer Price Index for all urban consumers (CPI-U) and a revised Consumer Price Index for urban wage earners and clerical workers (CPI-W). The CPI-U is the one most widely quoted, and it is published regularly in *The Wall Street Journal*.

increased 72.7% since the base period 1982–1984. The 50-year time series of the CPI from 1950 to 2000 is shown in Figure 17.1. Note how the CPI measure reflects the sharp inflationary behavior of the economy in the late 1970s and early 1980s.

Producer Price Index

The PPI is designed as a measure of price changes for domestic goods; imports are not included.

The **Producer Price Index (PPI),** also published monthly by the U.S. Bureau of Labor Statistics, measures the monthly changes in prices in primary markets in the United States. The PPI is based on prices for the first transaction of each product in nonretail markets. All commodities sold in commercial transactions in these markets are represented. The survey covers raw, manufactured, and processed goods at each level of processing and includes the output of industries classified as manufacturing, agriculture, forestry, fishing, mining, gas and electricity, and public utilities. One of the common uses of this index is as a leading indicator of the future trend of consumer prices and the cost of living. An increase in the PPI reflects producer price increases that will eventually be passed on to the consumer through higher retail prices.

Weights for the various items in the PPI are based on the value of shipments. The weighted average of price relatives is calculated by the Laspeyres method. The August 2000 PPI, computed with a 1982 base index of 100, was 138.1.

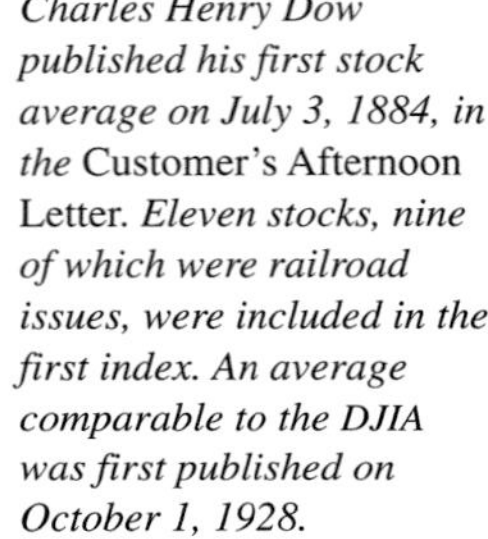
Charles Henry Dow published his first stock average on July 3, 1884, in the Customer's Afternoon Letter. *Eleven stocks, nine of which were railroad issues, were included in the first index. An average comparable to the DJIA was first published on October 1, 1928.*

Dow Jones Averages

The **Dow Jones averages** are indexes designed to show price trends and movements on the New York Stock Exchange. The best known of the Dow Jones indexes is the Dow Jones Industrial Average (DJIA), which is based on common stock prices of 30 large companies. It is the sum of these stock prices divided by a number, which is revised from time to time to adjust for stock splits and switching of companies in the index. Unlike the other price in-

FIGURE 17.1 CONSUMER PRICE INDEX, 1950–2000 WITH BASE 1982–1984 = 100

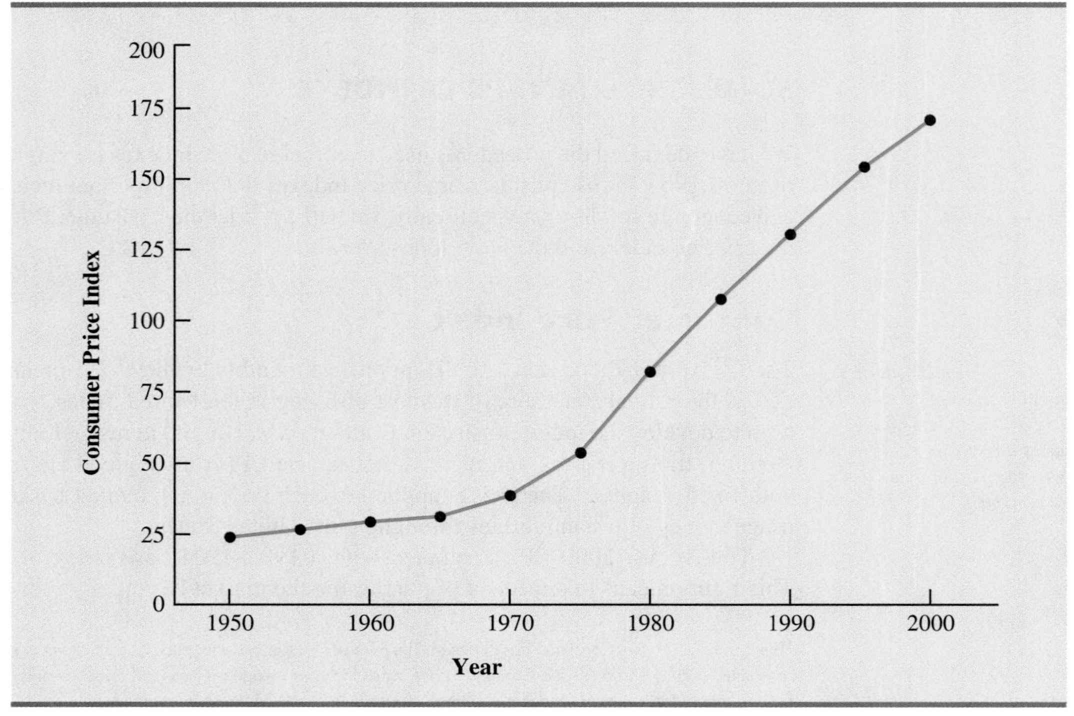

TABLE 17.7 THE 30 COMPANIES USED IN THE DOW JONES INDUSTRIAL AVERAGE (SEPTEMBER 2000)

Alcoa	Exxon Mobil	J. P. Morgan Chase
American Express	General Electric	McDonald's
AT&T	General Motors	Merck
Boeing	Hewlett-Packard	Microsoft
Caterpillar	Home Depot	Minnesota Mining
Citigroup	Honeywell Int'l	Philip Morris
Coca Cola	IBM	Procter & Gamble
Disney	Intel	SBC Communications
DuPont	International Paper	United Technologies
Eastman Kodak	Johnson & Johnson	Wal-Mart Stores

Source: Barron's, March 12, 2001.

dexes that we have studied, it is not expressed as a percentage of base-year prices. The specific firms used in September 2000 to compute the DJIA are listed in Table 17.7.

Other Dow Jones averages are computed for 20 transportation stocks and for 15 utility stocks. The Dow Jones averages are computed and published daily in *The Wall Street Journal* and other financial publications.

17.5 DEFLATING A SERIES BY PRICE INDEXES

Time series are deflated to remove the effects of inflation.

Many business and economic series reported over time, such as company sales, industry sales, and inventories, are measured in dollar amounts. These time series often show an increasing growth pattern over time, which is generally interpreted as indicating an increase in the physical volume associated with the activities. For example, a total dollar amount of inventory up by 10% might be interpreted to mean that the physical inventory is 10% larger. Such interpretations can be misleading if a time series is measured in terms of dollars, and the total dollar amount is a combination of both price and quantity changes. Hence, in periods when price changes are significant, the changes in the dollar amounts may not be indicative of quantity changes unless we are able to adjust the time series to eliminate the price change effect.

For example, from 1976 to 1980, the total amount of spending in the construction industry increased approximately 75%. That figure suggests excellent growth in construction activity. However, construction prices were increasing just as fast as—or sometimes even faster than—the 75% rate. In fact, while total construction spending was increasing, construction activity was staying relatively constant or, as in the case of new housing starts, decreasing. To interpret construction activity correctly for the 1976–1980 period, we must adjust the total spending series by a price index to remove the price increase effect. Whenever we remove the price increase effect from a time series, we say we are *deflating the series.*

In relation to personal income and wages, we often hear discussions about issues such as "real wages" or the "purchasing power" of wages. These concepts are based on the notion of deflating an hourly wage index. For example, Figure 17.2 shows the pattern of hourly wages of manufacturing workers for the period 1996–2000. We see a trend of wage increases from \$12.77 per hour to \$14.36 per hour. Should manufacturing workers be pleased with this growth in hourly wages? The answer depends on what has happened to the purchasing power of their wages. If we can compare the purchasing power of the \$12.77 hourly wage in 1996 with the purchasing power of the \$14.36 hourly wage in 2000, we will be better able to judge the relative improvement in wages.

FIGURE 17.2 ACTUAL HOURLY WAGES OF MANUFACTURING WORKERS

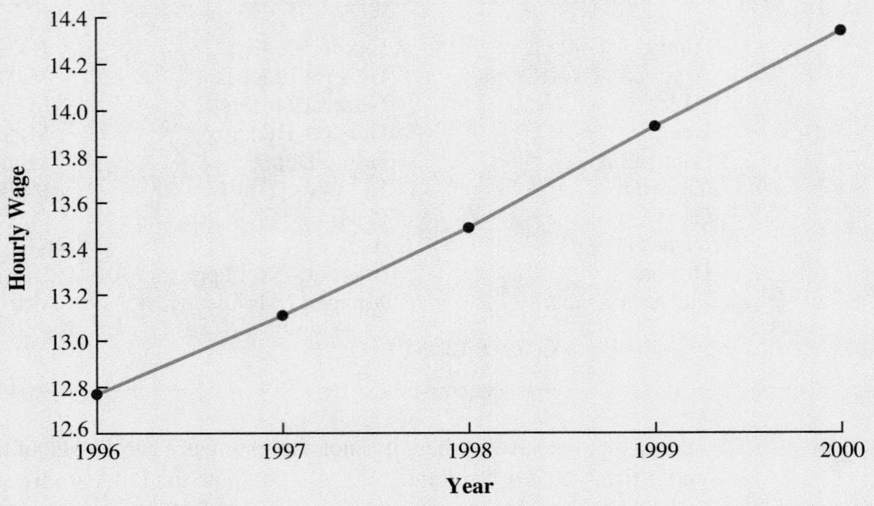

Table 17.8 reports both the hourly wage rate and the CPI for the period 1996–2000. With these data, we will show how the CPI can be used to deflate the index of hourly wages. The deflated series is found by dividing the hourly wage rate in each year by the corresponding value of the CPI and multiplying by 100. The deflated hourly wage index for manufacturing workers is given in Table 17.9; Figure 17.3 is a graph showing the deflated, or real, wages.

What does the deflated series of wages tell us about the real wages or purchasing power of workers during the 1996–2000 period? In terms of base period dollars (1982–1984 = 100), the hourly wage rate remained relatively flat over the period. After removing the inflationary effect we see that the purchasing power of the workers has changed only slightly. This effect is seen clearly in Figure 17.3. Thus, the advantage of using price indexes to deflate a series is that we have a clearer picture of the real dollar changes that are occurring.

Real wages are a better measure of purchasing power than actual wages. Indeed, many union contracts call for wages to be adjusted in accordance with changes in the cost of living.

This process of deflating a series measured over time has an important application in the computation of the Gross Domestic Product (GDP). The GDP is the total value of all goods and services produced in a given country. Obviously, over time the GDP will show gains that are in part due to price increases if the GDP is not deflated by a price index. There-

TABLE 17.8 HOURLY WAGES OF MANUFACTURING WORKERS AND CONSUMER PRICE INDEX: 1996–2000

Year	**Hourly Wage ($)**	**CPI (1982–1984 Base)**
1996	12.77	156.9
1997	13.11	160.5
1998	13.46	163.0
1999	13.93	166.6
2000	14.36	172.6

Source: Bureau of Labor Statistics

TABLE 17.9 DEFLATED SERIES OF HOURLY WAGES FOR MANUFACTURING WORKERS

Year	Deflated Hourly Wage
1996	($12.77/156.9)(100) = $8.14
1997	($13.11/160.5)(100) = $8.17
1998	($13.46/163.0)(100) = $8.26
1999	($13.93/166.6)(100) = $8.36
2000	($14.36/172.6)(100) = $8.32

FIGURE 17.3 REAL HOURLY WAGES OF MANUFACTURING WORKERS (1982–1984 = 100)

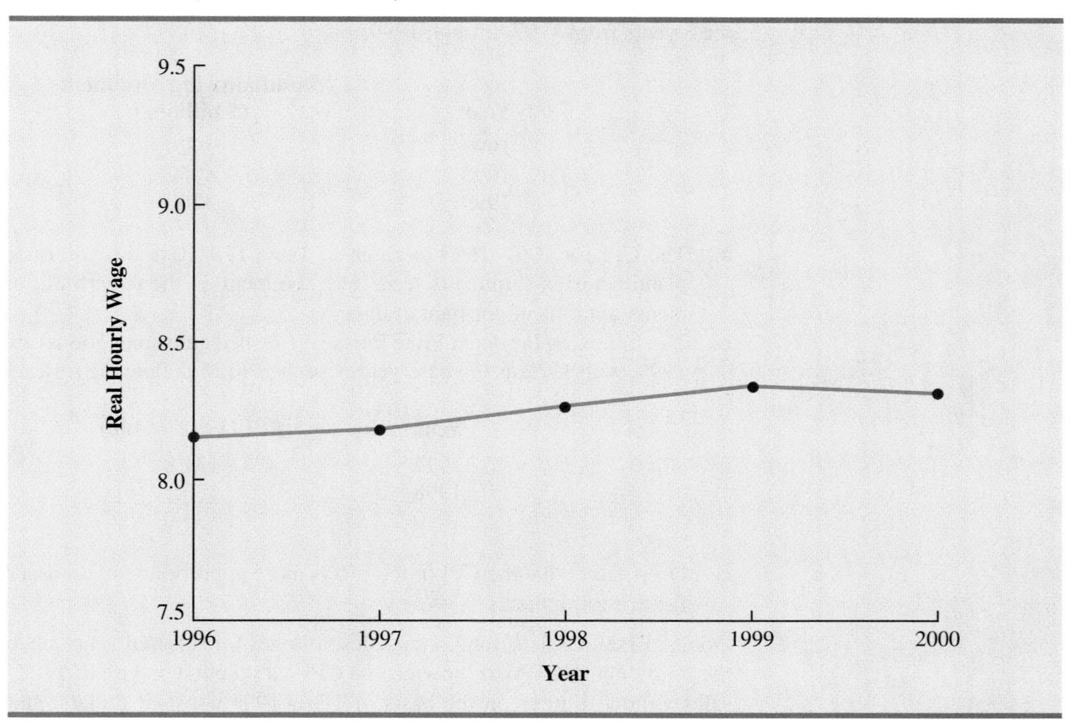

fore, to adjust the total value of goods and services to reflect actual changes in the volume of goods and services produced and sold, the GDP must be computed with a price index deflator. The process is similar to that discussed in the real wages computation.

EXERCISES

Applications

10. Average hourly wages for manufacturing workers in 1980 were $7.27; in 2000, they were $14.36. The CPI in 1980 was 82.4; in 2000 it was 172.6.

a. Deflate the hourly wage rates in 1980 and 2000 to find the real wage rates.

b. What is the percentage change in actual hourly wages from 1980 to 2000?

c. What is the percentage change in real wages from 1980 to 2000?

11. Average hourly wages for workers in service industries for the 5 years from 1996 through 2000 are reported here. Use the Consumer Price Index information in Table 17.8 to deflate the wages series. What has been the percentage increase in real wages and salaries from 1998 to 2000?

Year	Total Wages & Salaries ($ billions)
1996	11.76
1997	12.23
1998	12.84
1999	13.35
2000	13.82

Source: Bureau of Labor Statistics

12. The U.S. Bureau of the Census reported the following total manufacturing shipments for the 3 years from 1997 through 1999.

Year	Manufacturing Shipments ($ billions)
1997	3929
1998	4052
1999	4260

a. The CPI for 1997–1999 is given in Table 17.8. Use this information to deflate the manufacturing shipments series and comment on the pattern of manufacturers' shipments in terms of constant dollars.

b. The following Producer Price Indexes (finished consumer goods) are for 1997 through 1999, with 1982 as the base year. Use the PPI to deflate the series.

Year	PPI (1982 = 100)
1997	131.8
1998	130.7
1999	133.0

c. Do you feel that the CPI or the PPI is more appropriate to use as a deflator for manufacturing shipments?

13. Dooley Retail Outlets' total retail sales volumes for selected years since 1982 is shown in the following table. Also shown is the CPI with the index base of 1982–1984. Deflate the sales volume figures on the basis of 1982–1984 constant dollars, and comment on the firm's sales volumes in terms of deflated dollars.

Year	Retail Sales ($)	CPI (1982–1984 base)
1982	380,000	96.5
1987	520,000	113.6
1992	700,000	140.3
1997	870,000	160.5
2000	940,000	172.6

17.6 PRICE INDEXES: OTHER CONSIDERATIONS

In the preceding sections we described several methods used to compute price indexes, discussed the use of some important indexes, and presented a procedure for using price indexes to deflate a time series. Several other issues must be considered to enhance our understanding of how price indexes are constructed and how they are used. Some are discussed in this section.

Selection of Items

The primary purpose of a price index is to measure the price change over time for a specified class of items, products, and so on. Whenever the class of items is very large, the index cannot be based on all items in the class. Rather, a sample of representative items must be used. By collecting price and quantity information for the sampled items, we hope to obtain a good idea of the price behavior of all items that the index is representing. For example, in the Consumer Price Index the total number of items that might be considered in the population of normal purchase items for a consumer could be 2000 or more. However, the index is based on the price-quantity characteristics of just 400 items. The selection of the specific items in the index is not a trivial task. Surveys of user purchase patterns as well as good judgment go into the selection process. A simple random sample is not used to select the 400 items.

After the initial selection process, the group of items in the index must be periodically reviewed and revised whenever purchase patterns change. Thus, the issue of which items to include in an index must be resolved before an index can be developed and again before it is revised.

Selection of a Base Period

Most indexes are established with a base-period value of 100 at some specific time. All future values of the index are then related to the base-period value. But what base period is appropriate for an index? It is not an easy question, and the answer must be based on the judgment of the developer of the index.

Many of the indexes established by the United States government as of 2000 have a 1982 base period. As a general guideline, the base period should not be too far from the current period. For example, a Consumer Price Index with a 1945 base period would be difficult for most individuals to understand because of unfamiliarity with conditions in 1945. The base period for most indexes therefore is adjusted periodically to a more recent period of time. The CPI base period was changed from 1967 to the 1982–1984 average in 1988. The PPI currently uses 1982 as its base period (i.e., 1982 = 100).

Quality Changes

The purpose of a price index is to measure changes in prices over time. Ideally, price data are collected for the same set of items at several times, and then the index is computed. A basic assumption is that the prices are identified for the same items each period. A problem is encountered when a product changes in quality from one period to the next. For example, a manufacturer may alter the quality of a product by using less expensive materials, fewer features, and so on, from year to year. The price may go up in following years, but the price is for a lower quality product. Consequently, the price may actually go up more than is represented by the list price for the item. It is difficult, if not impossible, to adjust an index for decreases in the quality of an item.

A substantial quality improvement also may cause an increase in the price of a product. A portion of the price related to the quality improvement should be excluded from the index computation. However, adjusting an index for a price increase that is related to higher quality of an item is extremely difficult, if not impossible.

Although common practice is to ignore minor quality changes in develo[illegible] price index, major quality changes must be addressed because they can [illegible] uct description from period to period. If a product description is chan[illegible] must be modified to account for it; in some cases, the product might be [illegible] the index.

In some situations, however, a substantial improvement in quality is followed by a decrease in the price. This less typical situation has been the case with personal computers during the 1990s. Designers of the CPI are now making the proper adjustments in the CPI to reflect this situation.

17.7 QUANTITY INDEXES

In addition to the price indexes described in the preceding sections, other types of indexes are useful. In particular, one other application of index numbers is to measure changes in quantity levels over time. This type of index is called a **quantity index.**

Recall that in the development of the weighted aggregate price index in Section 17.2, to compute an index number for period t we needed data on unit prices at a base period (P_0) and period t (P_t). Equation (17.3) provided the weighted aggregate price index as

$$I_t = \frac{\Sigma P_{it} Q_i}{\Sigma P_{i0} Q_i}(100)$$

The numerator, $\Sigma P_{it} Q_i$, represents the total value of fixed quantities of the index items in period t. The denominator, $\Sigma P_{i0} Q_i$, represents the total value of the same fixed quantities of the index items in year 0.

Computation of a weighted aggregate quantity index is similar to that of a weighted aggregate price index. Quantities for each item are measured in the base period and period t, with Q_{i0} and Q_{it}, respectively, representing those quantities for item i. The quantities are then weighted by a fixed price, the value added, or some other factor. The "value added" to a product is the sales value minus the cost of purchased inputs. The formula for computing a weighted aggregate quantity index for period t is

$$I_t = \frac{\Sigma Q_{it} w_i}{\Sigma Q_{i0} w_i}(100) \tag{17.9}$$

In some quantity indexes the weight for item i is taken to be the base-period price (P_{i0}), in which case the weighted aggregate quantity index is

$$I_t = \frac{\Sigma Q_{it} P_{i0}}{\Sigma Q_{i0} P_{i0}}(100) \tag{17.10}$$

Quantity indexes can also be computed on the basis of weighted quantity relatives. One formula for this version of a quantity index follows.

$$I_t = \frac{\Sigma \frac{Q_{it}}{Q_{i0}}(Q_{i0} P_i)}{\Sigma Q_{i0} P_i}(100) \tag{17.11}$$

This formula is the quantity version of the weighted price relatives formula developed in Section 17.3 as in equation (17.8).

The **Index of Industrial Production,** developed by the Federal Reserve Board, is probably the best-known quantity index. It is reported monthly and the base period is 1992. The index is designed to measure changes in volume of production levels for a variety of manufacturing classifications in addition to mining and utilities. In August 2000 the index was 145.7.

EXERCISES

Methods

14. Data on quantities of three items sold in 1992 and 2000 are given here along with the sales prices of the items in 1992. Compute a weighted aggregate quantity index for 2000.

	Quantity Sold		
Item	**1992**	**2000**	**Price/Unit 1992 ($)**
A	350	300	18.00
B	220	400	4.90
C	730	850	15.00

Applications

15. A trucking firm handles four commodities for a particular distributor. Total shipments for the commodities in 1991 and 2000, as well as the 1991 prices, are reported in the following table.

	Shipments		
Commodity	**1991**	**2000**	**Price/Shipment 1991**
A	120	95	$1200
B	86	75	$1800
C	35	50	$2000
D	60	70	$1500

Develop a weighted aggregate quantity index with a 1991 base. Comment on the growth or decline in quantities over the 1991–2000 period.

16. An automobile dealer reports the 1989 and 2000 sales for three models in the following table. Compute quantity relatives and use them to develop a weighted aggregate quantity index for 2000 using the 2 years of data.

	Sales		
Model	**1989**	**2000**	**Mean Price per Sale (1989)**
Sedan	200	170	$15,200
Sport	100	80	$17,000
Wagon	75	60	$16,800

SUMMARY

Price and quantity indexes are important measures of changes in price and quantity levels within the business and economic environment. Price relatives are simply the ratio of the current unit price of an item to a base-period unit price multiplied by 100, with a value of 100 indicating no difference in the current- and base-period prices. Aggregate price indexes are created as a composite measure of the overall change in prices for a given group of items or products. Usually the items in an aggregate price index are weighted by their quantity of usage. A weighted aggregate price index can also be computed by weighting the price rela-tives by the usage quantities for the items in the index.

The Consumer Price Index and the Producer Price Index are both widely with 1982–1984 and 1982, respectively, as base years. The Dow Jones Indu is another widely quoted price index. It is a weighted sum of the prices of 30 co listed on the New York Stock Exchange. Unlike many other indexes, it is not stated as a percentage of some base-period value.

Often price indexes are used to deflate some other economic series reported over time. We saw how the CPI could be used to deflate hourly wages to obtain an index of real wages. Selection of the items to be included in the index, selection of a base period for the index, and adjustment for changes in quality are important additional considerations in the development of an index number. Quantity indexes were briefly discussed, and the Index of Industrial Production was mentioned as an important quantity index.

GLOSSARY

Price relative A price index for a given item that is computed by dividing a current unit price by a base-period unit price and multiplying the result by 100.

Aggregate price index A composite price index based on the prices of a group of items.

Weighted aggregate price index A composite price index in which the prices of the items in the composite are weighted by their relative importance.

Laspeyres index A weighted aggregate price index in which the weight for each item is its base-period quantity.

Paasche index A weighted aggregate price index in which the weight for each item is its current-period quantity.

Consumer Price Index A monthly price index that uses the price changes in a market basket of consumer goods and services to measure the changes in consumer prices over time.

Producer Price Index A monthly price index designed to measure changes in prices of goods sold in primary markets (i.e., first purchase of a commodity in nonretail markets).

Dow Jones averages Aggregate price indexes designed to show price trends and movements on the New York Stock Exchange.

Quantity index An index that is designed to measure changes in quantities over time.

Index of Industrial Production A quantity index that is designed to measure changes in the physical volume or production levels of industrial goods over time.

KEY FORMULAS

Price Relative in Period *t*

$$\frac{\text{Price in period } t}{\text{Base period price}}(100) \tag{17.1}$$

Unweighted Aggregate Price Index in Period *t*

$$I_t = \frac{\Sigma P_{it}}{\Sigma P_{i0}}(100) \tag{17.2}$$

Weighted Aggregate Price Index in Period *t*

$$I_t = \frac{\Sigma P_{it} Q_i}{\Sigma P_{i0} Q_i}(100) \tag{17.3}$$

Weighted Average of Price Relatives

$$I_t = \frac{\Sigma \frac{P_{it}}{P_{i0}}(100)w_i}{\Sigma w_i} \tag{17.6}$$

Weighting Factor for (17.6)

$$w_i = P_{i0}Q_i \tag{17.7}$$

Weighted Aggregate Quantity Index

$$I_t = \frac{\Sigma Q_{it}w_i}{\Sigma Q_{i0}w_i}(100) \tag{17.9}$$

SUPPLEMENTARY EXERCISES

17. The median sales prices for new single-family houses for the years 1996–1999 are as follows (*Statistical Abstract of the United States,* 2000).

Year	Price ($1000s)
1996	140.0
1997	146.0
1998	152.5
1999	160.0

a. Use 1996 as the base year and develop a price index for new single-family homes over this 4-year period.
b. Use 1997 as the base year and develop a price index for new single-family homes over this 4-year period.

18. Nickerson Manufacturing Company has the following data on quantities shipped and unit costs for each of its four products:

Products	Base-Period Quantities (1998)	Mean Shipping Cost per Unit ($)	
		1998	2001
A	2000	10.50	15.90
B	5000	16.25	32.00
C	6500	12.20	17.40
D	2500	20.00	35.50

a. Compute the price relative for each product.
b. Compute a weighted aggregate price index that reflects the shipping cost change o·· the four-year period.

19. Use the price data in Exercise 18 to compute a Paasche index for the shipp quantities are 4000, 3000, 7500, and 3000 for each of the four products.

20. Boran Stockbrokers, Inc., selects four stocks for the purpose of developing its own index of stock market behavior. Costs per share for a 1999 base period, January 2001, and March 2001 follow. Base-year quantities have been set on the basis of historical volumes for the four stocks.

			Cost per Share ($)		
Stock	**Industry**	**1999 Quantity**	**1999 Base**	**January 2001**	**March 2001**
A	Oil	100	31.50	32.75	32.50
B	Computer	150	65.00	59.00	57.50
C	Steel	75	40.00	42.00	39.50
D	Real Estate	50	18.00	16.50	13.75

Use the 1999 base period to compute the Boran index for January 2001 and March 2001. Comment on what the index tells you about what is happening in the stock market.

21. Compute the price relatives for the four stocks making up the Boran index in Exercise 20. Use the weighted aggregates of price relatives to compute the January 2001 and March 2001 Boran indexes.

22. Consider the following price relatives and quantity information for grain production in Iowa (*Statistical Abstract of the United States,* 1997).

Product	**1991 Quantities (millions of bushels)**	**Base Price per Bushel ($)**	**1991–1996 Price Relatives**
Corn	1427	2.30	113
Soybeans	350	5.51	123

What is the 1996 weighted aggregate price index for the Iowa grains?

23. Fresh fruit price and quantity data for the years 1988 and 1998 follow (*Statistical Abstract of the United States,* 1999). Quantity data reflect per capita consumption in pounds and prices are per pound.

Fruit	**1988 per Capita Consumption (pounds)**	**1988 Price ($/pound)**	**1998 Price ($/pound)**
Bananas	24.3	.41	.51
Apples	19.9	.71	.85
Oranges	13.9	.56	.61
Pears	3.2	.64	.98

a. Compute a price relative for each product.

b. Compute a weighted aggregate price index for fruit products. Comment on the change in fruit prices over the 10-year period.

24. Starting faculty salaries (9-month basis) for assistant professors of business administration at a major Midwestern university follow. Use the CPI to deflate the salary data to constant dollars. Comment on the trend in salaries in higher education as indicated by these data.

Year	**Starting Salary ($)**	**CPI (1982–1984 Base)**
1970	14,000	38.8
1975	17,500	53.8
1980	23,000	82.4
1985	37,000	107.6
1990	53,000	130.7
1995	65,000	152.4
2000	80,000	172.6

25. The 5-year historical prices per share for a particular stock and the Consumer Price Index with a 1982–1984 base period follow.

Year	Price per Share ($)	CPI (1982–1984 Base)
1996	51.00	156.9
1997	54.00	160.5
1998	58.00	163.0
1999	59.50	166.6
2000	59.00	172.6

Deflate the stock price series and comment on the investment aspects of this stock.

26. A major manufacturing company has reported the quantity and product value information for 1997 and 2001 in the table that follows. Compute a weighted aggregate quantity index for the data. Comment on what this quantity index means.

	Quantities		
Product	1997	2001	Values ($)
A	800	1200	30.00
B	600	500	20.00
C	200	500	25.00

Chapter 18

FORECASTING

CONTENTS

STATISTICS IN PRACTICE

NEVADA OCCUPATIONAL HEALTH CLINIC*

Sparks, Nevada

Nevada Occupational Health Clinic is a privately owned medical clinic in Sparks, Nevada. The clinic specializes in industrial medicine and has been in operation at the same site for more than 20 years. In the beginning of 1991, the clinic entered a rapid growth phase in which monthly billings increased from $57,000 to more than $300,000 in 26 months. The clinic was still undergoing dramatic growth when the main clinic building burned to the ground on April 6, 1993.

The clinic's insurance policy covered physical property and equipment as well as loss of income due to the interruption of regular business operations. Settling the property insurance claim was a relatively straightforward matter of determining the value of the physical property and equipment lost during the fire. However, determining the value of the income lost during the seven months that it took to rebuild the clinic was a complicated matter involving negotiations between the business owners and the insurance company. No preestablished rules could help calculate "what would have happened" to the clinic's billings if the fire had not occurred. To estimate the lost income, the clinic used a forecasting method to project the growth in business that would have been realized during the seven-month lost-business period. The actual history of billings prior to the fire provided the basis for a forecasting model with linear trend and seasonal components as discussed in this chapter. This forecasting model enabled the clinic to establish an accurate estimate of the loss, which eventually was accepted by the insurance company.

A 1993 fire closed the Nevada Occupational Health Clinic for seven months. © PhotoDisc, Inc.

*The authors are indebted to Bard Betz, Director of Operations, and Curtis Brauer, Executive Administrative Assistant, Nevada Occupational Health Clinic, for providing this Statistics in Practice.

An essential aspect of managing any organization is planning for the future. Indeed, the long-run success of an organization is closely related to how well management is able to anticipate the future and develop appropriate strategies. Good judgment, intuition, and an awareness of the state of the economy may give a manager a rough idea or "feeling" of what is likely to happen in the future. However, converting that feeling into a number that can be used as next quarter's sales volume or next year's raw material cost is difficult. The purpose of this chapter is to introduce several forecasting methods.

Most companies can forecast total demand for all products with errors of less than 5%. However, forecasting demand for individual products can result in significantly higher errors.

Suppose we have been asked to provide quarterly forecasts of the sales volume for a particular product during the coming one-year period. Production schedules, raw material purchasing, inventory policies, and sales quotas will all be affected by the quarterly forecasts we provide. Consequently, poor forecasts may result in poor planning and hence increased costs for the firm. How should we go about providing the quarterly sales volume forecasts?

We will certainly want to review the actual sales data for the product in past periods. Using these historical data, we can identify the general level of sales and any trend such as an increase or decrease in sales volume over time. A further review of the data might reveal a seasonal pattern such as peak sales occurring in the third quarter of each year and sales volume bottoming out during the first quarter. By reviewing historical data, we can often develop a better understanding of the pattern of past sales, leading to better predictions of future sales for the product.

A forecast is simply a prediction of what will happen in the future. Managers must learn to accept the fact that regardless of the technique used, they will not be able to develop perfect forecasts.

Historical sales form a time series. A **time series** is a set of observations on a variable measured at successive points in time or over successive periods of time. In this chapter, we will introduce several procedures for analyzing time series. The objective of such analyses is to provide good **forecasts** or predictions of future values of the time series.

Forecasting methods can be classified as quantitative or qualitative. Quantitative forecasting methods can be used when (1) past information about the variable being forecast is available, (2) the information can be quantified, and (3) a reasonable assumption is that the pattern of the past will continue into the future. In such cases, a forecast can be developed using a time series method or a causal method.

If the historical data are restricted to past values of the variable, the forecasting procedure is called a *time series method.* The objective of time series methods is to discover a pattern in the historical data and then extrapolate the pattern into the future; the forecast is based solely on past values of the variable and/or on past forecast errors. In this chapter we discuss three time series methods: smoothing (moving averages, weighted moving averages, and exponential smoothing), trend projection, and trend projection adjusted for seasonal influence.

Causal forecasting methods are based on the assumption that the variable we are forecasting has a cause-effect relationship with one or more other variables. In this chapter we discuss the use of regression analysis as a causal forecasting method. For instance, the sales volume for many products is influenced by advertising expenditures, so regression analysis may be used to develop an equation showing how these two variables are related. Then, once the advertising budget has been set for the next period, we could substitute this value into the equation to develop a prediction or forecast of the sales volume for that period. Note that if a time series method had been used to develop the forecast, advertising expenditures would not be considered; that is, a time series method would have based the forecast solely on past sales.

Qualitative methods generally involve the use of expert judgment to develop forecasts. For instance, a panel of experts might develop a consensus forecast of the prime rate for a year from now. An advantage of qualitative procedures is that they can be applied when the information on the variable being forecast cannot be quantified and when historical data are either not applicable or unavailable. Figure 18.1 provides an overview of the types of forecasting methods.

18.1 COMPONENTS OF A TIME SERIES

The pattern or behavior of the data in a time series has several components. The usual assumption is that four separate components—trend, cyclical, seasonal, and irregular—combine to provide specific values for the time series. Let us look more closely at each of these components.

Trend Component

In time series analysis, the measurements may be taken every hour, day, week, month, or year, or at any other regular interval.* Although time series data generally exhibit random fluctuations, the time series may still show gradual shifts or movements to relatively higher or lower values over a longer period of time. The gradual shifting of the time series is re-

*We limit our discussion to time series in which the values of the series are recorded at equal intervals. Cases in which the observations are not made at equal intervals are beyond the scope of this text.

FIGURE 18.1 OVERVIEW OF FORECASTING METHODS

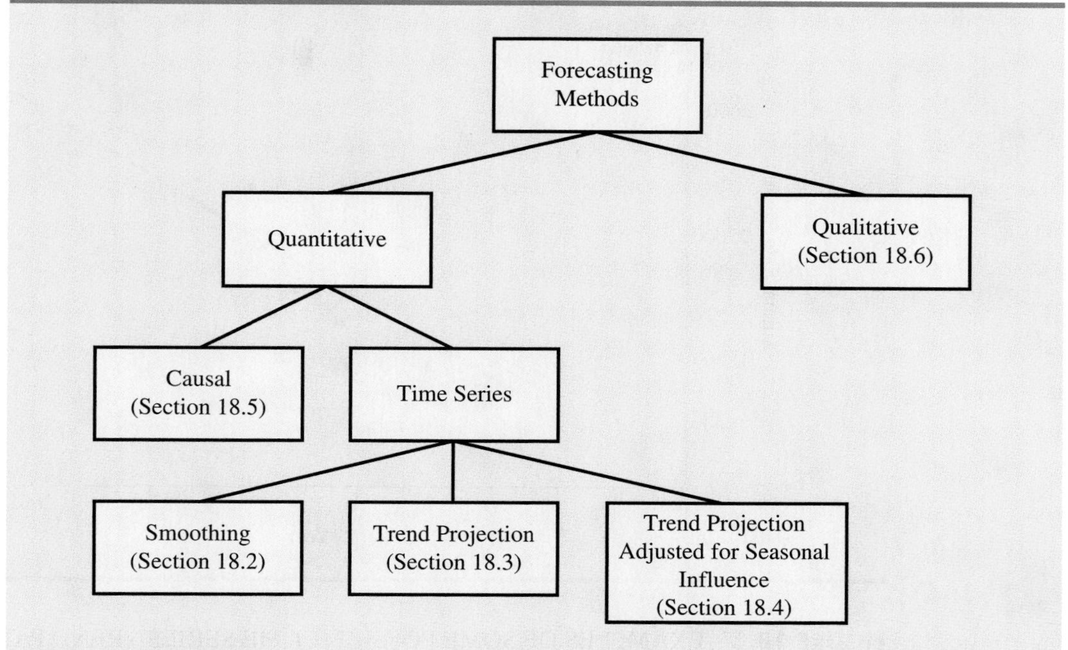

ferred to as the **trend** in the time series; this shifting or trend is usually the result of long-term factors such as changes in the population, demographic characteristics of the population, technology, and/or consumer preferences.

For example, a manufacturer of photographic equipment may see substantial month-to-month variability in the number of cameras sold. However, in reviewing the sales over the past 10 to 15 years, the manufacturer may find a gradual increase in the annual sales volume. Suppose the sales volume was approximately 17,000 cameras in 1991, 23,000 cameras in 1996, and 25,000 cameras in 2001. This gradual growth in sales over time shows an upward trend for the time series. Figure 18.2 shows a straight line that may be a good approximation of the trend in camera sales. Although the trend for camera sales appears to be linear and increasing over time, sometimes the trend in a time series can be described better by some other patterns.

Figure 18.3 shows some other possible time series trend patterns. Panel (A) shows a nonlinear trend; in this case, the time series indicates little growth initially, then a period of rapid growth, and finally a leveling off. This trend might be a good approximation of sales for a product from introduction through a growth period and into a period of market saturation. The linear decreasing trend in panel (B) is useful for a time series displaying a steady decrease over time. The horizontal line in panel (C) represents a time series that has no consistent increase or decrease over time and thus no trend.

Cyclical Component

Although a time series may exhibit a trend over long periods of time all future values of the time series will not fall exactly on the trend line. In fact, time series often show alternating sequences of points below and above the trend line. Any recurring sequence of points above

FIGURE 18.2 LINEAR TREND OF CAMERA SALES

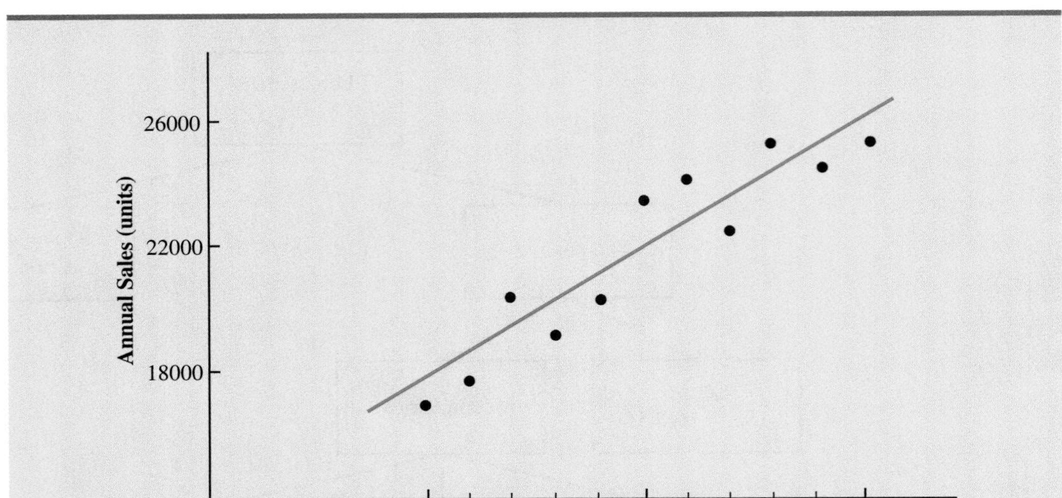

FIGURE 18.3 EXAMPLES OF SOME POSSIBLE TIME SERIES TREND PATTERNS

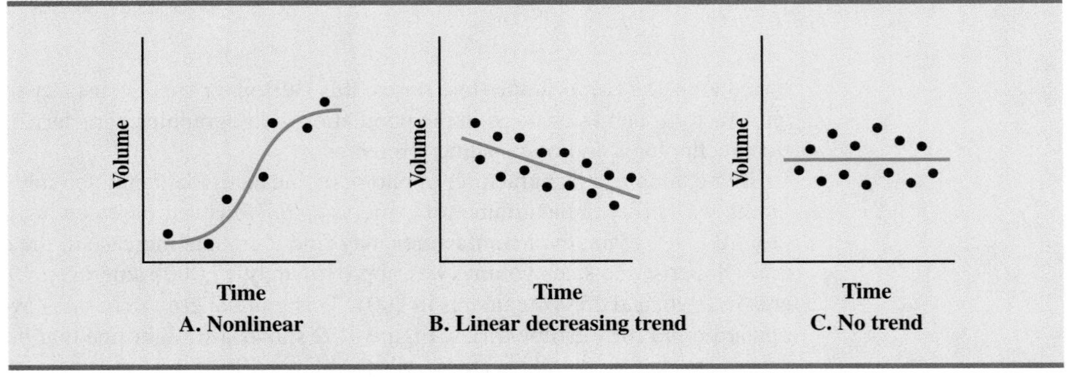

and below the trend line lasting more than one year can be attributed to the **cyclical component** of the time series. Figure 18.4 shows the graph of a time series with an obvious cyclical component. The observations are taken at intervals one year apart.

Many time series exhibit cyclical behavior with regular runs of observations below and above the trend line. Generally, this component of the time series is due to multiyear cyclical movements in the economy. For example, periods of moderate inflation followed by periods of rapid inflation can lead to time series that alternate below and above a generally increasing trend line (e.g., a time series for housing costs). Many time series in the early 1980s displayed this type of behavior.

Seasonal Component

Whereas the trend and cyclical components of a time series are identified by analyzing multiyear movements in historical data, many time series show a regular pattern over one-year periods. For example, a manufacturer of swimming pools expects low sales activity in the fall and winter months, with peak sales in the spring and summer months. Manufactur-

FIGURE 18.4 TREND AND CYCLICAL COMPONENTS OF A TIME SERIES WITH DATA POINTS ONE YEAR APART

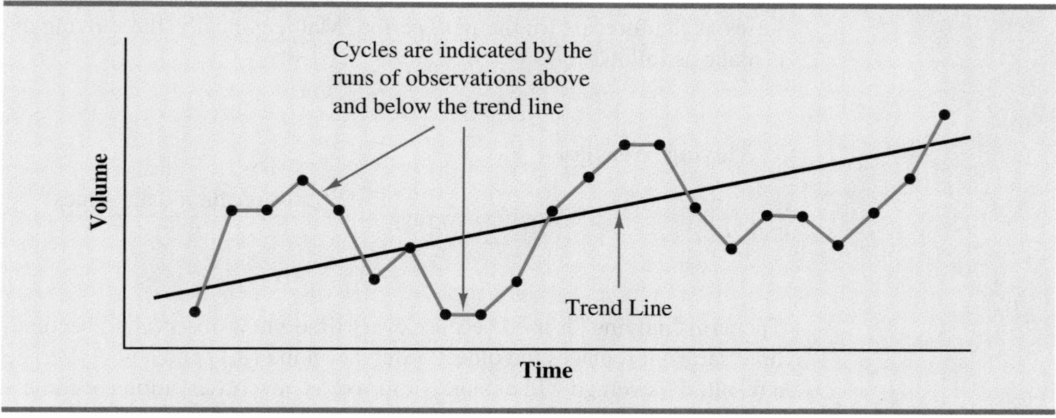

ers of snow removal equipment and heavy clothing, however, expect just the opposite yearly pattern. Not surprisingly, the component of the time series that represents the variability in the data due to seasonal influences is called the **seasonal component.** Although we generally think of seasonal movement in a time series as occurring within one year, the seasonal component can also be used to represent any regularly repeating pattern that is less than one year in duration. For example, daily traffic volume data show within-the-day "seasonal" behavior, with peak levels occurring during rush hours, moderate flow during the rest of the day and early evening, and light flow from midnight to early morning.

Irregular Component

The **irregular component** of the time series is the residual, or "catch-all," factor that accounts for the deviations of the actual time series values from those expected given the effects of the trend, cyclical, and seasonal components. The irregular component is caused by the short-term, unanticipated, and nonrecurring factors that affect the time series. Because this component accounts for the random variability in the time series, it is unpredictable. We cannot attempt to predict its impact on the time series.

18.2 SMOOTHING METHODS

Many manufacturing environments require forecasts for thousands of items weekly or monthly. Thus, in choosing a forecasting technique, simplicity and ease of use are important criteria. The data requirements for the techniques presented in this section are minimal, and the techniques are easy to use and understand.

In this section we discuss three forecasting methods: moving averages, weighted moving averages, and exponential smoothing. The objective of each of these methods is to "smooth out" the random fluctuations caused by the irregular component of the time series, therefore they are referred to as smoothing methods. Smoothing methods are appropriate for a stable time series—that is, one that exhibits no significant trend, cyclical, or seasonal effects—because they adapt well to changes in the level of the time series. However, without modification, they do not work as well when a significant trend, cyclical, and/or seasonal variation is present.

Smoothing methods are easy to use and generally provide a high level of accuracy for short-range forecasts, such as a forecast for the next time period. One of the methods, exponential smoothing, has minimal data requirements and thus is a good method to use when forecasts are required for large numbers of items.

Moving Averages

The **moving averages** method uses the average of the most recent n data values in the time series as the forecast for the next period. Mathematically, the moving average calculation is made as follows.

Moving Average

$$\text{Moving Average} = \frac{\Sigma(\text{most recent } n \text{ data values})}{n} \tag{18.1}$$

The term "moving" is used because every time a new observation becomes available for the time series, it replaces the oldest observation in (18.1) and a new average is computed. As a result, the average will change, or move, as new observations become available.

TABLE 18.1

GASOLINE SALES TIME SERIES

Week	Sales (1000s of gallons)
1	17
2	21
3	19
4	23
5	18
6	16
7	20
8	18
9	22
10	20
11	15
12	22

To illustrate the moving averages method, consider the 12 weeks of data in Table 18.1 and Figure 18.5. These data show the number of gallons of gasoline sold by a gasoline distributor in Bennington, Vermont, over the past 12 weeks. Figure 18.5 indicates that, although random variability is present, the time series appears to be stable over time. Hence, the smoothing methods of this section are applicable.

To use moving averages to forecast gasoline sales, we must first select the number of data values to be included in the moving average. As an example, let us compute forecasts using a 3-week moving average. The moving average calculation for the first 3 weeks of the gasoline sales time series is

$$\text{Moving Average (Weeks 1–3)} = \frac{17 + 21 + 19}{3} = 19$$

We then use this moving average as the forecast for week 4. Because the actual value observed in week 4 is 23, the forecast error in week 4 is $23 - 19 = 4$. In general, the error associated with any forecast is the difference between the observed value of the time series and the forecast.

The calculation for the second 3-week moving average is

$$\text{Moving Average (Weeks 2–4)} = \frac{21 + 19 + 23}{3} = 21$$

Hence, the forecast for week 5 is 21. The error associated with this forecast is $18 - 21 = -3$. Thus, the forecast error may be positive or negative depending on whether the forecast is too low or too high. A complete summary of the 3-week moving average calculations for the gasoline sales time series is provided in Table 18.2 and Figure 18.6.

Forecast accuracy is not the only consideration. Sometimes the most accurate method requires data on related time series that are difficult or costly to obtain. Trade-offs are often made between cost and forecast accuracy.

Forecast Accuracy. An important consideration in selecting a forecasting method is the accuracy of the forecast. Clearly, we want forecast errors to be small. The last two columns of Table 18.2, which contain the forecast errors and the squared forecast errors, can be used to develop a measure of accuracy.

For the gasoline sales time series, we can use the last column of Table 18.2 to compute the average of the sum of the squared errors. Doing so we obtain

$$\text{Average of the Sum of Squared Errors} = \frac{92}{9} = 10.22$$

FIGURE 18.5 GASOLINE SALES TIME SERIES

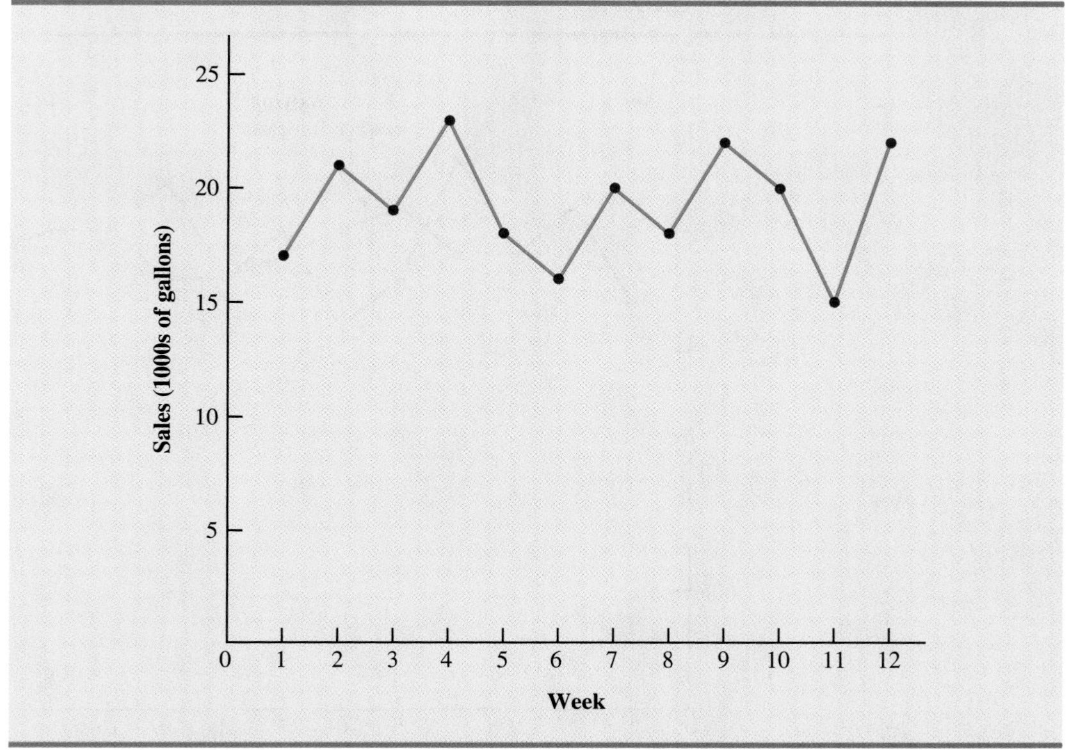

This average of the sum of squared errors is commonly referred to as the **mean squared error** (MSE). The MSE is an often-used measure of the accuracy of a forecasting method and is the one we use in this chapter.

As we indicated previously, to use the moving averages method, we must first select the number of data values to be included in the moving average. Not surprisingly, for a particular time series, moving averages of different lengths will differ in their ability to

TABLE 18.2 SUMMARY OF THREE-WEEK MOVING AVERAGE CALCULATIONS

Week	Time Series Value	Moving Average Forecast	Forecast Error	Squared Forecast Error
1	17			
2	21			
3	19			
4	23	19	4	16
5	18	21	−3	9
6	16	20	−4	16
7	20	19	1	1
8	18	18	0	0
9	22	18	4	16
10	20	20	0	0
11	15	20	−5	25
12	22	19	3	9
		Totals	0	92

FIGURE 18.6 GASOLINE SALES TIME SERIES AND THREE-WEEK MOVING AVERAGE FORECASTS

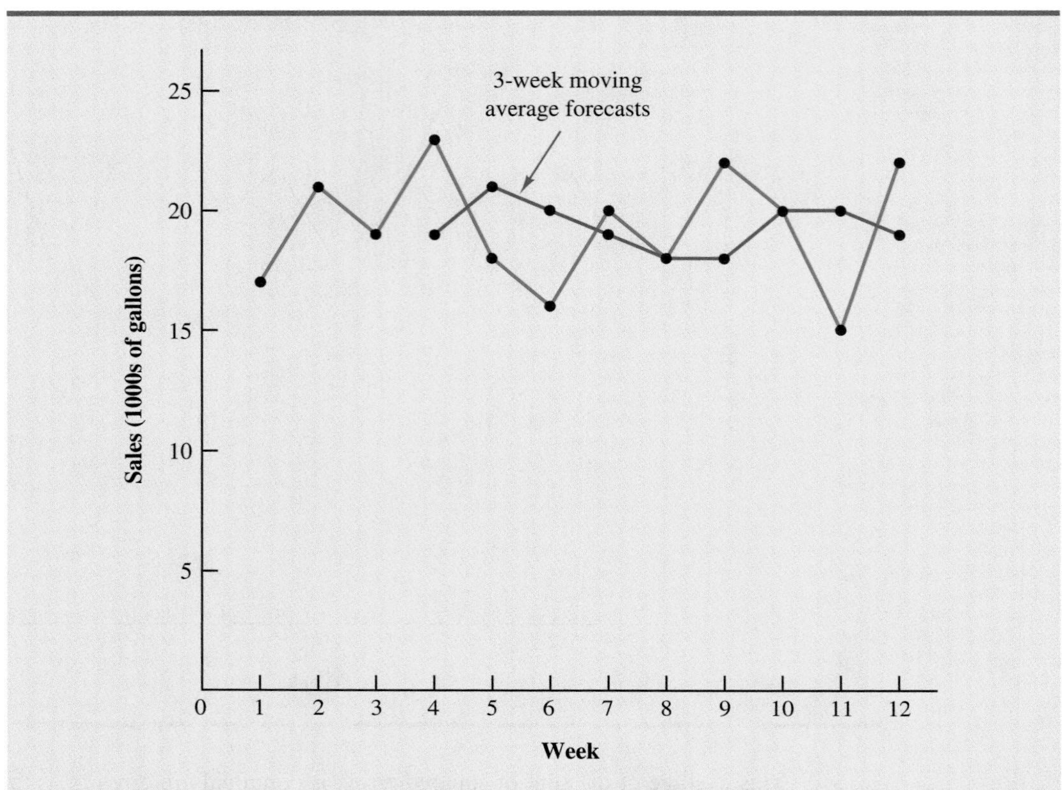

forecast the time series accurately. One possible approach to choosing the number of values to be included in the moving average is to use trial and error to identify the length that minimizes the MSE. Then, if we are willing to assume that the length that is best for the past will also be best for the future, we would forecast the next value in the time series by using the number of data values that minimized the MSE for the historical time series. Exercise 2 at the end of the section will ask you to consider 4-week and 5-week moving averages for the gasoline sales data. A comparison of the MSEs will indicate the number of weeks of data you may want to include in the moving average calculation.

Weighted Moving Averages

In the moving averages method, each observation in the moving average calculation receives the same weight. One variation, known as **weighted moving averages,** involves selecting a different weight for each data value and then computing a weighted average of the most recent n values as the forecast. In most cases, the most recent observation receives the most weight, and the weight decreases for older data values. For example, we can use the gasoline sales time series to illustrate the computation of a weighted 3-week moving average, with the most recent observation receiving a weight three times as great as that given the oldest observation, and the next oldest observation receiving a weight twice as great as the oldest. For week 4 the computation is:

$$\text{Forecast for Week 4} = \tfrac{1}{6}(17) + \tfrac{2}{6}(21) + \tfrac{3}{6}(19) = 19.33$$

Note that for the weighted moving average the sum of the weights is equal to 1. Actually the sum of the weights for the simple moving average also equalled 1: Each weight was 1/3. However, recall that the simple or unweighted moving average provided a forecast of 19.

Forecast Accuracy. To use the weighted moving averages method we must first select the number of data values to be included in the weighted moving average and then choose weights for each of the data values. In general, if we believe that the recent past is a better predictor of the future than the distant past, larger weights should be given to the more recent observations. However, when the time series is highly variable, selecting approximately equal weights for the data values may be best. Note that the only requirement in selecting the weights is that their sum must equal 1. To determine whether one particular combination of number of data values and weights provides a more accurate forecast than another combination, we will continue to use the MSE criterion as the measure of forecast accuracy. That is, if we assume that the combination that is best for the past will also be best for the future, we would use the combination of number of data values and weights that minimized MSE for the historical time series to forecast the next value in the time series.

Exponential Smoothing

Exponential smoothing is simple and has few data requirements, which makes it an inexpensive approach for firms that make many forecasts each period.

Exponential smoothing uses a weighted average of past time series values as the forecast; it is a special case of the weighted moving averages method in which we select only one weight—the weight for the most recent observation. The weights for the other data values are computed automatically and become smaller as the observations move farther into the past. The basic exponential smoothing model follows.

Exponential Smoothing Model

$$F_{t+1} = \alpha Y_t + (1 - \alpha)F_t \tag{18.2}$$

where

$$\begin{aligned} F_{t+1} &= \text{forecast of the time series for period } t + 1 \\ Y_t &= \text{actual value of the time series in period } t \\ F_t &= \text{forecast of the time series for period } t \\ \alpha &= \text{smoothing constant } (0 \le \alpha \le 1) \end{aligned}$$

Equation (18.2) shows that the forecast for period $t + 1$ is a weighted average of the actual value in period t and the forecast for period t; note in particular that the weight given to the actual value in period t is α and that the weight given to the forecast in period t is $1 - \alpha$. We can demonstrate that the exponential smoothing forecast for any period is also a weighted average of *all the previous actual values* for the time series with a time series consisting of three periods of data: Y_1, Y_2, and Y_3. To start the calculations, we let F_1 equal the actual value of the time series in period 1; that is, $F_1 = Y_1$. Hence, the forecast for period 2 is

$$\begin{aligned} F_2 &= \alpha Y_1 + (1 - \alpha)F_1 \\ &= \alpha Y_1 + (1 - \alpha)Y_1 \\ &= Y_1 \end{aligned}$$

Thus, the exponential smoothing forecast for period 2 is equal to the actual value of the time series in period 1.

The forecast for period 3 is

$$F_3 = \alpha Y_2 + (1 - \alpha)F_2 = \alpha Y_2 + (1 - \alpha)Y_1$$

Finally, substituting this expression for F_3 in the expression for F_4, we obtain

$$\begin{aligned} F_4 &= \alpha Y_3 + (1 - \alpha)F_3 \\ &= \alpha Y_3 + (1 - \alpha)[\alpha Y_2 + (1 - \alpha)Y_1] \\ &= \alpha Y_3 + \alpha(1 - \alpha)Y_2 + (1 - \alpha)^2 Y_1 \end{aligned}$$

The term exponential smoothing *comes from the exponential nature of the weighting scheme for the historical values.*

Hence, F_4 is a weighted average of the first three time series values. The sum of the coefficients, or weights, for Y_1, Y_2, and Y_3 equals one. A similar argument can be made to show that, in general, any forecast F_{t+1} is a weighted average of all the previous time series values.

Despite the fact that exponential smoothing provides a forecast that is a weighted average of all past observations, all past data do not need to be saved to compute the forecast for the next period. In fact, once the **smoothing constant** α has been selected, only two pieces of information are needed to compute the forecast. Equation (18.2) shows that with a given α we can compute the forecast for period $t + 1$ simply by knowing the actual and forecast time series values for period t—that is, Y_t and F_t.

To illustrate the exponential smoothing approach to forecasting, consider the gasoline sales time series in Table 18.1 and Figure 18.5. As indicated, the exponential smoothing forecast for period 2 is equal to the actual value of the time series in period 1. Thus, with $Y_1 = 17$, we will set $F_2 = 17$ to start the exponential smoothing computations. Referring to the time series data in Table 18.1, we find an actual time series value in period 2 of $Y_2 = 21$. Thus, period 2 has a forecast error of $21 - 17 = 4$.

Continuing with the exponential smoothing computations using a smoothing constant of $\alpha = .2$, we obtain the following forecast for period 3.

$$F_3 = .2Y_2 + .8F_2 = .2(21) + .8(17) = 17.8$$

Once the actual time series value in period 3, $Y_3 = 19$, is known, we can generate a forecast for period 4 as follows.

$$F_4 = .2Y_3 + .8F_3 = .2(19) + .8(17.8) = 18.04$$

By continuing the exponential smoothing calculations, we can determine the weekly forecast values and the corresponding weekly forecast errors, as shown in Table 18.3. Note that we have not shown an exponential smoothing forecast or the forecast error for period 1 because no forecast was made. For week 12, we have $Y_{12} = 22$ and $F_{12} = 18.48$. Can we use this information to generate a forecast for week 13 before the actual value of week 13 becomes known? Using the exponential smoothing model, we have

$$F_{13} = .2Y_{12} + .8F_{12} = .2(22) + .8(18.48) = 19.18$$

Thus, the exponential smoothing forecast of the amount sold in week 13 is 19.18, or 19,180 gallons of gasoline. With this forecast, the firm can make plans and decisions accordingly. The accuracy of the forecast will not be known until the end of week 13.

Figure 18.7 is the plot of the actual and forecast time series values. Note in particular how the forecasts "smooth out" the irregular fluctuations in the time series.

Forecast Accuracy. In the preceding exponential smoothing calculations, we used a smoothing constant of $\alpha = .2$. Although any value of α between 0 and 1 is acceptable, some

TABLE 18.3 SUMMARY OF THE EXPONENTIAL SMOOTHING FORECASTS AND FORECAST ERRORS FOR GASOLINE SALES WITH SMOOTHING CONSTANT $\alpha = .2$

Week (t)	Time Series Value (Y_t)	Exponential Smoothing Forecast (F_t)	Forecast Error ($Y_t - F_t$)
1	17		
2	21	17.00	4.00
3	19	17.80	1.20
4	23	18.04	4.96
5	18	19.03	−1.03
6	16	18.83	−2.83
7	20	18.26	1.74
8	18	18.61	−.61
9	22	18.49	3.51
10	20	19.19	.81
11	15	19.35	−4.35
12	22	18.48	3.52

FIGURE 18.7 ACTUAL AND FORECAST GASOLINE SALES TIME SERIES WITH SMOOTHING CONSTANT $\alpha = .2$

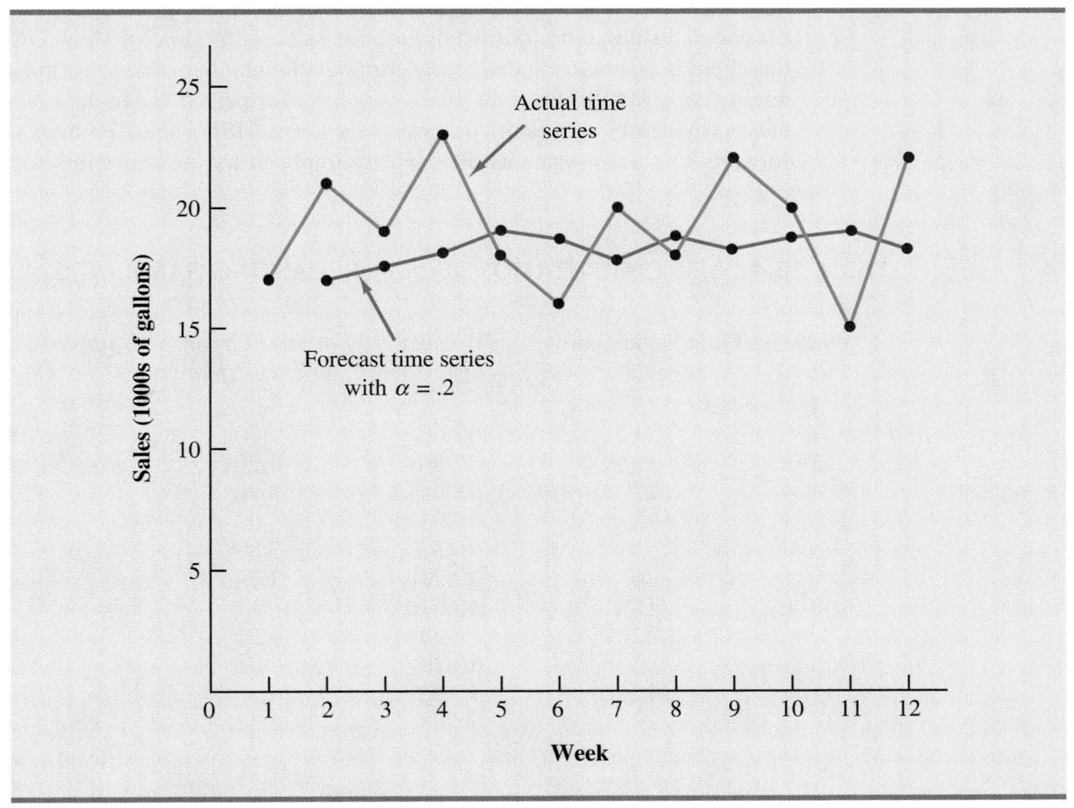

values will yield better forecasts than others. Insight into choosing a good value for α can be obtained by rewriting the basic exponential smoothing model as follows.

$$F_{t+1} = \alpha Y_t + (1 - \alpha)F_t$$
$$F_{t+1} = \alpha Y_t + F_t - \alpha F_t$$
$$F_{t+1} = \underset{\substack{\uparrow \\ \text{Forecast} \\ \text{in period } t}}{F_t} + \alpha\underbrace{(Y_t - F_t)}_{\substack{\text{Forecast error} \\ \text{in period } t}} \qquad \textbf{(18.3)}$$

Thus, the new forecast F_{t+1} is equal to the previous forecast F_t plus an adjustment, which is α times the most recent forecast error, $Y_t - F_t$. That is, the forecast in period $t + 1$ is obtained by adjusting the forecast in period t by a fraction of the forecast error. If the time series contains substantial random variability, a small value of the smoothing constant is preferred. The reason for this choice is that, because much of the forecast error is due to random variability, we do not want to overreact and adjust the forecasts too quickly. For a time series with relatively little random variability, larger values of the smoothing constant have the advantage of quickly adjusting the forecasts when forecasting errors occur and thus allowing the forecasts to react faster to changing conditions.

The criterion we will use to determine a desirable value for the smoothing constant α is the same as the criterion we proposed for determining the number of periods of data to include in the moving averages calculation. That is, we choose the value of α that minimizes the mean squared error (MSE). A summary of the MSE calculations for the exponential smoothing forecast of gasoline sales with $\alpha = .2$ is shown in Table 18.4. Note that there is one less squared error term than the number of time periods, because we had no past values with which to make a forecast for period 1. Would a different value of α have provided better results in terms of a lower MSE value? Perhaps the most straightforward way to answer this question is simply to try another value for α. We will then

TABLE 18.4 MSE COMPUTATIONS FOR FORECASTING GASOLINE SALES WITH $\alpha = .2$

Week (t)	Time Series Value (Y_t)	Forecast (F_t)	Forecast Error ($Y_t - F_t$)	Squared Forecast Error $(Y_t - F_t)^2$
1	17			
2	21	17.00	4.00	16.00
3	19	17.80	1.20	1.44
4	23	18.04	4.96	24.60
5	18	19.03	−1.03	1.06
6	16	18.83	−2.83	8.01
7	20	18.26	1.74	3.03
8	18	18.61	−.61	.37
9	22	18.49	3.51	12.32
10	20	19.19	.81	.66
11	15	19.35	−4.35	18.92
12	22	18.48	3.52	12.39
			Total	98.80

$$\text{MSE} = \frac{98.80}{11} = 8.98$$

TABLE 18.5 MSE COMPUTATIONS FOR FORECASTING GASOLINE SALES WITH $\alpha = .3$

Week (t)	Time Series Value (Y_t)	Forecast (F_t)	Forecast Error ($Y_t - F_t$)	Squared Forecast Error $(Y_t - F_t)^2$
1	17			
2	21	17.00	4.00	16.00
3	19	18.20	.80	.64
4	23	18.44	4.56	20.79
5	18	19.81	−1.81	3.28
6	16	19.27	−3.27	10.69
7	20	18.29	1.71	2.92
8	18	18.80	−.80	.64
9	22	18.56	3.44	11.83
10	20	19.59	.41	.17
11	15	19.71	−4.71	22.18
12	22	18.30	3.70	13.69
			Total	102.83

$$\text{MSE} = \frac{102.83}{11} = 9.35$$

compare its mean squared error with the MSE value of 8.98 obtained by using a smoothing constant of $\alpha = .2$.

The exponential smoothing results with $\alpha = .3$ are shown in Table 18.5. With MSE = 9.35, we see that for the current data set, a smoothing constant of $\alpha = .3$ results in less forecast accuracy than a smoothing constant of $\alpha = .2$. Thus, we would be inclined to prefer the original smoothing constant of $\alpha = .2$. Using a trial-and-error calculation with other values of α, we can find a "good" value for the smoothing constant. This value can be used in the exponential smoothing model to provide forecasts for the future. At a later date, after new time series observations have been obtained, we analyze the newly collected time series data to determine whether the smoothing constant should be revised to provide better forecasting results.

NOTES AND COMMENTS

1. Another measure of forecast accuracy is the *mean absolute deviation* (MAD). This measure is simply the average of the absolute values of all the forecast errors. Using the errors given in Table 18.2, we obtain

$$\text{MAD} = \frac{4 + 3 + 4 + 1 + 0 + 4 + 0 + 5 + 3}{9} = 2.67$$

One major difference between MSE and MAD is that the MSE measure is influenced much more by large forecast errors than by small errors (because for the MSE measure the errors are squared). The selection of the best measure of forecasting accuracy is not a simple matter. Indeed, forecasting experts often disagree as to which measure should be used. We use the MSE measure in this chapter.

2. Spreadsheet packages are an effective aid in choosing a good value of α for exponential smoothing and selecting weights for the weighted moving averages method. With the time series data and the forecasting formulas in the spreadsheets, you can experiment with different values of α (or moving average weights) and choose the value(s) of α providing the smallest MSE or MAD.

EXERCISES

Methods

1. Consider the following time series data.

Week	1	2	3	4	5	6
Value	8	13	15	17	16	9

   a. Develop a 3-week moving average for this time series. What is the forecast for week 7?
   b. Compute the MSE for the 3-week moving average.
   c. Use $\alpha = .2$ to compute the exponential smoothing values for the time series. What is the forecast for week 7?
   d. Compare the 3-week moving average forecast with the exponential smoothing forecast using $\alpha = .2$. Which appears to provide the better forecast?
   e. Use a smoothing constant of .4 to compute the exponential smoothing values. Does a smoothing constant of .2 or .4 appear to provide the better forecast? Explain.

2. Refer to the gasoline sales time series data in Table 18.1.
   a. Compute 4-week and 5-week moving averages for the time series.
   b. Compute the MSE for the 4-week and 5-week moving average forecasts.
   c. What appears to be the best number of weeks of past data to use in the moving average computation? Remember that the MSE for the 3-week moving average is 10.22.

3. Refer again to the gasoline sales time series data in Table 18.1.
   a. Using a weight of 1/2 for the most recent observation, 1/3 for the second most recent, and 1/6 for third most recent, compute a 3-week weighted moving average for the time series.
   b. Compute the MSE for the weighted moving average in part (a). Do you prefer this weighted moving average to the unweighted moving average? Remember that the MSE for the unweighted moving average is 10.22.
   c. Suppose you are allowed to choose any weights as long as they sum to one. Could you always find a set of weights that would make the MSE smaller for a weighted moving average than for an unweighted moving average? Why or why not?

4. With the gasoline time series data from Table 18.1, show the exponential smoothing forecasts using $\alpha = .1$. Applying the MSE criterion, would you prefer a smoothing constant of $\alpha = .1$ or $\alpha = .2$ for the gasoline sales time series?

5. With a smoothing constant of $\alpha = .2$, equation (18.2) shows that the forecast for the 13th week of the gasoline sales data from Table 18.1 is given by $F_{13} = .2Y_{12} + .8F_{12}$. However, the forecast for week 12 is given by $F_{12} = .2Y_{11} + .8F_{11}$. Thus, we could combine these two results to show that the forecast for the 13th week can be written

$$F_{13} = .2Y_{12} + .8(.2Y_{11} + .8F_{11}) = .2Y_{12} + .16Y_{11} + .64F_{11}$$

   a. Making use of the fact that $F_{11} = .2Y_{10} = .8F_{10}$ (and similarly for F_{10} and F_9), continue to expand the expression for F_{13} until it is written in terms of the past data values Y_{12}, Y_{11}, Y_{10}, Y_9, Y_8, and the forecast for period 8.
   b. Refer to the coefficients or weights for the past values Y_{12}, Y_{11}, Y_{10}, Y_9, Y_8; what observation can you make about how exponential smoothing weights past data values in arriving at new forecasts? Compare this weighting pattern with the weighting pattern of the moving averages method.

Applications

6. For the Hawkins Company, the monthly percentages of all shipments that were received on time over the past 12 months are 80, 82, 84, 83, 83, 84, 85, 84, 82, 83, 84, and 83.

a. Compare a 3-month moving average forecast with an exponential smoothing forecast for $\alpha = .2$. Which provides the better forecasts?
b. What is the forecast for next month?

7. Corporate triple A bond interest rates for 12 consecutive months follow.

9.5 9.3 9.4 9.6 9.8 9.7 9.8 10.5 9.9 9.7 9.6 9.6

a. Develop 3-month and 4-month moving averages for this time series. Does the 3-month or 4-month moving average provide the better forecasts? Explain.
b. What is the moving average forecast for the next month?

8. The values of Alabama building contracts ($ millions) for a 12-month period follow.

240 350 230 260 280 320 220 310 240 310 240 230

a. Compare a 3-month moving averages forecast with an exponential smoothing forecast using $\alpha = .2$. Which provides the better forecasts?
b. What is the forecast for the next month?

9. The following time series shows the sales of a particular product over the past 12 months.

Month	Sales	Month	Sales
1	105	7	145
2	135	8	140
3	120	9	100
4	105	10	80
5	90	11	100
6	120	12	110

a. Use $\alpha = .3$ to compute the exponential smoothing values for the time series.
b. Use a smoothing constant of .5 to compute the exponential smoothing values. Does a smoothing constant of .3 or .5 appear to provide the better forecasts?

10. Ten weeks of data on the Commodity Futures Index are 7.35, 7.40, 7.55, 7.56, 7.60, 7.52, 7.52, 7.70, 7.62, and 7.55.
a. Compute the exponential smoothing values for $\alpha = .2$.
b. Compute the exponential smoothing values for $\alpha = .3$.
c. Which exponential smoothing model provides the better forecasts? Forecast week 11.

11. The following data represent 15 quarters of manufacturing capacity utilization (in percentages).

Quarter/Year	Utilization (%)	Quarter/Year	Utilization (%)
1/1998	82.5	1/2000	78.8
2/1998	81.3	2/2000	78.7
3/1998	81.3	3/2000	78.4
4/1998	79.0	4/2000	80.0
1/1999	76.6	1/2001	80.7
2/1999	78.0	2/2001	80.7
3/1999	78.4	3/2001	80.8
4/1999	78.0		

a. Compute 3- and 4-quarter moving averages for this time series. Which moving average provides the better forecast for the fourth quarter of 2001?
b. Use smoothing constants of $\alpha = .4$ and $\alpha = .5$ to develop forecasts for the fourth quarter of 2001. Which smoothing constant provides the better forecast?
c. On the basis of the analyses in parts (a) and (b), which method—moving averages or exponential smoothing—provides the better forecast? Explain.

18.3 TREND PROJECTION

In this section we show how to forecast a time series that has a long-term linear trend. The type of time series for which the trend projection method is applicable shows a consistent increase or decrease over time; it is not stable so the smoothing methods described in the preceding section are not applicable.

Consider the time series for bicycle sales of a particular manufacturer over the past 10 years, as shown in Table 18.6 and Figure 18.8. Note that 21,600 bicycles were sold in year 1, 22,900 were sold in year 2, and so on. In year 10, the most recent year, 31,400 bicycles were sold. Although Figure 18.8 shows some up and down movement over the past 10 years, the time series seems to have an overall increasing or upward trend.

TABLE 18.6

BICYCLE SALES TIME SERIES

Year (t)	Sales (1000s) (Y_t)
1	21.6
2	22.9
3	25.5
4	21.9
5	23.9
6	27.5
7	31.5
8	29.7
9	28.6
10	31.4

We do not want the trend component of a time series to follow each and every up and down movement. Rather, the trend component should reflect the gradual shifting—in this case, growth—of the time series values. After we view the time series data in Table 18.6 and the graph in Figure 18.8, we might agree that a linear trend as shown in Figure 18.9 provides a reasonable description of the long-run movement in the series.

We use the bicycle sales data to illustrate the calculations involved in applying regression analysis to identify a linear trend. Recall that in the discussion of simple linear regression in Chapter 14, we described how the least squares method is used to find the best straight-line relationship between two variables. That is the methodology we will use to develop the trend line for the bicycle sales time series. Specifically, we will be using regression analysis to estimate the relationship between time and sales volume.

In Chapter 14 the estimated regression equation describing a straight-line relationship between an independent variable x and a dependent variable y was written

$$\hat{y} = b_0 + b_1 x \tag{18.4}$$

To emphasize the fact that, in forecasting, the independent variable is time, we will use t in equation (18.4) instead of x; in addition, we will use T_t in place of $\hat{y}$. Thus, for a

FIGURE 18.8 BICYCLE SALES TIME SERIES

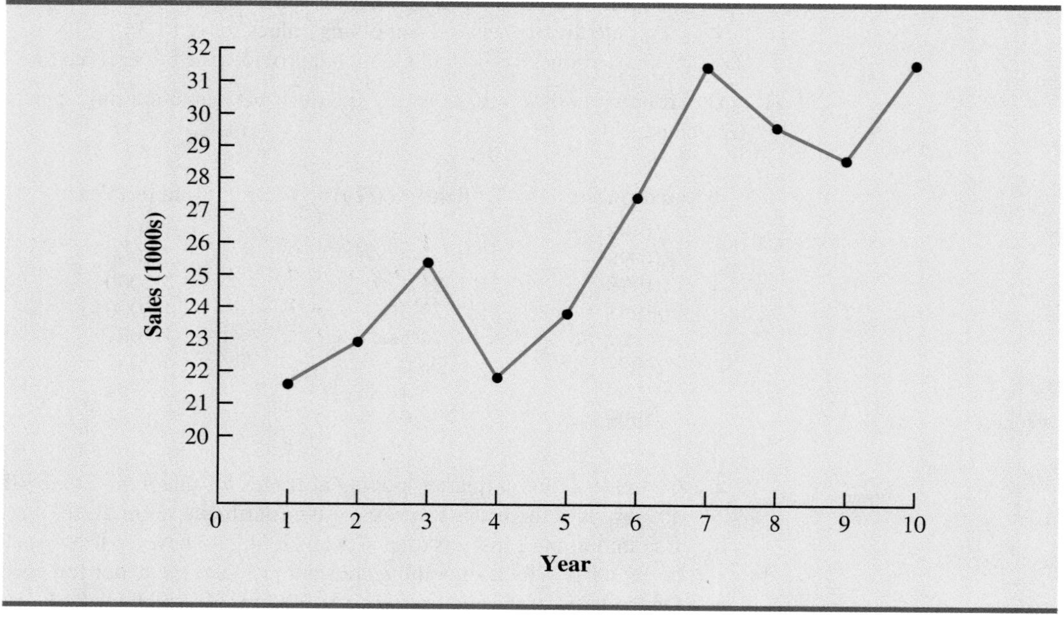

FIGURE 18.9 TREND REPRESENTED BY A LINEAR FUNCTION FOR BICYCLE SALES

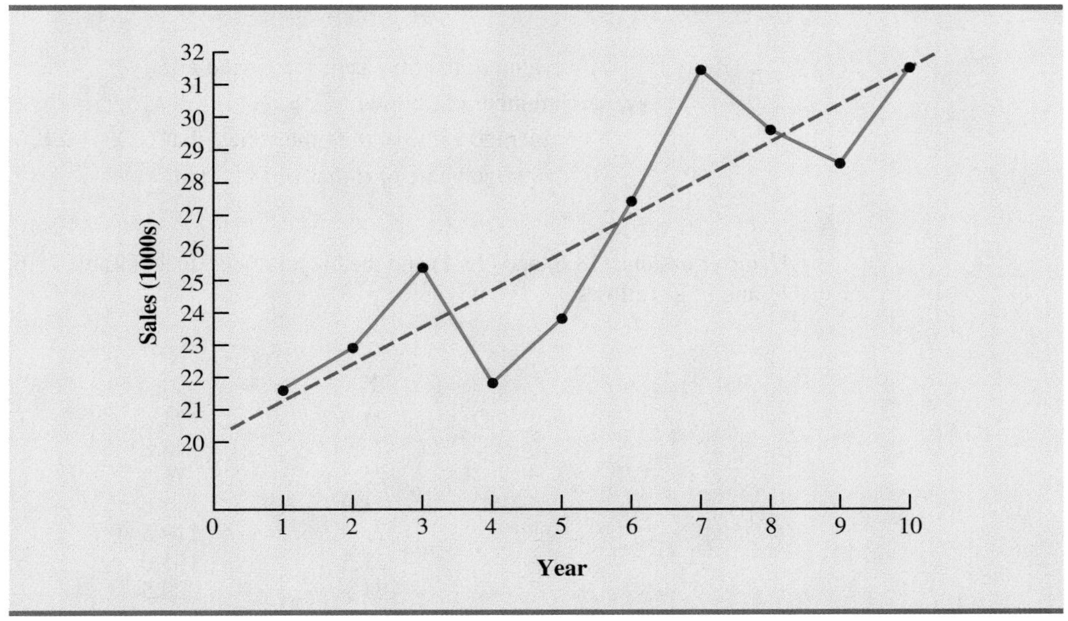

linear trend, the estimated sales volume expressed as a function of time can be written as follows.

Equation for Linear Trend

$$T_t = b_0 + b_1 t \tag{18.5}$$

where

T_t = trend value of the time series in period t
b_0 = intercept of the trend line
b_1 = slope of the trend line
t = time

In equation (18.5), we will let $t = 1$ for the time of the first observation on the time series data, $t = 2$ for the time of the second observation, and so on. Note that for the time series on bicycle sales, $t = 1$ corresponds to the oldest time series value and $t = 10$ corresponds to the most recent year's data. Formulas for computing the estimated regression coefficients (b_1 and b_0) in (18.5) are shown below.

Computing the Slope (b_1) and Intercept (b_0)

$$b_1 = \frac{\Sigma t Y_t - (\Sigma t \Sigma Y_t)/n}{\Sigma t^2 - (\Sigma t)^2/n} \tag{18.6}$$

$$b_0 = \bar{Y} - b_1 \bar{t} \tag{18.7}$$

where

Y_t = value of the time series in period t
n = number of periods
$\bar{Y}$ = average value of the time series; that is, $\bar{Y} = \Sigma Y_t / n$
$\bar{t}$ = average value of t; that is, $\bar{t} = \Sigma t / n$

Using equations (18.6) and (18.7) and the bicycle sales data of Table 18.6, we can compute b_0 and b_1 as follows:

	t	Y_t	tY_t	t^2
	1	21.6	21.6	1
	2	22.9	45.8	4
	3	25.5	76.5	9
	4	21.9	87.6	16
	5	23.9	119.5	25
	6	27.5	165.0	36
	7	31.5	220.5	49
	8	29.7	237.6	64
	9	28.6	257.4	81
	10	31.4	314.0	100
Totals	55	264.5	1545.5	385

$$\bar{t} = \frac{55}{10} = 5.5$$

$$\bar{Y} = \frac{264.5}{10} = 26.45$$

$$b_1 = \frac{1545.5 - (55)(264.5)/10}{385 - (55)^2/10} = 1.10$$

$$b_0 = 26.45 - 1.10(5.5) = 20.4$$

Therefore,

$$T_t = 20.4 + 1.1t \tag{18.8}$$

is the expression for the linear trend component for the bicycle sales time series.

Before the trend equation is used to develop a forecast, a statistical test of significance (see Chapter 14) should be conducted. In practice, such a test would be a routine part of fitting the trend line.

The slope of 1.1 indicates that over the past 10 years the firm has had an average growth in sales of about 1100 units per year. If we assume that the past 10-year trend in sales is a good indicator of the future, (18.8) can be used to project the trend component of the time series. For example, substituting $t = 11$ into (18.8) yields next year's trend projection, T_{11}.

$$T_{11} = 20.4 + 1.1(11) = 32.5$$

Thus, using the trend component only, we would forecast sales of 32,500 bicycles next year.

The use of a linear function to model the trend is common. However, as we discussed previously, sometimes time series have a curvilinear, or nonlinear, trend similar to those in

FIGURE 18.10 SOME POSSIBLE FORMS OF NONLINEAR TREND PATTERNS

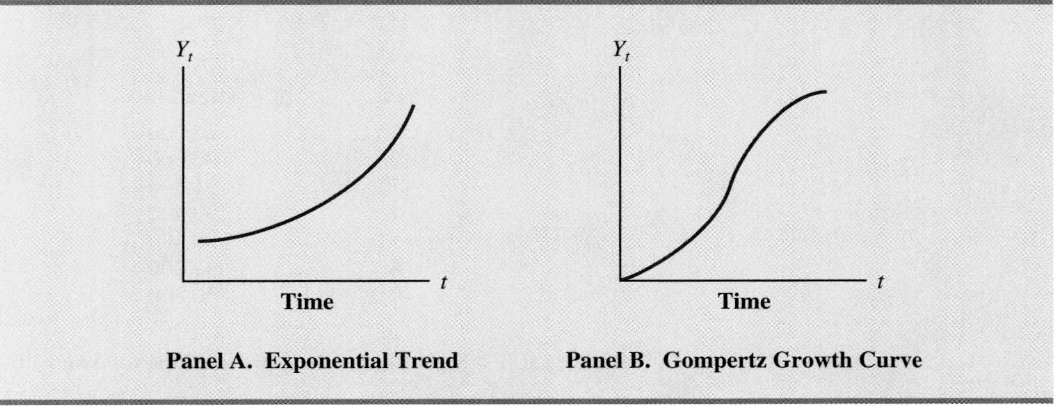

Figure 18.10. In Chapter 16 we discussed how regression analysis can be used to model curvilinear relationships of the type shown in panel A of Figure 18.10. More advanced texts discuss in detail how to develop regression models for more complex relationships, such as the one shown in panel B of Figure 18.10.

EXERCISES

Methods

12. Consider the following time series.

t	1	2	3	4	5
Y_t	6	11	9	14	15

Develop an equation for the linear trend component of this time series. What is the forecast for $t = 6$?

13. Consider the following time series.

t	1	2	3	4	5	6
Y_t	205	202	195	190	191	188

Develop an equation for the linear trend component for this time series. What is the forecast for $t = 7$?

Applications

14. The enrollment data (1000s) for a state college over the past 6 years are shown.

Year	1	2	3	4	5	6
Enrollment	20.5	20.2	19.5	19.0	19.1	18.8

Develop the equation for the linear trend component of this time series. Comment on what is happening to enrollment at this institution.

15. The following table gives average attendance figures for home football games at a major university for the past 7 years. Develop the equation for the linear trend component of this time series.

Year	Attendance
1	28,000
2	30,000
3	31,500
4	30,400
5	30,500
6	32,200
7	30,800

16. Automobile sales at B.J. Scott Motors, Inc., provided the following 10-year time series.

Year	Sales
1	400
2	390
3	320
4	340
5	270
6	260
7	300
8	320
9	340
10	370

Plot the time series and comment on the appropriateness of a linear trend. What type of functional form do you believe would be most appropriate for the trend pattern of this time series?

17. The president of a small manufacturing firm has been concerned about the continual increase in manufacturing costs over the past several years. The following figures provide a time series of the cost per unit for the firm's leading product over the past 8 years.

Year	Cost/Unit ($)
1	20.00
2	24.50
3	28.20
4	27.50
5	26.60
6	30.00
7	31.00
8	36.00

a. Show a graph of this time series. Does a linear trend appear to be present?

b. Develop the equation for the linear trend component of the time series. What is the average cost increase that the firm has been realizing per year?

18. Earnings per share for the Walgreen Company for a 10-year period follow.

.64 .73 .94 1.14 1.33 1.53 1.67 1.68 2.10 2.50

a. Use a linear trend projection to forecast this time series for the coming year.

b. What does this time series analysis tell you about the Walgreen Company? Do the historical data indicate that the Walgreen Company is a good investment?

19. In the late 1990s many firms began downsizing in order to reduce their costs. One of the results of these cost-cutting measures has been a decline in the percentage of private-industry jobs that are managerial. The following data show the percentage of females who are managers from 1990 to 1995 (*The Wall Street Journal Almanac,* 1998).

Year	1990	1991	1992	1993	1994	1995
Percentage	7.45	7.53	7.52	7.65	7.62	7.73

a. Develop a linear trend equation for this time series.
b. Use the trend equation to estimate the percentage of females who are managers for 1996 and 1997.

20. Gross revenue data ($ millions) for Regional Airlines for a 10-year period follow.

Year	Revenue	Year	Revenue
1	2428	6	4264
2	2951	7	4738
3	3533	8	4460
4	3618	9	5318
5	3616	10	6915

a. Develop a linear trend equation for this time series. Comment on what the equation tells about the gross revenue for Regional Airlines for the 10-year period.
b. Provide the forecasts for gross revenue for years 11 and 12.

21. ACT Networks, Inc., develops, markets, manufactures, and sells integrated wide-area network access products. The following are annual sales data from 1992 to 1997 (*Stock Investor Pro,* American Association of Individual Investors, August 31, 1997).

Year	Sales ($millions)
1992	5.4
1993	6.2
1994	12.7
1995	20.6
1996	28.4
1997	44.9

a. Develop a linear trend equation for this time series.
b. What is the firm's average increase in sales per year?
c. Use the trend equation to forecast sales for 1998.

18.4 TREND AND SEASONAL COMPONENTS

We have shown how to forecast a time series that has a trend component. In this section we extend the discussion by showing how to forecast a time series that has both trend and seasonal components.

Many situations in business and economics involve period-to-period comparisons. For instance, we might be interested to learn that unemployment is up 2% compared to last month, steel production is up 5% over last month, or that the production of electric power is down 3% from the previous month. Care must be exercised in using such information, however, because whenever a seasonal influence is present, such comparisons may be misleading. For instance, the fact that electric power consumption is down 3% from August to September might be only the seasonal effect associated with a decrease in the use of air conditioning and not because of a long-term decline in the use of electric power. Indeed, after adjusting for the seasonal effect, we might even find that the use of electric power has increased.

Removing the seasonal effect from a time series is known as deseasonalizing the time series. After we do so, period-to-period comparisons are more meaningful and can help identify whether a trend exists. The approach we take in this section is appropriate in situations when only seasonal effects are present or in situations when both seasonal and trend components are present. The first step is to compute seasonal indexes and use them to deseasonalize the data. Then, if a trend is apparent in the deseasonalized data, we use regression analysis on the deseasonalized data to estimate the trend component.

Multiplicative Model

In addition to a trend component (T) and a seasonal component (S), we will assume that the time series has an irregular component (I). The irregular component accounts for any random effects in the time series that cannot be explained by the trend and seasonal components. Using T_t, S_t, and I_t to identify the trend, seasonal, and irregular components at time t, we will assume that the time series value, denoted Y_t, can be described by the following **multiplicative time series model.**

$$Y_t = T_t \times S_t \times I_t \tag{18.9}$$

In this model, T_t is the trend measured in units of the item being forecast. However, the S_t and I_t components are measured in relative terms, with values above 1.00 indicating effects above the trend and values below 1.00 indicating effects below the trend.

We will illustrate the use of the multiplicative model with trend, seasonal, and irregular components by working with the quarterly data in Table 18.7 and Figure 18.11. These data show television set sales (in thousands of units) for a particular manufacturer over the past 4 years. We begin by showing how to identify the seasonal component of the time series.

TABLE 18.7

QUARTERLY DATA FOR TELEVISION SET SALES

Year	Quarter	Sales (1000s)
1	1	4.8
	2	4.1
	3	6.0
	4	6.5
2	1	5.8
	2	5.2
	3	6.8
	4	7.4
3	1	6.0
	2	5.6
	3	7.5
	4	7.8
4	1	6.3
	2	5.9
	3	8.0
	4	8.4

Calculating the Seasonal Indexes

Figure 18.11 indicates that sales are lowest in the second quarter of each year and increase in quarters 3 and 4. Thus, we conclude that a seasonal pattern exists for television set sales. We can begin the computational procedure used to identify each quarter's seasonal influence by computing a moving average to separate the combined seasonal and irregular components, S_t and I_t, from the trend component T_t.

To do so, we use one year of data in each calculation. Because we are working with a quarterly series, we will use four data values in each moving average. The moving average calculation for the first four quarters of the television set sales data is

$$\text{First Moving Average} = \frac{4.8 + 4.1 + 6.0 + 6.5}{4} = \frac{21.4}{4} = 5.35$$

Note that the moving average calculation for the first four quarters yields the average quarterly sales over year 1 of the time series. Continuing the moving average calculation, we next add the 5.8 value for the first quarter of year 2 and drop the 4.8 for the first quarter of year 1. Thus, the second moving average is

$$\text{Second Moving Average} = \frac{4.1 + 6.0 + 6.5 + 5.8}{4} = \frac{22.4}{4} = 5.60$$

Similarly, the third moving average calculation is $(6.0 + 6.5 + 5.8 + 5.2)/4 = 5.875$.

Before we proceed with the moving average calculations for the entire time series, we return to the first moving average calculation, which resulted in a value of 5.35. The 5.35 value represents an average quarterly sales volume (across all seasons) for year 1. As we

FIGURE 18.11 QUARTERLY TELEVISION SET SALES TIME SERIES

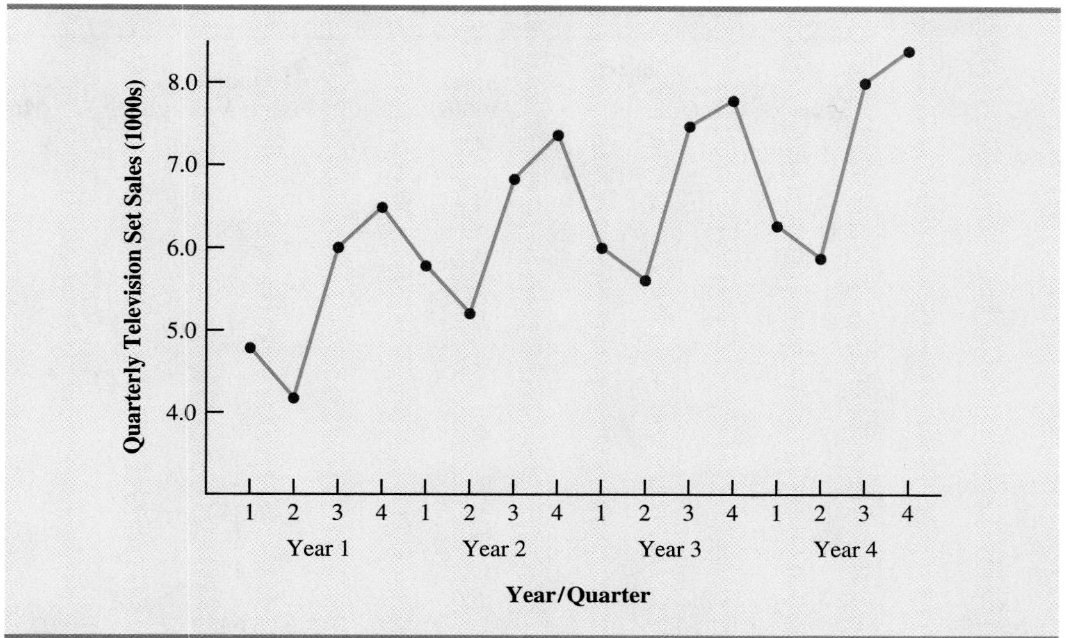

look back at the calculation of the 5.35 value, associating 5.35 with the "middle" quarter of the moving average group makes sense. Note, however, that we encounter some difficulty in identifying the middle quarter; with four quarters in the moving average there is no middle quarter. The 5.35 value corresponds to the last half of quarter 2 and the first half of quarter 3. Similarly, if we go to the next moving average value of 5.60, the middle corresponds to the last half of quarter 3 and the first half of quarter 4.

Recall that the reason for computing moving averages is to isolate the combined seasonal and irregular components. However, the moving average values we have computed do not correspond directly to the original quarters of the time series. We can resolve this difficulty by using the midpoints between successive moving average values. For example, if 5.35 corresponds to the first half of quarter 3 and 5.60 corresponds to the last half of quarter 3, we can use $(5.35 + 5.60)/2 = 5.475$ as the moving average value for quarter 3. Similarly, we associate a moving average value of $(5.60 + 5.875)/2 = 5.738$ with quarter 4. The result is a *centered moving average.* Table 18.8 shows a complete summary of the moving average calculations for the television set sales data.

If the number of data points in a moving average calculation is an odd number, the middle point will correspond to one of the periods in the time series. In such cases, we would not have to center the moving average values to correspond to a particular time period as we have done in the calculations in Table 18.8.

What do the centered moving averages in Table 18.8 tell us about this time series? Figure 18.12 is a plot of the actual time series values and the centered moving average values. Note particularly how the centered moving average values tend to "smooth out" both the seasonal and irregular fluctuations in the time series. The moving average values computed for 4 quarters of data do not include the fluctuations due to seasonal influences because the seasonal effect has been averaged out. Each point in the centered moving average represents the value of the time series as though there were no seasonal or irregular influence.

By dividing each time series observation by the corresponding centered moving average, we can identify the seasonal irregular effect in the time series. For example, the third quarter

TABLE 18.8 CENTERED MOVING AVERAGE CALCULATIONS FOR THE TELEVISION SET SALES TIME SERIES

Year	Quarter	Sales (1000s)	4-Quarter Moving Average	Centered Moving Average
1	1	4.8		
	2	4.1		
			5.350	
	3	6.0		5.475
			5.600	
	4	6.5		5.738
			5.875	
2	1	5.8		5.975
			6.075	
	2	5.2		6.188
			6.300	
	3	6.8		6.325
			6.350	
	4	7.4		6.400
			6.450	
3	1	6.0		6.538
			6.625	
	2	5.6		6.675
			6.725	
	3	7.5		6.763
			6.800	
	4	7.8		6.838
			6.875	
4	1	6.3		6.938
			7.000	
	2	5.9		7.075
			7.150	
	3	8.0		
	4	8.4		

of year 1 shows 6.0/5.475 = 1.096 as the combined seasonal irregular value. Table 18.9 summarizes the seasonal irregular values for the entire time series.

Consider the third quarter. The results from years 1, 2, and 3 show third-quarter values of 1.096, 1.075, and 1.109, respectively. Thus, in all cases, the seasonal irregular value appears to have an above-average influence in the third quarter. With the year-to-year fluctuations in the seasonal irregular value attributable primarily to the irregular component, we can average the computed values to eliminate the irregular influence and obtain an estimate of the third-quarter seasonal influence.

$$\text{Seasonal Effect of Third Quarter} = \frac{1.096 + 1.075 + 1.109}{3} = 1.09$$

We refer to 1.09 as the *seasonal index* for the third quarter. In Table 18.10 we summarize the calculations involved in computing the seasonal indexes for the television set sales time

FIGURE 18.12 QUARTERLY TELEVISION SET SALES TIME SERIES AND CENTERED MOVING AVERAGE

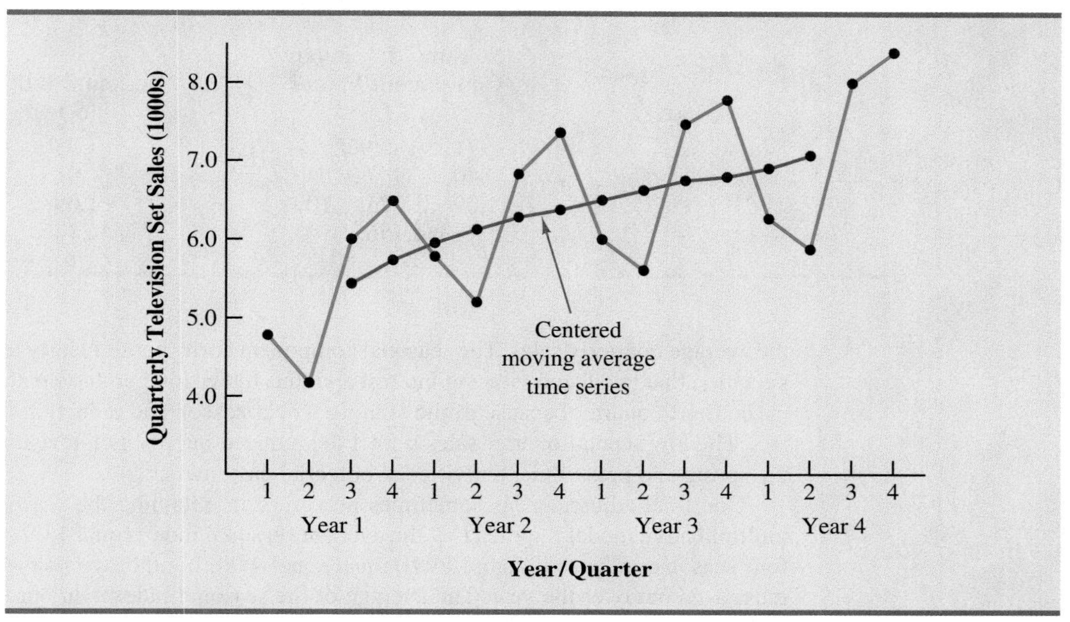

series. Thus, the seasonal indexes for the four quarters are: quarter 1, .93; quarter 2, .84; quarter 3, 1.09; and quarter 4, 1.14.

Interpretation of the values in Table 18.10 provides some observations about the seasonal component in television set sales. The best sales quarter is the fourth quarter, with sales averaging 14% above the average quarterly value. The worst, or slowest, sales quarter is the second quarter; its seasonal index of .84 shows that the sales average is 16% below

TABLE 18.9 SEASONAL IRREGULAR VALUES FOR THE TELEVISION SET SALES TIME SERIES

Year	Quarter	Sales (1000s)	Centered Moving Average	Seasonal Irregular Value
1	1	4.8		
	2	4.1		
	3	6.0	5.475	1.096
	4	6.5	5.738	1.133
2	1	5.8	5.975	.971
	2	5.2	6.188	.840
	3	6.8	6.325	1.075
	4	7.4	6.400	1.156
3	1	6.0	6.538	.918
	2	5.6	6.675	.839
	3	7.5	6.763	1.109
	4	7.8	6.838	1.141
4	1	6.3	6.938	.908
	2	5.9	7.075	.834
	3	8.0		
	4	8.4		

TABLE 18.10 SEASONAL INDEX CALCULATIONS FOR THE TELEVISION SET SALES TIME SERIES

Quarter	Seasonal Irregular Component Values ($S_t I_t$)	Seasonal Index (S_t)
1	.971, .918, .908	.93
2	.840, .839, .834	.84
3	1.096, 1.075, 1.109	1.09
4	1.133, 1.156, 1.141	1.14

the average quarterly sales. The seasonal component corresponds clearly to the intuitive expectation that television viewing interest and thus television purchase patterns tend to peak in the fourth quarter because of the coming winter season and reduction in outdoor activities. The low second-quarter sales reflect the reduced interest in television viewing due to the spring and presummer activities of potential customers.

One final adjustment is sometimes necessary in obtaining the seasonal indexes. The multiplicative model requires that the average seasonal index equal 1.00, so the sum of the four seasonal indexes in Table 18.10 must equal 4.00. In other words the seasonal effects must even out over the year. The average of the seasonal indexes in our example is equal to 1.00, and hence this type of adjustment is not necessary. In other cases, a slight adjustment may be necessary. To make the adjustment, multiply each seasonal index by the number of seasons divided by the sum of the unadjusted seasonal indexes. For instance, for quarterly data multiply each seasonal index by 4/(sum of the unadjusted seasonal indexes). Some of the exercises will require this adjustment to obtain the appropriate seasonal indexes.

Deseasonalizing the Time Series

The purpose of finding seasonal indexes is to remove the seasonal effects from a time series. This process is referred to as *deseasonalizing* the time series. Economic time series adjusted for seasonal variations (**deseasonalized time series**) are often reported in publications such as the *Survey of Current Business, The Wall Street Journal,* and *Business Week.* Using the notation of the multiplicative model, we have

With deseasonalized data, comparing sales in successive periods makes sense. With data that have not been deseasonalized, relevant comparisons can often be made between sales in the current period and sales in the same period one year ago.

$$Y_t = T_t \times S_t \times I_t$$

By dividing each time series observation by the corresponding seasonal index, we have removed the effect of season from the time series. The deseasonalized time series for television set sales is summarized in Table 18.11. A graph of the deseasonalized television set sales time series is shown in Figure 18.13.

Using the Deseasonalized Time Series to Identify Trend

Although the graph in Figure 18.13 shows some random up and down movement over the past 16 quarters, the time series seems to have an upward linear trend. To identify this trend, we will use the same procedure as in the preceding section; in this case, the data are quarterly deseasonalized sales values. Thus, for a linear trend, the estimated sales volume expressed as a function of time is

$$T_t = b_0 + b_1 t$$

FIGURE 18.13 DESEASONALIZED TELEVISION SET SALES TIME SERIES

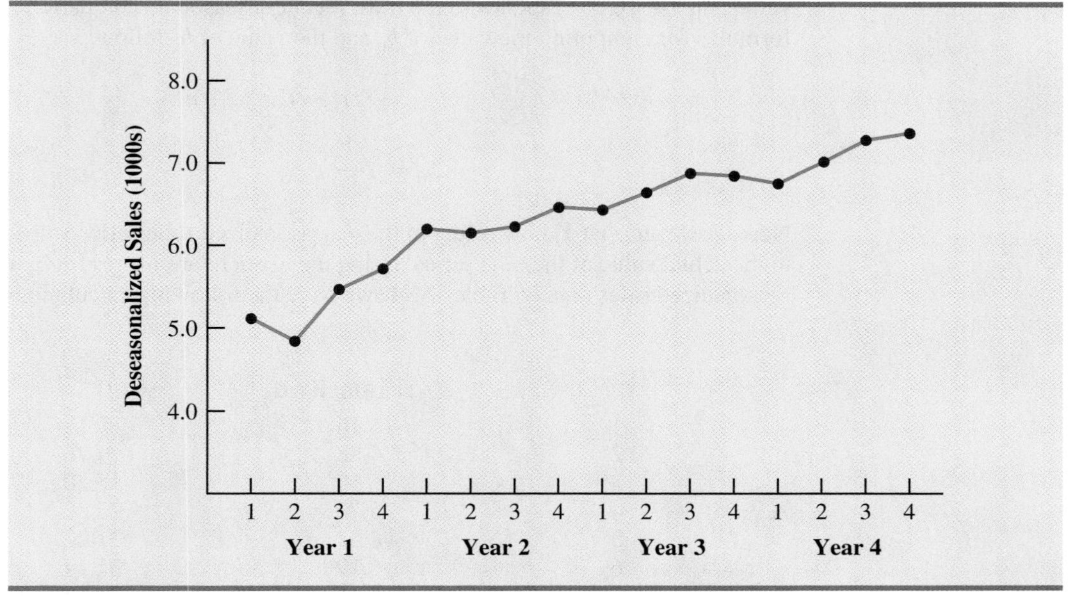

where

T_t = trend value for television set sales in period t
b_0 = intercept of the trend line
b_1 = slope of the trend line

As before, $t = 1$ corresponds to the time of the first observation for the time series, $t = 2$ corresponds to the time of the second observation, and so on. Thus, for the deseasonalized

TABLE 18.11 DESEASONALIZED VALUES FOR THE TELEVISION SET SALES TIME SERIES

Year	Quarter	Sales (1000s) (Y_t)	Seasonal Index (S_t)	Deseasonalized Sales ($Y_t/S_t = T_tI_t$)
1	1	4.8	.93	5.16
	2	4.1	.84	4.88
	3	6.0	1.09	5.50
	4	6.5	1.14	5.70
2	1	5.8	.93	6.24
	2	5.2	.84	6.19
	3	6.8	1.09	6.24
	4	7.4	1.14	6.49
3	1	6.0	.93	6.45
	2	5.6	.84	6.67
	3	7.5	1.09	6.88
	4	7.8	1.14	6.84
4	1	6.3	.93	6.77
	2	5.9	.84	7.02
	3	8.0	1.09	7.34
	4	8.4	1.14	7.37

television set sales time series, $t = 1$ corresponds to the first deseasonalized quarterly sales value and $t = 16$ corresponds to the most recent deseasonalized quarterly sales value. The formulas for computing the value of b_0 and the value of b_1 follow.

$$b_1 = \frac{\Sigma tY_t - (\Sigma t \Sigma Y_t)/n}{\Sigma t^2 - (\Sigma t)^2/n}$$

$$b_0 = \bar{Y} - b_1\bar{t}$$

Note, however, that Y_t now refers to the deseasonalized time series value at time t and not to the actual value of the time series. Using the given relationships for b_0 and b_1 and the deseasonalized sales data of Table 18.11, we have the following calculations.

	t	Y_t (Deseasonalized)	tY_t	t^2
	1	5.16	5.16	1
	2	4.88	9.76	4
	3	5.50	16.50	9
	4	5.70	22.80	16
	5	6.24	31.20	25
	6	6.19	37.14	36
	7	6.24	43.68	49
	8	6.49	51.92	64
	9	6.45	58.05	81
	10	6.67	66.70	100
	11	6.88	75.68	121
	12	6.84	82.08	144
	13	6.77	88.01	169
	14	7.02	98.28	196
	15	7.34	110.10	225
	16	7.37	117.92	256
Totals	136	101.74	914.98	1496

where

$$\bar{t} = \frac{136}{16} = 8.5$$

$$\bar{Y} = \frac{101.74}{16} = 6.359$$

$$b_1 = \frac{914.98 - (136)(101.74)/16}{1496 - (136)^2/16} = 0.148$$

$$b_0 = 6.359 - 0.148(8.5) = 5.101$$

Therefore,

$$T_t = 5.101 + 0.148t$$

is the expression for the linear trend component of the time series.

The slope of 0.148 indicates that over the past 16 quarters, the firm has had an average deseasonalized growth in sales of around 148 sets per quarter. If we assume that the past 16-quarter trend in sales data is a reasonably good indicator of the future, this equation can be used to project the trend component of the time series for future quarters. For example, substituting $t = 17$ into the equation yields next quarter's trend projection, T_{17}.

$$T_{17} = 5.101 + 0.148(17) = 7.617$$

Thus, the trend component yields a sales forecast of 7617 television sets for the next quarter. Similarly, the trend component produces sales forecasts of 7765, 7913, and 8061 television sets in quarters 18, 19, and 20, respectively.

Seasonal Adjustments

The final step in developing the forecast when both trend and seasonal components are present is to use the seasonal index to adjust the trend projection. Returning to the television set sales example, we have a trend projection for the next four quarters. Now we must adjust the forecast for the seasonal effect. The seasonal index for the first quarter of year 5 ($t = 17$) is 0.93, so we obtain the quarterly forecast by multiplying the forecast based on trend ($T_{17} = 7617$) by the seasonal index (0.93). Thus, the forecast for the next quarter is $7617(0.93) = 7084$. Table 18.12 gives the quarterly forecast for quarters 17 through 20. The high-volume fourth quarter has a 9190-unit forecast, and the low-volume second quarter has a 6523-unit forecast.

Models Based on Monthly Data

In the preceding television set sales example, we used quarterly data to illustrate the computation of seasonal indexes. However, many businesses use monthly rather than quarterly forecasts. In such cases, the procedures introduced in this section can be applied with minor modifications. First, a 12-month moving average replaces the 4-quarter moving average; second, 12 monthly seasonal indexes, rather than four quarterly seasonal indexes, must be computed. Other than these changes, the computational and forecasting procedures are identical.

Cyclical Component

Mathematically, the multiplicative model of (18.9) can be expanded to include a cyclical component.

$$Y_t = T_t \times C_t \times S_t \times I_t \tag{18.10}$$

The cyclical component, like the seasonal component, is expressed as a percentage of trend. As mentioned in Section 18.1, this component is attributable to multiyear cycles in the time series. It is analogous to the seasonal component, but over a longer period of time. However, because of the length of time involved, obtaining enough relevant data to estimate the cyclical component is often difficult. Another difficulty is that cycles usually vary in length. We leave further discussion of the cyclical component to texts on forecasting methods.

TABLE 18.12 QUARTERLY FORECASTS FOR THE TELEVISION SET SALES TIME SERIES

Year	Quarter	Trend Forecast	Seasonal Index (see Table 18.11)	Quarterly Forecast
5	1	7617	.93	(7617)(.93) = 7084
	2	7765	.84	(7765)(.84) = 6523
	3	7913	1.09	(7913)(1.09) = 8625
	4	8061	1.14	(8061)(1.14) = 9190

EXERCISES

Methods

22. Consider the following time series data.

Quarter	Year 1	Year 2	Year 3
1	4	6	7
2	2	3	6
3	3	5	6
4	5	7	8

a. Show the 4-quarter and centered moving average values for this time series.
b. Compute seasonal indexes for the 4 quarters.

Applications

23. The quarterly sales data (number of copies sold) for a college textbook over the past 3 years follow.

Quarter	Year 1	Year 2	Year 3
1	1690	1800	1850
2	940	900	1100
3	2625	2900	2930
4	2500	2360	2615

a. Show the 4-quarter and centered moving average values for this time series.
b. Compute seasonal indexes for the 4 quarters.
c. When does the textbook publisher have the largest seasonal index? Does this appear reasonable? Explain.

24. Identify the monthly seasonal indexes for the 3 years of expenses for a six-unit apartment house in southern Florida as given here. Use a 12-month moving average calculation.

	Expenses		
	Year 1	Year 2	Year 3
January	170	180	195
February	180	205	210
March	205	215	230
April	230	245	280
May	240	265	290
June	315	330	390
July	360	400	420
August	290	335	330
September	240	260	290
October	240	270	295
November	230	255	280
December	195	220	250

25. Air pollution control specialists in southern California monitor the amount of ozone, carbon dioxide, and nitrogen dioxide in the air on an hourly basis. The hourly time series data exhibit seasonality, with the levels of pollutants showing patterns over the hours in the day. On July 15, 16, and 17, the following levels of nitrogen dioxide were observed in the downtown area for the 12 hours from 6:00 A.M. to 6:00 P.M.

July 15:	25	28	35	50	60	60	40	35	30	25	25	20
July 16:	28	30	35	48	60	65	50	40	35	25	20	20
July 17:	35	42	45	70	72	75	60	45	40	25	25	25

a. Identify the hourly seasonal indexes for the 12 readings each day.

b. With the seasonal indexes from part (a), the data were deseasonalized; the trend equation developed for the deseasonalized data was $T_t = 32.983 + .3922t$. Using the trend component only, develop forecasts for the 12 hours for July 18.

c. Use the seasonal indexes from part (a) to adjust the trend forecasts developed in part (b).

26. Electric power consumption is measured in kilowatt-hours (kWh). The local utility company has an interrupt program whereby commercial customers that participate receive favorable rates but must agree to cut back consumption if the utility requests them to do so. Timko Products cut back consumption at 12:00 noon Thursday. To assess the savings, the utility must estimate Timko's usage without the interrupt. The period of interrupted service was from noon to 8:00 P.M. Data on electric power consumption for the previous 72 hours are available.

Time Period	**Monday**	**Tuesday**	**Wednesday**	**Thursday**
12–4 A.M.	—	19,281	31,209	27,330
4–8 A.M.	—	33,195	37,014	32,715
8–12 noon	—	99,516	119,968	152,465
12–4 P.M.	124,299	123,666	156,033	
4–8 P.M.	113,545	111,717	128,889	
8–12 midnight	41,300	48,112	73,923	

a. Is there a seasonal effect over the 24-hour period? Compute seasonal indexes for the six 4-hour periods.

b. Use trend adjusted for seasonal indexes to estimate Timko's normal usage over the period of interrupted service.

18.5 REGRESSION ANALYSIS

In the discussion of regression analysis in Chapters 14, 15, and 16, we showed how one or more independent variables could be used to predict the value of a single dependent variable. Looking at regression analysis as a forecasting tool, we can view the time series value that we want to forecast as the dependent variable. Hence, if we can identify a good set of related independent, or predictor, variables we may be able to develop an estimated regression equation for predicting or forecasting the time series.

The approach we used in Section 18.3 to fit a linear trend line to the bicycle sales time series is a special case of regression analysis. In that example, two variables—bicycle sales and time—were shown to be linearly related.* The inherent complexity of most real-world problems necessitates the consideration of more than one variable to predict the variable of interest. The statistical technique known as multiple regression analysis can be used in such situations.

Recall that to develop an estimated multiple regression equation, we need a sample of observations for the dependent variable and all independent variables. In time series analysis the n periods of time series data provide a sample of n observations on each variable that

*In a purely technical sense, the number of bicycles sold is not viewed as being related to time; instead, time is used as a surrogate for variables to which the number of bicycles sold is actually related but which are either unknown or too difficult or too costly to measure.

can be used in the analysis. For a function involving k independent variables, we use the following notation.

$$
\begin{aligned}
Y_t &= \text{value of the time series in period } t \\
x_{1t} &= \text{value of independent variable 1 in period } t \\
x_{2t} &= \text{value of independent variable 2 in period } t \\
&\ \ \vdots \\
x_{kt} &= \text{value of independent variable } k \text{ in period } t
\end{aligned}
$$

The n periods of data necessary to develop the estimated regression equation would appear as shown in the following table.

Period	Time Series (Y_t)	Value of Independent Variables						
		x_{1t}	x_{2t}	x_{3t}	.	.	.	x_{kt}
1	Y_1	x_{11}	x_{21}	x_{31}	.	.	.	x_{k1}
2	Y_2	x_{12}	x_{22}	x_{32}	.	.	.	x_{k2}
.	.	.	.	.	.	.	.	.
.	.	.	.	.	.	.	.	.
.	.	.	.	.	.	.	.	.
n	Y_n	x_{1n}	x_{2n}	x_{3n}	.	.	.	x_{kn}

As you might imagine, several choices are possible for the independent variables in a forecasting model. One possible choice for an independent variable is simply time. It was the choice made in Section 18.3 when we estimated the trend of the time series using a linear function of the independent variable time. Letting $x_{1t} = t$, we obtain an estimated regression equation of the form

$$\hat{Y}_t = b_0 + b_1 t$$

where $\hat{Y}_t$ is the estimate of the time series value Y_t and where b_0 and b_1 are the estimated regression coefficients. In a more complex model, additional terms could be added corresponding to time raised to other powers. For example, if $x_{2t} = t^2$ and $x_{3t} = t^3$, the estimated regression equation would become

$$
\begin{aligned}
\hat{Y}_t &= b_0 + b_1 x_{1t} + b_2 x_{2t} + b_3 x_{3t} \\
&= b_0 + b_1 t + b_2 t^2 + b_3 t^3
\end{aligned}
$$

Note that this model provides a forecast of a time series with curvilinear characteristics over time.

Other regression-based forecasting models have a mixture of economic and demographic independent variables. For example, in forecasting sales of refrigerators, we might select the following independent variables.

$$
\begin{aligned}
x_{1t} &= \text{price in period } t \\
x_{2t} &= \text{total industry sales in period } t - 1 \\
x_{3t} &= \text{number of building permits for new houses in period } t - 1 \\
x_{4t} &= \text{population forecast for period } t \\
x_{5t} &= \text{advertising budget for period } t
\end{aligned}
$$

According to the usual multiple regression procedure, an estimated regression equation with five independent variables would be used to develop forecasts.

Spyros Makridakis, a noted forecasting expert, has conducted research showing that simple techniques usually outperform more complex procedures for short-term forecasting. Using a more sophisticated and expensive procedure will not guarantee better forecasts.

Whether a regression approach provides a good forecast depends largely on how well we are able to identify and obtain data for independent variables that are closely related to the time series. Generally, during the development of an estimated regression equation, we will want to consider many possible sets of independent variables. Thus, part of the regression analysis procedure should be the selection of the set of independent variables that provides the best forecasting model.

In the chapter introduction we stated that the **causal forecasting models** use other time series related to the one being forecast in an effort to explain the cause of a time series' behavior. Regression analysis is the tool most often used in developing such causal models. The related time series become the independent variables, and the time series being forecast is the dependent variable.

In another type of regression-based forecasting model, the independent variables are all previous values of the same time series. For example, if the time series values are denoted $Y_1, Y_2, \ldots, Y_n$, then with a dependent variable Y_t, we might try to find an estimated regression equation relating Y_t to the most recent times series values Y_{t-1}, Y_{t-2}, and so on. With the three most recent periods as independent variables, the estimated regression equation would be

$$\hat{Y}_t = b_0 + b_1 Y_{t-1} + b_2 Y_{t-2} + b_3 Y_{t-3}$$

Regression models in which the independent variables are previous values of the time series are referred to as **autoregressive models.**

Finally, another regression-based forecasting approach incorporates a mixture of the independent variables previously discussed. For example, we might select a combination of time variables, some economic/demographic variables, and some previous values of the time series variable itself.

18.6 QUALITATIVE APPROACHES

If historical data are not available, managers must use a qualitative technique to develop forecasts. But the cost of using qualitative techniques can be high because of the time commitment required from the people involved.

In the preceding sections we discussed several types of quantitative forecasting methods. Most of those techniques require historical data on the variable of interest, so they cannot be applied when historical data are not available. Furthermore, even when such data are available, a significant change in environmental conditions affecting the time series may make the use of past data questionable in predicting future values of the time series. For example, a government-imposed gasoline rationing program would raise questions about the validity of a gasoline sales forecast based on historical data. Qualitative forecasting techniques afford an alternative in these and other cases.

Delphi Method

One of the most commonly used qualitative forecasting techniques is the **Delphi method,** originally developed by a research group at the Rand Corporation. It is an attempt to develop forecasts through "group consensus." In its usual application, the members of a panel of experts—all of whom are physically separated from and unknown to each other—are asked to respond to a series of questionnaires. The responses from the first questionnaire are tabulated and used to prepare a second questionnaire that contains information and opinions of the entire group. Each respondent is then asked to reconsider and possibly revise his or her previous response in light of the group information provided. This process continues until the coordinator feels that some degree of consensus has been reached. The goal of the

Delphi method is not to produce a single answer as output, but instead to produce a relatively narrow spread of opinions within which the majority of experts concur.

Expert Judgment

Empirical evidence and theoretical arguments suggest that between 5 and 20 experts should be used in judgmental forecasting. However, in situations involving exponential growth, judgmental forecasts may be inappropriate.

Qualitative forecasts often are based on the judgment of a single expert or represent the consensus of a group of experts. For example, each year a group of experts at Merrill Lynch gather to forecast the level of the Dow Jones Industrial Average and the prime rate for the next year. In doing so, the experts individually consider information that they believe will influence the stock market and interest rates; then they combine their conclusions into a forecast. No formal model is used, and no two experts are likely to consider the same information in the same way.

Expert judgment is a forecasting method that is commonly recommended when conditions in the past are not likely to hold in the future. Even though no formal quantitative model is used, expert judgment has provided good forecasts in many situations.

Scenario Writing

The qualitative procedure known as **scenario writing** consists of developing a conceptual scenario of the future based on a well-defined set of assumptions. Different sets of assumptions lead to different scenarios. The job of the decision maker is to decide how likely each scenario is and then to make decisions accordingly.

Intuitive Approaches

Subjective or *intuitive qualitative* approaches are based on the ability of the human mind to process a variety of information that, in most cases, is difficult to quantify. These techniques are often used in group work, wherein a committee or panel seeks to develop new ideas or solve complex problems through a series of "brainstorming sessions." In such sessions, individuals are freed from the usual group restrictions of peer pressure and criticism because they can present any idea or opinion without regard to its relevancy and, even more important, without fear of criticism.

SUMMARY

This chapter provided an introduction to the basic methods of time series analysis and forecasting. First, we showed that to explain the behavior of a time series, it is often helpful to think of the time series as consisting of four separate components: trend, cyclical, seasonal, and irregular. By isolating these components and measuring their apparent effect, one can forecast future values of the time series.

We discussed how smoothing methods can be used to forecast a time series that exhibits no significant trend, seasonal, or cyclical effect. The moving averages approach consists of computing an average of past data values and then using that average as the forecast for the next period. In the exponential smoothing method, a weighted average of past time series values is used to compute a forecast.

For time series that have only a long-term trend, we showed how regression analysis could be used to make trend projections. For time series in which both trend and seasonal influences are significant, we showed how to isolate the effects of the two factors and prepare better forecasts. Finally, regression analysis was described as a procedure for developing causal forecasting models. A causal forecasting model is one that relates the time series value (dependent variable) to other independent variables that are believed to explain (cause) the time series behavior.

Qualitative forecasting methods were discussed as approaches that could be used when little or no historical data are available. These methods are also considered most appropriate when the past pattern of the time series is not expected to continue into the future.

GLOSSARY

Time series A set of observations measured at successive points in time or over successive periods of time.

Forecast A projection or prediction of future values of a time series.

Trend The long-run shift or movement in the time series observable over several periods of time.

Cyclical component The component of the time series model that results in periodic above-trend and below-trend behavior of the time series lasting more than one year.

Seasonal component The component of the time series model that shows a periodic pattern over one year or less.

Irregular component The component of the time series model that reflects the random variation of the time series values beyond what can be explained by the trend, cyclical, and seasonal components.

Moving averages A method of forecasting or smoothing a time series by averaging each successive group of data points.

Mean squared error (MSE) An approach to measuring the accuracy of a forecasting model. This measure is the average of the sum of the squared differences between the forecast values and the actual time series values.

Weighted moving averages A method of forecasting or smoothing a time series by computing a weighted average of past data values. The sum of the weights must equal one.

Exponential smoothing A forecasting technique that uses a weighted average of past time series values to arrive at smoothed time series values that can be used as forecasts.

Smoothing constant A parameter of the exponential smoothing model that provides the weight given to the most recent time series value in the calculation of the forecast value.

Multiplicative time series model A model whereby the separate components of the time series are multiplied together to identify the actual time series value. When the four components of trend, cyclical, seasonal, and irregular are assumed present, we obtain $Y_t = T_t \times C_t \times S_t \times I_t$. When the cyclical component is not modeled, we obtain $Y_t = T_t \times S_t \times I_t$.

Deseasonalized time series A time series from which the effect of season has been removed by dividing each original time series observation by the corresponding seasonal index.

Causal forecasting methods Forecasting methods that relate a time series to other variables that are believed to explain or cause its behavior.

Autoregressive model A time series model whereby a regression relationship based on past time series values is used to predict the future time series values.

Delphi method A qualitative forecasting method that obtains forecasts through group consensus.

Scenario writing A qualitative forecasting method that consists of developing a conceptual scenario of the future based on a well-defined set of assumptions.

KEY FORMULAS

Moving Average

$$\text{Moving Average} = \frac{\Sigma(\text{most recent } n \text{ data values})}{n} \tag{18.1}$$

Exponential Smoothing Model

$$F_{t+1} = \alpha Y_t + (1 - \alpha)F_t \tag{18.2}$$

Equation for Linear Trend

$$T_t = b_0 + b_1 t \tag{18.5}$$

Multiplicative Time Series Model With Trend, Seasonal, and Irregular Components

$$Y_t = T_t \times S_t \times I_t \tag{18.9}$$

Multiplicative Time Series Model With Trend, Cyclical, Seasonal, and Irregular Components

$$Y_t = T_t \times C_t \times S_t \times I_t \tag{18.10}$$

SUPPLEMENTARY EXERCISES

27. Moving averages often are used to identify movements in stock prices. Weekly closing prices (in dollars per share) for Toys R Us for September 22, 1997, through December 8, 1997, follow (Prudential Securities, Inc.).

Week	Price ($)	Week	Price ($)
September 22	$34\frac{7}{8}$	November 3	$33\frac{5}{8}$
September 29	$35\frac{5}{8}$	November 10	$35\frac{1}{16}$
October 6	$34\frac{11}{16}$	November 17	$34\frac{1}{16}$
October 13	$33\frac{9}{16}$	November 24	$34\frac{1}{8}$
October 20	$32\frac{5}{8}$	December 1	$33\frac{1}{4}$
October 27	34	December 8	$32\frac{1}{16}$

a. Use a 3-month moving average to smooth the time series. Forecast the closing price for December 15, 1997.

b. Use a 3-month weighted moving average to smooth the time series. Use a weight of .4 for the most recent period, .4 for the next period back, and .2 for the third period back. Forecast the closing price for December 15, 1997.

c. Use exponential smoothing with a smoothing constant of $\alpha = .35$ to smooth the time series. Forecast the closing price for December 15, 1997.

d. Which of the three methods do you prefer? Why?

28. The Federal Election Commission maintains data showing the voting age population, the number of registered voters, and the turnout for federal elections. The following table shows the national voter turnout as a percentage of the voting age population from 1972 to 1996. (*The Wall Street Journal Almanac,* 1998)

Year	% Turnout	Year	% Turnout	Year	% Turnout
1972	55	1982	40	1990	37
1974	38	1984	53	1992	55
1976	54	1986	36	1994	39
1978	37	1988	50	1996	49
1980	53				

a. Use exponential smoothing to forecast this time series. Consider smoothing constants of $\alpha = .1$ and .2. What value of the smoothing constant provides the best forecast?

b. What is the forecast of the percentage turnout in 1998?

29. The percentage of individual investors' portfolios committed to stock depends on the state of the economy. As of April 1997, a typical portfolio consisted of cash (19%), stocks (30%), stock funds (37%), bonds (8%), and bond funds (6%) (*AAII Journal,* June 1997). The following table reports the percentage of stocks in a typical portfolio in nine quarters from 1995 to 1997.

Quarter	Stock %
1st—1995	29.8
2nd—1995	31.0
3rd—1995	29.9
4th—1995	30.1
1st—1996	32.2
2nd—1996	31.5
3rd—1996	32.0
4th—1996	31.9
1st—1997	30.0

a. Use exponential smoothing to forecast this time series. Consider smoothing constants of $\alpha = .2$, .3, and .4. What value of the smoothing constant provides the best forecast?

b. What is the forecast of the percentage of assets committed to stocks for the second quarter of 1997?

30. A chain of grocery stores noted the weekly demand (in cases) reported in the following table for a particular brand of automatic dishwasher detergent. Use exponential smoothing with $\alpha = .2$ to develop a forecast for week 11.

Week	Demand
1	22
2	18
3	23
4	21
5	17
6	24
7	20
8	19
9	18
10	21

31. United Dairies, Inc., supplies milk to several independent grocers throughout Dade County, Florida. Managers at United Dairies want to develop a forecast of the number of half-gallons of milk sold per week. Sales data for the past 12 weeks follow.

Week	Sales	Week	Sales
1	2750	7	3300
2	3100	8	3100
3	3250	9	2950
4	2800	10	3000
5	2900	11	3200
6	3050	12	3150

Using exponential smoothing with $\alpha = .4$, develop a forecast of demand for the 13th week.

32. The Garden Avenue Seven sells tapes of its musical performances. The following table reports sales (in units) for the past 18 months. The group's manager wants an accurate method for forecasting future sales.

Month	Sales	Month	Sales	Month	Sales
1	293	7	381	13	549
2	283	8	431	14	544
3	322	9	424	15	601
4	355	10	433	16	587
5	346	11	470	17	644
6	379	12	481	18	660

a. Use exponential smoothing with $\alpha = .3, .4$, and $.5$. Which value of α provides the best forecasts?
b. Use trend projection to provide a forecast. What is the value of MSE?
c. Which method of forecasting would you recommend to the manager? Why?

33. The Mayfair Department Store in Davenport, Iowa, is trying to determine the amount of sales lost while it was shut down during July and August because of damage caused by the Mississippi River flood. Sales data for January through June follow.

Month	Sales ($1000s)	Month	Sales ($1000s)
January	185.72	April	210.36
February	167.84	May	255.57
March	205.11	June	261.19

a. Use exponential smoothing, with $\alpha = .4$, to develop a forecast for July and August. (Hint: Use the forecast for July as the actual sales in July in developing the August forecast.) Comment on the use of exponential smoothing for forecasts more than one period into the future.
b. Use trend projection to forecast sales for July and August.
c. Mayfair's insurance company has proposed a settlement based on lost sales of $240,000 in July and August. Is this amount fair? If not, what amount would you recommend as a counteroffer?

34. Canton Supplies, Inc., is a service firm that employs approximately 100 individuals. Managers of Canton Supplies are concerned about meeting monthly cash obligations and want to develop a forecast of monthly cash requirements. Because of a recent change in operating policy, only the past seven months of data are considered to be relevant. With the following historical data, use trend projection to develop a forecast of cash requirements for each of the next two months.

Month	1	2	3	4	5	6	7
Cash Required ($1000s)	205	212	218	224	230	240	246

35. The following table reports the number of cable television subscribers (1000s) by year from 1980 to 1994 (*Television & Cable Factbook,* 1994).

Year	Number	Year	Number
1980	16,000	1988	44,000
1981	18,300	1989	47,500
1982	21,000	1990	50,000
1983	25,000	1991	51,000
1984	30,000	1992	53,000
1985	32,000	1993	55,000
1986	37,500	1994	57,000
1987	41,100		

a. Develop a linear trend equation for this time series. Comment on what the equation tells about the number of cable subscribers for the 15-year period.
b. Provide forecasts of the number of cable subscribers for 1995 and 1996.

36. The Costello Music Company has been in business for 5 years. During that time, sales of electric organs have increased from 12 units in the first year to 76 units in the most recent year. Fred Costello, the firm's owner, wants to develop a forecast of organ sales for the coming year. The historical data follow.

Year	1	2	3	4	5
Sales	12	28	34	50	76

a. Show a graph of this time series. Does a linear trend appear to be present?
b. Develop the equation for the linear trend component for the time series. What is the average increase in sales that the firm has been realizing per year?

37. Hudson Marine has been an authorized dealer for C&D marine radios for the past 7 years. The following table reports the number of radios sold each year.

Year	1	2	3	4	5	6	7
Number Sold	35	50	75	90	105	110	130

a. Show a graph of this time series. Does a linear trend appear to be present?
b. Develop the equation for the linear trend component of the time series.
c. Use the linear trend developed in part (b) to develop a forecast for annual sales in year 8.

38. The Motion Picture Association of America (MPAA) collects data that show the cost of making movies for MPAA members, including Disney, Paramount, Universal, Warner Brothers, MGM, Fox, Sony, and Turner. The following data show the average production costs in thousands of dollars from 1985 to 1996 for MPAA members (*The Wall Street Journal Almanac,* 1998).

Year	Average Production Costs ($1000s)	Year	Average Production Costs ($1000s)
1985	16,779	1991	26,136
1986	17,455	1992	28,858
1987	20,051	1993	29,910
1988	18,061	1994	34,288
1989	23,454	1995	36,370
1990	26,783	1996	39,836

a. Plot the time series and comment on the appropriateness of a linear trend.
b. Develop the equation for the linear trend component of this time series.
c. What is the average increase in production costs that MPAA members have been experiencing per year?
d. Use the trend equation to forecast the average production costs in 1997 and 1998.

39. The following table gives the number of cellular telephone subscribers (1000s) by year from 1986 to 1994 (*State of the Cellular Industry,* 1994).

Year	Number	Year	Number
1986	682	1990	5,283
1987	1,231	1991	7,557
1988	2,069	1992	11,033
1989	3,509	1993	16,009

Plot the time series and comment on the appropriateness of a linear trend. What type of functional form do you believe would be most appropriate for the trend pattern of this time series?

40. Refer to the Hudson Marine problem in Exercise 37. Suppose the quarterly sales values for the 7 years of historical data are as follow.

Year	Quarter 1	Quarter 2	Quarter 3	Quarter 4	Total Yearly Sales
1	6	15	10	4	35
2	10	18	15	7	50
3	14	26	23	12	75
4	19	28	25	18	90
5	22	34	28	21	105
6	24	36	30	20	110
7	28	40	35	27	130

a. Show the 4-quarter moving average values for this time series. Plot both the original time series and the moving average series on the same graph.
b. Compute the seasonal indexes for the four quarters.
c. When does Hudson Marine experience the largest seasonal effect? Does this seem reasonable? Explain.

41. Consider the Costello Music Company problem in Exercise 36. The quarterly sales data follow.

Year	Quarter 1	Quarter 2	Quarter 3	Quarter 4	Total Yearly Sales
1	4	2	1	5	12
2	6	4	4	14	28
3	10	3	5	16	34
4	12	9	7	22	50
5	18	10	13	35	76

a. Compute the seasonal indexes for the four quarters.
b. When does Costello Music experience the largest seasonal effect? Does this appear reasonable? Explain.

42. Refer to the Hudson Marine data in Exercise 40.
a. Deseasonalize the data and use the deseasonalized time series to identify the trend.
b. Use the results of part (a) to develop a quarterly forecast for next year based on trend.
c. Use the seasonal indexes developed in Exercise 40 to adjust the forecasts developed in part (b) to account for the effect of season.

43. Consider the Costello Music Company time series in Exercise 41.
a. Deseasonalize the data and use the deseasonalized time series to identify the trend.
b. Use the results of part (a) to develop a quarterly forecast for next year based on trend.
c. Use the seasonal indexes developed in Exercise 41 to adjust the forecasts developed in part (b) to account for the effect of season.

Case Problem 1 FORECASTING FOOD AND BEVERAGE SALES

The Vintage Restaurant is on Captiva Island, a resort community near Fort Myers, Florida. The restaurant, which is owned and operated by Karen Payne, has just completed its third year of operation. During that time, Karen has sought to establish a reputation for the restaurant as a high-quality dining establishment that specializes in fresh seafood. The efforts by Karen and her staff have proven successful, and her restaurant has become one of the best and fastest-growing restaurants on the island.

Karen has concluded that to plan for the growth of the restaurant in the future, she needs to develop a system that will enable her to forecast food and beverage sales by month for up to one year in advance. Karen has the following data ($1000s) on total food and beverage sales for the three years of operation.

Month	First Year	Second Year	Third Year
January	242	263	282
February	235	238	255
March	232	247	265
April	178	193	205
May	184	193	210
June	140	149	160
July	145	157	166
August	152	161	174
September	110	122	126
October	130	130	148
November	152	167	173
December	206	230	235

Managerial Report

Perform an analysis of the sales data for the Vintage Restaurant. Prepare a report for Karen that summarizes your findings, forecasts, and recommendations. Include:

1. A graph of the time series.
2. An analysis of the seasonality of the data. Indicate the seasonal indexes for each month, and comment on the high and low seasonal sales months. Do the seasonal indexes make intuitive sense? Discuss.
3. A forecast of sales for January through December of the fourth year.
4. Recommendations as to when the system that you have developed should be updated to account for new sales data.
5. Any detailed calculations of your analysis in the appendix of your report.

Assume that January sales for the fourth year turn out to be $295,000. What was your forecast error? If this is a large error, Karen may be puzzled about the difference between your forecast and the actual sales value. What can you do to resolve her uncertainty in the forecasting procedure?

Case Problem 2 FORECASTING LOST SALES

The Carlson Department Store suffered heavy damage when a hurricane struck on August 31, 2000. The store was closed for four months (September 2000 through December 2000), and Carlson is now involved in a dispute with its insurance company about the amount of lost sales during the time the store was closed. Two key issues must be resolved: (1) the amount of sales Carlson would have made if the hurricane had not struck and (2) whether Carlson is entitled to any compensation for excess sales due to increased business activity after the storm. More than $8 billion in federal disaster relief and insurance money came into the county, resulting in increased sales at department stores and numerous other businesses.

Table 18.13 gives Carlson's sales data for the 48 months preceding the storm. Table 18.14 reports total sales for the 48 months preceding the storm for all department stores in the county, as well as the total sales in the county for the four months the Carlson Department Store was closed. Carlson's managers have asked you to analyze these data and develop estimates of the lost sales at the Carlson Department Store for the months of September through December 2000. They also have asked you to determine whether a case can be

TABLE 18.13 SALES FOR CARLSON DEPARTMENT STORE, SEPTEMBER 1996 THROUGH AUGUST 2000 ($ MILLIONS)

Month	1996	1997	1998	1999	2000
January		1.45	2.31	2.31	2.56
February		1.80	1.89	1.99	2.28
March		2.03	2.02	2.42	2.69
April		1.99	2.23	2.45	2.48
May		2.32	2.39	2.57	2.73
June		2.20	2.14	2.42	2.37
July		2.13	2.27	2.40	2.31
August		2.43	2.21	2.50	2.23
September	1.71	1.90	1.89	2.09	
October	1.90	2.13	2.29	2.54	
November	2.74	2.56	2.83	2.97	
December	4.20	4.16	4.04	4.35	

made for excess storm-related sales during the same period. If such a case can be made, Carlson is entitled to compensation for excess sales it would have earned in addition to ordinary sales.

Managerial Report

Prepare a report for the managers of the Carlson Department Store that summarizes your findings, forecasts, and recommendations. Include:

1. An estimate of sales had there been no hurricane.
2. An estimate of countywide department store sales had there been no hurricane.
3. An estimate of lost sales for the Carlson Department Store for September through December 2000.

In addition, use the countywide actual department stores sales for September through December 2000 and the estimate in part (2) to make a case for or against excess storm-related sales.

TABLE 18.14 DEPARTMENT STORE SALES FOR THE COUNTY, SEPTEMBER 1996 THROUGH DECEMBER 2000 ($ MILLIONS)

Month	1996	1997	1998	1999	2000
January		46.8	46.8	43.8	48.0
February		48.0	48.6	45.6	51.6
March		60.0	59.4	57.6	57.6
April		57.6	58.2	53.4	58.2
May		61.8	60.6	56.4	60.0
June		58.2	55.2	52.8	57.0
July		56.4	51.0	54.0	57.6
August		63.0	58.8	60.6	61.8
September	55.8	57.6	49.8	47.4	69.0
October	56.4	53.4	54.6	54.6	75.0
November	71.4	71.4	65.4	67.8	85.2
December	117.6	114.0	102.0	100.2	121.8

Appendix 18.1 FORECASTING WITH MINITAB

In this appendix we show how Minitab can be used to develop forecasts using three forecasting methods: moving averages, exponential smoothing, and trend projection.

Moving Averages

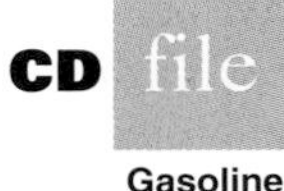

Gasoline

To show how Minitab can be used to develop forecasts using the moving averages method, we will develop a forecast for the gasoline sales time series in Table 18.1 and Figure 18.5. The sales data for the 12 weeks have been entered into column 2 of the worksheet. The following steps can be used to produce a 3-week moving average forecast for week 13.

Step 1. Select the **Stat** pull-down menu
Step 2. Choose **Time Series**
Step 3. Choose **Moving Average**
Step 4. When the Moving Average dialog box appears:
Enter C2 in the **Variable box**
Enter 3 in the **MA length** box
Select **Generate forecasts**
Enter 1 in the **Number of forecasts** box
Enter 12 in the **Starting from origin** box
Click **OK**

The 3-week moving average forecast for week 13 is shown in the session window. The mean square error of 10.22 is labeled as MSD in the Minitab output. Many other output options are available, including a summary table similar to Table 18.2 and graphical output similar to Figure 18.6.

Exponential Smoothing

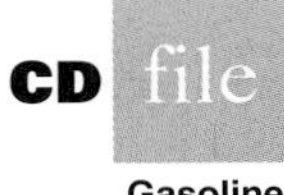

Gasoline

To show how Minitab can be used to develop an exponential smoothing forecast, we will again develop a forecast of sales in week 13 for the gasoline sales time series in Table 18.1 and Figure 18.5. The sales data for the 12 weeks have been entered into column 2 of the worksheet. The following steps can be used to produce a forecast for week 13 using a smoothing constant of $\alpha = .2$.

Step 1. Select the **Stat** pull-down menu
Step 2. Choose **Time Series**
Step 3. Choose **Single Exp Smoothing**
Step 4. When the Single Exponential Smoothing dialog box appears:
Enter C2 in the **Variable** box
Select the **Use** option for the Weight to Use in Smoothing
Enter 0.2 in the **Use** box
Select **Generate forecasts**
Enter 1 in the **Number of forecasts** box
Enter 12 in the **Starting from origin** box
Select **Options**
Step 5. When the Single Exponential Smoothing—Options dialog box appears:
Enter 1 in the **Use average of first** box
Click **OK**
Step 6. When the Single Exponential Smoothing dialog box appears:
Click **OK**

The exponential smoothing forecast for week 13 is shown in the session window. The mean square error is labeled as MSD in the Minitab output.* Many other output options are available, including a summary table similar to Table 18.3 and graphical output similar to Figure 18.7.

Trend Projection

Bicycle

To show how Minitab can be used for trend projection, we develop a forecast for the bicycle sales time series in Table 18.6 and Figure 18.8. The year numbers have been entered into column C1 and the sales data have been entered into column C2 of the worksheet. The following steps can be used to produce a forecast for week 13 using trend projection.

Step 1. Select the **Stat** pull-down menu
Step 2. Choose **Time Series**
Step 3. Choose **Trend Analysis**
Step 4. When the Trend Analysis dialog box appears:
Enter C2 in the **Variable** box
Choose **Linear** for the Model Type
Select **Generate forecasts**
Enter 1 in the **Number of forecasts** box
Enter 10 in the **Starting from origin** box
Click **OK**

The equation for linear trend and the forecast for the next period are shown in the session window.

Appendix 18.2 FORECASTING WITH EXCEL

In this appendix we show how Excel can be used to develop forecasts using three forecasting methods: moving averages, exponential smoothing, and trend projection.

Moving Averages

To show how Excel can be used to develop forecasts using the moving averages method, we will develop a forecast for the gasoline sales time series in Table 18.1 and Figure 18.5. The sales data for the 12 weeks have been entered into worksheet rows 2 through 13 of column B. The following steps can be used to produce a 3-week moving average.

Step 1. Select the **Tools** pull-down menu
Step 2. Choose **Data Analysis**
Step 3. Choose **Moving Average** from the list of Analysis Tools
Click **OK**
Step 4. When the Moving Average dialog box appears:
Enter B2:B13 in the **Input Range** box
Enter 3 in the **Interval** box
Enter C2 in the **Output Range** box
Click **OK**

The 3-week moving average forecasts will appear in column B of the worksheet. Forecasts for periods of other length can be computed easily by entering a different value in the **Interval** box.

---

*The value of MSD computed by Minitab is not the same as the value of MSE that appears in Table 18.4. Minitab uses a forecast of 17 for week 1 and computes MSD using all 12 time periods of data. In Section 18.2 we compute MSE using only the data for weeks 2 through 12, because we had no past values with which to make a forecast for period 1.

Exponential Smoothing

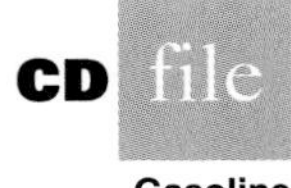

Gasoline

To show how Excel can be used for exponential smoothing, we again develop a forecast for the gasoline sales time series in Table 18.1 and Figure 18.5. The sales data for the 12 weeks have been entered into worksheet rows 2 through 13 of column B. The following steps can be used to produce a forecast using a smoothing constant of $\alpha = .2$.

Step 1. Select the **Tools** pull-down menu
Step 2. Choose **Data Analysis**
Step 3. Choose **Exponential Smoothing** from the list of Analysis Tools
Click **OK**
Step 4. When the Exponential Smoothing dialog box appears:
Enter B2:B13 in the **Input Range** box
Enter .8 in the **Damping factor** box
Enter C2 in the **Output Range** box
Click **OK**

The exponential smoothing forecasts will appear in column B of the worksheet. Note that the value we entered in the Damping factor box is $1 - \alpha$; forecasts for other smoothing constants can be computed easily by entering a different value for $1 - \alpha$ in the Damping factor box.

Trend Projection

CD file

Bicycle

To show how Excel can be used for trend projection, we develop a forecast for the bicycle sales time series in Table 18.6 and Figure 18.8. The data, with appropriate labels in row 1, have been entered into worksheet rows 1 through 11 of columns A and B. The following steps can be used to produce a forecast for year 11 by trend projection.

Step 1. Select an empty cell in the worksheet
Step 2. Select the **Insert** pull-down menu
Step 3. Choose **Function**
Step 4. When the Paste Function dialog box appears:
Choose **Statistical** in the Function Category box
Choose **Forecast** in the Function Name box
Click **OK**
Step 5. When the Forecast dialog box appears:
Enter 11 in the **x** box
Enter B2:B11 in the **Known y's** box
Enter A2:A11 in the **Known x's** box
Click **OK**

The forecast for year 11, in this case 32.5, will appear in the cell selected in Step 1.

Chapter 19

NONPARAMETRIC METHODS

CONTENTS

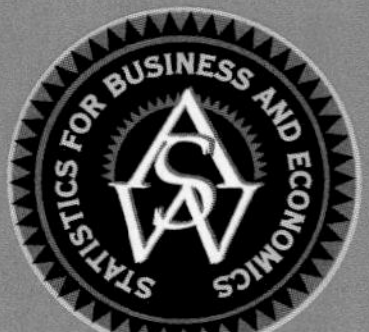

STATISTICS IN PRACTICE

WEST SHELL REALTORS*
Cincinnati, Ohio

West Shell Realtors was founded in 1958 with one office and a sales staff of three people. In 1964, the company began a long-term expansion program, with new offices added almost yearly. Over the years, West Shell grew to become one of the largest realtors in Greater Cincinnati, with offices in southwest Ohio, southeast Indiana, and northern Kentucky.

Statistical analysis helps real estate firms such as West Shell monitor sales performance. Monthly reports are generated for each of West Shell's offices as well as for the total company. Statistical summaries of total sales dollars, number of units sold, and median selling price per unit are essential in keeping both office managers and the company's top managers informed of progress and trouble spots in the organization.

In addition to monthly summaries of ongoing operations, the company uses statistical considerations to guide corporate plans and strategies. West Shell has implemented a strategy of planned expansion. Each time an expansion plan calls for the establishment of a new sales office, the company must address the question of office location. Selling prices of homes, turnover rates, and forecast sales volumes are the types of data used in evaluating and comparing alternative locations.

In one instance, West Shell identified two suburbs, Clifton and Roselawn, as prime candidates for a new office. A variety of factors were considered in comparing the two areas, including selling prices of homes. West Shell employed nonparametric statistical methods with small samples to help identify any differences in sales patterns for the two areas.

*The authors are indebted to Rodney Fightmaster of West Shell Realtors for providing this Statistics in Practice.

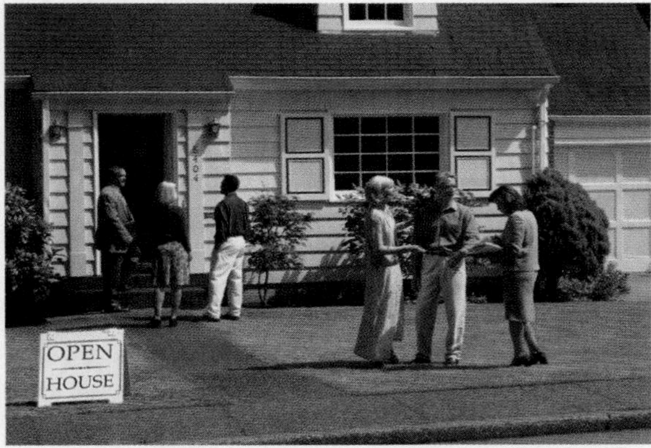

West Shell uses statistical analysis of home sales to remain competitive. © PhotoDisc, Inc.

Samples of 25 sales in the Clifton area and 18 sales in the Roselawn area were taken, and the Mann-Whitney-Wilcoxon rank-sum test was chosen as an appropriate statistical test of the difference in the pattern of selling prices. At the .05 level of significance, the Mann-Whitney-Wilcoxon test did not allow rejection of the null hypothesis that the two populations of selling prices were identical. Thus, West Shell was able to focus on criteria other than selling prices of homes in the site selection process.

In this chapter we will show how nonparametric statistical tests such as the Mann-Whitney-Wilcoxon test are applied. We will also discuss the proper interpretation of such tests.

The statistical methods presented thus far in the text are generally known as *parametric methods*. In this chapter we introduce several **nonparametric methods.** Such methods are often applicable in situations where the parametric methods of the preceding chapters are not. Nonparametric methods typically require less restrictive assumptions about the level of data measurement and fewer assumptions about the form of the probability distributions generating the sample data.

One consideration in determining whether a parametric or a nonparametric method is appropriate is the scale of measurement used to generate the data. All data are generated by one of four scales of measurement: nominal, ordinal, interval, and ratio. Hence, all statistical analyses are conducted with either nominal, ordinal, interval, or ratio data.

Let us define and provide examples of the four scales of measurement.

1. *Nominal scale.* The scale of measurement is nominal if the data are labels or categories used to define an attribute of an element. Nominal data may be numeric or nonnumeric.
   Examples. The exchange where a stock is listed (NYSE, NASDAQ, or AMEX) is nonnumeric nominal data. An individual's social security number is numeric nominal data.
2. *Ordinal scale.* The scale of measurement is ordinal if the data can be used to rank, or order, the observations. Ordinal data may be numeric or nonumeric.
   Examples. The measures small, medium, and large for the size of an item are nonnumeric ordinal data. The class ranks of individuals measured as 1, 2, 3, . . . are numeric ordinal data.
3. *Interval scale.* The scale of measurement is interval if the data have the properties of ordinal data and the interval between observations is expressed in terms of a fixed unit of measure. Interval data must be numeric.
   Examples. Measures of temperature are interval data. Suppose it is 70 degrees in one location and 40 degrees in another. We can rank the locations with respect to warmth; the first location is warmer than the second. The fixed unit of measure, a degree, enables us to say how much warmer it is at the first location: 30 degrees.
4. *Ratio scale.* The scale of measurement is ratio if the data have the properties of interval data and the ratio of measures is meaningful. Ratio data must be numeric.
   Examples. Variables such as distance, height, weight, and time are measured on a ratio scale. Temperature measures are not ratio data because there is no inherently defined zero point. For instance, the freezing point of water is 32 degrees on a Fahrenheit scale and zero degrees on a Celsius scale. Ratios are not meaningful with temperature data. For instance, it makes no sense to say that 80 degrees is twice as warm as 40 degrees.

In Chapter 1 we pointed out that nominal and ordinal scales provide qualitative data. Interval and ratio scales provide quantitative data.

If the level of data measurement is nominal or ordinal, computations of means, variances, and standard deviations are not meaningful. Thus, with these kinds of data, many of the statistical procedures discussed previously cannot be employed.

Most of the statistical methods referred to as parametric require the use of interval- or ratio-scaled data. With these levels of measurement arithmetic operations are meaningful, and means, variances, standard deviations, and so on can be computed, interpreted, and used in the analysis. With nominal or ordinal data, it is inappropriate to compute means, variances, and standard deviations; hence, parametric methods normally cannot be used. Nonparametric methods are often the only way to analyze such data and draw statistical conclusions.

Another consideration in determining whether a parametric method or a nonparametric method should be employed is the assumption about the population from which the data were obtained. For example, a parametric procedure for testing a hypothesis about the difference between the means of two populations was presented in Chapter 10. In the small-sample case, the *t* distribution can be used for this test if we are willing to assume that the populations are normally distributed with equal variances. If this assumption about the populations is not appropriate, the parametric method based on the use of the *t* distribution should not be used even if the data are interval or ratio scaled. However, nonparametric methods, which require no assumptions about the population probability distributions, are available for testing for differences between two populations. Because no probability distribution assumptions are required, nonparametric methods are often called **distribution-free methods.**

The statistical procedures based on small samples and the t distribution as discussed in Chapters 8, 9, and 10 are based on the assumption that the data are from a normal population. If this assumption is not appropriate, nonparametric methods may be used.

In general, for a statistical method to be classified as nonparametric, it must satisfy at least one of the following conditions.*

1. The method can be used with nominal data.
2. The method can be used with ordinal data.
3. The method can be used with interval or ratio data when no assumption can be made about the population probability distribution.

*See W. J. Conover, *Practical Nonparametric Statistics,* 3rd ed. (John Wiley & Sons, 1998).

If the level of data measurement is interval or ratio and if the necessary probability distribution assumptions for the population are appropriate, parametric methods provide more powerful or more discerning statistical procedures. In many cases where a nonparametric method as well as a parametric method can be applied, the nonparametric method is almost as good or almost as powerful as the parametric method. In cases where the data are nominal or ordinal or in cases where the assumptions required by parametric methods are inappropriate, only nonparametric methods are available. Because of the less restrictive data measurement requirements and the fewer assumptions needed about the population distribution, nonparametric methods are regarded as more generally applicable than parametric methods. The sign test, the Wilcoxon signed-rank test, the Mann-Whitney-Wilcoxon test, the Kruskal-Wallis test, and the Spearman rank correlation are the nonparametric methods presented in this chapter.

19.1 SIGN TEST

A common market-research application of the **sign test** involves using a sample of n potential customers to identify a preference for one of two brands of a product such as coffee, soft drinks, or detergents. The n expressions of preference are nominal data because the consumer simply names, or labels, a preference. Given these data, our objective is to determine whether a difference in preference exists between the two items being compared. As we will see, the sign test is a nonparametric statistical procedure for answering this question.

Small-Sample Case

The small-sample case for the sign test should be used whenever $n \leq 20$. Let us illustrate the use of the sign test for the small-sample case by considering a study conducted for Sun Coast Farms; Sun Coast produces an orange juice product marketed under the name Citrus Valley. A competitor of Sun Coast Farms has begun producing a new orange juice product known as Tropical Orange. In a study of consumer preferences for the two brands, 12 individuals were given unmarked samples of each product. The brand each individual tasted first was selected randomly. After tasting the two products, the individuals were asked to state a preference for one of the two brands. The purpose of the study is to determine whether consumers prefer one product over the other. Letting p indicate the proportion of the population of consumers favoring Citrus Valley, we want to test the following hypotheses.

$$H_0: p = .50$$
$$H_a: p \neq .50$$

If H_0 cannot be rejected, we will have no evidence indicating a difference in preference for the two brands of orange juice. However, if H_0 can be rejected, we can conclude that the consumer preferences are different for the two brands. In that case, the brand selected by the greater number of consumers can be considered the most preferred brand.

In the following discussion we will show how the small-sample version of the sign test can be used to test these hypotheses and draw a conclusion about consumer preferences. To record the preference data for the 12 individuals participating in the study, we use a plus sign if the individual expresses a preference for Citrus Valley and a minus sign if the individual expresses a preference for Tropical Orange. Because the data are recorded in terms of plus or minus signs, this nonparametric test is called the sign test.

Exact binomial probabilities are readily available when the sample size is less than or equal to 20. See Table 5 of Appendix B.

Under the assumption that H_0 is true ($p = .50$), the number of plus signs follows a binomial probability distribution with $p = .50$. With a sample size of $n = 12$, Table 5 in Appendix B shows the probabilities for the binomial probability distribution with $p = .50$ as displayed

TABLE 19.1

BINOMIAL PROBABILITIES WITH $n = 12$, $p = .50$

Number of Plus Signs	Probability
0	.0002
1	.0029
2	.0161
3	.0537
4	.1208
5	.1934
6	.2256
7	.1934
8	.1208
9	.0537
10	.0161
11	.0029
12	.0002

A shaded rejection region cannot be shown in Figure 19.1 because the sampling distribution is not continuous. The rejection region is defined by the spikes at 0, 1, 2, 10, 11, and 12.

in Table 19.1. Figure 19.1 is a graphical representation of this binomial probability distribution. It shows the probability of the number of plus signs under the assumption that H_0 is true and is therefore the appropriate sampling distribution for the hypothesis test. We use this sampling distribution to determine a rule for rejecting H_0; our approach is similar to the method we used to develop rejection rules for hypothesis testing in Chapter 9. For example, using $\alpha = .05$, we would place a rejection region or area of approximately .025 in each tail of the distribution in Figure 19.1. Starting at the lower end of the distribution, we see that the probability of obtaining zero, one, or two plus signs is $.0002 + .0029 + .0161 = .0192$. Note that we stop at 2 plus signs because adding the probability of three plus signs would make the area in the lower tail equal to $.0192 + .0537 = .0729$, which substantially exceeds the desired area of .025. At the upper end of the distribution, we find the same probability of .0192 corresponding to 10, 11, or 12 plus signs. Thus, the closest we can come to $\alpha = .05$ without exceeding it is $.0192 + .0192 = .0384$. We therefore adopt the following rejection rule.

Reject H_0 if the number of plus signs is less than 3 or greater than 9

The preference data obtained for the Sun Coast Farms example are reported in Table 19.2. Because only two plus signs are observed, the null hypothesis is rejected. The study provides evidence that consumer preference differs for the two brands of orange juice. We would advise Sun Coast Farms that consumers indicate a preference for the competitor's Tropical Orange brand.

Reject H_0 if p-value $< \alpha$ can also be used for nonparametric tests. With two plus signs, the p-value for this two-tailed test is $2(.0161 + .0029 + .0002) = .0384$. For a one-tailed test, the p-value would be found by summing probabilities in only one tail of the sampling distribution.

In the Sun Coast Farms example, all 12 individuals in the study were able to state a preference. In many situations, one or more individuals in the sample are not able to state a defi-

FIGURE 19.1 BINOMIAL PROBABILITIES FOR THE NUMBER OF PLUS SIGNS WHEN $n = 12$ AND $p = .50$

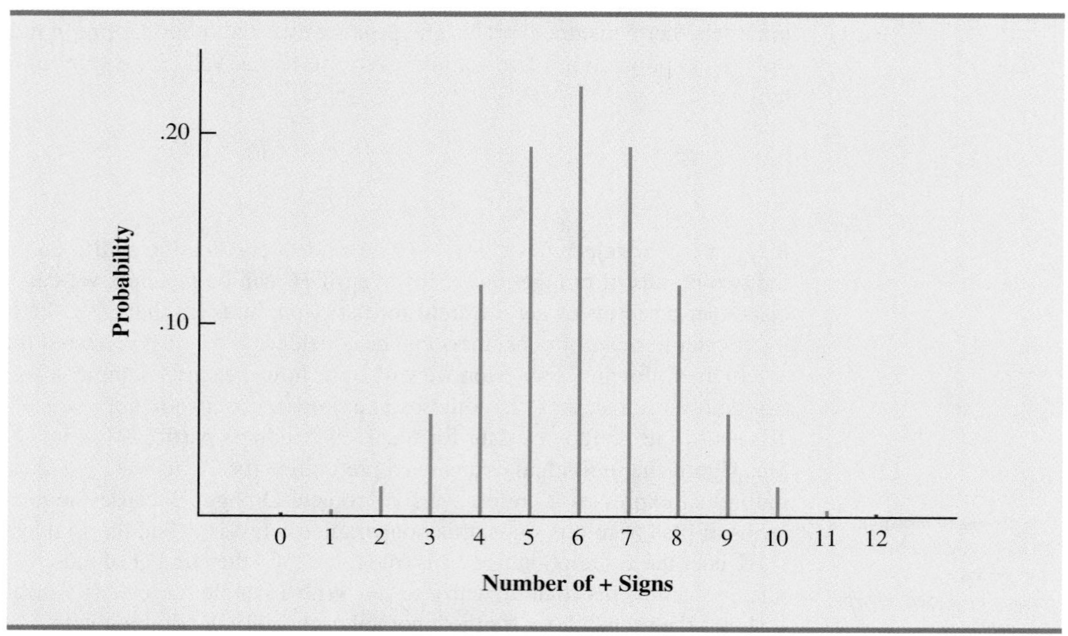

TABLE 19.2 PREFERENCE DATA FOR THE SUN COAST FARMS TASTE TEST

Individual	Brand Preference	Recorded Data
1	Tropical Orange	−
2	Tropical Orange	−
3	Citrus Valley	+
4	Tropical Orange	−
5	Tropical Orange	−
6	Tropical Orange	−
7	Tropical Orange	−
8	Tropical Orange	−
9	Citrus Valley	+
10	Tropical Orange	−
11	Tropical Orange	−
12	Tropical Orange	−

nite preference. In such cases, the individual's response of no preference can be dropped and the analysis conducted with a smaller sample size.

The binomial probability distribution as shown in Table 5 of Appendix B can be used to provide the decision rule for any sign test up to a sample size of $n = 20$. With the null hypothesis $p = .50$ and the sample size n, the decision rule can be established for any level of significance. In addition, by considering the probabilities in only the lower or upper tail of the binomial probability distribution, we can develop rejection rules for one-tailed tests. Appendix B does not provide binomial probability distribution tables for sample sizes greater than 20. In such cases, we can use the large-sample normal approximation of binomial probabilities to determine the appropriate rejection rule for the sign test.

Large-Sample Case

The large-sample sign test is equivalent to the test of a population proportion with $p = .5$ as presented in Chapter 9.

Using the null hypothesis H_0: $p = .50$ and a sample size of $n > 20$, the sampling distribution for the number of plus signs can be approximated by a normal probability distribution.

Normal Approximation of the Sampling Distribution of the Number of Plus Signs When No Preference Is Stated

$$\text{Mean: } \mu = .50n \tag{19.1}$$

$$\text{Standard Deviation: } \sigma = \sqrt{.25n} \tag{19.2}$$

Distribution form: approximately normal provided $n > 20$.

Let us consider an application of the sign test to political polling. A poll taken during a recent presidential election campaign asked 200 registered voters to rate the Democratic and Republican candidates in terms of best overall foreign policy. Results of the poll showed 72 rated the Democratic candidate higher, 103 rated the Republican candidate higher, and 25 indicated no difference between the candidates. Does the poll indicate a significant difference between the two candidates in terms of public opinion about their foreign policies?

Ties are handled by dropping the items from the analysis.

Using the sign test, we see that $n = 200 - 25 = 175$ individuals were able to indicate the candidate they believed had the best overall foreign policy. Using equations (19.1) and

(19.2), we find that the sampling distribution of the number of plus signs has the following properties.

$$\mu = .50n = .50(175) = 87.5$$
$$\sigma = \sqrt{.25n} = \sqrt{.25(175)} = 6.6$$

In addition, with $n = 175$ we can assume that the sampling distribution is approximately normal. This distribution is shown in Figure 19.2. Because the distribution is approximately normal, we can use the table of areas for the standard normal probability distribution to develop the rejection rule for the test. With $\alpha = .05$, the rejection rule for this two-tailed test can be written as follows.

$$\text{Reject } H_0 \text{ if } z < -1.96 \text{ or if } z > +1.96$$

Using the number of times the Democratic candidate received the higher foreign policy rating as the number of plus signs ($x = 72$), we have the following value of the test statistic.

$$z = \frac{x - \mu}{\sigma} = \frac{72 - 87.5}{6.6} = -2.35$$

If the analysis had used the number of times the Republican candidate was rated higher, $z = 2.35$ would have led us to the same conclusion.

Because $z = -2.35$ is less than -1.96, the hypothesis of no difference in foreign policy for the two candidates should be rejected at the .05 level of significance. With $z = -2.35$, the standard normal distribution can be used to show that the p-value is $2(.5000 - .4906) = .0188$. This study indicates that the candidates are perceived to have different foreign policy ratings.

Hypothesis Test About a Median

In Chapter 9 we described how hypothesis tests can be used to make an inference about a population mean. We now show how the sign test can be used to conduct hypothesis tests about a population median. Recall that the median splits a population in such a way that

FIGURE 19.2 PROBABILITY DISTRIBUTION OF THE NUMBER OF PLUS SIGNS FOR A SIGN TEST WITH $n = 175$

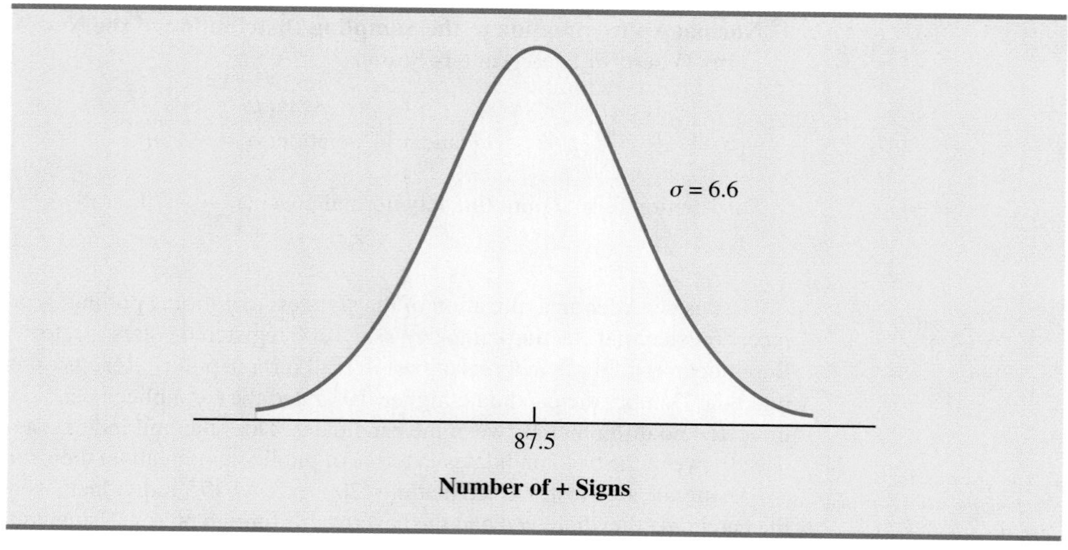

50% of the values are at the median or above and 50% are at the median or below. We can apply the sign test by using a plus sign whenever the data in the sample are above the hypothesized value of the median and a minus sign whenever the data in the sample are below the hypothesized value of the median. Any data exactly equal to the hypothesized value of the median should be discarded. The computations for the sign test are done in exactly the same way as before.

For example, the following hypothesis test is being conducted about the median price of new homes in St. Louis, Missouri.

$$H_0\text{: Median} = \$130{,}000$$
$$H_a\text{: Median} \neq \$130{,}000$$

In a sample of 62 new homes, 34 have prices above \$130,000, 26 have prices below \$130,000, and two have prices of exactly \$130,000.

Using (19.1) and (19.2) for the $n = 60$ homes with prices different from \$130,000, we obtain

$$\mu = .50n = .50(60) = 30$$
$$\sigma = \sqrt{.25n} = \sqrt{.25(60)} = 3.87$$

With $x = 34$ as the number of plus signs, the test statistic becomes

$$z = \frac{x - \mu}{\sigma} = \frac{34 - 30}{3.87} = 1.03$$

Using a two-tailed test and a level of significance of $\alpha = .05$, we reject H_0 if z is less than -1.96 or greater than $+1.96$. The test statistic is $z = 1.03$, therefore we cannot reject H_0. The p-value is $2(.5000 - .3485) = .303$. On the basis of these data, we are unable to reject the null hypothesis that the median selling price of a new home in St. Louis is \$130,000.

NOTES AND COMMENTS

The number of plus signs was used in the calculations to determine whether to reject the null hypothesis that $p = .5$. One could just as easily use the number of minus signs; the test result would be the same.

EXERCISES

Methods

1. The following table lists the preferences indicated by 10 individuals in taste tests involving two brands of a product.

Individual	Brand A versus Brand B	Individual	Brand A versus Brand B
1	+	6	+
2	+	7	−
3	+	8	+
4	−	9	−
5	+	10	+

With $\alpha = .05$, test for a significant difference in the preferences for the two brands. A plus indicates a preference for brand A over brand B.

2. The following hypothesis test is to be conducted.

$$H_0\text{: Median} \leq 150$$
$$H_a\text{: Median} > 150$$

A sample of size 30 yields 22 cases in which a value greater than 150 is obtained, three cases in which a value of exactly 150 is obtained, and five cases in which a value less than 150 is obtained. Use $\alpha = .01$ and conduct the hypothesis test.

Applications

3. Are stock splits beneficial to stockholders? SNL Securities studied stock splits in the banking industry over an 18-month period and found that stock splits tended to increase the value of an individual's stock holding. Assume that of a sample of 20 recent stock splits, 14 led to an increase in value, four led to a decrease in value, and two resulted in no change. Suppose a sign test is to be used to determine whether stock splits continue to be beneficial for holders of bank stocks.
   a. What are the null and alternative hypotheses?
   b. With $\alpha = .05$, what is the rejection rule?
   c. What is your conclusion?

4. A poll asked 1253 adults a series of questions about the state of the economy and their children's future. One question was, "Do you expect your children to have a better life than you have had, a worse life, or a life about as good as yours?" The responses were 34% better, 29% worse, 33% about the same, and 4% not sure. Use the sign test and a .05 level of significance to determine whether more adults feel their children will have a better future than feel their children will have a worse future. What is your conclusion?

5. *1996–1997 Neilson Media Research* identified *ER* and *Seinfeld* as the top-rated network television shows. Assume that in a local television preference poll, 410 individuals were asked to indicate their favorite network television show: 185 selected *ER*, 165 selected *Seinfeld*, and 60 selected another television show. Use a .05 level of significance to test the hypothesis that *ER* and *Seinfeld* have the same level of preference. What is your conclusion?

6. In 1996, competition in the consumer PC market was led by Packard Bell and Compaq (*USA Today,* July 21, 1997). A sample of 500 purchases showed 202 Packard Bell computers, 158 Compaq computers, and 140 other computers including IBM, Apple, and NEC. Use a .05 level of significance to test the hypothesis that Packard Bell and Compaq have the same share of the consumer PC market. What is your conclusion?

7. The median annual income of subscribers to *Barron's* magazine is \$131,000 (*www.barronsmag.com,* July 28, 2000). Assume a sample of 300 subscribers to *The Wall Street Journal* found 165 subscribers with an income over \$131,000 and 135 subscribers with an income under \$131,000. Can you conclude that there is any difference between the median incomes of the two subscriber groups? What is the *p*-value? Using $\alpha = .05$, what is your conclusion?

8. In a sample of 150 college basketball games, the home team won 98 games. Test to see whether the data support the claim of a home-team advantage in college basketball. Use a .05 level of significance. What is your conclusion?

9. The median number of part-time employees at fast-food restaurants in a particular city was known to be 15 last year. City officials think the use of part-time employees may be increasing. A sample of nine fast-food restaurants showed that more than 15 part-time employees worked at seven of the restaurants, one restaurant had exactly 15 part-time employees, and one had fewer than 15 part-time employees. Test at $\alpha = .05$ to see whether the median number of part-time employees has increased.

10. According to a national survey, the median annual income adults say would make their dreams come true is \$152,000. Suppose that of a sample of 225 individuals in Ohio, 122 individuals report that the amount of income needed to make their dreams come true is less than \$152,000, and 103 report that the amount needed is more than \$152,000. Test the null hypothesis that the median amount of annual income needed to make dreams come true in Ohio is \$152,000. Use $\alpha = .05$. What is your conclusion?

11. The U.S. Bureau of Labor Statistics reported the 1996 median weekly earnings for women in administrative and managerial positions to be \$585. A sample of administrative and managerial working women in the Chicago area provided the following data. Use the sample data to test H_0: median ≤ 585, H_a: median > 585 for the population of administrative and managerial working women in Chicago. Use a .05 level of significance. What is your conclusion?

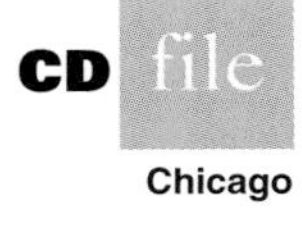

Chicago

622	516	631	498	715
571	494	525	664	721
657	692	551	580	649
706	597	518	725	635
548	604	671	607	487
583	702	622	714	693
600	721	662	633	681
624	551	632	544	485
655	721	669	677	609
656	562	721	489	582

19.2 WILCOXON SIGNED-RANK TEST

The **Wilcoxon signed-rank test** is the nonparametric alternative to the parametric matched-sample test presented in Chapter 10. In the matched-sample situation, each experimental unit generates two paired or matched observations, one from population 1 and one from population 2. The differences between the matched observations provide insight about the differences between the two populations.

The methodology of the parametric matched-sample analysis (the t test on paired differences) requires interval data and the assumption that the population of differences between the pairs of observations is *normally distributed.* With this assumption, the t distribution can be used to test the null hypothesis of no difference between population means. If the assumption of normally distributed differences is not appropriate, the nonparametric Wilcoxon signed-rank test can be used. We illustrate this nonparametric test by comparing the effectiveness of two production methods.

A manufacturing firm is attempting to determine whether two production methods differ in task completion time. A sample of 11 workers was selected, and each worker completed a production task using each of the production methods. The production method that each worker used first was selected randomly. Thus, each worker in the sample provided a pair of observations, as shown in Table 19.3. A positive difference in task completion times indicates that method 1 required more time, and a negative difference in times indicates that method 2 required more time. Do the data indicate that the methods are significantly different in terms of task completion times?

In effect, we have two populations of task completion times, one population associated with each method. The following hypotheses will be tested.

H_0: The populations are identical

H_a: The populations are not identical

TABLE 19.3 PRODUCTION TASK COMPLETION TIMES (MINUTES)

Worker	Method 1	Method 2	Difference
1	10.2	9.5	.7
2	9.6	9.8	−.2
3	9.2	8.8	.4
4	10.6	10.1	.5
5	9.9	10.3	−.4
6	10.2	9.3	.9
7	10.6	10.5	.1
8	10.0	10.0	.0
9	11.2	10.6	.6
10	10.7	10.2	.5
11	10.6	9.8	.8

If H_0 cannot be rejected, we will not have evidence to conclude that the task completion times differ for the two methods. However, if H_0 can be rejected, we will conclude that the two methods differ in task completion time.

The first step of the Wilcoxon signed-rank test requires a ranking of the *absolute value* of the differences between the two methods. We discard any differences of zero and then rank the remaining absolute differences from lowest to highest. Tied differences are assigned the average ranking of their positions in the combined data set. The ranking of the absolute values of differences is shown in the fourth column of Table 19.4. Note that the difference of zero for worker 8 is discarded from the rankings; then the smallest absolute difference of .1 is assigned the rank of 1. This ranking of absolute differences continues with the largest absolute difference of .9 assigned the rank of 10. The tied absolute differences for workers 3 and 5 are assigned the average rank of 3.5 and the tied absolute differences for workers 4 and 10 are assigned the average rank of 5.5.

Once the ranks of the absolute differences have been determined, the ranks are given the sign of the original difference in the data. For example, the .1 difference for worker 7,

TABLE 19.4 RANKING OF ABSOLUTE DIFFERENCES FOR THE PRODUCTION TASK COMPLETION TIME EXAMPLE

Worker	Difference	Absolute Value of Difference	Rank	Signed Rank
1	.7	.7	8	+ 8
2	−.2	.2	2	+ 2
3	.4	.4	3.5	+ 3.5
4	.5	.5	5.5	+ 5.5
5	−.4	.4	3.5	− 3.5
6	.9	.9	10	+10
7	.1	.1	1	+ 1
8	.0	.0	—	—
9	.6	.6	7	+ 7
10	.5	.5	5.5	+ 5.5
11	.8	.8	9	+ 9
			Sum of Signed Ranks	+44.0

which was assigned the rank of 1, is given the value of +1 because the observed difference between the two methods was positive. The .2 difference, which was assigned the rank of 2, is given the value of −2 because the observed difference between the two methods was negative for worker 2. The complete list of signed ranks, as well as their sum, is shown in the last column of Table 19.4.

Let us return to the original hypothesis of identical population task completion times for the two methods. If the populations representing task completion times for each of the two methods are identical, we would expect the positive ranks and the negative ranks to cancel each other, so that the sum of the signed rank values would be approximately zero. Thus, the test for significance under the Wilcoxon signed-rank test involves determining whether the computed sum of signed ranks (+44 in our example) is significantly different from zero.

Let T denote the sum of the signed-rank values in a Wilcoxon signed-rank test. It can be shown that if the two populations are identical and the number of matched pairs of data is 10 or more, the sampling distribution of T can be approximated by a normal probability distribution as follows.

Sampling Distribution of *T* for Identical Populations

$$\text{Mean: } \mu_T = 0 \tag{19.3}$$

$$\text{Standard deviation: } \sigma_T = \sqrt{\frac{n(n+1)(2n+1)}{6}} \tag{19.4}$$

Distribution form: approximately normal provided $n \geq 10$.

For the example, we have $n = 10$ after discarding the observation with the difference of zero (worker 8). Thus, using (19.4), we have

$$\sigma_T = \sqrt{\frac{10(11)(21)}{6}} = 19.62$$

Figure 19.3 is the sampling distribution of T under the assumption of identical populations.

The value of the test statistic z is:

$$z = \frac{T - \mu_T}{\sigma_T} = \frac{44 - 0}{19.62} = 2.24$$

FIGURE 19.3 SAMPLING DISTRIBUTION OF THE WILCOXON T FOR THE PRODUCTION TASK COMPLETION TIME EXAMPLE

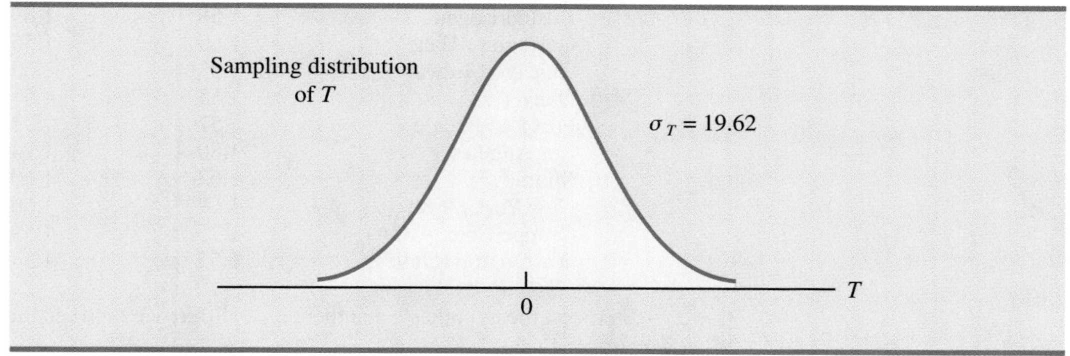

Testing the null hypothesis of no difference using a level of significance of $\alpha = .05$, we reject H_0 if $z < -1.96$ or if $z > 1.96$. With the value of $z = 2.24$, we reject H_0 and conclude that the two populations are not identical and that the methods differ in task completion time. Using $z = 2.24$, the p-value is $2(.5000 - .4875) = .025$. The fact that method 2 had the shorter completion times for eight of the 11 workers leads us to conclude that differences between the two populations indicate method 2 to be the better production method.

EXERCISES

Applications

12. Two fuel additives are being tested to determine their effect on miles per gallon for passenger cars. Test results for 12 cars follow; each car was tested with both fuel additives. Use $\alpha = .05$ and the Wilcoxon signed-rank test to see whether there is a significant difference in the additives.

	Additive			Additive	
Car	**1**	**2**	**Car**	**1**	**2**
1	20.12	18.05	7	16.16	17.20
2	23.56	21.77	8	18.55	14.98
3	22.03	22.57	9	21.87	20.03
4	19.15	17.06	10	24.23	21.15
5	21.23	21.22	11	23.21	22.78
6	24.77	23.80	12	25.02	23.70

13. A sample of 10 men was used in a study to test the effects of a relaxant on the time required to fall asleep for male adults. Data for 10 subjects showing the number of minutes required to fall asleep with and without the relaxant follow. Use a .05 level of significance to determine whether the relaxant reduces the time required to fall asleep. What is your conclusion?

Subject	**Without Relaxant**	**With Relaxant**	**Subject**	**Without Relaxant**	**With Relaxant**
1	15	10	6	7	5
2	12	10	7	8	10
3	22	12	8	10	7
4	8	11	9	14	11
5	10	9	10	9	6

14. Rental car gasoline prices per gallon were sampled at 10 major airports. Data for Avis and Budget car rental companies follow (*USA Today,* April 4, 2000).

Airport	**Avis**	**Budget**
Boston Logan	1.58	1.39
Chicago O'Hare	1.60	1.55
Chicago Midway	1.53	1.55
Denver	1.55	1.51
Fort Lauderdale	1.57	1.58
Los Angeles	1.80	1.74
Miami	1.62	1.60
New York (JFK)	1.69	1.60
Orange County, CA	1.75	1.59
Washington (Dulles)	1.55	1.54

Use $\alpha = .05$ to test the hypothesis that there is no difference between the two populations. What is your conclusion?

15. A test was conducted of two overnight mail delivery services. Two samples of identical deliveries were set up so that both delivery services were notified of the need for a delivery at the same time. The hours required to make each delivery follow. Do the data shown suggest a difference in the delivery times for the two services? Use a .05 level of significance for the test.

	Service	
Delivery	**1**	**2**
1	24.5	28.0
2	26.0	25.5
3	28.0	32.0
4	21.0	20.0
5	18.0	19.5
6	36.0	28.0
7	25.0	29.0
8	21.0	22.0
9	24.0	23.5
10	26.0	29.5
11	31.0	30.0

16. The 1997 price/earnings ratios for a sample of 12 stocks are shown in the following list (*Barron's*, December 8, 1997). Assume that a financial analyst has provided the estimated price/earnings ratio for 1998. Using a .05 level of significance, what is your conclusion about the differences between the price/earnings ratios for 1997 and 1998?

Stock	1997 P/E Ratio	1998 P/E Ratio (Est.)
Coca-Cola	40	32
Du Pont	24	22
Eastman Kodak	21	23
General Electric	30	23
General Mills	25	19
IBM	19	19
McDonald's	20	17
Merck	29	19
Motorola	35	20
Philip Morris	17	18
Walt Disney	33	27
Xerox	20	16

17. Ten test-market cities were selected as part of a market research study designed to evaluate the effectiveness of a particular advertising campaign. The sales dollars for each city were recorded for the week prior to the promotional program. Then the campaign was conducted for two weeks and new sales data were collected for the week immediately after the campaign. The two sets of sales data (in thousands of dollars) follow.

City	Precampaign Sales	Postcampaign Sales
Kansas City	130	160
Dayton	100	105
Cincinnati	120	140
Columbus	95	90
Cleveland	140	130
Indianapolis	80	82
Louisville	65	55
St. Louis	90	105
Pittsburgh	140	152
Peoria	125	140

Use $\alpha = .05$. What conclusion would you draw about the value of the advertising program?

19.3 MANN-WHITNEY-WILCOXON TEST

In this section we present another nonparametric method that can be used to determine whether a difference exists between two populations. This test, unlike the signed-rank test, is not based on a matched sample. Two independent samples, one from each population, are used. The test was developed jointly by Mann, Whitney, and Wilcoxon. It is sometimes called the *Mann-Whitney test* and sometimes the *Wilcoxon rank-sum test.* Both the Mann-Whitney and Wilcoxon versions of this test are equivalent; we refer to it as the **Mann-Whitney-Wilcoxon (MWW) test.**

Recall that in Chapter 10 we conducted a parametric test for the difference between the means of two populations. The following hypotheses were tested:

$$H_0\colon \mu_1 - \mu_2 = 0$$
$$H_a\colon \mu_1 - \mu_2 \neq 0$$

In the small-sample case, the hypothesis test required interval data and the assumption that both populations were normally distributed. Under these conditions, the t distribution was used to test for the difference between the means of the two populations.

The MWW test can be used without making the assumption that both populations are normally distributed.

The nonparametric MWW test does not require interval data or the assumption that both populations are normally distributed. The only requirement of the MWW test is that the measurement scale for the data is at least ordinal. Then, instead of testing for the difference between the means of the two populations, the MWW test determines whether the two populations are identical. The hypotheses for the MWW test are as follows.

H_0: The two populations are identical
H_a: The two populations are not identical

We demonstrate how the MWW test can be applied by first showing an application for the small-sample case.

Small-Sample Case

The small-sample case for the MWW test should be used whenever the sample sizes for both populations are less than or equal to 10. We illustrate the use of the MWW test for the small-sample case by considering the academic potential of students attending Johnston High School. The majority of students attending Johnston High School previously attended either Garfield Junior High School or Mulberry Junior High School. The question raised by school administrators was whether the population of students who had attended Garfield was identical to the population of students who had attended Mulberry in terms of academic potential. The following hypotheses were considered.

H_0: The two populations are identical in terms of academic potential
H_a: The two populations are not identical in terms of academic potential

Using high school records, Johnston High School administrators selected a random sample of four high school students who had attended Garfield Junior High and another random sample of five students who had attended Mulberry Junior High. The current high school class standing was recorded for each of the nine students used in the study. The ordinal class standings for the nine students are listed in Table 19.5.

The first step in the MWW procedure is to rank the *combined* data from the two samples from low to high. The lowest value (class standing 8) receives a rank of 1 and the highest value (class standing 202) receives a rank of 9. The ranking of the nine students is given in Table 19.6.

TABLE 19.5 HIGH SCHOOL CLASS STANDING DATA

Garfield Students		Mulberry Students	
Student	**Class Standing**	**Student**	**Class Standing**
Fields	8	Hart	70
Clark	52	Phipps	202
Jones	112	Kirkwood	144
Tibbs	21	Abbott	175
		Guest	146

TABLE 19.6 RANKING OF HIGH SCHOOL STUDENTS

Student	Class Standing	Combined Sample Rank	Student	Class Standing	Combined Sample Rank
Fields	8	1	Kirkwood	144	6
Tibbs	21	2	Guest	146	7
Clark	52	3	Abbott	175	8
Hart	70	4	Phipps	202	9
Jones	112	5			

The next step is to sum the ranks for each sample separately. This calculation is shown in Table 19.7. The MWW procedure can use the sum of the ranks for either sample. In the following discussion, we use the sum of the ranks for the sample of four students from Garfield. We denote this sum by the symbol T. Thus, for our example, $T = 11$.

Let us consider the properties of the sum of the ranks for the Garfield sample. With four students in the sample, Garfield could have the top four students in the study. If this were the case, $T = 1 + 2 + 3 + 4 = 10$ would be the smallest value possible for the rank sum T. Conversely, Garfield could have the bottom four students, in which case $T = 6 + 7 + 8 + 9 = 30$ would be the largest value possible for T. Hence, T for the Garfield sample must take a value between 10 and 30.

Note that values of T near 10 imply that Garfield has the significantly better, or higher ranking, students, whereas values of T near 30 imply that Garfield has the significantly weaker, or lower ranking, students. Thus, if the two populations of students were identical

TABLE 19.7 RANK SUMS FOR HIGH SCHOOL STUDENTS FROM EACH JUNIOR HIGH SCHOOL

	Garfield Students			Mulberry Students	
Student	**Class Standing**	**Sample Rank**	**Student**	**Class Standing**	**Sample Rank**
Fields	8	1	Hart	70	4
Clark	52	3	Phipps	202	9
Jones	112	5	Kirkwood	144	6
Tibbs	21	2	Abbott	175	8
			Guest	146	7
Sum of Ranks		11			34

in terms of academic potential, we would expect the value of T to be near the average of the two values, or $(10 + 30)/2 = 20$.

Critical values of the MWW T statistic are provided in Table 9 of Appendix B for cases in which both sample sizes are less than or equal to 10. In that table, n_1 refers to the sample size corresponding to the sample whose rank sum is being used in the test. The value of T_L is read directly from the table and the value of T_U is computed from (19.5).

$$T_U = n_1(n_1 + n_2 + 1) - T_L \tag{19.5}$$

Neither the value of T_L nor the value of T_U is in the rejection region. The null hypothesis of identical populations should be rejected only if T is strictly less than T_L or strictly greater than T_U.

For example, using Table 9 of Appendix B with a .05 level of significance, we see that the lower-tail critical value for the MWW statistic with $n_1 = 4$ (Garfield) and $n_2 = 5$ (Mulberry) is $T_L = 12$. The upper-tail critical value for the MWW statistic computed by using (19.5) is

$$T_U = 4(4 + 5 + 1) - 12 = 28$$

Thus, the MWW decision rule indicates that the null hypothesis of identical populations can be rejected if the sum of the ranks for the first sample (Garfield) is less than 12 or greater than 28. The rejection rule can be written as

$$\text{Reject } H_0 \text{ if } T < 12 \text{ or if } T > 28$$

If we had conducted the test with the rank sum of the Mulberry students, we would have had $n_1 = 5$, $n_2 = 4$, $T_L = 17$, $T_U = 33$, and $T = 34$. With $T > T_U$, we would have reached the same conclusion to reject H_0.

Referring to Table 19.7, we see that $T = 11$. Hence, the null hypothesis H_0 is rejected, and we can conclude that the population of students at Garfield differs from the population of students at Mulberry in terms of academic potential. The higher class ranking obtained by the sample of Garfield students suggests that Garfield students are better prepared for high school than the Mulberry students.

Large-Sample Case

When both sample sizes are greater than or equal to 10, a normal approximation of the distribution of T can be used to conduct the analysis for the MWW test. We illustrate the large-sample case by considering a situation at Third National Bank.

Third National Bank has two branch offices. Data collected from two independent simple random samples, one from each branch, are given in Table 19.8. Do the data indicate whether the populations of checking account balances at the two branch banks are identical?

The first step in the MWW test is to rank the *combined* data from the lowest to the highest values. Using the combined set of 22 observations in Table 19.8, we find the lowest data value of \$750 (sixth item of sample 2) and assign to it a rank of 1. Continuing the ranking gives us the following list.

Balance (\$)	Item	Assigned Rank
750	6th of sample 2	1
800	5th of sample 2	2
805	7th of sample 1	3
850	2nd of sample 2	4
.	.	.
.	.	.
.	.	.
1195	4th of sample 1	21
1200	3rd of sample 1	22

TABLE 19.8 ACCOUNT BALANCES FOR TWO BRANCHES OF THIRD NATIONAL BANK

Branch 1		Branch 2	
Account	**Balance ($)**	**Account**	**Balance ($)**
1	1095	1	885
2	955	2	850
3	1200	3	915
4	1195	4	950
5	925	5	800
6	950	6	750
7	805	7	865
8	945	8	1000
9	875	9	1050
10	1055	10	935
11	1025		
12	975		

In ranking the combined data, we may find that two or more data values are the same. In that case, the tied values are given the *average* ranking of their positions in the combined data set. For example, the balance of \$945 (eighth item of sample 1) will be assigned the rank of 11. However, the next two values in the data set are tied with values of \$950 (see the sixth item of sample 1 and the fourth item of sample 2). Because these two values will be considered for assigned ranks of 12 and 13, they are both assigned the rank of 12.5. At the next highest data value of \$955, we continue the ranking process by assigning \$955 the rank of 14. Table 19.9 is the entire data set with the assigned rank of each observation.

The next step in the MWW test is to sum the ranks for each sample. The sums are given in Table 19.9. The test procedure can be based on the sum of the ranks for either sample. We use the sum of the ranks for the sample from branch 1. Thus, for this example, $T = 169.5$.

TABLE 19.9 COMBINED RANKING OF THE DATA IN THE TWO SAMPLES FROM THIRD NATIONAL BANK

Branch 1			Branch 2		
Account	**Balance ($)**	**Rank**	**Account**	**Balance ($)**	**Rank**
1	1095	20	1	885	7
2	955	14	2	850	4
3	1200	22	3	915	8
4	1195	21	4	950	12.5
5	925	9	5	800	2
6	950	12.5	6	750	1
7	805	3	7	865	5
8	945	11	8	1000	16
9	875	6	9	1050	18
10	1055	19	10	935	10
11	1025	17		Sum of Ranks	83.5
12	975	15			
	Sum of Ranks	169.5			

Given that the sample sizes are $n_1 = 12$ and $n_2 = 10$, we can use the normal approximation to the sampling distribution of the rank sum T. The appropriate sampling distribution is given by the following expressions.

Sampling Distribution of T for Identical Populations

$$\text{Mean: } \mu_T = \tfrac{1}{2} n_1(n_1 + n_2 + 1) \quad (19.6)$$

$$\text{Standard Deviation: } \sigma_T = \sqrt{\tfrac{1}{12} n_1 n_2(n_1 + n_2 + 1)} \quad (19.7)$$

Distribution form: approximately normal provided $n_1 \geq 10$ and $n_2 \geq 10$.

For branch 1, we have

$$\mu_T = \tfrac{1}{2}\, 12(12 + 10 + 1) = 138$$
$$\sigma_T = \sqrt{\tfrac{1}{12}\, 12(10)(12 + 10 + 1)} = 15.17$$

Figure 19.4 is the sampling distribution of T. Following the usual hypothesis-testing procedure, we compute the test statistic z to determine whether the observed value of T appears to be from the sampling distribution of Figure 19.4. If T does not appear to be from that distribution, we will reject the null hypothesis and conclude that the populations are not identical. Computing the test statistic, we have

$$z = \frac{T - \mu_T}{\sigma_T} = \frac{169.5 - 138}{15.17} = 2.08$$

If we had used the rank sum for branch 2, we would have had $n_1 = 10$, $n_2 = 12$, $\mu_T = 115$, and $T = 83.5$. With $z = -2.08$ we would again reject H_0.

At a .05 level of significance, we know that, to reject H_0, z must be less than -1.96 or greater than $+1.96$. With $z = 2.08$, and the p-value $2(.5000 - .4812) = .0376$ less than $\alpha = .05$, we reject H_0. Thus, we conclude that the two populations are not identical. That is, the populations of account balances at the two branches are not the same.

In summary, the Mann-Whitney-Wilcoxon rank-sum test consists of the following steps to determine whether two independent random samples are selected from identical populations.

1. Rank the combined sample observations from lowest to highest, with tied values being assigned the average of the tied rankings.
2. Compute T, the sum of the ranks for the first sample.

FIGURE 19.4 SAMPLING DISTRIBUTION OF T FOR THE THIRD NATIONAL BANK EXAMPLE

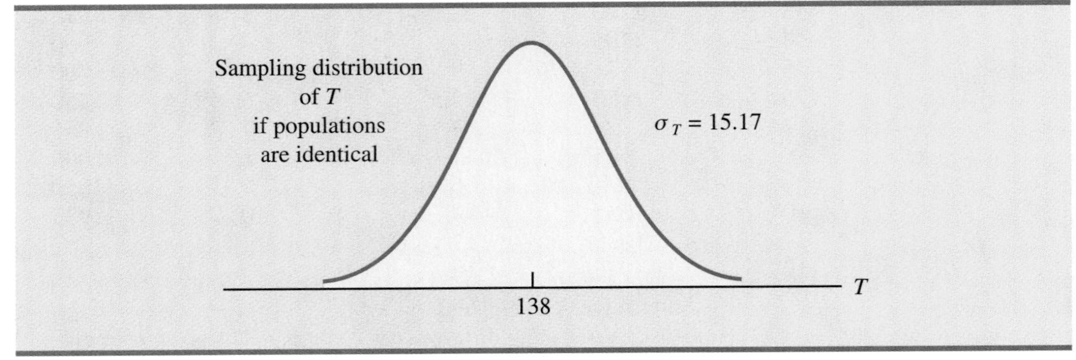

3. In the large-sample case, make the test for significant differences between the two populations by using the observed value of T and comparing it to the sampling distribution of T for identical populations as in equations (19.6) and (19.7). The value of the standardized test statistic z or the p-value will provide the basis for deciding whether to reject H_0. In the small-sample case, use Table 9 in Appendix B to find the critical values for the test.

NOTES AND COMMENTS

The nonparametric test discussed in this section is used to determine whether two populations are identical. Parametric statistical tests, such as the t test described in Chapter 10, test the equality of two population means. When we reject the hypothesis that the means are equal, we conclude that the populations differ in their means. When we reject the hypothesis that the populations are identical by using the MWW test, we cannot state how they differ. The populations could have different means, different variances, and/or different forms. Nonetheless, if we believe that the populations are the same in every aspect but the means, a rejection of H_0 by the nonparametric method implies that the means differ. The major advantages of the MWW test over the parametric t test are that it does not require any assumptions about the form of the probability distribution from which the measurements come and it can be used with ordinal data.

EXERCISES

Applications

18. Two fuel additives are being tested to determine their effect on gas mileage. Seven cars were tested with additive 1 and nine cars were tested with additive 2. The following data show the miles per gallon obtained with the two additives. Use $\alpha = .05$ and the MWW test to see whether there is a significant difference in gasoline mileage for the two additives.

Additive 1	Additive 2
17.3	18.7
18.4	17.8
19.1	21.3
16.7	21.0
18.2	22.1
18.6	18.7
17.5	19.8
	20.7
	20.2

19. Samples of starting annual salaries for individuals entering the public accounting and financial planning professions follow (*Fortune*, June 26, 1995). Annual salaries are shown in thousands of dollars.

Public Accountant	Financial Planner	Public Accountant	Financial Planner
25.2	24.0	30.0	28.6
33.8	24.2	25.9	24.7
31.3	28.1	34.5	28.9
33.2	30.9	31.7	26.8
29.2	26.9	26.9	23.9

a. Using a .05 level of significance, test the hypothesis that there is no difference between the starting annual salaries of public accountants and financial planners. What is your conclusion?

b. What are the sample mean annual salaries for the two professions?

20. The gap between the earnings of men and women with equal education is narrowing but has not closed (*USA Today,* September 15, 2000). Sample data for seven men and seven women with bachelor's degrees are as follows. Data are shown in thousands of dollars.

Men	30.6	75.5	45.2	62.2	38.2	49.9	55.3
Women	44.5	35.4	27.9	40.5	25.8	47.5	24.8

a. What is the median salary for men? For women?
b. Using $\alpha = .05$, conduct the hypothesis test for equal populations. What is your conclusion?

21. Mileage performance tests were conducted for two models of automobiles. Twelve automobiles of each model were selected randomly and a miles-per-gallon rating for each model was developed on the basis of 1000 miles of highway driving. The data follow.

Model 1		Model 2	
Automobile	**Miles per Gallon**	**Automobile**	**Miles per Gallon**
1	20.6	1	21.3
2	19.9	2	17.6
3	18.6	3	17.4
4	18.9	4	18.5
5	18.8	5	19.7
6	20.2	6	21.1
7	21.0	7	17.3
8	20.5	8	18.8
9	19.8	9	17.8
10	19.8	10	16.9
11	19.2	11	18.0
12	20.5	12	20.1

Use $\alpha = .10$ and test for a significant difference in the populations of miles-per-gallon ratings for the two models.

22. *Business Week* annually publishes statistics on the world's 1000 largest companies. A company's price/earnings (P/E) ratio is the company's current stock price divided by the latest 12 months' earnings per share. Listed in Table 19.10 are the P/E ratios for a sample of 10 Japanese and 12 U.S. companies (*Business Week,* July 11, 1994). Is the difference in P/E ratios between the two countries significant? Use the MWW test and $\alpha = .01$ to support your conclusion.

TABLE 19.10 P/E RATIOS FOR JAPANESE AND U.S. COMPANIES

Japan		United States	
Company	**P/E Ratio**	**Company**	**P/E Ratio**
Sumitomo Corp.	153	Gannet	19
Kinden	21	Motorola	24
Heiwa	18	Schlumberger	24
NCR Japan	125	Oracle Systems	43
Suzuki Motor	31	Gap	22
Fuji Bank	213	Winn-Dixie	14
Sumitomo Chemical	64	Ingersoll-Rand	21
Seibu Railway	666	American Electric Power	14
Shiseido	33	Hercules	21
Toho Gas	68	Times Mirror	38
		WellPoint Health	15
		Northern States Power	14

23. Police records show the following numbers of daily crime reports for a sample of days during the winter months and a sample of days during the summer months. Using a .05 level of significance, determine whether there is a significant difference between the winter and summer months in terms of the number of crime reports.

Winter	Summer
18	28
20	18
15	24
16	32
21	18
20	29
12	23
16	38
19	28
20	18

24. A certain brand of microwave oven was priced at 10 stores in Dallas and 13 stores in San Antonio. The data follow. Use a .05 level of significance and test whether prices for the microwave oven are the same in the two cities.

Dallas	San Antonio
445	460
489	451
405	435
485	479
439	475
449	445
436	429
420	434
430	410
405	422
	425
	459
	430

25. The National Association of Home Builders provided data on the cost of the most popular home remodeling projects (*USA Today,* June 17, 1997). Use the Mann-Whitney-Wilcoxon test to see whether it can be concluded that the cost of kitchen remodeling differs from the cost of master bedroom remodeling. Use a .05 level of significance.

Kitchen	Master Bedroom
25,200	18,000
17,400	22,900
22,800	26,400
21,900	24,800
19,700	26,900
23,000	17,800
19,700	24,600
16,900	21,000
21,800	
23,600	

19.4 KRUSKAL-WALLIS TEST

The MWW test in Section 19.3 can be used to test whether two populations are identical. It has been extended to the case of three or more populations by Kruskal and Wallis. The hypotheses for the Kruskal-Wallis test with $k \geq 3$ populations can be written as follows.

H_0: All populations are identical

H_a: Not all populations are identical

The Kruskal-Wallis test is based on the analysis of independent random samples from each of the k populations.

This test is an alternative to ANOVA in Chapter 13, which focused on the equality of the means of k populations.

In Chapter 13 we showed that analysis of variance (ANOVA) can be used to test for the equality of means among three or more populations. The ANOVA procedure requires interval or ratio data and the assumption that the k populations are normally distributed.

The nonparametric Kruskal-Wallis test can be used with ordinal data as well as with interval or ratio data. In addition, the Kruskal-Wallis test does not require the assumption of normally distributed populations. Hence, whenever the data from $k \geq 3$ populations are ordinal, or whenever the assumption of normally distributed populations is questionable, the Kruskal-Wallis test provides an alternate statistical procedure for testing whether the populations are identical. We demonstrate the Kruskal-Wallis test by using it in an employee selection application.

TABLE 19.11

PERFORMANCE EVALUATION RATINGS FOR 20 WILLIAMS EMPLOYEES

College A	College B	College C
25	60	50
70	20	70
60	30	60
85	15	80
95	40	90
90	35	70
80		75

Williams Manufacturing Company hires employees for its management staff from three local colleges. Recently, the company's personnel department has been collecting and reviewing annual performance ratings in an attempt to determine whether there are differences in performance among the managers hired from these colleges. Performance rating data are available from independent samples of seven employees from college A, six employees from college B, and seven employees from college C. These data are summarized in Table 19.11; the overall performance rating of each manager is given on a 0–100 scale, with 100 being the highest possible performance rating.

Suppose we want to test whether the three populations are identical in terms of performance evaluations. The Kruskal-Wallis test statistic, which is based on the sum of ranks for each of the samples, can be computed as follows.

Kruskal-Wallis Test Statistic

$$W = \left[\frac{12}{n_T(n_T + 1)} \sum_{i=1}^{k} \frac{R_i^2}{n_i}\right] - 3(n_T + 1) \tag{19.8}$$

where

k = the number of populations

n_i = the number of items in sample i

$n_T = \Sigma n_i$ = total number of items in all samples

R_i = sum of the ranks for sample i

Kruskal and Wallis were able to show that, under the null hypothesis in which the populations are identical, the sampling distribution of W can be approximated by a chi-square distribution with $k - 1$ degrees of freedom. This approximation is generally acceptable if each of the sample sizes is greater than or equal to five.

The Kruskal-Wallis test uses only the ordinal rank of the data.

To compute the W statistic for our example, we must first rank all 20 data items. The lowest data value of 15 from the college B sample receives a rank of 1, whereas the highest data value of 95 from the college A sample receives a rank of 20. The data values, their associated ranks, and the sum of the ranks for the three samples are given in Table 19.12. Note that we assign the average rank to tied items;* for example, the data values of 60, 70, 80, and 90 had ties.

The sample sizes are

$$n_1 = 7 \qquad n_2 = 6 \qquad n_3 = 7$$

and

$$n_T = \Sigma n_i = 7 + 6 + 7 = 20$$

We compute the W statistic by using equation (19.8).

$$W = \frac{12}{20(21)}\left[\frac{(95)^2}{7} + \frac{(27)^2}{6} + \frac{(88)^2}{7}\right] - 3(20 + 1) = 8.92$$

The computer procedures in Appendixes 11.1 and 11.2 show how Minitab and Excel can be used to compute p-value = .012.

The chi-square distribution table (Table 3 of Appendix B) shows that with $k - 1 = 2$ degrees of freedom and $\alpha = .05$ in the upper tail of the distribution, the critical chi-square value is $\chi^2 = 5.99147$. Because the test statistic $W = 8.92$ is greater than 5.99147, we reject the null hypothesis that the three populations are identical. As a result, we conclude that manager performance differs significantly depending on the college attended. Furthermore, because the performance ratings are lowest for college B, it would be reasonable for the company to either cut back recruiting from college B or at least evaluate its graduates more thoroughly.

TABLE 19.12 COMBINED RANKINGS FOR THE 20 WILLIAMS EMPLOYEES

College A	Rank	College B	Rank	College C	Rank
25	3	60	9	50	7
70	12	20	2	70	12
60	9	30	4	60	9
85	17	15	1	80	15.5
95	20	40	6	90	18.5
90	18.5	35	5	70	12
80	15.5			75	14
Sum of Ranks	95		27		88

NOTES AND COMMENTS

The Kruskal-Wallis procedure illustrated in the example began with the collection of interval-scaled data showing employee performance evaluation ratings. The procedure also would have worked had the data been the ordinal rankings of the 20 employees. In that case, the Kruskal-Wallis test could have been applied directly to the original data; the step of constructing the rank orderings from the performance evaluation ratings would have been omitted.

*If numerous tied ranks are observed, (19.8) must be modified; the modified formula is given in *Practical Nonparametric Statistics* by W. J. Conover.

EXERCISES

Methods

26. Three products received the following performance ratings by a panel of 15 consumers.

Product		
A	B	C
50	80	60
62	95	45
75	98	30
48	87	58
65	90	57

Use the Kruskal-Wallis test and $\alpha = .05$ to determine whether there is a significant difference in the performance ratings for the products.

27. Three admission test preparation programs are being evaluated. The scores obtained by a sample of 20 people who used the test preparation programs provided the following data. Use the Kruskal-Wallis test to determine whether there is a significant difference among the three test preparation programs. Use $\alpha = .01$.

Program		
A	B	C
540	450	600
400	540	630
490	400	580
530	410	490
490	480	590
610	370	620
	550	570

Applications

28. Forty-minute workouts of one of the following activities three days a week will lead to a loss of weight. The following sample data show the number of calories burned during 40-minute workouts for three different activities. Do these data indicate differences in the amount of calories burned for the three activities? Use a .05 level of significance. What is your conclusion?

Swimming	Tennis	Cycling
408	415	385
380	485	250
425	450	295
400	420	402
427	530	268

29. The miles-per-gallon data obtained from tests on three different automobiles are reported as follows. Use the Kruskal-Wallis test with $\alpha = .05$ to determine whether there is a significant difference in the gasoline mileage for the three automobiles.

Automobile		
A	B	C
19	19	24
21	20	26
20	22	23
19	21	25
21	23	27

30. A large corporation has been sending many of its first-level managers to an off-site supervisory skills course. Four different management development centers offer this course, and the corporation wants to determine whether they differ in the quality of training provided. A sample of 20 employees who have attended these programs has been chosen and the employees ranked in terms of supervisory skills. The results follow.

Course	Supervisory Skills Rank				
1	3	14	10	12	13
2	2	7	1	5	11
3	19	16	9	18	17
4	20	4	15	6	8

Note that the top-ranked supervisor attended course 2 and the lowest-ranked supervisor attended course 4. Use $\alpha = .05$ and test to see whether there is a significant difference in the training provided by the four programs.

31. The better-selling candies are high in calories. Assume that the following data show the calorie content from samples of M&Ms, Kit Kat, and Milky Way II. Test for significant differences in the calorie content of these three candies. At a .05 level of significance, what is your conclusion?

M&Ms	Kit Kat	Milky Way II
230	225	200
210	205	208
240	245	202
250	235	190
230	220	180

19.5 RANK CORRELATION

The Spearman rank-correlation coefficient is equal to the Pearson correlation coefficient applied to ordinal or rank data.

The correlation coefficient is a measure of the linear association between two variables for which interval or ratio data are available. In this section, we consider measures of association between two variables when only ordinal data are available. The **Spearman rank-correlation coefficient** r_s has been developed for this purpose.

Spearman Rank-Correlation Coefficient

$$r_s = 1 - \frac{6\Sigma d_i^2}{n(n^2 - 1)} \tag{19.9}$$

where

n = the number of items or individuals being ranked
x_i = the rank of item i with respect to one variable
y_i = the rank of item i with respect to a second variable
$d_i = x_i - y_i$

Let us illustrate the use of the Spearman rank-correlation coefficient with an example. A company wants to determine whether individuals who were expected at the time of employment to be better salespersons actually turn out to have better sales records. To investigate this question, the vice president in charge of personnel carefully reviewed the original job interview summaries, academic records, and letters of recommendation for 10 current

TABLE 19.13 SALES POTENTIAL AND ACTUAL TWO-YEAR SALES DATA FOR 10 SALESPEOPLE

Salesperson	Ranking of Potential	Two-Year Sales (units)	Ranking According to Two-Year Sales
A	2	400	1
B	4	360	3
C	7	300	5
D	1	295	6
E	6	280	7
F	3	350	4
G	10	200	10
H	9	260	8
I	8	220	9
J	5	385	2

members of the firm's salesforce. After the review, the vice president ranked the 10 individuals in terms of their potential for success, basing the assessment solely on the information available at the time of employment. Then a list was obtained of the number of units sold by each salesperson over the first two years. On the basis of actual sales performance, a second ranking of the 10 salespersons was carried out. Table 19.13 gives the relevant data and the two rankings. The statistical question is whether there is agreement between the ranking of potential at the time of employment and the ranking based on the actual sales performance over the first two years.

Let us compute the Spearman rank-correlation coefficient for the data in Table 19.13. The computations are summarized in Table 19.14. We see that the rank-correlation coefficient is a positive .73. The Spearman rank-correlation coefficient ranges from -1.0 to $+1.0$ and its interpretation is similar to that of the sample correlation coefficient in that positive values near 1.0 indicate a strong association between the rankings; as one rank increases, the other rank increases. Rank correlations near -1.0 indicate a strong negative association

TABLE 19.14 COMPUTATION OF THE SPEARMAN RANK-CORRELATION COEFFICIENT FOR SALES POTENTIAL AND SALES PERFORMANCE

Salesperson	x_i = Ranking of Potential	y_i = Ranking of Sales Performance	$d_i = x_i - y_i$	d_i^2
A	2	1	1	1
B	4	3	1	1
C	7	5	2	4
D	1	6	-5	25
E	6	7	-1	1
F	3	4	-1	1
G	10	10	0	0
H	9	8	1	1
I	8	9	-1	1
J	5	2	3	9
				$\Sigma d_i^2 = 44$

$$r_s = 1 - \frac{6\Sigma d_i^2}{n(n^2 - 1)} = 1 - \frac{6(44)}{10(100 - 1)} = .73$$

between the rankings; as one rank increases, the other rank decreases. The value $r_s = .73$ indicates a positive correlation between potential and actual performance. Individuals ranked high on potential tend to rank high on performance.

Test for Significant Rank Correlation

At this point, we have seen how sample results can be used to compute the sample rank-correlation coefficient. As with many other statistical procedures, we may want to use the sample results to make an inference about the population rank correlation ρ_s. To make an inference about the population rank correlation, we must test the following hypotheses.

$$H_0: \rho_s = 0$$
$$H_a: \rho_s \neq 0$$

Under the null hypothesis of no rank correlation ($\rho_s = 0$), the rankings are independent, and the sampling distribution of r_s is as follows.

Sampling Distribution of r_s

$$\text{Mean: } \mu_{r_s} = 0 \tag{19.10}$$

$$\text{Standard Deviation: } \sigma_{r_s} = \sqrt{\frac{1}{n-1}}$$

Distribution form: approximately normal provided $n \geq 10$.

The sample rank-correlation coefficient for sales potential and sales performance is $r_s = .73$. With this value, we can test for a significant rank correlation. From equation (19.10) we have $\mu_{r_s} = 0$ and from (19.11) we have $\sigma_{r_s} = \sqrt{1/(10-1)} = .33$. Using the test statistic, we have

$$z = \frac{r_s - \mu_{r_s}}{\sigma_{r_s}} = \frac{.73 - 0}{.33} = 2.21$$

At a .05 level of significance, we see that the null hypothesis of no correlation will be rejected if $z < -1.96$ or if $z > 1.96$. Because $z = 2.21$, and the p-value $= 2(.5000 - .4864) = .0272$ is less than $\alpha = .05$, we reject the hypothesis of no rank correlation. Thus, we can conclude that there is a significant rank correlation between sales potential and sales performance.

EXERCISES

Methods

SELF test

32. Consider the following set of rankings for a sample of 10 elements.

Element	x_i	y_i	Element	x_i	y_i
1	10	8	6	2	7
2	6	4	7	8	6
3	7	10	8	5	3
4	3	2	9	1	1
5	4	5	10	9	9

a. Compute the Spearman rank-correlation coefficient for the data.
b. Test for significant rank correlation using $\alpha = .05$ and state your conclusion.

33. Consider the following two sets of rankings for six items.

Case One			Case Two		
Item	**First Ranking**	**Second Ranking**	**Item**	**First Ranking**	**Second Ranking**
A	1	1	A	1	6
B	2	2	B	2	5
C	3	3	C	3	4
D	4	4	D	4	3
E	5	5	E	5	2
F	6	6	F	6	1

Note that in the first case the rankings are identical, whereas in the second case the rankings are exactly opposite. What value should you expect for the Spearman rank-correlation coefficient for each of these cases? Explain. Calculate the rank-correlation coefficient for each case.

Applications

34. For a sample of 11 states, the following table gives the ranks on pupil-teacher ratio (1 = lowest, 11 = highest) and expenditure per pupil (1 = highest, 11 = lowest).

	Rank			Rank	
State	**Pupil-Teacher Ratio**	**Expenditure per Pupil**	**State**	**Pupil-Teacher Ratio**	**Expenditure per Pupil**
Arizona	10	9	Massachusetts	1	1
Colorado	8	5	Nebraska	2	7
Florida	6	4	North Dakota	7	8
Idaho	11	2	South Dakota	5	10
Iowa	4	6	Washington	9	3
Louisiana	3	11			

At the $\alpha = .05$ level, does there appear to be a relationship between expenditure per pupil and pupil-teacher ratio?

35. A national study by Harris Interactive, Inc., evaluated the top Internet companies and their reputations (*The Wall Street Journal,* November 18, 1999). The following two lists show how 10 Internet companies ranked in terms of reputation and percentage of respondents who said they would purchase the company's stock. A positive rank correlation is anticipated because it seems reasonable to expect that a company with a higher reputation would be a more desirable purchase.

	Reputation	**Probable Purchase**
Microsoft	1	3
Intel	2	4
Dell	3	1
Lucent	4	2
Texas Instruments	5	9
Cisco Systems	6	5
Hewlett-Packard	7	10
IBM	8	6
Motorola	9	7
Yahoo	10	8

a. Compute the rank correlation between reputation and probable purchase.
b. Test for a significant positive rank correlation. What is the p-value?
c. Using $\alpha = .05$, what is your conclusion?

36. The 1996 rankings of a sample of professional golfers in both driving distance and putting follows (*Golf Digest,* January 1997). What is the rank correlation between driving distance and putting? Use a .10 level of significance.

Professional Golfer	Driving Distance	Putting
Fred Couples	1	5
David Duval	5	6
Ernie Els	4	10
Nick Faldo	9	2
Tom Lehman	6	7
Justin Leonard	10	3
Davis Love III	2	8
Phil Mickelson	3	9
Greg Norman	7	4
Mark O'Meara	8	1

37. A student organization surveyed both recent graduates and current students to obtain information on the quality of teaching at a particular university. An analysis of the responses provided the following teaching-ability rankings. Do the rankings given by the current students agree with the rankings given by the recent graduates? Use $\alpha = .10$ and test for a significant rank correlation.

	Ranking by	
Professor	**Current Students**	**Recent Graduates**
1	4	6
2	6	8
3	8	5
4	3	1
5	1	2
6	2	3
7	5	7
8	10	9
9	7	4
10	9	10

SUMMARY

In this chapter we have presented several statistical procedures that are classified as nonparametric methods. The parametric methods of the preceding chapters generally require interval or ratio data and often are based on assumptions about the population (for example, the assumption that the probability distribution is normal). Because nonparametric methods can be applied to nominal and ordinal data as well as interval and ratio data and do not require population distribution assumptions, they expand the class of problems that can be subjected to statistical analysis.

The sign test is a nonparametric procedure for identifying differences between two populations when the only data available are nominal data. In the small-sample case, the binominal probability distribution can be used to determine the critical values for the sign test; in the large-sample case, a normal approximation can be used. The Wilcoxon signed-rank test is a procedure for analyzing matched-sample data whenever interval- or ratio-scaled data are available for each matched pair. No assumptions are made about the population distribution. The Wilcoxon procedure tests the hypothesis that the two populations being considered are identical.

The Mann-Whitney-Wilcoxon test is a nonparametric method for testing for a difference between two populations based on two independent random samples. Tables were presented for the small-sample case, and a normal approximation was provided for the large-sample case. The Kruskal-Wallis test extends the Mann-Whitney-Wilcoxon test to the case of three or more populations. The Kruskal-Wallis test is the nonparametric analog of the parametric ANOVA test for differences among population means.

In the last section of this chapter we introduced the Spearman rank-correlation coefficient as a measure of association for two ordinal or rank-ordered sets of items.

GLOSSARY

Nonparametric methods Statistical methods that require few, if any, assumptions about the population probability distributions and the level of measurement. These methods can be applied when nominal or ordinal data are available.

Distribution-free methods Another name for nonparametric statistical methods that indicates the lack of assumptions about the population probability distribution.

Sign test A nonparametric statistical test for identifying differences between two populations based on the analysis of nominal data.

Wilcoxon signed-rank test A nonparametric statistical test for identifying differences between two populations based on the analysis of two matched or paired samples.

Mann-Whitney-Wilcoxon (MWW) test A nonparametric statistical test for identifying differences between two populations based on the analysis of two independent samples.

Kruskal-Wallis test A nonparametric test for identifying differences among three or more populations.

Spearman rank-correlation coefficient A correlation measure based on rank-ordered data for two variables.

KEY FORMULAS

Sign Test (Large-Sample Case)

$$\text{Mean: } \mu = .50n \tag{19.1}$$

$$\text{Standard Deviation: } \sigma = \sqrt{.25n} \tag{19.2}$$

Wilcoxon Signed-Rank Test

$$\text{Mean: } \mu_T = 0 \tag{19.3}$$

$$\text{Standard Deviation: } \sigma_T = \sqrt{\frac{n(n+1)(2n+1)}{6}} \tag{19.4}$$

Mann-Whitney-Wilcoxon Test (Large-Sample)

$$\text{Mean: } \mu_T = \tfrac{1}{2}\, n_1(n_1 + n_2 + 1) \tag{19.6}$$

$$\text{Standard Deviation: } \sigma_T = \sqrt{\tfrac{1}{12}\, n_1 n_2(n_1 + n_2 + 1)} \tag{19.7}$$

Kruskal-Wallis Test Statistic

$$W = \left[\frac{12}{n_T(n_T + 1)} \sum_{i=1}^{k} \frac{R_i^2}{n_i}\right] - 3(n_T + 1) \tag{19.8}$$

Spearman Rank-Correlation Coefficient

$$r_s = 1 - \frac{6\Sigma d_i^2}{n(n^2 - 1)} \tag{19.9}$$

SUPPLEMENTARY EXERCISES

38. The American Opinion Survey (*The Wall Street Journal,* March 4, 1997) asked the following question: Do you favor or oppose providing tax-funded vouchers or tax deductions to parents who send their children to private schools? Of the 2010 individuals surveyed, 905 favored the support, 1045 opposed the support, and 60 offered no opinion. Do the data indicate a significant difference in the preferences for the support for parents who send their children to private schools? Use a .05 level of significance.

39. The national median sales price of existing one-family homes is \$118,000 (*The Wall Street Journal Almanac*, 1998). Assume that the following data were obtained for sales of existing one-family homes in Houston and Boston.

	Greater than \$118,000	Equal to \$118,000	Less than \$118,000
Houston	11	2	32
Boston	27	1	13

a. Is the median resale price in Houston lower than the national median of \$118,000? Use a statistical test with $\alpha = .05$ to support your conclusion.

b. Is the median resale price in Boston higher than the national median of \$118,000? Use a statistical test with $\alpha = .05$ to support your conclusion.

40. Twelve homemakers were asked to estimate the retail selling price of two models of refrigerators. Their estimates of selling price are shown in the following table. Use these data and test at the .05 level of significance to determine whether there is a difference between the two models in terms of homemakers' perceptions of selling price.

Homemaker	Model 1	Model 2	Homemaker	Model 1	Model 2
1	\$650	\$900	7	\$700	\$ 890
2	760	720	8	690	920
3	740	690	9	900	1000
4	700	850	10	500	690
5	590	920	11	610	700
6	620	800	12	720	700

41. A study was designed to evaluate the weight-gain potential of a new poultry feed. A sample of 12 chickens was used in a 6-week study. The weight of each chicken was recorded before and after the 6-week test period. The differences between the before and after weights of the 12 chickens are 1.5, 1.2, −.2, .0, .5, .7, .8, 1.0, .0, .6, .2, −.01. A negative value indicates a weight loss during the test period, whereas .0 indicates no weight change over the period. Use a .05 level of significance to determine whether the new feed appears to provide a weight gain for the chickens.

42. The following data are product weights for items produced on two production lines. Test for a difference between the product weights for the two lines. Use $\alpha = .10$.

Production Line 1	Production Line 2
13.6	13.7
13.8	14.1
14.0	14.2
13.9	14.0
13.4	14.6
13.2	13.5
13.3	14.4
13.6	14.8
12.9	14.5
14.4	14.3
	15.0
	14.9

43. A client wants to determine whether there is a significant difference in the time required to complete a program evaluation with the three different methods that are in common use. The times (in hours) required for each of 18 evaluators to conduct a program evaluation follow.

Method 1	Method 2	Method 3
68	62	58
74	73	67
65	75	69
76	68	57
77	72	59
72	70	62

Use $\alpha = .05$ and test to see whether there is a significant difference in the time required by the three methods.

44. A sample of 20 engineers who have been with a company for three years has been rank-ordered with respect to managerial potential. Some of the engineers have attended the company's management-development course, others have attended an off-site management-development program at a local university, and the remainder have not attended any program. Use the following rankings and $\alpha = .025$ to test for a significant difference in the managerial potential of the three groups.

No Program	Company Program	Off-Site Program
16	12	7
9	20	1
10	17	4
15	19	2
11	6	3
13	18	8
	14	5

45. Course-evaluation ratings for four instructors follow. Use $\alpha = .05$ and the Kruskal-Wallis procedure to test for a significant difference in teaching abilities.

Instructor	Course-Evaluation Rating								
Black	88	80	79	68	96	69			
Jennings	87	78	82	85	99	99	85	94	
Swanson	88	76	68	82	85	82	84	83	81
Wilson	80	85	56	71	89	87			

46. A sample of 15 students received the following rankings on midterm and final examinations in a statistics course.

Rank		Rank		Rank	
Midterm	Final	Midterm	Final	Midterm	Final
1	4	6	2	11	14
2	7	7	5	12	15
3	1	8	12	13	11
4	3	9	6	14	10
5	8	10	9	15	13

Compute the Spearman rank-correlation coefficient for the data and test for a significant correlation with $\alpha = .10$.

STATISTICAL METHODS FOR QUALITY CONTROL

Chapter 20

CONTENTS

STATISTICS IN PRACTICE

DOW CHEMICAL U.S.A.*
Freeport, Texas

Dow Chemical U.S.A., Texas Operations, began in 1940 when The Dow Chemical Company purchased 800 acres of Texas land on the Gulf Coast to build a magnesium production facility. That original site has expanded to cover more than 5000 acres and is one of the largest petrochemical complexes in the world. Among the products from Texas Operations are magnesium, styrene, plastics, adhesives, solvent, glycol, and chlorine. Some products are made solely for use in other processes, but many end up as essential ingredients in products such as pharmaceuticals, toothpastes, dog food, water hoses, ice chests, milk cartons, garbage bags, shampoos, and furniture.

Dow's Texas Operations produces more than 30% of the world's magnesium, an extremely lightweight metal used in products ranging from tennis racquets to suitcases to "mag" wheels. The Magnesium Department was the first group in Texas Operations to train its technical people and managers in the use of statistical quality control. Some of the earliest successful applications of statistical quality control were in chemical processing.

In one application involving the operation of a drier, samples of the output were taken at periodic intervals; the average value for each sample was computed and recorded on a chart called an $\bar{x}$ chart. Such a chart enabled Dow analysts to monitor trends in the output that might indicate the process was not operating correctly. In one instance, analysts began to observe values for the sample mean that were not indicative of a process operating within its design limits. On further examination of the control chart and the operation itself, the analysts found that the variation could be traced to problems involving one operator. The $\bar{x}$ chart recorded after that operator was retrained showed a significant improvement in the process quality.

*The authors are indebted to Clifford B. Wilson, Magnesium Technical Manager, The Dow Chemical Company, for providing this Statistics in Practice.

Statistical quality control has enabled Dow Chemical U.S.A. to improve its processing methods and output. © Dan Guravich/CORBIS.

Dow Chemical has achieved quality improvements everywhere statistical quality control has been used. Documented savings of several hundred thousand dollars per year have been realized, and new applications are continually being discovered.

In this chapter we will show how an $\bar{x}$ chart such as the one used by Dow Chemical can be developed. Such charts are a part of statistical quality control known as statistical process control. We will also discuss methods of quality control for situations in which a decision to accept or reject a group of items is based on a sample.

The American Society for Quality Control (ASQC) defines *quality* as "the totality of features and characteristics of a product or service that bears on its ability to satisfy given needs." In other words, quality measures how well a product or service meets customer needs. Organizations recognize that to be competitive in today's global economy, they must strive for high levels of quality. As a result, an increased emphasis falls on methods for monitoring and maintaining quality.

Quality assurance refers to the entire system of policies, procedures, and guidelines established by an organization to achieve and maintain quality. Quality assurance consists of two principal functions: quality engineering and quality control. The objective of *quality engineering* is to include quality in the design of products and processes and to identify potential quality problems prior to production. **Quality control** consists of making a series of inspections and measurements to determine whether quality standards are being met. If quality standards are not being met, corrective and/or preventive action can be taken to achieve and maintain conformance. As we will show in this chapter, statistical techniques are extremely useful in quality control.

Traditional manufacturing approaches to quality control have been found to be less than satisfactory and are being replaced by improved managerial tools and techniques. Ironically, it was two U.S. consultants, Dr. W. Edwards Deming and Dr. Joseph Juran, who helped educate the Japanese in quality management.

After World War II, Dr. W. Edwards Deming became a consultant to Japanese industry; he is credited with being the person who convinced top managers in Japan to use the methods of statistical quality control.

Although quality is everybody's job, Deming stressed that quality must be led by managers. He developed a list of 14 points that he believed are the key responsibilities of managers. For instance, Deming stated that managers must cease dependence on mass inspection; must end the practice of awarding business solely on the basis of price; must seek continual improvement in all production processes and services; must foster a team-oriented environment; and must eliminate numerical goals, slogans, and work standards that prescribe numerical quotas. Perhaps most important, managers must create a work environment in which a commitment to quality and productivity is maintained at all times.

In 1987, the U.S. Congress enacted Public Law 107, the Malcolm Baldrige National Quality Improvement Act. The Baldrige Award is given annually to U.S. firms that excel in quality. This award, along with the perspectives of individuals like Dr. Deming and Dr. Juran, has helped top managers recognize that improving service quality and product quality is the most critical challenge facing their companies. Winners of the Malcolm Baldrige Award include Motorola, IBM, Xerox, and FedEx. In this chapter we present two statistical methods used in quality control. The first method, *statistical process control,* uses graphical displays known as *control charts* to monitor a production process; the goal is to determine whether the process can be continued or whether it should be adjusted to achieve a desired quality level. The second method, *acceptance sampling,* is used in situations where a decision to accept or reject a group of items must be based on the quality found in a sample.

20.1 STATISTICAL PROCESS CONTROL

In this section we consider quality control procedures for a production process whereby goods are manufactured continuously. On the basis of sampling and inspection of production output, a decision will be made to either continue the production process or adjust it to bring the items or goods being produced up to acceptable quality standards.

Continual improvement is one of the most important concepts of the total quality management movement. The most important use of a control chart is in improving the process.

Despite high standards of quality in manufacturing and production operations, machine tools will invariably wear out, vibrations will throw machine settings out of adjustment, purchased materials will be defective, and human operators will make mistakes. Any or all of these factors can result in poor quality output. Fortunately, procedures are available for monitoring production output so that poor quality can be detected early and the production process can be adjusted or corrected.

If the variation in the quality of the production output is due to **assignable causes** such as tools wearing out, incorrect machine settings, poor quality raw materials, or operator error, the process should be adjusted or corrected as soon as possible. Alternatively, if the variation is due to what are called **common causes**—that is, randomly occurring variations in materials, temperature, humidity, and so on, which the manufacturer cannot possibly control—the process does not need to be adjusted. The main objective of statistical process control is to determine whether variations in output are due to assignable causes or common causes.

Whenever assignable causes are detected, we conclude that the process is *out of control.* In that case, corrective action will be taken to bring the process back to an acceptable level of quality. However, if the variation in the output of a production process is due only to common causes, we conclude that the process is *in statistical control,* or simply *in control;* in such cases, no changes or adjustments are necessary.

Process control procedures are closely related to the hypothesis testing procedure discussed earlier in this text. In essence, control charts provide an ongoing test of the hypothesis that the process is in control.

The statistical procedures for process control are based on the hypothesis testing methodology presented in Chapter 9. The null hypothesis H_0 is formulated in terms of the production process being in control. The alternative hypothesis H_a is formulated in terms of the production process being out of control. Table 20.1 shows that correct decisions to continue an in-control process and adjust an out-of-control process are possible. However, as with other hypothesis testing procedures, both a Type I error (adjusting an in-control process) and a Type II error (allowing an out-of-control process to continue) are also possible.

Control Charts

Control charts that are based on data that can be measured on a continuous scale are called variables control charts. The $\bar{x}$ chart is a variables control chart.

A **control chart** provides a basis for deciding whether the variation in the output is due to common causes (in control) or assignable causes (out of control). Whenever an out-of-control situation is detected, adjustments and/or other corrective action will be taken to bring the process back into control.

Control charts can be classified by the type of data they contain. An **$\bar{x}$ chart** is used if the quality of the output is measured in terms of a variable such as length, weight, temperature, and so on. In that case, the decision to continue or to adjust the production process will be based on the mean value found in a sample of the output. To introduce some of the concepts common to all control charts, let us consider some specific features of an $\bar{x}$ chart.

Figure 20.1 shows the general structure of an $\bar{x}$ chart. The center line of the chart corresponds to the mean of the process when the process is *in control.* The vertical line identifies the scale of measurement for the variable of interest. Each time a sample is taken from the production process, a value of the sample mean $\bar{x}$ is computed and a data point showing the value of $\bar{x}$ is plotted on the control chart.

The two lines labeled UCL and LCL are important in determining whether the process is in control or out of control. The lines are called the *upper control limit* and the *lower control limit,* respectively. They are chosen so that when the process is in control, there will be a high probability that the value of $\bar{x}$ will be between the two control limits. Values outside the control limits provide strong statistical evidence that the process is out of control and corrective action should be taken.

Over time, more and more data points will be added to the control chart. The order of the data points will be from left to right as the process is sampled. In essence, every time a point is plotted on the control chart, we are carrying out a hypothesis test to determine whether the process is in control.

In addition to the $\bar{x}$ chart, other control charts can be used to monitor the range of the measurements in the sample (**R chart**), the proportion defective in the sample (**p chart**),

TABLE 20.1 THE OUTCOMES OF STATISTICAL PROCESS CONTROL

		State of Production Process	
		H_0 True Process in Control	**H_0 False Process Out of Control**
Decision	**Continue Process**	Correct decision	Type II error (allowing an out-of-control process to continue)
	Adjust Process	Type I error (adjusting an in-control process)	Correct decision

FIGURE 20.1 $\bar{x}$ CHART STRUCTURE

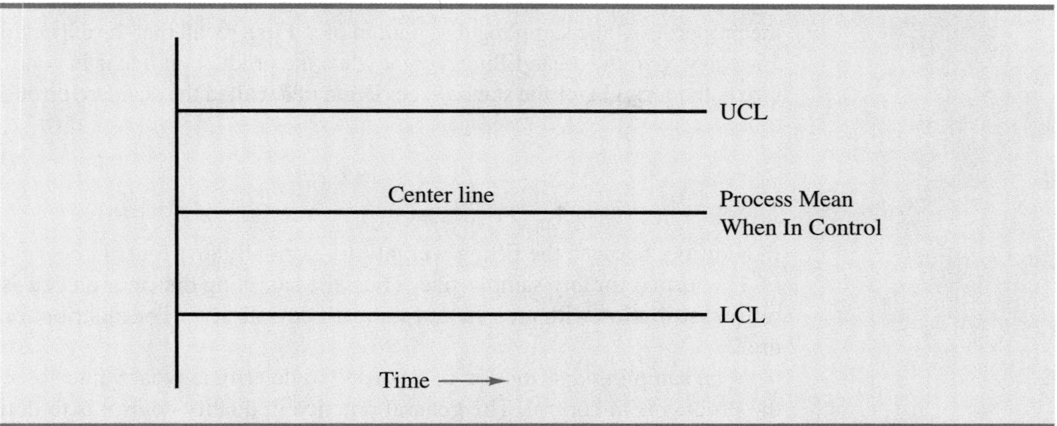

and the number of defective items in the sample (*np* **chart**). In each case, the control chart has an LCL, a center line, and a UCL similar to the $\bar{x}$ chart in Figure 20.1. The major difference among the charts is what the vertical axis measures; for instance, in a p chart the measurement scale denotes the proportion of defective items in the sample instead of the sample mean. In the following discussion, we will illustrate the construction and use of the $\bar{x}$ chart, R chart, p chart, and np chart.

$\bar{x}$ Chart: Process Mean and Standard Deviation Known

To illustrate the construction of an $\bar{x}$ chart, let us consider the situation at KJW Packaging. This company operates a production line where cartons of cereal are filled. Suppose KJW knows that when the process is operating correctly—and hence the system is in control—the mean filling weight is $\mu = 16.05$ ounces, and the process standard deviation is $\sigma = .10$ ounces. In addition, assume the filling weights are normally distributed. This distribution is shown in Figure 20.2.

FIGURE 20.2 DISTRIBUTION OF CEREAL-CARTON FILLING WEIGHTS

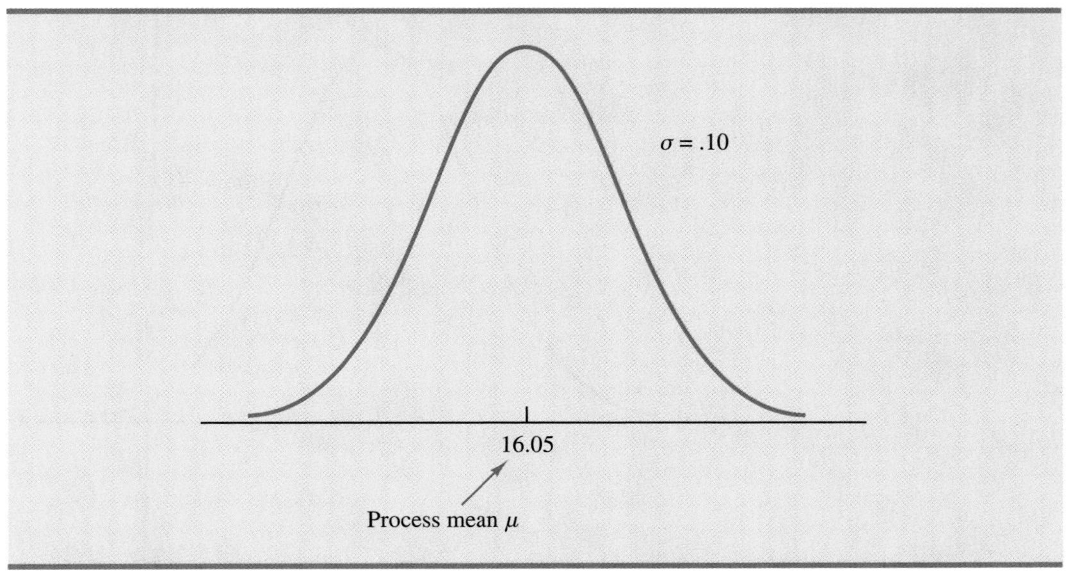

The sampling distribution of $\bar{x}$, as presented in Chapter 7, can be used to determine the variation that can be expected in $\bar{x}$ values for a process that is in control. Let us first briefly review the properties of the sampling distribution of $\bar{x}$. First, recall that the expected value or mean of $\bar{x}$ is equal to μ, the mean filling weight when the production line is in control. For samples of size n, the formula for the standard deviation of $\bar{x}$, called the standard error of the mean, is

$$\sigma_{\bar{x}} = \frac{\sigma}{\sqrt{n}} \tag{20.1}$$

In addition, because the filling weights are normally distributed, the sampling distribution of $\bar{x}$ is normal for any sample size. Thus, the sampling distribution of $\bar{x}$ is a normal probability distribution with mean μ and standard deviation $\sigma_{\bar{x}}$. This distribution is shown in Figure 20.3.

The sampling distribution of $\bar{x}$ is used to determine what values of $\bar{x}$ are reasonable if the process is in control. The general practice in quality control is to define as reasonable any value of $\bar{x}$ that is within 3 standard deviations above or below the mean value. Recall from the study of the normal probability distribution that approximately 99.7% of the values of a normally distributed random variable are within ± 3 standard deviations of its mean value. Thus, if a value of $\bar{x}$ is within the interval $\mu - 3\sigma_{\bar{x}}$ to $\mu + 3\sigma_{\bar{x}}$, we will assume that the process is in control. In summary, then, the control limits for an $\bar{x}$ chart are as follows.

Control Limits for an $\bar{x}$ Chart: Process Mean and Standard Deviation Known

$$\text{UCL} = \mu + 3\sigma_{\bar{x}} \tag{20.2}$$

$$\text{LCL} = \mu - 3\sigma_{\bar{x}} \tag{20.3}$$

Reconsider the KJW Packaging example with the process distribution of filling weights shown in Figure 20.2 and the sampling distribution $\bar{x}$ of shown in Figure 20.3. Assume that a quality control inspector periodically samples six cartons and uses the sample mean filling weight to determine whether the process is in control or out of control. Using (20.1),

FIGURE 20.3 SAMPLING DISTRIBUTION OF $\bar{x}$

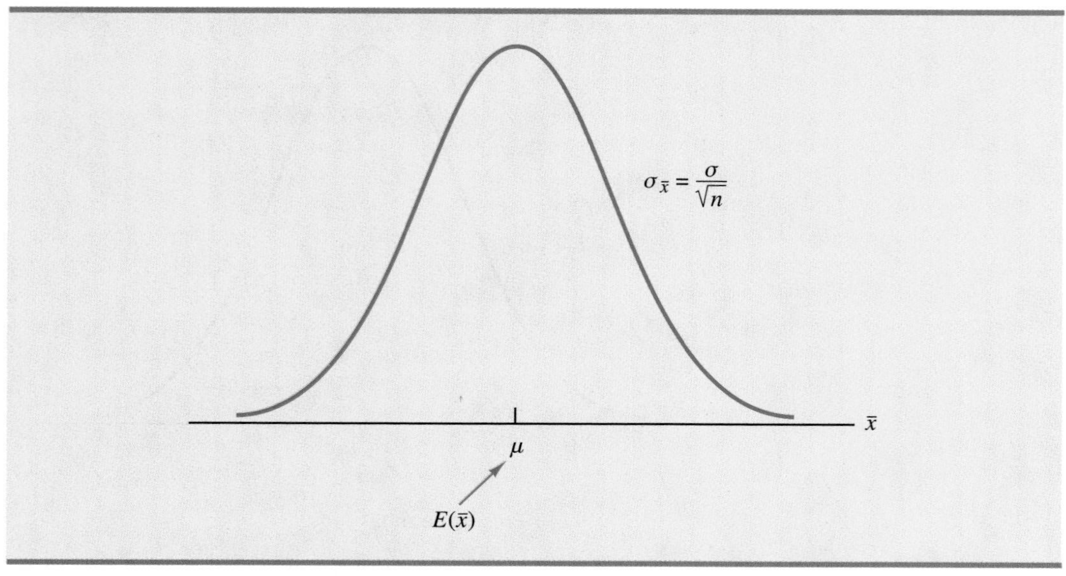

we find that the standard error of the mean is $\sigma_{\bar{x}} = \sigma/\sqrt{n} = .10/\sqrt{6} = .04$. Thus, with the process mean at 16.05, the control limits are UCL = 16.05 + 3(.04) = 16.17 and LCL = 16.05 − 3(.04) = 15.93. Figure 20.4 is the control chart with the results of 10 samples taken over a 10-hour period. For ease of reading, the sample numbers 1 through 10 are listed below the chart.

Note that the mean for the fifth sample in Figure 20.4 shows that the process is out of control. In other words, the fifth sample mean is below the LCL indicating that assignable causes of output variation are present and that underfilling is occurring. As a result, corrective action was taken at this point to bring the process back into control. The fact that the remaining points on the $\bar{x}$ chart are within the upper and lower control limits indicates that the corrective action was successful.

$\bar{x}$ Chart: Process Mean and Standard Deviation Unknown

In the KJW Packaging example, we showed how an $\bar{x}$ chart can be developed when the mean and standard deviation of the process are known. In most situations, the process mean and standard deviation must be estimated by using samples that are selected from the process when it is in control. For instance, KJW might select a random sample of five boxes each morning and five boxes each afternoon for 10 days of in-control operation. For each subgroup, or sample, the mean and standard deviation of the sample are computed. The overall averages of both the sample means and the sample standard deviations are used to construct control charts for both the process mean and the process standard deviation.

It is important to maintain control over both the mean and the variability of a process.

In practice, it is more common to monitor the variability of the process by using the range instead of the standard deviation because the range is easier to compute. The range can be used to provide good estimates of the process standard deviation; thus it can be used to construct upper and lower control limits for the $\bar{x}$ chart with little computational effort. To illustrate, let us consider the problem facing Jensen Computer Supplies, Inc.

Jensen Computer Supplies (JCS) manufactures 3.5-inch-diameter computer disks. Suppose random samples of five disks could be taken during the first hour of operation, during the second hour of operation, and so on, until 20 samples have been selected. Table 20.2 provides the diameter of each disk sampled as well as the mean $\bar{x}_j$ and range R_j for each of the samples.

FIGURE 20.4 THE $\bar{x}$ CHART FOR THE CEREAL-CARTON FILLING PROCESS

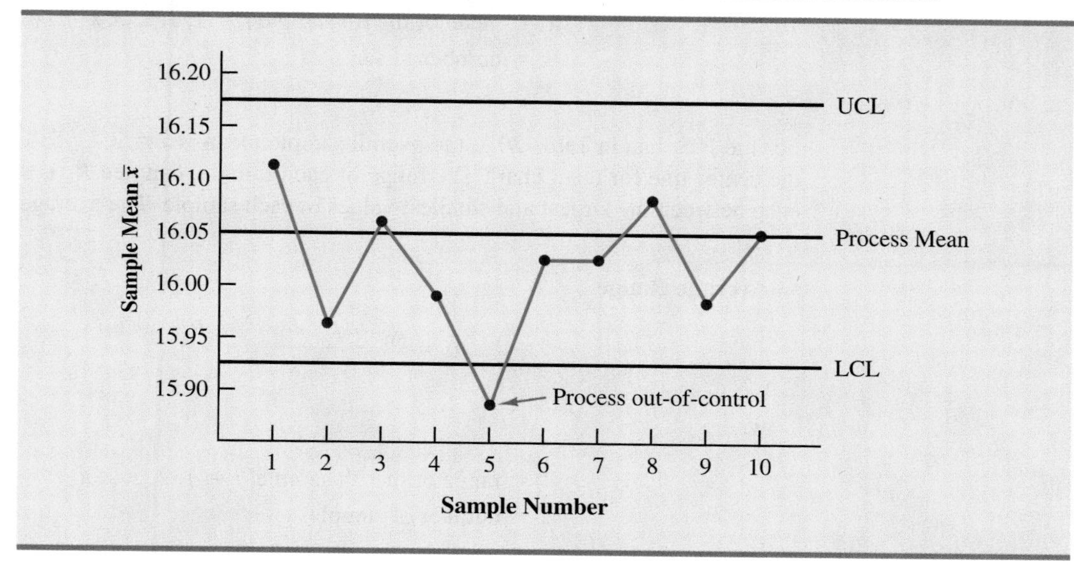

TABLE 20.2 DATA FOR THE JENSEN COMPUTER SUPPLIES PROBLEM

Jensen

Sample Number	Observations					Sample Mean $\bar{x}_j$	Sample Range R_j
1	3.5056	3.5086	3.5144	3.5009	3.5030	3.5065	.0135
2	3.4882	3.5085	3.4884	3.5250	3.5031	3.5026	.0368
3	3.4897	3.4898	3.4995	3.5130	3.4969	3.4978	.0233
4	3.5153	3.5120	3.4989	3.4900	3.4837	3.5000	.0316
5	3.5059	3.5113	3.5011	3.4773	3.4801	3.4951	.0340
6	3.4977	3.4961	3.5050	3.5014	3.5060	3.5012	.0099
7	3.4910	3.4913	3.4976	3.4831	3.5044	3.4935	.0213
8	3.4991	3.4853	3.4830	3.5083	3.5094	3.4970	.0264
9	3.5099	3.5162	3.5228	3.4958	3.5004	3.5090	.0270
10	3.4880	3.5015	3.5094	3.5102	3.5146	3.5047	.0266
11	3.4881	3.4887	3.5141	3.5175	3.4863	3.4989	.0312
12	3.5043	3.4867	3.4946	3.5018	3.4784	3.4932	.0259
13	3.5043	3.4769	3.4944	3.5014	3.4904	3.4935	.0274
14	3.5004	3.5030	3.5082	3.5045	3.5234	3.5079	.0230
15	3.4846	3.4938	3.5065	3.5089	3.5011	3.4990	.0243
16	3.5145	3.4832	3.5188	3.4935	3.4989	3.5018	.0356
17	3.5004	3.5042	3.4954	3.5020	3.4889	3.4982	.0153
18	3.4959	3.4823	3.4964	3.5082	3.4871	3.4940	.0259
19	3.4878	3.4864	3.4960	3.5070	3.4984	3.4951	.0206
20	3.4969	3.5144	3.5053	3.4985	3.4885	3.5007	.0259

The estimate of the process mean μ is given by the overall sample mean.

Overall Sample Mean

$$\bar{\bar{x}} = \frac{\bar{x}_1 + \bar{x}_2 + \cdots + \bar{x}_k}{k} \qquad \textbf{(20.4)}$$

where

$\bar{x}_j$ = mean of the jth sample $j = 1, 2, \ldots, k$
k = number of samples

For the JCS data in Table 20.2, the overall sample mean is $\bar{\bar{x}} = 3.4995$. This value will be the center line for the $\bar{x}$ chart. The range of each sample, denoted R_j, is simply the difference between the largest and smallest values in each sample. The average range follows.

Average Range

$$\bar{R} = \frac{R_1 + R_2 + \cdots + R_k}{k} \qquad \textbf{(20.5)}$$

where

R_j = range of the jth sample, $j = 1, 2, \ldots, k$
k = number of samples

For the JCS data in Table 20.2, the average range is $\bar{R} = .0253$.

In the preceding section we showed that the upper and lower control limits for the $\bar{x}$ chart are

$$\bar{x} \pm 3\frac{\sigma}{\sqrt{n}} \tag{20.6}$$

Hence, to construct the control limits for the $\bar{x}$ chart, we need to estimate σ, the standard deviation of the process. An estimate of σ can be developed by using the range data.

It can be shown that an estimator of the process standard deviation σ is the average range divided by d_2, a constant that depends on the sample size n. That is,

$$\text{Estimator of } \sigma = \frac{\bar{R}}{d_2} \tag{20.7}$$

The American Society for Testing and Materials Manual on Presentation of Data and Control Chart Analysis provides values for d_2 as shown in Table 11 of Appendix B. For instance, when $n = 5$, $d_2 = 2.326$, and the estimate of σ is the average range divided by 2.326. If we substitute $\bar{R}/d_2$ for σ in (20.6), we can write the control limits for the $\bar{x}$ chart as

$$\bar{\bar{x}} \pm 3\frac{\bar{R}/d_2}{\sqrt{n}} = \bar{\bar{x}} \pm \frac{3}{d_2\sqrt{n}}\bar{R} = \bar{\bar{x}} \pm A_2\bar{R} \tag{20.8}$$

Note that $A_2 = 3/(d_2\sqrt{n})$ is a constant that depends only on the sample size. Values for A_2 are provided in Table 11 in Appendix B. For $n = 5$, $A_2 = .577$; thus, the control limits for the $\bar{x}$ chart are

$$3.4995 \pm (.577)(.0253) = 3.4995 \pm .0146$$

Hence, UCL = 3.514 and LCL = 3.485.

Figure 20.5 shows the $\bar{x}$ chart for the Jensen Computer Supplies problem. We used the data in Table 20.2 and Minitab's control chart routine to construct the chart. The center line

FIGURE 20.5 $\bar{x}$ CHART FOR THE JENSEN COMPUTER SUPPLIES PROBLEM

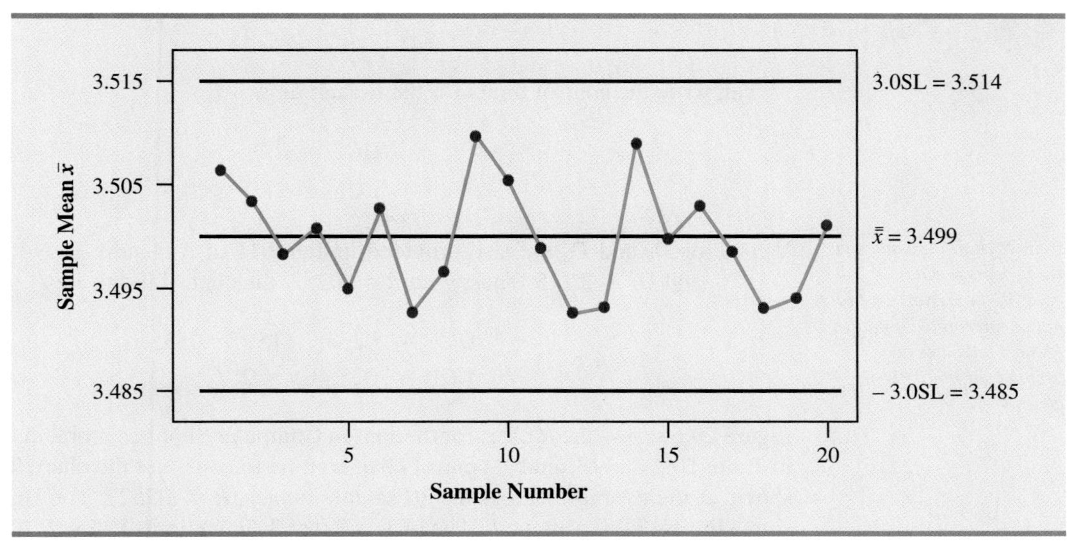

is shown at the overall sample mean $\bar{\bar{x}} = 3.499$. The upper control limit (UCL) is 3.514. Minitab uses the notation 3.0SL to indicate the UCL is 3 standard deviations or 3 "sigma limits" (SL) above $\bar{\bar{x}}$. The lower control (LCL) is 3.485, which is −3.0SL or 3 "sigma limits" below $\bar{\bar{x}}$. The $\bar{x}$ chart shows the 20 sample means plotted over time. Because all 20 sample means are within the control limits, the indication is that the mean of the Jensen manufacturing process is in control. This chart can now be used to monitor the process mean on an ongoing basis.

R Chart

Let us now consider a range chart (*R* chart) that can be used to control the variability of a process. To develop the *R* chart, we need to think of the range of a sample as a random variable with its own mean and standard deviation. The average range $\bar{R}$ provides an estimate of the mean of this random variable. Moreover, it can be shown that an estimate of the standard deviation of the range is

$$\hat{\sigma}_R = d_3 \frac{\bar{R}}{d_2} \tag{20.9}$$

where d_2 and d_3 are constants that depend on the sample size; values of d_2 and d_3 are also provided in Table 11 of Appendix B. Thus, the UCL for the *R* chart is given by

$$\bar{R} + 3\hat{\sigma}_R = \bar{R}\left(1 + 3\frac{d_3}{d_2}\right) \tag{20.10}$$

and the LCL is

$$\bar{R} - 3\hat{\sigma}_R = \bar{R}\left(1 - 3\frac{d_3}{d_2}\right) \tag{20.11}$$

If we let

$$D_4 = 1 + 3\frac{d_3}{d_2} \tag{20.12}$$

$$D_3 = 1 - 3\frac{d_3}{d_2} \tag{20.13}$$

we can write the control limits for the *R* chart as

$$\text{UCL} = \bar{R}D_4 \tag{20.14}$$

$$\text{LCL} = \bar{R}D_3 \tag{20.15}$$

If the R chart indicates that the process is out of control, the $\bar{x}$ chart should not be interpreted until the R chart indicates the process variability is in control.

Values for D_3 and D_4 are also provided in Table 11 of Appendix B. Note that for $n = 5$, $D_3 = 0$, and $D_4 = 2.115$. Thus, with $\bar{R} = .0253$, the control limits are

$$\text{UCL} = .0253(2.115) = .0535$$
$$\text{LCL} = .0253(0) = 0$$

Figure 20.6 shows the *R* chart for the Jensen Computer Supplies problem. We used the data in Table 20.2 and Minitab's control chart routine to construct the chart. The center line is shown at the overall mean of the 20 sample ranges, $\bar{R} = .02527$. The UCL is .05344 or 3 sigma limits (3.0SL) above $\bar{R}$. The LCL is 0.0 or 3 sigma limits below $\bar{R}$. The *R* chart shows

FIGURE 20.6 *R* CHART FOR THE JENSEN COMPUTER SUPPLIES PROBLEM

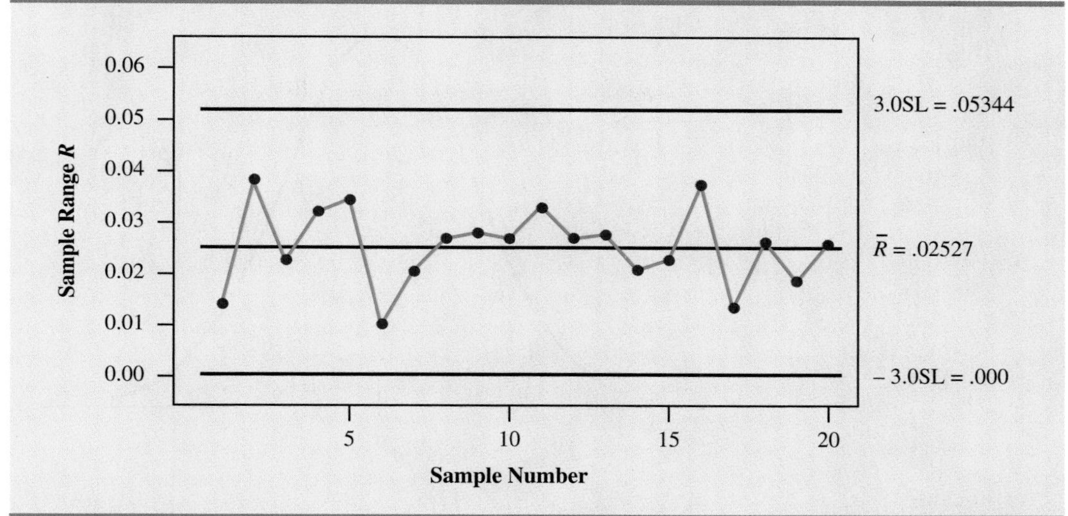

To monitor process variability, a sample standard deviation control chart (s chart) can be constructed instead of an R chart. If the sample size is 10 or less, the R chart and the s chart provide similar results. If the sample size is greater than 10, the s chart is generally preferred.

Control charts that are based on data indicating the presence of a defect or a number of defects are called attributes control charts. A p chart is an attributes control chart.

the 20 sample ranges plotted over time. Because all 20 sample ranges are within the control limits, we confirm that the process was in control during the sampling period.

p Chart

Let us consider the case in which the output quality is measured by either nondefective or defective items. The decision to continue or to adjust the production process will be based on $\bar{p}$, the proportion of defective items found in a sample. The control chart used for proportion-defective data is called a *p* chart.

To illustrate the construction of a *p* chart, consider the use of automated mail-sorting machines in a post office. These automated machines scan the zip codes on letters and divert each letter to its proper carrier route. Even when a machine is operating properly, some letters are diverted to incorrect routes. Assume that when a machine is operating correctly, or in a state of control, 3% of the letters are incorrectly diverted. Thus *p*, the proportion of letters incorrectly diverted when the process is in control, is .03.

The sampling distribution of $\bar{p}$, as presented in Chapter 7, can be used to determine the variation that can be expected in $\bar{p}$ values for a process that is in control. Recall that the expected value or mean of $\bar{p}$ is *p*, the proportion defective when the process is in control. With samples of size *n*, the formula for the standard deviation of $\bar{p}$, called the standard error of the proportion, is

$$\sigma_{\bar{p}} = \sqrt{\frac{p(1-p)}{n}} \tag{20.16}$$

We also learned in Chapter 7 that the sampling distribution of $\bar{p}$ can be approximated by a normal probability distribution whenever the sample size is large. With $\bar{p}$, the sample size can be considered large whenever the following two conditions are satisfied.

$$np \geq 5$$
$$n(1-p) \geq 5$$

In summary, whenever the sample size is large, the sampling distribution of $\bar{p}$ can be approximated by a normal probability distribution with mean *p* and standard deviation $\sigma_{\bar{p}}$. This distribution is shown in Figure 20.7.

FIGURE 20.7 SAMPLING DISTRIBUTION OF $\bar{p}$

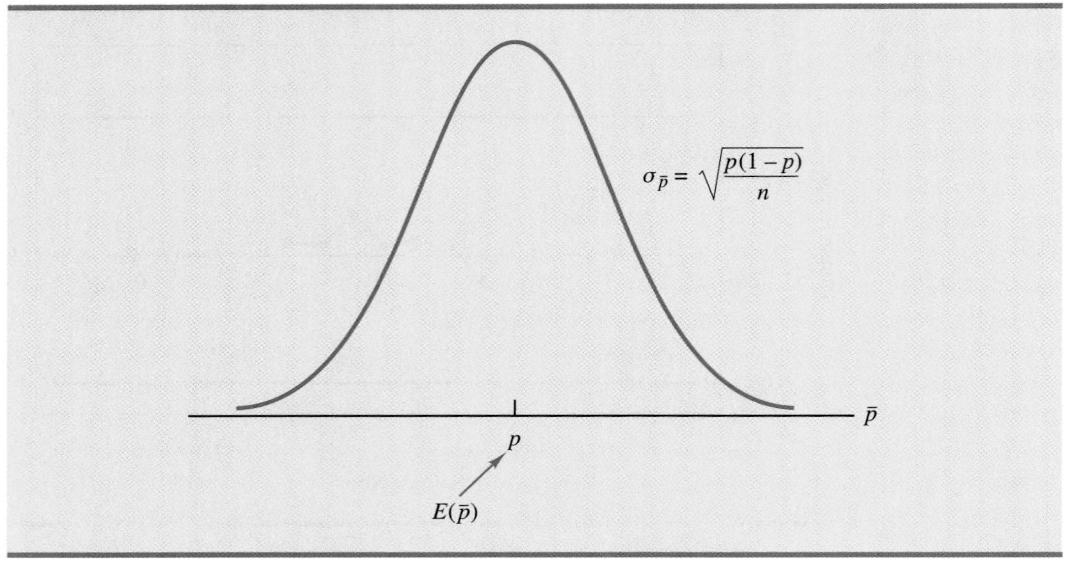

To establish control limits for a p chart, we follow the same procedure we used to establish control limits for an $\bar{x}$ chart. That is, the limits for the control chart are set at 3 standard deviations, or standard errors, above and below the proportion defective when the process is in control. Thus, we have the following control limits.

Control Limits for a p Chart

$$\text{UCL} = p + 3\sigma_{\bar{p}} \qquad \textbf{(20.17)}$$

$$\text{LCL} = p - 3\sigma_{\bar{p}} \qquad \textbf{(20.18)}$$

With $p = .03$ and samples of size $n = 200$, equation (20.16) shows that the standard error is

$$\sigma_{\bar{p}} = \sqrt{\frac{.03(1 - .03)}{200}} = .0121$$

Hence, the control limits are UCL $= .03 + 3(.0121) = .0663$, and LCL $= .03 - 3(.0121) = -.0063$. Because LCL is negative, LCL is set equal to zero in the control chart.

Figure 20.8 is the control chart for the mail-sorting process. The points plotted show the sample proportion defective found in samples of letters taken from the process. All points are within the control limits, providing no evidence to conclude that the sorting process is out of control. In fact, the p chart indicates that the process is in control and should continue to operate.

If the proportion of defective items for a process that is in control is not known, that value is first estimated by using sample data. Suppose, for example, that M different samples, each of size n, are selected from a process that is in control. The fraction or proportion of defective items in each sample is then determined. Treating all the data collected as one large sample, we can determine the average number of defective items for all the data; that value can then be used to provide an estimate of p, the proportion of defective items observed when the process is in control. Note that this estimate of p also enables us to estimate the standard error of the proportion; upper and lower control limits can then be established.

FIGURE 20.8 *p* CHART FOR THE PROPORTION DEFECTIVE IN A MAIL-SORTING PROCESS

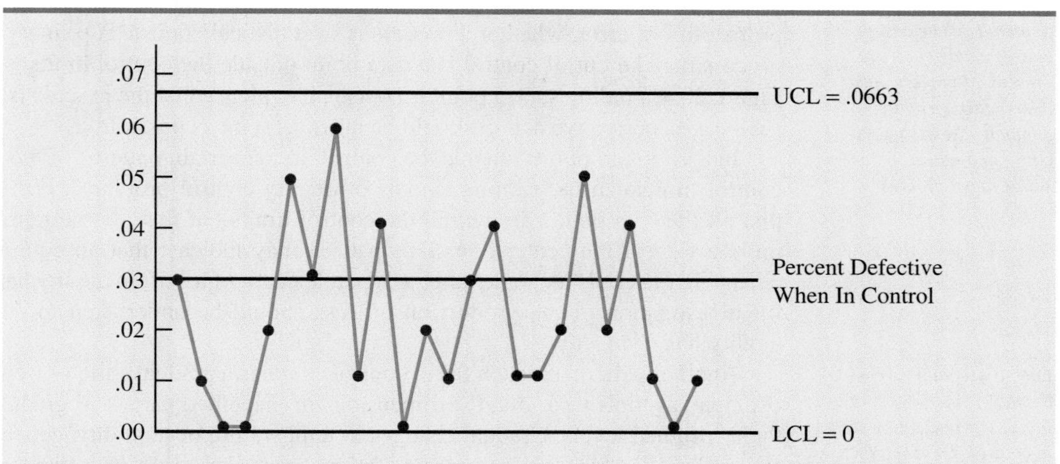

np Chart

An *np* chart is a control chart developed for the number of defective items in a sample. In this case, n is the sample size and p is the probability of observing a defective item when the process is in control. Whenever the sample size is large, that is when $np \geq 5$ and $n(1 - p) \geq 5$, the distribution of the number of defective items observed in a sample of size n can be approximated by a normal probability distribution with mean np and standard deviation $\sqrt{np(1-p)}$. Thus, for the mail-sorting example, with $n = 200$ and $p = .03$, the number of defective items observed in a sample of 200 letters can be approximated by a normal probability distribution with a mean of $200(.03) = 6$ and a standard deviation of $\sqrt{200(.03)(.97)} = 2.4125$.

The control limits for an *np* chart are set at 3 standard deviations above and below the expected number of defective items observed when the process is in control. Thus, we have the following control limits.

Control Limits for an *np* Chart

$$\text{UCL} = np + 3\sqrt{np(1-p)} \quad \textbf{(20.19)}$$

$$\text{LCL} = np - 3\sqrt{np(1-p)} \quad \textbf{(20.20)}$$

For the mail-sorting process example, with $p = .03$ and $n = 200$, the control limits are $\text{UCL} = 6 + 3(2.4125) = 13.2375$, and $\text{LCL} = 6 - 3(2.4125) = -1.2375$. When LCL is negative, LCL is set equal to zero in the control chart. Hence, if the number of letters diverted to incorrect routes is greater than 13, the process is concluded to be out of control.

The information provided by an *np* chart is equivalent to the information provided by the *p* chart; the only difference is that the *np* chart is a plot of the number of defective items observed whereas the *p* chart is a plot of the proportion of defective items observed. Thus, if we were to conclude that a particular process is out of control on the basis of a *p* chart, the process would also be concluded to be out of control on the basis of an *np* chart.

Interpretation of Control Charts

Control charts are designed to identify when assignable causes of variation are present. Managers must then authorize action to eliminate the assignable cause and return in-control status to the process.

The location and pattern of points in a control chart enable us to determine, with a small probability of error, whether a process is in statistical control. A primary indication that a process may be out of control is a data point outside the control limits, such as point 5 in Figure 20.4. Finding such a point is statistical evidence that the process is out of control; in such cases, corrective action should be taken as soon as possible.

In addition to points outside the control limits, certain patterns of the points within the control limits can be warning signals of quality control problems. For example, assume that all the data points are within the control limits but that a large number of points are on one side of the center line. This pattern may indicate that an equipment problem, a change in materials, or some other assignable cause of a shift in quality has occurred. Careful investigation of the production process should be undertaken to determine whether quality has changed.

Even if all points are within the upper and lower control limits, a process may not be in control. Trends in the sample data points or unusually long runs above or below the center line may also indicate out-of-control conditions.

Another pattern to watch for in control charts is a gradual shift, or trend, over time. For example, as tools wear out, the dimensions of machined parts will gradually deviate from their designed levels. Gradual changes in temperature or humidity, general equipment deterioration, dirt buildup, or operator fatigue may also result in a trend pattern in control charts. Six or seven points in a row that indicate either an increasing or decreasing trend should be cause for concern, even if the data points are all within the control limits. When such a pattern occurs, the process should be reviewed for possible changes or shifts in quality. Corrective action to bring the process back into control may be necessary.

NOTES AND COMMENTS

1. Because the control limits for the $\bar{x}$ chart depend on the value of the average range, these limits will not have much meaning unless the process variability is in control. In practice, the R chart is usually constructed before the $\bar{x}$ chart; if the R chart indicates that the process variability is in control, then the $\bar{x}$ chart is constructed. Minitab's Xbar-R option provides the $\bar{x}$ chart and the R chart simultaneously. The steps of this procedure are described in Appendix 20.1.
2. An np chart is used to monitor a process in terms of the number of defects. The Motorola Six Sigma Quality Level sets a goal of producing no more than 3.4 defects per million operations (*American Production and Inventory Control*, July 1991); this goal implies $p = .0000034$.

EXERCISES

Methods

1. A process that is in control has a mean of $\mu = 12.5$ and a standard deviation of $\sigma = .8$.
   a. Construct an $\bar{x}$ chart if samples of size four are to be used.
   b. Repeat part (a) for samples of size 8 and 16.
   c. What happens to the limits of the control chart as the sample size is increased? Discuss why this is reasonable.

2. Twenty-five samples, each of size five, were selected from a process that was in control. The sum of all the data collected was 677.5 pounds.
   a. What is an estimate of the process mean (in terms of pounds per unit) when the process is in control?
   b. Develop the control chart for this process if samples of size five will be used. Assume that the process standard deviation is .5 when the process is in control, and that the mean of the process is the estimate developed in part (a).

3. Twenty-five samples of 100 items each were inspected when a process was considered to be operating satisfactorily. In the 25 samples, a total of 135 items were found to be defective.
   a. What is an estimate of the proportion defective when the process is in control?
   b. What is the standard error of the proportion if samples of size 100 will be used for statistical process control?
   c. Compute the upper and lower control limits for the control chart.

SELF test

4. A process sampled 20 times with a sample of size eight resulted in $\bar{\bar{x}} = 28.5$ and $\bar{R} = 1.6$. Compute the upper and lower control limits for the $\bar{x}$ and R charts for this process.

Applications

5. Temperature is used to measure the output of a production process. When the process is in control, the mean of the process is $\mu = 128.5$ and the standard deviation is $\sigma = .4$.
   a. Construct an $\bar{x}$ chart if samples of size six are to be used.
   b. Is the process in control for a sample providing the following data?

      128.8 128.2 129.1 128.7 128.4 129.2

   c. Is the process in control for a sample providing the following data?

      129.3 128.7 128.6 129.2 129.5 129.0

6. A quality control process monitors the weight per carton of laundry detergent. Control limits are set at UCL = 20.12 ounces and LCL = 19.90 ounces. Samples of size five are used for the sampling and inspection process. What are the process mean and process standard deviation for the manufacturing operation?

7. The Goodman Tire and Rubber Company periodically tests its tires for tread wear under simulated road conditions. To study and control the manufacturing process, 20 samples, each containing three radial tires, were chosen from different shifts over several days of operation, with the following results. Assuming that these data were collected when the manufacturing process was believed to be operating in control, develop the R and $\bar{x}$ charts.

CD file

Tires

Sample		Tread Wear*	
1	31	42	28
2	26	18	35
3	25	30	34
4	17	25	21
5	38	29	35
6	41	42	36
7	21	17	29
8	32	26	28
9	41	34	33
10	29	17	30
11	26	31	40
12	23	19	25
13	17	24	32
14	43	35	17
15	18	25	29
16	30	42	31
17	28	36	32
18	40	29	31
19	18	29	28
20	22	34	26

*Hundredths of an inch

8. Over several weeks of normal, or in-control, operation, 20 samples of 150 packages each of synthetic-gut tennis strings were tested for breaking strength. A total of 141 packages of the 3000 tested failed to conform to the manufacturer's specifications.
   - **a.** What is an estimate of the process proportion defective when the system is in control?
   - **b.** Compute the upper and lower control limits for a p chart.
   - **c.** With the results of part (b), what conclusion should be made about the process if tests on a new sample of 150 packages find 12 defective? Do there appear to be assignable causes in this situation?
   - **d.** Compute the upper and lower control limits for an np chart.
   - **e.** Answer part (c) using the results of part (d).
   - **f.** Which control chart would be preferred in this situation? Explain.
9. An automotive industry supplier produces pistons for several models of automobiles. Twenty samples, each consisting of 200 pistons, were selected when the process was known to be operating correctly. The numbers of defective pistons found in the samples follow.

8	10	6	4	5	7	8	12	8	15
14	10	10	7	5	8	6	10	4	8

   - **a.** What is an estimate of the proportion defective for the piston manufacturing process when it is in control?
   - **b.** Construct a p chart for the manufacturing process, assuming each sample has 200 pistons.
   - **c.** With the results of part (b), what conclusion should be made if a sample of 200 has 20 defective pistons?
   - **d.** Compute the upper and lower control limits for an np chart.
   - **e.** Answer part (c) using the results of part (d).

20.2 ACCEPTANCE SAMPLING

In acceptance sampling, the items of interest can be incoming shipments of raw materials or purchased parts as well as finished goods from final assembly. Suppose we want to decide whether to accept or reject a group of items on the basis of specified quality characteristics. In quality control terminology, the group of items is a **lot,** and **acceptance sampling** is a statistical method that enables us to base the accept-reject decision on the inspection of a sample of items from the lot.

The general steps of acceptance sampling are shown in Figure 20.9. After a lot is received, a sample of items is selected for inspection. The results of the inspection are compared to specified quality characteristics. If the quality characteristics are satisfied, the lot is accepted and sent to production or shipped to customers. If the lot is rejected, managers must decide on its disposition. In some cases, the decision may be to keep the lot and remove the unacceptable or nonconforming items. In other cases, the lot may be returned to the supplier at the supplier's expense; the extra work and cost placed on the supplier can motivate the supplier to provide high-quality lots. Finally, if the rejected lot consists of finished goods, the goods must be scrapped or reworked to meet acceptable quality standards.

Acceptance sampling has the following advantages over 100% inspection:
1. *Usually less expensive*
2. *Less product damage due to less handling and testing*
3. *Fewer inspectors required*
4. *The only approach possible if destructive testing must be used*

The statistical procedure of acceptance sampling is based on the hypothesis testing methodology presented in Chapter 9. The null and alternative hypotheses are stated as follows.

$$H_0\text{: Good-quality lot}$$
$$H_a\text{: Poor-quality lot}$$

Table 20.3 shows the results of the hypothesis testing procedure. Note that correct decisions correspond to accepting a good-quality lot and rejecting a poor-quality lot. However, as with other hypothesis testing procedures, we need to be aware of the possibilities of making a Type I error (rejecting a good-quality lot) or a Type II error (accepting a poor-quality lot).

FIGURE 20.9 ACCEPTANCE SAMPLING PROCEDURE

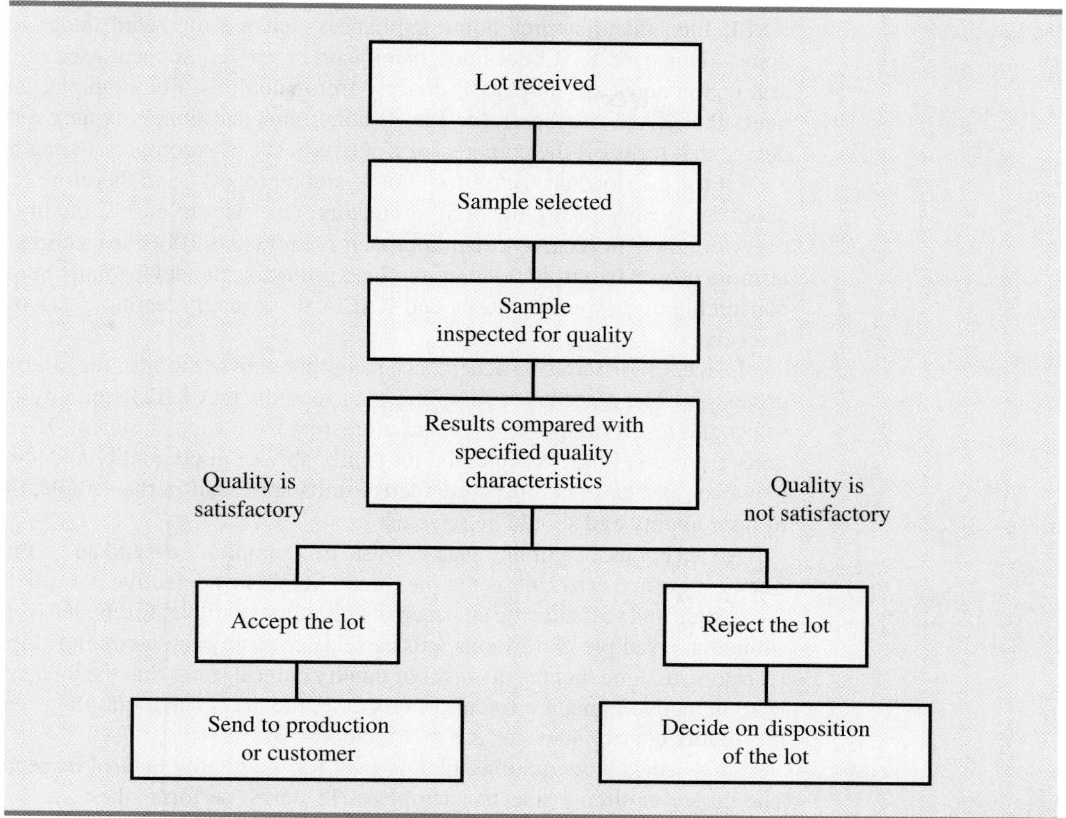

The probability of a Type I error creates a risk for the producer of the lot and is known as the **producer's risk.** For example, a producer's risk of .05 indicates a 5% chance that a good-quality lot will be erroneously rejected. The probability of a Type II error, on the other hand, creates a risk for the consumer of the lot and is known as the **consumer's risk.** For example, a consumer's risk of .10 means a 10% chance that a poor-quality lot will be erroneously accepted and thus used in production or shipped to the customer. Specific values for the producer's risk and the consumer's risk can be controlled by the person designing the acceptance sampling procedure. To illustrate how to assign risk values, let us consider the problem faced by KALI, Inc.

TABLE 20.3 THE OUTCOMES OF ACCEPTANCE SAMPLING

		State of the Lot	
		H_0 True Good-Quality Lot	H_0 False Poor-Quality Lot
Decision	**Accept the Lot**	Correct decision	Type II error (accepting a poor-quality lot)
	Reject the Lot	Type I error (rejecting a good-quality lot)	Correct decision

KALI, Inc.: An Example of Acceptance Sampling

KALI, Inc., manufactures home appliances that are marketed under a variety of trade names. However, KALI does not manufacture every component used in its products. Several components are purchased directly from suppliers. For example, one of the components that KALI purchases for use in home air conditioners is an overload protector, a device that turns off the compressor if it overheats. The compressor can be seriously damaged if the overload protector does not function properly, and therefore KALI is concerned about the quality of the overload protectors. One way to ensure quality would be to test every component received; that approach is known as 100% inspection. However, to determine proper functioning of an overload protector, the device must be subjected to time-consuming and expensive tests, and KALI cannot justify testing every overload protector it receives.

Instead, KALI uses an acceptance sampling plan to monitor the quality of the overload protectors. The acceptance sampling plan requires that KALI's quality control inspectors select and test a sample of overload protectors from each shipment. If very few defective units are found in the sample, the lot is probably of good quality and should be accepted. However, if a large number of defective units are found in the sample, the lot is probably of poor quality and should be rejected.

An acceptance sampling plan consists of a sample size n and an acceptance criterion c. The **acceptance criterion** is the maximum number of defective items that can be found in the sample and still indicate an acceptable lot. For example, for the KALI problem let us assume that a sample of 15 items will be selected from each incoming shipment or lot. Furthermore, assume that the manager of quality control states that the lot can be accepted only if no defective items are found. In this case, the acceptance sampling plan established by the quality control manager is $n = 15$ and $c = 0$.

This acceptance sampling plan is easy for the quality control inspector to implement. The inspector simply selects a sample of 15 items, performs the tests, and reaches a conclusion based on the following decision rule.

- *Accept the lot* if zero defective items are found.
- *Reject the lot* if one or more defective items are found.

Before implementing this acceptance sampling plan, the quality control manager wants to evaluate the risks or errors possible under the plan. The plan will be implemented only if both the producer's risk (Type I error) and the consumer's risk (Type II error) are controlled at reasonable levels.

Computing the Probability of Accepting a Lot

The key to analyzing both the producer's risk and the consumer's risk is a "What-if?" type of analysis. That is, we will assume that a lot has some known percentage of defective items and compute the probability of accepting the lot for a given sampling plan. By varying the assumed percentage of defective items, we can examine the effect of the sampling plan on both types of risks.

Let us begin by assuming that a large shipment of overload protectors has been received and that 5% of the overload protectors in the shipment are defective. For a shipment or lot with 5% of the items defective, what is the probability that the $n = 15$, $c = 0$ sampling plan will lead us to accept the lot? Because each overload protector tested will be either defective or nondefective and because the lot size is large, the number of defective items in a sample of 15 has a *binomial probability distribution.* The binomial probability function, which was presented in Chapter 5, follows.

Binomial Probability Function for Acceptance Sampling

$$f(x) = \frac{n!}{x!(n-x)!} p^x (1-p)^{(n-x)} \tag{20.21}$$

where

n = the sample size
p = the proportion of defective items in the lot
x = the number of defective items in the sample
$f(x)$ = the probability of x defective items in the sample

For the KALI acceptance sampling plan, $n = 15$; thus, for a lot with 5% defective ($p = .05$), we have

$$f(x) = \frac{15!}{x!(15-x)!} (.05)^x (1-.05)^{(15-x)} \tag{20.22}$$

Using (20.22), $f(0)$ will provide the probability that zero overload protectors will be defective and the lot will be accepted. In using (20.22), recall that $0! = 1$. Thus, the probability computation for $f(0)$ is

$$f(0) = \frac{15!}{0!(15-0)!} (.05)^0 (1-.05)^{(15-0)}$$

$$= \frac{15!}{0!(15)!} (.05)^0 (.95)^{15} = (.95)^{15} = .4633$$

We now know that the $n = 15$, $c = 0$ sampling plan has a .4633 probability of accepting a lot with 5% defective items. Hence, there must be a corresponding $1 - .4633 = .5367$ probability of rejecting a lot with 5% defective items.

Tables of binomial probabilities (see Table 5, Appendix B) can help reduce the computational effort in determining the probabilities of accepting lots. Selected binomial probabilities for $n = 15$ and $n = 20$ are listed in Table 20.4. Using this table, we can determine that if the lot contains 10% defective items, there is a .2059 probability that the $n = 15$, $c = 0$ sampling plan will indicate an acceptable lot. The probability that the $n = 15$, $c = 0$ sampling plan will lead to the acceptance of lots with 1%, 2%, 3%, . . . defective items is summarized in Table 20.5.

Using the probabilities in Table 20.5, a graph of the probability of accepting the lot versus the percent defective in the lot can be drawn as shown in Figure 20.10. This graph, or curve, is called the **operating characteristic (OC) curve** for the $n = 15$, $c = 0$ acceptance sampling plan.

Perhaps we should consider other sampling plans, ones with different sample sizes n and/or different acceptance criteria c. First consider the case in which the sample size remains $n = 15$ but the acceptance criterion increases from $c = 0$ to $c = 1$. That is, we will now accept the lot if zero or one defective component is found in the sample. For a lot with 5% defective items ($p = .05$), Table 20.4 shows that with $n = 15$ and $p = .05$, $f(0) = .4633$ and $f(1) = .3658$. Thus, there is a $.4633 + .3658 = .8291$ probability that the $n = 15$, $c = 1$ plan will lead to the acceptance of a lot with 5% defective items.

TABLE 20.4 SELECTED BINOMIAL PROBABILITIES FOR SAMPLES OF SIZE 15 AND 20

		p								
n	x	.01	.02	.03	.04	.05	.10	.15	.20	.25
15	0	.8601	.7386	.6333	.5421	.4633	.2059	.0874	.0352	.0134
	1	.1303	.2261	.2938	.3388	.3658	.3432	.2312	.1319	.0668
	2	.0092	.0323	.0636	.0988	.1348	.2669	.2856	.2309	.1559
	3	.0004	.0029	.0085	.0178	.0307	.1285	.2184	.2501	.2252
	4	.0000	.0002	.0008	.0022	.0049	.0428	.1156	.1876	.2252
	5	.0000	.0000	.0001	.0002	.0006	.0105	.0449	.1032	.1651
	6	.0000	.0000	.0000	.0000	.0000	.0019	.0132	.0430	.0917
	7	.0000	.0000	.0000	.0000	.0000	.0003	.0030	.0138	.0393
	8	.0000	.0000	.0000	.0000	.0000	.0000	.0005	.0035	.0131
	9	.0000	.0000	.0000	.0000	.0000	.0000	.0001	.0007	.0034
	10	.0000	.0000	.0000	.0000	.0000	.0000	.0000	.0001	.0007
20	0	.8179	.6676	.5438	.4420	.3585	.1216	.0388	.0115	.0032
	1	.1652	.2725	.3364	.3683	.3774	.2702	.1368	.0576	.0211
	2	.0159	.0528	.0988	.1458	.1887	.2852	.2293	.1369	.0669
	3	.0010	.0065	.0183	.0364	.0596	.1901	.2428	.2054	.1339
	4	.0000	.0006	.0024	.0065	.0133	.0898	.1821	.2182	.1897
	5	.0000	.0000	.0002	.0009	.0022	.0319	.1028	.1746	.2023
	6	.0000	.0000	.0000	.0001	.0003	.0089	.0454	.1091	.1686
	7	.0000	.0000	.0000	.0000	.0000	.0020	.0160	.0545	.1124
	8	.0000	.0000	.0000	.0000	.0000	.0004	.0046	.0222	.0609
	9	.0000	.0000	.0000	.0000	.0000	.0001	.0011	.0074	.0271
	10	.0000	.0000	.0000	.0000	.0000	.0000	.0002	.0020	.0099
	11	.0000	.0000	.0000	.0000	.0000	.0000	.0000	.0005	.0030
	12	.0000	.0000	.0000	.0000	.0000	.0000	.0000	.0001	.0008

Continuing these calculations we obtain Figure 20.11, which shows the operating characteristic curves for four alternative acceptance sampling plans for the KALI problem. Samples of size 15 and 20 are considered. Note that regardless of the proportion defective in the lot, the $n = 15$, $c = 1$ sampling plan provides the highest probabilities of accepting the lot. The $n = 20$, $c = 0$ sampling plan provides the lowest probabilities of accepting the lot; however, that plan also provides the highest probabilities of rejecting the lot.

TABLE 20.5 PROBABILITY OF ACCEPTING THE LOT FOR THE KALI PROBLEM WITH $n = 15$ and $c = 0$

Percent Defective in the Lot	Probability of Accepting the Lot
1	.8601
2	.7386
3	.6333
4	.5421
5	.4633
10	.2059
15	.0874
20	.0352
25	.0134

FIGURE 20.10 OPERATING CHARACTERISTIC CURVE FOR THE $n = 15$, $c = 0$ ACCEPTANCE SAMPLING PLAN

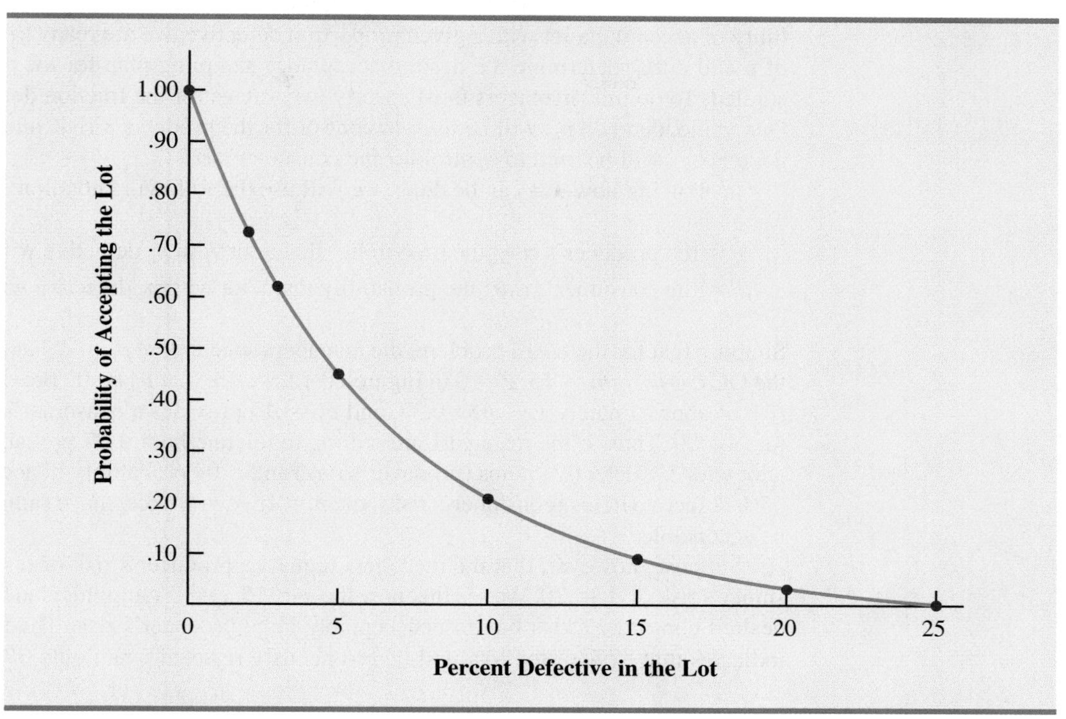

FIGURE 20.11 OPERATING CHARACTERISTIC CURVES FOR FOUR ACCEPTANCE SAMPLING PLANS

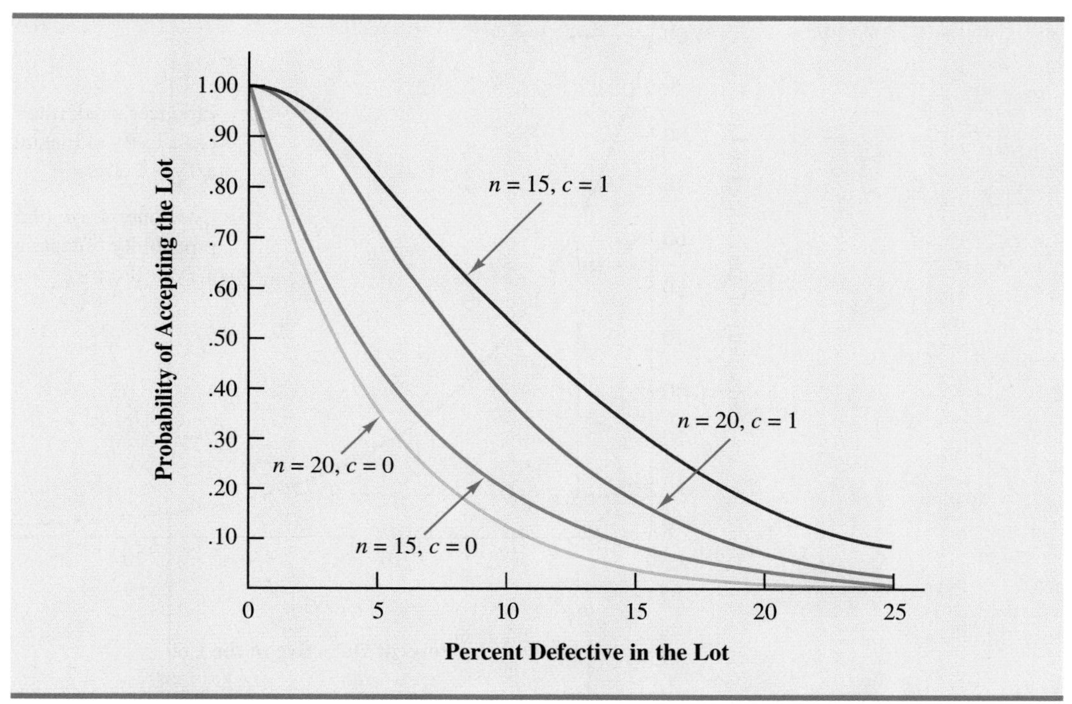

Selecting an Acceptance Sampling Plan

Now that we know how to use the binomial probability distribution to compute the probability of accepting a lot with a given proportion defective, we are ready to select the values of n and c that determine the desired acceptance sampling plan for the application being studied. To do this, managers must specify two values for the fraction defective in the lot. One value, denoted p_0, will be used to control for the producer's risk, and the other value, denoted p_1, will be used to control for the consumer's risk.

In showing how this can be done, we will use the following notation.

α = the producer's risk; the probability that a lot with p_0 defective will be rejected
β = the consumer's risk; the probability that a lot with p_1 defective will be rejected

Suppose that for the KALI problem, the managers specify that $p_0 = .03$ and $p_1 = .15$. From the OC curve for $n = 15$, $c = 0$ in Figure 20.12, we see that $p_0 = .03$ provides a producer's risk of approximately $1 - .63 = .37$, and $p_1 = .15$ provides a consumer's risk of approximately .09. Thus, if the managers are willing to tolerate both a .37 probability of rejecting a lot with 3% defective items (producer's risk) and a .09 probability of accepting a lot with 15% defective items (consumer's risk), the $n = 15$, $c = 0$ acceptance sampling plan would be acceptable.

Suppose, however, that the managers request a producer's risk of $\alpha = .10$ and a consumer's risk of $\beta = .20$. We see that now the $n = 15$, $c = 0$ sampling plan has a better-than-desired consumer's risk but an unacceptably large producer's risk. The fact that $\alpha = .37$ indicates that 37% of the lots will be erroneously rejected when only 3% of the items in

FIGURE 20.12 OPERATING CHARACTERISTIC CURVE FOR $n = 15$, $c = 0$ with $p_0 = .03$ AND $p_1 = .15$

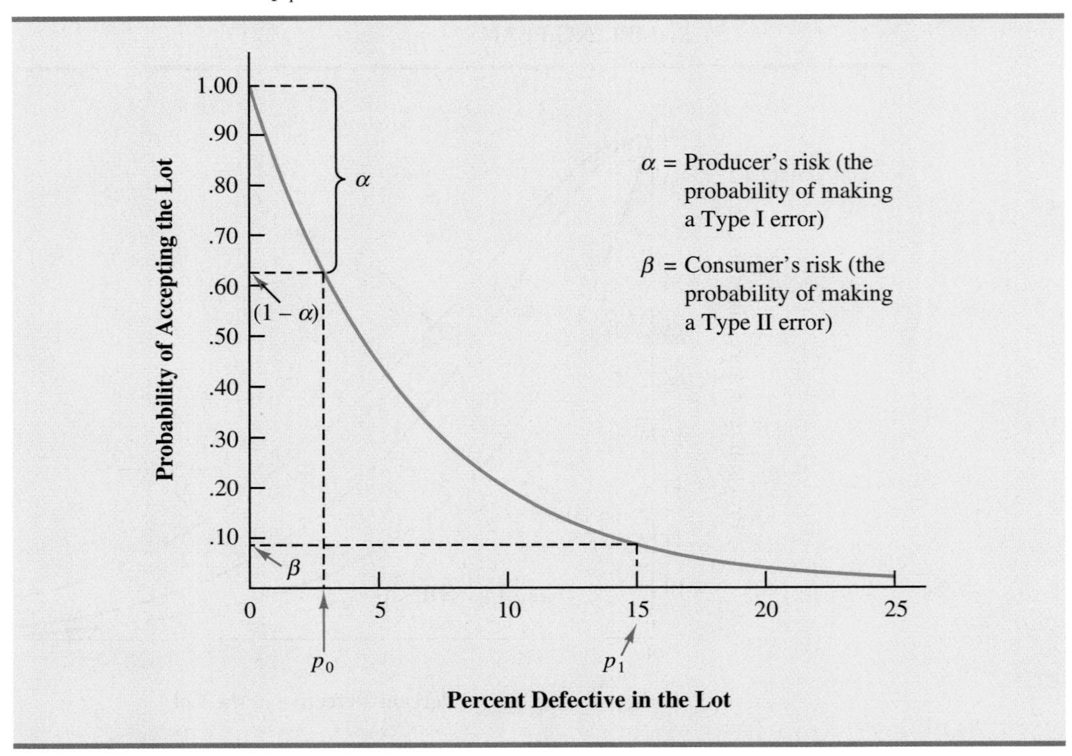

them are defective. The producer's risk is too high, and a different acceptance sampling plan should be considered.

Using $p_0 = .03$, $\alpha = .10$, $p_1 = .15$, and $\beta = .20$ in Figure 20.11 shows that the acceptance sampling plan with $n = 20$ and $c = 1$ comes closest to meeting both the producer's and the consumer's risk requirements. Exercise 13 at the end of this section will ask you to compute the producer's risk and the consumer's risk for the $n = 20$, $c = 1$ sampling plan.

As shown in this section, several computations and several operating characteristic curves may need to be considered to determine the sampling plan with the desired producer's and consumer's risk. Fortunately, tables of sampling plans are published. For example, the American Military Standard Table, MIL-STD-105D, provides information helpful in designing acceptance sampling plans. More advanced texts on quality control, such as those listed in the bibliography, describe the use of such tables. The advanced texts also discuss the role of sampling costs in determining the optimal sampling plan.

Multiple Sampling Plans

The acceptance sampling procedure that we have presented for the KALI problem is a *single-sample* plan. It is called a single-sample plan because only one sample or sampling stage is used. After the number of defective components in the sample is determined, a decision must be made to accept or reject the lot. An alternative to the single-sample plan is a **multiple sampling plan,** in which two or more stages of sampling are used. At each stage a decision is made among three possibilities: stop sampling and accept the lot, stop sampling and reject the lot, or continue sampling. Although more complex, multiple sampling plans often result in a smaller total sample size than single-sample plans with the same α and β probabilities.

The logic of a two-stage, or double-sample, plan is shown in Figure 20.13. Initially a sample of n_1 items is selected. If the number of defective components x_1 is less than or equal to c_1, accept the lot. If x_1 is greater than or equal to c_2, reject the lot. If x_1 is between c_1 and c_2 ($c_1 < x_1 < c_2$), select a second sample of n_2 items. Determine the combined, or total, number of defective components from the first sample (x_1) and the second sample (x_2). If $x_1 + x_2 \leq c_3$, accept the lot; otherwise reject the lot. The development of the double-sample plan is more difficult because the sample sizes n_1 and n_2 and the acceptance numbers c_1, c_2, and c_3 must meet both the producer's and consumer's risks desired.

NOTES AND COMMENTS

1. The use of the binomial probability distribution for acceptance sampling is based on the assumption of large lots. If the lot size is small, the hypergeometric probability distribution is the appropriate distribution. Experts in the field of quality control indicate that the Poisson distribution provides a good approximation for acceptance sampling when the sample size is at least 16, the lot size is at least 10 times the sample size, and p is less than .1. For larger sample sizes, the normal approximation to the binomial probability distribution can be used.
2. In the MIL-ST-105D sampling tables, p_0 is called the acceptable quality level (AQL). In some sampling tables, p_1 is called the lot tolerance percent defective (LTPD) or the rejectable quality level (RQL). Many of the published sampling plans also use quality indexes such as the indifference quality level (IQL) and the average outgoing quality limit (AOQL). The more advanced texts listed in the bibliography provide a complete discussion of these other indexes.
3. In this section we provided an introduction to *attributes sampling plans*. In these plans each item sampled is classified as nondefective or defective. In *variables sampling plans*, a sample is taken and a measurement of the quality characteristic is taken. For example, for gold jewelry a measurement of quality may be the amount of gold it contains. A simple statistic such as the average amount of gold in the sample jewelry is computed and compared with an allowable value to determine whether to accept or reject the lot.

FIGURE 20.13 A TWO-STAGE ACCEPTANCE SAMPLING PLAN

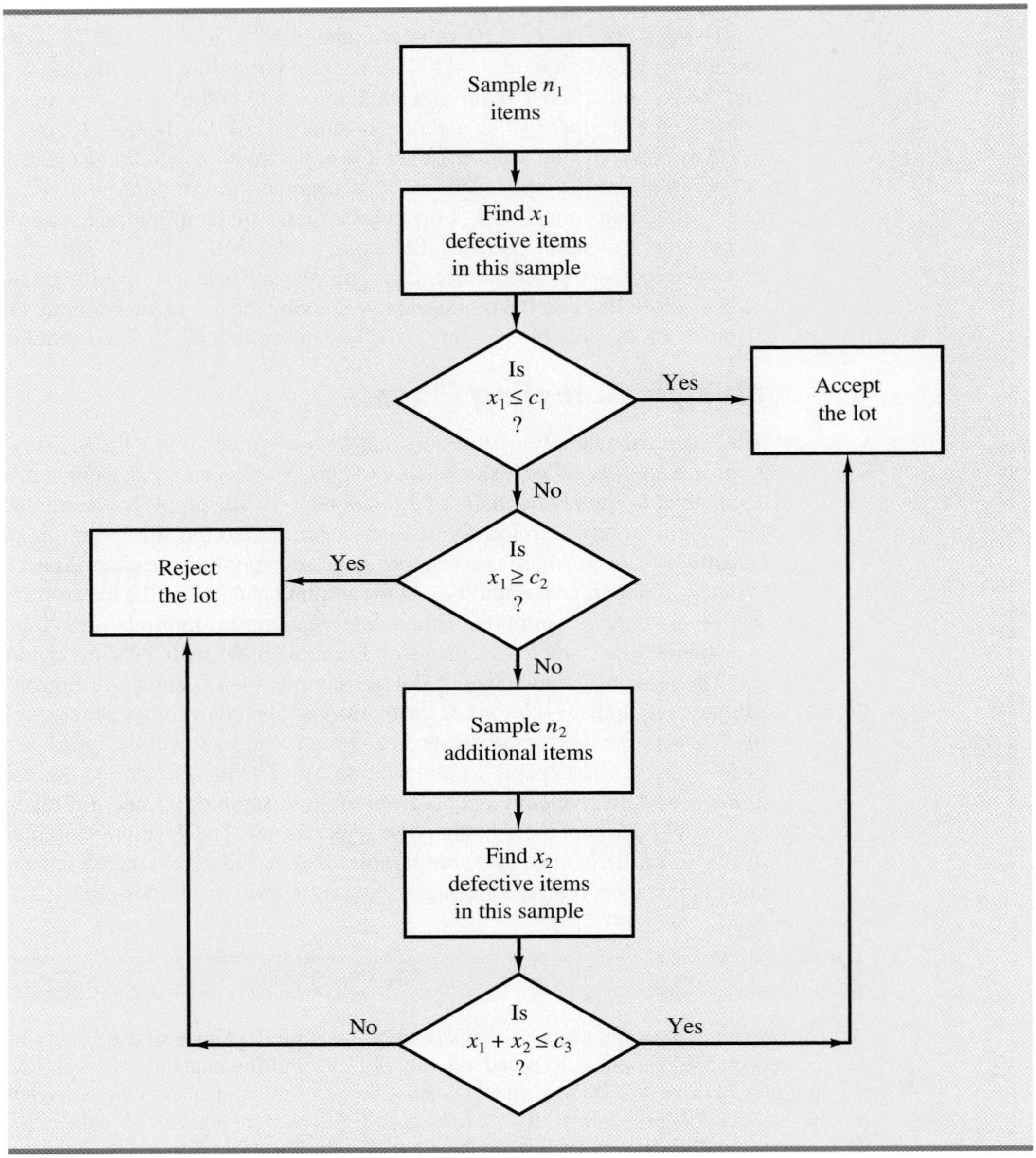

Sample n_1 items

Find x_1 defective items in this sample

Is $x_1 \le c_1$?

Yes → Accept the lot

No

Is $x_1 \ge c_2$?

Yes → Reject the lot

No

Sample n_2 additional items

Find x_2 defective items in this sample

Is $x_1 + x_2 \le c_3$?

Yes → Accept the lot

No → Reject the lot

EXERCISES

Methods

10. For an acceptance sampling plan with $n = 25$ and $c = 0$, find the probability of accepting a lot that has a defect rate of 2%. What is the probability of accepting the lot if the defect rate is 6%?

11. Consider an acceptance sampling plan with $n = 20$ and $c = 0$. Compute the producer's risk for each of the following cases.
    - **a.** The lot has a defect rate of 2%.
    - **b.** The lot has a defect rate of 6%.

12. Repeat Exercise 11 for the acceptance sampling plan with $n = 20$ and $c = 1$. What happens to the producer's risk as the acceptance number c is increased? Explain.

Applications

13. Refer to the KALI problem presented in this section. The quality control manager requested a producer's risk of .10 when p_0 was .03 and a consumer's risk of .20 when p_1 was .15. Consider the acceptance sampling plan based on a sample size of 20 and an acceptance number of 1. Answer the following questions.
    a. What is the producer's risk for the $n = 20$, $c = 1$ sampling plan?
    b. What is the consumer's risk for the $n = 20$, $c = 1$ sampling plan?
    c. Does the $n = 20$, $c = 1$ sampling plan satisfy the risks requested by the quality control manager? Discuss.
14. To inspect incoming shipments of raw materials, a manufacturer is considering samples of sizes 10, 15, and 20. Use the binomial probabilities from Table 5 of Appendix B to select a sampling plan that provides a producer's risk of $\alpha = .03$ when p_0 is .05 and a consumer's risk of $\beta = .12$ when p_1 is .30.
15. A domestic manufacturer of watches purchases quartz crystals from a Swiss firm. The crystals are shipped in lots of 1000. The acceptance sampling procedure uses 20 randomly selected crystals.
    a. Construct operating characteristic curves for acceptance numbers of 0, 1, and 2.
    b. If p_0 is .01 and $p_1 = .08$, what are the producer's and consumer's risks for each sampling plan in part (a)?

SUMMARY

In this chapter we discussed how statistical methods can be used to assist in the control of quality. We first presented the $\bar{x}$, R, p, and np control charts as graphical aids in monitoring process quality. Control limits are established for each chart; samples are selected periodically, and the data points plotted on the control chart. Data points outside the control limits indicate that the process is out of control and that corrective action should be taken. Patterns of data points within the control limits can also indicate potential quality control problems and suggest that corrective action may be warranted.

We also considered the technique known as acceptance sampling. With this procedure, a sample is selected and inspected. The number of defective items in the sample provides the basis for accepting or rejecting the lot. The sample size and the acceptance criterion can be adjusted to control both the producer's risk (Type I error) and the consumer's risk (Type II error).

GLOSSARY

Quality control A series of inspections and measurements that determine whether quality standards are being met.

Assignable causes Variations in process outputs that are due to factors such as machine tools wearing out, incorrect machine settings, poor-quality raw materials, operator error, and so on. Corrective action should be taken when assignable causes of output variation are detected.

Common causes Normal or natural variations in process outputs that are due purely to chance. No corrective action is necessary when output variations are due to common causes.

Control chart A graphical tool used to help determine whether a process is in control or out of control.

$\bar{x}$ chart A control chart used when the output of a process is measured in terms of the mean value of a variable such as a length, weight, temperature, and so on.

***R* chart** A control chart used when the output of a process is measured in terms of the range of a variable.

***p* chart** A control chart used when the output of a process is measured in terms of the proportion defective.

***np* chart** A control chart used to monitor the output of a process in terms of the number of defective items.

Lot A group of items such as incoming shipments of raw materials or purchased parts as well as finished goods from final assembly.

Acceptance sampling A statistical procedure in which the number of defective items found in a sample is used to determine whether a lot should be accepted or rejected.

Producer's risk The risk of rejecting a good-quality lot. This is the Type I error.

Consumer's risk The risk of accepting a poor-quality lot. This is the Type II error.

Acceptance criterion The maximum number of defective items that can be found in the sample and still allow acceptance of the lot.

Operating characteristic curve A graph showing the probability of accepting the lot as a function of the percentage defective in the lot. This curve can be used to help determine whether a particular acceptance sampling plan meets both the producer's and the consumer's risk requirements.

Multiple sampling plan A form of acceptance sampling in which more than one sample or stage is used. On the basis of the number of defective items found in a sample, a decision will be made to accept the lot, reject the lot, or continue sampling.

KEY FORMULAS

Standard Error of the Mean

$$\sigma_{\bar{x}} = \frac{\sigma}{\sqrt{n}} \tag{20.1}$$

Control Limits for an $\bar{x}$ Chart: Process Mean and Standard Deviation Known

$$\text{UCL} = \mu + 3\sigma_{\bar{x}} \tag{20.2}$$

$$\text{LCL} = \mu - 3\sigma_{\bar{x}} \tag{20.3}$$

Overall Sample Mean

$$\bar{\bar{x}} = \frac{\bar{x}_1 + \bar{x}_2 + \cdots + \bar{x}_k}{k} \tag{20.4}$$

Average Range

$$\bar{R} = \frac{R_1 + R_2 + \cdots + R_k}{k} \tag{20.5}$$

Control Limits for an $\bar{x}$ Chart: Process Mean and Standard Deviation Unknown

$$\bar{\bar{x}} \pm A_2\bar{R} \tag{20.8}$$

Control Limits for an *R* Chart

$$\text{UCL} = \bar{R}D_4 \tag{20.14}$$
$$\text{LCL} = \bar{R}D_3 \tag{20.15}$$

Standard Error of the Proportion

$$\sigma_{\bar{p}} = \sqrt{\frac{p(1 - p)}{n}} \tag{20.16}$$

Control Limits for a *p* Chart

$$\text{UCL} = p + 3\sigma_{\bar{p}} \tag{20.17}$$
$$\text{LCL} = p - 3\sigma_{\bar{p}} \tag{20.18}$$

Control Limits for an *np* Chart

$$\text{UCL} = np + 3\sqrt{np(1 - p)} \tag{20.19}$$
$$\text{LCL} = np - 3\sqrt{np(1 - p)} \tag{20.20}$$

Binomial Probability Function for Acceptance Sampling

$$f(x) = \frac{n!}{x!(n - x)!} p^x(1 - p)^{(n-x)} \tag{20.21}$$

SUPPLEMENTARY EXERCISES

16. Samples of size five provided the following 20 sample means for a production process that is believed to be in control.

95.72	95.24	95.18
95.44	95.46	95.32
95.40	95.44	95.08
95.50	95.80	95.22
95.56	95.22	95.04
95.72	94.82	95.46
95.60	95.78	

a. Based on these data, what is an estimate of the mean when the process is in control?
b. Assuming that the process standard deviation is $\sigma = .50$, develop a control chart for this production process. Assume that the mean of the process is the estimate developed in part (a).
c. Do any of the 20 sample means indicate that the process was out of control?

17. Product filling weights are normally distributed with a mean of 350 grams and a standard deviation of 15 grams.
a. Develop the control limits for samples of size 10, 20, and 30.
b. What happens to the control limits as the sample size is increased?
c. What happens when a Type I error is made?
d. What happens when a Type II error is made?
e. What is the probability of a Type I error for samples of size 10, 20, and 30?
f. What is the advantage of increasing the sample size for control chart purposes? What error probability is reduced as the sample size is increased?

18. Twenty-five samples of size five resulted in $\bar{\bar{x}} = 5.42$ and $\bar{R} = 2.0$. Compute control limits for the $\bar{x}$ and R charts, and estimate the standard deviation of the process.

19. The following are quality control data for a manufacturing process at Kensport Chemical Company. The data show the temperature in degrees centigrade at five points in time during a manufacturing cycle. The company is interested in using control charts to monitor the temperature of its manufacturing process. Construct the $\bar{x}$ chart and R chart. What conclusions can be made about the quality of the process?

Sample	$\bar{x}$	R	Sample	$\bar{x}$	R
1	95.72	1.0	11	95.80	.6
2	95.24	.9	12	95.22	.2
3	95.18	.8	13	95.56	1.3
4	95.44	.4	14	95.22	.5
5	95.46	.5	15	95.04	.8
6	95.32	1.1	16	95.72	1.1
7	95.40	.9	17	94.82	.6
8	95.44	.3	18	95.46	.5
9	95.08	.2	19	95.60	.4
10	95.50	.6	20	95.74	.6

20. The following were collected for the Master Blend Coffee production process. The data show the filling weights based on samples of 3-pound cans of coffee. Use these data to construct the $\bar{x}$ and R chart. What conclusions can be made about the quality of the production process?

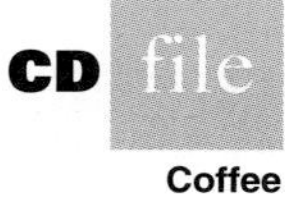

Coffee

	Observations				
Sample	**1**	**2**	**3**	**4**	**5**
1	3.05	3.08	3.07	3.11	3.11
2	3.13	3.07	3.05	3.10	3.10
3	3.06	3.04	3.12	3.11	3.10
4	3.09	3.08	3.09	3.09	3.07
5	3.10	3.06	3.06	3.07	3.08
6	3.08	3.10	3.13	3.03	3.06
7	3.06	3.06	3.08	3.10	3.08
8	3.11	3.08	3.07	3.07	3.07
9	3.09	3.09	3.08	3.07	3.09
10	3.06	3.11	3.07	3.09	3.07

21. Consider the following situations. Comment on whether the situation might cause concern about the quality of the process.

a. A p chart has LCL $= 0$ and UCL $= .068$. When the process is in control, the proportion defective is .033. Plot the following seven sample results: .035, .062, .055, .049, .058, .066, and .055. Discuss.

b. An $\bar{x}$ chart has LCL $= 22.2$ and UCL $= 24.5$. The mean is $\mu = 23.35$ when the process is in control. Plot the following seven sample results: 22.4, 22.6, 22.65, 23.2, 23.4, 23.85, and 24.1. Discuss.

22. Managers of 1200 different retail outlets make twice-a-month restocking orders from a central warehouse. Past experience has shown that 4% of the orders have one or more errors such as wrong item shipped, wrong quantity shipped, and item requested but not shipped. Random samples of 200 orders are selected monthly and checked for accuracy.

a. Construct a control chart for this situation.

b. Six months of data show the following numbers of orders with one or more errors: 10, 15, 6, 13, 8, and 17. Plot the data on the control chart. What does your plot indicate about the order process?

23. An $n = 10$, $c = 2$ acceptance sampling plan is being considered; assume that $p_0 = .05$ and $p_1 = .20$.

a. Compute both the producer's and the consumer's risk for this acceptance sampling plan.

b. Would either the producer, the consumer, or both be unhappy with the proposed sampling plan?

c. What change in the sampling plan, if any, would you recommend?

24. An acceptance sampling plan with $n = 15$ and $c = 1$ has been designed with a producer's risk of .075.
    a. Was the value of p_0 .01, .02, .03, .04, or .05? What does this value mean?
    b. What is the consumer's risk associated with this plan if p_1 is .25?

25. A manufacturer produces lots of a canned food product. Let p denote the proportion of the lots that do not meet the product quality specifications. An $n = 25$, $c = 0$ acceptance sampling plan will be used.
    a. Compute points on the operating characteristic curve when $p = .01, .03, .10$, and .20.
    b. Plot the operating characteristic curve.
    c. What is the probability that the acceptance sampling plan will reject a lot that has .01 defective?

26. Sometimes an acceptance sampling plan will be based on a large sample. In this case, the normal approximation to the binomial probability distribution can be used to compute the producer's and the consumer's risk associated with the plan. Referring to Chapter 6, we know that the normal distribution used to approximate binomial probabilities has a mean of np and a standard deviation of $\sqrt{np(1 - p)}$. Assume that an acceptance sampling plan is $n = 250$, $c = 10$.
    a. What is the producer's risk if p_0 is .02? As discussed in Chapter 6, a continuity correction factor should be used in this case. Thus, the probability of acceptance is based on the normal probability of the random variable being less than or equal to 10.5.
    b. What is the consumer's risk if p_1 is .08?
    c. What is an advantage of a large sample size for acceptance sampling? What is a disadvantage?

Appendix 20.1 CONTROL CHARTS WITH MINITAB

CD file Jensen

In this appendix we describe the steps required to generate Minitab control charts using the Jensen sample data shown in Table 20.2. The sample number appears in column C1. The first observation is in column C2, the second observation is in column C3, and so on. The following steps describe how to use Minitab to produce both the $\bar{x}$ chart and R chart simultaneously.

Step 1. Select the **Stat** pull-down menu
Step 2. Choose **Control Charts**
Step 3. Choose **Xbar-R**
Step 4. When the Xbar-R Chart dialog box appears:
    Select **Subgroups across rows of**
    Enter C2-C6 in the Subgroups across rows of box
    Select **Tests**
Step 5. When the Tests dialog box appears:
    Choose **One point more than 3 sigmas from center line***
    Click **OK**
Step 6. When the Xbar-R Chart dialog box appears
    Click **OK**

The $\bar{x}$ chart and the R chart will be shown together on the Minitab output. The choices available under Step 3 of the preceding Minitab procedure provide access to a variety of control chart options. For example, the $\bar{x}$ and the R chart can be selected separately. Additional options include the p chart, the np chart, and others.

*Minitab provides several additional tests for detecting special causes of variation and out-of-control conditions. The user may select several of these tests simultaneously.

Chapter 21

SAMPLE SURVEY

CONTENTS

STATISTICS IN PRACTICE

CINERGY*
Cincinnati, Ohio

Cinergy, formerly Cincinnati Gas & Electric Company (CG&E), is a public utility that provides gas and electric power to customers in the Greater Cincinnati area. To improve service to its customers, Cinergy continually strives to stay up-to-date with its customers' needs. Cinergy undertook a sample survey, the Building Characteristics Survey, to learn about the energy requirements of commercial buildings in its service area.

A variety of information concerning commercial buildings was sought, such as the floor space, number of employees, energy end-use, age of the building, type of building materials, and energy conservation measures. During preparations for the survey, Cinergy analysts determined that approximately 27,000 commercial buildings were in operation in the Cinergy service area. Based on available funds and the precision desired in the results, they recommended that a sample of 616 commercial buildings be surveyed.

The sample design chosen was stratified simple random sampling. Total electrical usage over the past year for each commercial building in the service area was available from company records, and because many of the building characteristics of interest (size, number of employees, etc.) were related to usage, it was the criteria used to divide the population of buildings into six strata.

The first stratum contained the commercial buildings for the 100 largest energy users; each building in this stratum was included in the sample. Although these buildings constituted only .2% of the population, they accounted for 14.4% of the total electrical usage. For the other strata, the number of buildings sampled was determined on the basis of obtaining the greatest precision possible per unit cost.

Statistical surveys enable Cinergy Company to determine the energy needs of its customers. © PhotoDisc, Inc.

A questionnaire was carefully developed and pretested before the actual survey was conducted. Data were collected through personal interviews. Completed surveys totaled 526 out of the sample of 616 commercial buildings. This response rate of 85.4% was considered to be excellent. Currently, Cinergy is using the survey results to improve the forecasts of energy demand and to improve service to its commercial customers.

In this chapter you will learn about the issues that statisticians consider in the design and execution of a sample survey such as the one conducted by Cinergy. Sample surveys are often used to develop profiles of a company's customers; they are also used by the government and other agencies to learn about various segments of the population.

*The authors are indebted to Jim Riddle of Cinergy for providing this Statistics in Practice.

21.1 TERMINOLOGY USED IN SAMPLE SURVEYS

In Chapter 1 we gave the following definitions of an element, a population, and a sample.

- An **element** is the entity on which data are collected.
- A **population** is the collection of all the elements of interest.
- A **sample** is a subset of the population.

To illustrate these concepts, consider the following situation. Dunning Microsystems, Inc. (DMI), a manufacturer of personal computers and peripherals, would like to collect data about the characteristics of individuals who have purchased a DMI personal computer. To obtain such data, a sample survey of DMI personal computer owners could be conducted. The *elements* in this sample survey would be individuals who have purchased a DMI personal computer. The *population* would be the collection of all people who have purchased a DMI personal computer, and the *sample* would be the subset of DMI personal computer owners who are surveyed.

In sample surveys it is necessary to distinguish between the target population and the sampled population. The **target population** is the population we want to make inferences about, while the **sampled population** is the population from which the sample is actually selected. It is important to understand that these two populations are not always the same. In the DMI example, the target population consists of all people who have purchased a DMI personal computer. The sampled population, however, might be all owners who have sent warranty registration cards back to DMI. Not every person who buys a DMI personal computer sends in the warranty card, so the sampled population would differ from the target population.

If inferences from a sample are to be valid, the sampled population must be representative of the target population.

Conclusions drawn from a sample survey apply only to the sampled population. Whether these conclusions can be extended to the target population depends on the judgment of the analyst. The key issue is whether the correspondence between the sampled population and the target population on the characteristics of interest is close enough to allow this extension.

Before sampling, the population must be divided into **sampling units.** In some cases, the sampling units are simply the elements. In other cases, the sampling units are groups of the elements. For example, suppose we want to survey certified professional engineers who are involved in the design of heating and air conditioning systems for commercial buildings. If a list of all professional engineers involved in such work were available, the sampling units would be the professional engineers we want to survey. If such a list is not available, we must find an alternative approach. A business telephone directory might provide a list of all engineering firms involved in the design of heating and air conditioning systems. Given this list, we could select a sample of the engineering firms to survey; then, for each firm surveyed, we might interview all the professional engineers. In this case, the engineering firms would be the sampling units and the engineers interviewed would be the elements.

A list of the sampling units for a particular study is called a **frame.** In the sample survey of professional engineers, the frame is defined as all engineering firms listed in the business telephone directory; the frame is not a list of all professional engineers because no such list is available. The choice of a particular frame and hence the definition of the sampling units is often determined by the availability and reliability of a list. In practice, the development of the frame can be one of the most difficult and important steps in conducting a sample survey.

21.2 TYPES OF SURVEYS AND SAMPLING METHODS

The three most common types of surveys are mail surveys, telephone surveys, and personal interview surveys; each of these types involves the design and administration of a questionnaire. Other types of surveys used to collect data do not involve questionnaires. For example, accounting firms are often hired to sample a company's inventory of goods to estimate the value of inventory on the company's balance sheet. In such surveys, someone simply counts the items and records the results.

In surveys that use questionnaires, the design of the questionnaire is critical. The designer must resist the temptation to include questions that *might* be of interest, because every question adds to the length of the questionnaire. Long questionnaires lead not only to respondent fatigue, but also to interviewer fatigue, especially in mail and telephone surveys.

Survey costs are lower for mail and telephone surveys. But with well-trained interviewers, higher response rates and longer questionnaires are possible with personal interviews. Given the amount of data collected on each element, personal interviews were the only way to collect the data for the Cinergy study in the Statistics in Practice feature in this chapter.

However, if personal interviews are to be used, a longer and more complex questionnaire is feasible. A large body of knowledge exists concerning the phrasing, sequencing, and grouping of questions for a questionnaire. These issues are discussed in more comprehensive books on survey sampling; several sources for this type of information are listed in the bibliography.

Sample surveys can also be classified in terms of the sampling method used. With **probabilistic sampling,** the probability of obtaining each possible sample can be computed; with a **nonprobabilistic sampling,** this probability is unknown. Nonprobabilistic sampling methods should not be used if the researcher wants to make statements about the precision of the estimates. In contrast, probabilistic sampling methods can be used to develop confidence intervals that provide bounds on the sampling error. In the following sections, four of the most popular probabilistic sampling methods are discussed: simple random sampling, stratified simple random sampling, cluster sampling, and systematic sampling.

Although statisticians prefer to use a probabilistic sampling method, nonprobabilistic sampling methods often are necessary. The advantages of nonprobabilistic sampling methods are their low expense and ease of implementation. The disadvantage is that statistically valid statements cannot be made about the precision of the estimates. Two of the more common nonprobabilistic methods are convenience sampling and judgment sampling.

With **convenience sampling,** the units included in the sample are chosen because of accessibility. For example, a professor conducting a research study at a university may ask student volunteers to participate in the study simply because they are in the professor's class; in this case, the sample of students is referred to as a convenience sample. In some situations, convenience sampling is the only practical approach. For example, to sample a shipment of oranges, an inspector might select oranges haphazardly from several crates since labeling each orange in the entire shipment to create a frame and employing a probabilistic sampling method would be impractical. Wildlife captures and volunteer panels for consumer research are other examples of convenience samples.

Although convenience sampling is a relatively easy approach to sample selection and data gathering, it is impossible to evaluate the "goodness" of the sample statistics obtained in terms of their ability to estimate the population parameters of interest. A convenience sample may provide good results or it may not; there is no statistically justified procedure for making any statistical inferences from the sample results. Nevertheless, at times some researchers apply a statistical method designed for a probability sample to the data gathered from a convenience sample. In doing so, the researcher may argue that the convenience sample can be treated as though it were a random sample in the sense that it is representative of the population. However, this argument should be questioned; one should be cautious in using a convenience sample to make statistical inferences about population parameters.

In using the nonprobability sampling technique referred to as **judgment sampling,** a person knowledgeable on the subject of the study selects sampling units that he or she feels are most representative of the population. Although judgment sampling is often a relatively easy way to select samples, users of the survey results must recognize that the quality of the results is dependent on the judgment of the person selecting the sample. Consequently, caution must be exercised in using judgment samples to make statistical inferences about a population parameter. In general, no statistical statements should be made about the precision of the results from a judgment sample.

With a nonprobabilistic sampling method, if methods can be employed to ensure that a representative sample is obtained, point estimates based on the sample can be useful. However, even then the precision of the results is unknown.

Both probabilistic and nonprobabilistic sampling methods can be used to select a sample. The advantage of nonprobabilistic methods is that they are generally inexpensive and easy to use. However, if it is necessary to provide statements about the precision of the estimates, a probabilistic sampling method must be used. Almost all large sample surveys employ probabilistic sampling methods.

21.3 SURVEY ERRORS

Two types of errors can occur in conducting a survey. One type, **sampling error**, is defined as the magnitude of the difference between the point estimator developed from the sample and the population parameter. In other words, sampling error is the error that occurs because not every element in the population is surveyed. The second type, **nonsampling error**, refers to all other types of errors that can occur when a survey is conducted, such as measurement error, interviewer error, and processing error. Although sampling error can occur only in a sample survey, nonsampling errors can occur in both a census and a sample survey.

Nonsampling Error

One of the most common types of nonsampling error occurs whenever we incorrectly measure the characteristic of interest. Measurement error can occur in a census or a sample survey. For either type of survey, the researcher must exercise care to ensure that any measuring instruments (e.g., the questionnaire) are properly calibrated and that the people who take the measurements are properly trained. Attention to detail is the best precaution in most situations.

Nonsampling error can be minimized by proper interviewer training, good questionnaire design, pretesting, and careful management of the process of coding and transferring the data to the computer.

Errors due to nonresponse are a concern to both the statistician responsible for designing the survey and the manager using the results. This type of nonsampling error occurs when data cannot be obtained for some of the units surveyed or when only partial data are obtained. The problem is most serious when a bias is created. For example, if interviews were conducted to assess women's attitudes toward working outside the home, making housecalls only during the daytime would create an obvious bias because women who work outside the home would be excluded from the sample.

In the 1990 U.S. Census, 25.9% of households did not respond. Census 2000 employed a sample survey of nonrespondents to estimate the characteristics of this portion of the population.

Nonsampling errors due to lack of respondent knowledge are common in technical surveys. For example, suppose building managers were surveyed to obtain detailed information about the types of ventilation systems used in office buildings. Managers of large office buildings may be especially knowledgeable about such systems because they may attend training seminars and have support staff to help keep them current. In contrast, managers of smaller office buildings may be less knowledgeable about such systems because of the wide variety of duties they must perform. This difference in knowledge can significantly affect the survey results.

Two other types of nonsampling error are selection error and processing error. Selection errors occur when an inappropriate item is included in the survey. Suppose a sample survey was designed to develop a profile of men with beards; if some interviewers interpreted the statement "men with beards" to include men with mustaches while other interviewers did not, the resulting data would be flawed. Processing errors occur whenever data are incorrectly recorded or incorrectly transferred from recording forms, such as from questionnaires to computer files.

Although some nonsampling errors will occur in most surveys, they can be minimized by careful planning. Care should be taken to ensure that the sampled population corresponds closely to the target population, that good questionnaire design principles are followed, that interviewers are well trained, and so on. The final report on a survey should include some discussion of the likely impact of nonsampling errors on the results.

Sampling Error

Recall the Dunning Microsystems (DMI) sample survey mentioned in Section 21.1. Suppose DMI wants to estimate the mean age of people who have purchased a DMI personal computer. If the entire population of DMI personal computer owners could be surveyed (a census) and nonsampling errors were not present, we could determine the mean age exactly. But what if less than 100% of the population of DMI owners can be surveyed? In this case, there will most likely be a difference between the sample mean and the population mean; the absolute value of this difference is the sampling error. In practice, it is not possible to know what the sampling error will be for any one particular sample because the population mean is unknown; however, it is possible to provide probability statements about the size of the sampling error.

Sampling error is minimized by proper choice of a sample design.

As stated, sampling error occurs because a sample, and not the entire population, is surveyed. Even though sampling error cannot be avoided, it can be controlled. Selecting an appropriate sampling method or design is one important way to control this type of error. In the following sections we will discuss four probabilistic sampling methods: simple random sampling, stratified simple random sampling, cluster sampling, and systematic sampling.

21.4 SIMPLE RANDOM SAMPLING

Recall the definition of simple random sampling from Chapter 7:

> A simple random sample of size n from a finite population of size N is a sample selected such that every possible sample of size n has the same probability of being selected.

To conduct a sample survey using **simple random sampling,** we begin by developing a frame or list of all elements in the sampled population. Then a selection procedure, based on the use of random numbers, is used to ensure that each element in the sampled population has the same probability of being selected. In this section we show how estimates of a population mean, total, and proportion are made for sample surveys that use simple random sampling.

Population Mean

In most sample surveys, the form of the probability distribution for the population is unknown. For example, in the DMI sample survey, management wants to estimate μ, the mean age of people who have purchased a DMI personal computer. Most likely, DMI would not know the form of the probability distribution of age for the population of all DMI owners. Not knowing the form of the probability distribution for a population is usually not a problem because the properties of the sampling distribution of $\bar{x}$, the point estimator of μ, depend primarily on the choice of the sample design.

In Chapter 7 we indicated that if a large ($n \geq 30$) simple random sample is selected, the central limit theorem enables us to conclude that the sampling distribution of $\bar{x}$ can be approximated by a normal probability distribution. In Chapter 8 we showed that for cases where the sampling distribution of $\bar{x}$ can be approximated by a normal probability distribution, an interval estimate of μ is given by

$$\bar{x} \pm z_{\alpha/2}\sigma_{\bar{x}} \tag{21.1}$$

where

$$\sigma_{\bar{x}} = \text{standard error of the mean.}$$

Recall that $1 - \alpha$ is the confidence coefficient, and $z_{\alpha/2}$ is the z value providing an area of $\alpha/2$ in the upper tail of the standard normal probability distribution. For example, for a 95% confidence interval, $z_{.025} = 1.96$. Note also that the standard error of the mean, $\sigma_{\bar{x}}$, is just the standard deviation of the sampling distribution of $\bar{x}$. In general, whenever we use the term *standard error* in this chapter, we will be referring to the standard deviation of the sampling distribution for the point estimator being considered.

When a simple random sample of size n is selected from a finite population of size N, an estimate of the standard error of the mean is

$$s_{\bar{x}} = \sqrt{\frac{N-n}{N}}\left(\frac{s}{\sqrt{n}}\right) \tag{21.2}$$

Using $s_{\bar{x}}$ as an estimate of $\sigma_{\bar{x}}$, the interval estimate of the population mean becomes

$$\bar{x} \pm z_{\alpha/2} s_{\bar{x}} \tag{21.3}$$

Use of $z = 2$ is based on the assumption that the sampling distribution is approximately normal; approximately 95% of the data for such a distribution lie within 2 standard deviations of the mean.

In a sample survey it is common practice to use a value of $z = 2$ when developing interval estimates. Thus, when simple random sampling is used, an approximate 95% confidence interval estimate of the population mean is given by the following expression.

Approximate 95% Confidence Interval Estimate of the Population Mean

$$\bar{x} \pm 2s_{\bar{x}} \tag{21.4}$$

As an example, consider the situation of the publisher of *Great Lakes Recreation,* a regional magazine specializing in articles on boating and fishing. The magazine currently has $N = 8000$ subscribers. A simple random sample of $n = 484$ subscribers shows mean annual income to be \$30,500 with a standard deviation of \$7040. An unbiased estimate of the mean annual income of all subscribers is given by $\bar{x} = \$30{,}500$. Using the sample results and (21.2), we obtain the following estimate of the standard error of the mean.

$$s_{\bar{x}} = \sqrt{\frac{8000-484}{8000}}\left(\frac{7040}{\sqrt{484}}\right) = 310$$

Therefore, using (21.4), we find that an approximate 95% confidence interval estimate of the mean annual income for the magazine subscribers is

$$30{,}500 \pm 2(310) = 30{,}500 \pm 620$$

or \$29,880 to \$31,120.

The preceding procedure can be used to compute an interval estimate for other population parameters such as the population total or the population proportion. In all cases where the sampling distribution of the point estimator can be approximated by a normal probability distribution, the approximate 95% confidence interval can be written as

Point Estimator ± 2(Estimate of the Standard Error of the Point Estimator)

In sample survey terminology, the ± value is often referred to as the bound on the sampling error. This is the same as the margin of error discussed in Chapter 8.

For example, in the *Great Lakes Recreation* sample survey, an estimate of the standard error of the point estimator is $s_{\bar{x}} = \$310$, and the **bound on the sampling error** is 2(\$310) = \$620.

Population Total

Consider the problem facing Northeast Electric and Gas (NEG). As part of an energy usage study, NEG needs to estimate the *total* square footage for the 500 public schools in its service area. We will denote the total square footage for the 500 schools as X; in other words,

X denotes the population total. Note that if μ, the mean square footage for the 500 public schools, were known, the value of X could be computed by multiplying N times μ. However, because μ is unknown, a point estimate of X is obtained by multiplying N times $\bar{x}$. We will denote the point estimator of X as $\hat{X}$.

Point Estimator of a Population Total

$$\hat{X} = N\bar{x} \tag{21.5}$$

An estimate of the standard error of this point estimator is given by

$$s_{\hat{X}} = Ns_{\bar{x}} \tag{21.6}$$

where

$$s_{\bar{x}} = \sqrt{\frac{N-n}{N}}\left(\frac{s}{\sqrt{n}}\right) \tag{21.7}$$

Note that (21.7) is the formula for the estimated standard error of the mean. With this standard error and (21.6), an approximate 95% confidence interval for the population total is given by the following expression.

Approximate 95% Confidence Interval Estimate of the Population Total

$$N\bar{x} \pm 2s_{\hat{X}} \tag{21.8}$$

Suppose that in the NEG study a simple random sample of $n = 50$ public schools is selected from the population of $N = 500$ schools; the sample mean is $\bar{x} = 22{,}000$ square feet and the sample standard deviation is $s = 4000$ square feet. Using (21.5), we find that the point estimator of the population total is

$$\hat{X} = (500)(22{,}000) = 11{,}000{,}000$$

Equation (21.7) can be used to compute an estimate of the standard error of the mean.

$$s_{\bar{x}} = \sqrt{\frac{500-50}{500}}\left(\frac{4000}{\sqrt{50}}\right) = 536.66$$

Then, using (21.6), we can obtain an estimate of the standard error of $\hat{X}$.

$$s_{\hat{X}} = (500)(536.66) = 268{,}330$$

Therefore, using (21.8), we find that an approximate 95% confidence interval estimate of the total square footage for the 500 public schools in NEG's service area is

$$11{,}000{,}000 \pm 2(268{,}330) = 11{,}000{,}000 \pm 536{,}660$$

or 10,463,340 to 11,536,660 square feet.

Population Proportion

The population proportion p is the fraction of the elements in the population with some characteristic of interest. In a market research study, for example, one might be interested in the proportion of consumers preferring a certain brand of product. The sample proportion $\bar{p}$ is an unbiased point estimator of the population proportion. An estimate of the standard error of the proportion is given by

$$s_{\bar{p}} = \sqrt{\left(\frac{N - n}{N}\right)\left(\frac{\bar{p}(1 - \bar{p})}{n - 1}\right)} \tag{21.9}$$

An approximate 95% confidence interval estimate of the population proportion is given by the following expression.

Approximate 95% Confidence Interval Estimate of the Population Proportion

$$\bar{p} \pm 2s_{\bar{p}} \tag{21.10}$$

As an illustration, suppose that in the Northeast Electric and Gas sampling problem, NEG would also like to estimate the proportion of the 500 public schools in its service area that use natural gas as fuel for heating. If 35 of the 50 sampled schools indicate the use of natural gas, the point estimate of the proportion of the 500 schools in the population that use natural gas is $\bar{p} = 35/50 = .70$. Using (21.9), we can compute an estimate of the standard error of the proportion.

$$s_{\bar{p}} = \sqrt{\left(\frac{500 - 50}{500}\right)\left(\frac{.7(1 - .7)}{50 - 1}\right)} = .0621$$

Therefore, using (21.10), we find that an approximate 95% confidence interval for the population proportion is

$$.7 \pm 2(.0621) = .7 \pm .1242$$

or .5758 to .8242.

As this example has shown, the width of the confidence interval can be rather large when one is estimating a population proportion. In general, large sample sizes are needed to obtain precise estimates of population proportions. A report of a sample survey of 529 mutual fund investors conducted by Louis Harris & Associates stated, "Results should be accurate to within 4.3 percentage points" (*Business Week,* August 15, 1994). This implies an approximate 95% confidence interval with width of .086. For large populations, samples of $n = 1200$ or more are often used.

Determining the Sample Size

An important consideration in sample design is the choice of sample size. The best choice usually involves a trade-off between cost and precision. Larger samples provide greater precision (tighter bounds on the sampling error), but are more costly. Often the budget for a study will dictate how large the sample can be. In other cases, the size of the sample must be large enough to provide a specified level of precision.

A common approach to choosing the sample size is to first specify the precision desired and then determine the smallest sample size providing that precision. In this context, the term *precision* refers to the size of the approximate confidence interval; smaller confidence intervals provide more precision. Because the size of the approximate confidence interval depends on the bound B on the sampling error, choosing a level of precision amounts to choosing a value for B. Let us see how this approach works in choosing the sample size necessary to estimate the population mean.

Equation (21.2) showed that the estimate of the standard error of the mean is

$$s_{\bar{x}} = \sqrt{\frac{N-n}{N}}\left(\frac{s}{\sqrt{n}}\right)$$

Recall that the bound on the sampling error is "2 times the estimate of the standard error of the point estimator." Thus,

$$B = 2\sqrt{\frac{N-n}{N}}\left(\frac{s}{\sqrt{n}}\right) \tag{21.11}$$

Solving (21.11) for n will provide a bound on the sampling error equal to B. Doing so yields

$$n = \frac{Ns^2}{N\left(\frac{B^2}{4}\right) + s^2} \tag{21.12}$$

Once a desired level of precision has been selected (by choosing a value for B), (21.12) can be used to find the value of n that will provide the desired level of precision. Using (21.12) to choose n for a practical study presents problems, however. In addition to specifying the desired bound on the sampling error B, one must know the sample variance s^2. But, s^2 will not be known until the sample is actually taken.

Cochran* suggests several ways to develop an estimate of s^2 in practice. Three of them are stated as the following:

1. Take the sample in two stages. Use the value of s^2 found in stage 1 in (21.12); the resulting value of n is what the size of the total sample must be. Then, select the number of additional units needed at stage 2 to provide the total sample size determined in stage 1.
2. Use the results of a pilot survey or pretest to estimate s^2.
3. Use information from a previous sample.

Let us now consider an example involving the estimate of the population mean for starting salaries of graduates of a particular university. Suppose there are $N = 5000$ graduates, and we want to develop an approximate 95% confidence interval with a width of at most \$1000. To provide such a confidence interval, $B = 500$. Before using (21.12) to determine the sample size, we need an estimate of s^2. Suppose a study of starting salaries conducted last year found that $s = \$3000$. We can use the data from this previous sample to estimate s^2. Using $B = 500$, $s = 3000$, and $N = 5000$, we can now use (21.12) to determine the sample size.

$$\begin{aligned} n &= \frac{5000(3000)^2}{5000\left(\frac{(500)^2}{4}\right) + (3000)^2} \\ &= 139.97 \end{aligned}$$

*William G. Cochran, *Sampling Techniques*, 3rd ed., Wiley, 1977.

Rounding up, we see that a sample size of 140 will provide an approximate 95% confidence interval of width \$1000. Keep in mind, however, that this calculation is based on the initial estimate of $s = \$3000$. If s turns out to be larger in this year's sample survey, the resulting approximate confidence interval will have a width greater than \$1000. Consequently, if cost considerations permit, the survey designer might choose a sample size of, say, 150 to provide added assurance that the final approximate 95% confidence interval will have a width less than \$1000.

The formula for determining the sample size necessary for estimating a population total with a bound B on the sampling error is as follows.

$$n = \frac{Ns^2}{\left(\frac{B^2}{4N}\right) + s^2} \tag{21.13}$$

In the previous example, we wanted to estimate the mean starting salary with a bound on the sampling error of $B = 500$. Suppose we are also interested in estimating the total salary of all 5000 graduates with a bound of \$2 million. We can use (21.13) with $B = 2{,}000{,}000$ to find the sample size needed to provide such a bound on the population total.

$$\begin{aligned} n &= \frac{5000(3000)^2}{\frac{(2{,}000{,}000)^2}{4(5000)} + (3000)^2} \\ &= 215.31 \end{aligned}$$

Rounding up, we see that a sample size of 216 is necessary to provide an approximate 95% confidence interval with a bound of \$2 million. We note here that if the same survey is expected to provide a bound of \$500 on the population mean and a bound of \$2 million on the population total, a sample size of at least 216 must be used. This size will provide a tighter bound than necessary on the population mean, while providing the minimum desired precision for the population total.

To choose the sample size for estimating a population proportion, we use a formula similar to the one for the population mean. We simply substitute $\bar{p}(1 - \bar{p})$ for s^2 in (21.12) to obtain

$$n = \frac{N\bar{p}(1 - \bar{p})}{N\left(\frac{B^2}{4}\right) + \bar{p}(1 - \bar{p})} \tag{21.14}$$

To use (21.14), we must specify the desired bound B and provide an estimate of $\bar{p}$. If a good estimate of $\bar{p}$ is not available, we can use $\bar{p} = .5$; it will ensure that the resulting approximate confidence interval will have a bound on the sampling error at least as small as desired.

EXERCISES

Methods

1. Simple random sampling has been used to obtain a sample of $n = 50$ elements from a population of $N = 800$. The sample mean was $\bar{x} = 215$, and the sample standard deviation was found to be $s = 20$.
   a. Estimate the population mean.
   b. Estimate the standard error of the mean.
   c. Develop an approximate 95% confidence interval for the population mean.

2. Simple random sampling has been used to obtain a sample of $n = 80$ elements from a population of $N = 400$. The sample mean was $\bar{x} = 75$, and the sample standard deviation was found to be $s = 8$.
   a. Estimate the population total.
   b. Estimate the standard error of the population total.
   c. Develop an approximate 95% confidence interval for the population total.
3. Simple random sampling has been used to obtain a sample of $n = 100$ elements from a population of $N = 1000$. The sample proportion was $\bar{p} = .30$.
   a. Estimate the population proportion.
   b. Estimate the standard error of the proportion.
   c. Develop an approximate 95% confidence interval for the population proportion.
4. A sample is to be taken to develop an approximate 95% confidence interval estimate of the population mean. The population consists of 450 elements, and a pilot study has resulted in $s = 70$. How large must the sample be if we want to develop an approximate 95% confidence interval with a width of 30?

Applications

5. In 1996, the Small Business Administration (SBA) granted 771 government-guaranteed loans to small businesses in North Carolina. (*The Wall Street Journal Almanac,* 1998). Suppose a sample of 50 small businesses showed an average loan of $149,670 with a standard deviation of $73,420 and that 18 of the businesses in the sample were manufacturing companies.
   a. Develop an approximate 95% confidence interval for the population mean value of a loan.
   b. Develop an approximate 95% confidence interval for the total value of all 771 SBA loans in North Carolina.
   c. In the sample, 18 of the 50 businesses were involved in manufacturing. Develop an approximate 95% confidence interval for the proportion of loans involving manufacturing companies.
6. A county in California had 724 corporate tax returns filed. The mean annual income reported was $161,220 with a standard deviation of $31,300. How large a sample will be necessary next year to develop an approximate 95% confidence interval for mean annual corporate earnings? The precision required is an interval width of no more than $5000.

21.5 STRATIFIED SIMPLE RANDOM SAMPLING

In stratified simple random sampling, the population is first divided into H groups, called strata. Then for stratum h a simple random sample of size n_h is selected. The data from the H simple random samples are combined to develop an estimate of a population parameter such as the population mean, total, or proportion.

If the variability within each stratum is smaller than the variability across the strata, a stratified simple random sample can lead to greater precision (narrower interval estimates of the population parameters). The basis for forming the various strata depends on the judgment of the designer of the sample. Depending on the application, a population might be stratified by department, location, age, product type, industry type, sales levels, and so on.

As an example, suppose the College of Business at Lakeside College wants to conduct a survey of this year's graduating class to learn about their starting salaries. There are five majors in the college: accounting, finance, information systems, marketing, and operations management. Of the $N = 1500$ students who graduated this year, there were $N_1 = 500$ accounting majors, $N_2 = 350$ finance majors, $N_3 = 200$ information systems majors, $N_4 = 300$ marketing majors, and $N_5 = 150$ operations management majors. Analysis of previous salary data suggests more variability in starting salaries across majors than within each major. As a result, a stratified simple random sample of $n = 180$ students is selected; 45 of the 180 students majored in accounting ($n_1 = 45$), 40 majored in finance ($n_2 = 40$), 30 majored in information systems ($n_3 = 30$), 35 majored in marketing ($n_4 = 35$), and 30 majored in operations management ($n_5 = 30$).

Population Mean

In stratified sampling an unbiased estimate of the population mean is obtained by computing a weighted average of the sample means for each stratum. The weights used are the fraction of the population in each stratum. The resulting point estimator, denoted $\bar{x}_{\text{st}}$, is defined as follows.

Point Estimator of the Population Mean

$$\bar{x}_{\text{st}} = \sum_{h=1}^{H} \left(\frac{N_h}{N}\right)\bar{x}_h \tag{21.15}$$

H = number of strata
$\bar{x}_h$ = sample mean for stratum h
N_h = number of elements in the population in stratum h
N = total number of elements in the population; $N = N_1 + N_2 + \cdots + N_H$

For stratified simple random sampling, the formula for computing an estimate of the standard error of the mean is a function of s_h, the sample standard deviation for stratum h.

$$s_{\bar{x}_{\text{st}}} = \sqrt{\frac{1}{N^2} \sum_{h=1}^{H} N_h(N_h - n_h)\frac{s_h^2}{n_h}} \tag{21.16}$$

Using these results, we see that an approximate 95% confidence interval estimate of the population mean is given by the following expression.

Approximate 95% Confidence Interval Estimate of the Population Mean

$$\bar{x}_{\text{st}} \pm 2s_{\bar{x}_{\text{st}}} \tag{21.17}$$

Suppose the survey of 180 graduates of the College of Business at Lakeland College provided the sample results shown in Table 21.1. The sample means for each major, or stratum, are \$35,000 for accounting, \$33,500 for finance, \$41,500 for information systems, \$32,000 for marketing, and \$36,000 for operations management. Using these results and (21.15), we can compute a point estimate of the population mean.

$$\bar{x}_{\text{st}} = \left(\frac{500}{1500}\right)(35{,}000) + \left(\frac{350}{1000}\right)(33{,}500) + \left(\frac{200}{1500}\right)(41{,}500)$$
$$+ \left(\frac{300}{1500}\right)(32{,}000) + \left(\frac{150}{1500}\right)(36{,}000) = 35{,}017$$

TABLE 21.1 LAKELAND COLLEGE SAMPLE SURVEY OF STARTING SALARIES OF GRADUATES

Major (h)	$\bar{x}_h$	s_h	N_h	n_h
Accounting	\$35,000	2000	500	45
Finance	\$33,500	1700	350	40
Information systems	\$41,500	2300	200	30
Marketing	\$32,000	1600	300	35
Operations management	\$36,000	2250	150	30

In Table 21.2 we show a portion of the calculations needed to estimate the standard error; note that

$$\sum_{h=1}^{5} N_h(N_h - n_h)\frac{s_h^2}{n_h} = 42{,}909{,}037{,}698$$

Thus,

$$s_{\bar{x}_{\text{st}}} = \sqrt{\left(\frac{1}{(1500)^2}\right)(42{,}909{,}037{,}698)} = \sqrt{19{,}070.68} = 138$$

Hence, using (21.17), we find that an approximate 95% confidence interval estimate of the population mean is 35,017 ± 2(138) = 35,017 ± 276, or \$34,741 to \$35,293.

TABLE 21.2 PARTIAL CALCULATIONS FOR THE ESTIMATE OF THE STANDARD ERROR OF THE MEAN FOR THE LAKELAND COLLEGE SAMPLE SURVEY

Major	h	$N_h(N_h - n_h)\frac{s_h^2}{n_h}$
Accounting	1	$500(500 - 45)\frac{(2000)^2}{45} = 20{,}222{,}222{,}222$
Finance	2	$350(350 - 40)\frac{(1700)^2}{40} = 7{,}839{,}125{,}000$
Information systems	3	$200(200 - 30)\frac{(2300)^2}{30} = 5{,}995{,}333{,}333$
Marketing	4	$300(300 - 35)\frac{(1600)^2}{35} = 5{,}814{,}857{,}143$
Operations management	5	$150(150 - 30)\frac{(2250)^2}{30} = 3{,}037{,}500{,}000$
		42,909,037,698

$$\sum_{h=1}^{5} N_h(N_h - n_h)\frac{s_h^2}{n_h}$$

Population Total

The point estimator of the population total (X) is obtained by multiplying N times $\bar{x}_{st}$.

Point Estimator of the Population Total

$$\hat{X} = N\bar{x}_{st} \tag{21.18}$$

An estimate of the standard error of this point estimator is

$$s_{\hat{X}} = Ns_{\bar{x}_{st}} \tag{21.19}$$

Thus, an approximate 95% confidence interval for the population total is given by the following expression.

Approximate 95% Confidence Interval Estimate of the Population Total

$$N\bar{x}_{st} \pm 2s_{\hat{X}} \tag{21.20}$$

Now suppose the College of Business at Lakeland College would also like to estimate the total earnings of the 1500 business graduates in order to estimate their impact on the economy. Using (21.18), we obtain an unbiased estimate of the total earnings.

$$\hat{X} = (1500)35{,}017 = 52{,}525{,}500$$

Using (21.19), we obtain an estimate of the standard error of the population total.

$$s_{\hat{X}} = 1500(138) = 207{,}000$$

Thus, using (21.20), we find that an approximate 95% confidence interval estimate of the total earnings of the 1500 graduates is 52,525,500 $\pm$ 2(207,000) = 52,525,500 $\pm$ 414,000 or \$52,111,500 to \$52,939,500.

Population Proportion

An unbiased estimate of the population proportion, p, for stratified simple random sampling is a weighted average of the proportions for each stratum. The weights used are the fraction of the population in each stratum. The resulting point estimator, denoted $\bar{p}_{st}$, is defined as follows.

Point Estimator of the Population Proportion

$$\bar{p}_{st} = \sum_{h=1}^{H} \left(\frac{N_h}{N}\right)\bar{p}_h \tag{21.21}$$

where

H = the number of strata
$\bar{p}_h$ = the sample proportion for stratum h
N_h = the number of elements in the population in stratum h
N = the total number of elements in the population: $N = N_1 + N_2 + \cdots + N_H$

An estimate of the standard error of $\bar{p}_{st}$ is given by

$$s_{\bar{p}_{st}} = \sqrt{\frac{1}{N^2}\sum_{h=1}^{H} N_h(N_h - n_h)\left[\frac{\bar{p}_h(1-\bar{p}_h)}{n_h - 1}\right]} \quad \textbf{(21.22)}$$

Thus, an approximate 95% confidence interval estimate of the population proportion is given by the following expression.

Approximate 95% Confidence Interval Estimate of the Population Proportion

$$\bar{p}_{st} \pm 2s_{\bar{p}_{st}} \quad \textbf{(21.23)}$$

In the Lakeland College survey, the college wants to know the proportion of graduates receiving a starting salary of \$36,000 or more. The results of the sample survey of 180 graduates show that 63 received starting salaries of \$36,000 or more and that 16 of the 63 majored in accounting, 3 majored in finance, 29 majored in information systems, 0 majored in marketing, and 15 majored in operations management.

Using (21.21), we can compute the point estimate of the proportion receiving starting salaries of \$36,000 or more.

$$\begin{aligned}\bar{p}_{st} &= \left(\frac{500}{1500}\right)\left(\frac{16}{45}\right) + \left(\frac{350}{1500}\right)\left(\frac{3}{40}\right) + \left(\frac{200}{1500}\right)\left(\frac{29}{30}\right) + \left(\frac{300}{1500}\right)\left(\frac{0}{35}\right) + \left(\frac{150}{1500}\right)\left(\frac{15}{30}\right)\\ &= .3149\end{aligned}$$

In Table 21.3 we show a portion of the calculations needed to estimate the standard error; note that

$$\sum_{h=1}^{5} N_h(N_h - n_h)\left[\frac{\bar{p}_h(1-\bar{p}_h)}{n_h - 1}\right] = 1570.6913$$

Thus,

$$\begin{aligned}s_{\bar{p}_{st}} &= \sqrt{\frac{1}{(1500)^2}(1570.6913)}\\ &= .0264\end{aligned}$$

Using (21.23), we find that an approximate 95% confidence interval for the proportion of graduates receiving starting salaries of \$36,000 or more is $.3149 \pm 2(.0264) = .3149 \pm .0528$, or .2621 to .3677.

Determining the Sample Size

With stratified simple random sampling we can think of choosing a sample size as a two-step process. First, a total sample size n must be chosen. Second, we must decide how to assign the sampled units to the various strata. Alternatively, we could first decide how large a sample to take in each stratum and then sum the stratum sample sizes to obtain the total sample size. It is often of interest to develop estimates of the mean, total, and proportion for the individual strata, therefore a combination of these two approaches is often employed. An overall sample size n and an allocation that will provide the necessary precision for the overall population parameter of interest are found. Then, if the sample sizes in some

TABLE 21.3 PARTIAL CALCULATIONS FOR THE ESTIMATE OF THE STANDARD ERROR OF $\bar{p}_{\text{st}}$ FOR THE LAKELAND COLLEGE SAMPLE SURVEY

Major	h	$N_h(N_h - n_h)\left[\dfrac{\bar{p}_h(1-\bar{p}_h)}{n_h - 1}\right]$
Accounting	1	$500(500-45)\left[\dfrac{(16/45)(29/45)}{45-1}\right] = 1184.7363$
Finance	2	$350(350-40)\left[\dfrac{(3/40)(37/40)}{40-1}\right] = 193.0048$
Information systems	3	$200(200-30)\left[\dfrac{(29/30)(1/30)}{30-1}\right] = 37.7778$
Marketing	4	$300(300-35)\left[\dfrac{(0/35)(35/35)}{35-1}\right] = 0.0000$
Operations management	5	$150(150-30)\left[\dfrac{(15/30)(15/30)}{30-1}\right] = 155.1724$
$\sum_{h=1}^{5} N_h(N_h - n_h)\left[\dfrac{\bar{p}_h(1-\bar{p}_h)}{n_h - 1}\right]$		1570.6913

of the strata are not large enough to provide the precision necessary for the estimates within the strata, the sample sizes for those strata are adjusted upward as necessary. In this section we discuss some of the issues pertinent to allocating the total sample to the various strata and present a method for choosing the total sample size and making the allocation.

The allocation task is to decide what fraction of the total sample should be assigned to each stratum. This fraction determines how large the simple random sample will be in each stratum. The factors considered most important in making the allocation are:

1. The number of elements in each stratum.
2. The variance of the elements within each stratum.
3. The cost of selecting elements within each stratum.

Generally, larger samples should be assigned to the larger strata and to the strata with larger variances. Conversely, to get the most information for a given cost, smaller samples should be allocated to the strata where the cost per unit of sampling is greatest.

The individual stratum variances often differ greatly. For example, suppose that in a particular study we are interested in determining the mean number of employees per building; because variability is greater in a stratum with larger buildings than in one with smaller buildings, a proportionately larger sample should be taken in such a stratum. The cost of selection can be an important consideration when significant interviewer travel between sampled units is necessary in some of the strata but not in others; this issue frequently arises when some of the strata involve rural areas and others involve cities.

In many surveys the cost per unit of sampling is approximately the same for each stratum (e.g., mail and telephone surveys); in such cases, the cost of sampling can be ignored in making the allocation. We present here the appropriate formulas for choosing the sample size and making the allocation in such cases. More advanced texts on sampling provide formulas for the case when sampling costs vary significantly across strata. The formulas we present in this section will minimize the total sampling cost for a given level of precision.

This method, known as *Neyman allocation,* allocates the total sample n to the various strata as follows.

$$n_h = n\left(\frac{N_h s_h}{\sum_{h=1}^{H} N_h s_h}\right) \tag{21.24}$$

Equation (21.24) shows that the number of units allocated to a stratum increases with the stratum size and standard deviation. Note that to make this allocation, we need to first determine the total sample size n. Given a specified level of precision B, we can use the following formulas to choose the total sample size when estimating the population mean and the population total.

Sample Size When Estimating the Population Mean

$$n = \frac{\left(\sum_{h=1}^{H} N_h s_h\right)^2}{N^2\left(\frac{B^2}{4}\right) + \sum_{h=1}^{H} N_h s_h^2} \tag{21.25}$$

Sample Size When Estimating the Population Total

$$n = \frac{\left(\sum_{h=1}^{H} N_h s_h\right)^2}{\frac{B^2}{4} + \sum_{h=1}^{H} N_h s_h^2} \tag{21.26}$$

As an example, suppose a Chevrolet dealer wants to survey the customers who have purchased a Corvette, Geo Prizm, or Cavalier to obtain information the dealer feels will be helpful in determining future advertising. In particular, suppose the dealer wants to estimate the mean monthly income for these customers with a bound on the sampling error of \$100. The dealer's 600 Corvette, Geo Prizm, and Cavalier customers have been divided into three strata: 100 Corvette owners, 200 Geo Prizm owners, and 300 Cavalier owners. A pilot survey was used to estimate the standard deviation in each stratum; the results are $s_1 = \$1300$, $s_2 = \$900$, and $s_3 = \$500$ for the Corvette, Geo Prizm, and Cavalier owners, respectively.

The first step in choosing a sample size for this survey is to use (21.25) to determine the total sample size necessary to provide a bound of $B = \$100$ on the estimate of the population mean. First, we compute

$$\sum_{h=1}^{3} N_h s_h = 100(1300) + 200(900) + 300(500) = 460{,}000$$

Next, we compute

$$\sum_{h=1}^{3} N_h s_h^2 = 100(1300)^2 + 200(900)^2 + 300(500)^2 = 406{,}000{,}000$$

Substituting these values into (21.25), we can determine the total sample size needed to provide a bound on the sampling error of $B = \$100$.

$$n = \frac{(460{,}000)^2}{\dfrac{(600)^2(100)^2}{4} + 406{,}000{,}000} = 162$$

Thus, a total sample size of 162 will provide the precision desired. To allocate the total sample to the three strata, we use (21.24).

$$n_1 = 162\left(\frac{100(1300)}{460{,}000}\right) = 46$$

$$n_2 = 162\left(\frac{200(900)}{460{,}000}\right) = 63$$

$$n_3 = 162\left(\frac{300(500)}{460{,}000}\right) = 53$$

We would therefore recommend sampling 46 Corvette owners, 63 Geo Prizm owners, and 53 Cavalier owners for a total sample size of 162 customers.

To determine the sample size when estimating a population proportion, we simply substitute $\sqrt{\bar{p}_h(1 - \bar{p}_h)}$ for s_h in (21.25); the result is

$$n = \frac{\left(\sum_{h=1}^{H} N_h \sqrt{\bar{p}_h(1 - \bar{p}_h)}\right)^2}{N^2\left(\dfrac{B^2}{4}\right) + \sum_{h=1}^{H} N_h \bar{p}_h(1 - \bar{p}_h)} \qquad \textbf{(21.27)}$$

Once the total sample size for the population proportion estimate has been determined, allocation to the various strata is again made by using (21.24) with $\sqrt{\bar{p}_h(1 - \bar{p}_h)}$ substituted for s_h.

NOTES AND COMMENTS

1. An advantage of stratified simple random sampling is that estimates of population parameters for each stratum are automatically available as a by-product of the sampling procedure. For example, besides obtaining an estimate of the average starting salary for all graduates in the Lakeland College sampling problem, we obtained an estimate of the average starting salary for each major. Because each of the starting salary estimates was based on a simple random sample from each stratum, the procedure for developing an approximate confidence interval estimate when a simple random sample is selected (see Equation (21.4)) can be used to compute an approximate 95% confidence interval estimate for the mean in each stratum. In a similar manner, interval estimates for the population total and the population proportion for each stratum can be developed by using (21.8) and (21.10), respectively.
2. Another type of allocation that is sometimes used with stratified simple random sampling is called *proportional allocation*. In this approach, the sample size allocated to each stratum is given by the following formula.

$$n_h = n\left(\frac{N_h}{N}\right) \qquad \textbf{(21.28)}$$

Proportional allocation is appropriate when the stratum variances are approximately equal and the cost per unit of sampling is about the same across strata. In the case where the stratum variances are equal, proportional allocation and the Neyman procedure result in the same allocation.

EXERCISES

Methods

7. A stratified simple random sample has been taken with the following results.

Stratum (h)	$\bar{x}_h$	s_h	$\bar{p}_h$	N_h	n_h
1	138	30	.50	200	20
2	103	25	.78	250	30
3	210	50	.21	100	25

a. Develop an estimate of the population mean for each stratum.
b. Develop an approximate 95% confidence interval for the population mean in each stratum.
c. Develop an approximate 95% confidence interval for the overall population mean.

8. Reconsider the sample results in Exercise 7.
a. Develop an estimate of the population total for each stratum.
b. Develop a point estimate of the total for all 550 elements in the population.
c. Develop an approximate 95% confidence interval for the population total.

9. Reconsider the sample results in Exercise 7.
a. Develop an approximate 95% confidence interval for the proportion in each stratum.
b. Develop a point estimate of the population proportion for the 550 elements in the population.
c. Estimate the standard error of the population proportion.
d. Develop an approximate 95% confidence interval for the population proportion.

10. A population has been divided into three strata with $N_1 = 300$, with $N_2 = 600$, and $N_3 = 500$. From a past survey, the following estimates for the standard deviations in the three strata are available: $s_1 = 150$, $s_2 = 75$, $s_3 = 100$.
a. Suppose an estimate of the population mean with a bound on the error of estimate of $B = 20$ is required. How large must the sample be? How many elements should be allocated to each stratum?
b. Suppose a bound of $B = 10$ is desired. How large must the sample be? How many elements should be allocated to each stratum?
c. Suppose an estimate of the population total with a bound of $B = 15{,}000$ is requested. How large must the sample be? How many elements should be allocated to each stratum?

Applications

11. A drug store chain has stores in four cities: 38 stores in Indianapolis, 45 in Louisville, 80 in St. Louis, and 70 in Memphis. Pharmacy sales in the four cities vary considerably because of the competition. The following sales data (in $1000s) are available from a sample survey. Each of the cities was considered a separate stratum, and a stratified simple random sample was taken.

Indianapolis	Louisville	St. Louis	Memphis
50.3	48.7	16.7	14.7
41.2	59.8	38.4	88.3
15.7	28.9	51.6	94.2
22.5	36.5	42.7	76.8
26.7	89.8	45.0	35.1
20.8	96.0	59.7	48.2
	77.2	80.0	57.9
	81.3	27.6	18.8
			22.0
			74.3

Continued

a. Estimate the mean sales for each city (stratum).
b. Develop an approximate 95% confidence interval for the mean sales in each city.
c. Estimate the proportion of stores with sales of \$50,000 or more.
d. Develop an approximate 95% confidence interval for the proportion of stores with sales of \$50,000 or more.

12. Reconsider the sample survey results in Exercise 11.
a. Estimate the population total for St. Louis.
b. Estimate the population total for Indianapolis.
c. Develop an approximate 95% confidence interval for mean pharmacy sales for the drug store chain.
d. Develop an approximate 95% confidence interval for total pharmacy sales for the drug store chain.

13. An accounting firm has a number of clients in the banking, insurance, and brokerage industries. There are $N_1 = 50$ banks, $N_2 = 38$ insurance companies, and $N_3 = 35$ brokerage firms. A marketing research firm has been hired to survey the accounting firm's clients in these three industries. The survey will ask a variety of questions about both the clients' businesses and their satisfaction with services provided by the accounting firm. Suppose an approximate 95% confidence interval is requested for the mean number of employees for the 123 clients with a bound on the error of estimation of $B = 30$.
a. Suppose a pilot study finds $s_1 = 80$, $s_2 = 150$, and $s_3 = 45$. Choose a total sample size, and explain how the sample size should be allocated to the three strata.
b. Suppose the pilot test is called into question and a decision is made to assume the stratum standard deviations are all equal to 100 in choosing the sample size. Choose a total sample size, and determine how many elements should be sampled in each stratum.

21.6 CLUSTER SAMPLING

Cluster sampling requires that the population be divided into N groups of elements called clusters such that each element in the population belongs to one and only one cluster. For example, suppose we want to survey registered voters in the state of Ohio. One approach would be to develop a frame consisting of all registered voters in the state of Ohio and then select a simple random sample of voters from this frame. Alternatively, in cluster sampling, we might choose to define the frame as the list of the $N = 88$ counties in the state (see Figure 21.1). In this approach, each county or cluster would consist of a group of registered voters, and each registered voter in the state would belong to one and only one cluster.

Suppose we select a simple random sample of $n = 12$ of the 88 counties. At this point, we could collect data for *all* registered voters in each of the 12 sampled clusters, an approach referred to as *single-stage cluster sampling,* or we could select a simple random sample of registered voters from each of the 12 sampled clusters, an approach referred to as *two-stage cluster sampling.* In either case, formulas are available for using the sample results to develop point and interval estimates of population parameters such as the population mean, total, or proportion. In this chapter, however, we consider only single-stage cluster sampling; more advanced texts on sampling present results for two-stage cluster sampling.

In the sense that both stratified and cluster sampling divide the population into groups of elements, the two sampling procedures are similar. The reasons for choosing cluster sampling, however, differ from the reasons for choosing stratified sampling. Cluster sampling tends to provide better results when the elements within the clusters are heterogeneous (not alike). In the ideal case, each cluster would be a small-scale version of the entire population. In this case, sampling a small number of clusters would provide good information about the characteristics of the entire population.

One of the primary applications of cluster sampling involves area sampling, where the clusters are counties, townships, city blocks, or other well-defined geographic sections of

FIGURE 21.1 COUNTIES OF THE STATE OF OHIO USED AS CLUSTERS OF REGISTERED VOTERS

All the elements in the population do not need to be listed. With cluster sampling, one needs only a list of the elements in the clusters sampled.

the population. Because data are collected from only a sample of the total geographic areas or clusters available, and the elements within the clusters are typically close to one another, significant savings in time and cost can be realized when a data collector or interviewer is sent to a sampled unit. As a result, even if a larger total sample size is required, cluster sampling may be less costly than either simple random sampling or stratified simple random sampling. In addition, cluster sampling can minimize the time and cost associated with developing the frame or list of elements to be sampled because cluster sampling does not require that a list of every element in the population be developed. One needs only a list of the elements in the clusters sampled.

To illustrate cluster sampling, let us consider a survey conducted by the CPA (certified public accountant) Society of the 12,000 practicing CPAs in a particular state. As part of the survey, the CPA Society collected information on income, gender, and factors related to the CPA's lifestyle. Because personal interviews were needed to obtain all the desired information, the CPA Society used a cluster sample to minimize the total travel and interviewing cost. The frame consisted of all CPA firms that were registered to practice accounting in the state. Suppose there are $N = 1000$ clusters, or CPA firms, registered to practice accounting in the state, and that a simple random sample of $n = 10$ CPA firms is to be selected.

In presenting the formulas for cluster sampling that are needed to develop approximate 95% confidence interval estimates of the population mean, total, and proportion, we will use the following notation.

$$\begin{aligned}
N &= \text{number of clusters in the population} \\
n &= \text{number of clusters selected in the sample} \\
M_i &= \text{number of elements in cluster } i \\
M &= \text{number of elements in the population; } M = M_1 + M_2 + \cdots + M_N \\
\bar{M} &= M/N = \text{average number of elements in a cluster} \\
x_i &= \text{total of all observations in cluster } i \\
a_i &= \text{number of observations in cluster } i \text{ with a certain characteristic}
\end{aligned}$$

For the CPA Society sample survey we have the following information.

$$\begin{aligned}
N &= 1000 \\
n &= 10 \\
M &= 12{,}000 \\
\bar{M} &= 12{,}000/1000 = 12
\end{aligned}$$

Table 21.4 shows the values of M_i and x_i for each of the sampled clusters as well as the number of female CPAs in the sampled firms (a_i).

Population Mean

The point estimator of the population mean obtained from cluster sampling is given by the following formula.

Point Estimator of the Population Mean

$$\bar{x}_c = \frac{\sum_{i=1}^{n} x_i}{\sum_{i=1}^{n} M_i} \tag{21.29}$$

TABLE 21.4 RESULTS OF CPA SAMPLE SURVEY

Firm (i)	CPAs (M_i)	Total Salary ($1000s) for Firm i (x_i)	Female CPAs (a_i)
1	8	384	2
2	25	1350	8
3	4	148	0
4	17	857	6
5	7	296	1
6	3	131	2
7	15	761	2
8	4	176	0
9	12	577	5
10	33	1880	9
Totals	128	6560	35

An estimate of the standard error of this point estimator is

$$s_{\bar{x}_c} = \sqrt{\left(\frac{N - n}{Nn\bar{M}^2}\right) \frac{\sum_{i=1}^{n}(x_i - \bar{x}_c M_i)^2}{n - 1}} \qquad \textbf{(21.30)}$$

Thus, the following expression gives an approximate 95% confidence interval estimate of the population mean.

Approximate 95% Confidence Interval Estimate of the Population Mean

$$\bar{x}_c \pm 2s_{\bar{x}_c} \qquad \textbf{(21.31)}$$

Using the data in Table 21.4, we obtain an estimate of the mean salary for practicing certified public accountants.

$$\bar{x}_c = \frac{6560}{128} = 51.250$$

The salary data in Table 21.4, listed in thousands of dollars, indicate an estimate of the mean salary for practicing certified public accountants in the state is $51,250.

In Table 21.5, we show a portion of the calculations needed to estimate the standard error; note that

$$\sum_{i=1}^{n}(x_i - \bar{x}_c M_i)^2 = 51{,}281.378$$

Thus,

$$s_{\bar{x}_c} = \sqrt{\left[\frac{1000 - 10}{(1000)(10)(12)^2}\right] \frac{51{,}281.378}{10 - 1}} = 1.979$$

TABLE 21.5 PARTIAL CALCULATIONS FOR THE ESTIMATE OF THE STANDARD ERROR OF THE MEAN FOR THE CPA SAMPLE SURVEY WHERE $\bar{x}_c = 51.250$

Firm (i)	M_i	x_i	$(x_i - 51.250M_i)^2$	
1	8	384	$[384 - 51.250(8)]^2 =$	676.000
2	25	1350	$[1350 - 51.250(25)]^2 =$	4,726.563
3	4	148	$[148 - 51.250(4)]^2 =$	3,249.000
4	17	857	$[857 - 51.250(17)]^2 =$	203.063
5	7	296	$[296 - 51.250(7)]^2 =$	3,937.563
6	3	131	$[131 - 51.250(3)]^2 =$	517.563
7	15	761	$[761 - 51.250(15)]^2 =$	60.063
8	4	176	$[176 - 51.250(4)]^2 =$	841.000
9	12	577	$[577 - 51.250(12)]^2 =$	1,444.000
10	33	1880	$[1880 - 51.250(33)]^2 =$	35,626.563
Totals	128	6560		51,281.378

$\sum_{i=1}^{n}(x_i - \bar{x}_c M_i)^2$ → 51,281.378

Hence, the standard error is \$1979. Using (21.31), we find that an approximate 95% confidence interval estimate for the mean annual salary is 51,250 ± 2(1979) = 51,250 ± 3958 or \$47,292 to \$55,208.

Population Total

The point estimator of the population total X is obtained by multiplying M times $\bar{x}_c$.

Point Estimator of the Population Total

$$\hat{X} = M\bar{x}_c \tag{21.32}$$

An estimate of the standard error of this point estimator is

$$s_{\hat{X}} = M s_{\bar{x}_c} \tag{21.33}$$

Thus, an approximate 95% confidence interval estimate for the population total is given by the following expression.

Approximate 95% Confidence Interval Estimate of the Population Total

$$M\bar{x}_c \pm 2s_{\hat{X}} \tag{21.34}$$

For the CPA sample survey,

$$\hat{X} = M\bar{x}_c = 12{,}000(51{,}250) = \$615{,}000{,}000$$
$$s_{\hat{X}} = M s_{\bar{x}_c} = 12{,}000(1979) = \$23{,}748{,}000$$

Thus, using (21.34), we find that an approximate 95% confidence interval is \$615,000,000 ± 2(\$23,748,000) = \$615,000,000 ± \$47,496,000 or \$567,504,000 to \$662,496,000.

Population Proportion

The point estimator of the population proportion obtained from cluster sampling follows.

Point Estimator of the Population Proportion

$$\bar{p}_c = \frac{\sum_{i=1}^{n} a_i}{\sum_{i=1}^{n} M_i} \tag{21.35}$$

where

a_i = number of elements in cluster i with the characteristic of interest

An estimate of the standard error of this point estimator is

$$s_{\bar{p}_c} = \sqrt{\left(\frac{N-n}{Nn\bar{M}^2}\right)\frac{\sum_{i=1}^{n}(a_i - \bar{p}_c M_i)^2}{n-1}} \tag{21.36}$$

Thus, an approximate 95% confidence interval estimate for the population proportion is given by the following expression.

Approximate 95% Confidence Interval Estimate of the Population Proportion

$$\bar{p}_c \pm 2s_{\bar{p}_c} \tag{21.37}$$

For the CPA sample survey, we can use (21.35) and the data in Table 21.4 to develop an estimate of the number of practicing certified public accountants who are women.

$$\bar{p}_c = \frac{2 + 8 + \cdots + 9}{8 + 25 + \cdots + 33} = \frac{35}{128} = .2734$$

In Table 21.6 we show a portion of the calculations needed to estimate the standard error; note that

$$\sum_{i=1}^{n}(a_i - \bar{p}_c M_i)^2 = 15.2098$$

Thus,

$$s_{\bar{p}_c} = \sqrt{\left[\frac{1000 - 10}{(1000)(10)(12)^2}\right]\frac{15.2098}{10 - 1}} = .0341$$

TABLE 21.6 PARTIAL CALCULATIONS FOR THE ESTIMATE OF THE STANDARD ERROR OF $\bar{p}_c$ FOR THE CPA SAMPLE SURVEY WHERE $\bar{p}_c = .2734$

Firm (i)	M_i	a_i	$(a_i - .2734M_i)^2$	
1	8	2	$[2 - .2734(8)]^2 =$	.0350
2	25	8	$[8 - .2734(25)]^2 =$	1.3572
3	4	0	$[0 - .2734(4)]^2 =$	1.1960
4	17	6	$[6 - .2734(17)]^2 =$	1.8284
5	7	1	$[1 - .2734(7)]^2 =$	.8350
6	3	2	$[2 - .2734(3)]^2 =$	1.3919
7	15	2	$[2 - .2734(15)]^2 =$	4.4142
8	4	0	$[0 - .2734(4)]^2 =$	1.1960
9	12	5	$[5 - .2734(12)]^2 =$	2.9556
10	33	9	$[9 - .2734(33)]^2 =$	.0005
Totals	128	35		15.2098

$\sum_{i=1}^{n}(a_i - \bar{p}_c M_i)^2$ → 15.2098

Hence, using (21.37), we find that an approximate 95% confidence interval for the proportion of practicing CPAs who are women is $.2734 \pm 2(.0341) = .2734 \pm .0682$ or .2052 to .3416.

Determining the Sample Size

Once the clusters have been formed, the primary issue in choosing a sample size is selecting the number of clusters n. The procedure for cluster sampling is similar to that for other methods of sampling. An acceptable level of precision is specified by choosing a value for B, the bound on the sampling error. Then a formula is developed for finding the value of n that will provide the desired precision.

The average cluster size and the variance between clusters are key factors in deciding how many clusters to include in the sample. If the clusters are similar, the variance between them will be small and the number of clusters sampled can be smaller. Also, if the average number of elements per cluster is larger, the number of clusters sampled can be smaller. The formulas for making an exact determination of sample size are included in more advanced texts on sampling.

EXERCISES

Methods

14. A sample of four clusters is to be taken from a population with $N = 25$ clusters and $M = 300$ elements. The values of M_i, x_i, and a_i for each cluster in the sample follow.

Cluster (i)	M_i	x_i	a_i
1	7	95	1
2	18	325	6
3	15	190	6
4	10	140	2
Totals	50	750	15

a. Develop point estimates of the population mean, total, and proportion.
b. Estimate the standard error for the estimates in part (a).
c. Develop an approximate 95% confidence interval for the population mean.
d. Develop an approximate 95% confidence interval for the population total.
e. Develop an approximate 95% confidence interval for the population proportion.

15. A sample of six clusters is to be taken from a population with $N = 30$ clusters and $M = 600$ elements. The following table shows values of M_i, x_i, and a_i for each cluster in the sample.

Cluster (i)	M_i	x_i	a_i
1	35	3,500	3
2	15	965	0
3	12	960	1
4	23	2,070	4
5	20	1,100	3
6	25	1,805	2
Totals	130	10,400	13

a. Develop point estimates of the population mean, total, and proportion.
b. Develop an approximate 95% confidence interval for the population mean.
c. Develop an approximate 95% confidence interval for the population total.
d. Develop an approximate 95% confidence interval for the population proportion.

Applications

16. A public utility is conducting a survey of mechanical engineers to learn more about the factors influencing the choice of heating, ventilation, and air conditioning (HVAC) equipment for new commercial buildings. A total of 120 firms in the utility's service area are engaged in designing HVAC systems. The sampling plan is to use cluster sampling with each firm representing a cluster. For each firm in the sample, all of the mechanical engineers will be interviewed. Approximately 500 mechanical engineers are believed to be employed by the 120 firms. A sample of 10 firms was taken. Among other things, the age of each respondent was recorded as well as whether the respondent had attended the local university.

Cluster (i)	M_i	Total of Respondents' Ages	Number Attending Local University
1	12	520	8
2	1	33	0
3	2	70	1
4	1	29	1
5	6	270	3
6	3	129	2
7	2	102	0
8	1	48	1
9	9	337	7
10	13	462	12
Totals	50	2000	35

a. Estimate the mean age of mechanical engineers engaged in this type of work.

b. Estimate the proportion of mechanical engineers in the utility's service area who attended the local university.

c. Develop an approximate 95% confidence interval for the mean age of mechanical engineers designing HVAC systems for commercial buildings.

d. Develop an approximate 95% confidence interval for the proportion of mechanical engineers in the utility's service area who attended the local university.

17. A national real estate company has just acquired a smaller firm that has 150 offices and 6000 agents in Los Angeles and other parts of Southern California. The national firm has conducted a sample survey to learn about attitudes and other characteristics of its new employees. A sample of eight offices has been taken, and all of the agents at these offices have completed the questionnaire. Results of the survey for the eight offices follow.

Office	Agents	Average Age	College Graduates	Male Agents
1	17	37	3	4
2	35	32	14	12
3	26	36	8	7
4	66	30	38	28
5	43	41	18	12
6	12	52	2	6
7	48	35	20	17
8	57	44	25	26

a. Estimate the mean age of the agents.

b. Estimate the proportion of agents who are college graduates and the proportion who are male.

c. Develop an approximate 95% confidence interval for the mean age of the agents.

d. Develop an approximate 95% confidence interval for the proportion of agents who are college graduates.

e. Develop an approximate 95% confidence interval for the proportion of agents who are male.

21.7 SYSTEMATIC SAMPLING

Systematic sampling is often used as an alternative to simple random sampling. In some sampling situations, especially those with large populations, it can be time-consuming to select a simple random sample by first finding a random number and then counting or searching through the frame until the corresponding element is found. Systematic sampling offers an alternative to simple random sampling in such cases. For example, if a sample size of 50 from a population containing 5000 elements is desired, we might sample one element for every 5000/50 = 100 elements in the population. A systematic sample for this case would involve randomly selecting one of the first 100 elements from the frame. The remaining sample elements are identified by starting with the first sampled element and then selecting every 100th element that follows in the frame. In effect, the sample of 50 is identified by moving systematically through the population and identifying every 100th element after the first randomly selected element. The sample of 50 will often be easier to select in this manner than it would be if simple random sampling were used. The first element selected is a random choice, allowing for the assumption that a systematic sample has the properties of a simple random sample. This assumption is usually appropriate when the frame is a random ordering of the elements in the population.

SUMMARY

We have provided a brief introduction to the field of survey sampling in this chapter. The purpose of survey sampling is to collect data for the purpose of making estimates of population parameters such as the population mean, total, or proportion. Survey sampling as a method of data collection can be contrasted with conducting experiments to generate data. When survey sampling is used, the design of the sampling plan is of critical importance in determining which existing data will be collected. When experiments are employed, experimental design issues are of critical importance in determining which data will be generated, or created.

Two types of errors can occur with sample surveys: sampling error and nonsampling error. Sampling error is the error that occurs because a sample, and not the entire population, is used to estimate a population parameter. Nonsampling error refers to all the other types of errors that can occur, such as measurement, interviewer, nonresponse, and processing error. Nonsampling errors are controlled by proper questionnaire design, thorough training of interviewers, careful verification of data, and so on. Sampling errors are minimized by a proper choice of sample design and by selecting an appropriate sample size.

We discussed four commonly used sample designs in this chapter: simple random sampling, stratified simple random sampling, cluster sampling, and systematic sampling. The objective of sample design is to get the most precise estimates for the least cost. When the population can be divided into strata so that the elements within each stratum are relatively homogeneous, stratified simple random sampling will provide more precision (smaller approximate confidence intervals) than simple random sampling. When the elements can be grouped in clusters so that all the elements in a cluster are close together geographically, cluster sampling often reduces interviewer cost; in these situations, cluster sampling will often provide the most precision per dollar. Systematic random sampling was presented as an alternative to simple random sampling.

GLOSSARY

Element The entity on which data are collected.

Population The collection of all elements of interest.

Sample A subset of the population.

Target population The population about which inferences are made.

Sampled population The population from which the sample is taken.

Sampling unit The units selected for sampling. A sampling unit may include several elements.

Frame A list of the sampling units for a study. The sample is drawn by selecting units from the frame.

Probabilistic sampling Any method of sampling for which the probability of each possible sample can be computed.

Nonprobabilistic sampling Any method of sampling for which the probability of selecting a sample cannot be computed.

Convenience sampling A nonprobabilistic method of sampling whereby elements are selected on the basis of convenience.

Judgment sampling A nonprobabilistic method of sampling whereby element selection is based on the judgment of the person doing the study.

Sampling error The error that occurs because a sample, and not the entire population, is used to estimate a population parameter.

Nonsampling error All types of errors other than sampling error, such as measurement error, interviewer error, and processing error.

Simple random sample A sample selected in such a manner that each sample of size n has the same probability of being selected.

Bound on the sampling error A number added to and subtracted from a point estimate to create an approximate 95% confidence interval. It is given by two times the standard error of the point estimator.

Stratified simple random sampling A probabilistic method of selecting a sample in which the population is first divided into strata and a simple random sample is then taken from each stratum.

Cluster sampling A probabilistic method of sampling in which the population is first divided into clusters and then one or more clusters are selected for sampling. In single-stage cluster sampling, every element in each selected cluster is sampled; in two-stage cluster sampling, a sample of the elements in each selected cluster is collected.

Systematic sampling A method of choosing a sample by randomly selecting the first element and then selecting every kth element thereafter.

KEY FORMULAS

SIMPLE RANDOM SAMPLING

Interval Estimate of the Population Mean

$$\bar{x} \pm z_{\alpha/2}\sigma_{\bar{x}} \tag{21.1}$$

Estimate of the Standard Error of the Mean

$$s_{\bar{x}} = \sqrt{\frac{N - n}{N}}\left(\frac{s}{\sqrt{n}}\right) \tag{21.2}$$

Interval Estimate of the Population Mean

$$\bar{x} \pm z_{\alpha/2} s_{\bar{x}} \tag{21.3}$$

Approximate 95% Confidence Interval Estimate of the Population Mean

$$\bar{x} \pm 2 s_{\bar{x}} \tag{21.4}$$

Point Estimator of a Population Total

$$\hat{X} = N\bar{x} \tag{21.5}$$

Estimate of the Standard Error of $\hat{X}$

$$s_{\hat{X}} = N s_{\bar{x}} \tag{21.6}$$

Approximate 95% Confidence Interval Estimate of the Population Total

$$N\bar{x} \pm 2 s_{\hat{X}} \tag{21.8}$$

Estimate of the Standard Error of the Proportion

$$s_{\bar{p}} = \sqrt{\left(\frac{N - n}{N}\right)\left(\frac{\bar{p}(1 - \bar{p})}{n - 1}\right)} \tag{21.9}$$

Approximate 95% Confidence Interval Estimate of the Population Proportion

$$\bar{p} \pm 2 s_{\bar{p}} \tag{21.10}$$

Sample Size for an Estimate of the Population Mean

$$n = \frac{Ns^2}{N\left(\frac{B^2}{4}\right) + s^2} \tag{21.12}$$

Sample Size for an Estimate of the Population Total

$$n = \frac{Ns^2}{\left(\frac{B^2}{4N}\right) + s^2} \tag{21.13}$$

Sample Size for an Estimate of the Population Proportion

$$n = \frac{N\bar{p}(1 - \bar{p})}{N\left(\frac{B^2}{4}\right) + \bar{p}(1 - \bar{p})} \tag{21.14}$$

STRATIFIED SIMPLE RANDOM SAMPLING

Point Estimator of the Population Mean

$$\bar{x}_{\text{st}} = \sum_{h=1}^{H}\left(\frac{N_h}{N}\right)\bar{x}_h \tag{21.15}$$

Estimate of the Standard Error of the Mean

$$s_{\bar{x}_{st}} = \sqrt{\frac{1}{N^2}\sum_{h=1}^{H} N_h(N_h - n_h)\frac{s_h^2}{n_h}} \tag{21.16}$$

Approximate 95% Confidence Interval Estimate of the Population Mean

$$\bar{x}_{st} \pm 2s_{\bar{x}_{st}} \tag{21.17}$$

Point Estimator of the Population Total

$$\hat{X} = N\bar{x}_{st} \tag{21.18}$$

Estimate of the Standard Error of $\hat{X}$

$$s_{\hat{X}} = Ns_{\bar{x}_{st}} \tag{21.19}$$

Approximate 95% Confidence Interval Estimate of the Population Total

$$N\bar{x}_{st} \pm 2s_{\hat{X}} \tag{21.20}$$

Point Estimator of the Population Proportion

$$\bar{p}_{st} = \sum_{h=1}^{H}\left(\frac{N_h}{N}\right)\bar{p}_h \tag{21.21}$$

Estimate of the Standard Error of $\bar{p}_{st}$

$$s_{\bar{p}_{st}} = \sqrt{\frac{1}{N^2}\sum_{h=1}^{H} N_h(N_h - n_h)\left[\frac{\bar{p}_h(1 - \bar{p}_h)}{n_h - 1}\right]} \tag{21.22}$$

Approximate 95% Confidence Interval Estimate of the Population Proportion

$$\bar{p}_{st} \pm 2s_{\bar{p}_{st}} \tag{21.23}$$

Allocating the Total Sample n to the Strata: Neyman Allocation

$$n_h = n\left(\frac{N_h s_h}{\sum_{h=1}^{H} N_h s_h}\right) \tag{21.24}$$

Sample Size When Estimating the Population Mean

$$n = \frac{\left(\sum_{h=1}^{H} N_h s_h\right)^2}{N^2\left(\frac{B^2}{4}\right) + \sum_{h=1}^{H} N_h s_h^2} \tag{21.25}$$

Sample Size When Estimating the Population Total

$$n = \frac{\left(\sum_{h=1}^{H} N_h s_h\right)^2}{\frac{B^2}{4} + \sum_{h=1}^{H} N_h s_h^2} \tag{21.26}$$

Sample Size for the Estimate of the Population Proportion

$$n = \frac{\left(\sum_{h=1}^{H} N_h \sqrt{\bar{p}_h(1 - \bar{p}_h)}\right)^2}{N^2\left(\frac{B^2}{4}\right) + \sum_{h=1}^{H} N_h \bar{p}_h(1 - \bar{p}_h)} \tag{21.27}$$

Proportional Allocation of Sample n to the Strata

$$n_h = n\left(\frac{N_h}{N}\right) \tag{21.28}$$

CLUSTER SAMPLING

Point Estimator of the Population Mean

$$\bar{x}_c = \frac{\sum_{i=1}^{n} x_i}{\sum_{i=1}^{n} M_i} \tag{21.29}$$

Estimate of the Standard Error of the Mean

$$s_{\bar{x}_c} = \sqrt{\left(\frac{N - n}{Nn\bar{M}^2}\right)\frac{\sum_{i=1}^{n}(x_i - \bar{x}_c M_i)^2}{n - 1}} \tag{21.30}$$

Approximate 95% Confidence Interval Estimate of the Population Mean

$$\bar{x}_c \pm 2s_{\bar{x}_c} \tag{21.31}$$

Point Estimator of the Population Total

$$\hat{X} = M\bar{x}_c \tag{21.32}$$

Estimate of the Standard Error of $\hat{X}$

$$s_{\hat{X}} = Ms_{\bar{x}_c} \tag{21.33}$$

Approximate 95% Confidence Interval Estimate of the Population Total

$$M\bar{x}_c \pm 2s_{\hat{X}} \tag{21.34}$$

Point Estimator of the Population Proportion

$$\bar{p}_c = \frac{\sum_{i=1}^{n} a_i}{\sum_{i=1}^{n} M_i} \tag{21.35}$$

Estimate of the Standard Error of $\bar{p}_c$

$$s_{\bar{p}_c} = \sqrt{\left(\frac{N - n}{Nn\bar{M}^2}\right)\frac{\sum_{i=1}^{n}(a_i - \bar{p}_c M_i)^2}{n - 1}} \tag{21.36}$$

Approximate 95% Confidence Interval Estimate of the Population Proportion

$$\bar{p}_c \pm 2s_{\bar{p}_c} \tag{21.37}$$

SUPPLEMENTARY EXERCISES

18. To assess consumer acceptance of a new series of ads for Miller Lite Beer, Louis Harris conducted a nationwide poll of 363 adults who had seen the Miller Lite ads (*USA Today*, November 17, 1997). The following responses are based on that survey. (Note: Since the survey sampled a very small fraction of all adults, assume $(N - n)/N = 1$ in any formulas involving the standard error.)

a. Nineteen percent of all respondents indicated they liked the ads a lot. Develop a 95% confidence interval for the population proportion.

b. Thirty-one percent of the respondents disliked the new ads. Develop a 95% confidence interval for the population proportion.

c. Seventeen percent of the respondents felt the ads are very effective. Develop a 95% confidence interval for the proportion of adults who think the ads are very effective.

d. Louis Harris reported that the "margin of error is five percentage points." What does this statement mean and how do you think they arrived at this number?

e. How might nonsampling error bias the results of such a survey?

19. *The Wall Street Journal* conducted a survey of subscribers to its interactive edition. One question asked the 504 respondents whether they used a laptop computer when traveling; 55% said they did. Another question asked respondents whether they used an express or package service when traveling; 31% said they did. (*The Wall Street Journal Interactive Edition Subscriber Survey,* 2000).

a. Develop an estimate of the standard error of the proportion for the proportion who use a laptop computer.

b. Develop an estimate of the standard error of the proportion for the proportion who use an express or package service.

c. Are the estimates of the standard error the same in parts (a) and (b)? If they differ, explain why.

d. Develop an approximate 95% confidence interval for the proportion who use a laptop computer.

e. Develop an approximate 95% confidence interval for the proportion who use an express or package service.

20. A quality of life survey was conducted with employees of a manufacturing firm. Of the firm's 3000 employees, a sample of 300 were sent questionnaires. Two hundred usable questionnaires were obtained for a response rate of 67%.

a. The mean annual salary for the sample was $\bar{x} = \$23{,}200$ with $s = \$3000$. Develop an approximate 95% confidence interval for the mean annual salary of the population.

b. Using the information in part (a), develop an approximate 95% confidence interval for the total salary of all 3000 employees.

c. Seventy-three percent of respondents reported that they were "generally satisfied" with their job. Develop an approximate 95% confidence interval for the population proportion.

d. Comment on whether you think the results in part (c) might be biased. Would your opinion change if you knew the respondents were guaranteed anonymity?

21. A U.S. Senate Judiciary Committee report showed the number of homicides in each state. In Indiana, Ohio, and Kentucky, the number of homicides was, respectively, 380, 760, and 260. Suppose a stratified random sample with the following results was taken to learn more about the victims and the cause of death.

Stratum	Sample Size	Shootings	Beatings	Black Victims
Indiana	30	10	9	21
Ohio	45	19	12	34
Kentucky	25	7	11	15

Continued

a. Develop an approximate 95% confidence interval for the proportion of shooting deaths in Indiana.
b. Develop an estimate for the total number of shooting deaths in Ohio.
c. Develop an approximate 95% confidence interval for the proportion of shooting deaths in Ohio.
d. Develop an approximate 95% confidence interval for the proportion of shooting deaths across all three states.

22. Refer again to the data in Exercise 21.
a. Develop an estimate of the total number of deaths (in the three states) due to beatings.
b. Develop an approximate 95% confidence interval for the proportion of deaths across all three states due to beatings.
c. Develop an approximate 95% confidence interval for the proportion of victims who are black.
d. Develop an estimate of the total number of victims who are black.

23. A stratified simple random sample is to be taken of a bank's customers to learn about a variety of attitudinal and demographic issues. The stratification is to be based on savings account balances as of June 30, 2001. A frequency distribution showing the number of accounts in each stratum, together with the standard deviation of account balances by stratum, follows.

Stratum ($)	Accounts	Standard Deviation of Account Balances
0.00–1,000.00	3000	80
1,000.01–2,000.00	600	150
2,000.01–5,000.00	250	220
5,000.01–10,000.00	100	700
over 10,000.00	50	3000

a. Assuming the cost per unit sampled is approximately equal across strata, determine the total number of persons to include in the sample. Assume we want a bound on the error of estimate of the population mean for savings account balances of $B = \$20$.
b. Use the Neyman allocation procedure to determine the number to be sampled for each stratum.

24. A public agency is interested in learning more about the persons living in nursing homes in a particular city. A total of 100 nursing homes are caring for 4800 people in the city and a cluster sample of six homes has been taken. Each person in the six homes has been interviewed. A portion of the sample results follows.

Home	Residents	Average Age of Residents	Disabled Residents
1	14	61	12
2	7	74	2
3	96	78	30
4	23	69	8
5	71	73	10
6	29	84	22

a. Develop an estimate of the mean age of nursing home residents in this city.
b. Develop an approximate 95% confidence interval for the proportion of disabled persons in the city's nursing homes.
c. Estimate the total number of disabled persons residing in nursing homes in this city.

APPENDIXES

Appendix A: References and Bibliography

General

Bowerman, B. L., and R. T. O'Connell, *Applied Statistics: Improving Business Processes,* Irwin, 1996.

Freedman, D., R. Pisani, and R. Purves, Statistics, 3rd ed., W. W. Norton, 1997.

Hogg, R. V., and A. T. Craig, *Introduction to Mathematical Statistics,* 5th ed., Prentice-Hall, 1994.

Hogg, R. V., and E. A. Tanis, *Probability and Statistical Inference,* 6th ed., Prentice Hall, 2001.

Joiner, B. L., and B. F. Ryan, *Minitab Handbook,* Brooks/Cole, 2000.

Miller, I., and M. Miller, *John E. Freund's Mathematical Statistics,* Prentice Hall, 1998.

Roberts, H., *Data Analysis for Managers with Minitab,* Scientific Press, 1991.

Tanur, J. M., *Statistics: A Guide to the Unknown,* 4th ed., Brooks/Cole, 2002.

Tukey, J. W., *Exploratory Data Analysis,* Addison-Wesley, 1977.

Experimental Design

Cochran, W. G., and G. M. Cox, *Experimental Designs,* 2d ed., Wiley, 1992.

Hicks, C. R., and K. V. Turner, *Fundamental Concepts in the Design of Experiments,* 5th ed., Oxford University Press, 1999.

Montgomery, D. C., *Design and Analysis of Experiments,* 5th ed., Wiley, 2000.

Winer, B. J., K. M. Michels, and D. R. Brown, *Statistical Principles in Experimental Design,* 3rd ed., McGraw-Hill, 1991.

Forecasting

Bowerman, B. L., and R. T. O'Connell, *Forecasting and Time Series: An Applied Approach,* 3rd ed., Duxbury, 1993.

Box, G. E. P., G. C. Reinsel, and G. Jenkins, *Time Series Analysis: Forecasting and Control,* 3rd ed., Prentice-Hall, 1994.

Makridakis, S., S. C. Wheelwright, and R. J. Hyndman, *Forecasting: Methods and Applications,* 3rd ed., Wiley, 1997.

Index Numbers

U.S. Department of Commerce, *Survey of Current Business.*

U.S. Department of Labor, Bureau of Labor Statistics, *CPI Detailed Report.*

U.S. Department of Labor, *Producer Price Indexes.*

Nonparametric Methods

Conover, W. J., *Practical Nonparametric Statistics,* 3rd ed., Wiley, 1998.

Gibbons, J. D., and S. Chakraborti, *Nonparametric Statistical Inference,* 3rd ed., Marcel Dekker, 1992.

Siegel, S., and N. J. Castellan, *Nonparametric Statistics for the Behavioral Sciences,* 2d ed., McGraw-Hill, 1988.

Sprent, P., *Applied Non-Parametric Statistical Methods,* CRC, 1993.

Probability

Feller, W., *An Introduction to Probability Theory and Its Application,* Vol. I, 3rd ed., Wiley, 1968.

Hogg, R. V., and E. A. Tanis, *Probability and Statistical Inference,* 6th ed., Prentice-Hall, 2001.

Ross, S. M., *Introduction to Probability Models,* 7th ed., Academic Press, 2000.

Wackerly, D. D., W. Mendenhall, and R. L. Scheaffer, *Mathematical Statistics with Applications,* 5th ed., PWS, 1996.

Quality Control

Deming, W. E., *Quality, Productivity, and Competitive Position,* MIT, 1982.

Duncan, A. J., *Quality Control and Industrial Statistics,* 5th ed., Irwin, 1986.

Evans, J. R., and W. M. Lindsay, *The Management and Control of Quality,* 4th ed., South-Western, 1998.

Gryna, F. M., and I. M. Juran, *Quality Planning and Analysis: From Product Development Through Use,* 3rd ed., McGraw-Hill, 1993.

Ishikawa, K., *Introduction to Quality Control,* Kluwer Academic, 1991.

Montgomery, D. C., *Introduction to Statistical Quality Control,* 3rd ed., Wiley, 1996.

Regression Analysis

Belsley, D. A., *Conditioning Diagnostics: Collinearity and Weak Data in Regression,* Wiley, 1991.

Chatterjee, S., and B. Price, *Regression Analysis by Example,* 3rd ed., Wiley, 1999.

Draper, N. R., and H. Smith, *Applied Regression Analysis,* 3rd ed., Wiley, 1998.

Graybill, F. A., and H. Iyer, *Regression Analysis: Concepts and Applications,* Duxbury Press, 1994.

Kleinbaum, D. G., L. L. Kupper, and K. E. Muller, *Applied Regression Analysis and Other Multivariate Methods,* 3rd ed., Duxbury Press, 1997.

Kutner, M. H., C. J. Nachtschiem, W. Wasserman, and J. Neter, *Applied Linear Statistical Models,* 4th ed., Irwin, 1996.

Mendenhall, M., and T. Sincich, *A Second Course in Statistics: Regression Analysis,* 5th ed., Prentice Hall, 1996.

Myers, R. H., *Classical and Modern Regression with Applications,* 2d ed., PWS, 1990.

Wonnacott, T. H., and R. J. Wonnacott, *Regression: A Second Course in Statistics,* Krieger, 1986.

Sampling

Cochran, W. G., *Sampling Techniques,* 3rd ed., Wiley, 1977.

Deming, W. E., *Some Theory of Sampling,* Dover, 1984.

Hansen, M. H., W. N. Hurwitz, W. G. Madow, and M. N. Hanson, *Sample Survey Methods and Theory,* Wiley, 1993.

Kish, L., *Survey Sampling,* Wiley, 1995.

Levy, P. S., and S. Lemeshow, *Sampling of Populations: Methods and Applications,* 3rd ed., Wiley, 1999.

Scheaffer, R. L., W. Mendenhall, and L. Ott, *Elementary Survey Sampling,* 5th ed., Duxbury, 1996.

Appendix B: Tables

TABLE 1 STANDARD NORMAL DISTRIBUTION

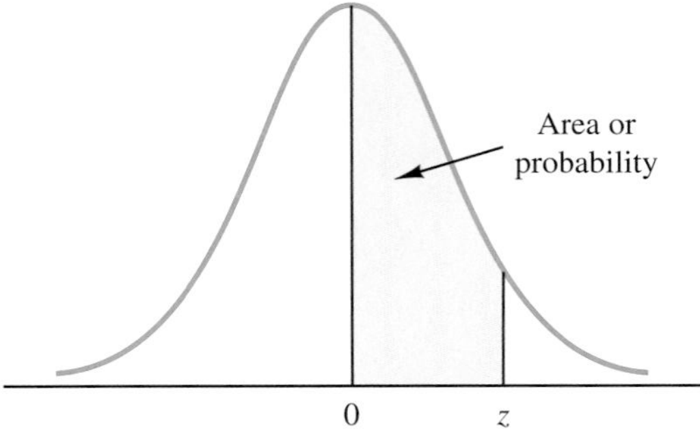

Entries in the table give the area under the curve between the mean and z standard deviations above the mean. For example, for $z = 1.25$ the area under the curve between the mean and z is .3944.

z	.00	.01	.02	.03	.04	.05	.06	.07	.08	.09
.0	.0000	.0040	.0080	.0120	.0160	.0199	.0239	.0279	.0319	.0359
.1	.0398	.0438	.0478	.0517	.0557	.0596	.0636	.0675	.0714	.0753
.2	.0793	.0832	.0871	.0910	.0948	.0987	.1026	.1064	.1103	.1141
.3	.1179	.1217	.1255	.1293	.1331	.1368	.1406	.1443	.1480	.1517
.4	.1554	.1591	.1628	.1664	.1700	.1736	.1772	.1808	.1844	.1879
.5	.1915	.1950	.1985	.2019	.2054	.2088	.2123	.2157	.2190	.2224
.6	.2257	.2291	.2324	.2357	.2389	.2422	.2454	.2486	.2518	.2549
.7	.2580	.2612	.2642	.2673	.2704	.2734	.2764	.2794	.2823	.2852
.8	.2881	.2910	.2939	.2967	.2995	.3023	.3051	.3078	.3106	.3133
.9	.3159	.3186	.3212	.3238	.3264	.3289	.3315	.3340	.3365	.3389
1.0	.3413	.3438	.3461	.3485	.3508	.3531	.3554	.3577	.3599	.3621
1.1	.3643	.3665	.3686	.3708	.3729	.3749	.3770	.3790	.3810	.3830
1.2	.3849	.3869	.3888	.3907	.3925	.3944	.3962	.3980	.3997	.4015
1.3	.4032	.4049	.4066	.4082	.4099	.4115	.4131	.4147	.4162	.4177
1.4	.4192	.4207	.4222	.4236	.4251	.4265	.4279	.4292	.4306	.4319
1.5	.4332	.4345	.4357	.4370	.4382	.4394	.4406	.4418	.4429	.4441
1.6	.4452	.4463	.4474	.4484	.4495	.4505	.4515	.4525	.4535	.4545
1.7	.4554	.4564	.4573	.4582	.4591	.4599	.4608	.4616	.4625	.4633
1.8	.4641	.4649	.4656	.4664	.4671	.4678	.4686	.4693	.4699	.4706
1.9	.4713	.4719	.4726	.4732	.4738	.4744	.4750	.4756	.4761	.4767
2.0	.4772	.4778	.4783	.4788	.4793	.4798	.4803	.4808	.4812	.4817
2.1	.4821	.4826	.4830	.4834	.4838	.4842	.4846	.4850	.4854	.4857
2.2	.4861	.4864	.4868	.4871	.4875	.4878	.4881	.4884	.4887	.4890
2.3	.4893	.4896	.4898	.4901	.4904	.4906	.4909	.4911	.4913	.4916
2.4	.4918	.4920	.4922	.4925	.4927	.4929	.4931	.4932	.4934	.4936
2.5	.4938	.4940	.4941	.4943	.4945	.4946	.4948	.4949	.4951	.4952
2.6	.4953	.4955	.4956	.4957	.4959	.4960	.4961	.4962	.4963	.4964
2.7	.4965	.4966	.4967	.4968	.4969	.4970	.4971	.4972	.4973	.4974
2.8	.4974	.4975	.4976	.4977	.4977	.4978	.4979	.4979	.4980	.4981
2.9	.4981	.4982	.4982	.4983	.4984	.4984	.4985	.4985	.4986	.4986
3.0	.4986	.4987	.4987	.4988	.4988	.4989	.4989	.4989	.4990	.4990

TABLE 2 *t* DISTRIBUTION

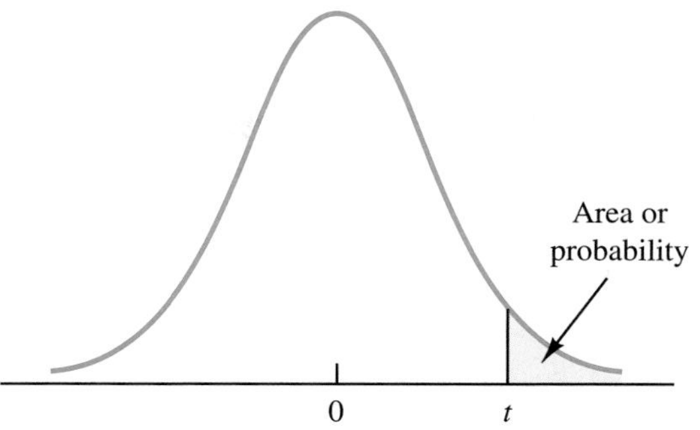

Entries in the table give t values for an area or probability in the upper tail of the t distribution. For example, with 10 degrees of freedom and a .05 area in the upper tail, $t_{.05} = 1.812$.

Degrees of Freedom	Area in Upper Tail .10	.05	.025	.01	.005
1	3.078	6.314	12.706	31.821	63.657
2	1.886	2.920	4.303	6.965	9.925
3	1.638	2.353	3.182	4.541	5.841
4	1.533	2.132	2.776	3.747	4.604
5	1.476	2.015	2.571	3.365	4.032
6	1.440	1.943	2.447	3.143	3.707
7	1.415	1.895	2.365	2.998	3.499
8	1.397	1.860	2.306	2.896	3.355
9	1.383	1.833	2.262	2.821	3.250
10	1.372	1.812	2.228	2.764	3.169
11	1.363	1.796	2.201	2.718	3.106
12	1.356	1.782	2.179	2.681	3.055
13	1.350	1.771	2.160	2.650	3.012
14	1.345	1.761	2.145	2.624	2.977
15	1.341	1.753	2.131	2.602	2.947
16	1.337	1.746	2.120	2.583	2.921
17	1.333	1.740	2.110	2.567	2.898
18	1.330	1.734	2.101	2.552	2.878
19	1.328	1.729	2.093	2.539	2.861
20	1.325	1.725	2.086	2.528	2.845
21	1.323	1.721	2.080	2.518	2.831
22	1.321	1.717	2.074	2.508	2.819
23	1.319	1.714	2.069	2.500	2.807
24	1.318	1.711	2.064	2.492	2.797
25	1.316	1.708	2.060	2.485	2.787
26	1.315	1.706	2.056	2.479	2.779
27	1.314	1.703	2.052	2.473	2.771
28	1.313	1.701	2.048	2.467	2.763
29	1.311	1.699	2.045	2.462	2.756
30	1.310	1.697	2.042	2.457	2.750
40	1.303	1.684	2.021	2.423	2.704
60	1.296	1.671	2.000	2.390	2.660
120	1.289	1.658	1.980	2.358	2.617
∞	1.282	1.645	1.960	2.326	2.576

TABLE 3 CHI-SQUARE DISTRIBUTION

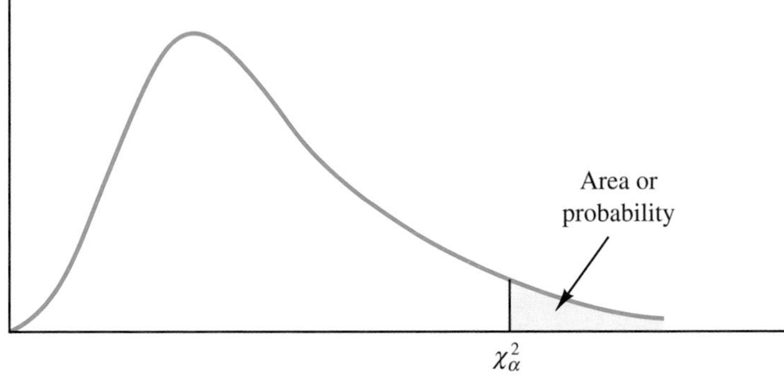

Entries in the table give χ^2_α values, where α is the area or probability in the upper tail of the chi-square distribution. For example, with 10 degrees of freedom and a .01 area in the upper tail, $\chi^2_{.01} = 23.2093$.

Degrees of Freedom	Area in Upper Tail									
	.995	.99	.975	.95	.90	.10	.05	.025	.01	.005
1	$392{,}704 \times 10^{-10}$	$157{,}088 \times 10^{-9}$	$982{,}069 \times 10^{-9}$	$393{,}214 \times 10^{-8}$	.0157908	2.70554	3.84146	5.02389	6.63490	7.87944
2	.0100251	.0201007	.0506356	.102587	.210720	4.60517	5.99147	7.37776	9.21034	10.5966
3	.0717212	.114832	.215795	.351846	.584375	6.25139	7.81473	9.34840	11.3449	12.8381
4	.206990	.297110	.484419	.710721	1.063623	7.77944	9.48773	11.1433	13.2767	14.8602
5	.411740	.554300	.831211	1.145476	1.61031	9.23635	11.0705	12.8325	15.0863	16.7496
6	.675727	.872085	1.237347	1.63539	2.20413	10.6446	12.5916	14.4494	16.8119	18.5476
7	.989265	1.239043	1.68987	2.16735	2.83311	12.0170	14.0671	16.0128	18.4753	20.2777
8	1.344419	1.646482	2.17973	2.73264	3.48954	13.3616	15.5073	17.5346	20.0902	21.9550
9	1.734926	2.087912	2.70039	3.32511	4.16816	14.6837	16.9190	19.0228	21.6660	23.5893
10	2.15585	2.55821	3.24697	3.94030	4.86518	15.9871	18.3070	20.4831	23.2093	25.1882
11	2.60321	3.05347	3.81575	4.57481	5.57779	17.2750	19.6751	21.9200	24.7250	26.7569
12	3.07382	3.57056	4.40379	5.22603	6.30380	18.5494	21.0261	23.3367	26.2170	28.2995
13	3.56503	4.10691	5.00874	5.89186	7.04150	19.8119	22.3621	24.7356	27.6883	29.8194
14	4.07468	4.66043	5.62872	6.57063	7.78953	21.0642	23.6848	26.1190	29.1413	31.3193
15	4.60094	5.22935	6.26214	7.26094	8.54675	22.3072	24.9958	27.4884	30.5779	32.8013
16	5.14224	5.81221	6.90766	7.96164	9.31223	23.5418	26.2962	28.8454	31.9999	34.2672
17	5.69724	6.40776	7.56418	8.67176	10.0852	24.7690	27.5871	30.1910	33.4087	35.7185
18	6.26481	7.01491	8.23075	9.39046	10.8649	25.9894	28.8693	31.5264	34.8053	37.1564
19	6.84398	7.63273	8.90655	10.1170	11.6509	27.2036	30.1435	32.8523	36.1908	38.5822

20	7.43386	8.26040	9.59083	10.8508	12.4426	28.4120	31.4104	34.1696	37.5662	39.9968
21	8.03366	8.89720	10.28293	11.5913	13.2396	29.6151	32.6705	35.4789	38.9321	41.4010
22	8.64272	9.54249	10.9823	12.3380	14.0415	30.8133	33.9244	36.7807	40.2894	42.7958
23	9.26042	10.19567	11.6885	13.0905	14.8479	32.0069	35.1725	38.0757	41.6384	44.1813
24	9.88623	10.8564	12.4011	13.8484	15.6587	33.1963	36.4151	39.3641	42.9798	45.5585
25	10.5197	11.5240	13.1197	14.6114	16.4734	34.3816	37.6525	40.6465	44.3141	46.9278
26	11.1603	12.1981	13.8439	15.3791	17.2919	35.5631	38.8852	41.9232	45.6417	48.2899
27	11.8076	12.8786	14.5733	16.1513	18.1138	36.7412	40.1133	43.1944	46.9630	49.6449
28	12.4613	13.5648	15.3079	16.9279	18.9392	37.9159	41.3372	44.4607	48.2782	50.9933
29	13.1211	14.2565	16.0471	17.7083	19.7677	39.0875	42.5569	45.7222	49.5879	52.3356
30	13.7867	14.9535	16.7908	18.4926	20.5992	40.2560	43.7729	46.9792	50.8922	53.6720
40	20.7065	22.1643	24.4331	26.5093	29.0505	51.8050	55.7585	59.3417	63.6907	66.7659
50	27.9907	29.7067	32.3574	34.7642	37.6886	63.1671	67.5048	71.4202	76.1539	79.4900
60	35.5346	37.4848	40.4817	43.1879	46.4589	74.3970	79.0819	83.2976	88.3794	91.9517
70	43.2752	45.4418	48.7576	51.7393	55.3290	85.5271	90.5312	95.0231	100.425	104.215
80	51.1720	53.5400	57.1532	60.3915	64.2778	96.5782	101.879	106.629	112.329	116.321
90	59.1963	61.7541	65.6466	69.1260	73.2912	107.565	113.145	118.136	124.116	128.299
100	67.3276	70.0648	74.2219	77.9295	82.3581	118.498	124.342	129.561	135.807	140.169

This table is reprinted by permission of Oxford University Press on behalf of The Biometrika Trustees from Table 8, Percentage Points of the χ^2 Distribution, by E. S. Pearson and H. O. Hartley, *Biometrika Tables for Statisticians,* Vol. 1, 3rd ed., 1966.

TABLE 4 *F* DISTRIBUTION

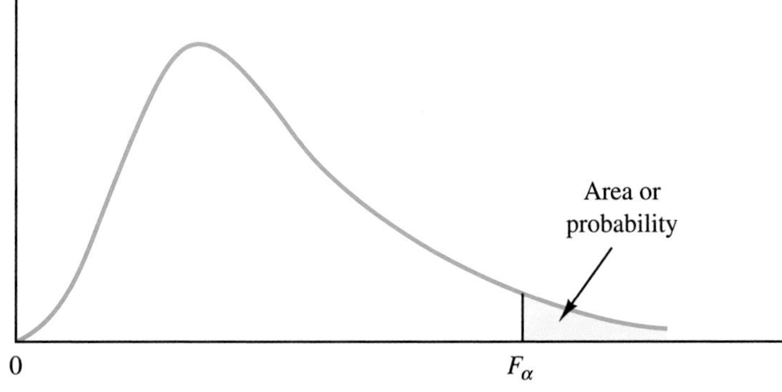

Entries in the table give F_α values, where α is the area or probability in the upper tail of the *F* distribution. For example, with 12 numerator degress of freedom, 15 denominator degrees of freedom, and a .05 area in the upper tail, $F_{.05} = 2.48$.

Table of $F_{.05}$ Values

Denominator Degrees of Freedom	Numerator Degrees of Freedom																		
	1	**2**	**3**	**4**	**5**	**6**	**7**	**8**	**9**	**10**	**12**	**15**	**20**	**24**	**30**	**40**	**60**	**120**	**∞**
1	161.4	199.5	215.7	224.6	230.2	234.0	236.8	238.9	240.5	241.9	243.9	245.9	248.0	249.1	250.1	251.1	252.2	253.3	254.3
2	18.51	19.00	19.16	19.25	19.30	19.33	19.35	19.37	19.38	19.40	19.41	19.43	19.45	19.45	19.46	19.47	19.48	19.49	19.50
3	10.13	9.55	9.28	9.12	9.01	8.94	8.89	8.85	8.81	8.79	8.74	8.70	8.66	8.64	8.62	8.59	8.57	8.55	8.53
4	7.71	6.94	6.59	6.39	6.26	6.16	6.09	6.04	6.00	5.96	5.91	5.86	5.80	5.77	5.75	5.72	5.69	5.66	5.63
5	6.61	5.79	5.41	5.19	5.05	4.95	4.88	4.82	4.77	4.74	4.68	4.62	4.56	4.53	4.50	4.46	4.43	4.40	4.36
6	5.99	5.14	4.76	4.53	4.39	4.28	4.21	4.15	4.10	4.06	4.00	3.94	3.87	3.84	3.81	3.77	3.74	3.70	3.67
7	5.59	4.74	4.35	4.12	3.97	3.87	3.79	3.73	3.68	3.64	3.57	3.51	3.44	3.41	3.38	3.34	3.30	3.27	3.23
8	5.32	4.46	4.07	3.84	3.69	3.58	3.50	3.44	3.39	3.35	3.28	3.22	3.15	3.12	3.08	3.04	3.01	2.97	2.93
9	5.12	4.26	3.86	3.63	3.48	3.37	3.29	3.23	3.18	3.14	3.07	3.01	2.94	2.90	2.86	2.83	2.79	2.75	2.71
10	4.96	4.10	3.71	3.48	3.33	3.22	3.14	3.07	3.02	2.98	2.91	2.85	2.77	2.74	2.70	2.66	2.62	2.58	2.54
11	4.84	3.98	3.59	3.36	3.20	3.09	3.01	2.95	2.90	2.85	2.79	2.72	2.65	2.61	2.57	2.53	2.49	2.45	2.40
12	4.75	3.89	3.49	3.26	3.11	3.00	2.91	2.85	2.80	2.75	2.69	2.62	2.54	2.51	2.47	2.43	2.38	2.34	2.30
13	4.67	3.81	3.41	3.18	3.03	2.92	2.83	2.77	2.71	2.67	2.60	2.53	2.46	2.42	2.38	2.34	2.30	2.25	2.21
14	4.60	3.74	3.34	3.11	2.96	2.85	2.76	2.70	2.65	2.60	2.53	2.46	2.39	2.35	2.31	2.27	2.22	2.18	2.13

15	4.54	3.68	3.29	3.06	2.90	2.79	2.71	2.64	2.59	2.54	2.48	2.40	2.33	2.29	2.25	2.20	2.16	2.11	2.07
16	4.49	3.63	3.24	3.01	2.85	2.74	2.66	2.59	2.54	2.49	2.42	2.35	2.28	2.24	2.19	2.15	2.11	2.06	2.01
17	4.45	3.59	3.20	2.96	2.81	2.70	2.61	2.55	2.49	2.45	2.38	2.31	2.23	2.19	2.15	2.10	2.06	2.01	1.96
18	4.41	3.55	3.16	2.93	2.77	2.66	2.58	2.51	2.46	2.41	2.34	2.27	2.19	2.15	2.11	2.06	2.02	1.97	1.92
19	4.38	3.52	3.13	2.90	2.74	2.63	2.54	2.48	2.42	2.38	2.31	2.23	2.16	2.11	2.07	2.03	1.98	1.93	1.88
20	4.35	3.49	3.10	2.87	2.71	2.60	2.51	2.45	2.39	2.35	2.28	2.20	2.12	2.08	2.04	1.99	1.95	1.90	1.84
21	4.32	3.47	3.07	2.84	2.68	2.57	2.49	2.42	2.37	2.32	2.25	2.18	2.10	2.05	2.01	1.96	1.92	1.87	1.81
22	4.30	3.44	3.05	2.82	2.66	2.55	2.46	2.40	2.34	2.30	2.23	2.15	2.07	2.03	1.98	1.94	1.89	1.84	1.78
23	4.28	3.42	3.03	2.80	2.64	2.53	2.44	2.37	2.32	2.27	2.20	2.13	2.05	2.01	1.96	1.91	1.86	1.81	1.76
24	4.26	3.40	3.01	2.78	2.62	2.51	2.42	2.36	2.30	2.25	2.18	2.11	2.03	1.98	1.94	1.89	1.84	1.79	1.73
25	4.24	3.39	2.99	2.76	2.60	2.49	2.40	2.34	2.28	2.24	2.16	2.09	2.01	1.96	1.92	1.87	1.82	1.77	1.71
26	4.23	3.37	2.98	2.74	2.59	2.47	2.39	2.32	2.27	2.22	2.15	2.07	1.99	1.95	1.90	1.85	1.80	1.75	1.69
27	4.21	3.35	2.96	2.73	2.57	2.46	2.37	2.31	2.25	2.20	2.13	2.06	1.97	1.93	1.88	1.84	1.79	1.73	1.67
28	4.20	3.34	2.95	2.71	2.56	2.45	2.36	2.29	2.24	2.19	2.12	2.04	1.96	1.91	1.87	1.82	1.77	1.71	1.65
29	4.18	3.33	2.93	2.70	2.55	2.43	2.35	2.28	2.22	2.18	2.10	2.03	1.94	1.90	1.85	1.81	1.75	1.70	1.64
30	4.17	3.32	2.92	2.69	2.53	2.42	2.33	2.27	2.21	2.16	2.09	2.01	1.93	1.89	1.84	1.79	1.74	1.68	1.62
40	4.08	3.23	2.84	2.61	2.45	2.34	2.25	2.18	2.12	2.08	2.00	1.92	1.84	1.79	1.74	1.69	1.64	1.58	1.51
60	4.00	3.15	2.76	2.53	2.37	2.25	2.17	2.10	2.04	1.99	1.92	1.84	1.75	1.70	1.65	1.59	1.53	1.47	1.39
120	3.92	3.07	2.68	2.45	2.29	2.17	2.09	2.02	1.96	1.91	1.83	1.75	1.66	1.61	1.55	1.50	1.43	1.35	1.25
∞	3.84	3.00	2.60	2.37	2.21	2.10	2.01	1.94	1.88	1.83	1.75	1.67	1.57	1.52	1.46	1.39	1.32	1.22	1.00

This table is reprinted by permission of Oxford University Press on behalf of The Biometrika Trustees from Table 18, Percentage Points of the *F* Distribution, by E. S. Pearson and H. O. Hartley, *Biometrika Tables for Statisticians,* Vol. 1, 3rd ed., 1966.

TABLE 4 *F* DISTRIBUTION (*Continued*)

Table of $F_{.025}$ Values

Denominator Degrees of Freedom	Numerator Degrees of Freedom 1	2	3	4	5	6	7	8	9	10	12	15	20	24	30	40	60	120	∞
1	647.8	799.5	864.2	899.6	921.8	937.1	948.2	956.7	963.3	968.6	976.7	984.9	993.1	997.2	1,001	1,006	1,010	1,014	1,018
2	38.51	39.00	39.17	39.25	39.30	39.33	39.36	39.37	39.39	39.40	39.41	39.43	39.45	39.46	39.46	39.47	39.48	39.49	39.50
3	17.44	16.04	15.44	15.10	14.88	14.73	14.62	14.54	14.47	14.42	14.34	14.25	14.17	14.12	14.08	14.04	13.99	13.95	13.90
4	12.22	10.65	9.98	9.60	9.36	9.20	9.07	8.98	8.90	8.84	8.75	8.66	8.56	8.51	8.46	8.41	8.36	8.31	8.26
5	10.01	8.43	7.76	7.39	7.15	6.98	6.85	6.76	6.68	6.62	6.52	6.43	6.33	6.28	6.23	6.18	6.12	6.07	6.02
6	8.81	7.26	6.60	6.23	5.99	5.82	5.70	5.60	5.52	5.46	5.37	5.27	5.17	5.12	5.07	5.01	4.96	4.90	4.85
7	8.07	6.54	5.89	5.52	5.29	5.21	4.99	4.90	4.82	4.76	4.67	4.57	4.47	4.42	4.36	4.31	4.25	4.20	4.14
8	7.57	6.06	5.42	5.05	4.82	4.65	4.53	4.43	4.36	4.30	4.20	4.10	4.00	3.95	3.89	3.84	3.78	3.73	3.67
9	7.21	5.71	5.08	4.72	4.48	4.32	4.20	4.10	4.03	3.96	3.87	3.77	3.67	3.61	3.56	3.51	3.45	3.39	3.33
10	6.94	5.46	4.83	4.47	4.24	4.07	3.95	3.85	3.78	3.72	3.62	3.52	3.42	3.37	3.31	3.26	3.20	3.14	3.08
11	6.72	5.26	4.63	4.28	4.04	3.88	3.76	3.66	3.59	3.53	3.43	3.33	3.23	3.17	3.12	3.06	3.00	2.94	2.88
12	6.55	5.10	4.47	4.12	3.89	3.73	3.61	3.51	3.44	3.37	3.28	3.18	3.07	3.02	2.96	2.91	2.85	2.79	2.72
13	6.41	4.97	4.35	4.00	3.77	3.60	3.48	3.39	3.31	3.25	3.15	3.05	2.95	2.89	2.84	2.78	2.72	2.66	2.60
14	6.30	4.86	4.24	3.89	3.66	3.50	3.38	3.29	3.21	3.15	3.05	2.95	2.84	2.79	2.73	2.67	2.61	2.55	2.49
15	6.20	4.77	4.15	3.80	3.58	3.41	3.29	3.20	3.12	3.06	2.96	2.86	2.76	2.70	2.64	2.59	2.52	2.46	2.40
16	6.12	4.69	4.08	3.73	3.50	3.34	3.22	3.12	3.05	2.99	2.89	2.79	2.68	2.63	2.57	2.51	2.45	2.38	2.32
17	6.04	4.62	4.01	3.66	3.44	3.28	3.16	3.06	2.98	2.92	2.82	2.72	2.62	2.56	2.50	2.44	2.38	2.32	2.25
18	5.98	4.56	3.95	3.61	3.38	3.22	3.10	3.01	2.93	2.87	2.77	2.67	2.56	2.50	2.44	2.38	2.32	2.26	2.19
19	5.92	4.51	3.90	3.56	3.33	3.17	3.05	2.96	2.88	2.82	2.72	2.62	2.51	2.45	2.39	2.33	2.27	2.20	2.13
20	5.87	4.46	3.86	3.51	3.29	3.13	3.01	2.91	2.84	2.77	2.68	2.57	2.46	2.41	2.35	2.29	2.22	2.16	2.09
21	5.83	4.42	3.82	3.48	3.25	3.09	2.97	2.87	2.80	2.73	2.64	2.53	2.42	2.37	2.31	2.25	2.18	2.11	2.04
22	5.79	4.38	3.78	3.44	3.22	3.05	2.93	2.84	2.76	2.70	2.60	2.50	2.39	2.33	2.27	2.21	2.14	2.08	2.00
23	5.75	4.35	3.75	3.41	3.18	3.02	2.90	2.81	2.73	2.67	2.57	2.47	2.36	2.30	2.24	2.18	2.11	2.04	1.97
24	5.72	4.32	3.72	3.38	3.15	2.99	2.87	2.78	2.70	2.64	2.54	2.44	2.33	2.27	2.21	2.15	2.08	2.01	1.94
25	5.69	4.29	3.69	3.35	3.13	2.97	2.85	2.75	2.68	2.61	2.51	2.41	2.30	2.24	2.18	2.12	2.05	1.98	1.91
26	5.66	4.27	3.67	3.33	3.10	2.94	2.82	2.73	2.65	2.59	2.49	2.39	2.28	2.22	2.16	2.09	2.03	1.95	1.88
27	5.63	4.24	3.65	3.31	3.08	2.92	2.80	2.71	2.63	2.57	2.47	2.36	2.25	2.19	2.13	2.07	2.00	1.93	1.85
28	5.61	4.22	3.63	3.29	3.06	2.90	2.78	2.69	2.61	2.55	2.45	2.34	2.23	2.17	2.11	2.05	1.98	1.91	1.83
29	5.59	4.20	3.61	3.27	3.04	2.88	2.76	2.67	2.59	2.53	2.43	2.32	2.21	2.15	2.09	2.03	1.96	1.89	1.81
30	5.57	4.18	3.59	3.25	3.03	2.87	2.75	2.65	2.57	2.51	2.41	2.31	2.20	2.14	2.07	2.01	1.94	1.87	1.79
40	5.42	4.05	3.46	3.13	2.90	2.74	2.62	2.53	2.45	2.39	2.29	2.18	2.07	2.01	1.94	1.88	1.80	1.72	1.64
60	5.29	3.93	3.34	3.01	2.79	2.63	2.51	2.41	2.33	2.27	2.17	2.06	1.94	1.88	1.82	1.74	1.67	1.58	1.48
120	5.15	3.80	3.23	2.89	2.67	2.52	2.39	2.30	2.22	2.16	2.05	1.94	1.82	1.76	1.69	1.61	1.53	1.43	1.31
∞	5.02	3.69	3.12	2.79	2.57	2.41	2.29	2.19	2.11	2.05	1.94	1.83	1.71	1.64	1.57	1.48	1.39	1.27	1.00

Table of $F_{.01}$ Values

Denominator Degrees of Freedom	Numerator Degrees of Freedom																		
	1	2	3	4	5	6	7	8	9	10	12	15	20	24	30	40	60	120	∞
1	4,052	4,999.5	5,403	5,625	5,764	5,859	5,928	5,982	6,022	6,056	6,106	6,157	6,209	6,235	6,261	6,287	6,313	6,339	6,366
2	98.50	99.00	99.17	99.25	99.30	99.33	99.36	99.37	99.39	99.40	99.42	99.43	99.45	99.46	99.47	99.47	99.48	99.49	99.50
3	34.12	30.82	29.46	28.71	28.24	27.91	27.67	27.49	27.35	27.23	27.05	26.87	26.69	26.60	26.50	26.41	26.32	26.22	26.13
4	21.20	18.00	16.69	15.98	15.52	15.21	14.98	14.80	14.66	14.55	14.37	14.20	14.02	13.93	13.84	13.75	13.65	13.56	13.46
5	16.26	13.27	12.06	11.39	10.97	10.67	10.46	10.29	10.16	10.05	9.89	9.72	9.55	9.47	9.38	9.29	9.20	9.11	9.06
6	13.75	10.92	9.78	9.15	8.75	8.47	8.26	8.10	7.98	7.87	7.72	7.56	7.40	7.31	7.23	7.14	7.06	6.97	6.88
7	12.25	9.55	8.45	7.85	7.46	7.19	6.99	6.84	6.72	6.62	6.47	6.31	6.16	6.07	5.99	5.91	5.82	5.74	5.65
8	11.26	8.65	7.59	7.01	6.63	6.37	6.18	6.03	5.91	5.81	5.67	5.52	5.36	5.28	5.20	5.12	5.03	4.95	4.86
9	10.56	8.02	6.99	6.42	6.06	5.80	5.61	5.47	5.35	5.26	5.11	4.96	4.81	4.73	4.65	4.57	4.48	4.40	4.31
10	10.04	7.56	6.55	5.99	5.64	5.39	5.20	5.06	4.94	4.85	4.71	4.56	4.41	4.33	4.25	4.17	4.08	4.00	3.91
11	9.65	7.21	6.22	5.67	5.32	5.07	4.89	4.74	4.63	4.54	4.40	4.25	4.10	4.02	3.94	3.86	3.78	3.69	3.60
12	9.33	6.93	5.95	5.41	5.06	4.82	4.64	4.50	4.39	4.30	4.16	4.01	3.86	3.78	3.70	3.62	3.54	3.45	3.36
13	9.07	6.70	5.74	5.21	4.86	4.62	4.44	4.30	4.19	4.10	3.96	3.82	3.66	3.59	3.51	3.43	3.34	3.25	3.17
14	8.86	6.51	5.56	5.04	4.69	4.46	4.28	4.14	4.03	3.94	3.80	3.66	3.51	3.43	3.35	3.27	3.18	3.09	3.00
15	8.68	6.36	5.42	4.89	4.56	4.32	4.14	4.00	3.89	3.80	3.67	3.52	3.37	3.29	3.21	3.13	3.05	2.96	2.87
16	8.53	6.23	5.29	4.77	4.44	4.20	4.03	3.89	3.78	3.69	3.55	3.41	3.26	3.18	3.10	3.02	2.93	2.84	2.75
17	8.40	6.11	5.18	4.67	4.34	4.10	3.93	3.79	3.68	3.59	3.46	3.31	3.16	3.08	3.00	2.92	2.83	2.75	2.65
18	8.29	6.01	5.09	4.58	4.25	4.01	3.84	3.71	3.60	3.51	3.37	3.23	3.08	3.00	2.92	2.84	2.75	2.66	2.57
19	8.18	5.93	5.01	4.50	4.17	3.94	3.77	3.63	3.52	3.43	3.30	3.15	3.00	2.92	2.84	2.76	2.67	2.58	2.49
20	8.10	5.85	4.94	4.43	4.10	3.87	3.70	3.56	3.46	3.37	3.23	3.09	2.94	2.86	2.78	2.69	2.61	2.52	2.42
21	8.02	5.78	4.87	4.37	4.04	3.81	3.64	3.51	3.40	3.31	3.17	3.03	2.88	2.80	2.72	2.64	2.55	2.46	2.36
22	7.95	5.72	4.82	4.31	3.99	3.76	3.59	3.45	3.35	3.26	3.12	2.98	2.83	2.75	2.67	2.58	2.50	2.40	2.31
23	7.88	5.66	4.76	4.26	3.94	3.71	3.54	3.41	3.30	3.21	3.07	2.93	2.78	2.70	2.62	2.54	2.45	2.35	2.26
24	7.82	5.61	4.72	4.22	3.90	3.67	3.50	3.36	3.26	3.17	3.03	2.89	2.74	2.66	2.58	2.49	2.40	2.31	2.21
25	7.77	5.57	4.68	4.18	3.85	3.63	3.46	3.32	3.22	3.13	2.99	2.85	2.70	2.62	2.54	2.45	2.36	2.27	2.17
26	7.72	5.53	4.64	4.14	3.82	3.59	3.42	3.29	3.18	3.09	2.96	2.81	2.66	2.58	2.50	2.42	2.33	2.23	2.13
27	7.68	5.49	4.60	4.11	3.78	3.56	3.39	3.26	3.15	3.06	2.93	2.78	2.63	2.55	2.47	2.38	2.29	2.20	2.10
28	7.64	5.45	4.57	4.07	3.75	3.53	3.36	3.23	3.12	3.03	2.90	2.75	2.60	2.52	2.44	2.35	2.26	2.17	2.06
29	7.60	5.42	4.54	4.04	3.73	3.50	3.33	3.20	3.09	3.00	2.87	2.73	2.57	2.49	2.41	2.33	2.23	2.14	2.03
30	7.56	5.39	4.51	4.02	3.70	3.47	3.30	3.17	3.07	2.98	2.84	2.70	2.55	2.47	2.39	2.30	2.21	2.11	2.01
40	7.31	5.18	4.31	3.83	3.51	3.29	3.12	2.99	2.89	2.80	2.66	2.52	2.37	2.29	2.20	2.11	2.02	1.92	1.80
60	7.08	4.98	4.13	3.65	3.34	3.12	2.95	2.82	2.72	2.63	2.50	2.35	2.20	2.12	2.03	1.94	1.84	1.73	1.60
120	6.85	4.79	3.95	3.48	3.17	2.96	2.79	2.66	2.56	2.47	2.34	2.19	2.03	1.95	1.86	1.76	1.66	1.53	1.38
∞	6.63	4.61	3.78	3.32	3.02	2.80	2.64	2.51	2.41	2.32	2.18	2.04	1.88	1.79	1.70	1.59	1.47	1.32	1.00

TABLE 5 BINOMIAL PROBABILITIES

Entries in the table give the probability of x successes in n trials of a binomial experiment, where p is the probability of a success on one trial. For example, with six trials and $p = .05$, the probability of two successes is .0305.

						p				
n	*x*	.01	.02	.03	.04	.05	.06	.07	.08	.09
2	0	.9801	.9604	.9409	.9216	.9025	.8836	.8649	.8464	.8281
	1	.0198	.0392	.0582	.0768	.0950	.1128	.1302	.1472	.1638
	2	.0001	.0004	.0009	.0016	.0025	.0036	.0049	.0064	.0081
3	0	.9703	.9412	.9127	.8847	.8574	.8306	.8044	.7787	.7536
	1	.0294	.0576	.0847	.1106	.1354	.1590	.1816	.2031	.2236
	2	.0003	.0012	.0026	.0046	.0071	.0102	.0137	.0177	.0221
	3	.0000	.0000	.0000	.0001	.0001	.0002	.0003	.0005	.0007
4	0	.9606	.9224	.8853	.8493	.8145	.7807	.7481	.7164	.6857
	1	.0388	.0753	.1095	.1416	.1715	.1993	.2252	.2492	.2713
	2	.0006	.0023	.0051	.0088	.0135	.0191	.0254	.0325	.0402
	3	.0000	.0000	.0001	.0002	.0005	.0008	.0013	.0019	.0027
	4	.0000	.0000	.0000	.0000	.0000	.0000	.0000	.0000	.0001
5	0	.9510	.9039	.8587	.8154	.7738	.7339	.6957	.6591	.6240
	1	.0480	.0922	.1328	.1699	.2036	.2342	.2618	.2866	.3086
	2	.0010	.0038	.0082	.0142	.0214	.0299	.0394	.0498	.0610
	3	.0000	.0001	.0003	.0006	.0011	.0019	.0030	.0043	.0060
	4	.0000	.0000	.0000	.0000	.0000	.0001	.0001	.0002	.0003
	5	.0000	.0000	.0000	.0000	.0000	.0000	.0000	.0000	.0000
6	0	.9415	.8858	.8330	.7828	.7351	.6899	.6470	.6064	.5679
	1	.0571	.1085	.1546	.1957	.2321	.2642	.2922	.3164	.3370
	2	.0014	.0055	.0120	.0204	.0305	.0422	.0550	.0688	.0833
	3	.0000	.0002	.0005	.0011	.0021	.0036	.0055	.0080	.0110
	4	.0000	.0000	.0000	.0000	.0001	.0002	.0003	.0005	.0008
	5	.0000	.0000	.0000	.0000	.0000	.0000	.0000	.0000	.0000
	6	.0000	.0000	.0000	.0000	.0000	.0000	.0000	.0000	.0000
7	0	.9321	.8681	.8080	.7514	.6983	.6485	.6017	.5578	.5168
	1	.0659	.1240	.1749	.2192	.2573	.2897	.3170	.3396	.3578
	2	.0020	.0076	.0162	.0274	.0406	.0555	.0716	.0886	.1061
	3	.0000	.0003	.0008	.0019	.0036	.0059	.0090	.0128	.0175
	4	.0000	.0000	.0000	.0001	.0002	.0004	.0007	.0011	.0017
	5	.0000	.0000	.0000	.0000	.0000	.0000	.0000	.0001	.0001
	6	.0000	.0000	.0000	.0000	.0000	.0000	.0000	.0000	.0000
	7	.0000	.0000	.0000	.0000	.0000	.0000	.0000	.0000	.0000
8	0	.9227	.8508	.7837	.7214	.6634	.6096	.5596	.5132	.4703
	1	.0746	.1389	.1939	.2405	.2793	.3113	.3370	.3570	.3721
	2	.0026	.0099	.0210	.0351	.0515	.0695	.0888	.1087	.1288
	3	.0001	.0004	.0013	.0029	.0054	.0089	.0134	.0189	.0255
	4	.0000	.0000	.0001	.0002	.0004	.0007	.0013	.0021	.0031
	5	.0000	.0000	.0000	.0000	.0000	.0000	.0001	.0001	.0002
	6	.0000	.0000	.0000	.0000	.0000	.0000	.0000	.0000	.0000
	7	.0000	.0000	.0000	.0000	.0000	.0000	.0000	.0000	.0000
	8	.0000	.0000	.0000	.0000	.0000	.0000	.0000	.0000	.0000

TABLE 5 BINOMIAL PROBABILITIES (*Continued*)

		p								
n	*x*	**.01**	**.02**	**.03**	**.04**	**.05**	**.06**	**.07**	**.08**	**.09**
9	0	.9135	.8337	.7602	.6925	.6302	.5730	.5204	.4722	.4279
	1	.0830	.1531	.2116	.2597	.2985	.3292	.3525	.3695	.3809
	2	.0034	.0125	.0262	.0433	.0629	.0840	.1061	.1285	.1507
	3	.0001	.0006	.0019	.0042	.0077	.0125	.0186	.0261	.0348
	4	.0000	.0000	.0001	.0003	.0006	.0012	.0021	.0034	.0052
	5	.0000	.0000	.0000	.0000	.0000	.0001	.0002	.0003	.0005
	6	.0000	.0000	.0000	.0000	.0000	.0000	.0000	.0000	.0000
	7	.0000	.0000	.0000	.0000	.0000	.0000	.0000	.0000	.0000
	8	.0000	.0000	.0000	.0000	.0000	.0000	.0000	.0000	.0000
	9	.0000	.0000	.0000	.0000	.0000	.0000	.0000	.0000	.0000
10	0	.9044	.8171	.7374	.6648	.5987	.5386	.4840	.4344	.3894
	1	.0914	.1667	.2281	.2770	.3151	.3438	.3643	.3777	.3851
	2	.0042	.0153	.0317	.0519	.0746	.0988	.1234	.1478	.1714
	3	.0001	.0008	.0026	.0058	.0105	.0168	.0248	.0343	.0452
	4	.0000	.0000	.0001	.0004	.0010	.0019	.0033	.0052	.0078
	5	.0000	.0000	.0000	.0000	.0001	.0001	.0003	.0005	.0009
	6	.0000	.0000	.0000	.0000	.0000	.0000	.0000	.0000	.0001
	7	.0000	.0000	.0000	.0000	.0000	.0000	.0000	.0000	.0000
	8	.0000	.0000	.0000	.0000	.0000	.0000	.0000	.0000	.0001
	9	.0000	.0000	.0000	.0000	.0000	.0000	.0000	.0000	.0001
	10	.0000	.0000	.0000	.0000	.0000	.0000	.0000	.0000	.0001
12	0	.8864	.7847	.6938	.6127	.5404	.4759	.4186	.3677	.3225
	1	.1074	.1922	.2575	.3064	.3413	.3645	.3781	.3837	.3827
	2	.0060	.0216	.0438	.0702	.0988	.1280	.1565	.1835	.2082
	3	.0002	.0015	.0045	.0098	.0173	.0272	.0393	.0532	.0686
	4	.0000	.0001	.0003	.0009	.0021	.0039	.0067	.0104	.0153
	5	.0000	.0000	.0000	.0001	.0002	.0004	.0008	.0014	.0024
	6	.0000	.0000	.0000	.0000	.0000	.0000	.0001	.0001	.0003
	7	.0000	.0000	.0000	.0000	.0000	.0000	.0000	.0000	.0000
	8	.0000	.0000	.0000	.0000	.0000	.0000	.0000	.0000	.0000
	9	.0000	.0000	.0000	.0000	.0000	.0000	.0000	.0000	.0000
	10	.0000	.0000	.0000	.0000	.0000	.0000	.0000	.0000	.0000
	11	.0000	.0000	.0000	.0000	.0000	.0000	.0000	.0000	.0000
	12	.0000	.0000	.0000	.0000	.0000	.0000	.0000	.0000	.0000
15	0	.8601	.7386	.6333	.5421	.4633	.3953	.3367	.2863	.2430
	1	.1303	.2261	.2938	.3388	.3658	.3785	.3801	.3734	.3605
	2	.0092	.0323	.0636	.0988	.1348	.1691	.2003	.2273	.2496
	3	.0004	.0029	.0085	.0178	.0307	.0468	.0653	.0857	.1070
	4	.0000	.0002	.0008	.0022	.0049	.0090	.0148	.0223	.0317
	5	.0000	.0000	.0001	.0002	.0006	.0013	.0024	.0043	.0069
	6	.0000	.0000	.0000	.0000	.0000	.0001	.0003	.0006	.0011
	7	.0000	.0000	.0000	.0000	.0000	.0000	.0000	.0001	.0001
	8	.0000	.0000	.0000	.0000	.0000	.0000	.0000	.0000	.0000
	9	.0000	.0000	.0000	.0000	.0000	.0000	.0000	.0000	.0000
	10	.0000	.0000	.0000	.0000	.0000	.0000	.0000	.0000	.0000
	11	.0000	.0000	.0000	.0000	.0000	.0000	.0000	.0000	.0000
	12	.0000	.0000	.0000	.0000	.0000	.0000	.0000	.0000	.0000
	13	.0000	.0000	.0000	.0000	.0000	.0000	.0000	.0000	.0000
	14	.0000	.0000	.0000	.0000	.0000	.0000	.0000	.0000	.0000
	15	.0000	.0000	.0000	.0000	.0000	.0000	.0000	.0000	.0000

TABLE 5 BINOMIAL PROBABILITIES (*Continued*)

		p								
n	*x*	**.01**	**.02**	**.03**	**.04**	**.05**	**.06**	**.07**	**.08**	**.09**
18	0	.8345	.6951	.5780	.4796	.3972	.3283	.2708	.2229	.1831
	1	.1517	.2554	.3217	.3597	.3763	.3772	.3669	.3489	.3260
	2	.0130	.0443	.0846	.1274	.1683	.2047	.2348	.2579	.2741
	3	.0007	.0048	.0140	.0283	.0473	.0697	.0942	.1196	.1446
	4	.0000	.0004	.0016	.0044	.0093	.0167	.0266	.0390	.0536
	5	.0000	.0000	.0001	.0005	.0014	.0030	.0056	.0095	.0148
	6	.0000	.0000	.0000	.0000	.0002	.0004	.0009	.0018	.0032
	7	.0000	.0000	.0000	.0000	.0000	.0000	.0001	.0003	.0005
	8	.0000	.0000	.0000	.0000	.0000	.0000	.0000	.0000	.0001
	9	.0000	.0000	.0000	.0000	.0000	.0000	.0000	.0000	.0000
	10	.0000	.0000	.0000	.0000	.0000	.0000	.0000	.0000	.0000
	11	.0000	.0000	.0000	.0000	.0000	.0000	.0000	.0000	.0000
	12	.0000	.0000	.0000	.0000	.0000	.0000	.0000	.0000	.0000
	13	.0000	.0000	.0000	.0000	.0000	.0000	.0000	.0000	.0000
	14	.0000	.0000	.0000	.0000	.0000	.0000	.0000	.0000	.0000
	15	.0000	.0000	.0000	.0000	.0000	.0000	.0000	.0000	.0000
	16	.0000	.0000	.0000	.0000	.0000	.0000	.0000	.0000	.0000
	17	.0000	.0000	.0000	.0000	.0000	.0000	.0000	.0000	.0000
	18	.0000	.0000	.0000	.0000	.0000	.0000	.0000	.0000	.0000
20	0	.8179	.6676	.5438	.4420	.3585	.2901	.2342	.1887	.1516
	1	.1652	.2725	.3364	.3683	.3774	.3703	.3526	.3282	.3000
	2	.0159	.0528	.0988	.1458	.1887	.2246	.2521	.2711	.2818
	3	.0010	.0065	.0183	.0364	.0596	.0860	.1139	.1414	.1672
	4	.0000	.0006	.0024	.0065	.0133	.0233	.0364	.0523	.0703
	5	.0000	.0000	.0002	.0009	.0022	.0048	.0088	.0145	.0222
	6	.0000	.0000	.0000	.0001	.0003	.0008	.0017	.0032	.0055
	7	.0000	.0000	.0000	.0000	.0000	.0001	.0002	.0005	.0011
	8	.0000	.0000	.0000	.0000	.0000	.0000	.0000	.0001	.0002
	9	.0000	.0000	.0000	.0000	.0000	.0000	.0000	.0000	.0000
	10	.0000	.0000	.0000	.0000	.0000	.0000	.0000	.0000	.0000
	11	.0000	.0000	.0000	.0000	.0000	.0000	.0000	.0000	.0000
	12	.0000	.0000	.0000	.0000	.0000	.0000	.0000	.0000	.0000
	13	.0000	.0000	.0000	.0000	.0000	.0000	.0000	.0000	.0000
	14	.0000	.0000	.0000	.0000	.0000	.0000	.0000	.0000	.0000
	15	.0000	.0000	.0000	.0000	.0000	.0000	.0000	.0000	.0000
	16	.0000	.0000	.0000	.0000	.0000	.0000	.0000	.0000	.0000
	17	.0000	.0000	.0000	.0000	.0000	.0000	.0000	.0000	.0000
	18	.0000	.0000	.0000	.0000	.0000	.0000	.0000	.0000	.0000
	19	.0000	.0000	.0000	.0000	.0000	.0000	.0000	.0000	.0000
	20	.0000	.0000	.0000	.0000	.0000	.0000	.0000	.0000	.0000

TABLE 5 BINOMIAL PROBABILITIES (*Continued*)

		p								
n	*x*	.10	.15	.20	.25	.30	.35	.40	.45	.50
2	0	.8100	.7225	.6400	.5625	.4900	.4225	.3600	.3025	.2500
	1	.1800	.2550	.3200	.3750	.4200	.4550	.4800	.4950	.5000
	2	.0100	.0225	.0400	.0625	.0900	.1225	.1600	.2025	.2500
3	0	.7290	.6141	.5120	.4219	.3430	.2746	.2160	.1664	.1250
	1	.2430	.3251	.3840	.4219	.4410	.4436	.4320	.4084	.3750
	2	.0270	.0574	.0960	.1406	.1890	.2389	.2880	.3341	.3750
	3	.0010	.0034	.0080	.0156	.0270	.0429	.0640	.0911	.1250
4	0	.6561	.5220	.4096	.3164	.2401	.1785	.1296	.0915	.0625
	1	.2916	.3685	.4096	.4219	.4116	.3845	.3456	.2995	.2500
	2	.0486	.0975	.1536	.2109	.2646	.3105	.3456	.3675	.3750
	3	.0036	.0115	.0256	.0469	.0756	.1115	.1536	.2005	.2500
	4	.0001	.0005	.0016	.0039	.0081	.0150	.0256	.0410	.0625
5	0	.5905	.4437	.3277	.2373	.1681	.1160	.0778	.0503	.0312
	1	.3280	.3915	.4096	.3955	.3602	.3124	.2592	.2059	.1562
	2	.0729	.1382	.2048	.2637	.3087	.3364	.3456	.3369	.3125
	3	.0081	.0244	.0512	.0879	.1323	.1811	.2304	.2757	.3125
	4	.0004	.0022	.0064	.0146	.0284	.0488	.0768	.1128	.1562
	5	.0000	.0001	.0003	.0010	.0024	.0053	.0102	.0185	.0312
6	0	.5314	.3771	.2621	.1780	.1176	.0754	.0467	.0277	.0156
	1	.3543	.3993	.3932	.3560	.3025	.2437	.1866	.1359	.0938
	2	.0984	.1762	.2458	.2966	.3241	.3280	.3110	.2780	.2344
	3	.0146	.0415	.0819	.1318	.1852	.2355	.2765	.3032	.3125
	4	.0012	.0055	.0154	.0330	.0595	.0951	.1382	.1861	.2344
	5	.0001	.0004	.0015	.0044	.0102	.0205	.0369	.0609	.0938
	6	.0000	.0000	.0001	.0002	.0007	.0018	.0041	.0083	.0156
7	0	.4783	.3206	.2097	.1335	.0824	.0490	.0280	.0152	.0078
	1	.3720	.3960	.3670	.3115	.2471	.1848	.1306	.0872	.0547
	2	.1240	.2097	.2753	.3115	.3177	.2985	.2613	.2140	.1641
	3	.0230	.0617	.1147	.1730	.2269	.2679	.2903	.2918	.2734
	4	.0026	.0109	.0287	.0577	.0972	.1442	.1935	.2388	.2734
	5	.0002	.0012	.0043	.0115	.0250	.0466	.0774	.1172	.1641
	6	.0000	.0001	.0004	.0013	.0036	.0084	.0172	.0320	.0547
	7	.0000	.0000	.0000	.0001	.0002	.0006	.0016	.0037	.0078
8	0	.4305	.2725	.1678	.1001	.0576	.0319	.0168	.0084	.0039
	1	.3826	.3847	.3355	.2670	.1977	.1373	.0896	.0548	.0312
	2	.1488	.2376	.2936	.3115	.2965	.2587	.2090	.1569	.1094
	3	.0331	.0839	.1468	.2076	.2541	.2786	.2787	.2568	.2188
	4	.0046	.0185	.0459	.0865	.1361	.1875	.2322	.2627	.2734
	5	.0004	.0026	.0092	.0231	.0467	.0808	.1239	.1719	.2188
	6	.0000	.0002	.0011	.0038	.0100	.0217	.0413	.0703	.1094
	7	.0000	.0000	.0001	.0004	.0012	.0033	.0079	.0164	.0312
	8	.0000	.0000	.0000	.0000	.0001	.0002	.0007	.0017	.0039

TABLE 5 BINOMIAL PROBABILITIES (*Continued*)

		p								
n	*x*	.10	.15	.20	.25	.30	.35	.40	.45	.50
9	0	.3874	.2316	.1342	.0751	.0404	.0207	.0101	.0046	.0020
	1	.3874	.3679	.3020	.2253	.1556	.1004	.0605	.0339	.0176
	2	.1722	.2597	.3020	.3003	.2668	.2162	.1612	.1110	.0703
	3	.0446	.1069	.1762	.2336	.2668	.2716	.2508	.2119	.1641
	4	.0074	.0283	.0661	.1168	.1715	.2194	.2508	.2600	.2461
	5	.0008	.0050	.0165	.0389	.0735	.1181	.1672	.2128	.2461
	6	.0001	.0006	.0028	.0087	.0210	.0424	.0743	.1160	.1641
	7	.0000	.0000	.0003	.0012	.0039	.0098	.0212	.0407	.0703
	8	.0000	.0000	.0000	.0001	.0004	.0013	.0035	.0083	.0176
	9	.0000	.0000	.0000	.0000	.0000	.0001	.0003	.0008	.0020
10	0	.3487	.1969	.1074	.0563	.0282	.0135	.0060	.0025	.0010
	1	.3874	.3474	.2684	.1877	.1211	.0725	.0403	.0207	.0098
	2	.1937	.2759	.3020	.2816	.2335	.1757	.1209	.0763	.0439
	3	.0574	.1298	.2013	.2503	.2668	.2522	.2150	.1665	.1172
	4	.0112	.0401	.0881	.1460	.2001	.2377	.2508	.2384	.2051
	5	.0015	.0085	.0264	.0584	.1029	.1536	.2007	.2340	.2461
	6	.0001	.0012	.0055	.0162	.0368	.0689	.1115	.1596	.2051
	7	.0000	.0001	.0008	.0031	.0090	.0212	.0425	.0746	.1172
	8	.0000	.0000	.0001	.0004	.0014	.0043	.0106	.0229	.0439
	9	.0000	.0000	.0000	.0000	.0001	.0005	.0016	.0042	.0098
	10	.0000	.0000	.0000	.0000	.0000	.0000	.0001	.0003	.0010
12	0	.2824	.1422	.0687	.0317	.0138	.0057	.0022	.0008	.0002
	1	.3766	.3012	.2062	.1267	.0712	.0368	.0174	.0075	.0029
	2	.2301	.2924	.2835	.2323	.1678	.1088	.0639	.0339	.0161
	3	.0853	.1720	.2362	.2581	.2397	.1954	.1419	.0923	.0537
	4	.0213	.0683	.1329	.1936	.2311	.2367	.2128	.1700	.1208
	5	.0038	.0193	.0532	.1032	.1585	.2039	.2270	.2225	.1934
	6	.0005	.0040	.0155	.0401	.0792	.1281	.1766	.2124	.2256
	7	.0000	.0006	.0033	.0115	.0291	.0591	.1009	.1489	.1934
	8	.0000	.0001	.0005	.0024	.0078	.0199	.0420	.0762	.1208
	9	.0000	.0000	.0001	.0004	.0015	.0048	.0125	.0277	.0537
	10	.0000	.0000	.0000	.0000	.0002	.0008	.0025	.0068	.0161
	11	.0000	.0000	.0000	.0000	.0000	.0001	.0003	.0010	.0029
	12	.0000	.0000	.0000	.0000	.0000	.0000	.0000	.0001	.0002
15	0	.2059	.0874	.0352	.0134	.0047	.0016	.0005	.0001	.0000
	1	.3432	.2312	.1319	.0668	.0305	.0126	.0047	.0016	.0005
	2	.2669	.2856	.2309	.1559	.0916	.0476	.0219	.0090	.0032
	3	.1285	.2184	.2501	.2252	.1700	.1110	.0634	.0318	.0139
	4	.0428	.1156	.1876	.2252	.2186	.1792	.1268	.0780	.0417
	5	.0105	.0449	.1032	.1651	.2061	.2123	.1859	.1404	.0916
	6	.0019	.0132	.0430	.0917	.1472	.1906	.2066	.1914	.1527
	7	.0003	.0030	.0138	.0393	.0811	.1319	.1771	.2013	.1964
	8	.0000	.0005	.0035	.0131	.0348	.0710	.1181	.1647	.1964
	9	.0000	.0001	.0007	.0034	.0016	.0298	.0612	.1048	.1527
	10	.0000	.0000	.0001	.0007	.0030	.0096	.0245	.0515	.0916
	11	.0000	.0000	.0000	.0001	.0006	.0024	.0074	.0191	.0417
	12	.0000	.0000	.0000	.0000	.0001	.0004	.0016	.0052	.0139
	13	.0000	.0000	.0000	.0000	.0000	.0001	.0003	.0010	.0032
	14	.0000	.0000	.0000	.0000	.0000	.0000	.0000	.0001	.0005
	15	.0000	.0000	.0000	.0000	.0000	.0000	.0000	.0000	.0000

TABLE 5 BINOMIAL PROBABILITIES (*Continued*)

		p								
n	*x*	**.10**	**.15**	**.20**	**.25**	**.30**	**.35**	**.40**	**.45**	**.50**
18	0	.1501	.0536	.0180	.0056	.0016	.0004	.0001	.0000	.0000
	1	.3002	.1704	.0811	.0338	.0126	.0042	.0012	.0003	.0001
	2	.2835	.2556	.1723	.0958	.0458	.0190	.0069	.0022	.0006
	3	.1680	.2406	.2297	.1704	.1046	.0547	.0246	.0095	.0031
	4	.0700	.1592	.2153	.2130	.1681	.1104	.0614	.0291	.0117
	5	.0218	.0787	.1507	.1988	.2017	.1664	.1146	.0666	.0327
	6	.0052	.0301	.0816	.1436	.1873	.1941	.1655	.1181	.0708
	7	.0010	.0091	.0350	.0820	.1376	.1792	.1892	.1657	.1214
	8	.0002	.0022	.0120	.0376	.0811	.1327	.1734	.1864	.1669
	9	.0000	.0004	.0033	.0139	.0386	.0794	.1284	.1694	.1855
	10	.0000	.0001	.0008	.0042	.0149	.0385	.0771	.1248	.1669
	11	.0000	.0000	.0001	.0010	.0046	.0151	.0374	.0742	.1214
	12	.0000	.0000	.0000	.0002	.0012	.0047	.0145	.0354	.0708
	13	.0000	.0000	.0000	.0000	.0002	.0012	.0045	.0134	.0327
	14	.0000	.0000	.0000	.0000	.0000	.0002	.0011	.0039	.0117
	15	.0000	.0000	.0000	.0000	.0000	.0000	.0002	.0009	.0031
	16	.0000	.0000	.0000	.0000	.0000	.0000	.0000	.0001	.0006
	17	.0000	.0000	.0000	.0000	.0000	.0000	.0000	.0000	.0001
	18	.0000	.0000	.0000	.0000	.0000	.0000	.0000	.0000	.0000
20	0	.1216	.0388	.0115	.0032	.0008	.0002	.0000	.0000	.0000
	1	.2702	.1368	.0576	.0211	.0068	.0020	.0005	.0001	.0000
	2	.2852	.2293	.1369	.0669	.0278	.0100	.0031	.0008	.0002
	3	.1901	.2428	.2054	.1339	.0716	.0323	.0123	.0040	.0011
	4	.0898	.1821	.2182	.1897	.1304	.0738	.0350	.0139	.0046
	5	.0319	.1028	.1746	.2023	.1789	.1272	.0746	.0365	.0148
	6	.0089	.0454	.1091	.1686	.1916	.1712	.1244	.0746	.0370
	7	.0020	.0160	.0545	.1124	.1643	.1844	.1659	.1221	.0739
	8	.0004	.0046	.0222	.0609	.1144	.1614	.1797	.1623	.1201
	9	.0001	.0011	.0074	.0271	.0654	.1158	.1597	.1771	.1602
	10	.0000	.0002	.0020	.0099	.0308	.0686	.1171	.1593	.1762
	11	.0000	.0000	.0005	.0030	.0120	.0336	.0710	.1185	.1602
	12	.0000	.0000	.0001	.0008	.0039	.0136	.0355	.0727	.1201
	13	.0000	.0000	.0000	.0002	.0010	.0045	.0146	.0366	.0739
	14	.0000	.0000	.0000	.0000	.0002	.0012	.0049	.0150	.0370
	15	.0000	.0000	.0000	.0000	.0000	.0003	.0013	.0049	.0148
	16	.0000	.0000	.0000	.0000	.0000	.0000	.0003	.0013	.0046
	17	.0000	.0000	.0000	.0000	.0000	.0000	.0000	.0002	.0011
	18	.0000	.0000	.0000	.0000	.0000	.0000	.0000	.0000	.0002
	19	.0000	.0000	.0000	.0000	.0000	.0000	.0000	.0000	.0000
	20	.0000	.0000	.0000	.0000	.0000	.0000	.0000	.0000	.0000

TABLE 6 VALUES OF $e^{-\mu}$

μ	$e^{-\mu}$	μ	$e^{-\mu}$	μ	$e^{-\mu}$
.00	1.0000				
.05	.9512	2.05	.1287	4.05	.0174
.10	.9048	2.10	.1225	4.10	.0166
.15	.8607	2.15	.1165	4.15	.0158
.20	.8187	2.20	.1108	4.20	.0150
.25	.7788	2.25	.1054	4.25	.0143
.30	.7408	2.30	.1003	4.30	.0136
.35	.7047	2.35	.0954	4.35	.0129
.40	.6703	2.40	.0907	4.40	.0123
.45	.6376	2.45	.0863	4.45	.0117
.50	.6065	2.50	.0821	4.50	.0111
.55	.5769	2.55	.0781	4.55	.0106
.60	.5488	2.60	.0743	4.60	.0101
.65	.5220	2.65	.0707	4.65	.0096
.70	.4966	2.70	.0672	4.70	.0091
.75	.4724	2.75	.0639	4.75	.0087
.80	.4493	2.80	.0608	4.80	.0082
.85	.4274	2.85	.0578	4.85	.0078
.90	.4066	2.90	.0550	4.90	.0074
.95	.3867	2.95	.0523	4.95	.0071
1.00	.3679	3.00	.0498	5.00	.0067
1.05	.3499	3.05	.0474	6.00	.0025
1.10	.3329	3.10	.0450	7.00	.0009
1.15	.3166	3.15	.0429	8.00	.000335
1.20	.3012	3.20	.0408	9.00	.000123
				10.00	.000045
1.25	.2865	3.25	.0388		
1.30	.2725	3.30	.0369		
1.35	.2592	3.35	.0351		
1.40	.2466	3.40	.0334		
1.45	.2346	3.45	.0317		
1.50	.2231	3.50	.0302		
1.55	.2122	3.55	.0287		
1.60	.2019	3.60	.0273		
1.65	.1920	3.65	.0260		
1.70	.1827	3.70	.0247		
1.75	.1738	3.75	.0235		
1.80	.1653	3.80	.0224		
1.85	.1572	3.85	.0213		
1.90	.1496	3.90	.0202		
1.95	.1423	3.95	.0193		
2.00	.1353	4.00	.0183		

TABLE 7 POISSON PROBABILITIES

Entries in the table give the probability of x occurrences for a Poisson process with a mean μ. For example, when $\mu = 2.5$, the probability of four occurrences is .1336.

	μ									
x	**0.1**	**0.2**	**0.3**	**0.4**	**0.5**	**0.6**	**0.7**	**0.8**	**0.9**	**1.0**
0	.9048	.8187	.7408	.6703	.6065	.5488	.4966	.4493	.4066	.3679
1	.0905	.1637	.2222	.2681	.3033	.3293	.3476	.3595	.3659	.3679
2	.0045	.0164	.0333	.0536	.0758	.0988	.1217	.1438	.1647	.1839
3	.0002	.0011	.0033	.0072	.0126	.0198	.0284	.0383	.0494	.0613
4	.0000	.0001	.0002	.0007	.0016	.0030	.0050	.0077	.0111	.0153
5	.0000	.0000	.0000	.0001	.0002	.0004	.0007	.0012	.0020	.0031
6	.0000	.0000	.0000	.0000	.0000	.0000	.0001	.0002	.0003	.0005
7	.0000	.0000	.0000	.0000	.0000	.0000	.0000	.0000	.0000	.0001

	μ									
x	**1.1**	**1.2**	**1.3**	**1.4**	**1.5**	**1.6**	**1.7**	**1.8**	**1.9**	**2.0**
0	.3329	.3012	.2725	.2466	.2231	.2019	.1827	.1653	.1496	.1353
1	.3662	.3614	.3543	.3452	.3347	.3230	.3106	.2975	.2842	.2707
2	.2014	.2169	.2303	.2417	.2510	.2584	.2640	.2678	.2700	.2707
3	.0738	.0867	.0998	.1128	.1255	.1378	.1496	.1607	.1710	.1804
4	.0203	.0260	.0324	.0395	.0471	.0551	.0636	.0723	.0812	.0902
5	.0045	.0062	.0084	.0111	.0141	.0176	.0216	.0260	.0309	.0361
6	.0008	.0012	.0018	.0026	.0035	.0047	.0061	.0078	.0098	.0120
7	.0001	.0002	.0003	.0005	.0008	.0011	.0015	.0020	.0027	.0034
8	.0000	.0000	.0001	.0001	.0001	.0002	.0003	.0005	.0006	.0009
9	.0000	.0000	.0000	.0000	.0000	.0000	.0001	.0001	.0001	.0002

	μ									
x	**2.1**	**2.2**	**2.3**	**2.4**	**2.5**	**2.6**	**2.7**	**2.8**	**2.9**	**3.0**
0	.1225	.1108	.1003	.0907	.0821	.0743	.0672	.0608	.0550	.0498
1	.2572	.2438	.2306	.2177	.2052	.1931	.1815	.1703	.1596	.1494
2	.2700	.2681	.2652	.2613	.2565	.2510	.2450	.2384	.2314	.2240
3	.1890	.1966	.2033	.2090	.2138	.2176	.2205	.2225	.2237	.2240
4	.0992	.1082	.1169	.1254	.1336	.1414	.1488	.1557	.1622	.1680
5	.0417	.0476	.0538	.0602	.0668	.0735	.0804	.0872	.0940	.1008
6	.0146	.0174	.0206	.0241	.0278	.0319	.0362	.0407	.0455	.0504
7	.0044	.0055	.0068	.0083	.0099	.0118	.0139	.0163	.0188	.0216
8	.0011	.0015	.0019	.0025	.0031	.0038	.0047	.0057	.0068	.0081
9	.0003	.0004	.0005	.0007	.0009	.0011	.0014	.0018	.0022	.0027
10	.0001	.0001	.0001	.0002	.0002	.0003	.0004	.0005	.0006	.0008
11	.0000	.0000	.0000	.0000	.0000	.0001	.0001	.0001	.0002	.0002
12	.0000	.0000	.0000	.0000	.0000	.0000	.0000	.0000	.0000	.0001

TABLE 7 POISSON PROBABILITIES (*Continued*)

	μ									
x	**3.1**	**3.2**	**3.3**	**3.4**	**3.5**	**3.6**	**3.7**	**3.8**	**3.9**	**4.0**
0	.0450	.0408	.0369	.0344	.0302	.0273	.0247	.0224	.0202	.0183
1	.1397	.1304	.1217	.1135	.1057	.0984	.0915	.0850	.0789	.0733
2	.2165	.2087	.2008	.1929	.1850	.1771	.1692	.1615	.1539	.1465
3	.2237	.2226	.2209	.2186	.2158	.2125	.2087	.2046	.2001	.1954
4	.1734	.1781	.1823	.1858	.1888	.1912	.1931	.1944	.1951	.1954
5	.1075	.1140	.1203	.1264	.1322	.1377	.1429	.1477	.1522	.1563
6	.0555	.0608	.0662	.0716	.0771	.0826	.0881	.0936	.0989	.1042
7	.0246	.0278	.0312	.0348	.0385	.0425	.0466	.0508	.0551	.0595
8	.0095	.0111	.0129	.0148	.0169	.0191	.0215	.0241	.0269	.0298
9	.0033	.0040	.0047	.0056	.0066	.0076	.0089	.0102	.0116	.0132
10	.0010	.0013	.0016	.0019	.0023	.0028	.0033	.0039	.0045	.0053
11	.0003	.0004	.0005	.0006	.0007	.0009	.0011	.0013	.0016	.0019
12	.0001	.0001	.0001	.0002	.0002	.0003	.0003	.0004	.0005	.0006
13	.0000	.0000	.0000	.0000	.0001	.0001	.0001	.0001	.0002	.0002
14	.0000	.0000	.0000	.0000	.0000	.0000	.0000	.0000	.0000	.0001

	μ									
x	**4.1**	**4.2**	**4.3**	**4.4**	**4.5**	**4.6**	**4.7**	**4.8**	**4.9**	**5.0**
0	.0166	.0150	.0136	.0123	.0111	.0101	.0091	.0082	.0074	.0067
1	.0679	.0630	.0583	.0540	.0500	.0462	.0427	.0395	.0365	.0337
2	.1393	.1323	.1254	.1188	.1125	.1063	.1005	.0948	.0894	.0842
3	.1904	.1852	.1798	.1743	.1687	.1631	.1574	.1517	.1460	.1404
4	.1951	.1944	.1933	.1917	.1898	.1875	.1849	.1820	.1789	.1755
5	.1600	.1633	.1662	.1687	.1708	.1725	.1738	.1747	.1753	.1755
6	.1093	.1143	.1191	.1237	.1281	.1323	.1362	.1398	.1432	.1462
7	.0640	.0686	.0732	.0778	.0824	.0869	.0914	.0959	.1002	.1044
8	.0328	.0360	.0393	.0428	.0463	.0500	.0537	.0575	.0614	.0653
9	.0150	.0168	.0188	.0209	.0232	.0255	.0280	.0307	.0334	.0363
10	.0061	.0071	.0081	.0092	.0104	.0118	.0132	.0147	.0164	.0181
11	.0023	.0027	.0032	.0037	.0043	.0049	.0056	.0064	.0073	.0082
12	.0008	.0009	.0011	.0014	.0016	.0019	.0022	.0026	.0030	.0034
13	.0002	.0003	.0004	.0005	.0006	.0007	.0008	.0009	.0011	.0013
14	.0001	.0001	.0001	.0001	.0002	.0002	.0003	.0003	.0004	.0005
15	.0000	.0000	.0000	.0000	.0001	.0001	.0001	.0001	.0001	.0002

	μ									
x	**5.1**	**5.2**	**5.3**	**5.4**	**5.5**	**5.6**	**5.7**	**5.8**	**5.9**	**6.0**
0	.0061	.0055	.0050	.0045	.0041	.0037	.0033	.0030	.0027	.0025
1	.0311	.0287	.0265	.0244	.0225	.0207	.0191	.0176	.0162	.0149
2	.0793	.0746	.0701	.0659	.0618	.0580	.0544	.0509	.0477	.0446
3	.1348	.1293	.1239	.1185	.1133	.1082	.1033	.0985	.0938	.0892
4	.1719	.1681	.1641	.1600	.1558	.1515	.1472	.1428	.1383	.1339

TABLE 7 POISSON PROBABILITIES (*Continued*)

5	.1753	.1748	.1740	.1728	.1714	.1697	.1678	.1656	.1632	.1606
6	.1490	.1515	.1537	.1555	.1571	.1587	.1594	.1601	.1605	.1606
7	.1086	.1125	.1163	.1200	.1234	.1267	.1298	.1326	.1353	.1377
8	.0692	.0731	.0771	.0810	.0849	.0887	.0925	.0962	.0998	.1033
9	.0392	.0423	.0454	.0486	.0519	.0552	.0586	.0620	.0654	.0688
10	.0200	.0220	.0241	.0262	.0285	.0309	.0334	.0359	.0386	.0413
11	.0093	.0104	.0116	.0129	.0143	.0157	.0173	.0190	.0207	.0225
12	.0039	.0045	.0051	.0058	.0065	.0073	.0082	.0092	.0102	.0113
13	.0015	.0018	.0021	.0024	.0028	.0032	.0036	.0041	.0046	.0052
14	.0006	.0007	.0008	.0009	.0011	.0013	.0015	.0017	.0019	.0022
15	.0002	.0002	.0003	.0003	.0004	.0005	.0006	.0007	.0008	.0009
16	.0001	.0001	.0001	.0001	.0001	.0002	.0002	.0002	.0003	.0003
17	.0000	.0000	.0000	.0000	.0000	.0001	.0001	.0001	.0001	.0001

	μ									
x	**6.1**	**6.2**	**6.3**	**6.4**	**6.5**	**6.6**	**6.7**	**6.8**	**6.9**	**7.0**
0	.0022	.0020	.0018	.0017	.0015	.0014	.0012	.0011	.0010	.0009
1	.0137	.0126	.0116	.0106	.0098	.0090	.0082	.0076	.0070	.0064
2	.0417	.0390	.0364	.0340	.0318	.0296	.0276	.0258	.0240	.0223
3	.0848	.0806	.0765	.0726	.0688	.0652	.0617	.0584	.0552	.0521
4	.1294	.1249	.1205	.1162	.1118	.1076	.1034	.0992	.0952	.0912
5	.1579	.1549	.1519	.1487	.1454	.1420	.1385	.1349	.1314	.1277
6	.1605	.1601	.1595	.1586	.1575	.1562	.1546	.1529	.1511	.1490
7	.1399	.1418	.1435	.1450	.1462	.1472	.1480	.1486	.1489	.1490
8	.1066	.1099	.1130	.1160	.1188	.1215	.1240	.1263	.1284	.1304
9	.0723	.0757	.0791	.0825	.0858	.0891	.0923	.0954	.0985	.1014
10	.0441	.0469	.0498	.0528	.0558	.0588	.0618	.0649	.0679	.0710
11	.0245	.0265	.0285	.0307	.0330	.0353	.0377	.0401	.0426	.0452
12	.0124	.0137	.0150	.0164	.0179	.0194	.0210	.0227	.0245	.0264
13	.0058	.0065	.0073	.0081	.0089	.0098	.0108	.0119	.0130	.0142
14	.0025	.0029	.0033	.0037	.0041	.0046	.0052	.0058	.0064	.0071
15	.0010	.0012	.0014	.0016	.0018	.0020	.0023	.0026	.0029	.0033
16	.0004	.0005	.0005	.0006	.0007	.0008	.0010	.0011	.0013	.0014
17	.0001	.0002	.0002	.0002	.0003	.0003	.0004	.0004	.0005	.0006
18	.0000	.0001	.0001	.0001	.0001	.0001	.0001	.0002	.0002	.0002
19	.0000	.0000	.0000	.0000	.0000	.0000	.0000	.0001	.0001	.0001

	μ									
x	**7.1**	**7.2**	**7.3**	**7.4**	**7.5**	**7.6**	**7.7**	**7.8**	**7.9**	**8.0**
0	.0008	.0007	.0007	.0006	.0006	.0005	.0005	.0004	.0004	.0003
1	.0059	.0054	.0049	.0045	.0041	.0038	.0035	.0032	.0029	.0027
2	.0208	.0194	.0180	.0167	.0156	.0145	.0134	.0125	.0116	.0107
3	.0492	.0464	.0438	.0413	.0389	.0366	.0345	.0324	.0305	.0286
4	.0874	.0836	.0799	.0764	.0729	.0696	.0663	.0632	.0602	.0573

TABLE 7 POISSON PROBABILITIES (*Continued*)

	μ									
x	**7.1**	**7.2**	**7.3**	**7.4**	**7.5**	**7.6**	**7.7**	**7.8**	**7.9**	**8.0**
5	.1241	.1204	.1167	.1130	.1094	.1057	.1021	.0986	.0951	.0916
6	.1468	.1445	.1420	.1394	.1367	.1339	.1311	.1282	.1252	.1221
7	.1489	.1486	.1481	.1474	.1465	.1454	.1442	.1428	.1413	.1396
8	.1321	.1337	.1351	.1363	.1373	.1382	.1388	.1392	.1395	.1396
9	.1042	.1070	.1096	.1121	.1144	.1167	.1187	.1207	.1224	.1241
10	.0740	.0770	.0800	.0829	.0858	.0887	.0914	.0941	.0967	.0993
11	.0478	.0504	.0531	.0558	.0585	.0613	.0640	.0667	.0695	.0722
12	.0283	.0303	.0323	.0344	.0366	.0388	.0411	.0434	.0457	.0481
13	.0154	.0168	.0181	.0196	.0211	.0227	.0243	.0260	.0278	.0296
14	.0078	.0086	.0095	.0104	.0113	.0123	.0134	.0145	.0157	.0169
15	.0037	.0041	.0046	.0051	.0057	.0062	.0069	.0075	.0083	.0090
16	.0016	.0019	.0021	.0024	.0026	.0030	.0033	.0037	.0041	.0045
17	.0007	.0008	.0009	.0010	.0012	.0013	.0015	.0017	.0019	.0021
18	.0003	.0003	.0004	.0004	.0005	.0006	.0006	.0007	.0008	.0009
19	.0001	.0001	.0001	.0002	.0002	.0002	.0003	.0003	.0003	.0004
20	.0000	.0000	.0001	.0001	.0001	.0001	.0001	.0001	.0001	.0002
21	.0000	.0000	.0000	.0000	.0000	.0000	.0000	.0000	.0001	.0001

	μ									
x	**8.1**	**8.2**	**8.3**	**8.4**	**8.5**	**8.6**	**8.7**	**8.8**	**8.9**	**9.0**
0	.0003	.0003	.0002	.0002	.0002	.0002	.0002	.0002	.0001	.0001
1	.0025	.0023	.0021	.0019	.0017	.0016	.0014	.0013	.0012	.0011
2	.0100	.0092	.0086	.0079	.0074	.0068	.0063	.0058	.0054	.0050
3	.0269	.0252	.0237	.0222	.0208	.0195	.0183	.0171	.0160	.0150
4	.0544	.0517	.0491	.0466	.0443	.0420	.0398	.0377	.0357	.0337
5	.0882	.0849	.0816	.0784	.0752	.0722	.0692	.0663	.0635	.0607
6	.1191	.1160	.1128	.1097	.1066	.1034	.1003	.0972	.0941	.0911
7	.1378	.1358	.1338	.1317	.1294	.1271	.1247	.1222	.1197	.1171
8	.1395	.1392	.1388	.1382	.1375	.1366	.1356	.1344	.1332	.1318
9	.1256	.1269	.1280	.1290	.1299	.1306	.1311	.1315	.1317	.1318
10	.1017	.1040	.1063	.1084	.1104	.1123	.1140	.1157	.1172	.1186
11	.0749	.0776	.0802	.0828	.0853	.0878	.0902	.0925	.0948	.0970
12	.0505	.0530	.0555	.0579	.0604	.0629	.0654	.0679	.0703	.0728
13	.0315	.0334	.0354	.0374	.0395	.0416	.0438	.0459	.0481	.0504
14	.0182	.0196	.0210	.0225	.0240	.0256	.0272	.0289	.0306	.0324
15	.0098	.0107	.0116	.0126	.0136	.0147	.0158	.0169	.0182	.1094
16	.0050	.0055	.0060	.0066	.0072	.0079	.0086	.0093	.0101	.0109
17	.0024	.0026	.0029	.0033	.0036	.0040	.0044	.0048	.0053	.0058
18	.0011	.0012	.0014	.0015	.0017	.0019	.0021	.0024	.0026	.0029
19	.0005	.0005	.0006	.0007	.0008	.0009	.0010	.0011	.0012	.0014
20	.0002	.0002	.0002	.0003	.0003	.0004	.0004	.0005	.0005	.0006
21	.0001	.0001	.0001	.0001	.0001	.0002	.0002	.0002	.0002	.0003
22	.0000	.0000	.0000	.0000	.0001	.0001	.0001	.0001	.0001	.0001

TABLE 7 POISSON PROBABILITIES (*Continued*)

	μ									
x	9.1	9.2	9.3	9.4	9.5	9.6	9.7	9.8	9.9	10
0	.0001	.0001	.0001	.0001	.0001	.0001	.0001	.0001	.0001	.0000
1	.0010	.0009	.0009	.0008	.0007	.0007	.0006	.0005	.0005	.0005
2	.0046	.0043	.0040	.0037	.0034	.0031	.0029	.0027	.0025	.0023
3	.0140	.0131	.0123	.0115	.0107	.0100	.0093	.0087	.0081	.0076
4	.0319	.0302	.0285	.0269	.0254	.0240	.0226	.0213	.0201	.0189
5	.0581	.0555	.0530	.0506	.0483	.0460	.0439	.0418	.0398	.0378
6	.0881	.0851	.0822	.0793	.0764	.0736	.0709	.0682	.0656	.0631
7	.1145	.1118	.1091	.1064	.1037	.1010	.0982	.0955	.0928	.0901
8	.1302	.1286	.1269	.1251	.1232	.1212	.1191	.1170	.1148	.1126
9	.1317	.1315	.1311	.1306	.1300	.1293	.1284	.1274	.1263	.1251
10	.1198	.1210	.1219	.1228	.1235	.1241	.1245	.1249	.1250	.1251
11	.0991	.1012	.1031	.1049	.1067	.1083	.1098	.1112	.1125	.1137
12	.0752	.0776	.0799	.0822	.0844	.0866	.0888	.0908	.0928	.0948
13	.0526	.0549	.0572	.0594	.0617	.0640	.0662	.0685	.0707	.0729
14	.0342	.0361	.0380	.0399	.0419	.0439	.0459	.0479	.0500	.0521
15	.0208	.0221	.0235	.0250	.0265	.0281	.0297	.0313	.0330	.0347
16	.0118	.0127	.0137	.0147	.0157	.0168	.0180	.0192	.0204	.0217
17	.0063	.0069	.0075	.0081	.0088	.0095	.0103	.0111	.0119	.0128
18	.0032	.0035	.0039	.0042	.0046	.0051	.0055	.0060	.0065	.0071
19	.0015	.0017	.0019	.0021	.0023	.0026	.0028	.0031	.0034	.0037
20	.0007	.0008	.0009	.0010	.0011	.0012	.0014	.0015	.0017	.0019
21	.0003	.0003	.0004	.0004	.0005	.0006	.0006	.0007	.0008	.0009
22	.0001	.0001	.0002	.0002	.0002	.0002	.0003	.0003	.0004	.0004
23	.0000	.0001	.0001	.0001	.0001	.0001	.0001	.0001	.0002	.0002
24	.0000	.0000	.0000	.0000	.0000	.0000	.0000	.0001	.0001	.0001

	μ									
x	11	12	13	14	15	16	17	18	19	20
0	.0000	.0000	.0000	.0000	.0000	.0000	.0000	.0000	.0000	.0000
1	.0002	.0001	.0000	.0000	.0000	.0000	.0000	.0000	.0000	.0000
2	.0010	.0004	.0002	.0001	.0000	.0000	.0000	.0000	.0000	.0000
3	.0037	.0018	.0008	.0004	.0002	.0001	.0000	.0000	.0000	.0000
4	.0102	.0053	.0027	.0013	.0006	.0003	.0001	.0001	.0000	.0000
5	.0224	.0127	.0070	.0037	.0019	.0010	.0005	.0002	.0001	.0001
6	.0411	.0255	.0152	.0087	.0048	.0026	.0014	.0007	.0004	.0002
7	.0646	.0437	.0281	.0174	.0104	.0060	.0034	.0018	.0010	.0005
8	.0888	.0655	.0457	.0304	.0194	.0120	.0072	.0042	.0024	.0013
9	.1085	.0874	.0661	.0473	.0324	.0213	.0135	.0083	.0050	.0029
10	.1194	.1048	.0859	.0663	.0486	.0341	.0230	.0150	.0095	.0058
11	.1194	.1144	.1015	.0844	.0663	.0496	.0355	.0245	.0164	.0106
12	.1094	.1144	.1099	.0984	.0829	.0661	.0504	.0368	.0259	.0176
13	.0926	.1056	.1099	.1060	.0956	.0814	.0658	.0509	.0378	.0271
14	.0728	.0905	.1021	.1060	.1024	.0930	.0800	.0655	.0514	.0387

TABLE 7 POISSON PROBABILITIES (*Continued*)

	μ									
x	**11**	**12**	**13**	**14**	**15**	**16**	**17**	**18**	**19**	**20**
15	.0534	.0724	.0885	.0989	.1024	.0992	.0906	.0786	.0650	.0516
16	.0367	.0543	.0719	.0866	.0960	.0992	.0963	.0884	.0772	.0646
17	.0237	.0383	.0550	.0713	.0847	.0934	.0963	.0936	.0863	.0760
18	.0145	.0256	.0397	.0554	.0706	.0830	.0909	.0936	.0911	.0844
19	.0084	.0161	.0272	.0409	.0557	.0699	.0814	.0887	.0911	.0888
20	.0046	.0097	.0177	.0286	.0418	.0559	.0692	.0798	.0866	.0888
21	.0024	.0055	.0109	.0191	.0299	.0426	.0560	.0684	.0783	.0846
22	.0012	.0030	.0065	.0121	.0204	.0310	.0433	.0560	.0676	.0769
23	.0006	.0016	.0037	.0074	.0133	.0216	.0320	.0438	.0559	.0669
24	.0003	.0008	.0020	.0043	.0083	.0144	.0226	.0328	.0442	.0557
25	.0001	.0004	.0010	.0024	.0050	.0092	.0154	.0237	.0336	.0446
26	.0000	.0002	.0005	.0013	.0029	.0057	.0101	.0164	.0246	.0343
27	.0000	.0001	.0002	.0007	.0016	.0034	.0063	.0109	.0173	.0254
28	.0000	.0000	.0001	.0003	.0009	.0019	.0038	.0070	.0117	.0181
29	.0000	.0000	.0001	.0002	.0004	.0011	.0023	.0044	.0077	.0125
30	.0000	.0000	.0000	.0001	.0002	.0006	.0013	.0026	.0049	.0083
31	.0000	.0000	.0000	.0000	.0001	.0003	.0007	.0015	.0030	.0054
32	.0000	.0000	.0000	.0000	.0001	.0001	.0004	.0009	.0018	.0034
33	.0000	.0000	.0000	.0000	.0000	.0001	.0002	.0005	.0010	.0020
34	.0000	.0000	.0000	.0000	.0000	.0000	.0001	.0002	.0006	.0012
35	.0000	.0000	.0000	.0000	.0000	.0000	.0000	.0001	.0003	.0007
36	.0000	.0000	.0000	.0000	.0000	.0000	.0000	.0001	.0002	.0004
37	.0000	.0000	.0000	.0000	.0000	.0000	.0000	.0000	.0001	.0002
38	.0000	.0000	.0000	.0000	.0000	.0000	.0000	.0000	.0000	.0001
39	.0000	.0000	.0000	.0000	.0000	.0000	.0000	.0000	.0000	.0001

TABLE 8 CRITICAL VALUES FOR THE DURBIN-WATSON TEST FOR AUTOCORRELATION

Entries in the table give the critical values for a one-tailed Durbin-Watson test for autocorrelation. For a two-tailed test, the level of significance is doubled.

Significance Points of d_L and d_U: $\alpha = .05$
Number of Independent Variables

	k	1		2		3		4		5
n	d_L	d_U	d_L	d_U	d_L	d_U	d_L	d_U	d_L	d_U
15	1.08	1.36	0.95	1.54	0.82	1.75	0.69	1.97	0.56	2.21
16	1.10	1.37	0.98	1.54	0.86	1.73	0.74	1.93	0.62	2.15
17	1.13	1.38	1.02	1.54	0.90	1.71	0.78	1.90	0.67	2.10
18	1.16	1.39	1.05	1.53	0.93	1.69	0.82	1.87	0.71	2.06
19	1.18	1.40	1.08	1.53	0.97	1.68	0.86	1.85	0.75	2.02
20	1.20	1.41	1.10	1.54	1.00	1.68	0.90	1.83	0.79	1.99
21	1.22	1.42	1.13	1.54	1.03	1.67	0.93	1.81	0.83	1.96
22	1.24	1.43	1.15	1.54	1.05	1.66	0.96	1.80	0.86	1.94
23	1.26	1.44	1.17	1.54	1.08	1.66	0.99	1.79	0.90	1.92
24	1.27	1.45	1.19	1.55	1.10	1.66	1.01	1.78	0.93	1.90
25	1.29	1.45	1.21	1.55	1.12	1.66	1.04	1.77	0.95	1.89
26	1.30	1.46	1.22	1.55	1.14	1.65	1.06	1.76	0.98	1.88
27	1.32	1.47	1.24	1.56	1.16	1.65	1.08	1.76	1.01	1.86
28	1.33	1.48	1.26	1.56	1.18	1.65	1.10	1.75	1.03	1.85
29	1.34	1.48	1.27	1.56	1.20	1.65	1.12	1.74	1.05	1.84
30	1.35	1.49	1.28	1.57	1.21	1.65	1.14	1.74	1.07	1.83
31	1.36	1.50	1.30	1.57	1.23	1.65	1.16	1.74	1.09	1.83
32	1.37	1.50	1.31	1.57	1.24	1.65	1.18	1.73	1.11	1.82
33	1.38	1.51	1.32	1.58	1.26	1.65	1.19	1.73	1.13	1.81
34	1.39	1.51	1.33	1.58	1.27	1.65	1.21	1.73	1.15	1.81
35	1.40	1.52	1.34	1.58	1.28	1.65	1.22	1.73	1.16	1.80
36	1.41	1.52	1.35	1.59	1.29	1.65	1.24	1.73	1.18	1.80
37	1.42	1.53	1.36	1.59	1.31	1.66	1.25	1.72	1.19	1.80
38	1.43	1.54	1.37	1.59	1.32	1.66	1.26	1.72	1.21	1.79
39	1.43	1.54	1.38	1.60	1.33	1.66	1.27	1.72	1.22	1.79
40	1.44	1.54	1.39	1.60	1.34	1.66	1.29	1.72	1.23	1.79
45	1.48	1.57	1.43	1.62	1.38	1.67	1.34	1.72	1.29	1.78
50	1.50	1.59	1.46	1.63	1.42	1.67	1.38	1.72	1.34	1.77
55	1.53	1.60	1.49	1.64	1.45	1.68	1.41	1.72	1.38	1.77
60	1.55	1.62	1.51	1.65	1.48	1.69	1.44	1.73	1.41	1.77
65	1.57	1.63	1.54	1.66	1.50	1.70	1.47	1.73	1.44	1.77
70	1.58	1.64	1.55	1.67	1.52	1.70	1.49	1.74	1.46	1.77
75	1.60	1.65	1.57	1.68	1.54	1.71	1.51	1.74	1.49	1.77
80	1.61	1.66	1.59	1.69	1.56	1.72	1.53	1.74	1.51	1.77
85	1.62	1.67	1.60	1.70	1.57	1.72	1.55	1.75	1.52	1.77
90	1.63	1.68	1.61	1.70	1.59	1.73	1.57	1.75	1.54	1.78
95	1.64	1.69	1.62	1.71	1.60	1.73	1.58	1.75	1.56	1.78
100	1.65	1.69	1.63	1.72	1.61	1.74	1.59	1.76	1.57	1.78

TABLE 8 CRITICAL VALUES FOR THE DURBIN-WATSON TEST FOR AUTOCORRELATION (*Continued*)

Significance Points of d_L and d_U: $\alpha = .025$
Number of Independent Variables

k	1		2		3		4		5	
n	d_L	d_U	d_L	d_U	d_L	d_U	d_L	d_U	d_L	d_U
15	0.95	1.23	0.83	1.40	0.71	1.61	0.59	1.84	0.48	2.09
16	0.98	1.24	0.86	1.40	0.75	1.59	0.64	1.80	0.53	2.03
17	1.01	1.25	0.90	1.40	0.79	1.58	0.68	1.77	0.57	1.98
18	1.03	1.26	0.93	1.40	0.82	1.56	0.72	1.74	0.62	1.93
19	1.06	1.28	0.96	1.41	0.86	1.55	0.76	1.72	0.66	1.90
20	1.08	1.28	0.99	1.41	0.89	1.55	0.79	1.70	0.70	1.87
21	1.10	1.30	1.01	1.41	0.92	1.54	0.83	1.69	0.73	1.84
22	1.12	1.31	1.04	1.42	0.95	1.54	0.86	1.68	0.77	1.82
23	1.14	1.32	1.06	1.42	0.97	1.54	0.89	1.67	0.80	1.80
24	1.16	1.33	1.08	1.43	1.00	1.54	0.91	1.66	0.83	1.79
25	1.18	1.34	1.10	1.43	1.02	1.54	0.94	1.65	0.86	1.77
26	1.19	1.35	1.12	1.44	1.04	1.54	0.96	1.65	0.88	1.76
27	1.21	1.36	1.13	1.44	1.06	1.54	0.99	1.64	0.91	1.75
28	1.22	1.37	1.15	1.45	1.08	1.54	1.01	1.64	0.93	1.74
29	1.24	1.38	1.17	1.45	1.10	1.54	1.03	1.63	0.96	1.73
30	1.25	1.38	1.18	1.46	1.12	1.54	1.05	1.63	0.98	1.73
31	1.26	1.39	1.20	1.47	1.13	1.55	1.07	1.63	1.00	1.72
32	1.27	1.40	1.21	1.47	1.15	1.55	1.08	1.63	1.02	1.71
33	1.28	1.41	1.22	1.48	1.16	1.55	1.10	1.63	1.04	1.71
34	1.29	1.41	1.24	1.48	1.17	1.55	1.12	1.63	1.06	1.70
35	1.30	1.42	1.25	1.48	1.19	1.55	1.13	1.63	1.07	1.70
36	1.31	1.43	1.26	1.49	1.20	1.56	1.15	1.63	1.09	1.70
37	1.32	1.43	1.27	1.49	1.21	1.56	1.16	1.62	1.10	1.70
38	1.33	1.44	1.28	1.50	1.23	1.56	1.17	1.62	1.12	1.70
39	1.34	1.44	1.29	1.50	1.24	1.56	1.19	1.63	1.13	1.69
40	1.35	1.45	1.30	1.51	1.25	1.57	1.20	1.63	1.15	1.69
45	1.39	1.48	1.34	1.53	1.30	1.58	1.25	1.63	1.21	1.69
50	1.42	1.50	1.38	1.54	1.34	1.59	1.30	1.64	1.26	1.69
55	1.45	1.52	1.41	1.56	1.37	1.60	1.33	1.64	1.30	1.69
60	1.47	1.54	1.44	1.57	1.40	1.61	1.37	1.65	1.33	1.69
65	1.49	1.55	1.46	1.59	1.43	1.62	1.40	1.66	1.36	1.69
70	1.51	1.57	1.48	1.60	1.45	1.63	1.42	1.66	1.39	1.70
75	1.53	1.58	1.50	1.61	1.47	1.64	1.45	1.67	1.42	1.70
80	1.54	1.59	1.52	1.62	1.49	1.65	1.47	1.67	1.44	1.70
85	1.56	1.60	1.53	1.63	1.51	1.65	1.49	1.68	1.46	1.71
90	1.57	1.61	1.55	1.64	1.53	1.66	1.50	1.69	1.48	1.71
95	1.58	1.62	1.56	1.65	1.54	1.67	1.52	1.69	1.50	1.71
100	1.59	1.63	1.57	1.65	1.55	1.67	1.53	1.70	1.51	1.72

TABLE 8 CRITICAL VALUES FOR THE DURBIN-WATSON TEST FOR AUTOCORRELATION (*Continued*)

Significance Points of d_L and d_U: $\alpha = .01$
Number of Independent Variables

k	1		2		3		4		5	
n	d_L	d_U	d_L	d_U	d_L	d_U	d_L	d_U	d_L	d_U
15	0.81	1.07	0.70	1.25	0.59	1.46	0.49	1.70	0.39	1.96
16	0.84	1.09	0.74	1.25	0.63	1.44	0.53	1.66	0.44	1.90
17	0.87	1.10	0.77	1.25	0.67	1.43	0.57	1.63	0.48	1.85
18	0.90	1.12	0.80	1.26	0.71	1.42	0.61	1.60	0.52	1.80
19	0.93	1.13	0.83	1.26	0.74	1.41	0.65	1.58	0.56	1.77
20	0.95	1.15	0.86	1.27	0.77	1.41	0.68	1.57	0.60	1.74
21	0.97	1.16	0.89	1.27	0.80	1.41	0.72	1.55	0.63	1.71
22	1.00	1.17	0.91	1.28	0.83	1.40	0.75	1.54	0.66	1.69
23	1.02	1.19	0.94	1.29	0.86	1.40	0.77	1.53	0.70	1.67
24	1.04	1.20	0.96	1.30	0.88	1.41	0.80	1.53	0.72	1.66
25	1.05	1.21	0.98	1.30	0.90	1.41	0.83	1.52	0.75	1.65
26	1.07	1.22	1.00	1.31	0.93	1.41	0.85	1.52	0.78	1.64
27	1.09	1.23	1.02	1.32	0.95	1.41	0.88	1.51	0.81	1.63
28	1.10	1.24	1.04	1.32	0.97	1.41	0.90	1.51	0.83	1.62
29	1.12	1.25	1.05	1.33	0.99	1.42	0.92	1.51	0.85	1.61
30	1.13	1.26	1.07	1.34	1.01	1.42	0.94	1.51	0.88	1.61
31	1.15	1.27	1.08	1.34	1.02	1.42	0.96	1.51	0.90	1.60
32	1.16	1.28	1.10	1.35	1.04	1.43	0.98	1.51	0.92	1.60
33	1.17	1.29	1.11	1.36	1.05	1.43	1.00	1.51	0.94	1.59
34	1.18	1.30	1.13	1.36	1.07	1.43	1.01	1.51	0.95	1.59
35	1.19	1.31	1.14	1.37	1.08	1.44	1.03	1.51	0.97	1.59
36	1.21	1.32	1.15	1.38	1.10	1.44	1.04	1.51	0.99	1.59
37	1.22	1.32	1.16	1.38	1.11	1.45	1.06	1.51	1.00	1.59
38	1.23	1.33	1.18	1.39	1.12	1.45	1.07	1.52	1.02	1.58
39	1.24	1.34	1.19	1.39	1.14	1.45	1.09	1.52	1.03	1.58
40	1.25	1.34	1.20	1.40	1.15	1.46	1.10	1.52	1.05	1.58
45	1.29	1.38	1.24	1.42	1.20	1.48	1.16	1.53	1.11	1.58
50	1.32	1.40	1.28	1.45	1.24	1.49	1.20	1.54	1.16	1.59
55	1.36	1.43	1.32	1.47	1.28	1.51	1.25	1.55	1.21	1.59
60	1.38	1.45	1.35	1.48	1.32	1.52	1.28	1.56	1.25	1.60
65	1.41	1.47	1.38	1.50	1.35	1.53	1.31	1.57	1.28	1.61
70	1.43	1.49	1.40	1.52	1.37	1.55	1.34	1.58	1.31	1.61
75	1.45	1.50	1.42	1.53	1.39	1.56	1.37	1.59	1.34	1.62
80	1.47	1.52	1.44	1.54	1.42	1.57	1.39	1.60	1.36	1.62
85	1.48	1.53	1.46	1.55	1.43	1.58	1.41	1.60	1.39	1.63
90	1.50	1.54	1.47	1.56	1.45	1.59	1.43	1.61	1.41	1.64
95	1.51	1.55	1.49	1.57	1.47	1.60	1.45	1.62	1.42	1.64
100	1.52	1.56	1.50	1.58	1.48	1.60	1.46	1.63	1.44	1.65

TABLE 9 T_L VALUES FOR THE MANN-WHITNEY-WILCOXON TEST

Reject the hypothesis of identical populations if the sum of the ranks for the n_1 items is *less* than the value T_L shown in the following table or if the sum of the ranks for the n_1 items is *greater* than the value T_U where

$$T_U = n_1(n_1 + n_2 + 1) - T_L$$

$\alpha = .01$		n_2 = 2	3	4	5	6	7	8	9	10
	2	3	3	3	4	4	4	5	5	5
	3	6	7	7	8	9	9	10	11	11
	4	10	11	12	13	14	15	16	17	18
	5	16	17	18	20	21	22	24	25	27
n_1	6	22	24	25	27	29	30	32	34	36
	7	29	31	33	35	37	40	42	44	46
	8	38	40	42	45	47	50	52	55	57
	9	47	50	52	55	58	61	64	67	70
	10	57	60	63	67	70	73	76	80	83

$\alpha = .05$		n_2 = 2	3	4	5	6	7	8	9	10
	2	3	3	3	3	3	3	4	4	4
	3	6	6	6	7	8	8	9	9	10
	4	10	10	11	12	13	14	15	15	16
	5	15	16	17	18	19	21	22	23	24
n_1	6	21	23	24	25	27	28	30	32	33
	7	28	30	32	34	35	37	39	41	43
	8	37	39	41	43	45	47	50	52	54
	9	46	48	50	53	56	58	61	63	66
	10	56	59	61	64	67	70	73	76	79

TABLE 10 CRITICAL VALUES OF THE STUDENTIZED RANGE DISTRIBUTION

$\alpha = .05$

Degrees of Freedom	Number of Populations 2	3	4	5	6	7	8	9	10	11	12	13	14	15	16	17	18	19	20
1	18.0	27.0	32.8	37.1	40.4	43.1	45.4	47.4	49.1	50.6	52.0	53.2	54.3	55.4	56.3	57.2	58.0	58.8	59.6
2	6.08	8.33	9.80	10.9	11.7	12.4	13.0	13.5	14.0	14.4	14.7	15.1	15.4	15.7	15.9	16.1	16.4	16.6	16.8
3	4.50	5.91	6.82	7.50	8.04	8.48	8.85	9.18	9.46	9.72	9.95	10.2	10.3	10.5	10.7	10.8	11.0	11.1	11.2
4	3.93	5.04	5.76	6.29	6.71	7.05	7.35	7.60	7.83	8.03	8.21	8.37	8.52	8.66	8.79	8.91	9.03	9.13	9.23
5	3.64	4.60	5.22	5.67	6.03	6.33	6.58	6.80	6.99	7.17	7.32	7.47	7.60	7.72	7.83	7.93	8.03	8.12	8.21
6	3.46	4.34	4.90	5.30	5.63	5.90	6.12	6.32	6.49	6.65	6.79	6.92	7.03	7.14	7.24	7.34	7.43	7.51	7.59
7	3.34	4.16	4.68	5.06	5.36	5.61	5.82	6.00	6.16	6.30	6.43	6.55	6.66	6.76	6.85	6.94	7.02	7.10	7.17
8	3.26	4.04	4.53	4.89	5.17	5.40	5.60	5.77	5.92	6.05	6.18	6.29	6.39	6.48	6.57	6.65	6.73	6.80	6.87
9	3.20	3.95	4.41	4.76	5.02	5.24	5.43	5.59	5.74	5.87	5.98	6.09	6.19	6.28	6.36	6.44	6.51	6.58	6.64
10	3.15	3.88	4.33	4.65	4.91	5.12	5.30	5.46	5.60	5.72	5.83	5.93	6.03	6.11	6.19	6.27	6.34	6.40	6.47
11	3.11	3.82	4.26	4.57	4.82	5.03	5.20	5.35	5.49	5.61	5.71	5.81	5.90	5.98	6.06	6.13	6.20	6.27	6.33
12	3.08	3.77	4.20	4.51	4.75	4.95	5.12	5.27	5.39	5.51	5.61	5.71	5.80	5.88	5.95	6.02	6.09	6.15	6.21
13	3.06	3.73	4.15	4.45	4.69	4.88	5.05	5.19	5.32	5.43	5.53	5.63	5.71	5.79	5.86	5.93	5.99	6.05	6.11
14	3.03	3.70	4.11	4.41	4.64	4.83	4.99	5.13	5.25	5.36	5.46	5.55	5.64	5.71	5.79	5.85	5.91	5.97	6.03
15	3.01	3.67	4.08	4.37	4.59	4.78	4.94	5.08	5.20	5.31	5.40	5.49	5.57	5.65	5.72	5.78	5.85	5.90	5.96
16	3.00	3.65	4.05	4.33	4.56	4.74	4.90	5.03	5.15	5.26	5.35	5.44	5.52	5.59	5.66	5.73	5.79	5.84	5.90
17	2.98	3.63	4.02	4.30	4.52	4.70	4.86	4.99	5.11	5.21	5.31	5.39	5.47	5.54	5.61	5.67	5.73	5.79	5.84
18	2.97	3.61	4.00	4.28	4.49	4.67	4.82	4.96	5.07	5.17	5.27	5.35	5.43	5.50	5.57	5.63	5.69	5.74	5.79
19	2.96	3.59	3.98	4.25	4.47	4.65	4.79	4.92	5.04	5.14	5.23	5.31	5.39	5.46	5.53	5.59	5.65	5.70	5.75
20	2.95	3.58	3.96	4.23	4.45	4.62	4.77	4.90	5.01	5.11	5.20	5.28	5.36	5.43	5.49	5.55	5.61	5.66	5.71
24	2.92	3.53	3.90	4.17	4.37	4.54	4.68	4.81	4.92	5.01	5.10	5.18	5.25	5.32	5.38	5.44	5.49	5.55	5.59
30	2.89	3.49	3.85	4.10	4.30	4.46	4.60	4.72	4.82	4.92	5.00	5.08	5.15	5.21	5.27	5.33	5.38	5.43	5.47
40	2.86	3.44	3.79	4.04	4.23	4.39	4.52	4.63	4.73	4.82	4.90	4.98	5.04	5.11	5.16	5.22	5.27	5.31	5.36
60	2.83	3.40	3.74	3.98	4.16	4.31	4.44	4.55	4.65	4.73	4.81	4.88	4.94	5.00	5.06	5.11	5.15	5.20	5.24
120	2.80	3.36	3.68	3.92	4.10	4.24	4.36	4.47	4.56	4.64	4.71	4.78	4.84	4.90	4.95	5.00	5.04	5.09	5.13
∞	2.77	3.31	3.63	3.86	4.03	4.17	4.29	4.39	4.47	4.55	4.62	4.68	4.74	4.80	4.85	4.89	4.93	4.97	5.01

TABLE 10 CRITICAL VALUES OF THE STUDENTIZED RANGE DISTRIBUTION (*Continued*)

$\alpha = .01$

Degrees of Freedom	Number of Populations																		
	2	3	4	5	6	7	8	9	10	11	12	13	14	15	16	17	18	19	20
1	90.0	135.0	164.0	186.0	202.0	216.0	227.0	237.0	246.0	253.0	260.0	266.0	272.0	277.0	282.0	286.0	290.0	294.0	298.0
2	14.0	19.0	22.3	24.7	26.6	28.2	29.5	30.7	31.7	32.6	33.4	34.1	34.8	35.4	36.0	36.5	37.0	37.5	37.9
3	8.26	10.6	12.2	13.3	14.2	15.0	15.6	16.2	16.7	17.1	17.5	17.9	18.2	18.5	18.8	19.1	19.3	19.5	19.8
4	6.51	8.12	9.17	9.96	10.6	11.1	11.5	11.9	12.3	12.6	12.8	13.1	13.3	13.5	13.7	13.9	14.1	14.2	14.4
5	5.70	6.97	7.80	8.42	8.91	9.32	9.67	9.97	10.2	10.5	10.7	10.9	11.1	11.2	11.4	11.6	11.7	11.8	11.9
6	5.24	6.33	7.03	7.56	7.97	8.32	8.61	8.87	9.10	9.30	9.49	9.65	9.81	9.95	10.1	10.2	10.3	10.4	10.5
7	4.95	5.92	6.54	7.01	7.37	7.68	7.94	8.17	8.37	8.55	8.71	8.86	9.00	9.12	9.24	9.35	9.46	9.55	9.65
8	4.74	5.63	6.20	6.63	6.96	7.24	7.47	7.68	7.87	8.03	8.18	8.31	8.44	8.55	8.66	8.76	8.85	8.94	9.03
9	4.60	5.43	5.96	6.35	6.66	6.91	7.13	7.32	7.49	7.65	7.78	7.91	8.03	8.13	8.23	8.32	8.41	8.49	8.57
10	4.48	5.27	5.77	6.14	6.43	6.67	6.87	7.05	7.21	7.36	7.48	7.60	7.71	7.81	7.91	7.99	8.07	8.15	8.22
11	4.39	5.14	5.62	5.97	6.25	6.48	6.67	6.84	6.99	7.13	7.25	7.36	7.46	7.56	7.65	7.73	7.81	7.88	7.95
12	4.32	5.04	5.50	5.84	6.10	6.32	6.51	6.67	6.81	6.94	7.06	7.17	7.26	7.36	7.44	7.52	7.59	7.66	7.73
13	4.26	4.96	5.40	5.73	5.98	6.19	6.37	6.53	6.67	6.79	6.90	7.01	7.10	7.19	7.27	7.34	7.42	7.48	7.55
14	4.21	4.89	5.32	5.63	5.88	6.08	6.26	6.41	6.54	6.66	6.77	6.87	6.96	7.05	7.12	7.20	7.27	7.33	7.39
15	4.17	4.83	5.25	5.56	5.80	5.99	6.16	6.31	6.44	6.55	6.66	6.76	6.84	6.93	7.00	7.07	7.14	7.20	7.26
16	4.13	4.78	5.19	5.49	5.72	5.92	6.08	6.22	6.35	6.46	6.56	6.66	6.74	6.82	6.90	6.97	7.03	7.09	7.15
17	4.10	4.74	5.14	5.43	5.66	5.85	6.01	6.15	6.27	6.38	6.48	6.57	6.66	6.73	6.80	6.87	6.94	7.00	7.05
18	4.07	4.70	5.09	5.38	5.60	5.79	5.94	6.08	6.20	6.31	6.41	6.50	6.58	6.65	6.72	6.79	6.85	6.91	6.96
19	4.05	4.67	5.05	5.33	5.55	5.73	5.89	6.02	6.14	6.25	6.34	6.43	6.51	6.58	6.65	6.72	6.78	6.84	6.89
20	4.02	4.64	5.02	5.29	5.51	5.69	5.84	5.97	6.09	6.19	6.29	6.37	6.45	6.52	6.59	6.65	6.71	6.76	6.82
24	3.96	4.54	4.91	5.17	5.37	5.54	5.69	5.81	5.92	6.02	6.11	6.19	6.26	6.33	6.39	6.45	6.51	6.56	6.61
30	3.89	4.45	4.80	5.05	5.24	5.40	5.54	5.65	5.76	5.85	5.93	6.01	6.08	6.14	6.20	6.26	6.31	6.36	6.41
40	3.82	4.37	4.70	4.93	5.11	5.27	5.39	5.50	5.60	5.69	5.77	5.84	5.90	5.96	6.02	6.07	6.12	6.17	6.21
60	3.76	4.28	4.60	4.82	4.99	5.13	5.25	5.36	5.45	5.53	5.60	5.67	5.73	5.79	5.84	5.89	5.93	5.98	6.02
120	3.70	4.20	4.50	4.71	4.87	5.01	5.12	5.21	5.30	5.38	5.44	5.51	5.56	5.61	5.66	5.71	5.75	5.79	5.83
∞	3.64	4.12	4.40	4.60	4.76	4.88	4.99	5.08	5.16	5.23	5.29	5.35	5.40	5.45	5.49	5.54	5.57	5.61	5.65

TABLE 11 FACTORS FOR $\bar{x}$ AND R CONTROL CHARTS

Observations in Sample, n	d_2	A_2	d_3	D_3	D_4
2	1.128	1.880	0.853	0	3.267
3	1.693	1.023	0.888	0	2.574
4	2.059	0.729	0.880	0	2.282
5	2.326	0.577	0.864	0	2.114
6	2.534	0.483	0.848	0	2.004
7	2.704	0.419	0.833	0.076	1.924
8	2.847	0.373	0.820	0.136	1.864
9	2.970	0.337	0.808	0.184	1.816
10	3.078	0.308	0.797	0.223	1.777
11	3.173	0.285	0.787	0.256	1.744
12	3.258	0.266	0.778	0.283	1.717
13	3.336	0.249	0.770	0.307	1.693
14	3.407	0.235	0.763	0.328	1.672
15	3.472	0.223	0.756	0.347	1.653
16	3.532	0.212	0.750	0.363	1.637
17	3.588	0.203	0.744	0.378	1.622
18	3.640	0.194	0.739	0.391	1.608
19	3.689	0.187	0.734	0.403	1.597
20	3.735	0.180	0.729	0.415	1.585
21	3.778	0.173	0.724	0.425	1.575
22	3.819	0.167	0.720	0.434	1.566
23	3.858	0.162	0.716	0.443	1.557
24	3.895	0.157	0.712	0.451	1.548
25	3.931	0.153	0.708	0.459	1.541

Adapted from Table 27 of ASTM STP 15D, *ASTM Manual on Presentation of Data and Control Chart Analysis.*

Appendix C: Summation Notation

Summations

Definition

$$\sum_{i=1}^{n} x_i = x_1 + x_2 + \cdots + x_n \tag{C.1}$$

Example for $x_1 = 5, x_2 = 8, x_3 = 14$;

$$\begin{aligned}\sum_{i=1}^{3} x_i &= x_1 + x_2 + x_3\\ &= 5 + 8 + 14\\ &= 27\end{aligned}$$

Result 1

For a constant c:

$$\sum_{i=1}^{n} c = \underbrace{(c + c + \cdots + c)}_{n \text{ times}} = nc \tag{C.2}$$

Example for $c = 5, n = 10$:

$$\sum_{i=1}^{10} 5 = 10(5) = 50$$

Example for $c = \bar{x}$

$$\sum_{i=1}^{n} \bar{x} = n\bar{x}$$

Result 2

$$\begin{aligned}\sum_{i=1}^{n} cx_i &= cx_1 + cx_2 + \cdots + cx_n\\ &= c(x_1 + x_2 + \cdots + x_n) = c\sum_{i=1}^{n} x_i\end{aligned} \tag{C.3}$$

Example for $x_1 = 5, x_2 = 8, x_3 = 14, c = 2$:

$$\sum_{i=1}^{3} 2x_i = 2\sum_{i=1}^{3} x_i = 2(27) = 54$$

Result 3

$$\sum_{i=1}^{n} (ax_i + by_i) = a\sum_{i=1}^{n} x_i + b\sum_{i=1}^{n} y_i \tag{C.4}$$

Example for $x_1 = 5, x_2 = 8, x_3 = 14, a = 2, y_1 = 7, y_2 = 3, y_3 = 8, b = 4$:

$$\begin{aligned}\sum_{i=1}^{3}(2x_i + 4y_i) &= 2\sum_{i=1}^{3}x_i + 4\sum_{i=1}^{3}y_i \\ &= 2(27) + 4(18) \\ &= 54 + 72 \\ &= 126\end{aligned}$$

Double Summations

Consider the following data involving the variable x_{ij}, where i is the subscript denoting the row position and j is the subscript denoting the column position:

		Column		
		1	**2**	**3**
Row	**1**	$x_{11} = 10$	$x_{12} = 8$	$x_{13} = 6$
	2	$x_{21} = 7$	$x_{22} = 4$	$x_{23} = 12$

Definition

$$\begin{aligned}\sum_{i=1}^{n}\sum_{j=1}^{m}x_{ij} &= (x_{11} + x_{12} + \cdots + x_{1m}) + (x_{21} + x_{22} + \cdots + x_{2m}) \\ &\quad + (x_{31} + x_{32} + \cdots + x_{3m}) + \cdots + (x_{n1} + x_{n2} + \cdots + x_{nm})\end{aligned} \tag{C.5}$$

Example:

$$\begin{aligned}\sum_{i=1}^{2}\sum_{j=1}^{3}x_{ij} &= x_{11} + x_{12} + x_{13} + x_{21} + x_{22} + x_{23} \\ &= 10 + 8 + 6 + 7 + 4 + 12 \\ &= 47\end{aligned}$$

Definition

$$\sum_{i=1}^{n}x_{ij} = x_{1j} + x_{2j} + \cdots + x_{nj} \tag{C.6}$$

Example:

$$\begin{aligned}\sum_{i=1}^{2}x_{i2} &= x_{12} + x_{22} \\ &= 8 + 4 \\ &= 12\end{aligned}$$

Shorthand Notation

Sometimes when a summation is for all values of the subscript, we use the following shorthand notations:

$$\sum_{i=1}^{n}x_i = \sum x_i \tag{C.7}$$

$$\sum_{i=1}^{n}\sum_{j=1}^{m}x_{ij} = \sum\sum x_{ij} \tag{C.8}$$

$$\sum_{i=1}^{n}x_{ij} = \sum_{i}x_{ij} \tag{C.9}$$

Appendix D: Answers to Even-Numbered Exercises

Chapter 1

2. a. 9
b. 4
c. Qualitative: country and room rate
Quantitative: number of rooms and overall score
d. Country is nominal; room rate is ordinal; number of rooms and overall score are ratio

4. a. 10
b. *Fortune* 500 largest U.S. Corporations
c. $14,227.59 million
d. $14,227.59 million

6. Questions a, c, and d are quantitative
Questions b and e are qualitative

8. a. 2013
b. Qualitative
c. Percentages
d. 563 or 564

10. a. Quantitative; ratio
b. Qualitative; nominal
c. Qualitative; ordinal
d. Qualitative; nominal
e. Quantitative; ratio

12. a. All visitors to Hawaii
b. Yes
c. Questions 1 and 4 provide quantitative data
Questions 2 and 3 provide qualitative data

14. a. 4
b. All are quantitative
c. Time series

16. a. Product taste tests and test marketing
b. Specially designed statistical studies

18. a. 40%
b. Qualitative

20. a. 56% and $387,325
b. 3.73
c. $387,325

22. a. All adult viewers reached by the station
b. Viewers contacted in the telephone survey
c. Sample

24. a. Correct
b. Incorrect
c. Correct
d. Incorrect
e. Incorrect

Chapter 2

2. a. .20
b. 40
c/d.

Class	Frequency	Percent Frequency
A	44	22
B	36	18
C	80	40
D	40	20
Total	200	100

4. a. Qualitative
b.

TV Show	Frequency	Percent Frequency
Millionaire	24	48
Frasier	15	30
Chicago Hope	7	14
Charmed	4	8
Total	50	100

d. Millionaire has the largest market share; Frasier is second

6. a.

Book	Frequency	Percent Frequency
7 Habits	10	16.66
Millionaire	16	26.67
Motley	9	15.00
Dad	13	21.67
WSJ Guide	6	10.00
Other	6	10.00
Total	60	100.00

b. First 5: *Millionaire, Dad, Motley, 7 Habits, WSJ Guide*
c. 48.33%

8. a.

Position	Frequency	Relative Frequency
P	17	.309
H	4	.073
1	5	.091
2	4	.073
3	2	.036
S	5	.091
L	6	.109
C	5	.091
R	7	.127
Totals	55	1.000

b. Pitcher
c. 3rd base
d. Rightfield
e. Infielders 16 to outfielders 18

10. a. Quality classifications
b.

Rating	Frequency	Relative Frequency
Poor	2	.03
Fair	4	.07
Good	12	.20
Very good	24	.40
Excellent	18	.30
Totals	60	1.00

12.

Class	Cumulative Frequency	Cumulative Relative Frequency
≤19	10	.20
≤29	24	.48
≤39	41	.82
≤49	48	.96
≤59	50	1.00

14. b/c.

Class	Frequency	Percent Frequency
6.0–7.9	4	20
8.0–9.9	2	10
10.0–11.9	8	40
12.0–13.9	3	15
14.0–15.9	3	15
Totals	20	100

16. a.

Stock Price ($)	Frequency	Relative Frequency	Percent Frequency
10.00–19.99	10	.40	40
20.00–29.99	4	.16	16
30.00–39.99	6	.24	24
40.00–49.99	2	.08	8
50.00–59.99	1	.04	4
60.00–69.99	2	.08	8
Total	25	1.00	100

b.

Earnings per Share ($)	Frequency	Relative Frequency	Percent Frequency
−3.00 to −2.01	2	.08	8
−2.00 to −1.01	0	.00	0
−1.00 to −0.01	2	.08	8
0.00 to 0.99	9	.36	36
1.00 to 1.99	9	.36	36
2.00 to 2.99	3	.12	12
Total	25	1.00	100

18. a. Lowest salary: $93,000
Highest salary: $178,000
b.

Salary ($1000s)	Frequency	Relative Frequency	Percent Frequency
91–105	4	0.08	8
106–120	5	0.10	10
121–135	11	0.22	22
136–150	18	0.36	36
151–165	9	0.18	18
166–180	3	0.06	6
Total	50	1.00	100

c. 20/50
d. 24%

20. a. 48.9%
b. 43.4%
c. 51.1%
d. 97.075 million
e. 46.6125 million

22.

```
5 | 7 8
6 | 4 5 8
7 | 0 2 2 5 5 6 8
8 | 0 2 3 5
```

24. Leaf Unit = 10

11	6
12	0 2
13	0 6 7
14	2 2 7
15	5
16	0 2 8
17	0 2 3

26. Leaf Unit = .1

0	4 7 8 9 9
1	1 2 9
2	0 0 1 3 5 5 6 8
3	4 9
4	8
5	
6	
7	1

28. a.

0	5 8
1	1 1 3 3 4 4
1	5 6 7 8 9 9
2	2 3 3 3 5 5
2	6 8
3	
3	6 7 7 9
4	0
4	7 8
5	
5	
6	0

b.

2000 P/E Forecast	Frequency	Relative Frequency
5–9	2	6.7
10–14	6	20.0
15–19	6	20.0
20–24	6	20.0
25–29	2	6.7
30–34	0	0.0
35–39	4	13.3
40–44	1	3.3
45–49	2	6.7
50–54	0	0.0
55–59	0	0.0
60–64	1	3.3
Total	30	100.0

30. b. Negative relationship

32. a.

Sales/ Margins/ ROE	EPS Rating 0–19	20–39	40–59	60–79	80–100	Total
A				1	8	9
B		1	4	5	2	12
C	1		1	2	3	7
D	3	1		1		5
E		2	1			3
Total	4	4	6	9	13	36

b.

Sales/ Margins/ ROE	EPS Rating 0–19	20–39	40–59	60–79	80–100	Total
A				11.11	88.89	100
B		8.33	33.33	41.67	16.67	100
C	14.29		14.29	28.57	42.86	100
D	60.00	20.00		20.00		100
E		66.67	33.33			100

Higher EPS ratings seem to be associated with higher ratings on Sales/Margins/ROE

34. No apparent relationship

36. a.

Vehicle	Frequency	Percent Frequency
F-Series	17	34
Silverado	12	24
Taurus	8	16
Camry	7	14
Accord	6	12
	50	100

b. Ford F-Series Pickup and Chevrolet Silverado

38. a.

Movie	Frequency	Percent Frequency
Blair Witch Project	159	36.0
Phantom Menace	89	20.2
Beloved	85	19.3
Primary Colors	57	12.9
Truman Show	51	11.6
Total	441	100.0

c. 56.2%

40.

Closing Price	Freq.	Rel. Freq.	Cum. Freq.	Cum. Rel. Freq.
0–9$\frac{7}{8}$	9	.225	9	.225
10–19$\frac{7}{8}$	10	.250	19	.475
20–29$\frac{7}{8}$	5	.125	24	.600
30–39$\frac{7}{8}$	11	.275	35	.875
40–49$\frac{7}{8}$	2	.050	37	.925
50–59$\frac{7}{8}$	2	.050	39	.975
60–69$\frac{7}{8}$	0	.000	39	.975
70–79$\frac{7}{8}$	1	.025	40	1.000
Totals	40	1.000		

42.

Income ($)	Frequency	Relative Frequency
18,000–21,999	13	0.255
22,000–25,999	20	0.392
26,000–29,999	12	0.235
30,000–33,999	4	0.078
34,000–37,999	2	0.039
Total	51	1.000

44. a. High Temperature

Stem	Leaves
3	
4	
5	7
6	1 4 4 4 4 6 8
7	3 5 7 9
8	0 1 1 4 6
9	0 2 3

b. Low Temperature

Stem	Leaves
3	9
4	3 6 8
5	0 0 0 2 4 4 5 5 7 9
6	1 8
7	2 4 5 5
8	
9	

c. The range of low temperatures is below the range of high temperatures

d. 8 cities

e.

Temperature	Frequency: High Temp.	Frequency: Low Temp.
30–39	0	1
40–49	0	3
50–59	1	10
60–69	7	2
70–79	4	4
80–89	5	0
90–99	3	0
Total	20	20

46. a.

Occupation	Satisfaction Score 30–39	40–49	50–59	60–69	70–79	80–89	Total
Cabinetmaker			2	4	3	1	10
Lawyer	1	5	2	1	1		10
Physical Therapist			5	2	1	2	10
Systems Analyst		2	1	4	3		10
Total	1	7	10	11	8	3	40

b.

Occupation	Satisfaction Score 30–39	40–49	50–59	60–69	70–79	80–89	Total
Cabinetmaker			20	40	30	10	100
Lawyer	10	50	20	10	10		100
Physical Therapist			50	20	10	20	100
Systems Analyst		20	10	40	30		100

c. Cabinetmakers seem to have the highest job satisfaction scores; lawyers seem to be the lowest

48. b.

Year	Freq.
1973 or before	247
1974–79	54
1980–86	82
1987–91	121
Total	504

Fuel	Freq.
Elect.	149
Nat. Gas	317
Oil	17
Propane	7
Other	14
Total	504

50. a. Crosstabulation of market value and profit

Market Value ($1000s)	Profit ($1000s) 0–300	300–600	600–900	900–1200	Total
0–8000	23	4			27
8000–16,000	4	4	2	2	12
16,000–24,000		2	1	1	4
24,000–32,000		1	2	1	4
32,000–40,000		2	1		3
Total	27	13	6	4	50

b. Crosstabulation of row percentages

Market Value ($1000s)	Profit ($1000s) 0–300	300–600	600–900	900–1200	Total
0–8000	85.19	14.81	0.00	0.00	100
8000–16,000	33.33	33.33	16.67	16.67	100
16,000–24,000	0.00	50.00	25.00	25.00	100
24,000–32,000	0.00	25.00	50.00	25.00	100
32,000–40,000	0.00	66.67	33.33	0.00	100

c. A positive relationship is indicated between profit and market value; as profit goes up, market value goes up

52. b. A positive relationship is demonstrated between market value and stockholders' equity

Chapter 3

2. 16, 16.5

4. 59.727, 57, 53

6. a. 91.45, 87.5, 120
b. 66.3, 69.5, 70
c. 500 @ $50
d. Yes

8. a. 38.75, 29
b. 38.5
c. 29.5, 47.5
d. 31

10. a. 48.33, 49; do not report a mode
b. 45, 55

12. *City:* mean = 15.58, median = 15.9, mode = 15.3
Country: mean = 18.92, median = 18.7, mode = 18.6 and 19.4

14. a. $639
b. 98.8 pictures
c. 110.2 minutes

16. 16, 4

18. a. 22
b. 75.2
c. 8.67
d. 4.87

20. a. Range = 32, IQR = 10
b. 92.75, 9.63

22. *Dawson:* range = 2, s = .67
Clark: range = 8, s = 2.58

24. a. 161, 92.5; 56, 19.5
b. 2705.38, 52.01; 290.85, 17.05
c. 56.38, 25.56
d. Greater for 500 @ $50

26. *Quarter-milers:* s = .056, Coef. of Var. = 5.8
Milers: s = .130, Coef. of Var. = 2.9

28. a. 95%
b. Almost all
c. 68%

30. .20, 1.50, 0, −.50, −2.20

32. a. 34%
b. 81.5%
c. 16%

34. a. −.95
b. 3.90
c. Labor cost in part (b) is an outlier

36. a. 100; 13.88 or approximately 14
b. 16%
c. 11.1 and 10.77; no outliers

38. 15, 22.5, 26, 29, 34

40. 5, 8, 10, 15, 18

42. a. 5, 9.6, 14.5, 19.2, 52.7
b. Limits: −4.8, 33.6
c. 41.6 is an outlier
52.7 is an outlier

44. a. 105.79, 52.7
b. 15.7, 78.3
c. Silicon Graphics, Toys R Us
d. 26.73; much greater

46. a. 37.48, 23.67
b. 7.91, 51.92
c. Limits: −58.11, 117.94
Russia and Turkey are outliers

48. b. There appears to be a linear relationship between x and y
c. $s_{xy} = 26.5$
d. $r_{xy} = .69$

50. −.91; negative relationship

52. a. .92
b. Strong positive linear relationship

54. a. 3.69
b. 3.175

56. a. 2.50
b. Yes

58. 10.74, 25.63, 5.06

60. **a.** 138.52, 129, 0
b. No, much more
c. 95, 169
d. 467, 74
e. 9271.01, 96.29
f. Yes, the $467 value

62. **a.** 18.57, 16.5
b. 53.49, 7.31
c. Quantex
d. 1.15
e. −.90
f. No, according to *z*-scores

64. 7195.5; 7019; 7,165,941; 2676.93

66. **a.** Public: 32; auto: 32
b. Public: 4.64; auto: 1.83
c. Auto has less variability

68. **a.** 400, 624, 836, 999, 1278
c. *Limits:* 61.5, 1561.5
No outliers

70. **b.** .9856; strong positive relationship

72. **b.** .75

74. **a.** 817
b. 833

76. 51.5, 227.37, 15.08

Chapter 4

2. 20 ways

4. **b.** (H,H,H), (H,H,T), (H,T,H), (H,T,T), (T,H,H), (T,H,T), (T,T,H), (T,T,T)
c. ⅛

6. .40, .26, .34; relative frequency method

8. **a.** 4: Commission Positive—Council Approves, Commission Positive—Council Disapproves Commission Negative—Council Approves, Commission Negative—Council Disapproves

10. **a.** .60
b. .09
c. .78
d. 86
e. $1.53 billion

12. **a.** 1,906,884
b. 1/1,906,884
c. 1/80,089,128

14. **a.** ¼
b. ½
c. ¾

16. **a.** 36
c. ⅙
d. 5⁄18
e. No; *P*(odd) = *P*(even) = $\frac{1}{2}$
f. Classical

18. **a.** *P*(0) = .05
b. *P*(4 or 5) = .20
c. *P*(0, 1, or 2) = .55

20. **a.** .112
b. .086
c. .49

22. **a.** .40, .40, .60
b. .80, yes
c. $A^c = (E_3, E_4, E_5)$; $C^c = (E_1, E_4)$; $P(A^c) = .60$; $P(C^c) = .40$
d. (E_1, E_2, E_5); .60
e. .80

24. .43

26. **a.** .30, .23
b. .17
c. .64

28. **a.** .698
b. .302

30. **a.** .67
b. .80
c. No

32. **a.**

	Single	Married	Total
Under 30	.55	.10	.65
30 or Over	.20	.15	.35
Total	.75	.25	1.0

b. Higher probability of under 30
c. Higher probability of single
d. .55
e. .8462
f. No

34. **a.** .44
b. .15
c. .0225
d. .0025
e. .136
f. .106

36. **a.**

	Satisfaction Score					
Occupation	**Under 50**	**50–59**	**60–69**	**70–79**	**80–89**	**Totals**
Cabinetmaker	.000	.050	.100	.075	.025	.250
Lawyer	.150	.050	.025	.025	.000	.250
Physical Therapist	.000	.125	.050	.025	.050	.250
Systems Analyst	.050	.025	.100	.075	.000	.250
Totals	.200	.250	.275	.200	.075	1.000

b. .075

c. .20
d. .25
e. .15
f. .60
g. .275

38. a. 52/190 = .2737
b. .0125
c. .3684

40. a. .10, .20, .09
b. .51
c. .26, .51, .23

42. a. .21
b. Yes

44. .6754

46. a. .68
b. 52
c. 10

48. a. .61
b. 18–34 and 65+
c. .30

50. a. 76
b. .24

52. b. .2022
c. .4618
d. .4005

54. a. .49
b. .44
c. .54
d. No
e. Yes

56. a. .25
b. .125
c. .0125
d. .10
e. No

58. 3.44%

60. a. .0625
b. .0132
c. Three

Chapter 5

2. a. x = time in minutes to assemble product
b. Any positive value: $x > 0$
c. Continuous

4. $x = 0, 1, 2, \ldots, 12$

6. a. 0, 1, 2, . . . , 20; discrete
b. 0, 1, 2, . . . ; discrete
c. 0, 1, 2, . . . , 50; discrete
d. $0 \le x \le 8$; continuous
e. $x > 0$; continuous

8. a.

x	1	2	3	4
$f(x)$	.15	.25	.40	.20

c. $f(x) \ge 0, \Sigma f(x) = 1$

10. a.

x	1	2	3	4	5
$f(x)$	.05	.09	.03	.42	.41

b.

x	1	2	3	4	5
$f(x)$	.04	.10	.12	.46	.28

c. .83
d. .28
e. Senior executives more satisfied

12. a. Yes
b. .65

14. a. .05
b. .70
c. .40

16. a. 5.20
b. 4.56, 2.14

18. a. $E(x) = 2.3$, same
b. $\text{Var}(x) = 1.23, \sigma = 1.11$

20. a. 166
b. −94; concern is to protect against the expense of a big accident

22. a. 445
b. 1250 loss

24. a. Medium: 145; large: 140
b. Medium: 2725; large: 12,400

26. a. $f(0) = .3487$
b. $f(2) = .1937$
c. .9298
d. .6513
e. 1
f. $\sigma^2 = .9000, \sigma = .9487$

28. a. .3292
b. .6422
c. .0182

30. a. Probability of a defective part must be .03 for each trial; trials must be independent
c. 2
d.

Number of defects	0	1	2
Probability	.9409	.0582	.0009

32. a. .90
b. .99
c. .999
d. Yes

34. a. .0634
b. .0634
c. .9729

38. a. $f(x) = \dfrac{3^x e^{-3}}{x!}$

b. .2241
c. .1494
d. .8008

40. a. .1952
b. .1048
c. .0183
d. .0907

42. a. .1465
b. 1
c. .8647

44. a. $\mu = 1.25$
b. .2865
c. .3581
d. .3554

46. a. 50
b. .067
c. .4667
d. .30

48. a. .50
b. .3333

50. a. .01
b. .07
c. .92
d. .07

52. a. .5333
b. .6667
c. .7778
d. $n = 7$

54. a. $f(x) \geq 0$ and $\Sigma f(x) = 1$
b. 3.64, .6704
c. Appears overvalued

56. a. .0364
b. .4420
c. 48
d. 6.7882

58. a. .9510
b. .0480
c. .0490

60. a. 328
b. 13.91
c. 13.91

62. .1912

64. a. .2240
b. .5767

66. a. .5333
b. .1333
c. .3333

Chapter 6

2. b. .50
c. .60
d. 15
e. 8.33

4. b. .50
c. .30
d. .40

6. a. .40
b. .64
c. .68

10. a. .3413
b. .4332
c. .4772
d. .4938

12. a. .2967
b. .4418
c. .3300
d. .5910
e. .8849
f. .2388

14. a. $z = 1.96$
b. $z = .61$
c. $z = 1.12$
d. $z = .44$

16. a. $z = 2.33$
b. $z = 1.96$
c. $z = 1.645$
d. $z = 1.28$

18. a. .2451
b. .1170
c. 69.48 minutes or more

20. a. .025
b. 5.16%
c. 33.72 or more

22. a. .4194
b. $517.44 or more
c. .0166

24. a. 902.75, 114.185
b. .1841
c. .1977
d. 1,091 million

26. a. .5276
b. .3935
c. .4724
d. .1341

28. a. .3935
b. .2231
c. .3834

30. a. 50 hours
b. .3935
c. .1353

32. a. $f(x) = 30e^{-30x}$
b. .0821
c. .7135

34. **a.** \$63,000
b. \$43,800 or less
c. 12.92%
d. \$87,675

36. **a.** \$220.33
b. .5999
c. \$1656.78

38. **a.** .0228
b. \$50

40. **a.** 38.3%
b. 3.59% better, 96.41% worse
c. 38.21%

42. $\mu = 19.23$ ounces

44. **a.** 4 hours
b. $1/4\, e^{-x/4}$
c. .7788
d. .1353

46. **a.** 2 minutes
b. .2212
c. .3935
d. .0821

Chapter 7

2. 22, 147, 229, 289

4. **a.** IBM, Microsoft, Intel, GE, AT&T
b. 252

6. 2782, 493, 825, 1807, 289

8. Washington, Clemson, Oklahoma, Colorado, USC, and Wisconsin

10. Finite, infinite, infinite, infinite, finite

12. **a.** .50
b. .3667

14. **a.** .19
b. .32
c. .79

16. .80

18. **a.** 200
b. 5
c. Normal with $E(\bar{x}) = 200$ and $\sigma_{\bar{x}} = 5$
d. The probability distribution of $\bar{x}$

20. 3.54, 2.50, 2.04, 1.77
$\sigma_{\bar{x}}$ decreases as n increases

22. **a.** Only for $n = 30$ and $n = 40$
b. $n = 30$; normal with $E(\bar{x}) = 400$ and $\sigma_{\bar{x}} = 9.13$
$n = 40$; normal with $E(\bar{x}) = 400$ and $\sigma_{\bar{x}} = 7.91$

24. **a.** Normal with $E(\bar{x}) = 51{,}800$ and $\sigma_{\bar{x}} = 516.40$
b. $\sigma_{\bar{x}}$ decreases to 365.15
c. $\sigma_{\bar{x}}$ decreases as n increases

26. **a.** Normal with $E(\bar{x}) = 1.20$ and $\sigma_{\bar{x}} = .014$
b. .8414
c. .5224

28. **a.** .5036, .6212, .7888, .9232, .9876
b. Higher probability within ±\$250

30. **a.** Normal with $E(\bar{x}) = 166{,}500$ and $\sigma_{\bar{x}} = 4200$
b. .9826
c. .7660, .4514, .1896
d. Increase the sample size

32. **a.** $n/N = .01$; no
b. 1.29, 1.30; little difference
c. .8764

34. **a.** .6156
b. .8530

36. **a.** .6156
b. .7814
c. .9488
d. .9942
e. Higher probability with larger n

38. **a.** Normal with $E(\bar{p}) = .76$ and $\sigma_{\bar{p}} = .0214$
b. .8384
c. .9452

40. **a.** Normal with $E(\bar{p}) = .25$ and $\sigma_{\bar{p}} = .0306$
b. .6730
c. .8968

42. **a.** Normal with $E(\bar{p}) = .15$ and $\sigma_{\bar{p}} = .0505$
b. .4448
c. .8389

44. 112, 145, 73, 324, 293, 875, 318, 618

46. **a.** Normal with $E(\bar{x}) = 31.5$ and $\sigma_{\bar{x}} = 1.70$
b. .4448
c. .9232

48. **a.** 8.49
b. .5000
c. .4448
d. .5934

50. **a.** 625
b. .7888

52. **a.** Normal with $E(\bar{p}) = .74$ and $\sigma_{\bar{p}} = .031$
b. .8030
c. .4778

54. **a.** .9606
b. .0495

56. **a.** 48
b. Normal, $E(\bar{p}) = .25$, $\sigma_{\bar{p}} = .0625$
c. .2119

Chapter 8

2. **a.** 30.60 to 33.40
b. 30.34 to 33.66
c. 29.81 to 34.19

4. 62

6. 362.80 to 375.20

8. **a.** 11,769 to 12,231
b. 11,725 to 12,275
c. 11,638 to 12,362
d. Width increases to be more confident

10. 7.25 to 8.25

12. **a.** 3.8
b. .81
c. 2.99 to 4.61

14. **a.** 1.734
b. −1.321
c. 3.365
d. −1.761 and 1.761
e. −2.048 and 2.048

16. **a.** 15.97 to 18.53
b. 15.71 to 18.79
c. 15.14 to 19.36

18. **a.** 1.58
b. .1474
c. 1.49 to 1.67

20. **a.** 21.15 to 23.65
b. 21.12 to 23.68
c. Intervals are essentially the same

22. **a.** 6.86
b. 6.54 to 7.18

24. **a.** 9
b. 35
c. 78

26. **a.** 340
b. 1358
c. 8487

28. **a.** 53
b. 75
c. 129
d. Must increase n

30. 59

32. **a.** .6733 to .7267
b. .6682 to .7318

34. 1068

36. **a.** .4393
b. .3870 to .4916

38. **a.** .0430
b. .2170 to .3030
c. 822

40. **a.** .2505
b. .0266

42. **a.** .0442
b. 601, 1068, 2401, 9604

44. **a.** 2009
b. 47,991 to 52,009

46. **a.** 49.8
b. 15.99
c. 47.58 to 52.02

48. **a.** 13.2
b. 7.8
c. 7.62 to 18.78
d. Wide interval; larger n desirable

50. 37

52. 176

54. **a.** .5420
b. .0508
c. .4912 to .5928

56. **a.** 1267
b. 1509

58. **a.** .68
b. .6391 to .7209

60. **a.** .3101
b. .2898 to .3304
c. 8219; No, this sample size is unnecessarily large

Chapter 9

2. **a.** H_0: $\mu \leq 14$
H_a: $\mu > 14$

4. **a.** H_0: $\mu \geq 220$
H_a: $\mu < 220$

6. **a.** H_0: $\mu \leq 1$
H_a: $\mu > 1$
b. Claiming $\mu > 1$ when it is not true
c. Claiming $\mu \leq 1$ when it is not true

8. **a.** H_0: $\mu \geq 220$
H_a: $\mu < 220$
b. Claiming $\mu < 220$ when it is not true
c. Claiming $\mu \geq 220$ when it is not true

10. **a.** Reject H_0 if $z > 2.05$
b. 1.36
c. .0869
d. Do not reject H_0

12. **a.** .0344; reject H_0
b. .3264; do not reject H_0
c. .0668; do not reject H_0
d. Approximately 0; reject H_0

14. **a.** Reject if $z > 2.33$
b. 3.11
c. Reject H_0

16. **a.** H_0: $\mu \geq 13$
H_a: $\mu < 13$
b. Reject if $z < -2.33$
c. −2.88
d. Reject H_0

18. a. $H_0: \mu \leq 5.72$
$H_a: \mu > 5.72$
b. 2.12
c. .0170
d. Reject H_0

20. a. $H_0: \mu \leq 37{,}000$
$H_a: \mu > 37{,}000$
b. 1.47
c. .0708
d. Do not reject H_0

22. a. Reject H_0 if $z < -2.33$ or $z > 2.33$
b. 1.13
c. .2584
d. Do not reject H_0

24. a. .0718; do not reject H_0
b. .6528; do not reject H_0
c. .0404; reject H_0
d. Approximately 0; reject H_0
e. .3174; do not reject H_0

26. a. Reject H_0 if $z < -1.96$ or if $z > 1.96$
b. -1.71
c. Do not reject H_0

28. a. $H_0: \mu = 1075$
$H_a: \mu \neq 1075$
b. $z = 1.43$
c. .1528
d. Do not reject H_0

30. a. $H_0: \mu = 26{,}133$
$H_a: \mu \neq 26{,}133$
b. -2.09
c. .0366
d. Reject H_0

32. a. 14.66 or less
b. Reject H_0

34. a. 18
b. 1.41
c. Reject H_0 if $t < -2.571$ or $t > 2.571$
d. -3.47
e. Reject H_0

36. a. .01; reject H_0
b. .10; do not reject H_0
c. Between .025 and .05; reject H_0
d. Greater than .10; do not reject H_0
e. Approximately 0; reject H_0

38. a. Reject H_0 if $t < -2.064$ or $t > 2.064$
b. -1.90
c. Do not reject H_0
d. Between .05 and .10

40. a. $H_0: \mu = 4000$
$H_a: \mu \neq 4000$
b. Reject H_0 if $t < -2.160$ or if $t > 2.160$
c. 1.63
d. Do not reject H_0
e. Between .10 and .20

42. a. $H_0: \mu \leq 2$
$H_a: \mu > 2$
b. Reject H_0 if $t > 1.833$
c. 2.4
d. .516
e. 2.45
f. Reject H_0
g. Between .01 and .025

44. a. Reject H_0 if $z < -1.96$ or $z > 1.96$
b. -1.25
c. .2112
d. Do not reject H_0

46. a. $H_0: p \leq .40$
$H_a: p > .40$
b. Reject H_0 if $z > 1.645$
c. 1.99
d. Reject H_0

48. a. .57
b. 3.13
c. Less than .001
d. Reject H_0
e. Yes

50. a. .6381
b. 2.83
c. .0046
d. Reject H_0

52. a. -1.20
b. .1151
c. Do not reject H_0

54. a. $H_0: p \geq .047$
$H_a: p < .047$
b. .0296
c. -2.82
d. .0024
e. Reject H_0

56. a. .2912
b. Type II error
c. .0031

58. a. Concluding $\mu \leq 15$ when it is not true
b. .2676
c. .0179

60. a. Concluding $\mu = 28$ when it is not true
b. .0853, .6179, .6179, .0853
c. .9147

62. .1151, .0015
Increasing n reduces β

64. 214

66. 109

68. 324

70. a. H_0: $\mu \leq 45{,}250$
H_a: $\mu > 45{,}250$
b. 2.71
c. .0034
d. Reject H_0

72. $z = 2.26$, p-value $= .0119$
Reject H_0

76. a. .0143
b. Reject H_0

78. a. .6502
b. $-.98$
c. .3270
d. Do not reject H_0

80. a. Show $p < .50$
b. $z = -6.62$; reject H_0

82. a. .825
b. 2.17
c. .03
d. Reject H_0

84. a. 352
b. -2.64
c. .0041
d. Reject H_0

86. 219

Chapter 10

2. a. 2.4
b. 5.27
c. .09 to 4.71

4. a. .60
b. .56 to .64

6. $-.51$ to 1.27

8. a. 1200
b. 438 to 1962
c. Populations normal with equal variances

10. a. 2.18
b. 4.41
c. .71 to 3.65

12. a. $z = -1.53$; do not reject H_0
b. .1260

14. a. H_0: $\mu_1 - \mu_2 = 0$
H_a: $\mu_1 - \mu_2 \neq 0$
b. Reject if $z < -1.96$ or $z > 1.96$
c. 2.18
d. Reject H_0
e. .0292

16. $z = 4.99$, p-value ≈ 0
Reject H_0

18. a. 1.08
b. .2802
c. Do not reject H_0

20. a. H_0: $\mu_1 - \mu_2 \leq 0$
H_a: $\mu_1 - \mu_2 > 0$
b. Reject H_0 if $t > 1.711$
c. 2.07
d. Reject H_0
e. Approximately .025

22. a. 3, -1, 3, 5, 3, 0, 1
b. 2
c. 2.0282
d. 2
e. .07 to 3.93

24. .16 to .35

26. $t = 1.63$; do not reject H_0

28. a. $t = 7.34$; reject H_0
b. 4.96 to 9.21

30. a. .12
b. .0586 to .1814
c. .0469 to .1931

32. .0009 to .1391

34. a. .2206
b. .1788 to .2624

36. $z = 3.94$; reject H_0

38. $z = 2.28$; p-value $= .0226$
Reject H_0

40. a. 4354 to 5646

42. a. H_0: $\mu_1 - \mu_2 \leq 0$
H_a: $\mu_1 - \mu_2 > 0$
b. $z = .59$; do not reject H_0
c. .2776

44. $t = 2.29$; reject H_0

46. a. H_0: $p_1 - p_2 \leq 0$
H_a: $p_1 - p_2 > 0$
b. $z = 4.70$; p-value ≈ 0
c. Reject H_0

48. .0174; reject H_0

Chapter 11

2. a. 15.76 to 46.95
b. 14.46 to 53.33
c. 3.8 to 7.3

4. a. .22 to .71
b. .47 to .84

6. a. 13.30
b. 7.97 to 38.23

8. a. .00845
b. .0919
c. .0042 to .0244
d. .0651 to .1561

10. $\chi^2 = 206.22$; reject H_0

12. a. .8106
b. $\chi^2 = 9.49$; do not reject H_0

14. $F = 2.42$; reject H_0

16. $F = 2.63$; reject H_0

18. $F = 3.54$; do not reject H_0

20. $F = 5.29$; reject H_0

22. a. $F = 4$; reject H_0
b. Drive carefully on wet pavement

24. 10.72 to 24.68

26. a. $\chi^2 = 27.44$; reject H_0
b. .00012 to .00042

28. $\chi^2 = 31.5$; reject H_0

30. a. 15
b. 6.25 to 11.13

32. $F = 1.39$; do not reject H_0

34. $F = 2.08$; reject H_0

Chapter 12

2. $\chi^2 = 15.33$, $\chi^2_{.05} = 7.81473$; reject H_0

4. $\chi^2 = 29.51$, $\chi^2_{.05} = 11.07$; reject H_0
Percentage figures have changed

6. $\chi^2 = 10.69$, $\chi^2_{.10} = 9.24$; reject H_0

8. $\chi^2 = 16.31$, $\chi^2_{.01} = 11.34$; reject H_0

10. $\chi^2 = 19.78$, $\chi^2_{.05} = 9.49$; reject H_0

12. a. $\chi^2 = 7.36$, $\chi^2_{.05} = 5.99$; reject H_0
b. Domestic 47.2%

14. $\chi^2 = 13.43$, $\chi^2_{.05} = 11.07$; reject H_0

16. a. $\chi^2 = 7.85$, $\chi^2_{.05} = 7.81$; reject H_0
b. Pharmaceutical, 98.6%

18. $\chi^2 = 13.42$, $\chi^2_{.01} = 13.28$; reject H_0

20. $\chi^2 = 9.03$, $\chi^2_{.05} = 7.81$; reject H_0

22. $\chi^2 = 4.30$, $\chi^2_{.05} = 5.99$; do not reject H_0

24. $\chi^2 = 2.8$, $\chi^2_{.10} = 6.25$; do not reject H_0

26. $\chi^2 = 8.04$, $\chi^2_{.05} = 7.81$; reject H_0

28. $\chi^2 = 31.43$, $\chi^2_{.05} = 7.81$; reject H_0

30. $\chi^2 = 42.53$, $\chi^2_{.05} = 9.49$; reject H_0

32. $\chi^2 = 23.37$, $\chi^2_{.05} = 7.81$; reject H_0

34. a. $\chi^2 = 12.86$, $\chi^2_{.01} = 9.21$; reject H_0
b. 66.7, 30.3, 2.9
54.0, 42.0, 4.0

36. $\chi^2 = 6.20$, $\chi^2_{.05} = 12.59$; do not reject H_0

38. $\chi^2 = 7.78$, $\chi^2_{.05} = 7.81$; do not reject H_0

Chapter 13

2. a. MSTR = 268
b. MSE = 92
c. Cannot reject H_0 because $F = 2.91 < F_{.05} = 4.26$
d.

Source of Variation	Sum of Squares	Degrees of Freedom	Mean Square	F
Treatments	536	2	268	2.91
Error	828	9	92	
Total	1364	11		

4. b. Reject H_0 because $F = 80 > F_{.05} = 2.76$

6. Reject H_0 because $F = 10.63 > F_{.05} = 4.26$

8. Significant difference;
$F = 7.00 > F_{.05} = 3.68$

10. Significant difference;
p-value = .015

12. No significant difference; p-value $= 0.403 > \alpha = .05$

14. -8.54 to -1.46

16. a. Significant difference; $F = 19.86 > F_{.05} = 3.10$
b. Significant difference; 2.3 > LSD = 1.19

18. 1 and 2: significant difference (LSD = 3.38)
1 and 3: significant difference (LSD = 3.17)
2 and 3: no significant difference (LSD = 3.51)

20. a.

Source of Variation	Sum of Squares	Degrees of Freedom	Mean Square	F
Treatments	1488	2	744	5.50
Error	2030	15	135.3	
Total	3518	17		

b. Significant difference between A and C

22. a. H_0: $\mu_1 = \mu_2 = \mu_3 = \mu_4 = \mu_5$
H_a: Not all the population means are equal
b. Reject H_0 because $F = 14.07 > 2.69$

24. Significant difference; $F = 43.99$ exceeds the critical value, which is between 3.15 and 3.23

26. b. Significant difference; $F = 9.87 > F_{.05} = 3.35$

28. Not significant; $F = 1.78 < F_{.05} = 3.89$

30. Not significant; $F = 2.54 < F_{.05} = 3.24$

32. Means are all different (LSD = 2.53)

34. Significant; $F = 6.60 > F_{.05} = 4.46$

36. Significant; $F = 12.60 > F_{.05} = 3.07$

38. Significant; $F = 7.12 > F_{.05} = 3.26$

40. Significant difference
$F = 22.46 > F_{.05} = 2.96$

42. Factor A is significant because $F = 3.72 > F_{.05} = 3.01$
Factor B is significant because $F = 4.94 > F_{.05} = 3.40$
Interaction is significant because $F = 12.52 > F_{.05} = 2.51$

44. No significant effect due to the loading and unloading method, the type of ride, or interaction

46. Factor A is not significant
Factor B is significant
Interaction is significant

48. Significant difference;
p-value $= .000 < \alpha = .05$

50. Significant difference

52. Significant difference

54. Significant; $F = 7.23 > F_{.05} = 4.26$

56. Not significant; $F = 1.48 < F_{.05} = 3.35$

58. Significant; $F = 5.19 > F_{.05} = 4.26$

60. Significant; $F = 6.99 > F_{.05} = 4.46$

62. Significant difference; p-value $= .000 < \alpha = .05$

64. Type of machine is significant; type of loading system and interaction are not significant

Chapter 14

2. b. There appears to be a linear relationship between x and y
d. $\hat{y} = 30.33 - 1.88x$
e. 19.05

4. b. There appears to be a linear relationship between x and y
d. $\hat{y} = -240.5 + 5.5x$
e. 106 pounds

6. c. $\hat{y} = 6.02 - .07x$
e. .42

8. a. $\hat{y} = 107.13 + 3.07x$
c. 153.2

10. b. $\hat{y} = 51.82 + .145x$
c. 84.4

12. c. $\hat{y} = 1293 + .3165x$
d. 25,031

14. b. $\hat{y} = 49.63 + 2.455x$
c. 69.3%

16. a. SSE = 6.3325, SST = 114.80, SSR = 108.47
b. $r^2 = .945$
c. $r = -.9721$

18. a. SSE = 85,135.14, SST = 335,000, SSR = 249,864.86
b. $r^2 = .746$
c. $r = +.8637$

20. a. $\hat{y} = -48.11 + 2.3325x$
b. $r^2 = .82$
c. $173,500

22. a. $\hat{y} = -1.02 + .13x$
b. $r^2 = .3631$
c. $r = +.6026$

24. a. 2.11
b. 1.453
c. .262
d. Significant
$t = -7.18 < -t_{.05} = -3.182$
e. Significant
$F = 51.41 > F_{.05} = 10.13$

26. a. Significant
$t = 3.43 > t_{.025} = 2.776$
b. Significant
$F = 11.74 > F_{.05} = 7.71$

28. They are related because $F = 20.17 > F_{.05} = 4.67$

30. No significant relationship

32. a. 1.11
b. 7.07 to 14.13
c. 2.32
d. 3.22 to 17.98

34. Confidence interval: $-.4$ to 4.98
Prediction interval: -2.27 to 7.31

36. a. 80.859
b. 78.58 to 83.14
c. 72.92 to 88.80

38. a. $5046.67
b. $3815.10 to $6278.24
c. Not out of line

40. a. 9
b. $\hat{y} = 20.0 + 7.21x$
c. 1.3626
d. Significant relationship because $F = 28 > F_{.05} = 5.59$
e. $380,500

42. a. $\hat{y} = 80.0 + 50.0x$
b. 30
c. Significant relationship because $F = 83.17 > F_{.05} = 4.20$
d. $680,000

44. b. Yes
c. $\hat{y} = 37.1 - .779x$
d. Significant relationship; p-value = 0.003
e. $r^2 = .434$; not a good fit
f. $12.28 to $22.91
g. $17.49 to $39.05

46. a. $\hat{y} = 2.32 + .64x$
b. No; the variance does not appear to be the same for all values of x

48. b. Yes

50. a. Yes; $x = 135$, $y = 145$ may be an outlier
b. Yes
c. Yes

52. a. $\hat{y} = 4.09 + .196x$
b. Minitab identifies observation 1 as having a large standardized residual; we would treat observation 1 as an outlier

54. a. $\hat{y} = 707 + .00482x$
b. Observation 6 is an influential observation

58. a. $\hat{y} = 9.26 + .711x$
b. Significant relationship
c. $r^2 = .744$; good fit
d. \$13.53

60. a. Market beta = .95
b. Significant relationship
c. $r^2 = .470$; not a good fit
d. Woolworth has a higher risk

62. a. $\hat{y} = 10.5 + .953x$
b. Significant relationship; p-value = .000
c. \$2874 to \$4952
d. Yes

64. a. Negative linear relationship
b. $\hat{y} = 8.10 - .344x$
c. Significant relationship; p-value = .002
d. $r^2 = .711$; reasonably good fit
e. 5.2 to 7.6 days

66. a. $\hat{y} = 5.85 + .830x$
b. Significant relationship; p-value = .000
c. 84.65 points
d. 65.35 to 103.96

Chapter 15

2. a. $\hat{y} = 45.06 + 1.94x_1$; $\hat{y} = 132.36$
b. $\hat{y} = 85.22 + 4.32x_2$; $\hat{y} = 150.02$
c. $\hat{y} = -18.37 + 2.01x_1 + 4.74x_2$; $\hat{y} = 143.18$

4. a. \$255,000

6. a. Speed = 49.8 + .015 Weight
b. Speed = 80.5 − .00312 Weight + .105 Horsepwr

8. a. Return = 247 − 32.8 Safety + 34.6 ExpRatio
b. 70.2

10. a. Revenue = 33.3 + 7.98 Cars
b. Increase of 1000 cars will increase revenue by \$7.98 million
c. Revenue = 106 + 8.94 Cars − .191 Locations

12. a. .926
b. .905
c. Yes

14. a. .75
b. .68

16. a. No, $R^2 = .311$
b. Multiple regression analysis

18. a. $R^2 = .942$, $R_a^2 = .932$
b. The fit is good

20. a. Significant; p-value = .000
b. Significant; p-value = .000
c. Significant; p-value = .000

22. a. SSE = 4000, $s^2 = 571.43$, MSR = 6000
b. Significant; $F = 10.50 > F_{.05} = 4.74$

24. a. Reject H_0: $\beta_1 = \beta_2 = 0$; p-value = .000
b. Weight: Cannot reject H_0: $\beta_1 = 0$; p-value = .386
Horsepwr: Reject H_0: $\beta_2 = 0$; p-value = .000

26. a. Significant; p-value = .000
b. Significant; p-value = .000
c. Not significant; p-value = .087

28. a. 132.16 to 154.15
b. 111.15 to 175.17

30. a. Speed = 72.6 + .0968 Horsepwr; estimate of speed is 101.29
b. 99.49 to 103.09
c. 94.594 to 107.986

32. a. $x_2 = 0$ if level 1; $x_2 = 1$ if level 2
$E(y) = \beta_0 + \beta_1 x_1 + \beta_2 x_2$
b. $E(y) = \beta_0 + \beta_1 x_1$
c. $E(y) = \beta_0 + \beta_1 x_1 + \beta_2$
d. $\beta_2 = E(y \mid \text{level 2}) - E(y \mid \text{level 1})$

34. a. \$15,300
b. \$56,100
c. \$41,600

36. a. $\hat{y}$ = 1.86 + 0.291 Months + 1.10 Type − 0.609 Person
b. Significant; p-value = $.002 < \alpha = .05$
c. Person is not significant

38. a. $\hat{y}$ = −91.8 + 1.08 Age + .252 Pressure + 8.74 Smoker
b. Significant; p-value = $.01 < \alpha = .05$
c. 95% prediction interval is 21.35 to 47.19 or a probability of .2135 to .4719; quit smoking and begin some type of treatment to reduce his blood pressure

40. a. $\hat{y} = -53.3 + 3.11x$
b. −1.40, −.15, 1.36, .47, −1.39; no
c. .38, .28, .22, .20, .98; no
d. .60, .00, .26, .03, 11.09; yes, the fifth observation

42. b. Unusual trend
c. No outliers
d. Observation 2 is an influential observation

44. b. 3.19

46. a. 3.04, 3.61, 5.08
b. Both are significant
d. .91

48. b. Significant; $F = 22.79 > F_{.05} = 5.79$
c. $R_a^2 = .861$; good fit
d. Both are significant

50. a. Speed = 97.6 + .0693 Price − .00082 Weight + .0590 Horsepwr − 2.48 Zero60
b. Significant relationship

c. Price and Weight are not significant
d. Speed = 103 + .0558 Horsepwr − 3.19 Zero60
e. Unusual trend
f. Observation 2 is an outlier
g. Observation 12 is an influential observation

52. a. Resale% = 38.8 + .000766 Price
b. Not a good fit; R-Sq = 36.7%
c. Resale% = 42.6 + 9.09 Type 1 + 7.92 Type 2 + .000341 Price
where Type 1 = 1 if a full-size pickup and Type 2 = 1 if a sport utility vehicle
d. Significant relationship; p-value corresponding to F = 14.79 = .000 < α = .05

Chapter 16

2. a. $\hat{y} = 9.32 + .424x$; p-value = .117 indicates that the relationship between x and y is not significant
b. $\hat{y} = -8.10 + 2.41x - .0480x^2$
$R_a^2 = .932$; a good fit
c. 20.965

4. a. $\hat{y} = 943 + 8.71x$
b. Significant; p-value = .005 < α = .01

6. b. No, the relationship appears to be curvilinear
c. Several possible models; e.g.,
$\hat{y} = 2.90 - .185x + .00351x^2$

8. a. It appears that a simple linear regression model is not appropriate
b. 2005% = 17.1 + 3.15 1999% − .0445 1999%Sq
c. Log2000% = 1.17 + .449 Log1999%
d. Part (b); higher percentage of variability is explained

10. a. Significant; $F = 49.52 > 4.24$
b. Significant; $F = 48.3 > 3.42$
c. Significant; $t = -4.46 < -2.069$
d. x_2 can be dropped

12. a. $\hat{y}$ = 170 + 6.61 TeamInt
b. $\hat{y}$ = 280 + 5.18 TeamInt − .0037 Rushing − 3.92 OpponInt
c. Addition of the two independent variables is not significant

14. a. $\hat{y}$ = −111 + 1.32 Age + .296 Pressure
b. $\hat{y}$ = −123 + 1.51 Age + .448 Pressure + 8.87 Smoker − .00276 AgePress
c. Significant

16. a. %College = −26.6 + .0970 SATScore
b. %College = −26.93 + .084 SATScore + .204 %TakeSAT
c. Same as part (b)

18. a. Green%
b. ScoreAvg = 58.2 − .00996 Distance − .152 Green% + .869 Putts
c. Yes
d. 72.65

20. $\hat{y}$ = −91.8 + 1.08 Age + .252 Pressure + 8.74 Smoker

22. d = 1.60; test is inconclusive

24.

x_1	x_2	Treatment
0	0	1
1	0	2
0	1	3

$x_3 = 0$ if block 1; $x_3 = 1$ if block 2
$E(y) = \beta_0 + \beta_1 x_1 + \beta_2 x_2 + \beta_3 x_3$

26. a.

D_1	D_2	Manufacturer
0	0	1
1	0	2
0	1	3

$E(y) = \beta_0 + \beta_1 D_1 + \beta_2 D_2$
b. $\hat{y} = 23.0 + 5.00D_1 - 2.00D_2$
c. $H_0: \beta_1 = \beta_2 = 0$
d. Mean time is not the same for each manufacturer; p-value = .004

28. Significant difference between the two analyzers

30. a. Let ExS denote the interaction between expense ratio and safety rating
Perform% = 23.3 + 222 Expense% − 28.9 ExS
b. R-Sq (adj) = 65.3%
c. 25.8 or approximately 26%

32. a. AUDELAY = 63.0 + 11.1 INDUS; no significant positive autcorrelation

34. Significant differences between comfort levels for the three types of browsers

Chapter 17

2. a. 32%
b. $8.14

4. $I_{2001} = 105$

6. $I = 125$

8. $I = 105$; portfolio is up 5%

10. a. 1980 wages: $8.82
2000 wages: $8.32
b. 97.5% increase
c. 5.7% decrease

12. a. 2448, 2486, 2557
Manufacturing shipments are increasing slightly in constant dollars
b. 2981, 3100, 3203
c. PPI

14. $I = 110$

16. $I = 83$

18. a. 151, 197, 143, 178
b. $I = 170$

20. $I_{\text{Jan}} = 96, I_{\text{Mar}} = 92$

22. $I = 117$

24. \$36,082; \$32,528; \$27,913; \$34,387; \$40,551; \$42,651; \$46,350

26. $I = 143$; quantity is up 43%

Chapter 18

2. a.

Week	4-Week	5-Week
10	19.00	18.80
11	20.00	19.20
12	18.75	19.00

b. 9.65, 7.41
c. 5-week

4. Weeks 10, 11, and 12: 18.48, 18.63, 18.27
MSE = 9.25; $\alpha = .2$ is better

6. a. MSE (3-month) = 1.24
MSE ($\alpha = .2$) = 3.55
Use 3-month moving averages
b. 83.3

8. a.

Month	3-Month	$\alpha = .2$
10	256.67	265.51
11	286.67	274.41
12	263.33	267.53

Using only the errors for months 4 to 12 for both, $\alpha = .2$ is better
b. 260

10. c. Use $\alpha = .3$; $F_{11} = 7.57$

12. $T_t = 4.7 - 2.1t$; 17.3

14. $T_t = 20.7466 - .3514t$

16. Consider a nonlinear trend

18. a. $T_t = .365 + .193t$; \$2.49
b. EPS increasing by an average of \$.193 per year

20. a. $T_t = 1997.6 + 397.545t$
b. $T_{11} = 6371$, $T_{12} = 6768$

22. a. Four-quarter moving average: 3.50, 4.00, 4.25, 4.75, 5.25, 5.50, 6.25, 6.50, 6.75
Centered moving average: 3.750, 4.125, 4.500, 5.000, 5.375, 5.875, 6.375, 6.625
b. Adjusted seasonal indexes: 1.2050, 0.7463, 0.8675, 1.1912
Note: adjustment = 0.9912

24. Adjusted seasonal indexes: 0.707, 0.777, 0.827, 0.966, 1.016, 1.305, 1.494, 1.225, 0.976, 0.986, 0.936, 0.787
Note: adjustment = 0.996

26. a. Yes
b. 12 − 4: 166,761.13
4 − 8: 146,052.99

28. a. .2 is better
b. 46.1

30. 20.26

32. a. $\alpha = .5$
b. $T_t = 244.778 + 22.088t$
c. Trend projection; smaller MSE

34. $T_8 = 252.28$, $T_9 = 259.10$

36. a. Yes
b. $T_t = 25 + 15t$

38. a. Linear trend appears to be appropriate
b. $T_t = 12{,}899.98 + 2092.066t$
c. \$2,092,066
d. 1997: \$40,096,838
1998: \$42,188,904

40. b. Adjusted seasonal indexes: 0.899, 1.362, 1.118, 0.621
Note: adjustment = 1.0101
c. Quarter 2; seems reasonable

42. a. $T_t = 6.329 + 1.055t$
b. 36.92, 37.98, 39.03, 40.09
c. 33.23, 51.65, 43.71, 24.86

Chapter 19

2. $z = 3.27$; reject H_0

4. $z = 3.15$; reject H_0

6. $z = 2.32$; reject H_0

8. $z = 3.76$; reject H_0

10. $z = 1.27$; do not reject H_0

12. $z = 2.43$; reject H_0

14. $z = 2.29$; reject H_0
Prices differ

16. $z = 2.62$; reject H_0

18. $T = 34$; reject H_0

20. $T = 36$; reject H_0

22. $z = 2.77$; reject H_0
P/E ratios differ

24. $z = -.25$; do not reject H_0

26. $W = 10.22$; reject H_0

28. $W = 9.26$; reject H_0

30. $W = 8.03$; reject H_0

32. a. .68
b. $z = 2.06$; reject H_0

34. $z = .72$; do not reject H_0

36. $r_s = -.709$; $z = -2.13$; reject H_0

38. $z = 3.17$; reject H_0

40. a. $z = -3.20$; reject H_0
Houston is below national median
b. $z = 2.21$; reject H_0
Philadelphia is above national median

42. $z = -2.97$; reject H_0

44. $W = 12.61$; reject H_0

46. $r_s = .76$; $z = 2.84$; reject H_0

Chapter 20

2. a. 5.42
b. UCL = 6.09, LCL = 4.75

4.

	R Chart	$\bar{x}$ Chart
UCL	2.98	29.10
LCL	.22	27.90

6. 20.01, .082

8. a. .0470
b. UCL = .0989, LCL = 2.0049 (use LCL = 0)
c. $\bar{p} = .08$; in control
d. UCL = 14.826, LCL = −0.726
Process is out of control if more than 14 defective
e. In control with 12 defective
f. *np* chart

10. $p = .02$; $f(0) = .6035$
$p = .06$; $f(0) = .2129$

12. $p_0 = .02$; producer's risk = .0599
$p_0 = .06$; producer's risk = .3396
Producer's risk decreases as the acceptance number *c* is increased

14. $n = 20$, $c = 3$

16. a. 95.4
b. UCL = 96.07, LCL = 94.73
c. No

18.

	R Chart	$\bar{x}$ Chart
UCL	4.23	6.57
LCL	0	4.27

Estimate of standard deviation = .86

20.

	R Chart	$\bar{x}$ Chart
UCL	.1121	3.112
LCL	0	3.051

22. a. UCL = .0817, LCL = −.0017 (use LCL = 0)

24. a. .03
b. $\beta = .0802$

26. a. Producer's risk = .0064
b. Consumer's risk = .0136
c. *Advantage:* excellent control
Disadvantage: cost

Chapter 21

2. a. 30,000
b. 320
c. 29,360 to 30,640

4. 73

6. 337

8. a. stratum 1: 27,600
stratum 2: 25,750
stratum 3: 21,000
b. 74,350
c. 70,599.88 to 78,100.12

10. a. $n = 93$, $n_1 = 30$, $n_2 = 30$, $n_3 = 33$
b. $n = 306$, $n_1 = 98$, $n_2 = 98$
$n_3 = 109$
c. $n = 275$, $n_1 = 88$, $n_2 = 88$, $n_3 = 98$

12. a. \$3,617,000
b. \$1,122,265
c. \$41,066 to \$56,499
d. \$9,568,261 to \$13,164,197

14. a. 15, 4500, .30
b. 1.4708, 441.24, .0484
c. 12.0584 to 17.9416
d. 3617.52 to 5382.48
e. .2032 to .3968

16. a. 40
b. .70
c. 35.8634 to 44.1366
d. .5234 to .8766

18. a. .1488 to .2312
b. .2615 to .3585
c. .1306 to .2094

20. a. \$22,790 to \$23,610
b. \$68,370,366 to \$70,829,634
c. .6692 to .7908

22. a. 431
b. .2175 to .3983
c. .6230 to .8002
d. 996

24. a. 75.275
b. .198 to .502
c. 1680

Appendix E: Solutions to Self-Test Exercises

Chapter 1

2. a. 9
b. 4
c. Country and room rate are qualitative variables; number of rooms and the overall score are quantitative variables
d. Country is nominal; room rate is ordinal; number of rooms and overall score are ratio

3. a. Average number of rooms = 808/9 = 89.78 or approximately 90 rooms
b. 2 of 9 are located in England; approximately 22%
c. 4 of 9 have a room rate of $$; approximately 44%

4. a. 10
b. *Fortune* 500 largest U.S. industrial corporations
c. Average revenue = $142,275.9/10 = $14,227.59 million
d. Using the sample average, statistical inference would let us estimate the average revenue for the population of 500 corporations as $14,227.59 million

13. a. Quantitative
b. Time series with 7 observations
c. Number of riverboat casinos
d. Time series shows a rapid increase; an increase would be expected in 1998, but it appears that the rate of increase is slowing

Chapter 2

3. a. $360° \times 58/120 = 174°$
b. $360° \times 42/120 = 126°$
c.

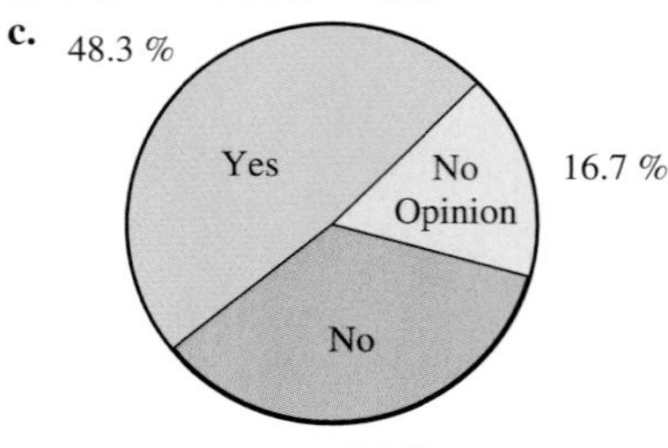

d.

7.

Rating	Frequency	Relative Frequency
Outstanding	19	.38
Very good	13	.26
Good	10	.20
Average	6	.12
Poor	2	.04

Management should be pleased with these results: 64% of the ratings are very good to outstanding, and 84% of the ratings are good or better; comparing these ratings to previous results will show whether the restaurant is making improvements in its customers' ratings of food quality

12.

Class	Cumulative Frequency	Cumulative Relative Frequency
≤19	10	.20
≤29	24	.48
≤39	41	.82
≤49	48	.96
≤59	50	1.00

15. a/b.

Waiting Time	Frequency	Relative Frequency
0–4	4	.20
5–9	8	.40
10–14	5	.25
15–19	2	.10
20–24	1	.05
Totals	20	1.00

c/d.

Waiting Time	Cumulative Frequency	Cumulative Relative Frequency
≤4	4	.20
≤9	12	.60
≤14	17	.85
≤19	19	.95
≤24	20	1.00

e. 12/20 = .60

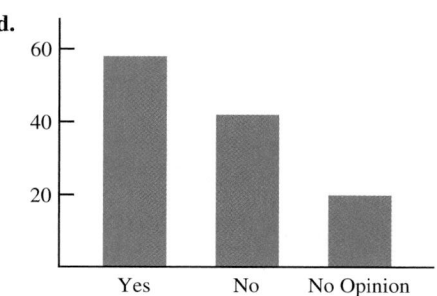

23. Leaf unit = .1

6	3
7	5 5 7
8	1 3 4 8
9	3 6
10	0 4 5
11	3

25.

9	8 9
10	2 4 6 6
11	4 5 7 8 8 9
12	2 4 5 7
13	1 2
14	4
15	1

29. a.

		y: 1	2	Total
x	A	5	0	5
	B	11	2	13
	C	2	10	12
	Total	18	12	30

b.

		y: 1	2	Total
x	A	100.0	0.0	100.0
	B	84.6	15.4	100.0
	C	16.7	83.3	100.0

c.

		y: 1	2
x	A	27.8	0.0
	B	61.1	16.7
	C	11.1	83.3
	Total	100.0	100.0

d. *A* values are always in $y = 1$

B values are most often in $y = 1$

C values are most often in $y = 2$

32. a.

Sales/ Margins/ ROE	EPS Rating 0–19	20–39	40–59	60–79	80–100	Total
A				1	8	9
B		1	4	5	2	12
C	1		1	2	3	7
D	3	1		1		5
E		2	1			3
Total	4	4	6	9	13	36

b.

Sales/ Margins/ ROE	EPS Rating 0–19	20–39	40–59	60–79	80–100	Total
A				11.11	88.89	100
B		8.33	33.33	41.67	16.67	100
C	14.29		14.29	28.57	42.86	100
D	60.00	20.00		20.00		100
E		66.67	33.33			100

Higher EPS ratings seem to be associated with higher ratings on Sales/Margins/ROE; of those companies with an "A" rating on Sales/Margins/ROE, 88.89% of them had an EPS Rating of 80 or higher; of the eight companies with a "D" or "E" rating on Sales/Margins/ROE, only 1 had an EPS rating above 60

Chapter 3

3. Arrange data in order: 15, 20, 25, 25, 27, 28, 30, 34

$i = \frac{20}{100}(8) = 1.6$; round up to position 2

20th percentile = 20

$i = \frac{25}{100}(8) = 2$; use positions 2 and 3

$$25\text{th percentile} = \frac{20 + 25}{2} = 22.5$$

$i = \frac{65}{100}(8) = 5.2$; round up to position 6

65th percentile = 28

$i = \frac{75}{100}(8) = 6$; use positions 6 and 7

$$75\text{th percentile} = \frac{28 + 30}{2} = 29$$

8. a. $\bar{x} = \frac{\Sigma x_i}{n} = \frac{775}{20} = 38.75$

Mode = 29 (appears three times)

b. Data in order: 22, 24, 29, 29, 29, 30, 31, 31, 32, 37, 40, 41, 44, 44, 46, 49, 50, 52, 57, 58

Median (10th and 11th positions)

$$\frac{37 + 40}{2} = 38.5$$

At home workers are slightly younger

c. $i = \frac{25}{100}(20) = 5$; use positions 5 and 6

$$Q_1 = \frac{29 + 30}{2} = 29.5$$

$i = \frac{75}{100}(20) = 15$; use positions 15 and 16

$$Q_3 = \frac{46 + 49}{2} = 47.5$$

d. $i = \frac{32}{100}(20) = 6.4$; round up to position 7

32nd percentile = 31

At least 32% of the people are 31 or younger

17. Range = 34 − 15 = 19

Arrange data in order: 15, 20, 25, 25, 27, 28, 30, 34

$$i = \frac{25}{100}(8) = 2;\ Q_1 = \frac{20 + 25}{2} = 22.5$$

$$i = \frac{75}{100}(8) = 6;\ Q_3 = \frac{28 + 30}{2} = 29$$

$IQR = Q_3 - Q_1 = 29 - 22.5 = 6.5$

$$\bar{x} = \frac{\Sigma x_i}{n} = \frac{204}{8} = 25.5$$

x_i	$(x_i - \bar{x})$	$(x_i - \bar{x})^2$
27	1.5	2.25
25	−.5	.25
20	−5.5	30.25
15	−10.5	110.25
30	4.5	20.25
34	8.5	72.25
28	2.5	6.25
25	−.5	.25
		242.00

$$s^2 = \frac{\Sigma(x_i - \bar{x})^2}{n - 1} = \frac{242}{8 - 1} = 34.57$$

$$s = \sqrt{34.57} = 5.88$$

18. a. Range = 190 − 168 = 22

b. $\bar{x} = \frac{\Sigma x_i}{n} = \frac{1068}{6} = 178$

$$s^2 = \frac{\Sigma(x_i - \bar{x})^2}{n - 1}$$

$$= \frac{4^2 + (-10)^2 + 6^2 + 12^2 + (-8)^2 + (-4)^2}{6 - 1}$$

$$= \frac{376}{5} = 75.2$$

c. $s = \sqrt{75.2} = 8.67$

d. $\frac{s}{\bar{x}}(100) = \frac{8.67}{178}(100) = 4.87$

27. Chebyshev's theorem: *at least* $(1 - 1/z^2)$

a. $z = \frac{40 - 30}{5} = 2;\ (1 - 1/2^2) = .75$

b. $z = \frac{45 - 30}{5} = 3;\ (1 - 1/3^2) = .89$

c. $z = \frac{38 - 30}{5} = 1.6;\ (1 - 1/1.6^2) = .61$

d. $z = \frac{42 - 30}{5} = 2.4;\ (1 - 1/2.4^2) = .83$

e. $z = \frac{48 - 30}{5} = 3.6;\ (1 - 1/3.6^2) = .92$

40. Arrange data in order: 5, 6, 8, 10, 10, 12, 15, 16, 18

$i = \frac{25}{100}(9) = 2.25$; round up to position 3

$Q_1 = 8$

Median (5th position) = 10

$i = \frac{75}{100}(9) = 6.75$; round up to position 7

$Q_3 = 15$

5-number summary: 5, 8, 10, 15, 18

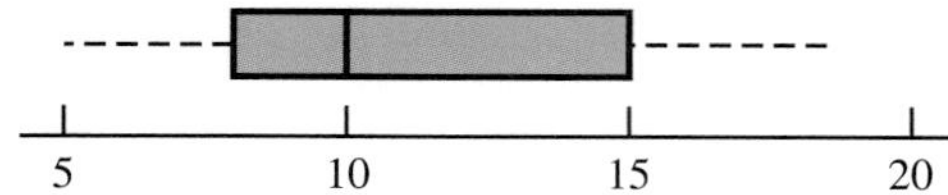

43. a. Arrange data in order low to high

$i = \frac{25}{100}(21) = 5.25$; round up to 6th position

$Q_1 = 1872$

Median (11th position) = 4019

$i = \frac{75}{100}(21) = 15.75$; round up to 16th position

$Q_3 = 8305$

5-number summary: 608, 1872, 4019, 8305, 14,138

b. $IQR = Q_3 - Q_1 = 8305 - 1872 = 6433$

Lower limit: 1872 − 1.5(6433) = −7777

Upper limit: 8305 + 1.5(6433) = 17,955

c. No; data are within limits

d. 41,138 > 27,604; 41,138 would be an outlier; data value should be reviewed and corrected

e.

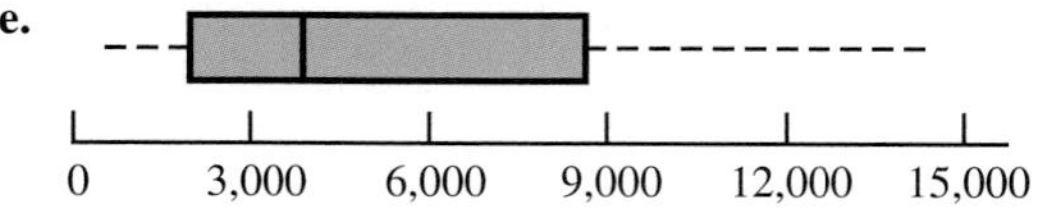

47. b. There appears to be a negative linear relationship between x and y

c.

x_i	y_i	$x_i - \bar{x}$	$y_i - \bar{y}$	$(x_i - \bar{x})(y_i - \bar{y})$
4	50	−4	4	−16
6	50	−2	4	−8
11	40	3	−6	−18
3	60	−5	14	−70
16	30	8	−16	−128
40	230	0	0	−240

$\bar{x} = 8; \bar{y} = 46$

$$s_{xy} = \frac{\Sigma(x_i - \bar{x})(y_i - \bar{y})}{n - 1} = \frac{-240}{4} = -60$$

The sample covariance indicates a negative linear association between x and y

d. $r_{xy} = \dfrac{s_{xy}}{s_x s_y} = \dfrac{-60}{(5.43)(11.40)} = -.97$

The sample correlation coefficient of −.97 is indicative of a strong negative linear relationship

55. a.

f_i	M_i	$f_i M_i$
4	5	20
7	10	70
9	15	135
5	20	100
25		325

$$\bar{x} = \frac{\Sigma f_i M_i}{n} = \frac{325}{25} = 13$$

b.

f_i	M_i	$(M_i - \bar{x})$	$(M_i - \bar{x})^2$	$f_i(M_i - \bar{x})^2$
4	5	−8	64	256
7	10	−3	9	63
9	15	2	4	36
5	20	7	49	245
25				600

$$s^2 = \frac{\Sigma f_i(M_i - \bar{x})^2}{n - 1} = \frac{600}{25 - 1} = 25$$

$$s = \sqrt{25} = 5$$

56. a.

Grade x_i	Weight w_i
4(A)	9
3(B)	15
2(C)	33
1(D)	3
0(F)	0
	60 credit hours

$$\bar{x} = \frac{\Sigma w_i x_i}{\Sigma w_i} = \frac{9(4) + 15(3) + 33(2) + 3(1)}{9 + 15 + 33 + 3} = \frac{150}{60} = 2.5$$

b. Yes

Chapter 4

2. $\binom{6}{3} = \dfrac{6!}{3!3!} = \dfrac{6\cdot5\cdot4\cdot3\cdot2\cdot1}{(3\cdot2\cdot1)(3\cdot2\cdot1)} = 20$

ABC	ACE	BCD	BEF
ABD	ACF	BCE	CDE
ABE	ADE	BCF	CDF
ABF	ADF	BDE	CEF
ACD	AEF	BDF	DEF

6. $P(E_1) = .40, P(E_2) = .26, P(E_3) = .34$
The relative frequency method was used

9. $\binom{50}{4} = \dfrac{50!}{4!46!} = \dfrac{50\cdot49\cdot48\cdot47}{4\cdot3\cdot2\cdot1} = 230{,}300$

10. a. Use the relative frequency approach
$P(\text{California}) = 1434/2374 = .60$

b. Number not from 4 states
$= 2374 - 1434 - 390 - 217 - 112$
$= 221$
$P(\text{Not from 4 states}) = 221/2374 = .09$

c. $P(\text{Not in early stages}) = 1 - .22 = .78$

d. Estimate of number of Massachusetts' companies in early stage of development = (.22)390 ≈ 86

e. If we assume the size of the awards did not differ by state, we can multiply the probability an award went to Colorado by the total venture funds disbursed to get an estimate
Est. of Colo. funds = (112/2374)($32.4)
= $1.53 billion
Authors' Note: The actual amount going to Colorado was $1.74 billion

15. a. S = (ace of clubs, ace of diamonds, ace of hearts, ace of spades)

b. S = (2 of clubs, 3 of clubs, . . . , 10 of clubs, J of clubs, Q of clubs, K of clubs, A of clubs)

c. There are 12; jack, queen, or king in each of the four suits

d. *For (a)*: 4/52 = 1/13 = .08
For (b): 13/52 = 1/4 = .25
For (c): 12/52 = .23

17. a. (4, 6), (4, 7), (4, 8)

b. .05 + .10 + .15 = .30

c. (2, 8), (3, 8), (4, 8)

d. .05 + .05 + .15 = .25

e. .15

23. a. $P(A) = P(E_1) + P(E_4) + P(E_6)$
$= .05 + .25 + .10 = .40$

$P(B) = P(E_2) + P(E_4) + P(E_7)$
$= .20 + .25 + .05 = .50$

$P(C) = P(E_2) + P(E_3) + P(E_5) + P(E_7)$
$= .20 + .20 + .15 + .05 = .60$

b. $A \cup B = \{E_1, E_2, E_4, E_6, E_7\}$
$P(A \cup B) = P(E_1) + P(E_2) + P(E_4) + P(E_6) + P(E_7)$
$= .05 + .20 + .25 + .10 + .05$
$= .65$

c. $A \cap B = \{E_4\}$, $P(A \cap B) = P(E_4) = .25$
d. Yes, they are mutually exclusive
e. $B^c = \{E_1, E_3, E_5, E_6\}$

$P(B^c) = P(E_1) + P(E_3) + P(E_5) + P(E_6)$
$= .05 + .20 + .15 + .10$
$= .50$

28. Let B = rented a car for business reasons
P = rented a car for personal reasons

a. $P(B \cup P) = P(B) + P(P) - P(B \cap P)$
$= .540 + .458 - .300$
$= .698$

b. $P(\text{Neither}) = 1 - .698 = .302$

30. a. $P(A \mid B) = \dfrac{P(A \cap B)}{P(B)} = \dfrac{.40}{.60} = .6667$

b. $P(B \mid A) = \dfrac{P(A \cap B)}{P(A)} = \dfrac{.40}{.50} = .80$

c. No, because $P(A \mid B) \neq P(A)$

33. a.

	Reason for Applying			
	Quality	**Cost/ Convenience**	**Other**	**Total**
Full-time	.218	.204	.039	.461
Part-time	.208	.307	.024	.539
Total	.426	.511	.063	1.000

b. A student is most likely to cite cost or convenience as the first reason (probability = .511); school quality is the first reason cited by the second largest number of students (probability = .426)
c. $P(\text{quality} \mid \text{full-time}) = .218/.461 = .473$
d. $P(\text{quality} \mid \text{part-time}) = .208/.539 = .386$
e. For independence, we must have $P(A)P(B) = P(A \cap B)$; from the table

$P(A \cap B) = .218$, $P(A) = .461$, $P(B) = .426$
$P(A)P(B) = (.461)(.426) = .196$

Because $P(A)P(B) \neq P(A \cap B)$, the events are not independent

39. a. Yes, because $P(A_1 \cap A_2) = 0$
b. $P(A_1 \cap B) = P(A_1)P(B \mid A_1) = .40(.20) = .08$
$P(A_2 \cap B) = P(A_2)P(B \mid A_2) = .60(.05) = .03$
c. $P(B) = P(A_1 \cap B) + P(A_2 \cap B) = .08 + .03 = .11$
d. $P(A_1 \mid B) = \dfrac{.08}{.11} = .7273$

$P(A_2 \mid B) = \dfrac{.03}{.11} = .2727$

42. M = missed payment
D_1 = customer defaults
D_2 = customer does not default
$P(D_1) = .05$, $P(D_2) = .95$, $P(M \mid D_2) = .2$, $P(M \mid D_1) = 1$

a. $$P(D_1 \mid M) = \frac{P(D_1)P(M \mid D_1)}{P(D_1)P(M \mid D_1) + P(D_2)P(M \mid D_2)} = \frac{(.05)(1)}{(.05)(1) + (.95)(.2)} = \frac{.05}{.24} = .21$$

b. Yes, the probability of default is greater than .20

Chapter 5

1. a. Head, Head (H, H)
Head, Tail (H, T)
Tail, Head (T, H)
Tail, Tail (T, T)
b. x = number of heads on two coin tosses
c.

Outcome	Values of x
(H, H)	2
(H, T)	1
(T, H)	1
(T, T)	0

d. Discrete; it may assume 3 values: 0, 1, and 2

3. Let: Y = position is offered
N = position is not offered

a. $S = \{(Y, Y, Y), (Y, Y, N), (Y, N, Y), (Y, N, N), (N, Y, Y), (N, Y, N), (N, N, Y), (N, N, N)\}$
b. Let N = number of offers made; N is a discrete random variable
c.

Experimental Outcome	(Y, Y, Y)	(Y, Y, N)	(Y, N, Y)	(Y, N, N)	(N, Y, Y)	(N, Y, N)	(N, N, Y)	(N, N, N)
Value of N	3	2	2	1	2	1	1	0

7. a. $f(x) \geq 0$ for all values of x
$\Sigma f(x) = 1$; therefore, it is a proper probability distribution
b. Probability $x = 30$ is $f(30) = .25$
c. Probability $x \leq 25$ is $f(20) + f(25) = .20 + .15 = .35$
d. Probability $x > 30$ is $f(35) = .40$

8. a.

x	$f(x)$
1	3/20 = .15
2	5/20 = .25
3	8/20 = .40
4	4/20 = .20
	Total 1.00

b.

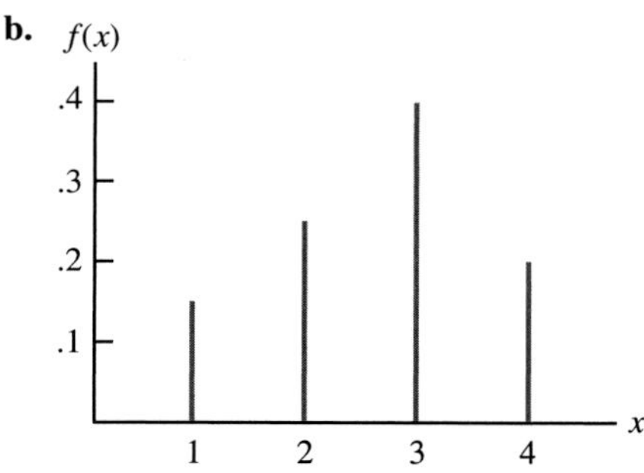

c. $f(x) \geq 0$ for $x = 1, 2, 3, 4$
$\Sigma f(x) = 1$

16. a.

y	$f(y)$	$yf(y)$
2	.20	.40
4	.30	1.20
7	.40	2.80
8	.10	.80
Totals	1.00	5.20

$E(y) = \mu = 5.20$

b.

y	$y - \mu$	$(y - \mu)^2$	$f(y)$	$(y - \mu)^2 f(y)$
2	−3.20	10.24	.20	2.048
4	−1.20	1.44	.30	.432
7	1.80	3.24	.40	1.296
8	2.80	7.84	.10	.784
			Total	4.560

$\text{Var}(y) = 4.56$
$\sigma = \sqrt{4.56} = 2.14$

18. a/b.

x	$f(x)$	$xf(x)$	$(x - \mu)$	$(x - \mu)^2$	$(x - \mu)^2 f(x)$
0	.01	.00	−2.3	5.29	.0529
1	.23	.23	−1.3	1.69	.3887
2	.41	.82	−0.3	0.09	.0369
3	.20	.60	0.7	0.49	.0980
4	.10	.40	1.7	2.89	.2890
5	.05	.25	2.7	7.29	.3645
	$E(x) = 2.30$				$\text{Var}(x) = 1.2300$
					$\sigma = 1.11$

The expected value, $E(x) = 2.3$, of the probability distribution is the same as the average reported in the *1997 Statistical Abstract of the United States*
$\text{Var}(x) = 1.23$ television sets squared
$\sigma = \sqrt{1.23} = 1.11$ television sets

25. a.

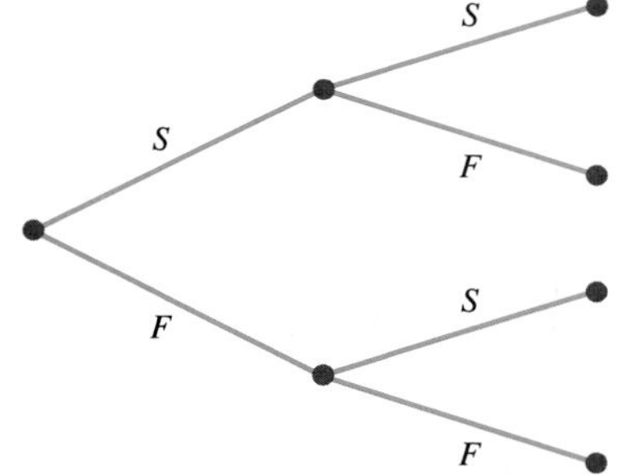

b. $f(1) = \binom{2}{1}(.4)^1(.6)^1 = \frac{2!}{1!1!}(.4)(.6) = .48$

c. $f(0) = \binom{2}{0}(.4)^0(.6)^2 = \frac{2!}{0!2!}(1)(.36) = .36$

d. $f(2) = \binom{2}{2}(.4)^2(.6)^0 = \frac{2!}{2!0!}(.16)(.1) = .16$

e. $P(x \geq 1) = f(1) + f(2) = .48 + .16 = .64$

f. $E(x) = np = 2(.4) = .8$
$\text{Var}(x) = np(1 - p) = 2(.4)(.6) = .48$
$\sigma = \sqrt{.48} = .6928$

30. a. Probability of a defective part being produced must be .03 for each trial; trials must be independent

b. Let D = defective
G = not defective

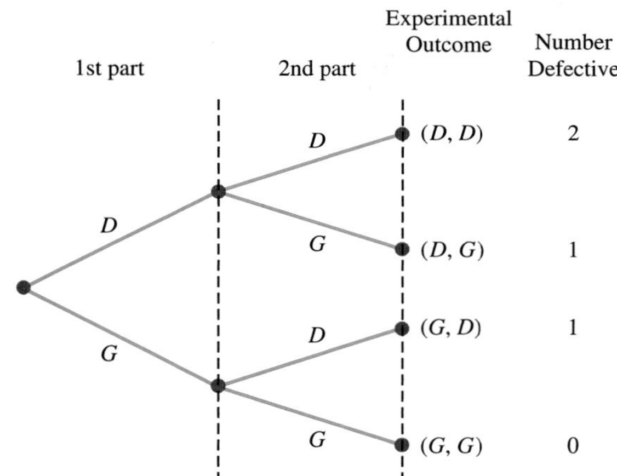

c. Two outcomes result in exactly one defect

d. $P(\text{no defects}) = (.97)(.97) = .9409$
$P(1 \text{ defect}) = 2(.03)(.97) = .0582$
$P(2 \text{ defects}) = (.03)(.03) = .0009$

39. a. $f(x) = \dfrac{2^x e^{-2}}{x!}$

b. $\mu = 6$ for 3 time periods

c. $f(x) = \dfrac{6^x e^{-6}}{x!}$

d. $f(2) = \dfrac{2^2 e^{-2}}{2!} = \dfrac{4(.1353)}{2} = .2706$

e. $f(6) = \dfrac{6^6 e^{-6}}{6!} = .1606$

f. $f(5) = \dfrac{4^5 e^{-4}}{5!} = .1563$

40. a. $\mu = 48(5/60) = 4$

$f(3) = \dfrac{4^3 e^{-4}}{3!} = \dfrac{(64)(.0183)}{6} = .1952$

b. $\mu = 48(15/60) = 12$

$f(10) = \dfrac{12^{10} e^{-12}}{10!} = .1048$

c. $\mu = 48(5/60) = 4$; one can expect 4 callers to be waiting after 5 minutes

$f(0) = \dfrac{4^0 e^{-4}}{0!} = .0183$; the probability none will be waiting after 5 minutes is .0183

d. $\mu = 48(3/60) = 2.4$

$f(0) = \dfrac{2.4^0 e^{-2.4}}{0!} = .0907$; the probability of no interruptions in 3 minutes is .0907

46. a. $f(1) = \dfrac{\binom{3}{1}\binom{10-3}{4-1}}{\binom{10}{4}} = \dfrac{\left(\frac{3!}{1!2!}\right)\left(\frac{7!}{3!4!}\right)}{\frac{10!}{4!6!}}$

$= \dfrac{(3)(35)}{210} = .50$

b. $f(2) = \dfrac{\binom{3}{2}\binom{10-3}{2-2}}{\binom{10}{2}} = \dfrac{(3)(1)}{45} = .067$

50. $N = 60, n = 10$

a. $r = 20, x = 0$

$f(0) = \dfrac{\binom{20}{0}\binom{40}{10}}{\binom{60}{10}} = \dfrac{(1)\left(\frac{40!}{10!30!}\right)}{\frac{60!}{10!50!}}$

$= \left(\dfrac{40!}{10!30!}\right)\left(\dfrac{10!50!}{60!}\right)$

$= \dfrac{40 \cdot 39 \cdot 38 \cdot 37 \cdot 36 \cdot 35 \cdot 34 \cdot 33 \cdot 32 \cdot 31}{60 \cdot 59 \cdot 58 \cdot 57 \cdot 56 \cdot 55 \cdot 54 \cdot 53 \cdot 52 \cdot 51}$

$\approx .01$

b. $r = 20, x = 1$

$f(1) = \dfrac{\binom{20}{1}\binom{40}{9}}{\binom{60}{10}} = 20\left(\dfrac{40!}{9!31!}\right)\left(\dfrac{10!50!}{60!}\right)$

$\approx .07$

c. $1 - f(0) - f(1) = 1 - .08 = .92$

d. Same as the probability one will be from Hawaii; in part (b) it was equal to approximately .07

Chapter 6

1. a.

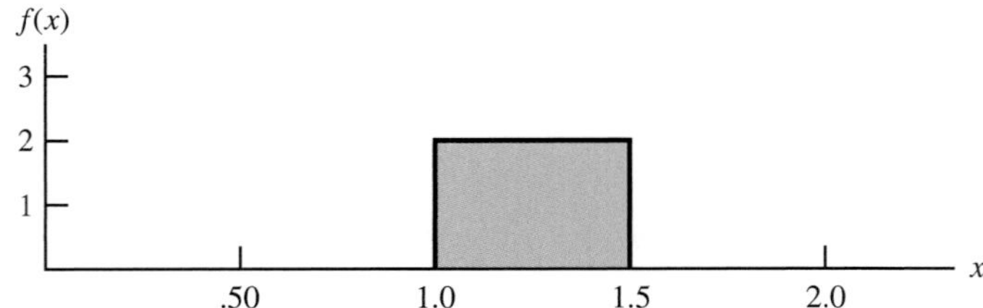

b. $P(x = 1.25) = 0$; the probability of any single point is zero since the area under the curve above any single point is zero

c. $P(1.0 \le x \le 1.25) = 2(.25) = .50$

d. $P(1.20 < x < 1.5) = 2(.30) = .60$

4. a.

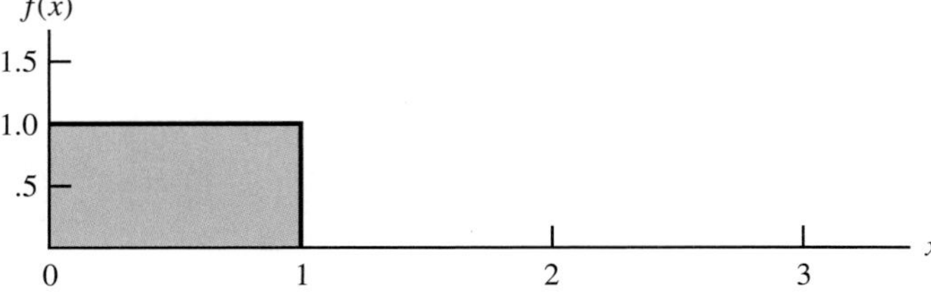

b. $P(.25 < x < .75) = 1(.50) = .50$

c. $P(x \leq .30) = 1(.30) = .30$
d. $P(x > .60) = 1(.40) = .40$

13. a. $.6879 - .0239 = .6640$
b. $.8888 - .6985 = .1903$
c. $.9599 - .8508 = .1091$

15. a. Look in the table for an area of $.5000 - .2119 = .2881$; $z = .80$ cuts off an area of .2119 in the upper tail; thus, for an area of .2119 in the lower tail, $z = -.80$
b. Look in the table for an area of $.9030/2 = .4515$; $z = 1.66$;
c. Look in the table for an area of $.2052/2 = .1026$; $z = .26$
d. Look in the table for an area of .4948; $z = 2.56$
e. Look in the table for an area of .1915; because the value we are seeking is below the mean, the z value must be negative; thus, $z = -.50$

18. a. Find $P(x \geq 60)$

$$\text{At } x = 60,\ z = \frac{60 - 49}{16} = \frac{11}{16} = .69$$

$P(x < 60) = .7549$

$P(x \geq 60) = 1 - P(x < 60) = 1 - .7549 = .2451$

b. Find $P(x \leq 30)$

$$\text{At } x = 30,\ z = \frac{30 - 49}{16} = -1.19$$

$$P(x \leq 30) = .5000 - .3830 = .1170$$

c. Find z-score so that $P(z \geq z\text{-score}) = .10$
A z-score of 1.28 cuts off 10% in upper tail; now, solve for corresponding value of x

$$1.28 = \frac{x - 49}{16}$$

$$x = 49 + (16)(1.28) = 69.48$$

So, 10% of subscribers spend 69.48 minutes or more reading *The Wall Street Journal*

27. a. $P(x \leq x_0) = 1 - e^{-x_0/3}$
b. $P(x \leq 2) = 1 - e^{-2/3} = 1 - .5134 = .4866$
c. $P(x \geq 3) = 1 - P(x \leq 3) = 1 - (1 - e^{-3/3}) = e^{-1} = .3679$
d. $P(x \leq 5) = 1 - e^{-5/3} = 1 - .1889 = .8111$
e. $P(2 \leq x \leq 5) = P(x \leq 5) - P(x \leq 2) = .8111 - .4866 = .3245$

29. a.

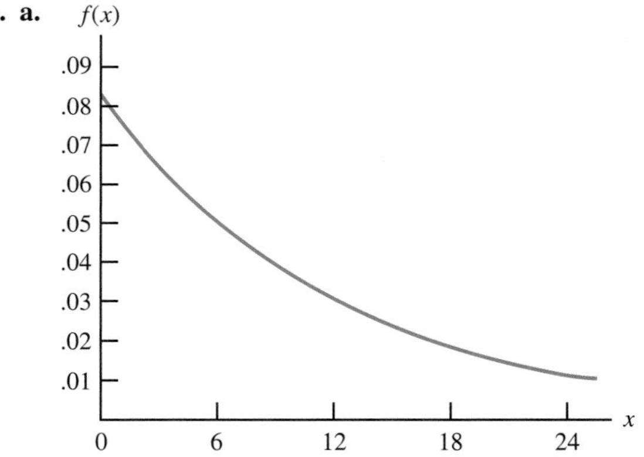

b. $P(x \leq 12) = 1 - e^{-12/12} = 1 - .3679 = .6321$
c. $P(x \leq 6) = 1 - e^{-6/12} = 1 - .6065 = .3935$
d. $P(x \geq 30) = 1 - P(x < 30) = 1 - (1 - e^{-30/12}) = .0821$

Chapter 7

1. a. AB, AC, AD, AE, BC, BD, BE, CD, CE, DE
b. With 10 samples, each has a 1/10 probability
c. E and C because 8 and 0 do not apply; 5 identifies E; 7 does not apply; 5 is skipped because E is already in the sample; 3 identifies C; 2 is not needed because the sample of size 2 is complete

3. 459, 147, 385, 113, 340, 401, 215, 2, 33, 348

11. a. $\bar{x} = \frac{\Sigma x_i}{n} = \frac{54}{6} = 9$

b. $s = \sqrt{\frac{\Sigma(x_i - \bar{x})^2}{n - 1}}$

$$\Sigma(x_i - \bar{x})^2 = (-4)^2 + (-1)^2 + 1^2 + (-2)^2 + 1^2 + 5^2 = 48$$

$$s = \sqrt{\frac{48}{6 - 1}} = 3.1$$

13. a. $\bar{x} = \frac{\Sigma x_i}{n} = \frac{465}{5} = 93$

b.

	x_i	$(x_i - \bar{x})$	$(x_i - \bar{x})^2$
	94	+1	1
	100	+7	49
	85	−8	64
	94	+1	1
	92	−1	1
Totals	465	0	116

$$s = \sqrt{\frac{\Sigma(x_i - \bar{x})^2}{n - 1}} = \sqrt{\frac{116}{4}} = 5.39$$

19. a. The sampling distribution is normal with:

$$E(\bar{x}) = \mu = 200$$

$$\sigma_{\bar{x}} = \frac{\sigma}{\sqrt{n}} = \frac{50}{\sqrt{100}} = 5$$

For $+5$, $(\bar{x} - \mu) = 5$,

$$z = \frac{\bar{x} - \mu}{\sigma_{\bar{x}}} = \frac{5}{5} = 1$$

$$\text{Area} = .3413 \times 2 = .6826$$

b. For ± 10, $(\bar{x} - \mu) = 10$,

$$z = \frac{\bar{x} - \mu}{\sigma_{\bar{x}}} = \frac{10}{5} = 2$$

$$\text{Area} = .4772 \times 2 = .9544$$

25. a.

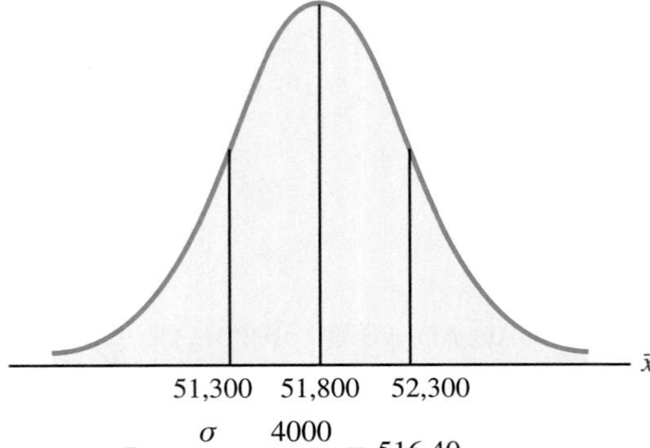

$$\sigma_{\bar{x}} = \frac{\sigma}{\sqrt{n}} = \frac{4000}{\sqrt{60}} = 516.40$$

$$z = \frac{52{,}300 - 51{,}800}{516.40} = +.97$$

Area = .3340 × 2 = .6680

b. $$\sigma_{\bar{x}} = \frac{\sigma}{\sqrt{n}} = \frac{4000}{\sqrt{120}} = 365.15$$

$$z = \frac{52{,}300 - 51{,}800}{365.15} = +1.37$$

Area = .4147 × 2 = .8294

34. a. $E(\bar{p}) = .40$

$$\sigma_{\bar{p}} = \sqrt{\frac{p(1-p)}{n}} = \sqrt{\frac{(.40)(.60)}{200}} = .0346$$

$$z = \frac{\bar{p} - p}{\sigma_{\bar{p}}} = \frac{.03}{.0346} = .87$$

Area = .3078 × 2 = .6156

b. $$z = \frac{\bar{p} - p}{\sigma_{\bar{p}}} = \frac{.05}{.0346} = 1.45$$

Area = .4265 × 2 = .8530

37. a.

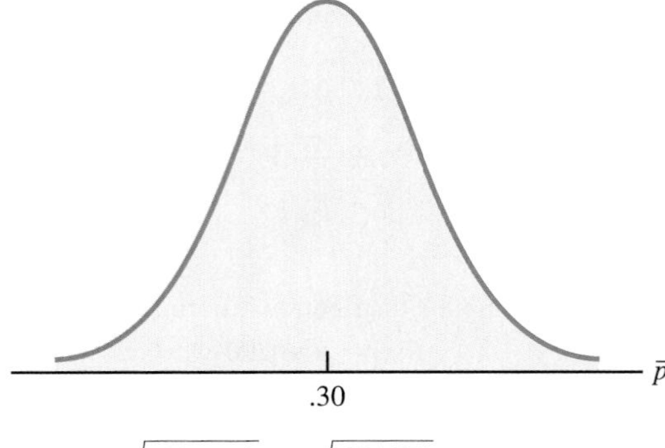

$$\sigma_{\bar{p}} = \sqrt{\frac{p(1-p)}{n}} = \sqrt{\frac{.30(.70)}{100}} = .0458$$

The normal distribution is appropriate because $np = 100(.30) = 30$ and $n(1-p) = 100(.70) = 70$ are both greater than 5

b. $P(.20 \le \bar{p} \le .40) = ?$

$$z = \frac{.40 - .30}{.0458} = 2.18$$

Area = .4854 × 2 = .9708

c. $P(.25 \le \bar{p} \le .35) = ?$

$$z = \frac{.35 - .30}{.0458} = 1.09$$

Area = .3621 × 2 = .7242

Chapter 8

2. Use $\bar{x} \pm z_{\alpha/2}(\sigma/\sqrt{n})$ with the sample standard deviation s used to estimate σ

a. $32 \pm 1.645\ (6/\sqrt{50})$
32 ± 1.4; 30.6 to 33.4

b. $32 \pm 1.96(6/\sqrt{50})$
32 ± 1.66; 30.34 to 33.66

c. $32 \pm 2.76(6/\sqrt{50})$
32 ± 2.19; 29.81 to 34.19

5. a. $1.96\sigma/\sqrt{n} = 1.96\ (5.00/\sqrt{49}) = 1.40$

b. $24.80 \pm .1.40$; 23.40 to 26.20

15. a. $$\bar{x} = \frac{\Sigma x_i}{n} = \frac{80}{8} = 10$$

b. $$s = \sqrt{\frac{\Sigma(x_i - \bar{x})^2}{n-1}} = \sqrt{\frac{84}{8-1}} = 3.46$$

c. With 7 degrees of freedom, $t_{.025} = 2.365$

$$\bar{x} \pm t_{.025}\frac{s}{\sqrt{n}}$$

$$10 \pm 2.365\frac{3.46}{\sqrt{8}}$$

10 ± 2.90; 7.10 to 12.90

17. At 90%, $80 \pm t_{.05}(s/\sqrt{n})$ with degrees of freedom = 17
$t_{.05} = 1.740$
$80 \pm 1.740(10/\sqrt{18})$
80 ± 4.10; 75.90 to 84.10

At 95%, $80 \pm t_{.025}(10/\sqrt{18})$ with degrees of freedom = 17
$t_{.025} = 2.110$
$80 \pm 2.110\ (10/\sqrt{18})$
80 ± 4.97; 75.03 to 84.97

24. a. Planning value of $\sigma = \frac{\text{Range}}{4} = \frac{36}{4} = 9$

b. $$n = \frac{z_{.025}^2\sigma^2}{E^2} = \frac{(1.96)^2(9)^2}{(3)^2} = 34.57; \text{ use } n = 35$$

c. $$n = \frac{(1.96)^2(9)^2}{(2)^2} = 77.79; \text{ use } n = 78$$

25. a. Use $n = \frac{z_{\alpha/2}^2\sigma^2}{E^2}$

$$n = \frac{(1.96)^2(6.82)^2}{(1.5)^2} = 79.41; \text{ use } n = 80$$

b. $$n = \frac{(1.645)^2(6.82)^2}{(2)^2} = 31.47; \text{ use } n = 32$$

31. a. $$\bar{p} = \frac{100}{400} = .25$$

b. $$\sqrt{\frac{\bar{p}(1-\bar{p})}{n}} = \sqrt{\frac{.25(.75)}{400}} = .0217$$

c. $\bar{p} \pm z_{.025}\sqrt{\frac{\bar{p}(1-\bar{p})}{n}}$

$.25 \pm 1.96(.0217)$

$.25 \pm .0424$; .2076 to .2924

35. a. $\bar{p} = 562/814 = .6904$

b. $1.645\sqrt{\frac{.6904(1-.6904)}{814}} = .0267$

c. $.6904 \pm .0267$; .6637 to .7171

39. a. $n = \frac{1.96^2 p(1-p)}{E^2}$

$n = \frac{1.96^2(.33)(.67)}{(.03)^2} = 943.75$; use $n = 944$

b. $n = \frac{2.576^2(.33)(.67)}{(.03)^2} = 1630.19$; use $n = 1631$

Chapter 9

2. a. H_0: $\mu \leq 14$
H_a: $\mu > 14$

b. No evidence that the new plan increases sales

c. The research hypothesis $\mu > 14$ is supported; the new plan increases sales

5. a. Rejecting H_0: $\mu \leq 8.6$ when it is true

b. Accepting H_0: $\mu \leq 8.6$ when it is false

10. a. $z = 2.05$
Reject H_0 if $z > 2.05$

b. $z = \frac{x - \mu}{s/\sqrt{n}} = \frac{16.5 - 15}{7/\sqrt{40}} = 1.36$

c. Area for $z = 1.36 = .4131$
p-value $= .5000 - .4131 = .0869$

d. Do not reject H_0

13. a. H_0: $\mu \geq 1056$
H_a: $\mu < 1056$

b. Reject H_0 if $z < -1.645$

c. $z = \frac{\bar{x} - \mu}{s/\sqrt{n}} = \frac{910 - 1056}{1600/\sqrt{400}} = -1.83$

d. Reject H_0; conclude $\mu < 1056$

e. p-value $= .5000 - .4664 = .0336$

22. a. Reject H_0 if $z < -2.33$ or $z > 2.33$

b. $z = \frac{\bar{x} - \mu}{\sigma/\sqrt{n}} = \frac{14.2 - 15}{5/\sqrt{50}} = 1.13$

c. p-value $= 2(.5000 - .3708) = .2584$

d. Do not reject H_0

25. a. Reject H_0 if $z < -1.96$ or $z > 1.96$

b. $z = \frac{\bar{x} - \mu_0}{s/\sqrt{n}} = \frac{38.5 - 39.2}{4.8/\sqrt{112}} = -1.54$

c. Do not reject H_0

d. p-value $= 2(.5000 - .4382) = .1236$

34. a. $\bar{x} = \frac{\Sigma x_i}{n} = \frac{108}{6} = 18$

b. $s = \sqrt{\frac{\Sigma(x_i - \bar{x})}{n-1}} = \sqrt{\frac{10}{6-1}} = 1.41$

c. Reject H_0 if $t < -2.571$ or $t > 2.571$

d. $t = \frac{\bar{x} - \mu}{s/\sqrt{n}} = \frac{18 - 20}{1.41/\sqrt{6}} = -3.47$

e. Reject H_0; conclude H_a is true

37. a. H_0: $\mu = 3.00$
H_a: $\mu \neq 3.00$

b. Reject H_0 if $t < -2.262$ or if $t > 2.262$

c. $\bar{x} = \Sigma x_i/n = \frac{28}{10} = 2.80$

d. $s = \sqrt{\frac{\Sigma(x_i - \bar{x})^2}{n-1}} = .70$

e. $t = \frac{\bar{x} - \mu_0}{s/\sqrt{n}} = \frac{2.80 - 3.00}{.70/\sqrt{10}} = -.90$

f. Do not reject H_0

g. $t_{.10} = 1.383$; p-value greater than $2(.10) = .20$

44. a. Reject H_0 if $z < -1.96$ or $z > 1.96$

b. $\sigma_{\bar{p}} = \sqrt{\frac{.20(.80)}{400}} = .02$

$z = \frac{\bar{p} - p}{\sigma_{\bar{p}}} = \frac{.175 - .20}{.02} = -1.25$

c. p-value $= 2(.5000 - .3944) = .2112$

d. Do not reject H_0

47. a. Reject H_0 if $z < -1.645$

b. $\bar{p} = 52/100 = .52$

$\sigma_{\bar{p}} = \sqrt{\frac{p(1-p)}{n}} = \sqrt{\frac{.64(1-.64)}{100}} = .0480$

$z = \frac{\bar{p} - p}{\sigma_{\bar{p}}} = \frac{.52 - .64}{.0480} = -2.50$

c. Reject H_0; conclude less than 64% agree

d. p-value $= .5000 - .4938 = .0062$

56.

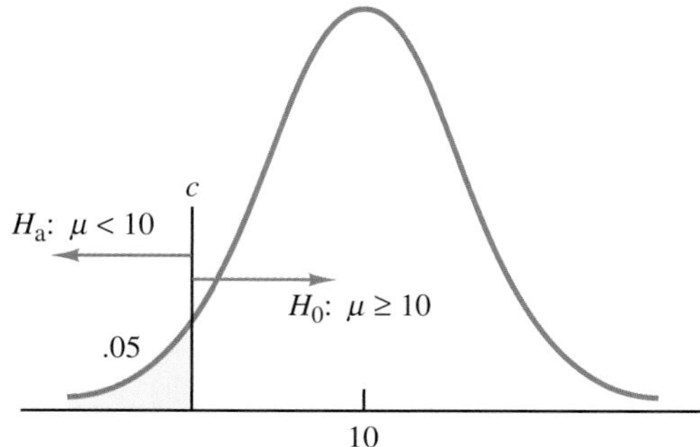

$c = 10 - 1.645(5/\sqrt{120}) = 9.25$

Reject H_0 if $\bar{x} < 9.25$

a. When $\mu = 9$,

$z = \frac{9.25 - 9}{5/\sqrt{120}} = .55$

$P(H_0) = (.5000 - .2088) = .2912$

b. Type II error

c. When $\mu = 8$,

$$z = \frac{9.25 - 8}{5/\sqrt{120}} = 2.74$$

$$\beta = (.5000 - .4969) = .0031$$

59. a. $H_0: \mu \geq 25$
$H_a: \mu < 25$
Reject H_0 if $z < -2.05$

$$z = \frac{\bar{x} - \mu_0}{\sigma/\sqrt{n}} = \frac{\bar{x} - 25}{3/\sqrt{30}} = -2.05$$

Solve for $\bar{x} = 23.88$
Decision Rule: Accept H_0 if $\bar{x} \geq 23.88$
Reject H_0 if $\bar{x} < 23.88$

b. For $\mu = 23$,

$$z = \frac{23.88 - 23}{3/\sqrt{30}} = 1.61$$

$$\beta = .5000 - .4463 = .0537$$

c. For $\mu = 24$,

$$z = \frac{23.88 - 24}{3/\sqrt{30}} = -.22$$

$$\beta = .5000 - .0871 = .5871$$

d. The Type II error cannot be made in this case; note that when $\mu = 25.5$, H_0 is true: the Type II error can only be made when H_0 is false

64. $$n = \frac{(z_\alpha - z_\beta)\sigma^2}{(\mu_0 - \mu_a)^2} = \frac{(1.645 + 1.28)^2(5)^2}{(10 - 9)^2} = 214$$

67. At $\mu_0 = 400$, $\alpha = .02$; $z_{.02} = 2.05$
At $\mu_a = 385$, $\beta = .10$; $z_{.10} = 1.28$
With $\sigma = 30$,

$$n = \frac{(z_\alpha + z_\beta)^2\sigma^2}{(\mu_0 - \mu_a)^2} = \frac{(2.05 + 1.28)^2(30)^2}{(400 - 385)^2} = 44.4 \text{ or } 45$$

Chapter 10

1. a. $\bar{x}_1 - \bar{x}_2 = 13.6 - 11.6 = 2$

b. $$s_{\bar{x}_1 - \bar{x}_2} = \sqrt{\frac{s_1^2}{n_1} + \frac{s_2^2}{n_2}} = \sqrt{\frac{(2.2)^2}{50} + \frac{3^2}{35}} = .595$$

$2 \pm 1.645(.595)$
$2 \pm .98$ or 1.02 to 2.98

c. $2 \pm 1.96(.595)$
2 ± 1.17, or .83 to 3.17

8. a. $\bar{x}_1 - \bar{x}_2 = 15{,}700 - 14{,}500 = 1200$

b. Pooled variance

$$s^2 = \frac{7(700)^2 + 11(850)^2}{18} = 632{,}083$$

$$s_{\bar{x}_1 - \bar{x}_2} = \sqrt{632{,}083\left(\frac{1}{8} + \frac{1}{12}\right)} = 362.88$$

With 18 degrees of freedom $t_{.025} = 2.101$,
$1200 \pm 2.101(362.88)$
1200 ± 762, or 438 to 1962

c. Populations are normally distributed with equal variances

11. a. $$s_{\bar{x}_1 - \bar{x}_2} = \sqrt{\frac{s_1^2}{n_1} + \frac{s_2^2}{n_2}} = \sqrt{\frac{(5.2)^2}{40} + \frac{6^2}{50}} = 1.18$$

$$z = \frac{(\bar{x}_1 - \bar{x}_2) - (\mu_1 - \mu_2)}{s_{\bar{x}_1 - \bar{x}_2}}$$

$$= \frac{(25.2 - 22.8)}{1.18} = 2.03$$

Reject H_0 if $z > 1.645$; therefore reject H_0; conclude H_a is true and $\mu_1 > \mu_2$

b. p-value $= .5000 - .4788 = .0212$

15. $H_0: \mu_1 - \mu_2 = 0$
$H_a: \mu_1 - \mu_2 \neq 0$
Reject H_0 if $z < -1.96$ or if $z > 1.96$

$$z = \frac{(\bar{x}_1 - \bar{x}_2) - 0}{\sqrt{\sigma_1^2/n_1 + \sigma_2^2/n_2}} = \frac{40 - 35}{\sqrt{(9)^2/36 + (10)^2/49}}$$

$= 2.41$

p-value $= 2(.5000 - .4920) = .0160$

Reject H_0; customers at the two stores differ in terms of mean ages

21. a. 1, 2, 0, 0, 2

b. $$\bar{d} = \frac{\Sigma d_i}{n} = \frac{5}{5} = 1$$

c. $$s_d = \sqrt{\frac{\Sigma(d_i - \bar{d})^2}{n - 1}} = \sqrt{\frac{4}{5 - 1}} = 1$$

d. With 4 degrees of freedom, $t_{.05} = 2.132$; reject H_0 if $t > 2.132$

$$t = \frac{\bar{d} - \mu_d}{s_d/\sqrt{n}} = \frac{1 - 0}{1/\sqrt{5}} = 2.24$$

Reject H_0; conclude $\mu_d > 0$

23. d = rating after − rating before

$H_0: \mu_d \leq 0$
$H_a: \mu_d > 0$
With 7 degrees of freedom, reject H_0 if $t > 1.895$; when $\bar{d} = .63$ and $s_d = 1.3025$,

$$t = \frac{\bar{d} - \mu_d}{s_d/\sqrt{n}} = \frac{.63 - 0}{1.3025/\sqrt{8}} = 1.36$$

Do not reject H_0; we cannot conclude that seeing the commercial improves the potential to purchase

31. a. $$\bar{p} = \frac{n_1\bar{p}_1 + n_2\bar{p}_2}{n_1 + n_2} = \frac{200(.22) + 300(.16)}{200 + 300} = .184$$

$$s_{\bar{p}_1 - \bar{p}_2} = \sqrt{(.184)(.816)\left(\frac{1}{200} + \frac{1}{300}\right)} = .0354$$

Reject H_0 if $z > 1.645$

$$z = \frac{(.22 - .16) - 0}{.0354} = 1.69$$

Reject H_0

b. p-value $= (.5000 - .4545) = .0455$

Chapter 11

2. $s^2 = 25$

a. With 19 degrees of freedom, $\chi^2_{.05} = 30.1435$ and $\chi^2_{.95} = 10.1170$

$$\frac{19(25)}{30.1435} \le \sigma^2 \le \frac{19(25)}{10.1170}$$

$$15.76 \le \sigma^2 \le 46.95$$

b. With 19 degrees of freedom, $\chi^2_{.025} = 32.8523$ and $\chi^2_{.975} = 8.90655$

$$\frac{19(25)}{32.8523} \le \sigma^2 \le \frac{19(25)}{8.90655}$$

$$14.46 \le \sigma^2 \le 53.33$$

c. $3.8 \le \sigma \le 7.3$

9. H_0: $\sigma^2 \le .0004$
H_a: $\sigma^2 > .0004$
$n = 30$

$$\chi^2_{.05} = 42.5569 \text{ (29 degrees of freedom)}$$

$$\chi^2 = \frac{(29)(.0005)}{.0004} = 36.25$$

Do not reject H_0; the product specification does not appear to be violated

15. We recommend placing the larger sample variance in the numerator; with $\alpha = .05$, $F_{.025,20,24} = 2.33$: reject H_0 if $F > 2.33$

$$F = \frac{8.2}{4.0} = 2.05; \text{ do not reject } H_0$$

Or, had we used the lower tail F value,

$$F_{.025,20,24} = \frac{1}{F_{.025,24,20}} = \frac{1}{2.46} = .41$$

$$F = \frac{4.0}{8.2} = .49$$

$$F > .41; \text{ do not reject } H_0$$

17. a. Let: σ_1^2 = variance in repair costs (4-year-old automobiles)

σ_2^2 = variance in repair costs (2-year-old automobiles)

H_0: $\sigma_1^2 \le \sigma_2^2$
H_a: $\sigma_1^2 > \sigma_2^2$

b. $s_1^2 = (170)^2 = 28{,}900$
$s_2^2 = (100)^2 = 10{,}000$

$$F = \frac{s_1^2}{s_2^2} = \frac{28{,}900}{10{,}000} = 2.89$$

$F_{.01,24,24} = 2.66$

Reject H_0; conclude that automobiles 4 years old have a larger variance in annual repair costs compared to automobiles 2 years old; this conclusion is expected due to the fact that older automobiles are more likely to have some expensive repairs that lead to greater variance in the annual repair costs

Chapter 12

1. Expected frequencies: $e_1 = 200(.40) = 80$
$e_2 = 200(.40) = 80$
$e_3 = 200(.20) = 40$

Actual frequencies: $f_1 = 60, f_2 = 120, f_3 = 20$

$$\chi^2 = \frac{(60-80)^2}{80} + \frac{(120-80)^2}{80} + \frac{(20-40)^2}{40}$$

$$= \frac{400}{80} + \frac{1600}{80} + \frac{400}{40}$$

$$= 5 + 20 + 10 = 35$$

$\chi^2_{.01} = 9.21034$, with $k - 1 = 3 - 1 = 2$ degrees of freedom

Because $\chi^2 = 35 > 9.21034$, reject the null hypothesis; that is, the population proportions are not as stated in the null hypothesis

3. H_0: $p_{ABC} = .29, p_{CBS} = .28, p_{NBC} = .25, p_{IND} = .18$
H_a: The proportions are not
$p_{ABC} = .29, p_{CBS} = .28, p_{NBC} = .25, p_{IND} = .18$

Expected frequencies: $300(.29) = 87$, $300(.28) = 84$
$300(.25) = 75$, $300(.18) = 54$
$e_1 = 87, e_2 = 84, e_3 = 75, e_4 = 54$

Actual frequencies: $f_1 = 95, f_2 = 70, f_3 = 89, f_4 = 46$

$\chi^2_{.05} = 7.81$ (3 degrees of freedom)

$$\chi^2 = \frac{(95-87)^2}{87} + \frac{(70-84)^2}{84} + \frac{(89-75)^2}{75} + \frac{(46-54)^2}{54} = 6.87$$

Do not reject H_0; there is no significant change in the viewing audience proportions

9. H_0: The column variable is independent of the row variable

H_a: The column variable is not independent of the row variable

Expected frequencies:

	A	B	C
P	28.5	39.9	45.6
Q	21.5	30.1	34.4

$$\chi^2 = \frac{(20-28.5)^2}{28.5} + \frac{(44-39.9)^2}{39.9} + \frac{(50-46.5)^2}{45.6} + \frac{(30-21.5)^2}{21.5} + \frac{(26-30.1)^2}{30.1} + \frac{(30-34.4)^2}{34.4}$$

$$= 7.86$$

$\chi^2_{.025} = 7.37776$, with $(2-1)(3-1) = 2$ degrees of freedom

Because $\chi^2 = 7.86 > 7.37776$, reject H_0; that is, conclude that the column variable is not independent of the row variable

11. H_0: Type of ticket purchased is independent of the type of flight

H_a: Type of ticket purchased is not independent of the type of flight

Expected Frequencies:

$e_{11} = 35.59$ $e_{12} = 15.41$

$e_{21} = 150.73$ $e_{22} = 65.27$

$e_{31} = 455.68$ $e_{32} = 197.32$

Ticket	Flight	Observed Frequency (f_i)	Expected Frequency (e_i)	$(f_i - e_i)^2/e_i$
First	Domestic	29	35.59	1.22
First	International	22	15.41	2.82
Business	Domestic	95	150.73	20.61
Business	International	121	65.27	47.59
Full-fare	Domestic	518	455.68	8.52
Full-fare	International	135	197.32	19.68
Totals		920		100.43

$\chi^2_{.05} = 5.99$ with $(3 - 1)(2 - 1) = 2$ degrees of freedom

Because $100.43 > 5.99$ we reject H_0; type of ticket purchased is not independent of the type of flight

20. First estimate μ from the sample data (sample size = 120)

$$\mu = \frac{0(39) + 1(30) + 2(30) + 3(18) + 4(3)}{120}$$

$$= \frac{156}{120} = 1.3$$

Therefore, we use Poisson probabilities with $\mu = 1.3$ to compute expected frequencies

x	Observed Frequency	Poisson Probability	Expected Frequency	Difference ($f_i - e_i$)
0	39	.2725	32.700	6.300
1	30	.3543	42.516	−12.516
2	30	.2303	27.636	2.364
3	18	.0998	11.976	6.024
4 or more	3	.0430	5.160	−2.160

$$\chi^2 = \frac{(6.300)^2}{32.700} + \frac{(-12.516)^2}{42.516} + \frac{(2.364)^2}{27.636} + \frac{(6.024)^2}{11.976} + \frac{(-2.160)^2}{5.160} = 9.0348$$

$\chi^2_{.05} = 7.81473$ with $5 - 1 - 1 = 3$ degrees of freedom

Because $\chi^2 = 9.0348 > 7.81473$, reject H_0; that is, conclude that the data do not follow a Poisson probability distribution

21. With $n = 30$ we will use six classes with $16\frac{2}{3}\%$ of the probability associated with each class

$$\bar{x} = 22.80, \; s = 6.2665$$

The z values that create 6 intervals, each with probability .1667 are −.98, −.43, 0, .43, .98

z	Cutoff value of x
−.98	22.8 − .98(6.2665) = 16.66
−.43	22.8 − .43(6.2665) = 20.11
0	22.8 + .00(6.2665) = 22.80
.43	22.8 + .43(6.2665) = 25.49
.98	22.8 + .98(6.2665) = 28.94

Interval	Observed Frequency	Expected Frequency	Difference
less than 16.66	3	5	−2
16.66–20.11	7	5	2
20.11–22.80	5	5	0
22.80–25.49	7	5	2
25.49–28.94	3	5	−2
28.94 and up	5	5	0

$$\chi^2 = \frac{(-2)^2}{5} + \frac{(2)^2}{5} + \frac{(0)^2}{5} + \frac{(2)^2}{5} + \frac{(-2)^2}{5} + \frac{(0)^2}{5}$$

$$= \frac{16}{5} = 3.20$$

$\chi^2_{.025} = 9.34840$ with $6 - 2 - 1 = 3$ degrees of freedom

Because $\chi^2 = 3.20 \le 9.34840$, do not reject H_0

The claim that the data come from a normal distribution cannot be rejected

Chapter 13

1. a. $\bar{\bar{x}} = (30 + 45 + 36)/3 = 37$

$$\text{SSTR} = \sum_{j=1}^{k} n_j(\bar{x}_j - \bar{\bar{x}})^2$$

$$= 5(30 - 37)^2 + 5(45 - 37)^2 + 5(36 - 37)^2$$

$$= 570$$

$$\text{MSTR} = \frac{\text{SSTR}}{k - 1} = \frac{570}{2} = 285$$

b. $$\text{SSE} = \sum_{j=1}^{k} (n_j - 1)s_j^2$$

$$= 4(6) + 4(4) + 4(6.5) = 66$$

$$\text{MSE} = \frac{\text{SSE}}{n_T - k} = \frac{66}{15 - 3} = 5.5$$

c. $$F = \frac{\text{MSTR}}{\text{MSE}} = \frac{285}{5.5} = 51.82$$

$F_{.05} = 3.89$ (2 degrees of freedom numerator and 12 denominator)

Because $F = 51.82 > F_{.05} = 3.89$, we reject the null hypothesis that the means of the three populations are equal

d.

Source of Variation	Sum of Squares	Degrees of Freedom	Mean Square	F
Treatments	570	2	285	51.82
Error	66	12	5.5	
Total	636	14		

6.

	Mfg 1	Mfg 2	Mfg 3
Sample mean	23	28	21
Sample variance	6.67	4.67	3.33

$$\bar{\bar{x}} = (23 + 28 + 21)/3 = 24$$

$$\text{SSTR} = \sum_{j=1}^{k} n_j(\bar{x}_j - \bar{\bar{x}})^2 = 4(23 - 24)^2 + 4(28 - 24)^2 + 4(21 - 24)^2 = 104$$

$$\text{MSTR} = \frac{\text{SSTR}}{k - 1} = \frac{104}{2} = 52$$

$$\text{SSE} = \sum_{j=1}^{k} (n_j - 1)s_j^2 = 3(6.67) + 3(4.67) + 3(3.33) = 44.01$$

$$\text{MSE} = \frac{\text{SSE}}{n_T - k} = \frac{44.01}{12 - 3} = 4.89$$

$$F = \frac{\text{MSTR}}{\text{MSE}} = \frac{52}{4.89} = 10.63$$

$F_{.05} = 4.26$ (2 degrees of freedom numerator and 9 denominator)

Because $F = 10.63 > F_{.05} = 4.26$, we reject the null hypothesis that the mean time needed to mix a batch of material is the same for each manufacturer

11. a. $\text{LSD} = t_{\alpha/2}\sqrt{\text{MSE}\left(\frac{1}{n_i} + \frac{1}{n_j}\right)}$

$$= t_{.025}\sqrt{5.5\left(\frac{1}{5} + \frac{1}{5}\right)}$$

$$= 2.179\sqrt{2.2} = 3.23$$

$|\bar{x}_1 - \bar{x}_2| = |30 - 45| = 15 >$ LSD; significant difference
$|\bar{x}_1 - \bar{x}_3| = |30 - 36| = 6 >$ LSD; significant difference
$|\bar{x}_2 - \bar{x}_3| = |45 - 36| = 9 >$ LSD; significant difference

b. $\bar{x}_1 - \bar{x}_2 \pm t_{\alpha/2}\sqrt{\text{MSE}\left(\frac{1}{n_1} + \frac{1}{n_2}\right)}$

$$(30 - 45) \pm 2.179\sqrt{5.5\left(\frac{1}{n_1} + \frac{1}{n_2}\right)}$$

$$-15 \pm 3.23 = -18.23 \text{ to } -11.77$$

13. $\text{LSD} = t_{\alpha/2}\sqrt{\text{MSE}\left(\frac{1}{n_1} + \frac{1}{n_3}\right)}$

$$= t_{.025}\sqrt{4.89\left(\frac{1}{4} + \frac{1}{4}\right)}$$

$$= 2.262\sqrt{2.45} = 3.54$$

Because $|\bar{x}_1 - \bar{x}_3| = |23 - 21| = 2 < 3.54$, there does not appear to be any significant difference between the means of populations 1 and 3

14. $\bar{x}_1 - \bar{x}_2 \pm \text{LSD}$
$23 - 28 \pm 3.54$
$-5 \pm 3.54 = -8.54$ to -1.46

19. a. $\bar{\bar{x}} = (156 + 142 + 134)/3 = 144$

$$\text{SSTR} = \sum_{j=1}^{k} n_j(\bar{x}_j - \bar{\bar{x}})^2 = 6(156 - 144)^2 + 6(142 - 144)^2 + 6(134 - 144)^2 = 1488$$

b. $\text{MSTR} = \frac{\text{SSTR}}{k - 1} = \frac{1488}{2} = 744$

c. $s_1^2 = 164.4, \quad s_2^2 = 131.2, \quad s_3^2 = 110.4$

$$\text{SSE} = \sum_{j=1}^{k} (n_j - 1)s_j^2 = 5(164.4) + 5(131.2) + 5(110.4) = 2030$$

d. $\text{MSE} = \frac{\text{SSE}}{n_T - k} = \frac{2030}{18 - 3} = 135.3$

e. $F = \frac{\text{MSTR}}{\text{MSE}} = \frac{744}{135.3} = 5.50$

$F_{.05} = 3.68$ (2 degrees of freedom numerator and 15 denominator)

Because $F = 5.50 > F_{.05} = 3.68$, we reject the hypothesis that the means for the three treatments are equal

34. ***Treatment Means***
$\bar{x}_{\cdot 1} = 13.6, \quad \bar{x}_{\cdot 2} = 11.0, \quad \bar{x}_{\cdot 3} = 10.6$
Block Means
$\bar{x}_{1\cdot} = 9,\ \bar{x}_{2\cdot} = 7.67,\ \bar{x}_{3\cdot} = 15.67,\ \bar{x}_{4\cdot} = 18.67,\ \bar{x}_{5\cdot} = 7.67$
Overall Mean
$\bar{\bar{x}} = 176/15 = 11.73$

Step 1

$$\text{SST} = \sum_i \sum_j (x_{ij} - \bar{\bar{x}})^2 = (10 - 11.73)^2 + (9 - 11.73)^2 + \cdots + (8 - 11.73)^2 = 354.93$$

Step 2

$$\text{SSTR} = b\sum_j (\bar{x}_{\cdot j} - \bar{\bar{x}})^2 = 5[(13.6 - 11.73)^2 + (11.0 - 11.73)^2 + (10.6 - 11.73)^2] = 26.53$$

Step 3

$$\text{SSBL} = k \sum_j (\bar{x}_{j\cdot} - \bar{\bar{x}})^2$$
$$= 3[(9 - 11.73)^2 + (7.67 - 11.73)^2 + (15.67 - 11.73)^2 + (18.67 - 11.73)^2 + (7.67 - 11.73)^2] = 312.32$$

Step 4

$$\text{SSE} = \text{SST} - \text{SSTR} - \text{SSBL}$$
$$= 354.93 - 26.53 - 312.32 = 16.08$$

Source of Variation	Sum of Squares	Degrees of Freedom	Mean Square	F
Treatments	26.53	2	13.27	6.60
Blocks	312.32	4	78.08	
Error	16.08	8	2.01	
Total	354.93	14		

$F_{.05} = 4.46$ (2 numerator degrees of freedom and 8 denominator)

Because $F = 6.60 > F_{.05} = 4.46$, we reject the null hypothesis that the means of the three treatments are equal

41. See Table E13.41

Step 1

$$\text{SST} = \sum_i \sum_j \sum_k (x_{ijk} - \bar{\bar{x}})^2$$
$$= (135 - 111)^2 + (165 - 111)^2 + \cdots + (136 - 111)^2 = 9028$$

Step 2

$$\text{SSA} = br \sum_i (\bar{x}_{i\cdot} - \bar{\bar{x}})^2$$
$$= 3(2)[(104 - 111)^2 + (118 - 111)^2] = 588$$

Step 3

$$\text{SSB} = ar \sum_j (\bar{x}_{\cdot j} - \bar{\bar{x}})^2$$
$$= 2(2)[(130 - 111)^2 + (97 - 111)^2 + (106 - 111)^2]$$
$$= 2328$$

Step 4

$$\text{SSAB} = r \sum_i \sum_j (\bar{x}_{ij} - \bar{x}_{i\cdot} - \bar{x}_{\cdot j} + \bar{\bar{x}})^2$$
$$= 2[(150 - 104 - 130 + 111)^2 + (78 - 104 - 97 + 111)^2 + \cdots + (128 - 118 - 106 + 111)^2] = 4392$$

Step 5

$$\text{SSE} = \text{SST} - \text{SSA} - \text{SSB} - \text{SSAB}$$
$$= 9028 - 588 - 2328 - 4392 = 1720$$

Source of Variation	Sum of Squares	Degrees of Freedom	Mean Square	F
Factor A	588	1	588	2.05
Factor B	2328	2	1164	4.06
Interaction	4392	2	2196	7.66
Error	1720	6	286.67	
Total	9028	11		

$F_{.05} = 5.99$ (1 degree of freedom numerator and 6 denominator)

$F_{.05} = 5.14$ (2 degrees of freedom numerator and 6 denominator)

Because $F = 2.05 < F_{.05} = 5.99$, factor A is not significant
Because $F = 4.06 < F_{.05} = 5.14$, factor B is not significant
Because $F = 7.66 > F_{.05} = 5.14$, interaction is significant

Chapter 14

1. a.

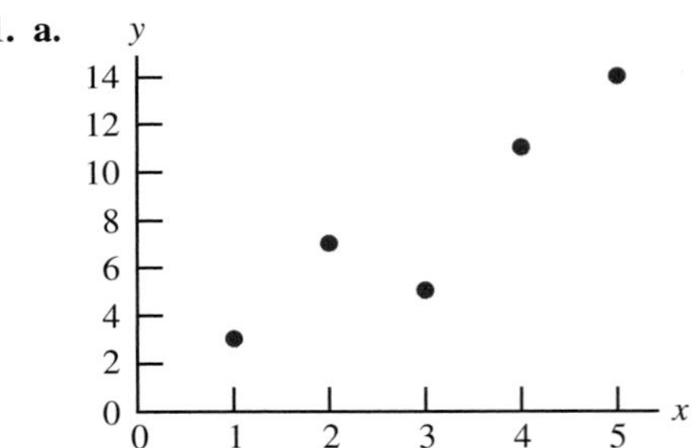

b. There appears to be a linear relationship between x and y

c. Many different straight lines can be drawn to provide a linear approximation of the relationship between x and y; in part (d) we will determine the equation of a straight line that "best" represents the relationship according to the least squares criterion

d. Summations needed to compute the slope and y-intercept:
$\Sigma x_i = 15$, $\Sigma y_i = 40$, $\Sigma(x_i - \bar{x})(y_i - \bar{y}) = 26$, $\Sigma(x_i - \bar{x})^2 = 10$

TABLE E13.41

		Factor B Level 1	Factor B Level 2	Factor B Level 3	Factor A Means
Factor A	Level 1	$\bar{x}_{11} = 150$	$\bar{x}_{12} = 78$	$\bar{x}_{13} = 84$	$\bar{x}_{1\cdot} = 104$
	Level 2	$\bar{x}_{21} = 110$	$\bar{x}_{22} = 116$	$\bar{x}_{23} = 128$	$\bar{x}_{2\cdot} = 118$
Factor B Means		$\bar{x}_{\cdot 1} = 130$	$\bar{x}_{\cdot 2} = 97$	$\bar{x}_{\cdot 3} = 106$	$\bar{\bar{x}} = 111$

$$b_1 = \frac{\Sigma(x_i - \bar{x})(y_i - \bar{y})}{\Sigma(x_i - \bar{x})^2} = \frac{26}{10} = 2.6$$

$b_0 = \bar{y} - b_1\bar{x} = 8 - (2.6)(3) = 0.2$

$\hat{y} = 0.2 - 2.6x$

e. $\hat{y} = .2 + 2.6x = .2 + 2.6(4) = 10.6$

4. a.

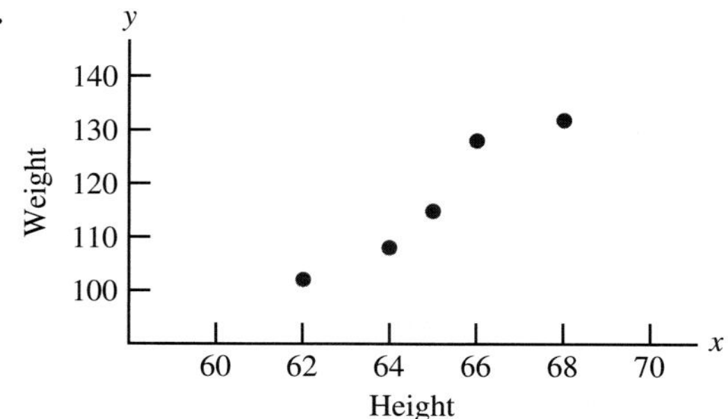

b. It indicates there may be a linear relationship between the variables

c. Many different straight lines can be drawn to provide a linear approximation of the relationship between x and y; in part (d) we will determine the equation of a straight line that "best" represents the relationship according to the least squares criterion

d. Summations needed to compute the slope and y-intercept:

$\Sigma x_i = 325, \quad \Sigma y_i = 585, \quad \Sigma(x_i - \bar{x})(y_i - \bar{y}) = 110,$
$\Sigma(x_i - \bar{x})^2 = 20$

$$b_1 = \frac{\Sigma(x_i - \bar{x})(y_i - \bar{y})}{\Sigma(x_i - \bar{x})^2} = \frac{110}{20} = 5.5$$

$b_0 = \bar{y} - b_1\bar{x} = 117 - (5.5)(65) = -240.5$

$\hat{y} = -240.5 + 5.5x$

e. $\hat{y} = -240.5 + 5.5(63) = 106$

The estimate of weight is 106 pounds

15. a. $\hat{y}_i = .2 + 2.6x_i$ and $\bar{y} = 8$

x_i	y_i	$\hat{y}_i$	$y_i - \hat{y}_i$	$(y_i - \hat{y}_i)^2$	$y_i - \bar{y}$	$(y_i - \bar{y})^2$
1	3	2.8	.2	.04	−5	25
2	7	5.4	1.6	2.56	−1	1
3	5	8.0	−3.0	9.00	−3	9
4	11	10.6	.4	.16	3	9
5	14	13.2	.8	.64	6	36
				SSE = 12.40		SST = 80

SSR = SST − SSE = 80 − 12.4 = 67.6

b. $r^2 = \frac{\text{SSR}}{\text{SST}} = \frac{67.6}{80} = .845$

The least squares line provided a good fit; 84.5% of the variability in y has been explained by the least squares line

c. $r = \sqrt{.845} = +.9192$

18. a. The estimated regression equation and the mean for the dependent variable:

$\hat{y} = 1790.5 + 581.1x, \quad \bar{y} = 3650$

The sum of squares due to error and the total sum of squares:

$\text{SSE} = \Sigma(y_i - \hat{y}_i)^2 = 85{,}135.14$

$\text{SST} = \Sigma(y_i - \bar{y})^2 = 335{,}000$

Thus, SSR = SST − SSE

= 335,000 − 85,135.14 = 249,864.86

b. $r^2 = \frac{\text{SSR}}{\text{SST}} = \frac{249{,}864.86}{335{,}000} = .746$

The least squares line accounted for 74.6% of the total sum of squares

c. $r = \sqrt{.746} = +.8637$

23. a. $s^2 = \text{MSE} = \frac{\text{SSE}}{n-2} = \frac{12.4}{3} = 4.133$

b. $s = \sqrt{\text{MSE}} = \sqrt{4.133} = 2.033$

c. $\Sigma(x_i - \bar{x})^2 = 10$

$$s_{b_1} = \frac{s}{\sqrt{\Sigma(x_i - \bar{x})^2}} = \frac{2.033}{\sqrt{10}} = .643$$

d. $t = \frac{b_1 - \beta_1}{s_{b_1}} = \frac{2.6 - 0}{.643} = 4.04$

$t_{.025} = 3.182$ (3 degrees of freedom)

Because $t = 4.04 > t_{.05} = 3.182$, we reject H_0: $\beta_1 = 0$

e. $\text{MSR} = \frac{\text{SSR}}{1} = 67.6$

$$F = \frac{\text{MSR}}{\text{MSE}} = \frac{67.6}{4.133} = 16.36$$

$F_{.05} = 10.13$ (1 degree of freedom numerator and 3 denominator)

Because $F = 16.36 > F_{.05} = 10.13$, we reject H_0: $\beta_1 = 0$

Source of Variation	Sum of Squares	Degrees of Freedom	Mean Square	F
Regression	67.6	1	67.6	16.36
Error	12.4	3	4.133	
Total	80	4		

26. a. $s^2 = \text{MSE} = \frac{\text{SSE}}{n-2} = \frac{85{,}135.14}{4} = 21{,}283.79$

$s = \sqrt{\text{MSE}} = \sqrt{21{,}283.79} = 145.89$

$\Sigma(x_i - \bar{x})^2 = .74$

$$s_{b_1} = \frac{s}{\sqrt{\Sigma(x_i - \bar{x})^2}} = \frac{145.89}{\sqrt{.74}} = 169.59$$

$$t = \frac{b_1 - \beta_1}{s_{b_1}} = \frac{581.08 - 0}{169.59} = 3.43$$

$t_{.025} = 2.776$ (4 degrees of freedom)

Because $t = 3.43 > t_{.025} = 2.776$, we reject H_0: $\beta_1 = 0$

b. $\text{MSR} = \dfrac{\text{SSR}}{1} = \dfrac{249{,}864.86}{1} = 249{,}864.86$

$$F = \frac{\text{MSR}}{\text{MSE}} = \frac{249{,}864.86}{21{,}283.79} = 11.74$$

$F_{.05} = 7.71$ (1 degree of freedom numerator and 4 denominator)

Because $F = 11.74 > F_{.05} = 7.71$, we reject $H_0{:}\ \beta_1 = 0$

c.

Source of Variation	Sum of Squares	Degrees of Freedom	Mean Square	F
Regression	29,864.86	1	29,864.86	11.74
Error	85,135.14	4	21,283.79	
Total	335,000	5		

32. a. $s = 2.033$

$\bar{x} = 3,\ \Sigma(x_i - \bar{x})^2 = 10$

$$s_{\hat{y}_p} = s\sqrt{\frac{1}{n} + \frac{(x_p - \bar{x})^2}{\Sigma(x_i - \bar{x})^2}}$$

$$= 2.033\sqrt{\frac{1}{5} + \frac{(4-3)^2}{10}} = 1.11$$

b. $\hat{y} = .2 + 2.6x = .2 + 2.6(4) = 10.6$

$\hat{y}_p \pm t_{\alpha/2} s_{\hat{y}_p}$

$10.6 \pm 3.182(1.11)$

10.6 ± 3.53, or 7.07 to 14.13

c. $$s_{\text{ind}} = s\sqrt{1 + \frac{1}{n} + \frac{(x_p - \bar{x})^2}{\Sigma(x_i - \bar{x})^2}}$$

$$= 2.033\sqrt{1 + \frac{1}{5} + \frac{(4-3)^2}{10}} = 2.32$$

d. $\hat{y}_p \pm t_{\alpha/2} s_{\text{ind}}$

$10.6 \pm 3.182(2.32)$

10.6 ± 7.38, or 3.22 to 17.98

35. a. $s = 145.89,\ \bar{x} = 3.2,\ \Sigma(x_i - \bar{x})^2 = .74$

$\hat{y} = 1790.5 + 581.1x = 1790.5 + 581.1(3)$

$= 3533.8$

$$s_{\hat{y}_p} = s\sqrt{\frac{1}{n} + \frac{(x_p - \bar{x})^2}{\Sigma(x_i - \bar{x})^2}}$$

$$= 145.89\sqrt{\frac{1}{6} + \frac{(3-3.2)^2}{.74}} = 68.54$$

$\hat{y}_p \pm t_{\alpha/2} s_{\hat{y}_p}$

$3533.8 \pm 2.776(68.54)$

3533.8 ± 190.27, or \$3343.53 to \$3724.07

b. $\hat{y} = 1790.5 + 581.1x = 1790.5 + 581.1(3)$

$= 3533.8$

$$s_{\text{ind}} = s\sqrt{1 + \frac{1}{n} + \frac{(x_p - \bar{x})^2}{\Sigma(x_i - \bar{x})^2}}$$

$$= 145.89\sqrt{1 + \frac{1}{6} + \frac{(3-3.2)^2}{.74}} = 161.19$$

$\hat{y}_p \pm t_{\alpha/2} s_{\text{ind}}$

$3533.8 \pm 2.776(161.19)$

3533.8 ± 447.46, or \$3086.34 to \$3981.26

40. a. 9

b. $\hat{y} = 20.0 + 7.21x$

c. 1.3626

d. $\text{SSE} = \text{SST} - \text{SSR} = 51{,}984.1 - 41{,}587.3 = 10{,}396.8$

$\text{MSE} = 10{,}396.8/7 = 1485.3$

$$F = \frac{\text{MSR}}{\text{MSE}} = \frac{41{,}587.3}{1485.3} = 28.0$$

$F_{.05} = 5.59$ (1 degree of freedom numerator and 7 denominator)

Because $F = 28 > F_{.05} = 5.59$, we reject $H_0{:}\ \beta_1 = 0$

e. $\hat{y} = 20.0 + 7.21(50) = 380.5$, or \$380,500

45. a. $\Sigma x_i = 14,\quad \Sigma y_i = 76,\quad \Sigma(x_i - \bar{x})(y_i - \bar{y}) = 200,$
$\Sigma(x_i - \bar{x})^2 = 126$

$$b_1 = \frac{\Sigma(x_i - \bar{x})(y_i - \bar{y})}{\Sigma(x_i - \bar{x})^2} = \frac{200}{126} = 1.5873$$

$b_0 = \bar{y} - b_1\bar{x} = 15.2 - (1.5873)(14) = -7.0222$

$\hat{y} = -7.02 + 1.59x$

b.

x_i	y_i	$\hat{y}_i$	$y_i - \hat{y}_i$
6	6	2.52	3.48
11	8	10.47	−2.47
15	12	16.83	−4.83
18	20	21.60	−1.60
20	30	24.78	5.22

c.

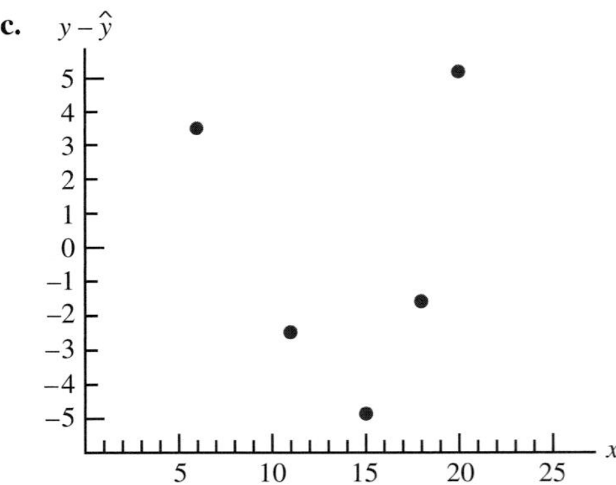

With only five observations, it is difficult to determine whether the assumptions are satisfied; however, the plot does suggest curvature in the residuals, which would indicate that the error term assumptions are not satisfied; the scatter diagram for these data also indicates that the underlying relationship between x and y may be curvilinear

d. $s^2 = 23.78$

$$h_i = \frac{1}{n} + \frac{(x_i - \bar{x})^2}{\Sigma(x_i - \bar{x})^2}$$

$$= \frac{1}{5} + \frac{(x_i - 14)^2}{126}$$

x_i	h_i	$s_{y_i - \hat{y}_i}$	$y_i - \hat{y}_i$	Standardized Residuals
6	.7079	2.64	3.48	1.32
11	.2714	4.16	−2.47	−.59
15	.2079	4.34	−4.83	−1.11
18	.3270	4.00	−1.60	−.40
20	.4857	3.50	5.22	1.49

e. The plot of the standardized residuals against $\hat{y}$ has the same shape as the original residual plot; as stated in part (c), the curvature observed indicates that the assumptions regarding the error term may not be satisfied

47. a. Let x = advertising expenditures and y = revenue

$\hat{y} = 29.4 + 1.55x$

b. SST = 1002, SSE = 310.28, SSR = 691.72

$$\text{MSR} = \frac{\text{SSR}}{1} = 691.72$$

$$\text{MSE} = \frac{\text{SSE}}{n-2} = \frac{310.28}{5} = 62.0554$$

$$F = \frac{\text{MSR}}{\text{MSE}} = \frac{691.72}{62.0554} = 11.15$$

$F_{.05} = 6.61$ (1 degree of freedom numerator and 5 denominator)

Because $F = 11.15 > F_{.05} = 6.61$, we conclude that the two variables are related

c.

x_i	y_i	$\hat{y}_i = 29.40 + 1.55x_i$	$y_i - \hat{y}_i$
1	19	30.95	−11.95
2	32	32.50	−.50
4	44	35.60	8.40
6	40	38.70	1.30
10	52	44.90	7.10
14	53	51.10	1.90
20	54	60.40	−6.40

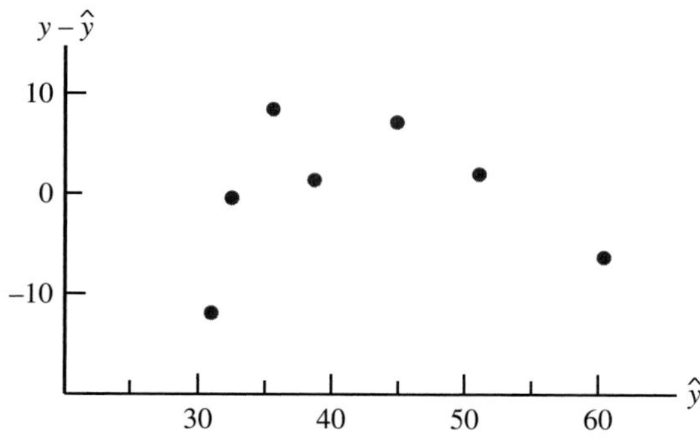

d. The residual plot leads us to question the assumption of a linear relationship between x and y; even though the relationship is significant at the $\alpha = .05$ level, it would be extremely dangerous to extrapolate beyond the range of the data

50. a. Using Minitab, we obtained the estimated regression equation $\hat{y} = 66.1 + .4023x$; a portion of the Minitab output is shown in Figure E14.50; the fitted values and standardized residuals are shown:

x_i	y_i	$\hat{y}_i$	Standardized Residuals
135	145	120.41	2.11
110	100	110.35	−1.08
130	120	118.40	.14
145	120	124.43	−.38
175	130	136.50	−.78
160	130	130.47	−.04
120	110	114.38	−.41

b.

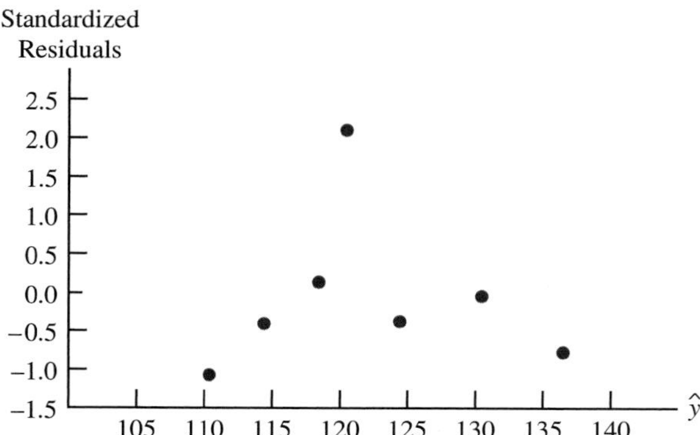

The standardized residual plot indicates that the observation $x = 135$, $y = 145$ may be an outlier; note that this observation has a standardized residual of 2.11

c. The scatter diagram is shown:

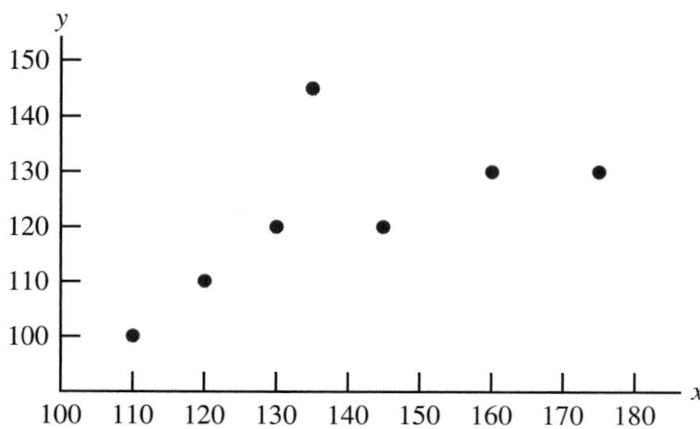

The scatter diagram also indicates that the observation $x = 135$, $y = 145$ may be an outlier; the implication is that for simple linear regression outliers can be identified by looking at the scatter diagram

52. a. A portion of the Minitab output is shown in Figure E14.52

b. Minitab identifies observation 1 as having a large standardized residual; thus, we would consider observation 1 to be an outlier

FIGURE E14.50

```
The regression equation is
Y = 66.1 + 0.402 X

Predictor        Coef       Stdev    t-ratio        p
Constant        66.10       32.06       2.06    0.094
X              0.4023      0.2276       1.77    0.137

s = 12.62       R-sq = 38.5%     R-sq(adj) = 26.1%

Analysis of Variance

SOURCE         DF          SS         MS        F        p
Regression      1       497.2      497.2     3.12    0.137
Error           5       795.7      159.1
Total           6      1292.9

Unusual Observations
Obs.        X        Y        Fit Stdev.Fit Residual   St.Resid
  1       135   145.00     120.42      4.87    24.58      2.11R
```

FIGURE E14.52

```
The regression equation is
Shipment = 4.09 + 0.196 Media$

Predictor        Coef       StDev          T        P
Constant        4.089       2.168       1.89    0.096
Media$        0.19552     0.03635       5.38    0.000

S = 5.044       R-Sq = 78.3%     R-Sq(adj) = 75.6%

Analysis of Variance
Source         DF          SS         MS          F        P
Regression      1      735.84     735.84      28.93    0.000
Error           8      203.51      25.44
Total           9      939.35

Unusual Observations
Obs    Media$   Shipment        Fit   StDev Fit    Residual     St Resid
  1       120      36.30      27.55        3.30        8.75        2.30R

R denotes an observation with a large standardized residual
```

Chapter 15

2. a. The estimated regression equation is
$\hat{y} = 45.06 + 1.94x_1$
An estimate of y when $x_1 = 45$ is
$\hat{y} = 45.06 + 1.94(45) = 132.36$

b. The estimated regression equation is
$\hat{y} = 85.22 + 4.32x_2$
An estimate of y when $x_2 = 15$ is
$\hat{y} = 85.22 + 4.32(15) = 150.02$

c. The estimated regression equation is
$\hat{y} = -18.37 + 2.01x_1 + 4.74x_2$

An estimate of y when $x_1 = 45$ and $x_2 = 15$ is
$\hat{y} = -18.37 + 2.01(45) + 4.74(15) = 143.18$

5. a. The Minitab output is shown in Figure E15.5a

b. The Minitab output is shown in Figure E15.5b

c. It is 1.60 in part (a) and 2.29 in part (b); in part (a) the coefficient is an estimate of the change in revenue due to a one-unit change in television advertising expenditures; in part (b) it represents an estimate of the change in revenue due to a one-unit change in television advertising expenditures when the amount of newspaper advertising is held constant

d. Revenue $= 83.2 + 2.29(3.5) + 1.30(1.8) = 93.56$ or \$93,560

12. a. $R^2 = \dfrac{\text{SSR}}{\text{SST}} = \dfrac{14{,}052.2}{15{,}182.9} = .926$

b. $R_a^2 = 1 - (1 - R^2)\dfrac{n - 1}{n - p - 1}$

$= 1 - (1 - .926)\dfrac{10 - 1}{10 - 2 - 1} = .905$

c. Yes; after adjusting for the number of independent variables in the model, we see that 90.5% of the variability in y has been accounted for

15. a. $R^2 = \dfrac{\text{SSR}}{\text{SST}} = \dfrac{23.435}{25.5} = .919$

$R_a^2 = 1 - (1 - R^2)\dfrac{n - 1}{n - p - 1}$

$= 1 - (1 - .919)\dfrac{8 - 1}{8 - 2 - 1} = .887$

FIGURE E15.5a

```
The regression equation is
Revenue = 88.6 + 1.60 TVAdv

Predictor      Coef       Stdev     t-ratio       p
Constant     88.638       1.582       56.02   0.000
TVAdv        1.6039      0.4778        3.36   0.015

s = 1.215      R-sq = 65.3%      R-sq(adj) = 59.5%

Analysis of Variance

SOURCE       DF        SS          MS          F        p
Regression    1    16.640      16.640      11.27    0.015
Error         6     8.860       1.477
Total         7    25.500
```

FIGURE E15.5b

```
The regression equation is
Revenue = 83.2 + 2.29 TVAdv + 1.30 NewsAdv

Predictor      Coef       Stdev     t-ratio       p
Constant     83.230       1.574       52.88   0.000
TVAdv        2.2902      0.3041        7.53   0.001
NewsAdv      1.3010      0.3207        4.06   0.010

s = 0.6426      R-sq = 91.9%     R-sq(adj) = 88.7%

Analysis of Variance

SOURCE       DF        SS          MS          F        p
Regression    2    23.435      11.718      28.38    0.002
Error         5     2.065       0.413
Total         7    25.500
```

b. Multiple regression analysis is preferred because both R^2 and R_a^2 show an increased percentage of the variability of y explained when both independent variables are used

19. a. $\text{MSR} = \dfrac{\text{SSR}}{p} = \dfrac{6216.375}{2} = 3108.188$

$\text{MSE} = \dfrac{\text{SSE}}{n-p-1} = \dfrac{507.75}{10-2-1} = 72.536$

b. $F = \dfrac{\text{MSR}}{\text{MSE}} = \dfrac{3108.188}{72.536} = 42.85$

$F_{.05} = 4.74$ (2 degrees of freedom numerator and 7 denominator)

Because $F = 42.85 > F_{.05} = 4.74$, the overall model is significant

c. $t = \dfrac{b_1}{s_{b_1}} = \dfrac{.5906}{.0813} = 7.26$

$t_{.025} = 2.365$ (7 degrees of freedom)

With $t = 7.26 > t_{.025} = 2.365$, β_1 is significant

d. $t = \dfrac{b_2}{s_{b_2}} = \dfrac{.4980}{.0567} = 8.78$

With $t = 8.78 > t_{.025} = 2.365$, β_2 is significant

23. a. $F = 28.38$

$F_{.01} = 13.27$ (2 degrees of freedom numerator and 1 denominator)

Because $F > F_{.01} = 13.27$, reject H_0

Alternatively, the p-value of .002 leads to the same conclusion

b. $t = 7.53$

$t_{.025} = 2.571$

Because $t > t_{.025} = 2.571$, β_1 is significant and x_1 should not be dropped from the model

c. $t = 4.06$

$t_{.025} = 2.571$

With $t > t_{.025} = 2.571$, β_2 is significant and x_2 should not be dropped from the model

28. a. Using Minitab, the 95% confidence interval is 132.16 to 154.15

b. Using Minitab, the 95% prediction interval is 111.15 to 175.17

29. a. See Minitab output in Figure E15.5b.

$\hat{y} = 83.230 + 2.2902(3.5) + 1.3010(1.8) = 93.588$ or $93,588

b. Using Minitab: 92.840 to 94.335, or $92,840 to $94,335

c. Using Minitab: 91.774 to 95.401, or $91,774 to $95,401

32. a. $E(y) = \beta_0 + \beta_1 x_1 + \beta_2 x_2$

where $x_2 = \begin{cases} 0 \text{ if level 1} \\ 1 \text{ if level 2} \end{cases}$

b. $E(y) = \beta_0 + \beta_1 x_1 + \beta_2(0) = \beta_0 + \beta_1 x_1$

c. $E(y) = \beta_0 + \beta_1 x_1 + \beta_2(1) = \beta_0 + \beta_1 x_1 + \beta_2$

d. $\beta_2 = E(y \mid \text{level 2}) - E(y \mid \text{level 1})$

β_1 is the change in $E(y)$ for a 1-unit change in x_1 holding x_2 constant

34. a. $15,300, because $b_3 = 15.3$

b. $\hat{y} = 10.1 - 4.2(2) + 6.8(8) + 15.3(0)$
$= 10.1 - 8.4 + 54.4$
$= 56.1$

Sales prediction: $56,100

c. $\hat{y} = 10.1 - 4.2(1) + 6.8(3) + 15.3(1)$
$= 10.1 - 4.2 + 20.4 + 15.3$
$= 41.6$

Sales prediction: $41,600

39. a. The Minitab output is shown in Figure E15.39

b. Using Minitab, we obtained the following values:

x_i	y_i	$\hat{y}_i$	Standardized Residual
1	3	2.8	.16
2	7	5.4	.94
3	5	8.0	−1.65
4	11	10.6	.24
5	14	13.2	.62

FIGURE E15.39

```
The regression equation is
Y = 0.20 + 2.60 X

Predictor      Coef      Stdev    t-ratio       p
Constant      0.200      2.132       0.09   0.931
X            2.6000     0.6429       4.04   0.027

s = 2.033     R-sq = 84.5%     R-sq(adj) = 79.3%

Analysis of Variance
SOURCE        DF         SS         MS        F        p
Regression     1     67.600     67.600    16.35    0.027
Error          3     12.400      4.133
Total          4     80.000
```

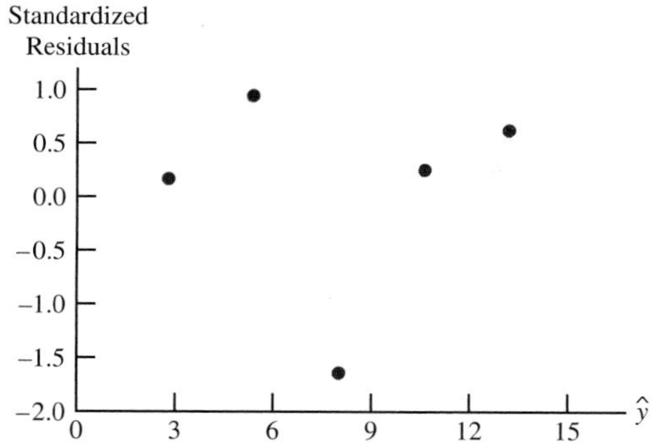

The point (3,5) does not appear to follow the trend of the remaining data; however, the value of the standardized residual for this point, −1.65, is not large enough for us to conclude that (3,5) is an outlier

c. Using Minitab, we obtained the following values:

x_i	y_i	Studentized Deleted Residual
1	3	.13
2	7	.92
3	5	−4.42
4	11	.19
5	14	.54

$t_{.025} = 4.303$ ($n - p - 2 = 5 - 1 - 2 = 2$ degrees of freedom)

Because the studentized deleted residual for (3,5) is $-4.42 < -4.303$, we conclude that the 3rd observation is an outlier

41. a. The Minitab output appears in Figure E15.5b; the estimated regression equation is

Revenue = 83.2 + 2.29 TVAdv + 1.30 NewsAdv

b. Using Minitab, we obtained the following values:

$\hat{y}_i$	Standardized Residual	$\hat{y}_i$	Standardized Residual
96.63	−1.62	94.39	1.10
90.41	−1.08	94.24	−.40
94.34	1.22	94.42	−1.12
92.21	−.37	93.35	1.08

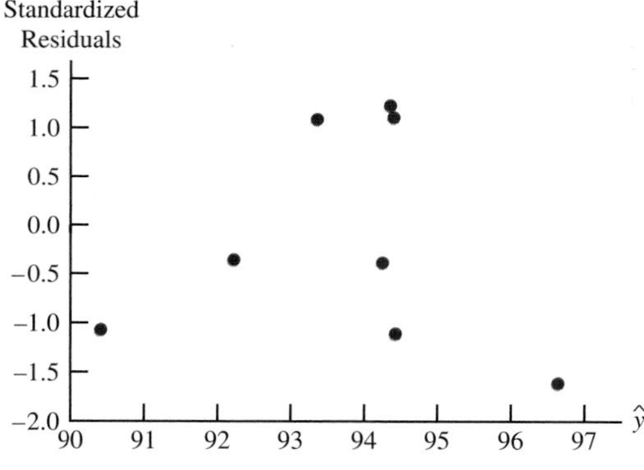

With relatively few observations, it is difficult to determine whether any of the assumptions regarding ϵ have been violated; for instance, an argument could be made that there does not appear to be any pattern in the plot; alternatively, an argument could be made that there is a curvilinear pattern in the plot

c. The values of the standardized residuals are greater than −2 and less than +2; thus, using this test, there are no outliers

As a further check for outliers, we used Minitab to compute the following studentized deleted residuals:

Observation	Studentized Deleted Residual	Observation	Studentized Deleted Residual
1	−2.11	5	1.13
2	−1.10	6	−.36
3	1.31	7	−1.16
4	−.33	8	1.10

$t_{.025} = 2.776$ ($n - p - 2 = 8 - 2 - 2 = 4$ degrees of freedom)

Because none of the studentized deleted residuals are less than −2.776 or greater than 2.776, we conclude that there are no outliers in the data

d. Using Minitab, we obtained the following values:

Observation	h_i	D_i
1	.63	1.52
2	.65	.70
3	.30	.22
4	.23	.01
5	.26	.14
6	.14	.01
7	.66	.81
8	.13	.06

The critical leverage value is

$$\frac{3(p+1)}{n} = \frac{3(2+1)}{8} = 1.125$$

Because none of the values exceed 1.125, we conclude that there are no influential observations; however, using Cook's distance measure, we see that $D_1 > 1$ (rule of thumb critical value); thus, we conclude that the first observation is influential

Final conclusion: observation 1 is an influential observation

Chapter 16

1. a. The Minitab output is shown in Figure E16.1a

b. The p-value corresponding to $F = 6.85$ is $.059 > \alpha = .05$, therefore the relationship is not significant

FIGURE E16.1a

```
The regression equation is
Y = - 6.8 + 1.23 X

Predictor        Coef        Stdev     t-ratio        p
Constant        -6.77        14.17       -0.48    0.658
X              1.2296       0.4697        2.62    0.059

s = 7.269        R-sq = 63.1%       R-sq(adj) = 53.9%

Analysis of Variance

SOURCE        DF          SS          MS          F        p
Regression     1      362.13      362.13       6.85    0.059
Error          4      211.37       52.84
Total          5      573.50
```

c.

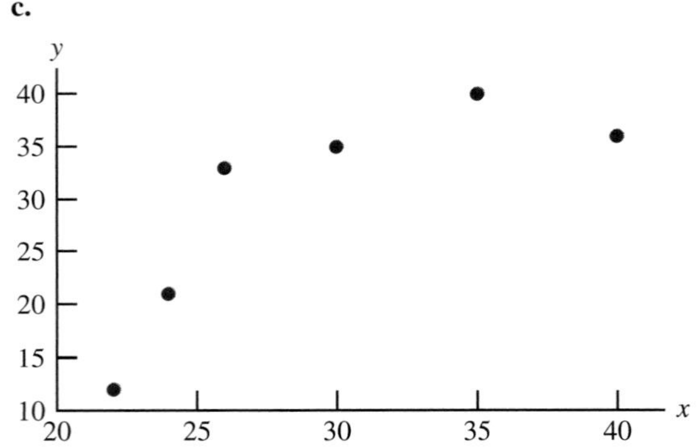

The scatter diagram suggests that a curvilinear relationship may be appropriate

d. The Minitab output is shown in Figure E16.1d

e. The p-value corresponding to $F = 25.68$ is $.013 < \alpha = .05$, therefore the relationship is significant

f. $\hat{y} = -168.88 + 12.187(25) - .17704(25)^2 = 25.145$

5. a. The Minitab output is shown in Figure E16.5a

b. The p-value corresponding to $F = 73.15$ is $.003 < \alpha = .01$, therefore the relationship is significant; we would reject H_0: $\beta_1 = \beta_2 = 0$

c. See Figure E16.5c

11. a. SSE = 1805 − 1760 = 45

$$F = \frac{\text{MSR}}{\text{MSE}} = \left(\frac{1760/4}{45/25}\right) = 244.44$$

$F_{.05} = 2.76$ (4 degrees of freedom numerator and 25 denominator)

Because $244.44 > 2.76$ we reject H_0; the relationship is significant

FIGURE E16.1d

```
The regression equation is
Y = - 169 + 12.2 X - 0.177 XSQ

Predictor        Coef        Stdev     t-ratio        p
Constant      -168.88        39.79       -4.74    0.024
X              12.187        2.663        4.58    0.020
XSQ          -0.17704      0.04290       -4.13    0.026

s = 3.248        R-sq = 94.5%       R-sq(adj) = 90.8%

Analysis of Variance

SOURCE        DF          SS          MS          F        p
Regression     2      541.85      270.92      25.68    0.013
Error          3       31.65       10.55
Total          5      573.50
```

FIGURE E16.5a

```
The regression equation is
Y = 433 + 37.4 X -0.383 XSQ

Predictor        Coef       Stdev     t-ratio        p
Constant        432.6       141.2        3.06    0.055
X              37.429       7.807        4.79    0.017
XSQ           -0.3829      0.1036       -3.70    0.034

s = 15.83        R-sq = 98.0%      R-sq(adj) = 96.7%

Analysis of Variance

SOURCE        DF          SS          MS         F        p
Regression     2       36643       18322     73.15    0.003
Error          3         751         250
Total          5       37395
```

FIGURE E16.5c

```
     Fit    Stdev.Fit            95% C.I.               95% P.I.
 1302.01         9.93    (1270.41, 1333.61)    (1242.55, 1361.47)
```

b. $SSE(x_1, x_2, x_3, x_4) = 45$

c. $SSE(x_2, x_3) = 1805 - 1705 = 100$

d. $F = \dfrac{(100 - 45)/2}{1.8} = 15.28 \qquad F_{.05} = 3.39$

Because $F = 15.28 > 3.39$, x_1 and x_2 are significant

12. a. The Minitab output is shown in Figure E16.12a

b. The Minitab output is shown in Figure E16.12b

c. $F = \dfrac{[\text{SSE(reduced)} - \text{SSE(full)}]/(\text{\# extra terms})}{\text{MSE(full)}}$

$= \dfrac{(23{,}157 - 14{,}317)/2}{1432} = 3.09$

$F_{.05} = 4.10$ (2 degrees of freedom numerator and 10 denominator)

Because $F = 3.09 < F_{.05} = 4.10$, the addition of the two independent variables is not significant

Note: Suppose that we consider adding only the number of interceptions made by the opponents; the corresponding Minitab output is shown in Figure E16.12c; in this case,

$F = \dfrac{(23{,}157 - 14{,}335)/1}{1303} = 6.77$

$F_{.05} = 4.84$ (1 degree of freedom numerator and 11 denominator)

Because $F = 6.77 > F_{.05} = 4.84$, the addition of the number of interceptions made by the opponents is significant

16. a. The Minitab output is shown in Figure E16.16a

b. Stepwise procedure (see Figure E16.16b)

c. Backward elimination procedure (see Figure E16.16c)

d. Best subsets regression (see Figure E16.16d)

21. a. The Minitab output is shown in Figure E16.21a

b. Residual plot as a function of the order in which the data are presented is shown; there does not appear to be any pattern indicative of positive autocorrelation

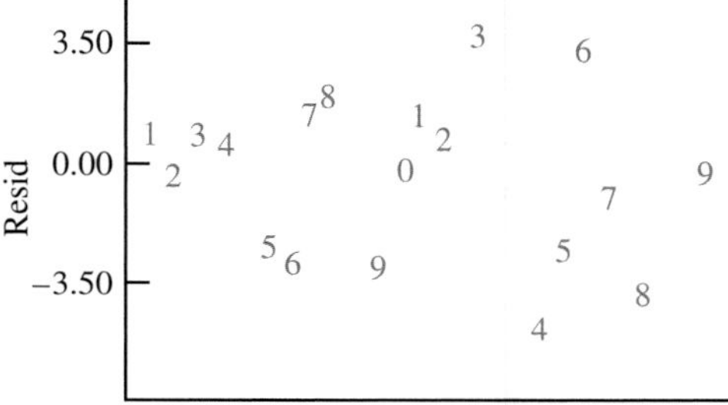

c. The Durban-Watson statistic (obtained from Minitab) is $d = 2.34$; at $\alpha = .05$, $d_L = 1.18$, and $d_U = 1.39$; because $d > d_U$, there is no significant positive autocorrelation

FIGURE E16.12a

```
The regression equation is
Points = 170 + 6.61 TeamInt

Predictor       Coef      Stdev    t-ratio       p
Constant      170.13      44.02       3.86   0.002
TeamInt        6.613      2.258       2.93   0.013

s = 43.93       R-sq = 41.7%     R-sq(adj) = 36.8%

Analysis of Variance

SOURCE       DF          SS          MS         F        p
Regression    1       16546       16546      8.57    0.013
Error        12       23157        1930
Total        13       39703

Unusual Observations
Obs. TeamInt    Points         Fit Stdev.Fit  Residual    St.Resid
 13     33.0     340.0       388.4      34.2     -48.4      -1.75 X

X denotes an obs. whose X value gives it large influence.
```

FIGURE E16.12b

```
The regression equation is
Points = 280 + 5.18 TeamInt - 0.0037 Rushing - 3.92 OpponInt

Predictor         Coef        Stdev    t-ratio        p
Constant        280.34        81.42       3.44    0.006
TeamInt          5.176        2.073       2.50    0.032
Rushing       -0.00373      0.03336      -0.11    0.913
OpponInt        -3.918        1.651      -2.37    0.039

s = 37.84       R-sq = 63.9%     R-sq(adj) = 53.1%

Analysis of Variance

SOURCE       DF          SS          MS         F        p
Regression    3       25386        8462      5.91    0.014
Error        10       14317        1432
Total        13       39703

SOURCE       DF      SEQ SS
Regression    1       16546
Error         1         776
Total         1        8064
```

FIGURE E16.12c

```
The regression equation is
Points = 274 + 5.23 TeamInt - 3.96 OpponInt

Predictor        Coef         Stdev    t-ratio         p
Constant       273.77         53.81       5.09     0.000
TeamInt         5.227         1.931       2.71     0.020
OpponInt       -3.965         1.524      -2.60     0.025

s = 36.10       R-sq = 63.9%     R-sq(adj) = 57.3%

Analysis of Variance

SOURCE         DF         SS          MS          F         p
Regression      2      25386       12684       9.73     0.004
Error          11      14335        1303
Total          13      39703

SOURCE         DF     SEQ SS
TeamInt         1      16546
OpponInt        1       8822
```

FIGURE E16.16a

```
The regression equation is
%College = -26.6 + 0.0970 SATScore

Predictor        Coef         Stdev    t-ratio         p
Constant       -26.61         37.22      -0.72     0.485
SATScore      0.09703       0.03734       2.60     0.019

s = 12.83      R-sq = 29.7%       R-sq(adj) = 25.3%

Analysis of Variance

SOURCE         DF         SS          MS          F         p
Regression      1     1110.8      1110.8       6.75     0.019
Error          16     2632.3       164.5
Total          17     3743.1
```

FIGURE E16.16b

```
STEP                1          2
CONSTANT       -26.61     -26.93

SATScore        0.097      0.084
t-RATIO          2.60       2.46

%TakeSAT                   0.204
t-RATIO                     2.21

s                12.8       11.5
R-sq            29.68      46.93
```

FIGURE E16.16c

STEP	1	2	3	4
CONSTANT	33.71	17.46	-32.47	-26.93
Size	-1.56	-1.39		
t-RATIO	-1.43	-1.42		
Spending	-0.0024	-0.0026	-0.0019	
t-RATIO	-1.47	-1.75	-1.31	
Salary	-0.00026			
t-RATIO	-0.40			
SATScore	0.077	0.081	0.095	0.084
t-RATIO	2.06	2.36	2.77	2.46
%TakeSAT	0.285	0.274	0.291	0.204
t-RATIO	2.47	2.53	2.60	2.21
S	11.2	10.9	11.2	11.5
R-sq	59.65	59.10	52.71	46.93

FIGURE E16.16d

Vars	R-sq	Adj. R-sq	S	Size	Spending	Salary	SATScore	%TakeSAT
1	29.7	25.3	12.826				X	
1	25.5	20.8	13.203					X
2	46.9	39.9	11.508				X	X
2	38.2	30.0	12.417	X			X	
3	52.7	42.6	11.244		X		X	X
3	49.5	38.7	11.618	X			X	X
4	59.1	46.5	10.852	X	X		X	X
4	52.8	38.3	11.660		X	X	X	X
5	59.6	42.8	11.219	X	X	X	X	X

23.

x_1	x_2	x_3	Treatment
0	0	0	A
1	0	0	B
0	1	0	C
0	0	1	D

$$E(y) = \beta_0 + \beta_1 x_1 + \beta_2 x_2 + \beta_3 x_3$$

26. a.

D_1	D_2	Manufacturer
0	0	1
1	0	2
0	1	3

$$E(y) = \beta_0 + \beta_1 D_1 + \beta_2 D_2$$

b. See Figure E16.26b

FIGURE E16.21a

```
The regression equation is
P/E = 6.51 + 0.569 %Profit

Predictor        Coef       Stdev    t-ratio        p
Constant        6.507       1.509       4.31    0.000
%Profit        0.5691      0.1281       4.44    0.000

s = 2.580       R-sq = 53.7%      R-sq(adj) = 51.0%

Analysis of Variance

SOURCE         DF         SS          MS          F         p
Regression      1     131.40      131.40      19.74     0.000
Error          17     113.14        6.66
Total          18     244.54
```

FIGURE E16.26b

```
The regression equation is
Time = 23.0 + 5.00 D1 - 2.00 D2

Predictor       Coef         Stdev    t-ratio        p
Constant      23.000         1.106      20.80    0.000
D1             5.000         1.563       3.20    0.011
D2            -2.000         1.563      -1.28    0.233

s = 2.211      R-sq = 70.3%       R-sq(adj) = 63.7%

Analysis of Variance

SOURCE         DF         SS          MS          F         p
Regression      2    104.000      52.000      10.64     0.004
Error           9     44.000       4.889
Total          11    148.000
```

c. H_0: $\beta_1 = \beta_2 = 0$

d. The p-value is .004 $< \alpha = .05$, therefore we conclude that the mean time to mix a batch of material is not the same for each manufacturer

Chapter 17

1. a.

Item	Price Relative
A	103 = (7.75/7.50)(100)
B	238 = (1500/630)(100)

b. $I_{2001} = \dfrac{7.75 + 1500.00}{7.50 + 630.00}(100) = \dfrac{1507.75}{637.50}(100) = 237$

c. $I_{2001} = \dfrac{7.75(1500) + 1500.00(2)}{7.50(1500) + 630.00(2)}(100)$

$= \dfrac{14{,}625.00}{12{,}510.00}(100) = 117$

d. $I_{2001} = \dfrac{7.75(1800) + 1500.00(1)}{7.50(1800) + 630.00(1)}(100)$

$= \dfrac{15{,}450.00}{14{,}130.00}(100) = 109$

3. a. Price relatives for A = (6.00/5.45)100 = 110

B = (5.95/5.60)100 = 106

C = (6.20/5.50)100 = 113

b. $I_{2001} = \dfrac{6.00 + 5.95 + 6.20}{5.45 + 5.60 + 5.50}(100) = 110$

c. $I_{2001} = \dfrac{6.00(150) + 5.95(200) + 6.20(120)}{5.45(150) + 5.60(200) + 5.50(120)}(100)$

$= 109$

9% increase over the two-year period

6.

Item	Price Relative	Base Period Price	Base Period Usage	Weight	Weighted Price Relative
A	150	22.00	20	440	66,000
B	90	5.00	50	250	22,500
C	120	14.00	40	560	67,200
			Totals	1250	155,700

$$I = \frac{155{,}700}{1250} = 125$$

7. a. Price relatives for A = (3.95/2.50)100 = 158
B = (9.90/8.75)100 = 113
C = (.95/.99)100 = 96

b.

Item	Price Relative	Base Price	Quantity	Weight $P_{i0}Q_i$	Weighted Price Relative
A	158	2.50	25	62.5	9,875
B	113	8.75	15	131.3	14,837
C	96	.99	60	59.4	5,702
			Totals	253.2	30,414

$$I = \frac{30{,}414}{253.2} = 120$$

Cost of raw materials is up 20% for the chemical

10. a. Deflated 1980 wages: $\dfrac{\$7.27}{82.4}(100) = \$8.82$

Deflated 1996 wages: $\dfrac{\$14.36}{172.6}(100) = \$8.32$

b. $\dfrac{14.36}{7.27}(100) = 197.5$; the percentange increase in actual wages is 97.5%

c. $\dfrac{8.32}{8.82}(100) = 94.3$; the change in real wages is a decrease of 5.7%

14. $I = \dfrac{300(18.00) + 400(4.90) + 850(15.00)}{350(18.00) + 220(4.90) + 730(15.00)}(100)$

$= \dfrac{20{,}110}{18{,}328}(100) = 110$

15. $I = \dfrac{95(1200) + 75(1800) + 50(2000) + 70(1500)}{120(1200) + 86(1800) + 35(2000) + 60(1500)}(100)$

$= 99$

Quantities are down slightly

Chapter 18

1. a.

Week	Time Series Value	Forecast	Forecast Error	Squared Forecast Error
1	8			
2	13			
3	15			
4	17	12	5	25
5	16	15	1	1
6	9	16	−7	49
			Total	75

Forecast for week 7 is (17 + 16 + 9)/3 = 14

b. MSE = 75/3 = 25

c.

Week (t)	Time Series Value (Y_t)	Forecast F_t	Forecast Error $Y_t - F_t$	Squared Error $(Y_t - F_t)^2$
1	8			
2	13	8.00	5.00	25.00
3	15	9.00	6.00	36.00
4	17	10.20	6.80	46.24
5	16	11.56	4.44	19.71
6	9	12.45	−3.45	11.90
Total				138.85

Forecast for week 7 is .2(9) + .8(12.45) = 11.76

d. For the $\alpha = .2$ exponential smoothing forecast

$$\text{MSE} = \frac{138.85}{5} = 27.77$$

Because the 3-week moving average has a smaller MSE, it appears to provide the better forecasts

e.

Week (t)	Time Series Value (Y_t)	Forecast F_t	Forecast Error $Y_t - F_t$	Squared Error $(Y_t - F_t)^2$
1	8			
2	13	8.0	5.0	25.00
3	15	10.0	5.0	25.00
4	17	12.0	5.0	25.00
5	16	14.0	2.0	4.00
6	9	14.8	−5.8	33.64
			Total	112.64

$$\text{MSE} = \frac{112.64}{5} = 22.52$$

A smoothing constant of .4 appears to provide the better forecasts; for week 7 the forecast using $\alpha = .4$ is .4(9) + .6(14.8) = 12.48

8. a.

Month	Time Series Value	3-Month Moving Average Forecast	$(\text{Error})^2$	$\alpha = .2$ Forecast	$(\text{Error})^2$
1	240				
2	350			240.00	12,100.00
3	230			262.00	1,024.00
4	260	273.33	177.69	255.60	19.36
5	280	280.00	0.00	256.48	553.19
6	320	256.67	4,010.69	261.18	3,459.79
7	220	286.67	4,444.89	272.95	2,803.70
8	310	273.33	1,344.69	262.36	2,269.57
9	240	283.33	1,877.49	271.89	1,016.97
10	310	256.67	2,844.09	265.51	1,979.36
11	240	286.67	2,178.09	274.41	1,184.05
12	230	263.33	1,110.89	267.53	1,408.50
Totals			17,988.52		27,818.49

MSE (3-month) = 17,988.52/9 = 1998.72

MSE (α = .2) = 27,818.49/11 = 2528.95

Based on the preceding MSE values, the 3-month moving average appears better; however, exponential smoothing was penalized by including month 2, which was difficult for any method to forecast. Using only the errors for months 4–12, the MSE for exponential smoothing is revised to

MSE(α = .2) = 14,694.49/9 = 1632.72

Thus, exponential smoothing was better considering months 4–12

b. Using exponential smoothing,

$$F_{13} = \alpha Y_{12} + (1 - \alpha)F_{12}$$
$$= .20(230) + .80(267.53) = 260$$

12. $\Sigma t = 15, \Sigma t^2 = 55, \Sigma Y_t = 55, \Sigma tY_t = 186$

$$b_1 = \frac{\Sigma tY_t - (\Sigma t \, \Sigma Y_t)/n}{\Sigma t^2 - (\Sigma t)^2/n}$$
$$= \frac{186 - (15)(55)/5}{55 - (15)^2/5} = 2.1$$
$$b_0 = \bar{Y} - b_1\bar{t} = 11 - 2.1(3) = 4.7$$
$$T_t = 4.7 + 2.1t$$
$$T_6 = 4.7 + 2.1(6) = 17.3$$

14. $\Sigma t = 21, \Sigma t^2 = 91, \Sigma Y_t = 117.1, \Sigma tY_t = 403.7$

$$b_1 = \frac{\Sigma tY_t - (\Sigma t \, \Sigma Y_t)/n}{\Sigma t^2 - (\Sigma t)^2/n}$$
$$= \frac{403.7 - (21)(117.1)/6}{91 - (21)^2/6} = -.3514$$
$$b_0 = \bar{Y} - b_1\bar{t} = 19.5167 - (-.3514)(3.5) = 20.7466$$
$$T_t = 20.7466 - .3514t$$

Enrollment appears to be decreasing by about 351 students per year

22. a.

Year	Quarter	Y_t	4-Quarter Moving Average	Centered Moving Average
1	1	4		
	2	2		
			3.50	
	3	3		3.750
			4.00	
	4	5		4.125
			4.25	
2	1	6		4.500
			4.75	
	2	3		5.000
			5.25	
	3	5		5.375
			5.50	
	4	7		5.875
			6.25	
3	1	7		6.375
			6.50	
	2	6		6.625
			6.75	
	3	6		
	4	8		

b.

Year	Quarter	Y_t	Centered Moving Average	Seasonal–Irregular Component
1	1	4		
	2	2		
	3	3	3.750	.8000
	4	5	4.125	1.2121
2	1	6	4.500	1.3333
	2	3	5.000	.6000
	3	5	5.375	.9302
	4	7	5.875	1.1915
3	1	7	6.375	1.0980
	2	6	6.625	.9057
	3	6		
	4	8		

Quarter	Seasonal–Irregular Component Values	Seasonal Index
1	1.3333, 1.0980	1.2157
2	.6000, .9057	.7529
3	.8000, .9302	.8651
4	1.2121, 1.1915	1.2018
	Total	4.0355

$$\text{Adjustment for seasonal index} = \frac{4}{4.0355} = .9912$$

Quarter	Adjusted Seasonal Index
1	1.2050
2	.7463
3	.8575
4	1.1912

Chapter 19

1. Binomial probabilities for $n = 10$, $p = .50$

x	Probability	x	Probability
0	.0010	6	.2051
1	.0098	7	.1172
2	.0439	8	.0439
3	.1172	9	.0098
4	.2051	10	.0010
5	.2461		

$P(0) + P(1) = .0108$; Adding $P(2)$ exceeds .025 required in the tail; therefore, reject H_0 if the number of plus signs is less than 2 or greater than 8;

Number of plus signs is 7

Do not reject H_0; conclude that there is no indication that a difference exists

2. $n = 27$ cases in which a value different from 150 is obtained

Use normal approximation with $\mu = np = .5(27) = 13.5$ and $\sigma = \sqrt{.25n} = \sqrt{.25(27)} = 2.6$

Use $x = 22$ as the number of plus signs and obtain the following test statistic:

$$z = \frac{x - \mu}{\sigma} = \frac{22 - 13.5}{2.6} = 3.27$$

With $\alpha = .01$, we reject if $z > 2.33$; because $z = 3.27 > 2.33$, reject H_0 and conclude the median is greater than 150

4. We need to determine the number of "better" responses and the number of "worse" responses; the sum of the two is the sample size used for the study

$$n = .34(1253) + .29(1253) = 789.4$$

Use the large-sample test using the normal distribution; this means the value of $n(n = 789.4)$ need not be integer. Use

$$\mu = .5n = .5(789.4) = 394.7$$
$$\sigma = \sqrt{.25n} = \sqrt{.25(789.4)} = 14.05$$

Let p = proportion of adults who feel children will have a better future

H_0: $p \leq .50$

H_a: $p > .50$

$$x = .34(1253) = 426.0$$

$$z = \frac{x - \mu}{\sigma} = \frac{426.0 - 394.7}{14.05} = 2.23$$

With $\alpha = .05$, we reject if $z > 1.645$; because $z = 2.23 > 1.645$, reject H_0 and conclude that more than half of the adults feel their children will have a better future

12. H_0: The populations are identical

H_a: The populations are not identical

Additive 1	Additive 2	Difference	Absolute Value	Rank	Signed Rank
20.12	18.05	2.07	2.07	9	+9
23.56	21.77	1.79	1.79	7	+7
22.03	22.57	−.54	.54	3	−3
19.15	17.06	2.09	2.09	10	+10
21.23	21.22	.01	.01	1	+1
24.77	23.80	.97	.97	4	+4
16.16	17.20	−1.04	1.04	5	−5
18.55	14.98	3.57	3.57	12	+12
21.87	20.03	1.84	1.84	8	+8
24.23	21.15	3.08	3.08	11	+11
23.21	22.78	.43	.43	2	+2
25.02	23.70	1.32	1.32	6	+6
					$T = 62$

$$\mu_T = 0$$

$$\sigma_T = \sqrt{\frac{n(n+1)(2n+1)}{6}} = \sqrt{\frac{12(13)(25)}{6}} = 25.5$$

$$z = \frac{T - \mu_T}{\sigma_T} = \frac{62 - 0}{25.5} = 1.83$$

Two-tailed test; reject H_0 if $z < -1.96$ or $z > 1.96$; because $z = 2.43 > 1.96$, reject H_0 and conclude that there is a significant difference between the additives

13.

Without Relaxant	With Relaxant	Difference	Rank of Absolute Difference	Signed Rank
15	10	5	9	+9
12	10	2	3	+3
22	12	10	10	+10
8	11	−3	6.5	−6.5
10	9	1	1	+1
7	5	2	3	+3
8	10	−2	3	−3
10	7	3	6.5	+6.5
14	11	3	6.5	+6.5
9	6	3	6.5	+6.5
				$T = 36$

$$\mu_T = 0$$

$$\sigma_T = \sqrt{\frac{n(n+1)(2n+1)}{6}} = \sqrt{\frac{10(11)(21)}{6}} = 19.62$$

$$z = \frac{T - \mu_T}{\sigma_T} = \frac{36}{19.62} = 1.83$$

One-tailed test; reject H_0 if $z > 1.645$

Reject H_0; there is a significant difference in favor of the relaxant

18. Rank the combined samples and find rank sum for each sample; this is a small-sample test because $n_1 = 7$ and $n_2 = 9$

Additive 1		Additive 2	
MPG	**Rank**	**MPG**	**Rank**
17.3	2	18.7	8.5
18.4	6	17.8	4
19.1	10	21.3	15
16.7	1	21.0	14
18.2	5	22.1	16
18.6	7	18.7	8.5
17.5	3	19.8	11
	34	20.7	13
		20.2	12
			102

$T = 34$

With $\alpha = .05$, $n_1 = 7$, and $n_2 = 9$

$T_L = 41$ and $T_U = 7(7 + 9 + 1) - 41 = 78$

Because $T = 34 < 41$, reject H_0 and conclude that there is a significant difference in gasoline mileage

19. a.

Public Accountant	Rank	Financial Planner	Rank
25.2	5	24.0	2
33.8	19	24.2	3
31.3	16	28.1	10
33.2	18	30.9	15
29.2	13	26.9	8.5
30.0	14	28.6	11
25.9	6	24.7	4
34.5	20	28.9	12
31.7	17	26.8	7
26.9	8.5	23.9	1
	136.5		73.5

$$\mu_T = \frac{1}{2}n_1(n_1 + n_2 + 1) = \frac{1}{2}(10)(10 + 10 + 1) = 105$$

$$\sigma_T = \sqrt{\frac{1}{12}n_1 n_2(n_1 + n_2 + 1)} = \sqrt{\frac{1}{12}(10)(10)(10 + 10 + 1)}$$

$= 13.23$

$T = 136.5$

Reject H_0 if $z < -1.645$ or if $z > 1.645$

$$z = \frac{136.5 - 105}{13.23} = 2.38$$

Reject H_0; salaries differ significantly for the two professions

b. Public Accountant $32,300
Financial Planner $26,700

26. Rankings:

Product A	Product B	Product C
4	11	7
8	14	2
10	15	1
3	12	6
9	13	5
34	65	21

$$W = \frac{12}{(15)(16)}\left[\frac{(34)^2}{5} + \frac{(65)^2}{5} + \frac{(21)^5}{5}\right] - 3(15 + 1)$$

$= 58.22 - 48 = 10.22$

$\chi^2_{.05} = 5.99147$ (2 degrees of freedom)

Reject H_0 and conclude the ratings for the products differ

28. Rankings:

Swimming	Tennis	Cycling
8	9	5
4	14	1
11	13	3
6	10	7
12	15	2
41	61	18

$$W = \frac{12}{15(15 + 1)}\left[\frac{41^2}{5} + \frac{61^2}{5} + \frac{18^2}{5}\right] - 3(15 + 1)$$

$= 9.26$

$\chi^2_{.05} = 5.99147$ (2 degrees of freedom)

Because $9.26 > 5.99147$, reject H_0 and conclude activities differ

32. a. $\Sigma d_i^2 = 52$

$$r_s = 1 - \frac{6\Sigma d_i^2}{n(n^2 - 1)} = 1 - \frac{6(52)}{10(99)} = .68$$

b. $$\sigma_{r_s} = \sqrt{\frac{1}{n - 1}} = \sqrt{\frac{1}{9}} = .33$$

$$z = \frac{r_s - 0}{\sigma_{r_s}} = \frac{.68}{.33} = 2.06$$

Reject H_0 if $z < -1.96$ or $z > 1.96$; because $z = 2.06 > 1.96$, reject H_0 and conclude that significant rank correlation exists

34. $\Sigma d_i^2 = 250$

$$r_s = 1 - \frac{6\Sigma d_i^2}{n(n^2 - 1)} = 1 - \frac{6(250)}{11(120)} = -.136$$

$$\sigma_{r_s} = \sqrt{\frac{1}{n - 1}} = \sqrt{\frac{1}{10}} = .32$$

$$z = \frac{r_s - 0}{\sigma_{r_s}} = \frac{-.136}{.32} = -.425$$

Reject H_0 if $z < -1.96$ or $z > 1.96$; because $z = -.425$, do not reject H_0; we cannot conclude that there is a significant relationship between the rankings

Chapter 20

4. *R chart:*

$UCL = \bar{R}D_4 = 1.6(1.864) = 2.98$

$LCL = \bar{R}D_3 = 1.6(.136) = .22$

$\bar{x}$ chart:

$UCL = \bar{\bar{x}} + A_2\bar{R} = 28.5 + .373(1.6) = 29.10$

$LCL = \bar{\bar{x}} - A_2\bar{R} = 28.5 - .373(1.6) = 27.90$

10. $f(0) = \dfrac{n!}{x!(n-x)!} p^x(1-p)^{n-x}$

When $p = .02$, the probability of accepting the lot is

$$f(0) = \frac{25!}{0!(25-0)!}(.02)^0(1-.02)^{25} = .6035$$

When $p = .06$, the probability of accepting the lot is

$$f(0) = \frac{25!}{0!(25-0)!}(.06)^0(1-.06)^{25} = .2129$$

Chapter 21

1. a. $\bar{x} = 215$ is an estimate of the population mean

b. $s_{\bar{x}} = \dfrac{20}{\sqrt{50}}\sqrt{\dfrac{800-50}{800}} = 2.7386$

c. $215 \pm 2(2.7386)$ or 209.5228 to 220.4772

5. a. $\bar{x} = 149{,}670$ and $s = 73{,}420$

$$s_{\bar{x}} = \sqrt{\left(\frac{771-50}{771}\right)}\frac{73{,}420}{\sqrt{50}} = 10{,}040.83$$

Approximate 95% confidence interval:
149,760 ± 2(10,040.83)
or
\$129,588.34 to \$169,751.66

b. $\hat{X} = N_{\bar{x}} = 771(149{,}670) = 115{,}395{,}570$
$s_{\hat{X}} = Ns_{\bar{x}} = 771(10{,}040.83) = 7{,}741{,}479.93$

Approximate 95% confidence interval:
115,395,570 ± 2(7,741,479.93)
or
\$99,912,810.14 to \$130,878,729.86

c. $\bar{p} = 18/50 = .36$ and

$$s_{\bar{p}} = \sqrt{\left(\frac{771-50}{771}\right)\frac{(.36)(.64)}{49}} = .0663$$

Approximate 95% confidence interval:
.36 ± 2(.0663)
or
.2274 to .4926

This interval is rather large; sample sizes must be large to obtain tight confidence intervals on a population proportion

7. a. Stratum *1*: $\bar{x}_1 = 138$
Stratum *2*: $\bar{x}_2 = 103$
Stratum *3*: $\bar{x}_3 = 210$

b. Stratum *1*

$$\bar{x}_1 = 138;\ s_{\bar{x}_1} = \left(\frac{30}{\sqrt{20}}\right)\sqrt{\frac{200-20}{200}} = 6.3640$$

Approximate 95% confidence interval:
138 ± 2(6.3640)
or 125.272 to 150.728

Stratum *2*

$$\bar{x}_2 = 103;\ s_{\bar{x}_2} = \left(\frac{25}{\sqrt{30}}\right)\sqrt{\frac{250-30}{250}} = 4.2817$$

Approximate 95% confidence interval:
103 ± 2(4.2817)
or 94.4366 to 111.5634

Stratum *3*

$$\bar{x}_3 = 210;\ s_{\bar{x}_3} = \left(\frac{50}{\sqrt{25}}\right)\sqrt{\frac{100-25}{100}} = 8.6603$$

Approximate 95% confidence interval:
210 ± 2(8.6603)
or 192.6794 to 227.3206

c. $\bar{x}_{st} = \left(\dfrac{200}{550}\right)138 + \left(\dfrac{250}{550}\right)103 + \left(\dfrac{100}{550}\right)210$
$= 50.1818 + 46.8182 + 38.1818 = 135.1818$

$$s_{\bar{x}_{st}} = \sqrt{\left(\frac{1}{(550)^2}\right)\left(200(180)\frac{(30)^2}{20} + 250(220)\frac{(25)^2}{30} + 100(75)\frac{(50)^2}{25}\right)}$$
$$= \sqrt{\left(\frac{1}{(550)^2}\right)3{,}515{,}833.3} = 3.4092$$

Approximate 95% confidence interval:
135.1818 ± 2(3.4092)
or 128.3634 to 142.0002

14. a. $\bar{x}_c = \dfrac{\Sigma x_i}{\Sigma M_i} = \dfrac{750}{50} = 15$

$\hat{X} = M\bar{x}_c = 300(15) = 4500$

$\bar{p}_c = \dfrac{\Sigma a_i}{\Sigma M_i} = \dfrac{15}{50} = .30$

b. $\Sigma(x_i - \bar{x}_c M_i)^2 = [95 - 15(7)]^2 + [325 - 15(18)]^2 + [190 - 15(15)]^2 + [140 - 15(10)]^2$
$= (-10)^2 + (55)^2 + (-35)^2 + (-10)^2$
$= 4450$

$$s_{\bar{x}_c} = \sqrt{\left(\frac{25-4}{(25)(4)(12)^2}\right)\left(\frac{4450}{3}\right)} = 1.4708$$

$s_{\hat{X}} = Ms_{\bar{x}_c} = 300(1.4708) = 441.24$

$\Sigma(a_i - \bar{p}_c M_i)^2 = [1 - .3(7)]^2 + [6 - .3(18)]^2 + [6 - .3(15)]^2 + [2 - .3(10)]^2$
$= (-1.1)^2 + (.6)^2 + (1.5)^2 + (-1)^2$
$= 4.82$

$$s_{\bar{p}_c} = \sqrt{\left(\frac{25-4}{(25)(4)(12)^2}\right)\left(\frac{4.82}{3}\right)} = .0484$$

c. Approximate 95% confidence interval for population mean:
15 ± 2(1.4708)
or 12.0584 to 17.9416

d. Approximate 95% confidence interval for population total:
4500 ± 2(441.24)
or 3617.52 to 5382.48

e. Approximate 95% confidence interval for population proportion:
.30 ± 2(.0484)
or .2032 to .3968

Index

D

E

F

G

Q

R

S

T

Chapter 1

Fortune	Exercises 4 and 5
Hotel	Exercises 2 and 3
Norris	Table 1.5
Shadow	Table 1.1

Chapter 2

ApTest	Table 2.10
Audit	Table 2.5
BwBooks	Exercise 6
Casino	Exercise 35
Computer	Exercise 21
Comstock	Exercise 40
Consolid	Case Problem
Crosstab	Exercise 29
Fortune	Table 2.18
HighLow	Exercise 44
IBD	Table 2.15
Income	Exercise 42
JobSat	Exercise 27
Nielsen	Exercise 4
OccupSat	Exercise 46
PEforcast	Exercise 28
Restaurant	Table 2.11
RevEmps	Exercise 47
Scatter	Exercise 30
Shadow	Exercise 41
SoftDrink	Table 2.1
Stereo	Table 2.14
Stocks	Table 2.9
Vehicles	Exercise 36
Wageweb	Exercise 18

Chapter 3

AcctSal	Exercise 5
Asian	Case Problem
Cameras	Exercise 14
Cities	Exercise 69
Consolid	Case Problem
Crime	Exercise 23
Discount	Exercises 6 & 24
DowS&P	Exercise 52
Eat	Exercise 60
Exam	Exercise 67
Health	Case Problem
HighLow	Exercise 53
Income	Exercise 65
Injury	Exercise 45
Mortgage	Exercise 63
Music	Exercise 7
NBA	Exercise 36
Notebook	Exercise 19
Options	Exercise 44
PCs	Exercise 51
ReEquity	Exercise 42
Retainer	Exercise 61
Salary	Table 3.1
Speakers	Exercise 37
Stereo	Table 3.7
WageWeb	Exercise 35
Websites	Exercise 9
World	Exercise 46

Chapter 4

Judge	Case Problem

Chapter 7

EAI	Section 7.1
MetAreas	Appendix 7.2, 7.3

Chapter 8

ActTemps	Exercise 46
Auto	Case Problem 3
Balance	Table 8.2
Bock	Case Problem 1
CJW	Section 8.1
FastFood	Exercise 12
GulfProp	Case Problem 2
Mail	Exercise 47
Miami	Exercise 11
Sleep	Exercise 22
Training	Table 8.4
TVtime	Exercise 19
Workers	Exercise 9

Chapter 9

BLS	Case Problem 2
Distance	Table 9.2
Heathrow	Section 9.5
Quality	Case Problem 1

Chapter 10

Airport	Exercise 6
Asia	Exercise 45
ExamData	Table 10.1
Golf	Case Problem
Matched	Table 10.4
Mutual	Exercise 42
PERatio	Exercise 28
Software	Table 10.3
TVRead	Exercise 27
Union	Exercise 10

Chapter 11

Bags	Exercise 19
DowJones	Exercise 21
Dullus	Section 11.2
MetroBus	Appendix 11.2
Training	Training
Travel	Exercise 25

Chapter 12

NYReform	Case Problem

Chapter 13

AgeExec	Exercise 49
AirTraf	Table 13.6
ArtDir	Exercise 52
Assembly	Exercise 56
AudJudg	Exercise 29
Browsing	Exercise 59
ChemTech	Table 13.3
Exer25	Exercise 25
GMAT	Table 13.11
IDSalary	Case Problem
ISP	Exercise 62
Medical1	Case Problem
Medical2	Case Problem
MktPerf	Exercise 55
MktValue	Exercise 10
NCP	Table 13.1
NFL	Exercise 53
Paint	Exercise 30
Resorts	Exercise 51
Salaries	Exercise 45
SatisJob	Exercise 50
Snow	Exercise 40
Stress	Exercise 9
Trucks	Exercise 48

Chapter 14

ADRs	Exercise 49
Alumni	Case Problem
Armand's	Table 14.1
Beer	Exercise 52
Casino	Exercise 11
CEO	Exercise 54
Cities	Exercise 20
DowS&P	Exercises 7 & 10
EmpRev	Exercise 12
HighLow	Exercise 61
HomeBldg	Exercise 43
Hotel	Exercise 14
IPO	Exercise 58
IRSAudit	Exercise 67
NAEP	Case Problem
Nielsen1	Exercise 39
Nielsen2	Exercise 53
OffRates	Exercise 44
Options	Exercise 59
PCs	Exercise 10
Safety	Case Problem

Chapter 15

Alumni	Case Problem
Auto1	Exercise 6
Auto2	Exercise 42
Auto3	Exercise 50
Butler	Table 15.1
CarRent	Exercise 10
Computer	Exercise 49
Consumer	Case Problem
Enquirer	Case Problem
Exer2	Exercise 2
Football	Exercise 37
ForFunds	Exercise 8
Johnson	Table 15.6
Mower	Exercise 7
NFL	Case Problem
Nielsen	Exercise 51
Repair	Exercise 35
Schools	Exercise 9
Showtime	Exercise 5
Stocks	Exercise 25
Stroke	Exercise 38
Trucks	Exercise 52

Chapter 16

Audit	Exercise 31
Beer	Exercise 7
Browsing	Exercise 34
Cravens	Table 16.5
Election	Exercise 35
Football	Exercise 12
GradRate	Case Problem
Internet	Exercise 8
Layoffs	Case Problem
LPGATour	Exercise 18
MPG	Table 16.4
MutFunds	Exercise 30
NFL	Exercise 15
PGATour	Case Problem
Reynolds	Table 16.1
Schools	Exercise 16
StkData	Exercise 19
Stroke	Exercise 14
Tyler	Table 16.2
WindGolf	Exercise 9

Chapter 18

Bicycle	Appendix 18.1, 18.2
Gasoline	Appendix 18.1, 18.2

Chapter 19

Chicago	Exercise 11

Chapter 20

Coffee	Exercise 20
Jensen	Table 20.2
Tires	Exercise 7